Conversion Factors

Mass
$1 \text{ g} = 10^{-3} \text{ kg} = 6.85 \times 10^{-5} \text{ slug}$
$1 \text{ kg} = 10^{3} \text{ g} = 6.85 \times 10^{-2} \text{ slug}$
$1 \text{ slug} = 1.46 \times 10^{4} \text{ g} = 14.6 \text{ kg}$
$1 \text{ u} = 1.66 \times 10^{-24} \text{ g} = 1.66 \times 10^{-27} \text{ kg}$
$1 \text{ metric ton} = 1000 \text{ kg}$

Length
$1 \text{ nm} = 10^{-9} \text{ m} = 10 \text{ Å}$
$1 \text{ cm} = 10^{-2} \text{ m} = 0.394 \text{ in.}$
$1 \text{ m} = 10^{-3} \text{ km} = 3.28 \text{ ft} = 39.4 \text{ in.}$
$1 \text{ km} = 10^{3} \text{ m} = 0.621 \text{ mi}$
$1 \text{ in.} = 2.54 \text{ cm} = 2.54 \times 10^{-2} \text{ m}$
$1 \text{ mi} = 5280 \text{ ft} = 1609 \text{ m} = 1.609 \text{ km}$

Area
$1 \text{ cm}^2 = 10^{-4} \text{ m}^2 = 0.1550 \text{ in.}^2 =$
$\quad 1.08 \times 10^{-3} \text{ ft}^2$
$1 \text{ m}^2 = 10^{4} \text{ cm}^2 = 10.76 \text{ ft}^2 = 1550 \text{ in.}^2$
$1 \text{ in.}^2 = 6.94 \times 10^{-3} \text{ ft}^2 = 6.45 \text{ cm}^2 =$
$\quad 6.45 \times 10^{-4} \text{ m}^2$
$1 \text{ ft}^2 = 144 \text{ in.}^2 = 9.29 \times 10^{-2} \text{ m}^2 = 929 \text{ cm}^3$

Volume
$1 \text{ cm}^3 = 10^{-6} \text{ m}^3 = 3.53 \times 10^{-5} \text{ ft}^3 =$
$\quad 6.10 \times 10^{-2} \text{ in.}^3$
$1 \text{ m}^3 = 10^{6} \text{ cm}^3 = 10^{3} \text{ L} = 35.3 \text{ ft}^3 =$
$\quad 6.10 \times 10^{4} \text{ in.}^3 = 264 \text{ gal}$
$1 \text{ liter} = 10^{3} \text{ cm}^3 = 10^{-3} \text{ m}^3 = 1.056 \text{ qt} =$
$\quad 0.264 \text{ gal}$
$1 \text{ in.}^3 = 5.79 \times 10^{-4} \text{ ft}^3 = 16.4 \text{ cm}^3 =$
$\quad 1.64 \times 10^{-5} \text{ m}^3$
$1 \text{ ft}^3 = 1728 \text{ in.}^3 = 7.48 \text{ gal} = 0.0283 \text{ m}^3 =$
$\quad 28.3 \text{ L}$
$1 \text{ qt} = 2 \text{ pt} = 946 \text{ cm}^3 = 0.946 \text{ L}$
$1 \text{ gal} = 4 \text{ qt} = 231 \text{ in.}^3 = 0.134 \text{ ft}^3 =$
$\quad 3.785 \text{ L}$

Time
$1 \text{ h} = 60 \text{ min} = 3600 \text{ s}$
$1 \text{ day} = 24 \text{ h} = 1440 \text{ min} = 8.64 \times 10^{4} \text{ s}$
$1 \text{ year} = 365 \text{ days} = 8.76 \times 10^{3} \text{ h} =$
$\quad 5.26 \times 10^{5} \text{ min} = 3.16 \times 10^{7} \text{ s}$

Angle
$1 \text{ rad} = 57.3°$
$\quad 1° = 0.0175 \text{ rad} \qquad 60° = \pi/3 \text{ rad}$
$\quad 15° = \pi/12 \text{ rad} \qquad 90° = \pi/2 \text{ rad}$
$\quad 30° = \pi/6 \text{ rad} \qquad 180° = \pi \text{ rad}$
$\quad 45° = \pi/4 \text{ rad} \qquad 360° = 2\pi \text{ rad}$
$1 \text{ rev/min} = \pi/30 \text{ rad/s} = 0.1047 \text{ rad/s}$

Speed
$1 \text{ m/s} = 3.60 \text{ km/h} = 3.28 \text{ ft/s} =$
$\quad 2.24 \text{ mi/h}$
$1 \text{ km/h} = 0.278 \text{ m/s} = 0.621 \text{ mi/h} =$
$\quad 0.911 \text{ ft/s}$
$1 \text{ ft/s} = 0.682 \text{ mi/h} = 0.305 \text{ m/s} =$
$\quad 1.10 \text{ km/h}$

$1 \text{ mi/h} = 1.467 \text{ ft/s} = 1.609 \text{ km/h} =$
$\quad 0.447 \text{ m/s}$
$60 \text{ mi/h} = 88 \text{ ft/s}$

Force
$1 \text{ N} = 10^{5} \text{ dynes} = 0.225 \text{ lb}$
$1 \text{ dyne} = 10^{-5} \text{ N} = 2.25 \times 10^{-6} \text{ lb}$
$1 \text{ lb} = 4.45 \times 10^{5} \text{ dynes} = 4.45 \text{ N}$
Equivalent weight of 1-kg mass
$\quad$ on Earth's surface $= 2.2 \text{ lb} = 9.8 \text{ N}$

Pressure
$1 \text{ Pa} (\text{N/m}^2) = 1.45 \times 10^{-4} \text{ lb/in.}^2 =$
$\quad 7.5 \times 10^{-3} \text{ torr} (\text{mm Hg}) =$
$\quad 10 \text{ dynes/cm}^2$
$1 \text{ torr} (\text{mm Hg}) = 133 \text{ Pa} (\text{N/m}^2) =$
$\quad 0.02 \text{ lb/in.}^2 = 1333 \text{ dynes/cm}^2$
$1 \text{ atm} = 14.7 \text{ lb/in.}^2 = 1.013 \times 10^{5} \text{ N/m}^2$
$\quad = 1.013 \times 10^{6} \text{ dynes/cm}^2 =$
$\quad 30 \text{ in. Hg} = 76 \text{ cm Hg}$
$1 \text{ bar} = 10^{6} \text{ dynes/cm}^2 = 10^{5} \text{ Pa}$
$1 \text{ millibar} = 10^{3} \text{ dynes/cm}^2 = 10^{2} \text{ Pa}$

Energy
$1 \text{ J} = 10^{7} \text{ ergs} = 0.738 \text{ ft·lb} = 0.239 \text{ cal} =$
$\quad 9.48 \times 10^{-4} \text{ Btu} = 6.24 \times 10^{18} \text{ eV}$
$1 \text{ kcal} = 4186 \text{ J} = 4.186 \times 10^{10} \text{ ergs} =$
$\quad 3.968 \text{ Btu}$
$1 \text{ Btu} = 1055 \text{ J} = 1.055 \times 10^{10} \text{ ergs} =$
$\quad 778 \text{ ft·lb} = 0.252 \text{ kcal}$
$1 \text{ cal} = 4.186 \text{ J} = 3.97 \times 10^{-3} \text{ Btu} =$
$\quad 3.09 \text{ ft·lb}$
$1 \text{ ft·lb} = 1.36 \text{ J} = 1.36 \times 10^{7} \text{ ergs} =$
$\quad 1.29 \times 10^{-3} \text{ Btu}$
$1 \text{ eV} = 1.60 \times 10^{-19} \text{ J} = 1.60 \times 10^{-12} \text{ ergs}$
$1 \text{ kWh} = 3.6 \times 10^{6} \text{ J}$

Power
$1 \text{ W} = 0.738 \text{ ft·lb/s} = 1.34 \times 10^{-3} \text{ hp} =$
$\quad 3.41 \text{ Btu/h}$
$1 \text{ ft·lb/s} = 1.36 \text{ W} = 1.82 \times 10^{-3} \text{ hp}$
$1 \text{ hp} = 550 \text{ ft·lb/s} = 745.7 \text{ W} =$
$\quad 2545 \text{ Btu/h}$

Mass-Energy Equivalents (at rest)
$1 \text{ u} = 1.66 \times 10^{-27} \text{ kg} \leftrightarrow 931.5 \text{ MeV}$
$1 \text{ electron mass} = 9.11 \times 10^{-31} \text{ kg} =$
$\quad 5.49 \times 10^{4} \text{ u} \leftrightarrow 0.511 \text{ MeV}$
$1 \text{ proton mass} = 1.672 \times 10^{-27} \text{ kg} =$
$\quad 1.007276 \text{ u} \leftrightarrow 938.28 \text{ MeV}$
$1 \text{ neutron mass} = 1.674 \times 10^{-27} \text{ kg} =$
$\quad 1.008665 \text{ u} \leftrightarrow 939.57 \text{ MeV}$

Temperature
$$T_F = \frac{9}{5} T_C + 32$$
$$T_C = \frac{5}{9}(T_F - 32)$$
$$T_K = T_C + 273.16$$

Physics

Third Edition

Jerry D. Wilson

Lander University

Anthony J. Buffa

California Polytechnic State University
San Luis Obispo

PRENTICE HALL
Upper Saddle River, NJ 07458

Library of Congress Cataloging-in-Publication Data

Wilson, Jerry D.

   College physics / Jerry D. Wilson, Anthony J. Buffa.—3rd ed.

     p. cm.

   Includes index.

   ISBN 0-13-398785-X

   1. Physics.   I. Buffa, Anthony J.   II. Title.

QC21.2.W548 1997

530—dc20
                                                  96–13252

                                                   CIP

Executive Editor: *Alison Reeves*

Editor in Chief: *Paul F. Corey*

Editorial Director: *Tim Bozik*

Assistant Vice President of Production and Manufacturing: *David W. Riccardi*

Executive Managing Editor: *Kathleen Schiaparelli*

Assistant Managing Editor: *Shari Toron*

Editor in Chief of ESM Development: *Ray Mullaney*

Development Editor: *Dan Schiller*

Project Management: *J. Carey Publishing Service*

Director of Marketing: *Kelly McDonald*

Manufacturing Manager: *Trudy Pisciotti*

Creative Director: *Paula Maylahn*

Art Director: *Heather Scott*

Art Manager: *Gus Vibal*

Artist: *Rolando Corujo*

Interior and Cover Designer: *Amy Rosen*

Cover Photographs: *Mountain Bike Racer Missy Giove / Carl Yarbrough Photography, Inc.*

Photo Editor: *Lorinda Morris-Nantz*

Photo Researcher: *Toby Zausner*

Editorial Assistant: *Pamela Holland-Moritz*

Assistant Editor: *Wendy Rivers*

 © 1997, 1994, 1990 by Prentice-Hall, Inc.

Simon & Schuster / A Viacom Company

Upper Saddle River, New Jersey 07458

Printed in the United States of America

10 9 8 7 6 5

ISBN 0-13-398785-X

Prentice-Hall International (UK) Limited, *London*

Prentice-Hall of Australia Pty. Limited, *Sydney*

Prentice-Hall Canada Inc., *Toronto*

Prentice-Hall Hispanoamericana, S.A., *Mexico*

Prentice-Hall of India Private Limited, *New Delhi*

Prentice-Hall of Japan, Inc., *Tokyo*

Simon & Schuster Asia Pte. Ltd., *Singapore*

Editora Prentice-Hall do Brasil, Ltda., *Rio de Janeiro*

About the Authors

Jerry D. Wilson   Jerry Wilson, a native of Ohio, is now Emeritus Professor of Physics and former Chair of the Division of Biological and Physical Sciences at Lander University in Greenwood, South Carolina. He received his B.S. degree from Ohio University, M.S. degree from Union College, and in 1970, a Ph.D. from Ohio University. He earned his M.S. degree while employed as a Materials Behavior Physicist by the General Electric Co.

As a doctoral graduate student, Professor Wilson held the faculty rank of Instructor and taught physical science courses. During this time, he co-authored a physical science text, which is now in its eighth edition. In conjunction with his teaching career, Professor Wilson has continued his writing and now has six titles in print that he has authored or co-authored. Recently retired from full-time teaching, he continues to write, including a weekly column, The Science Corner, for local newspapers.

With several competitive books available, one may wonder why another algebra-based physics text was written. Having taught introductory physics many times, I was well aware of the needs of students and the difficulties they have in mastering the subject. I decided to write a text that presented the basic physics principles in a clear and concise manner, with illustrative examples that help the major difficulty in learning physics: problem solving. Also, I wanted to write a text that is relevant so as to show students how physics applies in their everyday world—how things work and why things happen. Once the basics are learned, these follow naturally.

—Jerry Wilson

Anthony J. Buffa   Anthony Buffa received his B.S. degree in Physics from Rensselaer Polytechnic Institute and both his M.S. and Ph.D. degrees in Physics from the University of Illinois, Champaign-Urbana. In 1970, Professor Buffa joined the faculty at California Polytechnic State University, San Luis Obispo, where he is currently Professor of Physics, and has been a research associate with the Department of Physics Radioanalytical Facility since 1980.

Professor Buffa's main interest continues to be teaching. He has taught courses at Cal Poly ranging from introductory physical science to quantum mechanics, has helped in developing and revising laboratory experiments, and along with his colleagues has spent several summers teaching elementary physics to local teachers in a workshop sponsored by the NSF. With a strong interest in art and architecture, Dr. Buffa also spends a great deal of time developing his own artwork and sketches to be used in the teaching of introductory physics.

I try to emphasize to students how neat physics is, and I always make sure that significant components of my lectures, tests, and homework are conceptual and qualitative in nature. Only after the concepts are understood do we crunch numbers. This was the major reason that Jerry Wilson's book intrigued me: it had the same philosophy, was well-written, had simple but powerful artwork, and was concisely presented.

—Tony Buffa

Contents

About the Authors   xii
How to Use this Text   xiii
Preface   xix

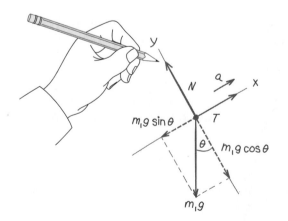

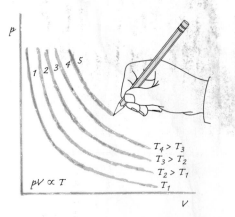

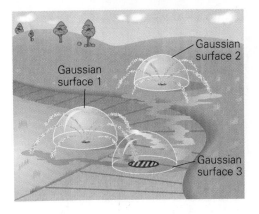

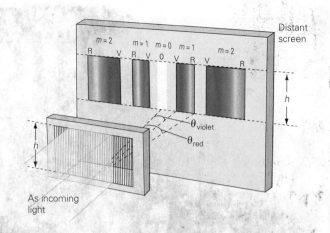

Demonstrations

Conceptual Examples

Applications (Insights in **boldface**)

How to Use this Text: A Guide for Students

Your instructor may select the text you use, but you are the one we are writing it for. We've tried to keep the material concise, focusing on the most important concepts, and yet rich with a variety of tools to help you understand physics as easily and as well as possible. The following few pages are designed to give you an overview of these tools and how you might best put them to use.

2 Kinematics: Description of Motion

2.1 Change of Position: Distance and Displacement

2.2 Speed and Velocity

2.3 Acceleration

2.4 Kinematic Equations

2.5 Free Fall

INSIGHT

■ Galileo Galilei and the Leaning Tower of Pisa

You can't look at a photo like this without mentally putting yourself into the picture. But how would you describe what you're experiencing? The sense of motion is so strong you can almost feel the air rushing by you. And yet, it's all an illusion! Motion takes place in time, but the photo can only "freeze" a single instant. How fast are you going? Are you speeding up as you plummet down the trail? Or are you brak- ing to slow down? You'll find that, without the dimension of time, you can hardly describe motion at all.

In this chapter we'll define motion and explore ways of de- scribing it. We'll also learn to analyze changes in motion— speeding up, slowing down, stopping. Along the way, we'll deal with a particularly interesting case of accelerated motion: free fall under the influence only of gravity.

30

Have you ever felt as though many of your courses and texts just presented a long list of unre- lated facts? **Chapter outlines** give you a quick "map" of the chapter to help you see the big picture and how all the pieces relate to each other.

What is physics to you? Does it seem like an abstract subject, unrelated to the world around you? In reading this text, we hope you'll come to see that, on the contrary, physics is exact- ly the study of the world around you. **Chapter opening photos** such as this one, along with extended captions, empha- size the connection between your world and the physics concepts that will be developed in the chapter ahead.

17.4 Electric Power

Objectives: To be able to (a) define electric power, (b) calculate the power delivery of simple electric circuits, and (c) explain joule heating and its significance.

How many times have you been reading a textbook and stopped to ask yourself, "What's the point?" or "Where is this leading?" **Integrated learning objectives** at the beginning of every chapter section help you focus on the most important con- cepts in the section ahead.

Conceptual Understanding is an Essential Problem-Solving Skill

If you understand why something works, you'll be able to use that knowledge over and over again, in many different situations. Throughout this book, we've stressed the fact that **conceptual understanding**—the "why"—is the basis for mastering a variety of problem-solving tools.

Suggested Problem-Solving Procedure

Say it in words

1. *Read the problem carefully and analyze it. Write down the given data and what it is you are to find.* Some data may not be given explicitly in numerical form. For example, if a car "starts from rest," its initial speed is zero ($v_o = 0$). In some instances, you may be expected to know certain quantities or to look them up in tables.

Say it in pictures

2. *Draw a diagram as an aid in visualizing and analyzing the physical situation of the problem where appropriate.* (This may not be necessary in every case, but it is usually helpful and can never do any harm.)

Say it in equations

3. *Determine what principle(s) and equation(s) are applicable to this situation, and how they can be used to get from the information given to what is to be found.* Keep in mind that many problems cannot be solved simply by plugging all of the given data into one equation; you may have to devise a strategy involving several steps.

Simplify the equations

4. *Simplify mathematical expressions as much as possible through algebraic manipulation before inserting actual values.* Trigonometric relationships (summarized in Appendix I) can also be used to simplify equations sometimes. The less calculation you do, the less likely you are to make a mistake— so *don't put the numbers in until you have to.*

Check the units

5. *Check units before doing calculations.* Make unit conversions if necessary so that all units are in the same system and quantities with the same dimensions have the same units (preferably standard units). This avoids mixed units and is helpful in unit analysis. (Unit checking and conversions are often done in Step 1 when writing the data.)

Insert numbers and calculate; check significant figures

6. *Substitute given quantities into equation(s) and perform calculations. Report the result with proper units and number of significant figures.*

Check the answer: is it reasonable?

7. *Consider whether the result is reasonable.* Does the answer have [appro]priate magnitude? (This means, is it "in the right ballpark"?) [For exam]ple, if a person's calculated mass turns out to be 2.30×10^2 kg, [it] should be questioned, since 230 kg corresponds to a weight of [...]

So, how do you get started? What generally is involved in solving physics problems? An extensive section in Chapter 1 (pp. 19–23) provides you with **a framework for thinking about problem solving.** Think of this seven step process as a set of general guidelines that can be applied to most problems.

To see the general problem solving procedures in action, work through the following three examples that show sample solutions with each step re-stated (and indicated by the number in red).

EXAMPLE 1.8 ■ FINDING THE AREA OF A RECTANGLE: PROBLEM-SOLVING PROCEDURE

Two students measure the lengths of adjacent sides of their dorm room. One reports 15 ft, 8 in., and the other reports 4.25 m. What is the area of the room in square meters?

Solution.

1. Adjacent sides of a room give its length and width, so we may write

   Given: Length = l = 15 ft, 8 in.   *Find:* Area (in m²)
   Width = w = 4.25 m

2. Sketch a diagram to help visualize the situation (•Fig. 1.10).

3 and 4. For this simple situation, the required equation is well known. The area (A) of a rectangle is $A = l \times w$, both of which are given.

5. It is obvious that a unit change is necessary. Let's first convert the length measurement to inches and then inches to meters:

$$15\ \text{ft} + 8\ \text{in.} = \left(15\ \text{ft} \times \frac{12\ \text{in.}}{\not{f}}\right) + 8\ \text{in.} = 188\ \text{in.}$$

and

$$188\ \text{in.} \times \frac{2.54\ \text{cm}}{\not{in.}} = 478\ \text{cm} = 4.78\ \text{m}$$

Notice how easy it is to convert units in the decimal metric system (cm to m). You should perform the conversion explicitly if necessary, using the conversion factor (1 m/100 cm).

6. Now perform the calculation.

   $A = l \times w = 4.78\ \text{m} \times 4.25\ \text{m}$
   $= 20.315\ \text{m}^2 = 20.3\ \text{m}^2$ (computed value rounded to 3 sf—why?)

7. The answer appears reasonable. Since $1\ \text{m} \approx 3\ \text{ft}$, the dorm room would be about 13 ft by 14 ft, which is about right (but, as always, too small). Suppose you had inadvertently punched 47.8 instead of 4.78 on your calculator. The result would be $A = 47.8\ \text{m} \times 4.25\ \text{m} = 203\ \text{m}^2$. A room with an area of about 200 m² would have dimensions of about 10 m by 20 m, which is roughly 30 ft by 60 ft. Since this is not the size of a typical dorm room, the magnitude of the result should make you suspect there may be an error.

Follow-up Exercise. The dimensions of a textbook are 0.22 m × 0.26 m × 4.0 cm. What volume in a book bag would the book take up? Give the answer in both m³ and cm³. *(Answers may be found in the Answers to Follow-up Exercises section at the back of the book.)*

PROBLEM-SOLVING STRATEGIES: "BALLPARK" AND ORDER-OF-MAGNITUDE CALCULATIONS

Calculations such as we have performed above are sometimes known as "ballpark" or "back of the envelope" calculations. In performing such calculations we know that, because the data are only approximate (they are not known precisely, or they are being rounded off to simple whole numbers for the purpose of rapid calculation), the answer can't be taken as very accurate (see Section 1.6). However, it is often precise enough for our needs. In the example given above, some of the data were unavoidable approximations (the glacier obviously isn't a simple rectangular object), while others, such as the density of ice, were deliberately approximated to simplify calculation. Nevertheless, this method gave us a good idea of the relative momenta and kinetic energies of the three objects.

Physicists often find it convenient to use a formalized version of the "ballpark" approach, commonly known as *order-of-magnitude calculation*. In this method, all values are rounded off to the nearest "order of magnitude," or power of 10: 10^2, 10^3, 10^4, etc. Since it is very easy to multiply and divide powers of 10, this shortcut makes calculation very fast. However, although sometimes just the powers of 10 can be used in a calculation, it is usually a good idea to retain one significant figure of the prefix. This is particularly true when quantities are squared or raised to higher powers, as in the kinetic energy calculations. For instance, $v_s = 4 \times 10^2$ m/s or $\approx 10^2$ m/s. Then $v_s{}^2 = (4 \times 10^2\ \text{m/s})^2 = 16 \times 10^4$ $\text{m}^2/\text{s}^2 \approx 10^5\ \text{m}^2/\text{s}^2$. However, if we had just used 10^2 m/s, we would have obtained 10^4 m^2/s^2.

In order-of-magnitude calculations, we can generally expect the answer to be correct to the nearest order of magnitude. In other words, if we get an answer of 1000, the true value might really be 750 or 1200, but we can be pretty confident that it is closer to 1000 than to 100 or to 10,000.

Example:

A light-year—a unit widely used in astronomy—is the distance light travels in one year in a vacuum. The speed of light is $c = 3.00 \times 10^8$ m/s. Approximately how many meters are there in a light-year?

Solution:

From the Conversion Factors Table inside the front cover (or simple calculation), we know that there are 3.16×10^7 s in one year. Using the relationship, distance = speed × time,

$$d = ct = (3 \times 10^8\ \text{m/s})(3 \times 10^7\ \text{s}) \approx 10^{16}\ \text{m}$$

Note that the product of the powers of 10 gives 10^{15}, but $3 \times 3 = 9$, so the prefixes give about another order of magnitude. Thus $(10 \times 10^{15}) = 10^{16}$ is the order of magnitude of the number of meters in a light-year.

While the general seven-step problem solving procedure can be applied to most problems, more situation-specific **problem solving hints and strategies** are provided in boxes throughout the text.

Visualization is one of the most important problem solving tools in physics. In many cases, if you can make a sketch of problem, you can probably solve it. You noted that we recommended "saying it in pictures" as the second step in general problem solving. **"Learn by Drawing"** features throughout the book show you how visualization can provide key insights into a variety of physical situations.

Conceptual examples reinforce the importance of understanding basic principles and give you practice in reasoning about physical situations.

Worked examples talk you through sample problems with fully explained step-by-step solution strategies. All worked examples close with a follow-up exercise, allowing you to test your understanding of the relevant concepts by tackling a closely related problem. Answers are in the back of the book so you can check your answer and make sure you're on the right track.

LEARN BY DRAWING

Leaning on Isotherms

When you are analyzing the various thermodynamic processes discussed in this chapter, it is sometimes hard to keep track of the heat flow (Q), work (W), and internal energy change (ΔU), together with their correct signs. One trick that can help with this bookkeeping is to superimpose a series of isotherms on the pV plot you are working with (as in Figures 12.2–12.5). This is generally useful even if the situation you are studying does not involve an isothermal process. It is also a good way of reinforcing your insight into what is happening physically in these processes.

Before starting, recall the two important properties of an isothermal process. Both of them follow directly from the fact that an isothermal process is, by definition, one in which the temperature remains constant.

1. In an isothermal process ΔU is zero. (Why?)
2. Since T is constant, the ideal gas law (Eq. 10.3) tells us that in an isothermal process pV must also be constant: $pV = k$. You may recall from algebra that this is the equation of a hyperbola, and on a pV diagram, each isotherm is indeed a hyperbola. The farther from the axes the hyperbola is, the higher the temperature that it represents (Fig. 1).

To take advantage of these properties, follow these steps:

- Sketch a set of isotherms for a series of different temperatures on the pV plot (Fig. 1).
- Then sketch the actual process or processes you are analyzing—for example, those shown in Fig. 2 and Fig. 3. (Take a moment to remind yourself what the names mean: isomet = constant volume, isobar = constant pressure, and adiabat = no heat flow.)
- Finally, use the first law of thermodynamics, $Q = \Delta U + W$, to calculate the final results, including signs. Usually, you will be able to find the sign of W by remembering that W is simply the area under the pV

curve for the process represented. The sign of ΔT will be clear from the isotherms that you have added; they will serve as a temperature scale, with progressively higher temperatures represented by isotherms at greater distances from the axes. A rise in T implies a positive value for ΔU, since the internal energy of the gas increases with its temperature. You will thus have all you need to determine the sign of Q, and thus whether heat flows into the system ($Q > 0$) or out of it ($Q < 0$).

The following two examples indicate the power of this approach. In Fig. 2, we are asked to determine whether heat flows into or out of the gas during an isobaric expansion. Expansion implies positive work ($p\Delta V > 0$). But what is the direction of the heat flow (or is it zero)? Sketching the isobar (with an arrow pointing to the right to indicate expansion), we see that it crosses from lower-temperature isotherms to higher-temperature ones. Hence there is a temperature increase, and ΔU must be positive. Since the gas has both gained internal energy and done positive work, both terms on the right-hand side of the first law equation are positive, so it follows that Q must be positive: $Q = (\Delta U + p\Delta V) > 0$. Positive heat flow means "into the gas," so we have our answer.

An isometric process involving a pressure drop is similarly analyzed in Fig. 3. Again, starting with lightly-drawn isotherms on the pV axes, we plot a vertical line (constant volume) with an arrow pointing downward to show pressure reduction. Since there is no volume change, the gas does no work. However, we can see that its temperature drops (why?), so its internal energy must decrease. This means that ΔU is negative, hence Q must be negative ($Q = \Delta U + p\Delta V = \Delta U < 0$). Thus heat flows out of the gas.

As an exercise, try analyzing an adiabatic processes in this way to see if you can graphically determine where the work comes from if there is no heat flow into or out of the gas.

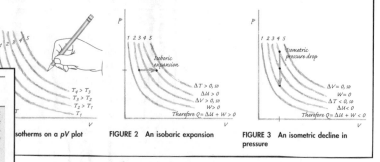

$T_4 > T_3$
$T_3 > T_2$
$T_2 > T_1$
T_1

isotherms on a pV plot

Isoboric expansion
$\Delta T > 0$, so
$\Delta U > 0$
$\Delta V > 0$, so
$W > 0$
Therefore $Q = \Delta U + W > 0$

FIGURE 2 An isobaric expansion

Isometric pressure drop
$\Delta V = 0$, so
$W = 0$
$\Delta T < 0$, so
$\Delta U < 0$
Therefore $Q = \Delta U + W < 0$

FIGURE 3 An isometric decline in pressure

12.2   The First Law of Thermodynamics   **391**

CONCEPTUAL EXAMPLE 4.9 ■ A JUMBLED JUGGLER: NEWTON'S THIRD LAW

A juggler carrying three heavy balls wants to cross a weak bridge that will support only his weight and two of the balls. Relying on his trade, he decides that to keep the bridge from collapsing and to save time, he will cross the bridge juggling the balls so that one is in the air at all times. Does he make it across safely? (a) yes (b) no *Clearly establish the reasoning and physical principle(s) used in determining your answer before checking below. That is, why did you select your answer?*

Reasoning and Answer. The juggler has apparently not studied Newton's laws. When throwing a ball upward, a force must be applied that is *greater* than the weight of the ball. (Why?) So, the juggler exerts an upward force on the ball, and the ball exerts an equal and opposite force on the juggler (Newton's third law). This adds a force *greater* than the weight of a third ball to the downward force already acting on the bridge (the juggler's weight and that of the other two balls). Down the bridge would go, so the answer is (b).

We can confirm this conclusion by analyzing the situation algebraically. The juggler's weight with three balls is $F = (M + 3m)g$. When juggling with one ball in the air, the weight is less, only $(M + 2m)g$. However, the force needed to accelerate the third ball upward is $mg + ma$. (Can you see why?) Thus, on the initial upward throw, the total downward force on the bridge would be $F' = (M + 2m)g + (mg + ma)$, or $F' = (M + 3m)g + ma$; and afterward, with one ball in the air, $F' = (M + m)g + (mg + ma)$, or $F' = (M + 2m)g + ma$. So it is clear that the juggler is no better off trying to juggle the third ball than just carrying it. Either way, he will get wet.

Follow-up Exercise(s). (a) Would the juggler be able to continue juggling while he was falling? (b) A canary is in a cage with an open wire floor that is suspended a sensitive scale. The weight of the cage and bird is noted when the bird is on a p When the yellow bird is flying about in the cage, is the scale reading different: how? (*Reasoning and Answers may be found in the Answers to Follow-up Exercises sect the back of the book.*)

EXAMPLE 8.5 ■ STACK THEM UP! CENTER OF GRAVITY

Uniform, identical bricks 20 cm long are stacked so that 4.0 cm of a brick extends beyond the brick beneath, as shown in •Fig. 8.12a. How many bricks can be stacked in this way before the stack falls over?

Solution. The stack will fall over when its center of mass is no longer above its base of support, which is the bottom brick. All of the bricks have the same mass, and the center of mass of each is located at its midpoint.

Taking the origin to be at the center of the bottom brick, the horizontal coordinate of the center of mass (or center of gravity) for the first two bricks in the stack is given by Eq. 6.19, where $m_1 = m_2 = m$ and x_2 is the displacement of the second brick:

$$X_{CM_2} = \frac{mx_1 + mx_2}{m + m}$$

$$= \frac{m(x_1 + x_2)}{2m} = \frac{x_1 + x_2}{2} = \frac{0 + 4.0\ cm}{2} = 2.0\ cm$$

Note that the masses of the bricks cancel out (since they are all the same). For three bricks,

xiv

How Does Physics Apply to the World Around You?

Applications are an important part of this text; we hope these features will spark your curiosity and answer some of your questions about how things work.

Insight boxes throughout the text show how physical principles discussed in the text apply to a wide variety of real-world situations, devices, and topics. A complete list of the Insight sections and other applications can be found on page xi.

Insight   Space Colonies and Artificial Gravity

Space colonies, long a staple of science fiction stories and films, are now actually being planned. However, humans cannot live for long periods of time in a "zero-gravity" environment (see Exercise 71) without detrimental physical effects.

FIGURE 1   Space colony and artificial gravity
It has been suggested that a space colony could be housed in a huge, rotating wheel as in this artist's conception. The rotation would supply the "artificial gravity" for the colonists.

It has been suggested that this problem could be avoided if the colony were housed in a huge rotating wheel, which would supply "artificial gravity" (Fig. 1).

As you know, centripetal force is necessary to keep an object in rotational circular motion. On the rotating Earth, that force is supplied by gravity, and we refer to it as weight. Because of our weight, we exert a force on the ground, and the reaction force (by Newton's third law) upward on our feet gives us the feeling of "having our feet on solid ground." In a rotating space colony, the situation is somewhat reversed. The rotating colony would supply the centripetal force on the inhabitants, which would be perceived as a weight sensation on the soles of our feet, or artificial gravity (Fig. 2a). Rotation at the proper speed would produce a simulation of normal gravity ($g = 9.80$ m/s^2) within the colony wheel. Note that in the colonists' world, "down" would be outward, toward the periphery of the space station, and "up" would always be inward, toward the axis of rotation.

The inhabitants would experience normal gravitational effects. For example, in Fig. 2a, suppose that the person simply let go of the ball. Viewed from outside the rotating wheel, you would say the ball has a tangential velocity v as a result of the rotation, and would go off in a straight line (Newton's first law) until it hits the side of the wheel (as indicated by the dashed line in the figure). However, the inhabitant rotates with the wheel, and from that perspective, the ball merely falls to the floor as it would on Earth. Think about it.

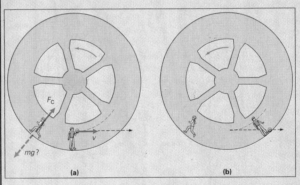

(a)   (b)

FIGURE 2   Rotating space colony
(a) In the frame of reference of someone in a rotating space colony, the centripetal force on the person would be perceived as weight sensation, or artificial gravity. Rotation at the proper speed would simulate normal gravity. From the point of view of an outside observer, a dropped ball would follow a tangential straight-line path. (b) A colonist, on the other hand, would observe the ball to fall downward as in a normal gravitational situation.

DEMONSTRATION 1 ■ Newton's First Law and Inertia

According to Newton's first law, an object remains at rest or in motion with a constant velocity unless acted upon by an unbalanced force.

(a) A pen is at rest on an embroidery hoop on top of a bottle.

(b) The hoop is struck sharply and accelerates horizontally. Because the friction between the pen and hoop is small and acts only for a short time, the pen does not move appreciably in the horizontal direction. However, there is now an unbalanced force acting vertically upon it—gravity.

(c) The pen falls into the bottle.

College Physics 3/e has seventeen sequential photographs of physics **demonstrations** in action—and in context of the chapter topic—to help you see and understand the physics being discussed.

Review Your Understanding, Practice Your Skills

The end of chapter review material gives you a brief summary of what you should know before going on to the exercises. Note that while we list the most **important terms, concepts,** and **equations,** we refer you back to the chapter discussion for explanations—again to encourage understanding over memorization.

Chapter Review

Important Terms

linear momentum 168
total linear momentum 168
conservation of linear
  momentum 174
impulse 178

impulse-momentum theorem 178
elastic collision 181
inelastic collision 181
completely inelastic collision 183
center of mass (CM) 188

center of gravity (CG) 191
jet propulsion 192
reverse thrust 194

Important Concepts

- Linear momentum of a particle is a vector and is defined as the product of mass and velocity.
- The total linear momentum of a system is the vector sum of the momenta of the individual particles.
- In the absence of a net external force, the total linear momentum is conserved.
- The impulse-momentum theorem relates impulse to the change in momentum.

- A collision is elastic if the total kinetic energy is conserved. Momentum is conserved in both elastic and inelastic collisions. In a completely inelastic collision, objects stick together after impact.
- The center of mass is the point at which all of the mass of an object or system may be considered to be concentrated. (The center of gravity is the point where all the weight may be considered to be concentrated.)

Important Equations

Linear Momentum:
$$\mathbf{p} = m\mathbf{v} \tag{6.1}$$

Total Linear Momentum of a System:
$$\mathbf{P} = \mathbf{p}_1 + \mathbf{p}_2 + \mathbf{p}_3 + \cdots = \Sigma_i \mathbf{p}_i \tag{6.2}$$

Newton's Second Law in Terms of Momentum:
$$\mathbf{F} = \frac{\Delta \mathbf{p}}{\Delta t} \tag{6.3}$$

Impulse-Momentum Theorem:
$$\text{Impulse} = \overline{\mathbf{F}}\,\Delta t = \Delta \mathbf{p} = m\mathbf{v} - m\mathbf{v}_\text{o} \tag{6.6}$$

Conditions for an Elastic Collision:
$$\begin{aligned} \mathbf{P}_\text{f} &= \mathbf{P}_\text{i} \\ K_\text{f} &= K_\text{i} \end{aligned} \tag{6.8}$$

Conditions for an Inelastic Collision:
$$\begin{aligned} \mathbf{P}_\text{f} &= \mathbf{P}_\text{i}, \\ K_\text{f} &< K_\text{i} \end{aligned} \tag{6.9}$$

Final Velocities in Head-On Two-Body Elastic Collisions: ($v_{2_\text{o}} = 0$):
$$v_1 = \left(\frac{m_1 - m_2}{m_1 + m_2}\right) v_{1_\text{o}} \tag{6.15}$$

$$v_2 = \left(\frac{2m_1}{m_1 + m_2}\right) v_{1_\text{o}} \tag{6.16}$$

Coordinate of the Center of Mass (using signs for directions):
$$X_{\text{CM}} = \frac{\Sigma_i m_i x_i}{M} \tag{6.19}$$

Exercises

Each section of **exercises** begins with review and conceptual questions that let you check your understanding of important information and concepts before proceeding to the quantitative problems.

6.1 Linear Momentum

1. Linear momentum has units of (a) N/m, (b) kg·m/s, (c) N/s, (d) all of these.

2. Momentum is (a) always conserved, (b) a scalar quantity, (c) a vector quantity that can be resolved into components, (d) unrelated to force.

3. A fan boat of the type used in swampy and marshy areas is shown in •Fig. 6.24. Explain the principle of its propulsion using momentum conservation.

4. Two objects have the same momentum. (a) Will they also have the same kinetic energy? (b) What can you say definitely about their motion?

•**FIGURE 6.24 Fan propulsion**
See Exercise 3.

68 ■■■ A stationary submerged submarine tracks an approaching submarine with sonar. A short pulse of ultrasound with a frequency of 400 kHz is sent toward the sub and returns 10 s later with an observed frequency of 412 kHz. (a) What is the speed of the approaching sub? Take the speed of sound in seawater to be 1200 m/s. [*Hint:* The ultrasound received by the moving sub is Doppler-shifted, and the moving sub acts as a moving source of sound with this shifted frequency. That is, the stationary sub receives a reflected pulse that is doubly Doppler-shifted.] (b) What is the approximate range of the approaching sub at the time of reflection?

14.6 Musical Instruments and Sound Characteristics

69 The human ear can best hear tones at (a) 1000 Hz, (b) 4000 Hz, (c) 6000 Hz, (d) any frequency

70 The quality of sound depends on its (a) waveform, (b) frequency, (c) speed, (d) intensity.

71 (a) After a snowfall, why does it seem particularly quiet? (b) Why do sounds in empty rooms sound hollow? (c) Why do people's voices sound fuller or richer when they sing in the shower?

72 The first three natural frequencies of an organ pipe are 136 Hz, 408 Hz, and 680 Hz. (a) Is the pipe an open or closed one? (b) Taking the speed of sound in air to be 340 m/s, find the length of the pipe.

73 Physically, why aren't the frets on a guitar evenly spaced?

74 It is possible for an open and a closed organ pipe of the same length to produce notes of the same frequency? Justify your answer.

75 ■ A closed organ pipe has a fundamental frequency of 528 Hz (a C note) at room temperature. What is the fundamental frequency of the pipe when the temperature is 0°C?

76 ■■ An open organ pipe has a length of 0.75 m. What would be the length of a closed organ pipe whose third harmonic ($m = 3$) is the same as the fundamental frequency of the open pipe?

77 ■■ A closed organ pipe has a length of 0.800 m. At room temperature, what are the frequencies of (a) the second harmonic and (b) the third harmonic?

78 ■■ An open organ pipe and a closed organ pipe both have lengths of 0.52 m at 20°C. What is the fundamental frequency of each pipe?

79 ■■ Suppose the pipes in Exercise 77 were at 10°C. Would the change in the fundamental frequency be the same for both pipes? Justify your answer. (Is there

another effect that might play a role here? Explain.)

80 ■■ An organ pipe has a fundamental frequency of 340 Hz in air at 0°C. When filled with another gas at the same temperature, the fundamental frequency is 992 Hz. What is the gas?

81 ■■ Show that the general equation for the Doppler effect for a moving source and a moving observer is given by

$$f_o = f_s \left(\frac{v \pm v_o}{v \pm v_s} \right)$$

using the sign convention of Eqs. 14.12 and 14.15.

82 ■■ A fire engine travels at a speed of 90 km/h with its siren emitting sound at a frequency of 500 Hz. What is the frequency heard by a passenger in a car traveling at 65 km/h in the opposite direction to the fire engine, (a) approaching it and (b) moving away from it? [*Hint:* See Exercise 67 and take the speed of sound to be 354 m/s (a hot day).]

83 ■■ A tuning fork with a frequency of 440 Hz is held above a resonance tube partially filled with water. Assuming that the speed of sound in air is 342 m/s, for what heights of the air column will resonances occur?

84 ■■■ A closed organ pipe is filled with helium. The pipe has a fundamental frequency of 660 Hz in air at 0°C. What is the fundamental frequency with the helium?

Additional Exercises

85 A jet flies at a speed of Mach 2.0. What is the half-angle of the conical shock wave formed by the aircraft?

86 Two sources of 440-Hz tones are located 6.97 m and 8.90 m from an observation point. If the air temperature that day is 15°C, how do the waves interfere at the point?

87 Derive Eqs. 14.18 and 14.19.

88 How long does it take sound to travel 3.5 km in air if the temperature is 30°C?

89 If a person standing 30.0 m from a 550-Hz point source walks 5.0 m toward the source, by what factor does the sound intensity change?

90 The value of the speed of sound in air at 100°C is listed in Table 14.1 as 387 m/s. Is this the value given by Eq. 14.1? If not, explain.

91 The speed of sound in air for normal *environmental* temperatures to a good approximation is $v = 331 + 0.6T_C$ m/s (Eq 14.1). Yet, a better approximation is given by $v = (331)\sqrt{1 + (T_C/273)}$ m/s. Show that this approximation is justified. [*Hint:* Consider a binomial expansion, $(1 \pm x)^n = 1 \pm nx + n(n-1)x^2/2! \pm \ldots$

Exercises **473**

Preface

The third edition of *College Physics* has undergone a major refinement process and has benefited in several key areas from Tony Buffa's participation as the new coauthor. Its approach to the teaching of physics, however, remains unchanged. We continue to believe there are two things that any introductory physics course must accomplish, regardless of its approach, emphasis, or what else it may try to do: impart an understanding of the basic physics principles, and enable a student to solve a variety of reasonable problems in topics presented in the text material.

These goals are linked. An understanding of physical principles is of limited use if it does not enable students to solve problems. Physics is a problem-solving science—and in the real world, students will be evaluated on their ability to produce correct answers on final exams or on the MCAT. Yet few people would consider that learning to solve problems by rote is the same thing as learning physics. Knowing and doing, insight and skill, must go hand in hand.

Any deficiency in meeting the first goal is likely to be obvious. Test scores quickly get the attention of both test takers and test graders. Low grades demoralize instructors while discouraging students who, quite understandably, conclude that physics is "too hard" for any but the phenomenally gifted. Deficiencies in meeting the second goal tend to be more subtle. Research in physics education has shown that a surprising number of students who do learn to solve typical problems well enough to pass examinations do so without ever arriving at a real understanding of the most elementary physical concepts. Such students often get high marks on exams, yet when asked to answer simple, qualitative questions designed to test their grasp of basic principles, they betray a surprising lack of insight. Simply put, they can solve quantitative problems and get the right answer, but they do not know why it is right.

Achieving Our Goals— The Features of the Third Edition

Most of the specific features of the text can be understood in light of these goals.

Conceptual Basis. We believe that giving students a more secure grasp of the principles will almost invariably enhance their problem-solving abilities. Central to this belief is a new approach to the development of problem-solving skills that stresses the understanding of basic concepts as the essential foundation, rather than mechanical and rote use of formulas. Throughout the writing of the book, we have organized discussions and incorporated pedagogical tools to help make sure conceptual insight drives the development of practical skills.

Concise Coverage. To maintain a sharp focus on essential concepts, a book should be concise, with a minimum of superfluous material. We have therefore kept the book brief with a strong emphasis on the basics. Topics of marginal interest have been avoided, along with ones that present formal or mathematical difficulties for students. Similarly, we have not wasted space deriving relationships when they shed no additional light on the principle involved. It is usually more important for students in a course such as this to understand what a relationship means and how it can be used rather than the mathematical or analytical techniques employed to derive it.

Integrated Learning Objectives. New to the third edition, specific learning objectives, placed at the beginning of each chapter section, help students structure their reading and facilitate review.

Suggested Problem-Solving Procedure. An extensive section (Section 1.7) provides a framework for thinking about problem solving. This section includes:

— an overview of problem-solving strategies;
— a 7-step procedure that is general enough to be applicable to most problems in physics but concrete enough to be easily used;
— three examples that illustrate the solution process, showing how the general procedure is applied in practice.

Problem-Solving Strategies and Hints. The initial treatment of problem solving is followed up throughout the text with an abundance of suggestions, tips, cautions, shortcuts, and useful techniques for solving specific kinds of problems. These help the students to apply general principles to specific contexts, as well as to avoid common pitfalls and misunderstandings.

"Learn by Drawing" features. New to the third edition, "Learn by Drawing" help students to use visualization as an aid in understanding concepts and solving problems.

Conceptual Examples. *College Physics* was the first physics text to include conceptual examples in addition to quantitative ones. We are glad to know this feature was so well received and have more than doubled their number in the third edition (from 22 to 48). Now highlighted for emphasis, these examples ask students to think about a physical situation and choose the correct prediction based on an understanding of relevant principles. The discussion that follows (Reasoning and Answer) explains clearly how the correct answer can be identified as well as why the others are wrong.

Follow-Up Exercises. Also well received in the second edition were conceptual follow-up exercises at the end of each conceptual example. The third edition now provides these with *all* worked examples, whether conceptual or quantitative. (Answers are given at the back of the book.)

More Explanation in Examples. Too many example solutions in other texts rely on formulas such as "From Equation 6.7 we have . . ." We have tried to make the solutions to in-text examples as clear, patient, and detailed as possible. The aim is not merely to show the student what equations to use but to explain the strategy being employed and the role of each step in the overall plan. Students are encouraged to learn the *why* of each step along with the *how*. This technique will make it easier for the student to apply the techniques being demonstrated to other problems that are not identical in structure.

Interactive Examples and Exercises. Many of the in-text examples and exercises can be dynamically simulated using either of two text-specific software programs with built-in "player" or "runtime" engines. The *Interactive Physics Player* for *College Physics 3/e* from Knowledge Revolution focuses on the mechanics section of the book and offers over 40 simulations while the *Physics Explorer Runtime* package for *College Physics 3/e* from Logal software has over 120 simulations covering mechanics, thermodynamics, electricity and magnetism, and light and optics. Both provide students with special insights into problem solving and allow them to perform their own experiments by varying parameters.

Integration of Conceptual and Quantitative Exercises. In order to help break down the artificial and ultimately counterproductive barrier between conceptual questions and quantitative problems, we have eliminated these categories in the end-of-chapter exercises. Instead, each section begins with a series of multiple choice and short answer questions that provide content review, test conceptual understanding, and ask students to reason from principles. The aim is to show students that the same kind of conceptual insight is required regardless of whether the desired answer involves words, equations, or numbers.

Insights. Applications are both intrinsically interesting and pedagogically useful. They satisfy the student's curiosity about the role of physics in the real world while reinforcing material presented in the text. The book includes 50 boxed Insight features, dealing with such diverse topics as The Greenhouse Effect; Automobile Air Bags; Fiber Optics; Magnetic Resonance Imaging; and Observing Black Holes.

Demonstrations. Photo sequences of 17 physics demonstrations bring physical principles to life, helping students understand that the information and equations on the page describe real-world phenomena.

Chapter Review. The end-of-chapter review material has been redesigned to support the dual aims of enhancing conceptual understanding and developing problem-solving skills. Each chapter review is now made up of three parts:

- Important Terms: A listing of the key terms introduced in the chapter that students should be able to define and explain, with page references.
- Important Concepts: A new summary of the key ideas of each chapter.
- Relationships for review: A listing of the major laws and mathematical relationships introduced, cross-referenced to the chapter. Specific applicability and limiting conditions are clearly stated for each expression.

Exercises. Each chapter ends with a wealth of exercises, organized by chapter section and ranked by general level of difficulty. In addition, the Exercises offer the following special features to help students refine both their conceptual understanding and their problem-solving skills:

—Each chapter integrates conceptual and quantitative exercises.
—Each chapter includes a supplemental section of *Additional Exercises* drawn from all sections of the chapter.
—Each section includes at least one set of paired problems that deal with similar situations. The first is solved in the Study Guide and the second allows the student to practice the same method (with the answer given at the back of the book).
—Answers to odd-numbered problems are at the back of the book.

Also New to the Third Edition

Revised Section on Electricity and Magnetism. Much of the section on Electricity and Magnetism (Chapters 15-21) has been rewritten, providing a clearer and more complete presentation of the concepts.

—The first chapter on electricity (formerly Chapter 15) has now been divided into two chapters: Chapter 15, "Electric Charge, Forces, and Fields," and Chapter 16, "Electric Potential, Energy, and Capacitance." Students now are given a slower and more thorough introduction to the basic principles of electricity.
—In an optional section, Chapter 15 now offers a qualitative, highly intuitive introduction to Gauss's Law.
—Unique "Learn by Drawing" boxes in Chapter 16, "Field Lines and Equipotentials," and Chapter 18, "Kirchhoff Plots," provide graphic reinforcement of important concepts.

The Absolutely Zero Tolerance for Errors Club (The AZTECs). Headed by Tony Buffa, this group of three accuracy checkers, along with the author of the *Instructor's Solutions Manual*, Bo Lou, and senior author Jerry Wilson, individually and independently worked all end-of-chapter exercises. Results were then collated and any discrepancies resolved by a "team" discussion. Then all data in the chapters, as well as the answers at the back of the book, were checked and rechecked in first and second pages. In addition, three other physics teachers reviewed the first pages for accuracy, paying special attention to the solutions for worked examples and answers to the follow-up exercises. Finally, two more instructors gave the second pages a final reading, again working with the authors to correct any remaining errors or ambiguities. While it is probably humanly impossible to produce a physics text with absolutely no errors, that was our goal and we worked hard to make the book as error-free as possible.

Supplements

The pedagogical value of the text is enhanced by a variety of supplements, developed to address the needs of both students and instructors.

For the Instructor

Annotated Instructor's Edition (0-13-505082-0)   The margins of the Annotated Instructor's Edition (AIE) contain an abundance of suggestions for classroom *demonstrations and activities*, along with *teaching tips* (points to emphasize, discussion suggestions, and common misunderstandings to avoid). In addition, the AIE contains:

- Marginal icons that identify each figure reproduced as a transparency in the Transparency Pack.
- Answers to end-of-chapter Exercises (following each exercise).
- Notes indicating an applicable video demonstration from the *Physics You Can See* video tape.

Instructor's Solutions Manual (0-13-505090-1)   Prepared by a new author, Bo Lou of Ferris State University, the *Instructor's Solutions Manual* supplies answers with complete, worked out solutions to all end-of-chapter exercises. Each solution has been checked for accuracy by a minimum of five instructors.

Test Item File (0-13-505124-X)   Fully revised by David Curott of the University of North Alabama, the *Test Item File* now offers approximately 2000 questions—a 42% increase over the last edition—including several new conceptual questions per chapter.

Prentice Hall Custom Test (Windows: 0-13-505173-8; MAC: 0-13-505157-6)   Based on the powerful testing technology developed by Engineering Software Associates, Inc. (ESA), the PH Custom Test allows instructors to create and tailor exams to their own needs. With the Online Testing Program, exams can also be administered online and data can then be automatically transferred for evaluation. A comprehensive desk reference guide is included, along with on-line assistance.

Transparency Pack (0-13-505132-0)   Contains 150 full-color acetates of text illustrations useful for class lectures. *Available upon adoption of the text.*

Physics You Can See Video Demonstrations (0-205-12393-7)   Each 2–5 minute segments demonstrates a classical physics experiment. Includes eleven segments such as "Coin & Feather" (*acceleration due to gravity*); "Monkey & Gun" (*rate of vertical free fall*); "Swivel Hips" (*force pairs*); and "Collapse a Can" (*atmospheric pressure*).

Presentation Manager CD-ROM (0-13-751629-0)   This new CD-ROM contains all the text art and videos from the *Physics You Can See* video tape as well as additional lab and demonstration videos and animations from the *Interactive Journey Through Physics CD-ROM,* also available from Prentice Hall (see below).

For the Student

Student Study Guide and Solutions Manual (0-13-505116-9) Significantly revised and expanded by Bo Lou of Ferris State University, the *Student Study Guide and Solutions Manual* presents chapter-by chapter reviews, chapter summaries, key terms, additional worked problems, and solutions to selected problems.

The New York Times "Themes of the Times" Program This innovative program, made possible through an exclusive partnership between Prentice Hall and *The New York Times*, is designed to bring current and relevant applications into the classroom. Through this program, adopters of *College Physics* 3/e are eligible to receive our unique "mini-newspapers," which bring together a collection of the latest and best physics articles from the highly respected pages of *The New York Times*. Updated annually. *Available free to qualified adopters up to the quantity of texts purchased. Contact your local representative for ordering.*

Physics Explorer Run-Time Software (Windows: 0-13-351719-5; MAC: 0-13-351701-2) Simulates problems and examples directly from the text with a wide range of experiments from among ten different models: One Body; Two Body; Gravity; Harmonic Motion; Waves; Ripple Tank; Diffraction; One Body Electrodynamics; AC/DC Circuits; and Electrostatics. Users can change values and parameters of their experiments, and observe the outcomes graphically. Extension Activities add additional scope, and built-in multiple choice Concept Checks assess understanding.

Interactive Physics II Player Software (Windows: 0-13-713736-2; MAC: 0-13-713422-3) This version of the highly acclaimed Interactive Physics II software animates and allows experimentation with select problems and examples directly from the text. Users can manipulate values, substitute variables, and observe how they affect the outcome through tools such as graphs and vectors.

Arco's *Physics for the MCAT* (0-13-505165-7) Adapted specifically for use with the text, this guide gives students planning to attend medical school the in-depth guidance needed for success on the physics section of the Medical College Admission Test (MCAT). Marginal annotations directly correlate material with *College Physics* 3/e.

Other Related Multimedia Materials

Interactive Physics II Player Workbook (Windows: 0-13-667312-0; MAC: 0-13-477670-4) Written by Cindy Schwarz of Vassar College, this highly interactive workbook/software package contains simulation projects of varying difficulty. Each contains a physics review, simulation details, hints, explanation of results, math help, and a self-test.

Interactive Mechanics (Windows: 0-13-192477-X; MAC: 0-13-261710-2) Written by Alejandro Garcia of San Jose State University, this is a supplement for any algebra-based or calculus-based physics course. Each 1 -1/2 hour experiment is similar to a traditional laboratory experiment in mechanics (such as a block sliding down an inclined plane), with computer simulations replacing lab equipment. Projects are self-contained with questions and problems provided for each. *Students must have access to Interactive Physics Software to run the simulations.*

Physics Explorer Student Edition (Windows: 0-13-351719-5; MAC: 0-13-351701-2) Fully interactive, text-independent software allows students to build their own experiments from the ground up. Experiments are built around the same ten models used in the *Run-Time Explorer*.

Interactive Journey Through Physics CD-ROM This highly interactive CD-ROM can be used as a stand-alone supplement, for any introductory physics course, or as a general reference tool. Through simulation, animation, video, and interactive problem solving, students can visualize difficult physics concepts in ways not available through the traditional lecture, lab, and text. Covers mechanics, electricity and magnetism, thermodynamics, and light and optics. The numerous analysis tools are easily navigated through a user-friendly interface.

Acknowledgments

We would like to acknowledge the generous assistance we received from many people during the preparation of this edition. First, our sincere thanks go to Professor Bo Lou of Ferris State University for his vital contributions and meticulous, conscientious help with checking problem solutions and answers, preparing the *Instructor's Solutions Manual* as well as the answer keys for the back of the book, and preparing a major revision of the *Student Study Guide and Solutions Manual*. We are similarly grateful to Dave Curott of the University of North Alabama for preparation of the *Test Item File*, as well as for his participation as an accuracy checker for all solutions to end-of-chapter exercises.

Indeed, all the members of AZTEC—Bo Lou, Dave Curott, William McCorkle (West Liberty State College) and Bela Karvaly (Lenoir Community College)—as well as the first and second page proof reviewers—Lattie Collins, Dennis Suchecki, Frederick Liebrand, Ronald Mowery, and especially Daryl Pedigo—deserve more than a special thanks for their tireless, timely, and extremely thorough review of all material in the book for scientific accuracy.

Dozens of other colleagues listed below helped us with reviews of the second edition to help make plans for the third edition, as well as reviews of manuscript as it was developed. We are indebted to them, as their thoughtful and constructive suggestions benefited the book greatly.

At Prentice Hall, the editorial staff continued to be particularly helpful. We are especially grateful to Dan Schiller, Senior Development Editor, for everything —his cheerful competence, experienced hand, insight, creativity, extreme attention to detail, and even his queries, all of which have helped make this book one of the most carefully crafted introductory physics texts available. Jennifer Carey, Project Manager, and Shari Toron, Assistant Managing Editor, kept the whole complex endeavor moving forward, while designers Amy Rosen and Heather Scott and artist Rolando Corujo made sure the ultimate physical presentation would be both visually engaging and clean and easy to use. We would also like to thank Kelly McDonald, Executive Marketing Manager, for her cheerful enthusiasm; Wendy Rivers for her work on the supplements; Alison Reeves, Executive Editor, and Pam Holland-Moritz, Editorial Assistant, for their help in coordinating all of these facets; and Paul Corey, Editor in Chief, for his support and encouragement.

In addition, I, (Tony Buffa), would like to especially acknowledge several people who were instrumental in my work as coauthor on this text. First, my coauthor Jerry Wilson deserves much thanks for his confidence in me. I sincerely hope I have brought the kinds of innovation and ideas he had in mind when he agreed to take me on board for this edition. As a skilled veteran writer, he was always there for me if I needed help. I also wish to acknowledge many fruitful discussions (and arguments!) with several Cal Poly colleagues of mine: Professors Joseph Boone, Ronald Brown, and Theodore Foster. They were a tremendous resource for help on several difficult points. At Prentice Hall I am deeply indebted to Tim Bozik and Ray Henderson for initially having enough confidence in me to recommend me to Jerry as a potential coauthor. Lastly, my wife Connie and daughters Jeanne and Julie deserve special mention. By unselfishly taking on some of my family responsibilities, they provided me with the element crucial to writing—time to think. Their constant encouragement and love was, and is, very much appreciated.

Finally, both of us would like to urge anyone using the book—student or instructor—to pass on to us any suggestions that you have for its improvement. We look forward to hearing from you.

Reviewers of the Third Edition

William Achor
West Maryland College

Arthur Alt
College of Great Falls

Zaven Altounian
McGill University

Frederick Anderson
University of Vermont

Charles Bacon
Ferris State University

Ali Badakhshan
University of Northern Iowa

Hugo Borja
Macomb Community College

Michael E. Browne
University of Idaho

David Bushnell
Northern Illinois University

Lyle Campbell
Oklahoma Christian University

Lowell Christensen
American River College

Lawrence Coleman
*University of California,
Davis*

Lattie F. Collins
East Tennessee State University

David Cordes
*Belleville Area Community
College*

William Dabby
Edison Community College

Donald Day
Montgomery College

Arnold Feldman
University of Hawaii

Donald Foster
Wichita State University

Frank Gaev
ITT-Ft. Lauderdale, Florida

Victor Keh
ITT-West Covina, California

James Kettler
Ohio University, Eastern Campus

Phillip Laroe
Carroll College

Bruce Layton
Mississippi Gulf Coast College

Federic Liebrand
Walla Walla College

Bryan Long
*Columbia State Community
College*

Michael LoPresto
Henry Ford Community College

Gary Motta
Lassen College

Ronald Mowrey
*Harrisburg Area Community
College*

Daryl Pedigo
Austin Community College

T.A.K.P. Pillai
*University of Wisconsin,
La Crosse*

Darden Powers
Baylor University

Earl Prohofsky
Purdue University

W. Steve Quon
Ventura College

Michael Ram
SUNY-Buffalo

William Rolnick
Wayne State University

Roy Rubins
University of Texas, Arlington

Anne Schmiedekamp
*Pennsylvania State University,
Ogontz Campus*

Mark Semon
Bates College

Thomas Sills
Wilbur Wright College

Michael Simon
*Housatonic Community
Technical College*

Dennis W. Suchecki
San Diego Mesa College

Frederick Thomas
Sinclair Community College

Arthur J. Ward
Nashville State Technical Institute

Arthur Wiggins
Oakland Community College

Reviewers of the Previous Edition

Charles Bacon
Ferris State College

William Berres
Wayne State University

Bennet Brabson
Indiana University

Michael Browne
University of Idaho

Philip A. Chute
*University of Wisconsin-
Eau Claire*

Lattie F. Collins
East Tennessee State University

David M. Cordes
*Belleville Area Community
College*

James R. Crawford
Southwest Texas State University

J.P. Davidson
University of Kansas

Richard Delaney
College of Aeronautics

Rober J. Foley
University of Wisconsin-Stout

Donald R. Franceschetti
Memphis State University

Simon George
California State-Long Beach

Barry Gilbert
Rhode Island College

Richard Grahm
Ricks College

Tom J. Gray
University of Nebraska

Douglas Al Harrington
Northeastern State University

Al Hilgendorf
University of Wisconsin-Stout

Joseph M. Hoffman
Frostburg State University

Omar Ahmad Karim
*University of North Carolina-
Wilmington*

S.D. Kaviani
El Camino College

Rubin Laudan
Oregon State University

Bruce A. Layton
*Mississippi Gulf Coast
Community College*

Robert March
University of Wisconsin

John D. McCullen
University of Arizona

J. Ronald Mowrey
*Harrisburg Area Community
College*

R. Daryl Pedigo
Austin Community College

T.A.K. Pillai
University of Wisconsin-La Crosse

Darden Powers
Baylor University

Donald S. Presel
*University of Massachusetts-
Dartmouth*

E.W. Prohofsky
Purdue University

Dan R. Quisenberry
Mercer University

George Rainey
*California State Polytechnic
University*

William Riley
Ohio State University

R.S. Rubins
*The University of Texas at
Arlington*

Sid Rudolph
University of Utah

Ray Sears
University of North Texas

Frederick J. Thomas
Sinclair Community College

Pieter B. Visscher
University of Alabama

John C. Wells
Tennessee Technical University

Dean Zollman
Kansas State University

1 Units and Problem Solving

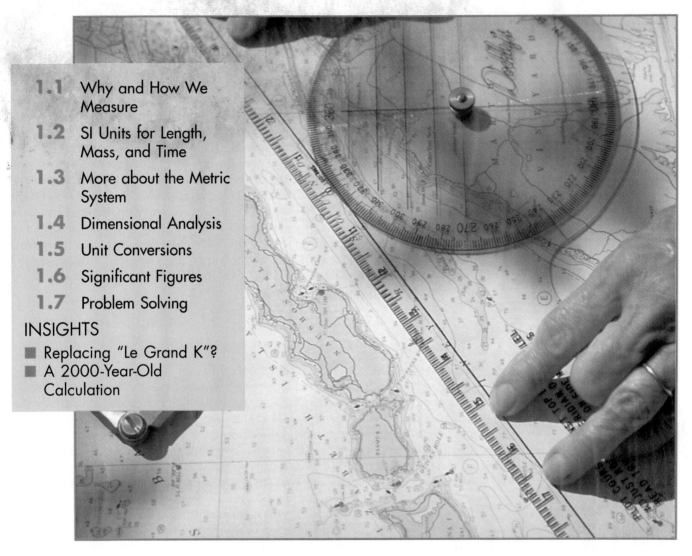

Not everyone has had the experience of laying out a course on a marine chart, as in the photo above. But most people have done something quite similar with a road map—plotting the shortest distance between two cities, measuring it, and estimating how long the trip will take. You might be planning a trip of 500 miles; traveling at an average speed of 50 mi/h, the trip would then take about 10 hours. Knowing your car's fuel economy, you might also estimate the amount of gas you expect to need and what it is likely to cost. (At 25 mpg and $1.25 per gallon, how much should you be prepared to spend?)

Measurement and problem solving are not confined to science—they are part of our lives. But they play a particularly central role in our attempts to describe and understand the physical world, as we shall see in this chapter.

1.1 Why and How We Measure

Objective: To be able to distinguish standard units and systems of units.

Imagine that someone is giving you directions to her house. Would you find it helpful to be told, "Drive along Elm Street for a little while and turn right at one of the lights. Then keep going for quite a long way." Or suppose that you were baking a cake. How would you like to use a recipe that said, "Beat up some eggs, add a little sugar and some butter and a *whole lot* of flour, and bake for a while in a pretty hot oven." Similarly, would you want to deal with a bank that sent you a statement at the end of the month saying, "You still have some money left in your account. Not a great deal, though."

Measurement is important to all of us. It is one of the concrete ways in which we deal with our world. This is particularly true in physics. *Physics is concerned with the description and understanding of nature*, and measurement is one of its most important tools.

There are certainly ways of describing the physical world that do not involve measurement. For instance, we might talk about the color of a flower or a dress. But color perception is subjective: it may vary from one person to another. Indeed, many people suffer from color blindness and cannot tell certain colors apart. Light can also be described in terms of wavelengths and frequencies. Different wavelengths are associated with different colors because of the physiological response of our eyes to light. But unlike the sensations or perceptions of color, the wavelengths can be measured. They are the same for everyone. In other words, they are *objective*. *Physics attempts to describe nature in an objective way through measurements.*

Standard Units

Measurements are expressed in unit values, or units. As you are probably aware, a large variety of units are used to express measured values. Some of the earliest units of measurement, such as the foot, were originally referenced to parts of the human body. (Even today, the hand is still used as a unit to measure the heights of horses.) If a unit becomes officially accepted, it is called a **standard unit**. Traditionally, a government or international body establishes standard units.

A group of standard units and their combinations is called a **system of units**. Two major systems of units are in use today—the metric system and the British system. The latter is still widely used in the United States but is gradually being phased out in favor of the metric system, which is used throughout most of the world.

Different units in the same system or units from different systems can be used to describe the same thing. For example, your height can be expressed in inches, feet, centimeters, meters—or even miles, for that matter (although this would not be a very *convenient* unit). It is always possible to convert from one unit to another, however, and this is sometimes necessary.

You may be surprised at how easy it is to determine very basic facts about our world through simple measurements and calculations that almost anyone can perform. Some 2200 years ago, a Greek scientist, using only some basic facts of geometry, was able to calculate the size of the Earth with considerable accuracy (see the Insight on p. 3).

Modern science may utilize more sophisticated mathematics and technology, but the approach is still the same. Much of our knowledge rests on a foundation of ingenious measurement and simple calculation. In a sense, mathematics is the language of physics. Principles can be expressed in words, but are stated most clearly—and usually most concisely—in mathematical terms. One equation is worth a thousand words, so to speak. But, in order to use equations to do calcu-

Standard unit

System of units

Insight A 2000-Year-Old Calculation: The Earth's Circumference

One of the first measurements of the circumference of the Earth was done sometime before 200 B.C. by the Greek scientist Eratosthenes (276–196 B.C.). His method, although quite simple, shows a great deal of insight and ingenuity. The basis for Eratosthenes' technique lies in the understanding that because the distance between the Earth and the Sun is so great compared to our planet's size, the Sun's rays arriving at the Earth are virtually parallel. Of course, the Sun would have to be at an infinite distance for the rays to be exactly parallel, but the divergence is so small that it is insignificant for practical purposes.

Eratosthenes lived in Alexandria, Egypt. He had learned that on the first day of summer at the town of Syene (now Aswan), some 5000 stadia to the south of Alexandria, the noon Sun was directly overhead. (A stadium, plural stadia, was a Greek unit of length believed to be equal to about $\frac{1}{6}$ km.) The Sun's position overhead was apparent because deep wells at Syene were lighted all the way to the bottom, and a vertical stick would cast no shadow at noon on that day.

However, at Alexandria, some 800 km to the north, Eratosthenes observed that a vertical stick *did* cast a noon-day shadow with a length $\frac{1}{8}$ that of the stick. This corresponds to an angle of 7.1° between the Sun's rays and the vertical (see Fig. 1 insert). Since lines extending through the well at Syene and through the vertical stick at Alexandria intersect at the Earth's center, the angular distance between these locations is also 7.1° (Fig. 1).

Then Eratosthenes reasoned that the 800 km arc length was to the 7.1° angle as the Earth's total circumference was to 360°, the number of degrees in a complete circle. That is, in ratio form,

$$\frac{\text{circumference}}{360°} = \frac{800 \text{ km}}{7.1°}$$

and

$$\text{circumference} = \frac{360°}{7.1°} \times 800 \text{ km} = 40,600 \text{ km}$$

The Earth's radius (R) can easily be calculated from the circumference (c), since $c = 2\pi R$:

$$R = \frac{c}{2\pi} = \frac{40,600 \text{ km}}{2\pi} = 6,460 \text{ km}$$

This is quite close to the modern value of 6,378 km for the Earth's equatorial radius (Appendix I). We cannot be certain of the accuracy of Eratosthenes' result because several different Greek stadia of various lengths were in use at that time. (You can see why it is important to have *standard* units.) Even so, his method was sound, and the circumference of the Earth was measured over 2000 years ago.

FIGURE 1 A 2000-year-old calculation
Observations on the incidence of sunlight allowed the Greek scientist Eratosthenes to calculate the circumference and radius of the Earth sometime before 200 B.C. (See Insight text for description.)

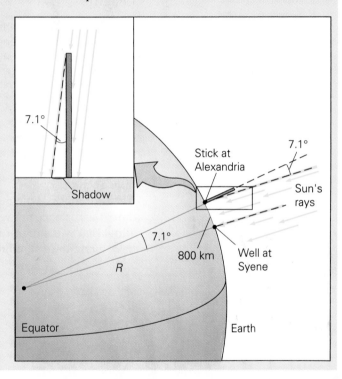

lations, you must have the proper background and understanding. Toward this end, you will begin your study of physics with a look at some basic quantities and units and an approach to problem solving. A firm understanding of these will benefit you throughout this entire course of study.

1.2 SI Units for Length, Mass, and Time

Objectives: To be able to **(a) describe the SI, and (b) specify the references for the three main base quantities of this system.**

Length, mass, and time are fundamental physical quantities that describe a great many objects and phenomena. In fact, the topics of mechanics (the study of motion and force) covered in the first part of this book require *only* these physical quantities. The system of units used by scientists to represent these and other quantities is based on the metric system.

Historically, the metric system was the outgrowth of proposals for a more uniform system of weights and measures in France during the seventeenth and eighteenth centuries. The modernized version of the metric system is called the **International System of Units**, officially abbreviated as **SI** (from the French *Système International des Unités*).

The SI includes *base quantities* and *derived quantities*, which are described by base units and derived units, respectively. **Base units** such as the meter and kilogram are easily represented by standards. Other quantities that may be expressed in terms of combinations of base units are called **derived units**. (As a familiar analogy, think of how we commonly measure the length of a trip in miles and the time in hours. We therefore express how fast we travel in the derived unit of miles/hour, which represents distance per unit of time, or length/time.)

One of the refinements of the SI was the adoption of new standard references for some base units, including those for length and time.

Length

Length is the base quantity used to measure distances or dimensions in space. We commonly say that length is the distance between two points. But the distance between any two points depends on how space is traversed, which may be in a straight or a curved line.

The SI unit of length is the **meter** (**m**). The meter was originally defined as 1/10,000,000 of the distance from the North Pole to the Equator along a meridian running through Paris (•Fig. 1.1a). A portion of this meridian between Dunkirk, France, and Barcelona, Spain, was surveyed to establish the standard length, which was assigned the name *metre* (from the Greek word *metron*, meaning "a measure"; the American spelling is *meter*). A meter is 39.37 inches—slightly longer than a yard.

The length of the meter was initially preserved in the form of a material standard: the distance between two marks on a metal bar (made of a platinum-iridium alloy) that was stored under controlled conditions. However, it is not desirable to have a reference standard that changes with external conditions, such as temper-

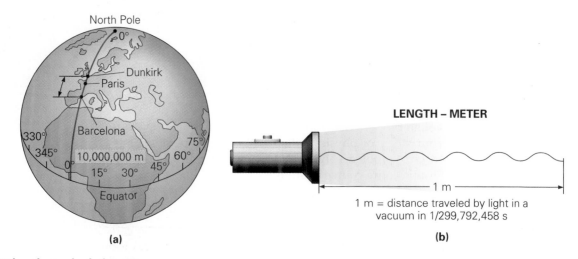

(a) **(b)**

•FIGURE 1.1 The SI length standard: the meter
(a) The meter was originally defined as 1/10,000,000 (or 10^{-7}) of the distance from the North Pole to the Equator along a meridian running through Paris, of which a portion was measured between Dunkirk and Barcelona. A metal bar (called the Meter of the Archives) was constructed as a standard. (b) The meter is currently defined in terms of the speed of light.

ature. Research provided more accurate standards. In 1983, the meter was redefined in terms of an unvarying property of light: the length of the path traveled by light in a vacuum during an interval of 1/299,792,458 of a second (Fig. 1.1b). Another way of saying this is that light travels 299,792,458 meters in a second, or that the speed of light in a vacuum is defined to be 299,792,458 meters per second. Note that the length standard is referenced to time, which is one of the most accurate measurements.

Mass

Mass is the base quantity used to describe amounts of matter. The more massive an object, the more matter it contains. (We will encounter more precise definitions of mass in Chapters 4 and 7.)

The SI unit of mass is the **kilogram** (**kg**). The kilogram was originally defined in terms of a specific volume of water, but is now referenced to a specific material standard: the mass of a prototype platinum-iridium cylinder kept at the International Bureau of Weights and Measures in Sèvres, France. The United States has a duplicate of the prototype cylinder, which serves as the standard mass for this country (•Fig. 1.2). (The kilogram may eventually be referenced to something less subject to time and chance than a piece of metal—see the Insight on p. 6.)

You may have noticed that the phrase "weights and measures" is generally used instead of "masses and measures." In the SI, mass is the base quantity, but in the familiar British system, weight is used instead to describe amounts of mass—for example, weight in pounds instead of mass in kilograms. The weight of an object is the gravitational attraction that the Earth exerts on an object. For example, when you weigh yourself on a scale, your weight is a measure of the downward gravitational force exerted on you by the Earth. We can use weight in this way because near the surface of the Earth, mass and weight are directly proportional to each other: they differ only by a particular constant.

But treating weight as the base quantity creates some problems. A base quantity is naturally most useful if its value is the same everywhere. This is the case with mass—an object has the same mass, or amount of matter, regardless of its location. *But this is not true of its weight.* For example, the weight of an object on the Moon is less than its weight on Earth. This is because the Moon is smaller (less massive) than the Earth, and the gravitational attraction (weight) for an object on the Moon is less than that on Earth. That is, an object with a given amount of mass has a particular weight on Earth, but on the Moon, the same amount of mass will weigh only about one-sixth as much. Similarly, the weight of an object would vary on the surfaces of different planets.

For now, keep in mind that *weight is related to mass, but they are not the same.* It is much more useful to take mass as the base quantity, as the SI does. Base quantities remain the same regardless of where they are measured, under normal or stationary conditions.* The distinction between mass and weight will be more fully explained in a later chapter. Our discussion until then will be chiefly concerned with mass.

Time

Time is a difficult concept to define. (Try it.) A common definition is that time is the forward flow of events. This is not so much a definition as an observation that time has never been known to run backwards (as it might appear to do when you view a film run backwards in a projector). Time is sometimes said to be a fourth dimension, accompanying the three dimensions of space. Thus, if something exists in space, it also exists in time. In any case, events can be used to mark time measurements. The events are analogous to the marks on a meterstick used for length measurements.

*We will see in Chapter 26 that when objects move at high speeds, approaching the speed of light, their masses increase, as the theory of relativity predicts.

MASS - KILOGRAM

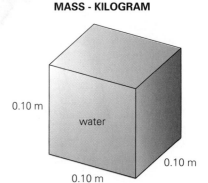

0.10 m

water

0.10 m

0.10 m

(a)

(b)

•FIGURE 1.2 **The SI mass standard: the kilogram**
(a) The kilogram was originally defined in terms of a specific volume of water, that of a cube 0.10 m on a side, thereby associating the mass standard with the length standard. The standard kilogram is now defined by a metal cylinder. **(b)** Prototype Kilogram No. 20 is the national standard for mass for the United States. It is kept under a double bell jar in a vault at the National Institute of Standards and Technology (NIST, formerly the National Bureau of Standards) in Washington, D.C.

The kilogram is the only SI base unit to be defined by a physical artifact or object. This is the platinum-iridium cylinder kept at the International Bureau of Weights and Measures in Sèvres, France. Nicknamed "Le Grand K," it is the *one and only* true kilogram; all others are copies. The other base units—length, time, and electric charge (current)—are defined in terms of constants of nature, such as the speed of light or atomic vibrations.

The mass of Le Grand K, for unknown reasons, has varied as much as 50 parts per billion over the last 100 years. The variation could be due to outgassing, absorption, or dirt accumulation and cleaning. Using a physical object as a standard has other disadvantages. The standard kilogram is difficult to reproduce. There is also the danger that it could be damaged or destroyed. As a result, there is an international effort to redefine the kilogram in such a way that it has an ab-

solute value, reproducible by national laboratories around the world.

The current research involves defining the kilogram in terms of Avogadro's number, which is the number of molecules in a particular amount (a "mole") of a substance. This number, which is equal to the number of molecules 22.4 L of a gas at a fixed temperature and pressure, is approximately 6.023×10^{23}. Basically, Avogadro's number relates macroscopic masses to atomic measurements.

Several complicated methods to measure Avogadro's number with sufficient accuracy are being explored. The uncertainty of an appropriate method must be very small—on the order of one part in a billion. It is estimated that it will take a decade or so before a new official definition not based on a physical object is adopted. Until then, Le Grand K will set the standard.

The SI unit of time is the **second (s)**. The solar "clock" was originally used to define the second. A solar day is the interval of time that elapses between two successive crossings of the same meridian by the Sun, at its highest point in the sky at that meridian (Fig. 1.3a). A second was fixed as 1/86,400 of this apparent solar day (1 day = 24 h = 1440 min = 86,400 s). However, the elliptical path of the Earth's motion around the Sun causes apparent solar days to vary in length.

As a more precise standard, an average, or mean, solar day was computed from the lengths of the apparent solar days during a solar year. In 1956, the second was referenced to this mean solar day. But the mean solar day is not exactly the same for each yearly period because of minor variations in the Earth's motions and a steady slowing of its rate of rotation due to tidal friction. So scientists kept looking for something better.

In 1967, an atomic standard was adopted as a more precise reference. Currently, the second is defined by the radiation of the cesium-133 atom. This particular cesium atom is used in an atomic "clock" that maintains our time standard. The atomic clock doesn't look much like an ordinary clock (Fig. 1.3b,c).

SI Base Units

The complete SI has seven base quantities and units. In addition to the meter, kilogram, and second for (1) length, (2) mass, and (3) time, there are the ampere (A) for (4) electric current (charge), the kelvin (K) for (5) temperature, the mole (mol) for (6) amount of substance, and the candela (cd) for (7) luminous intensity (Table 1.1). This is thought to be the smallest number of base quantities needed for a full description of everything observed or measured in nature. There are also two supplemental units for angular measure: the two-dimensional plane angle and the three-dimensional solid angle. There has not been general agreement about whether these geometric units are base or derived.

TABLE 1.1 The Seven Base Units of the SI	
Name of Unit (abbreviation)	*Property Measured*
meter (m)	length
kilogram (kg)	mass
second (s)	time
ampere (A)	electric current
kelvin (K)	temperature
mole (mol)	amount of substance
candela (cd)	luminous intensity

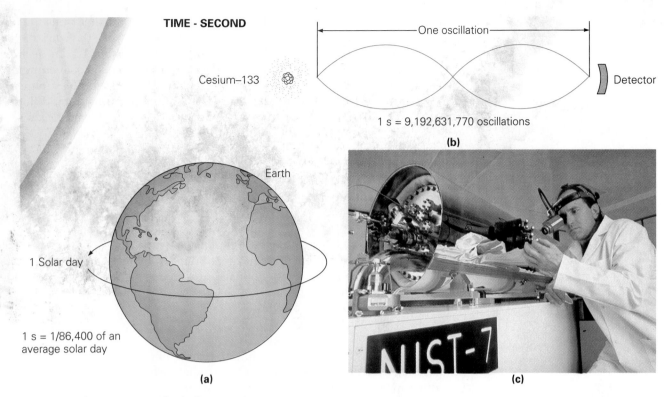

FIGURE 1.3   The SI time standard: the second
(a) The second was originally defined in terms of the average solar day. (b) It is currently defined in terms of the frequency of the radiation associated with an atomic transition. (c) This atomic "clock" is the primary frequency standard at the National Institute of Standards and Technology. The device keeps time with an accuracy of about three millionths of a second per year.

1.3   More about the Metric System

Objectives: To be able to use common (a) metric prefixes and (b) nonstandard metric units.

The metric system involving the standard units of length, mass, and time now incorporated in the SI was once called the **mks system** (for *m*eter-*k*ilogram-*s*econd). Another metric system that has been used in dealing with relatively small quantities is the **cgs system** (for *c*entimeter-*g*ram-*s*econd). In the United States, the system still generally in use is the British (or English) engineering system, in which the standard units of length, mass, and time are foot, slug, and second, respectively. You may not have heard of the slug, because, as we mentioned earlier, gravitational force (weight) is commonly used instead of mass—pounds instead of slugs—to describe quantities of matter. As a result, the British system is sometimes called the **fps system** (*f*oot-*p*ound-*s*econd).

The metric system is predominant throughout the world and is coming into increasing use in the United States. Primarily because of its mathematical simplicity, it is the preferred system of units for science and technology. SI units are used throughout most of this book. All quantities can be expressed in SI units. However, some units from other systems are accepted for limited use as a matter of practicality—for example, the time unit of hour and the temperature unit of degree Celsius. British units will sometimes be used in the early chapters for comparison purposes, since these units are still employed in everyday activities and for many practical applications.

The increasing worldwide use of the metric system means that you should be familiar with it. One of the greatest advantages of the metric system is that it is a decimal, or base-10, system. This means that larger or smaller units are obtained by multiplying or dividing a base unit by powers of 10. A list of the various multiples and corresponding prefixes is given in Table 1.2. In decimal measurements, the prefixes milli-, centi-, and kilo- are the ones most commonly used. The decimal characteristic makes it convenient to change measurements from one size metric unit to another. With the familiar British system, different factors must be used, such as 16 for converting pounds to ounces and 12 for converting feet to inches.

You are already familiar with one base-10 system—U.S. currency. Just as a meter can be divided into 10 decimeters, 100 centimeters, or 1000 millimeters, the "base unit" of the dollar can be broken down into 10 "decidollars" (dimes), 100 "centidollars" (cents), or 1000 "millidollars" (tenths of a cent, or mils, used in figuring property taxes and bond levies). Since the metric prefixes are all powers of 10, there are no metric analogs for quarters or nickels.

The official metric prefixes can help eliminate confusion. In the United States, a billion is a thousand million (10^9); in Great Britain, a billion is a million million (10^{12}). The use of metric prefixes eliminates any confusion, since giga- indicates 10^9 and tera- stands for 10^{12}.

Volume

In the SI, the standard unit of volume is the cubic meter (m^3)—the three-dimensional derived unit of the meter base unit. Since this is a rather large unit, it is often more convenient to use the nonstandard unit of volume (or capacity) of a cube 10 cm (centimeters), or 1 dm (decimeter), on a side (•Fig. 1.4). This volume was given the name *litre*, which is spelled **liter** (L) in the United States. The volume of a liter is 1000 cm^3 (10 cm × 10 cm × 10 cm). Since 1 L =

The liter

TABLE 1.2	Multiples and Prefixes for Metric Units*				
Multiple†	Prefix (and Abbreviation)	Pronunciation	Multiple†	Prefix (and Abbreviation)	Pronunciation
10^{24}	yotta- (Y)	yot'ta (*a* as in *a*bout)	10^{-1}	deci (d)	des'i (as in *deci*mal
10^{21}	zetta- (Z)	zet'ta (*a* as in *a*bout)	10^{-2}	centi- (c)	sen'ti (as in *senti*mental)
10^{18}	exa- (E)	ex'a (*a* as in *a*bout)	10^{-3}	milli- (m)	mil'li (as in *mili*tary)
10^{15}	peta- (P)	pet'a (as in *peta*l)	10^{-6}	micro- (μ)	mi'kro (as in *micro*phone)
10^{12}	tera- (T)	ter'a (as in *terra*ce)	10^{-9}	nano- (n)	nan'oh (*an* as in *ann*ual)
10^9	giga- (G)	ji'ga (*ji* as in *jiggle*, *a* as in *a*bout)	10^{-12}	pico- (p)	pe'ko (*peek-oh*)
10^6	mega- (M)	meg'a (as in *mega*phone)	10^{-15}	femto- (f)	fem'toe (*fem* as in *feminine*)
10^3	kilo- (k)	kil'o (as in *kilo*watt)	10^{-18}	atto- (a)	at'toe (as in *anatomy*)
10^2	hecto- (h)	hek'to (*heck-toe*)	10^{-21}	zepto- (z)	zep'toe (as in *zeppelin*)
10	deka- (da)	dek'a (*deck* plus *a* as in *a*bout)	10^{-24}	yocto- (y)	yock'toe (as in *sock*)

*For example, 1 gram (g) multiplied by 1000 (10^3) is 1 kilogram (kg), or multiplied by 1/1000 (10^{-3}) is 1 milligram (mg).

†The most commonly used prefixes are printed in color.

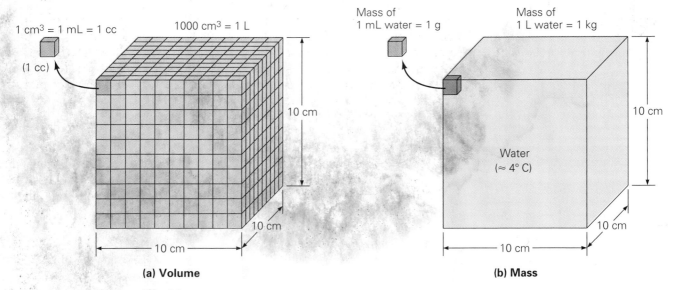

1 cm³ = 1 mL = 1 cc

(1 cc)

1000 cm³ = 1 L

10 cm

10 cm

10 cm

Mass of
1 mL water = 1 g

Mass of
1 L water = 1 kg

Water
(≈ 4° C)

10 cm

10 cm

10 cm

(a) Volume

(b) Mass

•**FIGURE 1.4   The liter and the kilogram**
Other metric units are derived from the meter. **(a)** A unit of volume (capacity) was taken to be the volume of a cube 1.0 dm (10 cm or 0.10 m) on a side and was given the name liter. **(b)** The mass of a liter of water (at its maximum density) was defined to be 1 kg. Note that the decimeter cube contains 1000 cm³, or 1000 mL. Thus, 1 cm³, or 1 mL, of water has a mass of 1 gram.

1000 mL (milliliters), it follows that $1\text{ mL} = 1\text{ cm}^3$. (Cubic centimeter is sometimes abbreviated cc, particularly in chemistry and biology.)

Recall that the standard unit of mass, the kilogram, was originally defined to be the mass of a cubic volume of water 10 cm or 0.10 m on a side, or the volume of one *liter* of water (at its maximum density, ≈ 4°C). The density of water does not vary greatly with temperature, so we can say with fairly good accuracy that *1 L of water has a mass of 1 kg*, an approximation that we will use throughout this book. Also, since 1 kg = 1000 g and 1 L = 1000 cm³, then 1 cm³ (or 1 mL) of water has a mass of 1 g.

You are probably more familiar with the liter than you think. The use of the liter is becoming quite common in some instances in the United States as •Fig. 1.5 indicates.

Note: Liter is sometimes abbreviated with a lowercase "ell" (l), but a capital "ell" (L) is preferred in the United States so that the abbreviation is less likely to be confused with the numeral one. (Isn't 1 L clearer than 1 l?)

EXAMPLE 1.1 ■ THE METRIC TON (OR TONNE): ANOTHER UNIT OF MASS

As we have seen, the metric unit of mass was originally related to the length standard, with a liter (1000 cm³) of water having a mass of 1 kg. The standard metric unit of volume is the m³ (cubic meter), and this volume of water was used to define a larger mass unit called the *metric ton* (or *tonne*, as it is sometimes spelled). A metric ton is equivalent to how many kilograms?

Solution.   Each liter of water has a mass of 1 kg, so we merely need to find how many liters there are in 1 m³. Since there are 100 cm in a meter, we can visualize a cubic meter as simply a cube with sides 100 cm in length. Therefore, a cubic meter (1 m³) has a volume of $10^2\text{ cm} \times 10^2\text{ cm} \times 10^2\text{ cm} = 10^6\text{ cm}^3$. Since 1 liter = 10^3 cm³, there must be $10^6\text{ cm}^3/10^3\text{ cm}^3 = 1000$ liters in 1 m³. Thus, 1 metric ton is equivalent to 1000 kg.

Note that this entire line of reasoning can be expressed very concisely in a single calculation:

$$\frac{1\text{ m}^3}{1\text{ L}} = \frac{100\text{ cm} \times 100\text{ cm} \times 100\text{ cm}}{10\text{ cm} \times 10\text{ cm} \times 10\text{ cm}} = 1000$$

Follow-up Exercise.   As presented later in the chapter, 1 kg has an equivalent weight of 2.2 lb. Using this relationship, how does the metric ton compare with the British ton? (*Answer may be found in the Answers to Follow-up Exercises section at the back of the book.*)

•**FIGURE 1.5   Two, three, one, and one-half liters**
The liter is now a common volume unit for soft drinks.

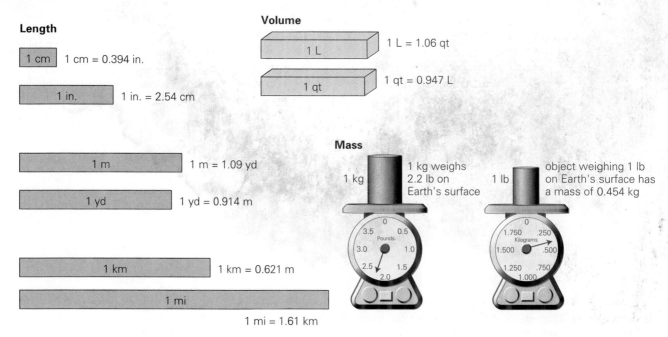

Length

1 cm   1 cm = 0.394 in.

1 in.   1 in. = 2.54 cm

1 m   1 m = 1.09 yd

1 yd   1 yd = 0.914 m

1 km   1 km = 0.621 m

1 mi

1 mi = 1.61 km

Volume

1 L   1 L = 1.06 qt

1 qt   1 qt = 0.947 L

Mass

1 kg   1 kg weighs 2.2 lb on Earth's surface

1 lb   object weighing 1 lb on Earth's surface has a mass of 0.454 kg

•**FIGURE 1.6 Comparison of some SI and British units**
The bars illustrate the relative magnitudes of each pair of units. (The comparison scales are different in each case.)

Because the metric system is coming into increasing use in this country, you may find it helpful to have an idea of how metric and British units compare. The relative sizes of some units are illustrated in •Fig. 1.6. The mathematical conversion from one unit to another will be discussed shortly.

1.4   Dimensional Analysis

Objectives: To be able to explain the advantages of and apply (a) dimensional analysis and (b) unit analysis.

The fundamental, or base, quantities used in physical descriptions are called *dimensions*. For example, length, mass, and time are dimensions. You could measure the distance between two points and express it in units of meters, centimeters, or feet. In any case, the quantity would have the dimension of length.

It is common to express dimensional quantities by bracketed symbols, such as [L], [M], and [T] for length, mass, and time, respectively. Derived quantities are combinations of dimensions; for example, velocity (v) has the dimensions [L]/[T] (think of miles/hour or meters/second), and volume (V) has the dimensions [L] × [L] × [L], or [L^3]. Addition and subtraction can only be done with quantities that have the same dimensions, for example, 10 s + 20 s = 30 s, or [T] + [T] = [T].

Dimensional analysis

Dimensional analysis is a procedure by which the dimensional consistency of any equation may be checked. You have used equations and know that an equation is a mathematical equality. Since physical quantities used in equations have dimensions, *the two sides of an equation must be equal not only in numerical value, but also in dimensions. And dimensions can be treated as algebraic quantities.* That is, they can be multiplied or divided. For example, suppose the quantities in an equation have the dimension of length: 3.0 m × 4.0 m = 12 m². This is [L] × [L] = [L²] dimensionally; thus both sides of the equation are *numerically* and *dimensionally* equal.

One of the major advantages of dimensional analysis is its usefulness for checking whether an equation that has been derived or is being used in solving a problem has the correct form. For example, suppose that you think you can solve a problem about the distance an object travels using the equation $x = at$, where x represents distance, or length, and has the dimension [L]; a represents acceleration, which, as we will see in Chapter 2, has the dimensions $[L]/[T^2]$; and t represents time and has the dimension [T]. (The dimensions and units of some common quantities are given in Table 1.3.) First, before trying to use numbers with the equation, you can check to see if it is dimensionally correct:

$$x = at$$

is dimensionally

$$[L] = \frac{[L]}{[T^2]} \times [T] \text{ or } [L] = \frac{[L]}{[T]}$$

which is obviously not true. Thus, $x = at$ *cannot* be a correct equation.

Dimensional analysis will tell you if an equation is incorrect, but a dimensionally consistent equation may *not* correctly express the real relationship of quantities. For example,

$$x = at^2$$

is dimensionally

$$[L] = \frac{[L]}{[T^2]} \times [T^2] \text{ or } [L] = [L]$$

This equation is dimensionally correct. But, as you will see in Chapter 2, it is not physically correct. The correct form of the equation—both dimensionally and physically—is $x = \frac{1}{2}at^2$. The fraction $\frac{1}{2}$ has no dimensions; it is a dimensionless number.

Using symbols for dimensions is fine, but in actual practice it is often more convenient to use units, which are often written into an equation along with the numerical values of the quantities. Units can also be treated as algebraic quantities, and like symbols can be canceled. Using units instead of symbols in dimensional analysis is called **unit analysis**. Obviously, if an equation is correct by unit analysis, it must be dimensionally correct. The following example demonstrates the use of unit analysis.

Unit analysis

TABLE 1.3 Dimensions and Units		
mass	[M]	kg
time	[T]	s
length	[L]	m
area	[L²]	m²
volume	[L³]	m³
velocity (*v*)	$\frac{[L]}{[T]}$	$\frac{m}{s}$
acceleration (*a* or *g*)	$\frac{[L]}{[T^2]}$	$\frac{m}{s^2}$

EXAMPLE 1.2 ■ CHECKING WITH UNITS: DIMENSIONAL ANALYSIS

A professor puts two equations on the board: (a) $v = v_o + at$ and (b) $x = v/2a$, where x is distance in meters (m), v and v_o are velocities in meters/second (m/s), a is acceleration in (meters/second)/second or meters/second² (m/s²), and t is time in seconds (s). Are the equations dimensionally correct? Use unit analysis to find out.

Solution.

(a) The equation is

$$v = v_o + at$$

Inserting units for the physical quantities gives

$$\frac{m}{s} = \frac{m}{s} + \left(\frac{m}{s^2} \times s\right) \quad \text{or} \quad \frac{m}{s} = \frac{m}{s} + \left(\frac{m}{s \times s} \times s\right)$$

Notice that units cancel like numbers in a fraction. Then, we have

$$\frac{m}{s} = \frac{m}{s} + \frac{m}{s} \quad \begin{bmatrix} \textit{Dimensionally} \\ \textit{correct} \end{bmatrix}$$

The equation is dimensionally correct, since the units on each side are meters per second. (The equation is also a correct relationship, as will be seen in the next chapter.)

(b) The equation

$$x = \frac{v}{2a}$$

with unit analysis is

$$m = \frac{\left(\dfrac{m}{s}\right)}{\left(\dfrac{m}{s^2}\right)} = \frac{m}{s} \times \frac{s^2}{m} \quad \text{or} \quad m = s \qquad \begin{bmatrix} \textit{Not dimensionally} \\ \textit{correct} \end{bmatrix}$$

Meters (m) cannot equal seconds (s), so in this case, the equation is dimensionally incorrect.

Follow-up Exercise. Is the equation $x = v^2/a$ dimensionally correct? What are the dimensions? *(Answer may be found in the Answers to Follow-up Exercises section at the back of the book.)*

Mixed Units

Unit analysis also allows you to check for mixed units. In general, in working problems, you should always use the same unit for a given dimension throughout a problem. For example, suppose that you wanted to find the floor area of a rectangular room in order to buy a new carpet to fit it. To compute the area, you would measure the lengths of two adjacent sides of the room and multiply them. Would you measure the sides in different units such as 10 ft × 3.0 m and express the area as 30 ft·m? Probably not. Such mixed units are not usually very useful.

Let's look at mixed units in an equation. Suppose that you used centimeters as the unit for x in this equation:

$$v^2 = v_o{}^2 + 2ax$$

and the units for the other quantities as in Example 1.2. Dimensionally, this would give

$$\left(\frac{m}{s}\right)^2 = \left(\frac{m}{s}\right)^2 + \left(\frac{m}{s^2} \times cm\right)$$

or

$$\frac{m^2}{s^2} = \frac{m^2}{s^2} + \frac{m \times cm}{s^2}$$

which is dimensionally correct. That is, it is equivalent to

$$\frac{[L^2]}{[T^2]} = \frac{[L^2]}{[T^2]} + \frac{[L^2]}{[T^2]}$$

But the units are mixed (m and cm). The terms on the right-hand side could not be added together without first changing one of the length units.

Determining the Units of Quantities

Another aspect of unit analysis that is very important in physics is the determination of the units of quantities from defining equations. For example, **density** (ρ) is defined by the equation

$$\rho = \frac{m}{V} \tag{1.1}$$

where m is mass and V is volume. (Density is the mass per unit volume and is a measure of the compactness of the mass of an object or substance.) What are the units of density? Since mass is measured in kilograms and volume in cubic meters in SI units, the defining equation

$$\rho = \frac{m}{V} \quad \left[\frac{\text{kg}}{\text{m}^3}\right]$$

gives the derived SI unit for density as kilograms per cubic meter (kg/m^3).

What are the units of π? Since the relationship between the circumference and the diameter of a circle is given by the equation $c = \pi d$, then $\pi = c/d$; and if length is measured in meters,

$$\pi = \frac{c}{d} \quad \left[\frac{\text{m}}{\text{m}}\right]$$

Thus, the constant π has no units because they cancel out. It is unitless or a dimensionless constant.

1.5   Unit Conversions

Objectives: To be able to (a) explain conversion factor relationships, and (b) apply them in converting units within a system, or from one system of units to another.

Because units in different systems, or even different units in the same system, can express the same quantity, it is sometimes necessary to convert the units of a quantity from one unit to another. For example, we may need to convert feet to yards or inches to centimeters. You already know how to do many unit conversions. For example, if a room is 12 feet long, what is its length in yards? Your immediate answer is 4 yards.

How did you do this? Well, you must have known a relationship between the units of foot and yard. That is, you know that 1 yd = 3 ft. This is what we call an *equivalence statement*. As we saw previously, the numerical values and units on both sides of an equation must be the same. In the instances of an equivalence statement, we commonly use an equal sign to indicate that 1 yd and 3 ft stand for the *same length*. The numbers are different because they stand for different *units* of length.

Mathematically to change units, we use **conversion factors**, which are simply equivalence statements expressed in the form of ratios; for example, 1 yd/3 ft or 3 ft/1 yd. (The 1 is often omitted in the denominators of such ratios for convenience, for example, 3 ft/yd.) To understand why such ratios are useful, note that dividing the expression 1 yd = 3 ft by 3 ft (or 3 ft = 1 yd by 1 yd) on both sides gives

Conversion factors

$$\frac{1 \text{ yd}}{3 \text{ ft}} = \frac{3 \text{ ft}}{3 \text{ ft}} = 1 \quad \text{or} \quad \frac{3 \text{ ft}}{1 \text{ yd}} = \frac{1 \text{ yd}}{1 \text{ yd}} = 1$$

As you can see from these examples, a conversion factor always has an actual value of 1—and you can multiply any quantity by 1 without changing its value, or size. Thus, a *conversion factor simply lets you express a quantity in terms of other units without changing its physical value or size*.

What is done in converting 12 feet to yards may be expressed mathematically as follows:

$$12 \text{ ft} \times \frac{1 \text{ yd}}{3 \text{ ft}} = 4 \text{ yd} \qquad \textit{(cancel feet)}$$

Using the appropriate conversion factor form, the feet cancel, as shown by the slash marks, giving the correct unit analysis, yd = yd.

Suppose you are asked to convert 12.0 inches to centimeters. You may not know the conversion factor in this case, but you can get it from a table (such as the one that appears inside the front cover of this book) that gives the needed relationships: 1 in. = 2.54 cm or 1 cm = 0.394 in. It makes no difference which of these equivalence statements is used. The question, once you have expressed the equivalence statement as a conversion factor, is whether to divide or multiply by that factor to make the conversion. *In doing unit conversions, you should take ad-*

vantage of unit analysis—that is, let the units choose the appropriate form of conversion factor, so to speak.

Note that the first of the equivalence statements, 1 in. = 2.54 cm, can give rise to two different forms of the conversion factor: 1 in./2.54 cm or 2.54 cm/in. Unit analysis tells you that the second form is the appropriate to multiply by:

$$12.0 \text{ in.} \times \frac{2.54 \text{ cm}}{1 \text{ in.}} = 30.5 \text{ cm}$$

What would be the result if you used the other form of the first conversion factor, 1 in./2.54 cm? Let's see:

$$12.0 \text{ in.} \times \frac{1 \text{ in.}}{2.54 \text{ cm}} = 4.72 \frac{\text{in.}^2}{\text{cm}}$$

This is dimensionally correct (why?), but it is not a conversion to centimeters. The units in.²/cm for length are nonstandard and confusing. To use this ratio properly, you must *divide* a quantity in inches by the conversion factor:

$$\frac{12.0 \text{ in.}}{\left(\dfrac{1 \text{ in.}}{2.54 \text{ cm}}\right)} = 30.5 \text{ cm}$$

(Compare this procedure with the first conversion of 12.0 in. above. Can you see why they are equivalent?)

Alternatively, you could multiply by a conversion factor derived from the second equivalence statement:

$$12.0 \text{ in.} \times \frac{1 \text{ cm}}{0.394 \text{ in.}} = 30.5 \text{ cm}$$

Or you could divide by the reciprocal form of this conversion factor:

$$\frac{12.0 \text{ in.}}{\left(\dfrac{0.394 \text{ in.}}{1 \text{ cm}}\right)} = 30.5 \text{ cm}$$

A few commonly used equivalence statements are not dimensionally correct; for example, 1 kg = 2.2 lb is used for conversions. The kilogram is a unit of mass and the pound is a unit of weight. What this means is that 1 kilogram is equivalent to 2.2 pounds; that is a 1-kilogram *mass* has a *weight* of 2.2 pounds, but only near the Earth's surface. Here mass and weight are directly proportional, differing only by a particular constant, which allows the use of the nondimensionally correct conversion factor of 1 kg/2.2 lb.

Note: 1 kg of mass has an equivalent weight of 2.2 lb near the surface of the Earth.

EXAMPLE 1.3 ■ CONVERTING UNITS: USE OF CONVERSION FACTORS

Do the following unit conversions: (a) a length of 15 meters to feet, (b) a month with 30 days to seconds, and (c) a speed of 50 miles/hour to meters/second.

Solution.

(a) From the conversion table, 1 m = 3.28 ft, so

$$15 \text{ m} \times \frac{3.28 \text{ ft}}{\text{m}} = 49 \text{ ft}$$

Another m-ft conversion is shown in ● Fig. 1.7. Is it correct? You should be able to do this one in your head.

(b) The conversion factor for days and seconds is available from the table (1 day = 86,400 s), but you may not always have a table handy, so instead use several better-known conversion factors to get the result:

$$30 \text{ days} \times \frac{24 \text{ h}}{\text{day}} \times \frac{60 \text{ min}}{\text{h}} \times \frac{60 \text{ s}}{\text{min}} = 2.6 \times 10^6 \text{ s}$$

Note how unit analysis checks the conversion factors for you. The rest is simple arithmetic.

(c) In this case, from the conversion table, 1 mi = 1609 m and 1 h = 3600 s. (The latter is easily computed.) These ratios are used to cancel the units that are to be changed, leaving behind the ones that are wanted:

$$\frac{50 \text{ mi}}{\text{h}} \times \frac{1609 \text{ m}}{\text{mi}} \times \frac{1 \text{ h}}{3600 \text{ s}} = 22 \text{ m/s}$$

Follow-up Exercise. (a) Convert 50 mi/h directly to m/s using a single conversion factor, and (b) show that this single conversion factor can be derived from those in part c of this Example. *(Answers may be found in the Answers to Follow-up Exercises section at the back of the book.)*

EXAMPLE 1.4 ■ CONVERTING UNITS OF AREA: CHOOSING THE CORRECT CONVERSION FACTOR

A hall bulletin board has an area of 2.5 m². What is this area in square centimeters (cm²)?

Solution. A common error in such conversions is the use of incorrect conversion factors. Because 1 m = 100 cm, it is sometimes assumed that 1 m² = 100 cm², which is *wrong*. The correct area conversion factor may be obtained directly from the correct linear conversion factor, 100 cm/1 m or 10^2 cm/1 m, by squaring it:

$$\left(\frac{10^2 \text{ cm}}{1 \text{ m}}\right)^2 = \frac{10^4 \text{ cm}^2}{1 \text{ m}^2}$$

Hence, 1 m² = 10^4 cm² (= 10,000 cm²). We can therefore write:

$$2.5 \text{ m}^2 \times \left(\frac{10^2 \text{ cm}}{1 \text{ m}}\right)^2 = 2.5 \text{ m}^2 \times \frac{10^4 \text{ cm}^2}{\text{m}^2} = 2.5 \times 10^4 \text{ cm}^2$$

Follow-up Exercise. How many cubic centimeters are there in one cubic meter? *(Answer may be found in the Answers to Follow-up Exercises section at the back of the book.)*

•**FIGURE 1.7 Converting units** Road signs show the elevation in both feet and meters, and distances in miles and kilometers.

CONCEPTUAL EXAMPLE 1.5 ■ SELECTING UNITS: UNITS AND MAGNITUDES

You want to express the fuel economy of a car so it will sound the most impressive. Which set of units would give the greatest numerical value? (a) mi/gal, (b) km/gal, (c) mi/L, (d) km/L. (Note: 1 mi = 1.61 km.) *Clearly establish the reasoning used in determining your answer before checking it below. That is, **why** did you select your answer?*

Reasoning and Answer. Since 1 mi is equivalent to 1.61 km, (b) would have a larger numerical value than (a), and (d) would have a larger value than (c). That is, 1 mi/gal = 1.61 km/gal and 1 mi/L = 1.61 km/L. Hence, the length unit would be km, and we must decide which volume unit would give the larger fuel economy value— km/gal or km/L. Since a liter is about a quart, a gallon is much larger than a liter (about 4 times larger). A gallon of gas would therefore run a car many more kilometers than a liter of gas. That is, 1 gal ≈ 4 L; so, for example, 40 km/gal ≈ 10 km/L. Thus the answer is (b).

Follow-up Exercise. (a) Which set of units would give the *smallest* numerical value? (b) Given that 1 L = 0.264 gal, justify your answer mathematically. *(Reasoning and answers may be found in the Answers to Follow-up Exercises section at the back of the book.)*

1.6 Significant Figures

Objectives: To be able to (a) determine the number of significant figures in a numerical value, and (b) report the proper number of significant figures after performing simple calculations.

Most of the time, you will be given numerical data when asked to solve a problem. In general, such data are either exact numbers or measured numbers (quantities). **Exact numbers** are those without any uncertainty or error. These include numbers such as the 100 used to calculate a percentage and the 2 in the equation $r = d/2$ relating the radius and diameter of a circle. **Measured numbers** are those obtained from measurement processes and so generally have some degree of uncertainty or error.

When calculations are done with measured numbers, the error of measurement is propagated, or carried along, by the mathematical operations. A question of how to report a result arises. For example, suppose that you are asked to find time (t) from the formula $x = vt$, and you are given that $x = 5.3$ m and $v = 1.67$ m/s. Then

$$t = \frac{x}{v} = \frac{5.3 \text{ m}}{1.67 \text{ m/s}} = ?$$

Doing the division operation on a hand calculator yields a result such as 3.173652695 (see •Fig. 1.8). How many figures, or digits, should you report in the answer?

The error or uncertainty of the result of a mathematical operation may be computed by statistical methods. A simpler, widely used procedure for making an estimate of this uncertainty involves the use of **significant figures** (sf), sometimes called significant digits. The degree of accuracy of a measured quantity depends on how finely divided the measuring scale of the instrument is. For example, you might measure the length of an object as 2.5 cm with one instrument and 2.54 cm with another; the second instrument provides more significant figures and a greater degree of accuracy. In general, *the number of significant figures of a numerical quantity is the number of reliably known digits it contains*. For a measured quantity, this is usually defined as all of the digits that can be read directly from the instrument used in making the measurement plus one uncertain digit that is obtained by estimating the fraction of the smallest division of the instrument's scale.

The quantities 2.5 cm and 2.54 cm have two and three significant figures, respectively. This is rather evident. However, some confusion may arise when a quantity contains one or more zeros. For example, how many significant figures does the quantity 0.0254 m have? What about 104.6 m? 2705.0 m? In such cases, we will use the following rules:

1. Zeros at the beginning of a number are not significant. They merely locate the decimal point.

    0.0254 m   three significant figures   (2, 5, 4)

2. Zeros within a number are significant.

    104.6 m   four significant figures   (1, 0, 4, 6)

3. Zeros at the end of a number after the decimal point are significant.

    2705.0 m   five significant figures   (2, 7, 0, 5, 0)

4. In whole numbers without a decimal point that end in one or more zeros (trailing zeros)—for example, 500 kg—the zeros may or may not be significant. In such cases, it is not clear which zeros serve only to locate the decimal point and which are actually part of the measurement. That is, if the first zero from the left (5<u>0</u>0 kg) was the estimated digit in the measurement, then only two digits are reliably known and there are only two significant figures. Similarly, if the last zero was the estimated digit

•FIGURE 1.8   Significant figures and insignificant figures
For the division operation of 5.3/1.67, a calculator with a floating decimal point gives many figures. A calculated quantity can be no more accurate than the least accurate quantity involved in the calculation, so this result should be rounded off to two significant figures, that is, 3.2.

Rules for determining the number of significant figures in quantities with zeros

($50\underline{0}$ kg), then there are three significant figures. This ambiguity may be removed by using scientific (powers-of-ten) notation:

$$5.0 \times 10^2 \text{ kg} \quad \text{two significant figures}$$

$$5.00 \times 10^2 \text{ kg} \quad \text{three significant figures}$$

This is helpful in expressing the results of calculations with the proper numbers of significant figures, as we shall see shortly. (A review of scientific notation may be found in Appendix I.)

Note: To avoid confusion with numbers having trailing zeroes used as given quantities in text examples and exercises, we will consider the trailing zeroes to be significant. For example, a time of 20 s should be assumed to have two significant figures, even if it is not written out as 2.0×10^1 s.

It is important to report the results of mathematical operations with the proper number of significant figures. The following general rule tells how to determine the number of significant figures in the result of a multiplication or division:

> The final result of a multiplication or division operation should have the same number of significant figures as the quantity with the least number of significant figures that was used in the calculation.

Multiplication and division with significant figures

What this means is that the result of a calculation can be no more accurate than the least accurate quantity used; that is, you cannot gain accuracy in performing mathematical operations. Thus, the result that should be reported for the division operation discussed at the beginning of this section is

$$\frac{\overset{(2\,sf)}{5.3 \text{ m}}}{\underset{(3\,sf)}{1.67 \text{ m/s}}} = 3.2 \text{ s} \quad (2\,sf)$$

The result is rounded off to two significant figures (see Fig. 1.8).

For multiple operations, rounding off to the proper number of significant figures should not be done at each step because rounding errors may accumulate. It is usually suggested that one or two insignificant figures be carried along, or if a calculator is being used, rounding off may be done on the final result of the multiple calculations.

The rules for rounding off numbers are as follows:

Rounding rules

1. If the next digit after the last significant figure is 5 or greater, the last significant figure is increased by 1. (For example, 2.136 becomes 2.14 rounded to 3 significant figures.)

2. If the next digit after the last significant figure is less than 5, the last significant figure is left unchanged. (For example, 2.132 becomes 2.13 rounded to 3 significant figures.)

EXAMPLE 1.6 ■ USING SIGNIFICANT FIGURES IN MULTIPLICATION AND DIVISION: ROUNDING APPLICATIONS

The following operations are performed and the results rounded off to the proper number of significant figures (sf).

Multiplication

$$\underset{\substack{(2\,sf) \quad (3\,sf)}}{2.4 \text{ m} \times 3.65 \text{ m}} \ (= 8.76 \text{ m}^2) = 8.8 \text{ m}^2 \quad \textit{(rounded to 2 sf)}$$

Division

$$\frac{\overset{(4\,sf)}{725.0 \text{ m}}}{\underset{(3\,sf)}{0.125 \text{ s}}} \; (= 5800 \text{ m/s}) = 5.80 \times 10^3 \text{ m/s} \qquad (\textit{represented with 3 sf})$$

Follow-up Exercise. Perform the following operations with answers expressed in standard powers-of-ten notation (one digit to the left of the decimal point) with the proper number of significant figures. (a) $(2.0 \times 10^5 \text{ kg})(0.035 \times 10^2 \text{ kg})$, and (b) $(148 \times 10^{-6} \text{ m})/(0.4906 \times 10^{-6} \text{ m})$ (*Answers may be found in the Answers to Follow-up Exercises section at the back of the book.*)

There is also a general rule for determining the number of significant figures for the results of addition and subtraction.

Addition and subtraction with significant figures

> The final result of the addition or subtraction of numbers should have the same number of decimal places as the quantity with the least number of decimal places that was used in the calculation.

This rule is applied by rounding off numbers to the least number of decimal places *before* adding or subtracting. The rounding off ensures that a result will have no more reliability than the quantity read from the measuring instrument with the least fine scale. Another way of looking at this is that the rounding off is determined by the quantity with the first doubtful or estimated figure from the left. (Why?)

EXAMPLE 1.7 ■ USING SIGNIFICANT FIGURES IN ADDITION AND SUBTRACTION: APPLICATION OF RULES

The following operations are performed by finding the number that has the first doubtful figure in a column (counting from the left) and rounding the other numbers to this column. (Units have been omitted for convenience.)

Addition
In the numbers to be added, note that 23.1 has the first doubtful figure in a column counting from the left, the 1. The other numbers are rounded to this column and the addition performed:

$$
\begin{array}{c}
23.1 \\
0.546 \\
\underline{1.45} \\
\end{array}
\xrightarrow{\textit{(rounding off)}}
\begin{array}{c}
23.1 \\
0.5 \\
\underline{1.5} \\
25.1 \\
\end{array}
$$

Subtraction
The same rounding procedure is used. Here, the 7 in the 157 is the first doubtful figure in a column from the left, and rounding the 5.5 to this column and subtracting,

$$
\begin{array}{r}
157 \\
- \quad 5.5 \\
\end{array}
\xrightarrow{\textit{(rounding off)}}
\begin{array}{r}
157 \\
- \quad 6 \\
\hline
151 \\
\end{array}
$$

(The decimal point, although omitted, is understood in aligning the numbers.)

Follow-up Exercise. Given the numbers 23.15, 0.546, and 1.058, (a) add the first two, and (b) subtract the last number from the first. (*Answers may be found in the Answers to Follow-up Exercises section at the back of the book.*)

Basically, the number of digits reported in a result depends on the number of digits in the given data. The rules for rounding will generally be observed for examples in this book. However, there will be exceptions that may make a difference, as explained in the following hint.

When working problems, you naturally strive to get the correct answer, and will probably want to check your answers against those listed in the Answers to Odd-numbered Exercises section in the back of the book. However, on occasion you may find that your answer differs from that given even though you have solved the problem correctly. There are several reasons why this could happen.

As stated previously, it is best to round off only the final result of a multi-part calculation, but this is not always convenient in elaborate calculations. Sometimes the results of intermediate steps are important in themselves and need to be rounded off to the appropriate number of digits as if each were a final answer. Similarly, text Examples in this book are often worked in steps for pedagogical reasons, to show the stages in the *reasoning* of the solution. The results obtained when intermediate steps are rounded off may differ slightly from those obtained when only the final answer is rounded.

Rounding differences may also occur when using conversion factors. For example, in changing 5.0 miles to kilometers with the conversion factors listed in the front of the book,

$$5.0 \text{ mi} \left(\frac{1.609 \text{ km}}{1 \text{ mi}} \right) = (8.045 \text{ km}) = 8.0 \text{ km} \qquad \textit{(2 significant figures)}$$

and

$$5.0 \text{ mi} \left(\frac{1 \text{ km}}{0.621 \text{ mi}} \right) = (8.051 \text{ km}) = 8.1 \text{ km} \qquad \textit{(2 significant figures)}$$

The difference arises because of rounding of the conversion factors. Actually, 1 km = 0.6214 mi, so 1 mi = (1/0.6214) km = 1.609269 km = 1.609 km. (Try repeating these conversions with the unrounded factors and see what you get.) To avoid rounding differences in conversions, we will generally use the multiplication form of a conversion factor, as in the first of the equations above, unless there is a convenient exact factor, such as 1 min/60 s.

Finally, slight differences in answers may occur when different methods are used to solve a problem, even though both may be legitimate. Keep in mind that when solving a problem (a general procedure for which is given in the following section), *if your answer differs from that in the text in only the last digit, the disparity is most likely the result of rounding difference or of an alternative method of solution being used.*

1.7 Problem Solving

Objectives: To be able to (a) establish a problem-solving procedure, and (b) apply it to typical problems.

An important aspect of physics is problem solving. In general, this involves the application of physical principles and equations to data from a particular situation in order to find some unknown or wanted quantity. There is no universal method for approaching a problem that will automatically produce a solution. A few general points, though, are worth keeping in mind:

- *Make sure you understand the problem.* The foundation of successful problem solving is having a thorough grasp of the problem. All too often a student will start working on a problem without really understanding the situation described or knowing what is to be found. (Basically, if you don't know exactly what you want, it is very difficult to find it!) Before starting to work a problem, make it a habit to list everything that is given or known (as we will do for most of the Examples in this text). Otherwise, it's all too easy to overlook a critical piece of information.

- To get from the given data to the ultimate solution, you may have to devise a strategy or plan. Many problems cannot be solved merely by finding one equation and "plugging in" the given quantities to get the solution. Often you will need to perform intermediate steps, each of which will bring you closer to the final answer. The process is a lot like crossing a stream by stepping from stone to stone—you need to keep in mind where you are and where you're going, and as you plan each step you must think ahead to the next one.

- Remember that equations are expressions of physical principles. Solving problems involves translating principles into the "language" of equations, but you should also be able to look at an equation and say what physical relationship it embodies. If you use an equation without knowing what it means, you *may* be lucky and get the right answer, but you will not be learning much physics—and you probably won't be able to repeat your success when confronted with a slightly different problem.

- A common cause of failure in problem solving is the *misapplication* of physical principles or equations. Principles and equations are generally subject to certain conditions or are limited in their applicability to physical situations.

- Many problems can be solved by more than one method. As with many things in life, however, there is often a hard way and an easy way. If you understand the problem and the relevant principles well, you can probably figure out how to solve it in the fastest and easiest way.

Although there is no magic formula for problem solving, there are some sound practices that can be very useful. The steps in the following procedure are intended to provide you with a framework that can be applied to solving most of the problems that you will encounter in this text. We will generally use these steps in dealing with the Example problems throughout this text. Additional helpful problem-solving hints will be given where appropriate in subsequent chapters.

Suggested Problem-Solving Procedure

Say it in words

1. *Read the problem carefully and analyze it. Write down the given data and what it is you are to find.* Some data may not be given explicitly in numerical form. For example, if a car "starts from rest," its initial speed is zero ($v_o = 0$). In some instances, you may be expected to know certain quantities or to look them up in tables.

Say it in pictures

2. *Draw a diagram as an aid in visualizing and analyzing the physical situation of the problem where appropriate.* (This may not be necessary in every case, but it is usually helpful and can never do any harm.)

Say it in equations

3. *Determine what principle(s) and equation(s) are applicable to this situation, and how they can be used to get from the information given to what is to be found.* Keep in mind that many problems cannot be solved simply by plugging all of the given data into one equation; you may have to devise a strategy involving several steps.

Simplify the equations

4. *Simplify mathematical expressions as much as possible through algebraic manipulation before inserting actual values.* Trigonometric relationships (summarized in Appendix I) can also be used to simplify equations sometimes. The less calculation you do, the less likely you are to make a mistake—so *don't put the numbers in until you have to.*

Check the units

5. *Check units before doing calculations.* Make unit conversions if necessary so that all units are in the same system and quantities with the same dimensions have the same units (preferably standard units). This avoids mixed units and is helpful in unit analysis. (Unit checking and conversions are often done in Step 1 when writing the data.)

Insert numbers and calculate; check significant figures

6. *Substitute given quantities into equation(s) and perform calculations. Report the result with proper units and number of significant figures.*

Check the answer: is it reasonable?

7. *Consider whether the result is reasonable.* Does the answer have an appropriate magnitude? (This means, is it "in the right ballpark"?) For example, if a person's calculated mass turns out to be 2.30×10^2 kg, the result should be questioned, since 230 kg corresponds to a weight of 506 lb.

•Figure 1.9 recapitulates these steps in the form of a flow chart. The following Examples illustrate the procedure. The steps are numbered to help you follow along.

EXAMPLE 1.8 ■ FINDING THE AREA OF A RECTANGLE: PROBLEM-SOLVING PROCEDURE

Two students measure the lengths of adjacent sides of their dorm room. One reports 15 ft, 8 in., and the other reports 4.25 m. What is the area of the room in square meters?

Solution.

1. Adjacent sides of a room give its length and width, so we may write

    Given: Length $= l = 15$ ft, 8 in.   **Find:** Area (in m^2)
       Width $= w = 4.25$ m

2. Sketch a diagram to help visualize the situation (•Fig. 1.10).

3 **and** 4. For this simple situation, the required equation is well known. The area (A) of a rectangle is $A = l \times w$, both of which are given.

5. It is obvious that a unit change is necessary. Let's first convert the length measurement to inches and then inches to meters:

$$15 \text{ ft} + 8 \text{ in.} = \left(15 \text{ ft} \times \frac{12 \text{ in.}}{\text{ft}}\right) + 8 \text{ in.} = 188 \text{ in.}$$

and

$$188 \text{ in.} \times \frac{2.54 \text{ cm}}{\text{in.}} = 478 \text{ cm} = 4.78 \text{ m}$$

Notice how easy it is to convert units in the decimal metric system (cm to m). You should perform the conversion explicitly if necessary, using the conversion factor (1 m/100 cm).

6. Now perform the calculation.

$$A = l \times w = 4.78 \text{ m} \times 4.25 \text{ m}$$
$$= 20.315 \text{ m}^2 = 20.3 \text{ m}^2 \quad \text{(computed value rounded to 3 sf—why?)}$$

7. The answer appears reasonable. Since $1 \text{ m} \approx 3 \text{ ft}$, the dorm room would be about 13 ft by 14 ft, which is about right (but, as always, too small). Suppose you had inadvertently punched 47.8 instead of 4.78 on your calculator. The result would be $A = 47.8 \text{ m} \times 4.25 \text{ m} = 203 \text{ m}^2$. A room with an area of about 200 m^2 would have dimensions of about 10 m by 20 m, which is roughly 30 ft by 60 ft. Since this is not the size of a typical dorm room, the magnitude of the result should make you suspect there may be an error.

Follow-up Exercise. The dimensions of a textbook are 0.22 m $\times$ 0.26 m $\times$ 4.0 cm. What volume in a book bag would the book take up? Give the answer in both m^3 and cm^3. (*Answers may be found in the Answers to Follow-up Exercises section at the back of the book.*)

Many problems will involve basic trigonometric functions. The major functions are given in the marginal note on p. 22; for other trigonometric relations, see Appendix I or the tables inside the back cover.

EXAMPLE 1.9 ■ FINDING THE LENGTH OF ONE SIDE OF A TRIANGLE: TRIGONOMETRY APPLICATION

A flower bed is laid out in the form of a triangle as shown in •Fig. 1.11. What is the length of the side of the bed that runs along the walkway?

Solution.

1 **and** 2. In some problems we are given diagrams with data. Here we have

    Given: $x = 5.4$ m   **Find:** r (long side of triangle)
       $\theta = 40°$

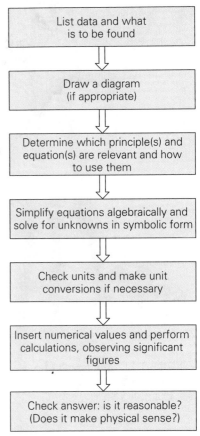

List data and what is to be found

Draw a diagram (if appropriate)

Determine which principle(s) and equation(s) are relevant and how to use them

Simplify equations algebraically and solve for unknowns in symbolic form

Check units and make unit conversions if necessary

Insert numerical values and perform calculations, observing significant figures

Check answer: is it reasonable? (Does it make physical sense?)

•FIGURE 1.9 A flow chart for the suggested problem-solving procedure.

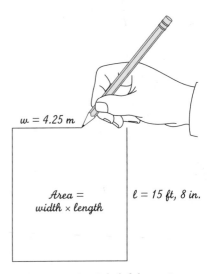

$w = 4.25$ m

Area = width × length

$l = 15$ ft, 8 in.

•FIGURE 1.10 A helpful step in problem solving
Drawing a diagram helps you to visualize and better understand the situation. See Example 1.8.

Trigonometric functions:

$$\cos \theta = \frac{x}{r} \left(\frac{\text{side adjacent}}{\text{hypotenuse}}\right)$$

$$\sin \theta = \frac{y}{r} \left(\frac{\text{side opposite}}{\text{hypotenuse}}\right)$$

$$\tan \theta = \frac{\sin \theta}{\cos \theta} = \frac{y}{x} \left(\frac{\text{side opposite}}{\text{side adjacent}}\right)$$

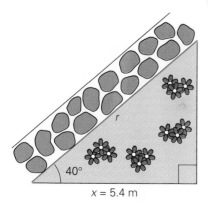

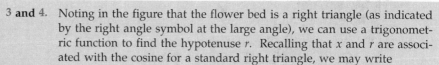

3 and 4. Noting in the figure that the flower bed is a right triangle (as indicated by the right angle symbol at the large angle), we can use a trigonometric function to find the hypotenuse r. Recalling that x and r are associated with the cosine for a standard right triangle, we may write

$$\cos \theta = \frac{x}{r} \qquad \text{or} \qquad r = \frac{x}{\cos \theta}$$

5 and 6. The units are OK. (Since x is in meters, r will come out in meters.) So put in the data:

$$r = \frac{x}{\cos \theta} = \frac{5.4 \text{ m}}{\cos 40°} = \frac{5.4 \text{ m}}{0.766} = 7.0 \text{ m}$$

7. The magnitude of r seems reasonable compared to the value of x.

Follow-up Exercise. To keep animals out of the flower bed in Fig. 1.11, the owner decides to fence in the perimeter of the bed. What will be the total length of the fence? (*Answer may be found in the Answers to Follow-up Exercises section at the back of the book.*)

•**FIGURE 1.11 The length of a triangle's side**
See Example 1.9.

It is understood that you do not need to explicitly write out the steps of problem-solving procedure each time. However, it is a good practice to run through them mentally when working a problem. In the examples in the following chapters, the problem-solving steps will not always be listed as was done here for illustration. Even so, you should be able to see the general pattern outlined above. In some instances, especially when introducing the application of a new principle or concept, problem-solving steps or hints will be listed to promote better understanding.

The main point of this section is that in solving problems you should have some systematic procedure for analyzing the situation and extracting the information wanted. The suggested problem-solving procedure given here is one option. Let's review it again.

EXAMPLE 1.10 ■ FINDING THE CAPACITY OF A CYLINDER: PROBLEM SOLVING IN THREE DIMENSIONS

A cylindrical container with a height of 28.5 cm and an inside diameter of 10.4 cm is filled with water. What is the mass of the water in kilograms?

Solution.

1. *Read the problem carefully and analyze it.*
   From the problem, we have

   Given: $h = 28.5$ cm (height)   **Find:** m (mass in kg)
   $d = 10.4$ cm (diameter)

2. *Draw a diagram as an aid in visualizing and analyzing the situation.*
   Make a sketch of the cylinder showing its dimensions, and adding any general information you might know about a cylinder. For example, the circular base has an area of $A = \pi r^2$, where r is the radius and $r = d/2$. (See •Fig. 1.12.)

3. *Determine what principle(s) and equation(s) are applicable and how to use them.*
   We are given the height and diameter of the cylinder, so we can find its volume, which is the volume of the water it contains. But how is volume related to mass, which is what we want? The link is density (ρ), which we saw earlier (Eq. 1.1) is mass per unit volume:

   $$\rho = \frac{m}{V} \qquad \text{or} \qquad m = \rho V$$

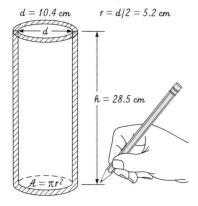

$d = 10.4$ cm $r = d/2 = 5.2$ cm

$h = 28.5$ cm

$A = \pi r^2$

•**FIGURE 1.12 The capacity of a cylinder**
Drawing a diagram helps in problem solving. See Example 1.10.

The density of water is not given. However, this can be looked up in a table, or you may remember from the definition of the kilogram that $\rho_{H_2O} = 1.00 \text{ g/cm}^3$ (at maximum density, which will be assumed here). So, knowing the volume and the density of water, we can find the mass.

4. *Simplify expressions algebraically if possible.*

Should you not remember the formula for the volume of a cylinder, your sketch can be helpful. The area of the circular end of the cylinder is $A = \pi r^2 = \pi d^2/4$ (since $r = d/2$), and the volume is this area times the height of the cylinder, or $V = Ah$. (Formulas for areas and volumes of common shapes can be found in Table 5 of Appendix I.)

We can therefore write an equation for the mass entirely in terms of known quantities:

$$m = \rho V = \rho Ah = \rho\left(\frac{\pi d^2}{4}\right)h$$

5. *Check units before doing calculations.*

The cylinder dimensions are in cm and the density in g/cm^3. Hence, the mass will come out in grams, which we can convert to kg as requested in the problem. (Another option would be to convert to meter and kilogram units first, but this would require more steps.)

Note: If the given data have mixed units, it is a good practice to convert to consistent units when writing down what is given. This avoids mistakes that might be made later if this step were overlooked.

6. *Substitute the given quantities into the equation and perform calculations, observing significant figures.*

Using the equation derived above and inserting the known quantities, we have

$$m = \rho V = \rho Ah = \rho\left(\frac{\pi d^2}{4}\right)h$$

$$= \left(\frac{1.00 \text{ g}}{\text{cm}^3}\right)\left[\frac{\pi(10.4 \text{ cm})^2}{4}\right](28.5 \text{ cm}) = 2.42 = 10^3 \text{ g}$$

$$= 2.42 \text{ kg}$$

7. *Consider whether the result is reasonable.*

It may be difficult sometimes to determine whether or not a result is reasonable because of unfamiliarity with units. Here, one might calculate that about 2.5 kg of water has a weight of about 5.0 lb (since 1 kg is equivalent to 2.2 lb or $\approx$ 2.0 lb). This seems to be a reasonable weight for this volume of water. (The dimensions are approximately those of a 2-L soda bottle.)

Follow-up Exercise. If the outside diameter of the cylinder in Fig. 1.12 is 10.5 cm, what is the total outside surface area of the sides and base of the cylinder? (*Answer may be found in the Answers to Follow-up Exercises section at the back of the book.*)

Chapter Review

Important Terms

You should be able to define and explain these chapter terms clearly.

standard unit   2
system of units   2
International System of Units
  (SI)   4
SI base units   4
SI derived units   4
meter (m)   4

kilogram (kg)   5
second (s)   6
mks system   7
cgs system   7
fps system   7
liter (L)   8
dimensional analysis   10

unit analysis   11
density (ρ)   12
conversion factor   13
exact number   16
measured number   16
significant figures (sf)   16

Important Concepts

- The SI units of length, mass, and time are the meter (m), kilogram (kg), and second (s) respectively.
- To a good approximation, a liter (L) of water has a mass of one kilogram.
- Dimensional analysis and/or unit analysis can be used to determine if an equation is correct, and unit analysis can be used to find the unit of a quantity.

- The number of significant figures of a numerical quantity is the number of reliably known digits it contains.
- Exercises should be worked with a consistent problem-solving procedure.

Important Equations

This equation is a mathematical statement of a concept or principle presented in the chapter. It will be needed in working some problem Exercises, so a good understanding is important. You should be able to identify the symbols and to explain their relationships before proceeding.

(The in-text equation reference number is given for convenience.)

Density:
$$\rho = \frac{m}{V} \left(\frac{mass}{volume} \right) \qquad (1.1)$$

Exercises

Throughout the text, many exercise sections will include "paired" exercises. These exercise pairs, identified with colored numbers, are intended to assist you in problem solving and learning. In such a pair, the first exercise is worked out in the Study Guide so that you can consult it should you need assistance in its solution. The second exercise is similar in nature and its answer is given at the back of the book.

1.2 Units

actual physical object.

1   The only SI standard represented by an artifact is the (a) meter (b) kilogram, (c) second, (d) electric charge.

2   Which of the following is *not* an SI base unit: (a) length, (b) mass, (c) weight, (d) time?

3   Are the following reasonable statements? (Justify your answers.) (a) It took 60 L of gasoline to fill up the car's tank. (b) The center on the basketball team has a height of 325 cm. (c) The area of a dorm room is 12 m².

4   ■ (a) In the British system, 16 oz = 1 pt and 16 oz = 1 lb. Is there something wrong here? Explain. (b) Here's an old one: a pound of feathers weighs more than a pound of gold. How can that be? (*Hint:* Look up ounce in the dictionary.) *TRUE*

5   ■ The metric system is a decimal (base-10) system and the British system is in part a duodecimal (base-12) system. Discuss the ramifications if our monetary system had a duodecimal base. What would be the possible values of our coins if this were the case?

1.4 Dimensional Analysis*

6   Both sides of an equation are equal in (a) numerical value, (b) units, (c) dimensions, (d) all of the preceding.

7   Unit analysis of an equation *cannot* tell you if (a) it is dimensionally correct, (b) the equation is correct, (c) the mathematics is correct, (d) both (b) and (c).

8   Can dimensional analysis tell you whether you have used the correct equation in solving a problem? Explain.

9   Must an equation be correct both by dimensional analysis and unit analysis? Explain.

10   Is the unit equation m·kg²·s/cm²·min·g·m² = g/km³ dimensionally correct? If so, explain how this can be with so many different units.

11   ■ Show that the equation $x = x_\text{o} + vt$, where v is velocity and x and x_o are lengths, is dimensionally correct.

12   ■ Use SI unit analysis to show that the equation $A = 4\pi r^2$, where A is the area and r is the radius of a sphere, is dimensionally correct.

13   ■ Is the equation $V = \pi d^3/4$, where V is the volume and d is the diameter of a sphere, dimensionally correct? (Use SI unit analysis to find out.)

14   ■■ The equation for the volume of a sphere is $V = 4\pi r^3/3$, where r is the radius of the sphere. Is the equation in Exercise 13 correct?

15   ■■ Show that $v^2 = v_\text{o}^2 - 2ax$ is dimensionally correct. (a is acceleration, v and v_o are velocities, and x is length.)

16   ■■ If $x = gt^2/2$, where x is length and t is time, is dimensionally correct, what are the SI units of the constant g?

*Dimensions and/or units of velocity and acceleration are given in the chapter.

17 ■■ Is the equation $v = v_o \sin \theta - gt$ dimensionally correct? Show by using SI unit analysis. (v and v_o are velocities, θ is an angle, t is time, and g is the same as in Exercise 16.)

18 ■■ Using unit analysis, determine whether the following equations are dimensionally correct (t is time, x is length, a is acceleration, and v is velocity): (a) $t = \sqrt{2x/a}$ and (b) $v = \frac{1}{2}(v_o + a)t$. (See Section 1.4 for units.)

19 ■■ Is the equation for the area of a trapezoid, $A = \frac{1}{2}a(b_1 + b_2)$, where a is the altitude and b_1 and b_2 are the bases, dimensionally correct? (See •Fig. 1.13.) If not, how should it be changed to correct it?

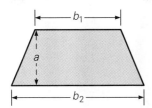

•**FIGURE 1.13 The area of a trapezoid**
See Exercise 19.

20 ■■ One student, using unit analysis, says that the equation $v = \sqrt{2ax}$ is dimensionally correct. Another says it isn't. With whom do you agree?

21 ■■ The formulas for a straight line, a parabola, and a hyperbola are commonly given as (a) $y = mx$, (b) $y = bx^2$, and (c) $y = c/x$, respectively, where m, b, and c are constants. What are the SI units of each of the constants if x and y are lengths?

22 ■■ In the equation $mgx = \frac{1}{2}kx^2$, the g is acceleration and k is a constant called the spring constant. What are the SI units of the spring constant?

23 ■■ The radius r of a circle with its center at the origin is given by $r = \sqrt{x^2 + y^2}$, where x and y are the Cartesian coordinates of a point on the circle. Based on this, a student concludes that the area A of the circle is $A = \pi\sqrt{x^2 + y^3}$. Is he correct? (Justify your answer.)

24 ■■■ The equation for the frequency (f) of a simple pendulum is $f = \dfrac{1}{2\pi}\sqrt{\dfrac{g}{L}}$, where L is the length of the pendulum string and g is the acceleration due to gravity. Frequency is commonly expressed in units of hertz (Hz). What is a hertz unit in terms of SI base units?

25 ■■■ Newton's second law of motion (Chapter 4) is expressed by the equation $F = ma$, where F represents force, m is mass, and a is acceleration. (a) The SI unit of force is, appropriately, called the newton (N). What are the units of the newton in terms of base quantities? (b) Another equation for force associated with uniform circular motion (Chapter 7) is $F = mv^2/r$, where v is velocity and r is the radius of the circular path. Does this equation give the same units for the newton?

26 ■■■ Newton's law of gravitation is expressed by the equation $F = Gm_1m_2/r^2$, where F is force, G is a constant (called the universal gravitational constant), m_1 and m_2 are masses, and r is a distance. (a) What are the SI base units of the constant G when the force is expressed in newtons? (See Exercise 25.) (b) What are the units of G in SI base quantities?

1.5 Unit Conversions*

27 A good way to ensure proper unit conversion is to (a) use another measurement instrument, (b) always work in one system of units, (c) use unit analysis, (d) use dimensional analysis.

28 A conversion factor written as 1 in. = 2.54 cm means that (a) 1 in. is equivalent to 2.54 cm, (b) this is a true equation, (c) 1 cm = 2.54 in. (d) none of the preceding.

29 The width of the nail of your little finger is about the size of a common metric unit. Which one?

30 ■ If you wanted to express your height as a larger number, which unit in each of the following pairs would you use: (a) meter or yard; (b) decimeter or foot; (c) centimeter or inch?

31 ■ What is the length in feet of (a) 1.00 km and (b) a 200-m free-style swimming event?

32 ■ Do the following conversions: (a) 25 m to feet, (b) 12.0 in. to centimeters, and (c) 7.0 days to seconds.

33 ■■ Participants in the Olympic light middleweight boxing event must have a minimum weight of 157 lb. The event is held in Europe, and one contender weighs in at 69.8 kg. Does he qualify?

34 ■■ The physics class at a particular school ordinarily lasts 60 minutes. The professor offers to change the class time to a microcentury. Do you think the students would accept? Justify your answer.

35 ■■ According to the *Guiness Book of Records*, the world's tallest person was Robert Wardlow who measured 8 ft, 11 in. and weighed 439 lb when he died in 1940. Translate these measurements for a metric *Book of Records*. Incidentally, Robert wore a size 37AA shoe, which is 47 cm in length. What is the length in inches?

36 ■■ A Boeing 777 jet has a length of 209 ft, 1 in., a wingspan of 199 ft, 11 in., and a fuselage diameter of 20 ft, 4 in. What are these dimensions in meters, and how big around is the fuselage?

37 ■■ Estimate the weight in pounds of a half gallon of skim milk.

*Conversion factors are listed inside the front cover.

38 ■■ An automobile speedometer is shown in ●Fig. 1.14. (a) What would the equivalent scale readings be in km/h? (b) What would the 55 mi/h speed limit be in km/h? How about a 65 mi/h speed limit?

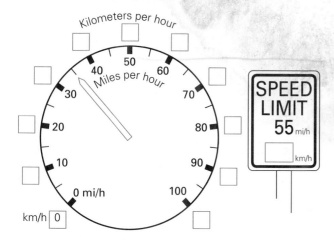

●**FIGURE 1.14 Speedometer readings**
See Exercise 38.

39 ■■ (a) Which holds more soda and how much more, a half-gallon bottle or a two-liter bottle? (b) Suppose that a 16-oz soda and a 500-mL soda sell for the same price. Which would you choose to get the most for your money and how much more (in milliliters) would you get? [*Hint*: 1 pint = 16 oz.]

40 ■■ A professor regularly buys 15 gal of gas, but the gas station has installed new pumps that deliver liters. How many liters of gas (rounded off to a whole number) should she ask for?

41 ■■ A student has a car that gets on the average 23.6 miles per gallon (mpg) of gasoline. She gets to spend a year in Europe and takes her car with her. (a) What should she expect the car's average gas mileage to be in km/L? (b) During the year there, she drove a total of 8000 km. With gas costing about $4.00/gal in Europe, how much did she spend on fuel? (Compute to the nearest cent.)

42 ■■ Some common product labels are shown in ●Fig. 1.15. From the units on the labels, form a set of conversion factors.

43 ■■ Another set of product labels is shown in ●Fig. 1.16, but there seem to be some inconsistencies. What is the problem and which are correct? [*Hint*: 1 lb = 0.4535924 kg; 1 L = 1.057618 qt.]

44 ■■ A particular student was 20 in. long when she was born. She is now 5 ft 4 in. and 18 years old. How many centimeters a year did she grow on the average?

●**FIGURE 1.15 Conversion factors**
See Exercise 42.

●**FIGURE 1.16 An extra gram and milliliter?**
See Exercise 43.

45 ■■ A basketball team from the United States has a center who is 6 ft 9 in. tall and weighs 210 lb. If the team plays exhibition games in Europe, what values will be listed for the center's height and mass on programs for fans there?

46 ■■ A 25-inch TV indicates the diagonal length of the TV tube. Assuming the face of the tube to be flat and rectangular, and that the diagonal makes an angle of 37° with the base of the tube, what is the area of the tube in (a) in.² and (b) cm²? (*Hint*: A diagram as suggested in the problem-solving procedure is helpful here.)

47 ■■ (a) A football field is 300 ft long and 160 ft wide. What are the field's dimensions in meters? (b) A football is 11 to 11¼ in. long. What is its length in centimeters?

48 ■■ Suppose that when the United States goes completely metric, the dimensions of a football field are established as 100 m by 54 m. Which would be larger, the metric football field or a current football field, and what would be the difference in the areas?

26   CHAPTER 1   Units and Problem Solving

49 ■■■ The width and length of a room are 3.2 yd and 4.0 yd. If the height of the room is 8.0 ft, what is the volume of the room in (a) cubic meters and (b) cubic feet?

50 ■■■ (a) The metal mercury has a density of 13.6 g/cm³. How much mercury would be required to fill a 0.225-L container? (b) What would be the weight of this amount of mercury in pounds?

51 ■■■ Engineers often express the density of a substance as a weight density, for example, in lb/ft³. (a) What is the weight density of water? (b) What is the weight of a gallon of water?

52 ■■■ In the Bible, Noah is instructed to build an ark 300 cubits long, 50.0 cubits wide, and 30.0 cubits high (see ●Fig. 1.17). A cubit was a unit of length based on the length of the forearm and equal to about half of a yard. (a) What would the dimensions of the ark be in meters? (b) What would its volume be in cubic meters? To approximate, assume that the ark was rectangular.

●**FIGURE 1.17 Noah and his ark**
See Exercise 52.

1.6 Significant Figures

53 Which of the following has the greatest number of significant figures? (a) 103.07, (b) 124.5, (c) 0.09916, (d) 5.408×10^5

54 In a multiplication and/or division operation involving the numbers 15,437; 201.08; and 408.0×10^5, the result should be rounded to how many places? (a) 3, (b) 4, (c) 5, (d) any number

55 ■ Express the length 50,570 μm (micrometers) in centimeters, decimeters, and meters to three significant figures.

56 ■ Using a meterstick, a student measures a length and reports it to be 23.8755 m. What is the smallest division on the meterstick scale?

57 ■ If a measured length is reported as 25.483 cm, could this length have been measured with an ordinary meterstick whose smallest division is millimeters? Discuss in terms of significant figures.

58 ■■ Determine the number of significant figures in the following measured numbers: (a) 1.007 m, (b) 8.03 cm, (c) 16.272 kg, (d) 0.015 μs (microseconds).

59 ■■ Express each of the numbers in Exercise 58 with two significant figures.

60 ■■ Which of the following quantities has three significant figures: (a) 305.0 cm, (b) 0.0500 mm, (c) 1.00081 kg, (d) 8.06×10^4 m²?

61 ■■ Express each of the following numbers with three significant figures: (a) 10.072 m, (b) 775.4 km, (c) 0.002549 kg, (d) 93,000,000 mi.

62 ■■ A circular flower bed has a radius of 4.25 m. Compute its (a) circumference and (b) area.

63 ■■ The parallel bases of a trapezoid are 7.75 cm and 4.6 cm, respectively, and the altitude is 3.25 cm. What is the area of the trapezoid? [*Hint:* See Exercise 19.]

64 ■■ The side of a cube is measured and its volume is reported to be 2.5×10^2 cm³. What was the measured length of the side of the cube?

65 ■■■ The outside dimensions of a cylindrical soda can are reported as 12.559 cm for the diameter and 5.62 cm for the height. What is the total outside area of the can?

66 ■■■ In doing a problem, a student adds 46.9 m and 5.72 m, and then subtracts 38 m from the result. What is the final answer?

67 ■■■ Work the following exercise by two procedures as directed, commenting on and explaining any difference in the answers. Use your calculator for the calculations. Compute $p = mv$, where $v = x/t$. (a) First compute v, and then p. (b) Compute $p = mx/t$ without an intermediate step. Given: $x = 8.5$ m, $t = 2.7$ s, and $m = 0.66$ kg.

1.7 Problem Solving

68 An important step in problem solving before actually mathematically solving an equation is (a) checking units, (b) checking significant figures, (c) consulting with a friend, (d) checking to see if the result is reasonable.

69 An important final step in problem solving before reporting an answer is (a) reading the problem again, (b) saving your calculations, (c) seeing if the answer is reasonable, (d) checking your results with another student.

70 After having calculated a result in liters, explain how you might determine if the result is reasonable.

71 ■ The mass of the Earth is 6.0×10^{24} kg, and it has a volume of 1.1×10^{21} m³. What is the Earth's average density?

72 ■ The metal mercury has a density of 13.6 g/cm³. What is the mass of a liter of mercury?

73 ■ Determine whether each of the following reported measurements has an appropriate magnitude. (a) A subcompact car has a mass of 2050 kg. (b) A person is 183 cm tall. (c) A runner runs a mile in 3.00×10^8 μs. (d) A normal highway driving speed is 25 m/s. (e) A dachshund has a tail 280 mm long and stands 5.0 dm off the ground. (f) A professor has a mass of 98 kg and a 50-cm waist.

74 ■■ Nutrition Facts labels now appear on most foods. An abbreviated label concerned with fat is shown in •Fig. 1.18. When burned in the body, each gram of fat supplies 9 calories. (A food calorie is really a kilocalorie. See Chapter 11.) (a) In that case, what percentage of the serving calories is supplied by fat? (b) You may notice that this doesn't agree with the listed Total Fat percentage in the figure. This is because the given % Daily Values are the percentages of the *maximum* recommended amounts of nutrients (in grams) contained in a 2000 calorie diet. What then is the maximum recommended amounts of total fat and saturated fat for a 2000 calorie diet?

Nutrition Facts
Serving Size: 1 can
Calories: 310

Amount Per Serving	% Daily Value*
Total Fat 18 g	28%
Saturated Fat 7g	35%

* Percent Daily Values are based on a 2,000 calorie diet.

•**FIGURE 1.18 Nutrition facts**
See Exercise 74

75 ■■ The thickness of the numbered pages of a text book is measured to be 3.75 cm. If the last page of the book is numbered 860, what is the average thickness of a page?

76 ■■ A light year is a unit of distance corresponding to the distance light can travel in a vacuum in 1 year. If the speed of light is 3.00×10^8 m/s, what is the length of a light year in kilometers?

77 ■■ For a 154-lb person, the body contains about 40 L of water and 5.6 L of blood. (a) What is the body's percent composition of water by weight? (b) What is the body's percent composition of blood by weight? (*Hint:* The density of blood is 1.05 g/cm³.)

78 ■■ Tony's Pizza Palace sells a 9.0-in. (diameter) pizza for $7.95 and a 12-in. pizza for $13.50. Which is the better buy?

79 ■■ In •Fig. 1.19, which black area is greater, the center circle or the outer ring?

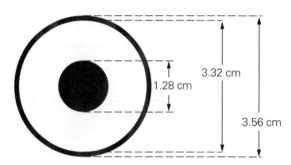

•**FIGURE 1.19 Which black area is greater?**
See Exercise 79.

80 ■■■ Cork, which floats in water, is very light with a density of 0.031 g/cm³. (a) Suppose you had a sphere of cork with a radius of 0.500 m. Would you be able to pick it up? Justify your answer by expressing the weight of the sphere in lb. (b) How about a cork sphere with a radius of 1.00 m? (What does this calculation demonstrate?)

81 ■■■ The Channel Tunnel or "Chunnel" that runs under the English Channel between Great Britain and France is 31 miles long. (There are actually three separate tunnels.) A shuttle train that carries passengers through the tunnel travels with an average speed of 75 mi/h. How long on the average does it take the shuttle to make a one-way trip through the Chunnel?

82 ■■■ When trains travel through the Chunnel (see Exercise 81), friction heats the air. Left unchecked, the air temperature in the tunnel would be unbearable, so a giant air-conditioning system was installed. In the cooling process, water is pumped through a 300-mi network of 24-in. diameter pipes. How many metric tons of water does the network contain?

Additional Exercises

83 The human heart beats on the average 70 times a minute (as indicated by the pulse rate). How many times does the heart beat on the average in a 70 year lifetime?

84 The salinity of the ocean is 3.5% by weight or mass. (That is, out of 100 g of sea water, 3.5 g is made up of mineral salts, mostly sodium chloride.) It is estimated that the total volume of water in the oceans and seas is about 1.37×10^9 km^3. If sea water has a density of 1.03×10^3 kg/m^3, what is the total mass of the minerals in the oceans and seas?

85 A right triangle has a base with a length of 10.5 cm and an altitude of 8.7 cm (•Fig. 1.20). What is the area of the triangle?

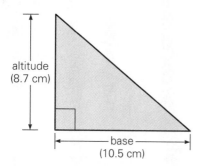

•**FIGURE 1.20 The area of a triangle**
See Exercise 85.

86 The general equation for a parabola is $y = ax^2 + bx + c$, where a, b, and c are constants. What are the units of each constant if y and x are in meters?

87 A rectangular block has the dimensions 4.85 cm, 6.5 cm, and 15.51 cm. What is the volume of the block in cubic centimeters?

88 Using unit analysis, show that the equation $t = x/(v_o + \frac{1}{2}at)$ (where v_o is velocity, a is acceleration, x is length, and t is time) is dimensionally correct.

89 A solid sphere has a radius of 12 cm. What is its surface area in (a) square centimeters and (b) square meters? (c) If it has a mass of 9.0 kg, what is its density in kg/m^3?

90 A cylindrical drinking glass has an inside diameter of 8.0 cm and a depth of 12 cm. If a person drinks a completely full glass of water, how much (in liters) will be consumed?

91 A fathom is a nautical measure of length or depth and is equal to 6 ft. A furlong is a unit of distance used in horse-racing and is equal to $\frac{1}{8}$ mi. What are the values of each of these distances expressed in meters using a metric prefix that allows the numerical part of the result to be as close as possible to 1?

92 The top of a rectangular table measures 1.245 m by 0.760 m. (a) What is the smallest division on the scale of the measurement instrument? (b) What is the area of the tabletop?

93 When computing the average speed of a cross-country runner, a student gets 25 m/s. Is this a reasonable result? Justify your answer.

94 The average density of the Moon is 3.36 g/cm^3, and it has a diameter of 2160 mi. What is the total mass of the Moon in kg? (Compare your answer to the value given inside the back cover.)

95 A thick-walled spherical metal shell has an inner diameter of 18.5 cm and an outer diameter of 24.6 cm. What is the volume occupied by the shell itself?

2 Kinematics: Description of Motion

Y̲OU can't look at a photo like this without mentally putting yourself into the picture. But how would you describe what you're experiencing? The sense of motion is so strong you can almost feel the air rushing by you. And yet, it's all an illusion! Motion takes place in time, but the photo can only "freeze" a single instant. How fast are you going? Are you speeding up as you plummet down the trail? Or are you braking to slow down? You'll find that, without the dimension of time, you can hardly describe motion at all.

In this chapter we'll define motion and explore ways of describing it. We'll also learn to analyze changes in motion—speeding up, slowing down, stopping. Along the way, we'll deal with a particularly interesting case of accelerated motion: free fall under the influence only of gravity.

The description of motion involves the representation of a restless world. Nothing is ever perfectly still. You may sit, apparently at rest, but your blood flows, and air moves into and out of your lungs. The air is composed of gas molecules moving at different speeds and in different directions. And, while you experience stillness, you, your chair, the building, and the air you breathe are all revolving through space with the Earth, part of a solar system in a spiraling galaxy in an expanding universe.

The branch of physics concerned with the study of motion and what produces and affects it is called **mechanics**. The roots of mechanics and of human interest in motion go back to early civilizations. The study of the motions of heavenly bodies, or celestial mechanics, grew out of the need to measure time and location. Several early Greek scientists, notably Aristotle, put forth theories of motion that were useful descriptions but were later proved to be incorrect. Our currently accepted concepts of motion were formulated in large part by Galileo (1564–1642) and Isaac Newton (1642–1727).

Mechanics is usually divided into two parts: kinematics and dynamics. **Kinematics** deals with the description of the motion of objects without consideration of what causes the motion. **Dynamics** analyzes the causes of motion. This chapter covers kinematics and reduces the description of motion to its simplest terms by considering the simple case of motion in a straight line, or linear motion, which is motion in one dimension (of space). Chapter 3 focuses on motion in two dimensions (which can easily be extended to three dimensions). Chapter 4 investigates dynamics to show what causes changes in motion.

2.1 Change of Position: Distance and Displacement

Objectives: To be able to (a) distinguish between scalars and vectors, and (b) define distance and displacement.

What is motion? This seems a simple question, but you might have some difficulty giving an immediate answer. After a little thought, you should conclude that **motion** involves the changing of position. Motion may be described in part by specifying *how far* something travels in changing position—that is, the distance it travels. **Distance** is simply the *total path length* traversed in moving from one location or point to another. For example, you may drive to school from your hometown and express the distance traveled in miles or kilometers. In general, the distance between two points depends on the path traveled (•Fig. 2.1).

Scalars and Vectors

In physics, quantities are distinguished as being scalars or vectors. Basically, the difference is that *direction* is associated with vectors but not with scalars. A **scalar quantity** is one with magnitude or size only. That is, a "scalar" (short for scalar quantity) has only a numerical value, such as 160 km or 100 mi. (Note that magnitudes include units.) Distance is a scalar quantity. Other examples of scalars are measurements or quantities such as 10 s, 3.0 kg, and 20°C.

A **vector quantity**, on the other hand, is one with both magnitude *and* direction. For example, when we describe an airplane as having flown 25 km north, we are giving a *vector* description. Other vector quantities include, as we shall shortly discover, velocity and acceleration. So, quite simply, a scalar quantity is one that does not have an associated direction, whereas vector quantities do.

Graphically, vectors are represented as arrows, with the direction of the arrow giving the direction of the vector (•Fig. 2.2). The length of the arrow may be drawn to scale and made proportional to the magnitude or numerical value of the vector. Note in the figure how one vector is drawn to be twice as long as the other.

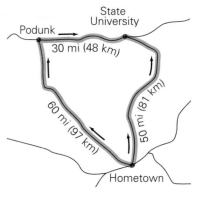

•FIGURE 2.1 Distance—total path length
In driving to school, one student may take the shortest route and travel a distance of 50 mi (81 km). Another student takes a longer route in order to visit a friend in Podunk before returning to school. The longer trip is in two segments; the distance traveled is the total path length, 60 mi + 30 mi = 90 mi (145 km).

Distance: How far

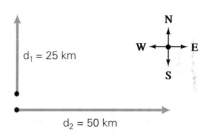

•FIGURE 2.2 Vector representation
A vector quantity is represented graphically by an arrow. The vector direction is indicated by the arrow direction, and the length of the arrow is proportional to the magnitude of the quantity.

Because of the property of direction, vectors are added and subtracted differently from scalars. Vector addition and subtraction will be considered in more detail in Chapter 3. Until then, we will generally limit our discussion of vectors to the description of motion in one dimension—that is, in a straight line.

Displacement

Displacement: How far, and in what direction

For straight-line, or linear, motion, it is convenient to specify position using the familiar Cartesian coordinate system with x and y axes at right angles. The straight-line path may be in any direction, but for simplicity we usually choose to orient the coordinate axes so that the motion is along one of them. We define **displacement** as the straight-line distance between two points, along with the direction. Hence, displacement is a vector quantity with both magnitude and direction from the initial position to the final position. A linear displacement, say along the x axis, is given by

$$\Delta x = x_2 - x_1 \tag{2.1}$$

where x_1 and x_2 are the initial and final positions, respectively. The Greek letter Δ (delta) is commonly used to represent a change. Here, the Δx means "the change (or difference) in x" and indicates a change in *position* along the x axis. In this form, the direction of the displacement is given by the sign ($+$ or $-$) associated with Δx.

Suppose a person moves in a straight line from the lockers toward the physics lab, 8.0 m away, as shown in •Fig. 2.3(a). With $x_1 = 1.0$ m and $x_2 = 9.0$ m, the displacement is

$$\Delta x = x_2 - x_1$$
$$= 9.0 \text{ m} - 1.0 \text{ m} = +8.0 \text{ m}$$

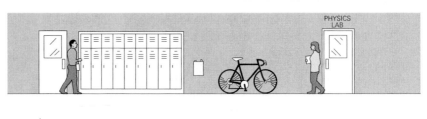

(a) Distance (magnitude or numerical value)

•**FIGURE 2.3 Distance (scalar) and displacement (vector)**
(a) The distance (straight-line path length) between the lockers and the physics lab is 8.0 m and is a scalar quantity. **(b)** The displacement of the person going toward the physics lab between these two points is +8.0 m (in the positive x direction) and is a vector quantity. **(c)** The displacement of the person going away from the physics lab between these two points is −8.0 m (in the negative x direction) and is a vector. Note how the sign gives the direction of the motion.

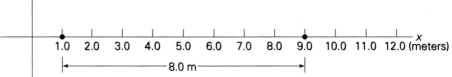

(b) Displacement (magnitude and direction)

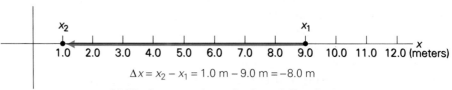

(c) Displacement (magnitude and direction)

The person's displacement (magnitude and direction) is then 8.0 m in the positive x direction, as indicated by the positive result. (The plus sign is often omitted, being taken as understood.) This displacement is represented by a vector arrow in •Fig. 2.3b.

Suppose the other person in the figure moves in the opposite direction, from $x_1 = 9.0$ m to $x_2 = 1.0$ m. In this case, the displacement is

$$\Delta x = x_2 - x_1$$
$$= 1.0 \text{ m} - 9.0 \text{ m} = -8.0 \text{ m}$$

The minus sign indicates that the direction of the displacement is in the negative x direction, and a representative vector arrow would point this way (•Fig. 2.3c).

Notice how vectors give more information than scalars. Both persons walked the same distance (8.0 m), but the displacements, with directional signs, tell you the direction of the motion of each, and hence that they walked in opposite directions.

A common designation for a vector quantity is a boldface symbol, e.g., $\Delta \mathbf{x}$, or, for a general vector, $\mathbf{A}$. The magnitude *and* direction of the vector must be expressed explicitly—for example, with numerical magnitudes and directional signs, as we have done for linear displacements. In our discussion of straight-line motion we will continue to use this magnitude-sign notation to describe vector quantities wherever appropriate.

Note: In the classroom, many instructors may use the arrow notation: $\vec{A}$. This notation should not be confused with the short line which denotes an average value. Using precise notation is well worth the extra time it takes. Poor or confusing notation can cause problems.

2.2 Speed and Velocity

Objectives: To be able to (a) define and calculate speed and velocity, and (b) perform graphical analyses of velocity.

Speed

When something is in motion, its position changes with time. That is, it moves a certain distance in a given time. Both length and time are therefore important quantities in describing motion. For example, imagine a car and a pedestrian moving down a street and traveling a distance (length) of one block. You would expect the car to travel faster, to cover the distance in a shorter time, than the person does. This can be expressed by using length and time to give the *rate at which distance is traveled*, or **speed**, for each.

Speed: How fast

Average speed is the distance traveled divided by the total time elapsed in traveling that distance:

$$\text{average speed} = \frac{\text{distance traveled}}{\text{total time to travel that distance}}$$

Definition of average speed

$$\bar{s} = \frac{\Delta d}{\Delta t} \qquad (2.2)$$

SI unit of speed: m/s

where the symbol d is used for distance, the actual path length. (A bar over a symbol, as $\bar{s}$ here, is a common way of indicating an average value.) The SI standard unit for speed is m/s (length/time), although km/h is used in many everyday applications. The British standard unit is ft/s, but we often use mi/h.

Since distance is a scalar quantity (as is time), speed is also a scalar. The distance does not have to be in a straight line (see Fig. 2.1). For example, you probably have computed the average speed for an automobile trip using the distance obtained from the starting and ending odometer readings. Suppose these readings were 17,455 km and 17,775 km, respectively, for a 4-hour trip. (We'll assume

• FIGURE 2.4   Instantaneous speed
The speedometer of a car gives the speed over a very short interval of time, so its reading approaches the instantaneous speed. Add the direction of the car's motion and you have a good approximation of the instantaneous velocity.

that you have a car with odometer readings in kilometers.) Subtracting the readings gives a traveled distance of 320 km, so the average speed for the trip is $\bar{s} = \Delta d/\Delta t = 320\ \text{km}/4.0\ \text{h} = 80\ \text{km/h}$ (or 50 mi/h).

Average speed gives a general description of motion over a time interval Δt. In the case of the auto trip with an average speed of 80 km/h, the car's speed wasn't *always* 80 km/h. With various stops and starts on the trip, the car must have been moving more slowly than the average speed part of the time. It therefore had to be moving more rapidly than the average speed part of the time. With an average speed you really don't know how fast the car was moving at any particular time during the trip. Similarly, the average test score for a class doesn't tell you the score of any particular student.

If the time interval (Δt) considered becomes smaller and smaller and approaches zero, the speed calculation gives an **instantaneous speed**. This is how fast something is moving *at a particular instant of time*. The speedometer of a car gives an approximate instantaneous speed. For example, the speedometer shown in • Fig. 2.4 indicates a speed of about 44 mi/h or 70 km/h. If the car travels with a constant speed (so that the speedometer reading does not change), then the average and instantaneous speeds will be equal. (Do you agree? Think of the average test score analogy.)

Velocity

As we have seen, speed, like the distance it incorporates, is a scalar quantity—it has magnitude only. Another quantity used to describe motion is velocity. Speed and velocity are often used synonomously, but the terms have different meanings in physics. Basically, speed is a scalar and velocity is a vector—it has both magnitude and direction. **Velocity** tells you how fast *and* in what direction. And just as we can speak of average and instantaneous speeds, we have average and instantaneous velocities, involving vector displacements. The **average velocity** is the displacement divided by the total travel time:

Velocity: How fast and in what direction

Definition of average velocity

$$\text{average velocity} = \frac{\text{displacement}}{\text{total travel time}}$$

$$\bar{\mathbf{v}} = \frac{\Delta \mathbf{x}}{\Delta t} = \frac{\mathbf{x} - \mathbf{x}_\text{o}}{t - t_\text{o}} \tag{2.3}$$

SI unit of velocity: m/s

The vector difference, $\Delta \mathbf{x} = \mathbf{x} - \mathbf{x}_\text{o}$, is simply the displacement between the initial and final positions. In Fig. 2.3, x_2 and x_1 denote the positions, but x and x_o are used in Eq. 2.3. These are the more general symbols for position and are more convenient because they involve the use of fewer subscripts. The subscripts on x_o and t_o stand for original and indicate that these quantities refer to original, or initial, position and time. The symbols x and t represent position and time at some arbitrary later time and are sometimes called the final position and time.

As discussed in Section 2.1, for motion in one dimension, it is convenient to use plus and minus signs (+ and −) to indicate the directions of displacements and velocities along the positive and negative axes—for example, $+x$ and $+\bar{v}$ (or simply x and $\bar{v}$) and $-x$ and $-\bar{v}$. Also, it is common to take $x_\text{o} = 0$ and $t_\text{o} = 0$. Then, $\Delta x = x$, $\Delta t = t$, and Eq. 2.3 becomes

Note: In Eq. 2.4, t stands for Δt and x stands for Δx.

$$\bar{v} = \frac{x}{t} \quad \text{or} \quad x = \bar{v}t \tag{2.4}$$

As can be readily seen from this equation, the standard units for velocity are the same as those for speed: m/s or ft/s.

You might be wondering whether there is a relationship between average speed and average velocity. A quick look at Fig. 2.3 will show you that *if* the motion is in one direction, the distance is equal to the magnitude of the displacement, and the average speed is the magnitude of the average velocity. However, be careful. This is not true if there is motion in both directions, as the following example shows.

EXAMPLE 2.1 ■ THERE AND BACK: AVERAGE SPEED VERSUS AVERAGE VELOCITY

A jogger jogs from one end to the other of a straight 300-m track (from point A to point B in •Fig. 2.5) in 2.50 min and then turns around and jogs 100 m back toward the starting point (to point C) in another 1.00 min. What are the jogger's average speeds and velocities in going (a) from A to B and (b) from A to C?

Solution. From the problem we have

Given: $x_B = +300$ m (from A to B)   *Find:* Average speeds and average
$x_C = -100$ m (from B to C)   velocities

$t_B = (2.50\text{ min})(60\text{ s/min}) = 150\text{ s}$ ⎱ (conversion to
$t_C = (1.00\text{ min})(60\text{ s/min}) = 60.0\text{ s}$ ⎰ standard units)

(a) The jogger's average speed in going from A to B is easily computed.

$$\bar{s}_B = \frac{\Delta d}{\Delta t}$$

$$= \frac{300\text{ m}}{150\text{ s}} = 2.00\text{ m/s}   \text{(a scalar)}$$

The average velocity in going from A to B with $x_o = 0$ is also easily found, but direction must be indicated. Using Eq. 2.4,

$$\bar{\mathbf{v}}_B = \frac{\mathbf{x}_B}{t_B}$$

$$= \frac{+300\text{ m}}{150\text{ s}} = +2.00\text{ m/s}   \text{(a vector)}$$

The positive direction is to the right in Fig. 2.5. Note that the average speed is equal to the magnitude of the average velocity in this instance.

(b) The average speed in going from A to C involves the *total* distance traveled, so

$$\bar{s}_C = \frac{\Delta d}{\Delta t}$$

$$= \frac{300\text{ m} + 100\text{ m}}{150\text{ s} + 60.0\text{ s}} = 1.90\text{ m/s}$$

where there are no directional signs. (Why?)

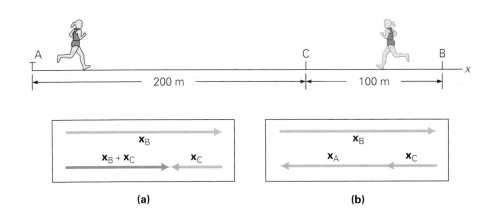

FIGURE 2.5 Average speed and velocity
(a) The jogger starts at A, jogs to B, then back to C. What are the average speeds and velocities for the segments of the lap? See Example 2.1. **(b)** If the jogger returns to the starting point (A), the total displacement, $\mathbf{x}_C + \mathbf{x}_B + \mathbf{x}_A$, is zero.

•**FIGURE 2.6   Back home again!**
Despite having covered nearly
110 m on the base paths (and
regardless of whether he is safe or
out), at the moment he slides
through the batter's box (his
original position) into home the
runner's displacement is zero—at
least if he is a right-handed hitter.
Therefore, no matter how fast he
ran the bases, his average velocity
for the round trip is also zero.
(What would you estimate his
displacement to be if he bats lefty?
Could you also estimate his
average velocity in this case?)

The average velocity, on the other hand, involves the sum of the vector displacements:

$$\bar{\mathbf{v}}_C = \frac{\mathbf{x}_B + \mathbf{x}_C}{t_B + t_C}$$

$$= \frac{+300 \text{ m} + (-100 \text{ m})}{150 \text{ s} + 60.0 \text{ s}} = +0.952 \text{ m/s}$$

Note that direction makes a difference; the average speed is *not* equal to the magnitude of the average velocity in this case. This is because the displacements add vectorially. Notice in Fig. 2.5a, which illustrates this simple case of vector addition, that adding $\mathbf{x}_B$ and $\mathbf{x}_C$ gives an effective or resultant displacement from the initial starting point (A) to the stopping point (C), or +200 m.

Suppose that next the jogger sprints from point C back to the starting point (A) in 0.500 min. What are the average speed and average velocity for the whole lap? The additional data are $x_A = -200$ m (Fig. 2.5) and $t_A = 30.0$ s. Then, for the round trip, the average speed is

$$\bar{s}_A = \frac{\Delta d}{\Delta t}$$

$$= \frac{300 \text{ m} + 100 \text{ m} + 200 \text{ m}}{150 \text{ s} + 60.0 \text{ s} + 30.0 \text{ s}} = 2.50 \text{ m/s}$$

The average velocity is

$$\bar{\mathbf{v}}_A = \frac{\mathbf{x}_B + \mathbf{x}_C + \mathbf{x}_A}{t_B + t_C + t_A}$$

$$= \frac{+300 \text{ m} + (-100 \text{ m}) + (-200 \text{ m})}{240 \text{ s}} = 0 \text{ m/s}$$

The average velocity in this case is zero!

The total displacement is measured from the initial starting point to the final stopping point. When the jogger comes back to the starting point, the displacement is zero and so, therefore, is the average velocity. This is true for any round trip (•Fig. 2.6). Notice in Fig. 2.5b that the (vector) addition of $\mathbf{x}_C + \mathbf{x}_A$ gives a combined or resultant vector that is equal and opposite (in direction) to $\mathbf{x}_B$. The sum of all three vectors is zero—they cancel each other, so to speak.

Follow-up Exercise.   (a) Example 2.1 shows that it is possible to have a zero average velocity. Is it possible for the average speed to be zero? (b) It took six years and 2.3 billion miles for the *Galileo* spacecraft to reach Jupiter in late 1995, where it went into orbit and sent a probe into the Jovian atmosphere. What was *Galileo's* average speed in mi/h for the trip to Jupiter? (*Answers may be found in the Answers to Follow-up Exercises section at the back of the book.*)

Instantaneous velocity: How fast and in what direction *right now*

As Example 2.1 shows, average velocity provides only a very general description of motion. One way to obtain a closer look at motion is to take smaller time segments, that is, to let the time of observation (Δt) become smaller. As with speed, when Δt approaches zero, we obtain the **instantaneous velocity**, which describes how fast something is moving and in what direction at a particular instant of time.

In some instances, the motion of an object may be uniform, which means that the velocity or speed (or both) is constant. For example, the car in •Fig. 2.7 has a uniform velocity (as well as a uniform speed). It travels the same distance in equal time intervals (50 km each hour), and the direction of its motion does not change.

Graphical Analysis

Graphical analysis is often helpful in understanding motion and its related quantities. For example, the car's motion may be represented on a plot of position versus time, or *x* versus *t*. As can be seen from Fig. 2.7, a straight line is obtained for a uniform, or constant, velocity on such a graph.

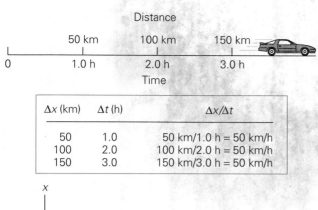

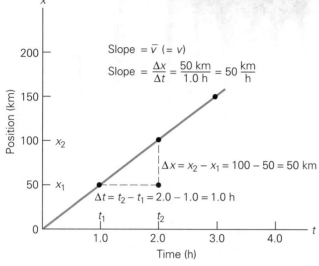

Uniform velocity

•**FIGURE 2.7   Uniform linear motion—constant velocity**
In uniform linear motion, an object travels with a constant velocity, covering the same distance in equal time intervals. Here, a car travels 50 km each hour. An x versus t plot is a straight line, since equal displacements are covered in equal times. The numerical value of the slope of the line is equal to the magnitude of the velocity and the sign of the slope gives its direction. (Average velocity equals instantaneous velocity in this case.)

Recall from Cartesian graphs of y versus x that the slope of a straight line is given by $\Delta y/\Delta x$. Here, with a plot of x versus t, the slope of the line, $\Delta x/\Delta t$, is equal to the average velocity ($\bar{v} = \Delta x/\Delta t$). For uniform motion, this is equal to the instantaneous velocity. That is, $\bar{v} = v$. (Why?) The numerical value of the slope is the magnitude of the velocity and the sign of the slope gives direction. A positive slope indicates that x increases with time, so the motion is in the positive x direction. (The plus sign is often omitted as being understood.)

Suppose that a plot of position versus time for a car's motion was a straight line with a negative slope, as in •Fig. 2.8. What does this indicate? As can be seen in the figure, the position (x) values get smaller with time, indicating that the car was traveling in uniform motion in the negative x direction.

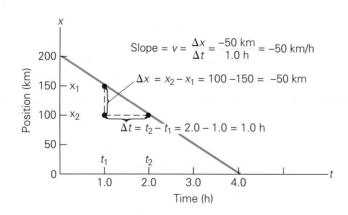

•**FIGURE 2.8   Position versus time graph for an object in uniform motion in the −x direction**
A straight line on an x versus t plot with a negative slope indicates uniform motion in the −x direction. Note that the object's location changes at a constant rate. At $t = 4.0$ h, the object is at the origin ($x = 0$). How does the graph look if the motion continues for $t > 4.0$ h?

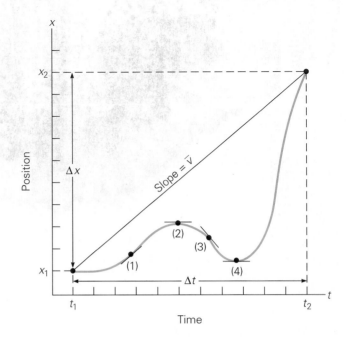

FIGURE 2.9 Position versus time graph for an object in nonuniform linear motion
For a nonuniform velocity, an x versus t plot is a curved line. The slope of the line between two positions is the average velocity between those positions, and the instantaneous velocity is the slope of a line tangent to the curve at any point. Four tangent lines are shown in the figure. Can you describe the object's motion in words?

In most instances the motion of an object is nonuniform, meaning that different distances are covered in equal intervals of time. An x versus t plot for such motion in one dimension is a curved line, as illustrated in ●Fig. 2.9. The average velocity for a particular interval of time is the slope of a straight line between the two points on the curve corresponding to the starting and ending times of the interval. In the figure, the average velocity for the total trip is the slope of the straight line joining the beginning and ending points of the curve (t_1 and t_2).

The instantaneous velocity is equal to the slope of a straight line tangent to the curve at a specific point. Several tangent lines are shown in Fig. 2.9. At (1), the slope is positive, and the motion is in the positive x direction. At (2), the slope of a horizontal tangent line is zero, so there is no motion. That is, the object has stopped ($v = 0$). At (3), the slope is negative, so the object is moving in the negative x direction. Thus, it stopped and changed direction at point (2). What is happening at point (4)?

Drawing various tangent lines along the curve, we see that their slopes vary, indicating that the instantaneous velocity is changing with time. An object in nonuniform motion speeds up and/or slows down. How motion with a change in velocity is described is the topic of the next section.

But, before speeding things up (or slowing them down), consider the following example.

CONCEPTUAL EXAMPLE 2.2 ■ A ROUND TRIP: USING DEFINITIONS

In driving the usual route to school, a student computes the average speed for the trip to be 20 km/h. On the return trip along the same route, there is less traffic and the average speed is 30 km/h. What is the average speed for the total trip? (a) less than 25 km/h, (b) 25 km/h, (c) greater than 25 km/h? *Clearly establish the reasoning used in determining your answer before checking it below. That is, **why** did you select your answer?*

Reasoning and Answer. It is tempting to choose 25 km/h as the correct answer, as this is the average of the two average speeds. However, this is not the case. The average of the average speeds would be 25 km/h only if the student spent *the same amount of time* traveling at each speed. For example, suppose you spent an hour traveling at each of the given speeds. Then it is easy to see that the average speed for the total trip would be 25 km/h: $\bar{s}$ = (20 km + 30 km)/2.0 h = 25 km/h.

But in this case, the *distances traveled* rather than the times traveled are the same. To travel the same distance, more time must be spent traveling at 20 km/h than at 30 km/h. Hence the total average is weighted toward the slower average speed, and the answer is (a).

You can confirm this by finding the actual average speed, using the definition of average speed: the total distance divided by the time to travel this distance. We are not given the distance, but let's say that it is d one way, for a total round-trip distance of $2d$. The times for each part of the trip are given by $t_1 = d/\bar{s}_1$ and $t_2 = d/\bar{s}_2$. Thus, the average speed for the total trip, by definition, is

$$\bar{s} = \frac{\text{total distance}}{\text{time to travel distance}} = \frac{2d}{t_1 + t_2} = \frac{2d}{d/\bar{s}_1 + d/\bar{s}_2} = \frac{2}{1/\bar{s}_2 + 1/\bar{s}_2}$$

where the d in the denominator was factored out and canceled. So,

$$\bar{s} = \frac{2}{1/\bar{s}_1 + 1/\bar{s}_2} = \frac{2}{1/20 + 1/30} = \frac{2}{3/60 + 2/60} = \frac{2}{5/60} = 120/5 = 24 \text{ km/h}$$

where the units were omitted in the calculation for convenience.

Follow-up Exercise. What average speed on the return trip would give an average of 25 km/h for the total trip? (*Reasoning and answer may be found in the Answers to Follow-up Exercises section at the back of the book.*)

2.3 Acceleration

Objectives: To be able to (a) explain the relationship between velocity and acceleration, and (b) perform graphical analyses of acceleration.

The basic description of motion involves the time rate of change of position, which we call velocity. Going one step further, we can consider how this *rate of change* changes. Suppose that something is moving at a constant velocity and then the velocity changes. Such a change in velocity is an *acceleration*. The gas pedal on an automobile is commonly called the accelerator. When you press down on the accelerator, the car speeds up; when you let up on the accelerator, the car slows down. In either case, there is a change in velocity with time. We define **acceleration** as the time rate of change of velocity.

Acceleration: A change in velocity

Analogous to average velocity is the **average acceleration**, or the change in velocity divided by the time taken to make the change:

$$\text{average acceleration} = \frac{\text{change in velocity}}{\text{time to make the change}}$$

$$\bar{\mathbf{a}} = \frac{\Delta \mathbf{v}}{\Delta t} = \frac{\mathbf{v} - \mathbf{v}_o}{t - t_o} \quad (2.5)$$

Definition of average acceleration

SI unit of acceleration: m/s^2

where $\mathbf{v}$ and $\mathbf{v}_o$ are instantaneous velocities—the velocities at times t and t_o. Here we use the boldface vector notation for, in general, the velocities may be in different directions. Since velocity is a vector quantity, so is acceleration. Analogous to instantaneous velocity is **instantaneous acceleration**, which is the acceleration at a particular instant of time.

Instantaneous acceleration: a change in velocity *right now*

The dimensions of acceleration are (length/time)/time (as is obvious from $\Delta v/\Delta t$). The SI units for acceleration are therefore (m/s)/s, or m/s·s, commonly written m/s^2 (and read as "meters per second squared"). In the British system, the units are ft/s^2.

Note: In compound units, multiplication is indicated by a dot.

Since velocity is a vector quantity, having both magnitude and direction, a change in velocity may involve either or both of these factors. An acceleration, therefore, may result from a change in *speed* (magnitude), a change in *direction*, or a change in *both*, as illustrated in •Fig. 2.10.

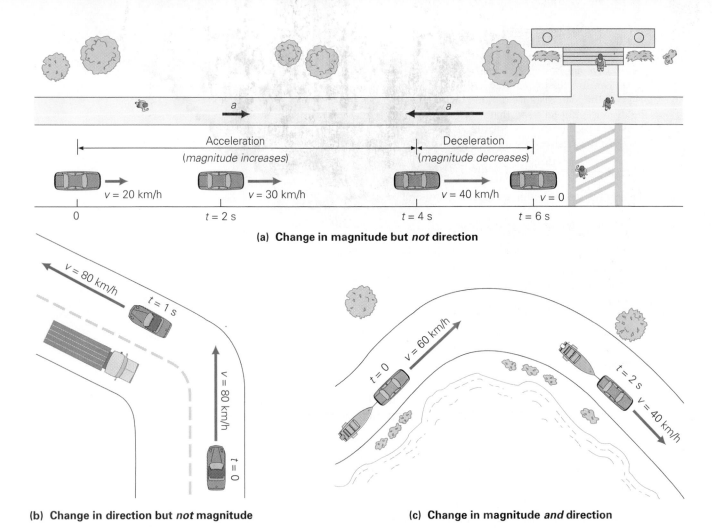

(a) **Change in magnitude but *not* direction**

(b) **Change in direction but *not* magnitude**

(c) **Change in magnitude *and* direction**

•**FIGURE 2.10   Acceleration—the time rate of change of velocity**
Since velocity is a vector quantity with magnitude and direction, an acceleration can occur when there is **(a)** a change in magnitude but not direction, **(b)** a change in direction but not magnitude, or **(c)** a change in both magnitude and direction.

For the special case of straight-line or linear motion, plus and minus signs will be used to indicate velocity directions, as was done for linear displacements. Then, Eq. 2.5 can be written as

$$\bar{a} = \frac{v - v_{\mathrm{o}}}{t} \tag{2.6}$$

where t_{o} is taken to be zero (v_{o} may not be zero, so it cannot generally be omitted).

EXAMPLE 2.3 ■ SLOWING IT DOWN: AVERAGE ACCELERATION

An automobile traveling on a straight road at 90 km/h slows down to 40 km/h in 5.0 s. What is its average acceleration?

Solution.   From the problem, we have the following data. [With the motion in a straight line, the instantaneous velocities are assumed to be in the positive direction, and conversions to standard units (km/s to m/s) are made right away since it is noted that the time is given in seconds. In general, one always works with acceleration in standard units.]

Given: $v_0 = (90 \text{ km/h})\left(\dfrac{0.278 \text{ m/s}}{1 \text{ km/h}}\right)$ *Find:* $\bar{a}$ (average acceleration)

$= 25$ m/s

$v = (40 \text{ km/h})\left(\dfrac{0.278 \text{ m/s}}{1 \text{ km/h}}\right)$

$= 11$ m/s

$t = 5.0$ s

Given the initial and final velocities and the time interval, the average acceleration may be found using Eq. 2.6.

$$\bar{a} = \frac{v - v_0}{t}$$

$$= \frac{11 \text{ m/s} - 25 \text{ m/s}}{5.0 \text{ s}} = -2.8 \text{ m/s}^2$$

The minus sign indicates the direction of the (vector) acceleration. In this case, the acceleration is opposite to the direction of the initial motion ($+v_0$), and it slows the car. Such an acceleration is sometimes called a *deceleration*, since the car is slowing.

Follow-up Exercise. Does a negative acceleration necessarily mean that a moving object is slowing down (decelerating) or its velocity is decreasing? (*Answer may be found in the Answers to Follow-up Exercises section at the back of the book.*)

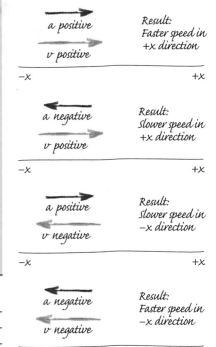
Although acceleration can vary with time, our study of motion will be restricted to constant accelerations for simplicity. (An important constant acceleration is the acceleration due to gravity near the Earth's surface, which will be considered in the next section.) Since for a constant acceleration the average is equal to the constant value ($\bar{a} = a$), the bar over the acceleration in Eq. 2.6 may be omitted. Thus, for a constant acceleration, the equation relating velocity, acceleration, and time is commonly written

$$v = v_0 + at \qquad \text{(constant acceleration only)} \qquad (2.7)$$

EXAMPLE 2.4 ■ FAST START, SLOW STOP: MOTION WITH CONSTANT ACCELERATION

A drag racer starting from rest accelerates in a straight line at a constant rate of 5.5 m/s^2 for 6.0 s. (a) What is the racer's velocity at the end of this time? (b) If a parachute deployed at this time causes the racer to slow down uniformly at a rate of 2.4 m/s^2, how long will it take the racer to come to a stop?

Solution. Notice that the racer first speeds up and then slows down, so we must pay close attention to the directional signs of the vector quantities. Taking the initial motion to be in the positive direction, we have,

Given: (a) $v_0 = 0$ (at rest)   *Find:* v_f (final velocity)
$a = 5.5$ m/s^2
$t = 6.0$ s

(b) $v_0 = v$ (from part a)   *Find:* t (time)
$v = 0$ (comes to stop)
$a = -2.4$ m/s^2 (opposite direction of v_0)

The data have been listed in two parts. This helps avoid confusion with symbols. Note that the final velocity (v) that is to be found in part (a) becomes the initial velocity (v_0) for part (b).

(a) To find v, we use Eq. 2.7 directly:

$$v = v_o + at = 0 + (5.5 \text{ m/s}^2)(6.0 \text{ s}) = 33 \text{ m/s}$$

(b) Here we want time, so solving Eq. 2.6 for t and using $v_o = 33$ m/s from part (a), we have

$$t = \frac{v - v_o}{a} = \frac{0 - 33 \text{ m/s}}{-2.4 \text{ m/s}^2} = 14 \text{ s}$$

Note that the time comes out positive, as it should. We start implicitly at zero time (taking $t_o = 0$ when the parachute is deployed) and time goes forward or in a "positive direction."

Follow-up Exercise. What is the racer's instantaneous velocity in this Example 10 seconds after the parachute is deployed? (*Answer may be found in the Answers to Follow-up Exercises section at the back of the book.*)

Motions with constant accelerations are easy to represent graphically using instantaneous velocity versus time. A v versus t plot is a straight line whose slope is equal to the acceleration, as illustrated in •Fig. 2.11. Note that Eq. 2.7 can be written $v = at + v_o$, which, as you may recognize, has the form of an equation for a straight line, $y = mx + b$ (slope m and intercept b). In Fig. 2.11(a), the motion is in the positive direction, and the accleration adds to the velocity for a time t, as

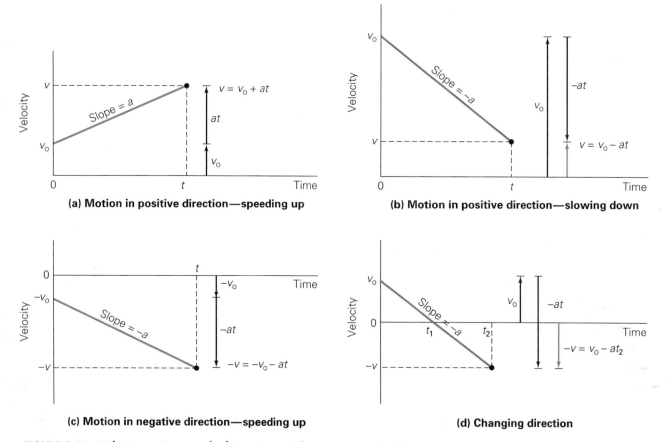

(a) Motion in positive direction—speeding up

(b) Motion in positive direction—slowing down

(c) Motion in negative direction—speeding up

(d) Changing direction

•**FIGURE 2.11   Velocity vs. time graphs for motions with constant accelerations**
The slope of a v versus t plot is the acceleration. **(a)** A positive slope indicates an increase in the velocity in the positive direction. The vertical arrows to the right indicate how the acceleration adds velocity to the initial velocity v_o. **(b)** A negative slope indicates a decrease in the initial velocity v_o, or a deceleration. **(c)** Here a negative slope indicates a negative acceleration, but the initial velocity is in the negative direction, $-v_o$, so the speed of the object increases in that direction. **(d)** The situation here is initially similar to that in (b), but ends by resembling that in (c). Can you explain what happened at time t_1?

illustrated by the vertical arrows at the right of the figure. In (b), the negative slope indicates a negative acceleration that produces a slowing down or deceleration. However, (c) illustrates how a negative acceleration can speed things up (for motion in the negative direction). The situation in (d) is slightly more complex. Can you explain what is happening there?

When an object moves with a constant acceleration, its velocity changes by the same amount in each time unit. For example, if the acceleration is 10 m/s^2 in the same direction as the initial velocity, the object's velocity increases by 10 m/s in each second. Suppose that the object has an initial velocity (v_o) of 20 m/s at $t_o = 0$. Then, for $t = 0, 1.0, 2.0, 3.0,$ and 4.0 s, the velocities are 20, 30, 40, 50, and 60 m/s, respectively. The average velocity over the 4-s interval is $\bar{v} = 40$ m/s. This average may be computed in the regular manner, or you may immediately recognize that the uniformly increasing series of numbers 20, 30, 40, 50, and 60 has an average value of 40 (the midway value of the series). Note that the average of the extreme (initial and final) values also gives the average of the series, that is, $(20 + 60)/2 = 40$. This is true in general: When the velocity changes at a uniform rate because of a constant acceleration, $\bar{v}$ will be the average of the initial and final velocities.

$$\bar{v} = \frac{v + v_o}{2} \qquad \text{(constant acceleration only)} \qquad (2.8)$$

EXAMPLE 2.5 ■ ON THE WATER: USING MULTIPLE EQUATIONS

A motorboat starting from rest on a lake accelerates in a straight line at a constant rate of 3.0 m/s^2 for 8.0 s. How far does the boat travel during this time?

Solution. Reading the problem and summarizing the given data and what is to be found, we have

Given: $v_o = 0$ **Find:** x (distance)
$a = 3.0 \text{ m/s}^2$
$t = 8.0$ s

(It is noted that all of the units are standard.)

In analyzing the problem, the reasoning might go something like this: To find x, we will need to use Eq. 2.4, $x = \bar{v}t$. This is the only equation we have for finding distance. (The average velocity $\bar{v}$ must be used because the velocity is changing and not constant.) With t given, the solution to the problem then involves finding $\bar{v}$. By Eq. 2.8, $\bar{v} = (v + v_o)/2$, so with $v_o = 0$, we need only find the final velocity v and the problem is solved. Eq. 2.7, $v = v_o + at$, enables us to calculate this from the given data. So, computing the required quantities in reverse order to the reasoning in solving the problem, we have:

The velocity of the boat at the end of 8.0 s is

$$v = v_o + at = 0 + (3.0 \text{ m/s}^2)(8.0 \text{ s}) = 24 \text{ m/s}$$

The average velocity over that time interval is

$$\bar{v} = \frac{v + v_o}{2} = \frac{24 \text{ m/s} + 0}{2} = 12 \text{ m/s}$$

Finally, the magnitude of the displacement, which in this case is the same as the distance traveled, is

$$x = \bar{v}t$$
$$= (12 \text{ m/s})(8.0 \text{ s}) = 96 \text{ m}$$

Follow-up Exercise. (Sneak preview.) In the next section, the following equation will be derived: $x = v_o t + \frac{1}{2}at^2$. Use the data in this Example to see if this equation gives the distance traveled. (*Answer may be found in the Answers to Follow-up Exercises section at the back of the book.*)

2.4 Kinematic Equations

Objectives: To be able to (a) explain the kinematic equations, and (b) apply them to physical situations.

The description of motion in one dimension with constant acceleration requires only three basic equations. From previous sections, these are

$$x = \bar{v}t \tag{2.4}$$

$$\bar{v} = \frac{v + v_o}{2} \qquad \text{(constant acceleration only)} \tag{2.8}$$

$$v = v_o + at \qquad \text{(constant acceleration only)} \tag{2.7}$$

(Keep in mind that the first equation is general and is not limited to situations where there is constant acceleration as the latter two are.)

However, as Example 2.5 showed, the description of motion in some instances requires multiple applications of these equations, which may not be obvious at first. It would be helpful if there were a way to reduce the number of operations in solving kinematic problems, and there is—combining equations algebraically.

For instance, combining the above equations involves first substituting for $\bar{v}$ from Eq. 2.8 into Eq. 2.4:

$$x = \bar{v}t = \left(\frac{v + v_o}{2}\right)t$$

Then substituting for v from Eq. 2.7 gives

$$x = \left(\frac{v + v_o}{2}\right)t = \left[\frac{(v_o + at) + v_o}{2}\right]t$$

Simplifying gives

$$x = v_o t + \tfrac{1}{2} at^2 \qquad \text{(constant acceleration only)} \tag{2.9}$$

Essentially, this series of steps was done in Example 2.5. This combined equation allows the distance traveled by the motorboat in that example to be computed directly.

$$x = v_o t + \tfrac{1}{2} at^2 = 0 + \tfrac{1}{2}(3.0 \text{ m/s}^2)(8.0 \text{ s})^2 = 96 \text{ m}$$

Much easier, isn't it?

Another possibility is to eliminate time (t), rather than the final velocity (v). As before, we first substitute the expression for $\bar{v}$ from Eq. 2.8 into Eq. 2.4, which gives us

$$x = \bar{v}t = \left(\frac{v + v_o}{2}\right)t$$

Then, writing Eq. 2.7 in the form $t = (v - v_o)/a$ gives us an expression for t that we can substitute into the previous equation, giving

$$x = \left(\frac{v + v_o}{2}\right)t = \left(\frac{v + v_o}{2}\right)\left(\frac{v - v_o}{a}\right)$$

Simplifying gives

$$v^2 = v_o^2 + 2ax \qquad \text{(constant acceleration only)} \tag{2.10}$$

Students in introductory physics courses are sometimes overwhelmed by the various kinematic equations—right in the second chapter! However, keep in mind that equations and mathematics are the "tools" of physics; and as any mechanic or carpenter will tell you, tools make your work easier as long as you are familiar with and know how to use them. The same is true with our physics "tools."

Basically, just equations

$$v = v_o + at \tag{2.7}$$
$$x = v_o t + \tfrac{1}{2} at^2 \tag{2.9}$$
$$v^2 = v_o^2 + 2ax \tag{2.10}$$

are used to solve the majority of kinematic problems. (Occasionally we are interested in average speed or velocity, but, as pointed out earlier, averages don't tell you a great deal.) Now, note that each of these three equations has four variables. As you know, three of these must be known in a particular equation to solve for the fourth, unknown variable. You should always try to understand and visualize a problem, but listing the data as in the Suggested Problem-Solving Procedure in Chapter 1 may help you to decide which equation to use, since you generally need to know three variables. Remember this as you go over the remaining examples in the chapter.

Also, don't overlook any *implied data*. For example, what does "dropped" in the following Example imply?

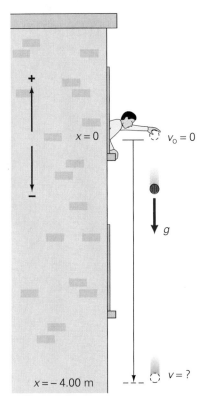

•FIGURE 2.12 **A dropped ball**
A sketch to help visualize Example 2.6.

EXAMPLE 2.6 ■ A DROPPED BALL: USING THE KINEMATIC EQUATIONS

A ball is dropped out of a window near the top of a building. If it accelerates toward the ground at a rate of 9.80 m/s², what is its velocity when it has fallen 4.00 m?

Solution. Here, we list the given data and make a sketch to help visualize the situation (•Fig. 2.12).

Given: $v_o = 0$ (dropped)   *Find:* v (velocity)
$a = -9.80$ m/s²
$x = -4.00$ m

Note that three variables are given: the statement that the ball is dropped means that the initial velocity is zero, $v_o = 0$. Upward is taken as the positive direction and downward as the negative direction. The magnitudes of a and x are then written with minus signs to indicate their directions. No directional sign is explicitly assigned to the velocity we wish to find. Its direction will be given in the result of the mathematical operations. Of course, in this case you should realize that the final velocity will be negative or in the downward direction.

The velocity may be obtained directly from Eq. 2.10:
$$v^2 = v_o^2 + 2ax$$
$$= 0 + 2(-9.80 \text{ m/s}^2)(-4.00 \text{ m}) = 78.4 \text{ m}^2/\text{s}^2$$
and
$$v = \pm\sqrt{78.4 \text{ m}^2/\text{s}^2} = -8.85 \text{ m/s}$$

where the negative root is chosen because the motion is downward (that is, in the negative direction) in this physical situation. (Keep in mind that when you take the square root of a quantity you always obtain both positive and negative roots.)

Using the combined forms of the equations can often save you a lot of steps and calculations.

Follow-up Exercise. In this Example, how long does it take for the ball to fall the 4.00 m? (*Answer may be found in the Answers to Follow-up Exercises section at the back of the book.*)

TABLE 2.1 Equations for Linear Motion with Constant Acceleration*	
$x = \bar{v}t$	(2.4)†
$\bar{v} = \dfrac{v + v_o}{2}$	(2.8)
$v = v_o + at$	(2.7)
$x = v_o t + \tfrac{1}{2}at^2$	(2.9)
$v^2 = v_o^2 + 2ax$	(2.10)

*It is assumed that $x_o = 0$ and the velocity is v_o at $t_o = 0$. The initial position and time, x_o and t_o, may be included for general cases, for example, $x - x_o = \bar{v}(t - t_o)$.

†Note that Eq. 2.4 is not limited to constant acceleration but applies generally.

For your convenience, the equations for linear motion with constant acceleration are summarized in Table 2.1 . Note that only the first three are basic equations; the last two are convenient combinations. Also, they are quite versatile.

CONCEPTUAL EXAMPLE 2.7 ■ TWO RACING CARS: THE EFFECT OF SQUARED QUANTITIES

During some time trials, race car A, starting from rest, accelerates uniformly along a straight, level track for a particular time interval. Similarly, car B accelerates at the same rate but for twice the time. Then, at the ends of their respective acceleration periods, which of these statement is true: (a) car A has traveled a greater distance, (b) car B has traveled twice as far as car A, (c) car B has traveled four times as far as car A, (d) both cars have traveled the same distance? *Clearly establish the reasoning and physical principle(s) used in determining your answer before checking it below. That is, **why** did you select your answer?*

Reasoning and Answer. It is given that $v_o = 0$ and the acceleration is the same for both cars. To find the distances traveled in a time t, you would use Eq. 2.9, which, with $v_o = 0$, becomes $x = \frac{1}{2}at^2$. The important thing to note here is that the distance increases as t^2. That is, if you double the time, the distance quadruples (increases by a factor of four).

In this example, car B accelerates twice as long as car A, or $t_B = 2t_A$, so car B travels four times as far as car A and the answer is (c). Expressed mathematically,

$$x_A = [\tfrac{1}{2}at_A^2] \quad \text{and} \quad x_B = \tfrac{1}{2}at_B^2 = \tfrac{1}{2}a(2t_A)^2 = \tfrac{1}{2}a(4t_A^2) = 4[\tfrac{1}{2}at_A^2] = 4x_A.$$

What if $t_B = 3t_A$?

Follow-up Exercise. How would the speeds of the cars compare at the ends of the acceleration periods? *(Reasoning and answer may be found in the Answers to Follow-up Exercises section at the back of the book.)*

EXAMPLE 2.8 ■ PUTTING ON THE BRAKES: VEHICLE STOPPING DISTANCE

The stopping distance for a vehicle after the brakes have been applied is an important factor in highway safety. This distance depends on the initial speed (v_o) and the braking capacity, or deceleration, $-a$, which is assumed to be constant. (Recall that the minus sign indicates that the acceleration is in the negative direction, which in this case is opposite that of the velocity. Thus the car slows to a stop.) Express the stopping distance x in terms of these quantities.

Solution. Here, we are working with variables, so we can represent quantities only in symbolic form.

Given: v_o *Find:* x (in terms of the given variables)
         $-a$ (opposite direction of v_o)
         $v = 0$ (car comes to stop)

Again, it is helpful to make a sketch of the situation, particularly when directional vector quantities are involved (see ●Fig. 2.13). Since Eq. 2.10 has the variables we want, it should allow us to find the stopping distance. Expressing the negative acceleration explicitly gives

$$v^2 = v_o^2 + 2(-a)x = v_o^2 - 2ax$$

Since the vehicle comes to a stop ($v = 0$), we can solve for x:

$$x = \frac{v_o^2}{2a}$$

●**FIGURE 2.13 Vehicle stopping distance**
A sketch to help visualize Example 2.8.

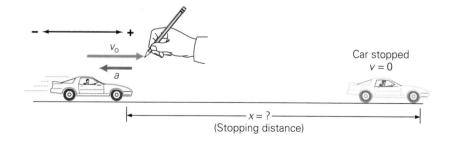

This gives us x expressed in terms of the vehicle's initial speed and stopping acceleration.

Notice that the stopping distance x is proportional to the *square* of the initial speed. Doubling the initial velocity therefore increases the stopping distance by a factor of 4 (for the same deceleration). That is, if the stopping distance is x_1 for an initial speed of v_1, then for a two-fold initial speed ($v_2 = 2v_1$), the stopping distance would be four-fold:

$$x_1 \propto v_1^2$$
$$x_2 \propto v_2^2 = (2v_1)^2 = 4v_1^2$$

so

$$x_2 = 4x_1$$

We can get the same result by using ratios:

$$\frac{x_2}{x_1} = \frac{v_2^2}{v_1^2} = \left(\frac{v_2}{v_1}\right)^2 = \left(\frac{2v_o}{v_o}\right)^2 = 4.$$

Do you think this is an important consideration in setting speed limits, for example, in school zones? (The driver's reaction time should also be considered. A method for approximating a person's reaction time is given in the next section.)

Follow-up Exercise. A driver traveling at 25 mi/h in a school zone can brake to an emergency stop in 8.0 ft. Under similar conditions, what would be the braking distance if the car were traveling at 40 mi/h? (*Answer may be found in the Answers to Follow-up Exercises section at the back of the book.*)

2.5 Free Fall

Objective: To be able to analyze free fall using the kinematic equations.

One of the more common cases of constant acceleration is the acceleration due to gravity near the Earth's surface. When an object is dropped, its initial velocity is zero (at the instant it is released), but at a later time while falling, it has a nonzero velocity. There has been a change in velocity and so, by definition, an acceleration. This **acceleration due to gravity** (g) has an approximate magnitude of

Acceleration due to gravity

$$g = 9.80 \text{ m/s}^2 \quad \textit{acceleration due to gravity}$$

or 980 cm/s^2 and is directed downward (toward the center of the Earth). In British units, the value of g is about 32 ft/s^2.

The values given here for g are only approximate because the acceleration due to gravity varies slightly at different locations as a result of differences in elevation and regional average mass density of the Earth. These small variations will be ignored in this book unless otherwise noted. (Gravitation is studied in more detail in Chapter 7.) Air resistance is another factor that affects the acceleration of a falling object. But for relatively dense objects and over the short distances of fall commonly encountered, air resistance produces only a small effect, which will also be ignored here for simplicity. (The frictional effect of air resistance will be considered in Chapter 4.)

*Objects in motion solely under the influence of gravity are said to be in **free fall**.* The words "free fall" bring to mind dropped objects that are moving downward under the influence of gravity ($g = 9.80 \text{ m/s}^2$ in the absence of air resistance). However, the term can be applied in general to any vertical motion under the influence of gravity. Objects released from rest or thrown upward or downward are in free fall once they are released. That is, after $t = 0$ (the time of release), only gravity is acting and influencing the motion. (Even when an object projected upward is traveling upward, *it is still accelerating downward.*) Thus, the set of equations for motion in one dimension (in Table 2.1) can be used to describe generalized free fall.

The acceleration due to gravity g is the *constant* acceleration for all free-falling objects, regardless of their mass or weight. It was once thought that heavier bodies fell faster than lighter bodies. This was part of Aristotle's theory of motion. You can easily observe that a coin falls faster than a sheet of paper when dropped simultaneously from the same height. But in this case air resistance plays a noticeable role. If the paper is crumpled into a compact ball, it gives the coin a better race. Similarly, a feather "floats" down much more slowly than a coin falls. However, in a near-vacuum, where there is negligible air resistance, the feather and the coin fall with the same acceleration—the acceleration due to gravity (•Fig. 2.14).

Astronaut David Scott performed a similar experiment on the Moon in 1971 by simultaneously dropping a feather and a hammer from the same height. Of course, he did not need a vacuum pump since the Moon has no atmosphere and no air resistance. The hammer and the feather reached the lunar surface together, but fell at a slower rate than on Earth. The acceleration due to gravity near the Moon's surface is approximately one-sixth of that near the Earth's surface ($g_m \approx g/6$).

Currently accepted ideas about the motion of falling bodies are due in large part to Galileo. He challenged Aristotle's theory and experimentally investigated the motion of objects. Legend has it that he studied the accelerations of falling bodies by dropping objects of different weights from the top of the Leaning Tower of Pisa (see the Insight on p. 49).

It is customary to use y to represent the vertical direction and to take upward as positive (as with the vertical y axis of Cartesian coordinates). Since the acceleration due to gravity is always downward, it is in the negative direction. This negative acceleration, $a = -g = -9.80 \text{ m/s}^2$, may be substituted each time into the equations of motion. However, the relationship $a = -g$ may be expressed explicitly in the equations for linear motion (see Table 2.1):

Note: A projected object while traveling upward ($+v$) is accelerating downward, or has an acceleration in the downward direction ($-g$).

$$y = \bar{v}t \tag{2.4'}$$

$$\bar{v} = \frac{v + v_o}{2} \tag{2.8'}$$

Free-fall equations with $-g$ expressed explicit

$$v = v_o - gt \tag{2.7'}$$

$$y = v_o t - \tfrac{1}{2}gt^2 \tag{2.9'}$$

$$v^2 = v_o^2 - 2gy \tag{2.10'}$$

Note: g here refers to the magnitude of acceleration due to gravity.

The origin ($y = 0$) of the reference frame is usually taken to be at the initial position of the object. Since upward is generally taken to be in the positive direction ($+y$ axis on a graph), writing $-g$ explicitly in the equations reminds you of direc-

•**FIGURE 2.14 Free fall and air resistance**
(a) When dropped simultaneously from the same height, a feather falls more slowly than a coin because of air resistance. But when both objects are dropped in an evacuated container with a good partial vacuum, where air resistance is negligible, the feather and the coin fall together with a constant acceleration. (b) An actual demonstration with multiflash photography. An apple and a feather are released simultaneously through a trap door into a large vacuum chamber, and they fall together—almost. Although the chamber has a partial vacuum, there is still some air resistance. (How can you tell?)

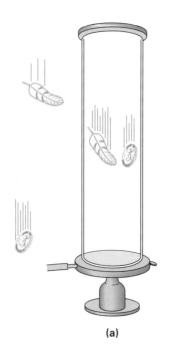

(a)

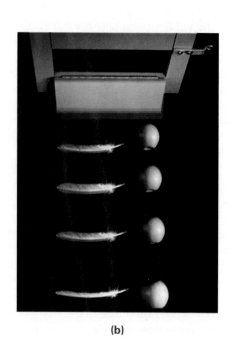

(b)

Galileo Galilei and the Leaning Tower of Pisa

Galileo Galilei (Fig. 1) was born in Pisa, Italy, in 1564 during the Renaissance. Today, he is known throughout the world by his first name and often referred to as the father of modern science or the father of modern mechanics and experimental physics, which attests to the magnitude of his scientific contributions.

One of Galileo's greatest contributions to science was the establishment of the scientific method, that is, investigation through experiment. In contrast, Aristotle's approach was based on logical deduction. By the scientific method, for a theory to be valid it must predict or agree with experimental results. If it doesn't, it is invalid or requires modification. Galileo said, "I think that in the discussion of natural problems we ought not to begin at the authority of places of Scripture, but at sensible experiments and necessary demonstrations."*

Probably the most popular and well-known legend about Galileo is that he performed experiments with falling bodies by dropping objects from the Leaning Tower of Pisa (see Fig. 2). There is some doubt as to whether Galileo actually did this, but there is little doubt that he questioned Aristotle's view on the motion of falling objects. In 1638, Galileo wrote:

> Aristotle says that an iron ball of one hundred pounds falling from a height of one hundred cubits reaches the ground before a one-pound ball has fallen a single cubit. I say that they arrive at the same time. You find, on making the experiment, that the larger outstrips the smaller by two finger-breadths, that is, when the larger has reached the ground, the other is short of it by two finger-breadths; now you would not hide behind these two fingers the ninety-nine cubits of Aristotle.•

This and other writings show that Galileo was aware of the effect of air resistance.

The experiments at the Tower of Pisa were supposed to have taken place around 1590. In his writings of about that time, Galileo mentions dropping objects from a high tower, but never specifically names the Tower of Pisa. A letter written to Galileo from another scientist in 1641 describes the dropping of a cannon ball and a musket ball from the Tower of Pisa. The first account of Galileo doing a similar experiment was written a dozen years after his death by Vincenzo Viviani, his last pupil and first biographer. Viviani related that the falling bodies "all moved at the same speed" and that Galileo demonstrated "this with repeated experiments from the height of the Campanile [Tower] of Pisa in the presence of the other teachers and philosophers, and the whole assembly of students." However, there is no other record of this event, which seems odd given that a crowd of people supposedly witnessed it.

It is not known whether Galileo told this story to Viviani in his declining years or Viviani created this picture of his former teacher.

The important point is that Galileo recognized (and probably experimentally showed) that free-falling objects fall at the

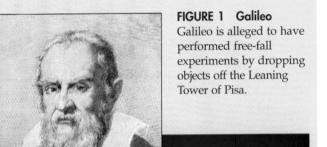

FIGURE 1 Galileo
Galileo is alleged to have performed free-fall experiments by dropping objects off the Leaning Tower of Pisa.

FIGURE 2 The Leaning Tower of Pisa
The tower, constructed as a belfry for the nearby cathedral, was built on shifting subsoil. Construction began in 1173, and the tower started to shift one way and then the other, before inclining in its present direction. Today, the tower leans about 4 meters (13 ft) from the vertical. It has been closed since 1990 and efforts are being made to stabilize the leaning.

same rate regardless of their mass or weight (see Fig. 2.14). Galileo gave no reason why all objects in free fall have the same acceleration, but Newton did, as you will learn in a later chapter.

*From *Growth of Biological Thought: Diversity, Evolution & Inheritance*, by F. Meyr (Cambridge, MA: Harvard University Press, 1982).
•This and the next quotation are both from *Aristotle, Galileo, and the Tower of Pisa*, by L. Cooper (Ithaca, NY: Cornell University Press, 1935).

tional differences, and the value of g is inserted as 9.80 m/s². However, the choice is arbitrary. The equations can be written with $a = g$, for example, $v = v_o + gt$, with the directional minus sign associated directly with g. In this case, a value of -9.80 m/s² must be substituted for g each time.

Note that you have to be explicit about the directions of vector quantities when using these equations. The displacement y and the velocities v and v_o may be positive or negative, depending on the direction of motion. The use of these equations and the sign convention is illustrated in the following Examples.

EXAMPLE 2.9 ■ A STONE THROWN DOWNWARD: THE KINEMATIC EQUATIONS REVISITED

A boy on a bridge throws a stone vertically downward toward the river below with an initial velocity of 14.7 m/s. If the stone hits the water 2.00 s later, what is the height of the bridge above the water?

Solution. As usual, we first read the problem and write down what is given and what is to be found.

Given: $v_o = -14.7$ m/s (downward taken as   *Find:* y (height)
         $t = 2.00$ s   the negative direction)
         g (= 9.80 m/s²)

Notice that g is now just a positive number, since the directional minus sign has already been put into the previous equations of motion. After a while, you will probably just write down the symbol g, since you will be familiar with its numerical value. This time, draw a sketch of your own to help analyze the situation.

Considering which equation(s) will provide the solution using the given data, it should become evident that the distance the stone travels in a time t is given directly by Eq. 2.9':

$$y = v_o t - \tfrac{1}{2} g t^2 = (-14.7 \text{ m/s})(2.00 \text{ s}) - \tfrac{1}{2} (9.80 \text{ m/s}^2)(2.00 \text{ s})^2$$

$$= -29.4 \text{ m} - 19.6 \text{ m} = -49.0 \text{ m}$$

The minus sign indicates that the displacement is downward, which agrees with what you know from the statement of the problem.

Follow-up Exercise. How much longer would it take for the stone to reach the river if the boy in this Example had dropped the ball rather than thrown it? (*Answer may be found in the Answers to Follow-up Exercises section at the back of the book.*)

EXAMPLE 2.10 ■ MEASURING REACTION TIME: FREE FALL

Reaction time is how long it takes a person to notice, think, and act in response to a situation—for example, the time between first observing and then responding to something happening on the road ahead while driving an automobile. Reaction time varies with the complexity of the situation (and the individual). In general, the largest part of a person's reaction time is spent thinking, but practice in dealing with a given situation can reduce this time.

A person's reaction time for a simple situation may be measured by having another person drop a ruler (without warning) through the thumb and forefinger as shown in •Fig. 2.15. The falling ruler is grasped by the first person as quickly as possible, and the length of the ruler below the top of the finger is noted. If the ruler descends 18.0 cm on the average before it is caught, what is the person's average reaction time?

Solution. Notice that only the distance of fall is given. However, we know a couple of other things, such as v_o and g, so

•**FIGURE 2.15 Reaction time**
A person's reaction time can be measured by having her (or him) grasp a dropped ruler. See Example 2.10.

Given: $y = -18.0 \text{ cm} = -0.180 \text{ m}$ **Find:** t (reaction time)
$v_o = 0$
$g \ (= 9.80 \text{ m/s}^2)$

(Note that the distance y has been converted directly to meters.) We can see that Eq. 2.9′ applies here:

$$y = v_o t - \tfrac{1}{2} g t^2$$

or

$$y = -\tfrac{1}{2} g t^2 \quad \text{(with } v_o = 0\text{)}$$

Solving for t gives

$$t = \sqrt{\frac{2y}{-g}}$$

$$= \sqrt{\frac{2(-0.180 \text{ m})}{-9.80 \text{ m/s}^2}} = 0.192 \text{ s}$$

Try this with a fellow student and measure your reaction time. Why do you think another person should drop the ruler rather than yourself?

Follow-up Exercise. A popular party trick is to substitute a crisp dollar bill for the ruler in Fig. 2.15, telling the person that he or she can have the dollar if caught. Is this a good deal? (The length of a dollar is 15.7 cm.) *(Answer may be found in the Answers to Follow-up Exercises section at the back of the book.)*

EXAMPLE 2.11 ■ FREE FALL UP AND DOWN: USING IMPLICIT DATA

A worker on a scaffold on a billboard throws a ball straight up. It has an initial velocity of 11.2 m/s when it leaves his hand at the top of the billboard (●Fig. 2.16). (a) What is the maximum height the ball reaches relative to the top of the billboard? (b) How long does it take to reach this height? (c) What is the position of the ball at $t = 2.00$ s?

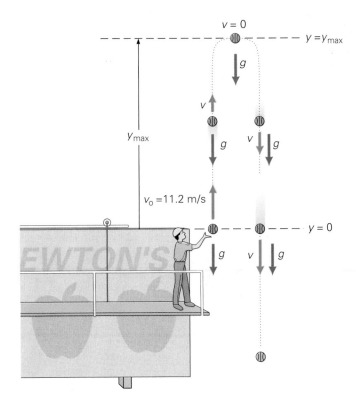

●FIGURE 2.16 **Free fall up and down**
(The ball is horizontally displaced for illustration.) Note the lengths of the velocity and acceleration vectors at different times. See Example 2.11.

Solution. It might appear that all that is given in the general problem is the initial velocity v_o. However, a couple of other things are implicitly "given" because they are understood. One is the acceleration g, and the other is the velocity at the maximum height where the ball stops. Here, in changing direction, the velocity of the ball is momentarily zero, so we have

Given: $v_o = 11.2 \text{ m/s}$
$\quad\quad\quad\; g \; (= 9.80 \text{ m/s}^2)$
$\quad\quad\quad\; v = 0 \; (\text{at } y_{max})$
$\quad\quad\quad\; t = 2.00 \text{ s (for part c)}$

Find: (a) y_{max} (maximum height)
$\quad\quad\quad$ (b) t_u (time upward)
$\quad\quad\quad$ (c) y (at $t = 2.00$ s)

(a) Notice that we reference the height ($y = 0$) to the top of the billboard. For this part of the problem, we need only be concerned with the upward motion—a ball is thrown upward and it stops at its maximum height y_{max}. With $v = 0$ at this height, y_{max} may be found directly from Eq. 2.10':

$$v^2 = 0 = v_o^2 - 2gy_{max}$$

and

$$y_{max} = \frac{v_o^2}{2g}$$
$$= \frac{(11.2 \text{ m/s})^2}{2(9.80 \text{ m/s}^2)} = 6.40 \text{ m}$$

relative to the top of billboard ($y = 0$, see figure).

(b) The time the ball travels upward is designated t_u. This is the time it takes to reach y_{max} where $v = 0$. Then, knowing v_o and v, the time t_u may be found directly from Eq. 2.7':

$$v = 0 = v_o - gt_u$$

and

$$t_u = \frac{v_o}{g}$$
$$= \frac{11.2 \text{ m/s}}{9.80 \text{ m/s}^2} = 1.14 \text{ s}$$

(c) The height of the ball at $t = 2.00$ s is given directly by Eq. 2.9':
$$y = v_o t - \tfrac{1}{2}gt^2$$
$$= (11.2 \text{ m/s})(2.00 \text{ s}) - \tfrac{1}{2}(9.80 \text{ m/s}^2)(2.00 \text{ s})^2 = 22.4 \text{ m} - 19.6 \text{ m} = 2.8 \text{ m}$$

Note that this is 2.8 m above, or measured upward from, the reference point ($y = 0$). The ball has reached its maximum height and is on its way back down.

Considered from another reference point, this situation is like dropping a ball from a height of y_{max} above the top of the billboard with $v_o = 0$ and asking how far it falls in a time $t = 2.00 \text{ s} - t_u = 2.00 \text{ s} - 1.14 \text{ s} = 0.86 \text{ s}$. The answer is
$$y = v_o t - \tfrac{1}{2}gt^2$$
$$= 0 - \tfrac{1}{2}(9.80 \text{ m/s}^2)(0.86 \text{ s})^2 = -3.6 \text{ m}$$

This is the same as the position found above but is measured with respect to the maximum height as the reference point; that is,
$$y_{max} - 3.6 \text{ m} = 6.4 \text{ m} - 3.6 \text{ m} = 2.8 \text{ m}$$

Follow-up Exercise. At what height does the ball in this Example have a speed of 5.00 m/s? (*Hint*: This occurs twice—once on the way up, and once on the way down.)
(*Answers may be found in the Answers to Follow-up Exercises section at the back of the book.*)

PROBLEM-SOLVING HINT

When working vertical projection problems involving motions up and down, it is often convenient to divide the problem into two parts and consider each separately. As seen in the previous example, for the upward part of the motion, the velocity is zero at the maximum height.

A zero quantity simplifies the calculations. Similarly, the downward part of the motion is analogous to that of an object dropped from a height, where the initial velocity is zero.

However, as the previous example shows, the appropriate equations may be used directly at any position or time of the motion. For instance, note in part (c) of Example 2.11 that the height was found directly for a time *after* the ball had reached the maximum height. The velocity of the ball at that time could also have been found directly from Eq. 2.7′, $v = v_o - gt$.

EXAMPLE 2.12 ■ VERTICAL FREE-FALL PROJECTIONS: SOME INTERESTING FACTS

Referring to the thrown ball in Example 2.11 and Fig. 2.16, (a) compare the travel time upward (t_u) with the travel time downward (t_d) required for the ball to return to its starting point. (b) Compare the initial velocity of the ball with the velocity it has when back at the starting point.

Solution. The given data in this case will include some of the results of Example 2.11. There, t_u and y_{max} were computed.

Given: $v_o = 11.2$ m/s   **Find:** (a) t_d, and compare with t_u
$t_u = 1.14$ s   (b) v (returning velocity),
$y_{max} = 6.40$ m   and compare with v_o

(a) The time to reach the maximum height is $t_u = 1.14$ s. The time for the return trip downward, t_d, is the time it takes the ball to fall 6.40 m (the maximum height) from rest ($v = 0$ at maximum height). Refer to Fig. 2.16. This situation is exactly the same as if the ball were dropped and we were asked to find the time it would take to fall 6.40 m. With $v_o = 0$ for this downward part of the motion, Eq. 2.9′ applies:

$$y = v_o t_d - \tfrac{1}{2}g t_d^2 = 0 - \tfrac{1}{2}g t_d^2 \quad \text{or} \quad y = -\tfrac{1}{2}g t_d^2$$

and

$$t_d = \sqrt{\frac{2y}{-g}}$$

$$= \sqrt{\frac{2(-6.40 \text{ m})}{-9.80 \text{ m/s}^2}} = 1.14 \text{ s}$$

Hence, $t_u = t_d$, or *the time of flight upward and the time to return to the starting point are the same* (in the absence of air resistance).

Notice that the total time for the ball to return to its starting point is given by

$$t = t_u + t_d = 1.14 \text{ s} + 1.14 \text{ s} = 2.28 \text{ s}$$

If we had been asked to compute this total time, it would not have been necessary to do this in two steps (time up and time down). Another way is to note that when the ball returns to its original starting point, it is again at $y = 0$, and then use Eq. 2.9′ to find the total time:

$$y = v_o t - \tfrac{1}{2}g t^2 = (v_o - \tfrac{1}{2}gt)t = 0$$

where v_o is the initial upward velocity. This expression involves the multiplication of two factors, so it is satisfied when either factor is zero. When $t = 0$, the ball is initially at $y = 0$. When $(v_o - \tfrac{1}{2}gt) = 0$, the ball is again at $y = 0$ on the return trip. From the latter,

$$v_o = \tfrac{1}{2}gt$$

and

$$t = \frac{2v_o}{g}$$

$$= \frac{2(11.2 \text{ m/s})}{9.80 \text{ m/s}^2} = 2.29 \text{ s}$$

Note that the answer is slightly different. This is an example of a rounding difference, which may occur when a problem is solved by different methods.

(b) The velocity of the ball when it returns to its starting point ($y = 0$) may be found using Eq. 2.10':

$$v^2 = v_o^2 - 2gy = v_o^2 - 0 \quad \text{or} \quad v^2 = v_o^2$$

Taking the square root of both sides, we have $v = \pm v_o$; that is, the expression has two roots, as mentioned previously. In this case,

$$v = -v_o = -11.2 \text{ m/s}$$

where the negative root was taken to indicate the proper downward direction. (The positive root, $v = +v_o$, gives the initial velocity, since this is also a velocity of the ball at $y = 0$.) Hence, the velocities are equal and opposite, and the ball returns to the starting point *with the same speed it had initially* (again, assuming no air resistance).

Notice that the returning velocity could also be found by considering the velocity of the ball at time t_d, or 1.14 s after falling from its maximum height. As for a dropped ball, $v_o = 0$, and Eq. 2.7' gives

$$v = v_o - gt$$
$$= 0 - (9.80 \text{ m/s}^2)(1.14 \text{ s}) = -11.2 \text{ m/s}$$

Keep in mind that, as the examples in this section show, there is often more than one approach to solving a problem.

Follow-up Exercise. For the situation in Fig. 2.16, the ball continues falling after $t = 2.28$ s. How would you find its position at some later time before it hits the ground? (*Answer may be found in the Answers to Follow-up Exercises section at the back of the book.*)

Chapter Review

Important Terms

You should be able to define and explain these chapter terms clearly.

mechanics 31	**displacement** 32	**instantaneous velocity** 36
kinematics 31	**speed** 33	**acceleration** 39
dynamics 31	**average speed** 33	**average acceleration** 39
motion 31	**instantaneous speed** 34	**instantaneous acceleration** 39
distance 31	**velocity** 34	**acceleration due to gravity** 47
scalar (quantity) 31	**average velocity** 34	**free fall** 47
vector (quantity) 31		

Important Concepts

- Motion is a change of position; it can be described in terms of the distance moved (a scalar) or the displacement (a vector).
- A scalar quantity has magnitude (value and units) only; a vector quantity has magnitude *and* direction.
- Speed (a scalar) is the time rate of change of distance, and velocity (a vector) is the time rate of change of displacement.

- Acceleration is the time rate of change of velocity, and hence a vector quantity.
- Most kinematic equations are limited to constant accelerations only.
- An object in free fall has a constant acceleration of $g = 9.80 \text{ m/s}^2$ (acceleration due to gravity) near the surface of the Earth.

Important Equations

These equations are mathematical statements of the concepts and principles presented in the chapter. They will be needed in working problem Exercises, so a good understanding is important. You should be able to identify the symbols and to explain the relationships and limitations before proceeding. (In-text equation reference numbers are given for convenience.)

Average speed: $$\bar{s} = \frac{\Delta d}{\Delta t}$$ (2.2)

Acceleration Due to Gravity:
$$g = 9.80 \text{ m/s}^2 = 980 \text{ cm/s}^2 \approx 32 \text{ ft/s}^2$$

Kinematic Equations for Linear Motion with Constant Acceleration

(with $x_o = 0$): $\qquad x = \bar{v}t$ (2.4)

(general, not limited to constant acceleration)

$$\bar{v} = \frac{v + v_o}{2}$$ (2.8)

$$v = v_o + at$$ (2.7)

$$x = v_o t + \tfrac{1}{2} at^2$$ (2.9)

$$v^2 = v_o^2 + 2ax$$ (2.10)

constant acceleration only

Kinematic Equations Applied to Free Fall
(with downward taken as the negative direction, $y_o = 0$, and $-g$ expressed explicitly):

$$y = \bar{v}t$$ (2.4′)

$$\bar{v} = \frac{v + v_o}{2}$$ (2.8′)

$$v = v_o - gt$$ (2.7′)

$$y = v_o t - \tfrac{1}{2}gt^2$$ (2.9′)

$$v^2 = v_o^2 - 2gy$$ (2.10′)

g constant

Exercises

2.1 Change of Position: Distance and Displacement

1 Distance is always (a) equal to the magnitude of the corresponding displacement, (b) less than or equal to the magnitude of the corresponding displacement, (c) greater than or equal to the magnitude of the corresponding displacement.

2 In specifying a position or a change in position, what is necessary?

3 Two people choose different reference points to specify an object's position. Does this affect their coordinate descriptions of the object? Explain.

4 Can a displacement from one point to another be zero, yet the distance involved in moving between these points be nonzero? How about the reverse situation? Explain.

5 You are told that a person walks 500 m. What can you safely say about the person's final position relative to the starting point?

2.2 Speed and Velocity

6 A scalar quantity has (a) magnitude, (b) direction, (c) both direction and magnitude.

7 For a constant velocity, the speed is (a) continually changing, (b) less than the magnitude of the velocity, (c) greater than the magnitude of the velocity, (d) equal to the magnitude of the velocity vector.

8 An object travels with a constant velocity. What is the relationship of its speed to the velocity?

9 A fellow student shows you a position versus time plot in which the straight-line graph segments form a square. He states that these data were obtained from a lab experiment. Would you believe him? Explain.

10 ■ A small airplane flies in a straight line at a speed of 125 km/h. How long does it take the plane to fly 300 km?

11 ■ A bus travels on an interstate highway at an average speed of 85 km/h. How far does the bus travel in 20 min on the average? Would this be the actual distance? Explain.

12 ■ A motorist drives 150 km from one city to another in 2.5 h, but makes the return trip in only 2.0 h. What are the average speeds for (a) each half of the round trip and (b) the total trip?

13 ■ A senior citizen walks 0.30 km in 10 min going around a shopping mall. What is her average speed in m/s? (b) If she wants to increase her average speed by 20% in walking a second lap, what would be the travel time in minutes for this case?

14 ■■ Is the average speed of the jogger in Example 2.1 going from A to C equal to the average of the average speeds in going from A to B and from B to C? Justify your answer.

15 ■■ A student throws a ball vertically upward such that it travels 9.0 m to its maximum height. If the ball is caught at the initial height 2.4 s after being thrown, (a) what is the ball's average speed? (b) its average velocity?

16 ■■ The distance of one lap around an oval dirt bike track is 1.25 km. If a rider going at a constant speed makes one lap in 1.10 min, what is the speed of the bike and rider in m/s? Is the velocity of the bike also constant? Explain.

17 ■■ Given that the speed of sound is 340 m/s, how much time will elapse between seeing a lightning flash and hearing the resulting thunder if the lightning strikes 3.50 km away? (The speed of light is 3.00×10^8 m/s, or about 186,000 mi/s, so the lightning flash is seen instantaneously.)

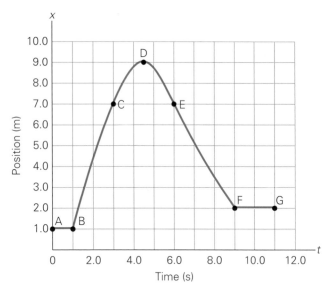

•FIGURE 2.17 Position vs. time
See Exercise 18.

18 ■■ A plot of position versus time is shown in •Fig. 2.17 for an object in linear motion. (a) What are the average velocities for the segments AB, BC, CD, DE, EF, FG, and BG? (b) State whether the motion is uniform or nonuniform in each case. (c) What is the instantaneous velocity at point D?

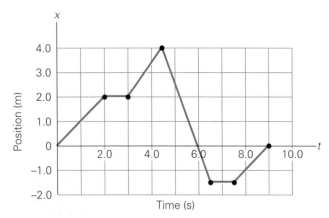

•FIGURE 2.18 Position vs. time
See Exercise 19.

19 ■■ In demonstrating a dance step, a person moves in one dimension as shown in •Fig. 2.18. What are (a) the average speed and (b) the average velocity for each phase of the motion? (c) What are the instantaneous velocities at $t = 1.0$ s, 2.5 s, 4.5 s, and 6.0 s? (d) What is

the average velocity for the interval between $t = 4.5$ s and $t = 9.0$ s? [*Hint:* Recall that the effective displacement is the displacement between the starting point and the ending point.]

20 ■■ A skydiver jumps out of a plane and falls 1000 m in 20.0 s before opening her parachute. Thereafter, the 500-m distance to the ground is traveled in 18.0 s. What is the skydiver's average velocity for the total trip?

21 ■■ The Indianapolis 500, a 500-mi auto race, was first run in 1911 in a time of 6 h, 42 min, and 8 s. In 1992, the race was run in a record time of 3 h, 7 min, and 12 s. (a) What were the average speeds for the Indy 500 for these years in mi/h? (Ignore significant figures.) (b) What was the percentage change in the average speed from 1911 to 1992?

22 ■■ According to the theory of continental drift and plate tectonics, the continents were once parts of a single giant supercontinent, called Pangaea, which broke up. Recent measurements of seafloor spreading along mid-oceanic ridges show the continental drift to be on the order of 4 cm per year. Assume that this rate has been constant throughout the past, and take the current distance between South America and Africa to be 5000 mi. Approximately how long ago did Pangaea break up?

23 ■■■ The first nonstop round-the-world flight was made in 1986 in a small aircraft named *Voyager*. The flight lasted 9 days, 3 min, and 44 seconds. (The plane carried 7011 lb of fuel—more than three times its own weight.) Assuming the flight was along a great circle, what was the plane's average speed in (a) km/h and (b) mi/h? (A great circle is one with a plane that passes through the center of the Earth and divides it into two equal halves. Do you know why this would be a preferred route?)

24 ■■■ A student driving home for the holidays starts at 8:00 a.m. to make the 675-km trip, practically all of which is on nonurban interstate highway. If she wants to arrive home no later than 3:00 p.m., what must her average speed be, at minimum? Will she have to exceed the 65 mi/h speed limit?

25 ■■■ Two runners approaching each other on a straight track have constant velocities of $+4.50$ m/s and -3.50 m/s, respectively, when they are 100 m apart. How long will it take for the runners to meet and at what position will this occur?

26 ■■■ In driving the usual route to school, a student computes his average speed to be 30 km/h. In a hurry to get home that afternoon, he wants to average 60 km/h for the *total* trip. What would his average speed for the return trip over the same route have to be in order to do this? (*Hint:* see Example 2.2, and be ready for a surprise.)

2.3 Acceleration

27 The gas pedal of an automobile is commonly referred to as the accelerator. Which of the following might also be called an accelerator: (a) the brakes, (b) the steering wheel, (c) both of the preceding?

28 On a v versus t plot for an initially moving object that receives a constant acceleration in the direction of motion, (a) the graph is a curved line, (b) the y-intercept is nonzero, (c) the slope of the line is necessarily positive, (d) the slope of the line is necessarily negative.

29 An object traveling with a constant velocity v_o experiences a constant acceleration in the same direction for a time t. Then, an acceleration of equal magnitude is experienced in the opposite direction for the same time t. What is the velocity after this?

30 What would be the general form of a v versus t plot for an object in linear motion with (a) a nonconstant increasing acceleration in the direction of a positive velocity, and (b) a nonconstant decreasing acceleration in the direction opposite a positive velocity? (c) Describe the motions of the two objects that have the v versus t plots shown in •Fig. 2.19.

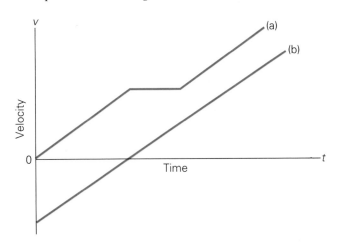

•**FIGURE 2.19 Description of motion**
See Exercise 30.

31 ■An automobile traveling at 30.0 km/h along a straight road accelerates to 55.0 km/h in 4.00 s. What is the magnitude of the average acceleration?

32 ■A car moving with a velocity of 24 m/s on a one-way street must be brought to a stop in 4.0 s. What is the required average acceleration?

33 ■If an automobile moving with a velocity of 40 km/h along a straight road is braked uniformly to rest in 5.0 s, by how much must the velocity change each second?

34 ■■ At a sports car rally, a car starting from rest accelerates uniformly at a rate of 9.0 m/s² over a straight-

line distance of 100 m. The time to beat in this event is 4.5 s. Does the driver do it? If not, what must the acceleration be to do so?

35 ■■ An object moves in the negative x direction with a speed of 2.5 m/s. If an acceleration of 0.50 m/s² is experienced in the positive x direction, after how long a time will the object have a positive velocity?

36 ■■ What is the acceleration for each graph segment in •Fig. 2.20? Describe the motion of the object over the total time interval.

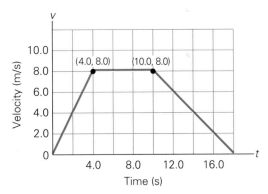

•**FIGURE 2.20 Velocity vs. time**
See Exercise 36.

37 ■■ •Figure 2.21 shows a plot of velocity versus time for an object in linear motion. (a) Compute the acceleration for each phase of motion. (b) Describe how the object moves during the last time segment.

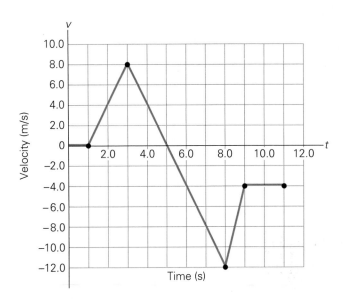

•**FIGURE 2.21 Velocity vs. time**
See Exercise 37.

38 ■■■ A train normally travels with a uniform speed of 72 km/h on a long stretch of straight, level track. On a particular day, the train must make a 2.0 min stop at a station along this track. If the train decelerates at a uni-

Exercises **57**

form rate of $1.0 \, \text{m/s}^2$ and accelerates at a rate of $0.50 \, \text{m/s}^2$, how much time is *lost* in stopping at the station?

2.4 Kinematic Equations

39 For a constant linear acceleration, the graph of position versus time would be (a) a horizontal straight line, (b) a straight line with a positive slope, (c) a straight line with a negative slope, (d) a curved line.

40 An object accelerates uniformly from rest for t seconds. The average speed for this time interval is (a) $\frac{1}{2}at$, (b) $\frac{1}{2}at^2$, (c) $2at$, (d) none of the preceding.

41 A classmate states that a negative acceleration always means that a moving object is decelerating. Is this true? Explain.

42 ■ A motorboat traveling on a straight course slows down uniformly from 70 km/h to 35 km/h in a distance of 50 m. What is the acceleration?

43 ■ A car initially at rest rolls down a hill with a uniform acceleration of $5.0 \, \text{m/s}^2$. How long will it take the car to travel 150 m, the distance to the bottom of the hill?

44 ■ A car accelerates uniformly from rest at a rate of $5.25 \, \text{m/s}^2$. (a) How far does the car travel in 7.00 s? (b) What is the speed of the car at that time?

45 ■■ The driver of a pickup truck going 100 km/h applies the brakes, giving the truck a uniform deceleration of $6.50 \, \text{m/s}^2$ while it travels 20.0 m. (a) What is the velocity of the truck in km/h at the end of this distance? (b) How much time has elapsed?

46 ■■ An experimental rocket car starting from rest reaches a speed of 560 km/h after a straight 400-m run on a level salt flat. Assuming the acceleration to be constant, what was the time of the run?

47 ■■ A rocket car travels with a constant velocity of 250 km/h on a salt flat. The driver gives the car a reverse thrust, and it experiences a continuous and constant deceleration of $8.25 \, \text{m/s}^2$. How much time elapses until the car is 175 m from the point where the reverse thrust is applied?

48 ■■ Large, ocean-going oil tankers may weigh as much as 640,000 tons. Such ships require 3.0 miles to stop from a top speed of 15 knots. Assuming that a ship slows down at a constant rate in a straight line, what would be the ship's acceleration in m/s^2 in such a case? (A knot is a nautical unit of speed, nautical mile/hour, and 1 nautical mile = 6067 ft = 1.15 mi. Our common mile (5280 ft) is properly called a statute mile.)

49 ■■ A rifle bullet with a muzzle velocity (speed) of 330 m/s is fired directly into a special, dense material that stops the bullet in 30 cm. Assuming the deceleration to be constant, what is its value?

50 ■■ A bullet traveling horizontally with a speed of 35.0 m/s hits a board perpendicular to the surface, passes through it, and emerges on the other side with a speed of 21.0 m/s. If the board is 4.00 cm thick, how long does the bullet take to pass through it?

51 ■■ A motorboat is traveling due north in calm water at a speed of 6.0 m/s. The operator puts the motor in reverse and the boat receives an acceleration of $-1.5 \, \text{m/s}^2$. (a) How long after the boat is put into reverse does it momentarily come to a halt? (b) What is the position of the boat when it momentarily comes to a halt relative to the point where the motor was put into reverse?

52 ■■ The speed limit in a school zone is 40 km/h (about 25 mi/h). A driver traveling at this speed sees a child run into the road 17 m ahead of his car. He applies the brakes, and the car decelerates at a uniform rate of $8.0 \, \text{m/s}^2$. If the driver's reaction time is 0.25 s, will the car stop before hitting the child?

53 ■■ Assuming a driver's reaction time of 0.25 s, construct a table of the stopping distances for an automobile traveling at initial velocities of 36 km/h, 72 km/h, and 90 km/h with uniform decelerations of (a) $8.0 \, \text{m/s}^2$ (on a dry pavement) and (b) $4.0 \, \text{m/s}^2$ (on a wet pavement). Also list the initial velocities in mi/h and the stopping distances in ft in parentheses.

54 ■■ An object moves in the positive x direction with a speed of 40 m/s. As it passes through the origin, it starts to experience a constant acceleration of $3.5 \, \text{m/s}^2$ in the negative x direction. How much time elapses before the object returns to the origin?

55 ■■ A jet aircraft being launched from an aircraft carrier is accelerated from rest along a 94 m track for 2.5 s. What is the launch speed of the plane?

56 ■■ Show that the area under the curve of a v versus t plot for a constant acceleration is equal to the displacement. Do this for the cases where (a) $a = 0$, (b) $v_0 = 0$, and (c) $v_0 \neq 0$. [*Hint:* The area of a triangle is $\frac{1}{2}ab$, or one-half the altitude times the base.]

57 ■■ Compute the distance traveled for the motion represented by Fig. 2.20.

58 ■■■ Figure 2.21 shows velocity versus time for an object in linear motion. (a) What are the instantaneous velocities at $t = 8.0$ s and $t = 11.0$ s? (b) Compute the final displacement of the object. (c) Compute the total distance the object travels.

59 ■■■ In our kinematic equations, it was assumed that $x_0 = 0$ and $t_0 = 0$ for simplicity and convenience. Rewrite the equations with x_0 and t_0 expressed explicitly. What is the difference between the x in Eq. 2.9 and the x in your rewritten equation?

60 ■■■ An object moves in the positive x direction with a constant acceleration. At $x = 5.0$ m, its speed is 10 m/s. In 2.5 s, the object is at $x = 65$ m. What is its acceleration?

61 ■■■ A car and a motorcycle at rest and 1000 m apart start toward each other at the same time on a level track. If the car accelerates at a uniform rate of 3.70 m/s^2 and the motorcycle accelerates at a uniform rate of 4.40 m/s^2, at what position will they pass each other relative to the car's starting point?

2.5 Free Fall

62 A dropped object in free fall (a) falls 9.8 m each second, (b) falls 9.8 m during the first second, (c) has an increase in speed of 9.8 m/s each second, (d) has an increase in acceleration of 9.8 m/s each second.

63 When an object is thrown vertically upward, neglecting air resistance, its (a) velocity changes nonuniformly, (b) maximum height is independent of the initial velocity, (c) travel time upward is slightly greater than its travel time downward, (d) speed on returning to its starting point is the same as its initial speed.

64 Given the data for a vertically upward projected object, you are asked to find the time the object is at height y. The equation $y = v_o t - \frac{1}{2}gt^2$ is solved using the quadratic formula and two roots are obtained. Does this mean there are two answers? Explain. [*Hint:* Do the roots of a quadratic equation necessarily have to be both positive and negative?]

65 For the motion of a dropped object in free fall, sketch the general forms of the graphs of (a) v versus t, and (b) y versus t.

66 ■ Sketch the general forms of the graphs of (a) velocity v versus t, and (b) y versus t for an object projected vertically upward.

67 ■ If a dropped object falls 19.6 m in 2.00 s, how far will it fall in 4.00 s?

68 ■■ A student drops a ball from the top of a tall building and it takes 2.8 s to reach the ground. (a) What was its speed in mi/h just before hitting the ground? (b) Find the height of the window above the ground from which the ball was dropped.

69 ■■ An object dropped from the top of a cliff takes 1.80 s to hit the water in the lake below. What is the height of the cliff above the water?

70 ■■ (a) The World Trade Center and the Empire State Building in the "Big Apple" have heights of about 417 m and 381 m, respectively. If objects were dropped from the top of each, assuming free fall, what would be the difference in time in their reaching the ground? (b) How does the time of fall compare for Chicago's Sears Tower, at 443 m?

71 ■■ With what speed must an object be projected vertically upward for it to reach a maximum height of 12.0 m above its starting point?

72 ■■ For an object dropped from rest, what is the distance it will fall *during* the second second (that is, between $t = 1.00$ s and $t = 2.00$ s)? Does this distance double during the fourth second of fall? Explain.

73 ■■ A spring-loaded gun shoots a 0.0050-kg bullet vertically upward with an initial velocity of 21 m/s. (a) What is the height of the bullet 4.0 s after firing? (b) At what times is the bullet 12 m above the muzzle of the gun?

74 ■■ The ceiling of a classroom is 3.75 m above the floor. A student tosses an apple vertically upward, releasing it 0.50 m above the floor. What is the maximum magnitude of the initial velocity that can be given to the apple if it is not to touch the ceiling?

75 ■■ A stone is thrown vertically downward with an initial speed of 12.4 m/s from a height of 65.0 m above the ground. (a) How far does the stone travel in 2.00 s? (b) What is its velocity when it hits the ground?

76 ■■ A ball is projected vertically downward with a speed of 3.00 m/s. (a) How far does the ball travel in 1.80 s? (b) What is the velocity of the ball at that time?

77 ■■ At what rate would a car have to accelerate from rest on a straight, level road to have the same speed in the same amount of time that an object dropped from a height of 20 m would have just before hitting the ground?

78 ■■ A ball is thrown upward with an initial speed of 6.0 m/s by someone on the top of a building 34 m tall who is leaning over the edge so that the ball will not strike the building on the return trip. (a) How far above the ground will the ball be at the end of 1.0 s? (b) What is the ball's velocity at that time? (c) When and with what speed will the ball strike the ground?

79 ■■ Test your algebra. (a) A ball thrown vertically upward is at a height h at time t_1 and again at time t_2. At what height, in terms of t_1 and t_2, does this occur? (b) With what speed, in terms of t_1 and t_2, is the ball projected upward?

80 ■■ Laura can jump a vertical distance of 0.85 m. (a) What is the total time she is off the ground? (b) With what velocity does she hit the ground?

81 ■■ One student throws a ball vertically upward with an initial speed of 9.8 m/s. Another student standing

5.0 m away starts running toward the ball on release and catches it at the same height. What was the student's acceleration? (Assume a uniform acceleration.)

82 ■■ Referring to the situation in Fig. 2.16, (a) what would be the position of the ball 3.00 s after it was thrown upward, and (b) if the top of the billboard is 20.0 m above the ground, what is the ball's total time of flight, and what is its velocity just before hitting the ground?

83 ■■ Draw plots of y versus t and v versus t for the following: (a) an object dropped from rest from a height of 30 m above the ground, and (b) an object projected vertically upward with an initial velocity of 34.3 m/s and its return to the same point. (c) Draw plots of a versus t for these motions.

84 ■■ A super ball is thrown vertically downward with an initial speed of 72.0 km/h from a height of 4.00 m. Assuming the ball rebounds with 95% of its impact speed, how high will the ball go? (Would it be 95.0% of the initial height?)

85 ■■ An aluminum ball with a mass of 4.0 kg and an iron ball of the same size with a mass of 11.6 kg are dropped simultaneously from a height of 49 m. (a) Neglecting air resistance, how long does it take the aluminum ball to fall to the ground? (b) How much later does the heavier iron ball strike the ground?

86 ■■■ A photographer in a helicopter ascending vertically at a constant rate of 12.50 m/s accidentally drops a camera when the helicopter is 60.0 m above the ground. (a) How long will it take the camera to reach the ground? (b) What will its speed be when it hits?

87 ■■■ The acceleration due to gravity on the Moon is one-sixth of that on Earth. (a) If an object were dropped from the same height on the Moon and on the Earth, how much longer (by what factor) would it take it to hit the surface of the Moon? (b) For a projectile with an initial velocity of 18.0 m/s upward, what would be the maximum height and the total time of flight on the Moon and on the Earth?

88 ■■■ A student at a window on the second floor of a dorm sees his math professor coming along the walkway beside the building. He drops a water balloon from 18.0 m above the ground when the prof is 1.00 m from the point directly beneath the window. If the prof is 170 cm tall and walks at a rate of 0.450 m/s, does the balloon hit his head? Does it hit him at all?

89 ■■■ A dropped object passes a window that is 1.35 m tall in 0.210 s. From what height above the top of the window was the object released?

90 ■■■ From a dormitory window 12.0 m above the ground, one student drops a ball, and at the same time, another student throws another ball vertically upward with an initial speed of 9.00 m/s. (a) Where is the dropped ball when the thrown ball reaches its maximum height? (b) How much time elapses between the balls' striking the ground?

Additional Exercises

91 In throwing an object vertically upward with a velocity of 7.25 m/s from the top of a tall building the thrower leans over the edge so that the object will not strike the building on the return trip. (a) What is the velocity of the object when it has traveled a total distance of 25.0 m? (b) How long does it take to travel this distance?

92 After landing, a jet plane comes uniformly to rest along a straight runway with an average velocity of -35.0 km/h. If this takes 7.00 s, what is the plane's acceleration?

93 A vertically moving projectile reaches a maximum height of 17.5 m above its starting position. (a) What was the projectile's initial velocity? (b) What is its height above the starting point at $t = 2.45$ s?

94 A drag racer traveling on a straight track with a speed of 200 km/h ejects a parachute and slows uniformly to a speed of 20 km/h in 12 s. (a) What is the racer's acceleration? (b) How far does it travel in the 12-s interval?

95 On a cross-country trip, a couple drives 500 mi in 10 h on the first day, 380 mi in 8.0 h on the second day, and 600 mi in 15 h on the third day. What was their average speed?

96 A projectile is given an initial velocity of 8.0 m/s directly upward. Use the quadratic formula (with $g = 10$ m/s^2 to simplify the calculations) to determine (a) the time when the projectile is 1.4 m above its starting point and (b) the time when it is 3.5 m above its starting point. (c) Suppose that the projectile is instead given an initial velocity of 8.0 m/s directly downward. At what time is the projectile a distance of 1.8 m below its starting point? (Comment on both roots of the quadratic equation in each case.)

97 A car and a motorcycle start from rest at the same time on a straight track, but the motorcycle is 25.0 m behind the car (•Fig. 2.22). The car accelerates at a uniform rate of 3.70 m/s^2, and the motorcycle at a uniform rate of 4.40 m/s^2. (a) How much time elapses until the motorcycle overtakes the car? (b) How far will each have traveled during that time? (c) How far ahead of the car will the motorcycle be 2.00 s later? (Both vehicles are still accelerating.)

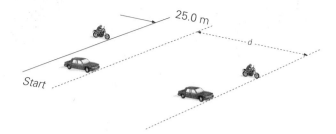

•**FIGURE 2.22   A tie race (drawing not to scale.)**
See Exercise 97.

98 A train on a straight, level track has an initial speed of 45.0 km/h. A uniform acceleration of 1.50 m/s^2 is applied while the train travels 200 m. (a) What is the speed of the train at the end of this distance? (b) How long did it take for the train to travel the 200 m?

99 A car going 85 km/h on a straight road is brought uniformly to a stop in 10 s. How far does the car travel during that time?

100 Two motorcyclists race against the clock on a 40-km cross-country route. The first cyclist travels the route with an average speed of 55 km/h. The second cyclist starts 3.5 min after the first but crosses the finish line at the same time. What is the average speed of the second cyclist?

101 An object initially at rest experiences an acceleration of 1.5 m/s^2 for 6.0 s and then travels at that constant velocity for another 8.0 s. What is the object's average velocity over the 14-s interval?

102 (This is an old one.) In order to find the depth of the water surface in a well, a person drops a stone from the top of the well and simultaneously starts a stopwatch. The watch is stopped when the splash is heard, giving a reading of 3.65 s. The speed of sound is 340 m/s. Find the depth of the water surface below the top of the well. Take the person's reaction time for stopping the watch to be 0.250 s.

103 When checking reaction times as described in Example 2.10, some students decide to calibrate the ruler in time units so they can obtain a quick approximation without having to do a calculation each time the test is made. They want to calibrate the ruler in time intervals of 20 ms. Show on a sketch of a ruler where the time gradations would be for the first 20 cm.

104 An arrow shot from a bow vertically upward has an initial velocity of 50.0 m/s. (a) What is the maximum height of the arrow above its launch point? (b) How long does it take for the arrow to return to its launch point? (Neglect air resistance and assume that the arrow travels in a straight line.)

105 Suppose that the boy on the bridge in Example 2.9 throws the stone vertically upward instead of downward. How long will the stone be in flight before it hits the river?

106 An object starting from rest and traveling in a straight line has velocities of 5.0 m/s, 10 m/s, 15 m/s, 20 m/s, and 25 m/s at the ends of the first, second, third, fourth, and fifth seconds, respectively. (a) What is the acceleration of the object? (b) What is its average velocity for the 5-s interval? (c) Sketch a graph of v versus t. (d) Sketch a graph of x versus t.

107 An automobile travels along a straight, level highway uniformly covering a distance of 88 ft each second. (a) What is the car's speed in mi/h? (b) What are the magnitudes of the car's average speed and average velocity in m/s? (c) What can you say about the car's instantaneous speed and velocity?

108 A circular track has a diameter of 0.50 km. A go-cart with a constant speed of 7.0 m/s makes two complete laps around the track. How long does it take the cart to complete the two laps?

109 On a walk in the country, a couple goes 1.80 km east along a straight road in 20.0 min and then 2.40 km directly north in 35.0 min. Answer the following using units of km/h. (a) What is the average velocity for each segment of their walk? (b) What is the average speed for the total distance walked? (c) If they take a straight-line path back to their original starting place in 25.0 min, what is the average speed and average velocity for the total trip?

110 An automobile is braked to a stop with a uniform deceleration in a time of t_s. Show that the distance traveled during this time is given by $x = (v_o^2/a) - \frac{1}{2}at_s^2$, where the positive direction is taken to be in the direction of the initial motion.

3 Motion in Two Dimensions

You *can* get from here to there! It's just a matter of knowing which way to head at the crossroads. But did you ever wonder why roads so often meet at right angles? There's a good reason. Living on the Earth's surface, we are used to describing locations in two dimensions, and one of the easiest ways to do this is by referring to a pair of mutually perpendicular axes. When you want to tell someone how to get to a particular place in the city, you often say, "Go uptown 4 blocks and then crosstown for three more." In the country, it's "Go south for 5 miles and then another half mile east." Either way,

though, you need to know how far to go in each of *two* different directions that are 90° apart.

You can use the same approach to describe motion—and the motion doesn't have to be in a straight line. As you will shortly see, we can use vectors, introduced in the last chapter, to describe motion in curved paths as well. Such analysis of *curvilinear* motion will eventually allow you to analyze the behavior of such things as batted balls, comets circling the Sun, and even electrons in atoms.

Curvilinear motion is quite easy to analyze using rectangular components of motion. Essentially, you break down, or *resolve*, the curved motion into rectangular (*x* and *y*) components and look at the straight-line motion in each dimension. To those components you can apply the kinematic equations introduced in Chapter 2. To find the position of an object moving in a curved path, for example, you simply find *x* and *y* coordinates at any time; then the object's position is given as the point (*x*, *y*).

Components of motion can be conveniently represented in vector notation: A displacement vector for example, has, or is made up of, the position components **x** and **y**. Similarly, a velocity vector has the velocity components $\mathbf{v}_x$ and $\mathbf{v}_y$; an acceleration vector has the acceleration components $\mathbf{a}_x$ and $\mathbf{a}_y$. In some instances, aspects of motion can be analyzed by adding vectors directly. Because every vector has both magnitude and direction, the scalar addition of numbers you learned in grade school does not apply. In this chapter, you'll learn how to add and subtract vectors, operations that take direction into account.

3.1 Components of Motion

Objectives: To be able to (a) analyze motion in terms of its components, and (b) apply the kinematic equations to components of motion.

An object moving in a straight line was considered in Chapter 2 to be moving along one of the Cartesian axes (*x* or *y*). But what if the motion is not along an axis? For example, consider the situation illustrated in •Fig. 3.1. Here, the balls are moving uniformly across a tabletop. The ball rolling in a straight line along the side of the table, designated as the *x* direction, is moving in one dimension. That is, its motion can be described with a single coordinate, *x*, as was done for motions in Chapter 2. Similarly, the motion of the ball rolling in the *y* direction can be described by a single *y* coordinate. However, both *x* and *y* coordinates are needed to describe the motion of the ball rolling diagonally across the table. We call this motion in two dimensions.

You might observe that if the diagonally-moving ball were the only object you had to consider, the *x* axis could be chosen in that direction and the motion reduced to one dimension. This is true, but once the coordinate axes are fixed, motions not along the axes must be described with two coordinates (*x*, *y*) or in two dimensions. Also, keep in mind that all motions in a plane (two dimensions) are not in straight lines. Think about the path of a ball you toss to another person.

In considering the motion of the ball moving diagonally across the table in Fig. 3.1a, one can think of the ball moving in the *x* and *y* directions simultaneously. That is, it has a velocity in the *x* direction ($+v_x$) *and* a velocity in the *y* direction ($+v_y$) at the same time. The combined velocities describe the actual motion of the ball. If the ball has a constant velocity (**v**) in a direction at an angle θ relative to the *x* axis, then the velocities in the *x* and *y* directions are obtained by resolving, or breaking down, the velocity vector into *components* in these directions, as shown in the pencil drawing in Fig. 3.1a. As may be seen from this drawing, the v_x and v_y components have magnitudes of

$$v_x = v \cos \theta \tag{3.1a}$$

$$v_y = v \sin \theta \tag{3.1b}$$

Magnitudes of velocity components in *x* and *y* directions

(Notice that $v = \sqrt{v_x^2 + v_y^2}$, so *v* is a combination of the *x* and *y* velocities.)

You are familiar with the use of two-dimensional length components in finding the *x* and *y* coordinates in a Cartesian system. For the ball rolling on the table,

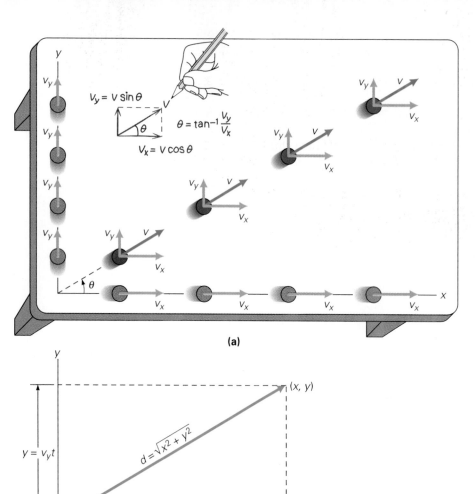

(a)

(b)

•FIGURE 3.1 Components of motion
(a) The velocity (and displacement) for uniform straight-line motion, may have x and y components (v_x and v_y as shown in the pencil drawing) because of the chosen orientation of the coordinate axes. Note that the velocity and displacement of the ball in the x direction are exactly the same as those that a ball rolling along the x axis with a uniform velocity of v_x would have. A comparable relationship holds true for the ball's motion in the y direction. Since the motion is uniform, the ratio v_y / v_x (and therefore θ) is constant. (b) The coordinates (x, y) of the ball's position and the distance d it has traveled from the origin may be found for any time t.

its position, (x, y), or the distance traveled from the origin in each of the component directions at time t, is given by

$$x = v_x t \qquad \text{\textit{Magnitude of}} \qquad (3.2a)$$
$$\text{\textit{displacement components}}$$
$$y = v_y t \qquad \text{\textit{(zero acceleration only)}} \qquad (3.2b)$$

just as for motion in the x and y directions separately. (Here, as usual, we assume that $x_o = y_o = 0$ at $t = 0$.) The ball's straight-line distance from the origin is then $d = \sqrt{x^2 + y^2}$ (Fig. 3.1b).

EXAMPLE 3.1. ■ ON A ROLL: USING COMPONENTS OF MOTION

If the diagonally-moving ball in Fig. 3.1 has a velocity of 0.50 m/s at an angle of 37° relative to the x axis, find how far it travels in 3.0 s using x and y components.

Solution. Organizing the data, we have

Given: $v = 0.50$ m/s   *Find:* d (distance)
$\theta = 37°$
$t = 3.0$ s

The distance traveled by the ball in terms of its x and y components is given by $d = \sqrt{x^2 + y^2}$. To find x and y as given by Eq. 3.2, we must first compute the velocity components v_x and v_y (Eq. 3.1).

$$v_x = v \cos 37° = (0.50 \text{ m/s})(0.80) = 0.40 \text{ m/s}$$
$$v_y = v \sin 37° = (0.50 \text{ m/s})(0.60) = 0.30 \text{ m/s}$$

Then, the component distances are

$$x = v_x t = (0.40 \text{ m/s})(3.0 \text{ s}) = 1.2 \text{ m}$$
$$y = v_y t = (0.30 \text{ m/s})(3.0 \text{ s}) = 0.90 \text{ m}$$

and the actual path distance is
$$d = \sqrt{x^2 + y^2} = \sqrt{(1.2 \text{ m})^2 + (0.90 \text{ m})^2} = 1.5 \text{ m}$$

Follow-up Exercise.   Suppose that a ball were rolling diagonally across a table with the same speed as in Fig. 3.1, *but* from the lower right corner, which is taken as the origin of the coordinate system, toward the upper left corner at an angle of $37°$ relative to the $-x$ axis. What would be the velocity components in this case? (*Answer may be found in the Answers to Follow-up Exercises section at the back of the book.*)

PROBLEM-SOLVING HINT

Note that for this simple case, the distance can also be obtained directly from $d = vt = (0.50 \text{ m/s})(3.0 \text{ s}) = 1.5$ m. However, we have solved this Example in a more general way to illustrate the use of components of motion. The direct solution would have been evident if the equations had been combined algebraically before calculation, that is,

$$x = v_x t = (v \cos \theta) t$$
$$y = v_y t = (v \sin \theta) t$$

and

$$d = \sqrt{x^2 + y^2} = \sqrt{(v \cos \theta)^2 t^2 + (v \sin \theta)^2 t^2} = \sqrt{v^2 t^2 (\cos^2 \theta + \sin^2 \theta)} = vt$$

Before embarking on the first solution strategy that occurs to you, pause for a moment to see whether there might be an easier or more direct way of approaching the problem.

Kinematic Equations for Components of Motion

The preceding Example involved two-dimensional motion in a plane. With a constant velocity (constant components v_x and v_y), the motion is in a straight line. The motion may also be accelerated. For motion in a plane *with a constant acceleration* having components a_x and a_y, the displacement (from the origin) and velocity components are given by the kinematic equations of Chapter 2 for the x and y directions:

$$x = v_{x_o} t + \tfrac{1}{2} a_x t^2 \qquad (3.3a)$$

$$y = v_{y_o} t + \tfrac{1}{2} a_y t^2 \qquad (3.3b)$$

(constant acceleration only)

$$v_x = v_{x_o} + a_x t \qquad (3.3c)$$

$$v_y = v_{y_o} + a_y t \qquad (3.3d)$$

Kinematic equations for displacement and velocity components

If an object is initially moving with a constant velocity and suddenly experiences an acceleration (a vector) in the direction of the velocity ($0°$) or opposite to it ($180°$), it would continue in a straight-line path, either speeding up or slowing down, respectively.

But motion can also be along a curved path. For the motion of an object to be *curvilinear*, that is, to vary from a straight-line path, an acceleration is required.

In this case, however, the acceleration vector must be at some angle to the velocity vector other than $0°$ or $180°$. Such a deflecting acceleration produces a curved path, for which the ratio of the velocity components varies with time. That is, the direction of the motion, $\theta = \tan^{-1}(v_y/v_x)$, varies with time because one or both of the velocity components do, so the motion is not in a straight line.

Consider a ball initially moving along the x axis, as illustrated in ●Fig. 3.2. Assume that starting at a time $t_o = 0$ it receives a constant acceleration a_y in the y direction. (You'll learn what produces an acceleration in the next chapter.) The magnitude of the x component of the ball's displacement is given by $x = v_x t$, since there is no acceleration in the x direction. Prior to t_o, the motion is in a straight line along the x axis. But at any time after t_o, there is a y displacement with a magnitude of $y = \frac{1}{2}a_y t^2$ (as given by Eq. 3.3b with $v_{y_o} = 0$). The result is a curved path for the ball.

Note that the length (magnitude) of the velocity component v_y changes with time. The total velocity vector (**v**) *at any time* is tangent to the curved path of the ball. It is at an angle θ relative to the x axis, given by $\theta = \tan^{-1}(v_y/v_x)$ (see Fig. 3.2), which now changes with time as we see in the following Example.

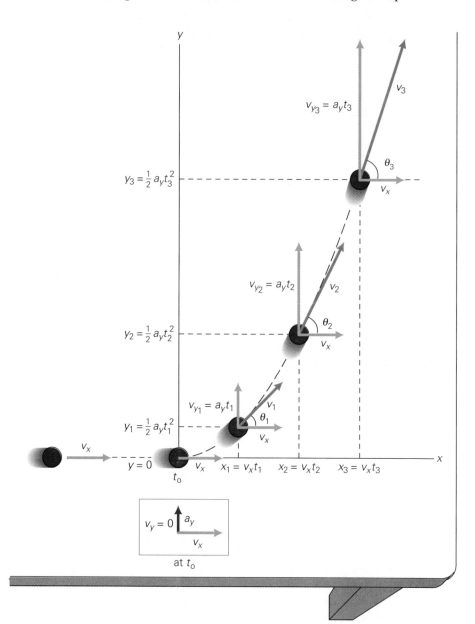

●**FIGURE 3.2 Curvilinear motion**
An acceleration not parallel to the instantaneous velocity produces a curved path. Here an acceleration (**a**$_y$) is applied at t_o to a ball initially moving with a constant velocity (**v**$_x$). The result is a curved path with the velocity components as shown. Notice how **v**$_y$ increases with time.

EXAMPLE 3.2 ■ A CURVING PATH: VECTOR COMPONENTS

Suppose that the ball in Fig. 3.2 has an initial velocity of 1.50 m/s along the x axis and starting at t_o receives an acceleration of 2.80 m/s² in the y direction. (a) What is the position of the ball 3.00 s after t_o? (b) What is the velocity of the ball at that time?

Solution. Referring to Fig. 3.2, we have the following:

Given: $v_{x_o} = v_x = 1.50$ m/s   **Find:** (a) (x, y) (position coordinates)
$v_{y_o} = 0$
$a_x = 0$
$a_y = 2.80$ m/s   (b) **v** (velocity)
$t = 3.00$ s

(a) At 3.00 s after $t_o = 0$, Eq. 3.3a and 3.3b tell us that the ball has traveled the following distances from the origin in the x and y directions:

$$x = v_{x_o}t + \tfrac{1}{2} a_x t^2 = (1.50 \text{ m/s})(3.00 \text{ s}) + 0 = 4.50 \text{ m}$$

$$y = v_{y_o}t + \tfrac{1}{2} a_y t^2 = 0 + \tfrac{1}{2}(2.80 \text{ m/s}^2)(3.00 \text{ s})^2 = 12.6 \text{ m}$$

Thus, its position is $(x, y) = (4.50$ m, 12.6 m). If you computed the distance, $d = \sqrt{x^2 + y^2}$, what would this be? (Note that it is not the actual distance the ball has traveled in 3.00 s, but the magnitude of the *displacement* or straight-line distance from the origin at $t = 3.00$ s.)

(b) The x component of the velocity is given by Eq. 3.3c:

$$v_x = v_{x_o} + a_x t = 1.50 \text{ m/s} + 0 = 1.50 \text{ m/s}$$

(This is constant since there is no acceleration in the x direction.) Similarly, the y component of the velocity is given by Eq. 3.3d:

$$v_y = v_{y_o} + a_y t = 0 + (2.80 \text{ m/s}^2)(3.00 \text{ s}) = 8.40 \text{ m/s}$$

The velocity therefore has a magnitude of

$$v = \sqrt{v_x^2 + v_y^2} = \sqrt{(1.50 \text{ m/s})^2 + (8.40 \text{ m/s})^2} = 8.53 \text{ m/s}$$

and its direction relative to the x axis is

$$\theta = \tan^{-1}\left(\frac{v_y}{v_x}\right) = \tan^{-1}\left(\frac{8.40 \text{ m/s}}{1.50 \text{ m/s}}\right) = 79.9°$$

Note: Don't confuse the direction of the velocity with the direction of the displacement from the origin. The direction of the velocity must always be tangent to the path.

Follow-up Exercise. Suppose that the ball in Example 3.2 also received an acceleration of 1.00 m/s² in the $+x$ direction at t_o. What would be the position of the ball 3.00 s after t_o in this case? *(Answer may be found in the Answers to Follow-up Exercises section at the back of the book.)*

3.2 Vector Addition and Subtraction

Objectives: To be able to add and subtract vectors (a) graphically and (b) analytically.

Many physical quantities are vectors. You have worked with a few related to motion (displacement, velocity, and acceleration) and will encounter more during this course of study. A very important technique in the analysis of many physical situations is the addition (and subtraction) of vectors. By adding or combining such quantities (**vector addition**), you can obtain the overall, or net, effect that occurs—the *resultant*, as we call the vector sum.

You have already been adding vectors. In Chapter 2, displacements were added to get the net displacement. In this chapter, vector components of motion were added to give net effects. Notice in the preceding Example 3.2 that the velocity components v_x and v_y were added to give the resultant velocity.

In this section, we will look at vector addition and subtraction in general, along with common vector notation. As you will learn, these operations are not the same as scalar or numerical addition and subtraction, with which you are already familiar. Vectors have magnitudes *and* directions, so different rules apply.

In general, there are geometric (graphical) methods and analytical (computational) methods of vector addition. The geometric methods are useful in helping to visualize the concepts of vector addition. Analytical methods are more commonly used, however, because they are faster and more precise. One thing that scalar addition and vector addition have in common is that the quantities being added must have the same units. You cannot meaningfully add a displacement vector to a velocity vector, just as you cannot add the scalars of distance and speed and obtain a result with any physical meaning.

Vector Addition: Geometric Methods

Triangle Method. To add two vectors, for example, to add **B** to **A** (that is, to find **A** + **B**) by the **triangle method**, you first draw **A** on a sheet of graph paper to some scale. For example, if **A** is a displacement in meters, a convenient scale is 1 cm:1 m, or 1 cm of vector length on the graph corresponds to 1 meter of displacement (•Fig. 3.3a). As shown in •Fig. 3.3b, the direction of the **A** vector is specified as being at an angle (θ_A) relative to a coordinate axis, usually the *x* axis.

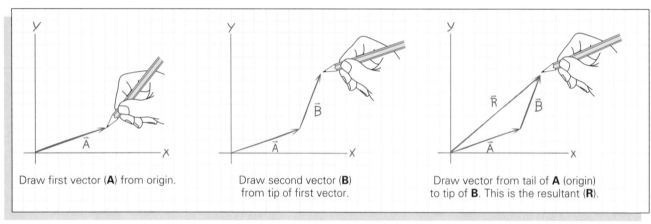

Draw first vector (**A**) from origin.

Draw second vector (**B**) from tip of first vector.

Draw vector from tail of **A** (origin) to tip of **B**. This is the resultant (**R**).

(a)

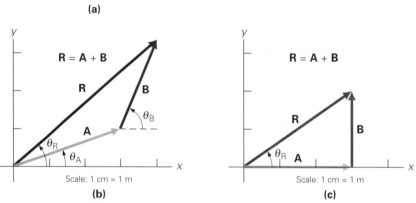

(b)

(c)

•**FIGURE 3.3 Triangle method of vector addition**
(a) The vectors **A** and **B** are placed tip to tail. The vector from the tail of **A** to the tip of **B**, forming the third side of the triangle, is the resultant, **R** = **A** + **B**. **(b)** When drawn to scale, the magnitude of **R** can be found by measuring its length and using the scale conversion, and the direction angle θ_R may be measured with a protractor. Analytical methods may also be used. For a non-right triangle as in (b), the laws of sines and cosines would be used. If the vector triangle is a right triangle as in **(c)**, **R** is easily obtained with the Pythagorean theorem and the direction angle is given by an inverse trigonometric function.

Next draw **B** with its tail starting at the tip of **A**. (Thus, this method is also called the *tip-to-tail method*.) The vector from the tail of **A** to the tip of **B** is then the vector sum, or the resultant of the two vectors: **R** = **A** + **B**.

If drawn to scale, the magnitude of **R** can be found by measuring its length and using the scale conversion. In such a graphical approach, the direction angle (θ_R) is measured with a protractor. Knowing the magnitudes and directions (θ's) of **A** and **B**, the magnitude and direction of **R** can also be found analytically using trigonometric methods. For the non-right triangle in •Fig. 3.3b, the laws of sines and cosines would be used (see Appendix I). The resultant of the vector right triangle in •Fig. 3.3c would be much easier to find using the Pythagorean theorem for the magnitude and an inverse trig function to find the direction angle. [Notice how **R** in this case is made up of x and y components (**A** and **B**).]

Parallelogram Method. Another graphical method of vector addition similar to the triangle method is the **parallelogram method**. In •Fig. 3.4, **A** and **B** are drawn tail to tail, and a parallelogram is formed as shown. The resultant **R** lies along the diagonal of the parallelogram. Drawing the diagram to scale with proper orientations, the magnitude and direction of **R** can be measured directly from the diagram as in the triangle method.

Notice that **B** could be moved over to the other side of the parallelogram, forming the **A** + **B** triangle (and demonstrating why the triangle and parallelogram methods are really equivalent). In general, a vector (arrow) can be moved around in vector addition methods. As long as you don't change its length (magnitude) or direction, you don't change the vector. In Fig. 3.4, this shifting of vector arrows shows that **A** + **B** = **B** + **A**—that is, the vectors can be added in either order.

Polygon Method. The tip-to-tail method can be extended to include the addition of any number of vectors. The method is then called the **polygon method** because the resulting graphical figure is a polygon. This is illustrated for four vectors in •Fig. 3.5, where **R** = **A** + **B** + **C** + **D**. Note that this addition is essentially three applications of the triangle method. The length and direction of the resultant could be found analytically by successive applications of the laws of sines and cosines, but an easier analytical method, the component method, will be described shortly. As in the parallelogram method, the four vectors (or any number of vectors) can be added in any order.

Vector Subtraction. Vector subtraction is a special case of vector addition.

$$\mathbf{A} - \mathbf{B} = \mathbf{A} + (-\mathbf{B})$$

That is, to subtract **B** from **A**, a *negative* **B** is added to **A**. In Chapter 2, you learned that a minus sign simply means that the direction of a vector is opposite to that of one with a plus sign, e.g., +x and −x. The same is true for a general vector.

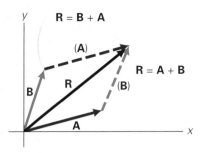

•**FIGURE 3.4 Parallelogram method of vector addition**
The diagonal of the parallelogram formed after **A** and **B** are drawn with their tails at the origin is the resultant, **R** = **A** + **B**. When drawn to scale, the magnitude and direction of **R** can be measured directly from the diagram as in the triangle method. Notice that the parallelogram method is equivalent to the triangle method. Shifting **B** to the right forms the same **A** + **B** vector triangle shown in Fig. 3.3b. Shifting **A** up has the same effect, with **R** = **B** + **A**.

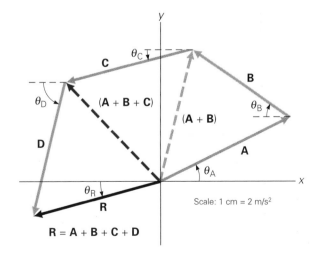

•**FIGURE 3.5 Polygon method of vector addition**
The vectors to be added are placed tip to tail. The resultant **R** is the vector from the tail of the first vector (**A**) to the tip of the last vector (**D**), which completes the polygon. This method is essentially multiple applications of the triangle method: (**A** + **B**) + **C**, and so on.

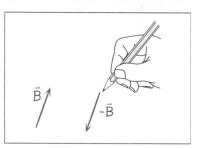

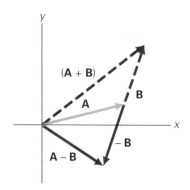

•FIGURE 3.6 Vector subtraction
Vector subtraction is a special case
of vector addition; that is, $\mathbf{A} - \mathbf{B} =$
$\mathbf{A} + (-\mathbf{B})$, where $-\mathbf{B}$ has the same
magnitude as $\mathbf{B}$ but is in the
opposite direction (see sketch).
Thus $\mathbf{A} + \mathbf{B}$ is *not* the same as
$\mathbf{A} - \mathbf{B}$. Can you show that $\mathbf{B} - \mathbf{A}$
$= -(\mathbf{A} - \mathbf{B})$ geometrically?

Note: The notation $\tan^{-1} x$ stands for
arctangent x, or the angle whose
tangent is x

The vector $-\mathbf{B}$ has the same magnitude as the vector $\mathbf{B}$, but is in the opposite direction (see •Fig. 3.6). The vector diagram in the figure provides a graphical representation of $\mathbf{A} - \mathbf{B}$.

EXAMPLE 3.3 ■ VECTOR SUBTRACTION: A SPECIAL CASE OF VECTOR ADDITION

Consider yourself in a car (A) traveling along a straight, level highway with a speed $v_A = 75$ km/h. Another car (B) travels at a different speed $v_B = 90$ km/h. Find the differences in the velocities, $\mathbf{v}_{BA} = \mathbf{v}_B - \mathbf{v}_A$ when (a) the other car travels in the same direction in front of you, and (b) the other car is approaching you traveling in the opposite direction.

Solution. Taking the direction in which you are traveling as positive, we have, where the signs indicate vector directions,

Given: $v_A = (+) 75$ km/h   *Find:* $\mathbf{v}_{BA} = \mathbf{v}_B - \mathbf{v}_A$
$\qquad$(a) $v_B = (+) 90$ km/h (same direction)
$\qquad$(b) $v_B = -90$ km/h (opposite direction)

(a) With both cars traveling in the same (+) direction,

$$\mathbf{v}_{BA} = \mathbf{v}_B - \mathbf{v}_A = 90 \text{ km/h} - (+75 \text{ km/h}) = +15 \text{ km/h}$$

(b) When traveling in opposite directions,

$$\mathbf{v}_{BA} = \mathbf{v}_B - \mathbf{v}_A = -90 \text{ km/h} - (+75 \text{ km/h}) = -165 \text{ km/h}$$

Here, $\mathbf{v}_{BA}$ is what is called a *relative* velocity—the velocity of B relative to A. In part (a), relative to you in car A, car B appears to travel at a speed of 15 km/h. However, when approaching you, car B appears to travel at 165 km/h. It's as though you consider yourself or your car's reference frame to be at rest. (We will explore the subject of relative velocity further in Section 3.3.)

Follow-up Exercise. To what are the given velocities in Example 3.3 referenced? *(Answer may be found in the Answers to Follow-up Exercises section at the back of the book.)*

Vector Components and the Analytical Component Method

Probably the most widely used analytical method for adding multiple vectors is the **component method**. It will be used again and again throughout the course of our study, so a basic understanding of the method is *essential*. Learn this section well.

Adding Rectangular Vector Components. By rectangular components, we mean those at right (90°) angles to each other and usually taken in the rectangular coordinate x and y directions. You have already had an introduction to the addition of such components in the discussion of the velocity components of motion. For the general case, suppose that $\mathbf{A}$ and $\mathbf{B}$, two vectors at right angles, are added, as illustrated in •Fig. 3.7a. The right angle makes things easy. The magnitude of $\mathbf{C}$ is given by the Pythagorean theorem:

$$C = \sqrt{A^2 + B^2} \qquad (3.4a)$$

The orientation of $\mathbf{C}$ to the x axis is given by the angle

$$\theta = \tan^{-1}\left(\frac{B}{A}\right) \qquad (3.4b)$$

This is how a resultant is expressed in **magnitude-angle form**.

Resolving a Vector into Rectangular Components; Unit Vectors. Resolving a vector into rectangular components is essentially the reverse of adding the components. Given a vector $\mathbf{C}$, •Fig. 3.7b illustrates how it may be resolved into x and y vector components, $\mathbf{C}_x$ and $\mathbf{C}_y$. You simply complete the vector triangle with x

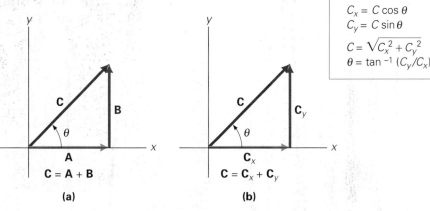

$$C_x = C \cos \theta$$
$$C_y = C \sin \theta$$
$$C = \sqrt{C_x^2 + C_y^2}$$
$$\theta = \tan^{-1}(C_y/C_x)$$

(a) $C = A + B$

(b) $C = C_x + C_y$

•**FIGURE 3.7   Vector components**
(a) The vectors **A** and **B** along the x and y axes, respectively, add to give **C**. **(b)** A vector **C** may be resolved into rectangular components C_x and C_y.

and y components. From the diagram, it can be seen that the magnitudes or vector lengths of these components are given by

$$C_x = C \cos \theta \qquad \text{magnitudes of} \qquad (3.5a)$$
$$C_y = C \sin \theta \qquad \text{components} \qquad (3.5b)$$

(similar to $v_x = v \cos \theta$ and $v_y = v \sin \theta$ in Example 3.1). The angle of direction for **C** can also be expressed in terms of the components, since $\tan \theta = C_y/C_x$, or

$$\theta = \tan^{-1}\left(\frac{C_y}{C_x}\right) \qquad \begin{array}{l}\text{direction of vector from} \\ \text{magnitudes of components}\end{array} \qquad (3.6)$$

A general notation for expressing the magnitude and direction of a vector involves the use of unit vectors. For example, as illustrated in •Fig. 3.8, a vector **A** may be written as **A** = A**a**. The numerical magnitude is represented by A, and **a** is called a **unit vector**. That is, it has a magnitude of unity, or one, and so simply indicates the vector's direction. For example, a velocity along the x axis might be written **v** = 4.0 m/s **x** (that is 4.0 m/s magnitude in the $+x$ direction).

Note in the figure how $-$**A** would be represented in this notation. Although the minus sign is sometimes put in front of the numerical magnitude, this is an absolute number; the minus actually goes with the unit vector; $-$**A** = $-A$**a** = $A(-$**a**).* That is, the unit vector is in the $-$**a** direction (opposite **a**). A velocity of **v** = -4.0 m/s **x** has a magnitude of 4.0 m/s in the $-x$ direction, that is, **v** = 4.0 m/s $(-$**x**).

This notation can be used to express explicitly the rectangular components of a vector. For example, the ball's displacement from the origin in Example 3.2 could be written **d** = (4.50 m) **x** + (12.6 m) **y**, where **x** and **y** are unit vectors in these directions. In some instances it may be more convenient to express a general vector in this unit vector **component form**:

$$\mathbf{C} = C_x\mathbf{x} + C_y\mathbf{y} \qquad (3.7)$$

Vector Addition Using Components.   The **analytical component method** of vector addition involves resolving the vectors into rectangular components and adding the components for each axis independently. This is illustrated graphically in •Fig. 3.9 for two vectors, $\mathbf{F}_1$ and $\mathbf{F}_2$. The sums of the x and y components of the vectors being added are then equal to the corresponding components of the resultant vector.

The same principle applies if you are given three (or more) vectors to add. You could find the resultant by applying the graphical tip-to-tail method as illustrated in •Fig. 3.10a. However, this involves drawing the vectors to scale and using a protractor to measure angles, which is time-consuming. Note that the magnitude

*The notation is sometimes written **A** = $|A|$**a** or $-$**A** = $-|A|$**a** to express the absolute value of the magnitude.

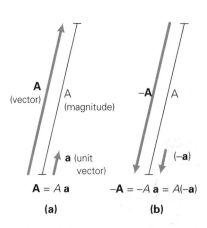

(a) $\mathbf{A} = A\,\mathbf{a}$

(b) $-\mathbf{A} = -A\,\mathbf{a} = A(-\mathbf{a})$

•**FIGURE 3.8   Unit vectors**
(a) A unit vector **a** has a magnitude of unity, or one, and so simply indicates a vector's direction. Written with the magnitude A, it represents the vector **A**, that is, **A** = A **a**. **(b)** For the vector $-$**A**, the unit vector is $-$**a**, and $-$**A** = $-A$ **a** = $A(-$**a**). The magnitude A is a positive number, i.e., $|A|$.

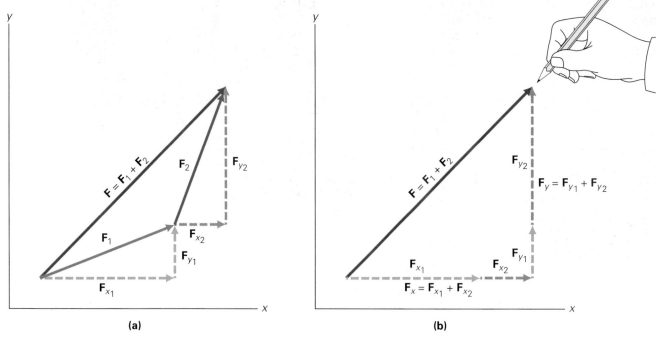

FIGURE 3.9 Component addition
(a) In adding vectors by the component method, each vector is first resolved into its x and y components. **(b)** The sums of the x and y components of $\mathbf{F}_1$ and $\mathbf{F}_2$ are the x and y components of the resultant $\mathbf{F}$; that is, $\mathbf{F}_x = \mathbf{F}_{x_1} + \mathbf{F}_{x_2}$ and $\mathbf{F}_y = \mathbf{F}_{y_1} + \mathbf{F}_{y_2}$.

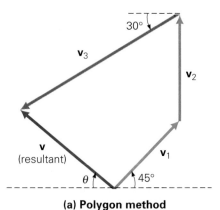

(a) Polygon method

FIGURE 3.10 Component method of vector addition
(a) Several vectors may be added graphically to find the resultant $\mathbf{v}$, but this is time-consuming. In the analytical component method **(b)** the vectors to be added are all first placed with their tails at the origin so that they may be easily resolved into rectangular components. **(c)** The respective summations of all the x components and all the y components are then added to give the resultant $\mathbf{v}$. Notice how the negative components subtract. See Example 3.4.

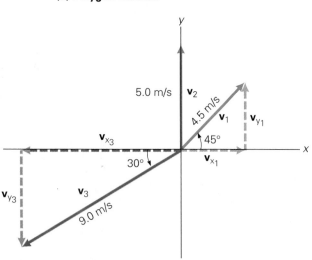

(b) Component method
(resolving components)

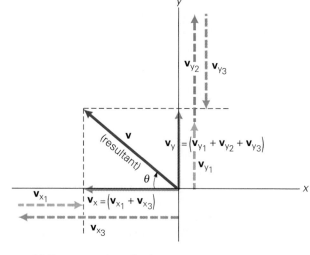

(c) Component method
(adding x and y components, shown as
offset dashed arrows, and finding resultant)

of the directional angle θ is not obvious in the figure and would have to be measured (or computed trigonometrically).

However, if you use the component method, you do not have to draw the vectors tip to tail. In fact, it is usually more convenient to put all of the tails together at the origin, as shown in •Fig. 3.10b. Also, the vectors do not have to be drawn to scale, since the sketch is just a visual aid in applying the analytical method.

In the component method, you simply resolve the vectors to be added into their x and y components, add the respective components, and recombine to find the resultant. This procedure is illustrated in •Fig. 3.10c. Notice that it is just linear vector addition in the component directions, and that components may add or subtract. A vector in the x or y direction has a "component" only in that direction (e.g., $\mathbf{v}_2$ in the y direction in the figure).

Notice also in Fig. 3.10c that the directional angle θ of the resultant is referred to the x axis, as are the individual vectors in Fig. 3.10b. *In adding vectors by the component method, we will reference all vectors to the nearest x axis, that is, the x axis or $-x$ axis.* As will be seen, this eliminates angles greater than 90° (as occurs when customarily measuring angles counterclockwise from the $+x$ axis) and the use of double angle formulas, e.g., $\cos(\theta + 90°)$. This greatly simplifies calculations. The recommended procedures for adding vectors analytically by the component method can be summarized as follows:

Procedures for Adding Vectors by the Component Method

1. Resolve the vectors to be added into their x and y components. Use the acute angles (those less than 90°) between the vectors and the x axis, and indicate the directions of the components by plus and minus signs. (See •Fig. 3.11.)

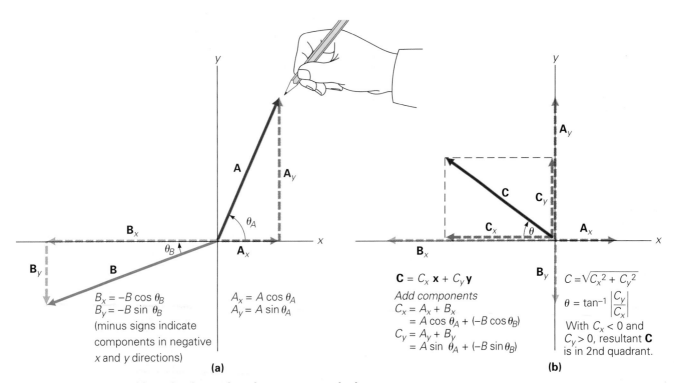

•**FIGURE 3.11 Vector addition by the analytical component method**
(a) Resolve the vectors into their x and y components. (b) Add all of the x components and all of the y components together vectorially to obtain the x and y components (C_x and C_y) of the resultant. Express the resultant in either component form or magnitude-angle form. See text for more detailed description.

2. Add all of the x components together and all of the y components together vectorially to obtain the x and y components of the resultant, or vector sum. (This is done algebraically using plus and minus signs.)

3. Express the resultant vector using:

   (a) the component form—e.g., $\mathbf{C} = C_x\mathbf{x} + C_y\mathbf{y}$—*or*
   (b) the magnitude-angle form.

   For the latter, find the magnitude of the resultant using the summed x and y components and the Pythagorean formula:

   $$C = \sqrt{C_x^2 + C_y^2}$$

   Find the angle of direction (relative to the x axis) by taking the arctangent of the *absolute value* of the ratio of the y and x components (that is, the positive value ignoring any minus signs):

   $$\theta = \tan^{-1}\left|\frac{C_y}{C_x}\right|$$

   Designate the quadrant in which the resultant lies. This is obtained from the signs of the summed components or a sketch of their addition using the triangle (or parallelogram) method (see Fig. 3.11). The angle θ is the angle between the resultant and the x axis in that quadrant.

Note: Taking the absolute value means to ignore minus signs (e.g., $|-3| = 3$). This is done to avoid negative values and angles greater than 90°.

EXAMPLE 3.4 ■ APPLYING THE ANALYTICAL COMPONENT METHOD: SEPARATING AND COMBINING X AND Y COMPONENTS

Let's apply the procedural steps of the component method to the addition of the vectors in Fig. 3.10b. The vectors with units of m/s represent velocities.

Solution.

1. The rectangular components of the vectors are shown in the figure.
2. Summing these components gives

   $$\mathbf{v} = v_x\mathbf{x} + v_y\mathbf{y} = (v_{x_1} + v_{x_2} + v_{x_3})\mathbf{x} + (v_{y_1} + v_{y_2} + v_{y_3})\mathbf{y}$$

   where

   $$v_x = v_{x_1} + v_{x_2} + v_{x_3} = v_1\cos 45° + v_2\cos 90° - v_3\cos 30°$$
   $$= (4.5\,\text{m/s})(0.707) + (5.0\,\text{m/s})(0) - (9.0\,\text{m/s})(0.866) = -4.6\,\text{m/s}$$
   $$v_y = v_{y_1} + v_{y_2} + v_{y_3} = v_1\sin 45° + v_2\sin 90° - v_3\sin 30°$$
   $$= (4.5\,\text{m/s})(0.707) + (5.0\,\text{m/s})(1) - (9.0\,\text{m/s})(0.50) = 3.7\,\text{m/s}$$

   In tabular form, the components are:

	x components		*y components*
v_{x_1}	$v_1\cos 45° = +3.2\,\text{m/s}$	v_{y_1}	$v_1\sin 45° = +3.2\,\text{m/s}$
v_{x_2}	$v_2\cos 90° = 0$	v_{y_2}	$v_2\sin 0° = +5.0\,\text{m/s}$
v_{x_3}	$v_3\cos 30° = -7.8\,\text{m/s}$	v_{y_3}	$v_3\sin 30° = -4.5\,\text{m/s}$
Sums: v_x	$= -4.6\,\text{m/s}$	v_y	$= +3.7\,\text{m/s}$

   The directions of the components are indicated by signs (the + sign usually omitted as being understood). In this case, v_2 has no x component. Note that in general for the analytical component method, the x components are cosine functions and the y components are sine functions.

3. In component form, the resultant vector is

   $$\mathbf{v} = (-4.6\,\text{m/s})\mathbf{x} + (3.7\,\text{m/s})\mathbf{y}$$

   In magnitude-angle form, the resultant velocity has a magnitude of

   $$v = \sqrt{v_x^2 + v_y^2} = \sqrt{(-4.6\,\text{m/s})^2 + (3.7\,\text{m/s})^2} = 5.9\,\text{m/s}$$

Since the x component is negative and the y component is positive, the resultant lies in the second quadrant at an angle of

$$\theta = \tan^{-1}\left|\frac{v_y}{v_x}\right| = \tan^{-1}\left(\frac{3.7 \text{ m/s}}{4.6 \text{ m/s}}\right) = 39°$$

above the negative x axis (see Fig. 3.10c).

Follow-up Exercise. Suppose in Example 3.4 there were an additional $v_4 = 4.6$ m/s $\mathbf{x}$ velocity vector. What would be the resultant in this case? (*Answer may be found in the Answers to Follow-up Exercises section at the back of the book.*)

Although this discussion is limited to motion in two dimensions (in a plane), the component method is easily extended to three dimensions. For a velocity in three dimensions, the vector has x, y, and z components: $\mathbf{v} = v_x \mathbf{x} + v_y \mathbf{y} + v_z \mathbf{z}$.

3.3 Relative Velocity

Objective: **To be able to determine relative velocities through vector addition and subtraction.**

Measurements must be made with respect to some reference. This is usually taken to be the origin of a coordinate system. The point you designate as the origin of a set of coordinate axes is arbitrary and entirely a matter of choice. For example, you may "attach" the coordinate system to the road or the ground and then measure the displacement or velocity of a car relative to these axes.

A situation may be analyzed from any frame of reference. For example, the origin of the coordinate axes may be attached to a car moving along a highway. In analyzing motion from another reference frame, you do not change the physical situation or what is taking place, only the point of view from which you choose to describe it. Hence, we say that motion is *relative* (to some reference frame), and we refer to **relative velocity**. Since velocity is a vector, vector addition and subtraction are helpful in determining relative velocities.

Relative Velocities in One Dimension

When the velocities are linear (along a straight line) in the same or opposite directions and all have the same reference (such as the Earth), relative velocities can be found simply by vector subtraction. As an illustration, consider cars moving with constant velocities along a straight, level highway, as in ●Fig. 3.12. The velocities shown in the figure are *relative to the Earth or ground*, as indicated by the reference set of coordinate axes in Fig. 3.12a. They are also relative to the stationary observers standing by the highway and sitting in the parked car A. That is, these observers see the cars moving with the velocities indicated. The relative velocity of two objects is given by the velocity (vector) difference. For example, the velocity of car B *relative to car A* is given by

$$\mathbf{v}_{BA} = \mathbf{v}_B - \mathbf{v}_A = +90 \text{ km/h} - 0 \text{ km/h} = +90 \text{ km/h}$$

Thus, a person sitting in car A would see car B move away at a speed of 90 km/h (in the positive x direction). For this linear case, the directions of the velocities are indicated by plus and minus signs (in addition to the minus sign in the formula).

Similarly, the velocity of car C relative to an observer in car A is

$$\mathbf{v}_{CA} = \mathbf{v}_C - \mathbf{v}_A = -60 \text{ km/h} - 0 \text{ km/h} = -60 \text{ km/h}.$$

The person in car A would see car C approaching with a speed of 60 km/h (in the negative x direction).

Note: Use subscripts carefully!
v_{AB} = velocity *of A relative to B*

•FIGURE 3.12 Relative velocity
The observed velocity of a car
depends on, or is relative to, the
frame of reference. The velocities
shown in **(a)** are relative to the
ground or to the parked car A. In
(b) the frame of reference is with
respect to car B, and the velocities
are those that the driver of car B
would observe. **(c)** These aircraft,
performing air-to-air refueling, are
normally described as traveling at
hundreds of km/h. To what frame
of reference do these velocities
refer? What is their velocity
relative to each other?

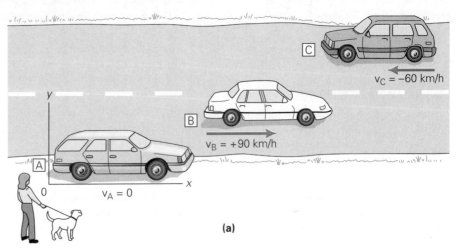

(a)

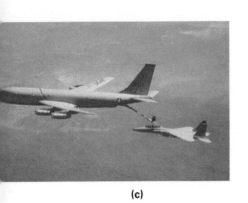

(c)

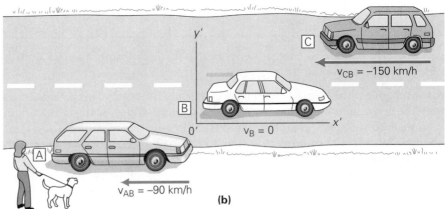

(b)

But suppose that you want to know the velocities of the other cars *relative to car B*, that is, from the point of view of an observer in car B, or relative to a set of coordinate axes with the origin fixed to car B (Fig. 3.12b). Relative to those axes, car B is not moving and acts as the fixed reference point. The other cars are moving relative to car B. The velocity of car C relative to car B is

$$\mathbf{v}_{CB} = \mathbf{v}_C - \mathbf{v}_B = -60 \text{ km/h} - (+90 \text{ km/h}) = -150 \text{ km/h}$$

Similarly, car A has a velocity relative to car B of

$$\mathbf{v}_{AB} = \mathbf{v}_A - \mathbf{v}_B = 0 \text{ km/h} - (+90 \text{ km/h}) = -90 \text{ km/h}$$

Notice that relative to B, the other cars are both moving in the negative x direction. That is, C is approaching B with a velocity of 150 km/h in the $-x$ direction and A appears to be receding from B with a velocity of 90 km/h in the $-x$ direction. (Imagine yourself in car B, and take that position as stationary. Car C would appear to be coming toward you at a high rate of speed, and car A would be getting farther and farther away, as though it were moving backward relative to you.) Note that, in general,

$$\mathbf{v}_{AB} = -\mathbf{v}_{BA}$$

What about the velocities relative to car C? From the point of view (or reference point) of car C, cars A and B would both appear to be moving in the positive x direction. For B relative to C,

$$\mathbf{v}_{BC} = \mathbf{v}_B - \mathbf{v}_C = 90 \text{ km/h} - (-60 \text{ km/h}) = +150 \text{ km/h}$$

Can you show that $\mathbf{v}_{AC} = +60 \text{ km/h}$?

In some instances, we may need to work with velocities that do not all have the same reference. In such cases, relative velocities can be found by means of vector addition. To solve problems of this kind, *it is essential to identify the velocity references with care.*

Let's look first at a one-dimensional (linear) example. Suppose that a straight moving walkway in a major airport moves with a velocity of $v_{wg} = +1.0$ m/s, where the subscripts indicate the velocity of the walkway (w) with respect to or relative to the ground (g). A passenger (p) on the walkway trying to make a flight connection walks with a velocity of $v_{pw} = +2.0$ m/s relative to the walkway (w). What is the passenger's velocity relative to an observer standing by the walkway (that is, relative to the ground)?

The velocity we are seeking, v_{pg} is given by

$$v_{pg} = v_{pw} + v_{wg} = +2.0 \text{ m/s} + 1.0 \text{ m/s} = +3.0 \text{ m/s}$$

Thus the stationary observer sees the passenger traveling with speed of 3.0 m/s down the walkway. (Make a sketch and show how the vectors add.)

PROBLEM-SOLVING HINT

Notice the pattern of the subscripts in this example. On the right side of the equation, the two inner subscripts are the same (w). The outer subscripts (p and g) are sequentially the same as those for the relative velocity on the left side of the equation. When adding relative velocities, always check to make sure that the subscripts have this relationship—it indicates that you have set up the equation correctly.

What if a passenger got on the walkway going in the opposite direction and walked with the same speed as that of the walkway? Now it is essential to indicate the direction in which the passenger is walking by means of a minus sign: $v_{pw} = -1.0$ m/s. In this case, relative to the stationary observer,

$$v_{pg} = v_{pw} + v_{wg} = -1.0 \text{ m/s} + 1.0 \text{ m/s} = 0$$

so the passenger is stationary with respect to the ground and the walkway acts as a treadmill. (Excellent physical exercise!)

Relative Velocities in Two Dimensions

Of course, velocities are not always in the same or opposite directions. However, knowing how to use rectangular components to add or subtract vectors, we can solve problems involving relative velocities in two dimensions, as the following Examples show.

EXAMPLE 3.5 ■ ACROSS AND DOWN THE RIVER: RELATIVE VELOCITY AND COMPONENTS OF MOTION

The current of a 500-m wide straight river has a flow rate of 2.55 km/h. A motorboat that travels with a constant speed of 8.00 km/h in still water crosses the river (•Fig. 3.13). (a) If the boat's bow points directly across the river toward the opposite shore, what is the velocity of the boat relative to the stationary observer at the bridge? (b) How far downstream will the boat's landing point be from the point directly opposite its starting point? (c) What is the distance traveled by the boat in crossing the river? (Assume that the boat comes instantaneously to rest on a sandy shore.)

Solution. Here again, identifying the data is important. As indicated in the figure, we take the river's flow velocity (v_{rs}, river to shore) to be in the x direction and the boat's velocity (v_{br}, boat to river) to be in the y direction. Note that the river's flow velocity is *relative to the shore*, and that the boat's velocity is *relative to the river* or water, as indicated by the subscripts. Listing the data, we have

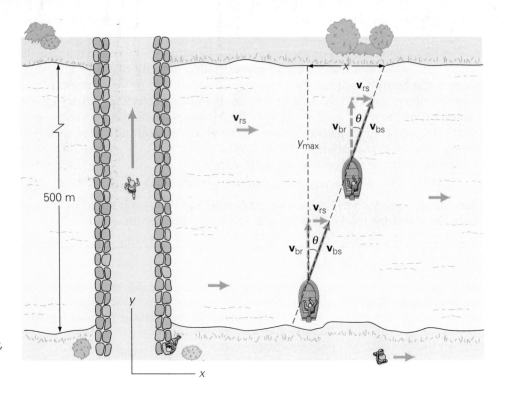

•FIGURE 3.13 Relative velocity and components of motion
As the boat moves across the river, it is carried downstream by the current. See Example 3.5.

Given: y_{max} = 500 m (river width)
$\mathbf{v}_{rs}$ = 2.55 km/h $\mathbf{x}$
   = 0.709 m/s $\mathbf{x}$
   (velocity of river *relative to shore*)
$\mathbf{v}_{br}$ = 8.00 km/h $\mathbf{y}$
   = 2.22 m/s $\mathbf{y}$
   (velocity of boat *relative to river*)

Find: (a) $\mathbf{v}_{bs}$ (velocity of boat *relative to shore*)
(b) x (distance downstream)
(c) d (distance traveled by boat)

(The conversion factor from km/h to m/s has been omitted for convenience.)

Notice that as the boat moves toward the opposite shore, it is also carried downstream by the current. These velocity components would be clearly apparent relative to the jogger crossing the bridge and to the person sauntering downstream in Fig. 3.13. If both stay even with the boat, the velocity of each will match one of the components of the boat's velocity. Since the velocity components are constant, the boat travels in a straight line diagonally across the river (much like the ball rolling across the table in Example 3.1).

(a) The velocity of the boat relative to the shore ($\mathbf{v}_{bs}$) is given by vector addition. In this case, we have

$$\mathbf{v}_{bs} = \mathbf{v}_{br} + \mathbf{v}_{rs}$$

Since the velocities are not along one axis as in the previous example of the cars, they cannot be added directly. Notice in Fig. 3.13 that the vectors form a right triangle, so we can apply the Pythagorean theorem to find the magnitude of $\mathbf{v}_{bs}$:

$$v_{bs} = \sqrt{(v_{br})^2 + (v_{rs})^2} = \sqrt{(2.22 \text{ m/s})^2 + (0.709 \text{ m/s})^2}$$
$$= 2.33 \text{ m/s}$$

This is in the direction defined by

$$\theta = \tan^{-1}\left(\frac{v_{rs}}{v_{br}}\right) = \tan^{-1}\left(\frac{0.709 \text{ m/s}}{2.22 \text{ m/s}}\right) = 17.7°$$

(b) To find the distance x that the current carries the boat downstream, we use components. Note that in the y direction, $y_{max} = v_{br}t$, and

$$t = \frac{y_{max}}{v_{br}} = \frac{500 \text{ m}}{2.22 \text{ m/s}} = 225 \text{ s}$$

which is the time it takes the boat to cross the river.

During this time, the boat is carried downstream by the current a distance of

$$x = v_{rs}t = (0.709 \text{ m/s})(225 \text{ s}) = 160 \text{ m}$$

(c) We could find the distance d traveled by the boat using x and y_{max} with the Pythagorean theorem again, but let's use the magnitude of the relative velocity and time:

$$d = v_{bs}t = (2.33 \text{ m/s})(225 \text{ s}) = 524 \text{ m}$$

Follow-up Exercise. In part (c), find the distance d by using the Pythagorean theorem. (The answer may be slightly different. Why?) *(Answer may be found in the Answers to Follow-up Exercises section at the back of the book.)*

EXAMPLE 3.6 ■ FLYING INTO THE WIND: RELATIVE VELOCITY

An airplane with an air speed of 200 km/h (its speed in still air) flies in a direction such that with a 50.0-km/h west wind blowing, it travels in a straight line northward. (Wind direction is specified by the direction *from* which the wind blows, so a west wind blows from west to east.) To maintain its course due north, the plane must fly at an angle into the wind as illustrated in •Fig. 3.14. What is the speed of the plane along its northward path?

Solution. As always, it is important to identify what the given velocities are relative to.

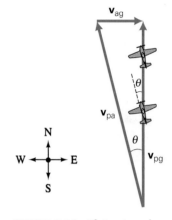

•**FIGURE 3.14 Flying into the wind** To fly directly north, the plane's heading must be west of north. See Example 3.6.

Given: $\mathbf{v}_{pa}$ = 200 km/h at angle θ (velocity of plane with respect to still air, i.e., air speed)

$\mathbf{v}_{ag}$ = 50 km/h east (velocity of air with respect to Earth or ground, i.e., wind speed)

plane flies due north with velocity $\mathbf{v}_{pg}$

Find: v_{pg} (ground speed of plane)

The speed of the plane with respect to the Earth or ground, v_{pg}, is called its ground speed, whereas v_{pa} is its air speed. Vectorially, the respective velocities are related by

$$\mathbf{v}_{pg} = \mathbf{v}_{pa} + \mathbf{v}_{ag}$$

As can be seen, if there were no wind blowing ($v_{ag} = 0$), the ground speed and air speed would be equal. However, a head wind (one blowing directly toward the plane) would cause a slower ground speed; and a tail wind, a faster ground speed. The situation is analogous to that of a boat going upstream and downstream.

The vector relationship applies to our situation. Here $\mathbf{v}_{pg}$ is the resultant of the other two vectors, which can be added by the triangle method. We use the Pythagorean theorem to find v_{pg}, noting that v_{pa} is the hypotenuse of the triangle.

$$v_{pg} = \sqrt{v_{pa}^2 - v_{ag}^2} = \sqrt{(200 \text{ km/h})^2 - (50.0 \text{ km/h})^2} = 194 \text{ km/h}$$

(Note that it was convenient to use the units of km/h, since the calculation did not involve any other standard units.)

Follow-up Exercise. What is the plane's heading (θ direction) in the preceding example in order to fly directly north? *(Answer may be found in the Answers to Follow-up Exercises section at the back of the book.)*

3.4 Projectile Motion

Objectives: To be able to analyze projectile motion to find (a) position, (b) time of flight, and (c) range.

A familiar example of two-dimensional, curvilinear motion is the motion of objects that are thrown or projected by some means. The motion of a stone thrown across a stream or a golf ball driven off a tee is **projectile motion.** A special case of projectile motion in one dimension occurs when an object is projected vertically upward. This case was treated in Chapter 2 in terms of free fall with air resistance neglected. We will generally neglect air resistance here also so that the only acceleration acting on a projectile is that due to gravity.

Projectile motion is easily analyzed using vector components. You simply break up the motion into its x and y components and treat them separately.

Note: Review Section 2.5 (free fall)

Horizontal Projections

It is worthwhile to first analyze the special case of the motion of an object projected horizontally, or parallel to a level surface. Suppose that you throw an object horizontally with an initial velocity v_{x_0} (•Fig. 3.15a). Projectile motion is analyzed beginning at the instant of release ($t = 0$). Once the object is released, the horizontal acceleration is zero ($a_x = 0$), so the horizontal velocity remains constant: $v_x = v_{x_0}$.

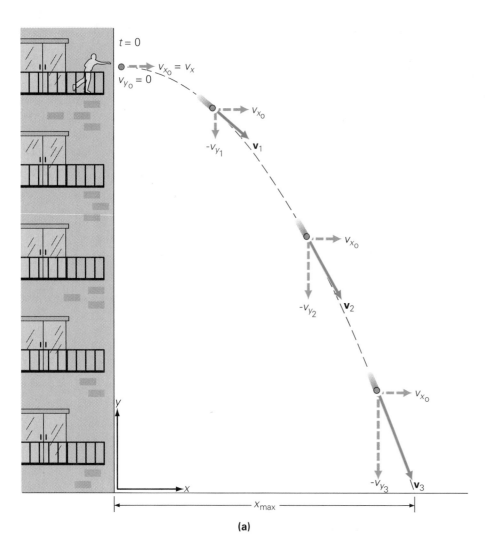

(a)

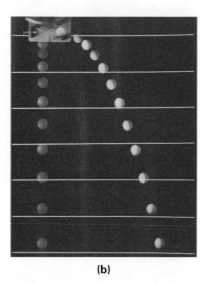

(b)

•**FIGURE 3.15 Horizontal projection**
(a) The velocity components of a projectile launched horizontally show that it travels to the right as it falls downward. **(b)** A multiflash photograph shows the paths of two golf balls. One was projected horizontally at the same time that the other was dropped. The horizontal lines are 15 cm apart, and the time interval between flashes was 1/30 s. You can see that the vertical motions of the balls are the same. Why? Can you describe the horizontal motion of the yellow ball?

According to the equation $x = v_x t$ (Eq. 3.2a), the projected object would continue to travel in the horizontal direction indefinitely. However, you know that this is not what really happens. As soon as the object is projected, it is in free fall in the vertical direction, with $v_{y_o} = 0$ (as though it were dropped) and $a_y = -g$. In other words, the projected object travels with a uniform velocity in the horizontal direction while *at the same time* undergoing acceleration in the downward direction under the influence of gravity. The result is a curved path, as illustrated in Fig. 3.15. (Compare the motions in Fig. 3.15 and Fig. 3.2. Do you see any similarities?). If there were no horizontal motion, the object would simply drop to the ground in a straight line, and, in fact, the time of flight of the projected object is *exactly the same as if it were falling vertically*.

Note the components of the velocity vector in Fig. 3.15a. The length of the horizontal component of the velocity vector remains the same, but the length of the vertical component increases with time. What is the instantaneous velocity at any point along the path? (Think in terms of vector addition, covered in the last section.) The photo in •Fig. 3.15b shows the actual motions of a horizontally projected golf ball and one that is simultaneously dropped from rest. The horizontal reference lines show that the balls fall vertically at the same rate. The only difference is that the horizontally projected ball also travels to the right as it falls.

EXAMPLE 3.7 ■ STARTING AT THE TOP: HORIZONTAL PROJECTION

Suppose that the ball in Fig. 3.15a is projected from a height of 25.0 m above the ground and is thrown with an initial horizontal velocity of 8.25 m/s. (a) How long is the ball in flight before striking the ground? (b) How far from the building does the ball strike the ground?

Solution. Writing the data with the origin chosen as the point from which the ball is thrown and downward taken as the negative direction, we have

Given: $y = -25.0$ m   **Find:** (a) t (time of flight)
$\quad\quad v_{x_o} = 8.25$ m/s   (b) x (horizontal distance)
$\quad\quad a_x = 0$
$\quad\quad v_{y_o} = 0$
$\quad\quad a_y = -g$

(a) As noted previously, the time of flight is the same as the time it takes for the ball to fall vertically to the ground. This may be found using the equation $y = v_{y_o}t - \frac{1}{2}gt^2$, in which the negative direction of g is expressed explicitly as was done in Chapter 2. With $v_{y_o} = 0$,

$$y = -\tfrac{1}{2}gt^2$$

and

$$t = \sqrt{\frac{2y}{-g}} = \sqrt{\frac{2(-25.0\ \text{m})}{-9.80\ \text{m/s}^2}} = 2.26\ \text{s}$$

(b) The ball travels in the x direction for the same amount of time it travels in the y direction (that is, 2.26 s), and (since there is no acceleration in the horizontal direction) with uniform velocity. Thus , with $a_x = 0$,

$$x = v_{x_o}t = (8.25\ \text{m/s})(2.26\ \text{s}) = 18.6\ \text{m}$$

Follow-up Exercise. What is the velocity (in component form) of the ball just before it strikes the ground? (*Answer may be found in the Answers to Follow-up Exercises section at the back of the book.*)

Projections at Arbitrary Angles

The general case of projectile motion involves an object projected at an arbitrary angle θ relative to the horizontal; for example, a golf ball struck by a club

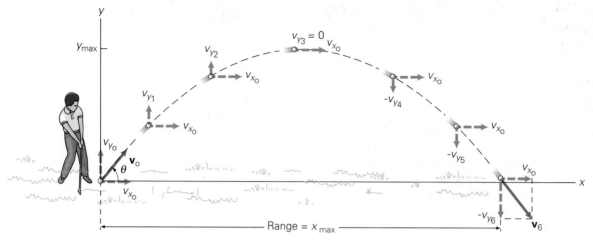

●FIGURE 3.16 Projection at an angle
The velocity components for the ball are shown for various times. Note that $v_y = 0$ at the top of the arc, or at y_{max}. The range (R) is the maximum horizontal distance, or x_{max}.

(●Fig. 3.16). During projectile motion, the object travels up and down while traveling horizontally with a constant velocity.

This motion is also analyzed using its components. As before, upward is taken as the positive direction and downward as the negative direction. The initial velocity v_o is first resolved into rectangular components:

$$v_{x_o} = v_o \cos \theta \qquad \qquad \qquad \text{(3.8a)}$$
$$v_{y_o} = v_o \sin \theta \qquad \qquad \qquad \text{(3.8b)}$$

initial velocity components, $(t = 0)$

Since there is no horizontal acceleration and gravity acts in the negative y direction, the x component of the velocity is constant and the y component varies with time (see Eq. 3.3):

$$v_x = v_{x_o} = v_o \cos \theta \qquad \qquad \text{(3.9a)}$$
$$v_y = v_{y_o} - gt = (v_o \sin \theta) - gt \qquad \text{(3.9b)}$$

velocity components

The components of the instantaneous velocity at various times are illustrated in Fig. 3.16. The instantaneous velocity is the sum of these components and is tangent to the curved path of the ball at any point.

Similarly, the displacement components are given by

$$x = v_{x_o}t = (v_o \cos \theta)t \qquad \qquad \text{(3.10a)}$$
$$y = v_{y_o}t - \tfrac{1}{2}gt^2 = (v_o \sin \theta)t - \tfrac{1}{2}gt^2 \qquad \text{(3.10b)}$$

displacement components

Note: A projectile launched at an angle to the horizontal follows a parabolic path

The curve produced by these equations, or the path of motion of the projectile, is called a **parabola**. The path of projectile motion is often referred to as a parabolic arc. Such arcs are commonly observed (●Fig. 3.17).

Note as in the case of horizontal projection, *time is the common feature shared by the components of motion.* Aspects of projectile motion that may be of interest in various situations include the time of flight, the maximum height reached, and the **range**, which is the maximum horizontal distance traveled.

Projectile range—the maximum horizontal distance traveled

EXAMPLE 3.8 ■ TEEING OFF: PROJECTION AT AN ANGLE

The golf ball in Fig. 3.16 leaves the tee with an initial velocity of 30.0 m/s at an angle of 37° to the horizontal. (a) What is the maximum height reached by the ball? (b) What is its range?

Solution.

Given: $v_o = 30.0$ m/s   **Find:** (a) y_{max}
$\qquad\quad \theta = 37°$ $\qquad\qquad\qquad$ (b) $R = x_{max}$
$\qquad\quad a_y = -g$

For illustration, let us compute v_{x_o} and v_{y_o} explicitly so as to be able to use simplified kinematic equations:

$$v_{x_o} = v_o \cos 37° = (30.0 \text{ m/s})(0.799) = 24.0 \text{ m/s}$$
$$v_{y_o} = v_o \sin 37° = (30.0 \text{ m/s})(0.602) = 18.1 \text{ m/s}$$

(a) Just as for an object thrown vertically upward, $v_y = 0$ at the maximum height (y_{max}). Thus, we can find the time to reach the maximum height (t_m) by using Eq. 3.9b with v_y set equal to zero:

$$v_y = 0 = v_{y_o} - gt_m$$

$$t_m = \frac{v_{y_o}}{g} = \frac{18.1 \text{ m/s}}{9.80 \text{ m/s}^2} = 1.85 \text{ s}$$

As in the case of vertical projection, the time in going up is equal to the time in coming down, so the total time of flight is $t = 2t_m$ (to return to the elevation from which the object was projected).

The maximum height (y_{max}) is then obtained by substituting t_m into Eq. 3.10b:

$$y_{max} = v_{y_o}t_m - \tfrac{1}{2}gt_m^2 = (18.1 \text{ m/s})(1.85 \text{ s}) - \tfrac{1}{2}(9.80 \text{ m/s}^2)(1.85 \text{ s})^2 = 16.7 \text{ m}$$

The maximum height could also be obtained directly from Eq. 2.10', $v_y^2 = v_{y_o}^2 - 2gy$, with $y = y_{max}$ and $v_y = 0$. However, the method of solution used here illustrates how the time of flight is obtained.

(b) The range (R) is equal to the horizontal distance traveled (x_{max}), which is easily found by substituting the total time of flight $t = 2t_m = 2(1.85 \text{ s}) = 3.70 \text{ s}$ into Eq. 3.10a:

$$R = x_{max} = v_x t = v_{x_o}(2t_m) = (24.0 \text{ m/s})(3.70 \text{ s}) = 88.8 \text{ m}$$

Follow-up Exercise. What is the velocity of the golf ball (in magnitude-angle form) just before striking the ground? (*Answer may be found in the Answers to Follow-up Exercises section at the back of the book.*)

•**FIGURE 3.17   Parabolic arcs** (a) Fireworks and (b) streams of water move in parabolic arcs, though air resistance may distort the trajectories to some extent.

EXAMPLE 3.9 ■ WHERE'S THE BALL? PROJECTILE COORDINATES

What is the location of the golf ball in the preceding example at $t = 3.00$ s?

Solution. At this time, the ball is still in flight and its location at that instant is given by its Cartesian coordinates (x, y). Using data from Example 3.8, we have:

$$x = v_{x_o}t = (24.0 \text{ m/s})(3.00 \text{ s}) = 72.0 \text{ m}$$
$$y = v_{y_o}t - \tfrac{1}{2}gt^2 = (18.1 \text{ m/s})(3.00 \text{ s}) - \tfrac{1}{2}(9.80 \text{ m/s}^2)(3.00 \text{ s})^2$$
$$= 10.2 \text{ m}$$

So the ball is at (x, y) = (72.0 m, 10.2 m) on its way to its maximum horizontal distance or range.

Follow-up Exercise. At what time would the golf ball have the same vertical height as at $t = 3.00$ s? (You should be able to do this in your head. *Hint:* Think symmetry in time.) (*Answer may be found in the Answers to Follow-up Exercises section at the back of the book.*)

The range of a projectile is an important consideration in various applications. This is particularly true in those sports where a maximum range is desired, such as when driving a golf ball or throwing a javelin.

At what angle should an object be projected so as to attain its maximum range? To answer this requires looking more closely at the equation used in Example 3.8 to calculate the range, $R = v_x t$. First let's look at the expressions for v_x and t. Since there is no acceleration in the horizontal direction, we know that

$$v_x = v_{x_o} = v_o \cos \theta$$

and $t = 2t_m$, where (as we found in Example 3.8)

$$t_m = \frac{v_{y_o}}{g} = \frac{v_o \sin \theta}{g}$$

Then,

$$R = v_x t = (v_o \cos \theta)(2t_m) = (v_o \cos \theta)\left(\frac{2v_o \sin \theta}{g}\right) = \frac{2v_o^2 \sin \theta \cos \theta}{g}$$

Using the trigonometric identity $\sin 2\theta = 2 \cos \theta \sin \theta$ (see Appendix I) we have

$$R = \frac{v_o^2 \sin 2\theta}{g} \qquad \begin{array}{l} \textit{projectile range, } x_{max} \\ \textit{(for } y_{initial} = y_{final}) \end{array} \qquad (3.11)$$

Note that the range depends on the magnitude of the initial velocity (or speed, v_o) and the angle of projection (θ); g is assumed to be constant. Keep in mind that this equation applies only to the special, but common, case of $y_{initial} = y_{final}$—that is, the landing point is at the same height as the launch point.

For a particular v_o, the range is a maximum (R_{max}) when $\sin 2\theta = 1$, since this is the maximum value of the sine function (which varies from 0 to 1). Thus

$$R_{max} = \frac{v_o^2}{g} \qquad (3.12)$$

Since, as we have seen, this maximum range is obtained when $\sin 2\theta = 1$, and $\sin 90° = 1$, we have

$$2\theta = 90°$$

or

$$\theta = 45°$$

for the maximum range for a given initial speed when the projectile returns to the elevation from which it was projected. At a greater or smaller angle, for a projectile with the same initial speed, the range will be less, as illustrated in ● Fig. 3.18. Also, the range is the same for angles equally above and below 45°, such as 30° and 60°.

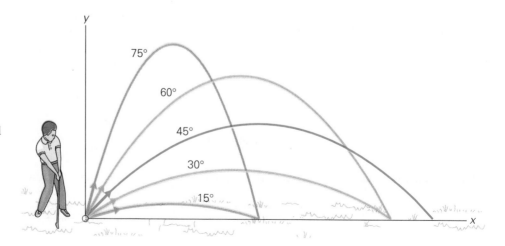

●**FIGURE 3.18 Range**
For a projectile with a given initial speed, the maximum range is ideally attained with a projection angle of 45°. For projection angles above and below 45°, the range is shorter, and equal for angles equally different from 45° (for example, 30° and 60°).

EXAMPLE 3.10 ■ TAKE YOUR CHOICE: CALCULATING PROJECTION ANGLES

A projectile has an initial launch speed of 18 m/s. If it is desired that it strike a target 31 m away at the same elevation, what should be the projection angle?

Solution. Listing the data and what is to be found, we have

Given: $v_o = 18$ m/s   **Find:** θ (projection angle)
        $R = 31$ m

The projection angle θ may be found using Eq. 3.11, $R = v_o^2 \sin 2\theta/g$. Solving for $\sin 2\theta$,

$$\sin 2\theta = \frac{Rg}{v_o^2} = \frac{(31\text{ m})(9.8\text{ m/s}^2)}{(18\text{ m/s})^2} = 0.94$$

Then

$$2\theta = \sin^{-1}(0.94) = 70°$$

and

$$\theta = \frac{70°}{2} = 35°$$

However, there is another solution. Recall that the sine has the same value for angles in the first and second quadrants. That is, $\sin \theta = \sin (180° - \theta)$, or for our double angle,

$$\sin 2\theta = \sin (180° - 2\theta) = \sin (180° - 70°) = \sin 110° = 0.94$$

Hence, we could also have

$$2\theta = 110°$$

and

$$\theta = 55°$$

So, there are *two* launch angles, 35° and 55°, that will give the same range. This is a general result: there will always be two initial angles that produce the same range except for $\theta = 45°$, which ideally produces the maximum range.

Follow-up Exercise. A projectile with an initial speed of 10 m/s is projected at an angle of 40° relative to the horizontal. At what angle should another projectile with an initial speed of 15 m/s be projected so it has the same range? (*Answer may be found in the Answers to Follow-up Exercises section at the back of the book.*)

Thus, to get the maximum range, a projectile should *ideally* be projected at an angle of 45°. However, air resistance has been neglected. In actual situations, such as when a ball or object is thrown or hit hard, this factor may have a significant effect. Air resistance reduces the speed of the projectile, thereby reducing the range. As a result, when air resistance is a factor the angle of projection for maximum range is less than 45°, which gives a greater initial horizontal velocity (●Fig. 3.19). Other factors, such as spin and wind, may affect the range of a projectile.

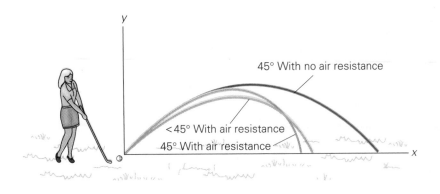

●**FIGURE 3.19 Air resistance and range**
When air resistance is a factor, the angle of projection for maximum range is less than 45°.

For example, backspin on a driven golf ball provides lift, and the projection angle for the maximum range may be considerably less than 45°.

One should keep in mind that the 45° maximum range requires that the initial velocity components be equal: $\tan^{-1} (v_{y_o}/v_{x_o}) = 45°$ and $\tan 45° = 1$, so $v_{y_o} = v_{x_o}$. However, this condition may not always be physically possible, as the following Example shows.

FIGURE 3.20 Hanging in there
Basketball players seem to "hang" in the air at the peak of their jump. Why is this? See the Follow-up Exercise, Example 3.11.

CONCEPTUAL EXAMPLE 3.11 ■ THE LONGEST JUMP: THEORY AND PRACTICE

In a long jump event, the jumper normally has a launch angle of (a) less than 45°, (b) equal to 45°, (c) greater than 45°? *Clearly establish the reasoning and physical principle(s) used in determining your answer before checking it below. That is, **why** did you select your answer?*

Reasoning and Answer. Air resistance is not a major factor here (although wind speed is taken into account for record settings in track and field events). Therefore, it would seem that, to achieve maximum range, the jumper would take off at an angle of 45°. But there is another physical consideration. Let's look more closely at the jumper's initial velocity components.

To maximize a long jump, the jumper runs as fast as possible and then pushes upward as strongly as possible to maximize the velocity components. The initial vertical velocity component ($\mathbf{v}_{y_o}$) depends on the upward push of the jumper's legs. The initial horizontal component ($\mathbf{v}_{x_o}$), on the other hand, depends mostly on the running speed toward the jump point. In general, a greater velocity can be achieved by running than by jumping, so $v_{x_o} > v_{y_o}$. Then, since $\theta = \tan^{-1} (v_{y_o}/v_{x_o})$, we have $\theta < 45°$, where $v_{y_o}/v_{x_o} < 1$ in this case. Hence, the answer is (a), and certainly could not be (c). A typical launch angle for a long jump is 20° to 25°. (Of course, a jumper could increase her launch angle closer to the ideal 45°, but she could do this only by decreasing her running speed, resulting in a decrease in range.)

Follow-up Exercise. Basketball players when jumping to score seem to be suspended momentarily or "hang" in the air (●Fig. 3.20). Explain the physics of this effect. (*Reasoning and answer may be found in the Answers to Follow-up Exercises section at the back of the book.*)

EXAMPLE 3.12 ■ A VIEW FROM THE BRIDGE: EXTENDED PROJECTILE RANGE

A stone thrown from a bridge 20 m above a river has an initial velocity of 12 m/s at an angle of 45° to the horizontal (●Fig. 3.21). (a) What is the range of the stone? (b) At what angle does the stone strike the water?

Solution.

Given: $y = -20$ m (river distance)   *Find:* (a) $R = x_{max}$ (range)
$\quad\quad\quad v_o = 12$ m/s   (b) θ (impact angle)
$\quad\quad\quad \theta = 45°$
$\quad\quad\quad a_y = -g$

Computing the velocity components:

$$v_{x_o} = v_o \cos 45° = (12 \text{ m/s})(0.707) = 8.5 \text{ m/s}$$
$$v_{y_o} = v_o \sin 45° = (12 \text{ m/s})(0.707) = 8.5 \text{ m/s}$$

(It is often convenient to compute these initially and treat them as given data.)

(a) There is more than one way to work this problem. Let's divide the range into two parts: (1) x_1, the symmetrical part of the path, which ends where the projectile returns to the same elevation as the starting point, and (2) x_2, the projectile's path below the launch point, such that $R = x_1 + x_2 = x_{max}$ (see figure).

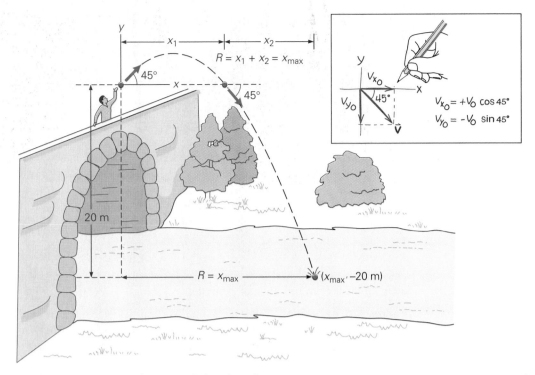

•FIGURE 3.21 Projectile range: landing point below launch point
See Example 3.12.

(1) The range of the first part of the motion, $x_1 = R_1$ may be found using Eq. 3.11. (This equation does not apply to the total range. Why?)

$$x_1 = R_1 = \frac{v_o^2 \sin 2\theta}{g} = \frac{(12 \text{ m/s})^2 \sin (2 \times 45°)}{9.8 \text{ m/s}^2} = 15 \text{ m}$$

(2) As with the initial and returning velocities for vertical projections (see Example 2.11), the magnitude of the velocity when the projectile returns to the elevation from which it started is the same as that of the initial velocity v_o. Also, the magnitude of the directional angle is the same, but below the x axis (Fig. 3.21). Looking only at the x_2 part of the motion, therefore, the situation is the same as for an object with a downward initial projection at an angle of 45° below the horizontal. The *"Given"* data for this part of the problem is then

Given: $y = -20 \text{ m}$ *Find:* x_2
         $v_{x_o} = 8.5 \text{ m/s}$
         $v_{y_o} = -8.5 \text{ m/s}$ (downward)
         $a_y = -g$

This part of the problem, too, can be solved in a couple of ways. We will do it in three steps. Time is the connecting feature, so let's first find the y component of the velocity with which the stone hits the water, use it to find the time of flight, and then use this information to find the horizontal distance the stone travels.

From Equation 2.10′,

$$v_y^2 = v_{y_o}^2 - 2gy$$

and

$$v_y = \sqrt{(-8.5 \text{ m/s})^2 - 2(9.8 \text{ m/s}^2)(-20 \text{ m})}$$
$$= -22 \text{ m/s}$$

where the negative root is the relevant one because we know that the y component of the velocity is in the downward direction.

Then, the time of flight (or fall) is found by solving $v_y = v_{y_o} - gt$ for t:

$$t = \frac{v_{y_o} - v_y}{g} = \frac{-8.5 \text{ m/s} - (-22 \text{ m/s})}{9.8 \text{ m/s}} = 1.4 \text{ s}$$

The horizontal distance for this part of the motion is then

$$x_2 = v_{x_o}t = (8.5 \text{ m/s})(1.4 \text{ s}) = 12 \text{ m}$$

Hence, the total range is

$$R = x_{\text{max}} = x_1 + x_2 = 15 \text{ m} + 12 \text{ m} = 27 \text{ m}$$

(b) We know that $v_x = v_{x_o} = 8.5$ m/s for the total flight, and it was found that the y component of the velocity on impact was $v_y = -22$ m/s. Hence, the velocity on impact is

$$\mathbf{v} = v_x\mathbf{x} + v_y\mathbf{y} = 8.5 \text{ m/s } \mathbf{x} - 22 \text{ m/s } \mathbf{y}$$

Relative to the river surface (x axis), this vector has an angle of

$$\theta = \tan^{-1}\left(\frac{v_y}{v_x}\right) = \tan^{-1}\left(\frac{-22 \text{ m/s}}{8.5 \text{ m/s}}\right) = -69°$$

where the minus sign indicates that $\mathbf{v}$ is directed below the horizontal. Note that this is quite different from the angle of projection.

Follow-up Exercise. What is the total vertical distance traveled by the stone before hitting the water? (*Answer may be found in the Answers to Follow-up Exercises section at the back of the book.*)

Example 3.12 shows that there are several approaches to solving problems involving projectile motion. There we used a two-step procedure to avoid dealing with a quadratic equation. But notice that the time for the motion of part (a) could also have been found from Eq. 3.10b,

$$y = v_{y_o}t - \tfrac{1}{2}gt^2 \quad \text{or} \quad -20 \text{ m} = (-8.5 \text{ m/s})t - \tfrac{1}{2}(9.8 \text{ m/s}^2)t^2$$

This is a quadratic equation ($at^2 + bt + c = 0$). You can use the quadratic formula (Appendix I), to find the roots, and in this case with both positive and negative roots, the positive root gives the time of fall. (Why?) Let's look more closely at this direct approach to the situation.

EXAMPLE 3.13 ■ EXAMPLE 3.12 REVISITED: USING A QUADRATIC EQUATION

Find the range of the thrown stone in Example 3.12 using the direct quadratic equation method outlined above.

Solution.

Given: (from Example 3.12)   *Find:* $R = x_{\text{max}}$
$y = -20$ m
$v_{y_o} = 8.5$ m/s ($t = 0$)
$v_{x_o} = 8.5$ m/s ($t = 0$)
$a_y = -g$

To find the range, we need to know the total time of flight in the horizontal direction. This is the same time as the time for the combined vertical motions (upward and downward). This *total* time of flight may be found from $y = v_{y_o}t - \tfrac{1}{2}gt^2$:

$$y = v_{y_o}t - \tfrac{1}{2}gt^2$$

or

$$-20 \text{ m} = (8.5 \text{ m/s})t - \tfrac{1}{2}(9.8 \text{ m/s}^2)t^2$$

Writing this in general quadratic equation form ($at^2 + bt + c = 0$), with the units omitted for convenience,

$$4.9t^2 - 8.5t - 20 = 0$$

Then, using the quadratic formula (Appendix I),

$$t = \frac{-b \pm \sqrt{b^2 - 4ac}}{2a}$$

$$= \frac{-(-8.5) \pm \sqrt{(-8.5)^2 - 4(4.9)(-20)}}{2(4.9)}$$

$$= \frac{8.5 \pm 21.5}{9.8} = 3.1 \text{ s}$$

where the positive root is taken. Therefore,

$$R = x_{max} = v_{x_o}t = (8.5 \text{ m/s})(3.1 \text{ s}) = 26 \text{ m}$$

Recall that the answer in Example 3.12 was $R = 27$ m. Is something wrong? Not really. This is an example of how rounding significant figures in problem steps can affect the final answer. If three significant figures were assumed for the data in Example 3.12, the result would be $R = 26.0$ m. (Try it and see.)

Note: Significant figures and rounding are discussed in Section 1.6.

Follow-up Exercise. Let's revisit Example 3.9 too. There it was found that the golf ball had a height of 10.2 m at $t = 3.00$ s. In that Follow-up Exercise, you were asked at what previous time the ball had the same vertical height (i.e., the same height on the way up). Use the quadratic formula to find the other time. (*Hint:* You'll get two positive roots in this case. What do they mean? *(Answer may be found in the Answers to Follow-up Exercises section at the back of the book.)*

Finally, for the situation in Fig. 3.21, it should be noted that a projection angle of 45° does not give the maximum range. As pointed out earlier, the 45°-maximum range angle applies only for a horizontal surface or where the landing point is at the same horizontal elevation as the launch point. For an object projected from a height above the ground (or above the landing point), the angle of projection for maximum range is less than 45°. (See Exercise 85.)

Chapter Review

Important Terms

components of motion   63
vector addition (subtraction)   67
triangle method   68
parallelogram method   69
polygon method   69
component method   70

magnitude-angle form   70
unit vector   71
component form   71
analytical component method
   71

relative velocity   75
projectile motion   80
parabola   82
range   82

Important Concepts

- Motion in two dimensions may be analyzed by considering linear component motion in each dimension individually.

- Vector subtraction is a special case of vector addition.

- Of the various methods of vector addition, the component method is the most powerful. A resultant vector may be expressed in the magnitude-angle form or the unit-vector component form.

- Relative velocity is a velocity expressed *relative* to a particular reference frame.

- Projectile motion is analyzed by considering horizontal and vertical components separately—constant velocity in the horizontal direction and an acceleration of g in the vertical direction.

- To get the maximum range, a projectile should ideally be projected at an angle of 45°. However, when air resistance is a factor, the angle of projection for the maximum range is less than 45°.

Important Equations

Components of Initial Velocity:

$$v_{x_o} = v_o \cos \theta \qquad (3.1a)$$

$$v_{y_o} = v_o \sin \theta \qquad (3.1b)$$

Components of Displacement (constant acceleration only):

$$x = v_{x_o}t + \tfrac{1}{2}a_x t^2 \qquad (3.3a)$$

$$y = v_{y_o}t + \tfrac{1}{2}a_y t^2 \qquad (3.3b)$$

Components of Velocity (constant acceleration only):

$$v_x = v_{x_o} + a_x t \qquad (3.3c)$$

$$v_y = v_{y_o} + a_y t \qquad (3.3d)$$

Component Method of Vector Addition:

$$\mathbf{v} = \mathbf{v}_1 + \mathbf{v}_2 + \mathbf{v}_3 + \cdots$$

$$v_x = \Sigma v_{x_i} \quad \text{and} \quad v_y = \Sigma v_{y_i}$$

$$\mathbf{v} = v_x \mathbf{x} + v_y \mathbf{y} \qquad component\ form \qquad (3.7)$$

$$\left. \begin{array}{l} v = \sqrt{v_x^2 + v_y^2} \\[2mm] \theta = \tan^{-1}\left|\dfrac{v_y}{v_x}\right| \end{array} \right\} magnitude\text{-}angle\ form \qquad \begin{array}{l}(3.4a)\\[4mm](3.4b)\end{array}$$

Projectile Range ($y_{initial} = y_{final}$ only):

$$R = \frac{v_o^2 \sin 2\theta}{g} \qquad (3.11)$$

Exercises*

3.1 Components of Motion

1 On Cartesian axes, the y component of a vector is generally associated with a (a) cosine, (b) sine, (c) tangent, (d) none of these.

2 For an object in curvilinear motion, (a) its rectangular velocity components are constant, (b) the y-velocity component is necessarily greater than the x-velocity component, (c) there is an acceleration nonparallel to the object's path, (d) the velocity and acceleration vectors must be at right angles (90°).

3 Describe the motion of an object that is initially traveling with a constant velocity and then receives an acceleration (a) in a direction parallel to the initial velocity, (b) in a direction perpendicular to the initial velocity, and (c) that is always perpendicular to the instantaneous velocity or direction of motion.

4 ■ (a) When you are sitting in a stationary car and it is raining (with no wind), the rain falls straight down relative to the car and the ground. But when you're driving, the rain appears to hit the windshield at an angle, which increases as the velocity of the car increases. Explain this effect in terms of velocity components. [*Hint:* Consider the velocity of the rain relative to that of the car.] (b) If the raindrops fall vertically with a speed of 10 m/s at an apparent angle of 25° to the vertical, what is the speed of the car?

5 ■ An object moves with a velocity of 7.5 m/s at an angle of 20° to the x axis. What are the x and y components of the velocity?

6 ■■ A student walks two blocks west and one block south. If the length of a block is 60 m, what displacement will bring the student back to the starting point?

7 ■■ A student strolls diagonally across a level rectangular campus plaza, covering the 50-m distance in 6.0 min. (a) If the diagonal route makes a 37° angle with the long side of the plaza, what would be the distance in walking half way around the outside of the plaza instead of the diagonal route? (b) If the outside route were also walked in 6.0 min at a constant speed, how much time would be spent on each leg?

8 ■■ A motor boat travels with a speed of 40.0 km/h in a straight path on a still lake. Suddenly, a strong, steady wind pushes the boat perpendicularly to its straight-line path with a speed of 15.0 km/h for 5.00 s. Relative to its position just when the wind started to blow, where is the boat located at the end of this time?

9 ■■ The y component of a velocity that has an angle of 27° to the x axis has a magnitude of 3.8 m/s. (a) What is the magnitude of the velocity? (b) What is the magnitude of the x component of the velocity?

10 ■■ The displacement vector of a moving object initially at the origin has a magnitude of 12.5 cm and is at an angle of 210° relative to the $+x$ axis at a particular instant. What are the coordinates of the object at that instant?

11 ■■ A ball rolls with a constant velocity of 1.50 m/s at an angle of 45° relative to the x axis in the 4th quadrant. Taking the ball to be at the origin at $t = 0$, what are its (x, y) coordinates 1.65 s later?

12 ■■ A ball rolling on a table has velocity with rectangular components $v_x = 0.60$ m/s and $v_y = 0.80$ m/s. What is the displacement of the ball in an interval of 2.5 s?

*Assume angles to be exact for significant figures.

13 ∎∎ A small plane takes off with a constant speed of 130 km/h at an angle of 37°. In 3.00 s, (a) how high is the plane above the ground, and (b) what horizontal distance has it traveled from the lift-off point?

14 ∎∎∎ A ball moves diagonally across a table top from corner to corner with a constant speed of 0.75 m/s. The table top is 3.0 m wide and 4.0 m long. Another ball, starting at the same time as the first one, moves along the longer edge of the table. What constant speed must the second ball have in order to reach the same corner at the same time as the ball traveling diagonally?

15 ∎∎∎ A particle moves with a speed of 2.5 m/s in the $+x$ direction. At the origin, it receives a continuing acceleration of 0.75 m/s² in the $+y$ direction. What is the position of the particle 4.0 s later?

16 ∎∎∎ An automobile travels 800 m up a straight highway that is inclined 15° to the horizontal with a constant speed of 60 km/h. An observer notes only the vertical motion of the car. What is the car's (a) vertical velocity magnitude and (b) vertical travel distance?

3.2 Vector Addition and Subtraction

17 A vector may be resolved into (a) only two components, (b) only three components, (c) any number of components, (d) a number of components that depends on the direction of the vector.

18 The resultant of $\mathbf{A} - \mathbf{B}$ is the same as (a) $\mathbf{B} - \mathbf{A}$, (b) $-\mathbf{A} + \mathbf{B}$, (c) $-(\mathbf{A} + \mathbf{B})$, (d) $-(\mathbf{B} - \mathbf{A})$

19 Find the rectangular components of a vector $\mathbf{v}$ that has a magnitude of 10.0 m/s and is oriented at an angle of 60° relative to the positive x axis.

20 ∎ Using the triangle method, show graphically that (a) $\mathbf{A} + \mathbf{B} = \mathbf{B} + \mathbf{A}$, and (b) if $\mathbf{A} - \mathbf{B} = \mathbf{C}$, then $\mathbf{A} = \mathbf{B} + \mathbf{C}$.

21 ∎ Is vector addition associative, that is, does $(\mathbf{A} + \mathbf{B}) + \mathbf{C} = \mathbf{A} + (\mathbf{B} + \mathbf{C})$? Justify your answer.

22 ∎ A vector $\mathbf{B}$ has an x component of -2.5 m and a y component of 4.2 m. What is $\mathbf{B}$ expressed in magnitude-angle form?

23 ∎ In Chapter 2, the displacement between two points on the x axis was written $x = x_2 - x_1$. Show what this means graphically in terms of vector subtraction.

24 ∎∎ For the two vectors $\mathbf{x}_1 = 20$ m $\mathbf{x}$ and $\mathbf{x}_2 = 15$ m $\mathbf{x}$, compute and show graphically (a) $\mathbf{x}_1 + \mathbf{x}_2$, (b) $\mathbf{x}_1 - \mathbf{x}_2$, and (c) $\mathbf{x}_2 - \mathbf{x}_1$.

25 ∎∎ For each of the following vectors, give a vector that when added to it yields a null vector (a vector with a magnitude of zero) Express the vector in the alternate component or magnitude-angle form: (a) $\mathbf{A} = 4.5$ cm,

40° first quadrant, (b) $\mathbf{B} = 2.0$ cm $\mathbf{x} - 4.0$ cm $\mathbf{y}$, (c) $\mathbf{C} = 8.0$ cm at an angle of 60° relative to the negative x axis in the second quadrant.

26 ∎∎ Find the resultant (or sum) of the three vectors $\mathbf{A} = 4.0$ m $\mathbf{x} + 2.0$ m $\mathbf{y}$, $\mathbf{B} = -6.0$ m $\mathbf{x} + 3.5$ m $\mathbf{y}$, and $\mathbf{C} = -5.5$ m $\mathbf{y}$.

27 ∎∎ (a) Find the resultant (or sum) of the vectors in •Fig. 3.22. (b) If $\mathbf{F}_1$ in the figure were at an angle of 27° with the x axis, instead of 37°, what would be the resultant (or sum) $\mathbf{A}$ and $\mathbf{B}$?

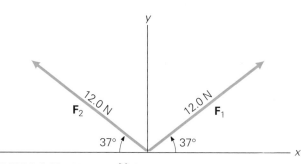

•**FIGURE 3.22 Vector addition**
See Exercises 27 and 28.

28 ∎∎ For the vectors in Fig. 3.22, determine (a) $\mathbf{F} = \mathbf{F}_1 - \mathbf{F}_2$, and (b) $\mathbf{F} = \mathbf{F}_2 - \mathbf{F}_1$. Do you think that the resultants will be the same? That is, does $\mathbf{F}_1 - \mathbf{F}_2 = \mathbf{F}_2 - \mathbf{F}_1$? (c) Is the following a true relationship? $(\mathbf{F}_1 - \mathbf{F}_2) \stackrel{?}{=} -(\mathbf{F}_2 - \mathbf{F}_1)$

29 ∎∎ An airplane takes off and flies directly north a distance of 50.0 km. Severe thunderstorms are reported in the flight path, so the pilot turns to a course of 40° east of north and flies 25.0 km in that direction, then takes a course of 20° west of north for 35.0 km. What is the plane's displacement from its starting point?

30 ∎∎ Two velocity vectors, $\mathbf{v}_1$ and $\mathbf{v}_2$, when added together have a resultant of $\mathbf{v}_3 = 2.0$ m/s $\mathbf{x} - 1.5$ m/s $\mathbf{y}$. If $\mathbf{v}_1 = 4.0$ m/s $\mathbf{x} - 3.0$ m/s $\mathbf{y}$, what is $\mathbf{v}_2$ in magnitude-angle form?

31 ∎∎ Find the resultant (or sum) of the three vectors in •Fig. 3.23. (Express in both magnitude-angle form and component notation.)

32 ∎∎ For the vectors in Fig. 3.23, find the magnitude and orientation of $\mathbf{v}_1 - \mathbf{v}_2 - \mathbf{v}_3$.

33 ∎∎ For the vectors in Fig. 3.23, find (a) $\mathbf{v}_1 + \mathbf{v}_2$, (b) $\mathbf{v}_1 + \mathbf{v}_3$, and (c) $\mathbf{v}_2 - \mathbf{v}_1$.

34 ∎∎ Two forces (vectors), $\mathbf{F}_1 = 3.0$ N $\mathbf{x} - 3.0$ N $\mathbf{y}$ and $\mathbf{F}_2 = -6.0$ N $\mathbf{x} + 4.5$ N $\mathbf{y}$, are applied to a particle. What third force $\mathbf{F}_3$ would make the net or resultant force on the particle zero?

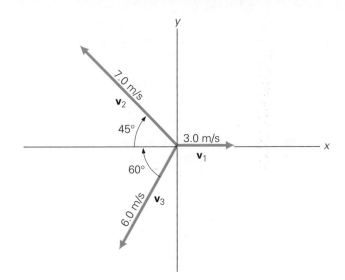

FIGURE 3.23   Vector addition
See Exercises 31, 32, and 33.

35   ■■ Two forces (vectors), $F_1 = 8.0$ N at an angle of 60° (first quadrant) and $F_2 = 5.5$ N at an angle of 45° (fourth quadrant), are applied to a particle at the origin. What third force F_3 would make the net or resultant force on the particle zero?

36   ■■ Five vectors with equal magnitude of 5.0 m have the following directional angles with respect to the x axis: 37° (1st and 2nd quadrants), 45° (3rd and 4th quadrants), and $-y$. What is the resultant of the vectors?

37   ■■ A student works four different problems involving the addition of different vectors F_1 and F_2. He states that the magnitudes of the four <u>resultants</u> are given by (a) $F_1 + F_2$, (b) $F_1 - F_2$, (c) $\sqrt{F_1^2 + F_2^2}$, and (d) $F_1 + F_2 = F_1 - F_2$. Is this possible? If so, describe the vectors in each case.

38   ■■ For three arbitrary vectors, F_1, F_2, and F_3, show graphically that the resultant F is the same regardless of the order in which the vectors are added. (How many possibilities are there?)

39   ■■ A block rests on an inclined plane. Its weight is a force directed vertically downward, as illustrated in

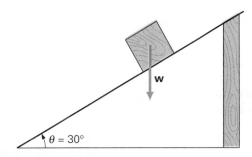

FIGURE 3.24   Block on an inclined plane
See Exercise 39.

Fig. 3.24. Find the components of the force along the surface of the plane and perpendicular to it.

40   ■■■ An amusement park roller coaster starts out initially on a level track 50.0 m long and then goes up a 25.0 m incline at an angle of 30° to the horizontal. It then goes down a 15.0 m ramp with an incline of 40° to the horizontal. What is the displacement of the roller coaster from its starting point when at the bottom of the ramp?

41   ■■■ A person walks from point A to point B as shown in Fig. 3.25. What is the person's displacement relative to A?

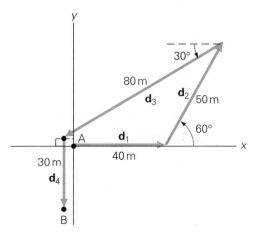

FIGURE 3.25   Walking along
See Exercise 41.

42   ■■■ A golfer's ball lies 2.00 m directly south of the hole on a level green. On the first putt, the ball travels a 3.00-m straight-line path at an angle of 5° east of north; and on the second putt, it travels a straight-line distance of 1.20 m at an angle of 6° south of west. What would be the displacement of a third putt that would put the ball in the hole?

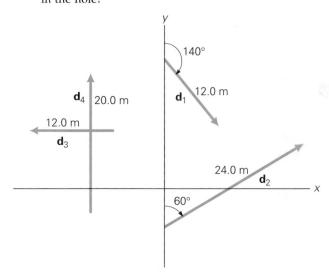

FIGURE 3.26   Vectors, vectors everywhere
See Exercises 44 and 45.

43 ■■■ Find the resultant of the displacement vectors $\mathbf{d}_1 = 3.0 \text{ m } \mathbf{x} + 3.0 \text{ m } \mathbf{y}$, $\mathbf{d}_2 = 2.5 \text{ m } \mathbf{x} - 6.0 \text{ m } \mathbf{y}$, and $\mathbf{d}_3 = -2.0 \text{ m } \mathbf{x} + 1.5 \text{ m } \mathbf{y}$; that is, find $\mathbf{d}_1 + \mathbf{d}_2 + \mathbf{d}_3$. (Express in both component notation and magnitude-angle form.)

44 ■■■ Find the resultant (or sum) of the four vectors in •Fig. 3.26.

45 ■■■ For the vectors in Fig. 3.26 find $(\mathbf{d}_2 - \mathbf{d}_4) + (-\mathbf{d}_1 + \mathbf{d}_3)$.

3.3 Relative Velocity

46 ■ While you are traveling in a car on a straight, level interstate highway at 100 km/h, another car passes; its speedometer reads 120 km/h. (a) What is your velocity relative to the other driver? (b) What are the velocities of both cars relative to a stationary observer standing by the road?

47 ■ A person riding in the back of a pickup truck traveling at 60 km/h on a straight, level road throws a ball with a speed of 20 km/h relative to the truck in the direction opposite its motion. (a) What is the velocity of the ball relative to a stationary observer by the side of the road? (b) What is the velocity of the ball relative to the driver of a car moving in the same direction as the truck at a speed of 90 km/h?

48 ■■ A swimmer maintains a speed of 0.15 m/s relative to the water when swimming directly toward the opposite shore of a straight river with a current that flows at 0.75 m/s. (a) How far downstream is the swimmer carried in 1.5 min? (b) What is the velocity of the swimmer relative to an observer on shore?

49 ■■ For the situation described in Example 3.5, (a) in what direction should the boat be headed in order to land at a point directly across the river from its starting point? (b) How long would such a crossing take?

50 ■■ A pontoon boat that can move at 3.50 m/s in still water crosses a straight river 200 m wide and arrives at a point 25.0 m upriver from the point directly across from its starting point. If the boat is directed upriver at an angle of 40.0° (relative to a line directly across the river), what is the speed of the current?

51 ■■ If the flow rate of the current in a straight river is greater than the speed of a boat in the water, the boat cannot make a trip *directly across* the river. Prove this statement.

52 ■■ A boat that travels at a speed of 6.75 m/s in still water goes directly across a river and back (•Fig. 3.27). The current flows at 0.50 m/s. (a) At what angle(s) must the boat be steered? (b) How long does it take to make a round trip? (Assume that the boat's speed is constant at all times and neglect turnaround time.)

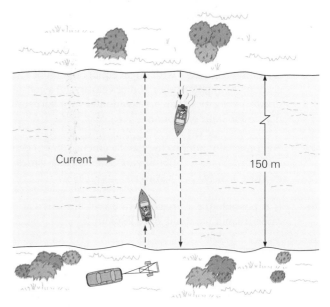

•**FIGURE 3.27   A boat race**
See Exercise 52.

53 ■■ The current of a 200-m wide straight river has a flow rate of 2.50 km/h. A motorboat with a speed of 30 km/h in still water crosses the river. (a) If the boat is pointed directly across the river toward the opposite shore, how far will it travel across *and* down the river in 9.00 s? (b) How far downstream will its landing point be from the point directly opposite its starting point? (Assume that the boat comes instantaneously to rest on a sandy shore.)

54 ■■ In Exercise 53, how long would it take the boat to travel *directly* across the river?

55 ■■ A passenger on a train traveling 90 km/h on a straight, level track notes that it takes another 180-m long, constantly moving train 4.0 s to pass by. What is the velocity of the passing train?

56 ■■ A moving walkway in an airport is 75 m long and moves with a speed of 0.30 m/s. A late passenger, after traveling 25 m standing on the walkway, starts to walk with a speed of 0.50 m/s. How long does it take for the passenger to travel the total walkway distance?

57 ■■ A shopper is carried up a moving escalator in 40 s. However, she can walk up the stationary escalator in 2.0 min. How long would it take her to walk up the moving escalator?

58 ■■■ Two cars travel on level streets that intersect at right angles to each other. Car A approaches the intersection from the east with a speed of 30 km/h. Car B has just passed through the intersection and is traveling 25 km/h northward. (a) What is the velocity of car B relative to car A? (b) What is the velocity of car A relative to car B?

3.4 Projectile Motion*

59 If air resistance is neglected, the motion of an object projected at an angle consists of a uniform downward acceleration combined with (a) an equal horizontal acceleration, (b) a uniform horizontal velocity, (c) a constant upward velocity, (d) an acceleration always perpendicular to the path of motion.

60 For projectile motion, the displacement component(s) (a) both vary as t, (b) both vary as t^2, (c) of x varies as t^2 and y as t, (d) of x varies as t and y as t^2.

61 Is it possible for a projectile to travel in a circular arc rather than a parabolic one? Explain.

62 Using a sketch of components, show that the velocity of a projectile at the same elevation as the initial projection ($y_{final} = y_{initial}$) has the same magnitude as the initial velocity and the same size directional angle, but directed below the horizontal.

63 ■ A ball with a horizontal speed of 1.25 m/s rolls off a bench with a height of 1.00 m. (a) How long will it take the ball to reach the floor? (b) How far from a point on the floor directly below the edge of the bench will it land?

64 ¿■ A spring-loaded gun can project a marble with an initial speed of 4.0 m/s. With the gun lying horizontally on a level table 1.0 m above the floor, what is the range of a marble fired from it?

65 ■ An electron leaves an electron gun horizontally in a TV set with a speed of 1.5×10^6 m/s. If the viewing screen is 35 cm away, how far will the electron fall or travel in the vertical direction before hitting the screen?

66 ■■ While warming up, a pitcher throws a ball horizontally with a speed of 23.5 m/s. How far will the ball have dropped from its straight-line path in a distance of 18.4 m (or 60.5 ft, the distance from the pitcher's mound to home plate)? If the ball were thrown from a height of 2.2 m, would it reach the plate?

67 ■■ A pitcher throws a fast ball horizontally with a speed of 140 km/h toward home plate 18.4 m away. (a) If the batter's combined reaction and swing times total 0.350 s, how long can the batter watch the ball after it has left the pitcher's hand before making a decision to swing? (A major reason why some players illegally used hollowed-out bats with cork or superball filling is to make the bat lighter, which decreases the swing time and allows more time to follow the ball and make a decision on swinging.) (b) How far does the ball drop from its original horizontal line in traveling to the plate?

68 ■■ One ball rolls on a table 1.0 m tall at a constant speed of 0.25 m/s, and another ball rolls on the floor directly under the first ball and with the same speed and direction. Will the balls collide after the first ball rolls off the table? If so, how far from the point directly below the edge of the table will they be when they strike each other?

69 ■■ A student throws a softball horizontally from a dorm window 15.0 m above the ground. Another student standing 10.0 m away from the dorm catches the ball at a height of 1.50 m above the ground. What is the initial velocity of the ball?

70 ■■ A package is to be dropped from an airplane to hit the ground at a designated spot near some campers. The airplane approaches the spot at an altitude of 0.500 km above level ground, moving horizontally with a constant velocity of 140 km/h. Having the designated point in sight, the pilot prepares to drop the package. (a) What should the angle between the horizontal and the pilot's line of sight be when the package is released? (b) What is the location of the plane when the package hits the ground?

71 ■■ A soccer player kicks a stationary ball, giving it a speed of 20.0 m/s at an angle of 15.0° to the horizontal. (a) What is the maximum height reached by the ball? (b) What is its range? (c) How could the range be increased?

72 ■■ A fellow student tells you that when a projectile is launched at an angle of 45° above the horizontal, its range is four times its maximum height. Do you agree? Justify your answer. (b) At what launch angle would the range and the maximum height be equal distances?

73 ■■ A javelin is thrown at angles of 35° and 60° to the horizontal from the same height with the same speed in each case. For which throw does the javelin go farther and how many times farther? (Assume that the landing place is at the same height as the launching.)

74 ■■ A mortar shell with a muzzle velocity of 125 m/s is fired at an angle of 35° above the horizontal. If the shell explodes 6.00 s after firing, what is its location?

75 ■■ A projectile demonstration consists of a spring-loaded "cannon" on a wheeled car that fires a metal ball vertically (●Fig. 3.28). Termed a "ballistics car," it is given a push and set in motion horizontally with a constant velocity. A pin is pulled with a string to launch the ball, which travels upward and then falls back into the moving cannon—every time if done properly. (a) Explain why the cannon always catches the ball under these conditions. (b) If the vertical initial speed of the ball is 5.0 m/s as the cannon moves horizontally with a speed of 0.75 m/s, how far from the launch point is the ball caught? (c) What would happen if the cannon were accelerating?

---

*Neglect air resistance unless stated otherwise.

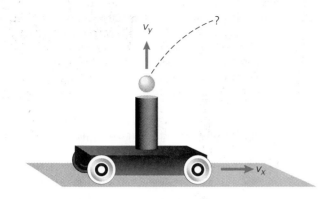

•FIGURE 3.28 A ballistics car
The car is used to demonstrate components of motion. See Exercise 75.

76 ■■ William Tell is said to have shot an apple off his son's head with an arrow. If the arrow was shot with an initial speed of 55 m/s and the boy was 15 m away, at what launch angle did Bill aim the arrow? (Assume that the arrow and apple are initially at the same height above the ground.)

77 ■■ This time William Tell is shooting at an apple hanging in a tree. The apple is a horizontal distance of 20.0 m away and at a height of 4.00 m above the ground. If the arrow is released from a height of 1.00 m above the ground and hits the apple 0.500 s later, what is its initial velocity?

78 ■■■ An astronaut on the Moon fires a projectile from a launcher sitting on a level surface so as to get the maximum range. If the launcher gives the projectile a muzzle velocity of 36 m/s, what is the range of the projectile? [*Hint:* Don't forget where the launcher is and gravitational effects.]

79 ■■■ A ditch 2.5 m wide crosses a trailbike path. An upward incline of 15° has been built up on the approach so that the top of the incline is level with the top of the ditch. What is the minimum speed at which trailbike must be moving to clear the ditch? (Add 1.4 m to the range for the back of the bike to clear the ditch safely.)

80 ■■■ A quarterback passes a football with a velocity of 50 ft/s at an angle of 40° to the horizontal toward an intended receiver 30 yd downfield. The pass is released 5.0 ft above the ground. Assume that the receiver is stationary and that he will catch the ball if it comes to him. Will the pass be completed?

81 ■■■ A golf ball is hit and receives an initial velocity of 40.0 m/s at an angle of 40.0° to the horizontal. (a) In the absence of air resistance, what would be the ball's range? (b) If a light wind gave the ball a retarding horizontal acceleration of 0.0560 m/s², what would be the range?

82 ■■■ Three identical balls are thrown or projected from a cliff at a particular height (*h*) above a horizontal plain. One ball is projected at an angle of 45° to the horizontal, the second ball is thrown horizontally, and the third is thrown at an angle of 30° below the horizontal (that is, downward). If all of the balls have the same initial speed, which one will have the greatest speed when it hits the ground? (Justify your answer mathematically.)

83 ■■■ A diver goes off a 4.0-m high diving board with a velocity of 8.0 m/s at an angle of 30° above the horizontal. (a) What is the maximum height of the diver above the water? (b) How far from a point directly below the edge of the board will the diver hit the water?

84 ■■■ The pin on a level, elevated golf green is 250 m horizontally from the tee and at an elevation of 45.0 m. A golfer hits a ball at an angle 10.0° greater than that of the hole above the tee and makes a hole-in-one! What was the initial speed of the ball? (b) Suppose the next golfer hits the ball toward the hole with the same speed, but at an angle of 10.5° greater than that of the hole above the tee. Where does the ball hit relative to the hole?

85 ■■■ Show that for an object projected from a height above the ground (or landing elevation), the angle of projection for maximum range is less than 45°. [*Hint:* Time of flight is obtained from $-y = v_{y_o}t - \frac{1}{2}gt^2$.]

Additional Exercises

86 Determine the initial velocity of the horizontally projected golf ball in Fig. 3.15(b).

87 The apparatus for a popular lecture demonstration is shown in •Fig. 3.29. The gun is aimed directly at the can, and when the gun is fired, the can is simultaneously released. This so-called monkey gun won't miss as long as the initial velocity of the bullet is sufficient to reach the falling target before it hits the floor. Verify this, using the figure. [*Hint:* Note that $y_o = x \tan \theta$.]

88 Two trains traveling with constant speeds of 90 km/h approach each other on straight, parallel tracks. If the trains are 5.0 km apart, how long will it take for them to meet?

89 A diver cleaves the water at an angle of 45°, and the water is 10.0 m deep. Assume that she maintains a constant underwater velocity with a magnitude of 0.850 m/s. Will she reach the bottom of the pool in 4.0 s?

90 A firefighter holds the nozzle of a hose a horizontal distance of 25.0 m from a flaming building. If the speed of the water coming from the nozzle is 20.0 m/s, (a) show that the stream will not reach a third-story window 11.0 m

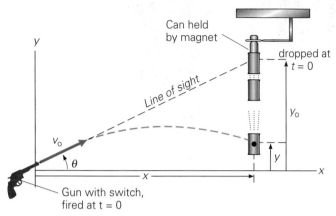

•FIGURE 3.29 A sure shot
See Exercise 87.

above the nozzle level for a projection angle of 45°. (b) A greater height can be achieved by directing the nozzle at a greater angle. Could the window be reached with a stream projection angle of 50°?

91 After being hit from a corner of the rink, a hockey puck has a constant velocity of 25.0 m/s at an angle of 40° to the side wall along the length of the rink. (Neglect friction.) (a) Using the wall as an axis, what are the components of the puck's velocity? (b) What are the displacement components for the puck 2.50 s after it is hit? (Let the direction along the wall be the x axis.)

92 In a movie, a monster climbs to the top of a building 30 m above the ground and hurls a boulder downward with a speed of 25 m/s at an angle of 45° below the horizontal. How far from the building does the boulder land?

93 The shells fired from an artillery piece have a muzzle speed of 1.80×10^2 m/s, and the target is at a horizontal distance of 3.00 km. (a) At what angle relative to the horizontal should the gun be aimed? (b) Could the gun hit a target 3.50 km away?

94 Show that the path of a projectile is a parabolic arc. The general form of the equation for a parabola is $y = ax - bx^2$, where a and b are constants. [*Hint:* Use the appropriate forms of Eq. 3.3a or 3.3b for displacements.]

95 A ball moving along the y axis with a speed of 2.5 m/s experiences an acceleration of 0.45 m/s² in the x direc-

tion when it is at $(x, y) = (0, 1.0$ m). What are (a) the velocity components and (b) the position of the ball 4.0 s after the acceleration is applied?

96 A ball is thrown horizontally from the top of a building at a height of 32.5 m above the ground and hits the level ground 56.0 m from the base of the building. (a) What is the initial velocity of the ball? (b) What is the velocity of the ball just before it hits the ground?

97 For projectile motion in general at an angle θ relative to a level surface, sketch graphs of (a) y versus t, (b) x versus t, (c) v_x versus t, (d) v_y versus t, and (e) v versus t.

98 A field goal is attempted with the football resting at the center of the field 40 yd from the goalposts. If the kicker gives the ball a velocity of 70.0 ft/s toward the goalposts at an angle of 45° to the horizontal, will the kick be good? (The crossbar of the goalposts is 10 ft above the ground, and the ball must be higher than this when it reaches the goalposts for the field goal to be good.)

99 A pickup truck travels on a straight, level highway with a speed of 90.0 km/h. If a person riding in the back of the truck throws an object toward the side of the highway with a speed of 10.5 m/s at an angle of 37.0° above the horizontal, where does the object land? Assume that the object is released 2.20 m above ground level and neglect air resistance.

100 (a) An object moving in a straight line with a speed of 10 m/s has a velocity with an x component of 6.0 m/s. In what direction is the object moving? (b) What is the value of the y component of the velocity?

101 At a track and field meet, the best long jump is measured as 8.20 m. The jumper took off at an angle of 37° to the horizontal. (a) What was the jumper's initial speed? (b) If there were another meet on the moon and the same jumper could attain only half of the initial earthly speed, what would be the maximum jump there? (Air resistance does not have to be neglected in part (b). Why?)

102 An airplane with a speed of 150 km/h heads directly north while a west wind blows with a constant speed of 30 km/h. How far does the plane travel during the two hours of flight?

4 Force and Motion

Here's a situation a lot of people can identify with. You don't have to understand any physics to know what's needed to get the car (or anything else) moving: a push or a pull. In other words, if the frustrated motorist (or the tow truck that he will soon call) can just apply enough *force*, the car will move.

But why does he need a tow truck? What's keeping the car stuck in the snow? A car's engine can generate plenty of force— so why doesn't he just put the car into reverse and back out?

However, in order for a car to move, another force is needed besides that exerted by the engine: *friction.* Here, the problem is most likely not enough friction between the tires and the snow.

This may seem odd to you, for we ordinarily think of friction as something that impedes or prevents motion. Yet without enough friction, cars wouldn't move—and you'd find that you couldn't walk either! In this chapter you'll learn why this isn't as paradoxical as it seems, as well as many other things about force and its relation to motion.

In Chapters 2 and 3 we learned how to analyze motion in terms of kinematics. Now we need to know more about the *dynamics* of motion—that is, what causes motion and changes in motion. This inquiry leads us to the concept of force.

If an object is at rest, a push or a shove may set it into motion, as when we push a sled or a stalled automobile. That is, a force is applied to the object. Similarly, an object in motion may be speeded up or slowed down by applying a force, as when we push a grocery cart or bring it to a stop. We can also change the direction of an object's motion by means of a *force*—think what happens when a batter hits a fastball over the center-field fence. In fact, we shall find that *all* changes in motion result from the action of a force or forces.

There are several types of forces. For example, there is the gravitational force that attracts people and objects to the Earth, giving rise to the force we call weight. There is also the common force of friction, which tends to oppose motion. For simplicity, frictional effects are sometimes ignored (as they were in Chapter 2 when air resistance was neglected in analyzing free fall). However, in many instances friction cannot be ignored and must be included in the analysis of motion. *In general, it is important to recognize a force acting on an object and to be able to calculate the effects of the force.*

The study of force and motion occupied many early scientists. It was Isaac Newton (1642–1727), the English scientist (•Fig. 4.1), who summarized the various relationships and principles into three statements, or laws, which not surprisingly are known as Newton's laws of motion. These laws sum up the concepts of dynamics. In this chapter, you'll learn what Newton had to say about force and motion.

4.1   The Concept of Force and Net Force

Objectives: To be able to (a) relate force and motion, and (b) explain what is meant by a net or unbalanced force.

Let's first take a closer look at the meaning of force. It is easy to give examples of forces, but how would you generally define this concept? An operational definition of force is based on observed effects. That is, a force is described in terms of what it does. From your own experience, you know that *forces can produce changes in motion.* A force can set a stationary object into motion. It can also speed up or slow down a moving object, or change the direction of its motion. In other words, a force can produce a change in velocity (speed and/or direction)—that is, an acceleration. Therefore, an observed change in motion, including motion starting from rest, is evidence of a force. This leads to a common definition of **force**:

A force is something capable of changing an object's state of motion.

The word "capable" is very significant here. It takes into account the fact that a force may be acting on an object, but its capability to produce a change in motion may be balanced or canceled by another force, or forces. The net effect is then zero. Thus, a force may not *necessarily* produce a change in motion. However, it follows that if a force acts *alone*, the object on which it acts *will* accelerate.

Since a force may produce an acceleration—a vector quantity—force is itself a vector quantity, with both magnitude and direction. When several forces act on an object, you will often be interested in their combined effect, or the net force. The **net force** is the vector sum, $\Sigma_i \mathbf{F}_i$, or resultant, of all the forces acting on an object or system. As illustrated in •Fig. 4.2a, the net force is zero when forces of equal magnitude act in opposite directions. Such forces are said to be *balanced forces*. A nonzero net force is referred to as an *unbalanced force*. In this case, *the situation can be analyzed as though only one force equal to the net force were acting.* An unbalanced or net force produces an acceleration. In some instances, an applied

•**FIGURE 4.1   Isaac Newton**
A modern portrait based on a contemporary original. Newton (1642–1727), one of the greatest scientific minds of all time, made fundamental contributions to mathematics, astronomy, and several branches of physics, including optics and mechanics. He formulated the laws of motion and universal gravitation (Chapter 7), and was one of the inventors of calculus. Some of his most profound work was done when he was in his middle twenties.

Inertia ∝ to mass.

Proportional

Note: In the notation, $\Sigma_i \mathbf{F}_i$, the Greek letter sigma means the "sum of" the individual forces, as indicated by the i subscripts, i.e., $\Sigma_i \mathbf{F}_i = \mathbf{F}_1 + \mathbf{F}_2 + \mathbf{F}_3 + \cdots$—a vector sum. The i subscripts are sometimes omitted as being understood, and one writes $\Sigma \mathbf{F}$.

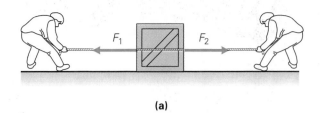

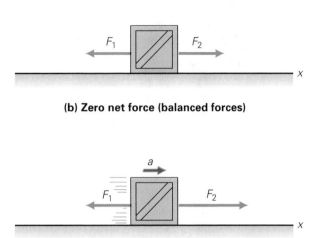

(a)

(b) Zero net force (balanced forces)

(c) Nonzero net force (unbalanced forces)

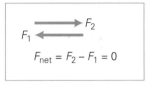

$$F_{net} = F_2 - F_1 = 0$$

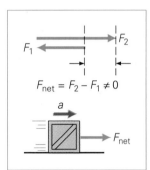

$$F_{net} = F_2 - F_1 \neq 0$$

•FIGURE 4.2   Net force
(a) Opposite forces are applied to a block. **(b)** If the forces are of equal magnitude, the vector resultant, or the net force acting on the block in the x direction, is zero. The forces acting on the block are said to be balanced. **(c)** If the forces are unequal in magnitude, the resultant is not zero. A nonzero net force, or an unbalanced force, then acts on the block, producing an acceleration (for example, setting it into motion if it was initially at rest).

unbalanced force may also deform an object, that is, change its size and/or shape (see Chapter 9). A deformation involves a change in motion for some part of an object, hence there is an acceleration.

Forces are sometimes divided into two types or classes. The more familiar of these consists of *contact forces*. Such forces arise because of physical contact between objects. For example, when you push on a door to open it or throw or kick a ball, you exert a contact force on the door or ball.

The other class of forces is called *action-at-a-distance forces*. Examples of these forces are gravity, the electrical force between two charges, and the magnetic force between two magnets. The Moon is attracted to the Earth and maintained in orbit by a gravitational force, but there seems to be nothing physically transmitting that force. Later in the text (Chapter 30), you will learn the modern view of how action-at-a-distance forces are transmitted.

Now, with our better understanding of the concept of force, let's see how force and motion are related through Newton's laws.

4.2   Inertia and Newton's First Law of Motion

Objectives: **To be able to (a) state and explain Newton's first law of motion, and (b) describe inertia and its relationship to mass.**

The groundwork for Newton's first law of motion was laid by Galileo. In his experimental investigations, Galileo dropped objects to observe motion under the influence of gravity. However, the relatively large acceleration due to gravity causes dropped objects to move quite fast and quite far in a short time. From the kinematic equations in Chapter 2, you can see that 3.0 s after being dropped (if we neglect air resistance), an object in free fall has a speed of about 29 m/s (64 mi/h) and has fallen a distance of 44 m (about 48 yd, or almost half the length of a football field). Thus, experimental measurements of free-fall distance versus time were particularly difficult to make with the instrumentation available in Galileo's time.

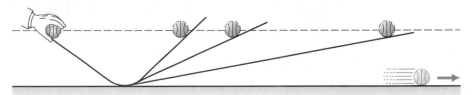

•FIGURE 4.4   A difference in inertia
The larger punching bag has more mass, and hence more inertia or resistance to a change in motion.

•FIGURE 4.3   Galileo's experiment
A ball rolls farther along the upward incline as the angle of incline is decreased. On a smooth, horizontal surface, the ball rolls a greater distance before coming to rest. How far would the ball travel on an ideal, perfectly smooth surface?

To slow things down so that he could study motion, Galileo used balls rolling on inclined planes. He allowed a ball to roll down one inclined plane and then up another with a different degree of incline (•Fig. 4.3). Galileo noted that the ball rolled to approximately the same height in every case, but it rolled farther in the horizontal direction when the angle of this incline was smaller. When allowed to roll onto a horizontal surface, the ball traveled a considerable distance and went even farther when the surface was made smoother. Galileo wondered how far the ball would travel if the horizontal surface could be made perfectly smooth (frictionless). Although this situation was impossible to attain experimentally, Galileo reasoned that in this ideal case with an infinitely long surface, the ball would continue to travel indefinitely with straight-line, uniform motion, since there would be nothing (no force) to cause its motion to change.

According to Aristotle's theory of motion, which had been accepted for about 1500 years prior to Galileo's time, the normal state of a body was to be at rest (with the exception of celestial bodies, which were naturally in motion). Aristotle probably observed that objects moving on a surface tend to slow down and come to rest, so this conclusion would have seemed logical to him. However, from his experiments, Galileo concluded that bodies in motion exhibit the behavior of maintaining that motion, and that if an object is initially at rest, it will remain so, unless something causes it to move.

Galileo called this tendency of an object to maintain its initial state of motion **inertia**. That is,

Definition of inertia

> Inertia is the natural tendency of an object to maintain a state of rest or to remain in uniform motion in a straight line (constant velocity).

For example, if you've ever tried to stop a slowly rolling automobile by pushing on it, you felt its resistance to a change in motion, to slowing down. Physicists describe the property of inertia in terms of observed behavior, as they do for all physical phenomena. A comparative example of inertia is illustrated in •Fig. 4.4. If two punching bags have the same density, the large one will have more mass and more inertia, as you would quickly notice when you tried to punch them.

Newton related the concept of inertia to mass. Originally, he called mass a quantity of matter, but later redefined it as follows:

Relationship of mass and inertia

> Mass is a measure of inertia.

That is, a massive object has more inertia, or resistance to a change in motion, than a less massive object does. For example, a car has more inertia than a bicycle.

Newton's first law of motion, sometimes called the law of inertia, summarizes these observations:

Newton's first law—the law of inertia

> In the absence of an unbalanced applied force, a body at rest remains at rest, and a body already in motion remains in motion with a constant velocity (constant speed and direction).

According to Newton's first law, an object remains at rest or in motion with a constant velocity unless acted upon by an unbalanced force.

(a) A pen is at rest on an embroidery hoop on top of a bottle.

(b) The hoop is struck sharply and accelerates horizontally. Because the friction between the pen and hoop is small and acts only for a short time, the pen does not move appreciably in the horizontal direction. However, there is now an unbalanced force acting vertically upon it—gravity.

(c) The pen falls into the bottle.

That is, if the net force acting on an object is zero, then its acceleration is zero. (See Demonstration 1.)

4.3 Newton's Second Law of Motion

Objectives: To be able to **(a) state and explain Newton's second law of motion** and **(b) apply it to physical situations.**

A change in motion, or an acceleration (change in speed and/or direction), is evidence of a force; and all experiments indicate that an acceleration is directly proportional to the applied net force, that is,

$$\mathbf{a} \propto \mathbf{F}_{net}$$

where the boldface symbols indicate vector quantities. For example, if you hit a ball twice as hard (applied twice as much force), you would expect the acceleration of the ball to be twice as great (in the direction of the force).

However, as Newton recognized, the inertia or mass of the object also plays a role. For a given force, the more massive the object, the less its acceleration will be; that is, the acceleration and mass (m) of an object are inversely proportional:

$$a \propto \frac{1}{m}$$

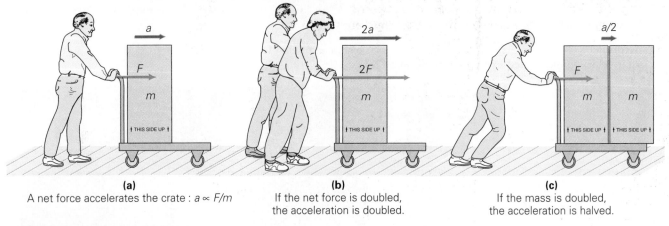

(a)
A net force accelerates the crate : $a \propto F/m$

(b)
If the net force is doubled, the acceleration is doubled.

(c)
If the mass is doubled, the acceleration is halved.

●**FIGURE 4.5 Newton's second law**
The relationships among force, acceleration, and mass shown here are expressed by Newton's second law of motion.

For example, if you hit two different balls with the same force, the less massive ball would acquire a greater acceleration.

Then combining these relationships, we have,

$$\mathbf{a} \propto \frac{\mathbf{F}_{net}}{m}$$

or in words,

Newton's second law—cause and effect

> The acceleration of an object is directly proportional to the *net* force acting on it and inversely proportional to its mass. The direction of the acceleration is in the direction of the applied net force.

●Figure 4.5 presents some illustrations of this principle. We can rewrite the preceding equation as

$$\mathbf{F}_{net} = \Sigma_i \mathbf{F}_i \propto m\mathbf{a}$$

The force has been expressed as $\mathbf{F}_{net} = \Sigma_i\mathbf{F}_i$ to emphasize that it is a net or unbalanced force—the vector sum ($\Sigma_i\mathbf{F}_i$) of the forces acting on the object—that produces an acceleration (**a**). It should be apparent that if the forces acting on an object are balanced, $\mathbf{F}_{net} = \Sigma_i\mathbf{F}_i = 0$, then there is no acceleration (**a** = 0). *The net force is commonly written as just* **F** *for convenience, with the understanding that this is the vector sum* ($\mathbf{F} = \Sigma_i\mathbf{F} = \mathbf{F}_{net}$). This simpler notation will be generally used hereafter; however, you may wish to write $\mathbf{F}_{net}$, or $\Sigma_i\mathbf{F}_i$, explicitly as a reminder.

A proportion may be expressed as an equation through the use of an appropriate *constant of proportionality*, usually symbolized k. Thus, $F \propto ma$ is expressed as $F = kma$. The k in $F = kma$ has been assigned a value of one by the definition of the SI unit of force, the newton (N). That is, $k = 1 \text{ N}/(\text{kg·m/s}^2)$, and therefore is omitted. **Newton's second law of motion** is thus commonly expressed:

$$\mathbf{F} = m\mathbf{a} \qquad \textit{Newton's second law} \qquad (4.1)$$

SI unit of force: kg·m/s^2, or newton (N)

Thus, if the net force acting on an object is zero, its acceleration is zero, and it remains at rest or in uniform motion, which is consistent with the first law. For a nonzero net force (an unbalanced force), the resulting acceleration is in the same direction as the force.*

*It may appear that Newton's first law is a special case of the second law, but not so. The first law *defines* what is called an *inertial reference system* (Section 26.1): one in which an isolated object, or one on which there is no net force, is stationary or moves with a constant velocity. If Newton's first law holds, then the second law in the form $\mathbf{F} = m\mathbf{a}$ applies to the system.

The SI unit of force is, appropriately, the **newton** (**N**). As Eq. 4.1 shows, the base units of the newton are kilograms times meters per second squared:

$$F = ma$$
$$N \equiv (kg)(m/s^2) = kg \cdot m/s^2$$

That is, a net force of 1 N gives a mass of 1 kg an acceleration of 1 m/s^2 (•Fig. 4.6). The British unit of force is the pound (lb): 1 lb is equivalent to about 4.5 N (actually 4.448 N). An average apple weighs about 1 N.

Weight

Equation 4.1 can be used to relate mass and weight. Recall from Chapter 1 that weight is the gravitational force of attraction that a celestial body exerts on an object. For us, this is the gravitational attraction of the Earth. Its effects are easily demonstrated: When you drop an object, it falls (accelerates) toward the Earth. Since there is only one force acting on it, we use the **weight** (*w*) for the net force *F*, and the acceleration due to gravity (*g*) for *a* in Eq. 4.1. We can therefore write in magnitude form

$$w = mg \qquad (4.2)$$

$$(F = ma)$$

The weight of 1.0 kg of mass is then $w = mg = (1.0 \text{ kg})(9.8 \text{ m/s}^2) = 9.8 \text{ N}$.

Thus, 1.0 kg of mass has a weight of approximately 9.8 N, or 2.2 lb. But, although weight and mass are simply related through Eq. 4.2, keep in mind that *mass is the fundamental property*. Mass doesn't depend on the value of *g*, but weight does. As pointed out previously, the acceleration due to gravity on the Moon is about $\frac{1}{6}$ that on Earth. The weight of an object on the Moon would thus be different from its weight on Earth; but its mass, which reflects the quantity of matter it contains, would be the same in both places.

Newton's second law also shows why all objects in free fall have the same acceleration. Consider, for example, two falling objects, one with twice the mass of the other. The object with twice as much mass would have twice as much weight, or two times as much gravitational force acting on it. But the more massive object also has twice the inertia, so twice as much force is needed to give it the same acceleration. Expressing this relationship mathematically, for the smaller mass (*m*) we can write $F/m = g$, and for the larger mass (2*m*), we have the same acceleration: $2F/2m = g$ (see •Fig. 4.7).

Newton's second law allows you to analyze dynamic situations. In using this equation, you should keep in mind that *F* is the *magnitude of the net force* and *m* is the *total mass of the system*. The boundaries defining a system may be real or imaginary. For example, a system might consist of all the gas molecules in a particular sealed vessel. But you might also define a system to be all the gas molecules in an arbitrary cubic meter of air. In studying dynamics we often have occasion to work with systems made up of one or more discrete masses: the Earth and Moon, for instance, or a series of blocks on a table top, or a tractor and wagon.

EXAMPLE 4.1 ■ NEWTON'S SECOND LAW: FINDING ACCELERATION

A tractor pulls a loaded wagon with a constant force of 440 N (•Fig. 4.8). If the total mass of the wagon and its contents is 275 kg, what is the wagon's acceleration? (Ignore any frictional forces.)

Solution. Listing the data from the Example, we have

Given: $F = 440 \text{ N}$ *Find:* *a* (acceleration)
$m = 275 \text{ kg}$

$a = \frac{F}{m}$ $F = m \cdot a$ $m = \frac{w}{g}$

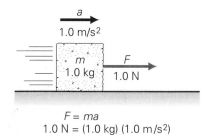

The newton (N), unit of force

$$F = ma$$
$$1.0 \text{ N} = (1.0 \text{ kg}) (1.0 \text{ m/s}^2)$$

•FIGURE 4.6   The newton (N)
A net force of 1.0 N acting on a mass of 1.0 kg produces an acceleration of 1.0 m/s^2.

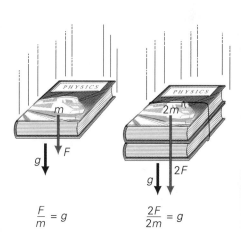

$$\frac{F}{m} = g \qquad \frac{2F}{2m} = g$$

•FIGURE 4.7   Newton's second law and free fall
In free fall, all objects fall with the same constant acceleration *g*. An object with twice the mass of another has twice as much gravitational force acting on it. But with twice the mass, it also has twice as much inertia, so twice as much force is needed to give it the same acceleration.

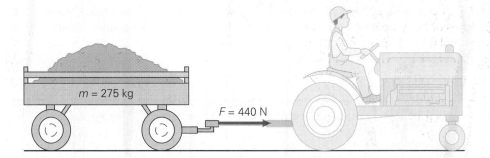

FIGURE 4.8 Force and acceleration
See Example 4.1.

It is noted that F is the net force in this case, and the acceleration is given by Eq. 4.1, $F = ma$. Solving for a,

$$a = \frac{F}{m} = \frac{440 \text{ N}}{275 \text{ kg}} = 1.60 \text{ m/s}^2$$

in the direction that the tractor is pulling.

Note that m was the *total* mass of the wagon and its contents. If the masses of wagon and contents had been given separately, say $m_1 = 75$ kg and $m_2 = 200$ kg, respectively, they would be added together in Newton's law, $F = ma = (m_1 + m_2)a$.

Of course, in reality there would be an opposing force of friction. Suppose there were an effective frictional force of $f = 140$ N. In this case, the net force would be the vector sum of the force exerted by the tractor and the frictional force, and the acceleration would be

$$a = \frac{F - f}{m} = \frac{440 \text{ N} - 140 \text{ N}}{275 \text{ kg}} = 1.09 \text{ m/s}^2$$

in the direction that the tractor is pulling.

With a constant net force, the acceleration is also constant, so the kinematic equations of Chapter 2 can be applied. Suppose the wagon started from rest ($v_o = 0$). Could you find how far it traveled in 4.00 s? Easily, using the appropriate kinematic equation:

$$x = v_o t + \tfrac{1}{2} at^2 = 0 + \tfrac{1}{2}(1.09 \text{ m/s}^2)(4.00 \text{ s})^2 = 8.72 \text{ m}$$

Follow-up Exercise. What is the change in velocity of the wagon in the preceding example during the 3rd second, that is between $t = 3.00$ s and $t = 4.00$ s, if it starts from rest? (*Answer may be found in the Answers to Follow-up Exercises section at the back of the book.*)

EXAMPLE 4.2 ■ NEWTON'S SECOND LAW: FINDING MASS

A student weighs 588 N. What is her mass?

Solution.

Given: $w = 588$ N   *Find:* m (mass)

Recall that weight is a (gravitational) force and Newton's second law, $F = ma$, can be written in the form $w = mg$ (Eq. 4.2), where g is the acceleration due to gravity (9.80 m/s²). Rearranging the equation, we have

$$m = \frac{w}{g} = \frac{588 \text{ N}}{9.80 \text{ m/s}^2} = 60.0 \text{ kg}$$

On the surface of the Earth, this is equivalent to 60.0 kg (2.2 lb/kg) = 132 lb. In countries, using the metric system, the kilogram unit of mass is used to express one's "weight," rather than a force unit as in the United States. It would be said that the student weighs 60.0 "kilos."

Follow-up Exercise. A person in Europe is a bit overweight and would like to lose 5.0 "kilos." What would be the equivalent pound loss? (*Answer may be found in the Answers to Follow-up Exercises section at the back of the book.*)

A dynamic system may consist of more than one discrete mass. In applications of Newton's second law, it is often advantageous, and sometimes necessary, to isolate a given mass within a system. This is possible because any part of a system can be treated as a discrete system to which the second law can be applied, as the following example shows.

EXAMPLE 4.3 ■ NEWTON'S SECOND LAW: ALL OR PART OF THE SYSTEM

Two masses, $m_1 = 2.5$ kg and $m_2 = 3.5$ kg, rest on a frictionless surface and are connected by a light string (●Fig. 4.9). A horizontal force of 12.0 N is applied to m_1 as shown in the figure. (a) What is the magnitude of the acceleration of the masses (the system)? (b) What is the magnitude of the force (T) in the string? [When a rope or string is stretched taut, it is said to be under *tension*, which is represented by the magnitude of the force acting at any point. In equilibrium, the force at the right end of the string has the same magnitude (T) as the force at the left end of the string.]

Solution. Carefully listing the data and what we want to find, we have

Given: $m_1 = 2.5$ kg   *Find:*   (a) a (acceleration)
         $m_2 = 3.5$ kg   (b) T (force)
         $F = 12.0$ N

Given an applied force that produces motion, the acceleration of the masses can be found from Newton's second law. In using the second law, it is important to keep in mind that it applies to the total system *or to any part of it*—that is, to the total mass ($m_1 + m_2$), or to m_1 individually, or to m_2 individually. However, *we must be sure to identify correctly the appropriate force or forces in each case.* The net force acting on the combined masses, for example, is not the same as the net force acting on m_2 considered separately.

(a) Drawing an imaginary line around m_1 and m_2, we see that the force acting on this system is F. Representing the total mass simply as m, we can thus write

$$a = \frac{F}{m} = \frac{F}{m_1 + m_2} = \frac{12.0 \text{ N}}{2.5 \text{ kg} + 3.5 \text{ kg}} = 2.0 \text{ m/s}^2$$

The acceleration is in the direction of the applied force, as the figure indicates. Note that m is the *total* mass of the system, or all the mass that is accelerated. (The mass of the string is small enough to be ignored.)

(b) Under tension a force is exerted *on* an object by flexible strings (or ropes or wires) and is directed along the string. Note in the figure that we are assuming the tension to be transmitted *undiminished* through the string. That is, the tension is the same

Note: When an object is described as "light," you can ignore its mass in analyzing the problem situation. That is, the mass is negligible relative to the other masses.

●**FIGURE 4.9 An accelerated system**
See Example 4.3.

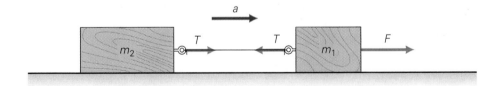

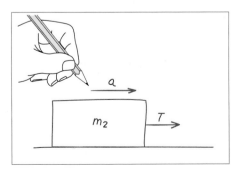

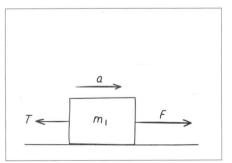

Free-body diagram for m_1

Isolating the masses

everywhere in the string. Thus, the magnitude of T acting on m_2 is the same as that acting on m_1. This is actually true only if the string has zero mass. Only such idealized *light* (negligible mass) strings or ropes will be considered in this book. (If the mass of the string were taken into account, the tensions acting on m_1 and m_2 would be different; the difference would be the net force required to accelerate the mass of the string.)

So, there is a force of magnitude T on each of the masses because of tension in the connecting string. The T forces on the masses are equal in magnitude and opposite. (We will learn more about such equal and opposite forces in the next section.) Note that the tension forces did not appear in part (a), where the total system of both masses was considered. In this case, the *internal* equal and opposite T forces cancel each other.

However, each mass may also be considered a separate system, to which Newton's second law applies. In these systems, the tension comes into play explicitly. Looking at the sketch of the isolated m_2 in Fig. 4.9, we see that the only force acting to accelerate this mass is T. Knowing the values of m_2 and a, the magnitude of this force is given directly by

$$T = m_2 a = (3.5 \text{ kg})(2.0 \text{ m/s}^2) = 7.0 \text{ N}$$

An isolated sketch of m_1 is also shown in Fig. 4.9, and the second law can equally well be applied to this mass to find T. Note the adjacent diagram showing all the forces acting on m_1. This is a simplified version of a **free-body diagram**, which represents an object as a "particle" or point mass and shows all the forces acting on it. Free-body diagrams, discussed more fully in the following Problem-Solving Strategy, are very helpful in analyzing situations involving several bodies and multiple forces. (In this simple case, the free-body diagram for m_2 would have only one vector and is not shown.)

From the free-body diagram we can easily see that we must add the forces vectorially to get the net force that produces the acceleration of m_1. That is,

$$F_{\text{net}} = F - T = m_1 a \qquad \textit{(direction of F taken as positive)}$$

Then, solving for T,

$$T = F - m_1 a$$
$$= 12.0 \text{ N} - (2.5 \text{ kg})(2.0 \text{ m/s}^2) = 12.0 \text{ N} - 5.0 \text{ N} = 7.0 \text{ N}$$

Follow-up Exercise. Suppose there were an additional horizontal force to the left of 3.0 N applied to m_2 in Fig. 4.9. What would be the tension in the connecting string in this case? (*Answer may be found in the Answers to Follow-up Exercises section at the back of the book.*)

PROBLEM-SOLVING STRATEGY: FREE-BODY DIAGRAMS

When working with problems in which two or more forces or components of force act on a body, it is convenient and instructive to draw a free-body diagram of the forces, as was done in the previous example. In such a diagram, we show all the forces acting on the body or object. If several bodies are involved, we may make a diagram for each body *separately*, showing all of the forces acting on each individual body.

In the figures that illustrate the physical situations, sometimes called *space diagrams*, force vectors may be drawn at different locations to indicate their points of application. However, since we are concerned only with linear motions, vectors in free-body diagrams may be shown emanating from a common point, which is chosen as the origin of x–y axes. One of the axes is generally chosen along the direction of the net force acting on a body, since that is the direction in which the body will accelerate. Also, it is often important to resolve force vectors into components, and properly chosen x–y axes simplify this.

In a sketch of a free-body diagram, the vector arrows do not have to be exactly to scale. However, it should be made apparent if there is a net force, and whether forces balance each other in a particular direction. When the forces aren't balanced, we know from Newton's second law that there must be an acceleration.

In summary, the general steps in constructing and using free-body diagrams are as follows. (Refer to the accompanying figure as you read.)

1. Sketch a space diagram (if one is not already available) and identify the forces acting on each body of the system.

2. Isolate the body for which the free-body diagram is to be constructed. Draw a set of Cartesian axes with the origin at a point through which the forces act and one of the axes along the direction of the body's acceleration. (This will be in the direction of the net force if there is one.)

3. Draw properly oriented force vectors on the diagram emanating from the origin of the axes. If there is an unbalanced force, assume a direction of acceleration and indicate with an acceleration vector.

4. Resolve any forces that are not directed along the x or y axes into x or y components. Use the free-body diagram to analyze the forces in terms of Newton's second law of motion. (Note: If the acceleration is in the direction opposite that selected, this will be indicated by an acceleration with an opposite sign in the solution. For example, if the motion is taken to be in the positive direction and it is actually in the opposite direction, the acceleration will be negative.)

Free-body diagrams are a particularly useful way of following one of the Suggested Problem-Solving Procedures in Chapter 1: Draw a diagram as an aid in visualizing and analyzing the physical situation of the problem. *Make it a practice to draw free-body diagrams for force problems, as is done in the following examples.*

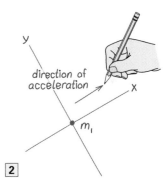

Space diagram

The Second Law in Component Form

Not only does Newton's second law hold for any part of a system, it also applies to the components of motion. A force may be expressed in component notation in two dimensions as follows:

$$\mathbf{F} = m\mathbf{a}$$

and

$$F_x\mathbf{x} + F_y\mathbf{y} = m(a_x\mathbf{x} + a_y\mathbf{y}) \tag{4.3a}$$
$$= ma_x\mathbf{x} + ma_y\mathbf{y}$$

so

$$F_x = ma_x \tag{4.3b}$$
$$F_y = ma_y$$

and Newton's second law applies to each component of motion. (Similarly, $F_z = ma_z$ in three dimensions.) As pointed out in the preceding Problem-Solving Strategy, free-body diagrams assist in resolving forces into rectangular components. An example of how the second law is applied to components follows.

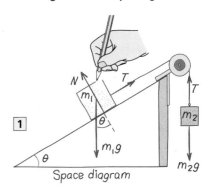

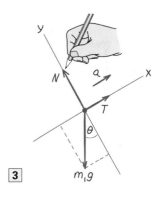

EXAMPLE 4.4 ■ NEWTON'S SECOND LAW: COMPONENTS OF FORCE

A force of 15.0 N is applied at an angle of 30° to the horizontal on a 0.750-kg block at rest on a frictionless surface as illustrated in •Fig. 4.10. (a) What is the magnitude of the resulting acceleration of the block? (b) If the force is applied for only 1.50 s, what happens after this?

Solution. First we write down the given data and what is to be found.

Given: $F = 15.0$ N   *Find:* (a) a (acceleration)
      $m = 0.750$ kg   (b) Describe the motion when
      $\theta = 30°$ the force is no longer applied
      $v_o = 0$
      $t = 1.50$ s

Then we draw a free-body diagram as in Fig. 4.10. Note that the forces act in the same directions as in the space diagram, for example, the force N is upward.

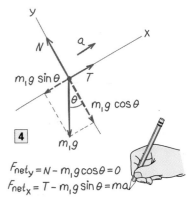

$F_{net_y} = N - m_1 g \cos\theta = 0$
$F_{net_x} = T - m_1 g \sin\theta = ma$

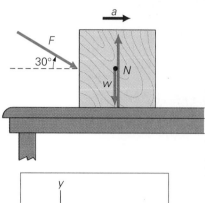

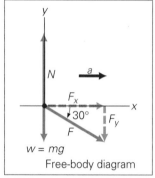

Free-body diagram

• **FIGURE 4.10** Newton's second law and components of force
See Example 4.4.

(a) The acceleration of the block is given by Newton's second law. We choose our axes so that **a** is in the $+x$ direction; this is the direction in which the block will move along the surface. We can see that only a component (F_x) of the applied force F acts in this direction. From the figure, we see that the component of F in the direction of motion is $F_x = F \cos \theta$. Applying Newton's law for this component,

$$F_x = F \cos 30° = ma_x$$

so

$$a_x = \frac{F \cos 30°}{m} = \frac{(15.0 \text{ N})(0.866)}{0.750 \text{ kg}} = 17.3 \text{ m/s}^2$$

This is the total acceleration of the block, since it does not accelerate in the y direction. Since $a_y = 0$, the sum of the forces in the y direction must then be zero. That is, the downward component of F acting on the block, F_y, and its downward weight force w must be balanced by the upward force N that the surface exerts on the block. (N is commonly called the *normal force* because it is normal, or perpendicular to the surface.) If this were not the case, then there would be a net force and an acceleration in the y direction. (The nature of the normal force will become clearer with further explanation in Sections 4.5 and 4.6.)

Summing the forces in the y direction with upward as positive,

$$N - F_y - w = 0$$

or

$$N - F \sin 30° - mg = 0$$

and

$$N = F \sin 30° + mg$$
$$= (15.0 \text{ N})(0.500) + (0.750 \text{ kg})(9.80 \text{ m/s}^2) = 14.9 \text{ N}$$

The surface then exerts a force of 14.9 N upward on the block, which balances the downward forces acting on it.

(b) When the force stops acting at the end of 1.50 s, the block will have a velocity in the x direction with a magnitude of

$$v_x = v_{x_0} + a_x t = 0 + (17.3 \text{ m/s}^2)(1.50 \text{ s}) = 26.0 \text{ m/s}$$

By Newton's first law, the block will continue to travel with this constant velocity until acted on by another force.

In checking the block's speed, you might notice that 26.0 m/s is about 94 km/h (or 58 mi/h), which is a bit unrealistic for a block on a surface. However, such ideal (frictionless) examples are used to make things simple for illustration purposes. (On a real surface, the frictional force would have to be taken into account; and after the applied force was removed, it would act to reduce the velocity, eventually decelerating the block to a halt.)

Follow-up Exercise. Suppose the block in Fig. 4.10 slides off the edge of the table with a horizontal velocity as found in part (b) of this Example. What would then be the net force on the block after it leaves the table? (*Answer may be found in the Answers to Follow-up Exercises section at the back of the book.*)

PROBLEM-SOLVING HINT

There is no fixed way to go about solving a problem. However, there are general strategies or procedures that are helpful in solving problems involving Newton's second law. Using our Suggested Problem-Solving Procedures introduced in Chapter 1, you might include the following for this case:

Note: Review pp. 20–22.

- Draw a free-body diagram for each individual body, showing all of the forces acting on that body.
- Depending on what is to be found, Newton's second law may be applied to the system as a whole (in which case internal forces cancel) or to a part of the system. Basically, *you want to obtain an equation containing the quantity for which you want to solve.* Review Example 4.3. (If there are two unknown quantities, application of Newton's law to two parts of the system may give you two equations and two unknowns. See Example 4.6 in next section.)
- Keep in mind that Newton's second law may be applied to components of motion, and forces may be resolved into components to do this. Review Example 4.4.

4.4 Applications of Newton's Second Law

Objective: To be able to apply Newton's second law, including the component form, to various situations.

The simple relationship expressed by Newton's second law, $\mathbf{F} = m\mathbf{a}$, allows the quantitative analysis of force and motion. This relationship can be thought of as a cause-and-effect one, with force being the cause and acceleration being the motional effect.

Note: Remember that $\mathbf{F}$ means the *net* force, or $\Sigma_i \mathbf{F}_i$.

In general, you will be concerned with applications involving constant forces. Constant forces result in constant accelerations for moving objects and allow the use of the kinematic equations from Chapter 2 in analyzing the motion. When there is a variable force, Newton's second law holds for the *instantaneous* force and acceleration, but the acceleration will vary with time. We will generally limit ourselves to constant accelerations and forces.

This section presents several examples of applications of Newton's second law so that you may become familiar with its use. This small but powerful equation will be used again and again throughout this text.

Newton's second law gives the acceleration resulting from an applied force, but it can also be used to find the force from motional effects, as the following example shows.

EXAMPLE 4.5 ■ A BRAKING CAR: FINDING A FORCE FROM MOTIONAL EFFECTS

A car traveling at 72.0 km/h along a straight level road is brought uniformly to a stop in a distance of 40.0 m. If the car weighs 8.80×10^3 N, what is the braking force?

Solution. In bringing the car to a stop, the braking force caused an acceleration (actually a deceleration), as illustrated in •Fig. 4.11. Listing what is given and wanted to find, we have

Given: $v_0 = 72.0$ km/h $= 20.0$ m/s $\qquad$ *Find:* F (braking force)
$\qquad\quad v = 0$
$\qquad\quad x = 40.0$ m
$\qquad\quad w = 8.80 \times 10^3$ N

We know that $F = ma$, so we can easily calculate F if we can find m and a. The car's mass m can be obtained from its given weight. The other given quantities should remind you of a kinematic equation from Chapter 2 from which the acceleration a may be found. Since the car is brought uniformly to a stop, the acceleration is constant and we may use Eq. 2.9, $v^2 = v_0^2 + 2ax$ to find a:

$$a = \frac{v^2 - v_0^2}{2x} = \frac{0 - (20.0 \text{ m})^2}{2(40.0 \text{ m})} = -5.00 \text{ m/s}^2$$

The minus sign indicates that the acceleration is opposite to v_0, as expected for a braking force, which slows the car.

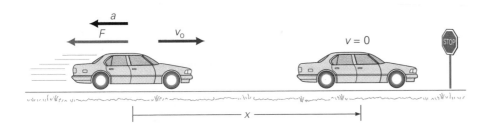

•**FIGURE 4.11 Finding force from motional effects**
See Example 4.5.

The mass of the car is obtained from the weight: $w = mg$ or $m = w/g$. Using this expression along with the acceleration obtained previously gives a braking force of

$$F = ma = \left(\frac{w}{g}\right)a$$

$$= \left(\frac{8.80 \times 10^3 \text{ N}}{9.80 \text{ m/s}^2}\right)(-5.00 \text{ m/s}^2) = -4.49 \times 10^3 \text{ N}$$

Follow-up Exercise. A 1000-kg car starting from rest travels in a straight line and uniformly reaches a speed of 10 m/s in 5.0 s. What is the magnitude of the net force that accelerates the car? (*Answer may be found in the Answers to Follow-up Exercises section at the back of the book.*)

EXAMPLE 4.6 ■ THE ATWOOD MACHINE: FINDING THE ACCELERATION OF A SYSTEM

The Atwood machine consists of two masses suspended from a fixed pulley, as shown in ●Fig. 4.12a. If $m_1 = 0.55$ kg and $m_2 = 0.80$ kg, what is the acceleration of the system? (Consider the pulley to be frictionless and the masses of the string and the pulley to be negligible.)

Solution. Listing the two given quantities, we have

Given: $m_1 = 0.55$ kg   *Find:* a (acceleration)
        $m_2 = 0.80$ kg

Since m_2 is greater than m_1, it is obvious that if released from rest, m_2 will fall and m_1 will rise with accelerations of the same magnitude in opposite directions. Looking at the two masses as separate systems (see the free-body diagrams), we are free to assign directional plus and minus signs arbitrarily. Here, the positive direction is taken to be the direction of acceleration for both masses. That is, for m_2, downward is taken to be the positive direction. The pulley is simply a direction changer; thus, the horizontal analog shown in ●Fig. 4.12b is equivalent to that in ●Fig. 4.12a, except for the changed directions of the forces and accelerations.

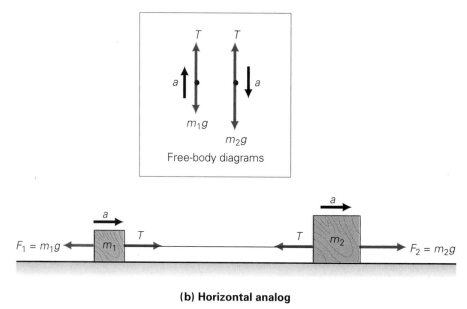

Free-body diagrams

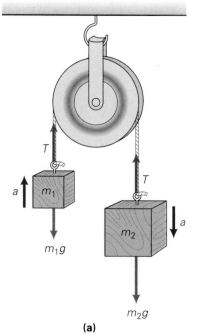

(a)

(b) Horizontal analog

●**FIGURE 4.12 The Atwood machine**
(a) A single, fixed pulley is simply a direction changer. **(b)** An equivalent horizontal analog. (The surface is assumed to be frictionless.) See Example 4.6.

Then, applying Newton's second law to the analog system as a whole (and thus ignoring the T forces, which are internal) gives

$$F_2 - F_1 = m_2 g - m_1 g$$
$$= (m_1 + m_2)a$$

$$\text{net force} = \text{total mass} \times \text{acceleration}$$

Solving for a gives

$$a = \frac{(m_2 - m_2)g}{m_1 + m_2}$$

$$= \frac{(0.80 \text{ kg} - 0.55 \text{ kg})(9.8 \text{ m/s}^2)}{0.55 \text{ kg} + 0.80 \text{ kg}} = 1.8 \text{ m/s}^2$$

Note that the forces resulting from the tension in the string cancel out and need not be considered in finding the solution. However, the problem may also be worked by applying Newton's second law to each isolated mass in the actual system on which the tension acts. (See the free-body diagrams in Fig. 4.12b.) This gives two equations (for a and T):

$$T - m_1 g = m_1 a$$
$$m_2 g - T = m_2 a$$

where the direction of the acceleration for each mass (upward for m_1 and downward for m_2) is taken as positive so as to avoid a minus sign for a. The magnitudes of the accelerations of the masses are equal. Eliminating T from the two equations gives the equation derived above:

$$m_2 g - m_1 g = m_2 a + m_1 a = (m_1 + m_2)a$$

Solving the problem this way gives you an equation containing T (actually, two of them as you see above) in case you need to find its value. For example, you might want to see whether it exceeds the tensile strength of the string (the force that would cause the string to break). Once the acceleration is found, you simply use it in either of the above two equations, solved for T. Using the first equation gives

$$T = m_1 a + m_1 g = m_1(a + g)$$
$$= (0.55 \text{ kg})(1.8 \text{ m/s}^2 + 9.8 \text{ m/s}^2) = 6.4 \text{ N}$$

Follow-up Exercise. Suppose there were a constant retarding force of friction, f, in the pulley of an Atwood machine when the masses were in motion. What would be the general (symbol) form of the acceleration in this case? (*Answers may be found in the Answers to Follow-up Exercises section at the back of the book.*)

Incidentally, the Atwood machine is named after the British scientist George Atwood (1746–1807), who used the arrangement to study motion and measure the value of g. Obviously, the masses can be chosen to minimize the acceleration, making it easier to measure the time of fall. The next Example concerns a variation of Atwood's machine, where one of the masses is on an inclined plane.

EXAMPLE 4.7 ■ THE ATWOOD MACHINE REVISITED: MOTION ON A FRICTIONLESS INCLINED PLANE

Two masses are connected by a light string running over a light pulley of negligible friction as illustrated in •Fig. 4.13. One mass ($m_1 = 5.0$ kg) is on a frictionless 20° inclined plane, and the other ($m_2 = 1.5$ kg) is freely suspended. What is the acceleration of the masses?

Solution. Following our usual procedure, we write

Given: $m_1 = 5.0$ kg   *Find:* a (acceleration)
        $m_2 = 1.5$ kg
        $\theta = 20°$

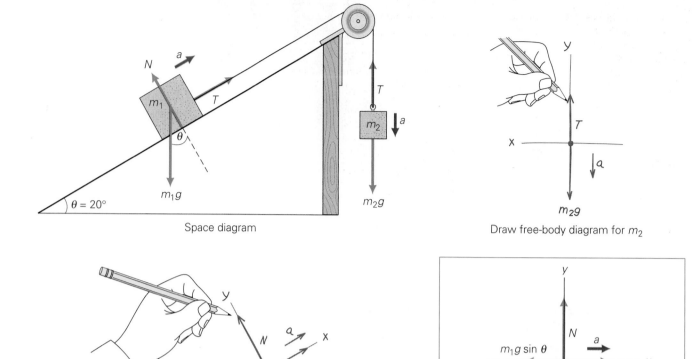

Space diagram

Draw free-body diagram for m_2

Draw free-body diagram for m_1

Free-body diagram for m_1
(reoriented and colored)

•FIGURE 4.13 Application of Newton's second law
See Example 4.7. (Drawing not to scale.)

To better visualize the forces involved, we isolate m_1 and m_2 and draw free-body diagrams for each mass. For mass m_1, there are three concurrent forces (forces acting through a common point). These are T, m_1g, and N, where T is the force in the string because of tension and N is the normal force of the table on the block. The forces are shown emanating from their common point of action. (Recall that a vector arrow can be moved as long as its direction is not changed.)

We will start by assuming that m_1 accelerates up the plane, which is taken to be in the x direction. (It makes no difference whether it is assumed that m_1 accelerates up or down the plane, as we shall see shortly.) Notice that m_1g (the weight) is broken down into components. The x component is in the assumed direction of acceleration, and the y component acts perpendicularly to the plane and is balanced by the normal force N. (There is no acceleration in the y direction, so there is no net force in this direction.)

Then, applying Newton's second law in component form to m_1, we have

$$N - m_1g \cos \theta = 0 \quad \textit{(no net force, forces cancel)}$$
$$T - m_1g \sin \theta = m_1a$$

And for m_2,

$$m_2g - T = m_2a$$

where the masses of the string and pulley have been neglected.

Adding the last two equations to eliminate T, we have

$$m_2g - m_1g \sin \theta = (m_1 + m_2)a$$

(net force = *total* mass × acceleration)

(Note that this is the equation that would be obtained by applying the second law to the system as a whole, in which the T forces cancel.)

Then, solving for a,

$$a = \frac{m_2g - m_1g \sin 20°}{m_1 + m_2}$$

$$= \frac{(1.5 \text{ kg})(9.8 \text{ m/s}^2) - (5.0 \text{ kg})(9.8 \text{ m/s}^2)(0.342)}{5.0 \text{ kg} + 1.5 \text{ kg}}$$

$$= -0.32 \text{ m/s}^2$$

The minus sign indicates that the acceleration is opposite to the assumed direction. That is, m_1 actually accelerates down the plane and m_2 accelerates upward. As this example shows, if you assume the acceleration to be in the wrong direction, the sign on the result will give you the correct direction anyway.

Could you find the tension force T in the string if you were asked to do so? How this could be done should be quite evident from the free-body diagram.

Follow-up Exercise. (a) In Example 4.7, what mass m_2 would cause m_1 to accelerate up the plane? (b) Keeping the masses the same as in the example, how should the angle of incline be adjusted so that m_1 would accelerate up the plane? (*Answers may be found in the Answers to Follow-up Exercises section at the back of the book.*)

Translational Equilibrium

Forces may act on an object without producing an acceleration. In such a case, with $a = 0$, we know from Newton's first law that

$$\Sigma \mathbf{F} = 0 \qquad (4.4)$$

That is, the vector sum of the forces, or the net force, is zero, so the object either remains at rest (as in •Fig. 4.14) or moves with a constant velocity. In such cases, objects are said to be in **translational equilibrium**. When at rest, an object is said to be in *static translational equilibrium*.

It follows that the sums of the rectangular components of the forces for an object in equilibrium are also zero. (Why?)

$$\begin{aligned} \Sigma \mathbf{F}_x &= 0 \qquad \textit{(translational} \qquad (4.5) \\ \Sigma \mathbf{F}_y &= 0 \qquad \textit{equilibrium only)} \end{aligned}$$

For three-dimensional problems, we would add $\Sigma \mathbf{F}_z = 0$. However, we will restrict our discussion of forces to two dimensions.

These equations give what is often referred to as the **condition for translational equilibrium.** (Another condition for rotational considerations will be given in Chapter 8.) Let's apply this translational equilibrium condition to a static equilibrium case.

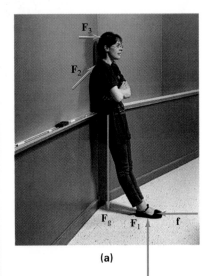

(a)

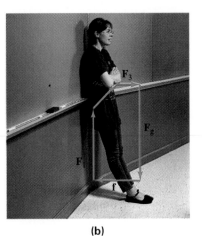

(b)

•FIGURE 4.14 **Many forces, no acceleration**
(a) At least five different external forces act on this physics teacher. Nevertheless, she experiences no acceleration. Why? (b) Adding the force vectors by the polygon method reveals that the vector sum of the forces is zero. The teacher is in static translational equilibrium. (She is also in static rotational equilibrium; we'll see why in Chapter 8.)

Note: In translational equilibrium, an object is at rest or moves with a constant velocity.

EXAMPLE 4.8 ■ A HANGING SIGN: STATIC TRANSLATIONAL EQUILIBRIUM

A 3.0-kg sign hangs in a hall in the Physics Department as shown in •Fig. 4.15a. What is the minimum tensile strength necessary for the cord that is used to hang the sign?

Solution.

Given: $m = 3.0 \text{ kg}$ *Find:* T_1 and T_2
        $\theta_1 = \theta_2 = 45°$ (tensions in the cord)

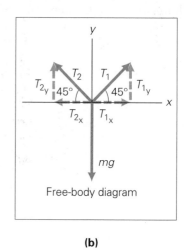

FIGURE 4.15 **Static translational equilibrium**
See Example 4.8.

$w = mg$

(a)

Free-body diagram

(b)

The minimum tensile strength of the cord is just the amount of tension the cord must be able to support without breaking. With a cord of sufficient strength, the sign will hang in static equilibrium. We want to find the values of T_1 and T_2 that will *just* support the weight (mg) of the sign.

The rectangular components of the tensions are shown in the free-body diagram (Fig. 4.15b). Then, applying the component conditions for static translational equilibrium (Eq. 4.5), we have

$$\Sigma F_x = T_{1_x} - T_{2_x} = 0$$

or

$$T_{1_x} = T_{2_x}$$

Hence, the magnitudes of the x components of the tensions are equal, as you might have expected. (They are the only forces in the x direction and they balance each other.) However, this gives no information about the magnitude of the tensions.

Since the angles θ_1 and θ_2 are equal, we also know that $T_{1_y} = T_{2_y}$. (Why?) Applying the other component condition,

$$\Sigma F_y = T_{1_y} + T_{2_y} - mg = 2T_{1_y} - mg$$
$$= 2(T_1 \sin 45°) - mg = 0$$

Solving for T_1,

$$T_1 = \frac{mg}{2 \sin 45°} = \frac{(3.0 \text{ kg})(9.8 \text{ m/s}^2)}{2(0.707)} = 21 \text{ N}$$

With the x and y tension components equal, $T_1 = T_2 = 21$ N, so the sign should be hung using a cord with a tensile strength of at least 21 N.

Follow-up Exercise. How could a heavier sign be hung in Example 4.8 using cord with a tensile strength of less than 21 N? (*Answer may be found in the Answers to Follow-up Exercises section at the back of the book.*)

4.5 Newton's Third Law of Motion

Objectives: To be able to (a) state and explain Newton's third law of motion, and (b) identify action–reaction force pairs.

Newton formulated a third law that is as far-reaching in its physical significance as the first two. For a simple introduction to the third law, consider the forces involved in seat belt safety. When the brakes are suddenly applied in a moving car, you continue to move forward (the frictional force on the seat of your pants is not enough to stop you). In doing so, you exert forces on the seat belt and shoulder strap. The belt and strap exert corresponding reaction forces on you, causing you to slow down with the car. If you haven't buckled up, you may keep on going (Newton's first law) until another applied force, such as that applied by the dashboard or windshield, slows you down.

We commonly think of single forces. However, Newton recognized that it is impossible to have a single force. He observed that in any application of force, there is always a mutual interaction, and forces always occur in pairs. An example given by Newton was this: If you press on a stone with a finger, then the finger is also pressed by, or receives a force from, the stone.

Newton termed the paired forces action and reaction, and **Newton's third law of motion** is

Newton's third law—action and reaction

For every force (action) there is an equal and opposite force (reaction).

In symbol notation,

$$\mathbf{F}_{12} = -\mathbf{F}_{21}$$

$\mathbf{F}_{12}$ is the force exerted *on* object 1 *by* object 2, and $-\mathbf{F}_{21}$ is the equal and opposite force exerted *on* object 2 *by* object 1. (The minus sign indicates the opposite direction.) *Which force is considered the action or the reaction is arbitrary.* F_{21} may be the reaction to F_{12} or vice versa.

The third law may seem to contradict the second law. If there are always equal and opposite forces, how can there be a nonzero net force? An important thing to remember about a force pair of the third law is that *the opposing forces do not act on the same object.* The second law is concerned with force(s) acting on a particular object (or system). The opposing forces of the third law act on different objects. Examples of this principle were encountered previously in our applications of Newton's second law. In both Examples 4.4 and 4.7, the normal force exerted by the surface *on the block* was the reaction force to the contact force of the block *on the surface.*

DEMONSTRATION 2 ■ Tension in a String: Action and Reaction Forces

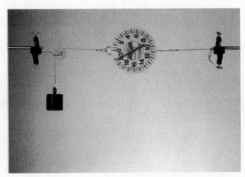

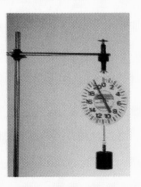

(a) Two suspended 2-kg masses are attached to opposite sides of a scale (calibrated in newtons). The total suspended weight is $w = mg = (4.0 \text{ kg})(9.8 \text{ m/s}^2) = 39.2 \text{ N}$, yet the scale reads about 20 N. Is something wrong with the scale?

(b) No, think of it in this manner. The tension in the string that is caused by the weight of one mass supplies the force that keeps the scale stationary, while the other mass stretches the scale spring giving a reading of 20 N [or $w = mg = (2.0 \text{ kg})(9.8 \text{ m/s}^2) = 19.6 \text{ N}$]. The reaction force can be equivalently supplied by a fixed support.

(c) A fixed pulley merely changes the direction of the force—the fixed support or reaction force can be vertical with the same effect. In all cases, the tension in the string is 19.6 N.

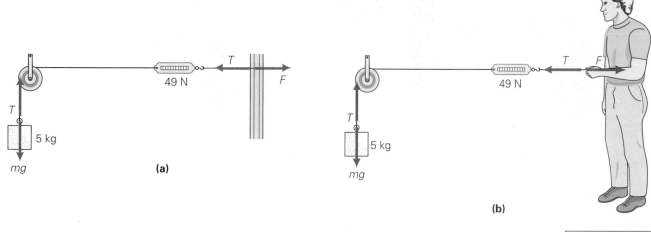

(a)

(b)

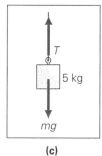

(c)

•FIGURE 4.16 Third law forces
(a) The string exerts a force on the wall (*T*) because of the suspended weight, and the wall exerts an equal and opposite force on the string (*F*), which may not be obvious. **(b)** However, replace the wall with yourself and the equal and opposite force becomes apparent. Note that the weight exerts a force on the string, and the string returns the favor, exerting an equal and opposite force on the weight. (The spring scale and string in parts (a) and (b) are assumed to be of negligible mass.) **(c)** If we look only at the forces *on the block*, we see that the force exerted by the string balances the gravitational force on the block (the block's weight). But these are *not* a third-law pair. (How can you tell?)

In most instances when you are applying Newton's second law, you need to consider only those forces acting on a single object. But the reaction forces of the third law are always there. For example, for the arrangement in •Fig. 4.16a, the suspended mass pulls on the string, and the string exerts an equal and opposite force on the mass. At the other end, the string exerts a force on the wall, and the wall exerts an equal and opposite force on the string, as you would realize if you were substituted for the wall (•Fig. 4.16b). Isolating the stationary, suspended mass (•Fig. 4.16c), you should see that there are two forces acting on it—the upward string force and the downward weight of the mass, which are equal and opposite, since the mass isn't accelerating. However, these are *not* a Newton's third law force pair. Why? For an interesting look at a situation similar to that in Fig. 4.16, see Demonstration 2 on page 115.

As another example, consider the situation in •Fig. 4.17a. Two third-law force pairs are acting when the person is holding the briefcase. There is a pair of contact forces: The person's hand exerts an upward force on the handle, and the handle exerts an equal downward force on the hand. That is, $F_{hand} = -F_{handle}$. This is an action/reaction pair, with the forces acting on *different* objects. The other third-law force pair consists of action-at-a-distance forces associated with gravitational attraction: The Earth attracts the briefcase (its weight), and the briefcase attracts the Earth.

Concentrating on the briefcase in isolation, we can see that only two of the four forces in Fig. 4.17a act on it—the upward force on the handle and the downward force of its weight. These are *not* a third-law force pair, however, because they act on the *same* object. Since the briefcase does not accelerate, these forces must be equal and opposite. Thus, the net force on the isolated stationary briefcase is zero, as required by Newton's first and second laws. When the person drops the briefcase (•Fig. 4.17b), there is then a nonzero net force (the unbalanced force of gravity) on the case, and it accelerates toward the Earth. But there is still a pair of third law forces *acting on different objects*—the briefcase and the Earth. (The Earth actually accelerates upward toward the briefcase as well, but the amount of the acceleration is infinitesimal. Can you explain why?)

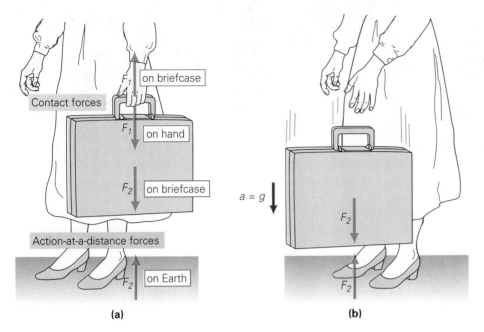

(a)

(b)

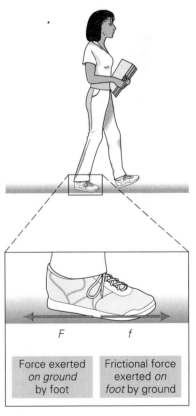

•**FIGURE 4.17 Force pairs of Newton's third law**
(a) When the person holds the briefcase, there are two force pairs: a contact pair (F_1) and an action-at-a-distance (gravity) pair (F_2). The net force acting on the briefcase is zero: The upward contact force balances the downward weight force. Note, however, that these are *not* a third-law pair. **(b)** When the briefcase is released, there is an unbalanced force acting on the case (its weight force), and it accelerates downward (at g for free fall).

CONCEPTUAL EXAMPLE 4.9 ■ A JUMBLED JUGGLER: NEWTON'S THIRD LAW

A juggler carrying three heavy balls wants to cross a weak bridge that will support only his weight and two of the balls. Relying on his trade, he decides that to keep the bridge from collapsing and to save time, he will cross the bridge juggling the balls so that one is in the air at all times. Does he make it across safely? (a) yes (b) no *Clearly establish the reasoning and physical principle(s) used in determining your answer before checking below. That is, **why** did you select your answer?*

Reasoning and Answer. The juggler has apparently not studied Newton's laws. When throwing a ball upward, a force must be applied that is *greater* than the weight of the ball. (Why?) So, the juggler exerts an upward force on the ball, and the ball exerts an equal and opposite force on the juggler (Newton's third law). This adds a force *greater* than the weight of a third ball to the downward force already acting on the bridge (the juggler's weight and that of the other two balls). Down the bridge would go, so the answer is (b).

We can confirm this conclusion by analyzing the situation algebraically. The juggler's weight with three balls is $F = (M + 3m)g$. When juggling with one ball in the air, the weight is less, only $(M + 2m)g$. However, the force needed to accelerate the third ball upward is $mg + ma$. (Can you see why?) Thus the total downward force on the bridge would be $F' = (M + 2m)g + (mg + ma)$, or $F' = (M + 3m)g + ma$. So it is clear that the juggler is actually worse off trying to juggle the third ball than just carrying it. Either way, however, he will get wet.

Follow-up Exercise(s). (a) Would the juggler be able to continue juggling while he was falling? (b) A canary is in a cage with an open wire floor that is suspended from a sensitive scale. The weight of the cage and bird is noted when the bird is on a perch. When the yellow bird is flying around in the cage, is the scale reading different? If so, how? *(Reasoning and Answers may be found in the Answers to Follow-up Exercises section at the back of the book.)*

4.6 Friction

Objectives: To be able to explain (a) the causes of friction, and (b) how it is described using coefficients of friction.

Friction refers to the ever-present resistance to motion that occurs whenever two materials, or media, are in contact with each other. This resistance occurs for

•**FIGURE 4.18 Friction and walking**
Note that the force of friction is shown in the direction of the walking motion. This may seem wrong at first glance, but it's not. The force of friction prevents the foot from slipping backwards while the other foot is brought forward.

(a)

(b)

•**FIGURE 4.19 Increasing and decreasing friction**
(a) To get a fast start, drag racers need to make sure that their wheels don't slip when the light goes on and they floor the accelerator. They therefore try to maximize the friction between their tires and the track by "burning in" the tires just before the start of the race. This is done by spinning the wheels with the brakes on until the tires are extremely hot. The rubber becomes so sticky that it almost welds itself to the road surface.
(b) Water serves as a good lubricant to reduce friction in rides like this one.

all types of media—solids, liquids, and gases—and is characterized as the **force of friction.** We have generally ignored all kinds of friction (including air resistance) in examples and problems for simplicity. Now that you know how to describe motion, we are ready to consider situations that are more realistic, in that the effects of friction are included.

In some real situations, we want to increase friction, for example by putting sand on an icy road or sidewalk to improve traction. This might seem contradictory, since an increase in friction presumably would increase the resistance to motion. However, consider the forces involved in walking as illustrated in •Fig. 4.18 on page 117. Without friction, the foot would slip backwards. (Think about walking on a slippery surface.) The force of friction prevents this, and sometimes needs to be increased on slippery surfaces (•Fig. 4.19a). In other situations, we try to reduce friction (•Fig. 4.19b). For instance, we lubricate moving machine parts to allow them to move more freely, lessen wear, and reduce expenditure of energy. Automobiles would not run without friction-reducing oils and greases.

This section is concerned chiefly with friction between solid surfaces. All surfaces are microscopically rough, no matter how smooth they appear or feel. It was originally thought that friction was primarily due to the mechanical interlocking of surface irregularities, or asperities (high spots). However, research has shown that the friction between the contacting surfaces of ordinary solids (metals in particular) is mostly due to local adhesion. When surfaces are pressed together, local welding or bonding occurs in a few small patches where the largest asperities make contact. To overcome this local adhesion, a force great enough to pull apart the bonded regions must be applied. Once contacting surfaces are in relative motion, another form of friction may result when the asperities of a harder material dig into a softer material, with a "plowing" effect.

Friction between solids is generally classified into three types: static, sliding (kinetic), and rolling. **Static friction** includes all cases in which the frictional force is sufficient to prevent relative motion between surfaces. **Sliding friction**, or **kinetic friction**, occurs when there is relative (sliding) motion at the interface of the surfaces in contact. **Rolling friction** occurs when one surface rotates as it moves over another surface but does not slip or slide at the point or area of contact. Rolling friction, such as occurs between a train wheel and a rail, is attributed to small local deformations in the contact region. This type of friction is somewhat difficult to analyze.

Frictional Forces and Coefficients of Friction

Here, we will consider the forces of friction on stationary and sliding objects. These are called the force of static friction and the force of kinetic (or sliding) friction. Experimentally, it is found that the force of friction depends on both the nature of the two surfaces and the load, or the force with which the surfaces are pressed together. For an object on a horizontal surface, this force is equal in magnitude to the object's weight. However, as shown in Fig. 4.13, on an inclined plane only a component of the weight force contributes to the load. Thus, to avoid confusion, you should remember that the force of friction is proportional to the normal force N, that is, $f \propto N$. The **normal force** always acts perpendicular to, and away from the surface. (It is the force exerted *by* the surface *on* the object.) In the absence of other perpendicular forces, the normal force is equal in magnitude to the component of the weight force acting perpendicular to the surface by the third law.

The force of static friction (f_s) between parallel surfaces in contact is in the direction that opposes the initiation of relative motion between the surfaces. The magnitude has different values such that

$$f_s \leq \mu_s N \quad \text{(static conditions)} \quad (4.6)$$

where μ_s is the **coefficient of static friction.** (Note that it is a dimensionless constant. Why?)

The less-than-or-equal-to sign (≤) indicates that the force of static friction may have different values or magnitudes up to some maximum. To understand this, look at the illustrations in •Fig. 4.20. In the first case, one person pushes on a file cabinet and it doesn't move. With no acceleration, the net force on the cabinet is zero, and $F - f_s = 0$, or $F = f_s$. Suppose that a second person also pushes and the file cabinet still doesn't budge. Then f_s must now be larger, since the applied force has been increased. Finally, if the applied force is made large enough to overcome the static friction, motion occurs. The greatest, or maximum, force of static friction is exerted just before the cabinet starts to slide (Fig. 4.20b), and for this case Eq. 4.6 can be written with an equal sign:

$$f_{s_{max}} = \mu_s N \qquad (4.7)$$

Once an object is in motion, or sliding, there is a force of kinetic friction (f_k) acting on it in the direction opposite to the direction of motion and having a magnitude of

$$f_k = \mu_k N \qquad \text{(sliding conditions)} \qquad (4.8)$$

Frictional forces that obey the general equation $f = \mu N$ are said to represent Coulomb friction, and the equation is sometimes called Coulomb's law of friction. The French scientist Charles A. de Coulomb studied friction during the latter half of the eighteenth century. Although Coulomb gets the credit, the relationship was actually formulated earlier by Leonardo da Vinci.

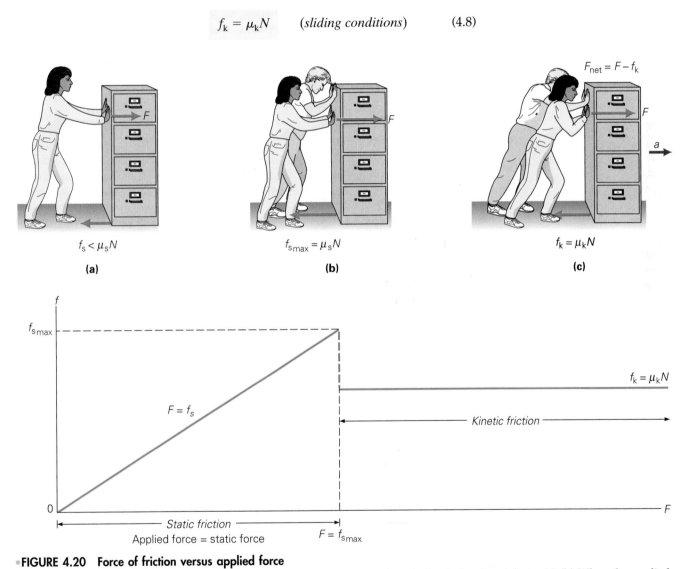

<figcaption>
•FIGURE 4.20 Force of friction versus applied force
(a) In the static region of the graph, as the applied force F increases, so does f_s: that is $f_s = F$ and $f_s < \mu_s N$. (b) When the applied force F exceeds $f_{s_{max}} = \mu_s N$, the heavy file cabinet is set into motion. (c) Once the cabinet is moving, the frictional force is decreased, since kinetic friction is less than static friction ($f_k < f_{s_{max}}$). Thus, if the applied force is maintained at $F = f_{s_{max}}$ there is a net force, and the cabinet is accelerated. For the cabinet to move with constant velocity, the applied force must be reduced to equal the kinetic friction force $f_k = \mu_k N$.
</figcaption>

where μ_k is the **coefficient of kinetic friction** (sometimes called the coefficient of sliding friction). Generally, the coefficient of kinetic friction is less than the coefficient of static friction ($\mu_k < \mu_s$) for two surfaces, which means that the force of kinetic friction is less than $f_{s_{max}}$ as illustrated in Fig. 4.20. The coefficients of friction between some common materials are listed in •Table 4.1.

Note that the force of static friction (f_s) exists in response to an applied force. The magnitude of f_s and its direction depend on the magnitude and direction of the applied force. Up to its maximum value, the force of static friction is equal in magnitude and opposite to the applied force (F), since there is no acceleration ($F - f_s = 0 = ma$). Thus, if the person in Fig. 4.20a pushed on the cabinet in the opposite direction, f_s would also be in the opposite direction. If there were no applied force F, then f_s would not exist. When F exceeds $f_{s_{max}}$, the block slides and kinetic friction comes into effect, with $f_k = \mu_k N$. If F is equal to f_k, the cabinet will slide with a constant velocity; and if F is greater than f_k, it will accelerate.

It has been experimentally determined that the coefficients of friction (and therefore the forces of friction) are nearly independent of the size of the contact area between metal surfaces. This means that the force of friction between a brick-shaped metal block and a metal surface is the same regardless of whether the block is lying on a larger side or a smaller side. The observation is not generally valid for other surfaces, such as wood, and does not apply to plastic or polymer surfaces. The lack of dependence of friction on contact area is related to pressure, which is force per unit of area ($p = F/A$). If the smaller side of a metal block has an area that is half as large as the area of a larger side, it will have, on the average, only half as many asperities for local welding as the larger side. However, the pressure producing the welding will be twice as great on the smaller side (the same weight acting over half the area).

Finally, you should keep in mind that although the equation $f = \mu N$ holds in general for frictional forces, friction may not be linear over a wide range. That is, μ is not always constant. For example, the coefficient of kinetic friction varies somewhat with the relative speed of the surfaces. However, for speeds up to several meters per second the coefficients are relatively constant. Thus, this discussion will neglect any variations due to speed, and the forces of static and kinetic friction will depend only on the load and the nature of the materials (as expressed in the given coefficients of friction).

TABLE 4.1 Approximate Values for Coefficients of Static and Kinetic Friction between Certain Surfaces		
Friction between Materials	μ_s	μ_k
aluminum on aluminum	1.90	1.40
glass on glass	0.94	0.35
rubber on concrete		
dry	1.20	0.85
wet	0.80	0.60
steel on aluminum	0.61	0.47
steel on steel		
dry	0.75	0.48
lubricated	0.12	0.07
Teflon on steel	0.04	0.04
Teflon on Teflon	0.04	0.04
waxed wood on snow	0.05	0.03
wood on wood	0.58	0.40
lubricated ball bearings	<0.01	<0.01
synovial joints (in humans at the ends of most long bones)	0.01	0.01

EXAMPLE 4.10 ■ PULLING A CRATE: STATIC AND KINETIC FORCES OF FRICTION

(a) If the coefficient of static friction between the 40.0-kg crate in ●Fig. 4.21 and the floor is 0.650, with what horizontal force must the worker pull to move the crate? (b) If the worker maintains that force once the crate starts to move and the coefficient of kinetic friction between the surfaces is 0.500, what is the magnitude of the acceleration of the crate?

Solution. Listing the given data and what we want to find, we have

Given: $m = 40.0 \text{ kg}$ *Find:*   (a) F (force necessary to move crate)
      $\mu_s = 0.650$ (b) a (acceleration)
      $\mu_k = 0.500$

(a) The crate will not move until the applied force F slightly exceeds the maximum static frictional force $f_{s_{max}}$. So we must find $f_{s_{max}}$ to see what force the worker must apply. The weight of the crate and the normal force are equal in magnitude in this case (see the free-body diagram in the figure), so the maximum force of static friction is

$$f_{s_{max}} = \mu_s N = \mu_s(mg)$$
$$= (0.650)(40.0 \text{ kg})(9.80 \text{ m/s}^2)$$
$$= 255 \text{ N} \quad \text{(about 57 lb)}$$

The crate moves if the applied force exceeds this maximum force.

(b) Now the crate is in motion and the worker maintains a constant applied force $F = f_{s_{max}} = 255 \text{ N}$. The force of kinetic friction f_k acts on the crate, but this is smaller than F because $\mu_k < \mu_s$. Hence, there is a net force, and the acceleration of the crate may be found by using Newton's second law:

$$F - f_k = F - \mu_k N = ma$$

or

$$a = \frac{F - \mu_k N}{m} = \frac{F - \mu_k(mg)}{m}$$
$$= \frac{255 \text{ N} - (0.500)(40.0 \text{ kg})(9.80 \text{ m/s}^2)}{40.0 \text{ kg}}$$
$$= 1.48 \text{ m/s}^2$$

Follow-up Exercise. On the average, by what factor does μ_s exceed μ_k for nonlubricated, metal-on-metal surfaces? (See Table 4.1.) *(Answer may be found in the Answers to Follow-up Exercises section at the back of the book.)*

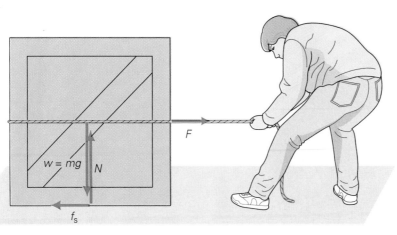

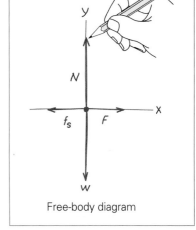

Free-body diagram

●**FIGURE 4.21** **Forces of static and kinetic friction**
See Example 4.10.

Let's look at the worker and the crate again, but this time assume that he applies the force at an angle (●Fig. 4.22).

EXAMPLE 4.11 ■ PULLING AT AN ANGLE: A CLOSER LOOK AT THE NORMAL FORCE

A worker pulling a crate applies a force at an angle of 30° to the horizontal as shown in Fig. 4.22. How large a force must he apply to move the crate? (Before looking at the solution, would you expect that the force needed in this case would be greater or smaller than in the preceding example?)

Solution. The data are the same as in Example 4.10, except that the force is applied at an angle.

Given: $\theta = 30°$ *Find:* F (force necessary to move crate)

In this case, the crate will move when the *horizontal component* of the applied force, $F \cos 30°$, slightly exceeds the maximum static friction force. So, we may write for the maximum friction:

$$F \cos 30° = f_{s_{max}} = \mu_s N$$

However, the magnitude of the normal force is *not* equal to that of the weight of the crate here because of the upward component of the applied force (see the free-body diagram in the figure). By the second law,

$$N + F \sin 30° - mg = 0$$

or

$$N = mg - F \sin 30°$$

In effect, the applied force partially supports the weight of the crate. Substituting this expression for N into the preceding equation gives

$$F \cos 30° = \mu_s(mg - F \sin 30°)$$

Solving for F gives

$$F = \frac{mg}{(\cos 30°/\mu_s) + \sin 30°}$$

$$= \frac{(40.0 \text{ kg})(9.80 \text{ m/s}^2)}{(0.866/0.650) + 0.500}$$

$$= 214 \text{ N} \quad \text{(about 48 lb)}$$

Thus, less applied force is needed in this case, reflecting the fact that the frictional force is less because of the reduced normal force.

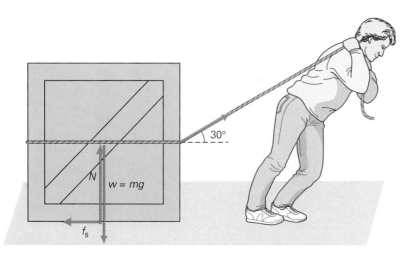

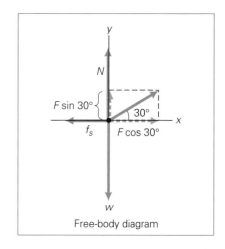

Free-body diagram

●**FIGURE 4.22 Normal force**
See Example 4.11.

Follow-up Exercise. Note that in this Example, applying the force at an angle produces two effects. As the angle between the applied force and the horizontal increases, the horizontal component of the applied force is reduced. However, the normal force also gets smaller, resulting in a lower $f_{s_{max}}$. Does one effect always outweigh the other? That is, does the applied force F necessary to move the crate *always decrease* with increasing angle? (*Hint*: Investigate F for different angles. For example, compute F for 20° and 50°. You already have a value for 30°. What do the results tell you?) (*Answer may be found in the Answers to Follow-up Exercises section at the back of the book.*)

CONCEPTUAL EXAMPLE 4.12 ■ PUSHING VERSUS PULLING: THE NORMAL FORCE ONE MORE TIME

Suppose that instead of pulling the crate as in Fig. 4.22, the worker pushes it, applying a force directed at an angle of 30° below the horizontal. In this case, the magnitude of the force needed to move the crate compared to that in Example 4.11 is (a) the same, (b) greater, (c) less. *Clearly establish the reasoning and physical principle(s) used in determining your answer before checking it below. That is,* **why** *did you select your answer?*

Reasoning and Answer. When the crate is pushed at a downward angle, only the horizontal component of the applied force is directly involved in producing motion, as in Example 4.11. But in this case, the vertical downward component of the applied force, $F \sin 30°$, *adds* to the load that the floor must support, so the magnitude of the normal force is increased: $N = mg + F \sin 30°$. Because N is greater, the frictional force $f_{s_{max}}$ is also greater. Thus a greater applied force is needed to move the crate, and the correct answer is (b).

Follow-up Exercise. Suppose the crate is moved by two workers, one pulling as in Example 4.11 and one pushing as in Example 4.12. What would be the magnitude of the normal force in this case? (*Reasoning and answer may be found in the Answers to Follow-up Exercises section at the back of the book.*)

Now let's look at a way to determine coefficients of friction.

EXAMPLE 4.13 ■ A SLIDING BLOCK: DETERMINING THE COEFFICIENT OF KINETIC FRICTION

A block slides with a constant velocity down a plane inclined at 37° to the horizontal (●Fig. 4.23). What is the coefficient of kinetic friction between the block and the plane?

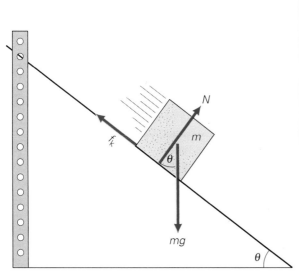

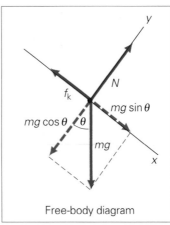

Free-body diagram

●FIGURE 4.23 **Coefficient of kinetic friction**
See Example 4.13.

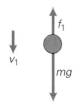

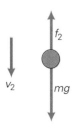

(a) As *v* increases, so does *f*.

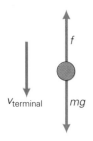

(b) When *f* = *mg*, the object falls with a constant (terminal) velocity.

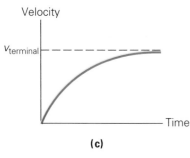

(c)

•**FIGURE 4.24 Air resistance and terminal velocity**
(a) As the speed of a falling object increases, so does the frictional force of air resistance. **(b)** When this force of friction equals the weight of the object, the net force is zero and the object falls with a constant (terminal) velocity. **(c)** A plot of velocity versus time showing these relationships.

Solution.

Given: $a = 0$ (because v is constant)   ***Find:*** μ_k (coefficient of kinetic friction)
$\theta = 37°$

Note that we are not given the mass of the block. We will not need it. Since the acceleration is zero, there is no net force on the block in either the x or y direction. So, noting the forces in the free-body diagram and using Newton's second law,

$$F_x = 0 = mg \sin\theta - f_k$$
$$F_y = 0 = N - mg \cos\theta$$

Rearranging the equations, we have

$$f_k = mg \sin\theta$$
$$N = mg \cos\theta$$

Then, since $f_k = \mu_k N$,

$$\mu_k = \frac{f_k}{N} = \frac{mg \sin\theta}{mg \cos\theta} = \tan\theta = \tan 37° = 0.75$$

Thus, adjusting the angle of incline until the velocity of the block sliding down the plane is constant allows μ_k to be determined experimentally from the angle of incline. From the preceding general result, we can write

$$\mu_k = \tan\theta$$

Suppose that μ_s between the plane and the block is 0.90. Can you determine the angle of incline at which the block will start to move down the plane? It will be an angle just slightly greater than that at which the component of the weight force down the plane equals the maximum force of static friction. That is,

$$mg \sin\theta = f_{s_{max}} = \mu_s N = \mu_s (mg \cos\theta)$$

and

$$\frac{\sin\theta}{\cos\theta} = \tan\theta = \mu_s$$

Thus,

$$\theta = \tan^{-1}\mu_s = \tan^{-1}(0.90) = 42°$$

Therefore, the block will move if the angle of incline exceeds 42°. Adjusting the angle of incline until the block just starts to slide down the plane is an experimental way of approximating μ_s. (This critical angle is called the *angle of repose.*)

Notice how the difference in angles in this Example points out the difference between static and kinetic friction. You must tilt the incline to 42° to start the block moving, but reduce it to 37° to keep the block's velocity constant.

Follow-up Exercise. What would be the dynamic situation if the plane in the preceding example were inclined at an angle greater than 37°? Would this be a convenient way to determine μ_k? (*Answer may be found in the Answers to Follow-up Exercises section at the back of the book.*)

Air Resistance

In analyses of free fall, you can generally ignore the effect of air resistance and still get valid approximations for objects falling relatively short distances. However, for longer distances, air resistance cannot be ignored.

Air resistance refers to the resistance force acting on an object moving through air. This arises because a moving object collides with air molecules. Therefore, air resistance depends on the object's shape and size (which determine the area exposed to collisions) as well as its speed. The larger the object and the faster it moves, the more collisions there will be with air molecules. (Air density is also a factor, but this can be assumed to be constant near the Earth's surface.) Since air resistance depends on speed, as a falling object accelerates under the influence of gravity, the retarding force of air resistance increases (•Fig. 4.24a). Eventually, the magnitude of the retarding force equals that of the object's weight force (•Fig. 4.24b), so that the net force on the object is zero. It then falls with a maximum constant velocity, which is called the **terminal velocity**.

This can be easily seen from Newton's second law. For the falling object, we have

$$F_{\text{net}} = ma$$

$$mg - f = ma$$

where downward has been taken as positive for convenience. Solving for a,

$$a = g - \frac{f}{m}$$

where a is the magnitude of the *instantaneous* acceleration.

Notice that the acceleration for a falling object when air resistance is included is less than g, that is, $a < g$ or $a < 9.8 \text{ m/s}^2$. As the object continues to fall, the force of air resistance f increases until $a = 0$ when $f = mg$ and $f - mg = 0$. The object then falls at its constant terminal velocity.

For a skydiver with an unopened parachute, the terminal velocity is about 200 km/h (about 125 mi/h). To reduce the terminal velocity, so that it can be reached sooner and the time of fall be extended, a skydiver will try to increase exposed body area to a maximum by assuming a spread-eagle position (•Fig. 4.25). Doing this takes advantage of the dependence of air resistance on the size and shape of the falling object. Once the parachute is open (giving a larger exposed area and a shape that catches the air), the additional air resistance slows the diver down to about 40 km/h (or 25 mi/h), which is preferable for landing.

•FIGURE 4.25 Terminal velocity
Sky divers assume a spread-eagle position in order to maximize air resistance. This causes them to reach terminal velocity more quickly and prolongs the time of fall.

CONCEPTUAL EXAMPLE 4.14 ■ RACE YOU DOWN: AIR RESISTANCE AND TERMINAL VELOCITY

From a high altitude, a balloonist simultaneously drops two balls of identical size but appreciably different weight. Assuming that both balls reach terminal velocity during the fall, (a) the heavier ball reaches terminal velocity first, (b) the balls reach terminal velocity at the same time, (c) the heavier ball hits the ground first, (d) the balls hit the ground at the same time. *Clearly establish the reasoning and physical principle(s) used in determining your answer before checking it below. That is, **why** did you select your answer?*

Reasoning and Answer. Terminal velocity is reached when the weight of a ball is balanced by the frictional air resistance. Both balls start to fall with the same acceleration, g, and their speeds and the retarding forces of air resistance increase at the same rate. The weight of the lighter ball will be balanced first, so (a) and (b) are incorrect. The lighter ball reaches terminal velocity ($a = 0$), but the heavier ball continues to accelerate and pulls ahead of the lighter ball. Hence, the heavier ball hits the ground first and the answer is (c), and (d) does not apply.

Follow-up Exercise. Suppose the heavier ball were much larger than the lighter ball. How might this affect the outcome? (*Reasoning and answer may be found in the Answers to Follow-up Exercises section at the back of the book.*)

Chapter Review

Important Terms

force   98
net (unbalanced) force   98
inertia   100

Newton's first law of motion
 (law of inertia)   100
Newton's second law of motion
  102

newton (unit)   103
weight   103
free-body diagram   106

Important Concepts

- A force is something capable of changing an object's state of motion. To produce a change in motion, there must be a net or unbalanced force.
- Newton's first law of motion is also called the law of inertia, where inertia is the natural tendency of an object to maintain its state of motion. It states that in the absence of a net applied force, a body at rest remains at rest and a body in motion remains in motion with constant velocity.
- Newton's second law relates the net force acting on an object or system to the (total) mass and the resulting acceleration. It defines the cause-and-effect relationship between force and acceleration ($\mathbf{F} \propto m\mathbf{a}$).
- Newton's third law states that for every force there is an equal and opposite reaction force. The opposing forces of a third-law pair always act on different objects.
- Friction is the resistance to motion that occurs between contacting surfaces.
- The frictional force between surfaces is characterized by coefficients of friction (μ), one for the static case and one for the kinetic (moving) case. In many cases, $f = \mu N$, where N is the normal force—the force perpendicular to the surface (the force exerted *by* the surface *on* the object) as a ratio of forces, μ is unitless.
- The force of air resistance on a falling object increases with increasing speed, until the object falls at a constant rate or terminal velocity.

Important Equations

Newton's Second Law:
$$\mathbf{F} = m\mathbf{a} \qquad (4.1)$$

Weight:
$$w = mg \qquad (4.2)$$

Component Form of Newton's Second Law:
$$\mathbf{F}_x\mathbf{x} + \mathbf{F}_y\mathbf{y} = ma_x\mathbf{x} + ma_y\mathbf{y} \qquad (4.3a)$$

Condition for Translational Equilibrium:
$$\Sigma\mathbf{F} = 0 \qquad (4.4)$$
or
$$\Sigma\mathbf{F}_x = 0 \quad \text{and} \quad \Sigma\mathbf{F}_y = 0 \qquad (4.5)$$

Force of Static Friction:
$$f_s \leq \mu_s N \qquad (4.6)$$
$$f_{s_{max}} = \mu_s N \qquad (4.7)$$

Force of Kinetic (Sliding) Friction:
$$f_k = \mu_k N \qquad (4.8)$$

Exercises

4.1 The Concept of Force and Net Force

4.2 Newton's First Law of Motion

1   The tendency of an object to maintain its state of motion is called (a) Newton's first law, (b) Galileo's principle, (c) mass, (d) inertia.

2   In the absence of a net force, an object will necessarily (a) be at rest, (b) be in motion with a constant velocity, (c) have zero acceleration, (d) none of these.

3   An object weighs 300 N on Earth and 50 N on the Moon. Does the object also have less inertia on the Moon?

4   In comparing the nearly equal weights of objects held in each hand, you tend to move them up and down several times. Why is this?

5   Is it easier to pick up and carry a heavy object on the Moon than on Earth? Explain.

6   Consider an air-bubble level sitting on a horizontal surface (•Fig. 4.26). Initially, the air bubble is in the middle of the horizontal glass tube. (a) If the level is released and it accelerates, which way would the bubble move? Which way would the bubble move if the force is removed and the level comes to rest? (b) Such a level is sometimes used as an "accelerometer" to indicate the *direction* of the acceleration of an applied force. Explain the principle involved. [*Hint:* Think about pushing a pan of water.]

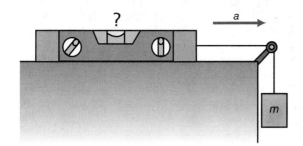

•FIGURE 4.26 An accelerometer
See Exercise 6.

7 As a follow-up to Exercise 6, consider the situation of a child holding a helium balloon in a closed car at rest. What would be observed when the car (a) accelerates from rest and (b) brakes to a stop. (The balloon does not touch the roof of the car.)

8 ■ Which has more inertia, 10 cm³ of gold or 20 cm³ of iron, and how many times more? (See Table 9.2.)

9 ■ You are told that Newton's first law applies to an object that has applied forces of $F_1 = 2.0\,N\,x - 3.5\,N\,y$ and $F_2 = -2.0\,N\,x + 3.0\,N\,y$ acting on it. What else do you know about the situation? Can you tell whether or not the object is at rest or in motion?

10 ■■ A particle of mass 2.5×10^{-6} kg moves in the $+x$ direction with constant velocity. If applied forces of $F_1 = 5.0$ N at an angle of 30° to the $+x$ axis (first quadrant) and $F_2 = 3.0$ N at 37° to the $-x$ axis (third quadrant) act on the particle, what other force, applied simultaneously, will allow it to maintain a constant velocity?

11 ■■ A 5.0-kg block at rest on a frictionless surface is acted on by forces $F_1 = 5.5$ N and $F_2 = 3.5$ N, as illustrated in •Fig. 4.27. What additional force will keep the block at rest?

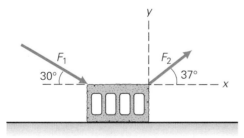

•FIGURE 4.27 Two applied forces
See Exercises 11 and 79.

12 ■■ A 1.5-kg object moves up the y axis with a constant speed. When it reaches the origin, the forces $F_1 = 5.0$ N at 37° relative to the positive x axis (first quadrant), $F_2 = 2.5$ N x, $F_3 = 3.5$ N at 45° relative to the negative x axis (third quadrant), and $F_4 = -1.5$ N y are applied to it. (a) Will the object maintain a path along the y axis? (b) If not, what simultaneously applied force will keep it moving along the y axis with a constant speed?

4.3 Newton's Second Law of Motion

4.4 Applications of Newton's Second Law

Note: In Exercises with strings and pulleys, "ideal conditions" means that the masses of the string(s) and pulley(s) should be neglected, as well as the friction of the pulley.

13 Newton's second law of motion relates the acceleration of an object acted upon by a net force as being (a) inversely proportional to its mass, (b) zero, (c) inversely proportional to the net force, (d) independent of the mass.

14 The newton unit of force is equivalent to (a) kg·m/s, (b) kg·m/s², (c) kg·m²/s, (d) none of these.

15 In general, this chapter considered forces that were applied to objects of constant mass. What would be the situation if mass were added or lost from a system while a force was being applied? Give examples of situations in which this might happen.

16 A sand hourglass sits on a delicate scale. The glass is turned over and the sand starts falling. Does the scale have a lighter reading during the time the sand is falling than when not? Explain.

17 ■ Determine the magnitude of the net force required to give a 4.50-kg object an acceleration of 1.50 m/s².

18 ■ A worker pushes on a crate, and it experiences a net force of 200 N. If the crate moves with an acceleration of 0.750 m/s², what is its weight?

19 ■ An ocean liner has a gross tonnage of 70,000 metric tons. What constant net force would give the liner an acceleration of 0.10 m/s²?

20 ■ A stalled 1800-kg automobile is towed by another car with a horizontal rope, and the cars accelerate at a rate of 0.25 m/s². Find the tension force in the rope.

21 ■■ A net force of 40 N acts on a container holding 2.0 L of water. Determine the resulting acceleration. (Neglect the mass of the container.)

22 ■■ At a party, 18 students lift a sports car. While holding the car off the ground, each student exerts an upward force of 415 N. (a) What is the mass of the car in kilograms? (b) What is its weight in pounds?

23 ■■ What force acts on a 0.75-kg object in free fall?

24 ■■ A horizontal force of 12.0 N acts on an object resting on a level, frictionless surface on the Moon, where the object has a weight of 98.0 N. (a) What is the magnitude of the acceleration of the object? (b) What would be the acceleration of the same object in a similar situation on Earth?

25 ■■ When a horizontal force of 300 N is applied to a 75.0-kg crate, it slides on a level floor opposed by a force of kinetic friction of 95.0 N. What is the acceleration of the crate?

26 ■■ A professor's car, which has a weight of 1.38×10^4 N, moves with a constant velocity of 55.0 km/h. Determine the unbalanced force acting on the car.

27 ■■ Two blocks of ice, weighing 80 N and 50 N, sit side by side in contact with each other on a horizontal surface. (a) If a constant horizontal force of 40 N is applied to one of the blocks in the direction of the other block, what is the resulting acceleration? (Neglect friction.) (b) If the force is applied to the block at an angle of 25° below the horizontal, what would be the acceleration?

28 ■■ A 640,000-metric ton ocean-going oil tanker traveling at a speed of 15 knots takes 5.0 km to come to a stop. What is the magnitude of the constant force required to do this? [A knot is a nautical mile/hour, and 1 knot = 1.15 (statute) mi/h.]

29 ■■ A stalled 1500-kg automobile is pushed toward a gas station by a man and a woman on a level road. The applied horizontal forces are 220 N for the woman and 335 N for the man. (a) If there is an effective force of friction of 400 N on the car as it moves, what is its acceleration? (b) Once the car is moving appreciably, what would be an appropriate combined applied force and why?

30 ■■ In an emergency stop to avoid an accident, a shoulder-strap seat belt holds a 60-kg passenger firmly in place. If the car was initially traveling at 90 km/h and came to a stop in 4.5 s along a straight, level road, what was the average force applied to the passenger by the seat belt?

31 ■■ A jet plane on a straight, level runway starts from rest and lifts off with an acceleration of 25 m/s². If the jet engine exerts a constant thrust of 2.0×10^6 N, what is the gross mass of the plane in metric tons?

32 ■■ A jet catapult on an aircraft carrier uniformly accelerates a plane with a mass of one metric ton uniformly from rest to a launch speed of 320 km/h in 2.00 s. What is the magnitude of the net force on the plane?

33 ■■ A 50-kg filing cabinet is to be lowered from an office building with a rope having a maximum tensile strength of 450 N. Is this possible? If so, explain and describe the motion.

34 ■■ In serving, a tennis player accelerates a 75-g tennis ball horizontally from rest to a speed of 45 m/s. Assuming the acceleration is uniform when the raquet is applied over a distance of 0.85 m, what is the magnitude of the force exerted on the ball by the raquet?

35 ■■ A 75-kg gymnast hangs vertically from a pair of parallel rings. (a) If the ropes supporting the rings are attached to the ceiling directly above, what is the tension in the ropes? (b) If the ropes are supported so that they make an angle of 45° with the ceiling, what is the tension in the ropes? (c) Analyze the tension as the angle becomes smaller and smaller. (Why are telephone and electric lines allowed to sag between the poles? Wouldn't it take less wire and be more economical if they were strung tightly?)

36 ■■ An Atwood machine (see Fig. 4.12) has suspended masses of 0.20 kg and 0.15 kg. What will be the acceleration of the smaller mass with ideal conditions?

37 ■■ For an ideal Atwood machine, m_2 (150 g) is released 1.0 m above the floor, with m_1 (140 g) resting on the floor. If it takes m_2 a time of 2.4 s to reach the floor, what is the experimental value of g?

38 ■■ Two boats pull a 75.0-kg water skier as illustrated in ●Fig. 4.28. (a) If each boat pulls with a force of 600 N and the skier travels with a constant velocity, what is the magnitude of the retarding force between the water and the skis? (b) Assuming the retarding force remains constant, if the boats each pull with a force of 700 N, what is the magnitude of the acceleration of the skier?

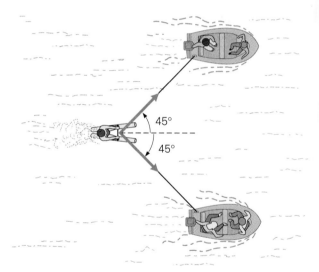

●FIGURE 4.28 Double tow
See Exercise 38.

39 ■■ (a) A 50-kg water skier is pulled on a lake by a boat with a horizontal force of 400 N. If the wind and current applies another force of 100 N at an angle of 30° relative to the straight-line direction of the boat, what is the skier's acceleration? (b) What is the acceleration if the wind-current force were in the opposite direction to that in part (a)?

40 ■■ For the arrangement in Fig. 4.13, with ideal conditions, what will be the acceleration of the system if $m_1 = 2.25$ kg, $m_2 = 1.65$ kg, and $\theta = 36.9°$?

41 ■■ Suppose that $\theta = 30°$ and $m_2 = 2.5$ kg for the arrangement in Fig. 4.13. What should m_1 be if it is to move with an acceleration of (a) 0.25 m/s² down the plane, or (b) 0.10 m/s² down the plane?

42 ■■ Three blocks are pulled along a frictionless surface by a horizontal force, as shown in •Fig. 4.29. (a) What is the acceleration of the system? (b) What are the tension forces in the light strings? (*Hint:* Can T_1 equal T_2? Investigate by drawing free-body diagrams of each block separately.)

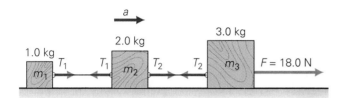

•**FIGURE 4.29 Three block system**
See Exercises 42 and 80.

43 ■■ Assume ideal conditions (no friction and masses of pulleys negligible) for the arrangement illustrated in •Fig. 4.30. What is the acceleration of the system if (a) $m_1 = 0.25$ kg, $m_2 = 0.50$ kg, and $m_3 = 0.25$ kg, and (b) $m_1 = 0.35$ kg, $m_2 = 0.15$ kg, and $m_3 = 0.50$ kg?

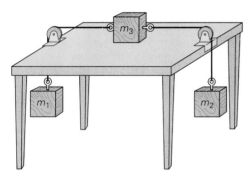

•**FIGURE 4.30 Which way will they accelerate?**
See Exercises 43, 83, and 84.

44 ■■ A horizontal net force of 40 N acting on a block on a frictionless level surface produces an acceleration of 2.5 m/s². A second block with a mass of 4.0 kg is dropped onto the first. What is the magnitude of the acceleration of the combination if the same force continues to act? (Assume that the second block does not slide on the first block.)

45 ■■■ One mass, $m_1 = 215$ g, of an ideal Atwood machine rests on the floor 1.10 m below the other mass, $m_2 = 255$ g. (a) If the masses are released from rest, how long does it take m_2 to reach the floor? (b) How high will mass m_1 ascend from the floor?

46 ■■■ A double Atwood machine is illustrated in •Fig. 4.31. Assuming ideal conditions and letting $m_3 = 4.0$ kg and $m_1 = m_2 = 3.0$ kg, (a) what is the magnitude of the acceleration of m_3? (b) What are the magnitudes of the tensions in the strings?

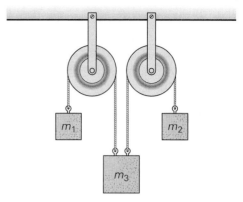

•**FIGURE 4.31 A double Atwood machine**
See Exercises 46, 47, 56, and 87.

47 ■■■ What is the magnitude of the acceleration of the system if the masses for a double Atwood machine similar to that in Fig. 4.31 are (a) $m_3 = 3.0$ kg, $m_2 = 4.0$ kg, and $m_1 = 3.0$ kg, or (b) $m_3 = 2.5$ kg, $m_2 = 3.5$ kg, and $m_1 = 4.0$ kg? (Assume ideal conditions.)

48 ■■■ A 70.0-kg person stands on a scale in an elevator. What weight in newtons does the scale read if the elevator is (a) at rest, (b) moving with an upward acceleration of 0.525 m/s², (c) moving with a downward acceleration of 0.525 m/s²?

49 ■■■ Two blocks are on a frictionless double inclined plane as illustrated in •Fig. 4.32. If one block (m_1) has a mass of 15 kg and the other (m_2) has a mass of 25 kg, what is the magnitude of the acceleration of the system? (Assume ideal conditions.)

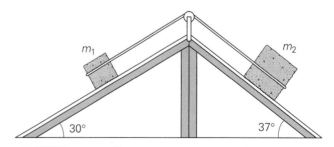

•**FIGURE 4.32 Double inclined plane**
See Exercises 49, 50, and 85.

50 ■■■ If the blocks on the double frictionless incline in Fig. 4.32 are at rest or move with a constant speed, (a) which block is more massive? (b) How many times more massive?

4.5 Newton's Third Law of Motion

51 The force pair of Newton's third law (a) consists of forces that are always opposite, but sometimes not equal, (b) always cancel each other when the second law is applied to a body, (c) always act on the same object, (d) consists of forces that are equal and opposite, but act on different objects.

52 One force of the action–reaction force pair (a) never produces an acceleration, (b) is always greater than the other, (c) may or may not produce a change in velocity, (d) none of these.

53 Is there something wrong with the following statement? A horse pulls forward on a wagon, and there is an equal and opposite force on the horse. The forces then cancel and there is no motion.

54 A person pushes on a block of wood that has been placed against a wall. Make a sketch, and analyze this situation in terms of Newton's third law.

55 Identify all of the forces in Fig. 4.9 in terms of Newton's third law. What forces are not explicitly illustrated in the figure?

56 Analyze the forces acting in the system of the double Atwood machine in Fig. 4.31 in terms of Newton's third law. Take the system to be in static equilibrium.

57 ■ Three blocks with masses of 1.0 kg, 2.0 kg, and 3.0 kg are stacked on a table with the smallest on top and the largest on the bottom. Make a sketch and analyze this system in terms of the force pairs of Newton's third law.

58 ■■ Two people, A and B, with masses of 50 kg and 60 kg, respectively, stand on a frictionless surface 10 m apart. B pulls on a connecting rope, giving A an acceleration of 0.92 m/s² towards B. (a) What is B's acceleration toward A? (b) If the pulling force is applied constantly, where will they meet?

59 ■■ A rifle weighs 50.0 N, and its barrel is 0.750 m long. It shoots a 25.0-g bullet, which leaves the barrel with a speed of 300 m/s (muzzle velocity) after being uniformly accelerated. What is the magnitude of the reaction force on the rifle?

4.6 Friction

Note: Neglect air resistance unless otherwise stated.

60 In general, the frictional force (a) is greater for smooth surfaces, (b) depends on slow sliding speeds, (c) is proportional to the load, (d) depends on the surface area.

61 The coefficient of kinetic friction (a) is usually greater than μ_s, (b) usually equals μ_s, (c) equals the applied force that exceeds the maximum static force, (d) is unitless.

62 In general, μ_k is less than μ_s. Can you suggest why?

63 Is it easier to push or pull a lawnmower? Explain.

64 (a) Why do front-wheel brakes on an automobile usually wear out first? (b) By equipping automobiles with wider brake bands or disks, safety would be increased because you could stop more quickly. Do you agree? Explain. Would there be any other advantages?

65 Suppose the slope conditions for the skier shown in •Fig. 4.33 are such that he travels with a constant velocity. Could you find the coefficient of kinetic friction between the snowy surface and the skis from the photo? If so, describe how this would be done.

•**FIGURE 4.33   A downslope run**
See Exercises 65 and 66.

66 (a) Explain why the skier in Fig. 4.33 is in a crouched position with arms tucked in. (b) Automobiles are streamlined so that they will get better gas mileage. Explain this in terms of air resistance.

67 (a) We commonly say that friction opposes motion. Yet, when we walk, the frictional force is *in the direction* of our motion (see Fig. 4.18). Is there an inconsistency here in terms of Newton's second law? Explain. (b) What effects would wind have on air resistance? [*Hint:* The wind can blow in different directions.]

68 ■ In moving a 35.0-kg desk from one side of a classroom to the other, a professor finds that a horizontal force of 275 N is necessary to set the desk in motion and a force of 195 N is necessary to keep it in motion with a constant speed. What are the coefficients of (a) static and (b) kinetic friction between the desk and the floor?

69 ■ A 40-kg crate is at rest on a level surface. If the coefficient of static friction between the crate and the surface is 0.69, what horizontal force is required to move the crate?

70 ■■ A 5.0-kg wooden block is placed on an adjustable wooden inclined plane. (a) What is the angle of incline above which the block will *start* to slide down the plane? (b) At what angle of incline will the block then slide down the plane at a constant speed?

71 ◼◼ A packing crate is placed on a 20° incline plane. If the coefficient of static friction between the crate and the plane is 0.65, will the crate slide down the plane? (Justify your answer.)

72 ◼◼ A glass paperweight with a weight of 3.5 N is placed on a pane of glass measuring 0.75 m × 0.75 m which is elevated at one end to act as an inclined plane. What is the elevation of the higher end of the pane when the paperweight slides down with a constant velocity once it is in motion?

73 ◼◼ A block with a mass of 2.0 kg and 10 cm on a side just slides down an inclined plane with a 30° angle of incline (●Fig. 4.34). Another block of the same height and same material has base dimensions of 20 cm × 10 cm, and thus a mass of 4.0 kg. (a) At what critical angle will the more massive block just start to slide down the plane? (b) Estimate the coefficient of static friction between the block and the plane.

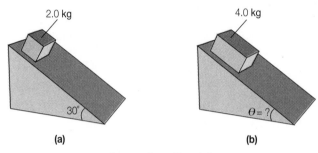

(a)   **(b)**

●**FIGURE 4.34 At what angle will it slide?**
See Exercise 73.

74 ◼◼ While being unloaded from an airplane, a 20.0-kg suitcase is placed on a flat ramp inclined at 37°. When released from rest, the suitcase accelerates down the ramp at 0.250 m/s². What is the coefficient of kinetic friction between the suitcase and the ramp?

75 ◼◼ Consider a system as in Fig. 4.13. (a) If the coefficient of static friction between $m_1 = 10$ kg and the plane surface is 0.80, what suspended mass (m_2) will just set m_1 into motion up in the plane? (b) What suspended mass is required to keep m_1 moving up the plane at a constant velocity if $\mu_k = 0.60$? (c) What suspended mass is required to keep m_1 moving down the plane with a constant velocity? (Assume ideal conditions for the string and pulley.)

76 ◼◼ A crate containing machine parts sits unrestrained on the back of a flatbed truck traveling along a straight road at a speed of 80 km/h. The driver applies a constant braking force and comes to a stop in a distance of 22 m. What is the minimum coefficient of static friction between the crate and the truck bed if the crate is not to slide forward?

77 ◼◼ A 1500-kg automobile travels at a speed of 90 km/h along a straight concrete highway. Faced with an emergency situation, the driver jams on the brakes and skids to a stop. What will the stopping distance be for (a) dry pavement and (b) wet pavement?

78 ◼◼ A school bus pulls into an intersection as a car approaches on an icy street at a speed of 35 km/h. Seeing the bus from 26 m away, the driver of the car locks the brakes causing the car to slide toward the intersection. If the coefficient of kinetic friction between the car's tires and the icy road is 0.25, does the car hit the bus? (You will learn in Chapter 8 why shorter stopping distances result from pumping the brakes.)

79 ◼◼ For the situation shown in Fig. 4.27, what is the minimum coefficient of static friction between the block and the surface that will keep the block from moving? ($F_1 = 5.0$ N, $F_2 = 4.0$ N, and $m = 5.0$ kg.)

80 ◼◼ For the system illustrated in Fig. 4.29, if $\mu_s = 0.45$ and $\mu_k = 0.35$ between the blocks and the surface, what applied forces will (a) set the blocks in motion and (b) move the blocks with a constant velocity?

81 ◼◼ A hockey player hits a puck with his stick, giving it an initial velocity of 5.0 m/s. If the puck slows uniformly and comes to rest in a distance of 20 m, what is the coefficient of kinetic friction between ice and puck?

82 ◼◼◼ A ramp 135 m long is to be built for a ski jump. If a skier starting from rest at the top is to have a speed of at least 24 m/s at the bottom, should the angle of the incline be greater than 20°? Justify your answer.

83 ◼◼◼ For the arrangement in Fig. 4.30, what is the minimum value of the coefficient of static friction between the block (m_3) and the table that would keep the system at rest if $m_1 = 0.25$ kg, $m_2 = 0.50$ kg, and $m_3 = 0.75$ kg? (Assume ideal conditions for the string and pulleys.)

84 ◼◼◼ If the coefficient of kinetic friction between the block and the table in Fig. 4.30 is 0.560, and $m_1 = 0.150$ kg and $m_2 = 0.250$ kg, (a) what should m_3 be if the system is to move with a constant speed? (b) If $m_3 = 0.100$ kg, what is the acceleration of the system? (Assume ideal conditions for the string and pulleys.)

85 ◼◼◼ For the double inclined plane illustrated in Fig. 4.32, if $m_1 = 1.5$ kg and $m_2 = 2.7$ kg, what is the minimum value of the coefficient of static friction between the blocks and the planes that will keep the system at rest? (Assume the same coefficient for both surfaces and ideal conditions for the string and pulleys.)

Additional Exercises

86 A force acts on a mass. If the force is tripled and the mass is halved, how is the acceleration affected? (Give a factor of change, e.g., 2 times as great or $\frac{1}{2}$ as great.)

87 If the masses for a double Atwood machine similar to the one shown in Fig. 4.31 are $m_1 = 2.0$ kg, $m_2 = 3.0$ kg, and $m_3 = 8.0$ kg, what is the magnitude of the acceleration of the system? (Assume ideal conditions.)

88 Does it require more force to start a block of steel moving on a horizontal steel surface or on a horizontal aluminum surface? How many times as much?

89 The maximum load that can safely be supported by a rope in an overhead hoist is 300 N. What is the maximum acceleration that can safely be given to a 15-kg object being hoisted vertically upward?

90 The coefficient of static friction between a 9.0-kg object and a horizontal surface is 0.45. Would a force of 35 N applied horizontally cause the object to move from rest?

91 An object with a mass of 2.0 kg travels with a constant velocity of 4.8 m/s northward. It is then acted upon by a force of 6.5 N in the direction of the motion and a force of 8.5 N to the south, which continue even after the mass comes momentarily to rest. (a) How far will the object travel before coming to rest? (b) What will be its position 1.5 s after coming momentarily to rest?

92 In the operation of a machine, a 5.0-kg steel part moves on a horizontal steel surface. The force applied to the part is downward at an angle of 30° from the horizontal. (a) What is the magnitude of the applied force required to set the part in motion if the surface is dry? (b) If the surface is lubricated, by what factor is the needed force reduced?

93 A skier on waxed skis coasts down a small slope with a constant velocity. What is the angle of the slope? [*Hint:* See Table 4.1.]

94 A crate weighing 9.80×10^3 N is pulled up a 37° incline by a pulley arrangement. If the coefficient of kinetic friction between the crate and the surface of the plane is 0.750, what is the magnitude of the applied force (parallel to the plane) required to move the crate with a constant velocity? (Assume ideal conditions for the pulley arrangement.)

95 For an Atwood machine with suspended masses of 0.30 kg and 0.40 kg, the acceleration of the masses is measured as 0.95 m/s². What is the effective force of friction for the system?

96 A loaded jet plane with a weight of 2.75×10^6 N is ready for take-off. If its engines supply 6.35×10^6 N of net thrust, how long a runway will the plane need to reach its minimum take-off speed of 285 km/h?

97 A 0.45-kg shuffleboard puck is given an initial velocity of 4.5 m/s down the playing surface. If the coefficient of sliding friction between the puck and the surface is 0.20, how far will the puck slide before coming to rest?

98 An object is acted on by $F_1 = 4.0$ N at an angle of +37° relative to the positive x axis and $F_2 = 6.0$ N along the negative x axis. (a) What single additional force will ensure that the particle has zero acceleration? (b) Describe the particle's motion in that case.

99 Three forces act on an object that is moving in a straight line with a constant speed. If two of the forces are $\mathbf{F}_1 = 2.5\,\text{N}\,\mathbf{x} - 1.5\,\text{N}\,\mathbf{y}$ and $\mathbf{F}_2 = -4.5\,\text{N}\,\mathbf{x} - 2.0\,\text{N}\,\mathbf{y}$, what is the third force?

100 In catching a baseball traveling horizontally with a speed of 15.0 m/s, a player moves the glove straight backward 20.0 cm from the time of contact to the time the ball comes to rest. If the ball has a mass of 0.0750 kg, what is the average force on the ball during that interval?

101 Two blocks initially at rest on a horizontal frictionless surface are each acted on by a constant force of 20 N. One block has a mass of 4.0 kg, and the other a mass of 5.0 kg. If the identical forces act on the blocks for 3.0 s, which block will have traveled the greater distance at the end of this time, and how far ahead of the other block will it be?

5 Work and Energy

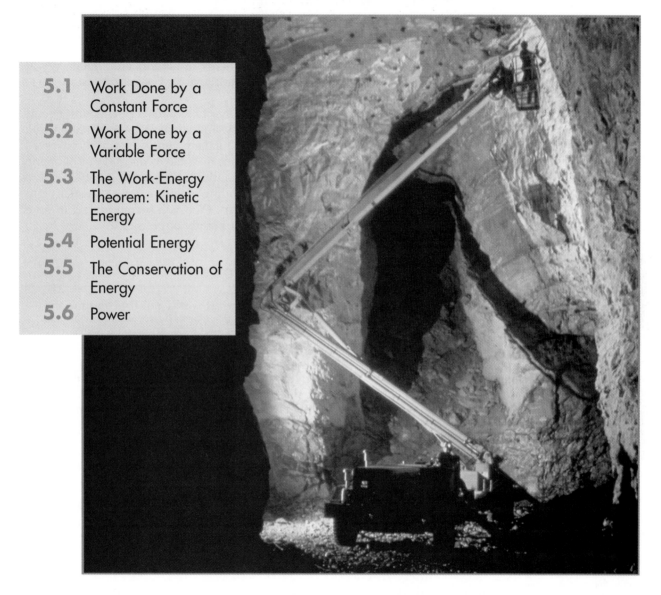

Work and energy can be somewhat slippery concepts. For the most part, our everyday sense of these words accords pretty well with their definitions in physics, but not always. You might think that this crane, for example, must be performing a good deal of work to support the miner and move him around, but that's not necessarily true from a physicist's point of view. Depending on how and where the bucket is moved, the crane might be doing no work at all. It might even be doing *negative* work—a concept that most people, unless they are intimately familiar with small children or

bureaucrats, can't easily imagine. A physicist would also say that the miner has slightly less energy down here than outside the mine—not because he prefers working on his lawn to mining zinc, but because he's a tiny bit closer to the center of the Earth.

In this chapter you'll learn how work and energy are defined in physics, and explore the close relationship between these very useful concepts. You'll also encounter one of the cornerstones of physics: the principal that energy of one kind can be transformed into another, or into work, but can never be lost or destroyed.

This chapter centers on two concepts that are important in both science and everyday life—*work* and *energy*. We commonly think of work as being associated with doing or accomplishing something. Because work makes us physically (and sometimes mentally) tired, we have invented machines and use them to decrease the amount of effort we expend personally. Energy, on the other hand, brings to mind the cost of fuel for transportation and heating, or perhaps the food that supplies the energy our bodies need to carry out life processes and to do work.

Although these notions do not really define work and energy, they point us in the right direction. As you might have guessed, and will learn in the chapter, work and energy are closely related. In physics, as in everyday life, when something possesses energy, it usually has the ability to do work. For example, water rushing through the sluices of a dam has energy of motion, and this energy allows it to do the work of driving a turbine or dynamo. Conversely, no work can be performed without energy.

Energy exists in various forms: There is mechanical energy, chemical energy, electrical energy, heat energy, nuclear energy, and so on. A transformation from one form to another may take place, but the total amount of energy is conserved, or always remains the same. This is the point that makes the concept of energy so useful. When a physically measurable quantity is conserved, it not only gives us an insight that leads to a better understanding of nature, but also usually provides another approach to practical problems. (You will be introduced to other conserved quantities during the course of our study of physics.)

5.1 Work Done by a Constant Force

Objectives: To be able to (a) define mechanical work, and (b) compute the work done in various situations.

The word work is commonly used in a variety of ways: We go to work; We work on projects; We work at our desks or on computers; We work problems. In physics, however, work has a very specific meaning. Mechanically, **work** involves force and displacement, and we use this word to describe quantitatively what is accomplished when a force moves an object through a distance. In the simplest case of a *constant* force:

> The work done by a constant force in moving an object is equal to the product of the magnitudes of the displacement and the component of the force parallel to the displacement.

Work—involves force and displacement

Work then involves moving an object through a distance. A force may be applied, as in •Fig. 5.1a, but if there is no motion (no displacement), then no work is done. For a constant force **F** acting in the same direction as the displacement **d** (•Fig. 5.1b), the work (*W*) is simply

$$W = Fd \tag{5.1}$$

The product of two vectors (force and displacement) in this case is a special type of vector multiplication and yields a scalar quantity equal to $Fd \cos \theta$. Thus, work is a scalar—it does not have direction. It can, however, be positive, zero, or negative, depending on the angle.

In general, work is done when a force moves an object through a distance and some *component* of the force is along the line of motion (•Fig. 5.1c). That is, if the force is at an angle θ to the object's displacement, then $F_{\parallel} = F \cos \theta$ is the component of force parallel to the displacement, and we have the more general equation

$$W = F_{\parallel}d = (F \cos \theta)d = Fd \cos \theta \tag{5.2}$$

SI unit of work: N·m or joule (J)

Note that if $\theta = 0°$, Eq. 5.2 reduces to Eq. 5.1. The perpendicular component of the force, $F_{\perp} = F \sin \theta$, does no work since there is no displacement in this direction.

Notice that θ is the angle *between* the force and displacement vectors. If the force is in the opposite direction to the displacement (for example, a braking force

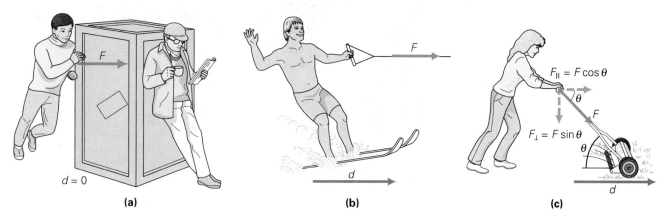

(a) (b) (c)

•FIGURE 5.1 **Work done by a constant force—the product of the magnitudes of the parallel component of force and the displacement**
(a) If there is no displacement, no work is done: $W = 0$. **(b)** For a constant force in the same direction as the displacement, $W = Fd$. **(c)** For a constant force at an angle to the displacement, $W = (F \cos \theta) \, d = Fd \cos \theta$.

that tends to slow down, or decelerate, an object) $\theta = 180°$ and $\cos 180° = -1$. Then the work is negative: $W = Fd \cos 180° = -Fd$.

Work is a scalar quantity. Since $W = Fd$, force is given in newtons, and displacement is given in meters, work has the SI unit of newton-meter (N·m). This unit is given the special name of **joule**:

$$Fd = W$$

$$\text{N·m} \equiv \text{joule (J)}$$

For example, the work done by a force of 25 N moving an object through a distance of 2.0 m is $W = Fd = (25 \text{ N})(2.0 \text{ m}) = 50 \text{ N·m}$ or 50 J.

From the equation, we also see that in the British system work would have the unit pound-foot. However, this is commonly written in reverse. The British standard unit for work is the **foot-pound** (**ft·lb**).

The name joule (J) (pronounced "jool") was given in honor of James Prescott Joule (1818–1889), a British scientist who investigated work and energy.

EXAMPLE 5.1 ■ APPLIED PSYCHOLOGY: MECHANICAL WORK

A student holds her psychology textbook, which has a mass of 1.5 kg, out of a second-story dormitory window until her arm is tired, and then releases it (•Fig. 5.2). (a) How much work is done on the book by the student in simply holding it out the window? (b) How much work will have been done by the force of gravity during the time in which the book falls 3.0 m?

Solution. Listing the data, we have

Given: $v_o = 0$ (initially at rest)   *Find:* (a) W (work holding)
$\quad\quad\quad m = 1.5$ kg   (b) W (work)
$\quad\quad\quad d = 3.0$ m

(a) Even though the student gets tired (because work is performed within the body to maintain muscles in a state of tension), she does *no* mechanical work *on the book* in merely holding it stationary. She exerts an upward force on the book (equal in magnitude to its weight), but the displacement is zero in this case ($d = 0$). Thus, $W = Fd = F \times 0 = 0$.

(b) While the book is falling, the net force acting on it is the force of gravity, which is equal in magnitude to the weight of the book, $F = w = mg$ (neglecting air resistance). The displacement is in the same direction as the force and has a magnitude of $d = 3.0$ m, so the work done by gravity is

$$W = Fd \cos 0° = (mg)d = (1.5 \text{ kg})(9.8 \text{ m/s}^2)(3.0 \text{ m}) = +44 \text{ J}$$

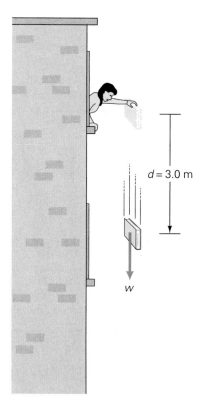

•FIGURE 5.2 **Mechanical work requires motion**
See Example 5.1.

Determining the Sign of Work.

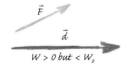

$W = W_0$

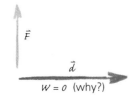

$W > 0$ but $< W_0$

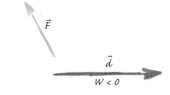

$W = 0$ (why?)

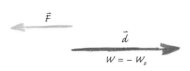

$W < 0$

$W = -W_0$

Follow-up Exercise. How much work is done on the book by gravity as it falls between heights of 3.0 m and 2.0 m? *(Answer may be found in the Answers to Follow-up Exercises section at the back of the book.)*

EXAMPLE 5.2 ■ YARD WORK: PARALLEL COMPONENT OF A FORCE

If the person in Fig. 5.1(c) pushes on the lawn mower with a constant force of 90.0 N at an angle of 40° to the horizontal, how much work does she do in pushing it a horizontal distance of 7.50 m?

Solution.

Given: $F = 90.0$ N   *Find:* W (work done by person)
$\theta = 40°$
$d = 7.50$ m

Here, the horizontal component of the applied force, $F \cos \theta$, is parallel to the displacement, so Eq. 5.2 applies.

$$W = Fd \cos 40° = (90.0 \text{ N})(7.50 \text{ m})(0.766) = +517 \text{ J}$$

Work is done by the horizontal component of the force, but the vertical component does no work.

Follow-up Exercise. When you wheel a wheelbarrow on a level surface, the force you apply has an upward component. Is work done by this component? *(Answer may be found in the Answers to Follow-up Exercises section at the back of the book.)*

We commonly specify what is doing work *on* what. For example, the force of gravity does work on a falling object, such as the book in Example 5.1. Also, when you lift an object, you do work *on* the object. We sometimes describe this as doing work *against* gravity because the force of gravity is in the direction opposite that of the applied lift force, and opposes it.

In both Examples 5.1 and 5.2, work was done by a single constant force. If more than one force acts on an object, the work done by each may be calculated separately. The total, or net, work is then the scalar sum of those quantities of work. Alternatively, you may find the vector sum of the forces and use this net force to compute the *total, or net, work*.

EXAMPLE 5.3 ■ TOTAL OR NET WORK

A 0.75-kg block slides down a 20° inclined plane with a uniform velocity (●Fig. 5.3). (a) How much work is done by the force of friction on the block as it slides the total length of the plane? (b) What is the net work done on the block? (c) Discuss the net work done if the angle of incline is adjusted so that the block accelerates down the plane.

Solution.   We list what is given, but equally importantly we want to know specifically what is to be found.

Given: $m = 0.75$ kg   *Find:*   (a) W_f (work done by friction)
$\theta = 20°$ (b) W_{net} (net work)
$L = 1.2$ m   (c) W (discuss work with block
(from figure)   accelerating)

(a)   Note from Fig. 5.3 that only two forces do work because there are only two forces parallel to the motion: f_k, the force of kinetic friction; and $mg \sin \theta$, the component of the block's weight acting down the plane. We first find the work done by the frictional force,

$$W_f = f_k d \cos 180° = -f_k d = -\mu_k N d$$

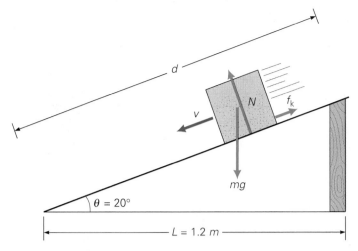

•**FIGURE 5.3 Total or net work**
See Example 5.3.

where $f_k = \mu_k N$. The angle 180° indicates that the force and displacement are in opposite directions. (It is common in such cases to write $W_f = -f_k d$ directly, since friction typically opposes motion.) The magnitude of the normal force N is equal to that of the component of the block's weight perpendicular to the surface (these are equal and opposite forces: why?):

$$N = mg \cos \theta$$

The distance d the block slides down the plane may be found using trigonometry. Note that $\cos \theta = L/d$, so

$$d = \frac{L}{\cos \theta}$$

But what is μ_k? It doesn't seem to be given (nor are the types of surfaces specified). This is an example in which a key piece of data is given to you indirectly in a problem. Note that the block slides down the plane with a uniform (constant) velocity. In Chapter 4, you learned that for uniform motion down a plane, $\mu_k = \tan \theta$. Thus,

$$W_f = -\mu_k N d = -(\tan \theta)(mg \cos \theta)\left(\frac{L}{\cos \theta}\right) = -mgL \tan 20°$$
$$= -(0.75 \text{ kg})(9.8 \text{ m/s}^2)(1.2 \text{ m})(0.364) = -3.2 \text{ J}$$

(b) The net work may be directly seen to be zero. Since the block moves with a constant velocity, the net force acting on it must be zero (by Newton's first or second law), and so the net work, $W_{net} = F_{net}d = 0$. Let's show this explicitly. We just computed the work done by the frictional force. The only other work is that done by the component of the gravitational force down the block, W_g.

$$W_g = F_\parallel d = (mg \sin \theta)\left(\frac{L}{\cos \theta}\right) = mgL \tan 20° = +3.2 \text{ J}$$

Then,

$$W_{net} = W_g + W_f = +3.2 \text{ J} - +3.2 \text{ J} = 0$$

Remember that work is a scalar quantity, so scalar addition is used to find net work.

(c) If the block accelerates down the plane, then from Newton's second law, $F = mg \sin \theta - f_k = ma$. The component of the gravitational force is greater than the opposing frictional force, so there is net work done on the block. You might be wondering what the effect of nonzero net work is. As you will learn shortly, it causes a change in the amount of energy an object has.

Follow-up Exercise.   In part (c) of this Example, is it possible for the frictional work to be greater than the gravitational work? (*Answers may be found in the Answers to Follow-up Exercises section at the back of the book.*)

Note: Recall the discussion of friction in Section 4.6.

Note how in part (a) of Example 5.3, the equation for W_f was easily simplified by using the algebraic expressions for N and d instead of computing these quantities initially. It is a good rule of thumb not to put numbers into an equation until you have to. Simplifying an equation through cancellation is easier with symbols, and saves computation time.

5.2 Work Done by a Variable Force

Objectives: **To be able to (a) differentiate work done by constant and variable forces, and (b) compute the work done by a spring force.**

The discussion in the preceding section was limited to work done by constant forces. In general, however, forces are variable; that is, they change with time and/or position. For example, a force applied to an object to overcome the force of static friction may be increased, until it exceeds $f_{s_{max}}$. However, the force of static friction does no work because there is no motion or displacement.

An example of variable force doing work in stretching a spring is illustrated in •Fig. 5.4. As a spring is stretched (or compressed) farther and farther, the restoring force of the spring gets greater and an increasing applied force is required. It is found that the applied force F is directly proportional to the change in length of the spring from its unstretched length. In equation form, this is expressed

$$F = k\Delta x = k(x - x_o)$$

or, if we chose $x_o = 0$,

$$F = kx$$

where x now represents the compressor or extensor of the spring from its unstretched length. As can be seen, the force varies with x. We describe this by saying that the *force is a function of position*.

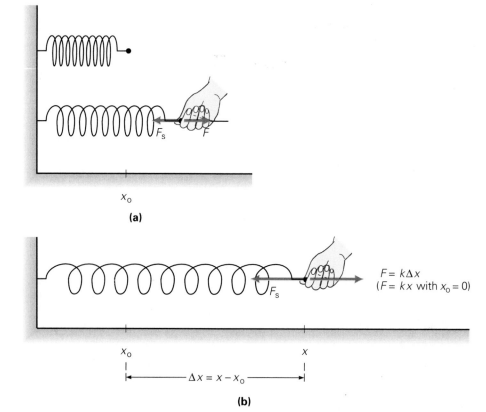

•**FIGURE 5.4 Spring force**
(a) An applied force F stretches the spring, and the spring exerts an equal and opposite force F_s on the hand. (b) The magnitude of the force depends on the change in the spring's length. This is often referenced to the position of a mass on the end of the spring.

The k in this equation is a constant of proportionality and is commonly called the **spring constant,** or **force constant.** The greater the value of k, the stiffer or stronger is the spring. As you should be able to prove to yourself, the units of k are newtons per meter (N/m).

k, the spring or force constant

The relationship expressed by Eq. 5.3 holds only for ideal springs. Real springs approximate this linear relationship between force and displacement within certain limits. For example, if a spring is stretched beyond a certain point, called its elastic limit, it will be pulled out of shape and $F = kx$ will no longer apply.

Note that a spring exerts an equal and opposite force,

$$F_s = -k\Delta x = -k(x - x_o)$$

or, with $x_o = 0$,

$$F_s = -kx \tag{5.3}$$

The minus sign indicates that this spring force is in the direction opposite to the displacement whether the spring is stretched or compressed. This equation is a form of what is known as *Hooke's law*, after Robert Hooke, a contemporary of Newton.

Our study of work will generally be limited to situations involving constant or average forces. As you learned in studying kinematics (Chapter 2), an average value may not be very useful except in special cases. One of these cases is the average velocity for a constant acceleration: $\bar{v} = (v + v_o)/2$. A special case of a variable force that can be analyzed without calculus is the spring force, where the spring constant is k. A plot of F versus x is shown in • Fig. 5.5 (which also shows the analogous case of the plot of v versus t for a constant acceleration). The slope of the line is equal to k, and F increases uniformly with x. The average force is then

$$\bar{F} = \frac{F + F_o}{2}$$

or, since $F_o = 0$ (why?)

$$\bar{F} = \frac{F}{2}$$

Thus, the work done in stretching or compressing an amount x the spring from its unstretched length is

$$W = \bar{F}x = \frac{Fx}{2}$$

Since $F = kx$, the work done is

$$W = \tfrac{1}{2}kx^2 \qquad \text{\textit{work done in stretching}} \tag{5.4}$$
$$\text{\textit{(or compressing) a spring}}$$

Note that the work is just the area under the curve in Fig. 5.5.

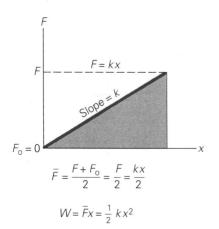

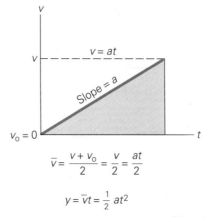

• FIGURE 5.5 **Work done by a uniformly variable force** The work done by a uniformly varying force of the form $F = kx$ is $W = \tfrac{1}{2}kx^2$. A plot of this special case of a variable force is graphically analogous to a plot of v versus t for a uniformly varying velocity starting at rest ($v_o = 0$): $v = at$. The work (W) and the distance (y) are equal to the areas under the respective lines.

5.2 Work Done by a Variable Force **139**

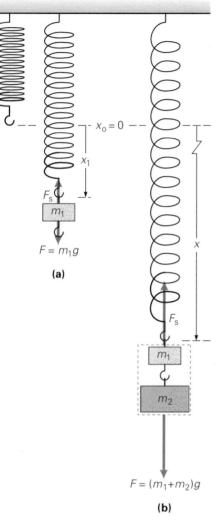

● FIGURE 5.6 Determining the spring constant and the work done in stretching a spring
See Example 5.4.

EXAMPLE 5.4 ■ DETERMINING THE SPRING CONSTANT

A 0.15-kg mass is suspended from a vertical spring and descends a distance of 4.6 cm, after which it hangs at rest (●Fig. 5.6). An additional 0.50-kg mass is then suspended from the first. What is the total extension of the spring? (Neglect the mass of the spring.)

Solution. The data given are as follows:

Given: $m_1 = 0.15 \text{ kg}$ *Find:* x (total stretch length)
$x_1 = 4.6 \text{ cm} = 0.046 \text{ m}$
$m_2 = 0.50 \text{ kg}$

The total stretch distance is given by $F = kx$, where F is the applied force or the weight of the mass suspended on the spring. However, the spring constant k, is not given. This may be found from the data pertaining to the suspension of m_1 and resulting displacement x_1. (This is a common method of determining spring constants.) As seen in Fig. 5.6a, the magnitudes of the weight force and the restoring spring force are equal, so we may write

$$F_s = m_1g = kx_1$$

and

$$k = \frac{m_1g}{x_1} = \frac{(0.15 \text{ kg})(9.8 \text{ m/s}^2)}{0.046 \text{ m}} = 32 \text{ N/m}$$

Then, knowing k, the total extension of the spring is found from the balanced force situation shown in Fig. 5.6b:

$$F = (m_1 + m_2)g = kx$$

Thus,

$$x = \frac{(m_1 + m_2)g}{k} = \frac{(0.15 \text{ kg} + 0.50 \text{ kg})(9.8 \text{ m/s}^2)}{32 \text{ N/m}}$$

$$= 0.20 \text{ m (or 20 cm)}$$

Follow-up Exercise. How much work is done in stretching the spring in the preceding example? *(Answer may be found in the Answers to Follow-up Exercises section at the back of the book.)*

PROBLEM-SOLVING HINT

The reference position x_o for the change in length of a spring is arbitrary and is usually chosen for convenience. *The important quantity in computing work is the position difference Δx, or the net change in the length of the spring.* As shown in ●Fig. 5.7 for a mass suspended on a spring, x_o can be referenced to the unloaded length of the spring or to the loaded position, which may be taken as zero for convenience. In the preceding Example, x_o was referenced to the end of the unloaded spring. Also, with the displacement only in one direction, downward was designated as the $+x$ direction to avoid minus signs.

When the net force on the suspended mass is zero, the mass is said to be at its equilibrium position (as in Fig. 5.7a with m_1 suspended). This position may also be taken as a zero reference ($x = x_o = 0$, Fig. 5.7b). The equilibrium position is a convenient reference point for a case in which the mass oscillates up and down on the spring. (We will describe this motion in a later chapter.) Notice that in general there are both positive and negative directions. Also, since the displacement is in the vertical direction, the x's are often replaced by y's.

5.3 The Work-Energy Theorem: Kinetic Energy

Objective: To be able to (a) explain the work-energy theorem, and (b) apply it in solving problems.

Now that we have an operational definition of work, we are ready to look at how work is related to energy. Energy is one of the most important concepts in science.

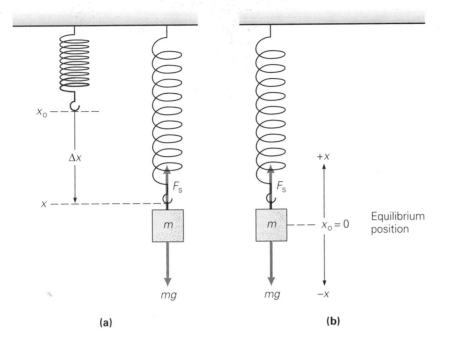

We describe it as a quantity possessed by objects or systems. Basically, work is something that is *done on* objects, whereas energy is something that objects *have*.

One form of energy that is closely associated with work is kinetic energy. (Another form of energy, potential energy, will be described in the next section.) Consider an object at rest on a frictionless surface. Let a horizontal force act on the object and set it in motion. Work is done *on* the object, but where does the work "go," so to speak? It goes into setting the object into motion or changing its *kinetic* conditions. Because of its motion, we say the object has energy—kinetic energy, which gives it the capability to do work.

Kinetic energy is often called the energy of motion. It is defined mathematically as one-half of the product of the mass and the square of the (instantaneous) speed of a moving object.

Kinetic energy—the energy of motion

$$K = \tfrac{1}{2}mv^2 \qquad kinetic\ energy \qquad (5.5)$$

SI unit of energy: joule (J)

For a constant net force doing work on a moving object, as illustrated in ●Fig. 5.8, the force does an amount of work, $W = Fx$. But what are the kinematic effects? The force causes the object to accelerate, and from Equation 2.9, $v^2 = v_o^2 + 2ax$,

$$a = \frac{v^2 - v_o^2}{2x}$$

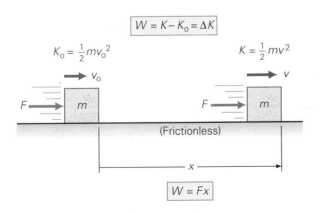

●FIGURE 5.8 **The relationship of work and kinetic energy**
The work done on a block in moving along a horizontal frictionless surface is equal to the change in its kinetic energy, $W = \Delta K$.

where v_o may or may not be zero. Writing the force in its Newton's second law form ($F = ma$) and then substituting the expression for a from the previous equation gives

$$F = ma = m\left(\frac{v^2 - v_o{}^2}{2x}\right)$$

Using this expression in the equation for work, we get

$$W = Fx = m\left(\frac{v^2 - v_o{}^2}{2x}\right)x$$
$$= \tfrac{1}{2}mv^2 - \tfrac{1}{2}mv_o{}^2$$

In terms of kinetic energy, then,

$$W = \tfrac{1}{2}mv^2 - \tfrac{1}{2}mv_o{}^2 = K - K_o = \Delta K$$

or

$$W_{net} = \Delta K \qquad (5.6)$$

Work-energy theorem

Work and energy

(a)

(b)

•**FIGURE 5.9 Kinetic energy and work**
(a) A moving object, such as a wrecking ball, possesses kinetic energy and so can do work. (b) A great deal of kinetic energy was converted to work when a large meteorite made this crater in Arizona, perhaps 10,000 years ago. The crater is 1200 m in diameter and 180 m deep.

This equation is called the **work-energy theorem** and relates the net work done on an object to the change in its kinetic energy. That is, *the net work done on a body by an external force is equal to the change in kinetic energy of the body.* Work and energy both have the units of joules, and are both *scalar* quantities.

Keep in mind that the work-energy theorem is true in general and not just for the special case considered in deriving Eq. 5.6. As a concrete example, recall that in Example 5.1 the force of gravity did +44 J of work on a book that fell from rest through a distance of $y = 3.0$ m. At that position and instant, the falling book had 44 J of kinetic energy. This can be easily shown. Since $v_o = 0$ in this case, $v^2 = 2gy$, and $gy = v^2/2$. Substituting this expression in the equation for the work done on the falling book, we get

$$W = Fd = mgy = \frac{mv^2}{2} = K$$

where $K_o = 0$. Thus the kinetic energy gained by the book is equal to the net work done on it: 44 J in this instance. (As an exercise, confirm this fact by calculating the speed of the book and evaluating its kinetic energy.)

As you can see from the work-energy theorem, kinetic energy has the same units as work. You should also note that, like work, kinetic energy is a scalar quantity.

What the work-energy theorem tells us is that when work is done, there is change or a transfer of energy. In general then, we might say that *work is a measure of the transfer of kinetic energy.* For example, a force doing work on an object that causes it to speed up gives rise to an increase in the object's kinetic energy. On the other hand, work done by the force of kinetic friction may cause a moving object to slow down and decrease its kinetic energy. So, for a change in or transfer of kinetic energy, there must be a net force and net work, as Eq. 5.6 tells us.

When an object is in motion, it possesses kinetic energy and has the capability to do work. For example, a moving automobile has kinetic energy and can do work in crumpling a fender in a fender-bender—not *useful* work in that case, but still work. Other examples are shown in •Fig. 5.9.

EXAMPLE 5.5 ■ A GAME OF SHUFFLEBOARD: THE WORK-ENERGY THEOREM

A shuffleboard player (•Fig. 5.10) pushes a 0.25-kg puck, initially at rest, in a way that causes a constant horizontal force of 6.0 N to act on it through a distance of 0.50 m. (Neglect friction.)

(a) What are the kinetic energy and the speed of the puck when the force is removed?
(b) How much work would be required to bring the puck to rest?

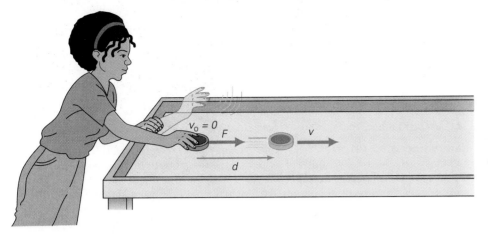

● FIGURE 5.10 Work and kinetic energy
See Example 5.5.

Solution. Listing the given data as usual, we have

Given: $m = 0.25$ kg   **Find:**   (a) K (kinetic energy)
        $F = 6.0$ N v (speed)
        $d = 0.50$ m   (b) W (work done in stopping puck)
        $v_o = 0$

(a) Since the speed is not known, we cannot compute the kinetic energy ($K = \frac{1}{2}mv^2$) directly. However, kinetic energy is related to *work* by the work-energy theorem. The work done on the puck is

$$W = Fd = (6.0 \text{ N})(0.50 \text{ m}) = +3.0 \text{ J}$$

Then, by the work-energy theorem,

$$W = \Delta K = K - K_o = 3.0 \text{ J}$$

But $K_o = \frac{1}{2}mv_o^2 = 0$, because $v_o = 0$, so

$$K = 3.0 \text{ J}$$

The speed may be found from the kinetic energy. Since $K = \frac{1}{2}mv^2$, we have

$$v = \sqrt{\frac{2K}{m}} = \sqrt{\frac{2(3.0 \text{ J})}{0.25 \text{ kg}}} = 4.9 \text{ m/s}$$

(b) As you might guess, the work required to bring the puck to rest is equal to its kinetic energy (the amount of energy that must be "removed" from the puck to stop its motion). To confirm this we essentially perform the reverse of the previous calculation, with $v_o = 4.9$ m/s and $v = 0$.

$$W = K - K_o = 0 - K_o = -\tfrac{1}{2}mv_o^2$$
$$= -\tfrac{1}{2}(0.25 \text{ kg})(4.9 \text{ m/s})^2 = -3.0 \text{ J}$$

The minus sign indicates that the puck loses energy as it slows down. The work is done *against* the motion of the puck; that is, the opposing force is in a direction opposite that of the motion. (In a real situation, the opposing force would be friction.)

Follow-up Exercise. Suppose the puck in this Example had twice the final speed. Would it then take twice as much work to stop the puck? (*Answer may be found in the Answers to Follow-up Exercises section at the back of the book.*)

PROBLEM-SOLVING HINT

Notice how work-energy considerations were used to find speed in Example 5.5. This could be done in another way. First, the acceleration could be found from $a = F/m$, and then the kinematic equation $v^2 = v_o^2 + 2ax$ could be used to find v (where $x = d$ in the equation). The point is that problems often can be solved in different ways, and finding the fastest and most efficient way is often the key to success. As our discussion of energy progresses, you will see how useful and powerful the notions of work and energy are, both as theoretical concepts and as practical tools for solving many kinds of problems.

CONCEPTUAL EXAMPLE 5.6 ■ KINETIC ENERGY: MASS VERSUS SPEED

In a football game, a 140-kg guard runs with a speed of 4.0 m/s and a 70-kg free safety moves at 8.0 m/s. Then, (a) both players have the same kinetic energy, (b) the safety has twice as much kinetic energy as the guard, (c) the guard has twice as much kinetic energy as the safety, (d) the safety has four times as much kinetic energy as the guard. *Clearly establish the reasoning and physical principle(s) used in determining your answer before checking it below. That is, **why** did you select your answer?*

Reasoning and Answer. The kinetic energy of a body depends on both its mass and its speed. You might think that, with half the mass but twice the speed, the safety would have the same kinetic energy as the guard, but this is not the case. Kinetic energy, $K = \frac{1}{2}mv^2$, is directly proportional to the mass, but proportional to the *square* of the speed. Thus, halving the mass decreases the kinetic energy by a factor of two; so if the two athletes had equal speeds, the safety would have half as much kinetic energy as the guard.

However, doubling the speed increases the kinetic energy, not by a factor of 2, but by a factor of 2^2 or 4. Thus, the safety, with half the mass but twice the speed, would have $\frac{1}{2} \times 4 = 2$ times as much kinetic energy as the guard, and the answer is (b).

Note that to answer this question it was not necessary actually to calculate the kinetic energy of each player. We can do so, however, to check our conclusions:

$$K_s = \tfrac{1}{2}m_s v_s^2 = \tfrac{1}{2}(70 \text{ kg})(8.0 \text{ m/s})^2 = 2.2 \times 10^3 \text{ J}$$

$$K_g = \tfrac{1}{2}m_g v_g^2 = \tfrac{1}{2}(140 \text{ kg})(4.0 \text{ m/s})^2 = 1.1 \times 10^3 \text{ J}$$

Thus we see explicitly that our answer was correct.

Follow-up Exercise. Suppose that the safety's speed were only 50 percent greater than the guard's, or 6.0 m/s. Which athlete would then have the greater kinetic energy, and how many times greater? *(Reasoning and answer may be found in the Answers to Follow-up Exercises section at the back of the book.)*

PROBLEM-SOLVING HINT

Note that the work-energy theorem relates the work done to the *change* in the kinetic energy. Often we have $v_o = 0$, and $K_o = 0$, so $W = \Delta K = K$. But take care! You *cannot* simply use the square of the change in velocity $(\Delta v)^2$ to calculate ΔK, as you might at first think. In terms of velocity, we have

$$W = \Delta K = K - K_o = \tfrac{1}{2}mv^2 - \tfrac{1}{2}mv_o^2 = \tfrac{1}{2}m(v^2 - v_o^2)$$

Note that $v^2 - v_o^2$ *not* the same as $(v - v_o)^2 = (\Delta v)^2$, since $(v - v_o)^2 = v^2 - 2vv_o + v_o^2$, so the work or change in kinetic energy is **not** equal to $\tfrac{1}{2}m(v - v_o)^2 = \tfrac{1}{2}m(\Delta v)^2$.

What this means is that you must compute the kinetic energy of an object at one point or time (using the instantaneous velocity to get the instantaneous kinetic energy) and also at another point or time. Then the quantities are subtracted to find the change in kinetic energy, or the work. Alternatively, you can find the difference of the *squares* of the velocities $(v^2 - v_o^2)$ first in computing the change.

CONCEPTUAL EXAMPLE 5.7 ■ AN ACCELERATING CAR: SPEED AND KINETIC ENERGY

A car traveling at 5.0 m/s speeds up to 10 m/s, an increase in kinetic energy that requires work W_1. Then, the speed is increased from 10 m/s to 15 m/s, requiring work W_2. How do the amounts of work compare? (a) $W_1 > W_2$, (b) $W_1 = W_2$, (c) $W_2 > W_1$. *Clearly establish the reasoning and physical principle(s) used in determining your answer before checking it below. That is, **why** did you select your answer?*

Reasoning and Answer. As noted above, the work-energy theorem relates the work done to the *change* in the kinetic energy. Since the speeds have the same increment in each case ($\Delta v = 5.0$ m/s), it might appear that (b) would be the answer. However, we need to keep in mind the relationships that we explored in the previous Example: the kinetic energy of an object is directly proportional to its mass, but proportional to the *square* of its speed. So we would expect that going from 5.0 m/s to 10 m/s, a change in speed of 100 percent, would entail a greater increase in kinetic energy than in going from 10 m/s to 15 m/s, which represents a smaller fractional change—only a 50 percent increase. Thus, we would expect the answer to be (c).

We can confirm this conclusion by calculating the actual changes in the kinetic energy in each of these cases:

$$W_1 = \tfrac{1}{2}m(v^2 - v_o^2) = \tfrac{1}{2}m[(10 \text{ m/s})^2 - (5.0 \text{ m/s})^2] = \tfrac{1}{2}m(75 \text{ m}^2/\text{s}^2)$$

Similarly,

$$W_2 = \tfrac{1}{2}m(v^2 - v_o^2) = \tfrac{1}{2}m[(15 \text{ m/s})^2 - (10 \text{ m/s})^2] = \tfrac{1}{2}m(125 \text{ m}^2/\text{s}^2)$$

Note that $W_2/W_1 = 125/75 = 1.67$, so W_2 is 67% larger than W_1.

Follow-up Exercise. Suppose the car is speeded up a third time, from 15 m/s to 20 m/s, a change requiring work W_3. How does the work done in this increment compare with W_2? Would you expect a similar percentage increase? *(Reasoning and answer may be found in the Answers to Follow-up Exercises section at the back of the book.)*

5.4 Potential Energy

Objectives: **To be able to (a) explain how potential energy depends on position, and (b) compute values of gravitational potential energy.**

An object in motion has kinetic energy. However, whether an object is in motion or not, it may have another form of energy—potential energy. As the name implies, an object having potential energy has the *potential* to do work. You can probably think of many examples: a compressed spring, a drawn bow, water held back by a dam, a wrecking ball poised to drop. In all such cases, the potential to do work derives from the position or configuration of bodies. The spring has energy because it is compressed, the bow because it is drawn, the water and the ball because they have been lifted above the surface of the earth (•Fig. 5.11). Consequently, **potential energy** (U) is often called the energy of position (and/or configuration).

In a sense, potential energy can be thought of as stored work. You have already seen an example of potential energy in Section 5.2 when work was done in compressing a spring. Recall that the work done in such a case is $W = \tfrac{1}{2}kx^2$ (with $x_o = 0$). Note that the amount of work done depends on the amount of compression (x). Because work is done, there is a *change* of position and a *change* in potential energy (ΔU), which is equal to the work done *by the applied force* in compressing the spring:

$$W = \Delta U = U - U_o = \tfrac{1}{2}kx^2 - \tfrac{1}{2}kx_o^2$$

Thus, with $x_o = 0$ and $U_o = 0$, as they are commonly taken to be, the *potential energy of a spring* is

$$U = \tfrac{1}{2}kx^2 \qquad (5.7)$$

SI unit of energy: joule (J)

[Since the potential energy has x^2 as one term, the previous problem-solving hint applies when $x_o \neq 0$; that is, $x^2 - x_o^2 \neq (x - x_o)^2$.]

Perhaps the most common type of potential energy is **gravitational potential energy**. In this case, position refers to the height of an object above some reference point such as the floor or the ground. Suppose that an object of mass m is lifted a distance Δh (•Fig. 5.12). Work is done against the force of gravity, and an applied force at least equal to the object's weight is necessary to lift it, $F = w = mg$. The work done in lifting is then equal to the change in potential energy. Expressed in equation form, since there is no overall change in kinetic energy we have

work = change in potential energy

(a)

(b)

•**FIGURE 5.11 Potential energy**
Potential energy has many forms. **(a)** Work must be done to bend the bow, giving it potential energy. That energy is converted into kinetic energy when the arrow is released. **(b)** Gravitational potential energy is converted into kinetic energy when an object falls. (Where did the gravitational potential energy of the water and the diver come from?)

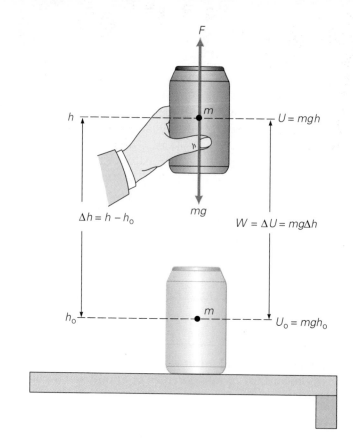

FIGURE 5.12 Gravitational potential energy
The work done in lifting an object is equal to the change in gravitational potential energy, $W = F\Delta h = mg(h - h_o)$.

or

$$W = F\Delta h = \Delta U = U - U_o = mgh - mgh_o$$

And, with the common choices of $h_o = 0$ and $U_o = 0$, the gravitational potential energy is

Gravitational potential energy

$$\boxed{U = mgh} \qquad \textit{gravitational potential energy} \qquad (5.8)$$

SI unit of energy: joule (J)

Because we commonly take upward to be the $+y$ direction, Eq. 5.8 is often written $U = mgy$. (Eq. 5.8 is the gravitational potential energy on or near the Earth's surface where g is considered to be constant. A more general form of gravitational potential energy is given in Section 7.5.)

As can be seen, a force and work are associated with a change in potential energy. Basically, the *change* in potential energy associated with a force (ΔU_F) is given by the *negative of the work* done by the force ($-W_F$) during the change, that is, $\Delta U_F = -W_F$. For example, when a spring is compressed, the spring force and displacement are in opposite directions. The work the force does is negative, so the change in potential energy is positive $[-(-W_F) = +\Delta U_F]$. Similarly, for an object in free fall downward, the gravitational force and displacement are in the same direction, so the force of gravity does positive work on the object, and the change in potential energy is negative (potential energy lost), since $-(+W_F) = -\Delta U_F$.

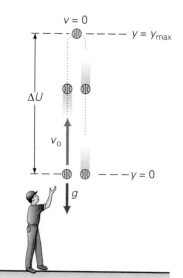

FIGURE 5.13 Kinetic and potential energies
See Example 5.8. (The ball is displaced sideways for clarity.)

EXAMPLE 5.8 ■ A THROWN BALL: KINETIC ENERGY AND GRAVITATIONAL POTENTIAL ENERGY

A 0.50-kg ball is thrown vertically upward with an initial velocity of 10 m/s (see Fig. 5.13). (a) What is the change in the ball's kinetic energy between the starting point and its maximum height? (b) What is the change in the ball's potential energy between its maximum height and the launch point? (Neglect air resistance.)

Solution. Studying Fig. 5.13 and listing the data, we have

Given: $m = 0.50$ kg   Find: (a) ΔK (change in kinetic energy)
$\qquad v_o = 10$ m/s $\qquad\qquad$ (b) ΔU (the change in potential energy
$\qquad a = g \qquad\qquad\qquad\qquad$ between y_o and y_{max})

(a) To find the *change* in kinetic energy, we first compute the kinetic energy at each point. We know the initial velocity v_o, and at the maximum height, $v = 0$, so $K = 0$. Thus,

$$\Delta K = K - K_o = 0 - K_o = -\tfrac{1}{2}mv_o^2 = -\tfrac{1}{2}(0.50 \text{ kg})(10 \text{ m/s})^2 = -25 \text{ J}$$

That is, the ball loses 25 J of kinetic energy as negative work is done on it by the force of gravity (force and displacement are in opposite directions).

(b) To find the change in potential energy we need to know the ball's height above its starting point when $v = 0$. Using the equation $v^2 = v_o^2 - 2gy$ to find y_{max},

$$y_{max} = \frac{v_o^2}{2g} = \frac{(10 \text{ m/s})^2}{2(9.8 \text{ m/s}^2)} = 5.1 \text{ m}$$

Then, with $y_o = 0$ and $U_o = 0$, $\Delta U = U - U_o = U - 0$, and

$$\Delta U = U = mgy_{max} = (0.50 \text{ kg})(9.8 \text{ m/s}^2)(5.1 \text{ m}) = +25 \text{ J}$$

The potential energy increases by 25 J, as might be expected. Notice that this is the change in potential energy with respect to the release point, which was taken as the zero reference point ($y_o = 0$).

Follow-up Exercise. In this Example, what are the total changes in the ball's kinetic and potential energies when it returns to the starting point? (*Answers may be found in the Answers to Follow-up Exercises section at the back of the book.*)

Zero Reference Point

An important point is illustrated in the preceding Example, namely, the choice of a zero reference point. Potential energy is the energy of *position*, and the potential energy at a particular position (U) is referenced to the potential energy at some other position (U_o). The reference position or point is arbitrary, as is the origin of a set of coordinate axes for analyzing a system. It is usually chosen with convenience in mind, for example, $y_o = 0$ or $h_o = 0$. The value of the potential energy at a particular position depends on the reference point used. However, the *difference or change in potential energy associated with two positions is the same regardless of the reference position.*

In Example 5.8 if ground level had been taken as the zero reference point, then U_o at the release point would not have been zero. However, U at the maximum height would have been greater and $\Delta U = U - U_o$ would have been the same. This concept is illustrated in •Fig. 5.14. Note that the potential energy can be negative. When an object has a negative potential energy, it is said to be in a potential energy *well*, which is analogous to being in an actual well. Work is needed to raise the object to a higher position in the well or to get it out of the well.

Also, for gravitational potential energy, the path by which an object is raised (or lowered) makes no difference (see •Fig. 5.15). That is, *the change in gravitational potential energy is independent of path*. As illustrated in the figure, an object raised to a height h has a change in potential energy of $\Delta U = mgh$ whether it is lifted vertically or moved along an inclined plane. This is because the force of gravity always acts downward and only vertical displacement (or vertical component) is involved in doing work against gravity and changing the potential energy.

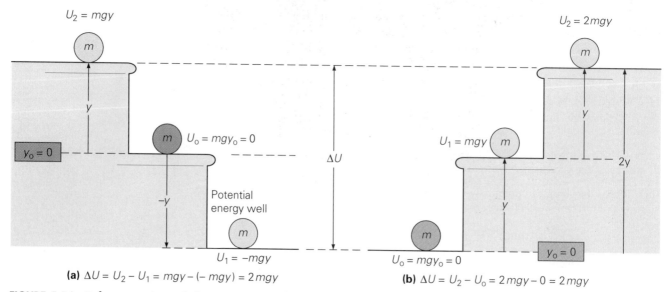

(a) $\Delta U = U_2 - U_1 = mgy - (-mgy) = 2mgy$

(b) $\Delta U = U_2 - U_0 = 2mgy - 0 = 2mgy$

•**FIGURE 5.14   Reference point and change in potential energy**
(a) The choice of a reference point (zero height) is arbitrary and may give rise to a negative potential energy. An object is said to be in a potential energy well in this case. **(b)** The well may be avoided by selecting a new zero reference. Note that the difference, or *change*, in potential energy (ΔU) associated with the two positions is the same regardless of the reference point. There is no physical difference even though there are two coordinate systems.

•**FIGURE 5.15   Path-independence of the change in gravitational potential energy**
When the crate is resting on the table, it has the same potential energy relative to the floor regardless of how it got there. Only the vertical component of the force that moves the crate does work against gravity in lifting it vertically or moving it up an inclined plane or ramp.

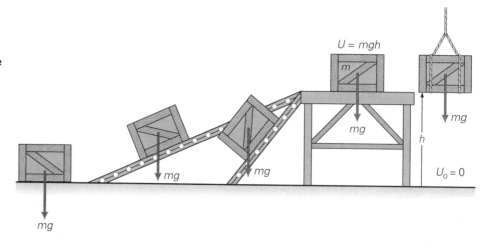

5.5   The Conservation of Energy

Objectives: To be able to (a) distinguish between conservative and nonconservative forces, and (b) explain their effects on the conservation of energy.

Conservation laws are the cornerstones of physics, both theoretically and practically. Most scientists would probably name the conservation of energy as the most profound and far-reaching of these important laws. When we say something is *conserved*, we mean that it is constant, or has a constant value. Because so many things continually change in physical processes, conserved quantities are extremely helpful in our attempts to understand and describe the universe. Of course, quantities are generally conserved only under special conditions. But even so, such quantities are given special status by being made the subject of conservation laws.

One of the most important conservation laws is that concerning conservation of energy. (You may have seen this coming in Example 5.8.) A familiar statement is that the total energy of the universe is conserved. This is true because the whole universe is taken to be a system. A system is defined as a definite quantity of matter enclosed by boundaries, either real or imaginary. In effect, the universe is the largest possible closed, or isolated, system we can imagine. Within a closed system, particles may interact with each other but have absolutely no interaction with anything outside. In general then, the amount of energy remains constant when no mechanical work is done on or by the system, and no energy is transmitted to or from the system (including thermal energy and radiation).

Thus, the **law of conservation of total energy** may be stated:

> The total energy of an isolated system is always conserved.

Conservation of total energy

Within such a system, energy may be converted from one form to another, but the total amount of all forms of energy is constant or unchanged.

Conservative and Nonconservative Forces

A general distinction among systems may be made by considering two categories of forces that may act within them: conservative and nonconservative forces. You have already been introduced to a couple of conservative forces: the force due to gravity and the spring force. A classic nonconservative force, friction, was considered in Chapter 4. Here's the distinction:

Note: Friction was discussed in Section 4.6.

> A force is said to be conservative if the work done by or against it in moving an object is independent of the object's path.

What this means is that the work done by a **conservative force** depends only on the initial and final positions of an object.

Conservative force—work independent of path

The concept of conservative and nonconservative forces is sometimes difficult to comprehend at first. Because this concept is so important in the conservation of energy, let's consider some illustrative examples to increase our understanding.

First, what does *independent of path* mean? An example was given in Fig. 5.15, where work was done against the conservative force of gravity. The figure illustrates that the work done in moving a crate to a table does not depend on the path, but only on the initial and final positions. The magnitude of the work done is equal to the change in potential energy, and in fact, the concept of potential energy is associated only with conservative forces. A change in potential energy may be defined in terms of the work done by a conservative force.

On the other hand work done against a **nonconservative force** such as friction *does* depend on path. Given the same frictional conditions, it would take more work (against friction) to move the crate up the longer ramp, so the work would depend on path—the longer the path, the more work done. It should therefore be evident that the *total* work done is not equal to a change in potential energy. Some work was done against a nonconservative force, in addition to the conservative gravitational force. In this case, the energy associated with the work done against friction would be converted to heat energy. Hence, in a sense, a conservative force allows you to conserve or store all of the energy as potential energy, whereas a nonconservative force does not.

To further highlight this idea, consider moving the crate around the table top. Here, work is done only against the nonconservative frictional force (why?). Certainly the work done in moving the crate between two points depends on the path taken. It would be greater, for example, if we moved the crate around the table before coming to the destination point than if we took a straight-line path.

Another approach to help you understand the distinction between conservative and nonconservative forces is through an equivalent statement of the previous definition:

> A force is conservative if the work done by or against it in moving an object through a round trip is zero.

Another way of describing a conservative force

Consider a book resting on a table. It has gravitational potential energy $U = mgh$, relative to some reference point $h_o = 0$. You could drop it on the floor, pick it up, and place it back at its original position; or you could pick it up from the table and carry it around with you all day, then place it back at its original position. Both are round trips, and the potential energy of the book is the same when it is returned to its original position. Thus, the change in its potential energy or the work done by the conservative force of gravity is $\Delta U = W = 0$. However, if you pushed the book around on the tabletop and eventually back to its original position, the work done against the nonconservative force of friction would depend on the path (the longer the path, the more work done), and the work done would not be stored but lost as heat and sound.

Notice that for the conservative gravitational force, the force and displacement are sometimes in the same direction (in which case positive work is done) and sometimes in opposite directions (in which case negative work is done against the force) during a round trip. Think of the simple case of the book falling to the floor and being placed back on the table. With positive and negative work, the total can be zero. However, for a nonconservative force of kinetic friction, which always opposes the motion or is in the opposite direction to the displacement, the total work done in a round trip can *never* be zero, and is always negative (energy lost).

Conservation of Total Mechanical Energy

The idea of a conservative force allows us to extend the conservation of energy to the special case of mechanical energy, which greatly helps us to better analyze many physical situations. The sum of the kinetic and potential energies is called the **total mechanical energy**:

$$\underset{\substack{total \\ mechanical \\ energy}}{E} = \underset{\substack{kinetic \\ energy}}{K} + \underset{\substack{potential \\ energy}}{U} \tag{5.9}$$

For a **conservative system** (one in which only conservative forces do work), the total mechanical energy is constant, or conserved; that is,

$$E = E_o$$
$$K + U = K_o + U_o$$

or

$$\tfrac{1}{2}mv^2 + U = \tfrac{1}{2}mv_o^2 + U_o \tag{5.10}$$

This is a mathematical statement of the **law of the conservation of mechanical energy**:

In a conservative system, the sum of the kinetic and potential energies is constant and equals the total mechanical energy of the system.

The kinetic and potential energies in a conservative system may change, but their sum is always constant. This is illustrated by the diagram in •Fig. 5.16a.

Notice for a conservative system, when work is done and energy is transferred, we have

$$\underset{initial\ energy}{K_o + U_o} = \underset{final\ energy}{K + U}$$

This equation can be rewritten as

$$K - K_o = -(U - U_o) \tag{5.11}$$

or

$$\Delta K = -\Delta U \qquad \text{(only when nonconservative forces do no work)}$$

What this expression tells us is that these quantities are related in see-saw fashion: If there is a decrease in potential energy $(-\Delta U)$, then the kinetic energy in-

Mechanical energy—kinetic plus potential

Conservation of mechanical energy

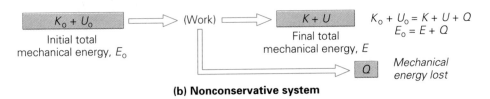

$K_0 + U_0$	$\Longrightarrow$ (Work) $\Longrightarrow$	$K + U$	$K_0 + U_0 = K + U$

Initial total mechanical energy, E_0 Final total mechanical energy, E $E_0 = E$

(a) Conservative system

| $K_0 + U_0$ | $\Longrightarrow$ (Work) $\Longrightarrow$ | $K + U$ | $K_0 + U_0 = K + U + Q$ |

Initial total mechanical energy, E_0 Final total mechanical energy, E $E_0 = E + Q$

Q Mechanical energy lost

(b) Nonconservative system

•**FIGURE 5.16 Conservative and nonconservative systems**
(a) The work done by a conservative force exchanges energy between kinetic and potential forms: that is, $\Delta K = -\Delta U$, or $K - K_0 = -(U - U_0)$. The total mechanical energy is conserved ($E = E_0$). **(b)** Part of the work done by a nonconservative force does not go into mechanical energy exchange, but is lost to friction or some other cause. The mechanical energy is not conserved in this case.

creases ($+\Delta K$) by an equal amount, and vice versa ($-\Delta K = \Delta U$). However, in a nonconservative system, some of the mechanical energy is lost (for example, to the heat of friction). The following examples illustrate the conservation of mechanical energy for conservative systems.

EXAMPLE 5.9 ■ LOOK OUT BELOW! CONSERVATION OF MECHANICAL ENERGY

A painter on a scaffold drops a 1.50-kg can of paint from a height of 6.00 m. (a) What is the kinetic energy of the can when it is at a height of 4.00 m? (b) With what speed will the can hit the ground? (Neglect air resistance.)

Solution. Listing what is given and what we are to find, we have

Given: $m = 1.50$ kg   *Find:*   (a) K (kinetic energy)
        $h_0 = 6.00$ m   (b) v (speed hitting the ground)
        $h = 4.00$ m
        ($v_0 = 0$)

(a) First, it is convenient to find the can's total mechanical energy, since this quantity is conserved while it is falling (why?). Initially, with $v_0 = 0$, the can's total mechanical energy is all potential. Taking the ground as the zero reference point, we have

$$E = K + U = 0 + mgh_0 = (1.50 \text{ kg})(9.80 \text{ m/s}^2)(6.00 \text{ m}) = 88.2 \text{ J}$$

The relation $E = K + U$ continues to hold while the can is falling, but now we know what E is. Rearranging the equation, we have $K = E - U$, and finding U at $h = 4.00$ m:

$$K = E - U = E - mgh = 88.2 \text{ J} - (1.50 \text{ kg})(9.80 \text{ m/s}^2)(4.00 \text{ m})$$
$$= 29.4 \text{ J}$$

Alternatively, we could have computed the change in (in this case, the loss of) potential energy, ΔU. This is equal to the negative of the change in (gain of) kinetic energy, ΔK (Eq. 5.11). Then,

$$\Delta K = -\Delta U$$
$$K - K_0 = -(U - U_0) = -(mgh - mgh_0)$$

and with $K_0 = 0$ (since $v_0 = 0$),

$$K = mgh(h_0 - h) = (1.50 \text{ kg})(9.8 \text{ m/s}^2)(6.00 \text{ m} - 4.00 \text{ m})$$
$$= 29.4 \text{ J}$$

(b) Just before the can strikes the ground ($h = 0$, $U = 0$), the total mechanical energy is all kinetic, or

$$E = K = \tfrac{1}{2}mv^2$$

Thus,

$$v = \sqrt{\frac{2E}{m}} = \sqrt{\frac{2(88.2 \text{ J})}{1.50 \text{ kg}}} = 10.8 \text{ m/s}$$

Basically, all of the potential energy of a free-falling object released from some height h is converted into kinetic energy just before it hits the ground, so

$$|\Delta K| = |\Delta U|$$
$$\tfrac{1}{2}mv^2 = mgh$$

or

$$v = \sqrt{2gh}$$

Note that the mass cancels and is not a consideration. This result is also obtained from the kinematic equation $v^2 = v_o^2 + 2gy$ (Eq. 2.9′), with $v_o = 0$ and $y = h$.

Follow-up Exercise. A fellow painter on the ground wishes to toss a paintbrush vertically upward a distance of 5.0 m to his partner on the scaffold. Use conservation of mechanical energy methods to determine the minimum speed which he must give to the brush. (*Answer may be found in the Answers to Follow-up Exercises section at the back of the book.*)

CONCEPTUAL EXAMPLE 5.10 ■ A MATTER OF DIRECTION? SPEED AND CONSERVATION OF ENERGY

Three balls of equal mass m are projected with the same speed in different directions as shown in ●Fig. 5.17. If air resistance is neglected, which ball would you expect to strike the ground with the greatest speed? (a) ball 1, (b) ball 2, (c) ball 3, (d) all strike with the same speed. *Clearly establish the reasoning and physical principle(s) used in determining your answer before checking it below. That is, **why** did you select your answer?*

Reasoning and Answer. All of the balls have the same initial kinetic energy, $K_o = \tfrac{1}{2}mv_o^2$. (Recall that energy is a scalar quantity, and the different directions of projections do not produce any difference in the kinetic energies.) Regardless of their trajectories, all of the balls ultimately descend a distance h relative to their common starting point, so they all lose the same amount of potential energy. (U is energy of *position*, and thus *independent* of path—see Fig. 5.15.)

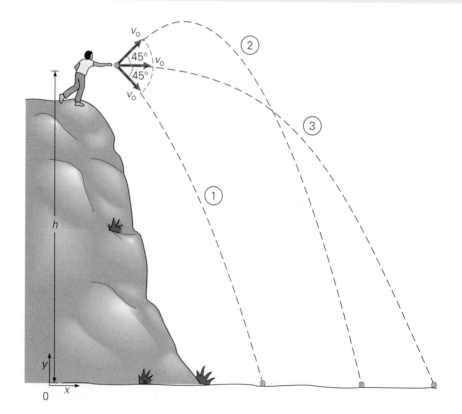

●**FIGURE 5.17 Speed and energy**
See Conceptual Example 5.10.

By the law of conservation of energy, the amount of potential energy each ball loses is equal to the amount of kinetic energy it gains. Since all balls start with the same amount of kinetic energy and gain the same amount of kinetic energy, all three will have equal kinetic energies just before striking the ground. This means that their speeds must be equal, so the answer is (d). Note that although balls 1 and 2 are projected at 45° angles, this is not a factor. Since the change in potential energy is independent of path, it is independent of the projection angle. The vertical distance between the starting point and the ground is the same (h) for projectiles at any angle.

Follow-up Exercise. Would the balls strike the ground with different speeds if their masses were different? *(Reasoning and answer may be found in the Answers to Follow-up Exercises section at the back of the book.)*

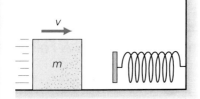

•FIGURE 5.18 Conservative force and the mechanical energy of a spring
See Example 5.11.

EXAMPLE 5.11 ■ CONSERVATIVE FORCES: MECHANICAL ENERGY OF A SPRING

A 0.30-kg mass sliding on a horizontal frictionless surface with a speed of 2.5 m/s, as depicted in •Fig. 5.18, strikes a light spring, which has a spring constant of 3.0×10^3 N/m. (a) What is the total mechanical energy of the system? (b) What is the kinetic energy (K_1) of the mass when the spring is compressed a distance $x_1 = 1.0$ cm? (Assume that no energy is lost in the course of the collision.)

Solution.

Given: $m = 0.30$ kg   *Find:*   (a) E (total mechanical energy)
    $v_o = 2.5$ m/s   (b) K_1 (kinetic energy)
    $k = 3.0 \times 10^3$ N/m
    $x_1 = 1.0$ cm $= 0.010$ m

(a) Before the mass makes contact with the spring, the total mechanical energy of the system is all in the form of kinetic energy, and

$$E = K_o = \tfrac{1}{2}mv_o^2 = \tfrac{1}{2}(0.30 \text{ kg})(2.5 \text{ m/s})^2 = 0.94 \text{ J}$$

Since the system is conservative, this is the total energy at any time.

(b) When the spring is compressed a distance x_1, it has potential energy $U_1 = \tfrac{1}{2}kx_1^2$, and

$$E = K_1 + U_1 = K_1 + \tfrac{1}{2}kx_1^2$$

Solving for K_1, we have

$$K_1 = E - \tfrac{1}{2}kx_1^2$$
$$= 0.94 \text{ J} - \tfrac{1}{2}(3.0 \times 10^3 \text{ N/m})(0.010 \text{ m})^2 = 0.94 \text{ J} - 0.15 \text{ J} = 0.79 \text{ J}$$

Follow-up Exercise. How far will the spring in the preceding example be compressed when the mass comes to a stop? *(Answer may be found in the Answers to Follow-up Exercises section at the back of the book.)*

Total Energy and Nonconservative Forces

In the preceding examples, the force of friction, which is probably the most common nonconservative force, was ignored. In general, both conservative and nonconservative forces do work on objects. However, as you know, when nonconservative forces do work, the mechanical energy is not conserved. Mechanical energy is "lost" through the work done by a nonconservative force such as friction.

It might be thought that we can no longer use an energy approach to analyze problems involving nonconservative forces, since mechanical energy can be lost or dissipated (•Fig. 5.19). However, in some instances we can use the conservation of total energy to find out how much energy was lost to the work done by a nonconservative force. Suppose an object initially has mechanical energy and that noncon-

•FIGURE 5.19 Nonconservative force and energy loss
Friction is a nonconservative force—when it is present, mechanical energy is not necessarily conserved. Can you tell from the photo what is happening to the work being done by the motor on the grinding wheel after it is converted into rotational kinetic energy?

• FIGURE 5.29 A one-horse open sleigh
See Exercise 88.

Additional Exercises

89 An electron ($m = 9.11 \times 10^{-31}$ kg) has 8.00×10^{-17} J of kinetic energy. What is the speed of the electron?

90 A large electric motor with an efficiency of 75% has a power output of 1.5 hp. If the motor is run steadily for 2.0 h and the cost of electricity is 12¢/kWh, how much does it cost to run the motor for this period?

91 A 28-kg child slides down a playground slide from a height of 3.0 m above the bottom of the slide. If her speed at the bottom is 2.5 m/s, how much energy is lost to friction?

92 An automobile is traveling at 90 km/h on level ground and then coasts up a hill. (a) Neglecting friction, how high above the level ground will the car go before stopping? (b) Assuming that 30% of the energy is lost to friction, how high will the car go?

93 A 1500-kg helicopter accelerates vertically upward from rest at a constant rate of 2.0 m/s² for a distance of 30 m. (a) What is the net work done on the helicopter? (b) How much work is done by the helicopter motor and rotor?

94 In planing a piece of wood 35 cm long, a carpenter applies a force of 40 N to a jack plane at a downward angle of 25° to the horizontal. How much work is done by the carpenter?

95 A 120-kg sleigh is pulled by one horse at a constant velocity for a distance of 0.75 km on a level snowy surface. The coefficient of kinetic friction between the sleigh runners and the snow is 0.25. (a) Calculate the work done by the horse. (b) Calculate the work done by friction.

96 A tractor pulls a wagon with a constant force of 700 N, giving the wagon a constant velocity of 20.0 km/h. (a) How much work is done by the tractor in 3.50 min? (b) What is the tractor's power output?

97 A sports car weighs a third as much as a large luxury car. (a) If the sports car is traveling at a speed of 90 km/h, at what speed would the larger car have to travel to have the same kinetic energy? (b) Suppose that the large car is traveling at a speed that gives it half the kinetic energy of the sports car. What is the speed of the large car?

98 A ball with a mass of 0.360 kg is dropped from a height of 1.20 m above the top of a fixed vertical spring, whose force constant is 350 N/m. (a) What is the maximum distance the spring is compressed by the ball? (Neglect energy loss due to the collision.) (b) What is the speed of the ball when the spring has been compressed 5.00 cm?

99 A roller coaster starts from rest at a point 45 m above the bottom of a dip. Neglecting friction, what will be the speed of the roller coaster at the top of the next slope, which is 30 m above the bottom of the dip?

100 How much work does a 1000-W motor do in 5.0 min if it has an efficiency of 90%?

101 On a snowy day when school has been canceled, a student on a sled starts from rest at a vertical height of 25 m above the horizontal base of the hill and slides down. If the sled and the student have a speed of 20 m/s at the bottom of the hill, is mechanical energy conserved? Justify your answer.

102 A constant horizontal force of 30 N moves a box along a rough surface with a constant speed. If the force does work at a rate of 50 W, (a) what is the box's speed? (b) How much work is done by the force in 2.5 s?

103 A 70-kg student who is late for a class runs up two flights of stairs whose combined vertical height is 7.0 m in 10 s (and then silently enters the classroom while the professor is writing on the blackboard). Compute the student's power output in doing work against gravity in (a) watts and (b) horsepower.

104 In Example 5.4, a spring is stretched a distance x_1 by a mass m_1, and then an additional mass m_2 stretches the spring an additional distance x_2, such that the total stretching distance $x = x_1 + x_2$. Show that the total work W done in stretching the spring is *not* equal to $W_1 + W_2$, that is $W \neq W_1 + W_2$ or $\frac{1}{2}kx^2 \neq \frac{1}{2}kx_1^2 + \frac{1}{2}kx_2^2$. How would you calculate W_2 explicitly?

105 A relaxed spring with a spring constant of 50 N/m is stretched 3.0 cm to x_1, and then stretched an additional 3.0 cm to x_2. What is the work done in stretching the spring from x_1 to x_2?

6 Momentum and Collisions

Tomorrow, the sportscasters may say that in this instant the momentum of the entire game changed. As a result of this clutch hit, one team gained momentum, while their opponents lost it. But regardless of the effect on the team, it's clear that the momentum of the *ball* must have changed dramatically in the instant before this photograph was taken. It was traveling from right to left, probably at a pretty good rate of speed—and thus with lots of momentum. But a collision with several pounds of hardwood—with plenty of momentum of its own—changed its trajectory in a fraction of a second. A fan might say that the batter turned it around; after studying Chapter 4, you might say that the force he applied gave it a large negative acceleration, reversing it's velocity vector (and probably increasing its magnitude some as well). Yet, if you summed up the momentum of the ball and bat just before the collision and just afterward, you'd discover that the total never changed! In this chapter, you'll find out why—and how that fact can help you to analyze physical processes in many different realms, from baseball to rocketry.

If you were bowling and the ball bounced off the pins and rolled back toward you, you would probably be very surprised. But why? What leads us to expect that the ball will send the pins flying and continue on its way, rather than rebounding? You might say that the momentum of the ball carries it onward even after the collision (and you would be right)—but what does that really mean? In this chapter, you will study the concept of *momentum* and learn how it is particularly useful in analyzing motion and collisions.

6.1 Linear Momentum

Objective: To be able to compute linear momentum and the components of momentum.

The term momentum may bring to mind a football player running down the field, knocking down players who are trying to stop him. Or you might have heard someone say that a team lost its momentum (and so lost the game). Such everyday usages give some insight into the meaning of momentum. They suggest the idea of mass in motion, and therefore of inertia. We tend to think of heavy or massive objects in motion as having a great deal of momentum, even if they move very slowly. However, according to the technical definition of momentum, a light object can have just as much momentum as a heavier one, and sometimes more.

Newton referred to what modern physicists term **linear momentum** as "the quantity of motion . . . arising from velocity and the quantity of matter conjointly." In other words, the momentum of a body is proportional to both its mass and its velocity. By definition,

The linear momentum of an object is the product of its mass and velocity.

$$\mathbf{p} = m\mathbf{v} \tag{6.1}$$

SI unit of momentum: kg·m/s

It is common to refer to linear momentum as simply momentum. From Eq. 6.1, you can see that the SI units for momentum are kg·m/s. Momentum is a vector quantity having the same direction as the velocity. Like velocity, it can be resolved into rectangular components: $\mathbf{p}_x = m\mathbf{v}_x$ and $\mathbf{p}_y = m\mathbf{v}_y$.

Equation 6.1 expresses the momentum of a single object or particle. For a system of more than one particle, the **total linear momentum** of the system is the vector sum of the momenta (plural of momentum) of the individual particles.

Total linear momentum—a vector sum

$$\mathbf{P} = \mathbf{p}_1 + \mathbf{p}_2 + \mathbf{p}_3 + \cdots = \Sigma_i \mathbf{p}_i \tag{6.2}$$

EXAMPLE 6.1 ■ MOMENTUM: MASS AND VELOCITY

A 100-kg football player runs straight down the field with a velocity of 4.0 m/s. A 1.0-kg artillery shell leaves the barrel of a gun with a muzzle velocity of 500 m/s. Which has the greater momentum (magnitude)?

Solution.   As usual, we first list the given data and what we are to find, using the subscripts *p* and *s* to refer to the player and shell, respectively.

Given: $m_p = 100 \, \text{kg}$ *Find:* p_p and p_s
  $v_p = 4.0 \, \text{m/s}$ (magnitudes of the momenta)
  $m_s = 1.0 \, \text{kg}$
  $v_s = 500 \, \text{m/s}$

The magnitude of the momentum of the player is

$$p_p = m_p v_p = (100 \, \text{kg})(4.0 \, \text{m/s}) = 4.0 \times 10^2 \, \text{kg·m/s}$$

and that of the shell is

$$p_s = m_s v_s = (1.0 \text{ kg})(500 \text{ m/s}) = 5.0 \times 10^2 \text{ kg·m/s}$$

Thus, the much lighter, or less massive, shell has the greater momentum. Keep in mind that the magnitude of momentum depends on *both* the mass and the magnitude of the velocity.

Follow-up Exercise. What would the football player's speed have to be in order for his momentum to have the same magnitude as the artillery shell's? Would this speed be possible? *(Answers may be found in the Answers to Follow-up Exercises section at the back of the book.)*

EXAMPLE 6.2 ■ MOMENTUM VERSUS KINETIC ENERGY: SOME BALLPARK COMPARISONS

(a) Consider the three objects shown in •Fig. 6.1—a .22-caliber bullet, a cruise ship, and a glacier. Assuming each to be moving at its normal speed, which do you think has the greatest linear momentum? The least? Which has the greatest kinetic energy? The least? (List your choices before going on.)

	Greatest	Least
Momentum	————	————
Kinetic energy	————	————

(b) Using the estimates given below as your data, calculate the magnitudes of the momenta and the kinetic energies of the three objects. How accurate were your original answers—any surprises?

Bullet: A typical .22-caliber bullet would have a weight of about 30 grains and a muzzle velocity of about 1300 ft/s. (A grain is an old British unit used for pharmaceuticals, such as 5-grain aspirin tablets, and bullets; 1lb = 7000 grains.)

Ship: A ship like the one shown would have a weight of about 70,000 tons and a speed of about 20 knots. (A knot is another old unit, still commonly used in nautical contexts; 1 knot = 1.15 mi/h.)

Glacier: The glacier might be 1 km wide, 10 km long, and 250 m deep, and might move at a rate of a meter per day. (Not surprisingly, there is much variation among glaciers. It is therefore obvious that these figures must involve more assumptions and rougher estimates than those for the bullet or ship. For example, we are assuming a uniform, rectangular cross-sectional area for the glacier. The depth is particularly difficult to estimate from a photograph; a minimum value is given by the fact that glaciers must be at least 50–60 m thick before they can "flow." Observed speeds range from a few centimeters to as much as 40 meters a day for valley glaciers like the one in the photograph. The value we have chosen is considered a typical one.)

Solution. Converting our data to metric units, we have:

Bullet:
$$m_b = 30 \text{ gr} \left(\frac{1 \text{ lb}}{7000 \text{ gr}}\right)\left(\frac{1 \text{ kg}}{2.2 \text{ lb}}\right) = 0.0019 \text{ kg}$$

$$v_b = 1300 \text{ ft/s} \left(\frac{0.305 \text{ m/s}}{\text{ft/s}}\right) = 397 \text{ m/s}$$

Ship:
$$m_s = 7.0 \times 10^4 \text{ ton} \left(\frac{2.0 \times 10^3 \text{ lb}}{\text{ton}}\right)\left(\frac{1 \text{ kg}}{2.2 \text{ lb}}\right) = 6.4 \times 10^7 \text{ kg}$$

$$v_s = 20 \text{ knots} \left(\frac{1.15 \text{ mi/h}}{\text{knot}}\right)\left(\frac{0.447 \text{ m/s}}{\text{mi/h}}\right) = 10 \text{ m/s}$$

Glacier:
width = 10^3 m, length = 10^4 m, depth = 2.5×10^2 m;

$$v_g = 1 \text{ m/day} \left(\frac{1 \text{ day}}{86,400 \text{ s}}\right) = 1.2 \times 10^{-5} \text{ m/s}$$

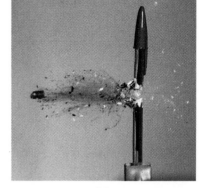

(a)

(b)

(c)

•**FIGURE 6.1 Three moving objects: a comparison of momenta and kinetic energies**
(a) A .22-caliber bullet shattering a ball point pen; **(b)** a cruise ship, **(c)** a glacier, Glacier Bay, Alaska. See Example 6.2.

We know the equations for kinetic energy, $K = \frac{1}{2}mv^2$ (Eq. 5.5), and momentum, $p = mv$ (Eq. 6.1). We have all the speeds and masses except for m_g, the mass of the glacier. To calculate this we need to know the density of ice, since $m = \rho V$ (Eq. 1.1). The density of ice is less than that of water, but the two are not very different, so we will use the density of water, $1.0 \times 10^3 \text{ kg/m}^3$, to simplify the calculations. (The actual density of ice is $0.92 \times 10^3 \text{ kg/m}^3$, but most of our other data in this Example are also approximations or estimates, so this shortcut should produce results that are good enough for our present purposes.)

$$m_g = \rho V = \rho(\ell \times \text{w} \times \text{d})$$
$$\approx (1.0 \times 10^3 \text{ kg/m}^3)(10^3 \text{ m})(10^4 \text{ m})(2.5 \times 10^2 \text{ m})$$
$$= 2.5 \times 10^{12} \text{ kg}$$

Then, calculating the magnitudes of the momenta of the objects, we have

Bullet: $p_b = m_b v_b = (0.0019 \text{ kg})(397 \text{ m/s}) = 0.75 \text{ kg·m/s}$

Ship: $p_s = m_s v_s = (6.4 \times 10^7 \text{ kg})(10 \text{ m/s}) = 6.4 \times 10^8 \text{ kg·m/s}$

Glacier: $p_g = m_g v_g = (2.5 \times 10^{12} \text{ kg})(1.2 \times 10^{-5} \text{ m/s}) = 3.0 \times 10^7 \text{ kg·m/s}$

Thus, the glacier has about 40 million times as much momentum as the bullet, and the ship nearly a billion times as much. These results are probably what you expected: the enormous differences in the masses of these objects outweigh the differences in their speeds. When we turn to the kinetic energies, however, the dependence on the *square* of the speed, which we have already called attention to in Example 5.6, produces results that might surprise you:

Bullet: $K_b = \frac{1}{2}m_b v_b^2 = \frac{1}{2}(0.0019 \text{ kg})(397 \text{ m/s})^2 = 1.5 \times 10^2 \text{ J}$

Ship: $K_s = \frac{1}{2}m_s v_s^2 = \frac{1}{2}(6.4 \times 10^7 \text{ kg})(10 \text{ m/s})^2 = 3.2 \times 10^9 \text{ J}$

Glacier $K_g = \frac{1}{2}m_g v_g^2 = \frac{1}{2}(2.5 \times 10^{12} \text{ kg})(1.2 \times 10^{-5} \text{ m/s})^2 = 1.8 \times 10^2 \text{ J}$

The bullet actually has about the same kinetic energy as the glacier—but the ship has more than 10 million times as much kinetic energy as either of them!

PROBLEM-SOLVING STRATEGIES: "BALLPARK" AND ORDER-OF-MAGNITUDE CALCULATIONS

Calculations such as we have performed above are sometimes known as "ballpark" or "back of the envelope" calculations. In performing such calculations we know that, because the data are only approximate (they are not known precisely, or they are being rounded off to simple whole numbers for the purpose of rapid calculation), the answer can't be taken as very accurate (see Section 1.6). However, it is often precise enough for our needs. In the example given above, some of the data were unavoidable approximations (the glacier obviously isn't a simple rectangular object), while others, such as the density of ice, were deliberately approximated to simplify calculation. Nevertheless, this method gave us a good idea of the relative momenta and kinetic energies of the three objects.

Physicists often find it convenient to use a formalized version of the "ballpark" approach, commonly known as *order-of-magnitude calculation*. In this method, all values are rounded off to the nearest "order of magnitude," or power of 10: 10^2, 10^3, 10^4, etc. Since it is very easy to multiply and divide powers of 10, this shortcut makes calculation very fast. However, although sometimes just the powers of 10 can be used in a calculation, it is usually a good idea to retain one significant figure of the prefix. This is particularly true when quantities are squared or raised to higher powers, as in the kinetic energy calculations. For instance, $v_s = 4 \times 10^2 \text{ m/s}$ or $\approx 10^2 \text{ m/s}$. Then $v_s^2 = (4 \times 10^2 \text{ m/s})^2 = 16 \times 10^4 \text{ m}^2/\text{s}^2 \approx 10^5 \text{ m}^2/\text{s}^2$. However, if we had just used 10^2 m/s, we would have obtained $10^4 \text{ m}^2/\text{s}^2$.

In order-of-magnitude calculations, we can generally expect the answer to be correct to the nearest order of magnitude. In other words, if we get an answer of 1000, the true value might really be 750 or 1200, but we can be pretty confident that it is closer to 1000 than to 100 or to 10,000.

Example:

A light-year—a unit widely used in astronomy—is the distance light travels in one year in a vacuum. The speed of light is $c = 3.00 \times 10^8$ m/s. Approximately how many meters are there in a light-year?

Solution:

From the Conversion Factors Table inside the front cover (or simple calculation), we know that there are 3.16×10^7 s in one year. Using the relationship, distance = speed × time,

$$d = ct = (3 \times 10^8 \text{ m/s})(3 \times 10^7 \text{ s}) \approx 10^{16} \text{ m}$$

Note that the product of the powers of 10 gives 10^{15}, but $3 \times 3 = 9$, so the prefixes give about another order of magnitude. Thus $(10 \times 10^{15}) = 10^{16}$ is the order of magnitude of the number of meters in a light-year. (The actual value is 9.48×10^{15} m.)

EXAMPLE 6.3 ■ TOTAL MOMENTUM: A VECTOR SUM

What is the total momentum for each of the systems of particles illustrated in •Fig. 6.2?

Solution.

Given: Magnitudes and directions
of momenta from Fig. 6.2

Find: (a) Total momentum (**P**)
(b) Total momentum (**P**)

(a) The total momentum of a system is the vector sum of the momenta of the individual particles, so

$$\mathbf{P} = \mathbf{p}_1 + \mathbf{p}_2 = (2.0 \text{ kg·m/s}) \mathbf{x} + (3.0 \text{ kg·m/s}) \mathbf{x} = (5.0 \text{ kg·m/s}) \mathbf{x} \quad (+x \text{ direction})$$

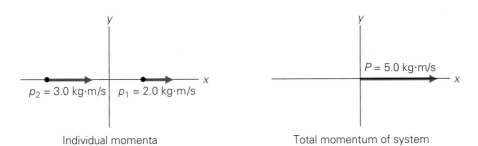

(a) **P** = **p**₁ + **p**₂

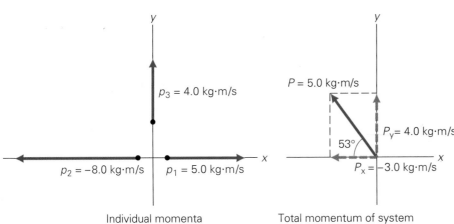

(b) **P** = **p**₁ + **p**₂ + **p**₃

•**FIGURE 6.2 Total momentum**
The total momentum of a system of particles is the vector sum of the individual momenta. See Example 6.3.

In this example, the momenta were along the coordinate axes and thus easily added. If the motion of one (or more) of the particles is not along an axis, its momentum may be broken up, or resolved, into rectangular components and added to the component sums—just as you learned to do with force components in Chapter 4. For example, suppose you were given a momentum vector with a magnitude of 5.0 kg·m/s at an angle of 53° relative to the negative x axis (like the total momentum vector in Fig. 6.2b). Then, its rectangular components are

Note: Review Example 4.4 and Fig. 4.11.

$$\mathbf{p}_x = -p \cos 53° \, \mathbf{x} = -(5.0 \text{ kg·m/s})(0.60) \, \mathbf{x} = -3.0 \text{ kg·m/s } \mathbf{x}$$
$$\mathbf{p}_y = p \sin 53° \, \mathbf{x} = (5.0 \text{ kg·m/s})(0.80) \, \mathbf{y} = 4.0 \text{ kg·m/s } \mathbf{y}$$

where the signs give the directions. Each momentum vector would be broken down thus and the rectangular components along the axis in each dimension added separately to find the vector sum.

Since momentum is a vector, a change in momentum can result from a change in magnitude and/or direction. Examples of changes in the momenta of particles because of changes of direction on collision are illustrated in •Fig. 6.3. In the figure, the magnitude of a particle's momentum is taken to be the same before and after collision (as indicated by the arrows of equal length). Fig. 6.3a illustrates a direct rebound—a 180° change in direction. Note that the change in momentum ($\Delta\mathbf{p}$) is the vector difference, and directional signs for the vectors are important. Fig. 6.3b shows a glancing collision, for which the change in momentum is given by the component differences.

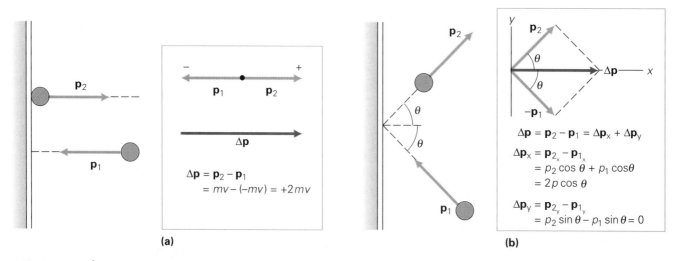

(a)   **(b)**

•**FIGURE 6.3 Change in momentum**
The change in momentum is given by the *difference* in the momentum vectors. **(a)** Here the vector sum is zero, but the vector *difference*, or change in momentum, is not. (The particles are displaced for convenience.) **(b)** The change in momentum is found by computing the change in the components.

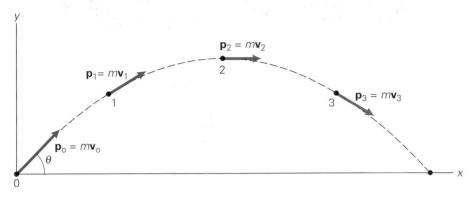

•FIGURE 6.4 **Change in the momentum of a projectile**
The total momentum vector of a projectile is tangential to its path (as is its velocity) and changes in magnitude and direction because of the action of an external force (gravity). The x component of the momentum is constant. (Why?)

Force and Momentum

As you know, a change in velocity (an acceleration) requires a force. Similarly, since momentum is directly related to velocity (as well as mass), a change in momentum also requires a force. In fact, Newton originally expressed his second law of motion in terms of momentum rather than acceleration. The force-momentum relationship may be seen by starting with $\mathbf{F} = m\mathbf{a}$ and using $\mathbf{a} = (\mathbf{v} - \mathbf{v}_o)/\Delta t$, where the mass is assumed to be constant:

$$\mathbf{F} = m\mathbf{a} = \frac{m(\mathbf{v} - \mathbf{v}_o)}{\Delta t} = \frac{m\mathbf{v} - m\mathbf{v}_o}{\Delta t} = \frac{\mathbf{p} - \mathbf{p}_o}{\Delta t} = \frac{\Delta\mathbf{p}}{\Delta t}$$

or

$$\mathbf{F} = \frac{\Delta\mathbf{p}}{\Delta t} \qquad (6.3)$$

Newton's second law of motion in terms of momentum

where $\mathbf{F}$ is the average force if the acceleration is not constant or the instantaneous force if Δt goes to zero.

Expressed in this form, Newton's second law is that *the net external force acting on an object is equal to the time rate of change of the object's momentum.* It is easily seen from the previous development that the equations $\mathbf{F} = m\mathbf{a}$ and $\mathbf{F} = \Delta\mathbf{p}/\Delta t$ are equivalent if the mass is constant. In some situations, however, the mass may vary. This will not be a consideration here in our discussion of particle collisions, but a special case will be given later in the chapter. The more general form of Newton's second law, is Equation 6.3, which is true even if the mass varies.

Just as the equation $\mathbf{F} = m\mathbf{a}$ indicates that an acceleration is evidence of a force, the equation $\mathbf{F} = \Delta\mathbf{p}/\Delta t$ indicates that *a change in momentum is evidence of a force.* For example, as illustrated in •Fig. 6.4, the momentum of a projectile is tangential to the projectile's parabolic path and changes in both magnitude and direction. The change in momentum indicates that there is a force acting on the projectile, which you know is the force of gravity. Changes in momentum were illustrated in Fig. 6.3. Can you identify the forces in these cases? Think in terms of Newton's third law.

6.2 The Conservation of Linear Momentum

Objectives: To be able to **(a)** explain the conditions for the conservation of linear momentum, and **(b)** apply it to physical situations.

Like energy, the momentum of a body or system is a conserved quantity under certain conditions, and this fact allows us to analyze a wide range of situations and solve many problems easily. The conservation of momentum is one of the most important principles in physics. In particular, it is used to analyze the collision of objects ranging from subatomic particles to automobiles in traffic accidents.

For the linear momentum of a particle or an object to be conserved (to remain constant with time), a condition must hold. This condition is apparent from the momentum form of Newton's second law (Eq. 6.3). If the net force acting on a particle is zero, that is,

$$\mathbf{F} = \frac{\Delta \mathbf{p}}{\Delta t} = 0$$

then,

$$\Delta \mathbf{p} = 0 = \mathbf{p} - \mathbf{p}_o$$

where $\mathbf{p}_o$ is the initial momentum and $\mathbf{p}$ is the momentum at some later time. Since they are equal, the momentum is conserved.

$$\mathbf{p} = \mathbf{p}_o \qquad (6.4)$$

or

$$m\mathbf{v} = m\mathbf{v}_o$$

Note that this is consistent with Newton's first law: an object remains at rest ($\mathbf{p} = 0$) or in motion with a *uniform* velocity (constant $\mathbf{p}$), unless acted on by a net external force.

The conservation of momentum may be extended easily to a system of particles by writing Newton's second law in terms of the sums (resultants) of the forces acting on the system and of the momenta of the particles: $\mathbf{F} = \Sigma \mathbf{F}_i$ and $\mathbf{P} = \Sigma \mathbf{p}_i = \Sigma m\mathbf{v}_i$. A more complicated argument similar to the development of Eq. 6.4 can be made for systems of particles. It turns out that if there is no net external force acting *on the system*, then $\mathbf{P} = \mathbf{P}_o$. This generalized condition is referred to as the law of **conservation of linear momentum**:

Conservation of momentum— no net external force

> The total linear momentum of a system is conserved if the net external force acting on the system is zero.

There are other ways to express this. For example, recall from Chapter 5 that an *isolated* system is one on which no external work is done. Therefore there is no net external force acting on such a system, so the total linear momentum of an isolated system is conserved.

Within a system, internal forces may be acting, for example, when particles collide. These are force pairs of Newton's third law and there is a good reason why such forces are not explicitly referred to in the condition for the conservation of momentum. By Newton's third law, these internal forces are equal and opposite and vectorially cancel each other. Thus, the net internal force of a system is zero.

Note: Third-law force pairs were discussed in Section 4.5.

An important point to understand, however, is that the momenta of *individual* particles or objects within a system may change. But in the absence of an external force, the *vector sum* of all the momenta remains the same. If the objects are initially at rest (total momentum is zero) and then set in motion as the result of internal forces, the total momentum must still add up to zero. An example of this is illustrated in •Fig. 6.5 and analyzed in Example 6.4. Objects in an isolated sys-

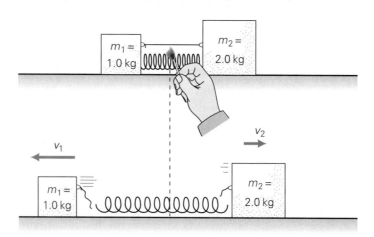

•**FIGURE 6.5 An internal force and the conservation of momentum**
The spring force is an internal force, so the momentum of the system is conserved. Example 6.4.

Demonstration 3 ■ Internal Forces (and Conservation of Momentum?)

(a) Getting ready to push off. Note that the forces are internal to the system.

(b) Play ball! Here a ball is used to exchange momentum between parts of the system.

In both cases, the initial total momentum is zero. What would happen after the actions (and reactions) begin? Actually there is an external force (what is it?), so the momentum is not conserved and the motion is limited.

tem may initially be in motion, and the motion may be changed by internal forces, but the total momentum after the changes must add up to the initial value, assuming friction is negligible. (See Demonstration 3.)

The conservation of momentum is often a powerful and convenient tool for analyzing situations involving motion. Its application is illustrated in the following examples.

EXAMPLE 6.4 ■ BEFORE AND AFTER: CONSERVATION OF MOMENTUM

Two masses, $m_1 = 1.0$ kg and $m_2 = 2.0$ kg, are held on either side of a light compressed spring by a light string joining them, as shown in Fig. 6.5. The string is burned (negligible external force), and the masses move apart on the frictionless surface, with m_1 having a velocity of 1.8 m/s to the left. What is the velocity of m_2?

Solution.   Listing the masses and speed given, we have

Given: $m_1 = 1.0$ kg   *Find:* v_2 (velocity—speed, and direction)
$m_2 = 2.0$ kg
$v_1 = 1.8$ m/s (left)

(The term *light* indicates that the masses of the spring and string can be ignored.)

Here the system consists of the two masses and the spring. Since the spring force is internal to the system, the momentum of the system is conserved. It should be apparent that the initial total momentum of the system ($\mathbf{P}_o$) is zero, as is the final momentum. Thus, we may write

$$\mathbf{P}_o = \mathbf{P} = 0 \quad \text{and} \quad \mathbf{P} = \mathbf{p}_1 + \mathbf{p}_2 = 0$$

(The spring does not come into the equations because its mass is negligible.) Thus,

$$\mathbf{p}_2 = -\mathbf{p}_1$$

which means that the momenta of m_1 and m_2 are equal and opposite. Using directional signs (with $+x$ to the right in the figure) gives

$$p_2 - p_1 = m_2 v_2 - m_1 v_1 = 0$$

or in terms of magnitudes

$$m_2 v_2 = m_1 v_1$$

and

$$v_2 = \left(\frac{m_1}{m_2}\right) v_1 = \left(\frac{1.0 \text{ kg}}{2.0 \text{ kg}}\right) (1.8 \text{ m/s}) = 0.90 \text{ m/s}$$

Thus, the velocity of m_2 is 0.90 m/s in the positive x direction, or to the right in the figure, as you might have expected since m_2 has twice the mass.

Follow-up Exercise.   (a) Suppose that the large block in Fig. 6.5 were attached to the Earth's surface, so that it could not move when the string was burned. Would momentum be conserved in this case? Explain. (b) Two girls, each having a mass of 50 kg, stand at rest on skateboards with negligible friction. The first tosses a 2.5-kg ball to the second. If the speed of the ball is 10 m/s, what is the speed of each girl after the ball is caught, and what is the momentum of the ball before it is tossed, while it is in the air, and after it is caught? *(Answers may be found in the Answers to follow-up Exercises section at the back of the book.)*

CONCEPTUAL EXAMPLE 6.5 ■ HOW FAR? CONSERVATION OF MOMENTUM WITH EXTERNAL FORCES

Consider the situation in Fig. 6.5 again. Let's assume there is friction and that the blocks are made from different materials such that the constant frictional forces acting on the moving blocks have the same magnitude. (Why would they have to be different materials?) Then, under these conditions, on coming to rest, (a) both blocks have traveled the same distance, (b) the lighter block has traveled twice the distance of the heavier one, (c) the lighter block has traveled four times the distance of the heavier one. *Clearly establish the reasoning and physical principle(s) used in determining your answer before checking it below. That is, **why** did you select your answer?*

Reasoning and Answer.   The spring force is internal and initially the frictional forces cancel within the system, so the linear momentum is conserved and we can get the initial speeds as in Example 6.4. However, let's solve this example algebraically rather than using numbers. (It's usually simpler to work with symbols.) To define the notation, we will say that $m_1 = m$ and $m_2 = 2m$, and specifically indicate the initial velocities, v_{1_o} and v_{2_o}, to distinguish them from the final velocities, $v_1 = v_2 = 0$. (Properly defined notation is important.) Then, using the conservation of momentum as above, we have

$$v_{1_o} = \left(\frac{m_2}{m_1}\right) v_{2_o} = \left(\frac{2m}{m}\right) v_{2_o} = 2 v_{2_o}$$

as shown previously with numbers.

With the same constant frictional decelerating force, each block has a constant deceleration, so we can use a kinematic equation from Chapter 2: $v^2 = v_o^2 + 2ax$. With $v = 0$, the magnitude of x is given by

$$v^2 = 0 = v_o^2 - 2ax \quad \text{and} \quad x = \frac{v_o^2}{2a}$$

The distance, then, is proportional to the *square* of the initial speed. So, with twice the initial speed, does the lighter block travel four times the distance of the heavier block, making the answer (c)? We had better look more closely. Note that the acceleration also appears in the expression for the distance. The frictional forces (f) on each block have the same magnitude, but since the masses of the blocks are different, so are the accelerations. Since after the spring releases, the net force on each block is friction, f, the acceleration of each is $a = f/m$, and the general expression for x becomes

$$x = \frac{m v_o^2}{2f}$$

Now we see that the distance is proportional not only to v^2 but also to m. Forming a ratio of x_1 and x_2, and using known ratio values,

$$\frac{x_1}{x_2} = \left(\frac{m_1}{m_2}\right)\left(\frac{v_{1_o}}{v_{2_o}}\right)^2 = (\tfrac{1}{2})(2)^2 = 2$$

Thus, $x_1 = 2x_2$, or the lighter block travels twice the distance of the heavier one, so the answer is (b).

Follow-up Exercise. If the magnitude of the constant frictional force on the lighter block were twice that on the heavier block, would the blocks then travel the same distance? Justify your answer. (*Reasoning and answer may be found in the Answers to Follow-up Exercises section at the back of the book.*)

EXAMPLE 6.6 ■ A GLANCING COLLISION: COMPONENTS OF MOMENTUM

A moving shuffleboard puck has a glancing collision with a stationary one of the same mass, as shown in •Fig. 6.6. Considering friction to be negligible, what are the speeds of the pucks after the collision?

Solution.

Given: $v_{1_o} = 0.95$ m/s   *Find:* v_1 and v_2 (speeds after collision)
$\theta_1 = 50°$
$\theta_2 = 40°$
$m_1 = m_2$

The collision forces of the two-puck system are internal forces, so the total momentum is conserved (the external force of friction is negligible for the very brief collision time). This means that the rectangular components of momentum are also conserved. That is, $\mathbf{P} = \mathbf{P}_x + \mathbf{P}_y$ is a constant, so $\mathbf{P}_x$ and $\mathbf{P}_y$ must be constant. For the x component of the momentum with signs indicating directions (positive axes shown in figure),

$$\overset{before}{} \quad \overset{after}{}$$
$$p_{x_o} = p_{x_1} + p_{x_2}$$

or

$$m_1 v_{1_o} = m_1 v_1 \cos 50° + m_2 v_2 \cos 40° \tag{1}$$

For the y component,

$$0 = p_{y_1} - p_{y_2}$$

or

$$0 = m_1 v_1 \sin 50° - m_2 v_2 \sin 40° \tag{2}$$

where the minus sign indicates that the component is in the negative y direction. With two equations and two unknowns, the rest is simple algebra. The masses cancel in Eq. 2 since they are equal ($m_1 = m_2$), and

$$v_2 = \left(\frac{\sin 50°}{\sin 40°}\right) v_1 = \left(\frac{0.766}{0.643}\right) v_1 = 1.19 v_1 \tag{3}$$

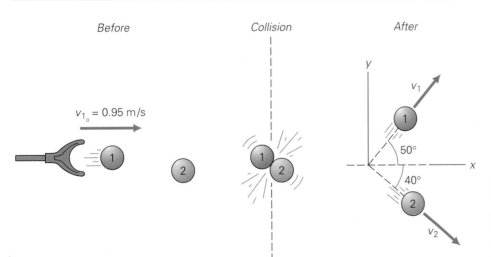

Before   Collision   After

$v_{1_o} = 0.95$ m/s

v_1

$50°$

$40°$

v_2

•FIGURE 6.6 **A glancing collision** Momentum is conserved in an isolated system. The motion in two dimensions may be analyzed in terms of the components of momentum which are also conserved. See Example 6.6.

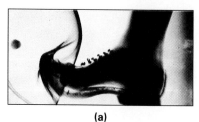

(a)

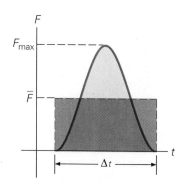

(b)

•**FIGURE 6.7 Collision impulse**
(a) Collision impulse causes the football to be deformed. **(b)** The impulse is the area under the curve of an F versus t graph. Note that the impulse force on the ball is not constant, but rises to a maximum.

Substituting for v_2 in Eq. 1 and canceling the masses (as above) gives

$$v_{1_o} = v_1 \cos 50° + (1.19v_1)(\cos 40°)$$

and

$$v_1 = \frac{v_{1_o}}{\cos 50° + (1.19)(\cos 40°)} = \frac{0.95 \text{ m/s}}{0.643 + (1.19)(0.766)} = 0.61 \text{ m/s}$$

Using this value in Eq. 3 gives

$$v_2 = 1.19v_1 = (1.19)(0.61 \text{ m/s}) = 0.73 \text{ m/s}$$

The directions of v_1 and v_2 are given by θ_1 and θ_2, respectively.

Follow-up Exercise.   (a) Is the kinetic energy conserved in the collision in this Example? Explain. (b) In such a collision, is it possible for both θ_1 and θ_2 to lie above the $+x$ axis? Why or why not? *(Answers may be found in the Answers to Follow-up Exercises section at the back of the book.)*

The conservation of momentum is one of the most important principles in physics. As mentioned previously, it is used to analyze the collisions of objects ranging from subatomic particles to automobiles in traffic accidents. In many instances, external forces may be acting on the objects, which means that the momentum is not conserved. But, as you will learn in the next section, the conservation of momentum often allows a good approximation *over the short time of a collision* because the internal forces, for which momentum is conserved, are much greater than the external forces. For example, external forces such as gravity and friction also act on colliding objects, but are often relatively small compared to the internal forces. Therefore, if the objects interact for only a brief time, the effects of the external forces may be negligible compared to those of the large internal forces.

6.3 Impulse

Objectives: To be able to relate (a) impulse and momentum, and (b) kinetic energy and momentum.

When two objects—such as a hammer and a nail, a golf club and a golf ball, or even two cars—collide, they exert large forces on one another for a short period of time (see •Fig. 6.7a). The force is not constant in this case. However, Newton's second law in momentum form is still useful for analyzing such situations using average values. Written in this form, the law states that the *average* force is equal to the time rate of change of momentum, $\bar{\mathbf{F}} = \Delta \mathbf{p}/\Delta t$ (Eq. 6.3). Rewriting the equation to express the change in momentum, we have

$$\bar{\mathbf{F}} \, \Delta t = \Delta \mathbf{p} = \mathbf{p} - \mathbf{p}_o \qquad (6.5)$$

The term $\bar{\mathbf{F}} \, \Delta t$ is known as the **impulse** of the force,

$$\text{Impulse} = \bar{\mathbf{F}} \, \Delta t = \Delta \mathbf{p} = m\mathbf{v} - m\mathbf{v}_o \qquad (6.6)$$

SI unit of impulse and momentum: N·s

Thus, *the impulse exerted on a body is equal to the change in the body's momentum.* This is referred to as the **impulse-momentum theorem**. Impulse obviously has units of newton·seconds (N·s), which are also units for momentum by Eq. 6.5 and Eq. 6.1 (kg·m/s = N·s).

In Chapter 5 you learned that by the work-energy theorem ($W = F \, \Delta x = \Delta K$), the area under an F (net force) versus x curve is equal to the net work or change in kinetic energy. Similarly, the area under an F versus t curve is equal to the impulse or the change in momentum (•Fig. 6.7b). An impulse force varies with time and is not a constant force, as Eq. 6.5 might suggest. However, in general, it is

•**FIGURE 6.8 Average impulse force**

The area under the average force line ($\bar{F} \, \Delta t$) is the same as the area under the F versus t curve, which is usually difficult to evaluate.

Note: Compare Fig. 6.7b with Fig. 5.5.

convenient to talk about the equivalent *constant* average force $\bar{F}$ acting over a time interval Δt to give the same impulse (same area under the force versus time curve) as shown in •Fig. 6.8.

EXAMPLE 6.7 ■ TEEING OFF: THE IMPULSE-MOMENTUM THEOREM

A golfer drives a 0.10-kg ball from an elevated tee, giving it an initial horizontal speed of 40 m/s (about 90 mi/h). The club and the ball are in contact for 1.0 ms (millisecond). What is the average force exerted by the club on the ball during this time?

Solution.

Given: $m = 0.10$ kg   *Find:* $\bar{F}$ (average force)
                $v = 40$ m/s
                $v_o = 0$
                $\Delta t = 1.0$ ms $= 1.0 \times 10^{-3}$ s

Notice that the mass and the initial and final velocities are given, so the change in momentum can be easily found. Then, the average force may be computed by the impulse-momentum theorem:

$$\bar{F} \Delta t = p - p_o = mv - mv_o$$

Thus,

$$\bar{F} = \frac{mv - mv_o}{\Delta t} = \frac{(0.10 \text{ kg})(40 \text{ m/s}) - 0}{1.0 \times 10^{-3} \text{ s}}$$

$$= 4000 \text{ N} \quad (\text{or } 900 \text{ lb})$$

The force is in the direction of the acceleration and is the *average* force. The instantaneous force is even greater near the midpoint of time interval of the collision (Δt in Fig. 6.8).

Follow-up Exercise.   Suppose the golfer in the example drives a ball with the same average force, but "follows through" on the swing so as to increase the contact time to 1.5 ms. What effect would this have on the initial horizontal speed of the drive? (*Answer may be found in the Answers to Follow-up Exercises section at the back of the book.*)

Example 6.7 illustrates the large internal forces that colliding objects can exert on one another during short contact times. However, in some instances the impulse may be manipulated to reduce the force. Suppose there is a fixed change in momentum in a given situation. Then, since $\Delta p = \bar{F} \Delta t$, if Δt could be made longer, the average impulse force $\bar{F}$ would be reduced.

You have probably tried to minimize the impulse force on occasion. For example, when jumping from a height onto a hard surface, you try not to land stiff-legged. The abrupt stop (small Δt) would apply a large impulse force to your leg bones and joints and could cause injury. Instead, you bend your knees as you land. The impulse is constant and vertically upward opposite your velocity in this case ($\bar{F} \Delta t = \Delta p = -mv_o$, with the final velocity zero). Thus increasing the time interval Δt makes the impulse force smaller. Similarly, in catching a fast-moving, hard ball, you quickly learn not to catch it with your arms rigid, but rather to move your hands with the ball. This increases the contact time and reduces the impulse force and the "sting." Another example in which the contact time is increased so as to decrease the impulse force is given in the Insight on p. 180.

In other instances, the *applied* impulse force may be relatively constant, and the contact time (Δt) may be deliberately increased to produce a greater impulse, and thus a greater change in momentum ($\bar{F} \Delta t = \Delta p$). This is the principle of "following through" in sports, for example, when hitting a ball with a bat or driving a golf ball. In the latter case (see •Fig. 6.9a), assuming that the golfer supplies the same average force with each swing, the longer the contact time, the greater will be the impulse or the momentum the ball receives. That is, with $\bar{F} \Delta t = mv$ (since

(a)

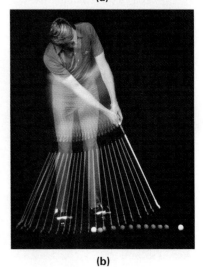

(b)

•FIGURE 6.9   **Increasing the contact time**
(a) A golfer follows through on a drive swing. One reason is to increase the contact time so that the ball receives greater impulse and momentum. (b) The follow-through on a long putt increases the contact time for greater impulse and momentum, but the main reason is for directional control. Notice that the putter is in contact with the ball for a time equivalent to about 4 flash intervals.

Insight   The Automobile Air Bag

A dark, rainy night—a car goes out of control and hits a big tree head-on! But the driver walks away with only minor injuries because he had his seatbelt buckled and had bought a car equipped with air bags. Air bags are now installed on most new cars. The bags, along with seatbelts, are safety devices designed to prevent injuries to passengers in the front seat in automobile collisions. (Backseat air bags are also available.)

When a car collides with something immovable such as a tree or a bridge abutment, or has a head-on collision with another vehicle, it stops almost instantaneously. If the front-seat passengers have not ``buckled up'' (and there are no air bags), they keep moving until acted upon by an external force (Newton's first law). For the driver, this force is supplied by the steering wheel and column; for the passenger, by the dashboard and/or windshield.

Even when everyone has buckled up, there can be injuries. Seatbelts absorb energy by stretching, and spread the force over a wide area to reduce the pressure. However, if a car is going fast enough and hits something truly immovable, there may be too much energy for the belts to absorb. This is where the air bag comes in. The bag inflates automatically on hard impact (Fig. 1), cushioning the driver (and front-seat passenger if both sides are equipped with air bags). In terms of impulse, the air bag increases the stopping contact time—the

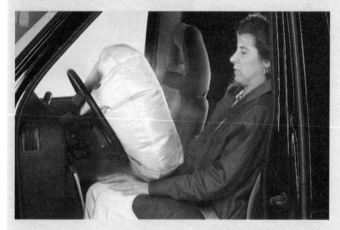

FIGURE 1 Impulse and safety   An automobile air bag increases the contact time, thereby decreasing the impulse force that could cause injury.

fraction of a second it takes your head to sink into the inflated bag is many times longer than the instant in which your head is stopped by the dashboard. A longer contact time means a reduced impact force, and thus much less likelihood of an injury. (Because the bag is large, the total impact force is also spread over a greater area of the body, so the force on any one part of the body is less.)

An interesting point is the inflating mechanism of an air bag. Think of how little time elapses between the front-end impact and the driver hitting the steering column in the collision of a fast-moving automobile. We say such a collision takes place instantaneously, yet during this time the air bag must be inflated! How is this done?

First, the air bag is equipped with sensors that detect the sharp deceleration associated with a head-on collision the instant it begins. If the deceleration exceeds the sensors' threshold settings, an electric current in an igniter in the air bag sets off a chemical explosion that generates gas to inflate the bag. The complete process from sensing to full inflation takes only on the order of 25 *thousandths* of a second (0.025 s).

The sensors' signals go first to a control unit, which determines whether a frontal collision is occurring rather than a system malfunction. (Accidental deployment could be dangerous as well as costly.) Typically, the control unit compares signals from two different sensors for collision verification. The unit is equipped with its own power source, since the car's battery and alternator are usually destroyed in a hard front-end collision. Sensing a collision, the control unit completes the circuit to the air bag igniter. It initiates a chemical reaction in a sodium azide (NaN_3) propellant. Gas (mostly nitrogen) is generated at an explosive rate, which inflates the air bag. The bag itself is made of thin nylon that is covered with cornstarch. The cornstarch acts as a lubricant to help the bag unfold smoothly on inflation.

Not only is sodium azide explosive, but it is also poisonous. Thus, one should not open a collapsed air bag, and if a car is junked, undeployed air bags must be disposed of properly. Other, nonpoisonous compounds are being developed for use as air bag inflators.

In general, air bags offer protection only if the occupants are thrown forward, because the bags are designed to deploy only in front-end collisions. They are of little use in side-impact crashes. Side air bags are now becoming available in some automobiles. But, whether or not the car is equipped with air bags, *all* passengers should wear their seat belts at *all* times.

$v_0 = 0$), the greater the value of Δt, the greater will be the final velocity of the ball. (This was illustrated in the preceding Follow-up Exercise.) As you learned in Section 3.3, a greater projection velocity gives a projectile a greater range. (What angle of projection should a golfer try to achieve when driving the ball on a level fairway?)

In some instances, a long follow-through that increases the contact time may be used to improve control (•Fig. 6.9b).

The word "impulse" implies that the impulse force acts only briefly or quickly (like an "impulsive" person), and this is true in many instances. However, the definition of impulse places no limit on the time interval over which the force may act. Technically, a comet at its closest approach to the Sun is involved in a collision, because in physics collision forces do not have to be contact forces. Basically, a **collision** is an interaction of objects in which there is an exchange of momentum and energy. As you might expect from the work-energy theorem and the impulse-momentum theorem, momentum and kinetic energy are directly related. A little algebraic manipulation of the formula for kinetic energy (Eq. 5.5) allows us to express kinetic energy in terms of the *magnitude* of momentum:

$$K = \tfrac{1}{2}mv^2 = \frac{(mv)^2}{2m} = \frac{p^2}{2m} \tag{6.7}$$

Collision—exchange of momentum and energy

Kinetic energy and momentum

Thus, kinetic energy and momentum are intimately related. The conservation of these quantities is an important consideration in collisions, as we shall see in the next section.

6.4 Elastic and Inelastic Collisions

Objective: To be able to describe the conditions on kinetic energy and momentum in elastic and inelastic collisions.

Taking a closer look at collisions in terms of the conservation of momentum is simpler if an isolated system is considered, such as a system of particles (or balls) involved in head-on collisions. For simplicity, we will consider only collisions in one dimension. Such collisions can also be analyzed in terms of the conservation of energy. On the basis of what happens to the total kinetic energy we can define two types of collisions: elastic and inelastic.

In an **elastic collision** (•Fig. 6.10a), the total kinetic energy is conserved. That is, the *total* kinetic energy of all the objects of the system after the collision is the same as their *total* kinetic energy before the collision:

Elastic collision—kinetic energy conserved

$$\begin{array}{ll} \text{total } K \text{ after} = \text{total } K \text{ before} & \textit{condition for an} \\ K_f = K_i & \textit{elastic collision} \end{array} \tag{6.8}$$

During the collision, some or all of the initial kinetic energy is temporarily converted to potential energy as the objects are deformed. But, after the maximum deformations, the objects elastically spring back to their original shapes, and the system regains all of its original kinetic energy. For example, two steel balls or two billiard balls may have a nearly elastic collision with each ball having the same shape afterwards as before; that is, there is no permanent deformation.

In an **inelastic collision** (•Fig. 6.10b), the total kinetic energy is *not* conserved. For example, one or more of the colliding objects may not spring back to its orig-

Note: In reality, only atoms and subatomic particles have truly elastic collisions, but some larger hard objects have nearly elastic collisions in which the kinetic energy is approximately conserved.

Inelastic collision—kinetic energy not conserved

(a)   (b)

•**FIGURE 6.10 Collisions**
(a) An elastic collision. (b) An inelastic collision.

inal shape, or heat may be generated. In such interactions, work is done by non-conservative forces (Section 5.5), such as friction, and some kinetic energy is lost.

$$\text{total } K \text{ after} < \text{total } K \text{ before} \qquad \textit{condition for an}$$
$$K_f < K_i \qquad \textit{inelastic collision} \qquad (6.9)$$

For example, a hollow aluminum ball that collides with a solid steel ball may be dented. Permanently deforming the ball takes work, and that work is done at the expense of the original kinetic energy of the system.

For isolated systems, momentum is conserved in both elastic and inelastic collisions. For an inelastic collision, only an amount of kinetic energy consistent with the conservation of momentum may be lost. It may seem strange that kinetic energy can be lost and momentum still be conserved, but this fact provides insight into the difference between scalar and vector quantities.

Energy and Momentum in Inelastic Collisions

To see how momentum can remain constant while the kinetic energy changes (decreases) in inelastic collisions, consider the examples illustrated in •Fig. 6.11. In the first case, two balls of equal mass ($m_1 = m_2$) approach each other with equal and opposite velocities ($v_{1_o} = -v_{2_o}$). Hence, the total momentum before the collision is vectorially zero, but the scalar kinetic energy is *not* zero. After the collision the balls are stuck together and stationary, so the total momentum is unchanged —still zero. Momentum is conserved because the forces of collision are internal to the system of the two balls; thus there is no net external force on the system. The total kinetic energy, however, has decreased to zero. In this case, the kinetic energy went into the work done in permanently deforming the balls. Some energy may also have gone into doing work against friction (producing heat) or been lost in some other way (for example, in producing sound).

It should be noted that the balls need not come to rest after collision. In a less inelastic collision, the balls may recoil in opposite directions at reduced but equal

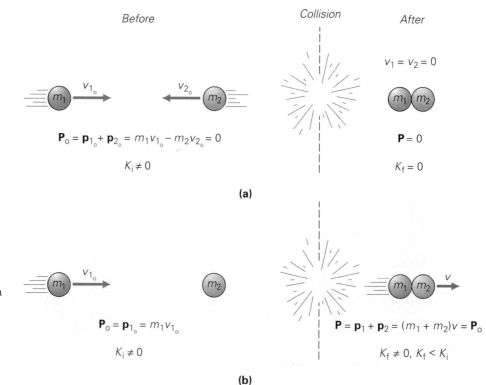

•**FIGURE 6.11 Inelastic collisions**
In inelastic collisions, momentum is conserved but kinetic energy is not. Collisions like the ones shown here, in which the objects stick together, are called completely or totally inelastic collisions. The maximum amount of kinetic energy lost is consistent with the conservation of momentum.

speeds. The momentum would be conserved (still equal to zero—why?), but the kinetic energy would again not be conserved.

In Fig. 6.11b, one ball is initially at rest as the other approaches. The balls stick together after collision, but are still in motion. Both of these cases are examples of a **completely inelastic collision**, in which the objects stick together and have the same velocity after colliding.

Assume that the balls in Fig. 6.11b have different masses. Since the momentum is conserved even in inelastic collisions,

before   after

$$m_1 v_o = (m_1 + m_2)v$$

and

$$v = \left(\frac{m_1}{m_1 + m_2}\right)v_o \qquad \begin{array}{l} m_2 \text{ initially at rest} \\ \text{(completely inelastic} \\ \text{collision only)} \end{array} \qquad (6.10)$$

Thus, v is less than v_o, since $m_1/(m_1 + m_2)$ must be less than 1. Now let us consider how much kinetic energy has been lost. Initially, $K_i = \frac{1}{2}m_1 v_o^2$, and finally, or after the collision,

$$K_f = \tfrac{1}{2}(m_1 + m_2)v^2.$$

Substituting for v from Eq. 6.10 and simplifying the result, we have

$$K_f = \tfrac{1}{2}(m_1 + m_2)\left(\frac{m_1 v_o}{m_1 + m_2}\right)^2 = \frac{\tfrac{1}{2}m_1^2 v_o^2}{m_1 + m_2} = \left(\frac{m_1}{m_1 + m_2}\right)\tfrac{1}{2}m_1 v_o^2 = \left(\frac{m_1}{m_1 + m_2}\right)K_i$$

and

$$\frac{K_f}{K_i} = \frac{m_1}{m_1 + m_2} \qquad \begin{array}{l} m_2 \text{ initially at rest} \\ \text{(completely inelastic} \\ \text{collision only)} \end{array} \qquad (6.11)$$

Eq. 6.11 gives the fractional amount of the initial kinetic energy that the system has after a completely inelastic collision. For example, if the masses of the balls are equal ($m_1 = m_2$), then $m_1/(m_1 + m_2) = \frac{1}{2}$ and $K_f/K_i = \frac{1}{2}$ or $K_f = K_i/2$. That is, only half of the initial kinetic energy is lost (consistent with the conservation of momentum).

Note that all the kinetic energy cannot be lost in this case no matter what the masses of the balls are. The momentum after collision cannot be zero, since it was not zero initially. Thus, the balls must be moving and must have at least some kinetic energy.

EXAMPLE 6.8 ■ STUCK TOGETHER: COMPLETELY INELASTIC COLLISION

A 1.0-kg ball with a speed of 4.5 m/s strikes a 2.0-kg stationary ball. If the collision is completely inelastic, (a) what are the speeds of the balls after the collision? (b) What percent of the initial kinetic energy do they have after the collision? (c) What is the total momentum after the collision?

Solution. Using the labeling as in the preceding discussion, we have

Given: $m_1 = 1.0$ kg *Find:* (a) v (speed after collision)
    $m_2 = 2.0$ kg
    $v_o = 4.5$ m/s   (b) $\dfrac{K_f}{K_i}$ (×100%)

             (c) $\mathbf{P}_f$ (total momentum after collision)

(a) The balls stick together and have the same speed after collision. This is given by Eq. 6.10:

$$v = \left(\frac{m_1}{m_1 + m_2}\right)v_o = \left(\frac{1.0 \text{ kg}}{1.0 \text{ kg} + 2.0 \text{ kg}}\right)(4.5 \text{ m/s})$$

$$= 1.5 \text{ m/s}$$

(b) The fractional part of the initial kinetic energy that the balls have after the completely inelastic collision is given by Eq. 6.11. Notice that this fraction, as given by the masses, is the same as that for the speeds (Eq. 6.10) in this special case. By inspection, we can write

$$\frac{K_f}{K_i} = \frac{m_1}{m_1 + m_2} = \frac{1.0 \text{ kg}}{1.0 \text{ kg} + 2.0 \text{ kg}} = \frac{1}{3} = 0.33 \ (\times 100\%) = 33\%$$

Let's show this explicitly:

$$\frac{K_f}{K_i} = \frac{\frac{1}{2}(m_1 + m_2)v^2}{\frac{1}{2}m_1 v_o^2} = \frac{\frac{1}{2}(1.0 \text{ kg} + 2.0 \text{ kg})(1.5 \text{ m/s})^2}{\frac{1}{2}(1.0 \text{ kg})(4.5 \text{ m/s})^2} = 0.33 \ (= 33\%)$$

Keep in mind that Eq. 6.11 applies *only* to *completely* inelastic collisions. For other types of collisions, the initial and final values of the kinetic energy must be computed explicitly.

(c) The momentum is conserved in all collisions (in the absence of external forces), so the total momentum after collision is the same as before, which is simply the momentum of the incident ball, with a magnitude of

$$P_f = p_{1_o} = m_1 v_o = (1.0 \text{ kg})(4.5 \text{ m/s}) = 4.5 \text{ kg·m/s}$$

and in the same direction.

Follow-up Exercise.   A small hard-metal ball of mass m collides with a larger, stationary soft-metal ball of mass M. A *minimum* amount of work W is required to make a dent in the larger ball. If the smaller ball initially has kinetic energy $K = W$, will the larger ball be dented in a completely inelastic collision? *(Answer may be found in the Answers to Follow-up Exercises section at the back of the book.)*

The amount of elasticity or inelasticity in collisions can be easily demonstrated by dropping various objects on to a hard surface and noting how high they rebound. For example, suppose you dropped a Superball, a baseball, and a piece of putty. Based on their rebound heights, how would you categorize the collisions?

Energy and Momentum in Elastic Collisions

For a general *elastic* collision of two objects, the condition is as follows:

$$\text{Conservation of } K: \quad \underbrace{\tfrac{1}{2}m_1 v_{1_o}^2 + \tfrac{1}{2}m_2 v_{2_o}^2}_{before} = \underbrace{\tfrac{1}{2}m_1 v_1^2 + \tfrac{1}{2}m_2 v_2^2}_{after} \qquad (6.12a)$$

Of course, linear momentum is conserved even if the collision is inelastic, thus we may also write

$$\text{Conservation of } \mathbf{P}: \quad m_1\mathbf{v}_{1_o} + m_2\mathbf{v}_{2_o} = m_1\mathbf{v}_1 + m_2\mathbf{v}_2 \qquad (6.12b)$$

Thus for elastic collisions, knowing the masses of the objects and the initial velocities allows you to find the final velocities (there are two equations and two unknowns).

A common collision situation is that in which one of the objects is initially stationary (Fig. 6.12). Assume that the motion of both of the balls after the collision is to the right, or in the positive direction, as shown in the figure. If this is not the case, the math will tell you. (How?)

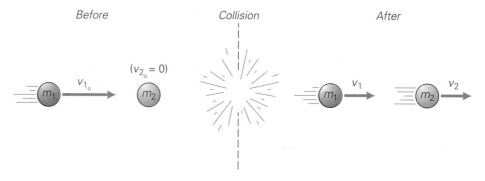

•FIGURE 6.12   **Elastic collision** For an elastic collision between two bodies, one of which is initially at rest, the velocities after the collision depend on the relative masses of the bodies.

Before   *Collision*   *After*

$(v_{2_o} = 0)$

v_{1_o}

v_1 v_2

This is a one-dimensional situation. The condition for an elastic, head-on collision in this situation is

$$\tfrac{1}{2}m_1 v_{1_o}^2 = \tfrac{1}{2}m_1 v_1^2 + \tfrac{1}{2}m_2 v_2^2 \tag{6.13a}$$

and conservation of linear momentum gives

$$m_1 v_{1_o} = m_1 v_1 + m_2 v_2 \tag{6.13b}$$

Rearranging these equations gives

$$m_1(v_{1_o}^2 - v_1^2) = m_2 v_2^2 \tag{1}$$

and

$$m_1(v_{1_o} - v_1) = m_2 v_2 \tag{2}$$

Then, using the relationship $x^2 - y^2 = (x + y)(x - y)$ and dividing the first equation by the second

$$\frac{m_1(v_{1_o} + v_1)(v_{1_o} - v_1) = m_2 v_2^2}{m_1(v_{1_o} - v_1) = m_2 v_2} \quad \frac{(\text{Eq. 1})}{(\text{Eq. 2})} \quad (divide)$$

we obtain

$$v_{1_o} + v_1 = v_2 \tag{6.14}$$

Eq. 6.14 can be used to eliminate v_1 or v_2 from either of the preceding two equations. Thus, each of these final velocities may be expressed in terms of the initial velocity (v_{1_o}):

$$v_1 = \left(\frac{m_1 - m_2}{m_1 + m_2}\right) v_{1_o} \qquad \text{(final velocities for} \tag{6.15}$$

$$v_2 = \left(\frac{2m_1}{m_1 + m_2}\right) v_{1_o} \qquad \begin{array}{l}\textit{elastic, head-on collision} \\ \textit{with } m_2 \textit{ initially stationary)}\end{array} \tag{6.16}$$

Note that the final velocities depend on mass *ratios* of the objects. Looking at Eq. 6.15, you can see that if m_1 is greater than m_2, then v_1 is positive, or in the same direction as the initial velocity of the incoming ball, as was assumed in Fig. 6.12. However, if m_2 is greater than m_1, then v_1 is negative, and the incoming ball recoils in the opposite direction after collision. Eq. 6.16 shows that v_2 is always in the same direction as the velocity of the incoming ball.

You can also get some general ideas about what happens after such a collision by considering three possible situations for the relative masses of the objects, illustrated in (•Fig. 6.13).

Case 1. $m_1 = m_2$ (Fig. 6.13a). From Eqs. 6.15 and 6.16,

$$v_1 = 0 \quad \text{and} \quad v_2 = v_{1_o}$$

Elastic head-on collision, $m_1 = m_2$

That is, if the masses of the colliding objects are equal, the objects simply exchange momentum and energy. The incoming ball is stopped on collision, and the originally stationary ball moves off with the same velocity that the incoming ball had, thus obviously conserving system kinetic energy and momentum.

Case 2. $m_1 \gg m_2$ (m_1 very much greater than m_2, Fig. 6.13b). In this case, m_2 can be ignored in the addition and subtraction with m_1 in Eqs. 6.15 and 6.16, and

$$v_1 \approx v_{1_o} \quad \text{and} \quad v_2 \approx 2v_{1_o}$$

Elastic head-on collision, $m_1 \gg m_2$

This tells you that if a very massive object collides with a stationary light object, the massive object is slowed down only slightly by the collision, and the light object is knocked away with a velocity almost twice that of the initial velocity of the massive object.

Case 3. $m_1 \ll m_2$ (m_1 very much less than m_2, Fig. 6.13c). Here, m_1 can be ignored in the addition and subtraction with m_2 in Eqs. 6.15 and 6.16, and

$$v_1 \approx -v_{1_o} \quad \text{and} \quad v_2 \approx 0$$

Elastic head-on collision, $m_1 \ll m_2$

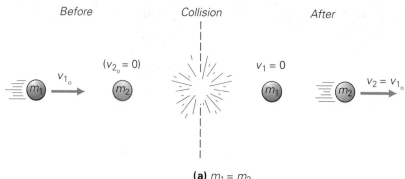

Before   Collision   After

(a) $m_1 = m_2$

•FIGURE 6.13 Special cases of head-on elastic collisions
(a) When a moving object collides elastically with a stationary object of equal mass, there is a complete exchange of momentum and energy. **(b)** When a very massive moving object collides elastically with a much less massive stationary object, the very massive object continues to move essentially as before, and the less massive object is given a velocity almost twice the initial velocity of the large mass. **(c)** When a moving object of small mass collides elastically with a very massive stationary object, the incoming object recoils in the opposite direction with approximately the same speed and the very massive object remains essentially stationary.

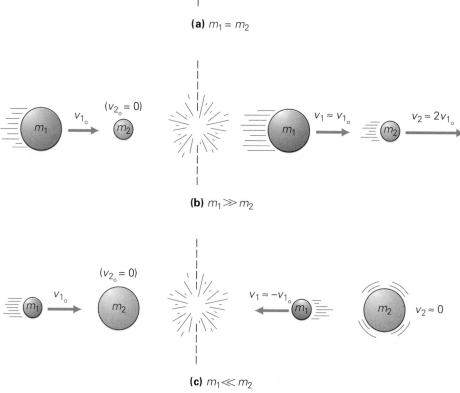

(b) $m_1 \gg m_2$

(c) $m_1 \ll m_2$

(In the second equation, the approximation $m_1/m_2 \approx 0$ is made.) Thus, if a light object collides with a massive stationary one, the massive object remains *almost* stationary, and the light object recoils backward with approximately the same speed it had before collision. An extreme case of this type is similar to a particle striking a solid, immovable wall (see Fig. 6.3). If the wall is anchored to the ground, then the Earth and wall recoil together, but this is totally unnoticeable (why?). Total momentum, however is still conserved. For the case in Fig. 6.13c, the massive ball must move a bit after the collision to conserve momentum.

EXAMPLE 6.9 ■ ELASTIC COLLISION: CONSERVATION OF MOMENTUM
AND KINETIC ENERGY

A 0.30-kg object with a speed of 2.0 m/s in the positive x direction has a head-on elastic collision with a stationary 0.70-kg object located at $x = 0$. What is the distance separating the objects 2.5 s after the collision?

Solution. Using the previous notation, we have

Given: $m_1 = 0.30$ kg   *Find:* $\Delta x = x_2 - x_1$ (separation distance)
           $v_{1_o} = 2.0$ m/s
           $m_2 = 0.70$ kg
           $t = 2.5$ s

From Eqs. 6.15 and 6.16, the velocities after collision are

$$v_1 = \left(\frac{m_1 - m_2}{m_1 + m_2}\right)v_{1_o} = \left(\frac{0.30\ \text{kg} - 0.70\ \text{kg}}{0.30\ \text{kg} + 0.70\ \text{kg}}\right)(2.0\ \text{m/s}) = -0.80\ \text{m/s}$$

$$v_2 = \left(\frac{2m_1}{m_1 + m_2}\right)v_{1_o} = \left[\frac{2(0.30\ \text{kg})}{0.30\ \text{kg} + 0.70\ \text{kg}}\right](2.0\ \text{m/s}) = 1.2\ \text{m/s}$$

Here m_1 is less than m_2, but not *so much* less that it could be ignored, as was done in part (c) of Fig. 6.13.

The objects are separating after collision and their positions are

$$x_1 = v_1 t = (-0.80\ \text{m/s})(2.5\ \text{s}) = -2.0\ \text{m}$$

$$x_2 = v_2 t = (1.2\ \text{m/s})(2.5\ \text{s}) = 3.0\ \text{m}$$

and

$$\Delta x = x_2 - x_1 = 3.0\ \text{m} - (-2.0\ \text{m}) = 5.0\ \text{m}$$

The objects are 5.0 m apart at that time.

Follow-up Exercise. Suppose the objects in this Example had equal masses. What would be the separation distance 2.5 s after collision in this case? *(Answer may be found in the Answers to Follow-up Exercises section at the back of the book.)*

EXAMPLE 6.10 ■ SPARE! CONSERVATION OF MOMENTUM AND KINETIC ENERGY

A 7.1-kg bowling ball with a speed of 6.0 m/s has a head-on elastic collision with a stationary 1.6-kg pin. (a) What is the velocity of each object after the collision? (b) What is the total momentum after the collision?

Solution. Listing the data with the same notation as before, we have

Given: $m_1 = 7.1$ kg $\qquad$ *Find:* (a) $\mathbf{v}_1$ and $\mathbf{v}_2$ (velocities after collision)
$\qquad\ v_{1_o} = 6.0$ m/s $\qquad\qquad\qquad$ (b) $\mathbf{P}$ (total momentum after collision)
$\qquad\ m_2 = 1.6$ kg

(a) The velocities are given directly by Eqs. 6.15 and 6.16, where the direction of the incoming ball is taken as positive.

$$v_1 = \left(\frac{m_1 - m_2}{m_1 + m_2}\right)v_{1_o} = \left(\frac{7.1\ \text{kg} - 1.6\ \text{kg}}{7.1\ \text{kg} + 1.6\ \text{kg}}\right)(6.0\ \text{m/s}) = 3.8\ \text{m/s}$$

$$v_2 = \left(\frac{2m_1}{m_1 + m_2}\right)v_{1_o} = \left[\frac{2(7.1\ \text{kg})}{(7.1\ \text{kg} + 1.6\ \text{kg})}\right](6.0\ \text{m/s}) = 9.8\ \text{m/s}$$

So both objects move in the same direction. Here m_1 is greater than m_2, but not *so much* greater that m_2 could be ignored, as was done in part (b) of Fig. 6.13.

(b) Again, momentum is conserved in elastic (and inelastic) collisions, so the total momentum afterwards is the same as that before the collision, which is that of the incident ball, with a magnitude of:

$$P_f = P_{1_o} = m_1 v_{1_o} = (7.1\ \text{kg})(6.0\ \text{m/s}) = 43\ \text{kg·m/s}$$

and in the same direction.

Follow-up Exercise. In general, for head-on, elastic collisions with one of the objects initially stationary, is it possible for the velocities to be the same after collision? Explain. *(Answer may be found in the Answers to follow-up Exercises section at the back of the book.)*

CONCEPTUAL EXAMPLE 6.11 ■ TWO IN, ONE OUT?

A novelty collision device, as shown in ●Fig. 6.14, consists of five identical metal balls. When one ball swings in, after multiple collisions, one ball swings out at the other end of the row of balls. When two balls swing in, two swing out; when three swing in, three swing out, and so on—always the same number out as in.

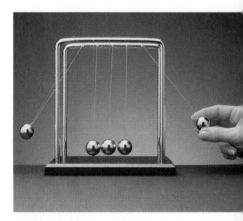

●**FIGURE 6.14 In and out**
See Example 6.11.

Suppose that two balls, each of mass m, swing in and collide with the next ball with a velocity v. Why doesn't one ball swing out at the other end with a velocity of $2v$? *Clearly establish the reasoning and physical principle(s) used in determining your answer before checking it below. That is, **how** did you arrive at your answer?*

Reasoning and Answer. The collisions along the horizontal row of balls are approximately elastic. Certainly two balls in and one out with twice the velocity doesn't violate the conservation of momentum: $(2m)v = m(2v)$. However, there's another condition if we assume elastic collisions—the conservation of kinetic energy. Let's check to see if this is so for this case.

$$K_o = K$$

$$\text{(before)} \qquad \text{(after)}$$
$$\tfrac{1}{2}(2m)v^2 \stackrel{?}{=} \tfrac{1}{2}m(2v)^2$$
$$mv^2 \neq 2mv^2$$

Hence, the kinetic energy would not be conserved if this happened, and what the equation is telling us is that this situation violates physical principles and does not occur. Note that there's a big violation—more energy out than in.

Follow-up Exercise. Suppose the first ball of mass m were replaced with a ball of mass $2m$. When this ball is pulled back and allowed to swing in, how many balls will swing out? (*Hint*: Think about the analogous situation in Fig. 6.13 and remember that the balls in the row are actually colliding. It may help to think of them as being separated.) (*Reasoning and answer may be found in the Answers to Follow-up Exercises section at the back of the book.*)

6.5 Center of Mass

Objectives: **To be able to (a) explain the concept of the center of mass and compute its location for simple systems, and (b) describe how the center of mass and center of gravity are related.**

The conservation of total momentum gives us a method for analyzing a "system of particles." Such a system may be virtually anything—a volume of gas, water in a container, or a baseball. Another important concept allows us to analyze the overall motion of a system of particles. It involves representing the whole system as a single particle. This concept will be introduced here and applied in more detail in the upcoming chapters.

If no net external force acts on a particle, we have seen that its linear momentum is constant. Similarly, if no net external force acts on a *system* of particles, the linear momentum of the system is constant. This similarity implies that a system of particles might be represented by an *equivalent* single particle. Moving rigid objects, such as balls, automobiles, and so forth, are essentially systems of particles and can be effectively represented by equivalent single particles when analyzing motion. Such representation is done through the concept of the **center of mass (CM)**.

The center of mass is the point at which all of the mass of an object or system may be considered to be concentrated, for the purposes of linear or translational motion only.

Even if a rigid object is rotating, the center of mass moves as though it were a particle (●Fig. 6.15). The center of mass is sometimes descriptively said to be at the balance point of a solid object. For example, if you balance a meterstick on your finger, the center of mass of the stick is located directly above your finger and all of the mass (or weight) seems to be concentrated there.

It turns out that an expression similar to Newton's second law for a single particle applies to a *system* when the center of mass is used:

$$\mathbf{F}_{\text{net external}} = M\mathbf{A}_{\text{CM}} \qquad (6.17)$$

where $\mathbf{F}$ is the net external force on the system, M is the total mass of the system or the sum of the masses of the particles of the system ($M = m_1 + m_2 + m_3 + \cdots + m_n$, where the system has n particles), and $\mathbf{A}_{\text{CM}}$ is the acceleration of the center of

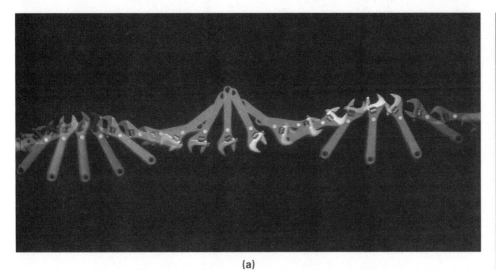

(a)

•**FIGURE 6.15 Center of mass**
(a) The center of mass of this sliding wrench moves as though it were a particle. Note the dot on the wrench that marks the center of mass. **(b)** Even after exploding, the center of mass of a fireworks projectile would follow a parabolic path in the absence of air resistance.

mass of the system. In words, Eq. 6.17 says that the *center of mass* of a system of particles moves as though all the mass of the system were concentrated there and acted on by the resultant of the external forces.

It follows that if the net external force on a system is zero, the total linear momentum of the center of mass is conserved (stays constant) because

$$\mathbf{F} = M\mathbf{A}_{CM} = M(\Delta\mathbf{V}_{CM}/\Delta t) = (\Delta M\mathbf{V}_{CM}/\Delta t) = (\Delta\mathbf{P}/\Delta t)$$

Thus the total momentum of the system, $\mathbf{P} = M\mathbf{V}_{CM}$, is constant. Since M is constant (why?), in this case $\mathbf{V}_{CM}$ is a constant. Thus the center of mass either moves with a constant velocity or remains at rest.

Although you may more readily visualize the center of mass of a solid object, the concept of the center of mass applies to any system of particles or objects, even a quantity of gas. For a system of n particles arranged in one dimension, along the x axis (•Fig. 6.16), the location of the center of mass is given by

$$\mathbf{X}_{CM} = \frac{m_1\mathbf{x}_1 + m_2\mathbf{x}_2 + m_3\mathbf{x}_3 + \cdots + m_n\mathbf{x}_n}{m_1 + m_2 + m_3 + \cdots + m_n} \tag{6.18}$$

That is, X_{CM} is the x coordinate of the center of mass of a system of particles. In shorthand notation (using signs to indicate vector directions),

$$X_{CM} = \frac{\Sigma_i m_i x_i}{M} \tag{6.19}$$

where Σ_i indicates the summation of the products $m_i x_i$ for i particles ($i = 1, 2, 3, \ldots, n$). If $\Sigma_i m_i x_i = 0$, then $X_{CM} = 0$, and the center of mass of the one-dimensional system is located at the origin.

Other coordinates of the center of mass for systems of particles are similarly defined. For a two-dimensional distribution of masses, the coordinates of the center of mass are (X_{CM}, Y_{CM}).

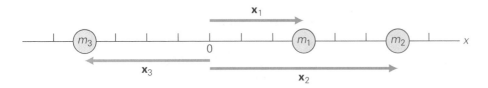

•**FIGURE 6.16 System of particles in one dimension**
Where is the system's center of mass? See Example 6.12.

EXAMPLE 6.12 ■ FINDING THE CENTER OF MASS: A SUMMATION PROCESS

Three masses, 2.0 kg, 3.0 kg, and 6.0 kg, are located at positions $(3.0, 0)$, $(6.0, 0)$, and $(-4.0, 0)$, respectively, in meters from the origin (Fig. 6.16). Where is the center of mass of this system?

Solution.

Given: $m_1 = 2.0$ kg   *Find:* X_{CM} (CM coordinate)
$m_2 = 3.0$ kg
$m_3 = 6.0$ kg
$x_1 = 3.0$ m
$x_2 = 6.0$ m
$x_3 = -4.0$ m

Then, simply performing the summation as indicated in Eq. 6.19,

$$X_{CM} = \frac{\Sigma_i m_i x_i}{M}$$

$$= \frac{(2.0 \text{ kg})(3.0 \text{ m}) + (3.0 \text{ kg})(6.0 \text{ m}) + (6.0 \text{ kg})(-4.0 \text{ m})}{2.0 \text{ kg} + 3.0 \text{ kg} + 6.0 \text{ kg}}$$

$$= 0$$

The center of mass is at the origin.

Follow-up Exercise.   Describe a single particle that could replace the system in this Example in order to analyze the translational part of the motion of the system as a whole. (*Answers may be found in the Answers to Follow-up Exercises section at the back of the book.*)

EXAMPLE 6.13 ■ A DUMBBELL: CENTER OF MASS REVISITED

A dumbbell (• Fig. 6.17) has a connecting bar of negligible mass. Find the location of the center of mass (a) if m_1 and m_2 are each 5.0 kg, and (b) if m_1 is 5.0 kg and m_2 is 10.0 kg.

Solution.

Given: (a) $m_1 = m_2 = 5.0$ kg   *Find:*   (a) (X_{CM}, Y_{CM}) (CM coordinates)
$x_1 = 0.20$ m   (b) (X_{CM}, Y_{CM})
$x_2 = 0.90$ m
$y_1 = y_2 = 0.10$ m

(b) $m_1 = 5.0$ kg
$m_2 = 10.0$ kg

Note that each mass is considered to be a particle located at the center of the sphere (its center of mass).

(a) Finding X_{CM} gives

$$X_{CM} = \frac{m_1 x_1 + m_2 x_2}{m_1 + m_2}$$

$$= \frac{(5.0 \text{ kg})(0.20 \text{ m}) + (5.0 \text{ kg})(0.90 \text{ m})}{5.0 \text{ kg} + 5.0 \text{ kg}} = 0.55 \text{ m}$$

Similarly, it is easy to find that $Y_{CM} = 0.10$ m. (You might have seen this right away, since each center of mass is at this height.) The center of mass of the dumbbell is then located at $(X_{CM}, Y_{CM}) = (0.55 \text{ m}, 0.10 \text{ m})$, or midway between the end masses.

(b) With $m_2 = 10.0$ kg,

$$X_{CM} = \frac{m_1 x_1 + m_2 x_2}{m_1 + m_2}$$

$$= \frac{(5.0 \text{ kg})(0.20 \text{ m}) + (10.0 \text{ kg})(0.90 \text{ m})}{5.0 \text{ kg} + 10.0 \text{ kg}} = 0.67 \text{ m}$$

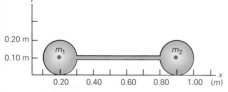

•**FIGURE 6.17   Location of the center of mass**
See Example 6.13.

In Example 6.13, when the value of one of the masses changed, the x coordinate of the center of mass changed. You might have expected the y coordinate to change also. However, the centers of the end masses were still at the same height, and Y_{CM} remained the same. To increase Y_{CM}, one or both of the end masses would have to be in a higher position.

Center of Gravity

As you know, mass and weight are directly related. Closely associated with the center of mass is the concept of the **center of gravity** (**CG**), the point where all of the weight of an object may be considered to be concentrated in representing the object as a particle. If the acceleration due to gravity is constant both in magnitude and direction over the extent of the object, Eq. 6.19 can be rewritten as

$$MgX_{CM} = \Sigma_i m_i g x_i \qquad (6.20)$$

Then, the weight, Mg, acts as if it were concentrated at X_{CM}, and the center of mass and the center of gravity coincide. As you may have noticed, the location of the center of gravity was implied in some previous figures where the vector arrows for weight ($\mathbf{w} = m\mathbf{g}$) were drawn from a point at or near the center of an object.

In some cases, the center of mass or the center of gravity of an object may be located by symmetry. For example, for a spherical object that is homogeneous (the mass is distributed evenly throughout), the center of mass is at the geometrical center (or center of symmetry). In Example 6.13a, where the end masses of the dumbbell were equal, it was probably apparent that the center of mass was midway between them.

The location of the center of mass of an irregularly shaped object is not so evident and is usually difficult to calculate (even with advanced mathematical methods that are beyond the scope of this book). In some instances, the center of mass may be located experimentally. For example, the center of mass of a flat, irregularly shaped object may be determined experimentally by suspending it freely from different points (•Fig. 6.18). A moment's thought should convince you that

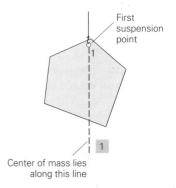

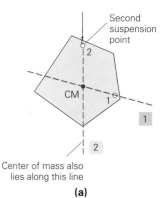

Center of mass also lies along this line

(a)

•**FIGURE 6.18 Location of the center of mass by suspension** (a) The center of mass of a flat, irregularly shaped object may be found by suspending the object from two or more points. The CM (and CG) lies on a vertical line under any point of suspension, so the intersection of two such lines marks its location midway through the thickness of the body. The sheet could be balanced horizontally at this point. Why? (b) The process is illustrated at left with a cutout map of the United States. Note that a plumb line dropped from any other point (third photo) does in fact pass through the CM as located in the first two photos.

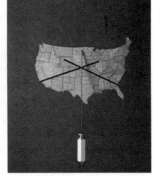

(b)

(a)

(b)

•**FIGURE 6.19 The center of mass may be located outside of a body** The center of mass may lie either inside or outside of a body, depending on its mass distribution. **(a)** For a uniform ring, the center of mass is at its center. **(b)** For an L-shaped object, if the mass distribution is uniform and the legs are of equal length, the center of mass lies on the diagonal between the legs.

•**FIGURE 6.20 Center of gravity** By arching her body, this high jumper can get over the bar even though her center of gravity passes beneath it.

the center of mass (or center of gravity) always lies vertically below the point of suspension. Since the center of mass is defined as the point at which all the mass of a body can be considered to be concentrated, this is analogous to a particle of mass suspended from a string. Suspending the object from two or more points and marking the vertical lines on which the center of mass must lie locates the center of mass as the intersection of the lines.

The center of mass of an object may lie outside the body of the object. Examples are shown in •Fig. 6.19. The center of mass of a homogeneous ring is at its center. The mass in any section of the ring is compensated by the mass in an equivalent section directly across the ring, and by symmetry the center of mass is at the center. For an L-shaped object with equal legs, the center of mass lies on a line that makes a 45° angle with both legs. Its location can easily be determined by suspending the L from a point on one of the legs and noting where a vertical line from that point intersects the diagonal line.

Keep in mind that the location of the center of mass or center of gravity of an object depends on the distribution of mass. Therefore, for a flexible object such as the human body, the position of the center of gravity changes as the object changes configuration (mass distribution). For example, when a person raises both arms overhead, his or her center of gravity is raised several centimeters. For a high jumper going over a bar, the center of gravity lies outside the arched body (•Fig. 6.20). In fact, the center of gravity passes *beneath* the bar. This is done purposefully because work must be done to raise the center of gravity, and only the jumper's body has to clear the bar, not the CG.

6.6 Jet Propulsion and Rockets

Objective: To apply the conservation of momentum in the explanation of jet propulsion and the operation of rockets.

The word "jet" is sometimes used to refer to a stream of liquid or gas emitted at a high speed, for example, a jet of water from a fountain or a jet of air from an automobile tire. **Jet propulsion** is the application of such jets to the production of motion. This usually brings to mind jet planes and rockets, but squids and octopuses propel themselves by squirting jets of water. You have probably tried the simple application of blowing up a balloon and releasing it. Lacking any guidance or rigid exhaust system, the balloon zig-zags around, driven by the escaping air. In terms of Newton's third law, the air is forced out by the contraction of the stretched balloon—that is, the balloon exerts a force on the air. Thus there must be an equal and opposite reaction force exerted by the air on the balloon. It is this force that propels the balloon on its erratic path.

Jet propulsion is explained by Newton's third law, and in the absence of external forces, the conservation of momentum also applies. You may understand this better by considering the recoil of a rifle, taking the rifle and the bullet as an isolated system (•Fig. 6.21). Initially, the total momentum of this system is zero. When the rifle is fired (by remote control to avoid external forces), the expansion of the gases from the exploding charge accelerates the bullet down the barrel. These gases push backward on the rifle as well, producing a recoil force (the "kick" experienced by the person firing the weapon). Since the initial momentum of the system is zero, and the force of the expanding gas is an internal force, the momenta of the bullet and of the rifle must be exactly equal and opposite at any instant. After the bullet leaves the barrel, there is no propelling force, so the bullet and the rifle move with constant velocities (unless acted upon by a net external force such as gravity or air resistance).

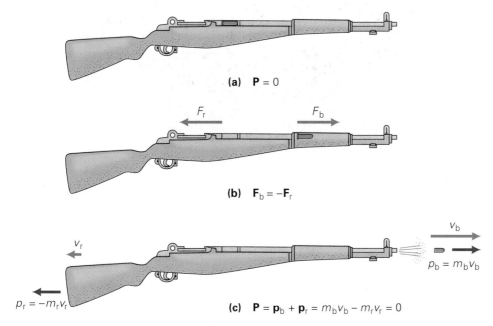

(a) $\mathbf{P} = 0$

(b) $\mathbf{F_b} = -\mathbf{F_r}$

F_r F_b

v_b

v_r

$p_b = m_b v_b$

$p_r = -m_r v_r$

(c) $\mathbf{P} = \mathbf{p_b} + \mathbf{p_r} = m_b v_b - m_r v_r = 0$

•**FIGURE 6.21 Conservation of momentum**
(a) Before it is fired, the total momentum of the rifle and bullet (as an isolated system) is zero. **(b)** During firing, there are equal and opposite internal forces, and the instantaneous total momentum of the rifle-bullet system remains zero (neglecting external forces, such as arise when a rifle is being held). **(c)** When the bullet leaves the barrel, the total momentum is still zero.

Similarly, the thrust for a rocket is created by exhausting the gas from burning fuel out the rear of the rocket. The expanding gas exerts a net force on the rocket that propels it in the forward direction (•Fig. 6.22a, b). The rocket exerts a reaction force on the gas, so that all of it is directed out the exhaust nozzle. If the rocket is at rest when the engines are turned on and there are no external forces (as in space, where friction is zero and gravitational forces are negligible), then the instantaneous momentum of the exhaust gas is equal and opposite to that of the rocket. The numerous exhaust gas molecules have small masses and high velocities, and the rocket has a much larger mass and a smaller velocity. (See Demonstration 4.)

Unlike a rifle firing a single shot, a rocket engine continually loses mass when burning fuel (it is more like a machine gun). Thus, the rocket is a system for which the mass is not constant. As the mass of the rocket decreases, it accelerates more easily. Multistage rockets take advantage of this fact. The hull of a burnt-out stage is jettisoned to give an in-flight reduction in mass (•Fig. 6.22b,c). The payload is actually a very small part of the initial mass of rockets for space flights.

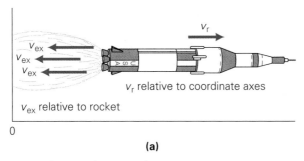

v_{ex}
v_{ex}
v_{ex}

v_r

v_r relative to coordinate axes

v_{ex} relative to rocket

0

(a)

•**FIGURE 6.22 Jet propulsion and mass reduction**
(a) A rocket burning fuel is continually losing mass, and so becomes easier to accelerate. The resulting force on the rocket (called the thrust) can be shown to depend on the product of the rate of change of its mass with time and the velocity of the exhaust gases: $(\Delta m/\Delta t)\mathbf{v_{ex}}$. Since the mass is decreasing, $(\Delta m/\Delta t)$ is negative and the thrust is opposite $\mathbf{v_{ex}}$. **(b)** The space shuttle makes use of a multistage rocket. Both the two booster rockets and the huge external fuel tank are jettisoned in flight. **(c)** The first and second stages of a Saturn V rocket are shown separating after 148 seconds of burn time.

(b)

(c)

DEMONSTRATION 4 ■ Jet Propulsion

(a) Ignition and blast off!

(b) Away we go—equal and opposite forces

A demonstration of Newton's third law and conservation of momentum. For a rocket or jet engine, thrust is obtained by burning fuel and exhausting the gas out the rear of the engine. Here a fire extinguisher exhausting CO_2 takes the place of the engine.

Suppose that the purpose of a space flight is to land a payload on the Moon. At some point on the journey, the gravitational attraction of the Moon will become greater than that of Earth, and the spacecraft will accelerate toward the Moon. A soft landing is desirable, so the spacecraft must be slowed down enough to go into orbit about the Moon. This is done by using the rocket engines to apply a **reverse thrust**, or braking thrust. The spacecraft is maneuvered through a 180° angle, or turned around, which is quite easy to do in space. The rocket engines are then fired, expelling the exhaust gas toward the Moon and supplying a braking action.

You have experienced a reverse thrust effect if you have flown in a commercial jet. In this instance, however, the craft is not turned around. Instead, after landing, the jet engines are revved up and a braking action can be felt. Ordinarily, revving up the engines accelerates the plane forward. The reverse thrust is accomplished by activating thrust reversers in the engines that deflect the exhaust gases forward (•Fig. 6.23). The gas experiences an impulse force and a change in momentum in the forward direction (see Fig. 6.3b), and the engine and the aircraft have an equal and opposite momentum change and braking impulse force.

Question: There are no end-of-chapter exercises for this section, so test your knowledge with this one: Astronauts use hand-held maneuvering devices (small rockets) to move around on space walks. Describe how these rockets would be used. Is there any danger on an untethered space walk?

•**FIGURE 6.23 Reverse thrust** Thrust reversers are activated on jet engines on landing to help slow the plane. The gas experiences an impulse force and a change in momentum in the forward direction, and the plane experiences an equal and opposite momentum change and a braking impulse force.

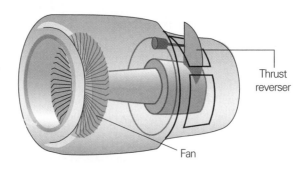

Normal operation

Thrust reverser activated

Chapter Review

Important Terms

linear momentum   168
total linear momentum   168
conservation of linear
  momentum   174
impulse   178

impulse-momentum theorem   178
elastic collision   181
inelastic collision   181
completely inelastic collision   183
center of mass (CM)   188

center of gravity (CG)   191
jet propulsion   192
reverse thrust   194

Important Concepts

- Linear momentum of a particle is a vector and is defined as the product of mass and velocity.
- The total linear momentum of a system is the vector sum of the momenta of the individual particles.
- In the absence of a net external force, the total linear momentum is conserved.
- The impulse-momentum theorem relates impulse to the change in momentum.

- A collision is elastic if the total kinetic energy is conserved. Momentum is conserved in both elastic and inelastic collisions. In a completely inelastic collision, objects stick together after impact.
- The center of mass is the point at which all of the mass of an object or system may be considered to be concentrated. (The center of gravity is the point where all the weight may be considered to be concentrated.)

Important Equations

Linear Momentum:
$$\mathbf{p} = m\mathbf{v} \tag{6.1}$$

Total Linear Momentum of a System:
$$\mathbf{P} = \mathbf{p}_1 + \mathbf{p}_2 + \mathbf{p}_3 + \cdots = \Sigma_i \mathbf{p}_i \tag{6.2}$$

Newton's Second Law in Terms of Momentum:
$$\mathbf{F} = \frac{\Delta \mathbf{p}}{\Delta t} \tag{6.3}$$

Impulse-Momentum Theorem:
$$\text{Impulse} = \overline{\mathbf{F}}\,\Delta t = \Delta \mathbf{p} = m\mathbf{v} - m\mathbf{v}_\text{o} \tag{6.6}$$

Conditions for an Elastic Collision:
$$\begin{aligned} \mathbf{P}_\text{f} &= \mathbf{P}_\text{i} \\ K_\text{f} &= K_\text{i} \end{aligned} \tag{6.8}$$

Conditions for an Inelastic Collision:
$$\begin{aligned} \mathbf{P}_\text{f} &= \mathbf{P}_\text{i}, \\ K_\text{f} &< K_\text{i} \end{aligned} \tag{6.9}$$

Final Velocities in Head-On Two-Body Elastic Collisions: ($v_{2_\text{o}} = 0$):
$$v_1 = \left(\frac{m_1 - m_2}{m_1 + m_2}\right) v_{1_\text{o}} \tag{6.15}$$

$$v_2 = \left(\frac{2m_1}{m_1 + m_2}\right) v_{1_\text{o}} \tag{6.16}$$

Coordinate of the Center of Mass (using signs for directions):
$$X_\text{CM} = \frac{\Sigma_i m_i x_i}{M} \tag{6.19}$$

Exercises

6.1 Linear Momentum

1   Linear momentum has units of (a) N/m, (b) kg·m/s, (c) N/s, (d) all of these.

2   Momentum is (a) always conserved, (b) a scalar quantity, (c) a vector quantity that can be resolved into components, (d) unrelated to force.

3   A fan boat of the type used in swampy and marshy areas is shown in •Fig. 6.24. Explain the principle of its propulsion using momentum conservation.

4   Two objects have the same momentum. (a) Will they also have the same kinetic energy? (b) What can you say definitely about their motion?

•**FIGURE 6.24   Fan propulsion**
See Exercise 3.

5 If you are riding in a car traveling at 90 km/h, what is your momentum relative to (a) the ground and (b) the car?

6 ■ Find the magnitude of the momentum of (a) a 0.50-kg ball traveling at 20 m/s and (b) a 1200-kg automobile traveling at 100 km/h.

7 ■ What is the magnitude of the linear momentum of a 7.1-kg bowling ball going down the alley with a speed of 20 m/s?

8 ■ The fastest time for the straight 200-m dash is about 19.8 s. What is the magnitude of the average momentum of a 65.5-kg runner who finishes the dash in that time?

9 ■■ After jumping vertically upward to make a slam-dunk, an 85-kg basketball player is essentially in free fall on the way back down. What is the magnitude of his time rate of change of momentum as he returns to the floor?

10 ■■ How fast would an automobile having a mass of 1000 kg have to be going to have the same linear momentum as a truck having a mass of 2.00 metric tons and traveling along a straight road with a constant speed of 30.0 km/h?

11 ■■ A 0.150-kg baseball traveling with a horizontal speed of 4.50 m/s is hit by a bat and then moves with a speed of 34.7 m/s in the opposite direction. What is the change in the ball's momentum?

12 ■■ If two electrons approach each other with speeds of 3.4×10^2 m/s and 4.5×10^2 m/s, respectively, what is the total momentum of the two-particle system? [*Hint:* Find the mass of an electron in the Table inside the back cover.]

13 ■■ If a 0.25-kg ball is dropped from a height of 10 m, what is the momentum of the ball (a) 0.75 s after being released and (b) just before it hits the ground?

14 ■■ A loaded tractor-trailer with a total mass of 5.0×10^3 kg traveling at 3.0 km/h coasts into a loading dock and collides, coming to a stop in 0.64 s. What is the magnitude of the average force exerted on the truck by the dock?

15 ■■ The spout of a water tower used to fill fire-fighting tank trucks is 6.0 m above the ground. If the spout is inadvertently left open and water runs out vertically at a rate of 4.0 L/s, what is the average force exerted by the water on the ground? (Assume $v_o = 0$.)

16 ■■ Taking the density of air to be 1.29 kg/m³, what is the magnitude of the momentum of a liter of air moving with a wind speed of (a) 30 km/h, and (b) 74 mi/h (the wind speed at which a tropical storm becomes a hurricane).

17 ■■ Two balls of equal mass (0.45 kg) approach the origin along the positive x and y axes at the same speed (3.3 m/s). (a) What is the total momentum of the system? (b) Will the balls necessarily collide at the origin?

What is the total momentum of the system after they have both passed through the origin?

18 ■■ A 0.20-kg billiard ball traveling at a speed of 15 m/s strikes the side rail of the table at an angle of 60° (•Fig. 6.25). If the ball rebounds at the same speed and angle, what is the change in its momentum?

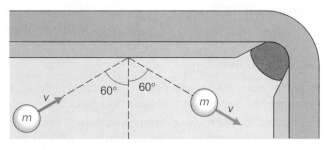

•FIGURE 6.25 Glancing collision
See Exercises 18, 19, and 60.

19 ■■ Suppose that the billiard ball in Fig. 6.25 approaches the rail at a speed of 15 m/s and an angle of 60° as shown, but rebounds at a speed of 10 m/s and an angle of 50°. What is the change in momentum in this case? [*Hint:* Use components.]

20 ■■■ Four particles have masses $m_1 = 10$ g, $m_2 = 15$ g, $m_3 = 13$ g, and $m_4 = 18$ g and velocities $\mathbf{v}_1 = (2.0$ m/s$)\,\mathbf{x}$, $\mathbf{v}_2 = (-0.50$ m/s$)\,\mathbf{x}$, $\mathbf{v}_3 = (0.70$ m/s$)\,\mathbf{y}$, and $\mathbf{v}_4 = (-1.9$ m/s$)\,\mathbf{y}$. What is the total momentum of the four-particle system?

21 ■■■ At a basketball game, a 120-lb cheerleader is tossed vertically upward with a speed of 4.50 m/s by another cheerleader. (a) What is the cheerleader's change in momentum from the time of release to just before being caught if she is caught at the same height? (b) Would there be any difference if she were caught 0.25 m below the point of release? If so, what is the change then?

6.2 The Conservation of Linear Momentum

22 The momentum of an object is conserved if (a) the force acting on the object is conservative, (b) there is a single, unbalanced internal force acting on the object, (c) the mechanical energy is conserved, (d) none of these.

23 Internal forces do not affect the conservation of momentum because (a) they cancel each other, (b) their effects are canceled by external forces, (c) they can never produce a change in velocity, (d) Newton's second law is not applicable to them.

24 What would be the difference in the physical effects of firing a blank cartridge and a regular cartridge (with a bullet) in a gun and why?

25 Imagine yourself standing in the middle of a frozen lake. The ice is so smooth that it is frictionless. How could you get to shore? (You couldn't walk. Why?)

26 ■ (a) A 4.0-metric ton truck traveling at 90 km/h collides with a stationary 1.0-metric ton car. If the vehicles remain joined together, what is their velocity immediately after collision? (b) What percentage of the initial kinetic energy is lost in the collision?

27 ■ A 70-kg man and his 40-kg daughter on skates stand together on a frozen lake. If they push apart and the father has a velocity of 0.50 m/s eastward, what is the velocity of the daughter? (Neglect friction.)

28 ■■ To get off a frozen, frictionless lake, a 70.0-kg person takes off a 0.150-kg shoe and throws it horizontally directly away from the shore with a speed of 2.00 m/s. If the person is 5.00 m from the shore, how long does it take for him to reach it?

29 ■■ An isolated 3.0-kg object moves along the y axis between the third and fourth quadrants at a speed of 2.0 m/s. When it reaches the origin, an internal explosion fractures the object into two pieces sending a 1.0-kg fragment in the positive x direction at a speed of 3.7 m/s. What is the velocity of the remainder of the object after the explosion? [*Hint:* Use components.]

30 ■■ An isolated 3.0-kg object initially at rest explodes and splits into three fragments. One fragment has a mass of 0.50 kg and flies off along the negative x axis at a speed of 2.8 m/s, and another has a mass of 1.3 kg and flies off along the negative y axis at a speed of 1.5 m/s. What is the speed and direction of the third fragment?

31 ■■ Suppose that the 3.0-kg object in Exercise 30 is initially traveling at a speed of 2.5 m/s in the positive x direction. What will the speed and direction of the third fragment be in this case?

32 ■■ Identical railroad freight cars hit each other and couple together. In each of the following cases, what are the velocities of the cars immediately after coupling? (a) A moving car approaches a stationary one with a velocity of +10 km/h. (b) Two cars approach each other with velocities of +20 km/h and −15 km/h, respectively. (c) Two cars travel in the same direction with velocities of +20 km/h and +15 km/h.

33 ■■ A 1600-kg empty hopper car rolls under a loading bin with a speed of 2.5 m/s, and a 3500-kg load is deposited in the car. What is the magnitude of the velocity of the car immediately after being loaded?

34 ■■■ For a movie scene, a 75-kg stunt man drops from a tree onto a 50-kg sled moving with a velocity of 6.0 m/s toward the shore on a frozen lake. (a) What is the speed of the sled after the stunt man is on board? (b) If the sled hits the bank and stops but the stunt man keeps on going, with what speed does he leave the sled? (Neglect friction.)

35 ■■■ A projectile that is fired from a gun has an initial velocity of 90.0 km/h at an angle of 60.0° to the horizontal. When the projectile is at the top of its trajectory, an internal explosion causes it to separate into two fragments that have equal masses. One of the fragments falls straight downward as though it has been released from rest. How far does the other fragment land from the gun?

36 ■■■ A ballistic pendulum is a device used to measure the velocity of a projectile, for example, the muzzle velocity of a rifle bullet. The projectile is shot horizontally into and becomes embedded in the bob of a pendulum as illustrated in ●Fig. 6.26. The pendulum swings upward to some height (h), which is measured. The masses of the block and the bullet are known. Using the laws of momentum and energy, show that the initial velocity of the projectile is given by $v_o = [(m + M)/m]\sqrt{2gh}$. (Neglect rotational considerations).

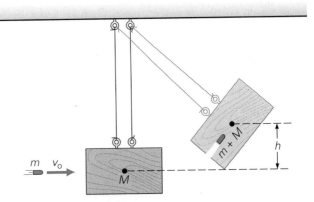

●FIGURE 6.26   A ballistic pendulum
See Exercises 36, 72, and 73.

6.3 Impulse

6.4 Elastic and Inelastic Collisions

37 Impulse is equal to (a) $F\Delta x$, (b) the change in kinetic energy, (c) the change in momentum, (d) $\Delta p/\Delta t$.

38 Which of the following is *not* conserved in an inelastic collision: (a) momentum, (b) mass, (c) kinetic energy, (d) total energy?

39 In a head-on elastic collision mass m_1 strikes a stationary mass m_2, there is a complete transfer of energy if (a) $m_1 = m_2$, (b) $m_1 \gg m_2$, (c) $m_1 \ll m_2$, (d) the masses stick together.

40 A ball with a large mass M and a ball with a small mass m have a head-on, elastic collision. Which ball receives the greater impulse? (a) The large mass ball M, (b) the small mass ball m, or (c) the impulses are equal.

41 Explain the difference for each of the following pairs of actions in terms of impulse: (a) a golfer's drive and chip shot, (b) a boxer's jab and knock-out punches, (c) a baseball player's bunting action and home-run swing.

42 Explain the principle behind (a) using styrofoam as packing to prevent objects from breaking and (b) padding dashboards in cars to prevent injurious bumps.

43 If $K = p^2/2m$, how can kinetic energy be lost in an inelastic collision, yet the total momentum still be conserved? Explain.

44 An inventor has the bright idea to make automobile bodies out of rubber so as to lower insurance costs for fender benders. Is this a good idea?

45 ■ A neutron (an electrically neutral subatomic particle) moving at a high speed collides elastically with a stationary (i) carbon atom and (ii) uranium atom. Compare what happens in each case. [*Hint*: Find the relative masses of these particles in Appendix V.]

46 ■ When tossed upward and hit horizontally by a batter, a 0.20-kg softball receives an impulse of 4.0 N·s. With what speed does the ball move away from the bat?

47 ■ An automobile with a linear momentum of 3.2×10^4 kg·m/s is brought to stop in 4.0 s. What is the magnitude of the average braking force?

48 ■ A pool player imparts an impulse of 3.0 N·s to a 0.25-kg cue ball with a cue stick. What is the speed of the ball after impact?

49 ■■ During a snowball fight, a 0.15-kg snowball traveling at a speed of 8.0 m/s hits a student in the back of the head. (a) What is the impulse? Is this an elastic collision? (b) If the contact time is 0.10 s, what is the average impulse force on the student's head?

50 ■■ A 0.45-kg volley ball comes over the net with a horizontal velocity of 2.0 m/s. One of the players on the front line jumps up and hits it back with a horizontal velocity of 8.0 m/s. What was the impulse imparted to the ball?

51 ■■ A basketball with a mass of 0.50 kg is thrown horizontally against a wall with a velocity of 15 m/s. If the ball rebounds with a velocity of 13 m/s, what is the impulse of the collision? (Was the collision elastic?)

52 ■■ A boy catches a 0.16-kg baseball coming directly toward him at a speed of 25 m/s in bare hands with his arms rigidly extended, and emits an audible "ouch!" because the ball stings his hands. He learns quickly to move his hands with the ball as he catches it. If the contact time for the collision is increased from 2.5 ms to 7.5 ms in this way, how do the magnitudes of the average impulse forces compare? (Assume that the ball has the same initial velocity for each catch.)

53 ■■ For a typical drive, the golf club and the ball are in contact for about 0.50 ms, and the ball leaves the tee with a speed of 70 m/s. What is the average force exerted by the club on the ball? Express your final answer in pounds. (The official weight of a golf ball is 1.620 oz.)

54 ■■ A 4.0-kg ball with a velocity of 4.0 m/s in the $+x$ direction has a head-on elastic collision with a stationary 2.0-kg ball. What are the velocities of the balls after the collision?

55 ■■ A ball with a mass of 100 g is traveling with a velocity of 50 cm/s in the $+x$ direction and collides head-on with a 5.0-kg ball that was at rest. Find the velocities of the balls after the collision, assuming that it is elastic.

56 ■■ A neutron (an electrically neutral subatomic particle) with a mass of 1.67×10^{-27} kg and traveling with a speed of 4.00×10^5 m/s collides elastically head-on with a stationary nucleus with a mass of 6.64×10^{-27} kg. (a) What are the velocities of the neutron and nucleus after collision? (b) What percentage of the neutron's initial kinetic energy is given to the nucleus?

57 ■■ For the apparatus in Fig. 6.14, show that one ball swinging in with speed v_o will not cause two balls to swing out with speed $v_o/2$.

58 ■■ A one-dimensional impulse force acts on a 2.0-kg object as diagrammed in the graph of ●Fig. 6.27. Find (a) the magnitude of the impulse given to the object, (b) the average force, and (c) the final velocity if the object had an initial velocity of -6.0 m/s.

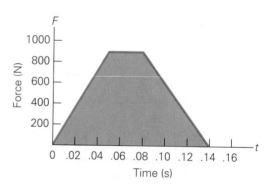

●**FIGURE 6.27 Force versus time graph**
See Exercise 58.

59 ■■ A 0.55-kg piece of putty is dropped from a height of 2.5 m above a flat surface. When the putty hits the surface, it comes to rest in 0.30 s. What is the average force exerted on the putty by the surface?

60 ■■ If the billiard ball in Fig. 6.25 is in contact with the rail for 0.010 s, what is the magnitude of the average force exerted on the ball? (See Exercise 18.)

61 ■■ A 4.0-metric ton truck traveling at 90 km/h plows into the back of a stalled 1500-kg automobile. The vehicles slide together 12 m before coming to a halt. As-suming the decelerating force to be constant, what is the magnitude of the force?

62 ■■ Two balls with masses of 2.0 kg and 6.0 kg travel to-ward each other at speeds of 12 m/s and 4.0 m/s, re-spectively. If the balls have a head-on, inelastic collision and the 2.0-kg ball recoils with a speed of 8.0 m/s, how much kinetic energy is lost in the collision?

63 ■■ Two balls with small and large masses (m and M) approach each other as shown in •Fig. 6.28. If they col-lide and stick together at the origin, what are the com-ponents of the velocity v' of the balls after collision in terms of the other parameters?

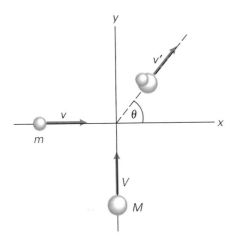

•**FIGURE 6.28 A completely inelastic collision**
See Exercise 63.

64 ■■ A 1500-kg car traveling east at 90.0 km/h and a 3000-kg minivan traveling south at 60.0 km/h collide at a perpendicular intersection. Assuming a completely inelastic collision, what is the velocity of the vehicles immediately after collision?

65 ■■ A cue ball traveling at 0.75 m/s hits the stationary 8-ball, which moves off with a speed of 0.25 m/s at an angle of 37° relative to the cue ball's initial direction. Assuming an inelastic collision, at what angle will the cue ball be deflected and what will be its speed?

66 ■■ A fellow student states that the total momentum of a 3-particle system ($m_1 = 0.25$ kg, $m_2 = 0.20$ kg, and $m_3 = 0.33$ kg) is initially zero, and that after an inelastic triple collision, he calculates the particles have velocities of 4.0 m/s in the $+x$ direction, 6.0 m/s at 120°, and 2.5 m/s at 230°, respectively. Do you agree with his calculations? If not, assuming the first two answers to be correct, what should be the momentum of the third particle?

67 ■■ In nuclear reactors, subatomic particles called neu-trons are slowed down by allowing them to collide with the atoms of a moderator material, such as carbon. (a) In a head-on, elastic collision with a carbon atom, what percentage energy is lost by a neutron? (b) If the neu-tron has an initial speed of 1.5×10^7 m/s, what will its speed be after collision?

68 ■■ A freight car with a mass of 2.0×10^4 kg rolls down an inclined track through a vertical distance of 3.3 m. At the bottom of the incline, on a level track, the car collides and couples with an identical freight car that was at rest. What percentage of the initial kinetic en-ergy is lost in the collision?

69 ■■■ A gondola car has a mass of 4.5 metric tons when empty. While it is moving at a speed of 2.0 m/s along a level track under a grain elevator, 7.5 metric tons of wheat are loaded into it from directly above. Is this an elastic collision? If not, where did the kinetic energy go?

70 ■■■ In an elastic, head-on collision with a stationary target particle, a moving particle recoils at $\frac{1}{3}$ of its inci-dent speed. (a) What is the ratio of the particle masses (m_1/m_2)? (b) What is the speed of the target particle af-ter the collision in terms of the initial speed of the in-coming particle?

71 ■■■ Show that the fraction of kinetic energy lost in the collision in Fig. 6.11b is equal to $m_2/(m_1 + m_2)$.

72 ■■■ Show that the fraction of kinetic energy lost in a ballistic pendulum collision (as in Fig. 6.26) is equal to $M/(m + M)$.

73 ■■■ A 10-g bullet is fired horizontally into and becomes embedded in a suspended block of wood whose mass is 0.890 kg (see Fig. 6.26). (a) What is the speed of the block with the embedded bullet immediately after the collision in terms of the initial speed (v_o)? (b) If the block with the embedded bullet swings upward and its cen-ter of mass is raised 0.40 m, what was the initial speed of the bullet? (c) Was the collision elastic? If not, what percentage of the initial kinetic energy was lost?

74 ■■■ A moving billiard ball collides with an identical stationary one, and the incoming ball is deflected at an angle of 45° from its original direction. Show that if the collision is elastic, both balls will have the same speed afterward and will move at a right angle (90°) relative to each other.

75 ■■■ (a) For an elastic, two-body head-on collision, show that in general $v_2 - v_1 = -(v_{2_o} - v_{1_o})$. That is, the relative speed of recession after the collision is the same as the relative speed of approach before it. (b) In gen-eral, a collision is either completely inelastic, com-pletely elastic, or somewhere in between. The degree of elasticity is sometimes expressed as the *coefficient of restitution* (e), which is defined as the ratio of the rela-tive velocities of recession and approach: $v_2 - v_1 = -e(v_{2_o} - v_{1_o})$. What are the values of e for an elastic col-lision and a completely inelastic collision?

76 ■■■ The coefficient of restitution (see Exercise 75) for steel colliding with steel is 0.95. If a steel ball is dropped

from a height h_o above a steel plate, to what height will the ball rebound?

6.5 Center of Mass

77 The center of mass of an object (a) always lies at the center of the object, (b) is at the location of the most massive particle in the object, (c) always lies within the object, (d) none of these.

78 The center of mass and center of gravity of an object coincide if (a) the object is flat, (b) there is a uniform mass distribution, (c) they both lie inside the object, (d) the acceleration due to gravity is constant.

79 A spacecraft is initially at rest in free space, and then its rocket engines are fired. Describe the motion of the center of mass of the system.

80 ■ (a) The center of mass of a system consisting of two 100-g particles is located at the origin. If one of the particles is at (0.45 m, 0) where is the other? (b) If the masses are moved so their center of mass is located at (0.25 cm, 0.15 cm), can you tell where the particles are located?

81 ■ The centers of a 4.0-kg sphere and a 7.5-kg sphere are separated by a distance of 1.5 m. Where is the center of mass of the system?

82 ■■ Find the center of mass of a system composed of three spherical objects with masses of 3.0 kg, 2.0 kg, and 4.0 kg and centers located at (−6.0 m, 0), (1.0 m, 0), and (3.0 m, 0), respectively.

83 ■■ For the system described in Exercise 82, where would a fourth sphere with a mass of 0.50 kg have to be located for the center of mass of the system to be at the origin?

84 ■■ (a) Find the center of mass of the Earth–Moon system. [*Hint:* Use data from Appendix I and consider the distance between the two to be measured from their centers.] (b) Where is that center of mass relative to the surface of the Earth?

85 ■■ Suppose that the mass of the dumbbell bar in Example 6.13 is not negligible, that the bar has a uniform mass of 1.0 kg. How does this affect the location of the center of mass of the dumbbell? (Consider both cases.)

86 ■■ A piece of uniform sheet metal measures 25 cm by 25 cm. If a circular piece with a radius of 5.0 cm is cut from the center of the sheet, where is the center of mass of the resulting shape?

87 ■■ Three particles, each with a mass of 0.25 kg, are located at (−4.0 m, 0), (2.0 m, 0), and (0, 3.0 m) and are acted on by forces $\mathbf{F}_1 = (-3.0 \text{ N}) \mathbf{y}$, $\mathbf{F}_2 = (5.0 \text{ N}) \mathbf{y}$, and $\mathbf{F}_3 = (4.0 \text{ N}) \mathbf{x}$, respectively. Find the acceleration (magnitude and direction) of the center of mass. [*Hint:* Consider the components of that acceleration.]

88 ■■ Locate the center of mass of the system shown in •Fig. 6.29 (a) if all of the masses are equal; (b) if $m_2 = m_4 = 2m_1 = 2m_3$; (c) if $m_1 = 1.0$ kg, $m_2 = 2.0$ kg, $m_3 = 3.0$ kg, and $m_4 = 4.0$ kg. (Coordinates are in meters.)

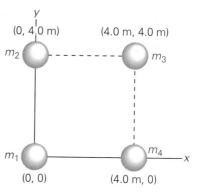

•**FIGURE 6.29 Where's the center of mass?**
See Exercise 88.

89 ■■ A system of two masses has a center of mass given by X_{CM_1}. Another system of three masses has a center of mass given by X_{CM_2}. Show that if all five masses are considered to be one system, the center of mass of that combined system is *not* $X_{CM} = X_{CM_1} + X_{CM_2}$.

90 ■■ A 100-kg astronaut (mass includes space gear) on a space walk is 5.0 m from a 4000-kg space capsule at the full length of her safety cord. To return to the capsule, she pulls herself along the cord. Where do the astronaut and capsule meet?

91 ■■■ Two skaters with masses of 90 kg and 60 kg, respectively, stand 8.0 m apart; each holds one end of a piece of rope. (a) If they pull themselves along the rope until they meet, how far does each skater travel? (Neglect friction.) (b) If only the 60-kg skater pulls along the rope until she meets her friend (who just holds onto the rope), how far does each skater travel?

92 ■■■ A 75.0-kg man stands in the far end of a 50.0-kg boat 100 m from the shore, as pictured in •Fig. 6.30. If he walks to the other end of the 6.00-m boat, how far is the man from the shore? Neglect friction and assume that the center of mass of the boat is at its center. [*Hint:* The answer

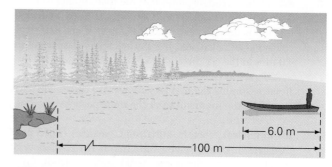

•**FIGURE 6.30 Walking toward shore**
See Exercise 92.

is *not* 94 m, because the boat moves as the man walks. Why? With no external force, the acceleration of the center of mass of the man-boat system is zero, and thus it is stationary. Calculate the location of the center of mass initially, and then apply to the final configuration.]

Additional Exercises

93 What is the total momentum of a system consisting of two particles, $m_1 = 0.015$ kg and $m_2 = 0.025$ kg, if (a) $v_1 = 8.5$ m/s and $v_2 = 9.0$ m/s in the positive x direction, and (b) $v_1 = -6.4$ m/s and $v_2 = 7.5$ m/s along the y axis?

94 Near the end of a chess game, each player has three pieces left: a knight (N), the queen (Q), and the king (K), as shown in •Fig. 6.31 (where B stands for black and W for white). If the queen has 3.0 times the mass of a knight and the king has 4.0 times the mass of a knight, find (a) the location (coordinates X_{CM} and Y_{CM}) of the CM of the black pieces, and (b) the location of the CM of the white pieces. (c) Find the location of the CM for *all* of the pieces, regardless of color. (Taking the lower left *corner* of the board as the origin of the coordinate system, express the coordinates in terms of d, the length of a side of the board's squares, assuming that each piece is located at the center of the square it occupies.)

95 In Exercise 94, the locations of the centers of mass of the black and white chess pieces were found in (a) and (b). Could these centers of mass be treated as the locations of point particles where the respective masses were considered concentrated to find the CM for all of the pieces? Check and find out.

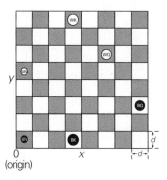

•**FIGURE 6.31 No checkmate yet**
See Exercises 94 and 95.

96 Two identical billiard balls approach each other at the same speed (2.0 m/s). At what speeds do they rebound after a head-on elastic collision?

97 A truck with a mass of 25 metric tons travels at a constant speed of 90 km/h. (a) What is the magnitude of the truck's instantaneous linear momentum? (b) What average force would be required to stop the truck in 8.0 s?

98 A 10-kg cannon ball in flight with a speed of 50 m/s in the $+x$ direction explodes and breaks into three pieces. A 2.5-kg piece goes off at an angle of $+30°$ to the x-axis with a speed of 60 m/s and a 4.5-kg piece goes off at an angle of $-45°$ to the x axis with a speed of 75 m/s. What is the momentum of the third piece?

99 A 15,000-N automobile travels at a speed of 45 km/h northward along a street, and a 7500-N sports car travels at a speed of 60 km/h eastward along an intersecting street. (a) If neither driver brakes and the cars collide at the intersection and stick together, what will the velocity of the cars be immediately after the collision? (b) What percentage of the initial kinetic energy will be lost in the collision?

100 A 100-g bullet is fired horizontally into a 14.9-kg block of wood resting on a horizontal surface, and the bullet becomes embedded in the block. If the muzzle velocity of the bullet is 250 m/s, what is the velocity of the block containing the embedded bullet immediately after the impact? (Neglect surface friction.)

101 Once in a while, there is a so-called grand alignment of planets; that is, all nine of the (known) planets are located along a straight line running through the Sun as viewed from above. Assuming that there is such an alignment, with all the planets on one side of the Sun, compute the approximate location of the center of mass of the nine planets at that time. (Obtain the necessary data from an introductory astronomy book or other reference.) Where is that center of mass relative to the surface of the Sun? When the planets move out of alignment, what effect does this have on the location of their center of mass?

102 A 2.5-kg block sliding on a frictionless horizontal surface with a constant velocity of 6.0 m/s approaches a stationary 6.5-kg block. (a) If the blocks have a completely inelastic collision, what is their velocity after it? (b) How much mechanical energy is lost in the completely inelastic collision?

103 A 90-kg astronaut is stranded in space at a point 12 m from his spaceship. In order to get back, he throws a 0.50-kg piece of equipment so that it moves at a speed of 4.0 m/s directly away from the spaceship. How long will it take him to reach the ship?

104 A 70-kg athlete achieves a height of 2.25 m in a standing high jump. Considering the jumper and the Earth as an isolated system, with what speed does the Earth initially move?

105 A loaded school bus with a total mass of 1.95 metric tons travels down a straight road with a speed of 60 km/h. The driver sees a speed limit sign reading 45 km/h and slows down to that speed. What is the change in the momentum of the bus?

106 A uniform, flat piece of metal is shaped like an equilateral triangle with sides that are 30 cm long. What are the coordinates of the center of mass in the xy plane if one apex is at the origin and one side along the y axis?

7 Circular Motion and Gravitation

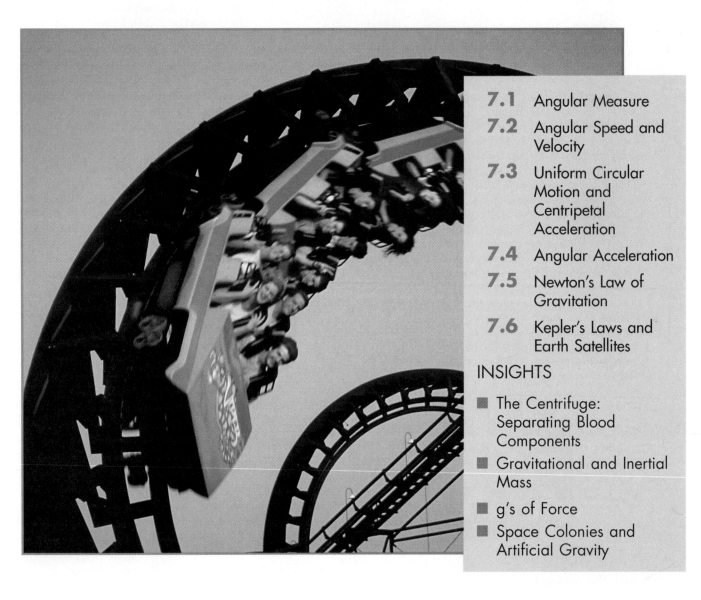

People often say that rides like this "defy gravity." Of course, you know that in reality gravity cannot be defied; it commands respect. There is nothing that will shield you from it, and no place in the universe where you can go to be entirely free of it. What, then, keeps these people in their seats, despite the tug of Earth's ever-present gravity? You may be surprised to find that if you can answer this question, you'll also be able to understand what keeps satellites from falling to Earth, and keeps the Earth in orbit around the Sun. These are some of the topics you'll explore in this chapter.

Circular motion is everywhere, from atoms to galaxies, from flagella to Ferris wheels. Two terms are frequently used to describe such motion. In general, we say that an object *rotates* when the axis of rotation lies within the body, and that it *revolves* when the axis is outside it. Thus, the Earth rotates on its axis and revolves about the Sun.

When a body rotates on its axis, all the particles of the body revolve—that is, they move in circular paths about the body's axis of rotation. For example, the particles that make up a compact disc all travel in circles about the hub of the CD player. In fact, as a "particle" on Earth, you are continually in circular motion about the Earth's rotational axis.

Circular motion is motion in two dimensions, and so can be described by rectangular components as used in Chapter 3. However, it is usually more convenient to describe circular motion in terms of angular quantities that will be introduced in this chapter. Being familiar with the description of circular motion will make the study of rotating rigid bodies much easier, as you will find in Chapter 8.

Gravity plays a large role in determining the motions of the planets, since it supplies the force necessary to maintain their nearly circular orbits. This chapter will consider Newton's law of gravitation, which describes this fundamental force, and will analyze planetary motion in terms of this and related basic laws. The same considerations will help you understand the motions of Earth satellites, of which there is one natural one (the Moon) and many artificial ones.

7.1 Angular Measure

Objectives: To be able to (a) define units of angular measure, and (b) show how angular measure is related to circular arc length.

Motion is described as a time rate of change of position. As you might guess, angular speed and velocity also involve a time rate of change of position, which is expressed by an angle. Consider a particle traveling in a circular path, as shown in •Fig. 7.1. At a particular instant, the particle's position (P) may be designated by the Cartesian coordinates x and y. However, the position may also be designated by the polar coordinates r and θ. The distance r extends from the origin, and the angle θ is commonly measured counterclockwise from the x axis. The transformation equations that relate one set of coordinates to the other are

$$x = r \cos \theta \tag{7.1a}$$

$$y = r \sin \theta \tag{7.1b}$$

as can be seen from the x and y components.

Note that r is the same for any point on a given circle. As a particle travels in a circle, the value of r is constant and only θ changes with time. Thus, circular motion can be described using only one polar coordinate (θ) that changes with time, instead of two Cartesian coordinates (x and y).

Analogous to linear displacement is **angular displacement**, which is given by

$$\Delta \theta = \theta - \theta_o \tag{7.2}$$

or simply $\Delta \theta = \theta$ when we choose $\theta_o = 0°$. A unit commonly used to express angular placement (or angles) is the degree (°); there are 360° in one complete circle, or revolution. Each degree is divided into 60 minutes and each minute into 60 seconds (these have no relationship to time units).

It is important to be able to relate the angular description of circular motion to the orbital or tangential description, that is, to relate the angular displacement to the arc length s. The arc length is the distance traveled along the circular path, and the angle θ is said to *subtend* (define) the arc length. A unit that is very convenient for relating angle to arc length is the **radian (rad)**, which is defined as the angle subtending an arc length (s) that is equal to the radius (r) (•Fig. 7.2a).

Polar coordinates, r and θ

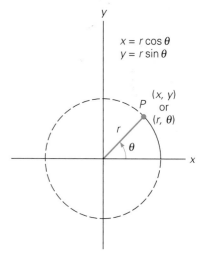

•FIGURE 7.1 **Polar coordinates**
A point, or location, may be described using polar coordinates instead of Cartesian coordinates— that is, as (r, θ) instead of (x, y). For a circle, θ is called the angular distance and r is the radial distance. The two types of coordinates are related by the transformation equations $x = r \cos \theta$ and $y = r \sin \theta$.

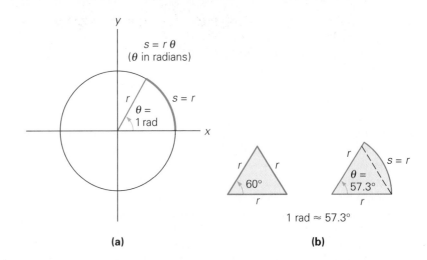

(a)   (b)

FIGURE 7.2 Radian measure
(a) Angular displacement may be measured either in degrees or in radians (rad). An angle θ is subtended by an arc length s. When $s = r$, the angle subtending s is defined to be 1 radian. More generally, $\theta = s/r$ with θ in radians. **(b)** A radian is approximately equal to 57.3°, and the circular sector marked off by this angle is similar to an equilateral triangle with sides r.

To get a general relationship between radians and degrees, let's consider the distance around a complete circle (360°). Basically, a radian is the angle subtending a circular arc equal in length to the radius. Thus for one full circle with $s = 2\pi r$ (the circumference), there is a total of $\theta = 2\pi r/r = 2\pi$ rad, or

$$2\pi \text{ rad} = 360°$$

This relationship can be used to obtain convenient conversions of common angles (Table 7.1). Thus,

$$1 \text{ rad} = 360°/2\pi = 57.3°$$

(to 3 significant figures). A radian is thus slightly smaller than one of the angles of an equilateral triangle, as •Fig. 7.2b shows. Notice in Table 7.1 how the angles in radians are expressed using π explicitly. This is done for convenience.

Similarly, the number of radians subtended by an arbitrary arc length s is equal to the number of radii that will go into s, or the number of radians $\theta = s/r$. Thus can write

$$s = r\theta \qquad (7.3)$$

which is an important relationship between the circular arc length s and its radius r. Notice that since $\theta = s/r$, the angle in radians is the ratio of two lengths. This means that a radian measure is a pure number—that is, it is dimensionless and so has no units.

TABLE 7.1 Equivalent Degree and Radian Measures	
Degrees	*Radians*
360°	2π
180°	π
90°	$\pi/2$
60°	$\pi/3$
57.3°	1
45°	$\pi/4$
30°	$\pi/6$

EXAMPLE 7.1 ■ FINDING ARC LENGTH: USING RADIAN MEASURE

A spectator standing at the center of a circular running track observes a runner start a practice race 256 m due east of her position (•Fig. 7.3). The runner runs on the same track to the finish line, which is located due north of the observer's position. What is the distance of the run?

Solution. Listing what is given and what is to be found, we have

Given: $r = 256$ m   *Find:* s (arc length)
        $\theta = 90° = \pi/2$ rad

Simply using Eq. 7.3 to find the arc length,

$$s = r\theta = (256 \text{ m})\left(\frac{\pi}{2}\right) = 402 \text{ m}$$

Note that the rad unit is omitted, but the equation is dimensionally correct. Why?

Follow-up Exercise. What would be the distance of one complete lap around the track in this Example? *(Answers may be found in the Answers to Follow-up Exercises section at the back of the book.)*

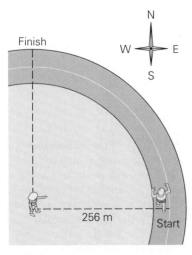

•**FIGURE 7.3 Arc length—easily found using radians**
See Example 7.1.

EXAMPLE 7.2 ■ HOW FAR AWAY? A USEFUL APPROXIMATION

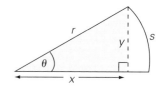

The Small Angle Approximation

A sailor measures the length of a distant tanker as an angular distance of 1° 9' (1 degree, 9 minutes) with a divided circle, as illustrated in •Fig. 7.4a. He knows from the shipping charts that the tanker is 150 m in length. Approximately how far away is the tanker?

Solution. To approximate the distance, we take the ship's length to be nearly equal to the arc length subtended by the measured angle. This is a good approximation for small angles. The data are then

Given: $\theta = 1° 9'$ *Find:* r (radial distance)
        $s = 150$ m

Equation 7.3 can be used to find r, but first we must convert θ to radians. Changing the minutes to degrees: $9'\left(\dfrac{1°}{60'}\right) = 0.15°$, and $\theta = 1.15°$. Then,

$$\theta = 1.15°\left(\frac{2\pi \text{ rad}}{360°}\right) = 0.0201 \text{ rad}$$

(Note that any of the equivalent relationships in Table 7.1 could be used for the conversion, such as π rad/180° or even 1 rad/57.3°.) Then,

$$r = \frac{s}{\theta} = \frac{150 \text{ m}}{0.0201} = 7.46 \times 10^3 \text{ m} = 7.46 \text{ km}$$

As noted, the distance r is an approximation, obtained by assuming that for small angles the arc length s and the straight-line chord length L are very nearly the same length (see •Fig. 7.4b). How good an approximation is this? To see, let's compute the perpendicular distance d to the ship. From the geometry, we have $\tan \theta/2 = L/2d$, and for small angles, $\tan \theta \approx \theta$ (see the Learn by Drawing on this page), so

$$d = \frac{L}{2 \tan \theta/2} \approx \frac{L}{2(\theta/2)} = \frac{150 \text{ m}}{0.0201} = 7.46 \times 10^3 \text{ m} = 7.46 \text{ km}$$

(Pretty good—the values are equal within the number of significant figures we're working with here.)

θ *not* small:

$$\theta \text{ (in rad)} = \frac{s}{r}$$

$$\sin \theta = \frac{y}{r} \quad \tan \theta = \frac{y}{x}$$

θ small:

$$y \longrightarrow s$$
$$x \longrightarrow r$$

$$\theta \text{ (in rad)} = \frac{s}{r} \approx \frac{y}{r} \approx \frac{y}{x}$$

$$\theta \text{ (in rad)} \approx \sin\theta \approx \tan\theta$$

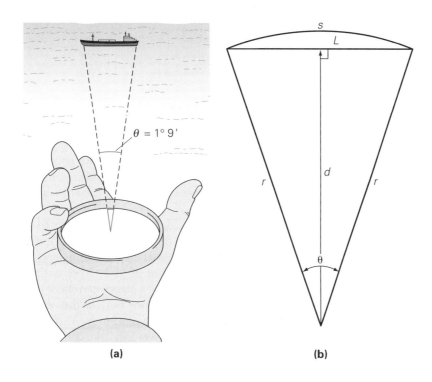

(a)   (b)

•FIGURE 7.4 Angular distance For small angles, the arc length is approximately a straight line, or the chord length. Knowing the length of the tanker, one can find how far away it is by measuring its angular size. See Example 7.2.

PROBLEM-SOLVING HINT

In computing trigonometric functions such as tan θ or sin θ, the angle may be expressed in degrees or radians, that is, sin 30° = sin (π/6 rad) = sin (0.524 rad) = 0.500. When finding trig functions with a calculator, note that there is usually a way to change the angle entry between "deg" and "rad" modes. Calculators commonly are set in the degree mode, so if you want to find the value of, say, sin (1.22 rad), first change to the "rad" mode and enter 1.22; sin (1.22 rad) = 0.940.

[Your calculator may have a third, "grad," mode. In case you're wondering about this, the grad is another, little-used angular unit. A grad is 1/100 of a right (90°) angle; that is, there are 100 grads in a right angle.]

7.2 Angular Speed and Velocity

Objectives: To be able to (a) describe and compute angular speed and velocity, and (b) explain their relationship to tangential speed.

The description of circular motion in angular form is analogous to the description of linear motion. In fact, you'll notice that the equations are almost mathematically identical, with different symbols being used to indicate that the quantities have different meanings. The lower-case Greek letter omega is used to represent **average angular speed** ($\overline{\omega}$), the angular displacement divided by the total time to travel the distance:

Angular speed

$$\overline{\omega} = \frac{\Delta\theta}{\Delta t} = \frac{\theta - \theta_o}{t - t_o} \qquad (7.4)$$

The units of angular speed are rad/s (actually s^{-1} since rad is unitless).

The *instantaneous angular speed* is given for an extremely small time interval (Δt approaches zero). It is common to take θ_o and t_o to be zero when possible, and we generally write

$$\overline{\omega} = \frac{\theta}{t} \qquad \text{or} \qquad \theta = \overline{\omega}t \qquad (7.5)$$

SI unit of angular distance: rad

As in the linear case, if the angular speed is *constant*, then $\overline{\omega} = \omega$.

Another common descriptive unit for angular speed is rpm (revolutions per minute); for example, the speed of an LP phonograph turntable is $33\frac{1}{3}$ rpm, and a CD (compact disk) rotates at a speed of 200–500 rpm (depending on location of the track). This nonstandard unit of rpm is readily converted to rad/s, since 1 revolution = 2π rad.

Angular velocity

The **average and instantaneous angular velocities** are analogous to their linear counterparts. Angular velocity is associated with angular displacement. Both are vectors, and so have direction; however, this may be specified in a special way. In one-dimensional, or linear, motion, a particle can go only in one direction or the other (+ or −), so the displacement and velocity vectors can have only these two directions. In the angular case, a particle moves one way or the other but the motion is along its *circular path*. Thus, the angular displacement and angular velocity vectors of a particle in circular motion can have only two directions, which correspond to going around the circular path with either increasing or decreasing angular displacement from θ_o ($\Delta\theta$). Let's focus on the angular velocity vector, ω. (The direction of the angular displacement will be the same as that of the angular velocity. Why?)

The direction of the angular velocity vector is given by a right-hand rule, illustrated in •Fig. 7.5a. When the fingers of your right hand are curled in the direction of circular motion, your extended thumb points in the direction of ω. Note that circular motion can be in only one of two circular *senses*, clockwise or counterclockwise. Plus and minus signs can be used to distinguish circular rotation directions relative to angular velocity vector. It is customary to take a counterclockwise rotation as positive (+) since the measurement of positive angular distance (and displacement) is conventionally done counterclockwise from the positive x axis.

QUESTION: Why not just designate the direction of the angular velocity vector to be either clockwise or counterclockwise?

ANSWER: Clockwise (cw) and counterclockwise (ccw) are really directional senses or indications rather than actual directions. These rotational senses are like right and left. Stand facing another person and ask if some object is on the right or left. (You'll disagree.) Similarly, if you held this book up toward a person facing you and rotated it, would it be rotating cw or ccw?

We can use cw and ccw to indicate rotational "directions" when specified relative to a reference, for example, the positive x axis as in the preceding discussion. Referring to Fig. 7.5, imagine yourself being first on one side of one of the rotating disks and then on the other. Which way is the disk rotating from each vantage point, cw or ccw? Then, apply the right-hand rule on both sides. You should find that the angular velocity vector direction is the same for both locations (because it is referenced to the right hand). Relative to this vector—for example, looking at the tip—there is no ambiguity in using + and − to indicate rotational senses or directions.

Relationship between Tangential and Angular Speeds

A particle moving in a circle has an instantaneous velocity tangential to its circular path. For a constant angular velocity and speed, the particle's orbital or **tangential speed** v (the magnitude of the tangential velocity) is also constant. How the angular and tangential speeds are related is revealed by starting with Eq. 7.3 ($s = r\theta$) and Eq. 7.5 ($\theta = \omega t$),

$$s = r\theta = r(\omega t)$$

The arc length, or distance, is also given by

$$s = vt$$

Combining the equations for s gives

$$v = r\omega \qquad \begin{array}{l} \textit{tangential speed} \\ \textit{for circular motion} \end{array} \qquad (7.6)$$

where ω is in rad/s. Equation 7.6 holds in general for instantaneous tangential and angular speeds for solid or rigid body rotation about a fixed axis.

Note that all the particles of an object rotating with constant angular velocity have the same angular speed, but the tangential speeds are different at different distances from the axis of rotation (•Fig. 7.6 and Demonstration 5).

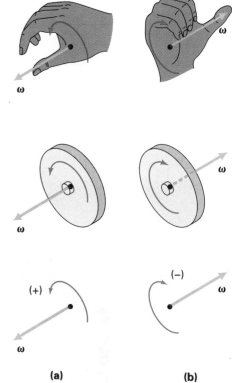

•**FIGURE 7.5 Angular velocity** The direction of the angular velocity vector for an object in rotational motion is given by a right-hand rule: when the fingers of the right hand are curled in the direction of the rotation, the extended thumb points in the direction of the angular velocity vector. Circular senses or directions are commonly indicated by plus and minus signs, as shown in parts (a) and (b), respectively.

EXAMPLE 7.3 ■ GOLDEN OLDIES: DIFFERENT RADII, DIFFERENT TANGENTIAL SPEEDS

On an old 45-rpm record, the beginning of the track is 8.0 cm from the center, and the end 5.0 cm from the center. What are (a) the angular speeds and (b) the tangential speeds of a point on the record at these distances when it is spinning at 45 rpm?

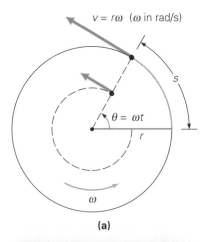

$v = r\omega$ (ω in rad/s)

s

$\theta = \omega t$

r

ω

(a)

(b)

•**FIGURE 7.6 Tangential and angular speeds**
(a) The tangential and angular speeds are related by $v = r\omega$, with ω in rad/s. Note that all of the particles of an object rotating about a fixed axis travel in circles. All the particles have the same angular speed (ω), but particles at different distances from the axis of rotation have different tangential speeds.
(b) Sparks from a grinding wheel provide a graphic illustration of instantaneous tangential velocity. (Can you explain why the paths curve slightly?)

Solution.

Given: $\omega = 45$ rpm
$r_1 = 8.0$ cm $= 0.080$ m
$r_2 = 5.0$ cm $= 0.050$ m

Find: (a) ω_1 and ω_2 (angular speeds)
(b) v_1 and v_2 (tangential speeds)

(a) It should be apparent that $\omega_1 = \omega_2$; that is, the record as a whole rotates with the same angular speed. All points on the record travel through 2π rad in the time it takes to make one revolution, so every point on the record has the same angular speed, 45 rpm (rev/min). This can be easily converted to rad/s:

$$\omega = \left(\frac{45 \text{ rev}}{\text{min}}\right)\left(\frac{2\pi \text{ rad}}{1 \text{ rev}}\right)\left(\frac{1 \text{ min}}{60 \text{ s}}\right) = 4.7 \text{ rad/s}$$

(A convenient conversion factor is 1 rev/min $= \pi/30$ rad/s. Can you see how this is obtained?)

(b) The tangential speeds at different locations on the record are different. All of the particles that make up the record go through one revolution in the same time. Therefore, the farther a point is from the center of the record, the longer its circular path will be, and its tangential speed must be greater, as Eq. 7.6 indicates (see also Fig. 7.6). Thus,

$$v_1 = r_1\omega = (0.080 \text{ m})(4.7 \text{ rad/s}) = 0.38 \text{ m/s}$$
$$v_2 = r_2\omega = (0.050 \text{ m})(4.7 \text{ rad/s}) = 0.24 \text{ m/s}$$

where the unitless rad is included in the angular quantities for clarity.

A point on the outer part of the record has a greater tangential speed than one on the inner part, as you might expect. The outer point must travel faster than the inner point to make one revolution in the same time.

Follow-up Exercise. (a) How long would it take for points at each of the radii in this Example to make one revolution? (b) Why on oval race tracks do inside and outside runners have different starting points, so that some runners start "ahead" of others? *(Answers may be found in the Answers to Follow-up Exercises section at the back of the book.)*

DEMONSTRATION 5 ■ Constant Angular Velocity, but not Tangential Velocity

This demonstration of uniform circular motion shows that tangential velocities are different at different radii. Neon bulbs are placed at different radii on a spinning wheel. The bulbs light up every 1/120 second for the same period of time. As seen in the photo, the bulbs at the greater radii have greater tangential speeds since they have longer lighted path lengths for the same time periods.

Period and Frequency

Some other quantities commonly used to describe circular motion are period and frequency. The **period** (T) is the time it takes for an object in circular motion to make one complete revolution, or cycle. For example, the period of revolution of the Earth about the Sun is 1 year, and the period of the Earth's axial rotation is 24 hours. The standard unit for the period is the second (s). Descriptively, the period is sometimes given in s/rev or s/cycle.

Closely related to the period is the **frequency** (f), which is the number of revolutions, or cycles, made in a given time, generally a second. For example, if a particle traveling uniformly in a circular orbit makes 5.0 revolutions in 1.0 s, the frequency (of revolution) is $f = 5.0$ rev/1.0 s $= 5.0$ rev/s, or 5.0 cycle/s (cps). Revolution and cycle are merely descriptive terms and not part of the frequency unit. The unit of frequency is 1/s, or s^{-1}, which is called the **hertz** (Hz) in the SI.

Since the descriptive units for frequency and period are inverses of one another (cycles/s and s/cycle) it follows that the two quantities are related by

$$f = \frac{1}{T} \qquad frequency\ and\ period \qquad (7.7)$$

SI unit of frequency: hertz (Hz, 1/s or s^{-1})

where the period is in seconds and the frequency, as noted, is in hertz or inverse seconds.

The frequency can also be related to the angular speed. For uniform circular motion, the orbital speed may be written as $v = 2\pi r/T$; that is, the distance traveled in one revolution divided by the time for one revolution (1 period). Similarly, for the angular case, since a distance of 2π rad is traveled in 1 period (by definition of the period), we have

$$\omega = \frac{2\pi}{T} = 2\pi f \qquad \begin{array}{l} angular\ speed\ in\ terms \\ of\ period\ and\ frequency \end{array} \qquad (7.8)$$

The hertz (Hz), a unit of frequency, is named for Heinrich Hertz (1857–1894), a German physicist and pioneering investigator of electromagnetic waves, which are also characterized by frequency.

Note: Frequency (f) and period (T) are inversely related.

EXAMPLE 7.4 ■ FREQUENCY AND PERIOD: AN INVERSE RELATIONSHIP

A compact disc (CD) rotates in a player at a constant speed of 200 rpm. What is the CD's (a) frequency and (b) period of revolution?

Solution. The angular speed is not in standard units, and so must be converted. Using the convenient conversion factor from Example 7.3 (1 rev/min = $\pi/30$ rad/s), we can write

Given: $\omega = 200$ rpm $\left(\dfrac{\pi/30 \text{ rad/s}}{\text{rpm}} \right)$ **Find:** (a) f (frequency)
                                                (b) T (period)

      $= 20.9$ rad/s

(a) Rearranging Eq. 7.8 and solving for f,

$$f = \frac{\omega}{2\pi} = \frac{20.9 \text{ rad/s}}{2\pi} = 3.33 \text{ Hz}$$

(The units of 2π are rad/cycle or revolution, so the result is in cycles/s or 1/s, which is hertz.)

(b) Eq. 7.8 could be used to find T, but Eq. 7.7 is a bit simpler,

$$T = \frac{1}{f} = \frac{1}{3.33 \text{ Hz}} = 0.300 \text{ s}.$$

Thus, it takes 0.300 s for the CD to make one revolution. (Notice that with Hz = 1/s, the equation is dimensionally correct.)

7.3 Uniform Circular Motion and Centripetal Acceleration

Objectives: To be able to (a) explain why there is a centripetal acceleration in constant or uniform circular motion, and (b) compute centripetal acceleration.

A simple, but important, type of circular motion is **uniform circular motion**, which occurs when an object moves at a constant speed in a circular path. An example of this is a car going around a circular track (•Fig. 7.7). The motion of the Moon around the Earth and of some electrons around the nucleus of an atom are approximated by uniform circular motion. Such motion is curvilinear, so you know from the discussion in Chapter 3 that there must be an acceleration. But what is its magnitude and direction?

Note: Review the discussion of curvilinear motion in Section 3.1.

Centripetal Acceleration

Obviously, the acceleration is not in the same direction as the instantaneous velocity (which is tangent to the circular path at any point). If it were, the object would speed up and the motion wouldn't be uniform. Recall that acceleration is the time rate of change of velocity, and velocity has both magnitude and direction. In uniform circular motion, the direction of the velocity is continually changing, which is a clue to the direction of the acceleration. (See Fig. 7.7.)

This is shown more explicitly in •Fig. 7.8. The velocity vectors at the beginning and end of a time interval give the change in velocity, or $\Delta \mathbf{v}$, via vector addition (or subtraction). The vector triangles for several instantaneous velocities are shown in the figure. All of the instantaneous velocity vectors have the same magnitude or length (constant speed), but differ in direction. Note that since $\Delta \mathbf{v}$ is not zero, there must be an acceleration ($\mathbf{a} = \Delta \mathbf{v}/\Delta t$).

Uniform circular motion—constant speed, but not constant velocity

As illustrated in the figure, as Δt (or $\Delta \theta$) becomes smaller, $\Delta \mathbf{v}$ points more toward the center of the circular path. As Δt approaches zero, the instantaneous change in the velocity, and therefore the acceleration, is exactly toward the center of the circle. As a result, the acceleration in uniform circular motion is called **centripetal acceleration**, which means center-seeking acceleration (from Latin *centri*, "center," and *petere*, "to fall toward" or "to seek").

Centripetal (center-seeking) acceleration: always directed toward the center of the circle.

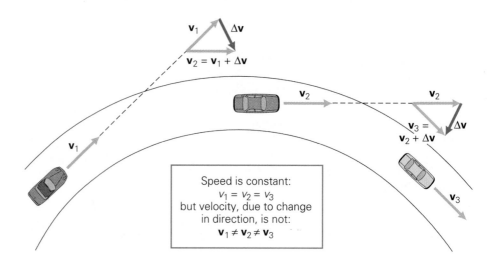

•**FIGURE 7.7 Uniform circular motion**
The speed of an object in uniform circular motion is constant, but its velocity changes because of changes in the direction of motion. Thus, there is an acceleration.

Speed is constant:
$v_1 = v_2 = v_3$
but velocity, due to change in direction, is not:
$\mathbf{v}_1 \neq \mathbf{v}_2 \neq \mathbf{v}_3$

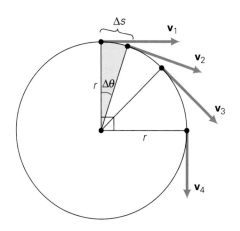

•FIGURE 7.8 **Analysis of centripetal acceleration**
The velocity vector of an object in uniform circular motion
is constantly changing direction. As Δt, the time interval
for $\Delta\theta$, is taken smaller and smaller, and approaches zero,
Δv (the change in the velocity, and therefore an
acceleration) is directed toward the center of the circle.
Analysis shows that the centripetal, or center-seeking,
acceleration has a magnitude of $a_c = v^2/r$.

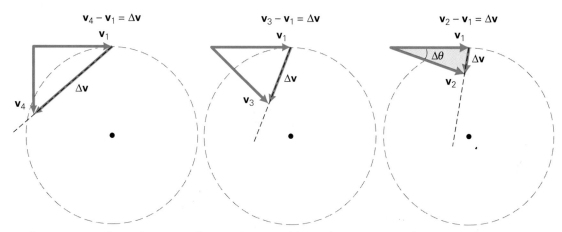

Without the centripetal acceleration, the motion would not be in a curved
path, but in a straight line. The centripetal acceleration must be directed radially
inward, that is, with no component in the direction of the (tangential) velocity, or
else the magnitude of that velocity would change (•Fig. 7.9). Note that for an object in uniform circular motion, the direction of the centripetal acceleration is continually changing. In terms of x and y components, a_x and a_y are not constant. Can
you describe how this differs from projectile motion?

The magnitude of the centripetal acceleration may be deduced from the small
shaded triangles in Fig. 7.8. (For very short time intervals, the arc length of Δs is almost a straight line—the chord.) These two triangles are similar because each has a
pair of equal sides surrounding the same angle $\Delta\theta$. (Note that the velocity vectors
have the same magnitude.) Thus, Δv is to v as Δs is to r, which may be written as

$$\frac{\Delta v}{v} \approx \frac{\Delta s}{r}$$

The arc length Δs is the distance traveled in time Δt, thus $\Delta s = v\,\Delta t$.

$$\frac{\Delta v}{v} \approx \frac{\Delta s}{r} = \frac{v\,\Delta t}{r}$$

and

$$\frac{\Delta v}{\Delta t} \approx \frac{v^2}{r}$$

Then, as Δt approaches zero this approximation becomes exact. The instantaneous
centripetal acceleration, $\Delta v/\Delta t = a_c$, thus has a magnitude

$$a_c = \frac{v^2}{r} \qquad \begin{array}{l}\textit{magnitude of}\\ \textit{centripetal acceleration}\end{array} \qquad (7.9)$$

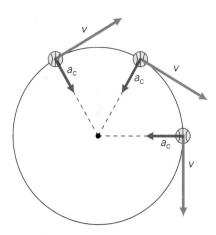

•FIGURE 7.9 **Centripetal
acceleration**
For an object in uniform circular
motion, the centripetal acceleration
is directed radially inward. There
is no acceleration component in the
tangential direction, otherwise, the
magnitude of the velocity
(tangential *speed*) would change.

Note: Centripetal acceleration depends
on tangential speed (v) and radius (r)

Using Eq. 7.6 ($v = r\omega$), the centripetal acceleration equation may be written in terms of the angular speed:

$$a_c = \frac{v^2}{r} = \frac{(r\omega)^2}{r} = r\omega^2 \qquad \begin{array}{l} \textit{centripetal acceleration} \\ \textit{in terms of angular speed} \end{array} \qquad (7.10)$$

EXAMPLE 7.5 ■ A SATELLITE IN ORBIT: CENTRIPETAL ACCELERATION

A space station is in a circular orbit about the Earth at an altitude h of 5.0×10^2 km. If the station makes one revolution every 95 min, what are its (a) orbital speed and (b) centripetal acceleration?

Solution.

Given: $\quad h = 5.0 \times 10^2$ km $\qquad\qquad$ *Find:* $\quad$ (a) v (tangential speed)
$\qquad\qquad = 5.0 \times 10^5$ m $\qquad\qquad\qquad\qquad$ (b) $\mathbf{a}_c$ (centripetal acceleration)
$\qquad\quad t = T = 95$ min $= 5.7 \times 10^3$ s

(a) The radius of the circular orbit is not h, but $R_E + h$, where R_E is the radius of the Earth, 6.4×10^6 m (see Table 6 in Appendix I):

$$r = R_E + h = (6.4 \times 10^6 \text{ m}) + (0.50 \times 10^6 \text{ m}) = 6.9 \times 10^6 \text{ m}$$

The station travels the circumference of its circular orbit ($2\pi r$) in the given time, so the tangential speed is

$$v = \frac{2\pi r}{T} = \frac{2\pi(6.9 \times 10^6 \text{ m})}{5.7 \times 10^3 \text{ s}} = 7.6 \times 10^3 \text{ m/s}$$

(b) Then the centripetal acceleration is (Eq. 7.9),

$$a_c = \frac{v^2}{r} = \frac{(7.6 \times 10^3 \text{ m/s})^2}{6.9 \times 10^6 \text{ m}} = 8.4 \text{ m/s}^2$$

directed toward the center of the orbit. Alternatively, we could have computed the angular speed, $\omega = 2\pi/T$ and used Eq. 7.10, $a_c = r\omega^2$.

As you might suspect, this centripetal acceleration is supplied by the Earth's gravitational force—it is the acceleration due to gravity at an altitude of 500 km.

Follow-up Exercise. (a) How does the acceleration found in this Example compare percentagewise with the acceleration due to gravity on Earth's surface? (b) Would an astronaut in the space station be "weightless"? (*Answers may be found in the Answers to Follow-up Exercises section at the back of the book.*)

A practical application of centripetal acceleration is discussed in the Insight on p. 213 and illustrated in the following Example.

EXAMPLE 7.6 ■ A CENTRIFUGE: CENTRIPETAL ACCELERATION

A laboratory centrifuge like that shown in •Fig. 7.10 operates at a rotational speed of 12,000 rpm. (a) What is the magnitude of the centripetal acceleration on a red blood cell at a radial distance of 8.00 cm from the centrifuge's axis of rotation? (b) How does this acceleration compare to g?

Solution. The data are as follows:

Given: $\quad \omega = 1.20 \times 10^4$ rpm $\qquad\qquad$ *Find:* $\quad$ (a) a_c
$\qquad\qquad\quad = 1.26 \times 10^3$ rad/s $\qquad\qquad\qquad\qquad$ (b) How a_c compares to g
$\qquad\qquad r = 8.00$ cm $= 0.0800$ m

(a) The centripetal acceleration is easily found using Eq. 7.10:

$$a_c = r\omega^2 = (0.0800 \text{ m})(1.26 \times 10^3 \text{ rad/s})^2 = 1.27 \times 10^5 \text{ m/s}^2$$

•**FIGURE 7.10 Centrifuge**
Centrifuges are used to separate particles of different sizes and densities suspended in liquids. For example, red and white blood cells can be separated from each other as well as the plasma that makes up the liquid portion of the blood.

Insight

The Centrifuge: Separating Blood Components

The centrifuge is a rotating machine used to separate particles of different sizes and densities suspended in a liquid (or a gas). For example, cream is separated from milk by centrifuging, and blood components are separated in centrifuges in medical laboratories (see Fig. 7.10).

Blood components will eventually settle toward the bottom of a tube in layers—a process called sedimentation—under the influence of normal gravity alone. The viscous drag of the plasma on the particles is analogous to (but much greater than) the air resistance that determines the terminal velocity of falling objects (Section 4.6). Red blood cells settle in the bottom layer of the plasma because they reach a greater terminal velocity than the white blood cells and platelets, and so get to the bottom sooner. However, gravitational sedimentation is generally very slow.

Since clinicians cannot afford to wait a long time to see the fractional volume of red cells in the blood, centrifugation is used to speed up the sedimentation process. Tubes are spun horizontally. The resistance of the fluid medium on the particles supplies the centripetal acceleration that keeps them moving in slowly widening circles as they settle toward the bottom of the tube. The bottom of the tube itself must exert a strong force on the contents as a whole, and must be strong enough so as not to break.

Laboratory centrifuges commonly operate at speeds sufficient to produce centripetal accelerations thousands of times larger than g (see Example 7.6). Since the principle of the centrifuge involves centripetal acceleration, perhaps "centripuge" would be a more descriptive name.

(b) Using the relationship $1\,g = 9.80\ \text{m/s}^2$ to express a_c in terms of g,

$$a_c = (1.27 \times 10^5\ \text{m/s}^2)\left(\frac{1\,g}{9.80\ \text{m/s}^2}\right) = 1.30 \times 10^4\,g\ (= 13{,}000\,g!)$$

Follow-up Exercise. What angular speed in rpm would give an acceleration of $1\,g$ at the radial distance in this Example? *(Answer may be found in the Answers to Follow-up Exercises section at the back of the book.)*

The concept of centripetal acceleration may be clearer if you look at the displacement components of uniform circular motion for a short time interval, Δt (•Fig. 7.11). The rectangular components in this case are directed tangentially and radially. An object travels a distance $v\Delta t$ tangent to the circle, and at the same time it is displaced a distance $\frac{1}{2}a_c(\Delta t)^2$ radially toward the center of the circle. As Δt approaches zero, a circular path is described. Because of its centripetal acceleration, the Moon in a sense is continually falling toward the Earth. Fortunately, it never gets here because it is traveling tangentially at the same time.

Suppose that an automobile moves into a level, circular curve. To negotiate the curve, the car must have a centripetal acceleration determined by the equation $a_c = v^2/r$. This acceleration is supplied by the force of friction between the tires and the road. Tires act outwardly on the road, and the reaction (friction) force to this is inward on the tires. However, this (static, why?) friction has a maximum limiting value. If the speed of the car is high enough, the friction will not be sufficient to supply the necessary centripetal acceleration, and the car will skid outward from the center of the curve. If the car moves onto a wet or icy spot, the friction between the tires and the road may be reduced, allowing the car to skid at an even lower speed.

CONCEPTUAL EXAMPLE 7.7 ■ BREAKING AWAY

A ball attached to a string is swung with uniform motion in a horizontal circle above a person's head (•Fig. 7.12a). If the string breaks, which of the trajectories shown in •Fig. 7.12b (viewed from above) would the ball follow? *Clearly establish the reasoning and physical principle(s) used in determining your answer before checking it below. That is, **why** did you select your answer?*

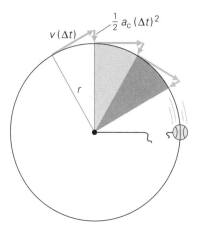

•**FIGURE 7.11 Displacement components, uniform circular motion**
In a short time, Δt, an object travels a distance $v\Delta t$ tangentially and a distance $\frac{1}{2}a_c(\Delta t)^2$ radially. Thus, the object is essentially falling toward the center of the circle compared to what it would do if let go. As Δt approaches zero, these components describe a circular path. (What happens if the string breaks? See Example 7.7.)

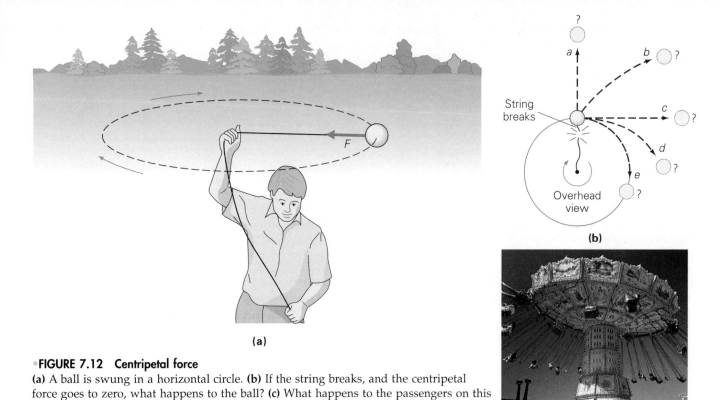

FIGURE 7.12 Centripetal force
(a) A ball is swung in a horizontal circle. (b) If the string breaks, and the centripetal force goes to zero, what happens to the ball? (c) What happens to the passengers on this amusement park ride as the angular velocity increases?

Reasoning and Answer. When the string breaks, the centripetal force goes to zero. Newton's first law states that if no force acts on an object in motion, the object will continue to move in a straight line. This rules out *b*, *d*, and *e*. It should be evident from Figs. 7.6, 7.8, 7.9, and 7.10 that at any instant (including the instant when the string breaks), the isolated ball has a horizontal, tangential velocity. The downward force of gravity acts on it, but this force affects only its vertical motion, which is not visible in Fig. 7.12b. The ball thus flies off tangentially and is essentially a horizontal projectile (with $v_{x_o} = v$, and $v_{y_o} = 0$, and $a_y = g$). Viewed from above, the ball would follow the path labeled *c*.

Follow-up Exercise. If you swing a ball in a horizontal circle about your head, can the string be exactly horizontal? (See Fig. 7.12a.) Explain your answer. *Hint: Analyze the forces acting on the ball. (Reasoning and answer may be found in the Answers to Follow-up Exercises section at the back of the book.)*

Centripetal Force

To provide an acceleration, there must be a net force, thus to produce a centripetal (inward) acceleration, we must have a **centripetal force** (net inward force). Expressing the magnitude of this force in terms of Newton's second law ($F = ma$) and inserting the expression for centripetal acceleration from Eq. 7.9, we can write

$$F_c = ma_c = \frac{mv^2}{r} \qquad \begin{array}{l}\textit{magnitude of}\\\textit{centripetal force}\end{array} \qquad (7.11)$$

The centripetal force, like the centripetal acceleration, is directed toward the center of the circular path.

Keep in mind that in general a force applied at an angle to the direction of motion of an object produces changes in the magnitude *and* direction of the velocity. However, when a force is continuously applied at an angle of 90° to the direction of motion (as is centripetal force), only the direction of the velocity changes. Also notice that because it is always perpendicular to the direction of motion, a centripetal force does no work (why?). Therefore, by the work-energy theorem, a centripetal force does not change the kinetic energy or speed of the object.

As we have seen in the two preceding Examples, a centripetal force can be provided by gravitational attraction or by a tension force in a string. In other situations, it could be a static frictional force, as •Fig. 7.13 and the following Examples show.

EXAMPLE 7.8 ■ WHERE THE RUBBER MEETS THE ROAD: FRICTION AND CENTRIPETAL FORCE

A car approaches a level, circular curve with a radius of 45.0 m. If the concrete pavement is dry, what is the maximum speed at which the car can negotiate the curve at a constant speed?

Solution. First writing down what is given and what is to be found, we have

Given: $r = 45.0$ m   *Find:* v (maximum speed)

To go around the curve at a particular speed, the car must have a centripetal acceleration and therefore a centripetal force must act on it. This force is supplied by static friction between the tires and the road. (The tires are not slipping or skidding relative to the road.) Recall from Chapter 4 that the maximum frictional force is given by $f_{s_{max}} = \mu_s N$, where here N is the magnitude of the normal force on the car and is equal to the weight of the car, mg, on the level road (why?). We may set this equal to the expression for centripetal force ($F_c = mv^2/r$) to find the maximum speed. To find $f_{s_{max}}$ we will need the coefficient of friction between rubber and concrete, and from Table 4.1, $\mu_s = 1.20$. Then, we can write

$$f_{s_{max}} = F_c$$

$$\mu_s N = \mu_s mg = \frac{mv^2}{r}$$

So

$$v = \sqrt{\mu_s rg}$$

$$= \sqrt{(1.20)(45.0 \text{ m})(9.80 \text{ m/s}^2)} = 23.0 \text{ m/s}$$

(about 83 km/h or 52 mi/h).

Follow-up Exercise. Would the centripetal force be the same for all vehicles in this Example? *(Answer may be found in the Answers to Follow-up Exercises section at the back of the book.)*

The proper safe speed for a highway curve is an important consideration. The coefficient of friction between tires and road may vary, depending on weather, road conditions, the design of the tires, the amount of tread wear, and so on. In designing a curved road, safety may be promoted by banking or inclining the roadway. This reduces the chances of skidding because the normal force exerted on the car by the road then has a component toward the center of the curve that reduces the need for friction. In fact, for a circular curve with a given banking angle and radius, there is one speed for which no friction is required at all. This condition is used in banking design (see Exercise 56).

Let's look at one more centripetal force example with two objects in uniform circular motion. It will help you understand the motions of satellites in circular orbits in a later section.

EXAMPLE 7.9 ■ STRUNG OUT: CENTRIPETAL FORCE AND NEWTON'S SECOND LAW

Suppose that two masses, $m_1 = 2.5$ kg and $m_2 = 3.5$ kg, are in uniform circular motion on a horizontal frictionless surface as illustrated in •Fig. 7.14, where $r_1 = 1.0$ m and $r_2 =$

•**FIGURE 7.13 Frictional centripetal force**
The centripetal force necessary for the skaters to round the curve is supplied by frictional force. If there were no friction between the skates and the ice, what would happen?

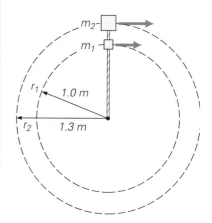

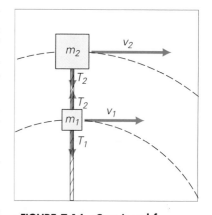

•**FIGURE 7.14 Centripetal force and Newton's second law**
See Example 7.9.

1.3 m. The forces acting on the masses are $T_2 = 2.9$ N and $T_1 = 4.5$ N. Find (a) the centripetal accelerations and (b) the magnitudes of the tangential velocities (tangential speeds) of the masses.

Solution.

Given: $r_1 = 1.0$ m and $r_2 = 1.3$ m *Find:* (a) $\mathbf{a}_{c_1}$ and $\mathbf{a}_{c_2}$ (centripetal accelerations)
$\quad\quad\quad m_1 = 2.5$ kg and $m_2 = 3.5$ kg $\quad\quad\quad$ (b) v_1 and v_2
$\quad\quad\quad T_2 = 2.9$ N
$\quad\quad\quad T_1 = 4.5$ N

By isolating m_2 in the figure, you can see that the centripetal force is provided by the tension in the string ($\mathbf{T}_2$ is the only force acting on m_2 toward the center of its circular path). Thus,

$$T_2 = m_2 a_{c_2}$$

and

$$a_{c_2} = \frac{T_2}{m_2} = \frac{2.9 \text{ N}}{3.5 \text{ kg}} = 0.83 \text{ m/s}^2$$

The acceleration is toward the center of the circle. You can find the tangential speed of m_2 from $a_c = v^2/r$:

$$v_2 = \sqrt{a_{c_2} r_2} = \sqrt{(0.83 \text{ m/s}^2)(1.3 \text{ m})} = 1.0 \text{ m/s}$$

The situation is a bit different for m_1. In this case there are two forces acting on m_1: the string tensions $\mathbf{T}_1$ and $-\mathbf{T}_2$. Also, you should realize that, by Newton's second law, in order to have a centripetal acceleration there must be a net force, which is given by the difference in the two tensions. Thus we have

$$F_{\text{net}_1} = +T_1 + (-T_2) = m_1 a_{c_1} = \frac{m_1 v_1^2}{r_1}$$

where the radial direction (toward the center of the circular path) is taken to be positive. Then

$$a_{c_1} = \frac{+T_1 - T_2}{m_1} = \frac{+4.5 \text{ N} + (-2.9 \text{ N})}{2.5 \text{ kg}} = 0.64 \text{ m/s}^2$$

and

$$v_1 = \sqrt{a_{c_1} r_1} = \sqrt{(0.64 \text{ m/s}^2)(1.0 \text{ m})} = 0.80 \text{ m/s}$$

Follow-up Exercise. Notice in this Example that the centripetal acceleration of m_2 is greater than that of m_1, yet $r_2 > r_1$, and $a_c \propto 1/r$. Is there something wrong here? Explain. *(Answer may be found in the Answers to Follow-up Exercises section at the back of the book.)*

7.4 Angular Acceleration

Objectives: To be able to (a) define angular acceleration, and (b) analyze rotational kinematics.

As you might have guessed, another type of acceleration in angular motion is angular acceleration. This is the time rate of change of angular velocity. In the case of circular motion, if there were an angular acceleration, the motion would not be uniform because the speed would be changing. Analogous to the linear case, the magnitude of the **average angular acceleration** ($\bar{\alpha}$) is

$$\bar{\alpha} = \frac{\Delta \omega}{\Delta t}$$

where the bar over the alpha indicates that it is an average. With $t_o = 0$ and if the angular acceleration is constant so that $\bar{\alpha} = \alpha$, we have

Angular acceleration—changes in angular velocity

$$\alpha = \frac{\omega - \omega_o}{t} \quad\quad \textit{(constant angular acceleration)}$$

SI unit of angular acceleration: rad/s^2

or

$$\omega = \omega_o + \alpha t \qquad (constant\ acceleration\ only) \qquad (7.12)$$

The standard units for angular acceleration are radians per second squared (rad/s^2).

No boldface symbols are used in Eq. 7.12 because, in general, plus and minus signs will be used to indicate angular directions, as described earlier. As in the case of linear motion, if the angular acceleration increases the angular velocity, both quantities have the same sign, meaning that their vector directions are the same (α is in the same direction as ω as given by the right-hand rule). If the angular acceleration decreases the angular velocity, the two quantities have opposite signs, meaning that their vectors are opposed (α is in the direction opposite to ω as given by the right-hand rule, or is an angular deceleration, so to speak).

EXAMPLE 7.10 ■ A ROTATING CD: ANGULAR ACCELERATION

A compact disc (CD) accelerates uniformly to its operational speed of 500 rpm in 3.50 s. (a) What is the angular acceleration of the CD during this time? (b) What is the angular acceleration after this time? (c) If the CD comes uniformly to a stop in 4.50 s, what is its angular acceleration?

Solution.

Given: $\omega_o = 0$

$\omega = 500\ rpm\left(\dfrac{\pi/30\ rad/s}{1\ rev/min}\right) = 52.4\ rad/s$

$t = 3.50\ s$ (starting up)
$t = 4.50\ s$ (coming to a stop)

Find: (a) α when $t < 3.50\ s$
(b) α when $t > 3.50\ s$
(c) α (in coming to a stop)

(a) Using Eq. 7.12,

$$\alpha = \frac{\omega - \omega_o}{t} = \frac{52.4\ rad/s - 0}{3.50\ s} = 15.0\ rad/s^2$$

in the direction of the angular velocity.

(b) After the CD reaches its operational speed, the angular velocity remains constant, so $\alpha = 0$.

(c) Again using Eq. 7.12, but this time with $\omega_o = 500$ rpm and $\omega = 0$,

$$\alpha = \frac{\omega - \omega_o}{t} = \frac{0 - 52.4\ rad/s}{4.50\ s} = -11.6\ rad/s^2$$

where the minus sign indicates that the angular acceleration is in the direction opposite that of the angular velocity

Follow-up Exercise. What are the directions of the ω and α vectors in part (a) of this Example if the CD rotates clockwise when viewed from above? (*Answer may be found in the Answers to Follow-up Exercises section at the back of the book.*)

As with arc length and angle ($s = r\theta$) and tangential and angular speeds ($v = r\omega$), there is a relationship between the tangential acceleration and the angular acceleration. The **tangential acceleration** is associated with the tangential speed changes, and hence continuously changes direction. The magnitudes of the tangential and angular accelerations are related by a factor of r. For circular motion with a constant radius r:

$$a_t = \frac{\Delta v}{\Delta t} = \frac{\Delta(r\omega)}{\Delta t} = \frac{r\Delta\omega}{\Delta t} = r\alpha$$

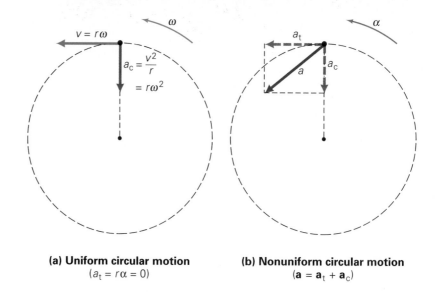

·FIGURE 7.15   Acceleration and circular motion
(a) In uniform circular motion, there is centripetal acceleration but no angular acceleration ($\alpha = 0$) or tangential acceleration ($a_t = r\alpha = 0$).
(b) In nonuniform circular motion, there are angular and tangential accelerations, and the total acceleration is the vector sum of the tangential and centripetal components.

(a) Uniform circular motion
$(a_t = r\alpha = 0)$

(b) Nonuniform circular motion
$(\mathbf{a} = \mathbf{a}_t + \mathbf{a}_c)$

so that

$$a_t = r\alpha \qquad \begin{array}{l}\textit{magnitude of}\\ \textit{tangential acceleration}\end{array} \qquad (7.13)$$

The tangential acceleration (a_t) is written with a subscript t to distinguish it from the radial, or *centripetal, acceleration* (a_c), which is necessary for circular motion.

For uniform circular motion, there is no angular acceleration ($\alpha = 0$) or tangential acceleration, as can be seen from Eq. 7.13. There is only centripetal acceleration (·Fig. 7.15a).

However, when there is an angular acceleration α (and therefore a tangential acceleration, $a_t = r\alpha$), there is a change in *both* the angular and the tangential velocities ($v = r\omega$). As a result, the centripetal acceleration, $a_c = v^2/r = r\omega^2$, must increase or decrease if the object is to maintain the same circular orbit (that is, if r is to stay the same). When there are both tangential and centripetal accelerations, the total instantaneous acceleration is their vector sum (·Fig. 7.15b). The tangential acceleration vector and the centripetal acceleration vector are perpendicular to each other at any instant, and the total acceleration is $\mathbf{a} = a_t\mathbf{t} + a_c\mathbf{r}$, where $\mathbf{t}$ and $\mathbf{r}$ are unit vectors directed tangentially and radially inward, respectively. You should be able to find the magnitude of $\mathbf{a}$ and the angle it makes relative to $\mathbf{a}_t$ using trigonometry (Fig. 7.15b).

By now, the direct correspondence between the linear and angular kinematic equations should be apparent to you. The other angular equations can be derived, as was done for the linear ones in Chapter 2. That development will not be shown; the set of angular equations with their linear counterparts for constant accelerations are listed in Table 7.2. A quick review of Chapter 2 (with a change of symbols) will show you how the angular equations are derived.

TABLE 7.2   Equations for Linear and Angular Motion with Constant Acceleration*

Linear	Angular	
$x = \bar{v}t$	$\theta = \bar{\omega}t$	(1)
$\bar{v} = \dfrac{v + v_o}{2}$	$\bar{\omega} = \dfrac{\omega + \omega_o}{2}$	(2)
$v = v_o + at$	$\omega = \omega_o + \alpha t$	(3)
$x = v_o t + \frac{1}{2}at^2$	$\theta = \omega_o t + \frac{1}{2}\alpha t^2$	(4)
$v^2 = v_o^2 + 2ax$	$\omega^2 = \omega_o^2 + 2\alpha\theta$	(5)

*For these equations, $x_o = 0$, $\theta_o = 0$, and $t_o = 0$. The first equation in each column is general, that is, is not limited to situations where the acceleration is constant.

EXAMPLE 7.11 ■ AROUND WE GO: ROTATIONAL KINEMATICS

A child on a merry-go-round selects a horse located 5.5 m from the central axis. When the ride begins, the merry-go-round accelerates at a rate of 0.069 rad/s² for 12 s and from then on maintains a constant angular speed. (a) How many revolutions does the child make before the constant operating speed is reached? (b) What is the constant angular speed of the ride and the child's tangential speed?

7.5 Newton's Law of Gravitation

Objectives: To be able to (a) describe Newton's law of gravitation and how it relates to the acceleration due to gravity, and (b) apply the general formulation of gravitational potential energy.

Another of Isaac Newton's many accomplishments was the formulation of what is called the **universal law of gravitation**. This law is very powerful and fundamental. Without it, for example, we would not understand the cause of tides or know how to put satellites into particular orbits around the Earth. This law allows us to analyze the motions of planets and comets, stars and galaxies. The word "universal" in the name indicates that we believe it to apply everywhere in the universe. (This term highlights the importance of the law, but for brevity it is common to refer simply to Newton's law of gravitation.)

Newton's law of gravitation in mathematical form gives a simple relationship for the gravitational interaction between two particles, or point masses, m_1 and m_2, separated by a distance r ($\bullet$Fig. 7.16). Basically, every particle in the universe has an attractive interaction with every other particle. The forces of mutual interaction are equal and opposite, a force pair as described by Newton's third law (Chapter 4).

The gravitational attraction or force (F) decreases as the distance (r) between two point masses increases; that is, the magnitude of the gravitational force and the distance separating the two particles are related as follows:

$$F \propto \frac{1}{r^2} \qquad (7.14)$$

(This type of relationship is called an *inverse square law*—F is inversely proportional to r^2.)

Newton's law also correctly postulates that the gravitational force or attraction of a body depends on its mass—the greater the mass, the greater the attraction. However, because gravity is a mutual interaction between masses, it should be directly proportional to both masses—i.e., to their product ($F \propto m_1 m_2$).

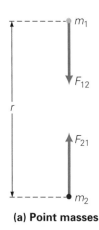

(a) Point masses

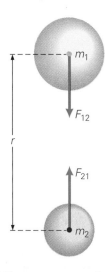

(b) Homogeneous spheres

$$F = \frac{Gm_1 m_2}{r^2}$$

$\bullet$**FIGURE 7.16 Universal law of gravitation**
(a) Any two particles or point masses are gravitationally attracted to each other with a force that has a magnitude given by Newton's universal law of gravitation.
(b) For homogeneous spheres, the masses may be considered to be concentrated at their centers.

Hence, Newton's law of gravitation has the form $F \propto m_1 m_2 / r^2$. Expressed as an equation, the magnitude of the mutually attractive gravitational forces between two masses is given by

Newton's Universal Law of Gravitation

$$F = \frac{G m_1 m_2}{r^2}$$

(7.15)

where G is a constant called the **universal gravitational constant**:

$$G = 6.67 \times 10^{-11} \text{ N} \cdot \text{m}^2/\text{kg}^2$$

This constant is often referred to as "big G" to distinguish it from "little g," the acceleration due to gravity. Note from Eq. 7.15 that F approaches zero only when r is infinitely large. Thus, the gravitational force has, or acts over, an infinite range.

You might wonder how Newton came to his conclusions about the force of gravity. Legend has it that his insight came after he observed an apple fall from a tree to the ground. Newton had been wondering what supplied the centripetal force to keep the Moon in orbit and might have had this thought: "If gravity attracts an apple toward the Earth, perhaps it also attracts the Moon, and the Moon is falling, or accelerating toward the Earth, under the influence of gravity" (see •Fig. 7.17).

Whether or not the legendary apple did the trick, Newton assumed that the Moon and the Earth were attracted to each other and could be treated as point masses, with their total masses concentrated at their centers. The inverse square relationship had been speculated on by some of his contemporaries. Newton's achievement was demonstrating that the relationship could be deduced using one of Johannes Kepler's laws of planetary motion (see Section 7.6).

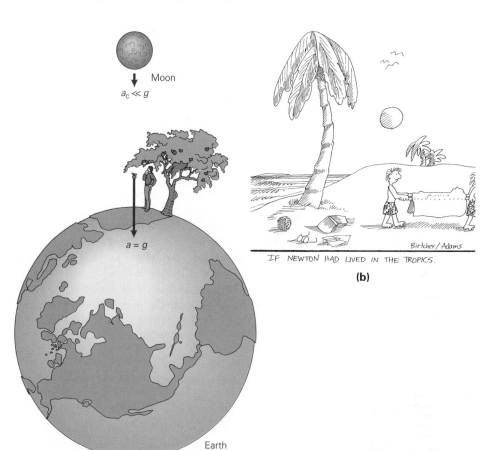

•**FIGURE 7.17 Gravitational insight?**
(a) Newton developed his law of gravitation while studying the orbital motion of the Moon. According to legend, his thinking was spurred when he observed an apple falling from a tree. He supposedly wondered whether the force causing the apple to accelerate toward the ground could extend to the Moon and cause it to fall toward Earth, that is, supply its orbital centripetal acceleration. **(b)** A popular version of the Newton legend—that he was actually hit on the head by a falling apple while sitting under the tree—inspired this cartoon.

Newton expressed Eq. 7.14 as a proportion ($F \propto m_1 m_2 / r^2$) because he did not know the value of G. It was not until 1798 (71 years after Newton's death) that the value of the universal gravitational constant was experimentally determined by an English physicist, Henry Cavendish. He used a sensitive balance to measure the gravitational force between separated spherical masses. Knowing F, r, and the m's from measurement, big G can be computed from Eq. 7.15.

As mentioned earlier, Newton considered the nearly spherical Earth and Moon to be point masses located at their respective centers. It took him some years, using mathematical methods he developed, to prove that this is the case for spherical, *homogeneous* objects.* The general concept is illustrated in •Fig. 7.18.

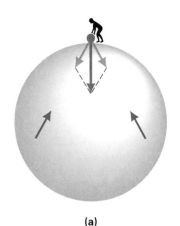

(a)

EXAMPLE 7.12 ■ GRAVITATIONAL ATTRACTION BETWEEN THE EARTH AND THE MOON: A CENTRIPETAL FORCE

Estimate the magnitude of the mutual gravitational force between the Earth and the Moon. (You can assume that the Earth and the Moon are homogeneous spheres.)

Solution. No data are given, so they must be available from references.

Given: (from Table 1 in Appendix I)   *Find:* F (gravitational force)
$M_E = 6.0 \times 10^{24}$ kg
$m_M = 7.4 \times 10^{22}$ kg
$r_{EM} = 3.8 \times 10^8$ m

The average distance from the Earth to the Moon (r_{em}) is taken to be the distance from the center of one to the center of the other, and using Eq. 7.15,

$$F = \frac{Gm_1 m_2}{r^2} = \frac{G M_E m_M}{r_{EM}^2}$$

$$= \frac{(6.67 \times 10^{-11} \text{ N·m}^2/\text{kg}^2)(6.0 \times 10^{24} \text{ kg})(7.4 \times 10^{22} \text{ kg})}{(3.8 \times 10^8 \text{ m})^2}$$

$$= 2.1 \times 10^{20} \text{ N}$$

This is the magnitude of the centripetal force that keeps the Moon revolving in its orbit around the Earth. It is a very large force, but our Moon is a very massive object, with a correspondingly large inertia to overcome.

Follow-up Exercise. With what acceleration is the Moon accelerating toward Earth? *(Answer may be found in the Answers to Follow-up Exercises section at the back of the book.)*

(b)

•**FIGURE 7.18 Uniform spherical masses**
(a) Gravity acts between any two particles. The resultant gravitational force exerted on an object outside a homogeneous sphere by two particles at symmetric locations within the sphere is directed toward the center of the sphere. **(b)** Because of the sphere's symmetry and uniform mass distribution, the net effect is as though all the mass of the sphere were concentrated as a particle at its center. For this special case, the gravitational center of force and center of mass coincide, but this is not generally true for other objects. (Only a few of the red arrows are shown because of space.)

An interesting relationship that is revealed when Newton's second law and gravitational attraction are used to measure mass is explained in the Insight on p. 222. This feature shows how things are questioned in science.

The acceleration due to gravity at a particular distance from a planet can also be investigated using Newton's second law of motion and his law of gravitation. The magnitude of the acceleration due to gravity, which we will generally write as a_g, is found by setting $m a_g$, the net force on the object, equal to the force of gravitational attraction due to a spherical mass M, at a distance r from its center

$$m a_g = \frac{GmM}{r^2}$$

Then, the acceleration due to gravity at any distance r from the planet's center is

$$a_g = \frac{GM}{r^2} \tag{7.16}$$

---

*For a homogeneous sphere, the equivalent point mass is located at the center of mass. However, this is a special case. The center of gravitational force and the center of mass of a configuration of particles or an object do not generally coincide (see Exercise 91).

Gravitational and Inertial Mass

To determine the mass of an object, you would find it easiest to measure its weight using a suitable scale. Then, in terms of weight at the Earth's surface, $w = mg$, or $m = w/g$, where the gravitational force is expressed using the acceleration due to gravity. But can the mass of an object be determined without reference to, or in the absence of, gravity? The answer is yes.

Newton's second law is $F = ma$. Therefore, by applying a force and measuring the resulting acceleration, you could determine the mass of an object. (This could be done even in the presence of gravity by applying a horizontal force to an object on a nearly frictionless horizontal surface.) This type of measurement involves the property of inertia. Thus, you would be measuring *inertial* mass by this procedure and *gravitational* mass by weighing.

Are the masses determined by these two methods the same, or are there two different types of mass? Comparisons have been made, and no significant difference has been found. Experimentally, the two types of measurements give results that differ by not much more than one part in a trillion (10^{12}).

Notice that a_g is proportional to $1/r^2$, so the farther away an object is from the planet, the smaller its acceleration due to gravity and the smaller the attractive force (ma_g) on the object. The force or weight vector is directed toward the center of the planet.

The preceding equation can be applied to the Moon or any planet. For example, taking the Earth to be a point mass M_E located at its center and R_E as its radius, we have the acceleration due to gravity (g) at the Earth's surface by setting the distance $r = R_E$:

$$a_{g_E} = g = \frac{GM_E}{R_E^{\,2}} \qquad (7.17)$$

This equation has several interesting implications. First, it reveals that taking g to be constant everywhere on the surface of the Earth involves assuming that the Earth has a homogeneous mass distribution and that the distance from the center of the Earth to any location on its surface is the same. Since these two assumptions are not true, taking g to be a constant is only an approximation, but one that works pretty well for most situations.

Also, you can see why the acceleration due to gravity is the same for all free-falling objects, that is, independent of the mass of the object. This mass of the object doesn't appear in Eq. 7.17, so all objects in free fall accelerate at the same rate.

Finally, if you're observant, you'll notice that Eq. 7.17 can be used to compute the mass of the Earth. All of the other quantities in the equation are measurable and their values are known, so M_E can readily be calculated. This is what Cavendish did after he determined the value of G experimentally. Then he also found the average density of the Earth. (What do you think this showed? Think of the difference between the Earth's crust and its metallic core. How does the density of a crustal rock compare to Cavendish's average density of the Earth?)

The acceleration due to gravity does vary with altitude. At a distance h above the Earth's surface, the acceleration is given by

$$a_g = \frac{GM_E}{(R_E + h)^2} \qquad (7.18)$$

PROBLEM-SOLVING HINT

When comparing accelerations due to gravity or gravitational forces, you will often find it convenient to work with ratios. For example, comparing a_g to g (Eqs. 7.16 and 7.17) for the Earth gives

$$\frac{a_g}{g} = \frac{GM_E/r^2}{GM_E/R_E^{\,2}} = \frac{R_E^{\,2}}{r^2} = \left(\frac{R_E}{r}\right)^2 \quad \text{or} \quad \frac{a_g}{g} = \left(\frac{R_E}{r}\right)^2$$

Note how the constants cancel out. Taking $r = R_E + h$ (Eq. 7.18), you can easily compute a_g/g, or the acceleration due to gravity at some altitude h above the Earth compared to g on the Earth's surface (9.80 m/s²).

Because R_E is very large compared to readily attainable everyday altitudes above the Earth's surface, the acceleration due to gravity does not decrease very rapidly as we ascend. At an altitude of 16 km (10 mi), $a_g/g = 0.99$, or a_g is still 99% of the value of g at the Earth's surface. At an altitude of 160 km (100 mi), a_g is 95% of g.

The value of g at the Earth's surface is referred to as the standard acceleration and is sometimes used as a unit. For example, when a spacecraft lifts off, astronauts are said to experience "several g's." This means that their acceleration is several times the standard acceleration, g. Since $g = w/m$, we can also think of g as the *gravitational force per unit mass*. Thus the term **g's of force** may be used for the force corresponding to a given acceleration. For example, in free fall on the surface of the Earth you experience an *acceleration* of 1 g = g, and the corresponding *force* that produces that 1 g is your weight: $F = m (1 \text{ g}) = mg = w$. If you experience a net force of 2 g's, this means the net force on you has a magnitude of twice your weight, and so on. (See the Insight below.)

Another aspect of the decrease of g with altitude concerns potential energy. In Chapter 5, you learned that $U = mgh$ for an object at a height h above some zero reference point, since g is essentially constant near the Earth's surface. This potential energy is equal to the work done in raising the object a distance h above the Earth's

Insight   "g's of Force"

To help better understand the non-standard but common use of the gravitational unit g as a unit of force, let's look at some examples. During the takeoff of a jet airliner, you experience an average horizontal force of about 0.20 g. This means that as the plane accelerates down the runway, the seat back exerts a horizontal force of about one-fifth of your weight against you. The reaction force (Newton's third law) is the force you exert against the seat, and you experience a feeling of being pushed back into the seat. On takeoff at an angle of 30°, this increases to about 0.70 g with a component of gravity helping push you into the seat.

A jet pilot diving in a circular arc is said to "pull" so many g's at the bottom of the arc. This tells the magnitude of the net force on the pilot, which supplies the centripetal acceleration necessary for circular motion. At about 10 g's, a pilot experiences a blackout, or loss of consciousness, because this force causes a lack of blood flow to the brain.

Similarly, an astronaut experiences as much as 8 g's of force on blast-off because of the upward acceleration (Fig. 1). This net force (experienced as apparent weight) is exerted on the astronaut by the seat which pushes on the astronaut with a force many times his or her weight. If an astronaut were standing upright, such a force would cause the body to accelerate away from its liquid blood. Thus the blood appears to go to the legs, where the blood vessels would be distended and the capillaries would rupture. Lack of blood circulation to the eyes and brain would cause temporary blindness and loss of consciousness. Consequently, astronauts are in a reclining position for blast-off.

You may have experienced several g's of force on a roller coaster. This occurs when the coaster is climbing after a dip. Usually, the force is limited to less than 3.5 g's. Otherwise, people become really frightened.

FIGURE 1   g's of force
Lt Col. John Stapp is shown before and during a high-speed rocket-propelled sled run. He reached Mach 1.7 (1.7 times the speed of sound or about 1200 mi/h) in just seconds and then quickly decelerated to rest.

surface in a *uniform* gravitational field. But what if the change in altitude is so large that g cannot be considered constant while work is done in moving an object? In this case, the equation $U = mgh$ doesn't apply. In general, it can be shown (using mathematical methods that are beyond the scope of this book) that the **gravitational potential energy** of two point masses separated by a distance r is given by

$$U = -\frac{Gm_1m_2}{r} \tag{7.19}$$

The minus sign in Eq. 7.19 arises from the choice of the zero reference point (the point where $U = 0$), which is $r = \infty$. In terms of the Earth and the mass m which is at an altitude h above the Earth's surface

Gravitational potential energy

$$U = -\frac{Gm_1m_2}{r} = -\frac{GmM_E}{R_E + h} \tag{7.20}$$

where r is the distance separating the Earth's center and the mass. What this means is that on Earth we are in a negative gravitational potential energy well (Fig. 7.19) that extends to infinity because the force of gravity has an infinite range. As h increases, so does U. That is, U becomes *less negative*, or gets closer to zero, corresponding to a higher position in the potential well. The same is true for the finite well approximation given by $U = mgh$, but over long distances, for which Eq. 7.20 applies, the change in U is not linear, but varies as $1/r$.)

Note: Potential energy wells are discussed in Section 5.4 (review Fig. 5.14).

Thus, when gravity does negative work (an object moves higher in the well) or gravity does positive work (an object falls lower in the well), there is a *change* in potential energy. As with finite potential energy wells, this change in energy is usually what's important in analyzing situations.

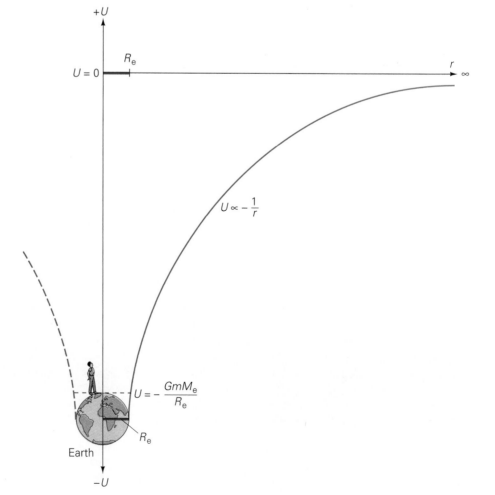

•**FIGURE 7.19 Gravitational potential energy well**
On Earth, we are in a negative gravitational potential energy well. As with an actual well or hole in the ground, work must be done against gravity to get higher in the well. The potential energy of an object increases as it moves higher in the well. This means that the value of U becomes less negative. The top of the Earth's gravitational well is at infinity, where the gravitational potential energy is, by choice, zero.

EXAMPLE 7.13 ■ DIFFERENT ORBITS: CHANGE IN GRAVITATIONAL POTENTIAL ENERGY

Two 50-kg satellites are in circular orbits about the Earth at altitudes of 1000 km (about 620 mi) and 35,000 km (about 22,000 mi). The lower one monitors particles about to enter the atmosphere, and the higher one, synchronous with the Earth's rotation, takes weather pictures from its stationary position with respect to the Earth's surface. What is the difference in the gravitational potential energies of the two satellites in their respective orbits?

Solution. Listing the data so that we can better see what's given, we have

Given: $m = 50$ kg   *Find:* ΔU (potential energy difference)
$h_1 = 1000$ km $= 1.0 \times 10^6$ m
$h_2 = 35{,}000$ km $= 35 \times 10^6$ m
$M_E = 6.0 \times 10^{24}$ kg (from table
               inside back
               cover)
$R_E = 6.4 \times 10^6$ m

The difference in the gravitational potential energy may be computed directly from Eq. 7.20. Keep in mind that the potential energy is the energy of position, so we compute the potential energies for each position or altitude and subtract. Thus we have

$$\Delta U = U_2 - U_1 = -\frac{GmM_E}{R_E + h_2} - \left(-\frac{GmM_E}{R_E + h_1}\right)$$

$$= GmM_e\left(\frac{1}{R_E + h_1} - \frac{1}{R_E + h_2}\right)$$

$$= (6.67 \times 10^{-11}\ \text{N·m}^2/\text{kg}^2)(50\ \text{kg})(6.0 \times 10^{24}\ \text{kg})$$

$$\times\left[\frac{1}{(6.4 \times 10^6\ \text{m}) + (1.0 \times 10^6\ \text{m})} - \frac{1}{(6.4 \times 10^6\ \text{m}) + (35 \times 10^6\ \text{m})}\right]$$

$$= +2.2 \times 10^9\ \text{J}$$

Notice that ΔU is positive indicating that m_2 is higher in the gravitational potential energy well.

Follow-up Exercise. (a) Suppose that the altitude of the higher satellite in this Example were doubled, to 70,000 km. Would the difference in the gravitational potential energies of the two satellites then be twice as great? Justify your answer. (b) Note that U_2 has a *smaller* negative value than U_1 (since $h_2 > h_1$). What does this mean?

Using the gravitational potential energy (Eq. 7.19) gives the equation for the total mechanical energy a different form than it had in Chapter 5. For example, the total mechanical energy of a mass m_1 moving near a stationary mass m_2 is

$$E = K + U = \tfrac{1}{2}m_1v^2 - \frac{Gm_1m_2}{r} \qquad (7.21)$$

(Note that the potential energy $U = -Gm_1m_2/r$ is negative in this case, and *not* written as *mgh*.) This equation and the principle of the conservation of energy can be applied to the Earth moving about the Sun, by neglecting other gravitational forces. The Earth's orbit is not quite circular, but slightly elliptical. At perihelion (the point of the Earth's closest approach to the Sun), the mutual gravitational potential energy is less (a larger negative number) than it is at aphelion (the point farthest from the Sun). Therefore, as can be seen from Eq. 7.21 in the form $\tfrac{1}{2}mv^2 = E + Gm_1m_2/r$, where E is constant, the Earth's kinetic energy and orbital speed are greatest at perihelion (the smallest value of r) and least at aphelion (the greatest value of r). Or, in general, the Earth's orbital speed is greater when it is nearer the Sun than when it is farther away.

Mutual gravitational potential energy also applies to a group, or configuration, of more than two masses. That is, there is gravitational potential energy due

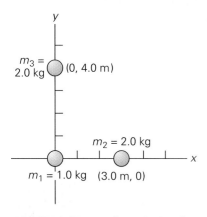

•FIGURE 7.20 **Total gravitational potential energy**
See Example 7.14.

to the masses being in a configuration. This is because work was done in bringing them together. Suppose that there is a single fixed mass m_1, and another mass m_2 is brought close to it from an infinite distance (where $U = 0$). The work done against the attractive force of gravity is negative (why?) and equal to the change in the mutual potential energy of the masses, which are now separated by a distance r_{12}, that is, $U_{12} = -Gm_1m_2/r_{12}$.

If a third mass m_3 is brought close to the other two fixed masses, there are then two forces of gravity acting on m_3, so $U_{13} = -Gm_1m_3/r_{13}$ and $U_{23} = -Gm_2m_3/r_{23}$. The total gravitational potential energy of the configuration is therefore

$$U = U_{12} + U_{13} + U_{23}$$
$$= -\frac{Gm_1m_2}{r_{12}} - \frac{Gm_1m_3}{r_{13}} - \frac{Gm_2m_3}{r_{23}} \tag{7.22}$$

A fourth mass could be brought in, but this development should be sufficient to suggest that the total gravitational potential energy of a configuration of particles is equal to the sum of the individual potential energies for all pairs of particles.

EXAMPLE 7.14 ■ TOTAL GRAVITATIONAL POTENTIAL ENERGY: ENERGY OF CONFIGURATION

Three masses are in a configuration as shown in •Fig. 7.20. What is their total gravitational potential energy?

Solution. From the figure we have,

Given: $m_1 = 1.0$ kg   *Find:* U (total gravitational potential)
$m_2 = 2.0$ kg
$m_3 = 2.0$ kg
$r_{12} = 3.0$ m
$r_{13} = 4.0$ m
$r_{23} = 5.0$ m (3-4-5-right triangle)

We can use Eq. 7.22 directly since only three masses are used in this example. (Note that Eq. 7.22 can be extended to four, five, . . . any number of masses.)

$$U = -\frac{Gm_1m_2}{r_{12}} - \frac{Gm_1m_3}{r_{13}} - \frac{Gm_2m_3}{r_{23}}$$
$$= (6.67 \times 10^{-11} \text{ N·m}^2/\text{kg}^2)$$
$$\times \left[-\frac{(1.0 \text{ kg})(2.0 \text{ kg})}{3.0 \text{ m}} - \frac{(1.0 \text{ kg})(2.0 \text{ kg})}{4.0 \text{ m}} - \frac{(2.0 \text{ kg})(2.0 \text{ kg})}{5.0 \text{ m}} \right]$$
$$= -1.3 \times 10^{-10} \text{ J}$$

Follow-up Exercise. Explain what the *negative* potential energy in this Example means physically. *(Answer may be found in the Answers to Follow-up Exercises section at the back of the book.)*

7.6 Kepler's Laws and Earth Satellites

Objectives: To be able to (a) state and explain Kepler's laws of planetary motion, and (b) describe the orbits and motions of satellites.

The force of gravity determines the motions of the planets and satellites and holds the solar system (and galaxy) together. A general description of planetary motion had been set forth shortly before Newton's time by the German astronomer and mathematician Johannes Kepler (1571–1630). Kepler was able to formulate three empirical laws from observational data gathered during a 20-year period by the Danish astronomer Tycho Brahe (1546–1601). Brahe's extensive observations of stars and planets were done without the benefit of a telescope, which had not yet been invented. But the observations Brahe made were more accurate than those of most of his contemporaries because of better instrumentation he had developed. As a result, he is considered to have been one of the greatest practical astronomers.

Kepler went to Prague to assist Brahe, who was the official mathematician at the court of the Holy Roman Emperor. Brahe died the next year and Kepler succeeded him, inheriting his records of the positions of the planets. Analyzing these data, Kepler announced the first two of his three laws in 1609 (the year Galileo built his first telescope). These laws were applied initially only to Mars. Kepler's third law came 10 years later.

Interestingly enough, Kepler's laws of planetary motion, which took him about 15 years to deduce from observed data, can now be derived theoretically with a page or two of calculations. These three laws apply not only to planets, but to any system composed of a body revolving about a larger body, to which the inverse square law of gravitation applies (for example, the Moon, artificial Earth satellites, and solar-bound comets obey these laws).

Kepler's first law (the law of orbits) says that

Kepler's first law: the law of orbits

Planets move in elliptical orbits with the Sun at one of the focal points.

An ellipse, shown in •Fig. 7.21a, has, in general, an oval shape, resembling a flattened circle. In fact, a circle is a special case of an ellipse in which the focal points, or *foci* (plural of focus), are at the same point (the center of the circle). Although the orbits of the planets are elliptical, most do not deviate very much from circles (Mercury and Pluto are notable exceptions; see Table 6 in Appendix I). For example, the difference between the perihelion and aphelion of the Earth (its closest and farthest distances from the Sun) is about 5 million km. This may sound like a lot, but it is only a little over 3 percent of 150 million km, which is the average distance between the Earth and the Sun.

Kepler's second law (the law of areas) says that

Kepler's second law: the law of areas

A line from the Sun to a planet sweeps out equal areas in equal lengths of time.

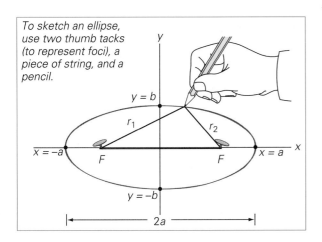

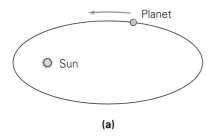

(a)

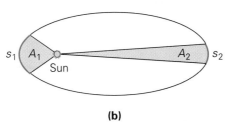

(b)

•FIGURE 7.21 Kepler's first and second laws of planetary motion (a) In general, an ellipse has an oval shape. The sum of the distances from the focal points to any point on the ellipse is constant: $r_1 + r_2 = 2a$, where $2a$ is the length of the line joining the two points on the ellipse at the greatest distance from its center, called the major axis. (The line joining the two points closest to the center is $2b$, the minor axis.) Planets revolve about the Sun in elliptical orbits for which the Sun is at one of the focal points and nothing is at the other. (b) A line joining the Sun and a planet sweeps out equal areas in equal times. Since $A_1 = A_2$, a planet travels faster along s_1 than along s_2.

This law is illustrated in •Fig. 7.21b. Since the time to travel the different orbital distances (s_1 and s_2) is the same, this law tells you that the orbital speed of a planet varies in different parts of its orbit. Because a planet's orbit is elliptical, its orbital speed is greater when it is closer to the Sun than when it is farther away. (This was deduced for the Earth using the conservation of energy in Section 7.5.)

Kepler's third law:
the law of periods

Kepler's third law (the law of periods) says that

> The square of the orbital period of a planet is directly proportional to the cube of the average distance of the planet from the Sun; that is, $T^2 \propto r^3$.

Kepler's third law is easily derived for the special case of a circular orbit, using Newton's law of gravitation. Since the centripetal force is supplied by the force of gravity, the expressions for these forces can be set equal:

$$\frac{m_\text{p}v^2}{r} = \frac{Gm_\text{p}M_\text{S}}{r^2}$$

$$\underset{\substack{\text{centripetal}\\\text{force}}}{} \qquad \underset{\substack{\text{gravitational}\\\text{force}}}{}$$

and

$$v = \sqrt{\frac{GM_\text{S}}{r}}$$

where m_p and M_s are the masses of the planet and the Sun, respectively, and v is the orbital speed. But, $v = 2\pi r/T$ (circumference/period), so

$$\sqrt{\frac{GM_\text{S}}{r}} = \frac{2\pi r}{T}$$

Squaring both sides and solving for T^2 gives

$$T^2 = \left(\frac{4\pi^2}{GM_\text{S}}\right)r^3$$

or

The law of periods

$$T^2 = Kr^3 \qquad (7.23)$$

The constant K for solar system planetary orbits is easily evaluated from orbital data (T and r) for the Earth: $K = 2.97 \times 10^{-19} \text{ s}^2/\text{m}^3$. (As an exercise, you might wish to convert K to the more useful units of y^2/km^3.) Knowing the value of K, notice from Eq. 7.23 that the mass of the Sun can be determined. (Note: this value of K does *not* apply to Earth satellites. Why?)

Earth Satellites

Note: For communications, many satellites are launched into a circular orbit above the equator at an altitude of about 36,000 km. Satellites there are *synchronous* to Earth's rotation. That is, they remain "fixed" over one point on the equator, and to an observer on Earth are always seen in the same position in the sky.

We are only about a half a century into the space age. Since the 1950s, numerous unmanned satellites have been put into orbit about the Earth, and now astronauts regularly spend days or weeks in orbiting space laboratories.

Putting a spacecraft into orbit about the Earth (or the Moon) is an extremely complex task. However, you can get a basic understanding of the problem from fundamental principles. First, suppose that a projectile could be given the initial speed required to take it just to the top of the Earth's potential energy well. At the exact top of the well, which is an infinite distance away ($r = \infty$), the potential energy is zero. By the conservation of energy and Eq. 7.19,

$$\overset{\textit{initial}}{K_\text{o} + U_\text{o}} = \overset{\textit{final}}{K + U}$$

$$\tfrac{1}{2}mv_\text{esc}^2 - \frac{GmM_\text{E}}{R_\text{E}} = 0 + 0$$

where v_{esc} is the **escape speed**: the initial speed needed to escape from the surface of a planet or moon. The final energy is zero since the projectile stops at the top of the well (at very large distances it is barely moving) and $U = 0$ there. Solving for v_{esc} gives

Escape speed

$$\tfrac{1}{2}mv_{esc}^2 = \frac{GmM_E}{R_E}$$

and

$$v_{esc} = \sqrt{\frac{2GM_E}{R_E}} \qquad (7.24)$$

Since $g = GM_E/R_E^2$ (Eq. 7.17), it is convenient to write

$$v_{esc} = \sqrt{2gR_E} \qquad (7.25)$$

Although derived for Earth, this equation may be used generally to find the escape speeds for other planets and our Moon.

EXAMPLE 7.15 ■ No Return: Escape Speed

What is the escape speed at the Earth's surface?

Solution. Equation 7.25 can be used with known data:

$$v_{esc} = \sqrt{2gR_E} = \sqrt{2(9.80 \text{ m/s}^2)(6.4 \times 10^6 \text{ m})} = 11 \times 10^3 \text{ m/s}$$
$$= 11 \text{ km/s} \quad \text{(about 7 mi/s)}$$

Thus, if a projectile or spacecraft could be given an initial upward speed of 11 km/s (about 40,000 km/h or 25,000 mi/h), it would leave the Earth and not fall back, or return. Note from the equation that the escape speed is independent of the mass of the spacecraft; this is the initial speed required to send *any* object to the top of the Earth's potential energy well. Such an initial speed is not attainable instantly at launch, but can be acheived in multi-stage steps. For example, in the 1970's probes to Jupiter (Pioneer series) left the Earth never to return and are currently at the limits of the Solar system.

Follow-up Exercise. How many times greater is the escape speed of the Earth than that of the Moon? *(Answers may be found in the Answers to Follow-up Exercises section at the back of the book.)*

It is clear that a tangential speed smaller than the escape speed is required to orbit a satellite. Consider the centripetal force for a satellite in circular orbit about the Earth. Since the centripetal force on the satellite is supplied by the gravitational attraction between the satellite and the Earth, we may again write:

$$F = \frac{mv^2}{r} = \frac{GmM_E}{r^2}$$

Then

$$v = \sqrt{\frac{GM_E}{r}} \qquad (7.26)$$

where $r = R_E + h$. For example, suppose that a satellite is in a circular orbit at an altitude of 500 km (about 300 mi); its tangential speed is

$$v = \sqrt{\frac{GM_E}{r}} = \sqrt{\frac{GM_E}{R_E + h}}$$

$$= \sqrt{\frac{(6.67 \times 10^{-11} \text{ N·m}^2/\text{kg}^2)(6.0 \times 10^{24} \text{ kg})}{(6.4 \times 10^6 \text{ m}) + (0.50 \times 10^6 \text{ m})}} = 7.6 \times 10^3 \text{ m/s}$$

$$= 7.6 \text{ km/s} \quad \text{(about 4.7 mi/s)}$$

This is about 27,000 km/h or 17,000 mi/h. As can be seen from Eq. 7.26, the required circular orbital speed *decreases* with altitude. That is, in terms of energy, for an increase in the potential energy U (higher altitude) there is a decrease in kinetic energy K (see Exercise 96).

In practice, a satellite is given a tangential speed by a component of the thrust from a rocket stage (•Fig. 7.22a). The inverse square relationship of Newton's law of gravitation means that the satellite orbits that are possible about a large mass are ellipses, of which a circular orbit is a special case. This is illustrated in •Fig. 7.22b for Earth, using the previously calculated values. If a satellite is not given a sufficient tangential speed, it will fall back to Earth (and possibly be burned up while falling through the atmosphere). If the tangential speed reaches the escape speed, the satellite will leave its orbit and go off into space.

Finally, the total energy of an orbiting satellite is

$$E = K + U = \tfrac{1}{2}mv^2 - \frac{GmM_E}{r} \tag{7.27}$$

Substituting the expression for v from Eq. 7.26 in the kinetic energy term gives

$$E = \frac{GmM_E}{2r} - \frac{GmM_E}{r}$$

Thus,

$$E = -\frac{GmM_E}{2r} \qquad \begin{array}{l}\textit{total energy of} \\ \textit{an orbiting satellite}\end{array} \tag{7.28}$$

Note that the total energy of the satellite is negative. To understand this, recall that on Earth we are in a negative potential energy well (see Fig. 17.19). More work is required to put a satellite into a higher orbit, where it has more energy because it is higher in the well. (The situation is analogous to throwing a ball upward from the bottom of a real well.) The total energy E increases as its *numerical value* becomes smaller—less negative—in going to a higher orbit toward the zero potential at the top of the well. That is, the farther the distance of a satellite from Earth, the greater its total energy.

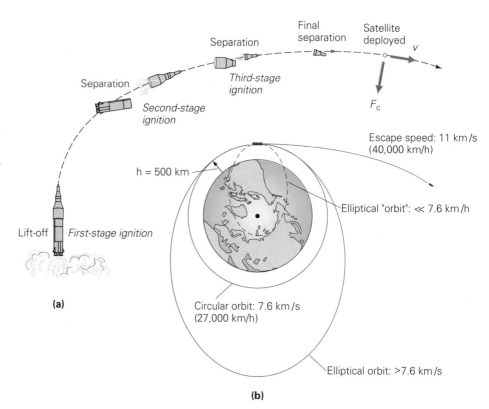

•FIGURE 7.22 **Satellite orbits**
(a) A satellite is put into orbit by giving it a tangential speed sufficient for maintaining an orbit at a particular altitude—the higher the orbit, the smaller the tangential speed. (b) At an altitude of 500 km, a tangential speed of 7.6 km/s is required for a circular orbit. With a greater tangential speed, the satellite would move out of the circular orbit. Since it would not have the escape speed, it would "fall" around the Earth in an elliptical orbit with the Earth's center at one focal point. A tangential speed less than 7.6 km/s would also give an elliptical path about the center of the Earth; but because the Earth is not a point mass, a certain minimum speed is needed to keep the satellite from striking the Earth's surface.

TABLE 7.3 Relationship of Radius, Speed, and Energy for Circular Orbital Motion		
	Increasing r (larger orbit)	Decreasing r (smaller orbit)
ω	decreases	increases
v	decreases	increases
K	decreases	increases
U	increases (smaller negative value)	decreases (larger negative value)
$E\ (=K + U)$	increases (smaller negative value)	decreases (larger negative value)

To help understand why we say that the total energy increases when its value becomes less negative, think of a change in energy from, say, 5.0 J to 10 J. Certainly this would be considered an increase in energy. Similarly, a change from -10 J to -5.0 J would be an increase in energy, even though the *absolute* value has decreased.

$$\Delta U = U - U_0 = -5.0 \text{ J} - (-10 \text{ J}) = +5.0 \text{ J}.$$

The relationship of speed and energy to orbital radius is summarized in Table 7.3.

Also note from the development of Eq. 7.28 that the kinetic energy of an orbiting satellite is equal to the absolute value of its total energy:

$$K = \frac{GmM_E}{2r} = |E| \qquad (7.29)$$

CONCEPTUAL EXAMPLE 7.16 ■ APPLYING THE BRAKES TO SPEED UP: CONSERVATION OF ENERGY

A spacecraft is in a circular orbit about the Earth, well outside the atmosphere. Describe what happens to the spacecraft if its retrorockets are fired. (Retrorockets point in the direction of the spacecraft's motion, and so produce a reverse thrust.) *Clearly establish the reasoning and physical principle(s) used in determining your answer before checking it below. That is, **how** did you arrive at your answer?*

Reasoning and Answer. The retrorockets applying a braking force to the spacecraft, *negative* work is done (because the force and displacement are in opposite directions), and the spacecraft loses some energy (the work-energy theorem). That is, the total energy of the spacecraft,

$$E = -\frac{GmM_E}{2r}$$

is reduced. If E decreases (that is, takes on a greater negative value), we can see from the above equation that r must also decrease. Thus, the spacecraft will spiral inward toward the Earth. From Eq. 7.29,

$$K = \tfrac{1}{2}mv^2 = \frac{GmM_E}{2r} \quad \text{or} \quad v^2 = \frac{GM_E}{r}$$

So, if r decreases, v increases, and the spacecraft actually speeds up!

Follow-up Exercise. Suppose the retrorockets of an orbiting spacecraft were fired until the craft was within the top layers of the Earth's atmosphere, where air resistance is appreciable. If the rockets were then shut off, what would happen to the spacecraft? (*Reasoning and answer may be found in the Answers to Follow-up Exercises section at the back of the book.*)

Note: The work-energy theorem is discussed in Section 5.3

CONCEPTUAL EXAMPLE 7.17 ■ FORWARD THRUST: A CHANGE IN ENERGY

The engines of a spacecraft in circular orbit about the Earth are fired so as to give the craft a *forward* thrust. How will the spacecraft's energy be changed after the firing, assuming another circular orbit is achieved? (a) K increased and U decreased, (b) K decreased and U increased, (c) both K and U increased, (d) the total energy decreased. *Clearly establish the reasoning and physical principle(s) used in determining your answer before checking it below. That is, **why** did you select your answer?*

Reasoning and Answer: You might think that a forward thrust would cause the spacecraft to speed up and increase its kinetic energy. However, an increase in v would require an increased centripetal force to maintain the spacecraft in orbit. This force is supplied by gravitational attraction, and so could increase only if r were reduced (i.e., if the spacecraft moved to a smaller orbit)—something that obviously would not result from the application of a forward thrust. (Draw a diagram to confirm this; in what direction would the thrust act with respect to the circular orbit?) Such a forward thrust (in contrast to a reverse thrust in the preceding Example) will send the craft into a larger rather than a smaller orbit, because the forward thrust does positive work on the spacecraft, causing it to rise higher in its potential energy well and increasing its potential energy. With an increase in r, the kinetic energy *decreases* (Eq. 7.29) while the total energy increases by becoming less negative (Eq. 7.28). Hence, the answer is (b). (In actuality, if the engine "burn" is brief the spacecraft would go into an elliptical orbit with its lowest altitude equal to the radius of the initial circular orbit.)

Follow-up Exercise: Imagine yourself in a spacecraft in a circular orbit, well behind a space station in the same radial orbit. Wishing to dock with the station, what maneuvers would you execute to catch up with it? *(Reasoning and answer may be found in the Answers to Follow-up Exercises section at the back of the book.)*

The advent of the space age and orbiting satellites has brought us the terms "weightlessness" and "zero gravity." These stem from the views of astronauts "floating" about in orbiting spacecraft (•Fig. 7.23a). However, the terms are misnomers. As mentioned earlier in the chapter, gravity is an infinite-range force, and Earth's gravity certainly acts on a spacecraft and astronauts, supplying the centripetal force necessary to keep them in orbit. Gravity is not zero, and there must be weight.

A better term to describe the floating effect of astronauts in orbiting spacecraft would be *"apparent weightlessness."* They "float" because both the astronauts and the spacecraft are centripetally accelerating (or falling) toward the Earth at the same rate. To help understand this effect, consider an analogous situation of a person standing on a scale in an elevator (•Fig. 7.23b). The "weight" measurement that the scale registers is actually the reaction force R of the scale on the person. In a non-accelerating elevator ($a = 0$), we have $R = mg = w$, and R is equal to the true weight of the individual. However, suppose the elevator is descending with an acceleration a, and $a < g$. As can be seen by the vector diagram in the figure,

$$mg - R = ma$$

and the *apparent* weight w' is

$$w' = R = m(g - a) < mg$$

where the downward direction is taken as positive in this instance. With a downward acceleration a, we see that R is less than mg, hence the scale indicates that the person weighs less than the true weight. Note that the *apparent acceleration* due to gravity is $g' = g - a$.

Now suppose the elevator were in free fall with $a = g$. As you can see, R and

(a)

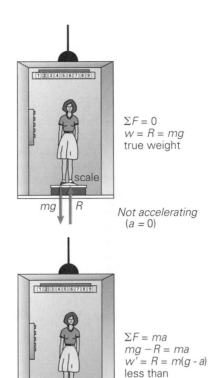

$\Sigma F = 0$
$w = R = mg$
true weight

Not accelerating
(a = 0)

$\Sigma F = ma$
$mg - R = ma$
$w' = R = m(g - a)$
less than
true weight

Descending with
acceleration a

$w' = R = 0$
"weightless"

Descending
with a = g

(b)

(c)

(d)

•**FIGURE 7.23 Apparent weightlessness**
(a) An astronaut "floats" in a spacecraft, seemingly in a "weightless" condition. **(b)** In a stationary elevator (top), a scale reads true weight. The weight reading is the reaction force R of the scale on the person. If the elevator is descending with an acceleration $a < g$ (middle), the reaction force and apparent weight are less than the true weight. If the elevator were in free fall, $a = g$, the reaction force and indicated weight would be zero since the scale would be falling as fast as the person. **(c)**, **(d)** "Weightlessness" on Earth.

the apparent weight w' would be zero. Essentially, the scale is accelerating or falling as the same rate of the person. The scale may indicate a "weightless" condition, but gravity still acts, as would be noted by the sudden stop at the bottom of the shaft.

(Note that you do not have to become an astronaut or be caught in a falling elevator to experience this condition. Some more familiar examples are shown in •Fig. 7.23(c) and (d).

Space has been called "the final frontier." Some day, instead of brief stays in Earth-orbiting spacecraft, we may have permanent space colonies (see the Insight on p. 234).

Insight | Space Colonies and Artificial Gravity

Space colonies, long a staple of science fiction stories and films, are now actually being planned. However, humans cannot live for long periods of time in a "zero-gravity" environment (see Exercise 71) without detrimental physical effects.

FIGURE 1 Space colony and artificial gravity
It has been suggested that a space colony could be housed in a huge, rotating wheel as in this artist's conception. The rotation would supply the "artificial gravity" for the colonists.

It has been suggested that this problem could be avoided if the colony were housed in a huge rotating wheel, which would supply "artificial gravity" (Fig. 1).

As you know, centripetal force is necessary to keep an object in rotational circular motion. On the rotating Earth, that force is supplied by gravity, and we refer to it as weight. Because of our weight, we exert a force on the ground, and the reaction force (by Newton's third law) upward on our feet gives us the feeling of "having our feet on solid ground." In a rotating space colony, the situation is somewhat reversed. The rotating colony would supply the centripetal force on the inhabitants, which would be perceived as a weight sensation on the soles of our feet, or artificial gravity (Fig. 2a). Rotation at the proper speed would produce a simulation of normal gravity ($g = 9.80$ m/s^2) within the colony wheel. Note that in the colonists' world, "down" would be outward, toward the periphery of the space station, and "up" would always be inward, toward the axis of rotation.

The inhabitants would experience normal gravitational effects. For example, in Fig. 2a, suppose that the person simply let go of the ball. Viewed from outside the rotating wheel, you would say the ball has a tangential velocity v as a result of the rotation, and would go off in a straight line (Newton's first law) until it hits the side of the wheel (as indicated by the dashed line in the figure). However, the inhabitant rotates with the wheel, and from that perspective, the ball merely falls to the floor as it would on Earth. Think about it.

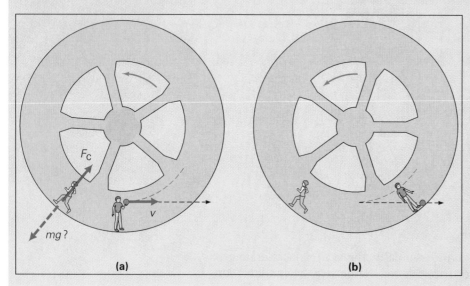

FIGURE 2 Rotating space colony
(a) In the frame of reference of someone in a rotating space colony, the centripetal force on the person would be perceived as weight sensation, or artificial gravity. Rotation at the proper speed would simulate normal gravity. From the point of view of an outside observer, a dropped ball would follow a tangential straight-line path. **(b)** A colonist, on the other hand, would observe the ball to fall downward as in a normal gravitational situation.

Chapter Review

Important Terms

angular distance   203
radian (rad)   203
average angular speed   206
angular velocity
   average   206
   instantaneous   206
tangential speed   207
period (*T*)   209
frequency (*f*)   209
hertz (Hz)   209

uniform circular motion   210
centripetal acceleration   210
centripetal force   214
average angular acceleration
   216
tangential acceleration   217
universal law of gravitation   219
universal gravitational constant,
   G   220
g's of force   223

gravitational potential energy
   224
Kepler's first law (the law of
   orbits)   227
Kepler' second law (the law of
   areas)   227
Kepler's third law (the law of
   periods)   228
escape speed   229

Important Concepts

- The radian (rad) is a measure of angle; 1 rad is the angle of a circle subtended by an arc length (*s*) equal to the radius (*r*).
- Tangential speed (*v*) and angular speed (*ω*) for circular motion are directly proportional, with the radius *r* being the constant of proportionality ($v = r\omega$).
- In uniform circular motion, a centripetal acceleration is required and is always directed toward the center of the circular path.
- A centripetal force (net force directed toward center of circle) is a requirement for circular motion.

- Angular acceleration (**α**) is the time rate of change of angular velocity.
- According to Newton's law of gravitation, every particle attracts every other particle in the universe with a force that is proprotional to both masses and inversely proportional to the square of the distance between them.
- An Earth satellite is in a negative potential energy well—the higher in the well, the greater the potential energy and the less the kinetic energy.

Important Equations

Arc Length (angle in radians):
$$s = r\theta \qquad (7.3)$$

Angular Kinematic Equations (see Table 7.2 for linear analogs):
$$\theta = \bar{\omega}t \qquad (7.5)$$

(general, not limited to constant acceleration)

$$\left.\begin{array}{l} \bar{\omega} = \dfrac{\omega + \omega_\text{o}}{2} \\[2mm] \omega = \omega_\text{o} + \alpha t \\[2mm] \theta = \omega_\text{o}t + \tfrac{1}{2}\alpha t^2 \\[2mm] \omega^2 = \omega_\text{o}{}^2 + 2\alpha\theta \end{array}\right\} \begin{array}{l}\text{constant}\\\text{acceleration}\\\text{only}\end{array}$$

(2, Table 7.2)
(7.12)
(4, Table 7.2)
(5, Table 7.2)

Tangential and Angular Speeds:
$$v = r\omega \qquad (7.6)$$

Angular Speed (with uniform circular motion):
$$\omega = \frac{2\pi}{T} = 2\pi f \qquad (7.8)$$

Frequency and Period:
$$f = \frac{1}{T} \qquad (7.7)$$

Centripetal Acceleration:
$$a_\text{c} = \frac{v^2}{r} = r\omega^2 \qquad (7.10)$$

Tangential and Angular Accelerations:
$$a_\text{t} = r\alpha \qquad (7.13)$$

Centripetal Force and Acceleration:
$$F_\text{c} = ma_\text{c} = \frac{mv^2}{r} \qquad (7.11)$$

Newton's Law of Gravitation:
$$F = \frac{Gm_1m_2}{r^2} \qquad (7.15)$$
$$G = 6.67 \times 10^{-11}\ \text{N·m}^2/\text{kg}^2$$

Acceleration Due to Gravity at an Altitude h:
$$a_\text{g} = \frac{GM_\text{E}}{(R_\text{E} + h)^2} \qquad (7.18)$$

Gravitational Potential Energy of Two Particles:
$$U = -\frac{Gm_1m_2}{r} \qquad (7.19)$$

Kepler's Law of Periods:

$$T^2 = Kr^3 \qquad (7.23)$$

(K depends on the mass of the object being orbited; for objects orbiting the Sun, $K = 2.97 \times 10^{-19}\, \text{s}^2/\text{m}^3$.)

Escape Speed:

$$v_{esc} = \sqrt{\frac{2GM_E}{r_E}} = \sqrt{2gR_E} \qquad (7.24)$$

Energy of Orbiting Satellite Orbiting Earth:

$$E = -\frac{GmM_E}{2r} \qquad (7.28)$$

$$K = |E| \qquad (7.29)$$

Exercises

7.1 Angular Measure

1 The radian unit is equivalent to (a) degree/time, (b) length, (c) length/length, (d) length/time.

2 A proper conversion factor for degrees to radians would be (a) $(\pi/7\, \text{rad})/270°$, (b) $(\pi/2\, \text{rad})/60°$, (c) $(3\pi\, \text{rad})/540°$, (d) $0.50\, \text{rad}/180°$.

3 The Cartesian coordinates of a point on a circle are $(\sqrt{2}\, \text{m}, \sqrt{2}\, \text{m})$. What are the polar coordinates (r, θ) of this point?

4 ■ The equation for a circle using Cartesian coordinates is $x^2 + y^2 = a^2$, where a is the radius of the circle. What is the equation of a circle using polar coordinates?

5 ■ Convert the following angles from degrees to radians. Report to 3 significant figures. (a) 10°, (b) 270°, (c) 45°, and (d) 450°.

6 ■ Convert the following angles from radians to degrees. Report to 3 significant figures. (a) $\pi/25$ rad, (b) $\pi/10$ rad, (c) 1.25 rad, and (d) 4π rad.

7 ■ What is the arc length subtended by an angle of $\pi/4$ rad on a circle with a radius of 6.0 cm?

8 ■■ A jogger on a circular track that has a radius of 0.250 km jogs a distance of 1.00 km. What angular distance does the jogger cover in (a) radians and (b) degrees?

9 ■■ If an arc length is twice the radius of a circle, what angle subtends this arc?

10 ■■ Assuming that the Earth's orbit around the Sun is circular, what is the approximate orbital distance in km the Earth travels in 4 months?

11 ■■ Using the ratio equation $s/\theta = c/360°$, where c is the circumference of a circle, (a) show that by definition 1 rad is equivalent to 57.3°. (b) Show that 2π rad is equivalent to 360°.

12 ■■ Two race cars start from rest and travel around a circular track whose diameter is 1.0 km. (a) One car develops engine trouble and stops after having traveled 300° around the track. What distance in km did it travel? (b) The other car makes four complete laps. How far did it travel?

13 ■■ In Europe, a large circular walking track with a diameter of 0.900 km is marked in angular distances in radians. An American tourist who walks 3.00 mi daily goes to the track and is confused. Help him out. How many radians should he walk?

14 ■■ (a) What is the angular width of a full Moon in degrees and radians as viewed from Earth? (b) What is the angular width of a full Earth in degrees and radians as viewed from the Moon? [*Hint:* Use data from Appendix I.]

15 ■■ To attend the 1996 Summer Olympics, a fan flies from Los Angeles (32° N, 119° W) to Atlanta (33° N, 84° W). Using your knowledge of angular measure, determine the approximate shortest flight distance in kilometers?

16 ■■■ Could a circular pie be cut such that all of the wedge-shaped pieces have an arc length along the outer crust equal to the pie's radius? If not, how many such pieces could you cut, and what would be the dimensions of the final piece?

17 ■■■ Electrical wire with a diameter of 0.75 cm is wound on a spool with a radius of 30 cm and a length of 24 cm. (a) Through how many radians must the spool be turned to wrap one even layer of wire? (b) What is the length of this wound wire?

18 ■■■ The Cartesian coordinates of a point on a circle with its center at the origin are (0.40 m, 0.30 m). What is the arc length measured counterclockwise on the circle from the positive x axis to this point?

7.2 Angular Speed and Velocity

19 Viewed from above, a rotating turntable is observed to rotate counterclockwise. The angular velocity vector then is (a) tangential to the turntable rim, (b) out of the plane of the turntable, (c) counterclockwise, (d) none of the preceding.

20 The frequency unit of hertz is equivalent to (a) that of the period, (b) cycle, (c) radian/s, (d) s^{-1}.

21 A turntable rotates clockwise as viewed from above. What is the direction of the angular velocity vector?

22 When clockwise or counterclockwise is used to describe rotational motion, why is a phrase such as "viewed from above" added?

23 ■ A race car makes two laps around a circular track in 3.0 min. What is the car's average angular speed?

24 ■ The blades of a helicopter rotor travel 14π rad each second. What is the angular speed of the rotor in rpm?

25 ■ A satellite in a circular orbit has a period of 10 hours. What is its frequency in revolutions per day?

26 ■ What is the period of revolution for (a) a $33\frac{1}{3}$ rpm phonograph record, (b) a 45-rpm record, and (c) an old-time 78-rpm record?

27 ■■ Determine which has the greater angular speed: particle A, which travels 160° in 2.00 s, or particle B, which travels 3π rad in 7.00 s.

28 ■■ A merry-go-round makes 24 revolutions in a 3.0-min ride. (a) What is the magnitude of its average angular velocity? (b) Assuming the merry-go-round to be operating with this angular velocity, what are the tangential speeds of persons 4.0 m and 5.0 m from the center or axis of rotation?

29 ■■ A runner running at a constant pace gets halfway around a circular track that has a diameter of 500 m in 2.5 min. What are the runner's (a) angular speed and (b) tangential speed?

30 ■■ For the second hand and the hour hand of a (running) clock, find (a) the period, (b) the frequency, (c) the angular speed, and (d) the angular velocity. Report to 2 significant figures.

31 ■■ A CD-ROM in a computer rotates at 500 rpm. How long will it take the disc to make one revolution?

32 ■■ Jupiter, the largest planet in our solar system, has a period of rotation of 10 hours and a period of revolution of 12 years. (a) What are its frequencies of rotation and revolution? (b) If the planet's diameter is 1.38×10^4 km and its mean distance from the Sun is 7.78×10^8 km, what are Jupiter's tangential speed at its equator and its angular speed of revolution? (Assume a circular orbit.)

33 ■■ A bicycle rider notes that the 26.0-inch diameter wheel makes 15.0 revolutions in a time of 8.50 s. (a) What is the angular speed of the wheel? (b) What distance in feet does the bicycle travel during this time?

34 ■■■ The driver of a motorboat sets ⸱ ties the wheel, making the boat trave⸱ speed of 12.5 m/s in a circle with a dia⸱ (a) Through what angular distance doe⸱ in 4.00 min? (b) What arc distance doe⸱ time?

35 ■■■ What are (a) the frequency of the Earth's rotation and (b) its angular *velocity*? (c) Express the tangential and angular speeds of a person at a latitude of 45° N (or S) as percentages of those of a person at the Equator.

7.3 Uniform Circular Motion and Centripetal Acceleration

36 In uniform circular motion, there is a (a) constant velocity, (b) constant angular velocity, (c) zero acceleration, (d) net tangential acceleration.

37 If the centripetal force on a particle in uniform circular motion is increased, (a) the motion will remain uniform, (b) the tangential speed will decrease, (c) the radius of the circular path will increase, (d) the tangential speed will increase and/or the radius will decrease.

38 A level curve on an interstate highway is a segment of a large circle and has a speed limit of 65 mi/h. A cloverleaf exit which feeds directly into an interstate highway is also a level, circular segment, but has a speed limit of only 30 mi/h. Why the difference?

39 The spin cycle of a washing machine is used to extract water from recently washed clothes. Explain the physical principle(s) involved here.

40 An apparatus as illustrated in •Fig. 7.24 is used to demonstrate forces in a rotating system. The floats are in jars of water. When the arm is rotated, which way will the floats move? (Compare with the accelerometer in Fig. 4.26.) Does it make a difference which way the arm is rotated?

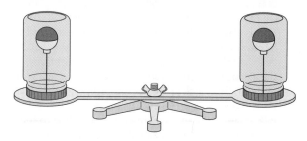

•**FIGURE 7.24 When set in motion, a rotating system**
See Exercise 40.

41 When rounding a curve in a fast-moving car, we experience a feeling of being thrown outward (•Fig. 7.25). It is sometimes said that this is because of an outward _entrifugal_ (center-fleeing) force. However, in terms of Newton's laws, this is called a pseudo or false force because it doesn't really exist. Analyze the situation in the figure and show that this is the case. [_Hint:_ Start with Newton's first law.]

•**FIGURE 7.25 A center-fleeing force?**
See Exercise 41.

42 ■ A ball is swung at a uniform speed in a horizontal circle with a radius of 1.25 m. If the ball has a centripetal acceleration of 1.25 m/s², what is its orbital speed?

43 ■ A race car goes around a level, circular track with a radius of 1.00 km at a speed of 120 km/h. What is the centripetal acceleration of the car?

44 ■■ The Moon revolves around the Earth in 29.5 days in a nearly circular orbit with a radius of 3.80×10^5 km. Assume that the Moon's motion is uniform. With what acceleration is it falling toward the Earth?

45 ■■ A car enters a circular curve with a radius of curvature of 0.400 km at a constant speed of 83.0 km/h. If the friction between the road and the car's tires can supply a centripetal acceleration of 1.25 m/s², does the car negotiate the curve smoothly? Justify your answer.

46 ■■ Compute the centripetal acceleration of a person on the rotating Earth (a) at the Equator, (b) at a latitude of 40°N, and (c) at the North Pole.

47 ■■ An electrically charged particle can travel in a circle in a uniform magnetic field (Chapter 19). An electron is observed to take a circular orbit with a radius of 30 cm traveling at a speed of 100 m/s. (a) What is the centripetal force acting on the electron resulting from the electromagnetic interaction? (b) How much work is done by this force in two electron revolutions?

48 ■■ Imagine that you swing a ball attached to the end of a string about your head at a constant speed in a horizontal circle with a radius of 1.50 m. If it takes 1.20 s for the object to make one revolution, (a) what is the magnitude of the tangential velocity of the object? (b) What centripetal acceleration are you imparting to the object via the string? (c) Is the string exactly horizontal?

49 ■■ Suppose the ball in Exercise 48 has a mass of 0.250 kg. If you supplied a tension force of 10.5 N to the string, what angle would the string make relative to the horizontal?

50 ■■ At a homecoming, one competition is to swing a bucket of water in a vertical circle without spilling any. If the distance from a person's shoulder to the center of mass of the bucket of water is 0.95 m, what is the minimum angular speed required to keep the water from coming out of the bucket at the top of the swing?

51 ■■ The Earth revolves around the Sun in a nearly circular orbit; its average distance from the Sun is about 1.5×10^8 km. Assume that the Earth's motion is uniform. (a) Compute the orbital speed in SI units. (Convert the answer to mi/h to get a more familiar measure of how fast you are traveling through space.) (b) What is the magnitude of the required centripetal acceleration for Earth to maintain this orbit?

52 ■■ Syncom satellites are satellites whose orbits are synchronized with the Earth's rotation. That is, they remain stationary above some location on the Earth's equator because the period of a satellite's revolution is the same as that of the Earth's rotation. The altitude of such satellites is 3.56×10^4 km. What is the centripetal acceleration of syncom satellite? What force produces this acceleration?

53 ■■ A jet pilot puts an aircraft into a vertical circular loop with a radius of 2.0 km at a constant speed of 700 km/h. (a) At the bottom of the loop, by what factor is the normal force on the pilot increased as compared to the value in level flight at the same speed? (b) What is the situation at the top of the loop?

54 ■■■ A block of mass m slides down an inclined plane into a loop-the-loop of radius R (•Fig. 7.26). (a) Neglecting friction, what is the minimum speed the block must have at the highest point of the loop to stay in the loop? [_Hint:_ What force must act on the block at the top of the loop to keep it on a circular path?] (b) At what vertical height on the inclined plane (in terms of the radius of the loop) must the block be released if it is to have the required minimum speed at the top of the loop?

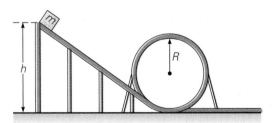

•**FIGURE 7.26 Loop-the-loop**
See Exercise 54.

55 ■■■ For a scene in a movie, a stunt driver drives a 1500-kg pickup truck with a length of 4.25 m around a circular curve with a radius of curvature of 0.333 km

(•Fig. 7.27). The truck is to curve off the road, jump across a gully, and land on the other side 2.96 m below and 10.0 m away. What is the minimum centripetal acceleration the truck must have going around the circular curve so that it will clear the gully and land on the other side?

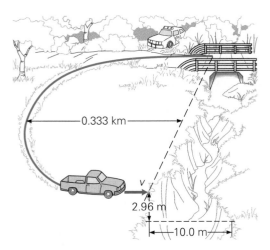

•**FIGURE 7.27 Over the gully**
See Exercise 55.

56 ■■■ In the design of a highway, curves are usually banked to improve safety (•Fig. 7.28).The horizontal component of the normal force toward the center of the curve reduces the need for friction to prevent skidding. In fact, for a circular curve banked at a given angle, there is one speed for which no frictional force is required — the centripetal force is supplied completely by the inward component of the normal force. This relationship is used in designing banked curves. (a) Show that the banking angle θ for which no friction is required for a car to safely negotiate a circular curve is given by $\tan \theta = v^2/gr$, where v is the car's speed and r is the radius of curvature. (b) How would the banking angle differ for a more massive truck? (c) Show how friction contributes to the situation.

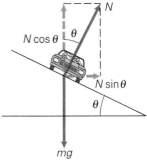

•**FIGURE 7.28 Banking safety**
See Exercise 56.

7.4 Angular Acceleration

57 The angular acceleration in circular motion (a) is equal in magnitude to the tangential acceleration divided by the

radius, (b) increases the angular velocity if in the same direction, (c) has units of s^{-2}, (d) all of the preceding.

58 ■ Suppose that the angular acceleration vector and the angular velocity vector in circular motion are in opposite directions. This means that (a) the centripetal acceleration will decrease, (b) there is no tangential acceleration, (c) the angular velocity will increase, (d) the vector sum of the accelerations is constant.

59 ■ A phonograph turntable set for 45 rpm is turned on and reaches its operating speed in 1.25 s. What is the average angular acceleration of the turntable? [*Hint:* Turntables rotate clockwise as viewed from above.]

60 ■ A particle initially at rest accelerates uniformly in a circular path at a rate of 1.5 rad/s² for 4.0 s. The circle has a radius of 30 cm. What are (a) the magnitude of the tangential acceleration during this time and (b) the tangential speed at the end of this time?

61 ■ A merry-go-round accelerating uniformly from rest achieves its operating speed of 2.5 rpm in four revolutions. What is the magnitude of its angular acceleration?

62 ■■ A $33\frac{1}{3}$-rpm record on a turntable uniformly reaches its operating speed in 2.45 s once the record player is turned on. (a) What is the angular distance traveled during this time? (b) What is the corresponding arc length in feet on the circumference of a 12-in. diameter record?

63 ■■ A bicycle being repaired is turned upside down and the wheel is rotated at a rate of 60 rpm. If the wheel slows uniformly to a stop in 15 s, how many revolutions does it make during this time?

64 ■■ A Ferris wheel with a diameter of 35.0 m starts from rest and achieves its maximum operational tangential speed of 2.20 m/s in a time of 15.0 s. (a) What is the magnitude of the wheel's angular acceleration? (b) What is the magnitude of the tangential acceleration after reaching the maximum operational speed?

65 ■■ The blades of a fan running at low speed turn at 250 rpm. When the fan is switched to high speed, the rotation rate increases uniformly to 350 rpm in 5.75 s. (a) What is the magnitude of the angular acceleration of the blades? (b) How many revolutions do the blades go through while the fan is accelerating?

66 ■■ A phonograph record initially moving at $33\frac{1}{3}$ rpm accelerates uniformly to 45 rpm in 2.5 s. What is the angular distance traveled by the record during this time?

67 ■■ A car on a circular track with a radius of 0.30 km accelerates from rest with a constant angular acceleration whose magnitude is 4.5×10^{-3} rad/s². (a) How long does it take the car to make one lap around the track? (b) What is the total acceleration of the car when it has completed half of a lap?

68 ■■■ A student investigating circular motion places a dime 10 cm from the center of a $33\frac{1}{3}$-rpm record on a turntable. Switching on the turntable, the student notes that the dime slides outward when the record is turning at 90% of its operating speed. (a) Why does the dime slide outward? (b) What is the coefficient of friction between the dime and the record?

7.5 Newton's Law of Gravitation

69 The gravitational force is (a) a linear function of distance, (b) an infinite-range force, (c) applicable only to our solar system, (d) varies because of G.

70 The acceleration due to gravity of an object on the Earth's surface (a) is a universal constant like G, (b) does not depend on the Earth's mass, (c) is directly proportional to the Earth's radius, (d) does not depend on the object's mass.

71 Astronauts in a spacecraft orbiting the Earth or out for a "spacewalk" (●Fig. 7.29) are seen to "float" in midair. This is sometimes referred to as *weightlessness* or *zero gravity (zero g)*. Are these terms correct? Explain why an astronaut appears to float in or near an orbiting spacecraft.

●**FIGURE 7.29 Out for a walk**
Why does this astronaut seem to "float"? See Exercise 71.

72 (a) Take a look at ●Fig. 7.30. If the cup were dropped, describe what would be observed as it fell. (b) A lighted candle is fixed vertically to the bottom of a tall beaker. If the beaker is dropped, is the candle affected? Explain.

●**FIGURE 7.30 Let it go**
See Exercise 72.

73 Give two ways to determine the mass of the Earth.

74 (a) A spring scale calibrated in kilograms is taken to the Moon to make measurements. Will the scale read correctly? Explain. (b) Will the scale reading be more or less at the Earth's Equator than at the North Pole for a given mass? Explain.

75 ■ Using Eq. 7.17, compute the value of g at the surface of the Earth.

76 ■ Compute the mass of the Sun using Eq. 7.23.

77 ■■ Approximate the gravitational force exerted on the Moon by the Earth and the Sun (a) during a solar eclipse, and (b) during a lunar eclipse.

78 ■■ (a) What is the acceleration due to gravity on the top of Mt. Everest? The summit is about 8.8 km above sea level. (Report to 3 significant figures.)

79 ■■ A 75-kg person weighs 735 N on Earth's surface. How far above the surface of the Earth would he have to go to "lose" 10% of his body weight?

80 ■■ (a) Two objects, each with a mass of 1.0 metric ton, are in contact with each other. How far apart are the centers of gravitational force, if there is a mutual gravitational attraction of 0.00010 N between them? (b) If the objects are homogeneous spheres of the same size, what is their minimum uniform density? [*Hint:* Remember that they are in contact.]

81 ■■ Four identical masses of 2.5 kg each are located at the corners of a square with 1.0-m sides. What is the net force on one of the masses?

82 ■■ For a spacecraft going directly from the Earth to the Moon, beyond what point will lunar gravity begin to dominate; that is, where will the lunar gravitational force be equal to the Earth's gravitational force? Are the astronauts on board truly weightless at this point?

83 ■■ Which is greater, the gravitational force exerted on the Earth by the Sun or by the Moon? (Compare these attractions by forming a ratio and giving a factor of how many times greater or smaller.)

84 ■■ In placing a 100-kg syncom satellite in orbit, what is the change in gravitational potential energy? (See Exercise 52.)

85 ■■ Plans are being made for a colony on Mars. Compute the approximate value of the acceleration due to gravity the colonists would experience on the Martian surface. (See Appendix III for data and compute explicitly.)

86 ■■ Assuming that we could get more earth (soil) from somewhere, estimate how thick would an added uniform outer layer on the Earth have to be to have $g = 10.0$ m/s² exactly? (Take the average density of the Earth to be 5.52 g/cm³ and $R_E = 6.40 \times 10^3$ km.)

87 ■■ A plane catapulted from an aircraft carrier goes from rest to a speed of 54 m/s in a distance of 50 m. How many g's of force does the pilot experience because of the catapult action?

88 ■■ Show that the value of the acceleration due to gravity on the Moon's surface is about one-sixth of that on the Earth's surface. [*Hint:* Use a ratio.]

89 ■■ Find the mutual gravitational potential energy for the configuration of masses shown in Fig. 6.29. (Use the mass values given in part c of the related Exercise.)

90 ■■■ (a) What is the mutual gravitational potential energy of the configuration shown in •Fig. 7.31 if all the masses are 1.0 kg? (b) What is the gravitational force per unit mass at the center of the configuration?

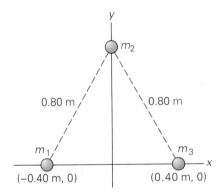

•FIGURE 7.31 **Gravitational potential, gravitational force, and center of mass**
See Exercises 90 and 91.

91 ■■■ For the configuration of 1.0-kg masses in Fig. 7.31, suppose that another 1.0-kg mass (m_4) is placed on the y axis the same distance below the origin as m_2 is above. (a) Find the net gravitational force on m_4. (b) Show that the result from part (a) is *not* the same as the gravitational force on m_4 calculated by considering all of the mass of the initial triangular configuration to be located at its center of mass. [*Hint:* Recall that a homogeneous sphere is a special case; in general, the center of mass of a body is not the same as the center of gravitational force.]

7.6 Kepler's Laws and Earth Satellites

92 A new planet is discovered and its period determined. The new planet's distance from the Sun could then be found using Kepler's (a) first law, (b) second law, (c) third law.

93 For a satellite to orbit the Earth, it must (a) have a tangential velocity, (b) be acted on by a centripetal force, (c) have greater gravitational potential energy than on Earth, (d) all of these.

94 (a) How much work does the centripetal force do on a satellite in circular orbit about the Earth in one revolution? (b) A person in a freely falling elevator thinks injury can be avoided by jumping upward just before the elevator strikes the floor. Would this work?

95 (a) In putting satellites into orbits, rockets are launched eastward. Why is this? (b) In the United States, satellites are launched from Florida. Why not from California, which generally has better weather conditions?

96 A shuttlecraft transporting a replacement crew to a space station in circular orbit about a planet is in a circular orbit at a lower altitude than that of the space station. (a) Explain how the shuttlecraft can change its orbit in order to dock with the space station. What effects would this change of orbit have on the shuttlecraft's energy and speed? (b) If the shuttlecraft moves to the altitude of the space station and establishes the same circular orbit, moving behind the space station, will it be able to overtake the space station and dock with it? Explain.

97 ■ An instrument package is projected vertically upward to collect data at the top of the Earth's atmosphere (at an altitude of about 800 km). (a) What initial speed is required at the Earth's surface for the package to reach this height? (b) What percentage of the escape speed is this?

98 ■ A 50-kg instrument package is put into a circular orbit about the Earth at an altitude of 1000 km. What is the orbital kinetic energy of the satellite?

99 ■■ Compute the constant K of Kepler's third law for (a) Earth and (b) Venus. (See Appendix I for data.)

100 ■■ Using a development similar to Kepler's law of periods for planets orbiting the Sun, find the required altitude of sychronous satellites above the Earth. (See Exercise 52.)

101 ■■ Venus has a rotational period of 243 days. What would be the altitude of a syncom satellite for this planet?

102 ■■ The asteroid belt that lies between Mars and Jupiter may be the debris of a planet that broke apart or did not form. The asteroid belt has a period of 5.0 years. How far from the Sun would this "fifth" planet have been?

103 ■■ Two identical satellites are in circular orbits around the Earth at altitudes of 500 km and 850 km. Which satellite has the (a) greater kinetic energy and (b) greater potential energy? (c) How many times greater is each of these values than that for the other satellite? (Take $R_E = 6.40 \times 10^3$ km.)

104 ■■ An ultracentrifuge operates at a speed of 500,000 rpm. (a) What is the centripetal acceleration on a virus

at a radial distance of 4.00 cm from the centrifuge's axis of rotation? (b) How does this acceleration compare with g, the acceleration due to gravity?

105 ■■■ In 1610, Galileo discovered four of the sixteen moons of Jupiter, the largest of which is Ganymede. This Jovian moon revolves around the planet in 7.16 days in a nearly circular orbit whose radius is about 1.07×10^6 km. Using these data, find the mass of Jupiter.

Additional Exercises

106 At sunset, the Sun has an angular width of about 0.50°. From the time the lower edge of the Sun just touches the horizon, about how long does it take in minutes to disappear?

107 Ocean tides are primarily produced by the gravitational attraction of the Moon. Explain how this attraction gives rise to two tidal "bulges" on opposite side of the Earth, resulting in two daily high tides and two daily low tides (•Figure 7.32). [*Hint:* Consider the inverse square relation of the distance and the gravitational force acting on the water on opposite sides of the Earth *and* on the Earth.] (You may want to go to the library for a little help on this one.)

(a)   (b)

•FIGURE 7.32 Tides
(a) Low tide and (b) high tide at the same location off the California coast. See Exercise 107.

108 A 60-kg girl prepares to swing out over a pond on a rope, as shown in •Fig. 7.33. What tension in the rope is required if she is to maintain a circular arc (a) when she starts her swing by stepping off the platform, (b) when she is 5.0 m above the pond surface, and (c) at the lowest point of the swinging motion? [*Hint:* Treat the girl and the rope as a simple pendulum.]

109 Suppose that the girl swinging on the rope in Fig. 7.33 (see Exercise 108) starts her swing at a lower point and has a tangential speed of 8.5 m/s when the rope is at an angle of 30° to the vertical. (a) What is her tangential acceleration at that time? (b) What is the magni-

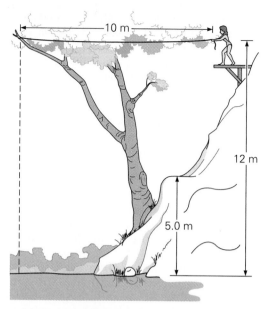

•FIGURE 7.33 A swinging time
See Exercises 108 and 109.

tude of the vector sum of the tangential and centripetal accelerations at that time? (c) From what height above the pond surface did she start her swing?

110 A daredevil stunt involves riding a motorcycle around the vertical inside wall of a cylindrical structure (•Fig. 7.34). Assume that the cylinder has a radius of 15 m and that the coefficient of static friction between the motorcycle tires and the wall is 1.1. (a) What is the minimum speed that will keep the motorcycle and the rider from sliding down the wall? Express this speed in km/h and mi/h. (b) Would the speed be different for a bigger (heavier) motorcycle? Explain.

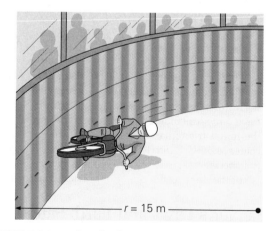

•FIGURE 7.34 A daredevil stunt
See Exercise 110.

111 A garden hose wound on a rotating storage cylinder has an outer layer at a distance of 0.45 m from the center of the cylinder. As this layer of wound hose is pulled out, the cylinder makes three revolutions. What is the length of the unwound hose?

112 The same side of the Moon always faces the Earth, because the period of the Moon's revolution around the Earth is the same as that of the Moon's rotation. Prove that these periods being equal means that the same hemisphere of the Moon is always visible from Earth.

113 Assuming that the Earth revolves about the Sun in uniform circular motion, find (a) the frequency of revolution in hertz and (b) the angular speed of revolution. (Use data given in Appendix I.)

114 What is the escape speed for a satellite in a circular orbit about the Earth at an altitude of 750 km?

115 An astronaut has a weight of 735 N on Earth. What would be the astronaut's weight in a spacecraft in a circular orbit at an altitude of 450 km?

116 The value of the Kepler constant is $K = 2.97 \times 10^{-19}$ s^2/m^3, and $T^2 = (2.97 \times 10^{-19} \ s^2/m^3)r^3$. With nonstandard units, K can be made equal to 1, and the equation may be written $T^2 = r^3$. Find the units of K that make it equal to 1, using Earth data. [*Hint:* The average, or mean, distance of the Earth from the Sun is used as a unit in astronomy and is called an astronomical unit (AU).]

117 Two point masses, $m_1 = 0.50$ kg and $m_2 = 0.75$ kg, are located at the positions (0, 0) and (0.60 m, 0), respectively. (a) What is the gravitational force on a third point mass, $m_3 = 0.25$ kg, located midway between the first two? (b) What is the gravitational force per unit mass midway between m_1 and m_2? (c) At what point on the x axis would the net force on m_3 be zero? Are there any other zero points?

118 Using Kepler's law of periods, show that the centripetal force acting on the Moon as it moves in its nearly circular orbit is an inverse square force. [*Hint:* Use the general form of the force and recall that $v = 2\pi r/T$.]

119 A pendulum swinging in a circular arc under the influence of gravity, as shown in •Fig. 7.35, has both centripetal and tangential components of acceleration. (a) If the pendulum bob has a speed of 2.7 m/s when the cord makes an angle of $\theta = 15°$ with the vertical, what are the magnitudes of the components at this time? (b) When is the centripetal acceleration a maximum? What is the value of the tangential acceleration at that time?

120 •Figure 7.36 depicts the Biblical story of David slaying Goliath with a stone hurled from a sling. Explain the physical principles involved in the use of this weapon. What potential advantage does it have over just throwing a stone? What factors might limit its effectiveness?

121 In Greek mythology, Atlas carried the entire Earth on his shoulders. (The statue of Atlas in •Figure 7.37 stands in New York's Rockefeller Center.) (a) How much work does Atlas do in supporting the Earth?

(b) Suppose Atlas decided to rearrange the solar system so as to relocate the Earth farther from the Sun—out to the orbit of Mars. What would be the change in Earth's gravitational potential energy relative to the Sun only? (Refer to the data in the Tables of Appendix III and the back endpapers, neglecting all bodies other than the Sun.)

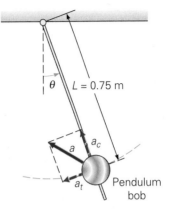

•**FIGURE 7.35   A swinging pendulum**
See Exercise 119.

•**FIGURE 7.36   David and Goliath**
Why not just throw the stone? See Exercise 120.

•**FIGURE 7.37   Unlucky Atlas**
Don't drop it! See Exercise 121.

8 Rotational Motion and Equilibrium

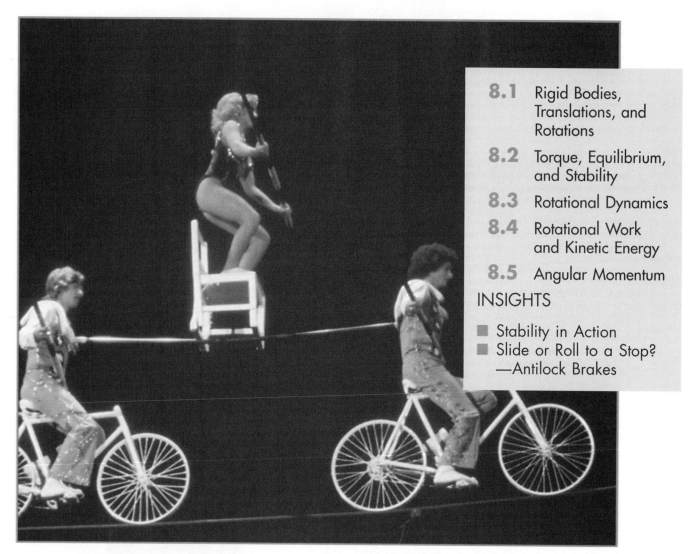

It's always a good idea to keep your equilibrium—but its more important in some situations than in others!

Looking at a photo like this, your first reaction is probably to wonder how these acrobats keep from falling. Presumably, the poles must help—but in what way? If you think a bit about the picture, however, and about what you have already learned, a less obvious but in some ways more interesting question may come to mind: why do we think they're in danger in the first place? After all, it's obvious that the wire is strong enough to support their weight. As long as this is the case, they would seem to be in equilibrium, as defined in Chapter 4. Why, then, do we get chills when we watch such a performance (especially if there is no net)? In this chapter you'll learn more about rotational motion; and, among other things, find that there's more to stability than you learned in Chapter 4.

Rotational motion is very important in physics because rotating objects are all around us: wheels on vehicles; gears and pulleys in machinery; planets in our solar system; and even many bones in the human body. (Can you think of a few bones that rotate in sockets?)

In Chapter 7, the circular motions of particles in rotating bodies were considered, but the description of the rotational motion of a body as a whole must take into account the motions of all its particles. As you may have noticed in previous chapters, the study of physics proceeds in a stepwise manner, going from simple, ideal descriptions to more complex, more realistic ones, which better describe actual physical events. This chapter takes another step into the real world of things, going beyond the circular motions of particles to the rotations of real objects. In addition, the conditions of equilibrium and stability are considered. In many situations we don't want things to rotate, or perhaps to move at all. (Think of a stepladder—or a bridge.)

Fortunately, the equations describing rotational motion can be written as almost direct analogs of those for translational (linear) motion. In Chapter 7, this similarity was pointed out for the kinematic equations. With the addition of equations describing rotational dynamics, you will be able to analyze the general motions of real objects.

8.1 Rigid Bodies, Translations, and Rotations

Objectives: To be able to (a) distinguish between pure translational and pure rotational motions of a rigid body, and (b) state the condition(s) for rolling without slipping.

It was convenient initially to consider motions of objects with the understanding that the motion of an object may be represented by a particle located at its center of mass. Rotation, or spinning, was not a consideration then because a point mass has no physical dimensions. Rotational motion becomes relevant when we analyze the motion of solid extended objects or rigid bodies, on which this chapter will focus.

> A *rigid body* is an object or system of particles in which the interparticle distances (distances between particles) are fixed and remain constant.

Definition of a rigid body

A quantity of liquid water is obviously not a rigid body, but the ice that would form if the water were frozen would be. The discussion of rigid body rotation is thus conveniently restricted to solids. Actually, the concept of a rigid body is an idealization. In reality, the particles (atoms and molecules) of a solid vibrate constantly. Also, solids can undergo elastic (and inelastic) deformations (Chapter 6). Even so, most solids can be considered to be rigid bodies for purposes of analyzing rotational motion.

A rigid body may be subject to either or both of two different types of motions: translational and rotational. Translational motion is basically the linear motion we studied in previous chapters. If an object has only **translational motion** (•Fig. 8.1a), every particle has the same instantaneous velocity, which means there is no rotation. (Why?)

An object may have only **rotational motion**, about a fixed axis (•Fig. 8.1b). In this case, all the particles of an object have the same instantaneous angular velocity and travel in circles about the axis of rotation. Although the axis of rotation of an object is commonly taken through its center of mass, this is not always the case. For example, you might pivot a meterstick through one end and have rotational motion about an axis through that end.

General rigid body motion is a combination of translational and rotational motions. When you throw a ball, the translational motion is described by the motion of its center of mass (as in projectile motion). But the ball may also spin, or rotate, and usually does.

Note: The words "rotation" and "revolution" are commonly used synonymously. In general, this book uses rotation when the axis of rotation goes through the body (the Earth's rotation on its axis) and revolution when the axis is outside the body (the revolution of the Earth about the Sun).

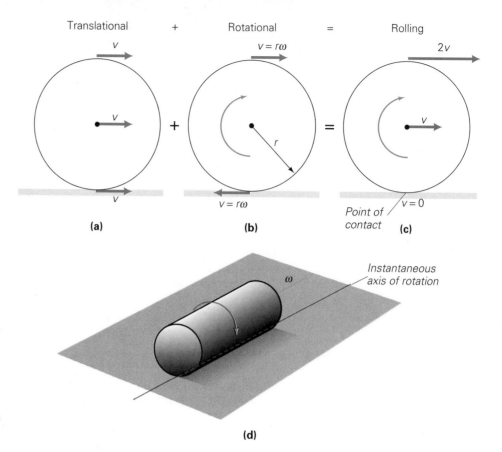

Translational + Rotational = Rolling

(a) **(b)** Point of contact **(c)**

(d)

• FIGURE 8.1 Rolling—a combination of translational and rotational motions
(a) In pure translational motion, all the particles of an object have the same instantaneous velocity. **(b)** For pure rotational motion, all the particles of an object have the same instantaneous angular velocity. **(c)** Rolling is a combination of translational and rotational motions. Summing the velocity vectors for these two motions shows that the point of contact (for a sphere) or the line of contact (for a cylinder) is instantaneously at rest. **(d)** The line of contact (or, for a sphere, a line through the point of contact) is called the instantaneous axis of rotation. Note that the center of mass of a rolling object on a level surface moves linearly and remains over the point or line of contact.

A common example of rigid body motion involving both translation and rotation is rolling, as illustrated in • Fig. 8.1c. The combined motion of any point or particle is given by the vector sum of its instantaneous velocity vectors. (Three points or particles are shown in the figure—top, middle, and bottom). At each instant, a rolling object rotates about an **instantaneous axis of rotation** through its point of contact with the surface (for a sphere) or along its line of contact with the surface (for a cylinder, see • Fig. 8.1d). The location of this axis changes with time. However, note in Fig. 8.1 that the point or line of contact of the body with the surface is instantaneously at rest (zero velocity), as can be seen from the vector addition of the combined motions at that point.

When an object rolls without slipping, the translational and rotational motions are related simply as discussed in the preceding chapter. For example, when a uniform ball (or cylinder) rolls in a straight line on a flat surface, it turns through an angle θ, and a point on the object which was initially in contact with the surface moves through an arc distance s (• Fig. 8.2). From Chapter 7, you know that $s = r\theta$. The center of mass of the ball is directly over the point of contact and moves a linear distance s. Then

$$v_{CM} = \frac{\Delta s}{\Delta t} = \frac{r\Delta\theta}{\Delta t} = r\omega$$

In terms of the speed of the center of mass and the angular speed, the *condition for rolling without slipping* is

$$v_{CM} = r\omega \qquad \text{condition for rolling without slipping} \qquad (8.1)$$

The condition is also expressed by $s = r\theta$, where s is the distance an object rolls, or the distance the center of mass moves. Carrying Eq. 8.1 one step further, we

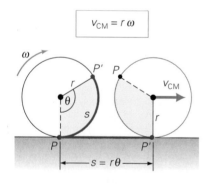

$$v_{CM} = r\omega$$

• FIGURE 8.2 Rolling without slipping
As an object rolls without slipping, the length of the arc between two points of contact on the circumference is equal to the linear distance traveled. This distance is $s = r\theta$. The speed of the center of mass is $v_{CM} = r\omega$.

can write an expression for the time rate of change of the velocity: $\Delta v / \Delta t = r\Delta\omega / \Delta t$. This yields an equation for *accelerated rolling without slipping*:

$$a_{\text{CM}} = \frac{\Delta v_{\text{CM}}}{\Delta t} = \frac{r\Delta\omega}{\Delta t} = r\alpha \qquad (8.1b)$$

8.2 Torque, Equilibrium, and Stability

Objectives: To be able to (a) define torque, (b) apply the conditions for mechanical equilibrium, and (c) describe the relationship between the location of the center of gravity and stability.

As with translational motion, a force is necessary to produce a change in rotational motion. But rotational motion is not always produced when a force acts on a rigid body. The motion or angular acceleration depends on *where* the force is applied. If a force acts through the axis of rotation, no rotation is produced (•Fig. 8.3a). But when the line of action of the force does not go through the axis of rotation, the body rotates (•Figs. 8.3b and c). The line of action of a force is simply a line extending through the force vector arrow, that is, the line along which the force acts.

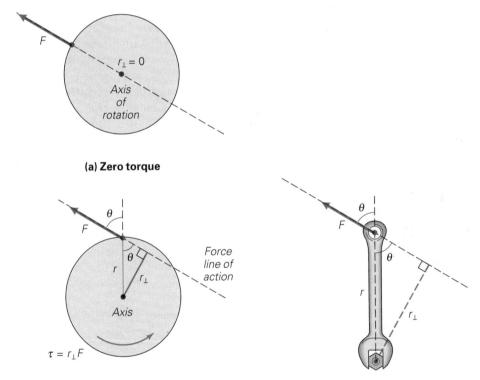

(a) Zero torque

(b) Counterclockwise torque

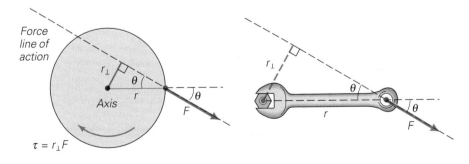

(c) Smaller clockwise torque

•**FIGURE 8.3   Torque and moment arm**
(a) When a force acts through the axis of rotation, $r_\perp = 0$ and $\tau = 0$. **(b)** The perpendicular distance $r_\perp$ from the axis of rotation to the line of action of a force is called the moment arm (or lever arm) and is equal to $r \sin \theta$. The torque, or twisting force, that produces rotational motion is given by $\tau = r_\perp F$. **(c)** The same force in the opposite direction with a smaller moment arm produces a smaller torque in the opposite direction.

As a practical example, think of applying a force to a heavy glass door that swings in and out. Where you apply the force makes a great difference in how easily the door opens, or rotates on its axis (through the hinges). Have you ever tried to open such a door and inadvertently pushed on the side near the hinges or axis of rotation?

Torque

The rate of change of rotation depends not only on the magnitude of the force, but also on the perpendicular distance of its line of action from the axis of rotation, $r_\perp$ (Fig. 8.3). Note from the figure that $r_\perp = r \sin \theta$, where r is the straight-line distance between the axis of rotation and the point at which the force acts, and θ is the angle between the line of r and the force vector $\mathbf{F}$. This perpendicular distance is called the **moment arm** or **lever arm.**

Torque: the product of a lever arm and a force

The product of the force and the lever arm is called **torque**, τ (from the Latin *torquere* meaning "to twist"). The magnitude of the torque is

$$\tau = r_\perp F = rF \sin \theta \qquad (8.2)$$

SI unit of torque: m·N

The SI units for torque are meters times newtons (m·N). These are the same as the units for work, $W = Fd$ (N·m, or J). However, we will write the units for torque as m·N to avoid confusion. Notice from Eq. 8.2 that $\tau = 0$ when $\theta = 0°$, or when the force acts through the axis of rotation. When $\theta = 90°$, the torque is maximum and the force acts perpendicularly to the lever arm, as when you push on a door to open it. Also note that $r_\perp F = rF_\perp$ (the product of the radial vector and the component of F perpendicular to that vector).

Torque in rotational motion may be thought of as the analog of force in translational motion. An unbalanced or net force changes translational motion, and an unbalanced or net torque changes rotational motion. The product of a force and moment arm, both vectors, torque is also a vector. Its direction is always perpendicular to the plane of the force and moment arm vectors and is given by a right-hand rule similar to that for angular velocity given in Section 7.2. If the fingers of the right hand are curled around the axis of rotation in the direction that the torque would produce a rotational (angular) acceleration, the extended thumb points in the direction of the torque. A sign convention, as in the case of linear motion, can be used to represent torque directions, as will be shown shortly.

EXAMPLE 8.1 ■ LIFTING AND HOLDING: MUSCLE TORQUE AT WORK

In our bodies, torques produced by the contraction of our muscles cause some bones to rotate at joints. For example, in lifting something with the forearm, a torque is applied by the biceps muscle on the lower arm (•Fig. 8.4). With the axis of rotation through the elbow joint and the muscle attached 4.0 cm from the joint, what are the magnitudes of the muscle torques for the cases in Fig. 8.4 if the muscle exerts a force of 600 N?

Solution. First we list the data given here and in the figure. This Example demonstrates an important point. Recall that θ is the angle *between* the radial vector $\mathbf{r}$ and the force $\mathbf{F}$.

Given: $r = 4.0 \text{ cm} = 0.040 \text{ m}$　　*Find:* (a) τ (muscle torque magnitude)
　　　　$F = 600 \text{ N}$　　　　　　　　(b) τ (muscle torque magnitude)
　　　　(a) $\theta = 30° + 90°$
　　　　(b) $\theta = 90°$

(a) In this case, $\mathbf{r}$ is directed along the forearm, so the angle between the $\mathbf{r}$ and $\mathbf{F}$ vectors is $\theta = 30° + 90°$. Thus, $r \sin \theta = r \sin (30° + 90°) = r \cos 30°$. (See Appendix I for this

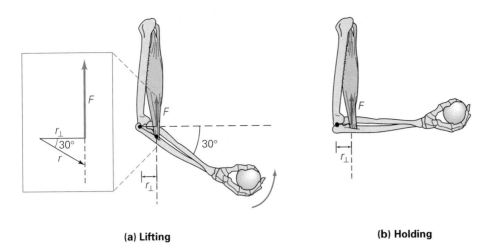

•FIGURE 8.4 Human torque
See Example 8.1.

(a) Lifting

(b) Holding

and other useful trigonometric relationships.) Or, just notice from the figure inset that $r_\perp = r \cos 30°$. Using Eq. 8.2, we have

$$\tau = r_\perp F = rF \sin (30° + 90°) = rF \cos 30°$$
$$= (0.040 \text{ m})(600 \text{ N}) \cos 30° = 21 \text{ m·N}$$

at that instant.

(b) Here, the lever arm and the line of action for the force are perpendicular ($\theta = 90°$), and $r_\perp = r \sin 90° = r$. Then,

$$\tau = r_\perp F = rF = (0.040 \text{ m})(600 \text{ N}) = 24 \text{ m·N}$$

Follow-up Exercise. In (a) of this Example, there must have been a net torque, since the ball was lifted by a rotation of the forearm. In (b), the ball is just being held and there is no rotational acceleration, so there is no net torque on the system. Identify the other torque(s) in each case. *(Answer may be found in the Answers to Follow-up Exercises section at the back of the book.)*

Before considering rotational dynamics with net torques and rotational motions, let's first take a look at the situation where the forces and torques acting on a body are balanced or in equilibrium.

Equilibrium

In general, equilibrium means that things are in balance or are stable. This definition applies in the mechanical sense to forces and torques. Unbalanced forces produce translational accelerations, but *balanced* forces produce the condition we call *translational equilibrium*. Similarly, unbalanced torques produce rotational accelerations, and *balanced* torques produce *rotational equilibrium*.

According to Newton's first law of motion, when the sum of the forces acting on a body is zero, it remains either at rest (static) or in motion with a constant velocity. In either case, the body is said to be in **translational equilibrium**. Stated another way, the *condition for translational equilibrium* is that the net force on a body is zero, $\Sigma \mathbf{F}_i = 0$. In other words, the vector sum of all forces equals zero: $\Sigma \mathbf{F}_i = \mathbf{F}_1 + \mathbf{F}_2 + \mathbf{F}_3 + \cdots = 0$. It should be apparent that this condition is satisfied for the situations illustrated in •Fig. 8.5a and b. Forces with lines of action through the same point are called **concurrent forces**, and when these vectorially add to zero as in (a) and (b), the body is in translational equilibrium.

But what about the situation pictured in •Fig. 8.5c? Here $\Sigma \mathbf{F}_i = 0$, but the opposing forces will cause the object to rotate, and it will clearly not be in a state of static equilibrium. (Such a pair of equal and opposite forces not having the same line of action is called a *couple*.) Thus, the condition $\Sigma \mathbf{F}_i = 0$ is a necessary, but *not sufficient*, condition for static equilibrium.

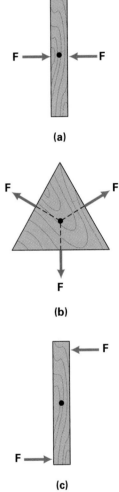

(a)

(b)

(c)

•**FIGURE 8.5 Equilibrium and forces** Forces with lines of action through the same point are said to be concurrent. The resultants of the concurrent forces acting on the objects in **(a)** and **(b)** are zero, and the objects are in equilibrium, because the net torque *and* net force are zero. In **(c)**, the object is in translational equilibrium, but it will undergo angular acceleration, hence is *not* in rotational equilibrium.

Since $\Sigma \mathbf{F}_i = 0$ is the condition for translational equilibrium, you might predict (and correctly so) that $\Sigma \tau_i = 0$ is the *condition for rotational equilibrium*. That is, if the sum of the *torques* acting on an object is zero, then the object is in **rotational equilibrium**—it is rotationally at rest or rotates with a constant angular velocity.

Thus we see that there are two equilibrium conditions. Taken together, they define what is called **mechanical equilibrium**.

A body is said to be in mechanical equilibrium when the conditions for both translational and rotational equilibrium are satisfied:

Conditions for mechanical equilibrium

$$\Sigma \mathbf{F}_i = 0 \qquad \textit{(for translational equilibrium)}$$
$$\Sigma \boldsymbol{\tau}_i = 0 \qquad \textit{(for rotational equilibrium)} \qquad (8.3)$$

A rigid body in mechanical equilibrium may be either at rest or moving with a constant linear and/or angular velocity. An example of the latter is an object rolling without slipping on a level surface with the center of mass having a constant velocity. Of course, this is an ideal condition because there is always some friction in reality. Of greater practical interest is **static equilibrium**, the condition that exists when a rigid body is at rest. There are many instances in which we do not want things to move, and this absence of motion can occur only if the equilibrium conditions are satisfied. It is particularly comforting to know, for example, that a bridge you are crossing is in static equilibrium, and not subject to translational or rotational motions.

Let's consider examples of static translational equilibrium and static rotational equilibrium separately, and then one in which both apply.

EXAMPLE 8.2 ■ TRANSLATIONAL STATIC EQUILIBRIUM: NO TRANSLATIONAL MOTION

A picture hangs motionless on a wall as shown in ●Fig. 8.6. If it has a mass of 5.0 kg, what are the tension forces in the wires?

Solution. Note that all the forces (tension and weight forces) are concurrent—that is, their lines of action pass through a common point, the nail. Because of this, the condition for rotational equilibrium ($\Sigma \tau_i = 0$) is automatically satisfied—with respect to an

●**FIGURE 8.6 Translational static equilibrium**
Since the picture hangs motionless on the wall, the sum of the forces acting on it must be zero. The forces are concurrent, with their lines of action passing through a common point at the nail. This is shown in the free body diagram, where all the forces have been represented as acting at this single point. (Note, however, that these are the forces acting on the *picture*, not on the nail.) See Example 8.2.

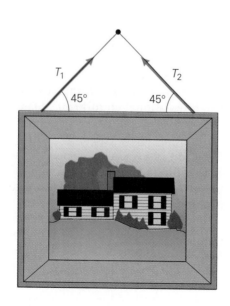

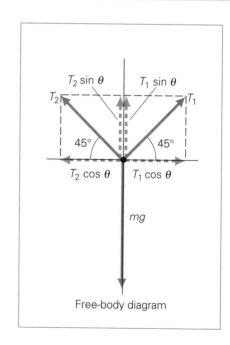

Free-body diagram

axis of rotation along the nail, the moment arms of the forces are zero. Thus, we need consider only translational equilibrium.

You will find it helpful to isolate the forces acting on the picture in a free-body diagram as was done in Chapter 4 for force problems (see Fig. 8.6). The diagram shows the concurrent forces acting through their common point. Note that we have moved all the force vectors to the common point, which is taken as the origin of the coordinate axes. The weight force mg acts downward.

With the system in static equilibrium, the net force is zero; that is, $\Sigma \mathbf{F}_i = 0$. Thus, the sums of the rectangular components are also zero: $\Sigma \mathbf{F}_{x_i} = 0$ and $\Sigma \mathbf{F}_{y_i} = 0$. Then

$$\Sigma \mathbf{F}_{x_i}: \quad T_1 \cos 45° - T_2 \cos 45° = 0$$

This just tells you that the x components of the forces are equal and opposite (and $T_1 = T_2 = T$). For the y components, with two upward components of $T \sin 45°$, we have

$$\Sigma \mathbf{F}_{y_i}: \quad T \sin 45° + T \sin 45° - mg = 0$$

or

$$2T \sin 45° - mg = 0$$

Thus,

$$T = \frac{mg}{2 \sin 45°} = \frac{(5.0 \text{ kg})(9.8 \text{ m/s}^2)}{2(0.707)} = 35 \text{ N}$$

Follow-up Exercise. Analyze the situation in Fig. 8.6 that would result if the wires were shortened so that the angles were decreased. Carry this to the limit where the angles approach zero. (*Answer may be found in the Answers to Follow-up Exercises section at the back of the book.*)

EXAMPLE 8.3 ■ ROTATIONAL STATIC EQUILIBRIUM: NO ROTATIONAL MOTION

Three masses are suspended from a meterstick as shown in •Fig. 8.7. How much mass must be suspended on the right side to have the system be in static equilibrium? (Neglect the mass of the meterstick.)

Solution. From the figure, we have

Given: $m_1 = 25$ g   *Find:* m_3 (unknown mass)
        $r_1 = 50$ cm
        $m_2 = 75$ g
        $r_2 = 30$ cm
        $r_3 = 35$ cm

The condition for translational equilibrium ($\Sigma \mathbf{F}_i = 0$) is automatically satisfied by the upward vertical reaction force (R) of the stand on the stick. Thus, $R - Mg = 0$ or $R = Mg$, where M is the total mass. This is true no matter what the total mass may be—that is, regardless of how much mass we add for m_3. However, unless the proper mass for m_3 is placed on the right side, the stick will experience a net torque and rotate.

Notice that the masses on the left side produce torques that tend to rotate the stick counterclockwise, and the mass on the right side produces a torque that tends to rotate it clockwise. As pointed out earlier, torque is a vector and directional difference is important. As in the linear case in which + and − were used to express opposite directions (e.g., +x and −x), we may designate torques as + and −, depending on the rotational motion they tend to produce. The rotational "directions" are taken as clockwise and counterclockwise around the axis of rotation, and we will use the sign convention shown in Fig. 8.7: A torque that tends to produce counterclockwise rotation will be taken as positive (+), and a torque that tends to produce a clockwise rotation will be taken as a negative (−).

This convention is arbitrary. The key thing to remember is that the magnitude of the sum of the counterclockwise torques must equal that of the clockwise torques to have rotational equilibrium. Thus, we apply the condition for rotational equilibrium by summing the torques about an axis. Let's take this to be through the center of the stick

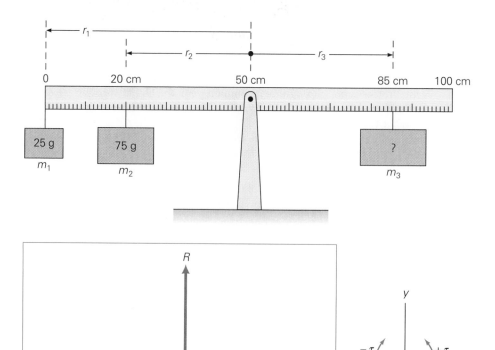

●FIGURE 8.7 Rotational static equilibrium
For the meterstick to be in rotational equilibrium, the sum of the torques acting about any selected axis must be zero. See Example 8.3.

at the 50-cm position, or point A in the Fig. 8.7 free-body diagram. Then, noting that R passes through the axis of rotation ($r = 0$) and produces no torque, we have

$$\Sigma \tau_i; \quad \tau_1 + \tau_2 + \tau_3 = + r_1 F_1 + r_2 F_2 - r_3 F_3 \qquad \text{(using our sign}$$
$$= r_1(m_1 g) + r_2(m_2 g) - r_3(m_3 g) = 0 \qquad \begin{array}{l}\textit{convention for}\\ \textit{torque vectors)}\end{array}$$

Noting that the g's cancel out, and solving for m_3, we have

$$m_3 = \frac{m_1 r_1 + m_2 r_2}{r_3}$$

$$= \frac{(25 \text{ g})(50 \text{ cm}) + (75 \text{ g})(30 \text{ cm})}{35 \text{ cm}} = 100 \text{ g}$$

where it was convenient not to convert to standard units. (The mass of the stick was neglected. If the stick is uniform, however, its mass will not affect the equilibrium as long as the pivot point is at the 50-cm mark. Why?)

Follow-up Exercise.   It should be noted that the axis of rotation could have been taken through any point along the stick. That is, if a system is in static rotational equilibrium, the condition $\Sigma \tau_i = 0$ holds for *any* axis of rotation. Demonstrate this for the system in this Example by taking the axis of rotation through the left end of the stick ($x = 0$). *(Answer may be found in the Answers to Follow-up Exercises section at the back of the book.)*

In general, the conditions for both translational and rotational equilibrium need to be written explicitly to solve the problem. The next Example is one such case.

EXAMPLE 8.4 ■ STATIC EQUILIBRIUM: NO TRANSLATION, NO ROTATION

A ladder with a mass of 15 kg rests against a smooth wall (●Fig. 8.8). A painter, who has a mass of 78 kg, stands on the ladder as shown in the figure. What frictional force must act on the bottom of the ladder to keep it from slipping?

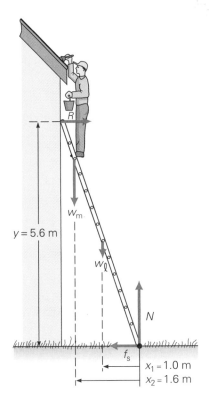

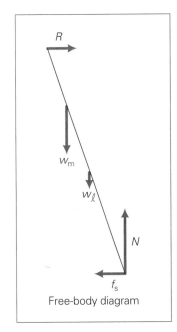

Free-body diagram

•**FIGURE 8.8 Static equilibrium**
For the painter's sake, the ladder
has to be in static equilibrium; that
is, both the sum of the forces and
the sum of the torques must be
zero. See Example 8.4.

Solution.

Given: $m_\ell = 15$ kg   *Find:* f_s (force of static friction)
       $m_m = 78$ kg
       Distances given in figure

Because the wall is smooth, there is negligible friction between it and the ladder, and
only the normal reaction force of the wall acts on the ladder at this point.

   In applying the conditions for static equilibrium, you are free to choose any axis of
rotation for the rotational condition. (The conditions must hold for any part of a system for static equilibrium; that is, there can't be motion in any part of the system.) Note
that choosing an axis at the end of the ladder where it touches the ground eliminates
the torques due to f_s and N since the moment arms are zero. Then you can write three
equations (using mg for w):

$$\Sigma \mathbf{F}_{x_i}: R - f_s = 0$$
$$\Sigma \mathbf{F}_{y_i}: N - m_m g - m_\ell g = 0$$

and

$$\Sigma \tau_i: +(m_\ell g)x_1 + (m_m g)x_2 + (-Ry) = 0$$

The weight of the ladder is considered to be concentrated at its center of gravity. Solving the third equation for R and substituting the given values for the masses and distances gives

$$R = \frac{(m_\ell g)x_1 + (m_m g)x_2}{y}$$

$$= \frac{(15 \text{ kg})(9.8 \text{ m/s}^2)(1.0 \text{ m}) + (78 \text{ kg})(9.8 \text{ m/s}^2)(1.6 \text{ m})}{5.6 \text{ m}}$$

$$= 2.4 \times 10^2 \text{ N}$$

From the first equation then,

$$f_s = R = 2.4 \times 10^2 \text{ N}$$

Follow-up Exercise.   In this Example, would the frictional force between the ladder and the ground (call it f_{s_1}) be the same if there were friction between the wall and
the ladder (call it f_{s_2})? Justify your answer. (*Answer may be found in the Answers to Follow-up Exercises section at the back of the book.*)

As may be seen from the preceding examples, a good procedure to follow in working problems involving static equilibrium is as follows:

1. Sketch a space diagram of the problem.
2. Draw a free-body diagram, showing and labeling all external forces, and normally resolving the forces into x and y components.
3. Apply the equilibrium conditions. Sum the forces. ($\Sigma \mathbf{F}_i = 0$, or usually in component form $\Sigma \mathbf{F}_{x_i} = 0$ and $\Sigma \mathbf{F}_{y_i} = 0$.) Sum the torques. ($\Sigma \tau_i = 0$), remembering to select an appropriate axis of rotation to reduce the number of terms as much as possible. Use $\pm$ sign conventions for both $\mathbf{F}$ and τ
4. Solve for the unknown quantities.

Stability and Center of Gravity

Note: Review Section 6.5

The equilibrium of a particle or a rigid body can also be described as being stable or unstable in a gravitational field. For rigid bodies, these categories of equilibria are conveniently analyzed in terms of the center of gravity. Recall from Chapter 6 that the **center of gravity** is the point at which all the weight of an object may be considered to be concentrated in representing it as a particle. When the acceleration due to gravity is constant, the center of gravity and the center of mass coincide.

Stable equilibrium—a restoring torque

If an object is in **stable equilibrium**, any small displacement results in a restoring force or torque, which tends to return the object to its original equilibrium position. As illustrated in •Fig. 8.9a, a ball in a bowl is in stable equilibrium.

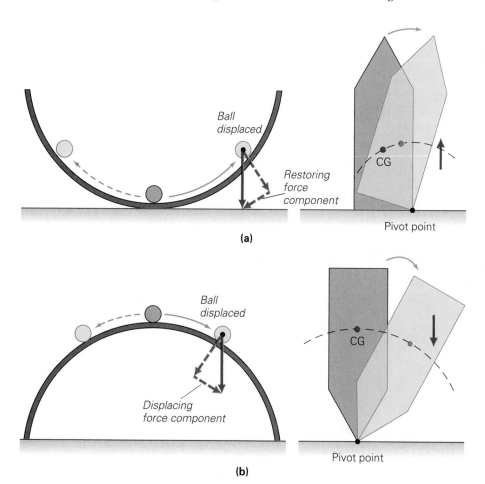

•FIGURE 8.9 Stable and unstable equilibrium
(a) When an object is in stable equilibrium, any small displacement from an equilibrium position results in a force or torque that tends to return the object to that position. A ball in a bowl (left) returns to the bottom after being displaced. Analogously, the center of gravity of an extended object (right) can be thought of as being in a potential energy bowl—a small displacement raises the CG, increasing the object's potential energy. **(b)** For an object in unstable equilibrium, any small displacement from its equilibrium position results in a force or torque that tends to take it farther away from that position. The ball on top of an overturned bowl (left) is in unstable equilibrium. For an extended object (right), the center of gravity can be thought of as being on an inverted potential energy bowl—a small displacement lowers the CG, decreasing the object's potential energy.

Analogously, the center of gravity of an extended body in stable equilibrium is essentially in a potential energy bowl. Any slight displacement raises its center of gravity, and a restoring force tends to return it to the position of minimum potential energy. This force is actually a torque that is due to a component of the weight force and that tends to rotate the object about a pivot point back to its original position.

For an object in **unstable equilibrium**, any small displacement from equilibrium results in a force or torque that tends to take the object farther away from its equilibrium position. This situation is illustrated in •Fig. 8.9b. Note that the center of gravity is at the top of an overturned, or inverted, potential energy bowl, that is, the potential energy is at a maximum in this case.

Unstable equilibrium—a
torque that topples

Small displacements or slight disturbances have profound effects on objects in unstable equilibrium—it doesn't take much to cause such an object to change its position. On the other hand, the angular displacement of an object in stable equilibrium can be quite substantial, and the object will still be restored to its equilibrium position. An object lying on a long side can be rotated through quite a distance and will still fall back to that original position. This is another way of expressing the *condition of stable equilibrium:*

> An object is in stable equilibrium as long as its center of gravity after a small displacement still lies above and inside its original base of support. That is, line of action of the weight force of the center of gravity intersects the original base of support.

Condition for stable
equilibrium

When this is the case, there will always be a restoring torque (•Fig. 8.10a). However, when the center of gravity or center of mass falls outside the base of support, over goes the object—because of a gravitational torque that rotates it away from its equilibrium position (•Fig. 8.10b).

Rigid bodies with wide bases and low centers of gravity are therefore more stable and less likely to tip over. This relationship is evident in the design of high-speed race cars, which have wide wheel bases and centers of gravity close to the ground (•Fig. 8.11). Also, the location of the center of gravity of the human body has an effect on certain physical abilities. For example, women can generally bend over and touch their toes or touch their palms to the floor more easily than can men, who often fall over trying. On the average, men have higher centers of gravity (larger shoulders) than do women (larger pelvises), so it is more likely that a man's center of gravity will be outside his base of support when he bends over. (See Demonstration 6.)

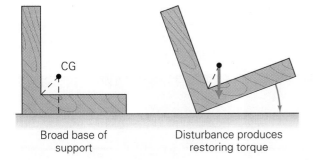
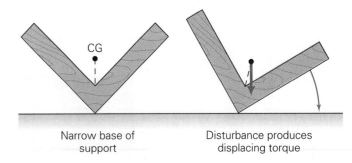

| (a) Stable Equilibrium | (b) Unstable Equilibrium |

•**FIGURE 8.10 Examples of stable and unstable equilibrium**
(a) When the center of gravity is above and inside an object's base of support, the object is in stable equilibrium (there is a restoring torque). Note how the line of action of the weight of the center of gravity intersects the original base of support after the displacement. **(b)** When the center of gravity lies outside the base of support or the line of action of the weight does not intersect the original base of support, the object is unstable (there is a displacing torque).

| (a) | (b) | (c) |

•FIGURE 8.11 Stable, stabler, stablest
(a) The acrobat's base of support is very narrow—the small area of head-to-head contact. As long as his center of gravity remains above this area, he is in equilibrium, but a displacement of only a few inches would probably be enough to topple him. (Why he is in a spread-eagle position will become clearer in the next section.) **(b)** This balancing rock in Idaho has a slightly larger base of support than the acrobat in (a); still, a displacement of its center of gravity by a few feet would probably send it crashing down. **(c)** Race cars are very stable because of their wide wheel bases and low centers of gravity.

Demonstration 6 ■ Stability and Center of Gravity

A demonstration to show that women generally have more stability than men—because of mass distribution and location of the center of gravity.

(a) An object is placed just beyond the fingertips of a man kneeling with elbows against his knees. Maintaining his balance, he can topple the object with his nose.

(b) With hands now clasped behind his back, the man's center of gravity shifts beyond his base of support (knees) as he bends over to topple the object, and he will topple instead.

(c) A woman can generally do it both ways.

(d) Because her center of gravity is positioned lower in the body, it does not move outside the base of support (past her knees) as she bends with hands clasped behind her back.

EXAMPLE 8.5 ■ STACK THEM UP! CENTER OF GRAVITY

Uniform, identical bricks 20 cm long are stacked so that 4.0 cm of a brick extends beyond the brick beneath, as shown in ▪Fig. 8.12a. How many bricks can be stacked in this way before the stack falls over?

Solution. The stack will fall over when its center of mass is no longer above its base of support, which is the bottom brick. All of the bricks have the same mass, and the center of mass of each is located at its midpoint.

Taking the origin to be at the center of the bottom brick, the horizontal coordinate of the center of mass (or center of gravity) for the first two bricks in the stack is given by Eq. 6.19, where $m_1 = m_2 = m$ and x_2 is the displacement of the second brick:

$$X_{CM_2} = \frac{mx_1 + mx_2}{m + m}$$

$$= \frac{m(x_1 + x_2)}{2m} = \frac{x_1 + x_2}{2} = \frac{0 + 4.0 \text{ cm}}{2} = 2.0 \text{ cm}$$

Note that the masses of the bricks cancel out (since they are all the same). For three bricks,

$$X_{CM_3} = \frac{m(x_1 + x_2 + x_3)}{3m} = \frac{0 + 4.0 \text{ cm} + 8.0 \text{ cm}}{3} = 4.0 \text{ cm}$$

For four bricks,

$$X_{CM_4} = \frac{m(x_1 + x_2 + x_3 + x_4)}{4m} = \frac{0 + 4.0 \text{ cm} + 8.0 \text{ cm} + 12.0 \text{ cm}}{4} = 6.0 \text{ cm}$$

and so on.

This series of results shows that the center of mass of the stack moves horizontally 2.0 cm for each brick added to the bottom one. For a stack of six bricks, the center of mass is 10 cm from the origin and directly over the edge of the bottom brick (2.0 cm × 5 *added* bricks = 10 cm, which is ½ of the length of the bottom brick), so the stack is in unstable equilibrium. The stack may not topple if the sixth brick is positioned very carefully, but it is doubtful that this could be done in practice. In any case, the seventh brick would definitely cause the stack to fall.

Follow-up Exercise. If the bricks in this Example were stacked so that alternately 4.0 cm and 6.0 cm extended beyond the brick beneath, how many bricks could be stacked in this manner before the stack toppled? *(Answer may be found in the Answers to Follow-up Exercises section at the back of the book.)*

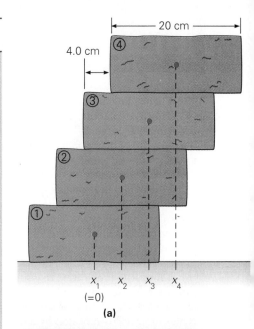

(b)

▪**FIGURE 8.12 Stack them up!**
(a) How many bricks can be stacked like this before the stack falls over? See Example 8.5.
(b) You may want to try similar experiments with books.

For another case of stability, see the chapter Insight on p. 258.

8.3 Rotational Dynamics

Objectives: To be able to (a) describe what is meant by the moment of inertia of a rigid body, and (b) apply the rotational form of Newton's second law to physical situations.

Moment of Inertia

Torque is the rotational analog of force in linear motion, and a net torque produces rotational motion. To analyze this relationship, consider a constant force acting on a particle of mass m. The magnitude of the torque on the particle is

$$\tau = r_\perp F = rF_\perp = rma_\perp = mr^2\alpha \qquad \text{torque on a particle} \qquad (8.4)$$

where $a_\perp = r\alpha$ is the tangential acceleration (a_t). For the rotation of a rigid body about a fixed axis, this equation can be applied to each particle and the results summed over the entire body (n particles) to find the total torque. Since, as you recall, all the

Torque on a particle

Insight   Stability in Action

When riding a bicycle and going around a curve or making a circular turn on a level surface, a rider instinctively leans into the curve. Why is this? One would think that leaning over, rather than remaining upright, is more likely to cause a spill. However, leaning really does increase stability—it's all a matter of torques.

When a bicycle or vehicle goes around a level circular curve, a centripetal force is needed, as we learned in Chapter 7. This is generally supplied by the force of static friction between the tires and the road. As illustrated in Fig. 1a, the reaction force **R** of the ground on the bicycle provides the required centripetal force **F**$_c$ to round the curve and the normal force **N**.

Suppose the rider tried to go around the curve with these forces while remaining upright as shown in the figure. Note that the line of action of **R** does not go through the system's center of gravity (indicated by a dot). Considering an axis of rotation through this point, there would be a counterclockwise torque that would tend to rotate the bicycle in such a way that the wheels would slide inward underneath the rider. However, if the rider leans inward at the proper angle (Fig. 1b), the line of action of **R** and the weight force both go through the center of gravity and there is no rotational instability (as the gentleman on the bicycle well knew).

(Being alert, you might observe there is still a torque on the rider. Indeed, when the rider leans into the curve, the weight force gives rise to a torque about an axis through the point of contact with the ground. This, along with the turning of the handle bars, causes the bicycle to turn. If the bicycle were not moving, there would be a rotation about this axis and the bicycle and rider would fall over.)

The need to lean into a curve is readily apparent in bicycle and motorcycle races on level tracks. Things can be made easier for the riders if tracks or roadways are banked to provide a natural lean (recall Section 7.3 and Exercise 7.56).

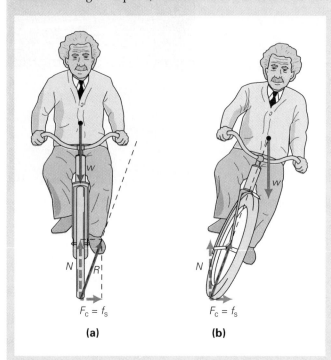

(a)   (b)

FIGURE 1   Leaning into a curve
When rounding a curve or making a turn, a bicycle rider must lean into the curve. (This rider could have told you why.)

particles of a rotating body have the same angular acceleration, we can simply add the individual torque magnitudes

$$\Sigma \tau = \tau_1 + \tau_2 + \tau_3 + \cdots + \tau_n$$

$$= m_1 r_1^2 \alpha + m_2 r_2^2 \alpha + m_3 r_3^2 \alpha + \cdots + m_n r_n^2 \alpha$$

$$= (m_1 r_1^2 + m_2 r_2^2 + m_3 r_3^2 + \cdots + m_n r_n^2)\alpha$$

$$\Sigma \tau = \left(\sum_{i=1}^{n} m_i r_i^2 \right) \alpha \tag{8.5}$$

But for a rigid body, the masses (m_i's) and the distances from the axis of rotation (r_i's) are constant. Therefore, the quantity in the parentheses is constant and called the **moment of inertia**, I:

$$I = \Sigma m_i r_i^2 \quad \textit{moment of inertia} \tag{8.6}$$

Definition of moment of inertia

SI unit of moment of inertia: $kg \cdot m^2$

We drop the summation sign and the magnitude of the torque is then

$$\tau = I\alpha \quad \textit{torque on a rigid body} \tag{8.7}$$

Rotational form of Newton's second law

This is the *rotational form of Newton's second law* ($\mathbf{F} = m\mathbf{a}$ and $\boldsymbol{\tau} = I\boldsymbol{\alpha}$, in vector form). Keep in mind that *net* forces and torques ($\mathbf{F}_{net}$ and $\boldsymbol{\tau}_{net}$) are necessary to produce accelerations, although this is not indicated explicitly in Eq. 8.7 τ means *net* torque.

As you might infer from a comparison of these two forms of Newton's law, the moment of inertia I is a measure of *rotational inertia*, or a body's tendency to resist change in its rotational motion. Although I is constant for a rigid body and is the rotational analog of mass, you must keep in mind that, unlike the mass of a particle, the moment of inertia is referenced to a particular axis and can have different values for different axes. The moment of inertia also depends on the mass distribution *relative* to the axis of rotation. It is easier (i.e., it takes less torque) to give an object an angular acceleration about some axes than about others. The following Example illustrates this point.

EXAMPLE 8.6 ■ ROTATIONAL INERTIA: MASS DISTRIBUTION AND AXIS OF ROTATION

Find the moment of inertia about the axis indicated for each of the simple one-dimensional dumbbell configurations in •Fig. 8.13 (Consider the mass of the connecting bar to be negligible and report to three significant figures.)

Solution.

Given: Values of m and r from the figure   *Find:* $I = \Sigma_i m_i r_i^2$

With $I = m_1 r_1^2 + m_2 r_2^2$,

(a) $I = (30 \text{ kg})(0.50 \text{ m})^2 + (30 \text{ kg})(0.50 \text{ m})^2 = 15 \text{ kg} \cdot \text{m}^2$

(b) $I = (40 \text{ kg})(0.50 \text{ m})^2 + (10 \text{ kg})(0.50 \text{ m})^2 = 12.5 \text{ kg} \cdot \text{m}^2$

(c) $I = (30 \text{ kg})(1.5 \text{ m})^2 + (30 \text{ kg})(1.5 \text{ m})^2 = 135 \text{ kg} \cdot \text{m}^2$

(d) $I = (30 \text{ kg})(0 \text{ m})^2 + (30 \text{ kg})(3.0 \text{ m})^2 = 270 \text{ kg} \cdot \text{m}^2$

(e) $I = (40 \text{ kg})(0 \text{ m})^2 + (10 \text{ kg})(3.0 \text{ m})^2 = 90 \text{ kg} \cdot \text{m}^2$

Example 8.6 clearly shows how the moment of inertia depends on the mass *and* its distribution relative to a particular axis of rotation. In general, the moment of inertia is larger the farther the mass is from the axis of rotation. This principle is important in the design of flywheels, which are used in automobiles to keep the engine running smoothly between cylinder firings. The mass of a flywheel is concentrated near the rim, giving a large moment of inertia, which resists changes in motion.

Follow-up Exercise. In parts (d) and (e) of this Example, would the moments of inertia be different if the axis of rotation were through m_2? Explain. *(Answer may be found in the Answers to Follow-up Exercises section at the back of the book.)*

CONCEPTUAL EXAMPLE 8.7 ■ BALANCING ACT: LOCATING THE CENTER OF GRAVITY

A rod with a movable ball, like that shown in •Fig. 8.14, is more easily balanced if the ball is in a higher position. This is because with the ball in a higher position, (a) the system has a higher center of gravity and more stability, (b) when the center of gravity is

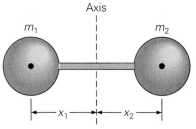

(a) $m_1 = m_2 = 30 \text{ kg}$
     $x_1 = x_2 = 0.50 \text{ m}$

(b) $m_1 = 40 \text{ kg}, m_2 = 10 \text{ kg}$
     $x_1 = x_2 = 0.50 \text{ m}$

(c) $m_1 = m_2 = 30 \text{ kg}$
     $x_1 = x_2 = 1.5 \text{ m}$

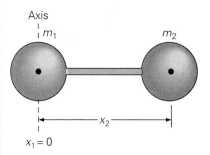

(d) $m_1 = m_2 = 30 \text{ kg}$
     $x_1 = 0, x_2 = 3.0 \text{ m}$

(e) $m_1 = 40 \text{ kg}, m_2 = 10 \text{ kg}$
     $x_1 = 0, x_2 = 3.0 \text{ m}$

•**FIGURE 8.13 Moment of inertia** The moment of inertia depends on the distribution of mass relative to a particular axis of rotation and, in general, has a different value when calculated about a different axis. This reflects the fact that objects are easier or more difficult to rotate about certain axes. See Example 8.6.

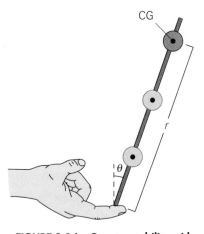

CG

•FIGURE 8.14 Greater stability with a higher center of gravity?
See Conceptual Example 8.7.

off the vertical there is less torque and smaller angular acceleration, (c) the moment of inertia about the axis of rotation is larger, (d) the center of gravity is closer to the axis of rotation. *Clearly establish the reasoning and physical principle(s) used in determining your answer before checking it below. That is, **why** did you select your answer?*

Reasoning and Answer.   With the ball at any position and the rod vertical, the system is in unstable equilibrium. We saw in Section 8.2 that rigid bodies with wide bases and *low* centers of gravity are more stable, so answer (a) isn't correct. Any slight movement will cause the rod to rotate about an axis through its point of contact. With the center of gravity (CG) at a higher position and off the vertical, there would be a greater lever arm (and thus a *greater* torque), so (b) too is incorrect. Also with the ball in a higher position, the center of gravity is *farther* from the axis of rotation, which eliminates (d). However, moving the CG farther from the axis of rotation has an interesting consequence—a greater moment of inertia or resistance to change in rotational motion.

With the ball in a higher position, as the rod starts to rotate there is a greater torque, but the increased moment of inertia produces an even greater resistance to rotational motion. With the rod rotationally accelerating more slowly, the person has more time to adjust his or her hand and bring the finger and the axis of rotation under the center of gravity. The torque is then zero and the rod is again in equilibrium, albeit unstable. Thus, the answer is (c).

We can understand this result more clearly if we consider the angular acceleration of the rod as it starts to tilt. Clearly, the smaller the angular acceleration, the more time is available to adjust one's hand under the rod to balance it. The angular acceleration of the rod is given by Eq. 8.7: $\alpha = \tau/I$. Now let us look at how the torque, τ, and moment of inertia, I, vary with r, the distance of the system's CG from the axis of rotation (through the balancing finger). Neglecting the mass of the rod and considering the CG to be simply that of the balls, $\tau = rF = rmg \sin \theta$ (Eq. 8.2) and $I = mr^2$ (Eq. 8.5). In other words, τ increases as r while I increases as r^2. Combining these relationships, we can see that α is proportional to $1/r$. Thus we have our paradoxical result: the farther the rod's CG from the axis of rotation, the more slowly it tilts and the easier it is to balance.

Follow-up Exercise.   When walking on a thin bar or rail, such as a railroad rail, you have probably found that it helps to hold your arms outstretched. Similarly, tightrope walkers often carry long poles, as in the chapter opening photo. How does this help a performer to maintain his or her balance? *(Reasoning and answer may be found in the Answers to Follow-up Exercises section at the back of the book.)*

The calculations for the moments of inertia of extended rigid bodies require math that is beyond the scope of this book. The results for some common shapes are given in •Fig. 8.15. The rotational axes are generally taken along axes of symmetry, that is, running through the center of mass so as to give a symmetrical mass distribution. An exception is the rod with an axis of rotation through one end (Fig. 8.15c). This axis is parallel to an axis of rotation through the center of mass of the rod (Fig. 8.15b). The moment of inertia about such a parallel axis is given by a useful theorem called the **parallel axis theorem**:

$$I = I_{CM} + Md^2 \qquad (8.8)$$

where I is the moment of inertia about an axis that is parallel to one through the center of mass, I_{CM}, and at a distance d from it, and M is the total mass of the body (•Fig. 8.16). For the axis through the end of the rod (Fig. 8.15c), the moment of inertia is obtained by applying the parallel axis theorem to the thin rod in Fig. 8.15b:

$$I = I_{cm} + Md^2 = \tfrac{1}{12}ML^2 + M\left(\frac{L}{2}\right)^2 = \tfrac{1}{12}ML^2 + \tfrac{1}{4}ML^2 = \tfrac{1}{3}ML^2$$

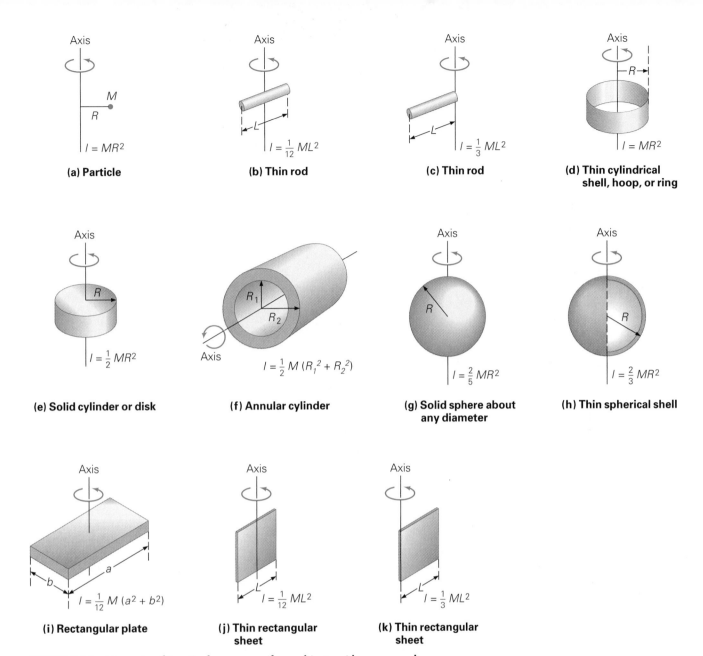

(a) Particle
$$I = MR^2$$

(b) Thin rod
$$I = \frac{1}{12} ML^2$$

(c) Thin rod
$$I = \frac{1}{3} ML^2$$

(d) Thin cylindrical shell, hoop, or ring
$$I = MR^2$$

(e) Solid cylinder or disk
$$I = \frac{1}{2} MR^2$$

(f) Annular cylinder
$$I = \frac{1}{2} M (R_1^2 + R_2^2)$$

(g) Solid sphere about any diameter
$$I = \frac{2}{5} MR^2$$

(h) Thin spherical shell
$$I = \frac{2}{3} MR^2$$

(i) Rectangular plate
$$I = \frac{1}{12} M (a^2 + b^2)$$

(j) Thin rectangular sheet
$$I = \frac{1}{12} ML^2$$

(k) Thin rectangular sheet
$$I = \frac{1}{3} ML^2$$

•**FIGURE 8.15** Moments of inertia for some uniform objects with common shapes

Applications of Rotational Dynamics

The rotational form of Newton's second law allows us to analyze dynamic rotational situations. The following examples illustrate how this is done. In such situations, it is very important to make certain that all the data are properly listed to help with the increasing number of variables.

EXAMPLE 8.8 ■ OPENING THE DOOR: TORQUE IN ACTION

A student opens a 12-kg door by applying a constant force of 40 N at a perpendicular distance of 0.90 m from the hinges (•Fig. 8.17). If the door is 2.0 m in height and 1.0 m wide, what is the magnitude of the angular acceleration of the door? (Assume that the door rotates freely on its hinges.)

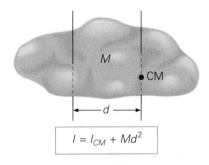

$$I = I_{CM} + Md^2$$

•**FIGURE 8.16 Parallel axis theorem**
The moment of inertia about an axis parallel to one through the center of mass of a body is given by $I = I_{CM} + Md^2$, where M is the total mass of the body and d is the distance between the two axes.

•FIGURE 8.17 **Torque in action**
See Example 8.8.

Solution. From the problem, we can list the following:

Given: $M = 12$ kg   *Find:* α (magnitude of angular
                $F = 40$ N   acceleration)
                $r_{\perp} = r = 0.90$ m
                $h = 2.0$ m   (door height)
                $w = 1.0$ m   (door width)

From the data and what is to be found, it should be apparent that we need to apply the rotational form of Newton's second law (Eq. 8.7), that is, $\tau = r_{\perp}F = I\alpha$. As can be seen, the problem boils down to finding the moment of inertia of the door.

Looking at Fig. 8.15, we see that part (k) applies to a door rotating on hinges, and $I = \frac{1}{3}ML^2$, where $L = w$, the width of the door. Essentially, the problem is now solved. We must simply do the calculations:

$$\tau = I\alpha$$

or

$$\alpha = \frac{\tau}{I} = \frac{r_{\perp}F}{\frac{1}{3}ML^2} = \frac{3rF}{Mw^2}$$

$$= \frac{3(0.90 \text{ m})(40 \text{ N})}{(12 \text{ kg})(1.0 \text{ m})^2}$$

$$= 9.0 \text{ rad/s}^2$$

Follow-up Exercise. In this Example, if the constant torque were applied through an angular distance of 45° and then removed, how long would it take the door to swing completely open (90°)? *(Answer may be found in the Answers to Follow-up Exercises section at the back of the book.)*

In problems involving pulleys in Chapter 4, the mass (and inertia) of the pulley was always neglected to simplify things. Now you know how to include this and can treat pulleys more realistically.

EXAMPLE 8.9 ■ PULLEYS HAVE MASS TOO: TAKING ACCOUNT OF PULLEY INERTIA

A block of mass m hangs from a string wrapped around a frictionless, disk-shaped pulley of mass M and radius R, as shown in •Fig. 8.18. If the block descends from rest under the influence of gravity, what is the magnitude of the linear acceleration of the block? (Neglect the mass of the string.)

Solution. The linear acceleration of the block depends on the angular acceleration of the pulley, so we look at the pulley system first. The pulley is a disk and thus has a moment of inertia of $I = \frac{1}{2}MR^2$ (Fig. 8.15e). It experiences a torque due to the tension force in the string (T). With $\tau = I\alpha$ (considering only the upper dashed box in Fig. 8.18),

$$\tau = r_{\perp}F = RT = I\alpha = \tfrac{1}{2}MR^2\alpha$$

so

$$\alpha = \frac{2T}{MR}$$

The linear acceleration of the block and the angular acceleration of the pulley are related by $a = R\alpha$ (why?), so

$$a = R\alpha = \frac{2T}{M} \tag{1}$$

But T is unknown. Looking at the descending mass (the lower dashed box) and summing the forces in the vertical direction gives

$$mg - T = ma$$

or

$$T = mg - ma \tag{2}$$

Using Eq. 2 to eliminate T from Eq. 1 gives

$$a = \frac{2T}{M} = \frac{2(mg - ma)}{M}$$

Solving for a,

$$a = \frac{2mg}{(2m + M)} \qquad (3)$$

Note that if $M \to 0$ (as for an ideal, massless pulley of previous chapters), then $I \to 0$, and $a = g$ (Eq. 3). Here, however, $M \neq 0$, so we have $a < g$. (Why?)

Follow-up Exercise. Pulleys can be analyzed even more accurately. In Example 8.9, friction was neglected, but practically a frictional torque (τ_f) exists and could be included. What would be the form of the angular acceleration in this case? Show your result to be dimensionally correct. *(Answer may be found in the Answers to Follow-up Exercises section at the back of the book.)*

Also in pulley problems, we neglect the mass of the string, which still gives a good approximation if the string is relatively light. Taking the mass of the string into account would give a continuously varying mass, and such a problem is beyond the scope of this book. Suppose you had masses suspended from each side of a pulley. Here, you'd have to compute the net torque. If you didn't know the values of the masses, or which way the pulley would rotate, then you could simply assume a direction. As in the linear case, if the result came out with the opposite sign, this would indicate that you had assumed the wrong direction.

PROBLEM-SOLVING HINT

For problems like those of Examples 8.8 and 8.9, dealing with coupled rotational and translational motions, keep in mind that the magnitudes of the accelerations are usually related by $a = r\alpha$, while $v = r\omega$ relates the magnitudes of the velocities at any instant of time. Applying Newton's second law (in rotational or linear form) to different parts of the system gives equations that can be combined using such relationships. Also, for rolling without slipping, $a = r\alpha$, and $v = r\omega$ relate the angular quantities to the linear motion of the center of mass.

Rotational problems sometimes give you surprising results, as the following Example shows.

CONCEPTUAL EXAMPLE 8.10 ■ FALLING FASTER THAN *g*? A ROTATIONAL PARADOX

A ball and cup demonstration is shown in •Fig. 8.19. A narrow, hinged board is propped up with a stick at a certain angle. In a depression on the upper end of the board is a steel ball, and a plastic cup is fixed to the board at a particular position nearer the hinge. When the stick is suddenly removed, (a) the board, cup, and ball all fall together, with the ball remaining in the depression, (b) the ball lands in the cup, (c) the ball lands to the right of the cup. *Clearly establish the reasoning and physical principle(s) used in determining your answer before checking it below. That is, why did you select your answer?*

Reasoning and Answer. This is not an easy question, but it's a lot of fun to grapple with. Though not purely conceptual, it nevertheless requires some conceptual insight. Most people would choose (a), but this answer is not generally correct. Generally (c) is not correct either, but for particular conditions (b) *is* correct—the ball actually lands in the cup. For this to happen, however, the board must be propped at a specific angle and the cup placed at a specific position.

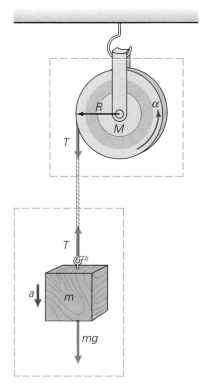

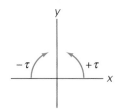

•FIGURE 8.18 Pulley with inertia Taking the mass, or rotational inertia, of a pulley into account allows a more realistic description of the motion. See Example 8.9.

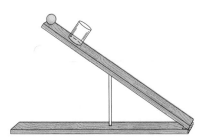

•FIGURE 8.19 A ball-and-cup trick See Conceptual Example 8.10.

You may be inclined to question this result. After all, you learned in Chapter 2 that all objects fall vertically at the same rate: g (the acceleration due to gravity). But for the ball to fall into the cup, the cup must beat the falling ball to the finish line (the base board). Note that the entire length of the falling board strikes the base board at the same instant—if the cup is to arrive before the ball, *all* points on the board must arrive before the ball. And since the end of the board starts the race even with the ball, it can't do this without having a vertical acceleration greater than g. How can this be?

The answer has to do with rotational motion. Keep in mind that the tangential acceleration of any point on the falling board is equal to the angular acceleration times the distance from the axis of rotation ($a = r\alpha$). Hence, points farther from the axis of rotation will exhibit larger tangential accelerations. The question is, does the end of the board have a *vertical* acceleration greater than g that will bring it to the base board before the ball? If so, then the cup attached to the board can be positioned so that the circular arc in which it falls brings it beneath the ball to catch it. Let's investigate this using the principles we have learned.

Let the length of the board be L and its mass M. Considering the center of mass to be at the midpoint of the board ($L/2$), then there is an *initial* torque on the board of $\tau = Fr_\perp = mg(L/2)\cos\theta$. Also, let's assume that the moment of inertia of the narrow board can be approximated by that of a thin rod, $I = ML^2/3$ (Fig. 8.15c). Hence,

$$\tau = I\alpha$$

or

$$Mg\left(\frac{L}{2}\cos\theta\right) = \left(\frac{ML^2}{3}\right)\alpha$$

and

$$\alpha = \frac{3g\cos\theta}{2L}$$

The tangential acceleration of the end of the board is then

$$a = r\alpha = L\alpha = \frac{3gL\cos\theta}{2L} = \frac{3g\cos\theta}{2}$$

The *vertical* component of this acceleration is

$$a_y = a\cos\theta = \frac{3g\cos^2\theta}{2}$$

For the end of the board to beat the ball, we must have

$$a_y > g$$

or

$$\frac{3g\cos^2\theta}{2} > g$$

which requires

$$\cos^2\theta > \tfrac{2}{3}$$

Solving for θ, this gives a maximum critical propping angle of about 35°. If the board is so set and the cup is at a particular position, the board accelerates faster than the ball and the ball falls into the cup.

Follow-up Exercise. (a) For a propping angle of 35°, at what position on the board must the cup be located? (b) Is the vertical component of the cup's acceleration greater than g? (*Reasoning and answers may be found in the Answers to Follow-up Exercises section at the back of the book.*)

Another application of rotational dynamics is the analysis of motion on an inclined plane. Previously, you worked with objects sliding down planes. Now you can let them roll.

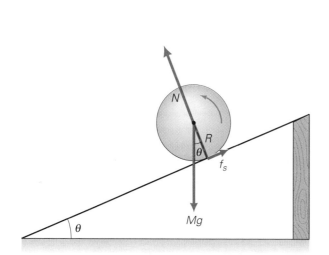

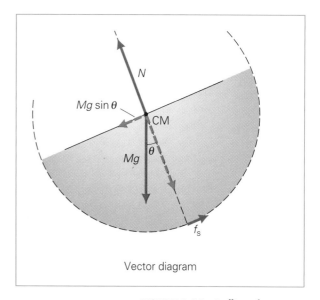

Vector diagram

●FIGURE 8.20 Rolling down an inclined plane

For a rigid ball rolling down a hard-surfaced inclined plane, the force of static friction acts on the ball in response to the parallel component of the weight force. See Example 8.11.

EXAMPLE 8.11 ■ ROLLING DOWN: TORQUE AND ACCELERATION

A solid, rigid spherical ball of mass M and radius R is released at the top of a hard-surfaced inclined plane. It rolls without slipping, with only static friction between it and the plane (●Fig. 8.20). What is the acceleration of the ball's center of mass?

Solution. Let's analyze this situation closely. Since the ball rolls without slipping, there is no relative motion between it and the plane at the point of contact (through which the instantaneous axis of rotation passes). At this point the force of static friction (f_s) acts on the ball. Without this force, the ball would slide down the plane. This force is perpendicular to the instantaneous axis of rotation, or directed up the plane.

You can analyze the motion using two different axes of rotation. First, let's consider an axis through the center of mass of the ball (the central dot in the figure) and perpendicular to the page so that we can see the effect of the frictional torque. The acceleration of the ball's center of mass is down the plane, and the ball's angular acceleration is about an axis through that center. Since the reaction force (the normal force N) and the weight force act through the axis of rotation (see the vector diagram in the figure), they produce no torques, and the angular acceleration is produced entirely by the frictional torque due to f_s.

Then applying Newton's second law, $\tau = I\alpha$, and using the moment of inertia about an axis through the ball's center of mass as found in Fig. 8.15g ($I = \frac{2}{5}MR^2$), we have

$$\tau = Rf_s = I\alpha = \left(\tfrac{2}{5}MR^2\right)\left(\frac{a}{R}\right)$$

where we substituted a/R for α. Then, solving for f_s,

$$f_s = \tfrac{2}{5}Ma$$

Looking at the translational motion of the center of mass and summing the forces acting parallel to the plane (see the vector diagram) gives

$$Mg\sin\theta - f_s = Ma$$

or

$$Mg\sin\theta - \tfrac{2}{5}Ma = Ma$$

Thus,

$$a = \tfrac{5}{7}g\sin\theta$$

Here is another situation where the acceleration does not depend on M or R. Therefore, uniform spheres of any size or density will roll down the same plane with the same acceleration (assuming no slippage).

Before leaving this Example, let's look at the situation from another perspective. Consider the rotational motion about the instantaneous axis of rotation, which runs through the point of contact of the sphere and the plane's surface. This point is instantaneously at rest, as noted earlier. As can be seen from Fig. 8.20, the normal force and the frictional force both act through this point, and so produce no torque. The torque on the ball from this perspective comes from the component of the weight force parallel to the plane, $Mg \sin \theta$ (see the vector diagram). The torque is then

$$\tau = rF_\perp = R(Mg \sin \theta) = I_i \alpha \qquad (1)$$

where I_i is the moment of inertia *about the instantaneous axis of rotation*. Note that this is *not* equal to the moment of inertia used above, $I_{CM} = \frac{2}{5}MR^2$, which is the moment of inertia about an axis *through the center of mass* of the ball. You can find the moment of inertia about the instantaneous axis using the parallel axis theorem.

$$I_i = I_{CM} + Md^2 = \frac{2}{5}MR^2 + MR^2 = \frac{7}{5}MR^2 \qquad (2)$$

Substituting for I_i from Eq. 2 into Eq. 1 gives

$$R(Mg \sin \theta) = I_i \alpha = (\tfrac{7}{5}MR^2)\left(\frac{a}{R}\right)$$

and

$$a = \tfrac{5}{7}g \sin \theta$$

This is the same as the acceleration found using the other axis of rotation (as it should be).

Follow-up Exercise. That uniform spheres of any size or density will roll down the same plane with the same acceleration is shown in this Example. Will different uniform objects with different shapes (for example, cylinders or hoops) roll down this incline with the same acceleration as a sphere? (*Answer may be found in the Answers to Follow-up Exercises section at the back of the book.*)

8.4 Rotational Work and Kinetic Energy

Objectives: To be able to discuss, explain, and use the rotational forms of (a) work, (b) kinetic energy, and (c) power.

This section gives the rotational analogs associated with work and kinetic energy for constant torques. Because their development is very similar to that given for their linear counterparts, detailed discussion is not needed.

Rotational Work. For rotational motion, $W = Fs$ for a single force acting tangentially along an arc length s:

$$W = Fs = F(r\theta) = \tau\theta$$

where θ is in radians. Thus,

Rotational work

$$W = \tau\theta \qquad (8.9)$$

In this book, both torque (τ) and angular displacement (θ) vectors are almost always along the fixed axis of rotation, so you will not need to be concerned about parallel components as you were for translational work. The torque and angular displacement may be in opposite directions, in which case the torque decreases, or slows down, the rotation of the body. Then the rotational work is negative, which is analogous to F and d being in opposite directions for translational motion.

Rotational Power. An expression for power, or the time rate of doing work, is easily obtained from Eq. 8.9:

Rotational power

$$P = \frac{W}{t} = \tau\left(\frac{\theta}{t}\right) = \tau\omega \qquad (8.10)$$

The Work-Energy Theorem and Kinetic Energy

The relationship between the net rotational work and change in rotational kinetic energy may be derived as follows, starting with the equation for net rotational work:

$$W = \tau\theta = I\alpha\theta$$

For a constant angular acceleration, $\omega^2 = \omega_0^2 + 2\alpha\theta$, and therefore

$$W = I\left(\frac{\omega^2 - \omega_0^2}{2}\right) = \tfrac{1}{2}I\omega^2 - \tfrac{1}{2}I\omega_0^2$$

Thus,

$$W = \tfrac{1}{2}I\omega^2 - \tfrac{1}{2}I\omega_0^2 = K - K_0 = \Delta K \qquad (8.11)$$

Rotation analog of work-energy theorem

where K is the **rotational kinetic energy**.

$$K = \tfrac{1}{2}I\omega^2 \qquad (8.12)$$

Rotational kinetic energy

Thus, the net rotational work is equal to the change in the rotational kinetic energy. Consequently, in order to change the *rotational* kinetic energy of an object, a torque must be applied, because Eq. 8.9 shows that rotational work involves a torque.

It is possible to derive the kinetic energy of a rotating rigid body directly. Summing the instantaneous tangential velocities and kinetic energies of the individual particles of the body gives

$$K = \tfrac{1}{2}\Sigma_i m_i v_i^2 = \tfrac{1}{2}(\Sigma_i m_i r_i^2)\omega^2 = \tfrac{1}{2}I\omega^2$$

where $v_i = r_i\omega$. Thus, Eq. 8.12 doesn't represent a new form of energy. It is simply another expression for kinetic energy, in a form that is more convenient for rigid body rotation.

A summary of the analogous equations for translational and rotational motion is given in Table 8.1. (The table contains momentum, which is discussed in the next section.)

When an object has both translational and rotational motions, its total kinetic energy may be divided into parts to reflect this. For example, for a cylinder rolling without slipping on a level surface, the motion is purely rotational relative to the instantaneous axis of rotation, which is instantaneously at rest (see Example 8.11). The kinetic energy is

$$K = \tfrac{1}{2}I_i\omega^2$$

TABLE 8.1	Translational and Rotational Quantities and Equations		
	Translational	*Rotational*	
Force:	$\mathbf{F}$	Torque (magnitude, moment of force):	$\tau = rF\sin\theta$
Mass (inertia):	m	Moment of inertia:	$I = \Sigma_i m_i r_i^2$
Newton's second law:	$\mathbf{F} = m\mathbf{a}$	Newton's second law:	$\boldsymbol{\tau} = I\boldsymbol{\alpha}$
Work:	$W = Fd$	Work:	$W = \tau\theta$
Power:	$P = Fv$	Power:	$P = \tau\omega$
Kinetic energy:	$K = \tfrac{1}{2}mv^2$	Kinetic energy:	$K = \tfrac{1}{2}I\omega^2$
Work-energy theorem:	$W = \tfrac{1}{2}mv^2 - \tfrac{1}{2}mv_0^2 = \Delta K$	Work-energy theorem:	$W = \tfrac{1}{2}I\omega^2 - \tfrac{1}{2}I\omega_0^2 = \Delta K$
Momentum:	$\mathbf{p} = m\mathbf{v}$	Momentum:	$\mathbf{L} = I\boldsymbol{\omega}$

where I_i is the moment of inertia about the instantaneous axis. This moment of inertia is given by the parallel axis theorem (Eq. 8.8): $I_i = I_{CM} + MR^2$, where R is the radius of the cylinder. Then

$$K = \tfrac{1}{2}I_i\omega^2 = \tfrac{1}{2}(I_{CM} + MR^2)\omega^2 = \tfrac{1}{2}I_{CM}\omega^2 + \tfrac{1}{2}MR^2\omega^2$$

But since there is no slipping, $R\omega = v_{CM}$, and

Rolling kinetic energy

$$K = \tfrac{1}{2}I_{CM}\omega^2 + \tfrac{1}{2}Mv_{CM}^2 \qquad (rolling, \qquad (8.13)$$
$$\underset{KE}{total} = \underset{KE}{rotational} + \underset{KE}{translational} \qquad no\ slipping)$$

Note: A rolling body has both translational kinetic energy and rotational kinetic energy.

Note that although a cylinder was used as an example here, this is a general result and applies to any object that is rolling without slipping. Thus, the total kinetic energy of such an object is the sum of two contributions: the translational kinetic energy of the center of mass and the rotational kinetic energy relative to a horizontal axis through the center of mass.

EXAMPLE 8.12 ■ DIVISION OF ENERGY: ROTATIONAL AND TRANSLATIONAL

A uniform solid 1.0-kg cylinder rolls without slipping on a flat surface at a speed of 1.8 m/s. (a) What is the total kinetic energy of the cylinder? (b) What percentage of this is rotational kinetic energy?

Solution.

Given: $M = 1.0$ kg
$v_{CM} = 1.8$ m/s
$I_{CM} = \tfrac{1}{2}MR^2$ (from Fig. 8.15e)

Find: (a) K (total kinetic energy)

(b) $\dfrac{K_r}{K}$ ($\times100\%$) (percentage of rotational energy)

(a) The cylinder rolls without slipping, so the condition $v_{CM} = R\omega$ applies. Then, the total kinetic energy is the sum of the rotational kinetic energy K_r and the translational kinetic energy of the center of mass K_{CM}:

$$= \tfrac{1}{2}(\tfrac{1}{2}MR^2)\left(\frac{v_{CM}}{R}\right)^2 + \tfrac{1}{2}Mv_{CM}^2 = \tfrac{1}{4}Mv_{CM}^2 + \tfrac{1}{2}Mv_{CM}^2$$

$$= \tfrac{3}{4}Mv_{CM}^2 = \tfrac{3}{4}(1.0\ \text{kg})(1.8\ \text{m/s})^2 = 2.4\ \text{J}$$

(b) The rotational kinetic energy K_r of the cylinder is the first term of the preceding equation, so, forming a ratio in symbol form,

$$\frac{K_r}{K} = \frac{\tfrac{1}{4}Mv_{CM}^2}{\tfrac{3}{4}Mv_{CM}^2} = \tfrac{1}{3}\ (\times100\%) \approx 33\%$$

Thus, the total kinetic energy of the cylinder is made up of rotational and translational parts, with one-third being rotational.

Note that the radius of the cylinder was not needed at all, and that neither the mass nor any numerical values were needed in part (b). *Don't* think that this exact division of energy is a general result, however. It is easy to show that the percentage is different for objects with different moments of inertia. For example, you should expect a rolling sphere to have a smaller percentage of rotational kinetic energy than a cylinder has because it has a smaller moment of inertia ($I = \tfrac{2}{5}mR^2$).

Potential energy can be brought into the act by applying the conservation of energy to an object rolling up or down an inclined plane. Try this in the Follow-up Exercise.

Follow-up Exercise. In this Example, suppose that the cylinder rolled up a 20° inclined plane without slipping. (a) At what vertical height on the plane does the cylinder stop? (b) To find the height in part (a), you probably equated the initial total kinetic energy to the final gravitational potential energy. That is, the total kinetic energy was reduced by the work done by gravity. However, a frictional force also acts (to prevent slipping—see Fig. 8.20). Is there not work done here too? *(Answers may be found in the Answers to Follow-up Exercises section at the back of the book.)*

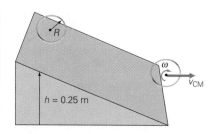

•**FIGURE 8.21 Rolling motion and energy**
When an object rolls down an inclined plane, potential energy is converted to translational *and* rotational kinetic energy. This makes the rolling slower than frictionless sliding. See Example 8.13.

EXAMPLE 8.13 ■ ROLLING DOWN VERSUS SLIDING DOWN: WHICH IS FASTER?

A uniform cylindrical hoop is released from rest at a height of 0.25 m near the top of an inclined plane (•Fig. 8.21). If the cylinder rolls down the plane without slipping and there is no energy loss due to friction, what is the linear speed of the cylinder's center of mass at the bottom of the incline?

Solution.

Given: $h = 0.25$ m
$\qquad I_{CM} = MR^2$ (from Fig. 8.15d)
Find: v_{CM} (speed of CM)

Since the total mechanical energy of the cylinder is conserved, you can write

$$E_o = E$$

or, taking $v = 0$ at the bottom of the incline, we have

$$\underset{\text{at rest}}{Mgh} = \underset{\text{at bottom of incline}}{\tfrac{1}{2}I_{CM}\omega^2 + \tfrac{1}{2}Mv_{CM}^2}$$

Using the condition $v_{CM} = R\omega$ gives

$$Mgh = \tfrac{1}{2}(MR^2)\left(\frac{v_{CM}}{R}\right)^2 + \tfrac{1}{2}Mv_{CM}^2 = Mv_{CM}^2$$

which, solved for v_{CM}, is

$$v_{CM} = \sqrt{gh} = \sqrt{(9.80 \text{ m/s}^2)(0.25 \text{ m})} = 1.6 \text{ m/s}$$

Again, not much numerical information was needed here. Note that the hoop rolls down from the same height as the solid cylinder rolled up in the Example 8.12 Follow-up Exercise, yet the speed of the hoop is less than that of the cylinder at the bottom of the incline. You should know why—differences in moment of inertia.

Follow-up Exercise. Suppose the inclined plane in this Example were frictionless and the hoop slid down the plane instead of rolling. How would the speed at the bottom compare in this case? *(Answer may be found in the Answers to Follow-up Exercises section at the back of the book.)*

As Example 8.13 shows, v_{CM} is independent of M and R. The masses and radii cancel out, so all objects of a particular shape (with the same formula for the moment of inertia) roll with the same speed, regardless of size or density. But the rolling speed does vary with the moment of inertia, which varies with the shape. Therefore, rigid bodies with different shapes roll with different speeds. For example, if you released a cylindrical hoop, a solid cylinder, and a uniform sphere at the same time from the top of an inclined plane, the sphere would win the race to the bottom, followed by the cylinder, with the hoop coming in last—every time!

You can try this experiment using a couple of food cans or other cylindrical containers—one full of some solid material (a rigid body) and one empty with the ends cut out—and a smooth ball of some sort. Remember that the masses and the radii make no difference. You might think that an annular cylinder (a hollow cylinder with inner and outer radii that vary appreciably—Fig. 8.15f) would be a possible front-runner, or front-roller, in such a race, but it wouldn't win. The rolling race down an incline is fixed even when you vary the masses and the radii. (See Exercise 70.)

Another aspect of rolling is discussed in the Insight on p. 270.

In an emergency situation, you may instinctively jam on the brakes of a car, trying to come to a quick stop, that is, to stop in the shortest distance. But with the wheels locked, the car skids, or slides, to a stop, often out of control. In this case, the force of sliding friction is acting on the wheels.

In order to prevent this, you learn to pump the brakes in order to roll rather than slide to a stop, particularly on wet or icy roads. Many newer automobiles have computerized antilock braking systems (ABS) that do this automatically. When the brakes are applied firmly and the car begins to slide, sensors in the wheels note this and a computer takes over control of the braking system. It momentarily releases the brakes and then varies the brake-fluid pressure with a pumping action (up to thirteen times per second!) so that the wheels will continue to roll without slipping.

In the absence of sliding, both rolling friction and static friction act. In many cases, however, the force of rolling friction is small and only static friction need be taken into account, as will be done here.

Does sliding instead of rolling make a big difference in the stopping distance for an automobile? It is easy to calculate the difference by assuming that rolling friction is negligible. Although the external force of static friction does no work to dissipate energy in slowing a car (this is done internally by friction on the brake pads), it does determine whether the wheels roll or slide.

In Example 2.7, a vehicle stopping distance was given by

$$x = \frac{v_o^2}{2a}$$

By Newton's second law, the net force in the horizontal direction is

$$F = f = \mu N = \mu mg = ma$$

and the stopping acceleration is then

$$a = \mu g$$

Thus,

$$x = \frac{v_o^2}{2\mu g} \qquad (1)$$

But, as was noted in Chapter 4, the coefficient of sliding (kinetic) friction is generally less than that of static friction; that is, $\mu_k < \mu_s$. The general difference between rolling and sliding stops can be seen by using the same initial velocity (v_o for both cases). Then, using Eq. 1 to form a ratio,

$$\frac{x_{roll}}{x_{slide}} = \frac{\mu_k}{\mu_s} \quad \text{or} \quad x_{roll} = \left(\frac{\mu_k}{\mu_s}\right) x_{slide}$$

The value of μ_k for rubber on wet concrete from Table 4.1 is 0.60, and the value of μ_s for these surfaces is 0.80. Using these values for a comparison of the stopping distances gives

$$x_{roll} = \left(\frac{0.60}{0.80}\right) x_{slide} = (0.75) x_{slide}$$

The car comes to a rolling stop in 75 percent of the distance required for a sliding stop, for example, 15 m instead of 20 m. Although this may vary for different conditions, it could be an important, perhaps life-saving, difference.

8.5   Angular Momentum

Objectives: To be able to (a) define angular momentum, and (b) apply the conservation of angular momentum to physical situations.

Another important quantity in rotational motion is angular momentum. Recall from Section 6.1 how linear momentum is associated with force. Similarly, angular momentum is associated with torque. As we have seen, torque is the product of a moment arm and a force. In a similar manner, **angular momentum (L)** is the product of a moment arm and linear momentum. For a particle of mass m, the magnitude of the linear momentum is $p = mv$ and $v = r\omega$. The magnitude of the angular momentum is

$$L = r_\perp p = mr_\perp v = mr_\perp^2 \omega \qquad (8.14)$$

where v is the tangential speed of the particle, $r_\perp$ the moment arm, and ω the angular speed. For circular motion, $r_\perp = r$, since $\mathbf{v}$ is perpendicular to $\mathbf{r}$. As can be seen from the equation, the units of L are kg·m²/s.

For a system of particles comprising a rigid body, the total magnitude of the angular momentum is

$$L = (\Sigma m_i r_i^2)\omega = I\omega \qquad (8.15)$$

which, in vector notation, is

Rigid body angular momentum

$$\mathbf{L} = I\boldsymbol{\omega} \qquad (8.16)$$

SI unit of angular momentum: kg·m²/s

Thus, **L** is in the direction of the angular velocity vector (**ω**) as given by the right-hand rule.

For linear motion, momentum is related to force by $\mathbf{F} = \Delta\mathbf{p}/\Delta t$. Angular momentum is analogously related to net torque, as can be easily shown:

$$\tau = I\alpha = \frac{I\Delta\omega}{\Delta t} = \frac{\Delta(I\omega)}{\Delta t} = \frac{\Delta\mathbf{L}}{\Delta t}$$

Note: Review Eq. 6.3, Section 6.1.

That is,

$$\tau = \frac{\Delta\mathbf{L}}{\Delta t} \qquad (8.17)$$

Net torque: the time rate of change of angular momentum

Thus, the *net* torque is equal to *the time rate of change of angular momentum*.

Conservation of Angular Momentum

Equation 8.17 was derived using $\tau = I\alpha$, which applies to a rigid system of particles or a rigid body having a constant moment of inertia. However, Eq. 8.17 is a general one that also applies to a nonrigid system of particles. In such a system, there may be a change in the mass distribution and a change in the moment of inertia. As a result, there may be an angular acceleration even in the absence of a torque. How can this be?

If the net torque on a system is zero, then, by Eq. 8.17, $\tau = \Delta\mathbf{L}/\Delta t = 0$, and

$$\Delta\mathbf{L} = \mathbf{L} - \mathbf{L}_o = I\omega - I_o\omega_o = 0$$

or

$$I\omega = I_o\omega_o \qquad (8.18)$$

Thus, the condition for the **conservation of angular momentum** is as follows:

> In the absence of an external, unbalanced torque, the total (vector) angular momentum of a system is conserved (remains constant).

Note: Angular momentum is conserved when the net torque is zero. This is the third conservation law in mechanics.

As with total linear momentum, the internal torques arising from internal forces cancel out.

For a rigid body with a constant moment of inertia (that is, $I = I_o$), the angular speed remains constant ($\omega = \omega_o$) in the absence of a net torque. But it is possible for the moment of inertia to change in some systems, giving rise to a change in the angular velocity, as the following example illustrates.

EXAMPLE 8.14 ■ PULL IT DOWN: CONSERVATION OF ANGULAR MOMENTUM

A small ball at the end of a string that passes through a tube is swung in a circle, as illustrated in ●Fig. 8.22. When the string is pulled downward through the tube, the angular speed of the ball increases. (a) Is this caused by a torque due to the pulling force? (b) If the ball is initially swung in a circle with a radius of 0.30 m at a speed of 2.8 m/s, what will be its tangential speed if the string is pulled down far enough to reduce the radius of the circle to 0.15 m?

Solution.

Given: $r_1 = 0.30$ m *Find:* (a) Cause of the increase in angular speed
$r_2 = 0.15$ m (b) v_2 (final tangential speed)
$v_1 = 2.8$ m/s

(a) The change in the angular velocity, or the angular acceleration, is not caused by a torque due to the pulling force. The force on the ball as transmitted by the string acts through the axis of rotation, and therefore the torque is zero. As the rotating

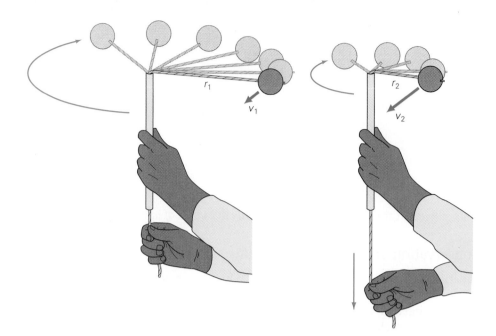

FIGURE 8.22 Conservation of angular momentum
When the string is pulled downward through the tube, the revolving ball speeds up. See Example 8.14.

portion of the string is shortened, the moment of inertia of the ball ($I = mr^2$, from Fig. 8.15a) decreases. This causes the ball to speed up, since in the absence of an external torque, the angular momentum of the ball is conserved.

(b) Since the angular momentum is conserved, we have

$$I_o \omega_o = I \omega$$

Then, using $I = mr^2$ and $\omega = v/r$ gives

$$m r_1 v_1 = m r_2 v_2$$

and

$$v_2 = \left(\frac{r_1}{r_2}\right) v_1 = \left(\frac{0.30 \text{ m}}{0.15 \text{ m}}\right) 2.8 \text{ m/s} = 5.6 \text{ m/s}$$

When the radial distance is shortened, the ball speeds up.

Follow-up Exercise. Let's look at the situation in this Example in terms of work and energy. If the vertical pulling force is 7.8 N, what is the final speed of the 0.10-kg ball? (*Answer may be found in the Answers to Follow-up Exercises section at the back of the book.*)

Example 8.14 should help you understand Kepler's law of areas (Chapter 7). A planet's angular momentum is essentially conserved, since the Sun exerts no torque on it (why?) and we can neglect the torques from other planets. Thus, when a planet is closer to the Sun in its elliptical orbit, and so has a shorter moment arm, its speed is greater by the conservation of angular momentum. Similarly, when an orbiting satellite's altitude varies during the course of its elliptical orbit, it speeds up or slows down in accordance with the same principle.

A popular lecture demonstration of the conservation of angular momentum is shown in •Fig. 8.23. A person sitting on a stool that rotates holds weights with his arms outstretched and is started slowly rotating. An external torque to start this rotation must be supplied by someone else, because the person on the stool cannot initiate the motion by himself. (Why not?) Once rotating, if the person brings his arms inward, the angular speed increases and he spins much faster. Extending his arms again slows him down. Can you explain this?

If L is constant, what happens to ω when I is made smaller by reducing r? The angular speed must increase to compensate and keep L constant. Ice skaters do dizzy toe spins by pulling in their arms to reduce their moments of inertia. Sim-

(a)

(b)

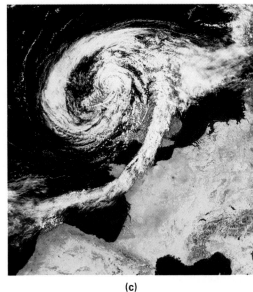
(c)

•FIGURE 8.23 **Moment of inertia change**
(a) Spinning slowly with masses in outstretched arms, the man's moment of inertia is relatively large (masses are at greater distances from the axis of rotation). Note that he is isolated, with no external torques (neglecting friction), so the angular momentum, $L = I\omega$, is conserved. (b) Pulling the arms inward decreases the moment of inertia (why?). Consequently ω must increase, and he goes into a dizzy spin. (c) The same principle helps to explain the violence of the winds that spiral around the center of a hurricane. As air rushes in toward the center of the storm, where the pressure is low, its rotational velocity must increase for angular momentum to be conserved.

ilarly, a diver spins during a high dive by tucking in, greatly decreasing his or her moment of inertia. The enormous wind speeds of tornadoes and hurricanes represent another example of the same effect (Fig. 8.23c).

Angular momentum, **L**, is a vector, and when it is conserved or constant, its magnitude *and* direction must remain unchanged. Thus, when no external torques act, the direction of **L** is fixed in space. This is the principle of the gyrocompass, as well as of passing a football accurately (•Fig. 8.24). The **L** vector of a spinning gyroscope in the compass is set in a particular direction (usually north). In the absence of external torques, the compass direction remains fixed, even though its carrier (an airplane or ship, for example) changes directions.

A football is normally passed with a spiraling rotation. This spin, or gyroscopic action, stabilizes the ball's motion in the direction of its spin axis. Similarly, rifle bullets are set spinning by the rifling in the barrel for directional stability.

The motion of a "sleeping" top is also illustrated in Fig. 8.24. A spinning top will stand straight up with its angular momentum vector fixed in space for some time. Note in the figure that the axis of rotation passes through the top's center of gravity, and thus there is no torque due to the top's weight.

However, you know that a sleeping top—or a gyroscope—eventually slows down because of frictional torques. In doing so, you have probably noticed how the spin axis revolves, or *precesses*, about the vertical axis. It revolves tilted over, so to speak (•Fig. 8.25). Since the gyroscope precesses, the angular momentum vector **L** is no longer constant in direction, indicating that a torque must be acting to give a change (Δ**L**) with time. As can be seen from the figure, the torque arises from the vertical component of the weight force, since the center of gravity no longer lies directly above the point of support or on the vertical axis of rotation. The instantaneous torque is such that the gyroscope's axis moves or precesses about the vertical axis.

In a similar manner, the Earth's rotational axis precesses. The Earth's rotational spin axis is tilted $23\frac{1}{2}°$ with respect to a line perpendicular to the plane of its revo-

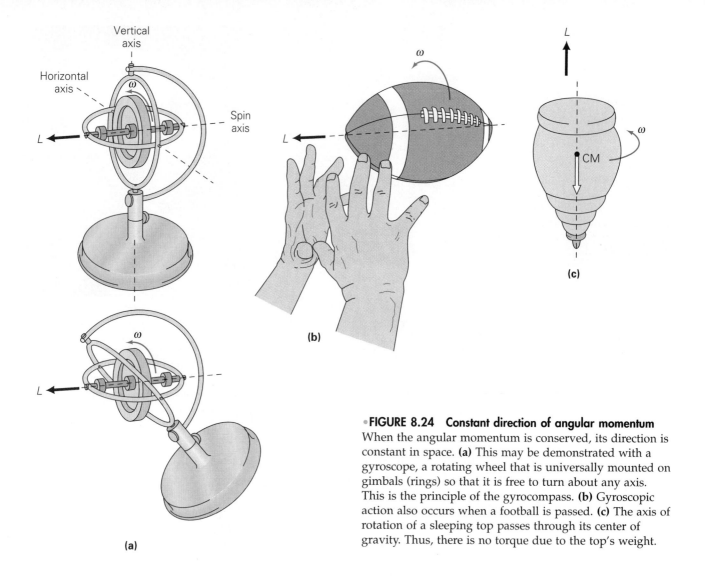

•FIGURE 8.24 Constant direction of angular momentum
When the angular momentum is conserved, its direction is
constant in space. **(a)** This may be demonstrated with a
gyroscope, a rotating wheel that is universally mounted on
gimbals (rings) so that it is free to turn about any axis.
This is the principle of the gyrocompass. **(b)** Gyroscopic
action also occurs when a football is passed. **(c)** The axis of
rotation of a sleeping top passes through its center of
gravity. Thus, there is no torque due to the top's weight.

lution about the Sun, and precesses about this line (Fig. 8.25b). The precession is due
to slight gravitational torques exerted on the Earth by the Sun and the Moon. The
torques arise because the Earth is not perfectly spherical—it has an equatorial bulge.

The period of the precession of the Earth's axis is about 26,000 years, so it has
little day-to-day effect. However, it does have an interesting long-term effect. Polaris will not always be (nor has it always been) the North Star, that is, the star
that the Earth's axis of rotation points toward. About 5000 years ago, Alpha
Draconis was the North Star, and 5000 years from now, it will be Alpha Cephei,
which is at an angular distance of about 68° away from Polaris on the circle described by the precession of the Earth's axis.

There are some other long-term torque effects on the Earth and the Moon.
Did you know that the Earth is slowing down and hence the days are getting
longer? Also, that the Moon is receding or getting farther away from the Earth?
This is primarily due to ocean tidal friction, which gives rise to a torque. As a result, the Earth's spin angular momentum, and therefore its rate of rotation, is
changing. The slowing rate of rotation causes the average day to be longer; this
century will be about 25 seconds longer than last. But this is an average rate. At
times, the Earth's rotation speeds up for relatively short periods. This is thought
to be associated with the rotational inertia of the liquid layer of the Earth's core
(see the Insight feature in Chapter 13).

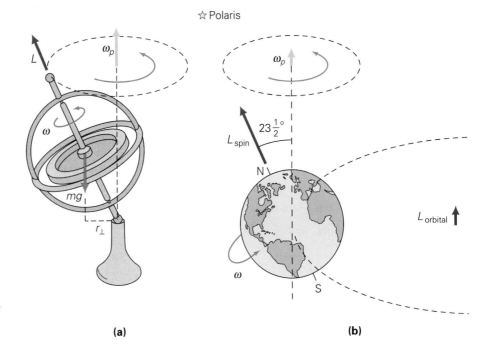

●FIGURE 8.25 Precession
An external torque causes a change in angular momentum. **(a)** For a spinning gyroscope, this change is directional, and the axis of rotation precesses about a vertical line. (The torque due to the weight force would point out of the page as drawn here, as would ΔL.) Note that although there is a torque that would topple a nonspinning gyroscope, a spinning gyroscope doesn't fall. **(b)** Similarly, the Earth's axis precesses because of gravitational torques caused by the Sun and the Moon. We don't notice this motion because the period of precession is about 26,000 years.

(a)   **(b)**

The tidal torque on the Earth results chiefly from the Moon's gravitational attraction, which is the main cause of ocean tides. This torque is *internal* to the Earth–Moon system, so the total angular momentum of that system is conserved. Since the Earth is losing angular momentum, the Moon must be gaining angular momentum to keep the total angular momentum constant. The Earth loses rotational (spin) angular momentum and the Moon gains orbital angular momentum. As a result, the Moon drifts slightly farther from Earth and its orbital speed decreases. The Moon moves away from the Earth at about 1 cm per lunar revolution, or lunar month (calculated from the rate of the slowing down of the Earth's rotation). Thus, the Moon moves in a slowly widening spiral.

Chapter Review

Important Terms

Important Concepts

- In pure translational motion, all the particles of an object have the same instantaneous velocity.
- In pure rotational motion (about a fixed axis), all the particles of an object have the same instantaneous angular velocity.

- Torque is the product of a force and a moment arm or lever arm.
- Mechanical equilibrium requires that the summation of the forces be zero (translational equilibrium) and the summation of the torques be zero (rotational equilibrium).

- An object is in stable equilibrium as long as its center of gravity, upon small displacement, lies above and inside its original base of support.
- Moment of inertia (I) is the rotational analog of mass

Important Equations

Condition for Rolling without Slipping:
$$v = r\omega \tag{8.1}$$
(or $s = r\theta$ or $a = r\alpha$)

Torque (magnitude):
$$\tau = r_\perp F = rF \sin \theta \tag{8.2}$$

Conditions for Translational and Rotational Mechanical Equilibrium:
$$\Sigma_i \mathbf{F}_i = 0 \quad \text{and} \quad \Sigma \tau_i = 0 \tag{8.3}$$

Torque on a Particle (magnitude):
$$\tau = r_\perp F = rF_\perp = rma_\perp = mr^2\alpha \tag{8.4}$$

Moment of Inertia:
$$I = \Sigma m_i r_i^2 \tag{8.6}$$

Rotational Form of Newton's Second Law (magnitude):
$$\tau = I\alpha \tag{8.7}$$

Parallel Axis Theorem:
$$I = I_{CM} + Md^2 \tag{8.8}$$

Rotational Work:
$$W = \tau\theta \tag{8.9}$$

- in the rotational form of Newton's second law, $\tau = I\alpha$.
- Angular momentum, the product of a moment arm and linear momentum, is conserved in the absence of an external, unbalanced torque.

Rotational Power:
$$P = \tau w \tag{8.10}$$

Work-Energy Theorem:
$$W = \Delta K = \tfrac{1}{2}I\omega^2 - \tfrac{1}{2}I\omega_o^2 \tag{8.11}$$

Rotational Kinetic Energy:
$$K = \tfrac{1}{2}I\omega^2 \tag{8.12}$$

Kinetic Energy of a Rolling Object:
$$K = \tfrac{1}{2}I_{CM}\omega^2 + \tfrac{1}{2}Mv_{CM}^2 \tag{8.13}$$

Angular Momentum of a Particle in Circular Motion (magnitude):
$$L = r_\perp p = mr_\perp v = mr_\perp^2\omega \tag{8.14}$$

Angular Momentum of a Rigid Body:
$$\mathbf{L} = I\omega \tag{8.16}$$

Torque in Terms of Angular Momentum:
$$\tau = \frac{\Delta \mathbf{L}}{\Delta t} \tag{8.17}$$

Conservation of Angular Momentum (with $\tau = 0$):
$$I\omega = I_o\omega_o \tag{8.18}$$

Exercises

8.1 Rigid Bodies, Translations, and Rotations

1 In pure rotational motion of a rigid body, (a) all the particles of the body have the same angular velocity, (b) all the particles of the body have the same tangential velocity, (c) the acceleration is always zero, (d) there are always two simultaneous axes of rotation.

2 The condition for rolling without slipping is (a) $a = r\omega^2$, (b) $v = r\omega$, (c) $F = ma$, (d) $a_c = v^2/r$.

3 Suppose someone in your physics class said that it is possible for a rigid body to have translational motion and rotational motion at the same time. Would you agree? Why?

4 For a rolling cylinder, what would happen if v were less than $R\omega$? Is it possible for v to be greater than $R\omega$? Explain.

5 ■ A wheel rolls uniformly on level ground without slipping. A piece of mud on the wheel flies off when at the 9-o'clock position (rear of wheel). Describe the subsequent motion of the mud.

6 ■ A rope goes over a circular pulley with a radius of 6.0 cm. If the pulley makes four revolutions without the rope slipping, what length of rope passes over the pulley?

7 ■ A circular disk with a radius of 0.15 m rolls without slipping on a level surface with an angular speed of 2.0 rad/s. (a) What is the linear speed of the center of mass of the disk? (b) What is the instantaneous tangential speed of the top of the disk?

8 ■ Show that $a = r\alpha$ is a condition for rolling without slipping.

9 ■■ A cylinder with a radius of 20 cm rolls with an angular speed of 2.4 rad/s. If the center of the cylinder moves 96 cm in 2.0 s, does the cylinder roll without slipping?

10 ■■ A ball with a radius of 15 cm rolls on a level surface with a translational speed of the center of mass of 0.25 m/s. What is the angular speed about the center of mass if the ball rolls without slipping?

11 ■■ A disk with a radius of 0.15 m rotates through 270° as it travels 0.71 m. Does the disk roll without slipping?

12 ■■ A friend tells you that he measured the translational speed of the center of mass of a sphere whose diameter is 0.182 m to be 0.410 m/s when the sphere was rolling without slipping with an angular speed of 4.98 rad/s. Is his measurement accurate?

13 ■■■ A cylinder with a diameter of 20 cm rolls on a level surface with an angular speed of 0.50 rad/s. If the cylinder experiences a uniform tangential acceleration of 0.018 m/s² without slipping until its angular speed is 1.25 rad/s, how many rotations does the cylinder go through during the acceleration time?

8.2 Torque, Equilibrium, and Stability

14 It is possible to have a net torque when (a) all forces act through the axis of rotation, (b) $\Sigma \mathbf{F}_i = 0$, (c) a body is in rotational equilibrium, (d) a body remains in unstable equilibrium.

15 If an object in unstable equilibrium is slightly displaced, (a) its potential energy will decrease, (b) the center of gravity is directly above the axis of rotation, (c) no gravitational work is done, (d) stable equilibrium follows.

16 (a) Why does a lower center of gravity give an object greater stability? (b) Does it make any difference how a tractor-trailer is loaded? [*Hint:* Think about an 18-wheeler going around a banked curve.]

17 Explain the balancing acts in •Fig. 8.26. Where are the centers of gravity?

•**FIGURE 8.26 Balancing acts**
See Exercise 17.

18 ■ How many different positions of stable equilibrium and unstable equilibrium are there for a cube? Consider each surface, edge, and corner to be a different position.

19 ■ The effects of torque and center of gravity on potential energy for slight displacements of bodies in stable and unstable equilibrium are discussed in the text. For a body in *neutral equilibrium*, a slight displacement produces no restoring or toppling torques, and the potential energy remains constant. Give some examples of bodies in neutral equilibrium.

20 ■ A student opens a glass door in the science building by pushing perpendicularly with a force of 50 N on a metal plate located 90 cm from the hinge side of the door. Another student opens the door by pushing perpendicularly at a distance of 60 cm from the hinge side. What is the magnitude of the force applied by the second student that produces the same torque as the first?

21 ■■ A 35-kg child sits on a uniform seesaw 2.0 m from the pivot point (or fulcrum). How far from the pivot point on the other side will her 30-kg playmate have to sit for the seesaw to be in equilibrium?

22 ■■ The pedals of a bicycle rotate in a circle with a diameter of 40 cm. What is the maximum torque a 55-kg rider can apply by putting all of her weight on a pedal?

23 ■■ A uniform meterstick pivoted at its center, as in Example 8.3, has a 100-g mass suspended at the 25.0-cm position. (a) At what position should a 75.0-g mass be suspended to put the system in equilibrium? (b) What mass would have to be suspended at the 90.0-cm position for the system to be in equilibrium?

24 ■■ Show that the balanced meterstick in Example 8.3 is in static rotational equilibrium about a horizontal axis through the zero end of the stick.

25 ■■ A force of 36 N is applied to a particle located 0.15 m from its axis of rotation. What is the magnitude of the torque about this axis if the angle between the direction of the applied force and the radius vector is (a) 60°, (b) 90°, and (c) 120°?

26 ■■ If $r = 0.25$ m in Exercise 25, what force would have to be applied to have the same torque in each case?

27 ■■ Telephone and electrical lines are intentionally allowed to sag between poles so that the tension will not be too great when something hits or sits on the line. Suppose that a line were stretched perfectly horizontally between two poles, which are 30 m apart. If a bird with a mass of 0.25 kg perched on the wire midway between the poles and the wire sagged 1.0 cm, what would the tension in the wire be?

28 ■■ In •Fig. 8.27, what is the force F_m supplied by the deltoid muscle so as to hold up the outstretched arm if the mass of the arm is 3.0 kg? (F_j is the joint force on the bone of the upper arm—the humerus.)

29 ■■ A variation of Russell traction (•Fig. 8.28) helps support the lower leg when in a cast. Suppose that the patient's leg and cast have a combined mass of 15.0 kg and m_1 is 4.50 kg. (a) What is the reaction force of the leg muscles to the traction? (b) What must m_2 be to keep the leg horizontal?

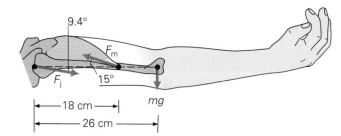

•FIGURE 8.27 Arm in static equilibrium
See Exercise 28.

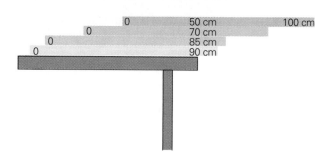

•FIGURE 8.29 Will they fall off?
See Exercise 33.

•FIGURE 8.28 Static traction
See Exercise 29.

30 ■■ A uniform meterstick weighing 5.0 N is pivoted so it can rotate about a horizontal axis through one end. If a 0.15-kg mass is suspended 75 cm from the pivoted end, what force must be applied to the free end of the meterstick to maintain it in a horizontal position?

31 ■■ A 10.0-kg solid uniform cube with 0.250-m sides rests on a level surface. What is the minimum amount of work necessary to put the cube in unstable equilibrium?

32 ■■ (a) How many uniform, identical textbooks with a width of 25 cm can be stacked on top of each other on a level surface without the stack falling over if each successive book is displaced 3.0 cm in the width direction relative to the next lower book? (b) If the books are 5.0 cm thick, what will be the height of the center of mass of the stack above the level surface?

33 ■■ If four metersticks were stacked on a table with 10-cm, 15-cm, 30-cm, and 50-cm segments over the edge, as shown in •Fig. 8.29, would the metersticks remain on the table?

34 ■■■ A uniform L-shaped object has a leg that is 0.50 m long and a shorter leg that is 0.30 m long (both measured along the outside edge). The L has a width of 6.0 cm and a thickness of 2.0 cm. Locate the center of gravity. [*Hint:* Divide the L into two rectangles and locate the center of gravity of each. Take long leg along *y* axis and short leg along *x* axis. Express mass in terms of density.]

35 ■■■ A 70-kg painter paints the side of a house while standing on a long board resting on a scaffold, as shown in •Fig. 8.30. If the board has a mass of 15 kg, how close to the end can the painter stand without tipping the board over?

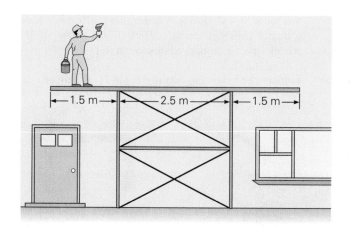

•FIGURE 8.30 Not too far!
See Exercises 35 and 36.

36 ■■■ Suppose that the board in Fig. 8.30 were suspended from vertical ropes attached to each end instead of resting on scaffolding. If the painter stood 1.5 m from one end of the board, what would the tensions in the ropes be? (See Exercise 35 for additional data.)

37 ■■■ An artist wishes to construct a birds-and-bees mobile, as shown in •Fig. 8.31. If the bee on the lower left has a mass of 0.10 kg and each vertical support string has a length of 30 cm, what are the masses of the other birds and bees? (Neglect the masses of the bars and strings.)

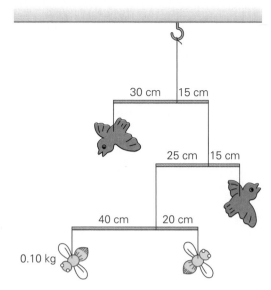

•**FIGURE 8.31 Birds and bees**
See Exercise 37.

8.3 Rotational Dynamics

38 The moment of inertia of a rigid body (a) depends on the axis of rotation, (b) cannot be zero, (c) depends on mass distribution, (d) all of the preceding.

39 Which of the following best describes the physical quantity called torque: (a) rotational analog of force, (b) energy due to rotation, (c) rate of change of linear momentum, (d) force that is tangent to a circle?

40 (a) Is the moment of inertia of a rigid body related in any way to the center of mass? Explain. (b) Can a moment of inertia have a negative value? If so, explain what this would mean.

41 Why does the moment of inertia of a rigid body have different values for different axes of rotation? What does this mean physically?

42 When a hard-boiled egg laying on a table is given a quick torque (spin), it will rise up and spin on one end like a top. A raw egg will not. Why the difference?

43 When playing ball, it is common to have a child "choke up" (hold closer to the bigger end) of the bat. How does this affect the moment of inertia?

44 A child stands near the edge of a rotating play ground merry-go-round (the hand-driven type). He then starts to walk toward the center of the merry-go-round. This results in a dangerous situation. Why?

45 The string of a stationary yo-yo is pulled as shown in •Fig. 8.32. (a) Which way does the yo-yo roll (without slipping), and why? [*Hint:* It may not be as you expect. Think about torque and axis of rotation. If you don't have a yo-yo, a spool of thread also works.] (b) Explain

what happens to the torque due to *F* if the string is pulled at a greater and greater angle to the horizontal.

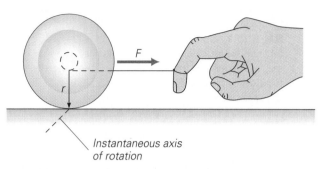

•**FIGURE 8.32 Pulling the yo-yo's string**
See Exercise 45.

46 (a) Why is it easier to balance a meterstick vertically on your finger than a pencil? (b) A softball, a volleyball, and a basketball are released at the same time from the top of an inclined plane. Give the results of the race down the plane.

47 ■ The release of vast amounts of carbon dioxide may give rise to an increase in the Earth's average temperature through the so-called greenhouse effect and cause a melting of the polar ice caps. If this occurred and the ocean level rose substantially, what effect would this have on the Earth's rotation and length of the day?

48 ■ A fixed 0.15-kg solid-disk pulley with a radius of 0.050 m is acted on by a net torque of 6.4 m·N. What is the angular acceleration of the pulley?

49 ■■ A light meterstick is loaded with masses of 2.0 kg and 4.0 kg at the 30-cm and 75-cm positions, respectively. (a) What is the moment of inertia about an axis through the 0-cm end of the meterstick? (b) What is the moment of inertia about an axis through the center of mass of the system? (c) Use the parallel axis theorem to find the moment of inertia about the axis through the 0-cm end of the stick, and compare the result with the result of part (a).

50 ■■ Using the parallel axis theorem, show that the moment of inertia for a thin rectangular sheet with an axis of rotation along one side is $I = \frac{1}{3}ML^2$ (see Fig. 8.15k).

51 ■■ Four 0.25-kg point masses connected by rods of negligible mass form a square rigid body with 0.10-m sides. Find the moment of inertia of the body about each of the following axes: (a) an axis perpendicular to the plane of the square and through its center, (b) an axis perpendicular to the plane and through one of the masses, (c) an axis in the plane and along one side through two of the masses, and (d) an axis in the plane running diagonally through two of the masses.

52 ■■ If the Earth rotated about a N–S axis tangent to the Earth at the Equator, how much more rotational inertia would it have?

53 ■■ A Ferris wheel accelerates from rest to an angular speed of 1.5 rad/s in 12 s. Approximating the Ferris wheel as a circular disk with a radius of 30 m and a mass of 2.0 metric tons, what is the net torque on the wheel?

54 ■■ A 15-kg uniform sphere with a radius of 15 cm rotates about an axis tangent to its surface at a rate of 3.0 rad/s. A constant torque of 10 m·N then increases the rotational speed to 7.5 rad/s. Through what angle does the sphere rotate while accelerating?

55 ■■ A disk-shaped 0.25-kg machine part with a radius of 6.0 cm rotates eccentrically about an axis that is normal to its flat surface and located ⅔ of the distance from the center to the circumference. (a) What torque is required to rotate the part about this off-center axis with an angular acceleration of 2.0 rad/s²? (b) How much greater is this than the torque required to rotate the part with the same angular acceleration about a parallel axis through the center?

56 ■■ Two masses are suspended from a pulley as shown in •Fig. 8.33 (the Atwood machine revisited, Chapter 4). The pulley itself has a mass of 0.20 kg, a radius of 0.15 m, and a constant torque of 0.35 m·N due to the friction between the rotating pulley and its axle. What is the magnitude of the acceleration of the suspended masses if $m_1 = 0.40$ kg and $m_2 = 0.80$ kg? (Neglect the mass of the string.)

57 ■■ A thin 1.0-m rod pivoted at one end falls (rotates) frictionlessly from a vertical position, starting from rest. What is the angular speed of the rod when it is horizontal? [*Hint:* Consider the center of mass and use the conservation of mechanical energy.]

58 ■■ For the system shown in •Fig. 8.34, $m_1 = 8.0$ kg, $m_2 = 3.0$ kg, $\theta = 30°$, and the radius and mass of the pulley are 0.10 m and 0.10 kg, respectively. (a) What is the acceleration of the masses? (Neglect friction and the string's mass.) (b) If the pulley has a constant frictional torque of 0.050 m·N when the system is in motion, what is the acceleration? [*Hint:* Isolate the forces. The tensions in the strings are different. Why?]

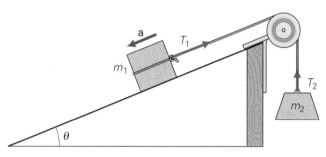

•**FIGURE 8.34   Inclined plane and pulley**
See Exercise 58.

59 ■■ What is the magnitude of the tangential force that must be applied to the rim of a 2.0-kg wheel that is disk-shaped and has a radius of 0.75 m in order to give it an angular acceleration of 4.8 rad/s²? (Neglect friction.)

60 ■■ A man starts his lawn mower by applying a constant tangential force of 150 N to the flywheel using a starter rope. The flywheel is a cylindrical ring with a mass of 0.30 kg and a diameter of 18 cm. What is the angular speed of the wheel after it has turned through 1 revolution? (Neglect friction and motor compression.)

61 ■■ A uniform sphere and a uniform cylinder with the same mass and radius roll side by side on a level surface without slipping and at the same velocity. If the sphere and the cylinder approach an inclined plane and roll up it without slipping, will they be at the same height on the plane when they come to a stop? If not, what will be the percent height difference?

62 ■■■ A uniform hoop rolls without slipping down a plane with an incline of 15° above the horizontal. What is the acceleration of the hoop's center of mass?

63 ■■■ For the ball rolling down the incline in Fig. 8.20, what would be the maximum angle of incline in terms of the coefficient of static friction for it to roll without slipping?

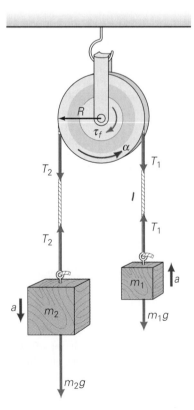

•**FIGURE 8.33   The Atwood machine revisited**
See Exercise 56.

64 ■■■ A uniform cylinder with a mass of 2.0 kg and a radius of 0.15 m is suspended by two strings wrapped around it (•Fig. 8.35). As the cylinder descends, the strings unwind from it. What is the acceleration of the center of mass of the cylinder? (Neglect the mass of the string.)

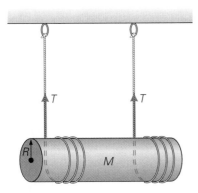

•FIGURE 8.35 **Unwinding**
See Exercise 64.

8.4 Rotational Work and Kinetic Energy

65 With $W = \tau\theta$, the unit(s) of rotational work is (a) watt, (b) N·m, (c) kg·rad/s², (d) N·rad.

66 The rotational kinetic energy of a body is equal to (a) $\frac{1}{2}Iv^2$, (b) I/τ, (c) $L\alpha$, (d) $L^2/2I$.

67 ■ A constant retarding torque of 20 m·N stops a rolling wheel whose diameter is 0.80 m in a distance of 15 m. How much work is done by the torque?

68 ■ A person opens a door by applying a force of 15 N perpendicular to it at a distance of 0.90 m from the hinges. The door is pushed wide open (to 90°) in 2.0 s. (a) How much work was done on the average? (b) What was the average power delivered?

69 ■ Use the conservation of mechanical energy to find the linear speed of the descending mass ($m = 1.0$ kg) of Fig. 8.18 after it has descended a vertical distance of 2.0 m from rest. (For the pulley, $M = 0.30$ kg and $R = 0.15$ m. Neglect friction and mass of the string.)

70 ■■ A sphere with a radius of 15 cm rolls on a level surface with a constant angular speed of 8.0 rad/s. To what height on a 30° inclined plane will the sphere roll before coming to rest? (Neglect frictional losses.)

71 ■■ A flywheel with a moment of inertia of 4.50 × 10² kg·m² rotates with a speed of 7500 rpm. How much work is required to bring it to rest? (b) If this is done uniformly in 1.5 min, how much power is expended?

72 ■■ A cylindrical hoop, a cylinder, and a sphere of equal radius and mass are released at the same time from the top of an inclined plane. Using conservation of mechanical energy, show that the sphere always gets to the bottom of the incline first with the faster speed and the hoop always arrives last with the slowest speed.

73 ■■ For the following objects, which are rolling without slipping, determine the rotational kinetic energy about the center of mass as a percentage of the total instantaneous kinetic energy: (a) solid sphere, (b) a thin spherical shell, (c) a thin cylindrical shell.

74 ■■ Which is greater and by what factor: the Earth's kinetic energy due to its rotation or its kinetic energy due to its revolution about the Sun? Assume a circular orbit. (Before computing, which do you think is greater and guess by how much.)

75 ■■ A 1.5-hp motor drives a machine disk having a mass of 4.5 kg and a diameter of 0.40 m from rest to a speed of 750 rpm. (a) What is the net work done by the motor? (b) How long does it take to reach the operating speed?

76 ■■ A solid 5.0-kg sphere with a diameter of 0.60 m starts from rest at a height of 1.2 m above the base of an inclined plane and rolls down under the influence of gravity. What is the linear speed of the sphere's center of mass just as it leaves the incline and rolls onto a horizontal surface? (Neglect friction.)

77 ■■ A 0.050-kg phonograph record with a radius of 0.15 m drops onto a turntable and is soon rotating at 33⅓ rpm. How much work must be supplied to get the record to rotate at this speed, and what supplies it?

78 ■■■ A pencil 20 cm long is initially held so that it stands vertically on its sharpened point on a table by a person with a finger on the eraser end. When the finger is removed, the pencil falls over onto the table. Assuming that the point does not slip, at what speed is the eraser end of the pencil moving when it strikes the horizontal surface? Consider the pencil to be a thin, uniform rod.

79 ■■■ A steel ball rolls down an incline into a loop-the-loop of radius R (•Fig. 8.36a). (a) What is the minimum speed the ball must have at the top of the loop in order to stay on the track? (b) At what vertical height (h) on the incline, in terms of the radius of the loop, must the ball be released in order for it to have the required minimum speed at the top of the loop? (Neglect frictional losses.) (c) •Fig. 8.36b shows a loop-the-loop for a roller coaster. What are the sensations of the riders if the roller coaster has the minimum speed or a greater speed at the top of the loop? [*Hint:* In case the speed is below the minimum, seat and shoulder straps hold the riders in.]

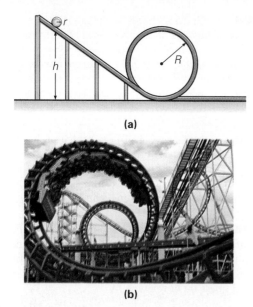

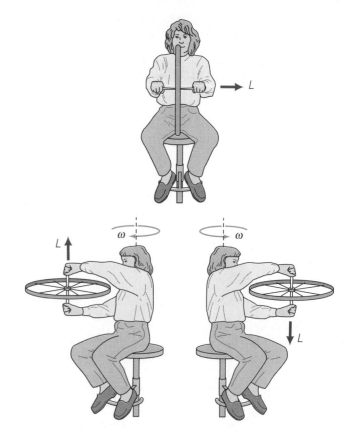

FIGURE 8.36 Loop-the-loop and rotational speed
See Exercise 79.

8.5 Angular Momentum

80 The units of angular momentum are (a) N·m, (b) kg·m/s², (c) m·N·s, (d) J·m.

81 Angular momentum is conserved when (a) a couple exists, (b) a body is in translational equilibrium, (c) **L** does not change with time, (d) none of the preceding.

82 A classroom demonstration is illustrated in •Fig. 8.37. A person on a rotating stool holds a rotating bicycle wheel (by handles attached to the wheel). When the wheel is held horizontally, she rotates one way (clockwise as viewed from above). When the wheel is turned over, she rotates in the opposite direction. Explain why this occurs. [*Hint:* Consider angular momentum vectors.]

83 An ice skater goes into a fast spin by tucking her arms in, and a ballerina does a pirouette with her arms over her head. Why are the arms put into these positions in each case?

84 Cats always seem to land on their feet when they fall, even if held upside down and dropped (•Fig. 8.38). While a cat is falling, there is no external torque and its center of mass falls as a particle. How can cats turn themselves over while falling?

85 For a non-sleeping top or gyroscope, the precessional motion is around the vertical axis. Explain why this is, and whether the precession is clockwise or counterclockwise (as viewed from above).

86 (a) Large helicopters have two overhead rotors that rotate in opposite directions (•Fig. 8.39), and small heli-

FIGURE 8.37 A double rotation
See Exercise 82.

copters have a single overhead rotor and an "antitorque" tail rotor. Explain the reason(s) for the differences. (b) Unlike helicopters, a single-engine airplane has only one propeller, or set of rotor blades. Discuss the stability of such an aircraft in light of part (a).

87 ■ A 0.010-kg ball in an arrangement like that shown in Fig. 8.22 swings initially in a circle having a 15-cm radius at an angular speed of 3π rad/s. If the force on the string is reduced enough that the ball swings in a circle with a radius of 25 cm, what is its angular speed?

88 ■ What is the angular momentum of a 2.0-g particle moving in a horizontal circle of 15-cm radius counterclockwise as viewed from above and with an angular speed of 6π rad/s? (Give magnitude and direction.)

89 ■■ The Earth's orbit about the Sun is slightly elliptical. The points on the orbit at maximum and minimum distances from the Sun are called the aphelion and the perihelion, respectively. (a) Show that the Earth's orbital speed is greater at the perihelion. (b) If $r_a = 1.52 \times 10^8$ km and $r_p = 1.47 \times 10^8$ km, how many times greater is the Earth's speed at the perihelion than its speed at the aphelion?

90 ■■ Compute the ratio of the Earth's orbital angular momentum and its rotational angular momentum. Are these momenta in the same direction?

•**FIGURE 8.39 Different rotors**
See Exercise 86.

•**FIGURE 8.38 One down, eight to go**
See Exercise 84.

91 ■■ The period of the Moon's rotation is the same as the period of its revolution: $29\frac{1}{2}$ days. What is the angular momentum for each motion? (Because the periods are equal, we see only one side of the Moon from Earth.)

92 ■■ A rotating spherical star in part of its life cycle expands to six times its normal volume. Assuming the mass to remain constant, how is the period of rotation affected?

93 ■■ A person on a rotating stool with arms outstretched as in Fig. 8.23a rotates at a speed of 0.75 rev/s. On bring-ing in his arms (Fig. 8.23b), the speed increases to 1.25 rev/s. By what factor does the moment of inertia change?

94 ■■ A skater has a moment of inertia of 100 kg·m² when his arms are outstretched and a moment of inertia of 75 kg·m² when his arms are tucked in close to his chest. If he starts to spin at an angular speed of 2.0 rps with his arms outstretched, what will his angular speed be when they are tucked in?

95 ■■ An ice skater doing a toe spin with outstretched arms has an angular speed of 4.0 rad/s. The skater then tucks in her arms, decreasing her moment of inertia by 7.5%. (a) What is the resulting angular speed? (b) By what factor does the skater's kinetic energy change? (Neglect any frictional effects.)

96 ■■■ In an ice show, two 70-kg skaters, both traveling at 8.0 m/s, approach each other along straight-line par-allel paths that are separated by a distance of 4.0 m. When directly opposite each other, the skaters grab the ends of a light rod 4.0 m in length. What is the initial rotational speed of the joined skaters? [*Hint:* Use the conservation of angular momentum.]

97 ■■■ A comet approaches the Sun as illustrated in •Fig. 8.40 and is deflected by the Sun's gravitational at-traction. This may be considered a collision, and b is called the impact parameter. Find the distance of clos-est approach (d) in terms of the impact parameter and the velocities (v_o at large distances and v_{max} at closest approach). Consider the radius of the Sun to be negli-gible compared to d. (As the figure shows, the tail of a comet always "points" away from the Sun.)

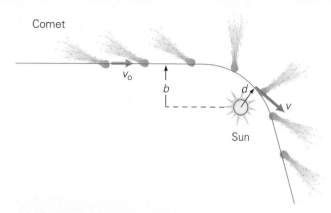

• FIGURE 8.40 A comet "collision"
See Exercise 97.

98 ■■■ A 0.50-kg kitten stands on the edge of a lazy Susan (a kind of turntable) that has a mass of 1.5 kg and
a radius of 0.30 m. Assume that the lazy Susan has frictionless bearings and is initially at rest. (a) What will
happen if the kitten starts walking around the edge of
the lazy Susan? (b) If the kitten walks at a speed of
0.25 m/s relative to the ground, what will the angular
speed of the lazy Susan be? (c) When the kitten has
walked completely around the edge and is back at its
starting point, will that point be above the same point
on the ground as it was at the start? If not, where is
the kitten relative to the starting ground point? (Speculate on what might happen if everyone on Earth suddenly started to run eastward. What effect might this
have on the length of a day?)

Additional Exercises

99 (a) For a bicycle rider making a circular turn on a level
surface as in Fig. 1 of the first Insight feature, show
that the proper angle of lean is given by $\theta = \tan^{-1} \mu_s$,
where μ_s is the coefficient of static friction between the
tires and the road. (b) If $\mu_s = 0.57$, what is the angle
of lean? (Does the angle increase with an increasing
coefficient? Explain.)

100 A solid 0.75-kg ball rolls without slipping on a level
surface. If the ball rolls in a straight line at a speed of
0.50 m/s, what is its kinetic energy?

101 Show that if two rigid bodies with the same axis of rotation are stuck together, the moment of inertia of the
combined rigid body about the common axis is equal
to the sum of the individual moments, or $I = I_1 + I_2$.

102 A spring is twisted 60° about its linear axis (axis of
symmetry) by a torque of 100 m·N. How much torsional, or rotational, energy is stored in the spring?
[Hint: Let the restoring torque of the spring be $\tau = k\theta$,
the rotational analog of $F = kx$ (Hooke's law—see
Chapter 5) with the minus sign omitted.]

103 Prove that a thin hoop or ring starting from rest will
roll more slowly down an inclined plane than an annular cylinder. (Remember that masses and radii do
not have to be taken into account.)

104 A bicycle wheel has a diameter of 0.80 m and a mass
of 2.0 kg. (a) If the bicycle travels with a linear speed
of 1.5 m/s, what is the angular speed of the wheel?
(b) What is the angular momentum of the wheel?
(Consider the wheel to be a thin ring.)

105 A piece of machinery with a moment of inertia of 2.6
kg·m² rotates with a constant angular speed of 4.0
rad/s when experiencing a frictional torque of 0.56 m·
N. What is the net torque acting on the piece?

106 The location of a person's center of gravity relative to
his or her height can be found using the arrangement
shown in •Fig. 8.41. The scales are initially adjusted
to read zero with the board alone. Locate the center of
gravity of the person relative to the horizontal dimension. Would you expect the location of the center
of gravity in other dimensions to be exactly at the midway points? Explain.

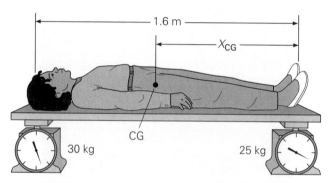

• FIGURE 8.41 Locating the center of gravity
See Exercise 106.

107 Using bricks identical to those in Example 8.5, you are
asked to create a stack of nine bricks (maximum) without its toppling, with each added brick displaced an
equal distance horizontally in the same direction.
(a) What is the maximum displacement that will allow this? (b) What is the height of the center of mass
of the stack if the uniform bricks are 8.0 cm thick?

108 The radius of gyration (k) is defined as the distance from
an axis of rotation at which all the mass (M) of an object would have to be concentrated to give the same
moment of inertia as would be calculated with the actual mass distribution, that is, $I = Mk^2$. This represents
an object as a rotating particle of mass M. For the axes
shown in Fig. 8.15, determine the radius of gyration
for (a) a uniform sphere, (b) a uniform cylinder, and
(c) a particle.

109   A 0.25-kg pulley has a diameter of 14 cm. Two masses, 0.10 kg and 0.15 kg, are suspended from the ends of a light string passing over the pulley. (a) Neglecting axle friction, what will the acceleration be when the masses are moving under the influence of gravity? (b) What is the magnitude of the combined kinetic frictional torque (due to both friction of the pulley bearings and friction between the string and the pulley) that would keep the masses moving at a constant speed? [*Hint:* Isolate the forces. The tensions in the strings are different. Why?]

110   A mass is suspended by two cords as shown in •Fig. 8.42. What are the tensions in the cords?

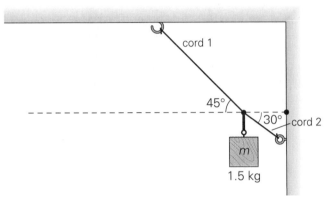

•**FIGURE 8.42   Strung-out equilibrium**
See Exercises 110 and 111.

111   If the cord attached to the vertical wall in Fig. 8.42 were horizontal (instead of at a 30° angle), what would the tensions in the cords be?

112   The outer wheels of an oversize tractor-trailer rig are 3.66 m apart. When the trailer is loaded, its center of gravity is equidistant from its sides and 3.58 m above the road surface. What banking angle of road will the parked trailer be in unstable equilibrium?

113   A thin rectangular plate has dimensions of 30 cm by 40 cm and a mass of 0.50 kg. For which axis of rotation does the plate have the greater rotational inertia, an axis tangent to a smaller side at its midpoint or an axis tangent to a larger side at its midpoint? How many times more? (Both axes are perpendicular to the plate surface.)

114   Show that $r_\perp F = rF_\perp$ for a torque.

9 Solids and Fluids

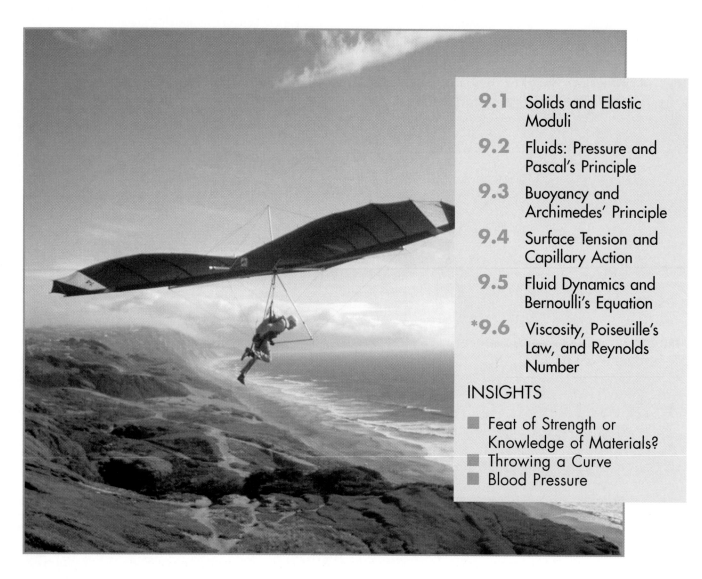

L ike the person in the photograph above, we exist at the intersection of three realms. We walk on the solid surface of the Earth, and in our daily lives use solid objects of all sorts, from scissors to computers. But we are surrounded by fluids—liquids and gases—on which we are completely dependent. Without the oxygen in the air that we breathe, we could not live for more than a few minutes; without the water that we drink, we could survive only for a few days at most. Indeed, we ourselves are not nearly as solid as we like to think. By far the most abundant substance in our bodies is water, and it is in the watery environment of our cells that all the chemical processes on which life depends take place.

In the pages that follow you will explore the three common phases of matter: solids, liquids, and gases. You'll learn how their properties differ, and why. This chapter pays particular attention to the behavior of fluids, shedding light on such questions as why icebergs and ocean liners float, why paper towels absorb spills, and what the 10W–40 on a can of motor oil means. Along the way, you'll also discover why the airborne person in the photo can neither float like a balloon nor fly like a hummingbird, yet can, with the aid of a suitably shaped piece of plastic, soar like an eagle.

On the basis of general physical distinctions, matter is commonly divided into three phases: solid, liquid, and gas. A solid has a definite shape and volume. A liquid has a definite volume, but assumes the shape of its container. A gas takes on the shape and volume of its container. Solids and liquids are sometimes called condensed matter. In this chapter, however, we will utilize a different classification scheme and consider matter in terms of solids and fluids. Liquids and gases are referred to collectively as fluids. A **fluid** is a substance that can flow, so liquids and gases qualify, but solids do not.

A simplistic description of solids is that they are made up of particles called atoms that are held rigidly together by interatomic forces. This concept of an ideal rigid body was used in Chapter 8 to describe rotational motion. Real solid bodies are not absolutely rigid and can be elastically deformed by external forces. Elasticity usually brings to mind a rubber band or spring that will resume its original dimensions even after being greatly deformed. In fact, all materials are elastic to some degree, even very hard steel. But, as you will learn, there's a limit to such deformation—an elastic limit.

Fluids, on the other hand, have little or no elastic response to force—a force merely causes an unconfined fluid to flow. Fluids are important in everyday life. You are surrounded by a fluid (air) and, for the most part, you are composed of fluids. Most of this chapter will focus on the behavior of fluids.

Because of their fluidity, liquids and gases have many properties in common, and it is convenient to study them together. Of course, there are some important differences as well: for example, liquids are not very compressible, whereas gases are easily compressed, as we will shortly see.

9.1 Solids and Elastic Moduli

Objectives: To be able to (a) distinguish between stress and strain, and (b) use elastic moduli to compute dimensional changes.

As stated previously, all solid materials are elastic to some degree. That is, a body that is slightly deformed by an applied force will return to its original dimensions or shape when the force is removed. The deformation may not be noticeable for many materials, but it's there.

You may be able to visualize why materials are elastic if you think in terms of the simplistic model of a solid in •Fig. 9.1. The atoms of the solid substance are imagined to be held together by springs. The elasticity of the springs represents the resilient nature of the interatomic forces. The springs resist permanent deformation, as do the forces between atoms.

The elastic properties of solids are commonly discussed in terms of stress and strain. **Stress** is a measure of the force causing a deformation. **Strain** is a relative measure of the deformation a stress causes. Quantitatively, *stress is the applied force per unit cross-sectional area:*

$$\text{stress} = \frac{F}{A} \tag{9.1}$$

SI unit of stress: N/m^2

Here F is the magnitude of the applied force normal (perpendicular) to the cross-sectional area. The equation shows that the SI units for stress are newtons per square meter (N/m^2).

As illustrated in •Fig. 9.2, a force applied to the ends of a rod gives rise to either an elongating tension, or *tensile stress*, or a *compressional stress*, depending on the direction of the force. In both these cases, the *tensile strain* is the ratio of the change in length to the original length:

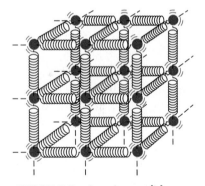

•**FIGURE 9.1 A springy solid**
The elastic nature of interatomic forces is indicated by simplistically representing them as springs, which similarly resist deformation.

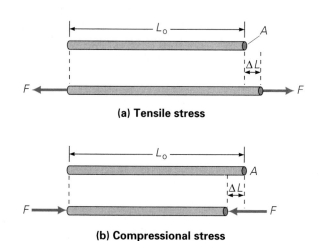

• FIGURE 9.2 **Tensile and compressional stress and strain** Tensile and compressional stresses are due to forces applied normally to the surface area of the ends of bodies, as shown here. **(a)** A tension, or tensile stress, tends to increase the length of an object. **(b)** A compressional stress tends to shorten the length.

(a) Tensile stress

(b) Compressional stress

$$\text{strain} = \frac{\text{change in length}}{\text{original length}} = \frac{\Delta L}{L_o} \qquad (9.2)$$

Strain is a unitless quantity

where $\Delta L = L - L_o$. Note that strain is a unitless quantity (length/length). It is the *fractional change* in length. For example, if the strain is 0.05, the material has changed in length by 5% of its original length.

As might be expected, the strain is proportional to the applied stress; that is, strain ∝ stress. For relatively small stresses, this is a direct proportion. The constant of proportionality, which depends on the nature of the material, is called the **elastic modulus**. Thus,

$$\text{stress} = \text{elastic modulus} \times \text{strain}$$

or

Elastic modulus

$$\text{elastic modulus} = \frac{\text{stress}}{\text{strain}} \qquad (9.3)$$

SI unit of elastic modulus: N/m^2

That is, the elastic modulus is the stress divided by the strain, or the ratio of stress to strain. The elastic modulus has the same units as stress. Why?

Three general types of elastic moduli (plural of modulus) are associated with stresses that produce changes in length, shape, or volume. These are called Young's modulus, shear modulus, and bulk modulus, respectively.

Change in Length: Young's Modulus

•Fig. 9.3 is a graph of the tensile stress versus the strain for a typical metal rod. The curve is a straight line up to a point called the *proportional limit*. Beyond this point, the strain begins to increase more rapidly to another critical point called the **elastic limit**. If the tension is removed at this point, the material will return to its original length. If the tension is applied beyond the elastic limit and then removed, the material will recover somewhat but will retain some permanent deformation.

The straight-line part of the graph shows a direct proportionality between stress and strain. This relationship, first formalized by Robert Hooke in 1678, is now known as *Hooke's law*. (It is the same general relationship as that given for a spring in Section 5.2—see Eq. 5.3.) The elastic modulus for a tension or a compression is called **Young's modulus** (Y):

Thomas Young (1773–1829) was a British physicist who investigated the mechanical properties of materials and optical phenomena.

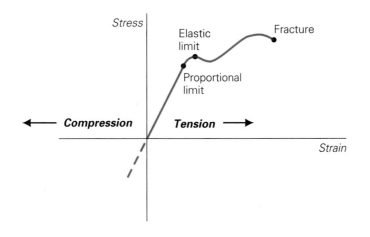

•FIGURE 9.3 **Stress versus strain**
A plot of stress versus strain for a
typical metal rod is a straight line
up to the proportional limit. Then
elastic deformation continues until
the elastic limit is reached. Beyond
that, the rod will be permanently
deformed and will eventually
fracture or break.

$$\frac{F}{A} = Y\left(\frac{\Delta L}{L_{o}}\right) \quad \text{or} \quad Y = \frac{F/A}{\Delta L/L_{o}} \qquad (9.4) \qquad \textbf{Young's modulus}$$

$$\underset{\text{stress}}{} \qquad \underset{\text{strain}}{}$$

SI unit of Young's modulus: N/m^2

The units for Young's modulus are the same as for stress (N/m^2) since the ratio of
lengths is unitless. Some typical values of Young's modulus are given in Table 9.1.

To obtain a conceptual or physical understanding of Young's modulus, let's
solve Eq. 9.4 for ΔL:

$$\Delta L = \left(\frac{FL_{o}}{A}\right)\frac{1}{Y} \quad \text{or} \quad \Delta L \propto \frac{1}{Y}$$

Hence we see that the larger the modulus of a material, the smaller its relative
change in length (with other parameters being equal).

TABLE 9.1	Elastic Moduli for Various Materials (in N/m^2)		
Substance	*Young's Modulus (Y)*	*Shear Modulus (S)*	*Bulk Modulus (B)*
Solids			
Aluminum	7.0×10^{10}	2.5×10^{10}	7.0×10^{10}
Brass	9.0×10^{10}	3.5×10^{10}	7.5×10^{10}
Copper	11×10^{10}	3.8×10^{10}	12×10^{10}
Glass	5.7×10^{10}	2.4×10^{10}	4.0×10^{10}
Iron	15×10^{10}	6.0×10^{10}	12×10^{10}
Steel	20×10^{10}	8.2×10^{10}	15×10^{10}
Liquids			
Alcohol, ethyl			1.0×10^{9}
Glycerin			4.5×10^{9}
Mercury			26×10^{9}
Water			2.2×10^{9}

EXAMPLE 9.1 ■ STRETCHING A WIRE: TENSILE STRESS AND YOUNG'S MODULUS

In a lab experiment, a total mass of 16 kg is suspended from a steel wire with a diameter of 0.10 cm. By what percent does the length of the wire increase?

Solution. Listing the data, along with Young's modulus for steel from Table 9.1, we
have

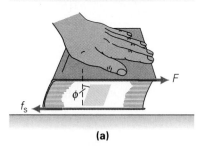

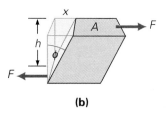

A shear stress is produced when a force is applied tangentially to a surface area. The strain is measured in terms of the relative displacement of the object's faces or the shear angle ϕ.

Given: $d = 0.10$ cm, so $r = 0.050$ cm $= 5.0 \times 10^{-4}$ m

$Y_{steel} = 20 \times 10^{10}$ N/m² (from Table 9.1)

Find: % increase in length

First we must recognize that $\Delta L/L_o$ is the *fractional* increase in length (which can readily be changed to a percentage). For example, if you had a spring with a length of 10 cm (L_o) and you stretched it 1.0 cm (ΔL), then $\Delta L/L_o = 1.0$ cm/10 cm $= 0.10$, and the length of the spring was increased by 10%.

The percent increase in length for the steel wire would be much less, and this can be found from Eq. 9.4:

$$F = mg = Y\left(\frac{\Delta L}{L_o}\right)A = Y\left(\frac{\Delta L}{L_o}\right)\pi r^2$$

Solving for $\Delta L/L_o$,

$$\frac{\Delta L}{L_o} = \frac{mg}{Y\pi r^2} = \frac{(16\,\text{kg})(9.8\,\text{m/s}^2)}{(20 \times 10^{10}\,\text{N/m}^2)\pi(5.0 \times 10^{-4}\text{m})^2} = 0.10 \times 10^{-2}\,(\times 100\%) = 0.10\%$$

Hence, the steel wire is stretched one-tenth of one percent.

Follow-up Exercise. The tensile or ultimate strength is the maximum stress a material can support before breaking or fracturing. If the tensile strength of the steel in the wire in this example is 4.9×10^8 N/m², how much mass could be suspended before it would break?

Change in Shape: Shear Modulus

Another way an elastic body can be deformed is by a *shear stress*. In this case, the deformation is due to an applied force that is tangential to the surface area (•Fig. 9.4). A change in shape results without a change in volume. The *shear strain* is given by x/h, where x is the relative displacement of the faces and h is the distance between them.

The shear strain is sometimes defined in terms of the *shear angle* (ϕ). As can be seen from the figure, $\tan\phi = x/h$. But the shear angle is usually quite small, so a good approximation is $\tan\phi \approx \phi \approx x/h$, where the angle ϕ is in radians. (If $\phi = 10°$, for example, there is only 1.0% difference between ϕ and $\tan\phi$.) The **shear modulus** (sometimes called the modulus of rigidity) is then

$$S = \frac{F/A}{x/h} \approx \frac{F/A}{\phi} \qquad (9.5)$$

SI unit of shear modulus: N/m²

As you may note in Table 9.1, the shear modulus is generally less than Young's modulus; S is approximately $Y/3$ for many materials, which indicates that there is a greater response to a shear stress than to a tensile stress. Also, note the inverse relationship of $\phi \propto 1/S$, similar to that pointed out previously for Young's modulus.

A shear stress may also be of the torsional type, resulting from the twisting action of a torque. For example, a torsional shear stress may shear off the head of a bolt that is being tightened.

Note in Table 9.1 that liquids do not have shear moduli (or Young's moduli). A shear stress cannot be effectively applied to a liquid or a gas. It is often said that *fluids cannot support a shear*. Why?

Change in Volume: Bulk Modulus

Suppose that a force directed inward acts over the entire surface of a body (•Fig. 9.5). Such a *volume stress* is often applied by pressure transmitted by a fluid (Section 9.2). An elastic material will be compressed by a volume stress; that is, it will show a change in volume but not in general shape in response to a pressure

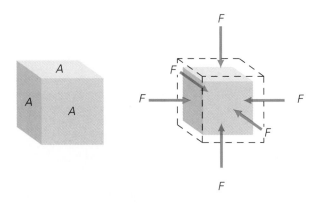

•FIGURE 9.5 **Volume stress and strain**
A volume stress is applied when a normal force acts over the entire surface area, as shown here for a cube of material. This most commonly occurs for gases. The resulting strain is a change in volume.

change. The change in pressure is equal to the volume stress, or $\Delta p = F/A$. The *volume strain* is the ratio of the volume change (ΔV) to the original volume (V_o). The **bulk modulus** (B) is then

$$B = \frac{F/A}{-\Delta V/V_o} = -\frac{\Delta p}{\Delta V/V_o} \qquad (9.6) \qquad \text{Bulk modulus}$$

SI unit of bulk modulus: N/m^2

The minus sign is introduced to make B a positive quantity, since $\Delta V = V - V_o$ is negative for an increase in external pressure. And, similar to the previous moduli relationships, $\Delta V \propto 1/B$.

Bulk moduli are listed for solids *and* liquids in Table 9.1. Gases have bulk moduli too, since they can be compressed. For a gas, it is common to talk about the reciprocal of the bulk modulus, which is called the *compressibility* (k):

$$k = \frac{1}{B} \qquad (9.7) \qquad \text{Compressibility}$$

The change in volume ΔV is thus directly proportional to the compressibility k.

Solids and liquids are relatively incompressible and have small values of compressibility. Gases, on the other hand, are easily compressed and have large compressibilities, which vary with pressure and temperature.

EXAMPLE 9.2 ■ COMPRESSING A LIQUID: VOLUME STRESS AND BULK MODULUS

By how much should the pressure on a liter of water be changed to compress it by 0.10%?

Solution. Listing the data, we have

Given: $\Delta V/V_o = 0.0010$ *Find:* $\Delta p = F/A$
        $V_o = 1.0 \text{ L} = 1000 \text{ cm}^3$
        $B_{H_2O} = 2.2 \times 10^9 \text{ N/m}^2$ (from Table 9.1)

Note that $-\Delta V/V_o$ is the *fractional* change in the volume, and for a compression of 0.10%, this is 0.0010 (no units, a ratio of volumes). With $V_o = 1000 \text{ cm}^3$, the volume reduction is

$$-\Delta V = (0.0010)V_o = (0.0010)(1000 \text{cm}^3) = 1.0 \text{ cm}^3$$

However, the change in volume is not needed. The fractional change may be used directly in Eq. 9.6 to find the pressure increase:

$$\Delta p = B\left(\frac{-\Delta V}{V_o}\right) = (2.2 \times 10^9 \text{ N/m}^2)(0.0010) = +2.2 \times 10^6 \text{ N/m}^2$$

(This increase is about 22 times normal atmospheric pressure!)

Follow-up Exercise. If an extra $1.0 \times 10^6 \text{ N/m}^2$ of pressure is applied to a half liter of water, what is the change in the water's volume?

Insight   Feat of Strength or Knowledge of Materials?

The breaking of wooden boards, concrete blocks, or similar materials with a bare hand or foot is an impressive demonstration often done by karate experts (Fig. 1). The physics of this feat can be analyzed in terms of the properties of the materials involved. In delivering the blow, the expert imparts a large impulse force (Chapter 6) to the top board or block, which bends under the pressure. (Note that the objects are struck midway between the end supports.) At the same time, the bones of the hand are being compressed as well. Fortunately, human bone can withstand more compressive force than can wood, tile, or concrete, which is why the expert's bones aren't damaged. (The ultimate compressive strength of bone is at least four times greater than that of concrete.*)

When the board or block is hit, the upper surface is compressed and the lower surface is elongated, or subjected to a tension force. Wood, tile, and concrete are weaker under tension than under compression. (The ultimate tensile strength of concrete is only about a twentieth of its ultimate compressive strength.) Thus, the board or block begins to crack at the bottom surface first. The crack propagates from the underside *toward* the hand, widens, and becomes a complete break. Thus the hand never actually "cuts through" the board or block.

The amount of force required to break a board or block in this way depends on several factors. For a wooden board, these are the type of wood, the width and thickness, and the distance between the end supports. Also, the edge of the hand must strike the board parallel to the grain of the wood. Similar considerations apply for a tile or a concrete block. Because tile and concrete are more rigid, more force is generally required to break these substances.

Some karate experts are able to break through a stack of boards or tiles and more than one concrete block. The force

FIGURE 1   Feat of strength or knowledge of materials? Breaking wooden boards, concrete blocks, or roofing tiles (as shown here) with a karate blow depends on both the physical strength of the expert and the strength of the material. Wood, concrete, and tile have different maximum tensile and compressional stresses, but the maximum compressional strength of bone is greater than any of these. (This photo, from a high-speed photography sequence, was made towards the end of the blow, when 3 of the 4 tiles were broken.)

required does not increase by a factor equal to the number of objects. High-speed photography shows that the hand makes contact with only one or two boards or tiles at the top of the stack. Then, as each object breaks, it collides with and breaks through the one below it.

Even though the properties of the materials seem to guarantee the success of this demonstration, you should not attempt the feat unless you are an expert and know what you're doing. The board or block might win.

*Ultimate strength is the maximum strength a material can stand before it breaks or fractures.

9.2   Fluids: Pressure and Pascal's Principle

Objectives: **To be able to (a) explain the pressure–depth relationship, and (b) state Pascal's principle and describe how it is used in practical applications.**

A force can be applied to a solid at a point of contact, but this won't work with a fluid, as we saw in the preceding discussion on bulk moduli. With fluids, a force must be applied over an area. Such an application of force is expressed in terms of **pressure**, or **the force per unit area**:

Pressure—force per unit area

$$p = \frac{F}{A} \tag{9.8}$$

SI unit of pressure: N/m^2 or pascal (Pa)

The force in this equation is understood to be acting normally (perpendicularly) to the surface area. F may be the perpendicular component of a force that acts at an angle to the surface (•Fig. 9.6). Pressure is a scalar quantity (with magnitude only), but the force producing it or the force produced within a fluid by a pressure does have direction and so is a vector.

Pressure has SI units of newtons per square meter (N/m²). This combined unit is given the special name **pascal (Pa)** in honor of the French scientist and philosopher Blaise Pascal (1623–1662):

$$1 \, Pa \equiv 1 \, N/m^2$$

(One of Pascal's major contributions has to do with fluids and will be discussed shortly.) In the British system, a common unit of pressure is lb/in.² (pounds per square inch, or psi). Other units, some of which will be introduced later, are used in special applications.

Pressure and Depth

If you have done any diving, you know that pressure increases with depth, because you have felt increased pressure on your eardrums. An opposite effect is commonly felt when flying in a plane or riding in a car going up a mountain. With increasing altitude, your ears may "pop" because of reduced air pressure.

How the pressure in a fluid varies with depth can be demonstrated by considering a container of liquid at rest, like the one in •Fig. 9.7. With the liquid at rest (in static equilibrium), all points at the same depth must experience the same pressure; otherwise, there would be a pressure difference and horizontal movement of the liquid. For the rectangular column shown in the figure, the force on the surface at the bottom of the container is equal to the weight of the liquid making up the column: $F = w = mg$. The mass of this column is its density times its volume: $m = \rho V$, where ρ, the Greek letter rho, is the density of the liquid. (Since the liquid is assumed incompressible, ρ is constant.) The volume of the liquid is equal to the height of the column times the area of its base: $V = Ah$. Thus we can write

$$F = mg = \rho V g = \rho g A h$$

With $p = F/A$, the extra pressure at a depth h (above the pressure at its top surface) due to the weight of the column is

$$p = \rho g h \qquad (9.9)$$

The pressure is the same everywhere on a horizontal plane at a depth h. Note that the total cross-sectional area of the bottom of the container can be taken as the base of a circular column with the same result.

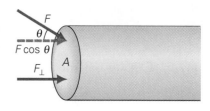

$$p = \frac{F_\perp}{A} = \frac{F \cos \theta}{A}$$

•**FIGURE 9.6 Pressure**
Pressure is usually written $p = F/A$, where it is understood that F is the force or component of force normal to the surface. In general, then, $p = (F \cos \theta)/A$.

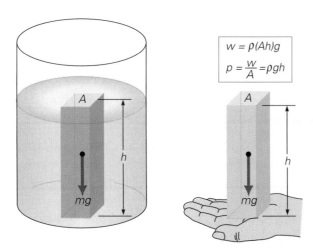

$$w = \rho(Ah)g$$
$$p = \frac{w}{A} = \rho g h$$

•**FIGURE 9.7 Pressure and depth**
The extra pressure at a depth h in a liquid is due to the weight of the liquid above: $p = \rho g h$, where ρ is the density of the liquid (assumed to be constant). This is shown here for an arbitrary column of liquid.

The derivation of Eq. 9.9 did not take into account pressure being applied to the open surface of the liquid. This adds to the pressure at a depth h to give a *total* pressure of

$$p = p_o + \rho g h \qquad \text{(constant density)} \qquad (9.10)$$

where p_o is the pressure applied to the liquid surface (that is, at $h = 0$).

For an open container, p_o is atmospheric pressure, or the weight (force) per area due to the gases in the atmosphere above the liquid's surface. The average atmospheric pressure at sea level is sometimes used as a unit, called an **atmosphere (atm)**:

$$1 \text{ atm} \equiv 101.325 \text{ kPa} = 1.01325 \times 10^5 \text{ N/m}^2 \approx 14.7 \text{ lb/in.}^2$$

How atmospheric pressure is measured will be described shortly.

EXAMPLE 9.3 ■ A SCUBA DIVER: PRESSURE AND FORCE

(a) What is the total pressure on the back of a scuba diver in a lake at a depth of 8.00 m?
(b) What is the force on the diver's back due to the water alone, taking the surface of the back to be a rectangle 60.0 cm by 50.0 cm?

Solution.

Given: $\quad h = 8.00$ m $\qquad\qquad$ *Find:* $\quad$ (a) p (total pressure)
$\qquad\qquad A = 0.600 \text{ m} \times 0.500 \text{ m}$ $\qquad\qquad\qquad$ (b) F (force due to water)
$\qquad\qquad\quad = 0.300 \text{ m}^2$
$\qquad\qquad \rho_{H_2O} = 1.00 \times 10^3 \text{ kg/m}^3$
$\qquad\qquad\quad$ (known from original
$\qquad\qquad\quad$ definition of the kilogram
$\qquad\qquad\quad$ or from Table 9.2 in the
$\qquad\qquad\quad$ next section)
$\qquad\qquad p_a = 1.01 \times 10^5 \text{ N/m}^2$

(a) The total pressure is the sum of the pressure due to the water and the atmospheric pressure (p_a). By Eq. 9.10, this is

$$\begin{aligned} p &= p_a + \rho g h \\ &= (1.01 \times 10^5 \text{ N/m}^2) + (1.00 \times 10^3 \text{ kg/m}^3)(9.80 \text{ m/s}^2)(8.00 \text{ m}) \\ &= (1.01 \times 10^5 \text{ N/m}^2) + (0.784 \times 10^5 \text{ N/m}^2) = 1.79 \times 10^5 \text{ N/m}^2 \text{ (or Pa)} \end{aligned}$$

(b) The pressure p_w due to the water alone is the $\rho g h$ portion of the preceding equation, so $p_w = 0.784 \times 10^5 \text{ N/m}^2$. Then, $p_w = F/A$, and

$$\begin{aligned} F = p_w A &= (0.784 \times 10^5 \text{ N/m}^2)(0.300 \text{ m}^2) \\ &= 2.35 \times 10^4 \text{ N} \quad \text{(or } 5.29 \times 10^3 \text{ lb—about 2.6 tons!)} \end{aligned}$$

Follow-up Exercise. You might question the answer to part (b) of this Example—how could the diver support such a force? To get a better idea of the forces our bodies can support, what would be the force on the diver's back at the surface from atmospheric pressure alone? How do you suppose our bodies are able to support such forces or pressures?

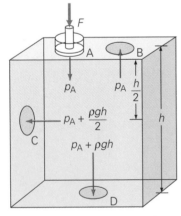

• **FIGURE 9.8 Pascal's principle**
The pressure applied at A is fully transmitted to all parts of the fluid and to the walls of the container. There is also pressure due to the weight of the fluid above at different depths.

Pascal's Principle

When the pressure (for example, air pressure) is increased on the entire open surface of an incompressible liquid at rest, the pressure at any point in the liquid or on the boundary surfaces increases by the same amount. The effect is the same if pressure is applied by means of a piston to any surface of an enclosed fluid (•Fig. 9.8). The transmission of pressure in fluids was studied by Blaise Pascal (after whom the SI pressure unit is named), and the observed effect is called **Pascal's principle**:

Pressure applied to an enclosed fluid is transmitted undiminished to every point in the fluid and to the walls of the container.

For an incompressible liquid, the pressure change is transmitted instantaneously. For a gas, a pressure change will generally be accompanied by a change in volume and/or temperature, but after equilibrium has been reestablished, Pascal's principle remains valid.

Common practical applications of Pascal's principle include the hydraulic braking systems used on automobiles. A force on the brake pedal transmits a force to the wheel brake cylinder. Similarly, hydraulic lifts and jacks are used to raise automobiles and other heavy objects (see •Fig. 9.9). Using Pascal's principle, we can show how such systems not only allow us to transmit force from one place to another, but also to multiply that force. The input pressure p_i, supplied by compressed air for a garage lift, for example, gives an input force F_i on a small piston area A_i. The full magnitude of the pressure is transmitted to the output piston, which has an area A_o. Since $p_i = p_o$,

$$\frac{F_i}{A_i} = \frac{F_o}{A_o}$$

and

$$F_o = \left(\frac{A_o}{A_i}\right)F_i \quad \text{force multiplication} \quad (9.11)$$

With A_o larger than A_i, then F_o will be larger than F_i. The input force is greatly multiplied.

EXAMPLE 9.4 ■ THE HYDRAULIC LIFT: PASCAL'S PRINCIPLE

A garage lift has input and lift pistons with diameters of 10 cm and 30 cm, respectively. The lift is used to hold up a car with a weight of 1.4×10^4 N. (a) What is the force on the input piston? (b) What pressure is applied to the input piston?

Solution.

Given: $d_i = 10$ cm   **Find:** (a) F_i (input force)
$\quad\quad\quad d_o = 30$ cm $\quad\quad\quad\quad$ (b) p_i (input pressure)
$\quad\quad\quad F_o = 1.4 \times 10^4$ N

(a) Rearranging Eq. 9.11 and using $A = \pi r^2 = \pi d^2/4$ for the circular piston ($r = d/2$) gives

$$F_i = \left(\frac{A_i}{A_o}\right)F_o = \left(\frac{\pi d_i^2/4}{\pi d_o^2/4}\right)F_o = \left(\frac{d_i}{d_o}\right)^2 F_o$$

$$F_i = \left(\frac{10 \text{ cm}}{30 \text{ cm}}\right)^2 F_o = \frac{F_o}{9} = \frac{1.4 \times 10^4 \text{ N}}{9} = 1.6 \times 10^3 \text{ N}$$

The input force is $\frac{1}{9}$ of the output force, or there was a force multiplication of 9 (that is, $F_o = 9F_i$)

(Note that we didn't really need to write the complete expressions for the areas at all. We know that the area of a circle is proportional to the square of its diameter. If the ratio of the piston diameters is 3 to 1, the ratio of their areas must therefore be 9 to 1, and we could use this ratio directly in Eq. 9.11.)

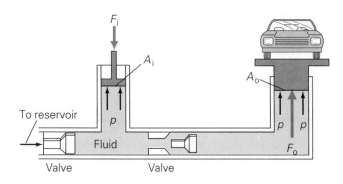

$$F_o = \left(\frac{A_o}{A_i}\right)F_i$$

•**FIGURE 9.9 The hydraulic lift** Because the input and output pressures are equal (Pascal's principle), a small input force gives a large output force, proportional to the ratio of the piston areas.

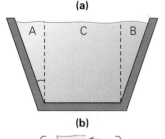

(a)

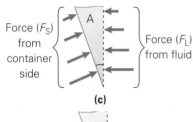

(b)

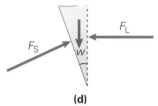

(c)

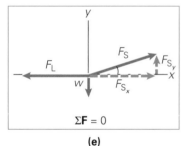

(d)

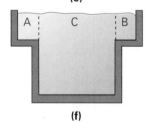

(e)

(f)

●**FIGURE 9.10 A hydrostatic paradox**
See Conceptual Example 9.5.
(a) Vessels with equal bases, filled to the same level. **(b)** Dividing the vessel into separate parts. Newton's laws apply to each part. **(c, d)** The forces on section A. **(e)** Free-body diagram of section A. **(f)** An analogous case; does this help?

(b) Then

$$p_i = \frac{F_i}{A_i} = \frac{F_i}{\pi r_i^2} = \frac{1.6 \times 10^3 \text{ N}}{\pi (0.050 \text{ m})^2}$$

$$= 2.0 \times 10^5 \text{ N/m}^2 \ (= 200 \text{ kPa})$$

This pressure is about 30 lb/in.², a common pressure used in automobile tires, and about twice atmospheric pressure (about 100 kPa or 15 lb/in.²).

Follow-up Exercise. Pascal's principle is used in shock absorbers found on automobiles and on the landing gear of airplanes. (The polished steel piston rods can be seen above the wheels on aircraft.) In these devices, a large force (the shock produced on hitting a bump in the road or an airport runway at high speed) must be reduced to a safe level. Basically, fluid is forced from a larger diameter cylinder into a smaller diameter cylinder (the reverse of the situation in Fig. 9.9). Suppose that the input piston of a shock absorber on a jet plane has a diameter of 8.0 cm. What would be the diameter of the output cylinder that would reduce the force by a factor of 10?

As the example shows, we can relate the force directly to the diameters of the pistons, $F_o = (d_o/d_i)^2 F_i$. By making $d_o \gg d_i$, we can get huge factors of force multiplication, as is typical for hydraulic presses, jacks, and earth-moving equipment. (The shiny input piston rods are often visible on front loaders and back hoes.) Inversely, we can get a force reduction by making $d_i > d_o$, as in the preceding Follow-up Exercise.

However, don't think that you are getting something for nothing with large force multiplications. Energy is still a factor, and this can never be multiplied by a machine. (Why not?) Looking at the work involved, and assuming the work output is equal to the work input, $W_o = W_i$ (an ideal condition—why?), we have from Eq. 5.1

$$F_o x_o = F_i x_i \quad \text{or} \quad F_o = \left(\frac{x_i}{x_o}\right) F_i$$

where x_o and x_i are the output and input distances moved by the respective pistons.

Thus, the output force can be much greater than the input force only if the input distance is much greater than the output distance. For example, if $F_o = 10F_i$, then $x_i = 10x_o$, and the input piston must travel ten times the distance of the output piston. We say that force is multiplied at the expense of distance. Specifically, with the preceding numbers, suppose the brake pads of a simple hydraulic braking system of a car move 5.0 mm when the brakes are applied. Then, the input distance of the pedal is $x_i = 10x_o = 10(5.0 \text{ mm}) = 50 \text{ mm}$—a greater input distance, but the brake pads are applied with 10 times the force applied to the pedal, $F_o = 10F_i$.

CONCEPTUAL EXAMPLE 9.5 ■ HYDROSTATIC PARADOX: PASCAL'S PRINCIPLE AND NEWTON'S THIRD LAW

Two containers with the same base area but different shapes are filled to the same level with water, as illustrated in ●Fig. 9.10a. Then, (a) the pressure on the bases is the same for both containers, (b) the pressure on the base of the container with sloping sides is greater, (c) the pressure on the base of the cylindrical container is greater. *Clearly establish the reasoning and the physical principle(s) used in determining your answer before checking it below. That is, **why** did you select your answer?*

Reasoning and Answer. It might be thought that we could simply take the weight of the liquid in each container and divide by the base (F/A) to get the pressure. However, this is valid only for cylindrical containers, as on the left in the figure. Sloping sides make a difference.

Looking at the sloping sides of the container on the right, Pascal's principle tells us that the pressure in a fluid is exerted in all directions. Hence, the liquid exerts a pressure on (and thus a force perpendicular to) the sides of the container. By Newton's third law, therefore, the sloping sides exert a force on the liquid. Since this force is perpendicular to the sloping sides, it has a vertical (upward) component, which we might expect to balance the downward weight of the liquid over the sloping sides. Hence, only the cylindrical column of liquid above the base contributes to the pressure there, and the answer is (a).

To confirm this supposition, consider the adjacent figure where the liquid is divided into sections (●Fig. 9.10b). (Recall from Chapter 4 that we can divide a system into discrete parts, and Newton's laws will apply to each part.) Considering section A as a separate system (●Fig. 9.10c and d), we see that the perpendicular force on the vertical boundary (from the liquid in section C) is balanced by the horizontal component of the force exerted on the liquid by the sloping side. Similarly, the weight of the liquid in this system is balanced by the upward component of this force, as shown in the free-body diagram (●Fig. 9.10e). Hence, the liquid over the sloping sides does not contribute to the pressure on the base of the container.

To make this conclusion more intuitively obvious, consider the similar case of a cylinder with a lip of greater radius, as shown in ●Fig. 9.10f. Here there would be little doubt that the pressure on the base would be due to the liquid in section C.

Follow-up Exercise. Suppose that you had a third container, with the same base area and water level, but with sides that sloped inward, so that the top was narrower than the base. Would the pressure on the base be the same, greater, or less than that in this Example?

Pressure Measurement

Pressure can be measured with a variety of mechanical devices that are often spring-loaded. Another type of instrument, called a manometer, uses a liquid—usually mercury. An *open-tube manometer* is illustrated in ●Fig. 9.11a. One end of the U-shaped tube is open to the atmosphere, and the other is connected to the container of gas whose pressure is to be measured. The liquid in the U-tube acts as a reservoir through which pressure is transmitted according to Pascal's principle.

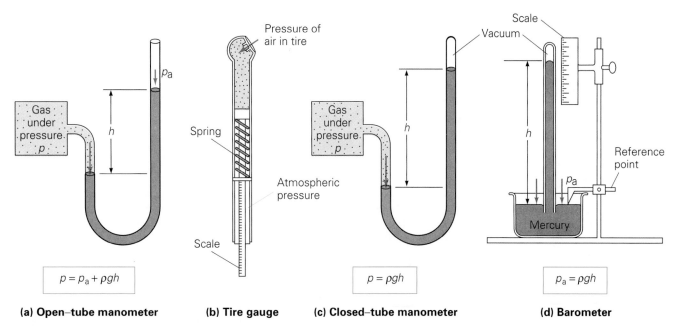

(a) **Open–tube manometer**   (b) **Tire gauge**   (c) **Closed–tube manometer**   (d) **Barometer**

●**FIGURE 9.11   Pressure measurement**
(a) For an open-tube manometer, the pressure of the gas in the container is balanced by the pressure due to the liquid column and atmospheric pressure acting on the open surface of the liquid. The absolute pressure of the gas equals the sum of the atmospheric pressure and ρgh, the gauge pressure. **(b)** A tire gauge measures gauge pressure, that is, the difference between the pressure in the tire and atmospheric pressure: $p_{\text{gauge}} = p - p_{\text{a}}$. Thus, if a tire gauge reads 200 kPa (30 lb/in.²), the actual pressure within the tire is 1 atm higher, or 300 kPa. **(c)** A closed-tube manometer is a closed system. It measures absolute pressure since here atmospheric pressure is not a consideration. **(d)** A barometer is a closed-tube manometer that is exposed to the atmosphere and thus reads atmospheric pressure.

The pressure of the gas (p) is balanced by the weight of the column of liquid (of height h measured from the lower surface level) and the atmospheric pressure (p_a) on the open liquid surface:

$$p = p_a + \rho g h \qquad (9.12)$$

The pressure p is called the **absolute pressure**, and $p - p_a = \rho g h$ is called the **gauge pressure**. The gauge pressure is the pressure that registers on many common pressure-measuring devices such as a tire pressure gauge (Fig. 9.11b). This instrument basically compares the pressure inside a tire with the pressure outside and records the pressure difference: $p_{gauge} = p - p_a$.

For the manometer illustrated in Fig. 9.11a, the absolute pressure of the gas is greater than the atmospheric pressure. Thus, the gauge pressure and the height (h) are positive ($p - p_a = \rho g h$). But if the absolute pressure of the gas in the container were equal to the atmospheric pressure, the gauge pressure would be zero. In this case, the column height would be zero ($h = 0$), or the liquid levels in the tubes would be at the same height. If the absolute pressure of the gas were less than atmospheric pressure (a partial vacuum), the gauge pressure and the column height would be negative.

If the open end of the U-shaped tube is evacuated and sealed, the device is a *closed-tube manometer* (Fig. 9.11c). Atmospheric pressure then has no effect on the liquid in the closed system, so the pressure registered by this manometer is equal to the absolute pressure: $p_{gauge} = \rho g h = p - 0 = p$. A practical application of a closed-tube manometer is the **barometer** (Fig. 9.11d), used to measure atmospheric pressure. Essentially, the container of gas whose pressure the barometer measures is the atmosphere.

The mercury barometer was invented by Evangelista Torricelli (1608–1647), who was Galileo's successor as professor of mathematics at the university in Florence. A tube filled with mercury is inverted into a reservoir. Some mercury runs out, but as much as is supported by the air pressure on the surface of the reservoir pool remains in the tube. (You may be wondering about the space in the tube above the liquid column. Since the tube was initially full, there is no air above the liquid surface. Some of the mercury will evaporate and the mercury vapor will exert some pressure on the surface, but this is taken to be negligibly small.) The atmospheric pressure is then equal to the pressure due to the weight of the column of mercury, or

Barometric pressure

$$p_a = \rho g h \qquad (9.13)$$

A *standard atmosphere* is defined as the pressure supporting a column of mercury exactly 76 cm in height at sea level and 0°C.

EXAMPLE 9.6 ■ STANDARD ATMOSPHERIC PRESSURE: CONVERTING TO PASCALS

If a standard atmosphere supports a column height of exactly 76 cm of mercury (chemical symbol Hg), what is the standard atmospheric pressure in pascals? (The density of mercury is 13.5951×10^3 kg/m^3 at 0°C, and $g = 9.80665$ m/s^2.)

Solution.

Given: $h = 76$ cm $= 0.76$ m (exact)   *Find:* p_a (atmospheric pressure)
   $\rho_{Hg} = 13.5951 \times 10^3$ kg/m^3
   $g = 9.80665$ m/s^2

Using Eq. 9.13

$$p_a = \rho g h = (13.5951 \times 10^3 \text{ kg/m}^3)(9.80665 \text{m/s}^2)(0.760000 \text{ m})$$
$$= (101{,}325 \text{ N/m}^2 = 1.01325 \times 10^5 \text{ Pa}) \quad \text{(or } 101.325 \text{ kPa)}$$

Follow-up Exercise. What would be the height of a barometer column for one standard atmosphere if water were used instead of mercury?

Changes in atmospheric pressure may be observed as changes in the height of the mercury column. Atmospheric pressure is commonly reported in terms of the height of the barometer column, and weather forecasters say that the barometer is rising or falling. That is, 1 standard atmosphere (atm) of pressure is expressed as

$$1 \text{ atm} = 76 \text{ cm Hg} = 760 \text{ mm Hg} = 29.92 \text{ in. Hg (about 30 in. Hg)}$$

In honor of Torricelli, a pressure supporting 1 mm of mercury is given the name *torr*:

$$1 \text{ mm Hg} \equiv 1 \text{ torr} \quad \text{and} \quad 1 \text{ atm} = 760 \text{ torr}$$

Because mercury is very toxic, it is sealed inside a barometer. A safer and less expensive device widely used to measure atmospheric pressure is the aneroid (without fluid) barometer. In an aneroid barometer, a sensitive metal diaphragm on an evacuated container (something like a drum head) responds to pressure changes, which are indicated on a dial. This is the kind of barometer you frequently find in homes in decorative wall mountings.

Since air is compressible, the atmospheric density and pressure are greatest at the Earth's surface and decrease with altitude. We live at the bottom of the atmosphere but don't notice its pressure very much in our ordinary daily activities. Remember that our bodies are composed largely of fluids, which exert a matching outward pressure. Indeed, the external pressure of the atmosphere is so important to our normal functioning that we take it with us wherever we can. The pressurized suits worn by astronauts in space or on the Moon are needed, not only to supply oxygen, but also to provide an external pressure similar to that on the Earth's surface (•Fig. 9.12).

9.3 Buoyancy and Archimedes' Principle

Objectives: To be able to (a) relate the buoyant force and Archimedes' principle, and (b) tell whether an object will float in a fluid based on relative densities.

When an object is placed in a fluid, it will either sink or float. This is most commonly observed with liquids, for example, objects floating or sinking in water. But the same effect occurs in gases. A falling object is sinking in the atmosphere, and other bodies float (•Fig. 9.13).

Things float because they are buoyant, or are buoyed up. For example, if you immerse a cork in water and release it, the cork will be buoyed up to the surface and float there. From your knowledge of forces, you know that such motion requires an upward net force on the object. That is, there must be an upward force acting on the object that is greater than its downward weight force. The forces are equal when the object floats, moves at constant velocity, or remains stationary. The upward force resulting from an object being wholly or partially immersed in a fluid is called the **buoyant force**.

How the buoyant force comes about can be seen by considering a buoyant object being held under the surface of a fluid, as shown in •Fig. 9.14a. The pressures on the upper and lower surfaces of the block are $p_1 = \rho_f g h_1$, and $p_2 = \rho_f g h_2$, where ρ_f is the density of the fluid. Thus, there is a pressure difference, $\Delta p = p_2 - p_1 = \rho_f g (h_2 - h_1)$, between the top and bottom of the block, which gives an upward force (the buoyant force), F_b. This force is balanced by the applied force and the weight of the block.

It is not difficult to derive an expression for the magnitude of the buoyant force. We know that pressure is force per unit area. Thus, if the top and bottom areas of the block are A, the magnitude of the net buoyant force in terms of the pressure difference is

$$F_b = p_2 A - p_1 A = (\Delta p)A = \rho_f g(h_2 - h_1)A.$$

•**FIGURE 9.12 Dressed for the occasion**
Astronauts need pressurized suits not only to supply oxygen but also to protect them from the vacuum of space, where there is no air pressure to oppose our body's internal pressure. What would happen otherwise?

Note: Another unit sometimes used in weather reports is the millibar (mb). By definition, $1 \text{ atm} = 1.01325 \times 10^5$ $\text{N/m}^2 = 1.01325 \text{ bar} = 1,013.25 \text{ mb}$. Normal atmospheric pressures are around 1000 mb.

•**FIGURE 9.13 Fluid buoyancy**
The air is a fluid in which objects such as these dirigibles float. The helium inside the blimps is lighter or less dense than the surrounding air. The blimps are supported by the resulting buoyant forces.

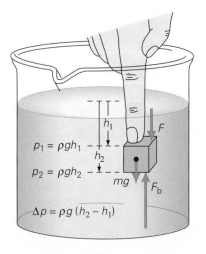

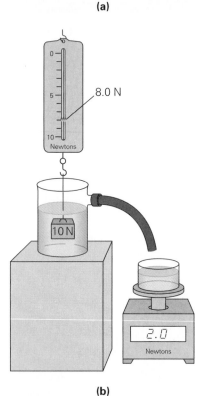

FIGURE 9.14 Buoyancy and Archimedes' principle
(a) A buoyant force arises from the pressure difference between different depths. The pressure on the bottom of the submerged block is greater than that on the top, so there is a (buoyant) force directed upward. (b) Archimedes' principle: The buoyant force on the object is equal to the weight of the volume of fluid it displaces. (The scale is set to read zero with an empty container.)

Since $(h_2 - h_1)A$ is the volume of the block, and hence the volume of fluid displaced by the block, V_f, we can write this expression as

$$F_b = \rho_f g V_f$$

But $\rho_f V_f$ is simply the mass of the fluid displaced by the block, m_f. Thus we can write the expression for the buoyant force $F_b = m_f g$, or the magnitude of the buoyant force is equal to the weight of the fluid displaced by the block (•Fig. 9.14b). This general result is known as **Archimedes' principle**:

> A body immersed wholly or partially in a fluid is buoyed up by a force equal in magnitude to the weight of the *volume of fluid* it displaces.

$$F_b = m_f g = \rho_f g V_f \qquad (9.14)$$

Archimedes (287–212 B.C.) was given the task of determining whether a crown made for a certain king was pure gold or contained some other, cheaper metal. Legend has it that the solution to the problem came to him when he was bathing, perhaps from seeing the water level rise when he got into the tub and experiencing the buoyant force on his limbs. In any case, it is said that he was so excited that he ran through the streets of the city shouting "Eureka!" (Greek for "I have found it"). Although Archimedes' solution to the problem involved density and volume (see Exercises 54 and 120), it presumably got him thinking about buoyancy.

EXAMPLE 9.7 ■ LIGHTER THAN AIR: BUOYANT FORCE

What is the buoyant force on a helium balloon with a radius of 30 cm in air if $\rho_{air} = 1.29$ kg/m³?

Solution.

Given: $r = 30$ cm $= 0.30$ m   *Find:* F_b (buoyant force)
           $\rho_{air} = 1.29$ kg/m³

The volume of the balloon is

$$V = \tfrac{4}{3}\pi r^3 = \tfrac{4}{3}\pi(0.30 \text{ m})^3 = 0.11 \text{ m}^3$$

Then, by Eq. 9.14, the weight of the air displaced by the balloon's volume, or the magnitude of the upward buoyant force, is

$$F_b = m_{air}g = (\rho_{air}V)g = (1.29 \text{ kg/m}^3)(0.11 \text{ m}^3)(9.8 \text{ m/s}^2) = 1.4 \text{ N}$$

Note that the buoyant force depends on the *density of the fluid* and *the volume of the body*. Shape makes no difference.

Follow-up Exercise. Would the buoyant force on the balloon in this example be greater or less if it were submerged in water, and by what factor? (Assume the same volume.)

CONCEPTUAL EXAMPLE 9.8 ■ WEIGHT AND BUOYANT FORCE: ARCHIMEDES' PRINCIPLE

An overflow container filled with water, like that shown on the left in Fig. 9.14b, sits on a scale that reads 40 N. (The water level is just below the exit tube in the side of the container.) A 5.0-N object that is less dense than water (e.g., a block of wood), is placed in the container. The water it displaces runs out the exit tube into another container that is not on the scale. Will the scale reading then be (a) exactly 45 N, (b) between 40 N and 45 N, (c) exactly 40 N, (d) less than 40 N? *Clearly establish the reasoning and physical principle(s) used in determining your answer before checking it below. That is, **why** did you select your answer?*

Reasoning and Answer. By Archimedes' principle, the block is buoyed upward with a force equal in magnitude to the weight of the water displaced. Since the block floats, the upward buoyant force must balance the weight of the block, and has a magnitude of 5.0 N. Thus a volume of water weighing 5.0 N is displaced from the container as a 5.0-N weight is added to the container. The scale still reads 40 N, so the answer is (c).

Note that the upward buoyant force and the block's weight force act *on the block*. The reaction force (pressure) of the block *on the water* is transmitted to the bottom of the container (Pascal's principle) and is registered on the scale.

Follow-up Exercise. Would the scale reading still be 40 N if the object had a density greater than that of water?

Buoyancy and Density

We commonly say that helium and hot-air balloons float because they are lighter than air. However, technically, they are *less dense than air*. An object's density will tell you whether it will sink or float in a fluid, as long as you also know the density of the fluid. Consider a solid uniform object totally immersed in a fluid. The weight of the object is

$$w_o = m_o g = \rho_o V_o g$$

The weight of the volume of fluid the object displaces, or the magnitude of the buoyant force, is

$$F_b = w_f = m_f g = \rho_f V_f g$$

But if the object is completely submerged, $V_o = V_f$, and dividing the second equation above by the first gives

$$\frac{F_b}{w_o} = \frac{\rho_f}{\rho_o} \quad \text{or} \quad F_b = \left(\frac{\rho_f}{\rho_o}\right) w_o \qquad (9.15)$$

Thus, if ρ_o is less than ρ_f, then F_b will be greater than w_o, and the object will be buoyed to the surface and float. If ρ_o is greater than ρ_f, then F_b will be less than w_o, and the object will sink. If ρ_o equals ρ_f, then F_b will be equal to w_o, and the object will remain in equilibrium at any submerged depth (as long as the density of the fluid is constant). If the object is not uniform, so that its density varies over its volume, then the density of the object is its average density.

These three conditions expressed in words are as follows:

An object will float in a fluid if the density of the object is less than the density of the fluid.

An object will sink in a fluid if the density of the object is greater than the density of the fluid. **Float or sink?**

An object will be in equilibrium at any submerged depth in a fluid if the densities of the object and the fluid are equal.

(See Demonstration 7.)

The densities of some solids and fluids are given in Table 9.2. A quick look will tell you if an object will float in a fluid, regardless of the shape or volume of the object. The conditions stated above also apply to a fluid in a fluid, provided the two are immiscible (do not mix).

In general, the densities of objects or fluids will be assumed to be uniform and constant in this book. (The density of the atmosphere varies with altitude but is relatively constant near the surface of the Earth.) However, in practical applications, as we have seen above, it is the *average* density of an object that often matters with regard to floating and sinking. An ocean liner is overall less dense than water, even though it is made of steel. Most of its volume is occupied by air, so the liner's average density is less than that of water. Similarly for the human body —most of us float in water.

In some instances, the overall density of an object may be purposefully varied. For example, a submarine submerges by flooding its tanks with sea water (called "taking on ballast"), which increases its average density. When the sub is to surface, the water is pumped out of the tanks, and the average density of the sub becomes less than that of the surrounding sea water.

DEMONSTRATION 7 ■ Buoyancy and Density

(a) Unopened cans of Coke are dropped into a container of water.

(b) The can of Classic Coke sinks and the can of Diet Coke floats.

A demonstration of buoyancy which shows that the overall density of a can of Diet Coke is less than that of water, while the density of a can of Classic Coke is greater.

Consider the following questions: Does one can have a greater volume of metal? higher gas pressure inside? more fluid volume? Do calories make a difference? You can investigate the possibilities yourself to determine the reason(s) for the different densities.

TABLE 9.2   Densities of Some Common Substances (in kg/m³)

Substance	Density (ρ)	Substance	Density (ρ)	Substance	Density (ρ)
Solids		*Liquids*		*Gases**	
Aluminum	2.7×10^3	Alcohol, ethyl	0.79×10^3	Air	1.29
Brass	8.7×10^3	Alcohol, methyl	0.82×10^3	Helium	0.18
Copper	8.9×10^3	Blood, whole	1.05×10^3	Hydrogen	0.090
Glass	2.6×10^3	Blood plasma	1.03×10^3	Oxygen	1.43
Gold	19.3×10^3	Gasoline	0.68×10^3	Water vapor (100°C)	0.63
Ice	0.92×10^3	Kerosene	0.82×10^3		
Iron	7.9×10^3	Mercury	13.6×10^3		
Lead	11.4×10^3	Sea water (4°C)	1.03×10^3		
Silver	10.5×10^3	Water, fresh (4°C)	1.00×10^3		
Steel	7.8×10^3				
Wood, oak	0.81×10^3				

*At 0°C and 1 atm unless otherwise specified.

EXAMPLE 9.9 ■ FLOAT OR SINK? COMPARISON OF DENSITIES

A uniform solid cube of material 10 cm on a side has a mass of 700 g. (a) Will the cube float in water? (b) If so, how much of its volume will be submerged?

Solution.

Given: $m = 700$ g $= 0.700$ kg
           $L = 10$ cm $= 0.10$ m
           $\rho_w = 1.00 \times 10^3$ kg/m³ (density of water from Table 9.2)

Find: (a) Whether the cube will float in water
           (b) The percentage of the volume submerged if it does float

It is sometimes convenient to work in cgs units in comparing quantities, particularly when working with ratios. For densities in g/cm^3, drop the "$\times 10^3$" from the values given in Table 9.2 for solids and liquids and add "$\times 10^{-3}$" for gases.

(a) The density of the cube material is

$$\rho_c = \frac{m}{V_c} = \frac{m}{L^3} = \frac{700 \text{ g}}{(10 \text{ cm})^3} = 0.70 \text{ g/cm}^3$$

Since ρ_c is less than ρ_w, the cube will float.

(b) The weight of the cube is $w_c = \rho_c g V_c$. When the cube is floating (in equilibrium), its weight is balanced by the buoyant force. That is, $F_b = \rho_w g V_w$, where V_w is the volume of water the submerged part of the cube displaces. Equating the expressions for weight and buoyant force gives

$$\rho_w g V_w = \rho_c g V_c$$

or

$$\frac{V_w}{V_c} = \frac{\rho_c}{\rho_w} = \frac{0.70 \text{ g/cm}^3}{1.00 \text{ g/cm}^3} = 0.70$$

Thus, $V_w = (0.70)V_c$, and 70% of the cube is submerged.

Follow-up Exercise. Most of an iceberg floating in the ocean (•Fig. 9.15) is submerged. What is seen is the proverbial "tip of the iceberg." What percentage of an iceberg's volume is seen above the surface? (Note: Icebergs are frozen *fresh* water.)

•**FIGURE 9.15 The tip of the iceberg** The vast majority of an iceberg's bulk is underneath the water. This large "berg" was photographed off the Antarctic Peninsula. (Does the submerged ice have to be directly below the exposed tip of the iceberg?)

CONCEPTUAL EXAMPLE 9.10 ■ A FLOATING CANDLE: ARCHIMEDES' PRINCIPLE AND BUOYANT FORCE

A dripless candle is weighted slightly on the bottom so that it floats upright in a container of water as illustrated in •Fig. 9.16. As the candle burns, (a) the portion of the candle above the water burns down and is extinguished, (b) the candle rises relative to the base of the container, (c) the candle sinks lower relative to the base of the container. (*Clearly establish the reasoning and physical principle(s) used in determining your answer before checking it below. That is, **why** did you select your answer?*)

Reasoning and Answer. To answer this, we need to look at the situation in terms of Archimedes' principle and buoyant force. Initially, the buoyant force balances the candle's original weight so that it floats at a certain level. As the candle burns, it gradually loses mass (weight). This gives rise to a net upward force, and the candle rises until the buoyant force and weight are again equal (with less candle submerged and a smaller buoyant force). Hence, the answer is not (a) or (c), but (b).

The candle continuously rises and stays lit. To better understand the process, it may be helpful to think of the candle as being burned in increments or segments and then extrapolate this to continuous burning.

Follow-up Exercise. A person sitting in a boat floating on a large swimming pool notes the water level on the side of the pool. He then drops several concrete blocks over the side of the boat into the water. Does the level of the pool change? If so, how?

•**FIGURE 9.16 Rise or sink?** See Conceptual Example 9.10.

You may have heard of a quantity called specific gravity, which is related to density. It is commonly used for liquids but also applies to solids. Basically, it is a comparison of the weight of a volume of a substance with the weight of an equal volume of water. The **specific gravity** of a substance is equal to the ratio of the density of the substance (ρ_s) to the density of water (ρ_w, at 4°C):

$$\text{sp. gr.} = \frac{\rho_s}{\rho_w}$$

Specific gravity

Because it is a ratio of densities, specific gravity has no units. Since in cgs units $\rho_w = 1.00 \text{ g/cm}^3$,

$$\text{sp. gr.} = \frac{\rho_s}{1.00} = \rho_s$$

That is, the specific gravity of a substance is equal to the numerical value of its density *in cgs units*. For example, if a liquid has a density of 1.5 g/cm^3, its specific gravity is 1.5, which tells you that it is 1.5 times denser than water. To get density values in gm/cm^3 simply divide the value in Table 9.2 by 10^3. The specific gravities of automobile coolant and battery electrolyte (both water solutions) are commonly measured so as to determine their relative concentrations of antifreeze and sulfuric acid, respectively.

9.4 Surface Tension and Capillary Action

Objectives: **To be able to (a) describe the source of surface tension and its effects, and (b) explain why liquids rise in small tubes by capillary action.**

A fluid will not support a shear stress, yet you have probably seen water striders walking on the surface of a pond (•Fig. 9.17a). Also, a razor blade or a needle placed carefully on the surface of water will float, even though steel is denser than water (about eight times denser according to Table 9.2). These things are possible because of an interesting property of liquids. A free liquid surface acts like a thin membrane that can be placed under a slight tension.

Surface Tension

The molecules of a liquid exert small attractive forces on each other. Even though molecules are electrically neutral overall, there is often some slight asymmetry of charge that gives rise to attractive forces between molecules (called van der Waals forces). Within a liquid, where any molecule is completely surrounded by other molecules, the net force is zero (•Fig. 9.17b). However, for molecules at the surface of the liquid, there is no attractive force acting from above the surface. (The effect of air molecules is small and considered negligible.) As a result, the mole-

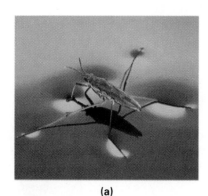

(a)

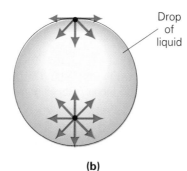

(b)

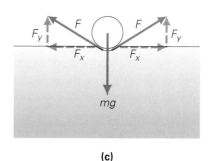

(c)

•**FIGURE 9.17 Surface tension**
(a) Insects like this water strider can walk on water because of surface tension, much as you might walk on a large trampoline. Note the depressions in the surface of the liquid where the legs touch it. **(b)** The net force on a molecule in the interior of a liquid is zero, because it is surrounded by other molecules. However, a molecule at the surface experiences a nonzero net force due to the attractive forces of the neighboring molecules just below the surface. **(c)** To form a surface depression, work must be done, since more interior molecules must be brought to the surface to increase the area. As a result, the surface area acts like a stretched elastic membrane, and the weight force of an object such as a needle is supported by the upward components of the surface tension. Insect legs make a similar depression, and the resulting upward force components allow the insect to walk on water.

(a)

(b)

•**FIGURE 9.18   Surface tension at work**
Because of surface tension, water droplets **(a)** and soap bubbles **(b)** tend to assume the shape that minimizes their surface area—that of a sphere.

TABLE 9.3 Surface Tensions for Some Liquids (in N/m)		
Liquid		*Surface tension* (γ)
Alcohol, ethyl	(20°C)	0.022
Blood, whole	(37°C)	0.058
Blood plasma	(37°C)	0.072
Mercury	(20°C)	0.45
Soapy water	(20°C)	0.025
Water	(0°C)	0.076
Water	(20°C)	0.073
Water	(100°C)	0.059

cules of the surface layer experience net forces due to the attraction of neighboring molecules just below the surface. This inward pull on the surface molecules causes the surface of the liquid to contract and to resist being stretched or broken, a property called **surface tension**.

If a sewing needle is carefully placed on the surface of a bowl of water, the surface acts like an elastic membrane under tension. There is a slight depression in the surface, and molecular forces along the depression are at an angle to the surface (•Fig. 9.17c). The vertical components of these forces balance the weight (mg) of the needle and it "floats" on the surface. Similarly surface tension supports the weight of a water strider.

The net effect of surface tension is to make the surface area of a liquid as small as possible. That is, a given volume of liquid tends to assume the shape that has the least surface area. As a result, drops of water and soap bubbles have spherical shapes because a sphere has the smallest surface area for a given volume (•Fig. 9.18). In forming a drop or bubble, surface tension pulls the molecules together to minimize the surface area.

Quantitatively, the surface tension (γ) in a liquid film is defined as the force per unit length acting along a line (for example, on a length of wire) when stretching the surface:

$$\gamma = \frac{F}{L} \tag{9.16}$$

SI unit of surface-tension: N/m

Surface tension

where the SI units for surface tension can be readily seen from the equation. Some surface tensions for liquids are given in Table 9.3. As might be expected, surface tension is highly dependent on temperature, since molecular motion increases with temperature.

An apparatus used to measure surface tension is shown in •Fig. 9.19. Basically, the device measures the force required to overcome the surface tension. For a circular wire loop, L is the length of the circumference, and $\gamma = F/2L$, since there are two film surfaces (one on each side of the wire).

Measuring surface tension

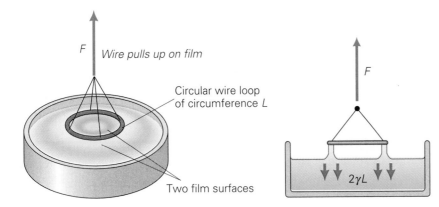

F | *Wire pulls up on film*

Circular wire loop of circumference L

Two film surfaces

F

$2\gamma L$

•**FIGURE 9.19   Measuring surface tension**
An experimental method of measuring the force required to just overcome surface tension. It is equal to $2\gamma L$, since there are two film surfaces.

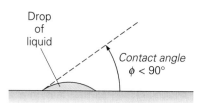

(a) Wetting condition

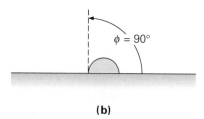

$\phi = 90°$

(b)

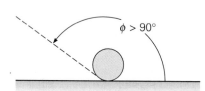

$\phi > 90°$

(c) Nonwetting condition

•**FIGURE 9.20   Contact angle**
A relative measure of adhesive and cohesive forces is given by the contact angle. **(a)** An angle ϕ less than 90° means that the liquid wets the surface, indicating that the adhesive forces are greater than the cohesive forces. **(b)** When ϕ is equal to 90°, the drop forms a hemisphere. **(c)** An angle ϕ greater than 90° means that the liquid does not wet the surface, indicating that the cohesive forces are greater than the adhesive forces.

TABLE 9.4 Contact Angles for Some Liquids on Solids	
Liquid-solid	*Contact angle (ϕ) (approximate)*
Alcohol–glass	0°
Kerosene–glass	26°
Mercury–glass	140°
Water–glass	0°
Water–silver	90°
Water–paraffin	107°

Another way of looking at surface tension is in terms of the work or energy needed to stretch the surface area. If a straight piece of wire of length L is used to stretch a surface by a parallel distance Δx, the work done against the surface tension is

$$W = F\Delta x = \gamma L\Delta x = \gamma\Delta A$$

since $F = \gamma L$ and $\Delta A = L\Delta x$ (the change in surface area). Thus,

$$\gamma = \frac{W}{\Delta A}$$

The surface tension, or force per unit length, is equivalently the work per unit change in the surface area, with units of J/m². (Is J/m² equivalent to N/m?)

Adhesion, Cohesion, and Capillary Action

Note the relatively small surface tension given in Table 9.3 for soapy water. Soaps and detergents have the effect of lowering the surface tension of water. Such substances are called *surfactants*. Without their help, the relatively high surface tension of plain water tends to prevent it from getting into small places, such as between the fibers of clothing. (You can also see from the table why warm water is generally used for cleaning.)

Soaps and detergents also act as *wetting agents*. Whether or not a liquid "wets" or adheres to a surface depends on the relative strengths of the adhesive and cohesive molecular forces. **Adhesive forces** are attractive forces between unlike molecules. **Cohesive forces** are attractive forces between like molecules. Cohesive forces hold a substance together *(cohesion)*, and adhesive forces hold different substances together *(adhesion)*.

If the adhesive forces between the molecules of the liquid and those of the surface are greater than the cohesive forces among the molecules of the liquid, the liquid wets the surface. On the other hand, if the cohesive forces are greater than the adhesive forces, the liquid does not wet the surface. Water beading up on a recently waxed car is a good example of the latter situation (see Fig. 9.18a). Water does not adhere well to waxes or oils on a surface. (This phenomenon, as well as surface tension, helps water striders to walk on water. The insects' feet are covered with a waxlike substance that prevents wetting.)

Although adhesive and cohesive forces are difficult to analyze, a relative measure of their effects is the *contact angle* (ϕ). This is the angle between the surface and a line drawn tangent to the liquid (•Fig. 9.20). Note from the figure that ϕ is less than 90° if the liquid wets the surface and greater than 90° if it does not. Water on clean glass has a contact angle of approximately 0° (it spreads out in a thin layer), and water on paraffin has a contact angle of 107° (Table 9.4). A little detergent on the paraffin will cause the water to spread out, or wet the surface more, and the contact angle will decrease. The cleansing action of soaps and detergents is due in large part to their enhancing of the capability of water to wet dirt particles so they can be washed away.

In a container, the free liquid surface curves upward (is concave) if the liquid wets the container wall and curves downward (is convex) if it does not (•Fig. 9.21). The curved shape of the liquid surface is referred to as the *meniscus* (from a Greek word meaning "crescent moon"). If a tube with a small diameter is positioned vertically with one end submerged in a liquid that wets the walls of the tube, the liquid in the tube will rise some distance above the surface of the surrounding liquid. This is called **capillary action** (or capillarity) and is a consequence of both surface tension and adhesion. Essentially, adhesion draws the water up the sides of the tube, and cohesion (surface tension) then pulls the column upward.

As might be expected, the height to which a liquid rises in a capillary tube depends on the diameter. (Capillary comes from a Latin word meaning "hairlike." Your smallest blood vessels are called capillaries and are so narrow that blood cells must pass through them in single file.) At equilibrium, the upward component of the surface tension force and the downward weight force of the liquid column must be equal in magnitude. The surface tension force is

$$F = \gamma L = \gamma(2\pi r)$$

where $L = 2\pi r$ since the liquid is in contact with the tube at all points around its circumference. The vertical component of this force has a magnitude of

$$F \cos \phi = \gamma(2\pi r)(\cos \phi)$$

The weight of the liquid column is given by

$$w = mg = \rho V g = \rho(\pi r^2 h)g$$

where mass in terms of density is $m = \rho V$ and the volume of the cylinder of liquid is $V = \pi r^2 h$. (Atmospheric pressure is not a consideration because it is the same on both surfaces.) Equating these force magnitudes ($F \cos \phi = w$) and solving for h gives

$$h = \frac{2\gamma \cos \phi}{\rho g r} \qquad (9.17)$$

A similar analysis for capillary depression, which occurs when the liquid does not wet the tube surface, gives the same equation. In this case, ϕ is greater than 90°, and h is negative.

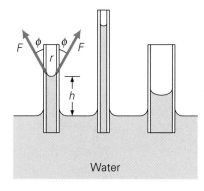

(a)

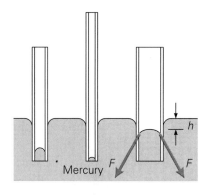

(b)

•**FIGURE 9.21 Capillary action**
(a) Liquids rise in small (capillary) tubes because of adhesion (wetting) and surface tension. The upward force (balanced by the weight of the liquid column) and the contact angle are shown in one tube. **(b)** If a liquid does not wet a capillary tube, there is a column depression.

EXAMPLE 9.11 ■ SOAP MAKES A DIFFERENCE: CAPILLARY ACTION

(a) How high will plain water at 20°C rise in a 1.0-mm-diameter glass capillary tube?

(b) Will soapy water rise more or less in the same capillary tube?

Solution.

Given: $d = 1.0 \text{ mm} = 1.0 \times 10^{-3} \text{ m}$,   *Find:*   (a) h_w (water height)
          so $r = 5.0 \times 10^{-4} \text{ m}$ (b) h_{sw} (soapy water
          $\rho_w = \rho_{sw} = 1.0 \times 10^3 \text{ kg/m}^3$ height compared
          (from Table 9.2)   to h_w)
          $\gamma_w = 0.073 \text{ N/m}$
          (from Table 9.3)
          $\gamma_{sw} = 0.025 \text{ N/m}$
          (from Table 9.3)

(a) From Table 9.4, the contact angle for water and glass is $\phi = 0°$. With $\cos \phi = \cos 0° = 1$, Eq. 9.17 becomes $h = 2\gamma/\rho g r$. For water,

$$h_w = \frac{2\gamma_w}{\rho_w g r}$$

$$= \frac{2(0.073 \text{ N/m})}{(1.0 \times 10^3 \text{ kg/m}^3)(9.8 \text{ m/s}^2)(5.0 \times 10^{-4} \text{ m})}$$

$$= 0.030 \text{ m (or 3.0 cm)}$$

(b) The height of the soapy water could be calculated using the same equation, but using a proportion saves some time.

$$h_{sw} = \left(\frac{\gamma_{sw}}{\gamma_w}\right)h_w = \left(\frac{0.025 \text{ N/m}}{0.073 \text{ N/m}}\right)h_w = (0.34)h_w$$

Hence, the rise of soapy water in the capillary is less than that of pure water—only about one-third as great, or $h_{sw} \approx 0.010 \text{ m (or 1.0 cm)}$.

Follow-up Exercise.   In part (a) of this Example, how would a change in temperature affect the capillary action?

Capillary action is important in liquid transport. Common examples are the absorption of water by paper towels, the wicking of oil in oil lamps, the distribution of water and nutrients in plants, and the holding of water in the soil. If it were not for the latter, the ground would be pretty dry down to the water table. Less desirable results of capillary action are the wetting of concrete blocks in building walls and the rise of water in cracks and joints.

9.5 Fluid Dynamics and Bernoulli's Equation

Objectives: **To be able to (a) identify the simplifications used in describing fluid flow, and (b) use the continuity equation and Bernoulli's equation to explain common effects.**

In general, fluid motion is difficult to analyze. For example, think of trying to describe the motion of a particle (a molecule) of water in a rushing stream. The overall motion of the stream may be apparent, but a mathematical description of the motion of some particle of it may be virtually impossible because of eddy currents (small whirlpool motions), the gushing of water over rocks, frictional drag on the stream bottom, and so on. A basic description of fluid flow is conveniently obtained by ignoring such complications and considering an *ideal fluid*. Actual fluid flow can then be approximated in reference to this more simple theoretical model.

In this simple approach to fluid dynamics, it is customary to consider four characteristics of an **ideal fluid**. In such a fluid, flow is *steady, irrotational, nonviscous*, and *incompressible*.

> *Steady flow* means that all the particles of a fluid have the same velocity as they pass a given point.

Steady flow might be called smooth or regular flow. The path of steady flow can be depicted in the form of **streamlines** (•Fig. 9.22). Every particle that passes a particular point moves along a streamline. That is, every particle moves along the same path (streamline) as particles that passed by earlier. Streamlines never cross. If they did, a particle would have alternate paths and abrupt changes in its velocity, in which case the flow would not be steady.

Steady flow requires low velocities. For example, it is approximated by the flow relative to a canoe that is gliding slowly through still water. When the flow velocity is high, eddies tend to appear, especially near boundaries, and the flow becomes turbulent. (The conditions for turbulent flow are discussed more fully at the end of the next section.)

Streamlines also indicate the relative magnitude of the fluid velocity. The velocity is greater where the streamlines are closer together. Notice this effect in Fig. 9.22a. The reason for this will be explained shortly.

> *Irrotational flow* means that a fluid element (a small volume of the fluid) has no net angular velocity, which eliminates the possibility of whirlpools and eddy currents (the flow is nonturbulent).

Consider the small paddle wheel in Fig. 9.22a. With a zero net torque, it does not rotate. Thus, the flow is irrotational.

> *Nonviscous flow* means that viscosity is neglected.

Viscosity refers to a fluid's internal friction, or resistance to flow. A truly nonviscous fluid would flow freely with no energy lost within it. Also, there would be no frictional drag between the fluid and the walls containing it. In reality, when a liquid flows through a pipe, the velocity is less near the walls because of frictional drag and is greater toward the center of the pipe. (Viscosity is discussed in more detail in Section 9.6.)

> *Incompressible flow* means that the fluid's density is constant.

Ideal fluid flow

Streamlines

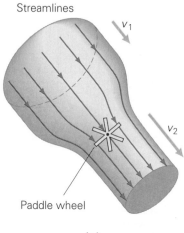

Paddle wheel

(a)

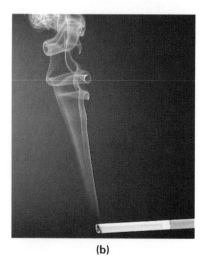

(b)

•**FIGURE 9.22 Streamline flow**
(a) Streamlines never cross and are closer together in regions of greater fluid velocity. The stationary paddle wheel indicates that the flow is irrotational, or without whirlpools and eddy currents.
(b) The smoke begins to rise in nearly streamline flow, but quickly becomes rotational and turbulent.

Liquids can usually be considered incompressible. Gases, on the other hand, are quite compressible. Sometimes, however, gases approximate incompressible flow—for example, air flowing relative to the wings of an airplane traveling at low speeds.

Theoretical or ideal fluid flow is not characteristic of most real situations. But the analysis of ideal flow provides results that approximate, or generally describe, a variety of applications. This analysis is not done in terms of Newton's laws but in terms of two basic principles: the conservation of mass and the conservation of energy.

Equation of Continuity

If there are no losses of fluid within a uniform tube, the mass of fluid flowing into the tube in a given time must be equal to the mass flowing out of the tube in the same time (by the conservation of mass). For example, in ●Fig. 9.23, the mass (Δm_1) entering the tube during a short time (Δt) is

$$\Delta m_1 = \rho_1 \, \Delta V_1 = \rho_1(A_1 v_1 \Delta t)$$

where A_1 is the cross-sectional area of the tube at the entrance, and in a time Δt, a fluid particle moves a distance equal to $v_1 \, \Delta t$. Similarly, the mass leaving the tube in the same time interval is

$$\Delta m_2 = \rho_2 \, \Delta V_2 = \rho_2(A_2 v_2 \Delta t)$$

Since the mass is conserved, $\Delta m_1 = \Delta m_2$, and

$$\rho_1 A_1 v_1 = \rho_2 A_2 v_2$$

or

$$\rho A v = \text{constant}$$

(9.18)   **Equation of continuity**

This general result is called the **equation of continuity**.

For an incompressible fluid, the density ρ is constant, so

$$A_1 v_1 = A_2 v_2$$

or

$$A v = \text{constant} \quad \textit{(for an incompressible fluid)}$$

(9.19)   **Flow rate equation**

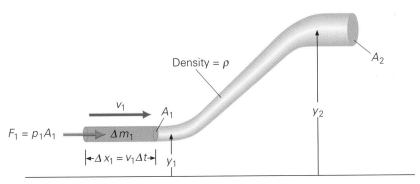

(a) Mass enters tube

(b) Mass exits tube

●**FIGURE 9.23   Flow continuity**
Ideal fluid flow can be described in terms of the conservation of mass by the equation of continuity. See the text for a description.

This is sometimes called the **flow rate equation**, since Av has the standard units of cubic meters per second (m^3/s, or volume/time). (In the British system, gal/min is often used.) Note that the flow rate equation shows that the fluid velocity is greater where the cross-sectional area of the tube is smaller. That is,

$$v_2 = \left(\frac{A_1}{A_2}\right)v_1$$

and v_2 is greater than v_1 if A_2 is less than A_1 (see •Fig. 9.24). (The streamlines will be closer together in a smaller tube.) This effect is evident in the common experience that the velocity of water is greater from a hose fitted with a nozzle than from the same hose without a nozzle.

The flow rate equation can be applied to the flow of blood in your body. Blood flows from the heart into the aorta. It then makes a circuit through the circulatory system, passing through arteries, arterioles (small arteries), capillaries, and back to the heart through veins. The velocity is lowest in the capillaries—a contradiction? No. The *total* area of the capillaries is much greater than that of the arteries or veins, so the flow rate equation holds.

Bernoulli's Equation

The conservation of energy or the general work-energy theorem leads to another relationship that has great generality for fluid flow. This relationship was first derived in 1738 by the Swiss mathematician Daniel Bernoulli (1700–1782) and is named for him.

Let's look again at the ideal fluid flowing in the tube in Fig. 9.23. Work is done by the external forces at the ends of the tube; F_1 positive work (in the same direction as the fluid's motion) and F_2 negative work (opposite to the fluid motion). The net work done on the system by these forces is then

$$W = F_1\Delta x_1 - F_2\Delta x_2 = (p_1 A_1)(v_1\Delta t) - (p_2 A_2)(v_2\Delta t)$$

The equation of continuity (Eq. 9.19) requires that $A_1 v_1 = A_2 v_2$, so we may write the work equation as

$$W = A_1 v_1\Delta t(p_1 - p_2)$$

Recall from the preceding derivation of the equation of continuity that $\Delta m_1 = \rho_1\Delta V_1 = \rho_1(A_1 v_1\Delta t)$, and $\Delta m_1 = \Delta m_2$, so we may write in general

$$W = \frac{\Delta m}{\rho}(p_1 - p_2)$$

The net work done on the system by the external forces (non-conservative work) must be equal to the change in total mechanical energy. That is, $W = \Delta E = \Delta K +$

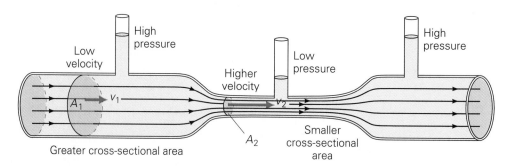

•**FIGURE 9.24 Flow rate and work-energy**
By the flow rate equation, Av = constant, and by the work-energy theorem, $W = \Delta K$. See the text for a description.

ΔU. Looking at the change in kinetic energy of an element of mass Δm, we have

$$\Delta K = \tfrac{1}{2}\Delta m\left(v_2^2 - v_1^2\right)$$

The corresponding change in gravitational potential energy is

$$\Delta U = \Delta mg(y_2 - y_1)$$

Thus,

$$W = \Delta K + \Delta U$$

$$\frac{\Delta m}{\rho}(p_1 - p_2) = \tfrac{1}{2}\Delta m(v_2^2 - v_1^2) + \Delta mg(y_2 - y_1)$$

Canceling the Δm's and rearranging gives the common form of **Bernoulli's equation**:

$$p_1 + \tfrac{1}{2}\rho v_1^2 + \rho gy_1 = p_2 + \tfrac{1}{2}\rho v_2^2 + \rho gy_2$$

or

$$p + \tfrac{1}{2}\rho v^2 + \rho gy = \text{constant} \tag{9.20}$$

Note that in working with a fluid, the terms in Bernoulli's equation are work or energy *per unit volume*. That is, $W = F\Delta x = p(A\Delta x) = p\Delta V$, and $p = W/\Delta V$ (work/volume). Similarly, with $\rho = m/V$, we have $\tfrac{1}{2}\rho v^2 = \tfrac{1}{2}mv^2/V$ (energy/volume, J/m^3) and $\rho gy = mgy/V$ (energy/volume).

Bernoulli's equation can be applied to many situations (see Demonstration 8). If there is horizontal flow ($y_1 = y_2$), then $p + \tfrac{1}{2}\rho v^2 = $ constant, which indicates that the pressure decreases if the fluid velocity increases (and vice versa). Chimneys and smokestacks are tall in order to take advantage of the more consistent and

Note: Compare the derivation of Eq. 5.10 in Section 5.5

Bernoulli effects

DEMONSTRATION 8 ■ Bernoulli-trapped Ball

(a) The demonstrator prepares to place the ball in an inverted funnel and to blow downward through the funnel.

(b) With air blowing through, the ball remains "trapped" in the funnel.

(c) When the demonstrator is out of breath, the ball falls.

How does blowing downward through a funnel keep a Ping Pong ball from falling? In this demonstration, seeing is believing in Bernoulli.

The air stream above and around the ball (along the wall of the funnel) is moving faster than the air below the ball, and the faster the air moves, the less pressure it exerts. Thus the greater pressure of the slower air below the ball holds it up. Without an air stream, the pressures are equalized (both atmospheric), and the ball falls.

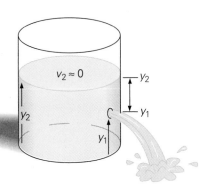

•FIGURE 9.25 **Fluid flow from a tank**
The flow rate is given by Bernoulli's equation. Example 9.12.

higher wind speeds at greater heights. The faster the wind blows over the top of a chimney, the lower the pressure, and the greater the pressure difference between the bottom and top of the chimney. Thus, the chimney draws better. Bernoulli's equation and the continuity equation (Av = constant) also tell you that if the cross-sectional area of a pipe is reduced, so that the velocity of the fluid passing through it is increased, the pressure is reduced.

Suppose a fluid is at rest ($v_2 = v_1 = 0$). Bernoulli's equation becomes

$$p_2 - p_1 = \rho g(y_1 - y_2)$$

This should look familiar; it is the pressure-depth relationship derived earlier (Eq. 9.10).

EXAMPLE 9.12 ■ FLOW RATE FROM A TANK: BERNOULLI'S EQUATION

Water escapes through a small hole in a tank as depicted in •Fig. 9.25. What is the initial flow rate of the water from the tank?

Solution. By Bernoulli's equation,

$$p_1 + \tfrac{1}{2}\rho v_1^2 + \rho g y_1 = p_2 + \tfrac{1}{2}\rho v_2^2 + \rho g y_2$$

where $y_2 - y_1$ is the height of the liquid surface above the hole. The atmospheric pressures acting on the open surface and at the hole, p_1 and p_2, respectively, are essentially equal and cancel from the equation, as does the density, giving

$$v_1^2 - v_2^2 = 2g(y_2 - y_1)$$

The equation of continuity says that $A_1v_1 = A_2v_2$, where A_2 is the cross-sectional area of the tank and A_1 is that of the hole. Since A_2 is much greater than A_1, then v_1 is much greater than v_2. So a good approximation is

$$v_1^2 = 2g(y_2 - y_1) \quad \text{or} \quad v_1 = \sqrt{2g(y_2 - y_1)}$$

The flow rate (volume/time) is then

$$\text{flow rate} = A_1v_1 = A_1\sqrt{2g(y_2 - y_1)}$$

Given the area of the hole and the height of the liquid above it, you can find the instantaneous speed of the water coming from the hole and the instantaneous flow rate.

Follow-up Exercise. What would be the percent change in the initial flow rate from the tank in this example if the diameter of the small circular hole were increased by 30.0%?

Another example of the Bernoulli effect (as it is sometimes called) is given in the Insight on p. 313.

*9.6   Viscosity, Poiseuille's Law, and Reynolds Number

Objectives: To be able to (a) define the coefficient of viscosity, (b) calculate flow rate using Poiseuille's law, and (c) interpret a fluid's Reynold's number with regard to turbulent flow.

Viscosity: resistance to flow

All real fluids have an internal resistance to flow, which is described as **viscosity**. Viscosity may be considered to be friction between the molecules of a fluid. In liquids, it is caused by short-range cohesive forces, and in gases, by collisions between molecules (see the discussion of air resistance in Section 4.6). The viscous drag for both liquids and gases depends on velocity and may be directly proportional to it in some cases. However, the relationship varies depending on conditions; for example, the drag is approximately proportional to v^2 or v^3 for turbulent flow.

Insight   Throwing a Curve

In baseball, a pitcher can cause a baseball to curve as it moves toward home plate by giving it an appropriate spin. Why a curve ball curves can be understood in terms of Bernoulli's equation and the viscous properties of air.

If air were an ideal fluid, the spin of a pitched ball would have no effect in changing the direction of its motion. (See Fig. 1a; the streamlines are those that would be seen by the ball or an observer moving with the ball.) However, because air is viscous, friction between it and the ball causes a thin (boundary) layer of air to be dragged around by the spinning ball. This is illustrated in Fig. 1b for a relatively slow airstream in which the flow is generally smooth. The speed of the ball relative to the air is thus greater on one side than on the other because the velocities are in the same direction on one side and in opposite directions on the other (Fig. 1b). By Bernoulli's equation, the low-velocity side (v') has a greater pressure than the high-velocity side (v''). Thus, the ball experiences a net force toward the low-pressure side and is deflected (curves).

However, baseballs are usually pitched very fast, and in this case, there is another effect. With a high air flow speed, the rotation of the ball causes the boundary layer to separate at different points on either side of the ball. This gives rise to a turbulent wake that is deflected in the spin direction (Fig. 2). As a result, there is a pressure difference and the ball is deflected or curves the opposite way. This is called the Magnus effect (after Gustav Magnus who observed the effect in 1852). From the standpoint of Newton's third law, the turbulent wake is deflected in one direction, and the reaction force deflects the ball in the opposite direction.

Note that the ball in Fig. 2 is rotating counterclockwise and is deflected to the left. To have the ball curve to the right would require a clockwise rotation. In general, the curve is in the direction that the front part of the ball is rotating. With the spin axis in the horizontal plane, the deflecting force may be either up or down, which can cause the ball not to drop as much under the influence of gravity, or to "sink" even faster. The spin axis can be oriented in any direction so as to produce a variety of effects with the deflecting force and gravity.

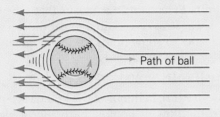

(a) If air were an ideal fluid, air flow would be the same on both sides.

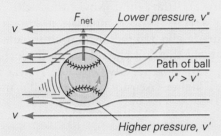

(b) Since air has viscosity, air flow is faster on one side of the ball, creating a pressure difference.

FIGURE 1   Bernoulli effect
(a) If air were an ideal fluid (without viscosity), or if no spin were applied, air flow would be the same on both sides of the ball. **(b)** Because air has viscosity, some air is dragged around with the ball, so the speed of the ball with respect to the air is greater on one side. Pressure on that side is lower, resulting in a net force in that direction (see text).

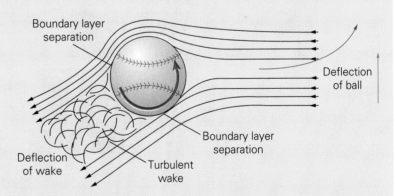

FIGURE 2   Magnus effect
The turbulent wake of a fast-moving ball gives rise to a pressure difference and the ball curves. In terms of Newton's third law, the wake is deflected in one direction and the ball is deflected in the opposite direction by the reaction force.

Internal friction causes fluid layers to move relative to each other in response to a shear stress (•Fig. 9.26a). This layered motion, called *laminar flow*, is characteristic of steady flow for viscous liquids at low velocities. At greater velocities, the flow becomes rotational, or turbulent. As Fig. 9.26a indicates, the magnitude of the shear stress can be used as a measure of viscosity. Viscosity is characterized by a *coefficient of viscosity*, η (the Greek letter eta), commonly referred to as simply the **viscosity**. If F/A is the shear stress needed to maintain laminar flow between two parallel planes separated by a distance h at a relative velocity of magnitude v, then

$$\eta = \left(\frac{F}{A}\right)\left(\frac{h}{v}\right) = \frac{Fh}{Av} \qquad (9.21)$$

SI unit of coefficient of viscosity: Pa·s

We can rearrange this equation slightly to give $\eta = (F/A)/(v/h)$. Since $v = \Delta x/\Delta t$, the expression (v/h) can be written as $(\Delta x/\Delta t)/h = (\Delta x/h)/\Delta t$, which is the rate of change of the shear strain. Thus we see that η is, in effect, the ratio of the shear stress to the rate of change of the shear strain. Essentially, this means that when the viscosity of a liquid is large, a lot more shear stress is required to get the liquid planes or layers to slide by each other. Viscosities of some fluids are given in Table 9.5.

Note that the SI unit for viscosity is pascal-seconds:

$$(N/m^2)(m/[m/s]) = (N/m^2)(s) = Pa·s$$

This combined unit is called the *poiseuille* (Pl), in honor of the French scientist Jean Poiseuille (1799–1869), who studied the flow of liquids, particularly blood. The cgs unit for viscosity is the *poise* (P). A smaller multiple, the centipoise (cP), is widely used because of its convenient size.

As might be expected, viscosity, and thus fluid flow, varies with temperature, which is evident from the old saying "slow as molasses in January." A familiar application is the viscosity grade given to motor oil used in automobiles. In winter, a low-viscosity, or relatively thin, oil should be used (such as SAE grade 10W or 20W) because it will flow more readily, particularly when the engine is cold at start-up. In summer, a higher-viscosity, or thicker, oil is used (SAE 30, 40, or even 50).

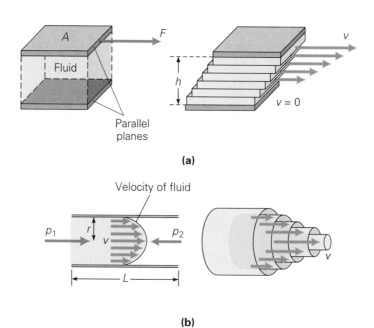

•FIGURE 9.26 Laminar flow (a) A shear stress causes fluid layers to move over each other in laminar flow. The shear force and the flow rate depend on the viscosity of the fluid. (b) For laminar flow through a pipe, the fluid velocity is less nearer the walls of the pipe because of frictional drag between the walls and the fluid.

TABLE 9.5	Viscosities of Various Fluids*		
		Viscosity (η)	
Fluid	Pl		cP
Liquids			
Alcohol, ethyl	1.2×10^{-3}		1.2
Blood, whole (37°C)	1.7×10^{-3}		1.7
Blood plasma (37°C)	2.5×10^{-3}		2.5
Glycerin	1.5		1.5×10^3
Mercury	1.55×10^{-3}		1.55
Oil, light machine	1.1		1.1×10^3
Water	1.00×10^{-3}		1.00
Gases			
Air	1.9×10^{-5}		1.9×10^{-2}
Oxygen	2.2×10^{-5}		2.2×10^{-2}

*At 20°C unless otherwise indicated. 1 poiseuille (Pl) = 10^3 centipoise (cP).

Of course, seasonal changes in the grade of motor oil are not necessary if you use the multigrade year-round oils. These contain additives called viscosity improvers, which are polymers whose molecules are long, coiled chains. A temperature increase causes the molecules to uncoil and intertwine with each other. Thus, the normal decrease in viscosity is counteracted. The action is reversed on cooling, and the oil maintains a relatively constant viscosity over a temperature range. Such motor oils are graded as, for example, SAE 10W-40 (or 10W-40 for short).

Note: SAE stands for Society of Automotive Engineers, an organization that designates the grades of motor oils based on flow rate or viscosity.

Poiseuille's Law

When a fluid flows through a pipe, there is frictional drag between the liquid and the walls, and the fluid velocity is greater toward the center of the pipe (•Fig. 9.26b). The average *flow rate Q* is given by

$$Q = A\bar{v} = \frac{\Delta V}{\Delta t} \qquad (9.22)$$

SI unit of flow rate: m^3/s

where ΔV is the volume passing a given point in the time Δt. The flow rate depends on properties of the fluid and dimensions of the pipe, as well as on the pressure difference between the ends of the pipe. Jean Poiseuille studied flow in pipes and tubes, assuming constant viscosity and steady or laminar flow. He derived a relationship for the flow rate, which is known as **Poiseuille's law**:

$$Q = \frac{\pi r^4 \Delta p}{8\eta L} \qquad (9.23)$$

Poiseuille's law—viscous flow

where r is the radius of the pipe and L is its length.

As should be expected, the flow rate (m^3/s) is inversely proportional to the viscosity and the length of the pipe. Also as expected, the flow rate is directly proportional to the pressure difference. Somewhat surprisingly, however, the flow rate is proportional to r^4, which makes it highly dependent on the radius of the tube, more so than might have been thought.

EXAMPLE 9.13 ■ FOURTH-POWER DEPENDENCE: POISEUILLE'S LAW

The radius of a length of pipe carrying a liquid is decreased by 5.0% because of deposits on the inner surface. By how much would the pressure difference between the ends of the constricted pipe have to be increased to maintain a constant flow rate?

Solution.

Given: $r_2 = 95\% \times r_1$, or   **Find:** Δp (pressure difference)
$r_2 = (0.95)r_1$
$Q_2 = Q_1$

With $Q_2 = Q_1$, by Eq. 9.23

$$\frac{Q_2}{Q_1} = \frac{r_2{}^4 \Delta p_2}{r_1{}^4 \Delta p_1}$$

and

$$\Delta p_2 = \left(\frac{r_1}{r_2}\right)^4 \Delta p_1 = \left(\frac{1}{0.95}\right)^4 \Delta p_1 = (1.23)\,\Delta p_1$$

The pressure difference must be increased by 23% to maintain the same flow rate.

Thus, with constricted arteries in the body, the blood pressure has to increase to maintain the normal flow of blood. As pointed out in the Insight on p. 317, this makes the heart work harder. A similar application of Poiseuille's law concerns administering gases in respiratory therapy for conditions involving obstructions in the air passages that reduce air flow to the lungs.

Follow-up Exercise. An engineer wants to have the same flow rate of water and light machine oil from pipes of the same length and with the same pressure head. Will the radii of the pipes be the same? If not, by what factor will they differ?

Reynolds Number

When the flow rate of a fluid exceeds a certain velocity, the flow ceases to be laminar and becomes turbulent. Poiseuille's law then no longer applies. Analyzing turbulent flow is a difficult task, but a value that tells when the onset of turbulence will occur has been determined experimentally. This is expressed in terms of a dimensionless number called the **Reynolds number** (R_n):

Reynolds number—turbulent flow

$$R_n = \frac{\rho \bar{v} d}{\eta} \qquad (9.24)$$

where ρ is the density of the fluid, $\bar{v}$ the average flow speed, d the diameter of a cylindrical pipe or tube, and η the viscosity.

In a pipe with smooth walls, the flow is laminar if R_n is below 2000. Turbulence generally sets in when R_n is about 2000 or more ($R_n \gtrsim 2000$). It is possible to have laminar flow if R_n is above 2000, but that flow will be very unstable. Any slight disturbance will cause it to become turbulent. Interestingly, flow rate is then often greater than if Poiseuille's law applied. Thus engineers deliberately design pipelines for turbulent flow, and much of the flow in our circulatory system is turbulent.

Insight   Blood Pressure

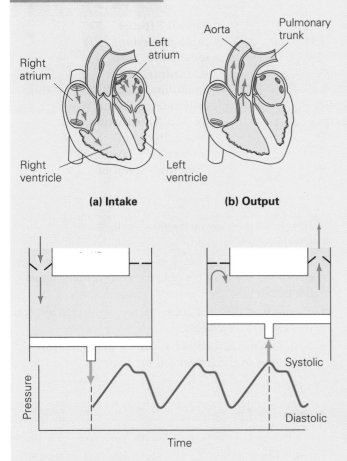

(a) Intake   **(b) Output**

Right atrium

Left atrium

Aorta

Pulmonary trunk

Right ventricle

Left ventricle

Pressure

Time

Systolic

Diastolic

FIGURE 1   The heart as a pump
The human heart is analogous to a mechanical force pump. Its pumping action, consisting of (a) intake and (b) output, gives rise to variations in blood pressure.

Basically, a pump is a machine that transfers mechanical energy to a fluid, causing it to flow. There are a wide variety of pumps, but one of interest to everyone is the human heart. The heart is a muscular pump that drives blood through the body's circulatory network of arteries, capillaries, and veins.

At the start of each pumping cycle, the heart's interior chambers enlarge and fill with freshly oxygenated blood arriving from the lungs (see Fig. 1). On contraction, the blood is forced out through the aorta into the arterial network. Smaller and smaller arteries branch off from the main ones, until the very small capillaries are reached. There food and oxygen being carried by the blood are exchanged with surrounding tissues, and wastes are picked up. The blood then flows into the veins to complete its circuit back to the heart.

The walls of the arteries have considerable elasticity and expand and contract with each pumping cycle. When the heart contracts, the blood pressure in the arteries increases. When the heart relaxes, the blood pressure decreases. The maximum pressure is called the systolic pressure, and the minimum pressure is called the diastolic pressure (after the two parts of the pumping cycle, the systole and diastole).

Taking a person's blood pressure involves measuring the pressure of the blood on the arterial walls. This is done with a sphygmomanometer (the Greek word *sphygmo* means "pulse"). An inflatable cuff is used to shut off the blood flow temporarily. The cuff pressure is slowly released, and the artery is monitored with a stethoscope (Fig. 2). A point is reached where blood is just forced through the constricted artery. This flow is turbulent and gives rise to a sound with each heartbeat. When the sound is first heard, the systolic pressure is noted, which is normally about 120 mm Hg. When the turbulent beats disappear because of smooth blood flow, the diastolic reading is taken. The pressure at this point is normally 80 mm Hg. Blood pressure is commonly reported by giving the systolic and diastolic pressures separated by a slash, for example, 120/80 (which is read as "120 over 80"). Normal blood pressure ranges between 100 and 140 for the systolic pressure and between 70 and 90 for the diastolic pressure.

High blood pressure is a common health problem. The elastic walls of the arteries expand under the hydraulic force of the blood pumped from the heart. Their elasticity may diminish with age, however. Fatty deposits (of cholesterols) can narrow and roughen the arterial passageways, impeding the blood flow and giving rise to a form of arteriosclerosis, or hardening of the arteries. Because of these defects, the driving pressure must increase to maintain a normal blood flow. The heart must work harder, which places a greater demand on its muscles. A relatively slight decrease in the effective cross-sectional area of a blood vessel has a rather large effect on the flow rate. (See the discussion of Poiseuille's law.)

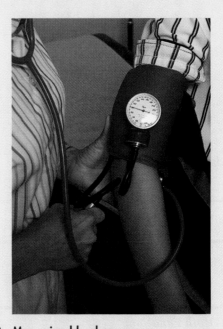

FIGURE 2   Measuring blood pressure
The pressure is indicated on the gauge in mm Hg or torr.

Chapter Review

Important Terms

Important Concepts

- Elastic modulus is the ratio of stress to strain.
- Young's modulus is the ratio of tensile stress to strain. (The stress may also be compressional.)
- Pressure applied to an enclosed fluid is transmitted undiminished to every point in the fluid and to the walls of the container (Pascal's principle).
- A body immersed wholly or partially in a fluid is buoyed up by a force equal in magnitude to the weight of the volume of fluid it displaces (Archimedes' principle).
- An object will float (sink) in a fluid if the average density of the object is less (greater) than the density of the fluid.
- Capillary action is a consequence of surface tension (cohesion) and adhesion.
- Bernoulli's equation is basically a mathematical statement of the conservation of energy for a fluid.

Important Equations

Stress:

$$\text{stress} = \frac{F}{A} \tag{9.1}$$

Strain (tensile):

$$\text{strain} = \frac{\Delta L}{L_\text{o}} = \frac{L - L_\text{o}}{L_\text{o}} \tag{9.2}$$

Young's Modulus:

$$Y = \frac{F/A}{\Delta L/L_\text{o}} \tag{9.4}$$

Shear Modulus:

$$S = \frac{F/A}{x/h} \approx \frac{F/A}{\phi} \tag{9.5}$$

Bulk Modulus:

$$B = \frac{F/A}{-\Delta V/V_\text{o}} = -\frac{\Delta p}{\Delta V/V_\text{o}} \tag{9.6}$$

Compressibility:

$$k = \frac{1}{B} \tag{9.7}$$

Pressure:

$$p = \frac{F}{A} \tag{9.8}$$

Pressure–Depth Equation:

$$p = p_\text{o} + \rho g h \tag{9.10}$$

Archimedes' Principle:

$$F_\text{b} = m_\text{f}g = \rho_\text{f}g V_\text{f} \tag{9.14}$$

Surface Tension:

$$\gamma = \frac{F}{L} \tag{9.16}$$

Capillary Height:

$$h = \frac{2\gamma \cos \phi}{\rho g r} \tag{9.17}$$

Equation of Continuity:

$$\rho_1 A_1 v_1 = \rho_2 A_2 v_2 \quad \text{or} \quad \rho A v = \text{constant} \tag{9.18}$$

Flow Rate Equation (for an incompressible fluid):

$$A_1 v_1 = A_2 v_2 \quad \text{or} \quad A v = \text{constant} \tag{9.19}$$

Bernoulli's Equation:

$$p_1 + \tfrac{1}{2}\rho v_1^2 + \rho g y_1 = p_2 + \tfrac{1}{2}\rho v_2^2 + \rho g y_2 \tag{9.20}$$

or

$$p + \tfrac{1}{2}\rho v^2 + \rho g y = \text{constant}$$

*Coefficient of Viscosity:

$$\eta = \frac{Fh}{Av} \tag{9.21}$$

*Flow Rate:

$$Q = A\bar{v} = \frac{\Delta V}{\Delta t} \tag{9.22}$$

$$Q = \frac{\pi r^4 \, \Delta p}{8\eta L} \qquad (9.23)$$

$$R_n = \frac{\rho \bar{v} d}{\eta} \qquad (9.24)$$

Exercises

9.1 Solids and Elastic Moduli

(Use as many significant figures as you need to show small changes.)

1 The pressure on an elastic body is described by (a) a modulus, (b) work, (c) stress, (d) strain.

2 Shear moduli exist for (a) solids, (b) liquids, (c) gases, (d) all of these.

3 One material has a greater Young's modulus than another. What does this tell you?

4 Why are scissors sometimes called shears? Is this a descriptive name?

5 Write the general form of Hooke's law and find the units of the "spring constant" for elastic deformation.

6 ■ A 100-kg mass is suspended using a cable with a diameter of 2.0 cm. What is the stress in the cable?

7 ■ A force of 500 N is applied at an angle of 37° to the surface of the end of a square bar. That surface is 4.0 cm on a side. What are (a) the compressional stress and (b) the shear stress on the bar?

8 ■ A carpenter applies a tangential force of 150 N to the upper surface of a block of wood. If the dimensions of the surface are 20 cm by 30 cm, what is the shear stress on the block?

9 ■■ A metal wire 1.0 mm in diameter and 2.0 m long hangs vertically with a 6.0-kg mass suspended from it. If the wire stretches 1.4 mm under the tension, what is the value of Young's modulus for the metal?

10 ■■ A steel rail in a railroad is 8.0 m long and has a cross-sectional area of 0.0025 m². No gap has been left between the rails, and on a hot day, the rail thermally expands 3.0 x 10⁻³ m. What is the force on the ends of the rail? (Is this a good reason to leave expansion gaps between the rails?)

11 ■■ A cable initially 130.00 cm long and 2.00 mm in diameter is stretched to a length of 130.26 cm by a force of 600 N. What is Young's modulus for the material of the cable?

12 ■■ A rectangular steel column (20.0 cm × 15.0 cm) supports a load of 12.0 metric tons. If the column was 4.00 m in length before being stressed, what is its loaded length?

13 ■■ A bimetallic rod as illustrated in •Fig. 9.27 is composed of brass and copper. If the rod is subjected to a compressive force of 5.00×10^4 N, which way will the rod bow or bend? (Justify your answer mathematically.)

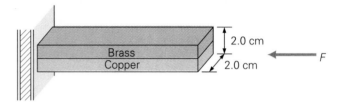

•**FIGURE 9.27 Bimetallic rod and mechanical stress**
See Exercise 13.

14 ■■ A shear force of 750 N is applied to one face of a cube of aluminum measuring 10 cm on a side. What is the displacement relative to the opposite face?

15 ■■ Two metal posts of aluminum and copper of the same size are subjected to equal shear stresses. Which post will show the larger deformation angle and by what factor?

16 ■■ A rectangular block of Jell-O with length, width, and height of 10 cm, 8.0 cm, and 4.0 cm, respectively, is subjected to a shear force of 0.40 N on its upper surface. If the top surface is displaced 0.30 mm relative to the bottom surface, what is the shear modulus of the gelatin?

17 ■■ Two metal plates are held together by two rivets with diameters of 0.40 cm. If the maximum shear stress a single rivet can withstand is 5.0×10^8 N/m², how much force must be applied parallel to the plates to shear off both rivets?

18 ■■ What pressure difference is required to compress a volume of mercury by 0.10%?

19 ■■ (a) Which of the liquids in Table 9.1 has the greatest compressibility? (b) For equal volumes of ethyl alcohol and water, which would require more pressure to be compressed by 0.10%, and how many times more?

20 ■■■ A brass cube 6.0 cm on a side is placed in a pressure chamber and subjected to a pressure of 1.2×10^7 N/m² on all its surfaces. What is the volume of the cube under this pressure?

21 ■■■ A 45-kg traffic light is suspended with two steel cables of equal length and radii of 0.50 cm. If the cables make an angle of 15° with the horizontal, what is the fractional increase in their length due to the weight of the light?

9.2 Fluids: Pressure and Pascal's Principle

22 For the pressure–depth relationship for a fluid ($p = \rho gh$), it is assumed (a) the pressure decreases with depth, (b) that a pressure difference depends on the reference point, (c) the fluid density is constant, (d) only liquids apply.

23 Two dams form artificial lakes of equal depth. However, one lake backs up 15 km behind the dam and the other backs up 50 km. What effect does the difference in length have on the pressures on the dams?

24 A water dispenser for pets has an inverted plastic bottle, as shown in •Fig. 9.28. (The water is dyed blue for contrast.) When a certain amount of water is drunk from the bowl, more water flows automatically from the bottle into the bowl. The bowl never overflows. Explain the operation of the dispenser. Does the height of the water in the bottle depend on the surface area of the water in the bowl?

•FIGURE 9.28   Pet barometer
See Exercise 24.

25 (a) Liquid storage cans, such as gasoline cans, generally have capped vent holes. What is the purpose of the vent, and what happens if you forget to remove the cap before pouring the liquid? (b) Explain how a medicine dropper works. (c) Explain how we breathe, i.e., inhalation and exhalation.

26 Automobile tires are inflated to about 30 lb/in.², whereas thin bicycle tires are inflated to 90–115 lb/in.² —three times or more pressure! Why is this?

27 Blood pressure is usually taken using the arm. However, suppose the pressure reading were taken on the calf of the leg of a standing person. Would there be a difference? Explain.

28 ■ Show that a water barometer would be impractical.

29 ■ A pure gold nugget with a volume of 15 cm³ is placed on a pan balance. What would be the volume of the brass weights needed to balance the nugget?

30 ■ Which has the greater volume and how many times greater, 11.6 kg of aluminum or 2.75 kg of lead?

31 ■ A 100-mL graduated cylinder is filled halfway with ethyl alcohol, and then the other half is filled with gasoline. What is the total mass of the combined liquids?

32 ■ A 75-kg athlete does a single handstand. If the area of the hand in contact with the floor is 125 cm², what pressure is exerted on the floor?

33 ■■ What is the pressure exerted on a 15° ski slope by a 90-kg skier going down it? (Assume that the area of contact of the skis with the snow is 0.40 m².)

34 ■■ A water tank sits on a hill above a town. If the water level in the tank is 20 m above the town's water plant, what is the static water pressure at the plant if the water is delivered through an 0.80-m diameter pipe?

35 ■■ The gauge pressure in both tires of a bicycle is 690 kPa. If the bicycle and the person riding it have a combined mass of 90.0 kg, what is the area of contact of *each* tire with the ground?

36 ■■ What is the fractional decrease in pressure when a barometer is raised 30 m to the top of a tall building? (Assume that the density of air remains constant.)

37 ■■ A student decides to compute the standard barometric reading on top of Mt. Everest (29,028 ft) by assuming the density of air has the same constant density as at sea level. Try this yourself. What does the result tell you?

38 ■■ In a sample of seawater taken from an oil spill region, it is found that an oil layer 4.0 cm thick floats on 55 cm of water. If the density of the oil is 0.75×10^3 kg/m³, what is the absolute pressure on the bottom of the container?

39 ■■ The pressure exerted by a person's lungs can be measured by having the person blow as hard as possible into one side of a manometer. If a person blowing into one side of an open-tube manometer produces a 70-cm difference in the heights of the columns of water in the manometer arms, what is the lung pressure?

40 ■■ In 1960, the U.S. Navy's bathyscaphe *Trieste* descended to a depth of 10,912 m (about 35,000 ft) in the Mariana Trench in the Pacific Ocean. (a) What was the pressure at that depth? (Assume that sea water is incompressible.) (b) What was the force on a circular observation window with a diameter of 15 cm?

41 ■■ Water is poured into one arm of an open U-tube containing mercury. What column height of water will cause the mercury level in the other arm to rise 2.5 cm?

42 ■■■ The output piston of a hydraulic garage lift has a cross-sectional area of $0.20 \ m^2$. (a) How much pressure on the input piston is required to support a car with a mass of 1.4 metric tons? (b) What force is applied to the input piston if it has a diameter of 5.0 cm? (c) The change in potential energy in lifting a load of mass m is $\Delta U = F_o h_o = mgh_o$, where h_o is the distance the output piston moves. Show that this is equal to the work done by the input piston, $W = F_i h_i$.

43 ■■■ A syringe with a plunger diameter of 2.0 cm is attached to a hypodermic needle with a diameter of 1.5 mm. What minimum force must be applied to the plunger to inject a fluid into a vein, where the blood pressure is 12 torr?

44 ■■■ An open-tube mercury manometer has a height difference between the columns of −10 cm. What are (a) the gauge pressure and (b) the absolute pressure?

45 ■■■ A hydraulic balance used to detect small mass changes is shown in •Fig. 9.29. If a mass of 0.25 g is placed on the balance platform, by how much will the height of the water in the smaller cylinder have changed when the balance comes to equilibrium?

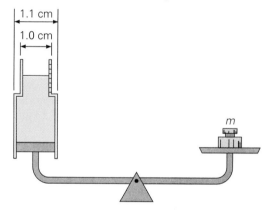

•FIGURE 9.29 A hydraulic balance
See Exercise 45.

9.3 Buoyancy and Archimedes' Principle

46 If a barometer were constructed using a liquid less dense than mercury, (a) a tube of larger diameter would be needed (b) the liquid column would be unchanged for a given pressure, (c) a longer tube may be necessary, (d) none of these.

47 If an object displaces an amount of liquid of greater weight than its own, the object will (a) float, (b) sink, (c) remain in equilibrium at any submerged position.

48 (a) What is the criterion for constructing a life jacket that will keep a person afloat? (b) Why is it so easy to float in the Great Salt Lake in Utah?

49 An ice cube floats in a glass of water. As the ice melts, how does the level of the water in the glass change? Would it make any difference if the ice cube were hollow? Explain.

50 ■ Ocean-going ships in port are loaded to the Plimsoll mark, which is a line indicating the maximum safe loading depth. However, in New Orleans, ships are loaded until the Plimsoll mark is somewhat below the water line. Why is this?

51 ■ A cube 8.5 cm on a side has a mass of 0.65 kg. Will the cube float in water?

52 ■■ A helium-filled balloon supports a load of 1 metric ton. (a) If the load includes the balloon's mass, what is its volume? (b) If the balloon were spherical, what would be its radius?

53 ■■ A submarine has a mass of 10,000 metric tons. How much water must be displaced for the sub to be in equilibrium just below the ocean surface?

54 ■■ Suppose that Archimedes found that the king's crown had a mass of 0.750 kg and a volume of $3.980 \times 10^{-5} \ m^3$. (a) How did he determine the volume? (b) Was the crown pure gold?

55 ■■ An 0.80-kg crown is submerged in water and its apparent weight is measured to be 7.30 N. Is the crown pure gold?

56 ■■ A rectangular boat as illustrated in •Fig. 9.30 is overloaded so that the water level is just 1.0 cm below the top of the side of the boat. What is the combined mass of the people and the boat?

•FIGURE 9.30 An overloaded boat
See Exercise 56.

57 ■■ An irregularly shaped piece of metal has a mass of 90 g in air. It is suspended from a scale and the scale reads 75 g when the piece is submerged in water. What are the volume and density of the piece of metal?

58 ■■ A flat-bottomed rectangular boat has a length of 4.0 m and a width of 1.5 m. If the load is 2000 kg (including the mass of the boat), how much of the boat will be submerged when it floats in a lake?

59 ■■ An ocean-going barge is 50.0 m long and 20.0 m wide and has a mass of 145 metric tons. Will the barge clear a reef 1.50 m below the surface of the water?

60 ■■ A block of iron quickly sinks in water, but ships constructed of iron float. A solid cube of iron 1.0 m on a side is made into sheets. To make an open cube that will not sink from these sheets, what should the minimum length of the sides be?

61 ■■ Plans are being made to bring back the zeppelin (a lighter-than-air airship, like the Goodyear blimp, that carries passengers and cargo—but filled with helium, not flammable hydrogen as was used in the ill-fated *Hindenburg*). One design calls for the ship to be 110 m long and have a total weight (without helium) of 30.0 metric tons. Assuming the ship's "envelope" to be cylindrical, what would its diameter be so as to lift the total weight?

62 ■■■ A girl floats in a lake with 97% of her body beneath the water. What are (a) her mass density and (b) her weight density?

63 ■■■ A bar of 24-karat (pure) gold is supposed to be completely solid. To check this, a scientist takes the bar into the lab, and using a scale finds that it has a mass of 1.00 kg in air and a measured mass of 0.93 kg when completely immersed in water. Is the bar solid?

9.4 Surface Tension and Capillary Action

64 Surface tension (a) depends on cohesion, (b) is a factor in capillary action, (c) causes liquids to form spherical drops, (d) all of the preceding.

65 Capillary action depends on (a) surface tension only, (b) cohesive forces only, (c) adhesive forces only, (d) all of the preceding.

66 Can you suggest (a) why the sand is wet well above the water line on a beach (assume there are no waves), and (b) why wet sand is firmer and easier to walk on than dry sand?

67 The speed of blood flow is greater in arteries than in capillaries. However, the flow rate equation ($Av =$ a constant) seems to predict that the speed should be greater in the smaller capillaries. Can you resolve this apparent inconsistency?

68 ■■ When sewing, why does a person often wet the end of the thread before trying to put it through the eye of a needle? How about putting water on a comb before combing one's wind-blown hair?

69 ■■ A vertical force of 0.014 N is required to lift a wire ring with a radius of 1.5 cm away from the surface of a liquid. What is the surface tension of the liquid? (Neglect the diameter of the wire.)

70 ■■ If a circular wire loop with a diameter of 3.0 cm is used to measure the surface tension of (a) soapy water and (b) mercury at room temperature (20°C), what would be the vertical force required to lift the ring from the liquid in each case?

71 ■■ To what height will water rise in a glass capillary tube with a diameter of 2.0 mm at room temperature?

72 ■■ In order to have water at 20°C rise a distance of 5.0 cm in a glass capillary tube, what would the diameter of the tube have to be?

73 ■■ (a) Would water or whole blood rise higher in a glass capillary tube having a diameter of 0.80 mm? (b) How many times greater would the height difference be? (Assume a contact angle of 0° for each, the water at 20°C, and the blood at normal body temperature.)

74 ■■ Given capillary tubes of the same diameter, will water or ethyl alcohol rise higher in the tubes, and how many times higher?

75 ■■ A glass capillary tube with a diameter of 1.0 mm is placed vertically in a dish of mercury. How far is the mercury in the tube depressed below the surface level in the dish?

76 ■■■ Water rises in plants through tiny capillaries (called xylem), which have diameters that range from about 0.01 mm to 0.30 mm. What is the maximum height to which water can rise in a xylem system under capillary action? (Obviously, this does not account for moisture getting to the tops of tall trees. Atmospheric pressure will support a column of water only about 10 m in height. It is believed that water is raised to great heights in trees by a negative pressure created by evaporation. Water evaporates from leaves, and the cohesive forces between water molecules pull up more molecules from below.)

9.5 Fluid Dynamics and Bernoulli's Equation

77 If the velocity at some point changes with time, the fluid flow is *not* (a) steady, (b) irrotational, (c) incompressible, (d) nonviscous.

78 According to Bernoulli's equation, if the pressure on a liquid is increased, (a) the flow velocity always increases, (b) the height of the liquid always increases, (c) both the flow velocity and the height of the liquid may increase, (d) none of these.

79 When a car is overtaken and passed by a fast-moving semitrailer truck, the driver of the car notices a force that tends to push the car toward the truck. What causes this? (This also occurs with ships, which is why they do not pass each other too closely.)

80 (a) The shape of an airplane wing causes the air to move faster over its upper surface than across its lower surface. Explain how this supplies lift to the plane, in terms of Bernoulli's equation. How about Newton's third law? (b) What supplies lift for a helicopter?

81 If you hold a narrow strip of paper in front of your mouth and blow over the top surface, the strip will rise. (Try it.) Explain why.

82 A stream of water from a faucet is narrower at the bottom than at the top. If allowed to fall a sufficient distance, the stream breaks into drops. Explain these effects.

83 Two common demonstrations of Bernoulli effects are shown in •Fig. 9.31. Explain these effects.

(a)

(b)

• **FIGURE 9.31 Bernoulli effects**
See Example 83.

84 ■ Show that the static pressure-depth equation can be derived from Bernoulli's equation.

85 ■ What is the flow rate of water that moves at an average speed of 2.8 m/s through a pipe with a diameter of 15 cm?

86 ■■ Fluid A flows three times as fast as fluid B through the same horizontal pipe. Which has the greater density, and how many times greater is it?

87 ■■ Water flows from a 2.0-cm-diameter pipe at a speed of 0.35 m/s. How long will it take to fill a 10-L container?

88 ■■ The speed of blood in a main artery, which has a diameter of 1.0 cm, is 4.5 cm/s. (a) What is the flow rate in the artery? (b) If the capillary system has a total cross-sectional area of 2500 cm², the average speed of blood through the capillaries is what percentage of that through the main artery? (c) Why is there a need for a low blood speed through the capillaries?

89 ■■ A room measures 3.0 m by 4.5 m by 6.0 m. If the heating and air conditioning ducts to and from the room have a diameter of 30 cm, what is the average flow rate of air necessary for all the air in the room to be exchanged every 10 min? (Assume that the density of the air is constant.)

90 ■■ The spout heights for the container shown in •Fig. 9.32 are 10 cm, 20 cm, 30 cm, and 40 cm. The water level is maintained at a height of 45 cm by an outside supply. (a) What is the speed of the water coming from each hole? (b) Which stream of water has the greatest range relative to the base of the container? Justify your answer.

• **FIGURE 9.32 Streams as projectiles**
See Exercise 90.

91 ■■■ In an industrial cooling process, water is circulated through a system. If water is pumped from the first floor through a 6.0-cm diameter pipe with a speed of 0.45 m/s under a pressure of 400 torr, what will be the pressure on the next floor 4.0 m above in a pipe with diameter of 2.0 cm?

92 ■■■ Water flows through a horizontal 7.0-cm-diameter pipe under a pressure of 6.0 Pa with a flow rate of 25 L/min. At one point, calcium deposits reduce the cross-sectional area of the pipe to 30 cm². What is the pressure at this point? (Consider the water to be an ideal fluid.)

Exercises **323**

93 ■■■ A Venturi meter can be used to measure the flow speed of a liquid. A simple one is shown in ●Fig. 9.33. Show that the flow speed for an ideal fluid is given by

$$v_1 = \sqrt{\frac{2g\Delta h}{(A_1^2/A_2^2) - 1}}$$

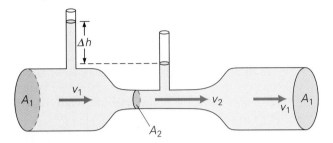

●**FIGURE 9.33   A flow speed meter**
See Exercise 93.

*9.6   Viscosity, Poiseuille's Law, and Reynolds Number

94 For a given flow rate of a viscous liquid, the required pressure difference varies as (a) r^2, (b) $1/r^2$, (c) r^4, (d) $1/r^4$.

95 The Reynolds number (a) is part of Poiseuille's law, (b) has units of poise, (c) gives an indication of turbulent flow, (d) is usually negative.

96 Show that the units for the coefficient of viscosity are N·s/m² or Pa·s.

97 Show that the Reynolds number is a dimensionless quantity.

98 ■■ A horizontal pipe with an inside diameter of 3.0 cm and a length of 6.0 m carries water, which has a flow rate of 40 L/min. What is the required pressure difference between the ends of the pipe?

99 ■■ What is the maximum flow rate of water for laminar flow in a 5.0-cm-diameter pipe?

100 ■■ A hospital patient receives a quick 500-cc blood transfusion through a needle with a length of 5.0 cm and an inner diameter of 1.0 mm. If the blood bag is suspended 0.85 m above the needle, how long does the transfusion take? (Neglect the viscosity of the blood flowing in the plastic tube between the bag and the needle.)

101 ■■ Blood goes through a vessel with a diameter of 1.5 cm. What would be the minimum average flow speed if the flow were turbulent?

102 ■■■ (a) Show that the average flow rate for turbulent flow is given by

$$Q = \frac{\pi d R_n \eta}{4\rho}$$

where d is the diameter of the pipe. (b) What is the flow rate of water through a 3.0-cm-diameter pipe?

Additional Exercises

103 How high will water rise in a 0.75-mm-diameter capillary tube made of silver?

104 A steel machine part has a volume of 0.36 m³. If it is suspended from a scale and immersed in gasoline, what will the scale read?

105 Ice will float in water, but sink in ethyl alcohol. In what proportion should these liquids be mixed so that an ice cube will be able to float at any depth?

106 A 60-kg woman balances herself on the heel of one of her high-heeled shoes. If the heel is a square with sides of 1.5 cm, what is the pressure exerted on the floor? (Express your answer in Pa, lb/in.², and atm.)

107 What is the force exerted on a person's body by the atmosphere if the surface area is 1.30 m²? (Express your answer in both newtons and pounds.)

108 A copper wire 100.00 cm long is stretched to a length of 100.02 cm when it supports a certain load. If an aluminum wire of the same diameter is used to support the same load, what should its initial length be if its stretched length is to be 100.02 cm also?

109 A scuba diver dives to a depth of 12 m in a lake. If the circular glass plate on the diver's face mask has a diameter of 18 cm, what is the force of the water only on it?

110 If water rises to a height of 4.5 cm in a glass capillary tube at 20°C, what is the diameter of the tube?

111 Oil is poured into the open side of an open tube manometer containing mercury. What is the density of the oil if a 7.0-cm column of mercury supports a 80-cm column of oil?

112 If the atmosphere had a constant density equal to its density at the Earth's surface, how high would it extend? What does this tell you?

113 The water level in a 15-cm-diameter cylinder like that shown in ●Fig. 9.34 is maintained at a constant height of 0.45 m. If the diameter of the spout pipe is 0.50 cm, how high is the vertical stream of water? (Assume the water to be an ideal fluid.)

114 A vertical steel beam with a rectangular cross-sectional area of 24 cm² is used to support a sagging floor in a building. If the beam supports a load of 10,000 N, by what percentage is the beam compressed?

FIGURE 9.34 How high a fountain?
See Exercise 113.

115 A container 100 cm deep is filled with water. If a small hole is punched in its side 25 cm from the top, with what initial speed will the water flow from the hole?

116 Show that specific gravity is the ratio of densities, given that by definition it is the ratio of the weight of a given volume of a substance and the weight of an equal volume of water.

117 The flow rate of water through a garden hose is 66 cm³/s, and the hose and nozzle have cross-sectional areas of 6.0 cm² and 1.0 cm², respectively. (a) If the nozzle is held 10 cm above the spigot, what are the flow speeds through the spigot and the nozzle? (b) What is the pressure difference between these points? (Consider the water to be an ideal fluid.)

118 Ancient stonemasons sometimes split huge blocks of rock by inserting wooden pegs into holes drilled in the rock and then pouring water on the pegs. Can you explain the physics that underlies this technique? [*Hint:* Think about sponges and paper towels.]

119 A crane lifts a rectangular iron bar whose dimensions are 0.25 m x 0.20 m x 10 m from the bottom of a lake. What is the minimum upward force the crane must supply when the bar is (a) in the water and (b) out of the water?

120 •Figure 9.35 shows a simple laboratory experiment. From your observations of the photographs, calculate (a) the volume and (b) the density of the suspended sphere. (Assume that the density of the sphere is uniform and that the liquid in the beaker is water.) (c) Would you be able to make the same determinations if the liquid in the beaker were mercury (see Table 9.2)? Explain.

FIGURE 9.35 Dunking a sphere
See Exercise 120.

10 Temperature

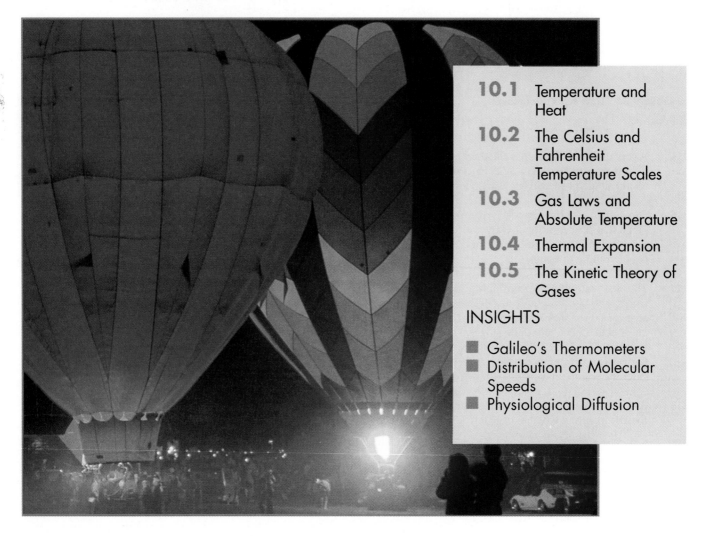

In the eighteenth century, long before the first blimp circled a football stadium to advertise tires, there were hot-air balloons. And today, these majestic airships still delight both those who ride in them and those who watch them floating serenely over the landscape. Part of their appeal must surely be their simplicity. Like sailboats, they are low-tech devices in a high-tech world. You can equip a balloon with the latest satellite-linked, computerized navigational system and attempt to fly across the Pacific, but the basic principles that keep you aloft were known and understood centuries ago. You can build a model of a hot-air balloon in your back yard, or even in the classroom.

But why must the air in a hot-air balloon be hot? More fundamentally, what do we mean by "hot"? How does hot air differ from cool air? In this chapter you'll find that answering such simple questions leads to an understanding of some far-reaching physical principles. You'll explore the nature of heat and the ways in which we measure temperature. You'll also encounter the gas laws, which explain not only the behavior of balloons but also more important phenomena, such as how our lungs supply us with the oxygen we need to live.

Temperature and heat are frequent subjects of conversation, but if you had to state what the words really mean, you might find yourself at a loss. Certainly these two concepts are related. We use thermometers of all sorts to record temperatures, which provide an objective equivalent for our sensory experience of hot and cold. And we all know that a temperature change generally results from the application or removal of heat. Temperature, therefore, is some sort of indication, or measure, of heat. But what is heat?

An early theory of heat considered it to be a fluidlike substance called caloric (from the Latin word *calor,* meaning "heat") that could be made to flow in and out of a body. Even though this theory has been abandoned, we still speak of heat as flowing from one object to another. Heat is now characterized as energy, and temperature and thermal properties are explained by considering the atomic and molecular behavior of substances. This and the next two chapters examine the nature of temperature and heat in terms of microscopic (molecular) theory and macroscopic observations.

10.1 Temperature and Heat

Objective: To be able to distinguish between temperature and heat.

A good way to begin studying thermal physics is with definitions of temperature and heat. **Temperature** is a relative measure, or indication, of hotness or coldness. A hot stove is said to have a high temperature, and an ice cube to have a low temperature. An object that has a higher temperature than another object does is said to be hotter. Note that hot and cold are relative terms, like tall and short. We can perceive temperature by touch. However, this temperature sense is somewhat unreliable, and its range is too limited to be useful for scientific purposes.

Heat is related to temperature and describes the process of energy transfer from one object to another. That is, **heat** is *energy that is transferred from one object to another because of a temperature difference.* Generally, heat transfer is the result of a temperature difference (though not always, as we shall find in Section 11.4). Thus, heat is energy in transit, so to speak. Once transferred, the energy becomes part of the total energy of the molecules of the object or system, its **internal energy**.

On a microscopic level, temperature is associated with molecular motion. In kinetic theory (Section 10.5), which treats gas molecules as point particles, temperature is a measure of the average random translational kinetic energy of the molecules. However, diatomic molecules and other real substances, besides having such translational "temperature" energy, also have kinetic energy due to linear vibration and rotation, as well as potential energy due to the attractive forces between molecules. These energies may not contribute directly to temperature, but are part of the internal energy, which is the sum total of all such energies (•Fig. 10.1).

However, a higher temperature does not necessarily mean that one system has a greater internal energy than another. For example, in a classroom on a cold day, the air temperature is relatively high compared to that of the outdoor air. But all that cold air outside the classroom has far more internal energy as a whole than does the warm air inside, simply because there is so much *more* of it. If this

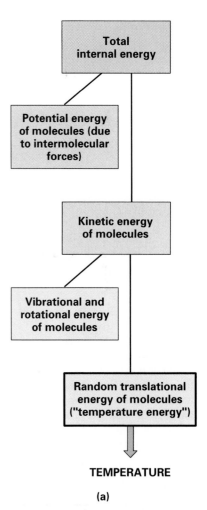

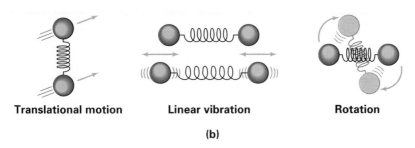

Translational motion   **Linear vibration**   **Rotation**

(b)

•**FIGURE 10.1 Molecular motions** **(a)** Temperature is associated with random translational motion; the internal energy of a system is the total energy. **(b)** A molecule may move as a whole in translational motion and/or have vibrational and rotational motions.

were not the case, heat pumps would not be practical (see Chapter 12). The internal energy of a system also depends on its mass, or the number of molecules in the system.

When heat is transferred between two objects, whether or not they are touching, they are said to be in *thermal contact*. When there is no longer a net heat transfer between objects in thermal contact, they are at the same temperature and are said to be in *thermal equilibrium*.

10.2 The Celsius and Fahrenheit Temperature Scales

Objectives: **To be able to (a) explain how a temperature scale is constructed, and (b) convert temperatures from one scale to another.**

A measure of temperature is obtained using a **thermometer**, a device constructed to make evident some property of a substance that changes with temperature. Fortunately, many physical properties of materials change sufficiently with temperature to be used as the bases for thermometers. By far the most obvious and commonly used property is **thermal expansion**, a change in the dimensions or volume of a substance that occurs when the temperature changes.

Almost all substances expand with increasing temperature, but to different extents. Most substances also contract with decreasing temperature. (Thermal expansion is used to mean both expansion and contraction; contraction is considered to be a negative expansion.) Because some metals expand more than others, a bimetallic strip (strips of two different metals bonded together) can be used to measure temperature changes. As heat is applied, the composite strip will bend away from the side made of the metal that expands more (•Fig. 10.2). Coils formed from such strips are used in dial thermometers and in common household thermostats (•Fig. 10.3).

A common thermometer is the liquid-in-glass type, which is based on the thermal expansion of a liquid. A liquid in a glass bulb expands into a glass stem, rising in a capillary bore. Mercury and alcohol (usually colored with a red dye to make it more visible) are the liquids used in most liquid-in-glass thermometers. These substances were chosen because of their relatively large thermal expansion and because they remain liquids over normal temperature ranges. (For a unique type of thermometer, see the Insight on p. 331.)

Thermometers are calibrated so that a numerical value may be assigned to a given temperature. For the definition of any standard scale or unit, two fixed reference points are needed. Since all substances change dimensions with temperature, an absolute reference for expansion is unavailable. However, the necessary fixed points may be correlated to physical phenomena that always occur at the same temperatures.

•**FIGURE 10.2 Thermal expansion** (a) A bimetallic strip is made of two strips of metal bonded together. **(b)** When such a strip is heated, it bends because of unequal expansions of the two metals. Here brass expands more than iron and the deflection is toward the iron. The deflection of the end of a strip could be used to measure temperature.

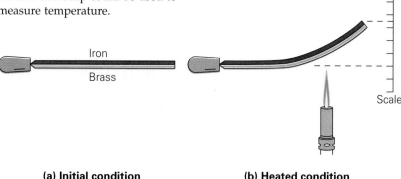

(a) Initial condition   **(b) Heated condition**

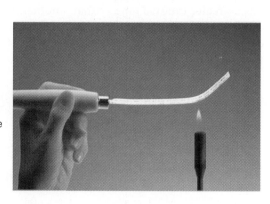

Insight   Galileo's Thermometers

Galileo, among his many other accomplishments, is generally credited with inventing one of the first thermometers. Several people worked on such thermometers about this time (around 1600), and it is not really clear who developed the first one. Called a *thermoscope*, it consisted of an air-filled glass bulb connected to a small-bore glass tube that extended into a container of colored liquid (Fig. 1). As the temperature of the air in the bulb changed, so did its pressure and volume ($pV \propto T$, Eq. 10.5), which affected the height of the column in the glass tube.

Although the thermoscope was relatively sensitive to variations in temperature, it was not a good measuring device because variations in atmospheric pressure on the liquid in the open container also affected the height of the column in the tube. (Recall the barometer in Section 9.2.) Our present-day liquid-in-glass thermometers are sealed so that atmospheric pressure has no effect on the reading.

Another novel thermometer is shown in Fig. 2. It is known as Galileo's buoyancy thermometer, and is cited as a further development of the thermoscope. Galileo actually had no direct involvement in the development of the buoyancy thermometer; his name was applied because of his popularity. Even so, it is an interesting instrument, with the temperature measurement being based on buoyancy and Archimedes' principle (Section 9.3). The density of the liquid in the column varies with temperature. This variation affects the buoyancies of the glass bulbs, which change heights accordingly. Calibrated temperatures are marked on disks attached to the bulbs. If one of the bulbs is in perfect equilibrium, so that it floats freely by itself in the middle of the column, then the value on its disk gives the temperature reading. If there is no freely floating bulb and the bulbs are grouped at the top and bottom of the column, then the temperature lies between the lowest value of the upper group and the highest value of the lower group. (Note that this is a Fahrenheit thermometer.)

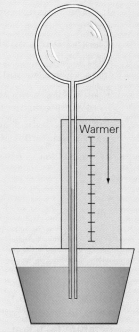

FIGURE 1   Galileo's thermoscope
An illustration of the thermoscope principle. Changes in the temperature of the air in the glass bulb produce variations in the height of the liquid column in the glass tube. With a measurement scale, the thermoscope becomes a thermometer.

FIGURE 2   Galileo's buoyancy thermometer
See Insight text for description. Can you tell what temperature the thermometer is registering?

(a)

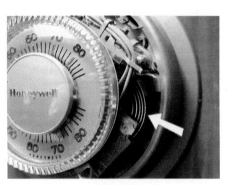

(b)

•**FIGURE 10.3   Bimetallic coil**
Bimetallic coils are used in dial thermometers **(a)** and household thermostats **(b)**. In a thermostat, the expansion and contraction of the coil are used to regulate the heating or cooling system, turning it off and on as the room temperature changes.

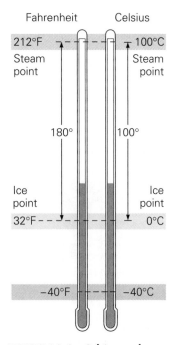

•FIGURE 10.4 Celsius and Fahrenheit temperature scales
Between the ice and steam fixed points, there are 100 degrees on the Celsius scale and 180 degrees on the Fahrenheit scale. Thus, a Celsius degree is 1.8 times larger than a Fahrenheit degree.

Note: For distinction, a particular temperature, e.g., T = 20°C, is written with °C (20 degrees Celsius), whereas a temperature interval, e.g., $\Delta T = 80°C - 60°C = 20$ C°, is written with C° (20 Celsius degrees).

The ice point and the steam point of water are two convenient fixed points. Also commonly known as the freezing and boiling points, these two points are the temperatures at which pure water freezes and boils under a pressure of 1 atm (standard pressure).

The two familiar temperature scales are the **Fahrenheit temperature scale** and the **Celsius temperature scale**. As shown in •Fig. 10.4, the ice and steam points have values of 32°F and 212°F, respectively, on the Fahrenheit scale and 0°C and 100°C on the Celsius scale. On the Fahrenheit scale, there are 180 equal intervals, or degrees, (F°), between the two reference points, and on the Celsius scale, there are 100 (C°). Therefore, a Celsius degree is larger than a Fahrenheit degree; since $180/100 = 9/5 = 1.8$, it is almost twice as large.

A relationship for converting between the two scales may be obtained from a graph of Fahrenheit temperature (T_F) versus Celsius temperature (T_C), like the one in •Fig. 10.5. The equation of the straight line (in general slope–intercept form, $y = mx + b$) is

$$T_F = \left(\frac{180}{100}\right)T_C + 32$$

and

$$T_F = \tfrac{9}{5}T_C + 32$$
or
$$T_F = 1.8T_C + 32$$ *Celsius to Fahrenheit conversion*   (10.1)

where $\tfrac{9}{5}$ or 1.8 is the slope of the line and 32 is the intercept on the vertical axis. Thus, to change from a Celsius temperature (T_C) to its equivalent Fahrenheit temperature (T_F), you simply multiply the Celsius reading by $\tfrac{9}{5}$ and add 32.

The equation may be solved for T_C to convert from Fahrenheit to Celsius:

$$T_C = \tfrac{5}{9}(T_F - 32)$$ *Fahrenheit to Celsius conversion*   (10.2)

EXAMPLE 10.1 ■ CONVERTING TEMPERATURE SCALE READINGS: FAHRENHEIT AND CELSIUS

What are (a) 20°C on the Fahrenheit scale, and (b) normal body temperature, 98.6°F on the Celsius scale?

Solution.   We wish to make the following conversions:

Given: (a) $T_C = 20\,°C$ *Find:*   (a) T_F
             (b) $T_F = 98.6\,°F$ (b) T_C

(a)   Eq. 10.1 is for changing Celsius readings to Fahrenheit:
$$T_F = \tfrac{9}{5}T_C + 32 = \tfrac{9}{5}(20) + 32 = 68°F$$
This temperature, often taken to be room temperature, is a good one to remember.

(b)   Eq. 10.2 changes Fahrenheit to Celsius:
$$T_C = \tfrac{5}{9}(T_F - 32) = \tfrac{5}{9}(98.6 - 32) = \tfrac{5}{9}(66.6) = 37.0°C$$
On the Celsius scale, normal body temperature has a whole-number value. Keep in mind that a Celsius degree is 1.8 times (almost twice) as large as a Fahrenheit degree, so a temperature elevation of several degrees on the Celsius scale makes a big difference. For example, a temperature of 40.0°C represents an elevation of 3.0 C° over normal body temperature. This is an increase of $3.0 \times 1.8 = 5.4°$ on the Fahrenheit scale, or a temperature of $98.6 + 5.4 = 104.0°F$.

Follow-up Exercise.   Convert the following temperatures: (a) −40°F to Celsius, and (b) −40°C to Fahrenheit.

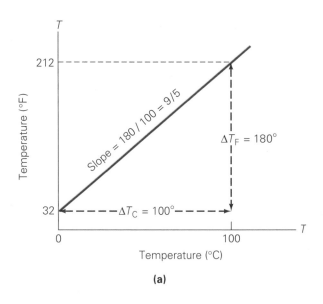

•FIGURE 10.5 Fahrenheit versus Celsius
(a) A plot of Fahrenheit temperature versus Celsius temperature gives a straight line of the general form $y = mx + b$, where $T_F = \frac{9}{5}T_C + 32$. **(b)** The temperature is given here in degrees Celsius. What is the equivalent temperature in degrees Fahrenheit?

PROBLEM-SOLVING HINT

Because of the similarity between Eqs. 10.1 and 10.2, it is easy to miswrite them. Since they are equivalent, you need to know only one of the equations, say Celsius to Fahrenheit (Eq. 10.1, $T_F = \frac{9}{5}T_C + 32$). Solving this for T_C algebraically gives Eq. 10.2. A good way to make sure that you have written the conversion equation correctly is to test it with a known temperature, such as the boiling point of water. For example, $T_C = 100°$, and

$$T_F = \frac{9}{5}T_C + 32 = \frac{9}{5}(100) + 32 = 212°F$$

Thus we know the equation is correct.

Liquid-in-glass thermometers are adequate for many temperature measurements, but problems arise when very accurate determinations are needed. A material may not expand uniformly over a wide temperature range. When calibrated to the ice and steam points, an alcohol thermometer and a mercury thermometer have the same readings at these points. But because of different expansion properties, the thermometers will not have exactly the same reading at an intermediate temperature, such as room temperature. For very sensitive temperature measurements and to define intermediate temperatures precisely, some other type of thermometer must be used. One such thermometer is discussed in the next section.

10.3 Gas Laws and Absolute Temperature

Objectives: To be able to (a) describe the ideal gas law, and (b) explain how it is used to determine absolute zero.

Different liquid-in-glass thermometers show slightly different readings for temperatures other than the fixed points because of differing expansion properties. A thermometer that uses a gas, however, gives the same readings regardless of which gas is used. Experiments show that all gases at very low densities exhibit the same expansion behavior.

The variables that describe the behavior of a given quantity (mass) of gas are pressure, volume, and temperature (p, V, and T). When temperature is held constant, the pressure and volume of a quantity of gas are related as follows:

$$pV = \text{constant} \quad \text{or} \quad p_1V_1 = p_2V_2 \qquad \text{(at constant temperature)} \qquad (10.3)$$

That is, the product of the pressure and volume is a constant. This relationship is known as **Boyle's law**, after Robert Boyle (1627–1691), the English chemist who discovered it. (See Demonstration 9.)

When the pressure is held constant, the volume of a quantity of gas is related to the *absolute* temperature (to be defined shortly):

$$\frac{V}{T} = \text{constant} \quad \text{or} \quad \frac{V_1}{T_1} = \frac{V_2}{T_2} \qquad \text{(at constant pressure)} \qquad (10.4)$$

That is, the ratio of the volume and the temperature is a constant. This relationship is known as **Charles' law**, after the French scientist Jacques Charles (1747–1823), who made early hot-air balloon flights and was therefore quite interested in the relationship between the volume and temperature of a gas. A popular demonstration of Charles' law is shown in •Fig. 10.6.

Low-density gases obey these laws, which may be combined into a single relationship. Notice that since $pV = \text{constant}$ and $V/T = \text{constant}$ for a given quantity of gas, then pV/T must also equal a constant. This relationship is known as the **ideal gas law**:

$$\frac{pV}{T} = \text{constant} \quad \text{or} \quad \frac{p_1V_1}{T_1} = \frac{p_2V_2}{T_2} \qquad \textit{ideal gas law}$$

That is, the ratio pV/T at time 1 is the same as at time 2, or at any other time, as long as the quantity of gas does not change.

This relationship can be written in a more general form that applies not just to a given quantity of a single gas but to any quantity of any dilute gas:

$$\frac{pV}{T} = Nk_B \quad \text{or} \quad pV = Nk_BT \qquad \textit{ideal gas law} \qquad (10.5)$$

•**FIGURE 10.6   Charles' law in action**
Demonstrations of the relationship between volume and temperature of a quantity of gas. A weighted balloon, initially at room temperature, is placed in a beaker of water. **(a)** When ice is placed in the beaker and the temperature lowered, the balloon's volume is reduced. **(b)** When the water is heated and the temperature is raised, the balloon's volume increases.

DEMONSTRATION 9 ■ Boyle's Shaving Cream

(a) A swirl of shaving cream is placed in a transparent vacuum chamber.

(b) A vacuum pump begins to evacuate the chamber.

(c) As the pressure is reduced, the swirl of cream grows in volume from the expansion of the air bubbles trapped in the cream.

(d) In a dramatic volume reduction, the shaving cream cannot stand up to the sudden increase in pressure when the chamber is vented to the atmosphere.

A demonstration that shows the inverse relationship of pressure (*p*) and volume (*V*) of Boyle's law, $p \propto 1/V$. Note: A similar effect occurs for a Boyle's marshmallow.

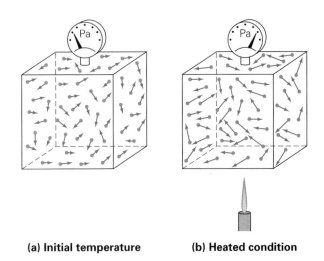

FIGURE 10.7 Constant-volume gas thermometer
Such a thermometer indicates temperature as a function of pressure, since for a low-density gas, $p \propto T$. **(a)** At some initial temperature, the pressure reading has a certain value. **(b)** When the gas thermometer is heated, the pressure (and temperature) reading is higher.

(a) Initial temperature   **(b) Heated condition**

Here N is the number of molecules in the sample of gas and k_B is a number known as Boltzmann's constant: $k_B = 1.38 \times 10^{-23}$ J/K. K stands for temperature on the Kelvin scale discussed below. (Can you show that the units are correct?) Note that the mass of the sample does not appear explicitly in Eq. 10.5. However, a given number of molecules of a gas is proportional to the total mass of the gas. The ideal gas law, sometimes called the *perfect gas law*, applies to all gases at low densities and describes fairly accurately the behavior of most gases at normal densities.

Absolute Zero and the Kelvin Temperature Scale

The important point for the purpose of this section is that the pressure and volume are directly proportional to temperature: $pV \propto T$. This relationship allows a gas to be used to measure temperature in a *constant-volume gas thermometer*. Holding the volume of a gas constant, which can be done easily in a rigid container (•Fig. 10.7), gives $p \propto T$. Thus, with a constant-volume gas thermometer, temperature is read in terms of pressure. A plot of pressure versus temperature gives a straight line in this case, as shown in •Fig. 10.8(a).

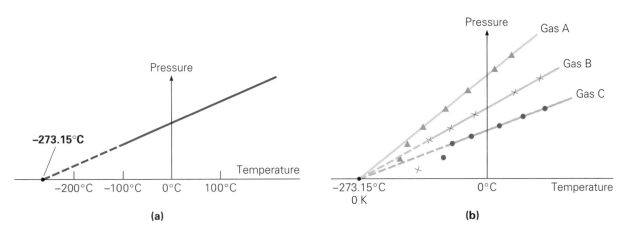

FIGURE 10.8 Pressure versus temperature
(a) A low-density gas kept at a constant volume gives a straight line on a p versus T graph: $p = (Nk_B/V)T$. When the line is extended to the zero pressure value, a temperature of $-273.15°C$ is obtained, which is taken to be absolute zero. **(b)** Extrapolation of lines for all low-density gases indicates the same absolute zero temperature. The actual behavior of gases deviates from the straight-line relationship at low temperatures because they start to liquefy.

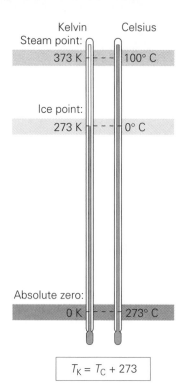

Kelvin   Celsius

Steam point:

373 K ╎╎ - - - ╎╎ 100° C

Ice point:

273 K ╎╎ - - - ╎╎ 0° C

Absolute zero:

0 K ╎╎ - - - ╎╎ 273° C

$$T_K = T_C + 273$$

•FIGURE 10.9 The absolute Kelvin temperature scale
The lowest temperature on the Kelvin scale (corresponding to −273.15°C) is absolute zero. A unit interval on the Kelvin scale, called a kelvin and abbreviated K, is equivalent to a temperature change of a Celsius degree; thus $T_K = T_C + 273.15$. (The constant is usually rounded to 273 for convenience. For example, a temperature of 0°C is equal to 273 kelvins.)

As can be seen in •Fig. 10.8(b), measurements of real gases (plotted data points) deviate from the values predicted by the ideal gas law at low temperatures. This is because the gases liquefy at low temperatures. However, the relationship is linear over a large temperature range, and it looks as though the pressure might reach zero with decreasing temperature if the gas continued to be a gas (ideal or perfect).

The absolute minimum temperature for an ideal gas may therefore be inferred by extrapolating, or extending the straight line to the axis, as in Fig. 10.8. This temperature is found to be −273.15°C and is designated as **absolute zero**. Absolute zero is believed to be the lower limit of temperature, but it has never been attained. In fact, there is a physical law that says it never can be (Chapter 12).

There is no known upper limit to temperature. The temperatures at the centers of some stars are estimated to be greater than 100 million degrees.

Absolute zero is the foundation of the **Kelvin temperature scale**, named after the British scientist Lord Kelvin.* On this scale, −273.15°C is taken as the zero point, that is, as 0 K (•Fig. 10.9). The size of the unit for Kelvin temperature is the same as the Celsius degree, so temperatures on these scales are related by

$$T_K = T_C + 273.15 \qquad (10.6)$$

where T_K is the temperature in **kelvins** (*not* degrees Kelvin). The kelvin is abbreviated as K (*not* °K). In many instances, a rounded value may be used in Eq. 10.6, that is, $T_K = T_C + 273$.

The absolute Kelvin scale is the official SI temperature scale; however, the Celsius scale is usually used for everyday temperature readings. The absolute temperature in kelvins is used primarily in scientific applications. *Caution:* Keep in mind that Kelvin temperatures *must* be used with the ideal gas law. It is a common mistake to use Celsius or Fahrenheit temperatures in that equation. Also note that there can be no negative temperatures on the Kelvin scale if absolute zero is the lowest possible temperature. That is, the Kelvin scale doesn't have an arbitrary zero temperature somewhere within the scale—0 K is absolute zero, period.

EXAMPLE 10.2 ■ DEEPEST FREEZE: ABSOLUTE ZERO ON THE FAHRENHEIT SCALE

What is absolute zero on the Fahrenheit scale?

Solution.

Given: $T_K = 0\,K$ *Find:* T_F

Temperatures on the Kelvin scale are related directly to Celsius temperatures by $T_K = T_C + 273.15$, so first you convert 0 K to a Celsius value:

$$T_C = T_K - 273.15 = 0 - 273.15 = -273.15°C$$

Then converting to Fahrenheit (Eq. 10.1) gives

$$T_F = \tfrac{9}{5}T_C + 32 = \tfrac{9}{5}(-273.15) + 32 = -459.67°F$$

Thus, absolute zero is about −460°F.

Follow-up Exercise.   There is an absolute temperature scale associated with the Fahrenheit temperature called the Rankine scale. A Rankine degree is the same size as a Fahrenheit degree and absolute zero is taken as 0°R (zero degrees Rankine). Write the conversion equations between (a) the Rankine and the Fahrenheit scales, (b) the Rankine and the Celsius scales, and (c) the Rankine and the Kelvin scales.

*Lord Kelvin, born William Thomson (1824–1907), developed devices to improve telegraphy and the compass and was involved in the laying of the first transatlantic cable. When he became a lord, it is said that he considered choosing Lord Cable or Lord Compass as his title, but decided on Lord Kelvin after a river that runs near the University of Glasgow in Scotland where he was a professor of physics for 50 years.

Initially, gas thermometers were calibrated using the ice and steam points. The Kelvin scale uses absolute zero and a fixed point adopted in 1954 by the International Committee on Weights and Measures. The second fixed point is the **triple point of water**, which represents a unique set of conditions where water coexists simultaneously in equilibrium as a solid, a liquid, and a gas. The conditions for the triple point are a pressure of 4.58 mm Hg, or about 610 Pa (760 mm Hg is normal atmospheric pressure), and a temperature taken to be 273.16 K. The kelvin is then defined as 1/273.16 of the temperature at the triple point of water.

Now let's use the ideal gas law, which requires absolute temperatures.

EXAMPLE 10.3 ■ THE IDEAL GAS LAW: USING ABSOLUTE TEMPERATURES

A quantity of low-density gas in a rigid container is initially at room temperature (20°C) and a particular pressure (p_1). If the gas is heated to a temperature of 60°C, by what factor does the pressure change?

Solution.

Given: $T_1 = 20°C$ **Find:** p_2/p_1 (pressure ratio or factor)
            $T_2 = 60°C$

Since we want the factor of pressure change, we write p_2/p_1 as a ratio. For example, if $p_2/p_1 = 2$, then $p_2 = 2p_1$, or the pressure would change (increase) by a factor of 2. The ratio should also indicate that we will make use of the ideal gas law in ratio form. The gas law requires *absolute* temperatures, so we first change the Celsius temperatures to kelvins:

$$T_1 = 20°C + 273 = 293 \text{ K}$$
$$T_2 = 60°C + 273 = 333 \text{ K}$$

where a rounded value of 273 was used in Eq. 10.6 for convenience. Then using the ideal gas law (Eq. 10.5) in the form $p_2V_2/T_2 = p_1V_1/T_1$, and since $V_1 = V_2$,

$$p_2 = \left(\frac{T_2}{T_1}\right)p_1 = \left(\frac{333 \text{ K}}{293 \text{ K}}\right)p_1 = 1.14p_1$$

Thus, p_2 is 1.14 times p_1; the pressure increases by a factor of 1.14, or by 14 percent. (What would the factor be if the Celsius temperatures were *incorrectly* used? It would be much larger, as you can readily see: 60°C/20°C = 3.)

Follow-up Exercise. If the gas in this Example is heated when at room temperature so that the pressure increases by a factor of 1.26, what is the final Celsius temperature?

Another Form of the Ideal Gas Law

The ideal gas law ($pV = Nk_BT$) is sometimes written in another form:

$$pV = nRT \qquad \textit{ideal gas law} \qquad (10.7)$$

with the constants nR rather than Nk_B used for convenience. Here, n is the number of moles (mol) of the gas, a quantity defined below, and R is called the *universal gas constant:*

$$R = 8.31 \text{ J/mol·K}$$

or
$$R = 0.0821 \text{ L·atm/mol·K}$$

The latter form is convenient because in working with gases, volumes are often measured in liters (L) and pressures in atmospheres (atm).

You may have learned in a chemistry course that a **mole** of a substance is the quantity that contains **Avogadro's number** (N_A) of molecules,

$$N_A = 6.02 \times 10^{23} \text{ molecules/mole}$$

Thus, the n and N in the two forms of the ideal gas law are related by $N = nN_A$. As in Eq. 10.5, neither the mass of gas in a particular sample nor the molecular

Note: N is the total number of molecules
N_A is Avogadro's number
$n = N/N_A$ is the number of moles

mass appears explicitly in this form of the ideal gas law; these quantities are reflected in the number of moles, n. From Eq. 10.7 it is easy to show that a mole of *any* gas occupies 22.4 L at 0°C and 1 atm. These conditions are known as standard temperature and pressure (STP).

Eq. 10.7 is a very practical form of the gas law, because we generally work with measured quantities or moles (n) of gases, rather than the number of molecules (N). To use Eq. 10.7, we need to know the number of moles of a gas. The weight of one mole of any substance is simply its formula mass expressed in grams. The formula mass is determined from the chemical formula and the atomic masses of the atoms (these are listed in Appendix IV and are commonly rounded to the nearest one-half). For example, water, H_2O, with two hydrogen atoms and one oxygen atom, has a formula mass of $2.0 + 16.0 = 18.0$; thus one mole of water has a mass of 18.0 grams. Similarly, the oxygen we breathe, O_2, has a formula mass of $2 \times 16.0 = 32.0$. Thus a mole of oxygen, 32 g, would occupy 22.4 L at STP.

EXAMPLE 10.4 ■ VOLUME OF A MOLE: APPLYING THE IDEAL GAS LAW

What volume in liters does one mole of helium occupy at one atmosphere of pressure and a temperature of (a) 0°C and (b) 20°C?

Solution.

Given: $n = 1.00$ mol   *Find:*   (a) V_1 (volume)
           $p = 1.00$ atm   (b) V_2
           (a) $T_1 = 0°C$
           (b) $T_2 = 20°C$
           $R = 0.0821$ L·atm/mol·K   (known)

Note that we use the liter·atm form of the universal gas constant. (Why?) Eq. 10.7 can be used to find the volumes, but first we must change the temperatures to kelvins. *Probably the most common error in working with the gas laws is forgetting to use absolute temperatures.*

(a) In kelvins, $T_1 = T_C + 273 = 0°C + 273 = 273$ K. Then, rearranging Eq. 10.7, we have

$$V_1 = \frac{nRT_1}{p} = \frac{(1.00 \text{ mol})(0.0821 \text{ L·atm/mol·K})(273 \text{ K})}{1 \text{ atm}} = 22.4 \text{ L}$$

You may have noticed that the gas was at STP, and so knew the answer already, since (as we saw previously) 22.4 L is the volume that one mole of *any* gas occupies at STP. This Example dramatizes the point. The gas in the Example was helium; but clearly the results would have been no different had the calculation involved one mole of O_2 or N_2 or any other gas.

(b) Converting T_2 to kelvins, $T_2 = T_C + 273 = 20°C + 273 = 293$ K. Then computing the volume at T_2,

$$V_2 = \frac{nRT_2}{p} = \frac{(1.00 \text{ mol})(0.0821 \text{ L·atm/mol·K})(293 \text{ K})}{1 \text{ atm}} = 24.1 \text{ L}$$

and the volume increases, as one might expect. Essentially, the gas is heated to 20°C at constant pressure.

Follow-up Exercise. A quantity of gas occupies 28.0 L at STP. How many molecules are there in this volume of gas?

Note: *Always* use Kelvin (absolute) temperatures with the ideal gas law.

10.4 Thermal Expansion

Objective: **To be able to calculate the thermal expansions of solids and liquids.**

Changes in the dimensions and volumes of materials are common thermal effects. As you learned earlier, thermal expansion provides a means of temperature mea-

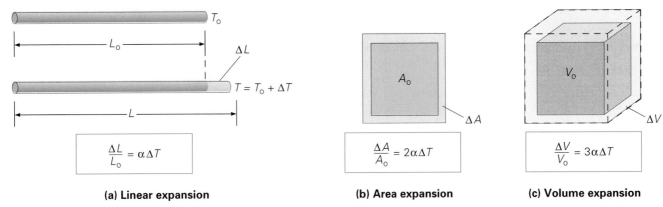

$$\frac{\Delta L}{L_o} = \alpha \Delta T$$

(a) Linear expansion

$$\frac{\Delta A}{A_o} = 2\alpha \Delta T$$

(b) Area expansion

$$\frac{\Delta V}{V_o} = 3\alpha \Delta T$$

(c) Volume expansion

•**FIGURE 10.10 Thermal expansion**
(a) Linear expansion is proportional to the temperature change; that is, the change in length ΔL is proportional to ΔT, and $\Delta L = \alpha L_o \Delta T$, where a is the thermal coefficient of linear expansion. **(b)** For isotropic expansion, the thermal coefficient of area expansion is approximately 2α. **(c)** The thermal coefficient of volume expansion for solids is about 3α.

surement. The thermal expansion of gases is generally described by the ideal gas law, and is very obvious. Less dramatic, but by no means less important, is the thermal expansion of solids and liquids.

Note: Solids are discussed in Section 9.1.

Thermal expansion results from a change in the average distance separating the atoms of a substance. The atoms are held together by bonding forces, which can be simplistically represented as springs in a simple model of a solid. The atoms vibrate back and forth, and with increased temperature (that is, more internal energy) become increasingly active and vibrate over greater distances. With wider vibrations in all dimensions, the solid expands as a whole.

The change in one dimension of a solid (length, width, or thickness) is called linear expansion. For small temperature changes, linear expansion is approximately proportional to ΔT, or $T - T_o$ (•Fig. 10.10a). The fractional change in length is $(L - L_o)/L_o$, or $\Delta L/L_o$, where L_o is the original length (at the initial temperature). This is related to the change in temperature by

$$\frac{\Delta L}{L_o} = \alpha \Delta T \quad \text{or} \quad \Delta L = \alpha L_o \Delta T \tag{10.8}$$

where α is the **thermal coefficient of linear expansion**. Note that the unit of α is inverse temperature: $1/C°$, or $C^{°-1}$. Values of α for some materials are given in Table 10.1.

A solid may have different coefficients of linear expansion for different directions, but for simplicity this book will assume that the same coefficient applies

TABLE 10.1 Values of Thermal Expansion Coefficients (in $C^{°-1}$) for Some Materials at 20°C			
Material	Coefficient of linear expansion (α)	Material	Coefficient of volume expansion (β)
Aluminum	24×10^{-6}	Alcohol, ethyl	1.1×10^{-4}
Brass	19×10^{-6}	Gasoline	9.5×10^{-4}
Brick or concrete	12×10^{-6}	Glycerin	4.9×10^{-4}
Copper	17×10^{-6}	Mercury	1.8×10^{-4}
Glass, window	9.0×10^{-6}	Water	2.1×10^{-4}
Glass, Pyrex	3.3×10^{-6}		
Gold	14×10^{-6}	Air (and most other gases) at 1 atm	3.5×10^{-3}
Ice	52×10^{-6}		
Iron and steel	12×10^{-6}		

Thermal Area Expansion

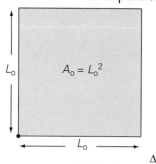

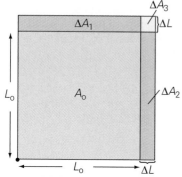

$\Delta A = \Delta A_1 + \Delta A_2 + \Delta A_3$

$\Delta A_1 = \Delta A_2 = L_0 \, \Delta L$

$\qquad = L_0 \, (\alpha L_0 \, \Delta T) = \alpha A_0 \Delta T$

Since ΔA_3 is very small
compared to ΔA_1 and ΔA_2,

$\Delta A \approx 2\alpha A_0 \Delta T$

to all directions (in other words, that solids show *isotropic* expansion). Also, the coefficient of expansion may vary slightly for different temperature ranges. Since this variation is negligible for most common applications, α will be considered to be constant and independent of temperature.

Eq. 10.8 may be rewritten to give the final length (L) after a temperature change:

$$\Delta L = \alpha L_0 \Delta T$$
$$L - L_0 = \alpha L_0 \Delta T$$
$$L = L_0 + \alpha L_0 \Delta T$$

or

$$L = L_0(1 + \alpha \Delta T) \qquad (10.9)$$

Since area (A) is length squared (L^2),

$$A = L^2 = L_0^2(1 + \alpha \Delta T)^2 = A_0(1 + 2\alpha\Delta T + \alpha^2\Delta T^2)$$

Because the values of α for solids are much less than 1 ($\sim 10^{-6}$, as shown in Table 10.1), the second-order term (containing α^2) may be dropped with negligible error. As a first-order approximation, then,

$$A = A_0(1 + 2\alpha\Delta T) \quad \text{or} \quad \frac{\Delta A}{A_0} = 2\alpha\Delta T \qquad (10.10)$$

Thus, the **thermal coefficient of area expansion** (•Fig. 10.10b) is twice as large as the coefficient of linear expansion (that is, it is equal to 2α).

Similarly, a first-order expression for volume expansion is

$$V = V_0(1 + 3\alpha\Delta T) \quad \text{or} \quad \frac{\Delta V}{V_0} = 3\alpha\Delta T \qquad (10.11)$$

The **thermal coefficient of volume expansion** (•Fig. 10.10c) is equal to 3α.

The following example illustrates that the equations for thermal expansions are approximations. Even though an equation is a description of a physical relationship, you should always keep in mind that it may be only an approximation of physical reality and/or may only apply in certain situations. Because calculations of thermal expansion involve such small numbers, you may assume that all quantities are exact, that is, that they have any number of significant figures you wish.

EXAMPLE 10.5 ■ A PARADOX OR AN APPROXIMATION? THERMAL EXPANSION AND CONTRACTION

An aluminum rod has a length of 1.0 m at 30°C. The temperature is decreased to 10°C and then raised back to 30°C again. Is the length of the rod still 1.0 m? (Ignore significant figures.)

Solution.

Given: $\quad L_0 = 1.0 \text{ m} = 100 \text{ cm}$ $\qquad\qquad$ *Find:* $\quad$ Final length of the rod
$\qquad\qquad T_0 = 30°C$
$\qquad\qquad T = 10°C$
$\qquad\qquad \Delta T_1 = T - T_0 = 10°C - 30°C = -20\ C°$
$\qquad\qquad \Delta T_2 = T_0 - T = 30°C - 10°C = 20\ C°$
$\qquad\qquad \alpha = 24 \times 10^{-6}\ C^{°-1} \quad \text{(from Table 10.1)}$

Intuitively, you know the rod should have the same length after its temperature is lowered and then raised by the same number of degrees. But what result does Eq. 10.9 give? With the first temperature change, the rod will contract to a length L_1:

$$L_1 = L_0(1 + \alpha\Delta T_1) = (100 \text{ cm})[1 + (24 \times 10^{-6}\ C^{°-1})(-20\ C°)]$$
$$= (100 \text{ cm})(1 - 0.000480) = 99.952 \text{ cm}$$

The minus from the negative temperature difference indicates a thermal contraction. Using the smaller units of centimeters better indicates the small effect. Any units of length can be used in the equation.

With the temperature rise, the rod with an initial length of L_1 will expand to a length L_2:

$$L_2 = L_1(1 + \alpha \Delta T_2) = (99.952 \text{ cm})[1 + (24 \times 10^{-6} \text{ C}^{\circ -1})(20 \text{ C}^\circ)]$$
$$= (99.952 \text{ cm})(1 + 0.00048) = 99.99998 \text{ cm}$$

Thus, the result obtained by applying the equation for linear expansion to this situation does *not* reflect reality—the rod would not be shorter when it came back to its starting temperature. If this were true, putting the rod through numerous cycles would make it shorter and shorter, and it would eventually disappear!

The problem arises because Eq. 10.9 is an approximation, and as a result is not symmetric with respect to the temperature change ΔT. This may be easily seen using Eq. 10.8. For a cooling, $\Delta L = L_o \alpha \Delta T$, whereas for heating, $\Delta L' = L_1 \alpha \Delta T$. Thus in cooling, ΔL is proportional to L_o, which is greater than L_1 to which the heating increase $\Delta L'$ is proportional. In other words, there is a change in basis or reference.

Follow-up Exercise. To help understand this Example, consider a monetary analogy. Suppose you invested an amount of money, and in one day your investment increased by 50%. Then, the next day you lost 50% of your money. The percentages of increase and decrease are equal. Would you have the same amount of money you started with? If not, how much of the original amount would you have?

The thermal expansion of materials is an important consideration in construction. Seams are put in concrete highways and sidewalks to allow room for expansion and prevent cracking. Expansion gaps in large bridge structures and between railroad rails are necessary to prevent damage. Similarly, expansion loops may be found in oil pipelines (•Fig. 10.11). The thermal expansion of steel beams and girders can produce tremendous pressures, as the following Example shows.

EXAMPLE 10.6 ■ TEMPERATURE RISING: THERMAL EXPANSION AND STRESS

A steel I-beam is 5.0 m long at a temperature of 20°C (68°F). On a hot day, the temperature rises to 40°C (104°F). (a) What is the change in the beam's length due to thermal expansion? (b) Suppose that the ends of the beam are initially in contact with rigid vertical supports. How much force will the expanded beam exert on the supports if it has a cross-sectional area of 60 cm²?

Solution.

Given: $L_o = 5.0 \text{ m}$ *Find:* (a) ΔL (change in length)
$T_o = 20°C$ (b) F (force)
$T = 40°C$
$\alpha = 12 \times 10^{-6} \text{ C}^{\circ -1}$
(from Table 10.1)

$$A = 60 \text{ cm}^2\left(\frac{1 \text{ m}}{100 \text{ cm}}\right)^2 = 6.0 \times 10^{-3} \text{ m}^2$$

(a) Using Eq. 10.8 to find the change in length with $\Delta T = T - T_o = 40°C - 20°C = 20 \text{ C}^\circ$,

$$\Delta L = L_o \alpha \Delta T = (5.0 \text{ m})(12 \times 10^{-6} \text{ C}^{\circ -1})(20 \text{ C}^\circ)$$
$$= 1.2 \times 10^{-3} \text{ m} = 1.2 \text{ mm}$$

This may not seem like much of an expansion, but it can give rise to a great deal of force if the beam is constrained and kept from expanding.

(b) By Newton's third law, the force that the beam exerts on its supports is equal to the force exerted by the supports to prevent the beam from expanding by a length ΔL.

(a)

(b)

•**FIGURE 10.11 Expansion gaps** (a) Expansion gaps can be found built into bridge roadways. They are designed to prevent contact stresses produced by thermal expansion. (b) These loops in oil pipelines serve a similar purpose. As hot oil passes through them, the pipes expand; loops take up the extra length.

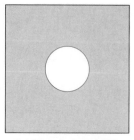

(a) Metal plate with hole

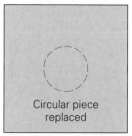

Circular piece replaced

(b) Metal plate without hole

•**FIGURE 10.12 A larger or smaller hole?**

See Conceptual Example 10.7.

This is the same as the force that would be required to compress the beam by that length. Using the Young's modulus form of Hooke's law (Chapter 9) with $Y = 20 \times 10^{10} \text{ N/m}^2$ (Table 9.1), the stress on the beam is

$$\frac{F}{A} = \frac{Y \Delta L}{L_{\text{o}}} = \frac{(20 \times 10^{10} \text{ N/m}^2)(1.2 \times 10^{-3} \text{ m})}{5.0 \text{ m}}$$

$$= 4.8 \times 10^7 \text{ N/m}^2$$

and the force is

$$F = (4.8 \times 10^7 \text{ N/m}^2)A = (4.8 \times 10^7 \text{ N/m}^2)(6.0 \times 10^{-3} \text{ m}^2)$$

$$= 2.9 \times 10^5 \text{ N} \quad \text{(about 65,000 lb or 32 tons!)}$$

Follow-up Exercise. Expansion gaps between identical steel I-beams laid end-to-end are specified to be 0.060% of the length of a beam at the installation temperature. With this specification, what is the temperature range for noncontact expansion?

CONCEPTUAL EXAMPLE 10.7 ■ LARGER OR SMALLER? AREA EXPANSION

A circular piece is cut from a flat metal sheet (•Fig. 10.12a). If the sheet is then placed in an oven and heated, the size of the hole will (a) become larger, (b) become smaller, (c) be unchanged. *Clearly establish the reasoning and physical principle(s) used in determining your answer before checking it below. That is, **why** did you select your answer?*

Reasoning and Answer. It is a common misconception to think that the area of the hole will shrink because of expansion of the metal around it. Think of the piece of metal removed from the hole, rather than of the hole itself. This piece would expand with increasing temperature. The metal in the heated sheet reacts as if the piece removed were still part of it. (Think of putting the piece of metal back into the hole and heating, as in •Fig. 10.12b. Or consider drawing a circle on an uncut metal sheet and heating it.) So, the answer is (a).

Follow-up Exercise. A circular ring of iron has tight-fitting metal bar inside it, across its diameter. If the arrangement is put into an oven and heated to a high temperature, would the circular ring be distorted or bent out of shape, or would it remain circular?

Liquids, like solids and gases, normally expand with increasing temperature. Because fluids (liquids and gases) have no definite shape, only volume expansion can be analyzed. The expression is

$$\frac{\Delta V}{V_{\text{o}}} = \beta \Delta T \quad \textit{fluid volume expansion} \tag{10.12}$$

where β is the coefficient of volume expansion for fluids. Note in Table 10.1 that the values of β for fluids are typically larger than the values of 3α for solids.

Water exhibits an anomalous volume expansion near its freezing point. The volume of a given amount of water decreases as it is cooled from room temperature, until its temperature reaches 4°C (•Fig. 10.13a). Below 4°C, the volume increases, and therefore the density decreases (•Fig. 10.13b). This means that water has a maximum density ($\rho = m/V$) at 4°C (actually 3.98°C).

When water freezes, the molecules form a hexagonal (six-sided) lattice pattern. (This is why snowflakes have hexagonal shapes.) It is the open structure of this lattice that gives water its almost unique property of expanding on freezing and being less dense as a solid than as a liquid. (This is why ice floats in water and frozen water pipes burst.) The variation of the density of water over the temperature range of 4°C to 0°C indicates that the open lattice structure is beginning to form at about 4°C, rather than arising instantaneously at the freezing point.

This property has an important environmental effect: Bodies of water, such as lakes and ponds, freeze at the top first. As a lake cools toward 4°C, water near

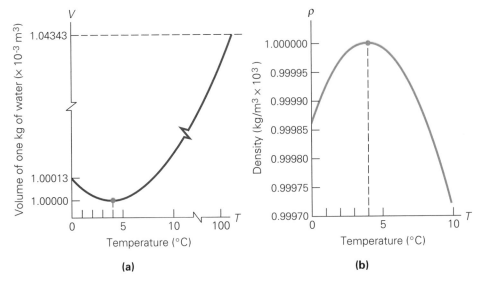

(a)   **(b)**

•**FIGURE 10.13 Thermal expansion of water**
Water exhibits nonlinear expansion behavior near its freezing point. **(a)** Above 4°C (actually 3.98°C), water expands with increasing temperature, but from 4°C down to 0°C, it expands with decreasing temperature. **(b)** As a result, water has its maximum density near 4°C.

the surface loses energy to the atmosphere, becomes denser, and sinks. The warmer, less dense water near the bottom rises. However, once the colder water on top reaches temperatures below 4°C, it becomes less dense and remains at the surface, where it freezes. If water did not have this property, lakes and ponds would freeze from the bottom up, which would destroy much of their animal and plant life (and would make ice skating a lot less popular). There would also be no oceanic ice caps at the polar regions. Rather, there would be a thick layer of ice at the bottom of the ocean, covered by a layer of water.

CONCEPTUAL EXAMPLE 10.8 ■ QUICK CHILL: TEMPERATURE AND DENSITY

Ice is put into a container of water at room temperature. For faster cooling, the ice (a) should be allowed to float naturally in the water or (b) be pushed to the bottom of the container with a stick and held there. *Clearly establish the reasoning and physical principle(s) used in determining your answer before checking it below. That is, **why** did you select your answer.*

Reasoning and Answer. When the ice melts, the water in the vicinity is cooled, and hence becomes denser (Fig. 10.13b). If the ice is permitted to float at the top, the denser water sinks and the warmer, less dense water rises. This mixing causes the water to cool quickly. If the ice were at the bottom of the container, however, the colder, denser water would remain there, and cooling would be slowed, so the answer is (a).

Follow-up Exercise. Suppose that the density versus temperature curve for water (Fig. 10.13b) were inverted, so that it dipped downward. What would this imply for the situation in this Example and the freezing of lakes?

10.5 The Kinetic Theory of Gases

**Objectives: To be able to (a) relate kinetic theory and temperature, and
(b) explain the process of diffusion and Graham's law.**

If atoms and molecules are viewed as colliding particles, the laws of mechanics can be applied to a quantity (system) of gas. We can then describe its characteristics in terms of molecular motion, pressure (force/area), energy, and so on. Because of the large number of particles involved, however, statistical analysis is necessary for such a description.

One of the major accomplishments of theoretical physics was to derive the ideal gas law from mechanical principles by interpreting the temperature in terms of the kinetic energy of the gas molecules. Theoretically, the molecules of an ideal gas are viewed as essentially point masses in random motion with relatively large distances separating them. As point masses, the molecules have no internal structure, so vibrational and rotational motions are not a consideration.

Note: Elastic collisions are discussed in Section 6.4.

The molecules make perfectly elastic collisions with each other and with the walls of the container. The forces between molecules are considered to act only over a short range, so the molecules interact with each other only during collisions. Since the molecules of an ideal gas interact only during collisions, such a gas would remain a gas at low temperatures and would not liquefy as real gases do. This gives the theoretical ideal gas its idealness. Colliding molecules lose and gain kinetic energy, with corresponding changes in their potential energy. This potential energy is ignored in summing the total energy of the system because molecules spend a negligible fraction of the time in collisions.

From Newton's laws of motion, the force on the walls of the container can be calculated from the change in momentum of the gas molecules when they collide with the walls (•Fig. 10.14). If this force is expressed in terms of pressure (force/area), the following equation is obtained (see Appendix II for derivation):

$$pV = \tfrac{1}{3}Nmv^2_{rms} \qquad (10.13)$$

where V is the volume of the container or gas, N the number of gas molecules in the closed container, and m the mass of a gas molecule. In this expression v_{rms} is the average speed, but a special kind of average. It is obtained by averaging the squares of the speeds and then taking the square root of the average—that is, $\sqrt{\overline{v^2}} = v_{rms}$. As a result, v_{rms} is called the root-mean-square (rms) speed.

Solving Eq. 10.5 for pV and equating that expression with Eq. 10.13 shows how the temperature came to be interpreted as a measure of the kinetic energy:

$$pV = Nk_BT = \tfrac{1}{3}Nmv^2_{rms}$$

or

$$\tfrac{1}{2}mv^2_{rms} = \tfrac{3}{2}k_BT \qquad (10.14)$$

Thus, the temperature of a gas (and that of the walls of the container or a thermometer bulb in thermal equilibrium with the gas) is directly proportional to its average random kinetic energy (per molecule): $\overline{K} = \tfrac{1}{2}mv^2_{rms} = \tfrac{3}{2}k_BT$. (Don't forget that T is the absolute temperature in kelvins.)

EXAMPLE 10.9 ■ MOLECULAR SPEED: RELATION TO ABSOLUTE TEMPERATURE

What is the average (rms) speed of a helium atom (He) in a helium balloon at room temperature? (Take the mass of the helium atom to be 6.65×10^{-27} kg.)

Solution.

Given: $m = 6.65 \times 10^{-27}$ kg   *Find:* v_{rms} (rms speed)
$T = 20°C$ (room temperature)
$k_B = 1.38 \times 10^{-23}$ J/K (known)

It should be evident that we will use Eq. 10.14, so we list k in the given quantities.

As we already know, we must change the Celsius temperature to kelvins, but note that k has units of J/K, which is a good reminder. So,

$$T_K = T_C + 273 = 20°C + 273 = 293 \text{ K}$$

where the 273.15 in Eq. 10.6 was rounded to 273 for convenience.

Rearranging Eq. 10.14, we have

$$v_{rms} = \sqrt{\frac{3k_BT}{m}} = \left[\frac{3(1.38 \times 10^{-23} \text{ J/K})(293 \text{ K})}{6.65 \times 10^{-27} \text{ kg}}\right]^{1/2}$$
$$= 1.35 \times 10^3 \text{ m/s} \;(=1.35 \text{ km/s})$$

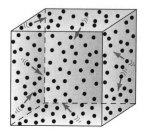

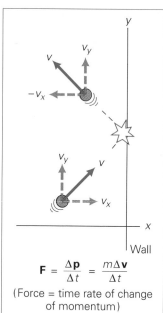

$$\mathbf{F} = \frac{\Delta \mathbf{p}}{\Delta t} = \frac{m\Delta \mathbf{v}}{\Delta t}$$

(Force = time rate of change of momentum)

•**FIGURE 10.14 Kinetic theory of gases**
The pressure a gas exerts on the walls of a container is due to the force resulting from the change in momentum of the gas molecules that collide with the wall. The force exerted by an individual molecule is equal to the time rate of change of momentum: $\mathbf{F} = \Delta \mathbf{p}/\Delta t = m\Delta \mathbf{v}/\Delta t$, where $\mathbf{p} = m\mathbf{v}$. The sum of the instantaneous normal components of the collision forces gives rise to the average pressure on the wall.

Follow-up Exercise. In this Example, if the temperature of the gas were increased by 10 C°, what would be the corresponding percent increase in the average (rms) speed?

For some interesting consequences related to molecular speeds, see the Insight on p. 345.

There are a couple of interesting points concerning Eq. 10.14. First, it predicts that at absolute zero ($T = 0$ K) all translational molecular motion of a gas would cease. According to classical theory, this would correspond to absolute zero energy. However, modern quantum theory says that there would still be some zero-point motion, and a corresponding minimum zero-point energy. Basically, absolute zero is the temperature at which all the energy that *can* be removed from an object has been removed.

Also, because the "particles" in an ideal gas interact only through collisions, as explained previously, the total average kinetic energy is the total internal energy of the gas. That is, its internal energy is all "temperature" energy as discussed in Section 10.1. With N molecules in a system, we can use Eq. 10.14, the energy per molecule, to write an expression for the total internal energy U,

$$U = N(\tfrac{1}{2}mv_{\text{rms}}^2) = \tfrac{3}{2}Nk_{\text{B}}T = \tfrac{3}{2}nRT \qquad \textit{(ideal gases only)} \qquad (10.15)$$

Total internal energy of an ideal gas

Note: Eq. 10.15 applies only to ideal gases or very dilute real gases.

where the latter expression is written in terms of the number of moles n, which is a more practical form.

Thus, we see that the internal energy of an ideal gas is directly proportional to its absolute temperature. This means that if the absolute temperature of a gas is doubled (by heat transfer), for example, from 200 K to 400 K, then its internal energy is also doubled. This does not apply to the Celsius and Fahrenheit temperatures, since their zero points are not referenced to zero energy. If a gas is at a temperature of 0°C (or 0°F) and its internal energy is doubled, the Celsius (or Fahrenheit) temperature is not doubled—doubling zero gives zero, which is not a new temperature. (What *would* its new temperature be?)

Diffusion

We depend on our sense of smell to detect odors, such as the smell of smoke from something burning. That you can smell something from a distance implies that molecules get from one place to another in the air, from the source to your nose. This process of random molecular mixing in which particular molecules move from a region where they are present in higher concentration to one where they are in lower concentration is called **diffusion**. Diffusion also occurs readily in liquids; think about what happens to a drop of ink in a glass of water (•Fig. 10.15). It even occurs to some degree in solids.

The rate of diffusion for a particular gas depends on the rms speed of its molecules. Even though gas molecules have large average speeds (Example 10.9), the molecules do not fly from one side of a room to the other. There are frequent collisions, and as a result the molecules "drift" rather slowly. For example, suppose someone opened a bottle of ammonia on the other side of a closed room. It would take some time for the ammonia to diffuse across the room for you to smell. (Much of the movement that people commonly attribute to diffusion is actually due to air currents.)

Gases can also diffuse through porous materials or permeable membranes. (This process is sometimes referred to as *effusion*.) Energetic molecules enter the material through the pores (openings), and, colliding with the pore walls, they

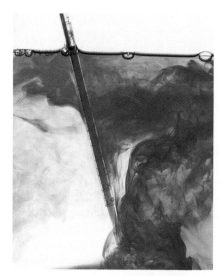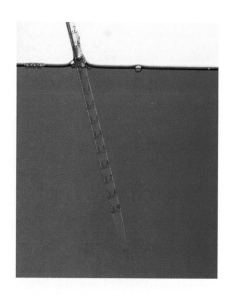

•FIGURE 10.15 Diffusion
Diffusion occurs in liquids. Random molecular motion would eventually distribute the dye throughout the water. Here there is some distribution due to mixing and the ink colors the water after a few minutes. The distribution would take some time with diffusion only.

slowly meander through the material. Such gaseous diffusion can be used to physically separate different gases from a mixture.

The kinetic theory of gases says that the average kinetic energy (per molecule) is proportional to the absolute temperature of a gas, $\frac{1}{2}mv_{rms}^2 = \frac{3}{2}k_BT$. So, on the average, the molecules of different gases (having different masses) move at different speeds at a given temperature. As you might expect, because they move faster, lighter gas molecules diffuse through the tiny openings of a porous material faster than heavier gas molecules. For example, at a particular temperature, molecules of oxygen (O_2) move faster on the average than the more massive molecules of carbon dioxide (CO_2). Because of this difference in molecular speed, oxygen can diffuse through a barrier faster than carbon dioxide can. Suppose that a mixture of equal volumes of oxygen and carbon dioxide is contained on one side of a porous barrier, as illustrated in •Fig. 10.16. After a while, some O_2 molecules and CO_2 molecules will have diffused through the barrier. Of course, the diffused gases would still be mixed. After another diffusion stage, however, the oxygen concentration would become even greater. Almost pure oxygen can be obtained by repeating the process many times.

Separation by gaseous diffusion is a key process in obtaining enriched uranium, which was used in the atomic bomb and in nuclear reactors that generate electricity (Chapter 30).

•FIGURE 10.16 Separation by gaseous diffusion
The molecules of both gases diffuse (or effuse) through the porous barrier. But, because oxygen molecules have the greater average speed, more of them pass through. Thus, there is a greater concentration of oxygen molecules on the other side of the barrier with time.

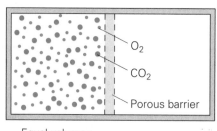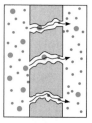

Equal volumes
of O_2 and CO_2

Diffusion
through barrier

Insight   Distribution of Molecular Speeds

The molecules in an ideal gas are assumed to be in random motion. Hence, some of the molecules have speeds less than the average speed and some have speeds that are greater than average. Using kinetic theory, it is possible to derive the distribution of molecular speeds in a gas. This was first done in 1859 by James Clerk Maxwell (1831–1879), who is best known for his contributions to electromagnetism (Chapter 20). The result, known as the *Maxwell distribution of speeds*, gives the relative number of molecules as a function of speed. A graph of such a distribution for an ideal gas is shown in Fig. 1.

Note that the root-mean-square value (v_{rms}) is different from the average value ($\bar{v}$). Also, because the curve is skewed to the right, neither of these values is at the peak or maximum of the curve, which represents the "most probable" speed, v_{mp}.

As you know, the average kinetic energy of the molecules is directly proportional to temperature, so the average molecular speed depends on temperature. In consequence, as the temperature increases, the curve for a particular gas shifts to the right, to higher speed values, as illustrated in Fig. 2.

From the characteristics of the distribution curves, an understanding of several important phenomena can be gained. For example, suppose that the speed distribution curves are plotted for two different real, dilute gases at the same temperature. As is evident from Eq. 10.14, the distribution of molecular speeds is in part a function of the molecules' mass ($v_{rms} = \sqrt{3k_B T/m}$). Hence, for a given temperature, the speed distribution curve for the lighter molecules would lie farther to the right, corresponding to higher speeds. This explains why lighter molecules, such as hydrogen and helium, escape more readily from the Earth's atmosphere than heavier molecules such as nitrogen and oxygen. (Recall the discussion of escape velocity in Section 7.6.) Does this give you a clue as to why the Moon and Mercury have no atmospheres?

The molecular speed distributions in liquids have general characteristics similar to those of gases. Hence, we see from Fig. 2 that some molecules in a liquid are very energetic (v_{mp}). Such fast-moving molecules may be able to penetrate the surface and break free of the liquid, even at room temperature. This process is known as *evaporation*. You can readily see that the higher the temperature, the more rapidly evaporation will take place. Because the most energetic molecules are the ones that leave the liquid, the average speed of those that remain behind is reduced, and the temperature of the liquid falls. Thus, evaporation is a cooling process (Section 11.3).

Finally, consider chemical reactions, which usually occur with the reactants in liquid solution. What causes a chemical reaction? Basically, molecules can interact chemically only when they have a certain minimum amount of kinetic energy, termed the *activation energy*. This is the energy needed to overcome the repulsion between the molecules (so that they can approach one another closely enough to react) and to break existing chemical bonds (so that new ones can be formed). There is thus an energy barrier, so to speak, that the reacting molecules must have sufficient energy to surmount. As with evaporation, the higher the temperature, the more molecules have the necessary energy (the shaded areas at the right of the curves in Fig. 2). Thus reactions proceed more rapidly as the temperature rises.

Because an increase in temperature results in both more frequent and more energetic molecular collisions, the effect of temperature on reaction rate is usually quite dramatic. The relationship is approximated by the *ten-degree rule* familiar to chemists: *a 10°C increase in temperature generally leads to a doubling of the reaction rate.*

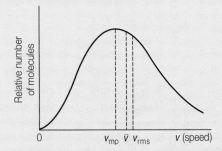

FIGURE 1   Distribution of molecular speeds for an ideal gas
Note that the most probable speed (v_{mp}) is not the same as either the average speed ($\bar{v}$) or the root-mean-square speed (v_{rms}).

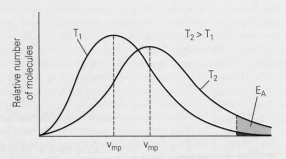

FIGURE 2   Distribution of molecular speeds at two different temperatures
When the temperature increases, the average kinetic energy of the gas molecules increases, and the curve representing the distribution of molecular speeds is shifted to the right.

Insight Physiological Diffusion

Diffusion plays a central role in many life processes. For example, consider a cell membrane in the lung. Such a membrane is permeable to a number of substances, any of which will diffuse through the membrane from a region where its concentration is high to a region where its concentration is low. Most importantly, the lung membrane is permeable to oxygen (O_2), and the transfer of O_2 across the membrane occurs because of a concentration gradient.

The blood carried to the lungs is low in O_2, having given up the oxygen to tissues for metabolism during its circulation through the body. The air in the lungs, on the other hand, is high in O_2 because there is a continual exchange of fresh air in the breathing process. As a result of this concentration difference, or gradient, O_2 diffuses from the lung volume into the blood as it flows through the lung tissue, and the blood leaving the lungs is high in O_2.

Exchanges between the blood and the tissues occur across capillary walls, and diffusion again is a major factor. The chemical composition of arterial blood is regulated to maintain the proper concentrations of particular solutes (substances dissolved in the blood solution), so that diffusion takes place in the appropriate directions across capillary walls. For example, as cells take up O_2 and glucose (blood sugar), the blood continuously brings in fresh supplies of the substances to maintain the concentration gradient needed for diffusion to the cells. The continual production of carbon dioxide (CO_2) and metabolic wastes in the cells produces concentration gradients in the opposite direction for these substances. They therefore diffuse out of the cells into the blood, to be carried away from the tissue by the circulatory system.

During periods of physical exertion, there is an increase in cellular activity. More O_2 is used up and more CO_2 is produced, thereby increasing the concentration gradients and the diffusion rates. With an increased demand for O_2, how do the lungs respond to provide this to the blood? As you might expect, the rate of diffusion depends on the surface area of the lung membrane and its thickness. Deeper breathing during exercise causes the aveoli (small air sacs in the lungs) to increase in volume. Such stretching increases the aveolar surface area and decreases the thickness of the membrane wall.

Also, the heart works harder during exercise, and the blood pressure is raised. The increased pressure forces open capillaries that are normally closed during rest or normal activity. As a result, total exchange area between the blood and cells is increased. Each of these changes help expedite the exchange of gases during exercise.

Now you can understand why people with emphysema (a disease involving the breakdown of aveoli walls) or pneumonia (characterized by fluid accumulations within or around the lungs) have difficulty providing enough oxygen to their tissues.

The rate of such diffusion depends on the average speed of the molecules ($R \propto v_{rms}$), which in turn depends, as we have seen, both on the temperature and on the mass of the gas molecules. Let's form a ratio and investigate:

$$\frac{R_1}{R_2} = \frac{v_{1_{rms}}}{v_{2_{rms}}}$$

where R_1 and R_2 are the diffusion rates (mass/time) of two gases, and the v_{rms}'s are the average rms speeds of the molecules. Then, by Eq. 10.14 for constant temperature (and pressure),

$$\frac{R_1}{R_2} = \frac{\sqrt{\frac{3k_BT}{m_1}}}{\sqrt{\frac{3k_BT}{m_2}}} = \frac{\sqrt{\frac{1}{m_1}}}{\sqrt{\frac{1}{m_2}}} = \sqrt{\frac{m_2}{m_1}} \qquad \begin{array}{l}\textit{(constant temperature} \\ \textit{and pressure)}\end{array} \qquad (10.16)$$

Thus, *the rate of porous diffusion of gas molecules is inversely proportional to the square root of the molecular mass.* This result is known as **Graham's law** and was experimentally determined by Thomas Graham (1805–1869), a Scottish chemist.

EXAMPLE 10.10 ■ DIFFUSION: GRAHAM'S LAW

Which gas will diffuse through a porous membrane faster, nitrogen (N_2) or oxygen (O_2), and how many times faster? (The molecular masses of N_2 and O_2 are about 4.65×10^{-26} kg and 5.31×10^{-26} kg, respectively.)

Solution.

Given: $m_N = 4.65 \times 10^{-26}$ kg   *Find:* Which gas diffuses faster,
$m_O = 5.31 \times 10^{-26}$ kg   and how many times faster

By Graham's law, the diffusion rate is inversely proportional to the molecular mass, and since nitrogen has the smaller molecular mass, it will have the greater diffusion rate and diffuse faster.

The ratio of the diffusion rates is given by Eq. 10.16, and

$$\frac{R_N}{R_O} = \sqrt{\frac{m_O}{m_N}} = \sqrt{\frac{5.31 \times 10^{-26} \text{ kg}}{4.65 \times 10^{-26} \text{ kg}}} = \sqrt{1.14} = 1.07$$

Thus nitrogen diffuses about 7 percent faster than oxygen: $R_N = (1.07)R_O$.

Follow-up Exercise. At what temperature would oxygen diffuse as fast as nitrogen does at room temperature?

Fluid diffusion is very important to organisms. In plant photosynthesis, carbon dioxide from the air diffuses into leaves, and oxygen and water vapor diffuse out. The diffusion of liquid water across a permeable membrane because of a concentration gradient is called **osmosis** and is vital in living cells. Osmotic diffusion is also important to kidney functioning: tubules in the kidneys concentrate waste matter from the blood in much the same way oxygen is removed from mixtures. (See the Insight on p. 346 for other examples of diffusion.)

Chapter Review

Important Terms

temperature   327
heat   327
internal energy   327
thermometer   328
thermal expansion   328
Fahrenheit temperature scale   329
Celsius temperature scale   329
Boyle's law   331
Charles' law   332

ideal (perfect) gas law   332
absolute zero   334
Kelvin temperature scale   334
kelvin   334
triple point of water   334
Avogadro's number   335
mole   335
thermal coefficient of linear
   expansion   337

thermal coefficient of area
   expansion   338
thermal coefficient of volume
   expansion   338
kinetic theory of gases   342
diffusion   343
Graham's law   346
osmosis   347

Important Concepts

- Heat is energy transferred from one object to another because of temperature differences. Once transferred, the energy becomes part of the internal energy of the object (or system).
- The ideal gas law relates pressure, volume, and absolute temperature.
- Absolute zero (0 K) corresponds to $-273.15°C$.

- Thermal coefficients of expansion relate the fractional change in dimension(s) to a change in temperature.
- According to kinetic theory, the absolute temperature of a gas is directly proportional to the average random kinetic energy per molecule.
- Graham's law relates the rate of porous diffusion to the square root of the molecular mass of a gas.

Important Equations

Celsius–Fahrenheit Conversion:

$$T_F = \tfrac{9}{5}T_C + 32 \tag{10.1}$$

$$T_C = \tfrac{5}{9}(T_F - 32) \tag{10.2}$$

Boyle's Law:

$$pV = \text{constant} \quad \text{or} \quad p_1V_1 = p_2V_2 \tag{10.3}$$

Charles' Law:

$$\frac{V}{T} = \text{constant} \quad \text{or} \quad \frac{V_1}{T_1} = \frac{V_2}{T_2} \tag{10.4}$$

Gas Law (always absolute

$$Nk_BT \quad \text{or} \quad \frac{p_1V_1}{T_1} = \frac{p_2V_2}{T_2} \qquad (10.5)$$

$$pV = nRT$$
$$k_B = 1.38 \times 10^{-23} \text{ J/K}$$
$$R = 0.0821 \text{ L·atm/mol·K}$$
$$= 8.31 \text{ J/mol·K}$$

Avogadro's number: $N_A = 6.02 \times 10^{23}$ molecules/mole

Kelvin–Celsius Conversion:

$$T_K = T_C + 273.15 \qquad (10.6)$$

Thermal Expansion of Solids:

linear: $\quad \dfrac{\Delta L}{L_o} = \alpha \Delta T \text{ or } L = L_o(1 + \alpha \Delta T) \quad (10.8, 10.9)$

area: $\quad \dfrac{\Delta A}{A_o} = 2\alpha \Delta T \quad \text{or} \quad A = A_o(1 + 2\alpha \Delta T) \qquad (10.10)$

volume: $\quad \dfrac{\Delta V}{V_o} = 3\alpha \Delta T \quad \text{or} \quad V = V_o(1 + 3\alpha \Delta T) \qquad (10.11)$

Thermal Volume Expansion of Fluids:

$$\frac{\Delta V}{V_o} = \beta \Delta T \qquad (10.12)$$

Results of Kinetic Theory of Gases:

$$pV = \tfrac{1}{3}Nmv_{rms}^2 \qquad (10.13)$$

$$\tfrac{1}{2}mv_{rms}^2 = \tfrac{3}{2}k_BT \qquad (10.14)$$

$$U = \tfrac{3}{2}Nk_BT = \tfrac{3}{2}nRT \qquad (10.15)$$

Diffusion Rate Ratio (Graham's Law):

$$\frac{R_1}{R_2} = \sqrt{\frac{m_2}{m_1}} \qquad \begin{array}{l}\textit{(constant temperature}\\ \textit{and pressure)}\end{array} \qquad (10.16)$$

Exercises*

10.2 The Celsius and Fahrenheit Temperature Scales

1 Which of the following is the hottest temperature: (a) 15°C, (b) 15°F, (c) 15 K?

2 Which of the following temperature scales has the smallest unit interval: (a) Celsius, (b) Fahrenheit, (c) Kelvin?

3 Heat always flows spontaneously from a body at a higher temperature to one at a lower temperature that is in thermal contact with it. Does it always flow from one with more internal energy to one with less internal energy? Explain.

4 What is the hottest (highest temperature) item in the home? (*Hint:* Think about this one and maybe a light will come on.)

5 ■ Suppose you didn't know that −40°C = −40°F (see Follow-up Exercise 10.1) and you were asked if the Celsius and Fahrenheit temperatures were ever equal. How could you find out algebraically?

6 ■ Convert the following to Celsius readings: (a) 1000°F, (b) 0°F, (c) −20°F, (d) −40°F.

7 ■ Convert the following to Fahrenheit readings: (a) 150°C, (b) 32°C, (c) −25°C, (d) −273°C.

8 ■■ A person running a fever has a body temperature of 39.4°C. What is the person's temperature on the Fahrenheit scale?

9 ■■ (a) Which is the lower temperature, −45°C or −45°F? (b) Which is the lower temperature, 200°C or 375°F?

10 ■■ The highest and lowest recorded air temperatures in the United States are 134°F (Death Valley, California, 1913) and −80°F (Prospect Creek, Alaska, 1971), respectively. What are these temperatures on the Celsius scale?

11 ■■ The highest and lowest recorded air temperatures in the world are 58°C (Libya, 1922) and −89°C (Antarctica, 1983). What are these temperatures on the Fahrenheit scale?

12 ■■ A cold front passing through a location causes an air temperature drop of 20 F° in a short time. What would be the corresponding change on the Celsius scale?

13 ■■■ (a) The largest temperature drop recorded in the United States in one day occurred in Browning, Montana, in 1916, with the temperature going from 7°C to −49°C. What is the corresponding *change* on the Fahrenheit scale? (b) The largest recorded temperature rise occurred in 2 min from −4°F to 45°F (Spearfish, South Dakota, 1943). What is the corresponding *change* on the Celsius scale?

14 ■■■ (a) On the plot of Fahrenheit temperature versus Celsius temperature in Fig. 10.5a, the y-intercept is 32°F. What would be the x-intercept value? (b) Suppose the graph were plotted in the opposite way (Celsius versus Fahrenheit).

*Assume all temperatures to be exact and neglect significant figures for small dimension changes.

10.3 Gas Laws and Absolute Temperature

15 The temperature used in the ideal gas law is (a) Celsius, (b) Fahrenheit, (c) Kelvin, (d) it doesn't matter.

16 When the temperature of a quantity of gas is increased, (a) the pressure must increase, (b) the volume must increase, (c) both the pressure and volume must increase, (d) none of the preceding.

17 The temperature of a quantity of gas is increased. (a) How is the density affected if the pressure is held constant? (b) How is the density affected if the volume is held constant?

18 A type of constant-volume gas thermometer is shown in •Fig. 10.17. Describe how it operates.

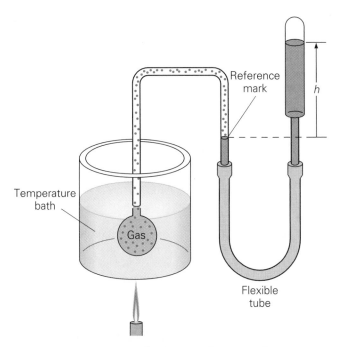

•**FIGURE 10.17 A type of constant-volume gas thermometer**
See Exercise 18.

19 Describe how a constant-pressure gas thermometer might be constructed.

20 In terms of the ideal gas law, what would a temperature of absolute zero imply? A negative absolute temperature?

21 ■ Convert the following temperatures to absolute temperatures in kelvins: (a) 0°C, (b) 100°C, (c) 20°C, and (d) −35°C.

22 ■ Convert the following temperatures to degrees Celsius: (a) 0 K, (b) 250 K, (c) 273 K, and (d) 325 K.

23 ■■ What are the absolute temperatures in kelvins for (a) −40°F and (b) −40°C?

24 ■■ Which is the higher temperature, 500°F or 500°K?

25 ■■ The surface temperature of the Sun is about 6000 K. (a) What is this temperature on the Fahrenheit and Celsius scales? (b) The surface temperature is sometimes reported to be 6000°C. Assuming that 6000 K is correct, what is the percentage error of this Celsius value?

26 ■■ What is the triple point temperature of water on the Celsius and Fahrenheit scales?

27 ■■ How many moles are there in (a) 40 g of water, (b) 245 g of H_2SO_4 (sulfuric acid), (c) 138 g of NO_2 (nitrogen dioxide), (d) 56 L of SO_2 (sulfur dioxide) at STP?

28 ■■ A constant-volume gas thermometer has a pressure of 1000 Pa at room temperature. If the pressure increases to 2000 Pa, what is the new Celsius temperature?

29 ■■ A constant-pressure gas thermometer is initially at a temperature of 25°C. If the volume of gas in the thermometer increases by 5.0%, what is the final Celsius temperature?

30 ■■ On a warm day (92°F), the air in a balloon occupies a volume of 0.20 m³ and exerts a pressure of 20.0 lb/in.² If the balloon is placed in a refrigerator and cooled to 32°F, the pressure decreases to 14.7 lb/in.² What is the volume of the balloon? (Assume that the air behaves as an ideal gas.)

31 ■■ A steel-belted radial automobile tire is inflated to 30.0 lb/in.² gauge pressure when the temperature is 61°F. Later in the day, the temperature rises to 100°F. Assuming the volume of the tire remains constant, what is the pressure at the elevated temperature? (*Hint*: remember that the ideal gas law uses absolute pressure.)

32 ■■ The temperature of a quantity of ideal gas is doubled and its volume is decreased by one-half. How is the pressure affected?

33 ■■ If 2.4 m³ of a gas initially at STP is compressed to 1.6 m³ and its temperature raised 30 C°, what is the final pressure?

34 ■■ How many molecules are there in (a) a liter and (b) 1.00 cm³ of a gas at STP?

35 ■■■ An automobile tire is inflated to 200 kPa gauge pressure when the air temperature is at the freezing point. It later warms up to 12.0°C and the air pressure in the tire is found to be 204 kPa. Does the volume of the tire change? If so, by what percentage? (Assume that the air is an ideal gas and that atmospheric pressure is constant.) (*Hint*: see Exercise 31.)

36 ■■■ A quantity of an ideal gas at 10°C occupies 4.0 L and has a pressure of 150 kPa. (a) What will the volume be if the temperature is kept constant and the pressure is decreased to 120 kPa? (b) What will the pressure be if the temperature is kept constant and the volume is compressed to 2.5 L? (c) What will the Celsius temperature be at a pressure of 120 kPa and 2.5 L?

37 ■■■ The geothermal gradient, the rate at which temperature increases with depth in the Earth's crust, is determined in deep mines and wells to be about 1 F°/150 ft. Assuming that this rate is uniform to any depth, what is the absolute temperature at the center of the Earth? Is this a reasonable assumption?

38 ■■■ Is there a temperature that has the same numerical value on the Kelvin and Fahrenheit scales? Justify your answer.

10.4 Thermal Expansion

39 The units of the thermal coefficient of linear expansion are (a) m/C°, (b) m²/C°, (c) m·C°, (d) C°⁻¹.

40 The thermal coefficient of volume expansion for a solid is (a) α, (b) 2α, (c) 3α, (d) α^3.

41 Consider a cube sitting on a bimetallic strip at room temperature, as depicted in •Fig. 10.18. What will happen if the cube is ice and (a) the upper strip aluminum and the lower strip brass, or (b) the upper strip iron and the lower strip copper? (c) If the two strips are brass and copper, which should be on top to keep a hot metal cube from falling off?

42 A demonstration of thermal expansion is shown in •Fig. 10.19. Initially, the ball goes through the ring made of the same metal. (a) When the ball is heated, it does not go through the ring. (b) If both the ball and the ring are heated, the ball again goes through the ring. Explain what is being demonstrated.

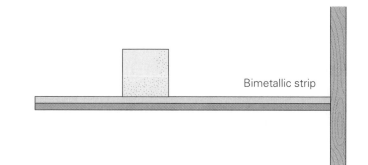

•FIGURE 10.18 Which way will the cube go?
See Exercise 41.

43 A solid metal disk rotates freely, so the conservation of angular momentum applies (Chapter 8). If the disk is heated while it is rotating, will there be any effect on the rate of rotation (the angular speed)?

44 ■ A steel meterstick is accurate at room temperature. If the meterstick is placed in a deep freeze at −5.0°C, what would be the stick's percent error because of thermal contraction?

45 ■ An aluminum flag pole is 20.00 m tall on a −15°C winter's day. How much higher will Old Glory fly from the pole on a hot 42°C summer's day?

46 ■■ A thin brass rod at room temperature (20°C) has a moment of inertia of $I = ML_o/12$ about an axis through its center and perpendicular to its length (see Fig. 8.15). To what temperature would the rod have to be heated to increase its moment of inertia by 0.10%?

47 ■■ A circular ring of iron has a tight-fitting iron bar across its diameter as illustrated in •Fig. 10.20. If the arrangement is put into an oven and heated to a high temperature, will the circular ring be distorted? Justify your answer.

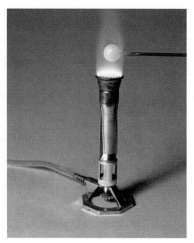
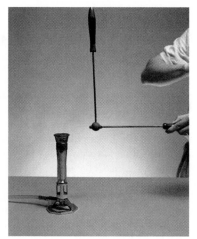

•Figure 10.19 Ball-and-ring expansion
See Exercise 42.

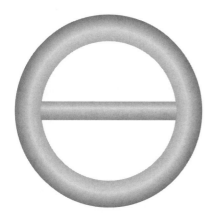

•FIGURE 10.20   Stress out of shape?
See Exercise 47.

48   ■■ A rectangular steel plate whose area is 0.060 m² is cooled from 350°C to room temperature. By what percentage does the area decrease?

49   ■■ What temperature change would cause a 0.10% increase in the volume of a quantity of water that was initially at room temperature?

50   ■■ A piece of copper tubing used in plumbing has a length of 60 cm and an inner diameter of 1.50 cm at room temperature. When hot water at 85°C flows through the tube, what are (a) its new length and (b) the change in its cross-sectional area? Does the latter affect the flow speed?

51   ■■ Concrete highway slabs are poured in lengths of 10.0 m. How wide should the expansion gaps between the slabs be to ensure that there will be no contact stress over a temperature range of −25°C to 45°C?

52   ■■ A circular piece with a diameter of 8.00 cm is cut from an aluminum sheet at room temperature. If the sheet is then placed in an oven and heated to 150°C, what will the area of the hole be?

53   ■■ A man's gold wedding ring has an inner diameter of 2.4 cm at room temperature. If the ring is dropped into boiling water, what is the change in the inner diameter of the ring?

54   ■■ A brass rod has a circular cross section with a radius of 0.500 cm. It fits into a circular hole in a copper sheet with a clearance of 0.0100 mm completely around it when both it and the sheet are at 20°C. (a) At what higher temperature will the clearance be zero? (b) Would such a tight fit be possible if the sheet were brass and the rod were copper?

55   ■■ Show that the density of a solid substance varies with temperature as $\rho = \rho_o(1 - 3\alpha\Delta T)$ or $\rho = \rho_o(1 - \beta\Delta T)$.

56   ■■ The density of water is 1.000×10^3 kg/m³ at 4°C. Calculate its density at 100°C (just before boiling).

57   ■■ A copper block has an internal spherical cavity with a diameter of 10 cm (•Fig. 10.21). The block is placed in an oven and heated from room temperature to 500 K. (a) Does the cavity get larger or smaller? (b) What is the change in the cavity's volume?

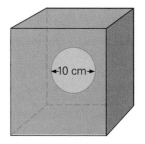

•Figure 10.21   A hole in a block
See Exercise 57.

58   ■■■ An aluminum cube is 10.0 cm on a side at room temperature. It is placed in an oven and heated to 300°C. (a) What is the cube's new volume? (b) Using this volume, compute the length of each edge of the cube. (c) Did each edge of the cube expand linearly when heated? (That is, could L have been found and the volume computed?)

59   ■■■ One morning when the temperature is 10°C, an employee at a rent-a-car company fills the 25-gal gas tank of an automobile to the top and then parks the car a short distance away. That afternoon, when the temperature is 20°C, gasoline drips from the tank onto the pavement. How much gas will be lost? (Neglect the expansion of the tank.)

60   ■■■ A Pyrex beaker that has a capacity of 1000 cm³ at 20°C contains 990 cm³ of mercury at that temperature. Is there some temperature at which the mercury will completely fill the beaker? Justify your answer. (Assume no mass loss by vaporization.)

10.5 The Kinetic Theory of Gases

61   In the kinetic theory of gases, the particles (a) experience gravitational attractions, (b) interact only through collisions, (c) move at very slow speeds, (d) have negligible masses.

62   If the temperature of a quantity of ideal gas is raised from 20°C to 40°C, its internal energy is (a) doubled, (b) quadrupled, (c) unchanged, (d) none of the preceding.

63   Equal volumes of helium gas (He) and neon (Ne) gas at the same temperature (and pressure) are on opposite sides of a porous membrane, as shown in •Fig. 10.22. Describe what happens after a period of time, and why.

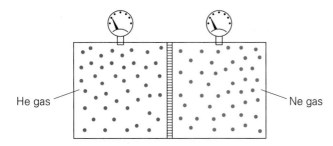

•FIGURE 10.22 What happens as time passes?
See Exercise 63.

64 An aqueous saline (salt) solution with a concentration of 0.85% is said to be isotonic. There is no osmosis through the cell membranes of red blood cells placed in an isotonic solution. Explain what would happen and why if red blood cells were placed in hypertonic and hypotonic solutions. [*Hint:* Think in terms of *water* concentrations.]

65 ■ What is the average kinetic energy per molecule of a low-density gas at (a) 20°C and (b) 100°C?

66 ■■ If the temperature of an ideal gas is raised from 25°C to 100°C, what will the percentage change in the average (rms) velocity of the gas molecules be?

67 ■■ (a) What is the average kinetic energy of molecules in a volume of gas at a temperature of 27°C? (b) What is the average (rms) speed of the molecules if the gas is helium? (A helium molecule is a single atom, which has a mass of 6.65×10^{-27} kg.)

68 ■■ What is the average speed of the molecules in low-density oxygen gas at room temperature? (The mass of an oxygen molecule, O_2, is 5.31×10^{-26} kg.)

69 ■■ At a given temperature, which would be greater, the rms speed of oxygen (O_2) or of ozone (O_3), and how many times greater?

70 ■■ What is the internal energy of 2.0 moles of ideal gas at room temperature?

71 ■■ A mole of ideal gas is heated from 100°C to 200°C. What is the change in the internal energy of the gas?

72 ■■ Heat is added to a quantity of an ideal gas that is initially at 25°C until its internal energy is doubled. What is the final Celsius temperature of the gas?

73 ■■ The temperature of an ideal gas is increased from 20°C to 40°C. What is the percentage increase in the internal energy of the gas?

74 ■■ At a given temperature, which has the faster diffusion rate, oxygen (O_2) or ozone (O_3), and how many times faster?

75 ■■ The rms speed of one gas is 2.5 times that of another at the same temperature. How do their diffusion rates vary?

76 ■■ A quantity of an ideal gas has a temperature of 0°C. An equal quantity of another ideal gas is twice as hot. What is its temperature?

Additional Exercises

77 Compute the final length of the rod in Example 10.5 with the thermal cycle reversed, that is, heating first and then cooling.

78 To conserve energy, thermostats in an office building are set at 78°F in the summer and 65°F in the winter. What would the settings be if the thermostats had a Celsius scale?

79 A copper wire has a length of 0.500 m at 20°C. If the temperature is increased to 100°C, what is the change in the wire's length?

80 Assuming that the interior temperature of the Sun decreases uniformly from the central core to the surface, what is the average Kelvin temperature gradient (K/km) if the core temperature is 15 million kelvins? (Use data given inside back cover.)

81 A mercury thermometer has a uniform capillary bore whose cross-sectional area is 0.012 mm². The volume of the mercury in the thermometer bulb at 10°C is 0.130 cm³. If the temperature is increased to 50°C, how much will the height of the mercury column in the capillary bore change? (Neglect the expansion of the mercury in the bore and of the glass of the thermometer.)

82 Show that a general expression for thermal stress is given by $F/A = Y\alpha\Delta T$, where Y is Young's modulus and α is the coefficient of linear expansion.

83 What temperature increase would produce a stress of 8.0×10^7 N/m² on a rigidly held steel beam?

84 Using the ideal gas law, express the coefficient of volume expansion in terms of temperature and pressure. (*Hint*: use different temperatures and pressures.)

85 An ideal gas occupies a container with a volume of 0.75 L at STP (standard temperature and pressure, 0°C and 1 atm). Find (a) the number of moles and (b) the number of molecules of the gas. (c) If the gas is carbon monoxide (CO), what is its mass?

86 A solid aluminum sphere has a diameter of 8.00 cm at room temperature. If the sphere is heated to 360°C, what is the change in its diameter?

87 (a) If the temperature drops by 10 C°, what is the corresponding temperature change on the Fahrenheit scale? (b) If the temperature rises by 10 F°, what is the corresponding change on the Celsius scale?

88 The core of the Sun is estimated to have a temperature of about 15 million kelvins. What is this temperature on the Fahrenheit and Celsius scales?

89 A new metal alloy is made into a 50-cm rod to determine its coefficient of linear expansion. After being heated from room temperature to 250°C, the rod is found to have a length of 50.044 cm. Is this a potentially valuable alloy? Justify your answer.

90 A square aluminum sheet 25 cm on a side is heated from room temperature to 300°C. What is the change in the length of a side?

91 Mercury has a density of 13.59 g/cm³ at room temperature. What is its density at 100°C?

92 The pressure on a low-density gas in a cylinder is kept constant as its temperature is increased from 10°C to 40°C. (a) Does the piston in the cylinder advance or recede? (b) What is the fractional change in the volume of the gas?

93 Show that the coefficient of volume expansion for solids is approximately equal to 3α.

94 In the troposphere (the lowest part of the atmosphere), the temperature decreases rather uniformly with altitude at a lapse rate of about 6.5 C°/km. What are the temperatures (a) near the top of the troposphere (which has an average thickness of 11 km) and (b) outside a commercial aircraft flying at a cruising altitude of 34,000 ft? (Assume that the ground temperature is normal room temperature.)

95 A mercury thermometer has a bulb volume of 0.200 cm³ and a capillary bore diameter of 0.065 mm. How far up the bore will the column of mercury move if the overall temperature of the thermometer is increased by 30 C°? (Neglect the expansion of the glass of the thermometer.)

96 Steel rails 7.50 m long are laid end to end on a cold day when the temperature is 0°C. The engineer on the project knows that the highest recorded summer temperature for that region is 40°C. If the engineer adds 20% to this value as a safety factor, what minimum expansion space should be left between the rails to avoid contact stress?

97 A quantity of an ideal gas initially at atmospheric pressure is maintained at a constant temperature while it is compressed to half of its volume. What is the final pressure of the gas?

98 Use algebraic formulas (not numbers) to demonstrate the discrepancy shown in Example 10.5.

11 Heat

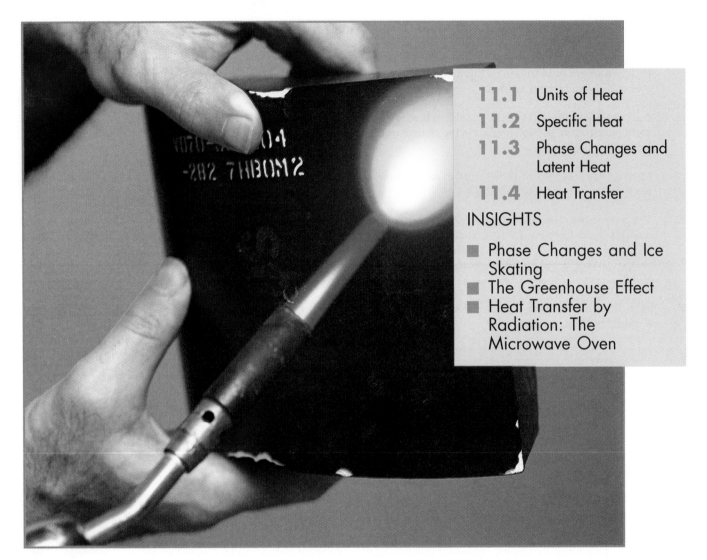

It's remarkable how much we know about the physical world solely from everyday experience—and often without any conscious recollection of having learned it. Probably everyone has some childhood memory of reaching for a pot on a stove and being warned: "Don't touch that—it's hot!" Because of such experiences—or perhaps from the results when we ignored such advice—most of us know to be very careful about handling anything that has recently been in contact with a flame or other source of heat. Yet how many of us can recall how and when we learned that, while the metal handle of a pot on the stove may be almost as hot as the bottom, a wooden pot handle is probably safe to touch? We know, in other words, not only that heat flows through objects, but that the rate of flow can vary

widely (as the photograph above shows in dramatic fashion). We also know that such direct physical contact isn't always necessary for heat to be transmitted: Think of touching the dashboard or the metal seat-belt buckle of your car after you have left it in direct sunlight for a few hours!

In this chapter you'll learn what heat is and how it is measured, and study the various processes by which heat passes from one body to another. Along the way, you'll find out why foam insulation is blown into attic walls and ceilings, and why the wind at the beach often reverses direction after the sun goes down. You'll also learn what happens when water boils or freezes, and why its temperature doesn't change during these processes even though a great deal of heat is flowing into or out of it.

Think for a moment how crucial heat is in our existence. Our bodies must precisely balance heat loss and heat gain to stay within the narrow temperature range necessary for life. On a larger scale, the average temperature of the Earth, so critical to our environment and the survival of the organisms that inhabit it, is maintained through a similar balance. We are warmed by energy that comes to us across a 150-million kilometer void. Each day, vast quantities of energy reach our planet's surface and atmosphere from the Sun, only to be radiated away into the cold of space.

These balances are delicate, and any disturbance can have serious consequences. In an individual, sickness can disrupt the balance, producing a chill or fever. Similarly, we worry that pollution of the atmosphere by "greenhouse" gases, the product of our industrial society, could give our whole planet a "fever" that could affect all of us.

Thus, heat, heat transfer, and phase changes not only affect the quality of our lives but are some of the most important topics in the study of physics. An understanding of them will allow you to explain many everyday things, as well as providing a basis for understanding the conversion of thermal energy into useful mechanical work.

11.1 Units of Heat

Objectives: To be able to (a) distinguish the various units of heat, and (b) define the mechanical equivalent of heat.

Like work, heat involves a transfer of energy. It is commonly thought that **heat** is a form of energy, but this is not true in the strictest sense. Rather, heat is the name we use to describe a type of energy *transfer*. When we refer to "heat" or "heat energy" in this book we mean the addition of internal energy to a body or system (or its removal).

Because heat is energy in transit, it is measurable as energy losses or gains and is described by standard energy units like any other quantity of energy. Recall that the SI standard unit of energy is the joule (J), or newton·meter (N·m). Thus, it is correct to say, for example, that 20 J of heat energy is transferred from one body to another. However, there are other commonly used units of heat. A chief one is the **kilocalorie (kcal)** (•Fig. 11.1a):

> One kilocalorie (kcal) is defined as the amount of heat needed to raise the temperature of 1 kg of water 1 C° (from 14.5°C to 15.5°C).

The kilocalorie

The kilocalorie is technically known as the 15° kilocalorie. The temperature range is specified because the energy needed to raise 1 kg of water by 1 C° varies slightly with temperature. It is a minimum in the temperature interval between 14.5°C and 15.5°C at a pressure of 1 atm. Over the temperature range in which water is a liquid, the variation is small and can be ignored for most purposes as will be done in our discussion.

$\Delta T = 1$ C°

1 kg
water

**(a) I kilocalorie (kcal)
or Calorie (Cal)**

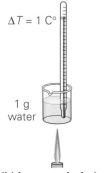

$\Delta T = 1$ C°

1 g
water

(b) I gram calorie (cal)

$\Delta T = 1$ F°

1 lb
water

(c) I British thermal unit (Btu)

•FIGURE 11.1 **Units of heat**
(a) A kilocalorie raises the temperature of 1 kg of water by 1 C°. **(b)** A calorie raises the temperature of 1 g of water by 1 C°. **(c)** A Btu raises the temperature of 1 lb of water by 1 F°. (Drawings not to scale.)

•FIGURE 11.2 It's a joule
In Australia, diet drinks are labeled as being "low joule." In Germany, the labeling is a bit more specific. The label reads: "Less than 4 kilojoules (1 kcal) in 0.3 Liter." How does this compare to our diet drinks?

Benjamin Thompson (1753–1814) was born in New England, but went to England at the time of the American Revolution. He later worked in Bavaria, where he was made a count. He took the name Count Rumford after his birthplace (now known as Concord, New Hampshire).

Because the kilocalorie is a larger multiple, the smaller **calorie (cal)**, is sometimes used (1 kcal = 1000 cal). *One calorie is the amount of heat needed to raise the temperature of 1 g of water by 1 C° (from 14.5°C to 15.5°C)* (•Fig. 11.1b).

A familiar use of the larger kilocalorie is for specifying the energy values of foods. In this context the word is usually shortened to Calorie (Cal). That is, people on diets really count kilocalories; for example, a piece of cake may contain 400 Cal (kcal), or 400,000 cal. A capital C is used to distinguish the larger kilogram-Calorie, or kilocalorie, from the smaller gram-calorie. They are sometimes referred to as "big Calorie" and "little calorie." (In some countries, the joule is used for food values—see •Fig. 11.2.)

A unit of heat that is commonly used in industry is the **British thermal unit (Btu)**. *One Btu is the amount of heat needed to raise the temperature of 1 lb of water by 1 F° (from 63°F to 64°F)* (•Fig. 11.1c). A Btu is more than 250 times larger than a calorie, but only about a fourth of a kilocalorie (1 Btu = 252 cal = 0.252 kcal). If you buy an air conditioner or an electric heater, you will find it rated in Btu's; for example, window air conditioners range from 4000 to 25,000 Btu's. This number is actually Btu's *per hour* and specifies how much heat the unit will transfer (or deliver in the case of a heater) during that time.

The Mechanical Equivalent of Heat

The current idea that heat is a transfer of energy is the result of work by many scientists on the relationship of heat and energy. Some early observations were made by Benjamin Thompson, Count Rumford, while he was supervising the boring of cannon barrels in Germany (see •Fig. 11.3). Rumford noticed that water put into the bore of the cannon to prevent overheating during drilling boiled away and had to be frequently replenished. He did several experiments, including ones in which he tried to detect "caloric fluid" by changes in the weights of heated substances. He eventually concluded that mechanical work was responsible for the heating of the water.

This conclusion was later proven quantitatively by James Joule, the English scientist after whom the SI unit for work and energy is named. Using the apparatus shown in •Fig. 11.4, Joule demonstrated that when a given amount of mechanical work was done, the water was heated, as indicated by an increase in temperature. He found that for every 4186 J of work done the temperature of the water rose 1 C° per kg or that 4186 J was equivalent to 1 kcal:

$$1 \text{ kcal} = 4186 \text{ J}$$

This relationship is called the **mechanical equivalent of heat** and provides a conversion factor between kilocalories and joules: 1 cal = 4.186 J.

EXAMPLE 11.1 ■ STIRRED, NOT SHAKEN: MECHANICAL EQUIVALENT OF HEAT

How much mechanical work in joules would have to be done in Joule's apparatus to raise the temperature of a liter of water from room temperature (20°C) to 25°C?

Solution. Listing the given data and what we want to find, we have

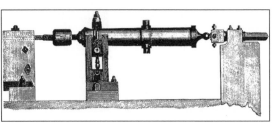

•FIGURE 11.3 Benjamin Thompson, Count Rumford (1753–1814) Engravings of the statesman and physicist Count Rumford and a cannon borer. Rumford's experience with the boring of cannon barrels led to his experiments on the nature of heat.

Given: $T_i = 20°C$ **Find:** W (work in joules)
           $T_f = 25°C$
           $V = 1.0 \, L$

where T_i and T_f are the initial and final temperatures, respectively.

From the definition of the kilocalorie (or calorie) we know how much heat (symbolized Q) it takes to raise the temperature of a certain amount of water one Celsius degree. Here, the amount or mass of water is given in terms of its volume. However, we know from Chapter 1 that 1.0 L of water has a mass of 1.0 kg. One kcal raises the temperature of one kg of water one Celsius degree. So, to raise the temperature of one kilogram of water by five Celsius degrees ($\Delta T = T_f - T_i = 25°C - 20°C = 5 \, C°$) would require 5.0 kcal, assuming no heat loss. Then,

$$Q = 5.0 \, \text{kcal}\left(\frac{4.186 \times 10^3 \, J}{1 \, \text{kcal}}\right) = 2.1 \times 10^4 \, J$$

heat   mechanical   mechanical
energy   equivalent   energy
           of heat

Follow-up Exercise.   To get an idea of the mechanical equivalent of 5.0 kcal of heat, to what height could a 100-kg (220 lb) object be lifted using this amount of energy?

11.2   Specific Heat

Objectives: To be able to (a) describe specific heat and its cause, and (b) explain how the specific heats of materials are obtained using calorimetry.

Recall from Chapter 10 that when heat is added to a substance, the energy may go to increase the random molecular motion, which results in a temperature change, and also to increase the potential energy associated with the molecular bonds. Different substances have different molecular configurations and bonding. Thus, if equal amounts of heat are added to equal masses of different substances, the resulting temperature changes will not generally be the same. For example, suppose that you had 1 kg of water and 1 kg of aluminum and added 1 kcal of heat to each. From the definition of the kilocalorie, you know that the temperature of the water would rise 1 C°. You would find that the temperature of the aluminum would increase by 4.5 C°. The reason for this difference is that more of the added energy goes into the nontranslational part of the internal energy of the water than of the aluminum.

The amount of heat (Q) required to change the temperature of a substance is proportional to the mass (m) of the substance and to the change in the temperature (ΔT). That is, $Q \propto m\Delta T$, or, in equation form,

$$Q = mc\Delta T \qquad (11.1)$$

Here $\Delta T = T_f - T_i$ is the temperature change, or the difference between the initial temperature (T_i) and final (T_f) temperature, and c is the *specific heat capacity*, or simply the **specific heat**. It is characteristic of, or *specific* for, a given substance and gives an indication of its internal molecular configuration and bonding.*

Writing Eq. 11.1 as $c = Q/m\Delta T$ shows us that the units of specific heat are J/kg·K (or kcal/kg·C°). The standard SI unit for specific heat is the one with the temperature in kelvins. However, the use of Celsius temperature in this unit (J/kg·C°) is commonly accepted because the size of a Celsius degree is the same as the size of a kelvin. Note that *the specific heat is the amount of energy re-*

Specific heat

---

*This may also be expressed in terms of *molar heat capacity*, where the mass is expressed in term of moles (Section 10.3) rather than kilograms. This is common in chemical analyses, particularly when working with gases.

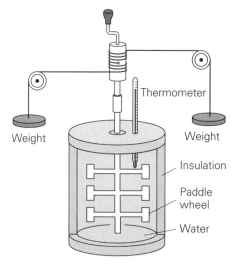

•FIGURE 11.4   Joule's apparatus for measuring the mechanical equivalent of heat
As the weights descend, the paddle wheels churn the water and mechanical energy or work is converted into heat energy. It was found that for every 4.186 J of work done, the temperature of the water rose 1 C° per gram, or that 4.186 J was equivalent to 1 calorie.

Note: Review Section 10.1 and Fig. 10.1

TABLE 11.1

TABLE 11.1 Specific Heats of Various Substances (20°C and 1 atm)

Substance	Specific heat (c)	
	J/kg·C°	kcal/kg·C° (or cal/g·C°)
Air (50°C)	1050	0.25
Alcohol, ethyl	2430	0.58
Aluminum	920	0.22
Copper	390	0.093
Glass	840	0.20
Ice (−5°C)	2100	0.50
Iron or steel	460	0.11
Lead	130	0.031
Mercury	140	0.033
Soil (average)	1050	0.25
Steam (110°C)	2010	0.48
Water (15°C)	4186	1.00
Wood (average)	1680	0.40

quired to raise the temperature of 1 kg of a substance by 1 C°. The specific heats of some common substances are given in Table 11.1. Specific heat depends somewhat on temperature (and pressure); but over normal temperature intervals, such variations can be neglected and c can be considered constant.

The greater the specific heat of a substance, the more energy must be transferred to it or taken from it to change the temperature of a given mass of it. That is, a substance with a greater specific heat has a greater heat capacity, or accepts or yields more heat for a given temperature change (and mass).

Water has a relatively large specific heat of 1.00 kcal/kg·C°. The value is exactly 1.00 because the definition of the kilocalorie states that 1 kcal raises the temperature of 1 kg of water by 1 C°.

You have been the victim of the large specific heat of water if you have ever burned your mouth on a baked potato or the cheese of a pizza. These foods have a high water content and a large heat capacity, so they don't cool off as quickly as some other foods do.

EXAMPLE 11.2 ■ WARMING UP: SPECIFIC HEAT

How much heat is required to raise the temperature of 0.20 kg of water from 15°C to 45°C?

Solution.

Given: $m = 0.20$ kg   *Find:* Q (heat)
       $\Delta T = T_f - T_i = 45°C - 15°C = 30$ C°
       $c = 4186$ J/kg·C°
          (from Table 11.1)

Equation 11.1 can be used directly with the quantities given:

$$Q = mc\Delta T = (0.20 \text{ kg})(4186 \text{ J/kg·C°})(30 \text{ C°}) = +2.5 \times 10^4 \text{ J}$$

+Q = heat added
−Q = heat removed

Note that when there is a temperature increase, ΔT is positive ($T_f > T_i$) and we designate Q as positive. This corresponds to energy being *added to* a system. Conversely, ΔT and Q are negative when energy is *removed from* a system.

Follow-up Exercise. How much heat (in joules) would have to be removed from 0.20 kg of water at 50°C in order to lower its temperature to 20°C?

EXAMPLE 11.3 ■ COOLING DOWN: SPECIFIC HEAT AGAIN

A half-liter of water at 30°C is cooled, with the removal of 15 kcal of heat. What is the final temperature of the water?

Solution. Because the amount of heat is given in kcal, let's use the kcal form of the specific heat for illustration.

Given: $m = 0.50$ kg   *Find:* T_f (final temperature)
(since 1 L of water
has a mass of 1 kg)
$T_i = 30$ C°
$Q = -15$ kcal
(negative because heat
energy is removed)
$c = 1.00$ kcal/kg·C°
(from Table 11.1)

Writing out the ΔT term of Eq. 11.1 gives
$$Q = mc\Delta T = mc(T_f - T_i)$$

Solving for T_f gives

$$T_f = \frac{Q}{mc} + T_i = \frac{-15 \text{ kcal}}{(0.50 \text{ kg})(1.00 \text{ kcal/kg·C°})} + 30°C = -30°C + 30°C = 0°C$$

The (liquid) water is at its freezing point; the removal of more heat would cause the water to freeze. However, Eq. 11.1 *does not apply* when the temperature change (ΔT) is an interval that includes a change of phase, as you will learn in the next section.

Follow-up Exercise. Suppose in this Example that 10 kcal of heat were initially removed, then 15 kcal added. What would be the final temperature of the water in this case?

EXAMPLE 11.4 ■ HEAT CAPACITY: DIFFERENT MATERIALS

Equal masses of aluminum (Al) and copper (Cu) are at the same temperature. Which will require the greater heat to raise its temperature by a given amount, and how many times greater is this than the heat that would have to be added to the other metal?

Solution.

Given: $m_{Al} = m_{Cu}$ *Find:* Which material requires the
$\Delta T_A = \Delta T_{Cu}$ greater value of Q and how many
$c_{Al} = 920$ J/kg·°C   times greater it is
$c_{Cu} = 390$ J/kg·°C
(from Table 11.1)

The first part is easy. Since aluminum has a larger specific heat, it has a larger heat capacity, and more heat is required to raise its temperature by a given amount.

The question "how many times greater" generally implies a ratio. That is, if we find $x = Q_{Al}/Q_{Cu}$, we can write $Q_{Al} = xQ_{Cu}$, or Q_{Al} is x times greater than Q_{Cu}. To find the ratio we use Eq. 11.1 for each metal and divide one by the other.

$$\frac{Q_{Al}}{Q_{Cu}} = \left(\frac{m_{Al}}{m_{Cu}}\right)\left(\frac{c_{Al}}{c_{Cu}}\right)\left(\frac{\Delta T_{Al}}{\Delta T_{Cu}}\right) = \frac{c_{Al}}{c_{Cu}}$$

where quantities that are given as being equal (mass and ΔT) have been grouped in parentheses so it can be easily seen that they cancel out.

Then,

$$Q_{Al} = \left(\frac{c_{Al}}{c_{Cu}}\right)Q_{Cu} = \left(\frac{920 \text{ J/kg·C°}}{360 \text{ J/kg·C°}}\right)Q_{Cu}$$

$$Q_{Al} = (2.56)Q_{Cu}$$

●FIGURE 11.5 **Calorimetry apparatus**
The calorimeter cup (center, with black insulating ring) goes into the larger container. The cover with the thermometer and stirrer is seen at the right. Metal shot or pieces of metal are heated in the small cup with handle that is inserted into the hole at the top of the steam generator on the tripod.

That is, 2.56 times more heat is required for the aluminum than for the copper.

Follow-up Exercise. As was mentioned, the specific heat of water is relatively large. To illustrate this point, let equal amounts of heat be added to equal masses of water and mercury initially at the same temperature. Which liquid has the greater temperature change, and how many times greater?

PROBLEM-SOLVING HINT

When the quantities given in a problem are equal, like the masses and temperature intervals in Example 11.4, it is a good indication that a ratio can be used to cancel out the equal quantities. Also keep in mind that phrases like "how many times more" imply a factor derived from a ratio.

Calorimetry

The specific heat of a substance is determined by measuring the quantities in Eq. 11.2 ($Q = mc\Delta T$) other than c. A simple laboratory apparatus for measuring specific heats is shown in ●Fig. 11.5. A substance of known mass and temperature is put into a quantity of water in a calorimeter. The water is at a different temperature, usually a lower one. The calorimeter is an insulated container that allows little heat loss (ideally, none). The principle of the conservation of energy is then applied to determine c. This procedure is sometimes called the *method of mixtures*. An example follows. Keep in mind that such heat exchange problems are just a matter of "thermal accounting." If something loses heat, something else must gain heat, or the conservation of energy would be violated.

EXAMPLE 11.5 ■ CALORIMETRY: THE METHOD OF MIXTURES

Students in a physics lab are to determine the specific heat of copper experimentally. They heat 0.150 kg of copper shot to 100°C and then carefully pour the hot shot into a calorimeter cup (Fig. 11.5) containing 0.200 kg of water at 20°C. The final temperature of the mixture in the cup is measured to be 25°C. If the aluminum cup has a mass of 0.037 kg, what is the specific heat of copper? (Assume that there is no heat loss to the surroundings.)

Solution. In calorimetry problems, it is important to identify and label all of the quantities so as to keep them straight. We will use subscripts m, w, and c to refer to the metal, water, and calorimeter cup, respectively, and the subscripts h, i, and f to refer to the temperatures of the hot metal shot, the water (and cup) initially at room temperature, and the final temperature of the system, respectively. With this notation, we have,

Given: $m_m = 0.150$ kg   *Find:* c_m (specific heat)
$m_w = 0.200$ kg
$c_w = 4186$ J/kg·C°
   (from Table 11.1)
$m_c = 0.037$ kg
$c_c = 920$ J/kg·C°
$T_h = 100°C, T_f = 25°C,$
   and $T_i = 20°C$

Then, assuming no heat is lost from the system, its energy is conserved and the magnitude of the heat lost by the metal $|Q_m|$ must equal that of the heat gained by the water and cup $|Q_{w+c}|$:

$$|\text{heat lost (by metal)}| = |\text{heat gained (by water and cup)}|$$
$$|Q_m| = |Q_{w+c}|$$

Substituting for these heats from Eq. 11.1 ($Q = mc\Delta T$) and using the absolute values of the ΔT's gives

$$m_m c_m (T_h - T_f) = m_w c_w (T_f - T_i) + m_c c_c (T_f - T_i)$$

where the metal initially at T_h cools to T_f and $|\Delta T_m| = (T_h - T_f)$; the water and the cup at T_i are heated to T_f and $\Delta T_{w+c} = (T_f - T_i)$.
Solving for c_m gives

$$c_m = \frac{(m_w c_w + m_c c_c)(T_f - T_i)}{m_m (T_h - T_f)}$$

$$= \frac{[(0.200 \text{ kg})(1.00 \text{ kcal/kg·C°}) + (0.037 \text{ kg})(0.22 \text{ kcal/kg·C°})] \times (25°C - 20°C)}{0.150 \text{ kg}(100°C - 25°C)}$$

$$= 0.093 \text{ kcal/kg·C°}$$

Follow-up Exercise.   In actual calorimetry experiments, some heat is lost from the system. What effect would this have on the experimentally-determined specific heat?

11.3   Phase Changes and Latent Heat

Objectives: To be able to (a) compare and contrast the three common phases of matter, and (b) relate latent heat to phase changes.

Matter normally exists in three *phases*: solid, liquid, and gas (see •Fig. 11.6). The phase that a substance is in depends on its internal energy (as indicated by its temperature) and the pressure on it. You probably think more readily of adding or removing heat to change the phase of a substance because most of your experience with phase changes has been at normal atmospheric pressure, which is relatively constant.

In the **solid phase**, molecules are held together by attractive forces, or bonds. (Simplistically, these bonds can be represented as springs, as was done in Section 9.1.) Adding heat causes increased motion about the molecular equilibrium positions. If enough heat is added to provide sufficient energy to break the intermolecular bonds, the solid undergoes a phase change and becomes a liquid. The temperature at which this occurs is called the **melting point**. The temperature at which a liquid becomes a solid is called the **freezing point**. In general, these temperatures are the same, but they can differ slightly.

Ice, table salt, and most metals are crystalline solids. That is, they have orderly molecular or atomic arrangements. Other substances, however, such as glass, are noncrystalline, or amorphous. Instead of melting at a particular temperature, these substances melt over a temperature range. This discussion will be concerned primarily with substances that have definite melting points.

In the **liquid phase**, molecules of a substance are relatively free to move, and a liquid assumes the shape of its container. In certain liquids, there may be some ordered structure, giving rise to so-called liquid crystals, such as are used in LCD's (liquid crystal displays) of calculators and clocks (Chapter 23). Adding heat increases the motion of the molecules of a liquid, and when they have enough energy to become separated by large distances (compared to their diameters), the liquid changes to the **gaseous phase**, or **vapor phase**. (The distinction between a gas and a vapor will be made shortly.) This change may occur slowly by the process of evaporation or rapidly at a particular temperature called the **boiling point**. The temperature at which a gas condenses and becomes a liquid is the **condensation point**.

Some solids, such as dry ice (solid carbon dioxide), moth balls, and certain air fresheners, change directly from the solid to the gaseous phase. This process is called **sublimation**. Like the rate of evaporation, the rate of sublimation increases with temperature. A phase change from a gas to a solid is called *deposition*. Frost, for example, is solidified water vapor deposited directly on grass, car windows, and other objects. Frost is not frozen dew, as is sometimes mistakenly assumed.

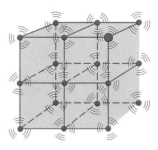

(a) Solid

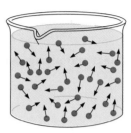

(b) Liquid

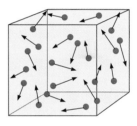

(c) Gas

•**FIGURE 11.6   Three phases of matter**
(a) The molecules of a solid are held together by bonds; consequently, a solid has definite shape and volume. **(b)** The molecules of a liquid can move more freely, so a liquid has a definite volume but assumes the shape of its container. **(c)** The molecules of a gas interact weakly and are separated by relatively large distances; thus, a gas has no definite shape or volume.

Note: Solid, liquid, and gas are sometimes referred to as states of matter, but the state of a system has a different meaning in physics, as you will learn in Chapter 12.

Latent Heat

In general, when heat energy is transferred to a substance, its temperature increases. However, when added (or removed) heat causes only a phase change, the temperature of the substance does *not* change. For example, if heat is added to a quantity of ice ($c = 2100$ J/kg·C° or 0.50 kcal/kg·C°) at -10°C, the temperature of the ice increases until it reaches its melting point of 0°C. At this point, the addition of more heat does not increase the temperature but causes the ice to melt, or change phase. (Of course, the heat must be added slowly so that the ice and melted water remain in thermal equilibrium.) Once the ice is melted, adding more heat will cause the temperature of the water to rise. A similar situation occurs during the liquid–gas phase change at the boiling point. Adding more heat to boiling water only causes more vaporization, not a temperature increase.

Note: Keep in mind that ice can be colder than 0°C.

Latent heat and phase changes

From the earlier description of the molecular nature of the different phases of matter, you can see that during a phase change the heat energy goes into the work of breaking bonds and separating molecules, rather than into increasing the temperature. The heat involved in a phase change is called the **latent heat** (L), and

$$Q = mL \tag{11.2}$$

where m is the mass of the substance. As you can see from this equation, the latent heat has units of joules per kilogram (J/kg) in the SI, or kcal/kg in other common units. The latent heat for a solid–liquid phase change is called the **latent heat of fusion** (L_f), and that for a liquid–gas phase change is called the **latent heat of vaporization** (L_v). These are often referred to as simply the heat of fusion and the heat of vaporization. The latent heats of some substances, along with their freezing and boiling points, are given in Table 11.2. (The latent heat for the less common solid–gas phase change is called the latent heat of sublimation and is symbolized by L_s.)

It is helpful to focus on the fusion and vaporization of water. A plot of temperature versus heat energy for a quantity of water is shown in •Fig. 11.7. Note that when heat is added (or removed) at the temperature of a phase change, 0°C or 100°C, the temperature remains constant. Once the phase change is complete, adding more heat causes the temperature to increase. Note in Fig. 11.7 that the slopes of the phase lines are not all the same, which indicates that the specific heats of the various phases are not all the same. (Why?)

For water, the latent heats of fusion and vaporization are

$$L_f = 3.33 \times 10^5 \text{ J/kg} \quad \text{(or about 80 kcal/kg)}$$
$$L_v = 22.6 \times 10^5 \text{ J/kg} \quad \text{(or about 540 kcal/kg)}$$
latent heats for water

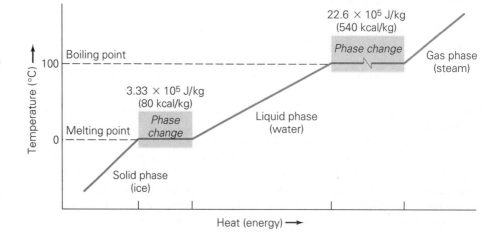

•**FIGURE 11.7 Temperature versus heat for water**
As heat is added to the various phases of water, the temperature increases. During a phase change, however, the heat energy does the work of separating the molecules and the temperature remains constant. Note the different slopes of the phase lines, which indicate different values of specific heat. (Not to scale.)

		L_f			L_v	
TABLE 11.2						

TABLE 11.2   Temperatures of Phase Changes and Latent Heats for Various Substances (1 atm)

Substance	Melting point	L_f J/kg	kcal/kg	Boiling point	L_v J/kg	kcal/kg
Alcohol, ethyl	−114°C	1.0×10^5	25	78°C	8.5×10^5	204
Gold	1063°C	0.645×10^5	15.4	2660°C	15.8×10^5	377
Helium*	—	—		−269°C	0.21×10^5	5
Lead	328°C	0.25×10^5	5.9	1744°C	8.67×10^5	207
Mercury	−39°C	0.12×10^5	2.8	357°C	2.7×10^5	65
Nitrogen	−210°C	0.26×10^5	6.1	−196°C	2.0×10^5	48
Oxygen	−219°C	0.14×10^5	3.3	−183°C	2.1×10^5	51
Tungsten	−3410°C	1.8×10^5	44	5900°C	48.2×10^5	1150
Water	0°C	3.33×10^5	80	100°C	22.6×10^5	540

*Not a solid at 1 atm pressure; melting point −272°C at 26 atm.

That is, 3.33×10^5 J (or 80 kcal) of heat are needed to melt 1 kg of ice at 0°C, and 22.6×10^5 J (or 540 kcal) of heat are needed to convert 1 kg of water to steam at 100°C (•Fig. 11.8). Note that the latent heat of vaporization is almost 7 times the latent heat of fusion. This indicates that more energy is needed to separate the molecules in going from water to steam than to break up the lattice structure in going from ice to water.

The word "latent" means "hidden," and its use in this context may be understood by considering a situation involving human skin. Since 540 kcal of heat energy are required to convert 1 kg of water into steam, the conservation of energy tells you that when 1 kg of steam condenses into water, 540 kcal of energy must be given up. As a result, burns from steam are usually more serious than those from boiling water. The condensing of the steam on the skin provides an additional 540 kcal/kg of heat that is seemingly hidden until contact.

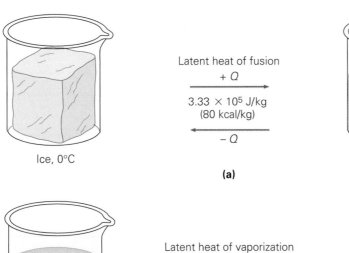

Latent heat of fusion
+ Q
3.33×10^5 J/kg
(80 kcal/kg)
− Q

Ice, 0°C

Water, 0°C

(a)

Latent heat of vaporization
+ Q
22.6×10^5 J/kg
(540 kcal/kg)
− Q

Water, 100°C

Steam, 100°C

(b)

•FIGURE 11.8 **Phase changes and latent heats**
(a) At 0°C, 80 kcal must be added to 1 kg of ice or removed from 1 kg of water to change its phase. **(b)** At 100°C, 540 kcal must be added to 1 kg of water or removed from 1 kg of steam to change its phase.

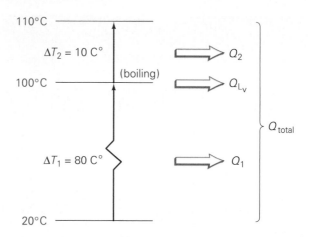

•**FIGURE 11.9 Steps in changing phases**
In going from water at 20°C to steam at 110°C, three heat-adding steps are involved. See Example 11.6.

EXAMPLE 11.6 ■ TEMPERATURE *AND* PHASE CHANGES: SPECIFIC HEAT *AND* LATENT HEAT

Heat is added to 0.500 kg of water at room temperature (20°C). How many joules of heat are required to change the water to steam at 110°C?

Solution.

Given: $m = 0.500$ kg $\qquad\qquad\qquad$ *Find:* Q (heat in joules)
$\quad\quad\quad T_i = 20°C, T_f = 110°C$
$\quad\quad\quad L_v = 22.6 \times 10^5$ J/kg (from Table 11.2)
$\quad\quad\quad\quad$ (specific heats from Table 11.1)

The temperature intervals and the heats required in the process are shown in •Fig. 11.9. First, the water is heated to the boiling point (100°C), where a phase change occurs, and then the steam is heated to 100°C. The Q's for the various steps are

(heating water) $\quad Q_1 = mc_w\Delta T_1 = (0.500 \text{ kg})(4186 \text{ J/kg·C°})(100°C - 20°C) = 1.67 \times 10^5$ J

(phase change) $\quad Q_{L_v} = mL_v = (0.500 \text{ kg})(22.6 \times 10^5 \text{ J/kg}) = 11.3 \times 10^5$ J

(heating steam) $\quad Q_2 = mc_s\Delta T_2 = (0.500 \text{ kg})(2010 \text{ J/kg·C°})(110°C - 100°C) = 0.101 \times 10^5$ J

Then $\qquad\qquad Q_{total} = \Sigma_i Q_i = (1.7 + 11.3 + 0.1) \times 10^5 \text{ J} = +13.1 \times 10^5$ J

Follow-up Exercise. How many joules of heat would have to be removed to condense 1.00 kg of steam at 110°C?

PROBLEM-SOLVING HINT

Note that you must compute the latent heat at each phase change. It is a common error to use the specific heat equation with a temperature interval that includes a phase change.

EXAMPLE 11.7 ■ WATER *AND* ICE: THERMAL EQUILIBRIUM

A 0.30-kg piece of ice at 0°C is placed in a liter of water at room temperature (20°C) in an insulated container. Assuming that no heat is lost to the container, what is the final temperature of the water?

Solution.

Given: $m_i = 0.30$ kg $\qquad\qquad\qquad$ *Find:* T_f (final temperature)
$\quad\quad\quad T_i = 0°C$
$\quad\quad\quad V_w = 1.0$ L, so $m_w = 1.0$ kg
$\quad\quad\quad T_w = 20°C$

The subscripts i and w refer to ice and water, respectively. You know that 1.0 L of water has a mass (m_w) of 1.0 kg and that the water supplies the heat to melt the ice. If all of the ice melts, you could then view the system as being two masses of water at different temperatures, which come to equilibrium at some intermediate temperature. Since there is no heat loss, it is tempting to write an expression equating the amount of heat lost by the water to the amounts of heat used to melt the ice and to warm up the ice water.

Note: This is simply a matter of "thermal accounting"—if something loses, something else gains.

But does all the ice melt? Let's work in kcal to avoid large numbers. To melt 0.30 kg of ice requires

$$Q_i = m_i L_f = (0.30 \text{ kg})(80 \text{ kcal/kg}) = 24 \text{ kcal}$$

Then, looking at the water, you must ask how much heat it can supply to the ice. The maximum amount would be that given up in lowering the temperature of the water to 0°C, which would be a temperature decrease of $\Delta T_w = -20$ C°. This maximum amount of heat would be

$$Q_w = m_w c_w \Delta T_w = (1.0 \text{ kg})(1.0 \text{ kcal/kg·C°})(-20 \text{ C°}) = -20 \text{ kcal}$$

Thus, all of the ice does *not* melt, since the water does not have enough energy to do this. The final temperature of the water is therefore 0°C.

The heat given up by the water (20 kcal) will melt a mass of ice equal to

$$m_i = \frac{Q_i}{L_f} = \frac{20 \text{ kcal}}{80 \text{ kcal/kg}} = 0.25 \text{ kg}$$

Thus, the final mixture would be 0.30 kg − 0.25 kg = 0.05 kg of ice in thermal equilibrium with 1.25 kg (or 1.25 L) of water at $T_f = 0°C$.

Follow-up Exercise. In this Example, what would the initial temperature of the water have to be to just melt all of the ice?

Information about phase changes is represented on graphs called **phase diagrams.** An example is the *p–T* (pressure–temperature) diagram for water shown in •Fig. 11.10. The curves are formed of the points (p, T), or the pressure–temperature combinations at which different phases are in equilibrium. For example, the point at 1 atm and 100°C corresponds to the normal boiling point, at which liquid water and steam are in equilibrium.

The *triple point* is the point at which all three phases coexist. This is the unique point used as a reference for the Kelvin scale (Chapter 10). The three curves branching out from this point separate the phase regions. Note that if you started heating a quantity of ice that was below 0°C at 1 atm, the plotted state of the system

Note: The Kelvin temperature scale is discussed in Section 10.3.

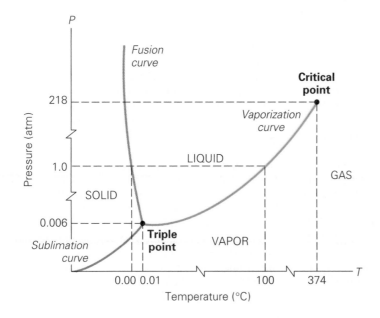

•**FIGURE 11.10 Phase diagram for water**
The curves branch out from the triple point, where water exists in all three phases. On the transition curves, it exists in two phases.

would first cross the fusion curve (along the horizontal dashed line extending from 1.0 atm in the graph) into the liquid phase region and, with continued heating, would cross the vaporization curve into the vapor phase region at 100°C.

"Vapor" is another term commonly used for "gas." Water vapor is water in the gas phase. (Vapor is also used in a nontechnical sense to mean visible droplets of water, such as condensed steam or clouds.) The distinction between vapor and gas is often made relative to the critical point at the end of the vaporization curve. At temperatures less than the critical temperature (374°C for water), a gas will change to a liquid if sufficient pressure is applied. If a gas is above its critical temperature, no amount of pressure will cause it to become a liquid. It becomes denser and denser with increasing pressure, but never quite becomes a liquid. A substance that is gaseous and has a temperature above its critical temperature is called a gas, and a substance that is gaseous but with a temperature below its critical temperature is known as a vapor.

As can be seen from Fig. 11.10, the boiling point of water decreases with decreasing pressure (the curve between liquid and vapor phases). In fact, a container of water in a vacuum chamber will boil at room temperature. The cooling effect of the boiling (the removal of latent heat) will eventually cause the remaining water to freeze if the vacuum is maintained. At high altitudes, where there is lower atmospheric pressure, the boiling point of water is lowered. For example, at Pike's Peak, Colorado, at an elevation of about 4300 m, the atmospheric pressure is about 600 torr, and water boils at about 94°C rather than at 100°C. The lower temperature lengthens the cooking time of food. A pressure cooker may be used to reduce the cooking time (at Pike's Peak or at sea level)—by increasing the pressure, a pressure cooker raises the boiling point.

Some interesting features of the water phase diagram are discussed in the Insight below.

Evaporation

The **evaporation** of water from an open container becomes evident only after a relatively long period of time. This phenomenon can be explained in terms of kinetic theory (Section 10.5). The molecules in a liquid are in motion, at different

Insight Phase Changes and Ice Skating

It is interesting to note from the phase diagram of water (Figure 11.10) that the fusion curve slopes upward to the left and the vaporization curve slopes upward to the right. These slopes indicate that the freezing point of water decreases slightly with increasing pressure and the boiling point increases with increasing pressure. The latter fact is what makes possible the operation of the pressure cooker, which cooks foods faster because boiling temperatures greater than 100°C can be obtained at pressures above atmospheric pressure.

Such a slope of the fusion curve is characteristic of only a very few substances that, like water, expand on freezing. It was once thought that the lowering of the freezing point of water by pressure provided the mechanism that made ice skating possible. Supposedly, the pressure of the narrow skate blade on the ice would lower the melting point below the ambient temperature. The skater would thus glide on a thin film of water, which quickly refroze when the pressure was re-

moved. However, from the phase diagram for water, it is found that a pressure of about 140 atm is needed to lower the freezing point from 0°C to −1°C. Only a very heavy person could produce a pressure on this order, and the outdoor ice skating temperature is usually well below −1°C.

Frictional heating between the skate blade and the ice does contribute to melting. For high-friction surfaces, such as skis on snow, this is the main mechanism. However, for ice skating there is another factor that contributes to the low coefficient of friction of ice. This is called *surface melting*, proposed by the English scientist Michael Faraday (1791–1867; see Chapter 20). He suggested that a thin layer of liquid normally exists on the surface of a solid even at temperatures well below the solid's melting point. Modern techniques have shown this to be the case for most solids. For ice, the thickness of the water film is about 40 nm near 0°C and about 0.50 nm near −35°C. (Recall that 1 nm = 10^{-9} m.)

speeds. A faster-moving molecule that is near the surface may momentarily leave the liquid. If its velocity is not too large, it will return to the liquid because of the attractive forces exerted by the other molecules. Occasionally, however, a molecule has a large enough velocity that it leaves the liquid entirely and becomes part of the air. The higher the temperature of the liquid, the more likely this is to occur. (Why?—see the Insight in Section 10.5 on distribution of molecular speeds.)

The escaping molecules take their energy with them. Since those molecules with greater than average energy are the ones most likely to escape, the energy and temperature of the remaining liquid will be reduced. Thus, *evaporation is a cooling process.* You have probably noticed this when drying off after a bath or shower. About 600 kcal are needed to evaporate 1 kg of water from the skin. This energy requirement can be estimated by considering a different process. The amount of heat per kilogram (Q/m) needed to raise the temperature of water from 35°C (approximately skin temperature) to 100°C ($\Delta T = 100°C - 35°C = 65$ C°) is $c\Delta T$ (since $Q = mc\Delta T$). Adding the latent heat of vaporization, which is also the heat per unit mass ($Q = mL_v$ and $L_v = Q/m$), gives

$$\frac{Q}{m} = c\Delta T + L_v = (1 \text{ kcal/kg·C°})(65\text{C°}) + 540 \text{ kcal/kg} = 605 \text{ kcal/kg}$$

Of course, evaporation doesn't involve heating to the boiling point, but this estimate gives the maximum possible cooling effect.

Although evaporation is a relatively slow process, it is often important in preventing our bodies from overheating. Usually, radiation and conduction are sufficient to maintain a rate of heat loss to the air that keeps us comfortable, given the temperature difference between our bodies and the surroundings. However, when the air gets hot and the temperature difference narrows (or disappears), we start to perspire. The evaporation of perspiration helps to cool our bodies. On a hot summer day, a person may stand in front of a fan and remark how cool the blowing air feels. But the fan is merely blowing hot air from one place to another. The air feels cool because its flow promotes evaporation, which removes heat energy. Of course, evaporation depends on the humidity (the amount of moisture already in the air). We feel less comfortable on hot humid days because evaporation is reduced.

11.4 Heat Transfer

Objectives: To be able to (a) describe the three methods of heat transfer, and (b) give practical and/or environmental examples of each.

Since heat is defined as energy in transit, how the transfer takes place is important. Heat moves from place to place (from a higher-temperature region to a lower-temperature region) by three mechanisms: conduction, convection, and radiation.

Conduction

You can keep a pot of coffee hot on a stove because heat is conducted through the coffee pot from the hot burner. The process of **conduction** is visualized as resulting from molecular interactions. Molecules in one part of a body at a higher temperature vibrate faster. They collide with and transfer some of their energy to less energetic molecules located toward the cooler part of the body. In this way, energy is conductively transferred from a higher-temperature region to a lower-temperature region—transfer as a result of a temperature difference.

Solids can be divided into two general categories: metals and nonmetals. Metals are good conductors of heat, or **thermal conductors.** Modern theory views metals as having a large number of electrons that are free to move around (not permanently bound to a particular molecule or atom). These free electrons are be-

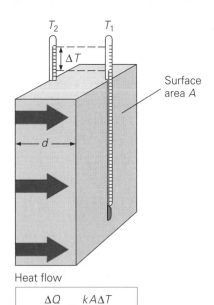

$$\frac{\Delta Q}{\Delta t} = \frac{kA\Delta T}{d}$$

•FIGURE 11.11   Thermal conduction
Heat conduction is characterized by the time rate of flow of heat ($\Delta Q/\Delta t$) in a material with a temperature difference ΔT. For a slab of material, $\Delta Q/\Delta t$ is directly proportional to the cross-sectional area (A) and the thermal conductivity of the material (k), and is inversely proportional to the slab thickness (d).

•FIGURE 11.12   Copper-bottomed
A layer of copper is often found on the bottoms of pots and saucepans. The high heat conductivity of this metal ensures the rapid and even spread of heat from the burner, reducing the likelihood of local "hot spots." (Contrast the chapter-opening photograph.)

lieved to be primarily responsible for the heat conduction of metals. Nonmetals, such as wood or cloth, have relatively few free electrons and are poor heat conductors. A poor heat conductor is called a **thermal insulator**.

In general, the ability of a substance to conduct heat depends on its phase. Gases are poor thermal conductors because their molecules are relatively far apart, and collisions are therefore infrequent. Liquids are better thermal conductors than gases are because their molecules are closer together and can interact more readily.

Heat conduction may be described quantitatively as the time rate of heat flow ($\Delta Q/\Delta t$) in a material for a given temperature difference (ΔT), as illustrated in •Fig. 11.11. It has been established through experiments that the rate of heat flow through a substance depends on the temperature difference between its boundaries. Heat conduction also depends on the size and shape of an object. Thus, the analysis of heat flow is generally done using a uniform slab of the material.

A moment's thought should convince you that the heat flow through a slab of material is directly proportional to its surface area (A) and inversely proportional to its thickness (d). That is,

$$\frac{\Delta Q}{\Delta t} \propto \frac{A\Delta T}{d}$$

The term $\Delta T/d$ is called *thermal gradient* (the change in temperature per unit of length). Using a constant of proportionality allows the relation to be written as an equation:

$$\frac{\Delta Q}{\Delta t} = \frac{kA\Delta T}{d} \qquad \textit{(conduction only)} \qquad (11.3)$$

The constant k is called the **thermal conductivity** and characterizes the heat-conducting ability of a material. The greater the value of k for a material, the more rapidly it will conduct heat. The units of k are J/m·s·C° (or kcal/m·s·C°). The thermal conductivities of various substances are listed in Table 11.3 on p. 370. These values actually vary slightly over different temperature ranges, but can be considered to be constant over normal temperature ranges and differences.

Compare the relatively large thermal conductivities of the good thermal conductors, the metals, with the relatively small thermal conductivities of some good thermal insulators, such as Styrofoam and wood. The large thermal conductivities of metals are attributable to "free" electrons that are not bound to a particular metal atom. This property is also important in electrical conductivity (Chapter 16). You may have noticed that some cooking pots have copper coatings on the bottoms (•Fig. 11.12). Being a good conductor of heat, the copper promotes the distribution of heat over the bottom of a pot for even cooking. Plastic foams, on the other hand, are good insulators, mainly because they contain pockets of air. Recall that gases are poor conductors, and note the low thermal conductivity of air in the table.

EXAMPLE 11.8 ■ INSULATION: CONTROLLING THERMAL CONDUCTIVITY

A room with a pine ceiling that measures 3.0 m × 5.0 m × 2.0 cm thick has a 6.0 cm layer of glass wool insulation above it (•Fig. 11.13a). On a cold day, the temperature inside the room is 20°C and the temperature in the attic is 8.0°C. Assuming that the temperatures remain constant, how much energy is saved in one hour by having installed the layer of insulation? Assume loses are due to conduction only.

Solution.   Computing some of the quantities and making conversions as we list the data, we have

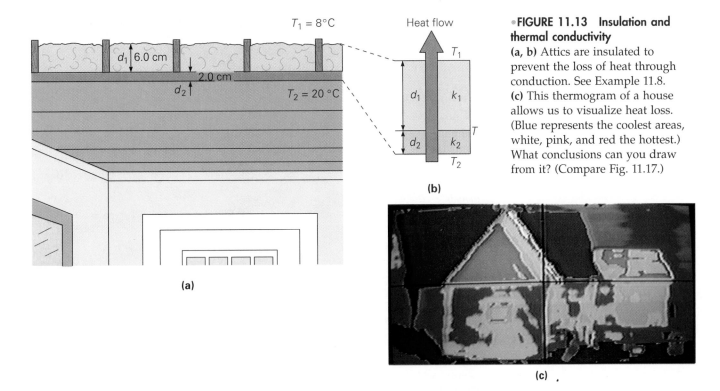

•FIGURE 11.13 **Insulation and thermal conductivity**
(a, b) Attics are insulated to prevent the loss of heat through conduction. See Example 11.8. **(c)** This thermogram of a house allows us to visualize heat loss. (Blue represents the coolest areas, white, pink, and red the hottest.) What conclusions can you draw from it? (Compare Fig. 11.17.)

*T*₁ = 8°C

Heat flow

*T*₁

2.0 cm

*T*₂ = 20 °C

*d*₁ 6.0 cm

*d*₂

*d*₁ *k*₁

*d*₂ *k*₂

T

*T*₂

(b)

(a)

(c)

Given: $A = 3.0 \text{ m} \times 5.0 \text{ m} = 15 \text{ m}^2$
    $d_1 = 6.0 \text{ cm} = 0.060 \text{ m}$
    $d_2 = 2.0 \text{ cm} = 0.020 \text{ m}$
    $\Delta T = T_2 - T_1 = 20°C - 8.0°C = 12 \text{ C}°$
    $\Delta t = 1.0 \text{ h} = 3.6 \times 10^3 \text{ s}$
    $k_1 = 0.042 \text{ J/m·s·C}° \text{ (glass wool)}$
    $k_2 = 0.12 \text{ J/m·s·C}° \text{ (wood)}$ (from Table 11.3)

Find: Energy saved in 1.0 h

(In working such problems with lots of given quantities, it is especially important to label all the data correctly.)

Then, to find the energy saved in one hour, we need to compute how much heat is conducted in this time with and without the layer of insulation. First, let's consider how much heat would be conducted in one hour through the wooden ceiling with no insulation present.

Equation 11.3 gives the rate of heat flow ($\Delta Q/\Delta t$). Since we know Δt, we can rearrange the equation to find ΔQ_c (heat conducted through the wooden ceiling).

$$\Delta Q_c = \left(\frac{k_2 A \Delta T}{d_2}\right)\Delta t$$

$$= \left[\frac{(0.12 \text{ J/m·s·C}°)(15 \text{ m}^2)(12 \text{ C}°)}{0.020 \text{ m}}\right](3.6 \times 10^3 \text{ s})$$

$$= 3.9 \times 10^6 \text{ J}$$

Now we need to find the heat conducted through the ceiling *and* the insulation layer together. Let *T* be the temperature at the interface of the materials, and T_2 and T_1 refer to the warmer and cooler temperatures, respectively (see •Fig. 11.13b). Then,

$$\frac{\Delta Q_1}{\Delta t} = \frac{k_1 A(T - T_1)}{d_1} \quad \text{and} \quad \frac{\Delta Q_2}{\Delta t} = \frac{k_2 A(T_2 - T)}{d_2}$$

The problem is that we don't know *T*. However, when the conduction is steady, the flow rates are the same for both, that is, $\Delta Q_1/\Delta t = \Delta Q_2/\Delta t$. This relationship enables us to eliminate *T*. We equate the two preceding equations, finding an expression for *T*, and then substitute this expression into either equation, which will give us a general expression for $\Delta Q/\Delta t$ for the combined layers.

$$\frac{\Delta Q_1}{\Delta T} = \frac{\Delta Q_2}{\Delta t}$$

TABLE 11.3 Thermal Conductivities of Some Substances

Substance	Thermal conductivity (k)	
	$J/m \cdot s \cdot C°$	$kcal/m \cdot s \cdot C°$
Metals		
Aluminum	240	5.7×10^{-2}
Copper	390	9.4×10^{-2}
Iron and steel	46	1.1×10^{-2}
Silver	420	10×10^{-2}
Liquids		
Transformer oil	0.18	4.2×10^{-5}
Water	0.57	14×10^{-5}
Gases		
Air	0.024	0.57×10^{-5}
Hydrogen	0.17	4.0×10^{-5}
Oxygen	0.024	0.58×10^{-5}
Other materials		
Brick	0.71	17×10^{-5}
Concrete	1.3	31×10^{-5}
Cotton	0.075	1.8×10^{-5}
Fiberboard	0.059	1.4×10^{-5}
Floor tile	0.67	16×10^{-5}
Glass (typical)	0.84	20×10^{-5}
Glass wool	0.042	1.0×10^{-5}
Human tissue (average)	0.20	5.0×10^{-5}
Ice	2.2	53×10^{-5}
Styrofoam	0.042	1.0×10^{-5}
Wood		
Wood, oak	0.15	3.5×10^{-5}
Wood, pine	0.12	2.8×10^{-5}
Vacuum	0	0

or

$$\frac{k_1 A(T - T_1)}{d_1} = \frac{k_2 A(T_2 - T)}{d_2}$$

The A's cancel, and solving for T gives

$$T = \frac{k_1 d_2 T_1 + k_2 d_1 T_2}{k_1 d_2 + k_2 d_1}$$

Substituting into either of the flow rate equations, and rearranging, we get (11.4*):

$$\frac{\Delta Q}{\Delta t} = \frac{A(T_2 - T_1)}{(d_1/k_1) + (d_2/k_2)} \tag{11.4*}$$

$$= \frac{(15 \text{ m}^2)(12 \text{ C}°)}{(0.060 \text{ m}/0.042 \text{ J/m} \cdot \text{s} \cdot \text{C}°) + (0.020 \text{ m}/0.12 \text{ J/m} \cdot \text{s} \cdot \text{C}°)}$$

$$= 1.1 \times 10^2 \text{ J/s}$$

and in $1.0 \text{ h} = 3.6 \times 10^3 \text{ s}$,

$$\Delta Q = (1.1 \times 10^2 \text{ J/s})\Delta t = (1.1 \times 10^2 \text{ J/s})(3.6 \times 10^3 \text{ s})$$

$$= 4.0 \times 10^5 \text{ J}$$

*This equation may be extended to any number of layers or slabs of materials by adding additional d/k terms to the denominator: $\Delta Q/\Delta t = A(T_2 - \Delta T_1)/\Sigma(d_i/k_i)$.

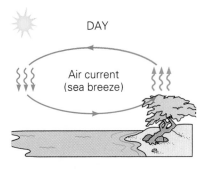

Land warmer than water

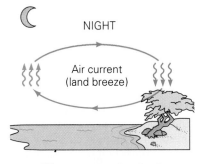

Water warmer than land

•**FIGURE 11.14 Convection cycles** During the day, natural convection cycles give rise to sea breezes near large bodies of water. At night, the pattern of circulation is reversed and land breezes blow. The temperature differences between land and water are the result of their different specific heats. Water has a much greater specific heat, so the land warms up more quickly during the day. At night, the land cools more quickly, while the water remains warmer because of its greater heat capacity.

Convection

In general, compared to solids, liquids and gases are not good thermal conductors. The mobility of molecules in fluids permits heat transfer by another process —convection. Heat transfer by **convection** involves mass transfer. For example, when cold water is in contact with a hot object, such as the bottom of a pot on a stove, the object transfers heat to the water by conduction, and the water carries the heat away with it by convection.

Natural convection cycles occur in liquids and gases. Such cycles are important in atmospheric processes, as illustrated in •Fig. 11.14. During the day, the ground heats up more quickly than do large bodies of water, as anyone who has been to the beach knows. This occurs both because the water has a greater specific heat and because mixing currents disperse the absorbed heat throughout the great volume of water. The air in contact with the warm ground is heated by conduction. That air expands, becoming less dense than in surrounding cooler air. As a result, the warm air rises (air currents) and other air moves horizontally (winds) to fill the space—creating a sea breeze near a large body of water. Cooler air descends, and a thermal convection cycle is set up, which transfers heat away from the land. At night, the ground loses its heat more quickly, and the water surface is warmer than the land. As a result, the cycle is reversed.

You can see convection currents in the air above a hot road surface in the summer and in transparent liquids, such as heated water in a glass container. This is because regions of different temperatures have different densities, which cause a bending, or refraction, of light (see Chapter 22).

Convection can also be forced, which means that the medium of heat transfer is moved mechanically. Note here that we can have heat transfer without a temperature difference. In fact, we can have energy transferred from a low-temperature region to a high-temperature region, as in the case of the forced convection of a refrigerator coolant removing energy from the inside of the refrigerator. (The circulating coolant carries heat energy from the inside of the refrigerator, and this heat is given up to the environment—see Section 12.4) Other common examples of forced convection systems are forced-air heating systems in homes (•Fig. 11.15), the human circulatory system, and the cooling system of an automobile engine. The human body does not use all of the energy obtained from food; a great deal is lost. [Keep in mind that there's usually a temperature difference between your body and the surroundings.] So that body temperature will stay normal, the internally generated heat energy is transferred close to the surface by blood circulation. From the skin, it is conducted to the air or lost by radiation (the other heat-transfer mechanism, to be discussed shortly).

Water or some other coolant is circulated (pumped) through most automobile cooling systems (some engines are air-cooled). The fluid medium carries heat to the radiator (a heat exchanger), where forced air flow produced by the fan carries it away. The radiator of an automobile is actually misnamed—most of the heat is transferred from it by convection rather than radiation.

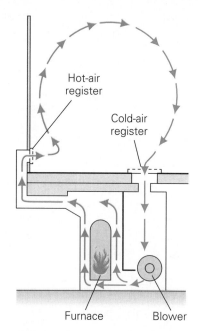

•FIGURE 11.15 Forced convection
Houses are commonly heated by forced convection. In older homes, natural convection was (and still is) used. Registers or gratings in the floors or walls allow heated air to enter and cooler air to return to the heat source.

Radiation

Conduction and convection require some material as a transport medium. The third mechanism for heat transfer needs no medium; it is called **radiation**, which refers to energy transfer by electromagnetic waves (Chapter 20). This is how heat is transferred to the Earth from the Sun through empty space. Visible light and other forms of electromagnetic radiation are commonly referred to as radiant energy.

You have experienced heat transfer by radiation if you've ever stood near an open fire (•Fig. 11.16). You can feel the heat on your exposed hands and face. This heat transfer is not due to convection or conduction, since heated air rises and air is a poor conductor. Visible radiation is emitted from the burning material, but most of the heating effect comes from the invisible **infrared radiation** emitted by the glowing embers or coals. You feel this radiation because it is absorbed by water molecules in your skin. (Body tissue is about 85% water.) The water molecule has an internal vibration whose frequency coincides with that of infrared radiation, which is therefore absorbed readily. [This is called *resonance absorption*. The electromagnetic wave drives the molecular vibration, and energy is transferred to the molecule, somewhat like pushing a swing; see Chapter 13.]

Infrared radiation is sometimes referred to as "heat rays." You may have noticed the red infrared lamps used to keep food warm in cafeterias. Heat transfer

•FIGURE 11.16 Heating by conduction, convection, and radiation
The hands of the person at left are warmed by the convection of rising hot air (and some radiation). The hand of the person at upper right is warmed by conduction. The hands of the person at lower right are warmed by radiation.

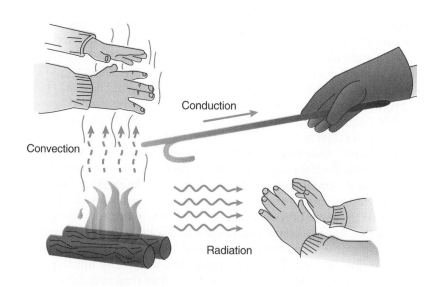

by infrared radiation is also important in maintaining our planet's warmth by a mechanism known as the *greenhouse effect*. This important environmental topic and another common example of heating by radiation absorption are discussed in the Insights on pp. 374 and 375.

Although infrared radiation is invisible to the human eye, it can be detected by other means. The frequency of the infrared radiation is proportional to the temperature of its source. This is the basis for infrared thermometers, which, using infrared detectors, can measure temperature remotely. Also, you can buy special infrared film for some cameras. A picture taken with this film will be an image consisting of contrasting light and dark areas corresponding to regions of higher and lower temperatures. Special instruments that apply such thermography are used in industry and medicine (•Fig. 11.17).

A new application of thermograms is for security. The system consists of an infrared camera and a computer that identifies people by means of the heat patterns emitted by the facial blood vessels. The camera takes a picture of the radiation from a person's face, which is compared with an earlier image stored in the computer memory. It is reported that the system can even distinguish between identical twins, whose facial features are slightly different. Also, changes in body temperature from weather conditions or a fever do not affect the identification, as the relative patterns of radiation remain the same.

The rate at which an object radiates energy has been found to be proportional to the fourth power of the absolute temperature (T^4). This is expressed in an equation known as **Stefan's law**:

$$P = \sigma A e T^4 \tag{11.5}$$

where P is the power radiated in watts (W) or joules/s (J/s). [The calorie is not generally used to measure radiation. Radiated power can be written $\Delta Q/\Delta t$ to indicate a rate of heat loss if desired.] The symbol σ (the Greek letter sigma) is called the *Stefan-Boltzmann constant*: $\sigma = 5.67 \times 10^{-8}$ W/m^2·K^4. The radiated power is also proportional to the surface area (A) of the object. The **emissivity** (e) is a number between 0 and 1 that is characteristic of the material (e is unitless). Dark surfaces have emissivities close to 1, and shiny surfaces have emissivities close to 0. The emissivity of human skin is about 0.70.

Dark surfaces are not only better emitters of radiation, they are also good absorbers. In general, *a good emitter is also a good absorber*. An ideal, or perfect, absorber (and emitter) is referred to as a **blackbody** ($e = 1.0$). Shiny surfaces are poor absorbers, since most of the incident radiation is reflected. This fact may be demonstrated easily as shown in •Fig. 11.18. (You should see why it is better to wear light-colored clothes in the summer and dark-colored clothes in the winter.)

When an object is in thermal equilibrium with its surroundings, its temperature is constant; thus, it must be emitting and absorbing radiation at the same rate. If the temperatures of the object and its surroundings are different, there must be a net flow of radiant energy. If an object is at a temperature T and its surroundings are at a temperature T_s, the net rate of energy loss or gain per unit time (power) is given by

$$P_{\text{net}} = \sigma A e (T_s^4 - T^4) \tag{11.6}$$

Note that if T_s is less than T, then P (or $\Delta Q/\Delta t$) will be negative, indicating a net energy loss. *Keep in mind that the temperatures used in calculating radiated power are the absolute temperatures in kelvins.*

Note that we can have heat transfer without a temperature difference. If $T_s = T$, there is a continuous exchange of radiant energy, but there is no *net* change of internal energy.

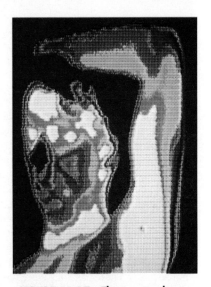

•FIGURE 11.17 Thermography
A thermogram of a man holding his arm above his head. The skin temperature varies about 1 C° for each color, from white (hottest) to black (coldest). Note the high temperature of the armpit, due to the proximity of blood vessels to the skin, and the cooler nose and hair.

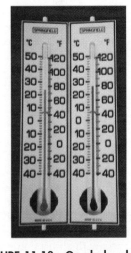

•FIGURE 11.18 Good absorber
Black objects are good absorbers. The bulb of the thermometer on the right has been painted black. Note the difference in the temperatures.

Insight   The Greenhouse Effect

You may have heard or read about the greenhouse effect in concern over global warming. The greenhouse effect helps regulate the Earth's long-term average temperature, which has been fairly constant. A portion of the solar radiation we receive reaches the Earth's surface and warms it. The Earth in turn reradiates energy in the form of infrared radiation. The balance between absorption and radiation is a major factor in stabilizing the Earth's temperature.

It is this balance that is affected by the concentration of "greenhouse gases"—primarily water vapor and carbon dioxide (CO_2)—in the atmosphere. As the infrared radiation passes through the atmosphere, some of it is absorbed by the greenhouse gases. These gases are selective absorbers; that is, they absorb radiation at certain wavelengths but not at others (Fig. 1).

If terrestrial infrared radiation is absorbed, the atmosphere warms, warming the Earth (heat transfer by radiation). This rise in surface temperature causes a shift in the wavelength of the emitted radiation. The wavelength is eventually shifted to a "window" in the absorption spectrum where little or no absorption takes place, and the terrestrial radiation passes through the atmosphere into space. Thus, the Earth loses energy and its surface temperature decreases. But with a temperature decrease, the terrestrial radiation shifts to a longer wavelength and is again absorbed by the greenhouse gases. We have a turning on and off, so to speak, similar to the action of a thermostat. Hence, the selective absorption of atmospheric gases play an important role in maintaining the Earth's average temperature

You may be wondering why this phenomenon is called the greenhouse effect. The reason is that the atmosphere functions like the glass in a greenhouse. That is, the absorption and transmission properties of glass are similar to those of the atmospheric greenhouse gases—in general, visible radiation is transmitted, but infrared radiation is selectively absorbed (Fig. 1). We have all observed the warming effect of sunlight passing through glass, for example, in a closed car on a sunny, cold day. Simi-larly, a greenhouse heats up by absorbing sunlight and trapping the reradiated infrared radiation. Thus, it is quite warm inside on a sunny day, even in winter. (Of course, the glass enclosure also keeps warm air from escaping upward, as it normally would. In practice, this elimination of heat loss by convection is the chief factor in maintaining an elevated temperature. The temperature of a greenhouse in the summer is controlled by painting the glass panels white so that some of the sunlight is reflected, and opening panels to let some hot air escape.)

The problem is that on Earth, human activities may accentuate the greenhouse warming. With all the combustion of fuels for heating and industrial processes, vast amounts of CO_2 and other greenhouse gases are vented into the atmosphere. There is concern that the result of this trend will be global warming: an increase in the Earth's average temperature that could dramatically affect the environment. For example, the climate in many parts of the globe might be altered, with effects on agricultural production and world food supplies that are very difficult to predict. It has also been suggested that a general rise in temperature might cause partial melting of the polar ice caps. Sea levels would rise, flooding low-lying regions and endangering coastal ports and population centers.

A less morbid, and interesting sidelight to the greenhouse effect concerns the transmission of radiation through glass. In an actual greenhouse, the visible portion of the sunlight is transmitted through the glass, but the short-wavelength ultraviolet radiation is absorbed. The ultraviolet spectrum from 280 nm to 400 nm is divided into two regions called UVA (320–400 nm) and UVB (280–320 nm). Ordinary window glass does not appreciably transmit UVB radiation, which is the primary cause of suntans and sunburns (see Section 20.4).

Hence, although a considerable warming effect occurs through glass from visible radiation, one does not receive a tan or severe sunburn through ordinary window glass because of UVB absorption. Some slight reddening may occur for sensitive skin from the UVA radiation that is transmitted.

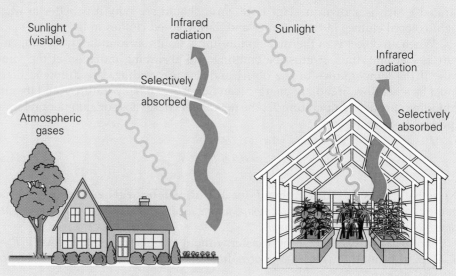

FIGURE 1   The greenhouse effect
The "greenhouse" gases of the atmosphere, particularly water vapor and carbon dioxide, are selective absorbers with absorption properties similar to those of glass used in greenhouses. Visible light is transmitted and heats the surface, while much of the infrared radiation that is re-emitted is absorbed.

Insight Heat Transfer by Radiation: The Microwave Oven

The microwave oven has quickly become a common kitchen appliance. It is both time-saving and energy-saving, since the oven doesn't have to be warmed up as does a conventional oven. The principle of operation of the microwave oven is heat transfer by radiation.

Microwaves are a form of electromagnetic radiation; they have a frequency range just below that of infrared radiation (Chapter 21). Like infrared radiation, microwaves are absorbed chiefly by water molecules (in a molecular resonance). In a microwave oven, the microwaves are generated electronically and distributed by reflection from a metal stirrer or fan

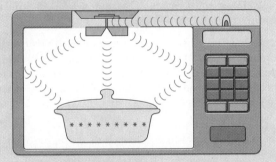

FIGURE 1   Heat transfer by radiation
In a microwave oven, microwaves (a type of electromagnetic radiation) are absorbed by molecules, chiefly those of water (and some fat), raising their temperature.

and the metal walls (Fig. 1). Because the walls reflect the radiant energy, they do not get hot.

Microwaves pass through plastic wrap, glass, or dishes made of other "microwave-safe" materials and are absorbed by water molecules in the food, producing rapid heating. (Metal utensils or objects cannot be used in microwave ovens because the microwaves can dislodge electrons from metals, causing sparking and possibly damage.) The microwaves do not penetrate the food completely but are absorbed near the surface. Heat is then conducted to the interior of the food, just as it is in conventional oven heating. This is why it is advisable to let large items or portions sit for a time after the microwave oven has shut off, so that they will be warmed or cooked throughout.

Since microwaves could be absorbed by water molecules in the skin, causing burns, microwave ovens have several important safety features. The door is tight-fitting so that microwaves cannot leak out. You will note that the glass in the door is fitted with a metal shield that has small holes through which food can be viewed without opening the door. Microwaves are reflected by this shield and prevented from coming through the glass (essentially, the waves are larger than the holes). Also, there is a mechanism that automatically shuts off the oven when the door is opened, so a person cannot get into the oven while it is running. In fact, the oven cannot be turned on when the door is open.

You may be hearing more about microwaves soon. Developmental work is being done on microwave clothes dryers,

---

EXAMPLE 11.10 ■ BODY HEAT: RADIANT HEAT TRANSFER

Suppose that your skin has an emissivity of 0.70 and its exposed area is 0.27 m². How much net energy will be radiated per second from this area if the air temperature is 20°C? Assume your skin temperature to be the same as normal body temperature, 37°C. (The average skin temperature is actually less. Why?)

Solution.

Given: $T_s = 20°C + 273 = 293$ K   *Find:* P_{net} (net power)
$T = 37°C + 273 = 310$ K
$e = 0.70$
$A = 0.27$ m²
$\sigma = 5.67 \times 10^{-8}$ W/m²·K⁴ (known)

We simply use Eq. 11.6 directly,

$$P_{net} = \sigma A e (T_s^4 - T^4)$$
$$= (5.67 \times 10^{-8} \text{ W/m}^2\text{·K}^4)(0.27 \text{ m}^2)(0.70)[(293 \text{ K})^4 - (310 \text{ K})^4]$$
$$= -20 \text{ W (or } -20 \text{ J/s)}$$

Since $P_{net} = \Delta Q/\Delta t$, 20 J of energy is radiated or *lost* (as indicated by the minus sign) each second.

Follow-up Exercise.   A piece of cake has 100 food Calories. If all of this energy were converted to body heat, how long would it supply the radiative heat loss computed in this Example?

Note that in Example 11.10 the fourth powers of the temperatures were found first and then their difference. It is *not* correct to find the temperature difference and then raise it to the fourth power: $T_s^4 - T^4 \neq (T_s - T)^4$.

Let's end this chapter with a thoughtful example.

CONCEPTUAL EXAMPLE 11.11 ■ REDUCING THE HEAT TRANSFER

Solar panels are used to collect solar energy to heat water, which may then be used for hot water or to heat a home at night. The panel boxes have black interiors (why?) through which the piping runs to carry the water, and the top side is covered with glass. However, ordinary glass absorbs most of the Sun's ultraviolet radiation (the type of radiation that causes us to become sunburned and tan, Section 19.4). This reduces the heating effect. Wouldn't it be better to leave the glass off the panel boxes? *Clearly establish the reasoning and physical principle(s) used in determining your answer before checking it below. That is, **how** did you arrive at your answer?*

Reasoning and Answer. Some energy is absorbed by the glass, but it serves a useful purpose and saves a lot more than it absorbs. As the black interior and piping of the panel box heat up, there could be heat loss by radiation (infrared) and convection. The glass prevents this through the "greenhouse effect." It absorbs much of the infrared radiation and keeps the convection inside the solar panel box. (See the Insight on the greenhouse effect, p. 374.)

Follow-up Exercise. Is there any practical reason for window drapes (other than privacy)?

Chapter Review

Important Terms

heat   355
kilocalorie (kcal)   355
calorie (cal)   356
British thermal unit (Btu)   356
mechanical equivalent of heat   356
specific heat   357
solid phase   361
melting point   361
freezing point   361

liquid phase   361
gaseous (vapor) phase   361
boiling point   361
condensation point   361
sublimation   361
latent heat   362
latent heat of fusion   362
latent heat of vaporization   362
phase diagrams   365
evaporation   366

conduction   367
thermal conductors   367
thermal insulators   368
thermal conductivity   368
convection   371
radiation   372
infrared radiation   372
Stefan's law   373
emissivity   373
blackbody   373

Important Concepts

- As a form of energy in transit, heat has the SI unit of joule (J). Common nonstandard units are the kilocalorie (kcal) and the British thermal unit (Btu).

- The mechanical equivalent of heat relates joules to kilocalories.

- The amount of heat required to raise the temperature of 1 kg of a substance by 1 C° is called the spe-

cific heat. The greater the specific heat of a substance, the greater the heat energy necessary to change the temperature of a given mass.

- Latent heat is the energy in J/kg associated with a phase change.

- Heat transfer is effected by three methods: conduction, convection, and radiation.

Important Equations

Mechanical Equivalent of Heat:

$$1 \text{ kcal} = 4186 \text{ J}$$
$$1 \text{ cal} = 4.186 \text{ J}$$

Specific Heat (capacity):

$$Q = mc\Delta T \text{ or } c = \frac{Q}{m\Delta T} \quad (11.1)$$

Latent Heat:

$$Q = mL \text{ or } L = \frac{Q}{m} \quad (11.2)$$

(for water)

$$L_f = 3.3 \times 10^5 \text{ J/kg (or 80 kcal/kg)}$$
$$L_v = 22.6 \times 10^5 \text{ J/kg (or 540 kcal/kg)}$$

Thermal Conduction:

$$\frac{\Delta Q}{\Delta t} = \frac{kA\Delta T}{d} \quad (11.3)$$

Stefan's Law:

$$P = \sigma A e T^4 \quad (11.5)$$

Radiant Power Loss or Gain:

$$P_{net} = \sigma A e (T_s^4 - T^4) \quad (11.6)$$

Exercises*

11.1 Units of Heat

1 The SI unit of heat energy is the (a) calorie, (b) kilocalorie, (c) Btu, (d) joule.

2 Which of the following is the largest unit of heat energy: (a) calorie, (b) kilocalorie, (c) Btu, (d) joule?

3 ■ A heater supplies 120 Btu of energy. What is this in joules?

4 ■ A person goes on a 1600-Cal diet to lose weight. What is the equivalent daily allowance in joules?

5 ■ A window air conditioner has a rating of 20,000 Btu (per hour). What would this rating be in watts?

6 ■ Suppose food energy values were listed in Btu. What would be the listed values of (a) a less-than-one-Calorie soft drink and (b) a 210-Calorie candy bar?

7 ■ What is the mechanical equivalent of heat in Btu?

11.2 Specific Heat

8 The amount of heat necessary to change the temperature of 1 kg of a substance by one degree Celsius is the (a) specific heat, (b) latent heat, (c) heat of combustion, (d) mechanical equivalent of heat.

9 Fifteen kilocalories would raise the temperature of three liters of water how many degrees? (a) 2 C°, (b) 3 C°, (c) 4 C°, (d) 5 C°.

10 Why is specific heat "specific"?

*Neglect heat losses in the exercises unless instructed otherwise and consider all temperatures to be exact.

11 When you swim in the ocean or a lake at night, the water may feel pleasantly warm even when the air is quite cool. Why?

12 Is it possible to have a negative specific heat? Explain.

13 ■ When 2100 J of heat is removed from 0.100 kg of a substance, its temperature is observed to decrease from 40°C to 10°C. What is the specific heat of the substance?

14 ■ A 5.0-g pellet of aluminum at room temperature gains 42 cal of heat. What is its final temperature?

15 ■ How many joules heat must be added to 5.0 L of water at room temperature to bring it to the boiling point?

16 ■■ How many minutes does it take an 800-W coffee pot to bring 1.00 L of water from room temperature to its boiling point? (Assume all the energy goes into heating the water.)

17 ■■ A 100-g piece of aluminum at 90.0°C is immersed in a liter of water at room temperature. Assuming no heat is lost to the surroundings or container, what is the temperature of the metal and water when they reach thermal equilibrium?

18 ■■ A 0.250-kg coffee cup at room temperature is filled with 250 cc of boiling coffee. The cup and the coffee come to thermal equilibrium at 80°C. If no heat is lost, what is the specific heat of the cup? [*Hint:* Consider the coffee to be essentially boiling water.]

19 ■■ A gallon of iced tea at 25°C is placed in a refrigerator to cool. How much energy must be removed from the tea to cool it to 10°C? (Take the specific heat of the sweetened tea to be 1.05 kcal/kg·C°.)

20 ■■ A waterfall is 75 m high. If all of the gravitational potential energy of the water were converted into heat energy, by how much would the temperature of the wa-

ter increase in going from the top to the bottom of the falls? [*Hint:* Consider a gram or kilogram of water going over the falls.]

21 ■■ A block of aluminum 10 cm on a side is cooled from 100°C to room temperature. If the energy removed from the aluminum block were added to a copper block of similar dimensions at room temperature, what would the final temperature of the copper block be?

22 ■■ A camper heats 30 L of water to boiling to take a bath. What volume of water from a 15°C stream must he add to have the bath water at 45°C? (Neglect any losses.)

23 ■■ How many kilograms of aluminum will experience the same temperature rise as 3.00 kg of copper when the same amount of heat is added to each?

24 ■■ If 50 g of ice at 0°C is added to 300 cm³ of water at 25°C in a 100-g aluminum calorimeter cup, what is the final temperature of the water?

25 ■■ In determining the specific heat of a new metal alloy, 150 g of the substance is heated to 400°C and then placed in a 200-g aluminum calorimeter cup containing 400 g of water at 10.0°C. If the final temperature of the mixture is 30.5°C, what is the specific heat of the alloy in SI units? (Ignore the calorimeter stirrer and thermometer.)

26 ■■■ In a calorimetry experiment, 0.50 kg of a metal at 100°C is added to 0.50 L of water at room temperature in an aluminum calorimeter cup. The cup has a mass of 250 g. (a) If the final temperature of the mixture is 25°C, what is the specific heat of the metal? (b) What would be the effect if some water splashed out of the cup when the metal was added?

27 ■■■ An electric immersion heater has a power rating of 1500 W. If the heater is placed in a liter of water at room temperature, how many minutes will it take to bring the water to a boil? (Assume that there is no heat loss.)

28 ■■■ At what average rate would heat have to be removed from 1.5 L of (a) water and (b) mercury to reduce the liquid's temperature from room temperature to its freezing point in 3.0 min?

11.3 Phase Changes and Latent Heat

29 The units of latent heat are (a) 1/C°, (b) J/kg·C°, (c) J/C°, (d) J/kg.

30 Latent heat is always (a) part of the specific heat, (b) the same as the mechanical equivalent of heat, (c) involved in a phase change.

31 Why do different substances have different freezing and boiling points? Would you expect the latent heats to be different for different substances? Explain.

32 (a) Ice cubes left in a frost-free refrigerator freezing compartment for a long time become smaller. Explain why. [*Hint:* Newer refrigerators have circulating fans.] (b) Could you cook a 3-minute egg in 2 minutes by turning up the heat of the boiling water? Explain.

33 (a) Automobile cooling systems operate under pressure. What is the purpose of this? What would happen if you removed the radiator cap shortly after turning off a hot engine? (*Don't do this*—it is quite dangerous!) (b) The directions for cooking a rice dish call for 2 cups of water, but there is a statement that at high altitudes the amount of water should be increased to $2\frac{1}{2}$ cups. Why is this?

34 ■ How much heat is required to boil away 1.00 L of water that is at 100°C?

35 ■ How much more heat is needed to convert 1.0 kg of ice at 0°C to steam at 100°C than to raise the temperature of 1.0 kg of water from 0°C to 100°C?

36 ■ How much heat must be added to 0.75 kg of lead at room temperature to cause it to melt?

37 ■ How many joules of energy are required to boil away 0.50 L of liquid nitrogen? (Take the density of liquid nitrogen to be 0.80×10^3 kg/m³.)

38 ■■ A mercury diffusion vacuum pump contains 0.015 kg of mercury vapor at a temperature of 630 K. Suppose that you wanted to condense the vapor. How much heat would have to be removed?

39 ■■ (a) A 0.250-kg piece of ice at −20°C is converted to steam at 115°C. How much heat in kcal must be supplied to do this? (b) To convert the steam back to ice at −5°C, how much heat would have to be removed?

40 ■■ How much ice (at 0°C) must be added to a liter of water at 100°C so as to end up with all liquid at room temperature?

41 ■■ A 0.60-kg piece of ice at −10°C is placed in 0.30 L of water at 50°C. How much liquid is there when the system reaches thermal equilibrium?

42 ■■ Heat is added to 40 mL of water and to 40 mL of ethyl alcohol, both initially at room temperature, at a rate of 0.27 kcal/s. How long will it take for each liquid to change completely to a gas?

43 ■■ Steam at 100°C is bubbled into 250 cm³ of water at room temperature in a calorimeter cup. How much steam will have been added when the water in the cup is at 60°C? (Ignore the effect of the cup.)

44 ■■ Ice is added to 0.75 L of tea at room temperature

to make iced tea. If enough ice is added to bring ice and liquid to thermal equilibrium, how much liquid is in the pitcher when this occurs?

45　■■ A half liter of water at 16°C is put into an ice cube tray. How much energy in kcal must be removed by the refrigerator to make it into ice at −8.0°C?

46　■■ If the ice cube tray in the preceding exercise is made of aluminum and has a mass of 0.250 kg, what would be the total heat removed from the tray of water in making ice cubes?

47　■■ Suppose 3.0 cm of rain is received over a region that has equivalent rectangular dimensions of 2.0 km × 3.0 km. How much energy in kcal was released when water vapor condensed to form this amount of rain? Is it a large amount of energy?

48　■■■ A kilogram of a substance gives a T versus Q graph as shown in ●Fig. 11.19. What are the (a) melting and boiling points? In SI notation, what are (b) the specific heats of the various phases, and (c) the latent heats?

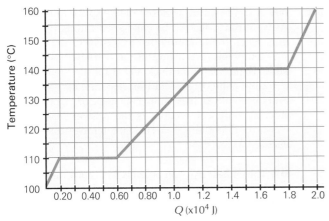

●FIGURE 11.19　Temperature versus heat
See Exercise 48.

49　■■■ In an experiment, a 150-g piece of ceramic superconductor material at room temperature is placed in liquid nitrogen (N_2) to cool. Assuming that this is done in a perfectly insulated flask, how many liters of liquid nitrogen will be boiled away in doing this? (Take the specific heat of the ceramic material to be the same as that of glass, and the density of liquid nitrogen to be 0.80×10^3 kg/m³.)

11.4 Heat Transfer

50　Solid thermal conductors commonly contain (a) free electrons, (b) air cavities, (c) metallic additives, (d) black bodies.

51　The warming of the atmosphere involves (a) conduction, (b) convection, (c) radiation, (d) all of these.

52　Underground water pipes sometimes freeze during extended cold spells. Why don't they freeze in a day or two?

53　(a) People blow on a spoonful of hot soup to cool it. Your breath is warm. How does blowing cause cooling? (b) On hot, humid days, ice can build up on the cooling coils of a window air conditioner. Does this help with the room cooling? Explain.

54　Why is the warning shown on the highway road sign in ●Fig. 11.20 so often necessary?

●FIGURE 11.20　A cold warning
See Exercise 54.

55　What is the purpose of the fins on the motorcycle radiator shown in ●Fig. 11.21?

●FIGURE 11.21　Radiator with fins
See Example 55.

56　Newton's law of cooling (Sir Isaac was a busy man) states that, in general, the heat loss from an object is proportional to the temperature *difference* between it and the surroundings: $\Delta Q / \Delta t = K \Delta T$, where K is a constant that includes losses by conduction, convection, and radiation. Apply this law to the following question: Would a cup of hot coffee stay hotter longer if you put cream in it right away or waited until you were ready to drink it, at a later time?

57 A Thermos bottle, shown in •Fig. 11.22, keeps cold beverages cold and hot ones hot. It consists of a double-walled, partially evacuated container with silvered walls (mirrored interior). The bottle is constructed to counteract all three mechanisms of heat transfer. Explain how.

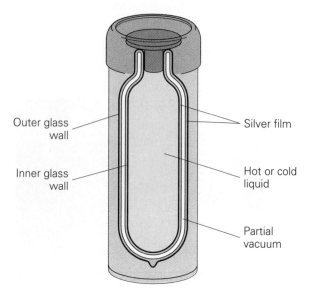

•**FIGURE 11.22 Thermal insulation**
The Thermos bottle is an application involving all three methods of heat transfer. See Exercise 57.

58 ■ How many times faster would heat be conducted from your bare feet by a tile floor than by an oak floor? (Assume that both floors have the same temperature and thickness.)

59 ■ If an object has an emissivity of 0.75 and an area of 0.20 m², how much energy does it radiate outward at room temperature?

60 ■ Of the metals listed in Table 11.3, for which two do the thermal conductivities form the greatest ratio?

61 ■■ The glass pane in a window has dimensions of 2.00 m × 1.50 m and is 4.00 mm thick. How much heat in joules will be conducted through the glass in one hour if there is a temperature difference of 2 C° between the inner and outer surfaces?

62 ■■ Assuming the human body has 4.0 cm layer of tissue and a surface area of 1.5 m², estimate the rate at which heat (in kcal/s) is conducted from inside the body to the surface if the skin temperature is 33°C. (Assume normal body temperature for the interior temperature. Also, neglect insulating clothes.)

63 ■■ A copper teakettle with a circular bottom that is 30 cm in diameter and has a uniform thickness of 2.5 mm sits on a burner whose temperature is 150°C. (a) If the teakettle is full of boiling water, what is the rate of heat conduction through its bottom? (b) Assuming that the heat from the burner is the only heat input, how much water is boiled away in 5.0 min? Is your answer reasonable? If not, explain why.

64 ■■ An aluminum bar and a copper bar of identical cross-sectional area have the same temperature difference between their ends and conduct heat at the same rate. Which bar is longer and how much longer?

65 ■■ The thermal insulation used in building is commonly rated in terms of its R-value, defined as L/k, where L is the thickness of the insulation in inches and k is the thermal conductivity. For example, 3.0 in. of foam plastic would have an R-value of 3.0 in./0.30 = 10, where, in British units, $k = 0.30$ Btu·in./ft²·h·F°. This value is expressed as R-10. (a) What does the R-value tell you; that is, how does thermal insulation vary with R-value? (b) What thicknesses of (1) fiberboard and (2) brick would give an R-value of R-10? (*Hint:* use known ratio for unit conversion.)

66 ■■ Pine wood 14 in. thick has an R-value of 19. What thickness of (a) glass wool and (b) fiber board would have the same R-value? (See Exercise 65.)

67 ■■ A large picture window measures 2.0 m by 3.0 m. At what rate will heat be conducted through the window when the room temperature is 20°C and the outside temperature is 0°C (a) if the window has a single pane of glass 4.0 mm thick and (b) if the window has a double pane of glass (a thermopane), each pane 2.0 mm thick with an intervening air space of 1.0 mm? Considering only conduction, is the answer reasonable? (Assume that there is a constant temperature difference.)

68 ■■ The emissivity of an object is 0.70. How many times greater would the outward radiation be from a similar black body at the same temperature?

69 ■■ A house wall is composed of a solid concrete block with outside brick veneer and faced on the inside with fiberboard as illustrated in •Fig. 11.23. If the outside temperature on a cold day is −5.0 C and it is room temperature on the inside, how much energy in joules is conducted through a wall with dimensions of 3.5 m × 5.0 m in one hour?

70 ■■ Suppose you wished to cut the heat loss through the wall in the preceding problem by 50% by installing insulation. What thickness of Styrofoam would have to be placed between the fiberboard and concrete block to do this?

71 ■■ If the temperature of an object at room temperature is increased from 20°C to 40°C, how are its (a) emissivity and (b) output radiation rate affected? (c) What are the effects if the object's temperature (room temperature) in kelvins is doubled?

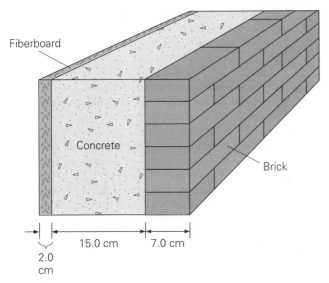

Fiberboard

Concrete

Brick

15.0 cm   7.0 cm

2.0 cm

•FIGURE 11.23 Thermal conductivity and heat loss
See Exercise 69.

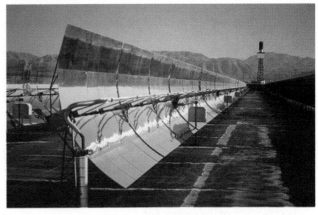

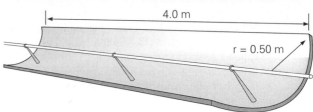

4.0 m

r = 0.50 m

•FIGURE 11.24 Solar collector and solar heating
See Exercise 74.

72 ■■ A metal sphere with a diameter of 15 cm has an emissivity of 0.58. If it is heated to a temperature of 300°C when the temperature of the surroundings is 20°C, approximately how much radiant energy does the sphere lose in 1.0 min? (Why is this value approximate without further heating?)

73 ■■ The average intensity ($I = P/A$, power/area) of solar radiation received at the top of the atmosphere is called the solar constant and is about 0.33 kcal/m²·s. How much radiant energy is emitted by the Sun each second? [*Hint:* The Earth intersects a cross-sectional area of the radiation at the Earth's average distance from the Sun. Think of the Sun's emitted energy as being spread evenly over a sphere with a radius equal to the distance of the Earth from the Sun.]

74 ■■■ Solar heating takes advantage of solar collectors such as the type shown in •Fig. 11.24. About 50% of the solar radiation received at the top of the atmosphere reaches the Earth. (The other part is reflected, scattered, absorbed, etc.) Assuming this is the case, how much heat energy would be collected, on the average, by the cylindrical collector shown in the figure during 8.0 h of collection?

75 ■■■ A steel cylinder with a radius of 5.0 cm and a length of 4.0 cm is placed in end-to-end thermal contact with a copper cylinder of the same dimensions. If the free ends of the two cylinders are maintained at constant temperatures of 95°C (steel) and 15°C (copper), how much heat will flow through the cylinders in 10 min?

76 ■■■ For the metal cylinders in Exercise 75, what is the temperature at the interface of the cylinders?

Additional Exercises

77 In a calorimetry experiment, 0.35-kg of aluminum shot at 100°C is carefully poured into 50 mL of water that has been chilled to 10°C. Neglecting any losses to the container (or otherwise), what is the final equilibrium temperature of the mixture?

78 A lamp filament radiates 100 W of power when the temperature of the surroundings is 20°C and 99.5 W of power when the temperature of the surroundings is 30°C. If the temperature of the filament is the same in each case, what is that temperature on the Celsius scale?

79 How much heat must be removed from 2.5 L of water at room temperature to form ice whose temperature is −8.0°C?

80 A student doing an experiment pours 150 g of heated copper shot into a 375-g aluminum calorimeter cup containing 200 mL of water at 25°C. The mixture (and the cup) comes to thermal equilibrium at 28°C. What was the initial temperature of the shot?

81 What is the Celsius temperature of a black body that radiates outwardly with an intensity of 462 W/m²?

82 A student mixes 1.0 L of water at 40°C with 1 L of ethyl alcohol at 20°C. Assuming that there is no heat loss to the container or the surroundings, what is the final temperature of the mixture? [*Hint:* See Table 11.1.]

83 A Styrofoam ice chest is filled with a block of ice at 0°C. The dimensions of the chest are 0.75 m × 0.40 m × 0.40 m, with walls 2.0 cm thick. (a) How much heat is conducted through the walls (including top and bottom) in 30 min if the outside temperature is 25°C? (Assume that the temperature difference remains constant.) (b) How much ice will melt in this time? (Assume that ice and water are in thermal equilibrium.) (c) Will water overflow from the chest?

84 After a barrel jump, a 65-kg ice skater traveling at 25 km/h glides to a stop. If 40% of the frictional heat generated by the skate blades goes into melting ice at 0°C, how much ice is momentarily melted? Where does the other 60% of the energy go?

85 A 0.030-kg lead bullet hits a steel plate and melts and splatters on impact. (This action has been photographed.) Assuming that the bullet receives 80% of the collision energy, at what minimum speed must it be traveling to melt on impact? (Assume that the bullet and the steel plate are initially at room temperature.)

86 An ice cube tray holds one-half liter of water at room temperature. If it is placed in a refrigerator freezer compartment at −5.0°C, how much heat has been removed from the water when it is ice at the freezer temperature?

87 Initially at room temperature, 0.50 kg of aluminum and 0.50 kg of iron are heated to 100°C. Which metal gains more heat and how much more?

88 A 0.025-kg lead bullet traveling with a speed of 250 m/s hits a 4.0-kg wooden block, penetrates, and stops inside the block. Assuming no heat loss, what is the temperature increase of the bullet and the block when they come to thermal equilibrium? (Assume that both had the same initial temperature.)

89 Equal amounts of heat are added to different quantities of copper and lead. The temperature of the copper increases by 10 C°, and the temperature of the lead by 5.0 C°. Which piece of metal has the greater mass and how much greater?

90 How much heat is required to convert 0.75 kg of ice at − 10°C to water at 20°C?

12 Thermodynamics

At first glance, this photo doesn't seem to make sense. The snow-covered landscape suggests that this is hardly the place where one would want to take a dip. Yet the bathers here seem to be enjoying themselves in comfort, with no shivering or blue lips. The mist rising from the surface points to the same conclusion: cold air, but warm water. (After reading the last chapter you should be able to explain why.)

In fact, these people are taking advantage of a lucky geological accident. At certain locations on Earth, hot springs from deep in the interior rise to the surface. In Yellowstone National Park, they produce boiling pools and geysers such as Old Faithful. On the coast of Iceland, they warm the ocean, creating the odd scene shown above—a tropical lagoon surrounded by glaciers.

But the uses of such hot springs extend beyond recreation. As you will learn in this chapter, wherever there is a temperature difference, the potential exists for obtaining useful work. This fact has not escaped the engineers of Iceland; The Blue Lagoon is the site, not only of a beach, but also of a geothermal power plant (see Figure 12.8) that draws on a renewable resource and produces no pollution. In this chapter you'll learn under what conditions, and with what efficiency, heat can be exploited to perform work, in machines as different as automobile engines and home freezers. You'll also find that the laws governing such energy conversions include some of the most general and far-reaching in all of physics.

As the word implies, **thermodynamics** deals with the transfer or the actions (dynamics) of heat (the Greek word for "heat" is *therme*). The development of thermodynamics started about 200 years ago and grew out of efforts to develop heat engines. The steam engine was one of the first of such devices, which convert heat energy to mechanical work. Steam engines in factories and locomotives powered the industrial revolution that changed the world. Although this course of study is primarily concerned with heat and work, thermodynamics is a broad and comprehensive science that includes a great deal more than heat engine theory.

Classical thermodynamics was not based on hypotheses about the structure of matter, but on experimental observations. However, it is possible to gain deeper insight into the principles of thermodynamics by applying modern molecular and kinetic theory and statistical mechanics, which deals with large numbers of particles. In this chapter, you will learn about the two general laws on which thermodynamics is based, as well as the concept of entropy.

12.1 Thermodynamic Systems, States, and Processes

Objectives: To be able to (a) define thermodynamic systems and states of systems, and (b) explain how processes affect such systems.

Thermodynamics is a field that makes use of many special terms, or everyday terms with special meanings, and you will find it useful to become familiar with some of them at the onset. The term **system**, as used in thermodynamics, refers to a definite quantity of matter enclosed by boundaries or surfaces, either real or imaginary. For example, a quantity of gas in the piston-cylinder of an engine has real boundaries. On the other hand, you may imagine boundaries enclosing a cubic meter of air in a room. The enclosing boundaries of a system need have no definite shape, nor do they have to enclose a fixed volume. For example, our previous system of gas could be compressed by advancing the piston.

Thermodynamic systems

Occasionally it is necessary to consider systems for which there is a transfer of matter across boundaries. However, for the most part, we will consider systems of constant mass. An important thermodynamic consideration is the interchange of energy between a system and its surroundings. This may occur through a transfer of heat and/or the performance of mechanical work. For example, if a balloon is heated, it expands and work is done against the surrounding atmospheric pressure (and in stretching the balloon). If heat transfer into or out of the system is impossible, the system is said to be a **thermally isolated system**). Work may be done on a thermally isolated system. For example, a thermally isolated balloon may be compressed by an external force or pressure, so that work is done on the system; and as we know, work results in a transfer of energy. If conditions are such that no energy exchange (that is, no interaction at all) can occur between a system and its surroundings, then the system is said to be a **completely isolated system** (or simply an *isolated system*).

When heat does enter or leave a system, it is absorbed from or given up to the surroundings or to theoretical heat reservoirs, which have constant temperatures. A **heat reservoir** is a system with an unlimited heat capacity. This implies that any quantity of heat may be withdrawn from or added to the reservoir without appreciably changing its temperature. By analogy, this is like taking a cup of water from or adding a cup of water to the ocean—you don't change its level appreciably.

State of a System

Just as there are kinematic equations to describe aspects of the motion of an object, there are **equations of state** to describe the conditions of thermodynamic systems. Such an equation expresses a mathematical relationship of the thermodynamic variables of a system. The ideal gas law, $pV = Nk_BT$, is a simple equation

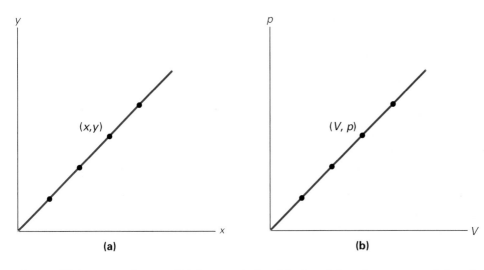

(a) On a common Cartesian graph, the coordinates (x, y) represent an individual point. **(b)** Similarly, on a p-V graph or diagram, the coordinates (V, p) represent a particular state of a system. It is common to refer to a p-V diagram because of the order of the coordinates in the ideal gas law (p, V, T). The graph is merely for illustration and is not that described by the ideal gas law equation of state.

of state. This expression establishes a relationship involving the pressure (p), volume (V), absolute temperature (T), and mass (N, the number of molecules) of a gas. These quantities are called *state variables*. Different states have different sets of values for these variables. A set of these variables that satisfies the ideal gas law specifies a state of a system of ideal gas completely. Of course, the system must be in thermal equilibrium and have a uniform temperature.

The state of a given mass of gas in a closed system can be specified by the variables p, V, and T. It is convenient to plot the states using the thermodynamic coordinates (p, V, T), much as we plot graphs with Cartesian coordinates (x, y, z). A general two-dimensional illustration of such a plot is shown in •Fig. 12.1. (This is not a p-V graph for an ideal gas. Several of these will be shown shortly.) Just as the coordinates (x, y) specify individual points on a Cartesian graph, the coordinates (V, p) specify individual *states* on the p-V graph or diagram.

In three dimensions, the states describe a surface, and each point (p, V, T) on the surface represents an equilibrium state. The three-dimensional p-V-T plot for carbon dioxide (a real gas) is shown in •Fig. 12.2. Note that the two-dimensional projections in the figure are phase diagrams, like those introduced in Chapter 11. The fusion curve between the solid and liquid phase regions on the p-T diagram

Note: Review Fig. 11.10

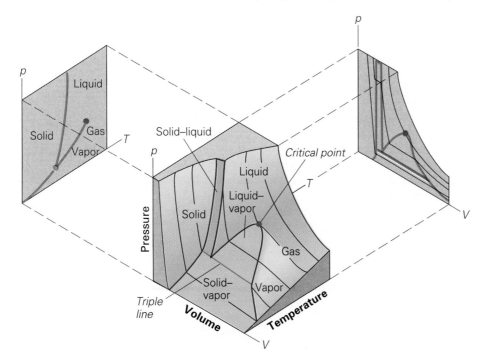

•FIGURE 12.2 The *pVT* surface for carbon dioxide
The thermodynamic states are points (p, V, T) on a three-dimensional pVT surface. The two-dimensional projections on the pT and pV planes illustrate the behavior of the substance at constant volume and constant temperature, respectively.

for carbon dioxide slopes slightly upward and to the right. (Recall that the slope of the fusion curve for water was upward and to the left.) As you can see from the pVT surface, when a quantity of carbon dioxide goes from liquid to solid, its volume is reduced; that is, it contracts on freezing (unlike water, which expands on freezing). The pVT plot of a substance is very helpful for understanding and predicting its behavior when the state variables change.

Processes

Note: A change of state (of a system) does not necessarily mean a change of phase, such as from liquid to solid. To avoid confusion, always refer to solid, liquid, and gas as *phases* rather than states of matter.

A **process** is a change in the state, or the thermodynamic coordinates, of a system. That is, when a system undergoes a process, the set of coordinates on the pVT plot describing it changes. For example, we go from state (p_1, V_1, T_1) to state (p_2, V_2, T_2) by a thermodynamic process. Processes are said to be either reversible or irreversible.

Suppose that a system of gas in equilibrium (known p, V, T values) is allowed to expand quickly. The state of the system will change rapidly and unpredictably, but will eventually return to equilibrium with another set of thermodynamic coordinates, in another state. On a graph, such as a p-V diagram, one could show the initial equilibrium point, or state, and the final state, but not what happened in between. Since the intermediate states changed so fast, there would be no data describing them. This is called an **irreversible process**, that is, one for which the intermediate steps are nonequilibrium states. "Irreversible" does not mean that the system can't be taken back to the initial state; it only means the process path can't be retraced, because of the nonequilibrium conditions.

If, however, the gas expands very, very slowly, passing from one known equilibrium state to the neighboring one and eventually arriving at the final state, then the process path between the initial and final states would be known. This is called a **reversible process**, that is, one whose path is known. In fact, a perfectly reversible process cannot be achieved. All real thermodynamic processes are irreversible to some degree because they follow complicated paths with many intermediate states. However, the concept of an ideal reversible process is useful.

These thermodynamic terms and ideas will now be applied in the laws of thermodynamics.

12.2 The First Law of Thermodynamics

Objectives: To be able to (a) explain the relationship of internal energy, heat, and work expressed by the first law, and (b) analyze various "iso-" processes.

The conservation of energy is considered to be valid for any system, and the **first law of thermodynamics** is simply a statement of the conservation of energy for thermodynamic systems. Heat, internal energy, and work are the quantities involved in a thermodynamic system. Suppose that some heat (Q) is added to a system. Where does it go? That it serves to increase the system's internal energy (ΔU) is certainly a possibility. Another possibility is that it could result in work (W) being done by the system. For example, when a heated gas expands, it does work on its surroundings (think of expanding air in a balloon).

Thus, it is possible for the added heat to go into internal energy or work, or both. Writing this as an equation gives a mathematical expression for the first law:

First law of thermodynamics: conservation of energy

$$Q = \Delta U + W \tag{12.1}$$

Our sign convention will be as follows: a positive value of heat ($+Q$) means that heat is *added to* the system and a positive value of work ($+W$) means that work is *done by* the system (for example, the work done by an expanding gas). Negative quantities mean that heat *is removed from the system* ($-Q$) and work *is done on the system* ($-W$), for example, when a gas is compressed.

Before going on with some examples of the first law, let's make certain we understand the relationship of U, Q, and W. Any given system in a particular state will have a certain amount of internal energy U. On the other hand, a system does not *have* certain amounts of heat or work. These are what *change* the state of the system, and generally change the internal energy. When heat is added to or removed from a system, or work is done on or by a system, *thermodynamic processes* occur that can change the system from one state to another, each having a particular internal energy U. That is, the internal energy depends only on the state of the system, and not what brought it there.

Hence, to find the change in internal energy, ΔU, in going from one state to another, we need to know only the internal energy of the states, and $\Delta U = U_2 - U_1$. Another way of saying this is that the change in the internal energy ΔU is independent of the process path, and depends only on the initial and final states. This is analogous to the change in gravitational potential energy (mgh) being independent of path (Section 5.4).

The first law can be applied to several processes for a closed system of an ideal gas in which one of the thermodynamic variables is kept constant. These processes have names beginning with iso- (from the Greek *isos*, meaning "equal"). Keep in mind that we are working with constant mass systems, unless otherwise stated.

Isobaric Process. A constant-pressure process is called an **isobaric process**. One for an ideal gas is illustrated in •Fig. 12.3. On a p-V diagram, the path of an isobaric process is along a horizontal line called an *isobar*. When heat is added to the gas in the cylinder, the ratio V/T must remain constant (from Eq. 10.5, $pV = Nk_BT$, so $V/T = Nk_B/p =$ a constant when p is constant). The heated gas expands; there is an increase in its volume. The temperature must therefore also increase proportionally, which means the internal energy of the gas increases. (Recall from Chapter 10 that kinetic theory says that the internal energy of an ideal gas is directly proportional to its absolute temperature.)

Note: Review Sections 10.1 and 10.5

Work is done by the gas as it expands, in moving the piston. From Section 5.1,

$$W = F\Delta x$$

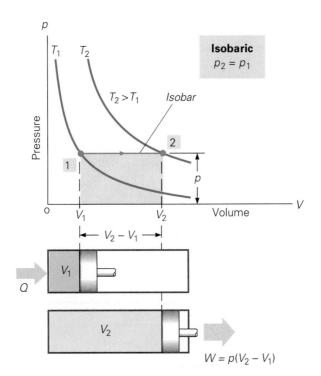

•**FIGURE 12.3 Isobaric (constant pressure) process**
The heat added to the gas in the frictionless piston goes into increasing the internal energy of the gas and doing work (the expanding gas moves the piston): $Q = \Delta U + W$. The work is equal to the area under the process path (from state 1 to state 2 here) on the p-V diagram.

In terms of pressure ($p = F/A$), the force may be written $F = pA$, where A is the area of the piston. Then, the previous equation becomes

$$W = pA\Delta x$$

But $A\Delta x$ is simply the change in the volume of the gas, $A\Delta x = \Delta V = V_2 - V_1$. Thus,

$$W = p\Delta V = p(V_2 - V_1) \qquad \textit{(for an isobaric process)} \qquad (12.2)$$

In Fig. 12.3, you can see that $p\Delta V$ is the area under the isobar on the p-V diagram. For a nonisobaric process (one in which the pressure changes), the work is also equal to the area under the line showing the process path. Thus, the work depends on the process path as well as on the initial and final states: there will be different areas under different paths.

On the other hand, since the internal energy of a quantity of an ideal gas depends only on its (absolute) temperature, a change in this internal energy is *independent* of the process path and depends only on the initial and final states, or the temperatures of these states ($\Delta U = U_2 - U_1 \propto T_2 - T_1$).

Since V_2 is greater than V_1 for an expanding gas, work is done by the system ($+W$). In terms of the first law, then,

$$Q = \Delta U + W = \Delta U + p\Delta V \qquad (12.3)$$

That is, the heat added to the system goes into both increasing the internal energy and into work done by the system. If the process were reversed and the gas were being compressed by an external force doing work *on* the system, all the quantities would be negative. Heat would go out of the system ($-Q$), and the internal energy, or the temperature, of the gas would decrease ($-\Delta U$).

Isometric Process. An **isometric process** (short for iso*volu*metric process) is a constant-volume process, sometimes called an isochoric process. As illustrated in •Fig. 12.4, the process path on a p-V diagram is along a vertical line, commonly called an *isomet*. No work is done ($W = p\Delta V = 0$, since $\Delta V = 0$), so all the added heat goes into increasing the internal energy, and therefore the temperature, of the gas. By the first law,

$$Q = \Delta U + W = \Delta U + 0$$

and

$$Q = \Delta U \qquad \textit{(for an isometric process)}$$

Isothermal Process. An **isothermal process** is a constant-temperature process (•Fig. 12.5). In this case, the process path is along an *isotherm*, or a line of constant temperature. Since $p = (Nk_BT)/V = \text{(constant)}/V$ for an isothermal process, an isotherm is a hyperbola on a p-V diagram (the general form of the equation for a hyperbola is $y = a/x$).

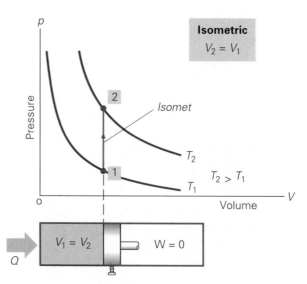

•**FIGURE 12.4 Isometric (constant volume) process**
All of the heat added to the gas goes into increasing the internal energy when the volume is held constant: $Q = \Delta U$. This causes an increase in temperature.

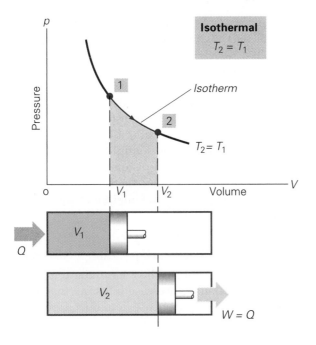

•FIGURE 12.5 Isothermal (constant temperature) process
All of the heat added to the gas goes into doing work (the expanding gas moves the piston): since $\Delta T = 0$ then $\Delta U = 0$, and $Q = W$. As always, the work is equal to the area under the process path on the p-V diagram.

In going from state 1 to state 2 in Fig. 12.5, heat is added to the system, and both the pressure and volume change in order to keep the temperature constant (pressure decreases and volume increases). The work done by the expanding system ($+W$) is again equal to the area under the process path.

For an isothermal process, the internal energy of the ideal gas remains constant ($\Delta U = 0$), because the temperature is constant. By the first law,

$$Q = \Delta U + W = 0 + W$$

and $\qquad\qquad Q = W \qquad$ *(for an isothermal process)*

Thus, for an ideal gas an isothermal process is one in which heat energy is converted to mechanical work (or vice versa for the reverse path).

Adiabatic Process. There's one more type of process in which a thermodynamic condition remains constant (•Fig. 12.6). An **adiabatic process** is one in which no

Adiabatic process: no heat transfer

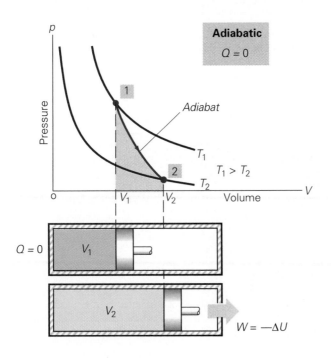

•FIGURE 12.6 Adiabatic (no heat transfer) process
In an adiabatic process, no heat is added or removed from the system: $Q = 0$. Work is done at the expense of the internal energy: $W = -\Delta U$. The pressure, volume, and temperature all change in the process.

heat is transferred into or out of the system; that is, $Q = 0$ (the Greek word *adiabatos* means "impassable.") The condition $Q = 0$ is satisfied for a thermally isolated system. Of course, this is an ideal or theoretical situation, since there is always some heat transfer in actual system processes. Under actual conditions, we can only approximate adiabatic processes. For example, processes that are nearly adiabatic can take place in systems that are not thermally isolated if they occur rapidly enough that there isn't time for much energy to be transferred into or out of the system.

As the system follows the process path, a curve called an *adiabat*, all three thermodynamic coordinates change. For example, suppose a quantity of ideal gas were compressed in a thermally isolated piston-cylinder. If the piston were suddenly released, the gas would expand (i.e., there would be changes in p and V). Work would be done at the expense of the internal energy of the gas, so its temperature would change. By the first law,

$$Q = 0 = \Delta U + W$$

and

$$W = -\Delta U \qquad \text{(for an adiabatic expansion)}$$

Thus, in an adiabatic expansion, work (the area under the process path) is done by the system with a corresponding decrease in its internal energy. The decrease in internal energy is evidenced by a decrease in temperature in going from state 1 to state 2.

CONCEPTUAL EXAMPLE 12.1 ■ BLOWING HOT AND COLD?

The air in your lungs is warm. This can be shown by putting your bare upper arm near your mouth and blowing air on your arm with your mouth opened widely. If you were to repeat this with your lips puckered, the air would feel (a) warmer, (b) cooler, (c) the same. *Clearly establish the reasoning and physical principle(s) used in determining your answer before checking it below. That is, **why** did you select your answer?*

Reasoning and Answer. We'll go from answer to reasoning on this one, since it is easy to experimentally determine the correct answer. Try it, and you might be surprised to find that the answer is (b).

The interesting thing is why this occurs. When you blow on your arm with mouth open, you get a gush of warm air as indicated by your temperature sense. However, when you blow with your lips puckered, the stream of air is compressed. Then, as it emerges, the air expands, doing work against the atmosphere. This work is done at the expense of the internal energy of the air, so the temperature of the air is decreased and it feels cooler. This takes place quickly and is a "quasi-adiabatic" process.

Follow-up Exercise. On a hot day, a motorist lets some air out of an overinflated tire. On touching the metal valve stem immediately thereafter, he finds that it feels cool. Why is this?

EXAMPLE 12.2 ■ ISOBARIC EXPANSION: CHANGE IN INTERNAL ENERGY

A quantity of an ideal gas has a volume of 22.4 L at STP (standard temperature and pressure). While absorbing 315 cal of heat from the surroundings, the gas expands isobarically to 32.4 L. What is the change in the internal energy of the gas?

Solution. Listing the data for inspection, along with appropriate conversions, we have

Given: $p_1 = p_2 = 1$ atm
$\qquad\qquad = 1.01 \times 10^5 \, \text{N/m}^2 \, \text{(Pa)}$
$\qquad V_1 = 22.4 \, \text{L} = 22.4 \times 10^{-3} \, \text{m}^3$
$\qquad V_2 = 32.4 \, \text{L} = 32.4 \times 10^{-3} \, \text{m}^3$
$\qquad T_1 = 0°\text{C} = 273 \, \text{K}$
$\qquad Q = 315 \, \text{cal} = 1.32 \times 10^3 \, \text{J}$

Find: ΔU (change in internal energy)

LEARN BY DRAWING

Leaning on Isotherms

When you are analyzing the various thermodynamic processes discussed in this chapter, it is sometimes hard to keep track of the heat flow (Q), work (W), and internal energy change (ΔU), together with their correct signs. One trick that can help with this bookkeeping is to superimpose a series of isotherms on the pV plot you are working with (as in Figures 12.2–12.5). This is generally useful even if the situation you are studying does not involve an isothermal process. It is also a good way of reinforcing your insight into what is happening physically in these processes.

Before starting, recall the two important properties of an isothermal process. Both of them follow directly from the fact that an isothermal process is, by definition, one in which the temperature remains constant.

1. In an isothermal process ΔU is zero. (Why?)
2. Since T is constant, the ideal gas law (Eq. 10.3) tells us that in an isothermal process pV must also be constant: $pV = k$. You may recall from algebra that this is the equation of a hyperbola, and on a pV diagram, each isotherm is indeed a hyperbola. The farther from the axes the hyperbola is, the higher the temperature that it represents (Fig. 1).

To take advantage of these properties, follow these steps:

- Sketch a set of isotherms for a series of different temperatures on the pV plot (Fig. 1).

- Then sketch the actual process or processes you are analyzing—for example, those shown in Fig. 2 and Fig. 3. (Take a moment to remind yourself what the names mean: isomet = constant volume, isobar = constant pressure, and adiabat = no heat flow.)

- Finally, use the first law of thermodynamics, $Q = \Delta U + W$, to calculate the final results, including signs. Usually, you will be able to find the sign of W by remembering that W is simply the area under the pV

curve for the process represented. The sign of ΔT will be clear from the isotherms that you have added; they will serve as a temperature scale, with progressively higher temperatures represented by isotherms at greater distances from the axes. A rise in T implies a positive value for ΔU, since the internal energy of the gas increases with its temperature. You will thus have all you need to determine the sign of Q, and thus whether heat flows into the system ($Q > 0$) or out of it ($Q < 0$).

The following two examples indicate the power of this approach. In Fig. 2, we are asked to determine whether heat flows into or out of the gas during an isobaric expansion. Expansion implies positive work ($p\Delta V > 0$). But what is the direction of the heat flow (or is it zero)? Sketching the isobar (with an arrow pointing to the right to indicate expansion), we see that it crosses from lower-temperature isotherms to higher-temperature ones. Hence there is a temperature increase, and ΔU must be positive. Since the gas has both gained internal energy and done positive work, both terms on the right-hand side of the first law equation are positive, so it follows that Q must be positive: $Q = (\Delta U + p\Delta V) > 0$. Positive heat flow means "into the gas," so we have our answer.

An isometric process involving a pressure drop is similarly analyzed in Fig. 3. Again, starting with lightly-drawn isotherms on the pV axes, we plot a vertical line (constant volume) with an arrow pointing downward to show pressure reduction. Since there is no volume change, the gas does no work. However, we can see that its temperature drops (why?), so its internal energy must decrease. This means that ΔU is negative, hence Q must be negative ($Q = \Delta U + p\Delta V = \Delta U < 0$). Thus heat flows out of the gas.

As an exercise, try analyzing an adiabatic processes in this way to see if you can graphically determine where the work comes from if there is no heat flow into or out of the gas.

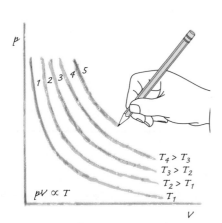

FIGURE 1 Isotherms on a pV plot

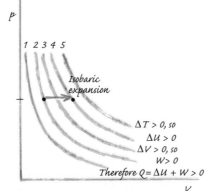

FIGURE 2 An isobaric expansion

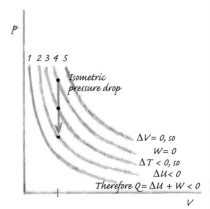

FIGURE 3 An isometric decline in pressure

As you might expect, the change in the internal energy is given by the first law: $\Delta U = Q - W$. We know Q, but we must first compute W, the work done by the expanding gas, to find ΔU. Using Eq. 12.2,

$$W = p\Delta V = p(V_2 - V_1)$$
$$= (1.01 \times 10^5 \text{ N/m}^2)[(32.4 \times 10^{-3} \text{ m}^3) - (22.4 \times 10^{-3} \text{ m}^3)]$$
$$= 1.01 \times 10^3 \text{ J}$$

Then, by the first law,

$$\Delta U = Q - W = (1.32 \times 10^3 \text{ J}) - (1.01 \times 10^3 \text{ J}) = 310 \text{ J}$$

Follow-up Exercise. In this Example, what is the equilibrium temperature in degrees Celsius of the gas after the expansion?

12.3 The Second Law of Thermodynamics and Entropy

Objectives: To be able to (a) state and explain the second law of thermodynamics in several forms, and (b) explain the concept of entropy.

Suppose that a piece of hot metal is placed in an insulated container of cool water. Heat will be transferred from the metal to the water, and the two will come to thermal equilibrium at some intermediate temperature. For a thermally isolated system, the total energy remains constant, in accordance with the first law of thermodynamics. Could there be a reverse process in which heat would be transferred from the cooler water to the hot metal? Everyone realizes that this would not happen naturally. But if it did, the total energy of the system would remain constant, and the energy transfer would not violate the first law.

Clearly, there must be another principle, not expressed in the first law of thermodynamics, that specifies the *direction* in which a process can take place. This principle is embodied in the **second law of thermodynamics**, which says that certain processes do not take place, or have never been observed to take place, even though they are consistent with the first law.

There are many equivalent statements of the second law, which are worded differently according to their application. One of these, which is applicable to the above situation, is

Heat will not flow spontaneously from a colder body to a warmer body.

Another common statement of the second law is this:

Heat energy cannot be transformed completely into mechanical work.

This statement applies to heat engines, which are discussed in more detail in the next section.

In general, the second law applies to all forms of energy. It is considered to be true because no one has ever found an exception to it (regardless of the form of its statement). If it were not valid, a perpetual motion machine could be built. Such a machine could transform heat from a reservoir completely into work and motion (mechanical energy), with no energy loss. The mechanical energy could then be transformed back into heat and be used to reheat the reservoir (again with no loss). Since the processes could be repeated indefinitely, the machine would run perpetually. All of the energy is accounted for, so this situation does not violate the first law. However, it is obvious that real machines are always less than 100% efficient, that the work output is always less than the energy input. Another statement of the second law is therefore:

It is impossible to construct an operational perpetual motion machine.

Many attempts have been made to construct perpetual motion machines, but there have been no known successes.

It would be convenient to have some way of expressing the direction of a process in terms of the thermodynamic properties of a system. One such property is temperature. In analyzing a conductive heat transfer process, you need to know the temperatures of the system and its surroundings. Knowing the temperature difference allows you to state the direction in which the heat transfer will spontaneously take place, into or out of the system (indicated by $+Q$ or $-Q$, respectively).

Note: Perpetual motion machines are distinguished as being of the first kind or the second kind. A perpetual motion machine of the first kind would violate the first law of thermodynamics, the law of the conservation of energy. That is, energy would be created, or the machine would have an efficiency *greater* than 100%. A perpetual motion machine of the second kind would not create energy but would violate the second law, which specifies the direction of spontaneous heat flow. In this case, 100% efficiency would be enough.

Entropy

A more general property that indicates the direction of a process was first described by Rudolf Clausius (1822–1888), a German physicist. This property is called **entropy**, a name coined by Clausius. Physically, entropy is a particularly rich, multifaceted concept. Its interpretations are diverse and intriguing. One might draw an analogy to the fable of the Blind Men and the Elephant, in which three blind men describe what each thinks an elephant to be after touching different parts. Thus, a discussion of entropy is likely to resemble a debate:

"Entropy is a measure of a system's ability to do useful work. As a system loses the ability to do work, its entropy increases."

"No, I say that entropy is what determines the direction of time. It's 'time's arrow' that points out the forward flow of events, thereby distinguishing the past from the future."

"You're both wrong. Entropy is really a measure of disorder. A system naturally moves toward a state of greater disorder or disarray. The more order, the less entropy you have."

As you see, entropy has several physical interpretations, and these are discussed more fully in the Insight on p. 394. Here, however, we will be concerned with the mathematical definition of entropy in terms of thermodynamic properties.

Entropy is related to temperature and heat. The change in entropy (ΔS) when an amount of heat (Q) is added to (or removed from) a system by a reversible process at a constant temperature is given by

$$\Delta S = \frac{Q}{T} \quad \begin{array}{l} \textit{change in entropy} \\ \textit{(constant temperature only)} \end{array} \quad (12.4)$$

where T is the Kelvin temperature and the units for entropy are joules per kelvin (J/K). [Note what would happen if you used the Celsius temperature in this equation for a phase change from ice to water at 0°C.].

If the temperature changes during the process, the change in entropy can be calculated using advanced mathematics. This discussion will be limited to isothermal processes or ones with relatively small temperature changes. For the latter, reasonable approximations of entropy changes may be obtained using average temperatures.

EXAMPLE 12.3 ■ Change in Entropy: An Isothermal Process

What is the change in entropy when 0.25 kg of ethyl alcohol ($L_v = 1.0 \times 10^5$ J/kg) vaporizes at its boiling point of 78°C?

Solution. From the problem, we have

Given: $m = 0.25$ kg   **Find:** ΔS (change in entropy)
$\quad\quad\quad T = 78°C + 273 = 351$ K
$\quad\quad\quad L_v = 1.0 \times 10^5$ J/kg

Notice that the temperature was converted directly to kelvins. Then, to compute the

Insight   Aspects of Entropy

In developing the concept of entropy, Clausius realized that when heat flows from one body to another, there is more involved than energy transfer. Energy could be transferred from either body to the other without violating the first law of thermodynamics, but heat will flow spontaneously only from a hotter body to a colder one. The temperature difference between the bodies that is necessary for heat flow corresponds to different "levels" of energy. When the bodies reach thermal equilibrium their temperatures (and thus their energy levels) are the same, and the heat flow ceases.

Even though there is a transfer of heat, the total energy of such an isolated two-body system is the same before and after the bodies come to equilibrium. But something is lost or reduced through the heat transfer—*the ability to do work*. When a temperature difference exists, there can be useful work output, as in a heat engine (Section 12.4). With no temperature difference, there is no thermal ability to do work. We can make an analogy here with different levels of gravitational potential energy represented by different heights. As the water in a stream flows down a mountainside, its descent can be made to do useful work (turning a mill wheel, for example). Once the water reaches sea level, however, it will fall no further and so cannot be made to do more work.

This loss of the ability to do useful work is one aspect of entropy. *As a system loses its ability to do useful work, its entropy increases.*

A closely related aspect of entropy is that it is a measure of disorder: *The more disordered the system, the greater is its entropy.* This is consistent with the previous discussion. Useful work can generally be extracted from highly ordered systems. As the degree of order decreases, so does the ability of the system to do work. Consider, for example, two adjacent chambers, one filled with gas under pressure and the other empty (evacuated). If a valve between the chambers is opened, the gas will expand into the empty space until its concentration is the same in both chambers. Such a system is more ordered when all of the gas is in one container. When the expansion occurs, the molecules are distributed randomly between two chambers, so the disorder (and thus the entropy) is increased. If you put a paddle wheel between the chambers, you could force the gas to perform work as it expanded into the second chamber. Once the gas has filled both chambers, however, and the system is in a state of maximum disorder, no more work can be extracted from it in this manner.

According to the second law of thermodynamics, the entropy of the universe increases in every natural process. This means that systems naturally move toward states of greater disorder or disarray. You know from experience that highly ordered systems—the neatly arranged items on the shelves of a stockroom, or the neatly piled papers on your desk—do not generally stay that way for long. As time passes, they tend to become disordered. (Eventually, they may become totally random, so that you can't predict where *anything* will be found.) Similarly, when ice cubes melt in a tray, the orderly crystalline structure of the ice is replaced by the more disorderly arrangement of molecules in the liquid.

It is important to realize that in applying the second law, one must take a very broad view. Within a given system, order can be created, restored, or even increased through the expenditure of energy. For example, you could work all afternoon tidying up the shelves of the stockroom or organizing your papers. And your freezer, using electrical energy, could perform the work necessary to refreeze the tray of ice cubes. As we shall see in a later Insight, however, such work merely purchases local order at the expense of greater disorder elsewhere. The entropy of the universe *as a whole* always increases.

Another interesting and important point about the second law is that it is based on probability. Consider the expansion of a quantity of gas described previously. You know that the expansion will not spontaneously reverse, with all the gas returning to the original container, thereby increasing the order. Statistically, the probability of this occurring is very small. However, statistics does not say that it is impossible, only that the likelihood is virtually zero. The validity of the second law is based not on the impossibility of certain processes (although they tend to be viewed as such), but on the fact that they have never been observed.

Like the simple process of gas expansion, more complex processes also have a natural direction specified by the second law. Suppose that you filmed a match being lit, a car rolling down a hill, or cream being poured into a cup of coffee. If the developed film were run backward, you would see the match "unburn," the car roll uphill, and the cream separate itself out from the coffee—processes that you would immediately recognize as unnatural. Such impossible processes would involve the creation of a more highly ordered state from a less highly ordered one. They would thus correspond to negative changes in the entropy of the universe.

The increase in entropy provides direction not only for processes but also for time. This is necessarily the case because all our measurements of time involve some physical process, whether it be the fall of sand in an hourglass, the swing of a pendulum, or the vibration of a quartz crystal. Thus, entropy is sometimes referred to as *time's arrow*. It points out the forward direction of the flow of events. Basically, entropy distinguishes the past from the future. Overall there will be more entropy in the future, and there was less entropy in the past.

As time goes on and natural processes occur, the total entropy of the universe increases, and differences in energy levels or temperatures are reduced. For example, if a system is at a high temperature, heat will naturally be transferred to its surroundings until thermal equilibrium is reached. In a universe of ongoing natural processes, there is a continual net entropy increase as heat is continually being transferred from systems at higher temperatures to those at lower temperatures. Over a long period of time, this will lead to the so-called *heat death of the universe:* the condition when the entropy of the universe has reached a maximum, and everything is at the same temperature. Not a very appealing future—but fortunately not too imminent either.

entropy change, we need to know the amount of heat involved in the process. This is a phase change and we are given the latent heat, so

$$Q = mL_v = (0.25 \text{ kg})(1.0 \times 10^5 \text{ J/kg}) = 2.5 \times 10^4 \text{ J}$$

Then

$$\Delta S = \frac{Q}{T} = \frac{+2.5 \times 10^4 \text{ J}}{351 \text{ K}} = +71 \text{ J/K}$$

Note that Q is positive because heat is added to the system. The change in entropy, then, is also positive, and the entropy of the alcohol increases.

Follow-up Exercise. What is the change in entropy when 1.00 kg of water freezes at 0°C?

EXAMPLE 12.4 ■ CHANGE IN ENTROPY: AN INCREASE OR A DECREASE?

A piece of metal at 24°C is placed in 1.00 L of water at 18°C. The thermally isolated system comes to equilibrium at a temperature of 20°C. Find the approximate change in the entropy of the system.

Solution. Using subscripts m and w for metal and water, and i and f for initial and final, respectively, we have

Given: $T_{m_i} = 24°C$ *Find:* ΔS (change in entropy of the
$T_{w_i} = 18°C$ system)
$m_w = 1.00$ kg
$T_f = 20°C$
$c_w = 4186$ J/kg·C°
     (from Table 11.2)

As in the preceding Example, we need to find the amount of heat (Q) to solve for ΔS. However, in this system there are two Q's: Q_w, the heat gained by the water and Q_m, the heat lost by the metal.

With $\Delta T_w = T_f - T_{w_i} = 20°C - 18°C = 2.0$ C°, the heat gained by the water is

$$Q_w = m_w c_w \Delta T = (1.00 \text{ kg})(4186 \text{ J/kg})(2 \text{ C°}) = 8.37 \times 10^3 \text{ J}$$

This is also the amount of heat *lost* by the metal, since by the conservation of energy, heat gained must equal heat lost. Therefore, $Q_m = -8.37 \times 10^3$ J, where the minus sign indicates a loss of heat.

In this case, we have small temperature changes, so we can approximate the entropy changes using average temperatures:

$$\overline{T}_w = \frac{T_f + T_{w_i}}{2} = \frac{20°C + 18°C}{2} = 19°C = 292 \text{ K}$$

$$\overline{T}_m = \frac{T_{m_i} + T_f}{2} = \frac{24°C + 20°C}{2} = 22°C = 295 \text{ K}$$

We can then use these average temperatures to compute the approximate entropy changes for both the water and the metal.

$$\Delta S_w = \frac{Q_w}{\overline{T}_w} = \frac{+8.37 \times 10^3 \text{ J}}{292 \text{ K}} = +28.7 \text{ J/K}$$

$$\Delta S_m = \frac{Q_m}{\overline{T}_m} = \frac{-8.37 \times 10^3 \text{ J}}{295 \text{ K}} = -28.4 \text{ J/K}$$

The change in the entropy of the whole system is the sum of these individual changes:

$$\Delta S = \Delta S_w + \Delta S_m = +28.7 \text{ J/K} - 28.4 \text{ J/K} = +0.3 \text{ J/K}$$

The entropy of the metal decreased because heat flowed out of it. The entropy of the water increased by a greater amount, so the total entropy change of the system was positive (an increase).

Follow-up Exercise. What would be the situation in this Example if the entropy change were zero ($\Delta S = 0$)?

Note: The entropy of a closed system increases for every natural process.

Note that the total entropy change for the system in Example 12.4 is positive. In general, the direction of any process is toward an increase in entropy. That is, *the entropy of an isolated system never decreases.*

Another way to state this observation about entropy is to say that *the entropy of an isolated system increases for every natural process* ($\Delta S > 0$). The water and metal in coming to an intermediate temperature in Example 12.4 are undergoing a natural process, so-called because this is what is always observed to occur.

If the metal in coming to thermal equilibrium with the water somehow spontaneously became hotter while the water became cooler, this would certainly be an *unnatural* process ($\Delta S < 0$). Similarly, water at room temperature in an isolated ice cube tray will not naturally turn into ice.

If a system is not isolated, there may be a decrease in the entropy of the system. For example, if an ice cube tray filled with water is put into a freezer compartment, the water will freeze, with a decrease in entropy. But there will be a larger increase in entropy somewhere else in the environment or the universe. Work has to be done or energy expended to bring about the transfer of heat involved in the change from water to ice. Thus, a general statement of the second law of thermodynamics in terms of entropy is:

The total entropy of the universe increases in every natural process.

Note: Energy can neither be created nor destroyed; entropy can be created but not destroyed.

The entropy of a system is a function of its state. Each state of a system has a particular value of entropy, and a change in entropy depends only on the initial and final states for a process, or $\Delta S = S_f - S_i$. (This is analogous to the change in internal energy for an ideal gas, $\Delta U = U_f - U_i$.) An entropy change can be represented on a *T-S* diagram, as shown in • Fig. 12.7. Just as the area under a curve on a *p-V* diagram for a gas is equal to the work ($W = p\Delta V$), the area under a *T-S* curve is equal to the change in the heat energy ($Q = T\Delta S$), which is easily seen for the isothermal process in the figure. Like work for a process with variable pressure, heat transferred in a process with variable temperature is also equal to the area under the *T-S* curve (which generally requires advanced mathematics to calculate).

There is a theoretical process for which the entropy is constant. Recall that for an adiabatic process $Q = 0$. Since in this case $\Delta S = Q/T = 0$, there is no change in entropy. Hence, an adiabatic process is a constant-entropy process, or an **isentropic process**. With the inclusion of an ideal adiabatic process, we may say that the entropy can only increase or remain constant ($\Delta S \geq 0$). How would an isentropic process appear on a *T-S* diagram?

•**FIGURE 12.7** *T-S* (temperature-entropy) diagrams
When the temperature (*T*) is plotted versus entropy (*S*), the area under the process path is equal to the heat (*Q*) transferred in the process. This is similar to work being equal to the area under a process path on a *p-V* diagram. **(a)** This *T-S* diagram is for an isothermal process. (How can you tell?) Note that the area under the path is that of a rectangle. **(b)** This *T-S* diagram is for a general process.

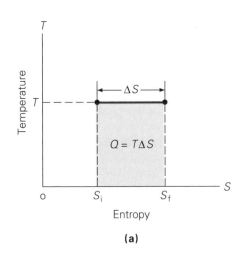

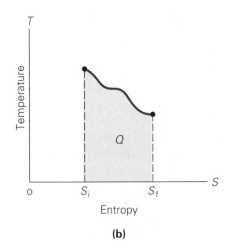

(a)   (b)

CONCEPTUAL EXAMPLE 12.5 ■ ENTROPY DECREASING?

Which of the following involves a *decrease* in the system's entropy: (a) a hot fudge sundae is left uneaten too long, so that the ice cream melts and the fudge solidifies, (b) a green plant combines water and carbon dioxide molecules in photosynthesis to make a larger, more complex molecule of sugar, (c) you drop your term paper on the way up the library stairs and find that, when you have gathered up all the pages, they are no longer in sequence, (d) all the perfume in an opened bottle evaporates and fills the room with scent, (e) a shortage of library personnel makes it increasingly difficult to find the books you want in their proper locations on the shelves? *Clearly establish the reasoning and physical principle(s) used in determining your answer before checking it below. That is, why did you select your answer?*

Reasoning and Answer. It is not difficult to see that cases (c) and (e) both involve a decrease in order, and thus an increase in entropy. The same is true, in a slightly less obvious way, of case (d). When the perfume molecules are no longer confined within the restricted space of the bottle but instead are distributed at random throughout a larger volume of air in the room, there is a loss of order—the positions of individual perfume molecules cannot be specified or predicted as accurately. Similarly in case (a), the system (fudge plus ice cream) is less highly ordered when the speeds of all the molecules are randomized than when they are divided into two distinct populations, one hot and one cold. (You may also recall that we showed in another way in Example 12.4 how the flow of heat from a warmer body to a colder one entails an increase in entropy.) Hence the answer is (b). In this instance a more complex, highly ordered structure was created from simpler components—a process that will not occur naturally without an input of energy and a larger increase in entropy somewhere else in the universe (see the Insight on p. 398).

Follow-up Exercise. You look through a microscope and find that your pet amoeba has died. Does this represent an increase or decrease in the total entropy of the amoeba? Of the universe?

•**FIGURE 12.8 Work from heat**
At the geothermal energy facility in the background, located in Iceland, natural heat is exploited to do useful work (generating electricity) without the need to "burn" fuel. Water heated by thermal processes beneath the Earth's surface serves as the hot reservoir. (What do you think is the cold reservoir?)

12.4 Heat Engines and Heat Pumps

Objectives: To be able to (a) explain the concept of a heat engine and compute thermal efficiency, and (b) explain the concept of a heat pump and compute coefficient of performance.

A **heat engine** is any device that converts heat energy to work. Since the second law says that perpetual motion machines are impossible, some of the heat supplied to a heat engine will necessarily be lost. In studying thermodynamics, there is no need to be concerned with the usual mechanical components of an engine, such as pistons, cylinders, and gears. For theoretical purposes, a heat engine is simply a device that takes heat from a high-temperature source (a hot reservoir), converts some of it to useful work, and transfers the rest to its surroundings (a cold, or low-temperature, reservoir). For example, most of the turbines that generate the electricity we use (Chapter 20) are heat engines, using heat from various sources: the burning of chemical fuels (oil, gas, or coal); nuclear reactions (Chapter 30); or the heat already present beneath the Earth's surface (•Figure 12.8). A generalized heat engine is usually represented as shown in •Fig. 12.9a.

Simply adding heat to a gas in a piston-cylinder arrangement will produce work in a single process. But since a continual output is usually wanted, practical heat engines usually operate in a cycle, or a series of processes, which brings the engine or system back to its original condition. Such cyclic heat engines include steam engines and internal combustion engines, such as automobile engines. An idealized, rectangular thermodynamic cycle is shown in •Fig. 12.9b. It consists of two isobars and two isomets. When these processes occur in the sequence indicated, the system goes through a cycle (1-2-3-4-1), returning to its orig-

Heat engine: heat → work

Insight | Life, Order, and the Second Law

The second law of thermodynamics is one of the founda-tions of physics and operates universally—the total en-tropy of the universe increases in every natural process. As we saw in the previous Insight, this means that systems naturally move toward states of greater disorder or disarray. Certainly life is a natural process. Yet clearly, there is an increase in or-der during the development of the human embryo into a ma-ture adult. Moreover, in virtually all life processes larger, more complex molecules and cellular structures are continuously being synthesized from simpler components. Does this mean that living things are exempt from the second law?

To answer this question, we must realize that cells are not closed systems. They are involved in continual exchanges of matter and energy with their surroundings. Living cells are constantly occupied in the exchange and conversion of energy from one form to another, which is governed by the first law of thermodynamics. For example, cells convert chemical po-tential energy into kinetic energy (movement) and electrical energy (the basis of nerve impulses).

Many cells, such as those of plants, can also convert sun-light (radiant energy) into chemical energy through photo-synthesis (Fig. 1). In all these processes, biological systems maintain or even increase their organization by causing a larger decrease in the order of their surroundings. The second law is not violated by a *local* decrease in entropy, as long as the *total* entropy of the universe increases in the process.

As an example of how cells can create order while still re-maining within the confines of the second law, consider how a cell uses chemical raw materials such as the sugar glucose ($C_6H_{12}O_6$). Cells use carbon atoms from glucose (and other or-ganic compounds) to build up more complex molecules—a decrease in entropy. But they get the energy for these reactions by "burning" a great many glucose molecules as well—oxi-dizing them to produce many smaller, simpler molecules of water (H_2O) and carbon dioxide (CO_2). This process involves an increase in entropy. For one thing, the product molecules, being less highly structured than glucose, have greater en-tropy. Moreover, heat is released by the oxidation process, rais-ing the entropy of the cell's surroundings. Thus a cell "pays" for its increase in order by increasing the disorder of the rest of the universe.

But, you may ask, where did the glucose molecule, with its complex structure and conveniently available chemical po-tential energy, come from in the first place? As we mentioned above, photosynthetic cells use the energy of sunlight to make glucose and related compounds. Thus all life on Earth is de-pendent on the Sun's energy—the plants that use it directly, the animals that eat the plants, the animals that eat those an-imals, and so on up the food chain.

We can look at the flow of energy through the living world in a more abstract and general way. As we will see in the Chap-ter 20 Insight on the greenhouse effect, life on Earth is de-pendent on receiving radiant energy in the form of sunlight and reradiating infrared energy back into space. In effect, the Sun is a high-temperature source and space is a low-tempera-ture sink. The flow of energy from one to another allows liv-ing things to decrease entropy locally, creating this island of order that we call the biosphere. Thus biological organisms are not exempt from the second law—they merely take ad-vantage of a loophole in it.

(a)

FIGURE 1 Solar collectors
(a) Human beings have developed devices to capture some of the energy of sunlight, although the technology is not yet in widespread use. **(b)** Green plants capture trillions of joules of solar energy every day, and have been doing so for billions of years. The chemical energy that they store in the form of sugars and other complex molecules is used by nearly all organisms on Earth (animals as well as plants) to maintain and increase the highly organized state that we call life.

(b)

inal condition. Recall that for an isobaric process, the work done is equal to the area under the isobar, $W = p\Delta V$, on a *p-V* diagram (see Fig. 12.3). Here, the work in the 1-2 process is positive and the work in the 3-4 process is negative ($-\Delta V$, a compression). Hence, the net work is the area of the rectangle formed by the iso-bars and isomets.

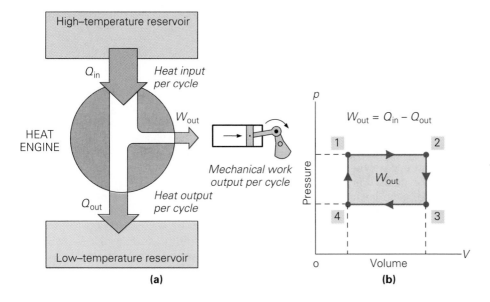

•FIGURE 12.9   Heat engine
(a) In this diagram of energy flow for a generalized cyclic heat engine, note that the width of the arrow representing Q_{in} is equal to the combined widths of the arrows representing W_{out} and Q_{out}, reflecting the conservation of work–energy. **(b)** A cyclic process consisting of two isobars and two isomets is shown here. The net work output per cycle is the area of the rectangle formed by the process paths.

EXAMPLE 12.6 ■ THE DRINKING BIRD: A HEAT ENGINE

An unusual example of a cyclic heat engine is a novelty item called a drinking bird, shown in •Fig. 12.10. The glass bird contains a volatile liquid (usually ether). When the absorbent material on the head and beak is wet, evaporation occurs. Heat is removed from the head, which lowers its temperature and internal vapor pressure. With a temperature difference, there is a pressure difference between the head and body, and the liquid rises to the head. This raises the bird's center of gravity above the pivot point, and a torque causes a forward rotation. The bird then rewets its beak in the glass of water.

The liquid from the body runs into the head, warming it. With the tube not completely filled with liquid and partially open, the pressures in the head and body equalize so that there is no longer a pressure difference. The liquid then drains back into the body. With the lowering of its center of gravity below the pivot point, the bird swings back to the vertical position, and the cycle begins again. The net effect is that heat is transferred from a high-temperature reservoir (the body) to a low-temperature reservoir (the head), with work being done (evidenced by the motion). Of course, gravity and the atmosphere also play roles in the operation of this specialized heat engine.

Follow-up Exercise.   Describe how atmospheric conditions would affect the operation of the drinking bird.

The most common cyclic heat engine is the internal combustion engine used in automobiles and for a variety of other applications. Such engines have more complicated cyclic processes than the one in Fig. 12.9, as you might imagine. (See the Insight on the Otto cycle on p. 400.) The cycles for automotive gasoline engines and diesel engines are similar, but there are other differences between diesel and gasoline engines. One is the type of fuel—the diesel engine runs on oil, not gasoline. Another is that the diesel engine lacks an ignition system or spark plugs. During the compression stroke, oil is injected into the cylinder. Combustion occurs spontaneously when the temperature of the air-fuel mixture gets high enough (because it is being compressed).

The higher compression and pressure present in a diesel engine give it a higher efficiency than a gasoline engine, about 40% as opposed to 30%. Thermal efficiency is used to rate heat engines. The **thermal efficiency** (ϵ_{th}) is

$$\epsilon_{th} = \frac{\text{work out}}{\text{heat in}} = \frac{W_{out}}{Q_{in}}$$

(12.5)

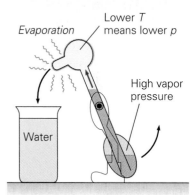

Lower T means lower p

Evaporation

High vapor pressure

Water

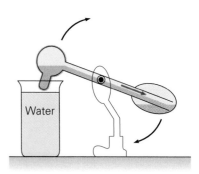

Water

•FIGURE 12.10   A heat engine?
See Example 12.6.

Insight · The Otto Cycle

The internal combustion engine is a cyclic heat engine employed in automobiles and many other devices. Two different cycles of operation are in general use: a two-stroke cycle and a four-stroke cycle. (The number of strokes tells how many up and down motions a piston makes in a cycle; for example, a two-stroke cycle has one upstroke and one downstroke per cycle.) Engines with a two-stroke cycle are used in motorcycles, outboard motors, chain saws, and some lawn mowers.

Most automobile engines have a four-stroke cycle. The steps in this cycle are shown in Fig. 1, along with a p-V diagram of the theoretical thermodynamic processes that the cycle approximates. The theoretical cycle is called the Otto cycle, for the German engineer Nickolas Otto (1832–1891), who built one of the first successful gasoline engines.

During the intake stroke (1–2), an isobaric expansion, the air–fuel mixture is admitted through the open valve. This mixture is adiabatically compressed on the compression stroke (2–3), which is followed by ignition (3–4) and an adiabatic expansion during the power stroke (4–5). Next is an isometric cooling of the system when the piston is at its lowest position (5–2). The final, exhaust stroke is along the isobaric leg of the Otto cycle (2–1).

A two-stroke engine does not have an isobaric leg on its theoretical cycle. The air–fuel mixture is added to the cylinder at the same time as the previously combusted gases are expelled, during the isometric cooling process (5–2). This results in some loss of fuel (with the exhaust gases), which is the major disadvantage of the two-stroke engine. An advantage is that there is one power stroke for every crankshaft revolution for a two-stroke engine, compared to only one power stroke for two crankshaft revolutions for a four-stroke engine.

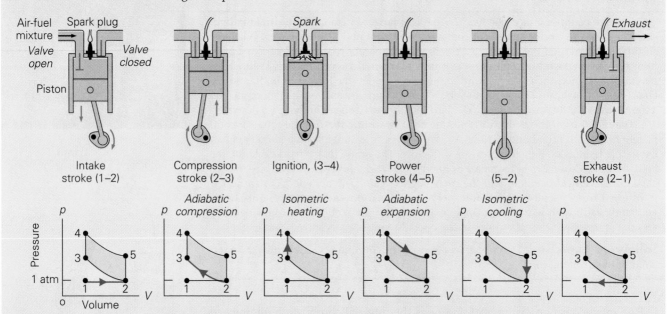

FIGURE 1   The four-stroke cycle for a heat engine
The process steps for the four-stroke Otto cycle. The piston moves up and down two times each cycle, making a total of four strokes per cycle.

The efficiency tells you what you get out for what you put in, so to speak. For one cycle of a cyclic heat engine, the work output per cycle is $W = Q_{in} - Q_{out}$ (see Fig. 12.9a), or, by the first law,

$$Q = \Delta U + W \quad \text{and} \quad Q_{in} - Q_{out} = 0 + W$$

where ΔU is zero, since the system returns to its original state in completing a cycle. Thus, taking $Q_{in} = Q_{hot}$ and $Q_{out} = Q_{cold}$, the thermal efficiency (per cycle) of a cyclic heat engine is

Thermal efficiency

$$\epsilon_{th} = \frac{W}{Q_{in}} = \frac{Q_{hot} - Q_{cold}}{Q_{hot}} = 1 - \frac{Q_{cold}}{Q_{hot}} \qquad (12.6)$$

In general, an engine is expected to have the same efficiency each cycle.

Like mechanical efficiency, thermal efficiency is a fraction and is commonly expressed as a percentage. From Eq. 12.6, you can see that a heat engine could have 100% efficiency if Q_{cold} were zero. This would mean that no heat energy would be lost and all the heat input would be converted to useful work, which is impossible, according to the second law of thermodynamics. In 1851, Lord Kelvin stated this observation, another form of the second law:

No heat engine operating in a cycle can convert its heat input completely to work.

The second law applied to heat engines

Practically, we see from Eq. 12.6 that to maximize the work output per cycle of a heat engine, we must work to minimize the ratio of Q_{cold}/Q_{hot}, which increases the efficiency.

EXAMPLE 12.7 ■ THERMAL EFFICIENCY: WHAT YOU GET OUT OF WHAT YOU PUT IN

With each cycle a heat engine removes 1500 J of heat energy from a high-temperature reservoir and exhausts 1000 J to a low-temperature reservoir. What is the engine's thermal efficiency?

Solution.

Given: $Q_{hot} = 1500$ J   **Find:** ϵ_{th} (thermal efficiency)
$\quad\quad\quad Q_{cold} = 1000$ J

We may use Eq. 12.6 directly, and

$$\epsilon_{th} = 1 - \frac{Q_{cold}}{Q_{hot}}$$

$$= 1 - \frac{1000 \text{ J}}{1500 \text{ J}} = 0.33 \, (\times 100\%) = 33\%$$

Follow-up Exercise. What is the work output per cycle of the engine in this Example?

Heat Pumps

The function performed by a heat pump is basically the reverse of that of a heat engine. That is, a **heat pump** is a device that transfers heat energy from a low-temperature reservoir to a high-temperature reservoir (Fig. 12.11). To do this, there must be work input, since the second law says that heat will not spontaneously flow from a cold body to a hot body. An example of a heat pump is an air conditioner.

Another familiar example of a heat pump is a refrigerator. With work input (from electrical energy), heat is transferred from inside the refrigerator (low-temperature reservoir) to the surroundings (high-temperature reservoir). Have you ever wondered how a refrigerator (or an air conditioner, another heat pump) operates? The knowledge you have gained about thermodynamic processes will allow you to understand the basic operation (see Fig. 12.12).

The refrigerant, or heat-transferring medium, is a substance with a relatively low boiling point. Ammonia (boiling point $-33.3°C$ at 1 atm), sulfur dioxide ($-10.1°C$), and Freon ($-29.8°C$) can be used as refrigerants. Freon is used in most domestic refrigerators. The other compounds were once used, and their smells were quite evident if the system happened to leak.

It is convenient to think of the system diagrammed in Fig. 12.12 as having high and low (temperature and pressure) sides. On the low side, heat is transferred to the evaporator coils from inside the refrigerator. The heat causes the refrigerant to boil and is carried away as latent heat of vaporization. The gaseous

Note: As used here, the term "heat pump" is general and does not refer specifically to the devices that are commonly used for heating and cooling of buildings. These will be discussed shortly.

Note: Freon, a chlorofluorocarbon (CFC), is involved in the depletion of the atmospheric ozone (O_3) layer that absorbs and protects us from harmful ultraviolet rays from the sun. Substitutes are now being developed and used as refrigerants as CFCs are phased out (see the Insight in Chapter 19).

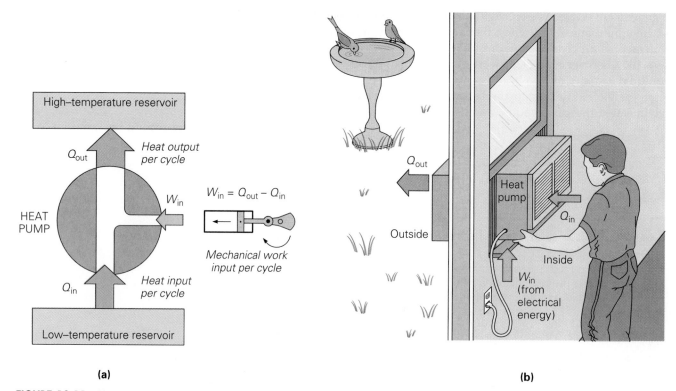

(a)

(b)

•FIGURE 12.11   Heat pump
(a) An energy flow diagram for a generalized cyclic heat pump. Note that the width of the arrow representing Q_{out}, the heat transferred *in*to the high-temperature reservoir is equal to the combined widths of the arrows representing W_{in} and Q_{in}, reflecting the conservation of work-energy. **(b)** An air conditioner is an example of a heat pump. With work input, it transfers heat (Q_{in}) out of or from a low-temperature reservoir (inside the house) to a high-temperature reservoir (the outside).

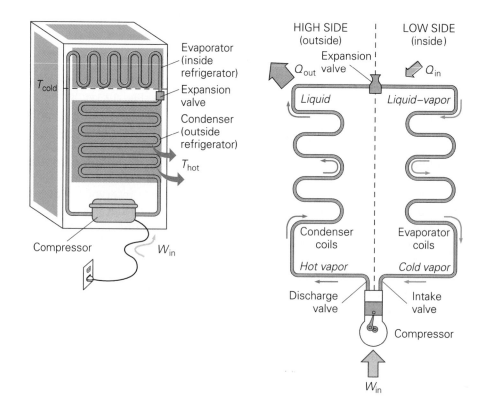

•FIGURE 12.12   Refrigerator operation
Heat (Q_{in}) is carried away from the interior by the refrigerant as latent heat. This heat energy and that of the work input (W_{in}) are discharged from the condenser to the surroundings (Q_{out}). See the text for a detailed description.

vapor is drawn into the compressor chamber on the downstroke of the piston and is compressed on the upstroke (work input by compressor motor). The compression increases the temperature of the gas, which is discharged from the compressor as a superheated vapor (on the high side). This vapor condenses in the cooler condenser unit, where circulating air (or water in larger units) carries away the latent heat of condensation and the heat of compression. The condensed liquid collects in a receiver.

An expansion valve maintains a balance between the high and low sides, thereby controlling the rate at which the refrigerant passes back to the low side. On being admitted to the low side, the liquid immediately boils and vaporizes because of the low pressure. This is a cooling process because the heat of vaporization is supplied by the internal energy of the liquid. The cooled refrigerant is drawn into the evaporator, and the cycle begins again.

In essence, a refrigerator (or air conditioner) pumps heat up a temperature gradient or "hill." (As an analogy, think of pumping water up an actual hill against the force of gravity.) The cooling efficiency of this operation depends on the amount of heat extracted from the cold temperature reservoir (the freezer compartment or the inside of a house), $Q_{in} = Q_{cold}$, and the work W needed to do so. Since a practical refrigerator operates in a cycle to provide a continuous removal of heat, $\Delta U = 0$ for the cycle. Then, by the conservation of energy (or first law), $Q_{cold} + W = Q_{hot}$, where $Q_{out} = Q_{hot}$ is the heat ejected to the high temperature reservoir or the outside.

The measure of a refrigerator or air conditioner's performance is defined differently from that of a heat engine, because of the difference in their functions. For these appliances, the efficiency is expressed in terms of the **coefficient of performance (cop)**. Since the purpose is to extract the most heat (Q_{cold}) per unit work input (W), the coefficient of performance for a refrigerator (or air conditioner), cop_{ref}, is expressed as their ratio:

$$cop_{ref} = \frac{Q_{in}}{W} = \frac{Q_{cold}}{Q_{hot} - Q_{cold}} \qquad (\textit{for refrigerator or air conditioner}) \quad (12.7)$$

where the conservation relationship given above was used to express the work in terms of heat.

Thus, the greater the cop, the better performance—more out for what you put in. For normal refrigerator operation, the work input is less than the heat removed, so the cop is greater than 1. The cop's of typical refrigerator cycles range from 3 to 5, depending on operating conditions. This means that the heat removed from the cold reservoir (the refrigerator freezer or the inside of a house) is 3-5 times the work needed to remove it.

Refrigerators and air conditioners are generically referred to as heat pumps because they basically "pump" heat uphill. However, the term *heat pump* is now more specifically applied to the common commercial devices used to cool homes and offices in the summer and to heat them in the winter. The summer operation is that of an air conditioner, as described above. That is, in this mode it cools the inside (of a house) and heats the outdoors. Operating in the heating mode, a heat pump heats the inside and cools the outdoors, usually by taking heat from the air. This may seem hard to believe, as it may be quite cold outside. However, keep in mind that the air has internal energy regardless of its temperature.

For a heat pump in the heating mode, the heat input is the major interest, so the cop is defined differently from that of a refrigerator or air conditioner. As you might guess, it is the ratio of Q_{out} to W (what you get out for what you put in):

$$cop_{hp} = \frac{Q_{out}}{W} = \frac{Q_{hot}}{Q_{hot} - Q_{cold}} \qquad (\textit{for heat pump in heating mode}) \quad (12.8)$$

where again $Q_{cold} + W = Q_{hot}$. Typical cop's for heat pumps range between 2 and 4, depending on operating conditions.

Compared to electrical heating, heat pumps are efficient. For each kilowatt-hour of electricity consumed, a heat pump typically provides ("pumps in") from one and one-half to three times as much heat as electrical baseboard units. However, if the outside temperature is very low, a heat pump may not be efficient enough to heat a building adequately. Then it must be supplemented by some conventional heating system, such as an electrical one. Some heat pumps use water from underground reservoirs, wells or in buried loops of pipe as a heat source and sink. These are more efficient than the ones that use the outside air because water has a larger specific heat than air, and the average temperature difference between the water and the inside air is smaller.

12.5 The Carnot Cycle and Ideal Heat Engines

Objectives: To be able to (a) explain how the Carnot cycle applies to heat engines, (b) compute the ideal Carnot efficiency, and (c) state the third law of thermodynamics.

Lord Kelvin's statement of the second law of thermodynamics says that any *cyclic* heat engine, regardless of its design, will always lose some heat energy. But how much heat must be lost? In other words, what is the maximum possible efficiency of a heat engine? In designing heat engines, engineers strive to make them as efficient as possible, but there must be some theoretical limit, and, according to the second law, it must be less than 100%.

Sadi Carnot (1796–1832), a French engineer, solved this problem. The first thing he considered was the thermodynamic cycle an ideal heat engine would use, that is, the most efficient cycle. A heat engine absorbs heat from a *constant* high-temperature reservoir and exhausts it to a *constant* low-temperature reservoir. These are ideally reversible isothermal processes and may be represented as two isotherms on a *p-V* diagram. But what are the processes that complete the cycle and make it the most efficient cycle? Carnot showed that these are reversible adiabatic processes, called adiabats when represented on a graph (•Fig. 12.13a).

Thus, the ideal **Carnot cycle** consists of two isotherms and two adiabats and is conveniently represented on a *T-S* diagram, where it forms a rectangle (•Fig. 12.13b). The area under the upper isotherm (1–2) is the heat added to the system from the high-temperature reservoir: $Q_{in} = Q_{hot} = T_{hot}\Delta S$. Similarly, the area under the lower isotherm (3–4) is the heat exhausted: $Q_{out} = Q_{cold} = T_{cold}\Delta S$. Here, Q_{in} and Q_{out} are heat transfers at *constant* temperatures (T_{hot} and T_{cold}). There is no heat transfer ($Q = 0$) in the adiabatic legs of the cycle. (Why?) This is not the case in practical heat-engine cycles where there is heat transfer in a leg of a cycle at *different* temperatures (not an isotherm). See Exercises 70 and 71.

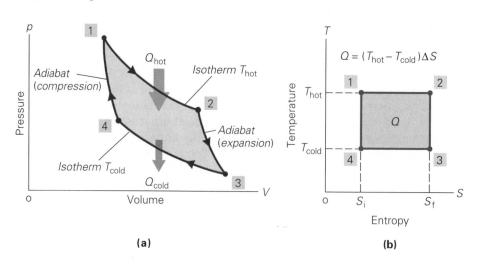

•**FIGURE 12.13 The Carnot cycle**
(a) The Carnot cycle consists of two isotherms and two adiabats. Heat is absorbed during the isothermal expansion and exhausted during the isothermal compression. **(b)** On a *T-S* diagram, the Carnot cycle forms a rectangle, the area of which is equal to *Q*.

The difference between these is the work output, which is equal to the area of the rectangle enclosed by the processes (the shaded area on the diagram):

$$W = Q = Q_{hot} - Q_{cold} = (T_{hot} - T_{cold})\Delta S$$

Since ΔS is the same for the areas under both isotherms, these expressions can be used to relate the temperatures and heats. That is, since

$$Q_{hot} = T_{hot}\Delta S \quad \text{and} \quad Q_{cold} = T_{cold}\Delta S$$

then

$$\frac{Q_{hot}}{T_{hot}} = \frac{Q_{cold}}{T_{cold}} \quad \text{or} \quad \frac{Q_{hot}}{Q_{cold}} = \frac{T_{hot}}{T_{cold}} \qquad (12.9)$$

This equation can be used to express the efficiency of an ideal heat engine in terms of temperature. From Eq. 12.6, this ideal **Carnot efficiency** (ϵ_C) is

$$\epsilon_C = 1 - \frac{Q_{cold}}{Q_{hot}} = 1 - \frac{T_{cold}}{T_{hot}}$$

or

$$\epsilon_C = 1 - \frac{T_{cold}}{T_{hot}} = \frac{T_{hot} - T_{cold}}{T_{hot}} \qquad (12.10)$$

Note: Keep in mind that the temperature in the Carnot efficiency is the absolute or Kelvin temperature.

where the fractional efficiency is often expressed as a percentage. Note that T_{cold} and T_{hot} are Kelvin temperatures.

The Carnot efficiency expresses the theoretical upper limit on thermodynamic efficiency for a cyclic heat engine, which can never be achieved. It corresponds to the theoretical mechanical advantage for a simple machine. A true Carnot engine cannot be built because the necessary reversible processes can only be approximated. Although reversible adiabatic processes can be approached, reversible isothermal processes are virtually impossible to approximate during the heat transfer processes in a real engine.

Note: The Carnot efficiency, which can never be attained, gives an ideal upper limit.

However, the Carnot efficiency does show that the greater the difference in the temperatures of the heat reservoirs, the greater the efficiency. For example, if T_{hot} is two times T_{cold}, or $T_{cold}/T_{hot} = \frac{1}{2} = 0.50$, the Carnot efficiency will be

$$\epsilon_C = 1 - \frac{T_{cold}}{T_{hot}} = 1 - 0.50 \; (\times 100\%) = 50\%$$

But if T_{hot} is four times T_{cold}, or $T_{cold}/T_{hot} = \frac{1}{4} = 0.25$, then

$$\epsilon_C = 1 - \frac{T_{cold}}{T_{hot}} = 1 - 0.25 \; (\times 100\%) = 75\%$$

EXAMPLE 12.8 ■ CARNOT EFFICIENCY: THE UNATTAINABLE UPPER LIMIT

An engineer is designing a cyclic heat engine to operate between temperatures of 150°C and 27°C. What is the maximum theoretical efficiency that can be achieved?

Solution.

Given: $T_{hot} = 150°C = 423 \text{ K}$ *Find:* ϵ_C (Carnot efficiency)
$ T_{cold} = 27°C = 300 \text{ K}$

Using Eq. 12.10

$$\epsilon_C = 1 - \frac{T_{cold}}{T_{hot}} = 1 - \frac{300 \text{ K}}{423 \text{ K}} = 0.291 \; (\times 100\%) = 29.1\%$$

This means that in actual operation the thermal efficiency of the heat engine will be less than 29.1%.

Follow-up Exercise. Suppose that the operating high temperature of the engine in this Example were increased to 200°C. What would be the change in the theoretical efficiency?

There are also Carnot cop's for refrigerators and heat pumps. (See Exercise 90.)

The Third Law of Thermodynamics

Another interesting conclusion can be drawn from the expression for the Carnot efficiency (Eq. 12.10). To have ϵ_C equal to 100%, T_{cold} would have to be (absolute) zero. Since a heat engine with 100% efficiency is impossible by the second law, we have the so-called **third law of thermodynamics**:

It is impossible to reach a temperature of absolute zero.

Absolute zero has never been observed experimentally. If it were, this would in effect violate the second law. In cryogenic (low-temperature) experiments, scientists have come close to absolute zero—to about 20 nK (2.0×10^{-8} K)—but have never reached it. Near absolute zero, reducing the temperature by an order of magnitude becomes more difficult at each step.

You may be wondering how such low temperatures are reached. There are two methods. One uses laser light to slow a beam of atoms. The light exerts a small force in the form of *radiation pressure*, probably best known for its role in the formation of comet tails (Section 20.4). Here, the radiation pressure acts as a retarding force. Recall that temperature is related to average kinetic energy, so reducing the speed of the atoms lowers the temperature.

The beam of slowed atoms is then passed through the convergence point of four laser beams. The bombarded atoms absorb light, and re-emit it at a higher frequency or energy (Section 26.3). Consequently, the atoms steadily lose energy; and as the power of the lasers is progressively lowered, the temperature of the atoms drops to a minimum.

The second method uses a magnetic "trap." A magnetic field (Section 19.1) is used to trap a large number of ionized atoms in a small volume (magnetic confinement, discussed in Section 30.3). As the magnetic field is gradually reduced, the fastest (or hottest) atoms escape the trap, leaving behind the slower (or colder) atoms. This is similar to an evaporative cooling process (Section 11.3). In this way, the temperature of the gaseous atoms is steadily lowered.

For both techniques, the temperature is determined by optical methods that measure the average atomic speed, and hence the average kinetic energy and temperature. (As you can see by the various references in the preceding discussion, physics builds on itself. That is, the more physical principles you learn, the easier it becomes to understand new phenomena.)

Chapter Review

Important Terms

thermodynamics   384
system   384
thermally isolated system   384
completely isolated system   384
heat reservoir   384
equations of state   384
process   386
irreversible process   386
reversible process   386

first law of thermodynamics   386
isobaric process   387
isometric process   388
isothermal process   388
adiabatic process   389
second law of thermodynamics   392
entropy   393
isentropic process   396

heat engine   397
thermal efficiency   399
heat pump   401
coefficient of performance (cop)   403
Carnot cycle   404
Carnot (ideal) efficiency   405
third law of thermodynamics   406

Important Concepts

• The state of a system is described by an equation of state. A process is a change in the state of a system. There are ideal reversible processes (changes through equilibrium states) and irreversible

processes (changes through nonequilibrium states).
• The first law of thermodynamics is a statement of the conservation of energy for a thermodynamic system.

- The second law of thermodynamics specifies whether or not a process can take place, or the direction of a process.
- The total entropy of the universe increases in every natural process.
- A heat engine is a device that converts heat energy into work. Thermal efficiency is the ratio of the work output and the heat input—what you get out for what you put in.
- A heat pump is a device that transfers heat energy from a low-temperature reservoir to a high-temperature reservoir. The coefficient of performance (cop) is the ratio of useful heat transferred to the work input required to do this.

Important Equations

First Law of Thermodynamics:
$$Q = \Delta U + W$$
$+ Q$, heat *added* to system
$- Q$, heat *removed* from system
$+ W$, work done *by* system   (12.1)
$- W$, work done *on* system

Work Done by an Expanding Gas (constant pressure):
$$W = p\Delta V = p(V_2 - V_1) \qquad (12.2)$$

Change in Entropy: (constant temperature)
$$\Delta S = \frac{Q}{T} \qquad (12.4)$$

Thermal Efficiency of a Heat Engine:
$$\epsilon_{th} = \frac{W}{Q_{in}} = \frac{Q_{hot} - Q_{cold}}{Q_{hot}} = 1 - \frac{Q_{cold}}{Q_{hot}} \qquad (12.6)$$

Coefficients of Performance:
$$cop_{ref} = \frac{Q_{in}}{W} = \frac{Q_{cold}}{Q_{hot} - Q_{cold}} \qquad (12.7)$$
(for refrigerator or air conditioner)
$$cop_{hp} = \frac{Q_{out}}{W} = \frac{Q_{hot}}{Q_{hot} - Q_{cold}} \qquad (12.8)$$
(for heat pump in heating mode)

Carnot Efficiency of an Ideal Heat Engine:
$$\epsilon_C = 1 - \frac{T_{cold}}{T_{hot}} = \frac{T_{hot} - T_{cold}}{T_{hot}} \qquad (12.10)$$

Exercises*

12.1 Thermodynamic Systems, States, and Processes

1   There may be an interchange of heat with the surroundings for (a) a thermally isolated system, (b) a completely isolated system, (c) a heat reservoir, (d) none of these.

2   Only initial and final states are known for irreversible processes on (a) *p-V* diagrams, (b) *p-T* diagrams, (c) *V-T* diagrams, (d) all of these.

3   Explain why the *p-V* graph in Fig. 12.1 is not that of an ideal gas.

4   On a *p-T* diagram, sketch the general paths for the following reversible processes for an ideal gas: (a) isothermal, (b) isobaric, (c) isometric, and (d) adiabatic.

5   On a *V-T* diagram, sketch the general paths for the following reversible processes for an ideal gas: (a) isothermal, (b) isobaric, (c) isometric, and (d) adiabatic.

6   On a *p-V* diagram, sketch a general cyclic process that consists of (a) an isothermal expansion, (b) an isobaric compression, and (c) an isometric process, in that order.

*Take temperatures to be exact and room temperature to be 20°C.

7   The *pVT* diagram for water is shown in •Fig. 12.14 (compare this with Fig. 11.10). (a) Explain the processes on the liquid–vapor and solid–vapor surfaces. (b) Why is there a triple-point line?

8   Explain how you can show from Fig. 12.14 that a quantity of water has a greater volume when it is frozen than when it is melted.

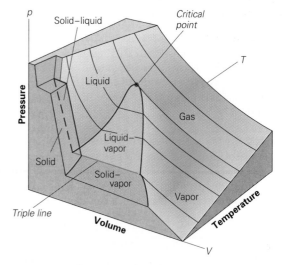

•FIGURE 12.14 The *pVT* surface for water
See Exercises 7, 8, and 99.

12.2 The First Law of Thermodynamics

9 According to the first law of thermodynamics, if work is done on a system, then (a) the internal energy of the system must change, (b) heat must be transferred from the system, (c) the internal energy of the system changes and/or heat is transferred from the system, (d) heat is transferred to the system.

10 If a system of ideal gas has heat added by an isothermal process, (a) work is done on the system, (b) the internal energy decreases, (c) the effect is the same as for an isometric process, (d) none of these.

11 (a) Why does the body of a hand pump become hot when a tire is pumped up? (Neglect friction. The plunger is usually lubricated to prevent air leakage.) (b) Why does the valve of a tire become cold when air is released from the tire?

12 ■ A rigid container contains 1 mole of nitrogen gas that slowly receives 3.0 kcal of heat. What is the change in the internal energy of the gas in joules?

13 ■ A quantity of ideal gas goes through a cyclic process and does 400 J of work. (a) Is heat added or removed from the system, and how much heat is involved? (b) Does the temperature of the gas increase or decrease?

14 ■■ A system of ideal gas is expanded adiabatically to 1.5 times its volume, while doing 500 J of work. (a) Did the temperature of the gas increase or decrease? (b) How much heat was transferred? (c) What was the change in the internal energy of the gas?

15 ■■ A quantity of ideal gas is taken through processes that are depicted as straight-line paths on a p-V diagram: from (1.0 atm, 1.0 m^3) to (4.0 atm, 1.0 m^3), then to (4.0 atm, 3.0 m^3), and back to (1.0 atm, 1.0 m^3). What is the net work done?

16 ■■ A system of gas at low density has an initial pressure of 1.65×10^4 Pa and occupies a volume of 0.20 m^3. The slow addition of 250 cal of heat to the system causes it to expand isobarically to a volume of 0.40 m^3. (a) How much work is done by the system in the process? (b) Did the internal energy of the system change? If so, by how much?

17 ■■ An Olympic weight lifter lifts 145 kg a vertical distance of 2.1 m. In doing so, his internal energy decreases by 6.0×10^4 J. Considering the weight lifter to be a thermodynamic system, how much heat in kcal flows and in what direction?

18 ■■ One mole of ideal gas at STP expands isobarically to 2.5 times its original volume. (a) Is work done by or on the system? How much work? (b) Does heat flow into or out of the system?

19 ■■ A system of ideal gas is taken through the reversible processes shown in ●Fig. 12.15. In terms of the state variables, (a) what is the change in the internal energy of the gas? (b) How much work is done by or on the gas? (c) What is the amount of heat transfer in the overall process?

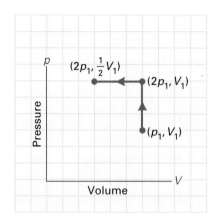

●**FIGURE 12.15   A p-V diagram for an ideal gas**
See Exercise 19

20 ■■ The temperature of 2.0 moles of ideal gas is increased from 150°C to 250°C using two different process paths. In the one, 2500 J of heat are added to the system, and in the other, 3000 J of heat are added. (a) What is the change in the internal energy of the gas in each case? (*Hint*: see Eq. 10.15.) (b) In which case is more work done and how much more?

21 ■■ A gram of water (1.00 cm^3) at 100°C is converted to 1671 cm^3 of steam at atmospheric pressure. What is the change in the internal energy of the system?

22 ■■ A quantity of gas undergoes the reversible changes illustrated in the p-V diagram in ●Fig. 12.16. How much work is done in each process?

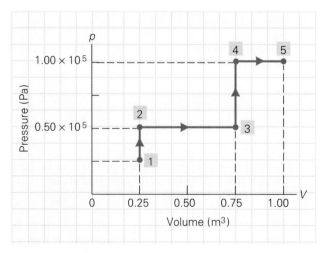

●**FIGURE 12.16   A p-V diagram and work**
See Exercises 22 and 23.

23 ■■ Suppose that after the final process shown in Fig. 12.16 (see Exercise 22) the pressure of the gas is decreased isometrically from 1.0×10^5 Pa to 0.70×10^5 Pa, and then the gas is compressed isobarically from $1.0 \ m^3$ to $0.80 \ m^3$. What is the total work done in all of these processes?

24 ■■ A quantity of gas is compressed as shown on the p-V diagram in ●Fig. 12.17. How much work is done on the system?

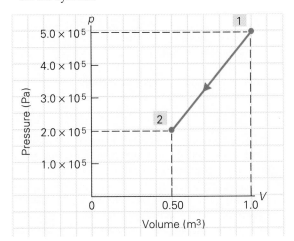

●**FIGURE 12.17 A variable p-V process and work**
See Exercises 24 and 25.

25 ■■ Suppose that after the process shown in Fig. 12.17 (see Exercise 24) the gas undergoes an isobaric expansion to $1.0 \ m^3$ and then the pressure is increased isometrically to 5.0×10^5 Pa. (a) How much work is done in each of these processes? (b) What is the total work done in taking the gas from its initial state [(1) in the figure] to its final state? (c) Is heat added or removed from the system? How much heat?

26 ■■■ One mole of ideal gas is taken through the cyclic process shown in ●Fig. 12.18. (a) Compute the work involved (W) for each of the four processes. (b) Find ΔU, W, and Q for the complete cyclic process. (c) What is T_3?

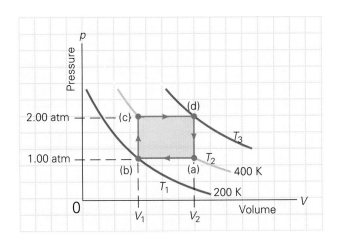

●**FIGURE 12.18 A cyclic process**
See Exercise 26.

12.3 The Second Law of Thermodynamics and Entropy

27 The second law of thermodynamics (a) gives an equation of state for a system, (b) applies only when the first law is satisfied, (c) precludes perpetual motion machines, (d) does not apply to a closed system.

28 Overall in a natural process, the change in the entropy of the universe is (a) negative, (b) positive, (c) zero, (d) the same as the change in the internal energy.

29 In Example 12.4, what would be the implication if the total change in entropy had been negative?

30 When a quantity of hot water is mixed with a quantity of cold water, the combined system comes to thermal equilibrium at some intermediate temperature. How does the entropy of the system change?

31 ■ What change in entropy is associated with the reversible phase change of 1.0 kg of ice to water at 0°C?

32 ■ What is the change in entropy when 0.50 kg of steam condenses to water at 100°C?

33 ■■ Which process has the greater change in entropy: 0.75 kg of water changing to ice at 0°C or 0.25 kg of water changing to steam at 100°C? Comment on the entropy changes in terms of order and disorder.

34 ■■ A quantity of an ideal gas initially at STP undergoes a reversible isothermal expansion and does 3.0×10^3 J of work on its surroundings in the process. What is the change in the entropy of the gas?

35 ■■ An ice tray is filled with 250 mL of water at 4.00°C from a cold-water tap and placed in a freezer, where the temperature is −6.00°C. Estimate the change in entropy that will have occurred when the water freezes and the ice comes to thermal equilibrium with its surroundings.

36 ■■ A certain system undergoes an isothermal process at 27°C, and there is an entropy change of 35 J/K. If 2.0×10^3 J of work is done on the system in the process, what is the change in the system's internal energy?

37 ■■ During a liquid-to-solid phase change of a quantity of a substance, the change in entropy is $−4.19 \times 10^3$ J/K. If 400 kcal of heat is removed in the process, what is the freezing point of the substance in degrees Celsius?

38 ■■ What is the change in entropy when 0.50 kg of mercury vapor ($L_v = 2.7 \times 10^5$ J/kg) condenses to a liquid at its boiling point of 357°C?

39 ■■ A 5.00-L quantity of water at 20.0°C is mixed with the same amount of water at 24.0°C. What is the approximate change in entropy in J/K?

40 ■■ Two heat reservoirs at temperatures of 200°C and 60°C are brought into thermal contact and 1.50×10^3 J of heat spontaneously flows from one to the other. What is the change in the entropy of the universe? What would a negative entropy change imply?

41 ■■ One mole of ideal gas goes through an isothermal compression at room temperature. If 7.5×10^3 J of work is done in compressing the gas, what is the change in entropy of the system?

42 ■■ A system of ideal gas at room temperature undergoes an isometric process with an internal energy decrease of 4.50×10^3 J to a final temperature of 16°C. What is the approximate change in entropy in J/K?

43 ■■ (a) How much heat is transferred in the processes shown on the T-S diagram in •Fig. 12.19? (b) What type of process takes place as the system goes from state b to c?

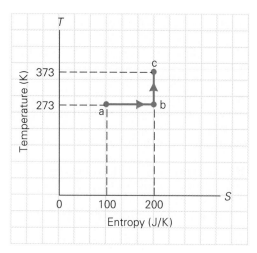

•**FIGURE 12.19 Entropy and heat**
See Exercises 43 and 44.

44 ■■ Suppose that the system described by the T-S diagram in Fig. 12.19 is returned to its original state (state a) by a reversible process that is depicted by a straight line from c to a. What is the change in entropy for the total cycle? How much heat is transferred in the cyclic process?

45 ■■■ A 50.0-g ice cube at 0°C is placed in 500 mL of water at room temperature. Estimate the change in entropy when all the ice has melted (a) for the ice, (b) for the water, and (c) for the universe.

12.4 Heat Engines and Heat Pumps

(Note: Consider efficiencies to be exact.)

46 For a cyclic heat engine, (a) $\epsilon_{th} > 1$, (b) $Q_{hot} = W$, (c) $\Delta U = W$, (d) $Q_{hot} > Q_{cold}$.

47 A heat pump (a) is rated by thermal efficiency, (b) requires work input, (c) is not consistent with the second law, (d) violates the first law.

48 (a) Is any external work done on the drinking bird in Example 12.6? How would the bird be represented in a generalized heat engine diagram? (b) Is leaving a refrigerator door open a practical way to air condition a room? Explain.

49 Lord Kelvin's statement of the second law as applied to heat engines refers to their operating *in a cycle*. Why is this phrase included?

50 In normal atmospheric convection cycles, colder air from a higher altitude is transferred to a lower, warmer level. Does this violate the second law? Explain.

51 ■ A heat engine with an efficiency of 40% does 800 J of work each cycle. How much heat is rejected to the surroundings (the low-temperature reservoir)?

52 ■ An internal combustion engine with a thermal efficiency of 15.0% produces 2.60×10^4 J of work each cycle. How much heat is rejected by the engine?

53 ■ The heat output of a particular engine is 7.5×10^3 J per cycle and the work output is 4.0×10^3 J. What is the thermal efficiency of the engine?

54 ■■ In each cycle, a heat engine absorbs 150 kcal of heat from a high-temperature reservoir and exhausts 100 kcal. (a) What is the engine's thermal efficiency? (b) How much of the heat input per cycle is converted to work?

55 ■■ A heat engine has a thermal efficiency of 25%. It absorbs 800 J from a high-temperature reservoir each cycle. (a) What is the work output of the engine? (b) How much of the input per cycle is not converted to work?

56 ■■ A steam engine does 4.5×10^3 J of useful work each cycle but loses 5.0×10^2 J to friction and exhausts 1.5 kcal of heat energy. What is the engine's efficiency?

57 ■■ A theoretical cyclic heat engine absorbs 1800 kcal of heat from a high-temperature reservoir and exhausts 600 kcal to a low-temperature reservoir. When run in reverse as a heat pump with the same reservoirs, 5870 J of work must be input to deliver 1800 kcal of heat energy to the high-temperature reservoir. (a) What is the engine's thermal efficiency? (b) Is the heat engine reversible in the sense that the work input and output are the same when the engine is run between the same reservoirs?

58 ■■ An engineer redesigns a heat engine and improves its thermal efficiency from 20% to 25%. (a) Does the ratio of the heat output to heat input increase or decrease, and why? (b) What is the change in the Q_{cold}/Q_{hot} ratio?

59 ■■ A gasoline engine consumes 7.8×10^4 kcal per hour. (a) What is the energy input during a period of 2.0 h? (b) If the engine delivers 25 kW of power during this time, what is its efficiency?

60 ■■ A refrigerator takes heat from its low-temperature reservoir at a rate of 1.5 kW when work is done at a rate of 2.5 kW. At what rate is heat given to the high-temperature reservoir?

61 ■■ A refrigerator with a cop of 2.2 removes 100 kcal of heat from its storage area each cycle. (a) How much heat is exhausted each cycle? (b) What is the total work input in joules for 10 cycles?

62 ■■ A heat pump removes 2.0×10^3 J of heat from the outdoors and delivers 3.5×10^3 J of heat to the inside of the house each cycle. (a) How much work does this require? (b) What is the cop of the heat pump?

63 ■■ An air conditioner has a cop of 1.75. What is the wattage input required for the unit to remove 2.50×10^5 kcal of heat in 20 minutes?

64 ■■ A heat engine has a thermal efficiency of 30.0%. If its heat input each cycle is supplied by the condensation of 8.00 kg of steam at 100°C, (a) what is the work output per cycle? (b) How much heat is rejected to the surroundings each cycle?

65 ■■ A heat pump uses an underground water reservoir as a heat source. It extracts 50 kcal each cycle, while doing 3.0×10^4 J of work. (a) How much heat is delivered to the inside of the house? (b) What is the temperature drop of the water when returned to the reservoir if 5.0 kg are used each cycle?

66 ■■■ A coal-fired power plant produces 900 MW of electricity and operates at an overall thermal efficiency of 35%. (a) What is the thermal input to the plant? (b) What is the rate of heat discharge from the plant? (c) Why is the water heated by the discharged heat cooled in a cooling tower before being discharged into a nearby river?

67 ■■■ Show that the change in entropy for a cycle of a heat engine is

$$\Delta S = \frac{Q_{cold}}{T_{cold}} - \frac{Q_{hot}}{T_{hot}}$$

12.5 The Carnot Cycle and Ideal Heat Engines

68 The Carnot cycle consists of (a) two isobars and two isotherms, (b) two isomets and two adiabats, (c) two adiabats and two isotherms, (d) four arbitrary processes which return the system to its initial state.

69 Which of the following temperature reservoir relationships would have the highest efficiency for a Carnot en-

gine? (a) $T_{cold} = (0.15)T_{hot}$, (b) $T_{cold} = (0.25)T_{hot}$, (c) $T_{cold} = (0.50)T_{hot}$, (d) $T_{cold} = (0.90)T_{hot}$

70 An idealized cycle for a two-stroke internal combustion engine is shown in •Fig. 12.20. It consists of an adiabatic expansion and compression (legs 1 and 3, respectively) and an isometric decompression and compression (legs 2 and 4, respectively). In the chapter it was stated that in a Carnot cycle, heat transfers were at *constant* temperatures. For the cycle in Fig. 12.20, (a) is heat taken in at a constant temperature? (b) Is heat rejected at a constant temperature? (c) Draw sketches that indicate the areas representing work (W, with signs) during expansion and compression, and for the net work.

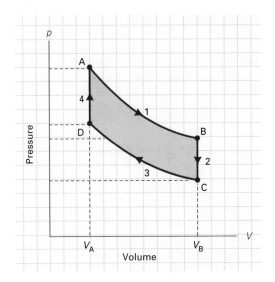

•**FIGURE 12.20 Non-Carnot cycle**
Two adiabats and two isomets. See Exercises 70 and 71. (Not to scale.)

71 In Fig. 12.20, legs 1 and 3 are adiabatic processes. Suppose these were isothermal processes instead. (a) Does this change make the cycle a Carnot cycle? Explain. (b) In which legs are there heat transfers? [Indicate directions by signs.] Are these tranfers at constant temperatures?

72 Automobile engines can either be air-cooled or water-cooled. Which type of engine would you expect to be more efficient and why?

73 ■■ The exhaust temperature of a Carnot engine is 95°C, and the engine has a heat output of 100 kcal per cycle at this temperature. If the engine has an efficiency of 30%, what is its heat input per cycle in joules?

74 ■■ An ideal heat engine with a Carnot efficiency of 35% takes in heat from a high-temperature reservoir at 147°C. What is the Celsius temperature of the low-temperature reservoir?

75 ■■ An ideal heat engine takes 6.5 kcal of heat per cycle from a high-temperature reservoir at 320°C and exhausts some of it to a low-temperature reservoir at 120°C. How much work is done by the engine?

76 ■■ A Carnot engine with an efficiency of 40% operates with a low-temperature reservoir at 50°C and exhausts 1200 J of heat each cycle. What is (a) the heat input per cycle and the (b) Celsius temperature of the high-temperature reservoir?

77 ■■ An ideal heat engine takes in heat from a reservoir at 327°C and has an efficiency of 30%. If the exhaust temperature did not vary and the efficiency increased to 40%, what would be the increase in the temperature of the hot reservoir?

78 ■■ An inventor claims to have developed a heat engine that, on each cycle, takes in 120 kcal of heat from a high-temperature reservoir at 400°C and exhausts 48 kcal to the surroundings at 125°C. Would you invest your money in the production of this engine? Explain.

79 ■■ Which has the greatest theoretical efficiency: a heat engine operating between reservoirs at 300°C and 100°C or one operating between reservoirs at 300 K and 100 K? What are the efficiencies in each case?

80 ■■ An engineer wants to run a heat engine with an efficiency of 40% between a high-temperature reservoir at 350°C and a low-temperature reservoir. Below what temperature must the low-temperature reservoir be for practical operation of the engine?

81 ■■ A heat engine operates at 45% of its ideal efficiency. If the temperature of the high-temperature and low-temperature reservoirs are 400°C and 100°C, respectively, what is the thermal efficiency of the engine?

82 ■■ Assume a heat engine has a thermal efficiency of 100%. Prove that this engine would also have a Carnot efficiency of 100%.

83 ■■ The Carnot efficiency shows that the greater the temperature difference of the reservoirs of a heat engine, the greater the efficiency. Suppose you had the choice of either raising the high-temperature reservoir a certain number of kelvins, or lowering the low-temperature reservoir the same number of kelvins. Which would you choose (assuming you wanted to increase the efficiency, of course)?

84 ■■ The exhaust temperature of a Carnot engine is 200°C. What must the high-temperature reservoir's temperature be for the engine to have an efficiency of 40.0%?

85 ■■ The working substance of a cyclic heat engine is 0.75 kg of an ideal gas. The cycle consists of two isobaric processes and two isometric processes as shown in •Fig. 12.21. What would be the efficiency of a Carnot engine operating with the same high-temperature and low-temperature reservoirs?

86 ■■ In each cycle, a Carnot heat engine takes 774 J of heat from a high-temperature reservoir and discharges 258 J to a low-temperature reservoir. (a) What is the

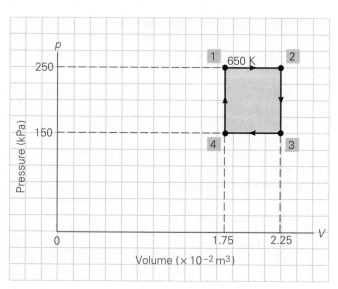

•**FIGURE 12.21 Thermal efficiency**
See Exercise 85.

Carnot efficiency of the engine? (b) How many times greater is the temperature of the high-temperature reservoir than that of the low-temperature reservoir?

87 ■■ A Carnot engine operating between reservoirs at 27°C and 227°C does 1500 J of work each cycle. (a) What is the efficiency of the engine? (b) What is the change in entropy each cycle? [*Hint:* See Exercise 67.]

88 ■■ A heat engine has an efficiency that is half that of a Carnot engine that operates between temperatures of 100°C and 375°C. If the real engine absorbs heat at a rate of 50 kW per cycle, at what rate is heat exhausted each cycle?

89 ■■■ The maximum temperature for the superheated steam used in a turbine for electrical generation is about 540° because of material limitations. (a) If the steam condenser operated at room temperature, what would the ideal efficiency be? (b) The actual efficiency is about 35–40%. What does this tell you?

90 ■■■ There is also a Carnot coefficient of performance (cop_C) for an ideal or Carnot heat pump. (a) Show that this is given by

$$cop_C = \frac{T_{hot}}{T_{hot} - T_{cold}}$$

(b) What does this tell you about the operating parameters of a heat pump? (Could you guess the equation for the Carnot cop for a refrigerator?)

91 ■■■ A salesperson tells you that a new air conditioner with a high cop removes 2.6×10^3 J each cycle from the inside of the refrigerator at a temperature of 5.0°C, and rejects 2.8×10^3 J into the 30°C surroundings. (a) What is the air conditioner's cop? (b) Would you buy one? Justify your answer.

92 ■■■ A Carnot heat engine has an efficiency of 50%. Suppose it could be run in reverse as a heat pump. What would be the Carnot cop of the pump? (See Exercise 90.)

93 ■■■ It has been proposed that the temperature difference in the ocean could be utilized to run a heat engine to generate electricity. In tropical regions, the water temperature is about 25°C at the surface and about 5°C at a low depth. (a) What would the maximum theoretical efficiency of such an engine be? (b) Do you think a heat engine with such a relatively low efficiency would be practical? Explain.

94 ■■■ An ideal heat pump is equivalent to a Carnot engine running in reverse. Show that the work required for an ideal heat pump is given by

$$W = Q_{cold}\left[\left(\frac{T_{hot}}{T_{cold}}\right) - 1\right]$$

Additional Exercises

95 A thermally isolated quantity of gas with an initial volume of 10 L is compressed isobarically at a pressure of 300 kPa. If 6.0×10^2 J of work is done on the system, what is the final volume of the gas?

96 For the two-step reversible process illustrated in •Fig. 12.22, the same amount of heat is added to and removed from the system. Prove that the total entropy (of the universe) increases in the process.

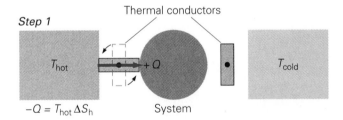

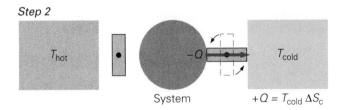

•**FIGURE 12.22 Does the entropy increase?**
See Exercise 96.

97 Use the general definition of efficiency of a heat engine, $\epsilon = W/Q_{in}$, to derive the Carnot efficiency. [*Hint:* Consider the cycle in terms of entropy, and see Fig. 12.13b.]

98 A gas is enclosed in a piston-cylinder arrangement with a diameter of 24.0 cm. Heat is slowly added to the system while the pressure is maintained at 1.00 atm. During the process, the piston moves 6.00 cm. (a) What type of process is this? (b) If the quantity of heat transferred to the system during the expansion is 0.100 kcal, what is the change in the internal energy of the gas?

99 Sketch the projection of the pVT surface for water in Fig. 12.14 on the pT plane and compare it with that for CO_2 in Fig. 12.1. What do the pT diagrams tell you about the freezing points of the substances as a function of temperature?

100 A heat engine operates between a reservoir at 30°C and one at 275°C. What is the maximum theoretical efficiency of the engine? Could this ever be achieved?

101 In a piston-cylinder arrangement 2.0 L of water at room temperature slowly receive 3.5 kcal of heat with the piston locked in place. (a) What is the change in the internal energy of the water? (b) What is the final temperature of the water? (Neglect pressure effects.)

102 Suppose that two automobiles with equal masses of 1.5×10^3 kg are both traveling at 60 km/h when they have a head-on, totally inelastic collision. Estimate the change in entropy of this process.

103 The energy or useful work lost as a result of a change in the entropy of the universe is given by $W = T_{cold}\Delta S_u$, where T_{cold} is the temperature of the cold reservoir for the process. (a) Show that for heat (Q) transferred from a high-temperature reservoir to a low-temperature reservoir $W = Q[1 - (T_{cold}/T_{hot})]$. (b) What is the quantity within the brackets?

104 A half-liter of ethyl alcohol at 18.0°C is mixed with a half-liter of water at 25.0°C. What is the approximate change in entropy once the mixture has reached thermal equilibrium? (Consider the system to be thermally isolated.)

105 A Carnot engine operates between temperature reservoirs at 250°C and 80°C. During each cycle, it absorbs 3.0×10^4 J of heat from the high-temperature reservoir. (a) What is the work output of this ideal engine? (b) How much heat does it reject to the low-temperature reservoir each cycle?

106 An ideal gas in a cylinder expands adiabatically while doing 2.5×10^3 J of work in moving the piston. What is the change in the internal energy of the gas? Does the temperature of the gas increase or decrease?

107 In a thermodynamic process, 2.50×10^3 J of heat is transferred to a system, and 6.00×10^2 J of work is done on the system. What is the change in the internal energy of the system?

108 During an isobaric process at 1 atm of pressure, a quantity of gas absorbs 125 cal of heat while expanding its volume from 1.0 L to 1.2 L. What is the change in the internal energy of the system?

13 Vibrations and Waves

13.1 Simple Harmonic Motion

13.2 Equations of Motion

13.3 Wave Motion

13.4 Wave Phenomena

13.5 Standing Waves and Resonance

INSIGHTS

■ Earthquakes and Seismology
■ Desirable and Undesirable Resonances

This is what a lot of people probably think of first when they hear the word "wave." We're all familiar with ocean waves, or their smaller relatives, the ripples that form on the surface of a lake or pond when something disturbs the surface. Yet in many ways, the waves that are most important to us, as well as most interesting to the physicist, are either invisible or don't look like waves. Sound, for example, is a wave. As a wave traveling through a fluid medium, it shares many properties with water waves. Perhaps most surprisingly, light is a wave—and not only visible light, but its invisible cousins that we know as radio waves, infrared and ultraviolet light, X-rays, and gamma rays. Whenever you peer through a microscope or put on a pair of glasses or look up at a rainbow, you are experiencing the consequences of the wave nature of light. Even par-

ticles turn out to have wave properties, though these become apparent only in the submicroscopic realm.

The specific kinds of waves mentioned above are discussed in more detail in subsequent chapters. Here, you will explore some more basic questions: what is a wave, and what properties do waves of *all* kinds have in common? You will find that to understand wave motion, it is necessary to study some other kinds of motion that at first might not seem closely related to waves. It turns out, however, that the motion of a weight bouncing up and down on a spring or the swinging of a pendulum in a grandfather clock provide important keys to an understanding of wave motion, whether it be that of an electron, a slinky, or the perfect wave that surfers dream about.

A vibration or oscillation involves back-and-forth motion. A swinging pendulum, for instance, oscillates back and forth, as does the dial on a bathroom scale as it comes to equilibrium to give a weight reading. Another example is a ball in a round-bottomed bowl. If the ball is displaced from its equilibrium position, it will roll back and forth, or oscillate, and finally come to rest at the equilibrium position, where its potential energy is lowest.

Recall from Chapter 8 that the ball must be oscillating about a point of stable equilibrium, for which there is a restoring force or torque. This is true in general for particles that undergo vibrating or oscillating motions. In a material medium, the restoring force is provided by intermolecular forces. If a molecule is disturbed, restoring forces exerted by its neighbors tend to return it to its original position, and it begins to oscillate. In so doing, it affects the adjacent particle, which is in turn set into oscillation.

In this situation, we can see that something happens at point A at time t_1, and this causes a similar happening at point B at t_2, and so on. This is referred to as propagation. (By analogy, think of how a rumor is propagated.) But what is propagated by the molecules in a material? A moment's thought should tell you it is energy. A single disturbance, such as you produce when you give the end of a stretched rope a quick shake, gives rise to what is referred to as a wave pulse. A continuous, repetitive disturbance gives rise to a continuous propagation of energy that we call wave motion.

In this chapter, you will learn how to describe vibrations and wave motions. This will be very important in the study of future topics, for wave phenomena are everywhere. The ripples on the surface of a pond or tank or the waves of the ocean are only the most familiar examples. Indeed, without waves, we would know very little about our world, for sound (Chapter 14) is a type of wave, and light (Chapter 20) is a type of wave. In fact, all electromagnetic radiations are waves—radio waves, microwaves, X-rays, and so on. In Chapter 28 you'll learn how even moving particles have wave-like properties. But first, we need to look at the basic descriptions of waves.

13.1 Simple Harmonic Motion

Objectives: To be able to **(a) describe simple harmonic motion, and (b) relate energy and speed in such motion.**

The motion of an oscillating particle depends on the restoring force that makes it go back and forth. It is convenient to begin to study such motion by considering the simplest type of force, which is one that is directly proportional to the displacement. One such force is the spring force, described by **Hooke's law,**

$$\mathbf{F} = -k\mathbf{x} \qquad (13.1)$$

where k is the spring constant. Recall from Chapter 5 that the minus sign indicates that the force is always in the opposite direction to the displacement ($\mathbf{F} = k(-\mathbf{x})$; that is, it always tends to restore the spring to its equilibrium position.

Suppose that a mass on a horizontal frictionless surface is connected to a spring as shown in •Fig. 13.1. When the mass is displaced to one side of its equilibrium position and released, it will move back and forth—that is, it will vibrate or oscillate. Here, we consider an oscillation or a vibration to be *periodic motion*, that is, a motion that repeats itself again and again along the same path. For linear oscillations, like those of a mass attached to a spring, the path may be back and forth or up and down. For the angular oscillation of a pendulum, the path is back and forth along a circular arc.

Motion under the influence of the type of force described by Hooke's law is called **simple harmonic motion (SHM)**, because the force is the simplest restoring force and because the motion can be described by harmonic functions (sines and

Note: Hooke's law is discussed in Section 5.2

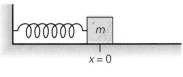

(a) Equilibrium

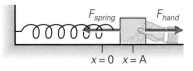

(b) $t = 0$ Just before release

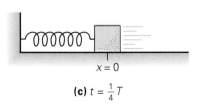

(c) $t = \frac{1}{4}T$

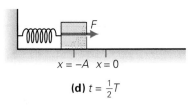

(d) $t = \frac{1}{2}T$

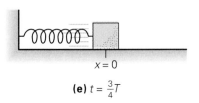

(e) $t = \frac{3}{4}T$

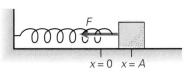

(f) $t = T$

FIGURE 13.1   Simple harmonic motion (SHM)
When displaced from its equilibrium position $x = 0$ and released **(a)**, **(b)**, the mass on a spring undergoes SHM (assuming no frictional losses). The time it takes to complete one cycle is the period of oscillation (T). At $t = T/4$ **(c)**, the mass is back at its equilibrium position; and at $t = T/2$ **(d)**, it is at $x = -A$. During the next half cycle, the motion is to the right; and at $t = T$, the mass is back at its initial ($t = 0$) starting position.

TABLE 13.1	Terms Used to Describe SHM

displacement—directed distance of an object ($\pm x$) from its equilibrium position.

amplitude (A)—magnitude of the maximum displacement or the maximum distance of an object from its equilibrium position.

period (T)—the time for one complete cycle of motion.

frequency (f)—the number of cycles per second (in Hz or 1/s, where $f = 1/T$).

cosines), as you will see later in the chapter. The directed distance of the object in SHM from its equilibrium position is its **displacement**. Note in the figure that the displacement can be either positive or negative ($+x$ or $-x$), which indicates direction. The maximum displacements are $+A$ and $-A$ (Fig. 13.1, parts b and d).

The magnitude of the maximum displacement, or the maximum distance of a mass or particle from its equilibrium position, is called the **amplitude** (A). It is a scalar quantity that expresses the distance for both extreme displacements.

Besides the amplitude, other important quantities for describing an oscillation are its period and frequency. The **period** (T) is the time for one complete cycle of motion. A cycle is a complete round trip, or motion through a complete oscillation. For example, if an object starts at $x = A$ (Fig. 13.1b), when it returns to position A (as in Fig. 13.1f), it will have completed one cycle in a time of one period. If an object were initially at $x = 0$ when disturbed, then its second return to this point would mark a cycle. In either case, the object would travel a distance of $4A$ in a cycle. Do you agree?

The **frequency** (f) is the number of cycles per second. The frequency and the period are related by

$$f = \frac{1}{T} \qquad \begin{array}{l} \textit{frequency} \\ \textit{and period} \end{array} \qquad (13.2)$$

The inverse relationship is reflected in the units: The period is the number of seconds per cycle, and the frequency is the number of cycles per second. For example, when $T = \frac{1}{2}$ s/cycle, $f = 2$ cycles/s.

The standard unit for frequency is the **hertz** (**Hz**), which is one cycle per second (cps).* From Eq. 13.2, frequency has the unit 1/s, or s^{-1}, since the period is a measure of time. Although cycle is not really a unit, you might find it convenient at times to express frequency in cycles/s to help with unit analysis. This is similar to the way the radian (rad) is used in the description of circular motion in Sections 7.1 and 7.2.

The preceding terms used to describe SHM are summarized in ●Table 13.1.

Energy and Speed of a Spring-Mass System in SHM

Recall from Chapter 5 that the potential energy stored in a stretched or compressed spring at a particular location is given by

$$U = \tfrac{1}{2}kx^2 \qquad (13.3)$$

The *change* in potential energy of a mass oscillating on a spring is related to the work done by the spring force, as illustrated and explained in ●Fig. 13.2. (Compare Fig. 5.5, Section 5.2.) A mass m oscillating on a spring has kinetic energy. Thus, the kinetic and potential energies together give the total mechanical energy of the system:

$$E = K + U = \tfrac{1}{2}mv^2 + \tfrac{1}{2}kx^2 \qquad (13.4)$$

*The unit is named for Heinrich Hertz (1857–1894), a German physicist and early investigator of electromagnetic waves.

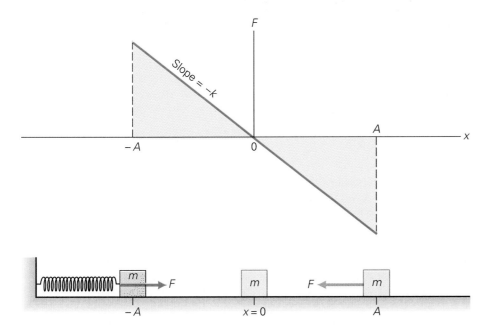

•FIGURE 13.2 Work and potential energy
The work done *by* the spring force on the oscillating mass is equal to the negative change in potential energy ($-\Delta U$). The magnitude of this change is equal to the area under the force (F) versus displacement (x) curve (where $\mathbf{F} = -k\mathbf{x}$). Note that the work can be both positive and negative. For example, in going from $x = 0$ to $x = A$, the work is negative (F and x in opposite directions), and U increases. But, going back from $x = A$ to $x = 0$, the work is positive (why?), and U decreases by the same amount. Thus, the overall ΔU is zero. (How about going from $x = 0$ to $x = -A$ and back—or half way back?)

When the mass is at one of its maximum displacements, $+A$ or $-A$, it is instantaneously at rest ($v = 0$). Thus, all the energy is potential energy at this time; that is,

$$E = \tfrac{1}{2}m(0)^2 + \tfrac{1}{2}kA^2 = \tfrac{1}{2}kA^2$$

or

$$E = \tfrac{1}{2}kA^2 \qquad \text{\textit{total energy in}} \atop \text{\textit{SHM of a spring}} \qquad (13.5)$$

where A is the magnitude of the maximum displacement or the amplitude. Neglecting any losses, the total energy is conserved. This is a special case of a more general result for SHM:

> The total energy of an object in simple harmonic motion is directly proportional to the square of the amplitude.

Equation 13.5 allows the velocity of a mass oscillating on a spring to be expressed as a function of position:

$$E = K + U \quad \text{or} \quad \tfrac{1}{2}kA^2 = \tfrac{1}{2}mv^2 + \tfrac{1}{2}kx^2$$

Then

$$v^2 = \frac{k}{m}(A^2 - x^2)$$

and

$$v = \pm\sqrt{\frac{k}{m}(A^2 - x^2)} \qquad (13.6) \qquad \textbf{Velocity of a particle in SHM}$$

where the $\pm$ indicates direction. Note that at $x = \pm A$, the velocity is zero and the mass is instantaneously at rest at its maximum displacement positions.

Also note that when the oscillating mass passes through the origin, or equilibrium position ($x = 0$), the potential energy is zero. At that instant, all the energy is kinetic and the mass is traveling at its maximum speed, v_{max}. (See Fig. 13.1c and e.) The energy expression for this case is

$$E = \tfrac{1}{2}kA^2 = \tfrac{1}{2}mv_{\text{max}}^2$$

and

$$v_{\text{max}} = \sqrt{\frac{k}{m}}(A) \qquad (13.7)$$

Note: This discussion will be limited to light springs, the mass of which can be considered negligible.

EXAMPLE 13.1 ■ A BLOCK AND A SPRING: SIMPLE HARMONIC MOTION

A block with a mass of 0.25 kg sitting on a frictionless surface is connected to a light spring that has a spring constant of 180 N/m (see Fig. 13.1). If the block is displaced 15 cm from its equilibrium position and released, what are (a) the total energy of the system, and (b) the speed of the block when it is 10 cm from its equilibrium position?

Solution. First we sort out and list the given data, as usual, along with determining what it is we are to find.

Given: $m = 0.25$ kg   *Find:*   (a) E (total energy)
        $k = 180$ N/m   (b) v (speed)
        $A = 15$ cm $= 0.15$ m
        $x = 10$ cm $= 0.10$ m

(a) The total energy is given directly by Eq. 13.5:
$$E = \tfrac{1}{2}kA^2 = \tfrac{1}{2}(180 \text{ N/m})(0.15 \text{ m})^2 = 2.0 \text{ J}$$

(b) The instantaneous speed of the block at a distance of 10 cm from the equilibrium position is given by Eq. 13.6 without directional signs:
$$v = \sqrt{\frac{k}{m}(A^2 - x^2)} = \sqrt{\frac{180 \text{ N/m}}{0.25 \text{ kg}}[(0.15 \text{ m})^2 - (0.10 \text{ m})^2]}$$
$$= \sqrt{9.0 \text{ m}^2/\text{s}^2}$$
$$= 3.0 \text{ m/s}$$

Follow-up Exercise. In part (b) of this Example, the block at $x = 10$ cm is at $\tfrac{2}{3}$ or 67% of its maximum displacement. Is its speed at that position therefore 67% of its maximum speed?

EXAMPLE 13.2 ■ THE SPRING CONSTANT: EXPERIMENTAL DETERMINATION

When a 0.50-kg mass is suspended from a spring, the spring stretches a distance of 10 cm (●Fig. 13.3a). (a) What is the spring constant of the spring? (b) The mass is then pulled down another 5.0 cm and released. What is the highest position of the oscillating mass?

Given: $m = 0.50$ kg   *Find:*   (a) k (spring constant)
        $y_o = 10$ cm $= 0.10$ m   (b) A (amplitude)
        $y = -5.0$ cm $= -0.050$ m

●FIGURE 13.3 **Determination of the spring constant**
(a) When a mass suspended on a spring is in equilibrium, the two forces on it cancel: $F_{sp} = w$, or $ky_o = mg$. Thus the spring constant k can easily be computed: $k = mg/y_o$.
(b) The zero reference point of SHM of a mass suspended on a spring is conveniently taken as the new equilibrium position, as the motion is symmetric about this point. See Example 13.2.

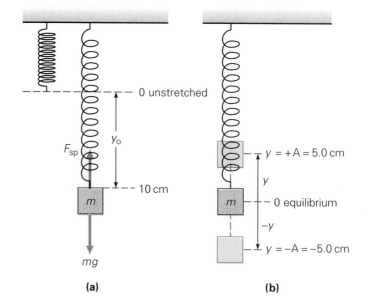

Since this oscillation is vertical, the displacement is designated by y, the coordinate commonly used for this direction.

(a) Suspending a mass on a spring is a common method of determining k. When the suspended mass and the stretched spring are in equilibrium (Fig. 13.3a), the weight of the mass and the spring force are equal and opposite. Then, equating their magnitudes,

$$F = w$$

or

$$ky_o = mg$$

Thus,

$$k = \frac{mg}{y_o} = \frac{(0.50 \text{ kg})(9.8 \text{ m/s}^2)}{0.10 \text{ m}} = 49 \text{ N/m}$$

(b) Once set into motion, the mass oscillates up and down through the equilibrium position. Since the motion is symmetric about this point, it is designated as the zero reference point of the oscillation (•Fig. 13.3b). The initial displacement is the negative amplitude ($-A$) of the oscillation, so the highest position of the mass will be 5.0 cm above the equilibrium position ($+A$).

It should be noted that as the mass oscillates up and down there is also a change in potential energy. Similar to a simple pendulum, there is a constant interchange of kinetic energy into potential energy, and vice versa. As a result, the total energy of the oscillating mass is still $E = \frac{1}{2}kA^2$.

Follow-up Exercise. What is the total energy of the oscillating system in this Example referenced to the equilibrium position? (Neglect gravitational potential energy.)

13.2 Equations of Motion

Objectives: To be able to (a) write the equation of motion for SHM, and (b) explain what is meant by phase and phase differences.

We refer to the **equation of motion** for an object or particle as that equation which gives its position as a function of time. For example, the equation of motion with a constant linear acceleration is $x = v_o t + \frac{1}{2}at^2$, where v_o is the initial velocity (Chapter 2). However, the acceleration is not constant for simple harmonic motion, so the kinematic equations of Chapter 2 do not apply to this case.

The equation of motion for an object in simple harmonic motion can be derived using a relationship between simple harmonic and uniform circular motions. SHM can be simulated by a component of uniform circular motion, as illustrated in •Fig. 13.4. As the illuminated object moves in uniform circular motion

•**FIGURE 13.4 Reference circle for horizontal motion**
(a) The shadow of the object in uniform circular motion has the same horizontal motion as the mass on the spring in simple harmonic motion. **(b)** The motion can thus be described by $x = A \cos \theta = A \cos \omega t$ (assuming $x = A$ at $t = 0$).

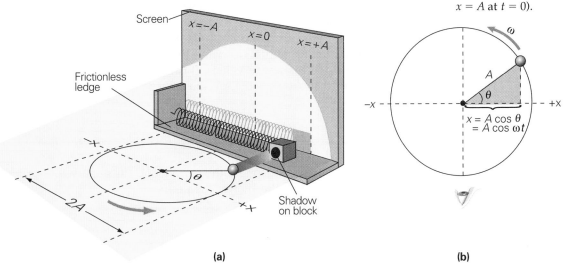

(a)

(b)

(with constant angular speed ω) in a horizontal plane, its shadow moves back and forth horizontally, following the same path as the mass on the spring, which is in simple harmonic motion. Since the shadow and the mass have the same position at any time, it follows that the equation of *horizontal* motion for the object in circular motion is the same as the equation of motion for horizontally oscillating mass on the spring.

From the reference circle in Fig. 13.4b the x coordinate (position) of the object is given by

$$x = A \cos \theta$$

But the object moves with a constant angular velocity with a magnitude ω, and in terms of angular distance θ, assuming $\theta = 0$ at $t = 0$, we have $\theta = \omega t$, so

$$x = A \cos \omega t \tag{13.8}$$

Note: See Eq. 7.5, Section 7.2.

The angular speed ω (in rad/s) is sometimes called the angular frequency, since $\omega = 2\pi f$, where f is the frequency of revolution or rotation (Section 7.2). Note in the figure that this is the same as the frequency of oscillation of the mass on the spring. Thus,

$$x = A \cos 2\pi f t = A \cos \frac{2\pi t}{T} \tag{13.9}$$

where $f = 1/T$.

Different forms of the equation of motion

Equations 13.8 and 13.9 give three equivalent forms of the equation of motion for a mass in simple harmonic motion. Any one of them may be used for convenience, depending on the known parameters. (Recall how the kinematic equations in Chapter 2 were expressed in different or combined forms for convenience.) For example, suppose you are given the time t in terms of the period T—say $t_o = 0$, $t_1 = T/2$, and $t_3 = T$—and asked to find the position of an object in SHM at these times. In this case, it is convenient to use Eq. 13.9, and

$$
\begin{aligned}
t_o &= 0 & x_o &= A \cos 2\pi(0)/T = A \cos 0 = A \\
t_1 &= \frac{T}{2} & x_1 &= A \cos 2\pi(T/2)/T = A \cos \pi = -A \\
t_2 &= T & x_2 &= A \cos 2\pi T/T = A \cos 2\pi = A
\end{aligned}
\tag{13.10}
$$

Hence, the results tell us that the object was initially at (released from) $x = A$. One-half period later it was at $x = -A$ or the opposite extreme of its oscillation; and at a time of one period (T), it was back where it started, which is to be expected since the motion is periodic.

The period of a mass oscillating on a spring can be expressed in terms of the system's parameters m and k. These are general parameters—that is, we can experimentally select different masses (m) and springs with different constants (k). Our algebraic equation of motion describes the motion of any combination once you know or select the numerical values. It can be shown using a reference circle as in Fig. 13.4b that the period for a mass oscillating in SHM on a spring is given by

$$T = 2\pi \sqrt{\frac{m}{k}} \qquad \textit{period of mass oscillating on a spring} \tag{13.11}$$

Thus, the greater the mass, the longer the period; and the greater the spring constant (the stiffer the spring), the shorter the period.

Since $f = 1/T$,

$$f = \frac{1}{2\pi} \sqrt{\frac{k}{m}} \qquad \textit{frequency of mass oscillating on a spring} \tag{13.12}$$

Thus, the greater the spring constant (the stiffer the spring), the faster and more frequently the spring vibrates, as you might expect.

Also, note that since $\omega = 2\pi f$, we may write

$$\omega = \sqrt{\frac{k}{m}} \qquad \text{\textit{anguluar frequency of mass}} \atop \text{\textit{oscillating on a spring}} \qquad (13.13)$$

As another example, a simple pendulum (with no energy losses) undergoes simple harmonic motion for small angles of oscillation. The equation for the period of a pendulum in simple harmonic motion is similar in form to that for a mass oscillating on a spring. It can be shown that the period of a simple pendulum oscillating through a small angle ($\theta \leq 10°$) is given to a good approximation by

$$T = 2\pi \sqrt{\frac{L}{g}} \qquad \text{\textit{period of a}} \atop \text{\textit{simple pendulum}} \qquad (13.14)$$

where L is the length of the pendulum and g is the acceleration due to gravity.

Note that the pendulum period is independent of the mass of the bob. Can you explain why? Think about what supplies the restoring force for the pendulum's oscillations. As you might expect, it is gravity. Hence, we would expect the acceleration (along with the velocity and period) to be independent of mass. That is, the gravitational force automatically provides the same acceleration to different bob masses of pendulums with the same length. The greater the mass of the bob, the greater the gravitational force on it, but also the greater its inertia. These effects cancel each other out, so that the acceleration doesn't vary with the mass. Similar effects occur in free fall (Chapter 2) and for blocks sliding and cylinders rolling (Chapters 4 and 8) down inclines. This is a general result when dealing with gravitational forces.

EXAMPLE 13.3 ■ AN OSCILLATING MASS: APPLYING THE EQUATION OF MOTION

A mass on a spring oscillates horizontally on a frictionless surface with an amplitude of 15 cm, a frequency of 0.20 Hz, and an equation of motion as given in Eq. 13.8. (a) What is the displacement of the mass at $t = 3.1$ s? (b) How many oscillations has it made since $t = 0$?

Solution.

Given: $A = 15$ cm $= 0.15$ m **Find:** (a) x (displacement)
$\qquad\quad f = 0.20$ Hz $\qquad\qquad\qquad$ (b) n (number of oscillations)
$\qquad\quad x = A \cos \omega t$ (Eq. 13.8)
$\qquad\quad t = 3.1$ s

(a) First, since we are given the frequency f, it is convenient to use the equation of motion in the form $x = A \cos 2\pi f t$ (Eq. 13.9), where $\omega = 2\pi f$. At $t = 0$, we have $x = A$, so initially the mass is released from its positive amplitude distance. Then, at $t = 3.1$ s,

$$x = A \cos 2\pi f t$$
$$= (0.15 \text{ m}) \cos \left[2\pi(0.20 \text{ s}^{-1})(3.1 \text{ s})\right] = (0.15 \text{ m}) \cos (3.9 \text{ rad}) = -0.11 \text{ m}$$

and the mass is 0.11 m on the other side of its equilibrium position.

(b) The number of oscillations (cycles) is equal to the product of the frequency (cycle/s) and the elapsed time (s), which are both given:

$$n = ft = (0.20 \text{ cycle/s})(3.1 \text{ s}) = 0.62 \text{ cycle}$$

Thus, at $t = 3.1$ s, the mass has traveled to its maximum negative displacement position (in 0.50 cycle) and is back at $x = -0.11$ m approaching its equilibrium position ($x = 0$).

Follow-up Exercise. Describe how the mass's postition varies with time in the course of its oscillation.

PROBLEM-SOLVING HINT

Note that in the part (a) calculation of Example 13.3, where we have cos (3.9), the angle is in radians, *not* degrees. Don't forget to set your calculator to radians (rather than degrees) when finding the value of a trigonometric function in equations for simple harmonic or circular motion.

EXAMPLE 13.4 ■ JEFFERSON'S PENDULUM: LENGTH AND PERIOD OF OSCILLATION

Thomas Jefferson once proposed that a standard reference for time be a simple pendulum with a period of one second. What would be the length of such a pendulum?

Solution.

Given: $T = 1.00$ s   *Find:* L (pendulum length)

The length of the simple pendulum can be found directly from Eq. 13.14. Squaring both sides of the equation and solving for L gives

$$L = \frac{T^2 g}{4\pi^2} = \frac{(1.00 \text{ s})^2 (9.80 \text{ m/s}^2)}{4\pi^2} = 0.248 \text{ m}$$

This standard was not adopted. (Can you think of any reasons?)

Follow-up Exercise. A simple pendulum can also be used to measure the acceleration due to gravity. Describe how this can be done experimentally.

CONCEPTUAL EXAMPLE 13.5 ■ GOING UP! A PENDULUM IN AN ELEVATOR

A simple pendulum attached to the ceiling of a stationary elevator has a period T. If the elevator accelerates upward and the period is measured, it will be (a) less than T, (b) greater than T, (c) equal to T. *Clearly establish the reasoning and physical principle(s) used in determining your answer before checking it below. That is,* **why** *did you select your answer?*

Reasoning and Answer. From Eq. 13.14, we see that what might vary here is the acceleration of the pendulum, since L is constant. In the stationary elevator, the period is inversely proportional to the square root of g. However, in an accelerating elevator, the effective or net acceleration is different.

This can be shown using an approach similar to that used in explaining apparent weightlessness in Chapter 7 (see Fig. 7.23). For a person on a scale in an upward accelerating elevator, we can write by Newton's second law, $R - mg = ma$, where R is the upward reaction force of the scale on the person, and is equal to the scale or weight reading. Then, the apparent weight w' is

$$w' = R = m(g + a)$$

So, in the upward accelerating elevator, the *effective* value of the acceleration due to gravity is $g' = g + a$. Putting this into Eq. 13.14, we see that the period of the pendulum in the upward accelerating elevator would be decreased, hence the answer is (a).

Follow-up Exercise. Suppose the elevator were in free fall. What would be the period of the pendulum in this case?

A vertical reference circle can also be used to describe simple harmonic motion, as illustrated in •Fig. 13.5 for a suspended mass oscillating on a spring. Suppose that the mass initially started its upward motion from its equilibrium position (having been given an upward push). In this case, the vertical displacement is given by $y = A \sin \theta$. A development similar to that for the horizontal case leads to this equation of motion:

$$y = A \sin \omega t$$

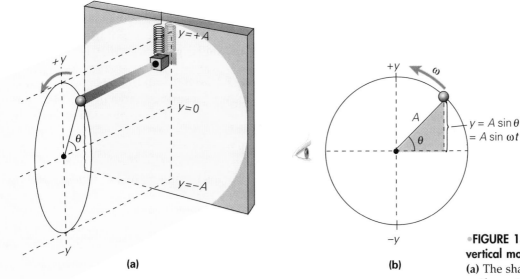

•FIGURE 13.5 Reference circle for vertical motion
(a) The shadow of the object in uniform circular motion has the same vertical motion as the mass oscillating on the spring in simple harmonic motion. (b) The motion can thus be described by $y = A \sin \theta = A \sin \omega t$ (assuming $y = 0$ at $t = 0$).

Notice initially at $t = 0$, the mass was at its equilibrium point ($y = 0$).

However, suppose the mass starts initially from its maximum positive displacement position. As shown in •Fig. 13.6, the resulting curve is a cosine, and

$$y = A \cos \omega t$$

Note that this equation correctly describes the initial conditions, that is, $y = +A$ at $t = 0$.

Thus, the equation of motion for an oscillating mass may be either a sine or a cosine function. Both of these functions are referred to as being *sinusoidal*. That is, simple harmonic motion is described by a sinusoidal function of time.

Initial Conditions and Phase

You may be wondering how to decide whether to use a sine or cosine function to describe a particular case of simple harmonic motion. In general, the form of the function is determined by the initial displacement of the mass, or the *initial condition* of the system. This initial condition is the value of the displacement at $t = 0$ and, along with the velocity, tells how the system is initially set into motion.

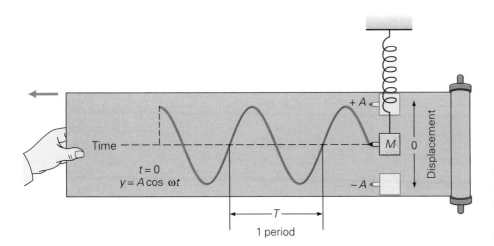

•FIGURE 13.6 Sinusoidal equation of motion

As time passes, the oscillating mass traces out a sinusoidal curve on the moving paper. In this case, $y = A \cos \omega t$.

Let's take four simple examples to illustrate this. As can be seen, if an object in SHM has an initial displacement of $y = 0$ at $t = 0$ and moves initially upward, the equation of motion is given by $y = A \sin \omega t$ (see curve in •Fig. 13.7a). Note that $y = A \cos \omega t$ does not satisfy the initial condition, that is, $y = A \cos \omega t = A \cos \omega(0) = A$, since $\cos 0 = 1$.

Suppose that the object is initially released ($t = 0$) from its positive amplitude position ($+A$), as in the case of the mass on a spring shown in Fig. 13.6. Here, the equation of motion is $y = A \cos \omega t$ (•Fig. 13.7b). This expression satisfies the initial condition: $y = A \cos \omega(0) = A$.

The other two simple cases are $y = 0$ at $t = 0$, with initial downward motion (for a mass on a spring), or in the negative direction (for horizontal SHM); and $y = -A$ at $t = 0$, meaning that the object is initially at its negative amplitude position. These motions are described by $y = -A \sin \omega t$ and $y = -A \cos \omega t$, respectively, as illustrated in •Fig. 13.7c and 13.7d.

Note in Fig. 13.7 that if the curves are completed to the horizontal axis (dashed lines in figure), they all have the same shape, but have been "shifted," so to speak.

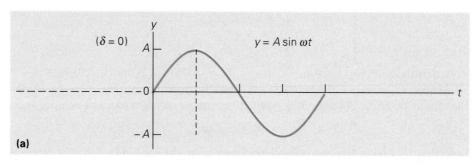

(a)

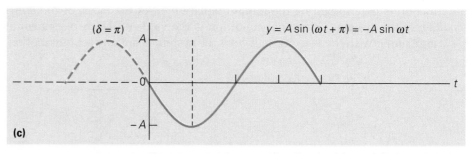

(b)

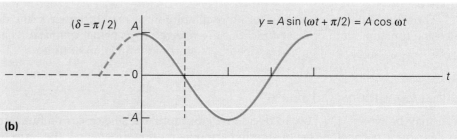

(c)

•FIGURE 13.7 **Phase differences**
With the general equation $y = A \sin (\omega t + \delta)$, motions are described by either sine or cosine terms for the phase constants shown. The initial displacement determines δ. Each curve is 90° (or $\pi/2$ rad) out of phase with the preceding one. Note that this is equivalent to shifting the curve a quarter of a cycle.

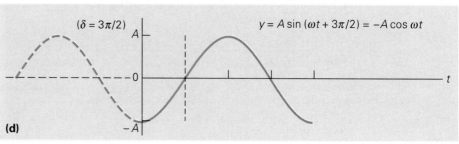

(d)

We describe this change as a shift in *phase*. For the general case, we may write the equation of simple harmonic motion as

$$y = A \sin (\omega t + \delta) \qquad (13.15)$$

where $(\omega t + \delta)$ is the *phase angle* and δ is the **phase constant**. The phase constant essentially matches the appropriate sinusoidal function to the motion.

The phase constants for the cases in Fig. 13.7 are given in the figure. For example, for $\delta = 90°$ (or $\pi/2$ rad), the equation of motion is $y = A \cos \omega t$, and a mass oscillating up and down on a spring would be initially released at the $+A$ position. You probably recognize that the curve for the $\delta = 90°$ is a cosine. This can be shown algebraically using the trigonometric double angle formula: $\sin (a + b) = \sin a \cos b + \cos a \sin b$. Then, with $\delta = 90°$,

$$y = A \sin (\omega t + 90°) = A\big[\sin \omega t \cos 90° + \cos \omega t \sin 90°\big] = A \cos \omega t$$

since $\cos 90° = 0$ and $\sin 90° = 1$.

The curves for $\delta = 0$ and $\delta = 90°$ (or $\pi/2$ rad) are said to be $90°$ out of phase, or shifted by a quarter cycle, with respect to one another. Notice in Fig. 13.7 that the curve for $\delta = 90°$ has essentially been shifted to the left by a quarter cycle ($90°$) from the curve for $\delta = 0$. The cases for $\delta = 180°$ (or π rad) and $\delta = 270°$ (or $3\pi/2$ rad) are each shifted an additional $90°$ out of phase.

Two masses oscillating with the same frequency in phase (having the same δ) will oscillate together. Two that oscillate completely out of phase ($180°$ difference in δ) will always be going in opposite directions or be at opposite maximum displacements.

For each of the four special cases in Fig. 13.7, the equation of motion is either a sine or cosine curve. If the initial displacement is not zero or $\pm A$, then δ is not a multiple of $90°$, and things get a bit more complicated. Eq. 13.15 still applies, but the double-angle formula above gives an equation of motion with both sine and cosine terms.

PROBLEM-SOLVING HINT

Keep in mind that the phase angle $(\omega t + \delta)$ is usually expressed in radians. For example, if $y = \sin [0.50t + \pi]$ then at $t = 20$ s, we have $y = \sin [(0.50)(20) + \pi] = \sin 13$, or the sine of 13 rad. If you are using a calculator for such computations, again make sure it is in the "rad" mode and not the "deg" mode.

See Demonstration 10 for a pictorial view of a sinusoidal oscillation and phase.

Velocity and Acceleration in SHM

Expressions for the velocity and acceleration of an object in SHM can be easily derived using energy and force considerations. We have already derived an expression (Eq. 13.6) for the velocity of a mass oscillating on a spring in terms of the displacement. We can now express the velocity as a function of time. For the case of vertical simple harmonic motion, with the displacement $y = A \sin \omega t$, we can rewrite Eq. 13.6 as

$$v = \sqrt{\frac{k}{m}(A^2 - y^2)} = \sqrt{\frac{k}{m}(A^2 - A^2 \sin^2 \omega t)} = A\sqrt{\frac{k}{m}}\sqrt{1 - \sin^2 \omega t}$$

Since $\omega = \sqrt{k/m}$ (Eq. 13.13) and $\cos \theta = \sqrt{1 - \sin^2\theta}$ (see Appendix I), we can write this expression as

$$v = A\omega \cos \omega t \qquad (13.16)$$

Note: Maximum velocity $v = \pm \omega A$.

DEMONSTRATION 10 ■ Simple Harmonic Motion (SHM) and Sinusoidal Oscillation

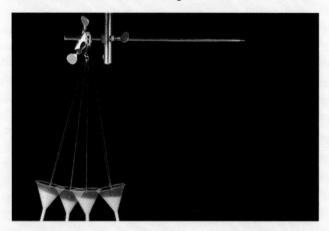

(a) A salt-filled funnel oscillates from a bi-filar suspension.

(b) The salt falls on a black-painted poster board that will be pulled in a direction perpendicular to the plane of the funnel's oscillation.

(c) Away we go.

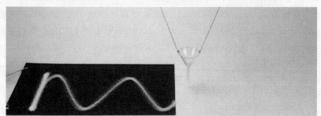

(d) The salt trail traces out a plot of displacement versus time, or $y = A \sin(\omega t + \delta)$. Note that in this case the phase constant is about $\delta = 90°$, and $y = A \cos \omega t$. (Why?)

A demonstration to show that SHM can be represented by a sinusoidal function. A "graph" of the function is generated with an analog of a strip chart recorder.

where the directional signs are given by the cosine function. Using Newton's second law to find the acceleration with the spring force $F = -ky$, we have

$$a = \frac{F}{m} = \frac{-ky}{m} = \frac{-k}{m} A \sin \omega t$$

and since $\omega = \sqrt{k/m}$,

$$a = -\omega^2 A \sin \omega t = -\omega^2 y \qquad (13.17)$$

Note that the functions for the velocity and acceleration are out of phase with that for the displacement (shown in Fig. 13.7). Since the velocity is 90° out of phase with the displacement, the velocity is greatest ($v = \pm\omega A$) when the oscillating mass is at (passing through) its zero equilibrium position. The acceleration is 180° out of phase with the displacement (as indicated by the minus sign on the right-hand side of Eq. 13.17). Therefore, the magnitude of the acceleration is a maximum ($a = \pm\omega^2 A$) when the displacement is a maximum, or when the mass is at an amplitude position. At any position except the equilibrium position, the directional sign of the acceleration is opposite that of the displacement, as it should be for an acceleration resulting from a restoring force. At the equilibrium position, the displacement and acceleration are both zero. (Can you see why?)

Also note that the acceleration for SHM is not constant with time. Hence, the kinematic equations of Chapter 2 cannot be used since they are for constant accelerations.

Note: Maximum acceleration $a = \pm\omega^2 A$.

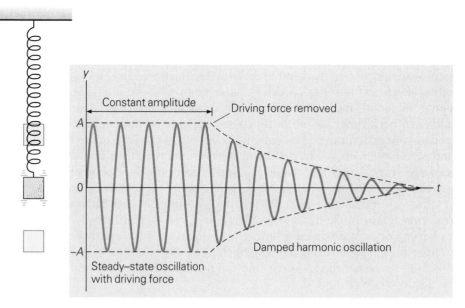

•FIGURE 13.8 Damped harmonic motion

•FIGURE 13.8 Damped harmonic motion
When a driving force adds energy to a system equal to its losses, the oscillation is steady with a constant amplitude. When the driving force is removed, the oscillations decay (that is, are damped), and the amplitude decreases exponentially with time.

Damped Harmonic Motion

Simple harmonic motion with a constant amplitude implies that there are no losses of energy, but in practical applications there are always some frictional losses. Therefore, to maintain a constant-amplitude motion, energy must be added to the system by some driving force (•Fig. 13.8). Without a driving force, the amplitude and the energy of an oscillator decrease with time, giving rise to **damped harmonic motion**. If the frictional force is proportional to the velocity of the oscillator, the motion changes over time as illustrated in Fig. 13.8. The time required for the oscillations to cease, or be damped out, depends on the magnitude of the damping force.

In many applications involving continuous periodic motion, damping is unwanted and necessitates energy input. However, in some instances, damping is desirable. For example, the dial in a spring-operated bathroom scale oscillates briefly before stopping at the weight. If not properly damped, these oscillations would continue for some time and you would have to wait before you could read your weight. Damping is also required for shock absorbers on automobiles and needle indicators on instruments measuring electrical quantities.

13.3 Wave Motion

Objectives: **To be able to (a) describe wave motion in terms of various parameters, and (b) identify different types of waves.**

The world is full of waves of various types—some examples are water waves, shock waves, sound waves, waves generated by earthquakes, and light waves. Any type of wave results from a disturbance. In this chapter we will be concerned with mechanical waves, or those that are propagated in some medium. (Light waves, which do not require a propagating medium, will be considered in more detail in later chapters.)

When a medium is disturbed, energy is imparted to it. Suppose that energy is added to a material mechanically, such as by a blow or (in the case of a gas) by compression. This sets some of the particles vibrating. Because the particles are linked by intermolecular forces, the oscillation of each particle affects that of its neighbors. The added energy propagates, or spreads, by means of interactions between the particles of the medium. An analogy for this process is shown in •Fig. 13.9, where the "particles" are dominoes. As each domino falls, it topples the one next to it. Thus energy is transferred from domino to domino, and the disturbance propagates through the medium.

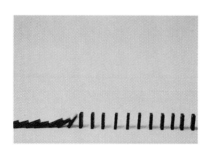

•FIGURE 13.9 Energy transfer
The propagation of a disturbance or a transfer of energy is seen in a row of falling dominos.

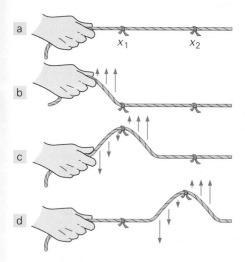

•FIGURE 13.10 Wave pulse
The hand disturbs the stretched rope in a quick up and down motion, and a wave pulse propagates along the rope. Note that the rope "particles" move up and down as the pulse passes. The energy in the pulse is *both* kinetic and potential (elastic and gravitational).

In this case, there is no restoring force between the dominoes, so they do not oscillate as do particles in a continuous material medium. Therefore the disturbance moves in space, but it does not repeat itself in time at any one location.

Similarly, if the end of a stretched rope is given a quick shake, the disturbance transfers energy from the hand to the rope, as illustrated in •Fig. 13.10. The forces acting between the rope "particles" cause them to move in response to the motion of the hand, and a *wave pulse* travels down the rope. Each "particle" goes up and then back down as the pulse passes by. This motion and that of the wave pulse propagation can be observed by tying pieces of ribbon onto the rope (at x_1 and x_2 in Fig. 13.10). As the disturbance passes point x_1, the ribbon rises and falls, as do the rope "particles." Later, the same occurs for the ribbon at x_2, which indicates that the disturbance energy is propagating or traveling along the rope.

In a continuous material medium, particles interact with their neighbors, and restoring forces cause them to oscillate when they are disturbed. Thus any disturbance not only propagates through space but may be repeated over and over in time at each place along the way. Such a regular, rhythmic disturbance in both time and space is called a **wave**, and the transfer of energy is said to take place by means of **wave motion**.

A continuous wave motion, or *periodic wave*, requires a constant disturbance from an oscillating source (•Fig. 13.11). In this case, the particles move up and down continuously. If the driving source is such that a constant amplitude is maintained (and the restoring force has the form of Hooke's law), the particle motion can be described as simple harmonic motion.

Such wave motion will have sinusoidal forms (sine or cosine) in both time and space. Being sinusoidal in space means that if you took a photograph of the wave at any instant (freezing it in time), you would see a sinusoidal waveform (such as one of the curves in Fig. 13.11). However, if you looked at a single point in space as a wave passed by, you would see a particle of the medium oscillating up and down sinusoidally with time, like the mass on a spring discussed in Section 13.2. (For example, imagine looking through a thin slit at the moving paper in Fig. 13.6. The wave trace would be seen rising and falling like a particle.)

Wave Characteristics

Specific characteristics of sinusoidal waves are used to describe them. As for a particle in simple harmonic motion, the *amplitude* (A) of a wave is the magnitude of the maximum displacement or the maximum distance from the particle equilibrium position (Fig. 13.11). This corresponds to the height of a crest or the depth of a trough. As in SHM, the total energy transported by a wave is proportional to the square of its amplitude ($E \propto A^2$).

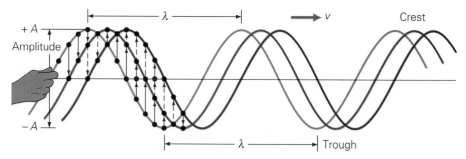

•FIGURE 13.11 Periodic wave
A continuous disturbance can set up a sinusoidal wave in stretched rope or string, and the wave travels down the rope with a wave speed *v*. Note that the rope "particles" oscillate vertically in simple harmonic motion. The distance between two successive points that are in phase (e.g., at two crests) on the waveform is the wavelength of the wave.

The distance between two successive crests (or troughs) is called the **wavelength** (λ) (Fig. 13.11). Actually, it is the distance between any two successive particles that are in phase (that is, at identical points on the wave form). The crest and trough positions are usually used for convenience. Note that a wavelength corresponds spatially to one cycle.

The *frequency* (f) of a wave is the number of cycles per second, that is, the number of complete waveforms, or wavelengths, that pass by a given point during a second. Keep in mind that the wave is moving. The period $T = 1/f$ is the time for one complete waveform (a wavelength) to pass by a given point.

Since a wave moves, it has a **wave speed** (or velocity if direction is specified). You should be able to convince yourself that the wave (or a particular point, such as a crest) travels a distance of one wavelength λ in a time of one period T. Then, since $v = d/t$,

$$v = \frac{\lambda}{T} = \lambda f \qquad \textit{wave speed} \qquad (13.18)$$

Note that the dimensions are correct (length/time). In general, the wave speed depends on the nature of the medium.

EXAMPLE 13.6 ■ DOCK OF THE BAY: FINDING WAVE SPEED

A person on a pier observes incoming waves that have a sinusoidal form with a distance of 1.6 m between the crests. If a wave laps against the pier every 4.0 s, what are (a) the frequency and (b) the speed of the waves?

Solution. The distance between crests is the wavelength, so we have

Given: $\lambda = 1.6$ m   *Find:* (a) f (frequency)
        $T = 4.0$ s   (b) v (wave speed)

(a) The lapping indicates the arrival of a wave crest, so 4.0 s is the period of the wave—the time it takes to travel one wavelength (the crest-to-crest distance). Then

$$f = \frac{1}{T} = \frac{1}{4.0 \text{ s}} = 0.25 \text{ s}^{-1} = 0.25 \text{ Hz}$$

(b) The frequency or the period can be used in Eq. 13.18 to find the wave speed:

$$v = \lambda f = (1.6 \text{ m})(0.25 \text{ s}^{-1}) = 0.40 \text{ m/s}$$

Alternatively,

$$v = \frac{\lambda}{T} = \frac{1.6 \text{ m}}{4.0 \text{ s}} = 0.40 \text{ m/s}$$

Follow-up Exercise. On another day, the person on the pier times the speed of a sinusoidal wave crest to be 0.25 m/s. (a) How far does the crest travel in 2.0 s? (b) Can you find the wavelength and frequency of the wave?

Types of Waves

Waves may be divided into two types based on the direction of the particles' oscillations relative to the wave velocity. A **transverse wave** is one for which the particle motion is perpendicular to the direction of the wave velocity. The wave produced in a stretched string (Fig. 13.11) is an example of a transverse wave, as is the wave shown in ∎Fig. 13.12a. A transverse wave is sometimes called a *shear wave* because the disturbance supplies a force that tends to shear the medium. Shear waves can propagate only in solids, since a liquid or a gas cannot support a shear. That is, a liquid or a gas does not have sufficient restoring forces between its particles to propagate a transverse wave.

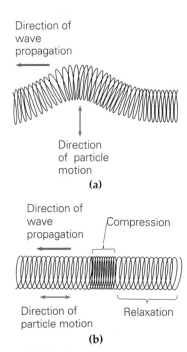

Direction of wave propagation

Direction of particle motion

(a)

Direction of wave propagation   Compression

Direction of particle motion   Relaxation

(b)

●**FIGURE 13.12 Transverse and longitudinal waves** (a) For a transverse wave, the particle motion is perpendicular to the direction of the wave velocity, as shown here in a spring for a wave moving to the left. (b) For a longitudinal wave, the particle motion is parallel to (or *along*) the direction of the wave velocity. Here a wave pulse also moves to the left.

Insight   Earthquakes and Seismology

The Earth's interior structure is still something of a mystery. The deepest mine shafts and drillings extend only a few kilometers into the Earth. Using waves to probe the Earth's structure is one way to investigate it further. Appropriately, waves generated by earthquakes have proved to be especially useful for this purpose. Seismology is the study of these so-called seismic waves.

FIGURE 1   The San Andreas Fault
A small section of the fault, which runs through the San Francisco Bay area as well as the less hospitable region of California shown here.

Earthquakes are caused by the sudden release of built-up stress along cracks and faults, such as the famous San Andreas Fault in California (Fig. 1). The geological theory of plate tectonics views the outer layer of the Earth as being a series of rigid plates, or huge slabs of rock, that are in very slow motion relative to one another. Stresses are continually being built up, particularly along boundaries between plates.

The energy from a stress-relieving disturbance propagates outward (seismic) waves. These are of two general types: surface waves and body waves. The surface waves, which move along the Earth's surface, account for most earthquake damage (Fig. 2). Body waves, as the name implies, travel through the Earth. There are both longitudinal and transverse vibrations. The compressional (longitudinal) waves are called P waves, and shear (transverse) waves are called S waves (see Fig. 3). The P and S stand for primary and secondary and indicate the waves' relative speeds (actually, their arrival times at monitoring stations). Primary waves travel through materials faster than do secondary waves and are detected first. An earthquake's rating on the Richter scale is related to the amplitude or energy of the seismic waves.

Seismic stations around the world monitor these waves with sensitive detecting instruments called seismographs (Fig. 4). From the data gathered, the paths of the waves through the Earth can be mapped, giving knowledge of the interior structure. The Earth's interior seems to be divided into three general regions: the crust, the mantle, and the core, which has

FIGURE 2   Bad vibrations
Earthquake damage caused by the major shock that struck Kobe, Japan in January 1995.

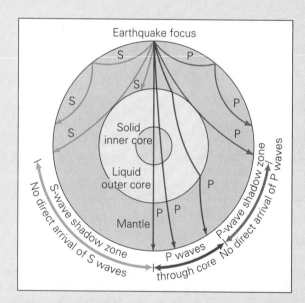

FIGURE 3   Earth waves
Earthquakes produce waves that travel through the Earth. Because transverse S waves are not detected on the opposite side of the Earth, scientists believe that part of the Earth's interior is a viscous liquid.

a solid inner part and a liquid outer part.* The locations of the boundaries of these regions are determined by the refraction, or bending, of the waves (see Section 13.4).

Longitudinal waves can travel through solids or liquids, but transverse waves can travel only through solids. When an earthquake occurs at a particular location, P waves are detected on the other side of the Earth and S waves are not (see Fig. 3). The absence of S waves in a shadow zone leads to the conclusion that the Earth must have a region near its center that is in the liquid phase. This region is a highly viscous metallic liquid, but definitely a liquid since it does not sup-

port a shear (transverse waves are not propagated). When the transmitted P waves enter and leave the liquid region, they are refracted (bent). This gives rise to a P wave shadow zone, which indicates that only the outer part of the core is liquid.

As you will learn in Chapter 18, the combination of a liquid outer core and rotation may be responsible for the Earth's magnetic field.

*The crust is in most places about 24–30 km (15–20 mi) thick; the mantle is 2900 km (1800 mi) thick; and the core has a radius of 3450 km (2150 mi). The solid inner core has a radius of about 1200 km (750 mi).

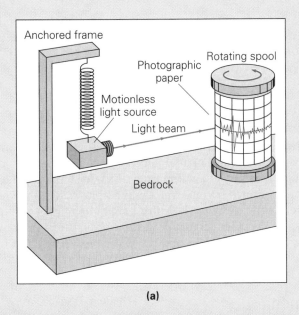

(a)

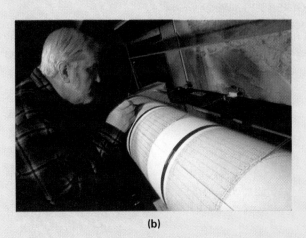

(b)

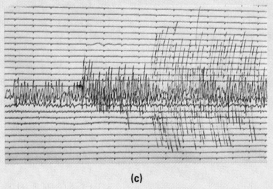

(c)

FIGURE 4   The seismograph
(a) The design of a simple seismograph. The device records the amplitude of the ground vibrations. The energy of the earthquake waves is proportional to the square of their amplitude. **(b)** A scientist of the U.S. Geological Survey monitors seismograph readings. **(c)** Seismograph trace from the destructive Loma Prieta, California, earthquake of 1989.

In a **longitudinal wave**, the particle oscillation is parallel to the direction of the wave velocity. A longitudinal wave may be produced in a stretched spring by moving the coils back and forth along the spring axis (Fig. 13.12b). Alternating pulses of compressions and relaxations move along the spring. A longitudinal wave is sometimes called a *compressional wave*.

Sound waves in air are another example of longitudinal waves. A periodic disturbance produces compressions in the air. The intervening relaxations are called rarefactions because the density of the air in these regions is reduced, or rarefied. Sound waves will be discussed in detail in Chapter 14.

Longitudinal waves can propagate in solids, liquids, and gases. All phases of matter can be compressed to some extent. The propagations of transverse and longitudinal waves in different media give information about the Earth's interior structure, as discussed in the Insight above.

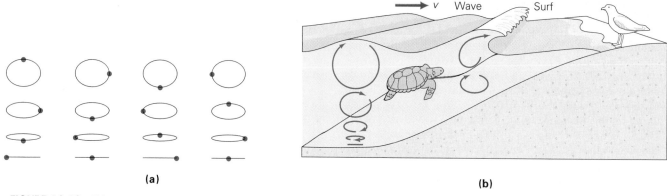

FIGURE 13.13 Water waves
Water waves are a combination of longitudinal and transverse motions. **(a)** At the surface, the water particles move in circles, but their motions become more longitudinal with depth. **(b)** When a wave approaches shore, the lower particles are forced into steeper paths until finally the wave breaks or falls over to form surf.

The sinusoidal profile of water waves might make you think that these are transverse waves. Actually, they reflect a combination of longitudinal and transverse motions (•Fig. 13.13). The particle motion may be nearly circular at the surface but becomes more elliptical with depth, eventually becoming longitudinal. A hundred meters or so below the surface of a large body of water, the wave disturbances have little effect. For example, a submarine at these depths is undisturbed by large waves on the ocean's surface. As a wave approaches shallower water near shore, the water particles have difficulty completing their elliptical paths. Finally, when the water becomes too shallow, the particles can no longer move through the bottom parts of their paths and the wave breaks. Its crest falls forward to form surf.

13.4   Wave Phenomena

Objective: To be able to explain various wave phenomena.

Among the phenomena common to all waves are interference, superposition, reflection, refraction, and diffraction.

Interference and Superposition

Strange as it may seem, when two or more waves meet, or pass through the same region of a medium, they pass through each other and proceed without being altered. While they are in the same region, the waves are said to be interfering.

What happens during interference, or what does the combined waveform look like? The relatively simple answer to this is given by the **principle of superposition**:

> At any time, the combined waveform of two or more interfering waves is given by the sum of the displacements of the individual waves at each point in the medium.

This principle is illustrated in •Fig. 13.14. The displacement of the combined waveform at any point is given by $y = y_1 + y_2$, where y_1 and y_2 are the displacements of the individual pulses at that point (directions are indicated by plus and minus signs). Interference, then, is the physical addition of waves.

In Fig. 13.14, the vertical displacements of the two pulses are in the same direction, and the amplitude of the combined waveform is greater than that of either pulse. This is called **constructive interference**. On the other hand, if one pulse

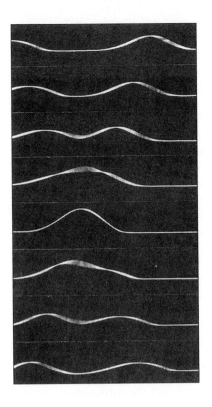

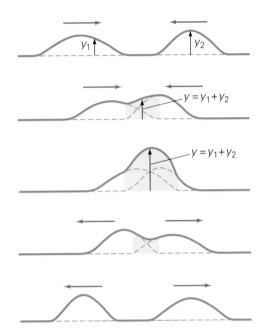

•FIGURE 13.14 Principle of superposition
When two waves meet, they interfere. The yellow tint marks the area where the two waves, moving in opposite directions, overlap and combine. The displacement at any point on the combined wave is equal to the sum of the displacements on the individual waves: $y = y_1 + y_2$.

has a negative displacement, the two pulses tend to cancel each other when they overlap, and the amplitude of the combined waveform is smaller than that of either pulse. This is called **destructive interference**.

Special cases of constructive and destructive interference are shown in •Fig. 13.15. These are shown for traveling wave pulses with the same width and amplitude. When these interfering waves are exactly in phase (crest coinciding with crest), the amplitude of the combined waveform is twice that of either individual wave. This is sometimes referred to as **total constructive interference**. When these interfering pulses are completely out of phase (180° difference, or crest coinciding with trough), the waveforms disappear; that is, the amplitude of the combined wave is zero. This is called **total destructive interference**.

The word "destructive" unfortunately tends to imply that the energy as well as the form of the waves is destroyed. This is not the case. At the point of total destructive interference, the wave energy is stored in the medium.

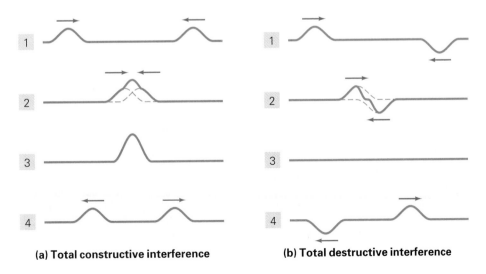

(a) Total constructive interference

(b) Total destructive interference

•FIGURE 13.15 Interference **(a)** When two waves of the same frequency and amplitude meet and are in phase, they interfere constructively (shown here for wave pulses). When the waves are exactly superimposed (3), total constructive interference occurs. **(b)** When the interfering waves are completely out of phase (180°) and are exactly superimposed (3), total destructive interference occurs.

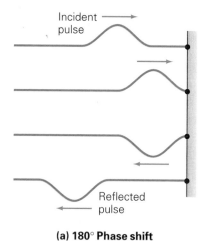

(a) 180° Phase shift

(b) No phase shift

•**FIGURE 13.16   Reflection**
(a) When a wave (pulse) in a string is reflected from a fixed boundary, the reflected wave is inverted, or undergoes a 180° phase shift. **(b)** If the string is free to move at the boundary, there is no phase shift of the reflected wave.

•**FIGURE 13.17   Diffraction**
Diffraction effects are greater when the opening (or object) is about the same size as or smaller than the wavelength of the waves, as shown here for water waves. **(a)** With an opening much larger than the wavelength of the plane water waves, diffraction is noticeable only near the edges. **(b)** With an opening about the same size as the wavelength of the waves, diffraction produces circular waves.

Reflection, Refraction, and Diffraction

Besides meeting other waves, waves can (and do) meet objects or a boundary with another medium. In such cases, several things may occur. One of these is reflection. **Reflection** occurs when a wave strikes an object or comes to a boundary of another medium and is at least partly diverted backward. An echo is an example of reflection (sound waves), and mirrors reflect light waves.

Two cases of reflection are illustrated in •Fig. 13.16. If the end of the string is fixed, the reflected pulse is inverted, or undergoes a 180° phase shift (Fig. 13.16a). This is because the pulse causes the string to exert an upward force on the wall, and the wall exerts an equal and opposite downward force on the string (by Newton's third law). Thus, the reflected pulse is downward, or inverted. If the end of the string is free to move, then the reflected pulse is not inverted (zero or no phase shift). This is illustrated in Fig. 13.16b, where the string is attached to a light ring that can move freely on a smooth pole. The ring is accelerated upward by the front portion of the incoming pulse and is then brought downward.

In general, when a wave strikes a boundary, some of its energy is reflected and some is transmitted or absorbed. When a wave crosses a boundary into another medium, its velocity changes because the new material has different characteristics. Entering the medium obliquely (at an angle), the transmitted wave moves in a direction different from that of the incident wave. This phenomenon is called **refraction**.

Diffraction refers to the bending of waves around an edge of an object. For example, if you stand along an outside wall of a building near the corner, you can hear people talking around the corner. Assuming there are no reflections or air motion (wind), this would not be possible if the sound waves traveled in a straight line.

In general, the effects of diffraction are evident only when the size of the diffracting object or opening is about the same as or smaller than the wavelength of the waves. This relationship can be readily observed in water waves. For example, water waves in a pond pass by blades of grass or reeds with little noticeable effect because the widths of these objects are much smaller than the wavelength of the waves. The dependence of diffraction on wavelength and size of the object is illustrated in •Fig. 13.17.

Reflection, refraction, and diffraction will be considered in more detail for light waves in Chapter 22.

13.5   Standing Waves and Resonance

Objective: To be able to (a) describe the formation and characteristics of standing waves, and (b) explain the phenomenon of resonance.

If you shake one end of a stretched rope, waves travel down it to the fixed end and are reflected back. The waves going down and back interfere. In most cases, the combined waveforms will have a changing, jumbled appearance. But if the rope is shaken at just the right frequency, a waveform appears to stand in place

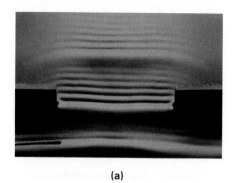

(a)

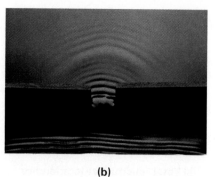

(b)

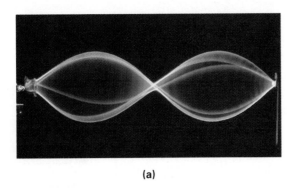

(a)

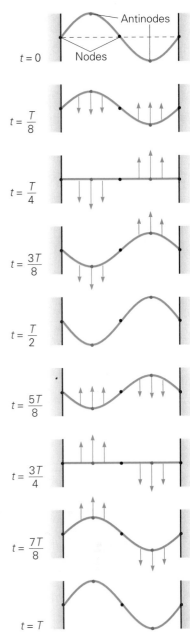

(b)

•**FIGURE 13.18 Standing waves**
(a) Standing waves are formed by interfering waves traveling in opposite directions. **(b)** Conditions of destructive and constructive interference recur as each wave travels a distance of $\lambda/4$ in a time of $t = T/4$. The motions of the rope particles are indicated by the arrows. This gives rise to standing waves with stationary nodes and maximum amplitude antinodes.

along the rope. Appropriately, this phenomenon is called a **standing wave** (see •Fig. 13.18a and Demonstration 11 on p. 438). This arises because of interference with the reflected waves, which have the same wavelength, amplitude, and speed.

Some points on the rope are stationary and are called **nodes**. At these points, the displacements of the interfering waves are always equal and opposite. Thus, by the principle of superposition, the interfering waves must cancel each other completely at these points, and the rope does not undergo any displacement. At all other points, the rope oscillates back and forth with the same frequency. The points of maximum amplitude, where constructive interference is greatest, are called **antinodes**. As you can see in •Fig. 13.18b, adjacent antinodes are separated by half of a wavelength ($\lambda/2$), or one loop; adjacent nodes are also separated by half of a wavelength. The wavelength is that of the interfering waves that produce the standing wave.

Standing waves can be generated in a rope by more than one driving frequency; the higher the frequency, the greater the number of oscillating half-wavelength loops in the rope. The frequencies at which large-amplitude standing waves are produced are called **natural frequencies**, or **resonant frequencies**. The standing wave patterns are called normal, or resonant, modes of vibration. In general, all things have one or more natural frequencies, which depend on such factors as mass, elasticity or restoring force, and geometry (boundary conditions). The natural frequencies of a system are sometimes called its characteristic frequencies.

A stretched string or rope can be analyzed to determine its natural frequencies (•Fig. 13.19). The boundary conditions are that the ends are fixed; thus, there must be a node at each end. The number of closed segments or loops of a standing wave that will fit between the nodes at the ends (along the length of the string) is equal to an integral number of half-wavelengths. Note that $L = \lambda_1/2$, $L = 2\lambda_2/2$, $L = 3\lambda_3/2$, $L = 4\lambda_4/2$, and so on. In general,

$$L = n\frac{\lambda_n}{2} \quad \text{or} \quad \lambda_n = \frac{2L}{n} \quad \text{(for } n = 1, 2, 3, \ldots)$$

The natural frequencies of oscillation, where v is the wave speed, are

$$f_n = \frac{v}{\lambda_n} = \frac{nv}{2L} \quad \text{for } n = 1, 2, 3, \ldots \qquad \begin{array}{l}\textit{natural frequencies for}\\ \textit{a stretched string}\end{array} \qquad (13.19)$$

The lowest natural frequency ($f_1 = v/2L$) is called the **fundamental frequency**. All of the other natural frequencies are integral multiples of the fundamental frequency: $f_n = nf_1$ ($n = 1, 2, 3, \ldots$). The set of frequencies $f_1, f_2 = 2f_1, f_3 = 3f_1$, and so on, is called a **harmonic series**: f_1 is the *first harmonic*, f_2 the *second harmonic*, and so on.

Strings fixed at each end are found in musical instruments such as violins and guitars. When such a string is excited, the resulting vibration will include several harmonics in addition to the fundamental frequency. The number of harmonics depends on how the string is excited, that is, plucked or bowed. In any case, it is the combination of harmonic frequencies that gives a particular instrument its characteristic sound quality. (More on this in Chapter 14.) As Eq. 13.19 shows, the

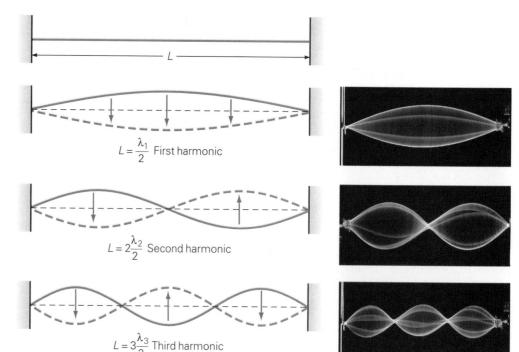

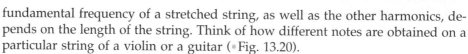

$L = \dfrac{\lambda_1}{2}$ First harmonic

$L = 2\dfrac{\lambda_2}{2}$ Second harmonic

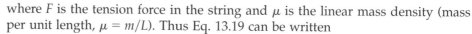

$L = 3\dfrac{\lambda_3}{2}$ Third harmonic

•FIGURE 13.19 Resonant frequencies
A stretched string can have standing waves only at certain frequencies. These correspond to the numbers of half-wavelength loops that will fit along the length of string between the nodes at the fixed ends.

fundamental frequency of a stretched string, as well as the other harmonics, depends on the length of the string. Think of how different notes are obtained on a particular string of a violin or a guitar (•Fig. 13.20).

Natural frequencies also depend on other parameters, such as mass and force, which affect the wave speed. For a stretched string, the wave speed (v) can be shown to be

$$v = \sqrt{\dfrac{F}{\mu}} \qquad (13.20)$$

where F is the tension force in the string and μ is the linear mass density (mass per unit length, $\mu = m/L$). Thus Eq. 13.19 can be written

$$f_n = \dfrac{nv}{2L} = \dfrac{n}{2L}\sqrt{\dfrac{F}{\mu}} \qquad \text{(for } n = 1, 2, 3, \ldots) \qquad (13.21)$$

Note that the greater the linear mass density of a string, the lower its natural frequencies are. As you may know, the low-note strings on a violin or guitar are thicker, or more massive, than the high-note strings.

•FIGURE 13.20 Fundamental frequencies
Performers on string instruments such as the violin or guitar use their fingers to stop or fret the strings. By pressing a string against the fingerboard the player alters the amount of its length that is free to vibrate. This changes the resonant frequency of the string and the pitch of the tone it produces.

EXAMPLE 13.7 ■ A PIANO STRING: FUNDAMENTAL FREQUENCY AND HARMONICS

A piano string with an effective length of 1.15 m and a mass of 20.0 g is under a tension of 6.30×10^3 N. (a) What is the fundamental frequency of the tone produced when the string is struck? (b) What are the frequencies of the next two harmonics?

Solution.

Given: $L = 1.15$
$m = 20.0 \text{ g} = 0.0200 \text{ kg}$
$F = 6.30 \times 10^3 \text{ N}$

Find: (a) f_1 (fundamental frequency)
(b) f_2 and f_3 (frequencies of next two harmonics)

(a) The linear mass density of the string is

$$\mu = \dfrac{m}{L} = \dfrac{0.0200 \text{ kg}}{1.15 \text{ m}} = 0.0174 \text{ kg/m}$$

Then, using Eq. 13.21,

$$f_1 = \frac{1}{2L} \sqrt{\frac{F}{\mu}} = \frac{1}{2(1.15 \text{ m})} \sqrt{\frac{6.3 \times 10^3 \text{ N}}{0.0174 \text{ kg/m}}} = 262 \text{ Hz}$$

This is approximately the frequency of middle C (C_4) on a piano (261.6 Hz).

(b) Since $f_2 = 2f_1$ and $f_3 = 3f_1$,

$$f_2 = 2f_1 = 2(262 \text{ Hz}) = 524 \text{ Hz}$$

and

$$f_3 = 3f_1 = 3(262 \text{ Hz}) = 786 \text{ Hz}$$

The second harmonic corresponds approximately to C_5 on a piano, since the frequency doubles with each octave (or every eighth white key).

Follow-up Exercise. A musical note is referenced to the fundamental vibrational frequency or first harmonic. In musical terms, the second harmonic is the first overtone, the third harmonic the second overtone, and so on. If an instrument has a third overtone with a frequency of 880 Hz, what is the frequency of the first overtone?

CONCEPTUAL EXAMPLE 13.8 ■ TUNING UP: RAISING THE PITCH OF A STRING

You wish to raise the pitch of a guitar string from the A note (220 Hz) below middle C to the A note (440 Hz) above middle C. Would you (a) loosen the string so that its tension is halved, (b) tighten the string so as to double the tension, (c) use another string of the same material with half the diameter at the same tension, (d) use another string of the same material with twice the diameter at the same tension? *Clearly establish the reasoning and physical principle(s) used in determining your answer before checking it below. That is,* **why** *did you select your answer?*

Reasoning and Answer. The fundamental frequency or pitch of a stretched string is given by Eq. 13.21:

$$f = \frac{1}{2L} \sqrt{\frac{F}{\mu}} \quad \text{(for } n = 1)$$

We can see from this equation that the frequency of the string is proportional to the *square root* of the tension force F, so loosening the string—that is, decreasing F—would not increase the frequency. Nor would doubling the tension double the frequency (because $\sqrt{2F} \neq 2\sqrt{F}$). Thus, neither (a) nor (b) is the correct answer.

The question then is, how does the frequency vary with the diameter of the string? Clearly, the greater the diameter of the string, the greater its mass per unit length (μ). Hence, a thinner string will vibrate with a higher frequency, and the answer is (c).

It is often helpful to confirm a conceptual answer using mathematics and equations. Let's do that here. We know that μ is the mass per unit length ($\mu = m/L$), and the mass can be written in terms of the volume density (ρ) and the geometry of the string, $m = \rho V$, where V is the volume of the string. The volume can be written in terms of its circular cross-sectional area (assumed constant) and the length of the string:

$$V = AL = \frac{\pi d^2 L}{4}$$

where d is the diameter of the string ($A = \pi r^2 = \pi d^2/4$).

Writing the frequency in terms of these parameters, we have

$$f = \frac{1}{2L} \sqrt{\frac{F}{\mu}} = \frac{1}{2L} \sqrt{\frac{FL}{m}} = \frac{1}{2L} \sqrt{\frac{4F}{\rho \pi d^2}}$$

So, with constant F and L (the active length of the string between the bridge and neck of a stringed instrument is constant), the frequency is inversely proportional to the diameter of the string,

$$f \propto \frac{1}{d}$$

Hence, using a string with half the diameter ($d_2 = d_1/2$) would double the frequency.

Forming a ratio shows this explicitly:

$$\frac{f_2}{f_1} = \frac{d_1}{d_2} \quad \text{or} \quad f_2 = \left(\frac{d_1}{d_2}\right)f_1$$

For $f_2 = 2f_1$ (that is, 440 Hz = 2 × 220 Hz), we must have $d_1/d_2 = 2$ or $d_2 = d_1/2$.

Follow-up Exercise. The pitch of a violin string is A below middle C (220 Hz). Could you tune this string to middle C (264 Hz)? If so, how?

•**FIGURE 13.21 Resonance in the playground**
The swing behaves like a pendulum in SHM. To transfer energy efficiently, you must time your pushes to its natural frequency.

When an oscillating system is driven at one of its natural, or resonant, frequencies, maximum energy transfer to the system occurs. The system is physically suited to a natural frequency, or "wants" to vibrate at it, so to speak. The condition of vibrating at a natural frequency is referred to as **resonance**.

A common example of a system in mechanical resonance is someone being pushed on a swing. Basically, a swing is a simple pendulum and has only one resonant frequency for a given length [$f = (1/2\pi)\sqrt{g/L}$]. If you push the swing with this frequency and in phase with its motion, the amplitude and energy increase (•Fig. 13.21). If you push at a slightly different frequency, the energy transfer is no longer a maximum. (What do you think happens if you push with the resonant frequency, but 180° out of phase with the swing's motion?)

Unlike a simple pendulum, a stretched string has many natural frequencies. Almost any driving frequency will cause a disturbance in the string. However, if the frequency of the driving force is not equal to one of the natural frequencies, the amplitude of the standing wave will be relatively small. When the frequency of the driving force is at one of the natural frequencies, more energy is transferred to the string, and the amplitude of the antinodes is relatively large.

Mechanical resonance is not the only type. When you tune a radio, you are changing the resonance frequency of an electrical circuit (Chapter 21) so that it will be driven by, or pick up, the frequency signal of the station you want. Another example of resonance is described in the Insight on p. 439.

DEMONSTRATION 11 ■ Flame standing wave pattern

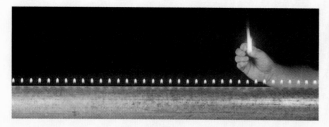

(a) Regularly spaced holes in a piece of downspout allow a line of uniformly sized flames when flammable gas is supplied to the pipe. One end of the pipe is closed by a metal plate and the other end is fitted with a rubber diaphragm and loudspeaker.

(b) When an audio oscillator driving the speaker is tuned to a resonant frequency of the pipe, standing sound (compressional) waves are created. The higher flames mark the locations of the antinodes, half a wavelength apart. [See *The Physics Teacher*, 17, 307 (1979).] Note: Some fast, contemporary music with good bass instead of an oscillator output produces a rapidly varying, dancing flame pattern.

Demonstration of a compressional standing wave: a sound wave in a closed tube. The high and low flames correspond to antinodes and nodes, respectively, in the gas column.

Insight   Desirable and Undesirable Resonances

When we hold a seashell to our ear, we say we hear the ocean, or a sound like that of the ocean. What causes this is a resonance effect—a desirable one because of the pleasant sound. Soft sounds enter the shell from the environment, and some of these are at the resonant frequencies of the shell. Standing waves are set up which can be heard when the shell is held close to the ear.

The mixture of sound frequencies is not harmonic because of the complicated shape of the shell's interior. As a result, the sound has a variety of frequencies—"white noise," so to speak (analogous to "white light," which is a mixture of all visible frequencies). The resulting sound is similar to that of the ocean.

FIGURE 1 Galloping Gertie
The collapse of the Tacoma Narrows Bridge on November 7, 1940, is captured in this frame from a movie camera.

When a large number of soldiers march over a small bridge, they are generally ordered to break step. The reason is that the marching frequency may correspond to one of the natural frequencies of the bridge and set it into resonant vibration, which could cause it to collapse. This actually occurred for a suspension bridge in England in 1831. The bridge was weak and in need of repair, but the resonance vibrations induced by the marching soldiers crossing the bridge helped it fail sooner—with some injuries.

Another incidence of bridge vibration wasn't due to marching soldiers, but to the driving force of the wind. On the morning of November 7, 1940, winds gusting to 40–45 mi/h started the main span of the Tacoma Narrows Bridge (in Washington state) vibrating. The bridge, 2800 ft (855 m) long and 39 ft (12 m) wide, had been opened to traffic only 4 months earlier.

During the first month of use, small transverse modes of vibration had been observed. But on November 7, special wind effects drove the bridge in near resonance, and the main span vibrated with a frequency of 36 vib/min and an amplitude of 1.5 ft.* At 10 a.m., the main span began to vibrate in a torsional (twisting) mode in two segments with a frequency of 14 vib/min. The wind continued to drive the bridge in resonance and the vibrational amplitude increased. Shortly after 11 a.m., the main span collapsed (Fig. 1).

"Galloping Gertie" (the nickname given to the bridge) was rebuilt using the same tower foundations. However, the new design made the structure stiffer to increase its resonant frequency so that high winds could not produce unwanted resonance.

*It is doubtful that the gusting of the wind set the bridge into vibration. The wind velocity was moderately steady and gust fluctuations are normally random. One explanation for the driving source of the oscillations involves the formation of vortices as the wind blew past the bridge. Vortices are like the eddies that form in the water at the end of the oars when a boat is rowed. The wind blowing over and under the bridge formed vortices that rotated in opposite directions. The formation and "shedding" of the vortices (like eddies coming off oars) would impart energy to the bridge, and if the frequency of this action were near a natural frequency, a standing wave would be set up.

Chapter Review

Important Terms

Hooke's law   415
simple harmonic motion (SHM)
 415
displacement   416
amplitude   416
period   416
frequency   416
hertz (Hz)   416
equation of motion   419
phase constant   425
damped harmonic motion   427
wave   428

wave motion   428
wavelength (λ)   429
wave speed   429
transverse wave   429
longitudinal wave   431
principle of superposition   432
constructive interference   432
destructive interference   433
total constructive interference
 433
total destructive interference
 433

reflection   434
refraction   434
diffraction   434
standing wave   434
node   435
antinode   435
natural (resonant) frequencies
 435
fundamental frequency   435
harmonic series   435
resonance   438

Important Concepts

- Simple harmonic motion (SHM) requires a restoring force directly proportional to the displacement, such as the spring force.
- In general, the total energy of an object in SHM is directly proportional to the square of the amplitude.
- An equation of motion gives an object's position as a function of time.
- The form of an equation of motion for a mass in SHM depends on the initial displacement or initial condition of the mass. This is expressed in terms of a phase constant.

- A wave is a disturbance in time and space; energy is transferred or propagated by wave motion.
- At any time, the combined waveform of two or more interfering waves is given by the sum of the displacements of the individual waves at each point in the medium.
- Standing waves result from the interference of waves of identical wavelength, amplitude, and speed traveling in opposite directions.

Important Equations

Hooke's Law:

$$\mathbf{F} = -k\mathbf{x} \tag{13.1}$$

Frequency and Period for SHM:

$$f = \frac{1}{T} \tag{13.2}$$

Total Energy of a Spring and Mass in SHM:

$$E = \tfrac{1}{2}kA^2 = \tfrac{1}{2}mv^2 + \tfrac{1}{2}kx^2 \tag{13.4–5}$$

Velocity of Oscillating Mass on a Spring:

$$v = \pm\sqrt{\frac{k}{m}(A^2 - x^2)} \tag{13.6}$$

Maximum Speed of Oscillating Mass on a Spring:

$$v_{\max} = \sqrt{\frac{k}{m}}(A) \tag{13.7}$$

Period of Mass Oscillating on a Spring:

$$T = 2\pi\sqrt{\frac{m}{k}} \tag{13.11}$$

Frequency of a Mass Oscillating on a Spring:

$$f = \frac{1}{2\pi}\sqrt{\frac{k}{m}} \tag{13.12}$$

Angular Frequency of a Mass Oscillating on a Spring:

$$\omega = 2\pi f = \sqrt{\frac{k}{m}} \tag{13.13}$$

Period of a Simple Pendulum (small-angle approximation):

$$T = 2\pi\sqrt{\frac{L}{g}} \tag{13.14}$$

Displacement of a Mass in SHM:

$$y = A\sin(\omega t + \delta) \tag{13.15}$$

(with $\delta = 0$)

$$y = A\sin\omega t = A\sin 2\pi f t = A\sin\frac{2\pi t}{T}$$

Velocity of a Mass in SHM ($\delta = 0$):

$$v = A\omega\cos\omega t \tag{13.16}$$

Acceleration of a Mass in SHM ($\delta = 0$):

$$a = -A\omega^2\sin\omega t = -\omega^2 y \tag{13.17}$$

Wave Speed:

$$v = \frac{\lambda}{T} = \lambda f \tag{13.18}$$

Natural Frequencies in a Stretched String:

$$f_n = \frac{nv}{2L} = \frac{n}{2L}\sqrt{\frac{F}{\mu}} \quad \text{(for } n = 1, 2, 3, \ldots\text{)} \tag{13.19–20}$$

Exercises

13.1 Simple Harmonic Motion

1 A particle in SHM (a) has variable amplitudes, (b) has a restoring force in the form of Hooke's law, (c) has a frequency directly proportional to its period, (d) may be represented graphically by $y = ax + b$.

2 The maximum kinetic energy of a particle in SHM is equal to (a) A, (b) A^2, (c) kA, (d) $kA^2/2$

3 If the amplitude of a mass in SHM is doubled, how are (a) the energy and (b) the maximum speed affected?

4 How is the speed of a mass in SHM affected as it approaches its equilibrium position? Explain.

5 ■ A particle is in SHM with an amplitude A. What is the total distance it travels in a time of $2T$?

6 ■ A mass oscillating on a spring completes a cycle every 0.025 s. What is the frequency of the oscillation?

7 ■ A particle in simple harmonic motion has a frequency of 40 Hz. What is the period of oscillation?

8 ■ The frequency of a simple harmonic oscillation is doubled from 0.25 Hz to 0.50 Hz. What is the change in the period of oscillation?

9 ■ For the oscillating mass in Fig. 13.1, describe its position and direction of motion at $t = 4T/3$.

10 ■■ Show that the total energy of a spring-mass system in simple harmonic motion is given by $\frac{1}{2} m\omega^2 A^2$.

11 ■■ Show that for a pendulum to oscillate with the same frequency as a mass on a spring, the pendulum's length must be $L = mg/k$.

12 ■■ What is the spring constant of a spring that stretches 6.0 cm when a 0.25-kg mass is suspended from it?

13 ■■ Atoms in a solid are in continual vibration due to thermal energy. At room temperature, the amplitude of the atomic vibrations is about 10^{-9} cm, and the frequency of oscillation is about 10^{12} Hz. (a) What is the approximate period of oscillation of an atom? (b) What is the magnitude of its maximum velocity?

14 ■■ A 0.350-kg mass resting on a horizontal frictionless surface is attached to a spring with a spring constant of 150 N/m. If the mass is pulled 0.100 m from its equilibrium position and released, what is the magnitude of force on the mass and its acceleration at (a) $t = 0$, (b) $x = 0.050$ m, and (c) $x = 0$?

15 ■■ In a lab experiment, you are given a spring with a spring constant of 12 N/m. What mass would you suspend on the spring so as to have an oscillation period of 0.91 when in SHM?

16 ■■ If the oscillating spring in Exercise 15 had an amplitude of 15 cm, (a) what would be its maximum speed and where would this occur? (b) What would be the speed at a half-amplitude position?

17 ■■ A 0.25-kg mass on a vertical spring oscillates with a frequency of 1.0 Hz. Suppose a 0.50-kg mass were instead suspended from the same spring. What would be the frequency of oscillation in this case?

18 ■■ A 0.20-kg mass is oscillating on a spring that has a spring constant of 40 N/m. The instantaneous speed of the mass is 0.95 m/s as it passes through its equilibrium position. What is the total energy of the system?

19 ■■ The oscillations of two spring-mass systems are graphed in •Fig. 13.22. If the mass for (a) is 4 times that of (b), which system has more energy and how many times more?

20 ■■ A 0.25-kg object suspended on a light spring is released from a position 15 cm above the stretched equilibrium position. The spring has a spring constant of 80 N/m. (a) What is the total spring energy of the system? (Neglect gravitational potential energy.) Does this energy depend on the mass of the object? Explain.

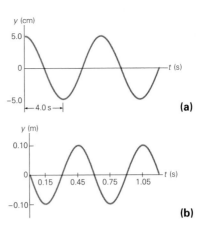

•**FIGURE 13.22   Wave energy and equation of motion**
See Exercises 19, 41, and 48.

21 ■■ What is the speed of the object in Exercise 20 when it is (a) 5.0 cm above its equilibrium position and (b) 5.0 cm below its equilibrium position? (c) What is the object's maximum speed, and where does this occur?

22 ■■ A mass of 0.25 kg oscillates horizontally on a frictionless surface on a light spring with a spring constant of 15 N/m. If the mass was initially displaced 6.0 cm from its equilibrium position, what is (a) its total energy, and (b) the difference between its maximum speed and its speed at $x = 2.0$ cm?

23 ■■■ A 75-kg circus performer jumps from a height of 5.0 m onto a trampoline and stretches it a depth of 0.30 m. Assume that the trampoline obeys Hooke's law. (a) How far will it stretch if the performer jumps from a height of 8.0 m? (b) How far will it be stretched when the performer stands still on it while taking a bow?

24 ■■■ A 0.250-kg ball is dropped from a height of 10.0 cm onto a spring as illustrated in •Fig. 13.23. If the spring has a spring constant of 60.0 N/m, (a) what distance will the spring be compressed? (b) On recoiling upward, how high will the ball go?

•**FIGURE 13.23   How far down?**
See Exercise 24.

13.2 Equations of Motion

25 An equation of motion for a particle in SHM (a) is always a cosine function, (b) reflects damping action, (c) is independent of the initial conditions, (d) may be represented by a sinusoidal function.

26 If pendulum A has a period twice that of pendulum B, then (a) $L_1 = 2L_2$, and $m_1 = m_2/2$, (b) $L_1 = 2L_2$ and $m_1 = 2m_2$, (c) $L_1 = 2L_2$ regardless of mass, (d) none of the preceding.

27 The apparatus in Fig. 13.6 demonstrates that the motion of a mass on a spring can be described by a sinusoidal function of time. How could this be demonstrated for a pendulum?

28 Could simple harmonic motion be described using a tangent function? Explain.

29 Would the period of a pendulum in a downward accelerating elevator be increased or decreased from that of the period in a stationary elevator? Explain.

30 ■ A mass of 0.25 kg oscillates in simple harmonic motion on a spring with a spring constant of 2.0×10^2 N/m. What are (a) the period and (b) the frequency of the oscillation?

31 ■ Write the general equation of motion for a mass that is on a horizontal frictionless surface and connected to a spring at equilibrium (a) if the mass is initially given a quick push away from the spring and (b) if the mass is pulled away from the spring and released.

32 ■ The displacement of an object is given by $y = (5.0 \text{ cm}) \sin 20\pi t$. What are (a) the amplitude, (b) the frequency, and (c) the period of oscillation?

33 ■■ Students use a pendulum with a length of 36.90 cm to measure the acceleration of gravity at the location of their school. If the period of the pendulum is found to be 1.220 s, what is the small-angle approximation value of g?

34 ■■ Make sketches showing two masses oscillating on springs when one is at equilibrium and they are (a) in phase, (b) 90° out of phase, (c) 180° out of phase, (d) 270° out of phase, and (e) 30° out of phase.

35 ■■ Compute the percentage difference between the angle θ in rad and the sine of θ for (a) $\theta = 10°$, (b) $\theta = 20°$, and (c) $\theta = 25°$. (Use four significant figures in % difference calculations, and report to two.)

36 ■■ The equation of motion for a 25-g mass in simple harmonic motion is given by $y = 8.0 \cos 10t$, where y is in centimeters. What are (a) the period and (b) the phase constant for this motion? (c) What is the total energy of the system?

37 ■■ The equation of motion of a particle in simple harmonic motion is given by $y = 10 \sin 0.50t$ (where y is in centimeters). What are the particle's (a) displacement, (b) velocity, and (c) acceleration at $t = 1.0$ s?

38 ■■ A 0.15-kg mass oscillates vertically in simple harmonic motion on a spring with a spring constant of 6.0 N/m. The spring was initially raised 8.0 cm. (a) Write the equation of motion. (b) Determine the phase constant. (c) What is the displacement of the mass at $t = 0.50$ s?

39 ■■ The simple harmonic motion of a 0.20-kg mass on a spring is described by $y = 20 \sin 2\pi t$ (where y is in centimeters). At $t = \frac{1}{8}$ s, what are the mass' (a) displacement and (b) velocity? (c) What is the total energy of the system?

40 ■■ A 0.30-kg mass on a spring oscillates with a frequency of 2.0 Hz. If initially released from a height of 10 cm, what will be (a) its position at $t = 1.8$ s, and (b) its maximum speed?

41 ■■ The motion of a particle is described by the curve in Fig. 13.22a. Write the equation of motion in three equivalent cosine forms (see Eqs. 13.8–13.10).

42 ■■ A 0.20-kg mass is suspended on a spring, which stretches a distance of 5.0 cm when at rest. The mass is then pulled down an additional distance of 10 cm and released. What are (a) the period of oscillation, (b) the equation of motion of the mass, (c) the total spring energy of the system, and (d) the displacement of the mass at $t = T/6$ s?

43 ■■ The equation of motion for a particle in simple harmonic motion is $x = 15 \sin 2\pi t$ (where x is in centimeters). At $t = 2.5$ s, what are the particle's (a) displacement, (b) velocity, and (c) acceleration?

44 ■■ A particle undergoes SHM with an amplitude A on a spring with a spring constant k. What is the particle's kinetic energy when its displacement is $-A/2$? (Express in terms of k and A.)

45 ■■ Two equal masses m_1 and m_2 oscillate on light springs, the second with a spring constant twice that of the first. Which mass will have the greater period and how many times greater?

46 ■■ A mass of 0.075 kg oscillates on a light spring, and another 0.075-kg mass oscillates as the bob of a simple pendulum in SHM. If the pendulum length is 0.30 m, what would be the spring constant of the spring if the periods of the oscillations are equal?

47 ■■ A 1.0-kg block on a horizontal frictionless surface is attached to a spring as in Fig. 13.1. The block is displaced 10 cm from its equilibrium position by a force of 150 N, and then released. (a) What is the spring constant of the spring? (b) What is its period of oscillation? (c) What is the force on the block at $t = \pi/2$ s?

48 ■■ The motion of a 0.35-kg mass oscillating on a light spring is described by the curve in Fig. 13.22b. (a) Write the equation for its displacement as a function of time. (b) What is the spring constant of the spring?

49 ■■■ The forces acting on a simple pendulum are shown in •Fig. 13.24. (a) Show that for the small-angle approximation (sin $\theta \approx \theta$) the force producing the motion has the same form as Hooke's law. (b) Show by analogy with a mass on a spring that the period of a simple pendulum is given by $T = 2\pi\sqrt{L/g}$.

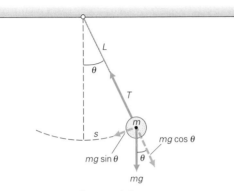

•**FIGURE 13.24   SHM of a pendulum**
See Exercise 49.

50 ■■■ A grandfather clock has a pendulum that is 75.00 cm long. It is accidentally broken, and when repaired the length is shorter by 2.0 mm. Considering the clocks pendulum to be a simple pendulum (a) will the repaired clock gain or lose time? (b) By how much will the repaired clock differ from the correct time (taken to be the time determined by the original pendulum in 24 h)? (c) If the pendulum were metal, would temperature make a difference in the timekeeping of the clock? Explain.

13.3   Wave Motion

51 Wave motion in a material medium involves (a) the propagation of a disturbance, (b) interparticle interactions, (c) the transfer of energy, (d) all of the preceding.

52 For a periodic wave propagating with a speed v, the following relationship holds: (a) $\lambda = v/f$, (b) $v = \lambda/f$, (c) $v = \lambda f^2$, (d) $f = \lambda/f$.

53 What type(s) of wave(s) will propagate in (a) solids, (b) liquids, and (c) gases?

54 ■ A longitudinal sound wave has a speed of 340 m/s in air. If this produces a tone having a frequency of 4.0×10^3 Hz, what is the wavelength?

55 ■ A transverse wave has a wavelength of 0.50 m and a frequency of 40 Hz. What is the wave speed?

56 ■ A student reading his physics book on a lake dock notices that the distance between two incoming wave crests is about 2.4 m, and then measures the time of ar-rival between wave crests to be 1.6 s. What is the approximate speed of the waves?

57 ■ Light waves travel in a vacuum at a speed of 300,000 km/s. The frequency of visible light is about 10^{14} Hz. What is the approximate wavelength of the light?

58 ■ Red light has a frequency of 4.3×10^{14} Hz. What is the wavelength of this light in a vacuum?

59 ■■ The range of sound audible to the human ear has frequencies from about 20 Hz to 20 kHz. If the speed of sound in air is 345 m/s, what are the limits of this audible range in wavelengths?

60 ■■ If you look on a radio dial, you will see that the AM frequencies range from 550 kHz to 1600 kHz, and the FM frequencies range from 88.0 MHz to 108 MHz. All of these radio waves travel at a speed of 3.00×10^8 m/s (speed of light). What are the ranges of (a) the AM band and (b) the FM band in terms of wavelengths?

61 ■■ A sonar generator on a submarine produces periodic ultrasonic waves at a frequency of 2.50 MHz. The wavelength of the waves in sea water is 4.80×10^{-4} m. When the generator is directed downward, an echo reflected from the ocean floor is received 16.7 s later. How deep is the ocean at that point? (Assume wavelength is constant at all depths.)

62 ■■ In watching a transverse wave go by, a person notes that 13 crests go by in a time of 3.0 s. If the distance between two successive crests is measured to be 0.75 m, what is the speed of the wave?

63 ■■ A wave traveling in the +x direction is shown in •Fig. 13.25a. The particle displacement at a particular location in the medium through which the wave travels is shown in Fig. 13.25b. (a) What is the amplitude of the traveling wave? (b) What is the wave speed?

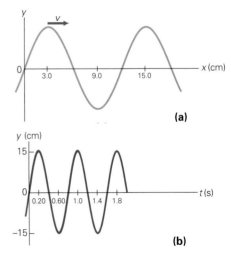

•**FIGURE 13.25   How high and how fast?**
See Exercise 63.

64 ■■ A wave with a frequency of 60 Hz has a speed of 12 m/s in a particular medium. (a) What is the wavelength? (b) If the wave is transmitted into another medium in which it is propagated at a speed of 20 m/s, by how much will the wavelength change? (The frequency remains the same.)

65 ■■ Assume that P and S waves (primary and secondary) from an earthquake with a focus near the Earth's surface travel through the Earth at average speeds of 8.0 km/s and 6.0 km/s, respectively. Assume that there is no deflection or refraction of the waves. (a) How long is the delay between the arrivals of successive waves at a seismic monitoring station located 90° in latitude from the epicenter of the quake? (b) Do the waves cross the boundary of the mantle? (c) How long does it take for the waves to arrive at a monitoring station on the opposite side of the Earth?

66 ■■■ The speed of longitudinal waves traveling in a long solid rod is given by $v = \sqrt{Y/\rho}$, where Y is Young's modulus and ρ is the density of the solid. If a disturbance has a frequency of 40 Hz, what is the wavelength of the waves it produces in (a) an aluminum rod and (b) a copper rod? [*Hint:* See Tables 9.1 and 9.2.]

67 ■■■ As noted in Exercise 66, the speed of longitudinal waves in a solid rod is given by $v = \sqrt{Y/\rho}$. Someone strikes a steel train rail with a hammer at a frequency of 0.50 Hz, and someone else puts his or her ear to the rail 1.0 km away. (a) How long after the first strike does the observer hear the sound? (b) What is the time interval between successive sound pulses heard by the observer? [*Hint:* See Tables 9.1 and 9.2]

68 ■■■ The general form of the equation of motion for a traveling wave, expressing both spatial and temporal dependence, is

$$y = A \sin(kx - \omega t)$$

where k is the *wave number* (*not* the spring constant) and is equal to $2\pi/\lambda$. (a) Show mathematically or graphically that this equation represents a wave traveling in the positive x direction. (b) Show that

$$y = A \sin(kx + \omega t)$$

represents a wave traveling in the negative x direction.

13.4 Wave Phenomena

69 When waves meet each other and interfere, the resultant waveform is determined by (a) reflection, (b) refraction, (c) diffraction, (d) superposition.

70 Refraction (a) involves constructive interference, (b) refers to a "bending" or change in direction at media interfaces, (c) is synonymous with diffraction, (d) occurs only for media or mechanical waves.

71 What is destroyed when destructive interference occurs?

13.5 Standing Waves and Resonance

72 For traveling waves to set up standing waves, it is necessary that the waves have the same (a) frequency, (b) amplitude, (c) speed, (d) all of the preceding.

73 When a stretching string or cord is oscillated at the third harmonic of its natural frequencies, then the standing wave in the string will exhibit (a) 3 wavelengths, (b) $\frac{1}{3}$ wavelength, (c) $\frac{3}{2}$ wavelengths, (d) 2 wavelengths.

74 A child's swing (a pendulum) has only one natural frequency f_o. Yet it can be driven or pushed smoothly at frequencies of $f_o/2$, $f_o/3$, $2f_o$, $3f_o$. How is this possible?

75 Give an example or two of physical objects that can vibrate in resonance with $L = \lambda/4$ or $3\lambda/4$. [*Hint:* Think of some linear, metal objects.]

76 ■ The fundamental frequency of a stretched string is 220 Hz. What is the frequency of (a) the third harmonic and (b) the fourth harmonic?

77 ■ A standing wave is formed in a stretched string that is 3.0 m long. What is the wavelength of (a) the first harmonic and (b) the third harmonic?

78 ■ A father pushes a 25 kg child on a swing. If the length of the swing is 3.0 m, how often does the father have to push to drive the system in resonance?

79 ■■ A piece of rubber tubing with a linear mass density of 0.125 kg/m is stretched by a force of 9.00 N. (a) What will be the wave speed along the tubing? (b) If the stretched tubing has a length of 10.0 m, what are its natural frequencies?

80 ■■ Find the first four harmonics for a string that is 2.0 m long, has a linear mass density of 2.5×10^{-2} kg/m, and is under a tension of 40 N.

81 ■■ Will a standing wave be formed in a 4.0-m length of stretched string that transmits waves with a speed of 12 m/s if it is driven at a frequency of (a) 15 Hz or (b) 20 Hz?

82 ■■ Two waves of equal amplitude and wavelength of 0.80 m travel in a string in opposite directions with a speed of 250 m/s. If the string is 2.0 m long, what is the order of the standing wave set up in the string?

83 ■■ Two stretched strings A and B have the same tension and linear mass density. Are any of the first six harmonics of the strings equal if the string lengths are (a) 1.0 m and 3.0 m or (b) 1.5 m and 2.0 m respectively?

84 ■■ A violin string is tuned to 440 Hz. When playing the instrument, the violinist puts a finger down on the string $\frac{1}{8}$ of the string length from the neck end. What is the frequency of the string when played like this?

85 ■■ A "whip" CB antenna on a car is 3.0 m long. When the car is moving along a highway, a standing wave is observed in the antenna that has 1.5 loop segments in it. What is the wavelength of the fundamental frequency of the antenna?

86 ■■■ A thin, flexible metal rod is 1.0 m long. It is clamped at one end to a table, and the other end can vibrate freely. What are the natural frequencies of the rod if the wave speed in the material is 3.5×10^3 m/s?

87 ■■■ In a common laboratory experiment on standing waves, the waves are produced in a stretched string by an electrical vibrator that oscillates at 60 Hz (•Fig. 13.26). The string runs over a pulley, and a hanger is suspended from the end. The tension in the string is varied by adding weights to the hanger. If the active length of string (the part that vibrates) is 1.5 m and this length of string has a mass of 0.10 g, what masses must be suspended to produce the first four harmonics in that length?

•**FIGURE 13.26 Standing waves in strings**
Twin vibrating strings with standing waves. This demonstration model allows variations in string tension and length (linear density) of the string. Also, the vibration frequency may be adjusted. See Exercise 87.

Additional Exercises

88 A 0.10-kg mass suspended on a spring is pulled to 8.0 cm below its equilibrium position and released. When the mass passes through the equilibrium position, it has a speed of 0.40 m/s. What is the speed of the mass when it is 3.0 cm from the equilibrium position?

89 Two objects oscillate on springs in simple harmonic motion. One has twice the mass of the other and oscillates with half the amplitude. The more massive object's spring has a spring constant that is two-thirds of that of the other spring. How do the total energies of the two systems compare?

90 A standing wave has nodes at $x = 0$ cm, $x = 6.0$ cm, $x = 12$ cm, and $x = 18$ cm. (a) What is the wavelength of the waves that are interfering to produce this standing wave? (b) At what positions are the antinodes?

91 A pendulum makes 6.0 complete cycles in a time of 10 s. What are the frequency and period of the pendulum's oscillation?

92 The motion of an object is described by $x = A \cos \omega t$. What are the general equations for the object's velocity and acceleration?

93 A steel piano wire is 60 cm long and has a mass of 3.0 g. If the tension in the wire is 550 N, what are the fundamental frequency and wavelength?

94 A 0.15-kg mass oscillates on a spring that has a spring constant of 500 N/m. (a) What is the energy of the system if the mass oscillates with an amplitude of 0.10 m? (b) What is the energy of the system if the 0.15-kg mass is replaced with a 0.30-kg mass that oscillates with the same amplitude?

95 A stretched string is observed to have four equal loops in a standing wave when driven at a frequency of 420 Hz. What driving frequency will set up a wave with two equal segments?

96 On a violin, a correctly tuned A string has frequency of 440 Hz. If an A string is found to produce a sharp note of 450 Hz when under a tension of 500 N, to what should the tension be adjusted to produce the correct frequency?

97 The amplitude of a damped harmonic oscillation is given as a function of time by $A = A_o e^{-t/\tau}$, where e is the base of the natural logarithms, A_o is the maximum amplitude at $t = 0$, and τ is a constant that depends on the damping forces. (a) How long does it take for the amplitude to be halved if $\tau = 12$ s? (b) How long does it take for the amplitude to go to zero?

98 A stretched string with a length of 1.6 m is driven by a variable frequency oscillator. It is observed that a standing wave with four closed loop segments occurs when the string is driven at 640 Hz. What frequency will produce a standing wave with two loop segments?

99 An object in simple harmonic motion is described by $y = 0.20 \sin 1.8\pi t$ (where y is in centimeters). What is the speed of the object at $t = 10$ s?

100 For an object in simple harmonic motion with a frequency of 3.0 Hz and an amplitude of 7.5 cm, what are the magnitudes of (a) the maximum displacement, (b) the maximum velocity, and (c) the maximum acceleration? (d) What is the object's displacement when each of these conditions occurs?

101 A mass resting on a horizontal frictionless surface is connected to a fixed spring. The mass is displaced 16 cm from its equilibrium position and released. At $t = 0.50$ s, the mass is 8.0 cm from its equilibrium position (and has not passed through it yet). What is the period of oscillation of the mass?

14 Sound

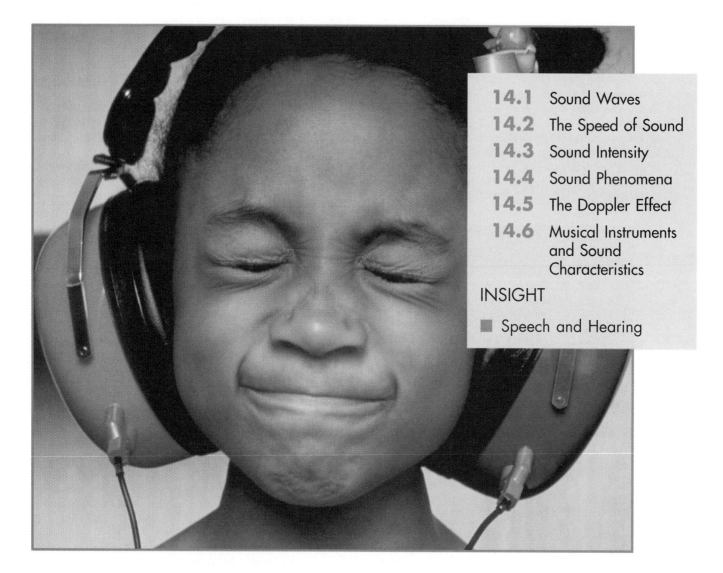

Good vibrations, clearly! But why should mere vibrations in the air have such an effect on us? Whatever the reason, it's clear that we owe a lot to sound waves. Not only do they provide us with one of our main sources of enjoyment in the form of music, they also bring us a wealth of vital information about our environment, from the chime of a doorbell to the warning shrill of a police siren to the song of a mockingbird. Indeed, they are the basis for our major form of communication: speech. They can also constitute a highly irritating distraction (noise). But sound waves become music, speech, or noise only when our ears perceive them. Physically, sound waves are just longitudinal waves that propagate in solids, liquids, and gases. Without a medium, there can be no sound: in a vacuum such as outer space, there would be utter silence.

This distinction between the sensory and physical meanings of sound gives you a way to answer the old philosophical question: If a tree falls in the forest where there is no one to hear it, does it make a sound? The answer depends on how sound is defined—it is no if we are thinking in terms of sensory hearing, but yes if we are considering only the physical waves.

Since sound waves are all around us most of the time, we are exposed to many interesting sound phenomena. You'll explore some of the most important of these in this chapter.

Objectives: To be able to (a) define sound, and (b) explain the sound
frequency spectrum.

For there to be sound waves, there must be a disturbance or vibrations in some
medium. This disturbance may be the clapping of hands, or the skidding of tires
as a car comes to a sudden stop. Underwater you can hear the click of rocks against
one another. If you put your ear to a thin wall, you can hear sounds from the
other side through it. **Sound waves** in gases and liquids (fluids) are primarily longitudinal waves. However, sound disturbances moving through solids can have
both longitudinal and transverse components. The intermolecular interactions in
solids are much stronger than in fluids, and allow transverse waves to propagate.

The characteristics of sound waves can be visualized by considering those
produced by a tuning fork (see •Fig. 14.1). A tuning fork is essentially a metal bar
bent into a U shape. The prongs, or tines, vibrate when struck. The fork vibrates
at its fundamental frequency (with an antinode at the end of each tine), so a single tone is heard. The vibrations disturb the air, producing alternating higher-den-
sity (compressed) regions called *condensations* and lower-density regions called
rarefactions. As the fork vibrates, these disturbances propagate outward, and a series of them can be described by a sinusoidal wave.

As the disturbances traveling through the air reach the ear, the eardrum (a
thin membrane) is set into vibration by the pressure variations. On the other side
of the eardrum, tiny bone structures (the hammer, anvil, and stirrup) carry the vibrations to the inner ear, where they are picked up by the auditory nerve (see the
Insight on p. 450).

Characteristics of the human ear limit the perception of sound. Only sound
waves with frequencies between about 20 Hz and 20 kHz (kilohertz) initiate nerve
impulses that are interpreted by the brain as sound. This frequency range is called
the **audible region** of the **sound frequency spectrum** (•Fig. 14.2). Frequencies
lower than 20 Hz are in the **infrasonic region**. The longitudinal waves generated
by earthquakes have infrasonic frequencies. Above 20 kHz is the **ultrasonic
region**. Ultrasonic waves can be generated by high-frequency vibrations in crystals.

Ultrasonic waves, or ultrasound, cannot be detected by humans but can be
by other animals. The audible region for dogs extends to about 45 kHz, so ultrasonic whistles can be used to call dogs without disturbing people. Cats and bats
have even higher ranges, up to about 70 kHz and 100 kHz, respectively.

There are many other practical applications of ultrasound. Since ultrasound
can travel for kilometers in water, it is used in sonar. Sonar is the ultrasound counterpart of radar, which uses radio waves for ranging and detection. Sound pulses
generated by the sonar apparatus are reflected by underwater objects, and the resulting echoes are picked up by a detector. The time required for a sound pulse
to make the round trip, together with the speed of sound in water, gives the distance of the reflecting object. In a similar manner, ultrasound is used in autofocus

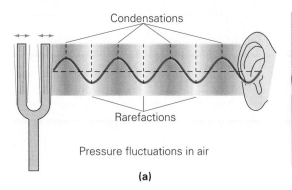

Condensations

Rarefactions

Pressure fluctuations in air

(a)

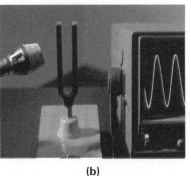

(b)

•**FIGURE 14.1 Vibrations make
waves**
(a) A vibrating tuning fork disturbs
the air, producing alternating high-
pressure regions (condensations)
and low-pressure regions
(rarefactions), which form sound
waves. **(b)** After being picked up
by a microphone, the pressure
variations are converted to
electrical signals. When these are
displayed on an oscilloscope, the
sinusoidal waveform is evident.

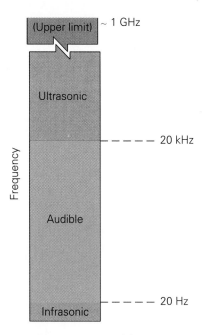

●FIGURE 14.2 Sound frequency spectrum
The audible region of sound for humans lies between about 20 Hz and 20 kHz. Below this is the infrasonic region, and above it is the ultrasonic region.

●FIGURE 14.3 Some uses of ultrasound
(a) This computer-enhanced false-color image, constructed from sonar signals, shows the wreck of the American sailing ship *Hamilton*. The schooner, sunk in Lake Ontario during the War of 1812, lies under 300 feet of water.
(b) Bats hunt flying insects with the aid of their own natural sonar systems. They emit pulses of ultrasonic waves, which lie within their audible region, and use the echoes reflected from their prey to guide their attack. (Notice the size of the animal's ears.)
(c) Ultrasound generated by crystal transducers, which convert electrical oscillations into mechanical vibrations, is transmitted through tissue and is reflected from internal structures. The reflected waves are detected by the transducers and the signals are used to construct an image, or echogram, as shown here for a well-developed fetus.

cameras. Distance measurement allows focus adjustments to be made. Sonar can be used not only for ranging but also to create images of the seafloor and objects resting on it, such as shipwrecks (●Fig. 14.3a).

Interestingly, sonar appeared in the animal kingdom long before it was developed by human engineers. Bats use a kind of natural sonar to navigate in and out of their caves and to locate and catch flying insects on their nocturnal hunting flights. The bats emit pulses of ultrasound and track their prey by means of the reflected echoes (●Fig. 14.3b). Certain species of cave-dwelling birds have evolved the same capability.

In medicine, ultrasound is used to examine internal tissues and organs that are nearly invisible to X-rays. Perhaps the best known medical application of ultrasound is its use to view a fetus without exposing it to the dangerous effects of X-rays (●Fig. 14.3c). Ultrasonic generators (transducers) made of quartz crystals produce high-frequency waves that are used to scan a designated region of the body from several angles. Reflections from the scanned areas are monitored, and a computer constructs an image from the reflected signals. Images are recorded several times each second. The series of images provides a "moving picture" of an internal structure, such as the heart of a fetus. A still shot, or echogram, is shown in the figure.

In industrial and home applications, ultrasonic baths are used to clean metal machine parts and jewelry. The high-frequency (short-wavelength) ultrasound vibrations loosen particles in otherwise inaccessible places.

Ultrasonic frequencies extend into the megahertz (MHz) range, but the sound frequency spectrum does not continue indefinitely. There is an upper limit of about 10^9 Hz, or 1 GHz (gigahertz), which is determined by the upper limit of the elasticity of materials.

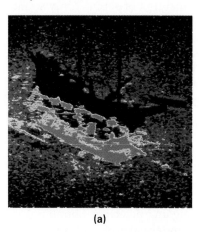

(a)

(b)

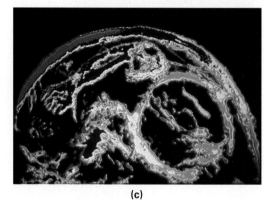

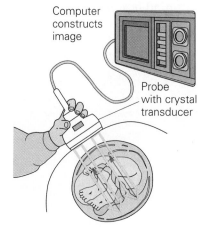

(c)

14.2 The Speed of Sound

Objectives: **To be able to (a) tell how the speed of sound differs in different media, and (b) describe the temperature dependence of the speed of sound in air.**

In general, the speed at which a disturbance moves through a medium depends on the elasticity of the medium and its density. For example, as you learned in Chapter 13, the wave speed in a stretched string is given by $v = \sqrt{F/\mu}$, where F is the tension in the string and μ is the linear mass density.

Similar expressions describe wave speeds in solids and liquids. Here, the elasticity is expressed in terms of moduli (Chapter 9). In general, the speeds of sound in solids and liquids are given by $v = \sqrt{Y/\rho}$ and $v = \sqrt{B/\rho}$, respectively, where Y is Young's modulus, B is the bulk modulus, and ρ is the density. The speed of sound in a gas is inversely proportional to the square root of the density, as for solids and liquids, but the equation is somewhat more complicated, and will not be presented here.

Note: Review Section 9.1.

Solids are generally more elastic than liquids, which in turn are more elastic than gases. In a highly elastic material, the restoring forces between the atoms or molecules cause a disturbance to propagate faster. Thus, the speed of sound is generally about 2–4 times faster in solids than in liquids and about 10–15 times faster in solids than in gases such as air (see Table 14.1).

Although not expressed explicitly in the preceding equations, the speed of sound also depends on the temperature of the medium. In air, for example, the speed of sound is 331 m/s (about 740 mi/h) at a temperature of 0°C. As the temperature increases, so do the speeds of the gas molecules. As a result, the molecules collide with each other more frequently, and a disturbance is transmitted more quickly. Thus the speed of sound in air increases with increasing temperature. For *normal environmental temperatures*, the speed of sound in air increases by about 0.6 m/s for each degree Celsius above 0°C. Thus, a good approximation of the speed of sound in air for a particular (environmental) temperature is given by

$$v = (331 + 0.6T_{\text{C}}) \text{ m/s} \qquad (14.1)$$

where T_{C} is the air temperature in degrees Celsius.* (Although not written explicitly, the units associated with the factor 0.6 are m/s·C°.)

Let's take a comparative look at the speeds of sound in different media.

EXAMPLE 14.1 ■ SOLID, LIQUID, GAS: THE SPEED OF SOUND IN DIFFERENT MEDIA

Find the values of the speeds of sound in (a) a solid copper rod, (b) liquid water, and (c) air at room temperature (20°C) from their material properties.

Solution. To find the speeds of sound in copper and water, we will need the moduli and densities. These are given in Tables 9.1 and 9.2.

Given: $Y_{\text{Cu}} = 11 \times 10^{10} \text{ N/m}^2$ *Find:*   (a) v_{Cu} 0(speed in copper)
$B_{\text{H}_2\text{O}} = 2.2 \times 10^9 \text{ N/m}^2$ (b) $v_{\text{H}_2\text{O}}$ (speed in water)
$\rho_{\text{Cu}} = 8.9 \times 10^3 \text{ kg/m}^3$ (c) v_{air} (speed in air)
$\rho_{\text{H}_2\text{O}} = 1.0 \times 10^3 \text{ kg/m}^3$
$T_{\text{C}} = 20°\text{C}$ (for air)

---

*A better approximation for these and higher temperatures is given by the expression

$$v = 331\sqrt{1 + \frac{T_{\text{C}}}{273}} \text{ m/s}$$

See v for air at 100°C in Table 14.1, which is outside the normal environmental temperature range.

TABLE 14.1   The Speed of Sound in Various Media (typical values)

Medium	Speed (m/s)
Solids	
Aluminum	5100
Copper	3500
Iron	4500
Glass	5200
Polystyrene	1850
Liquids	
Alcohol, ethyl	1125
Mercury	1400
Water	1500
Gases	
Air (0°C)	331
Air (100°C)	387
Helium (0°C)	965
Hydrogen (0°C)	1284
Oxygen (0°C)	316

Speech and Hearing

The Human Voice

Sound is one of our most important means of interpersonal communication, and our most important sound source is the human voice. Let's take a look at the anatomy and physics of the voice and of the ear, which serves as our receiver for sound.

The Human Voice

The energy for sounds associated with the human voice originates in the muscle action of the diaphragm, which forces air up from the lungs. To produce variations in sound, this steady stream of air must be periodically disturbed or "modulated." The fundamental modulating organ is the larynx (the "voice box"), across which are stretched membrane-like bands called the vocal cords. The opening and closing of the vocal cords modulates the air stream to produce sounds.

The effect of the vocal cords is similar to that of the reed or reeds of a woodwind instrument such as a clarinet or oboe. In these instruments the induced vibrations of the reeds convert a steady air flow into a periodic one. For the voice, the fundamental frequency of the modulated sound waves is determined by the tension of the vocal cords, which can be controlled voluntarily. The extent of this control is a primary factor in determining the basic frequency range of the speaking and singing voice.

The sound waves from the larynx are further modulated in the numerous resonance cavities in the throat, mouth, and nose, where standing waves are set up. Some of the vocal cavities can be altered by means of controllable structures, such as the tongue and lips, so as to produce a wide variety of sounds. (Sound out the vowel letters *a*, *e*, *i*, *o*, and *u*, and notice the positions of the tongue and lips.)

The waveforms of voice sounds are quite specific for individuals and provide a "voice print" that can be used for identification, just as fingerprints are. The validity of voice prints for legal identification, however, is still highly controversial.

Hearing

The anatomy of the human ear is illustrated in Fig. 1. Sound enters the outer ear and travels through the ear (or auditory) canal to the *eardrum* (the tympanum), which separates the outer ear from the middle ear. The eardrum is a membrane that vibrates in response to the impinging sound waves. The vibrations are transmitted through the middle ear, which contains an intricate set of connected bones, commonly called the *hammer* (malleus), *anvil* (incus), and *stirrup* (stapes), because of their resemblance to these objects.

The bones of the middle ear form a delicate set of levers with a force multiplication factor (mechanical advantage) of about 2. However, the amplification of the pressure in a wave is much greater because the area of the eardrum is about 20 times that of the *oval window*, the membrane-covered opening to the inner ear. (The stirrup is in contact with the membrane covering.) Thus, pressure is amplified by a factor of about 40.

The inner ear includes the semicircular canals, which are important in controlling balance, and the *cochlea*. It is in the cochlea that sound waves are translated into nerve impulses and that pitch or frequency discrimination is made. The cochlea consists of a series of liquid-filled tubes, coiled into a spiral shape that resembles a snail shell. Coiled up between two of the tubes is a structure called the *basilar membrane* that is supported and stiffened by reed-like *basilar fibers*. On the

---

(a) To find the speed of sound in a copper rod, we use the expression $v = \sqrt{Y/\rho}$ and
$$v_{\text{Cu}} = \sqrt{Y/\rho} = \sqrt{(11 \times 10^{10}\,\text{N/m}^2)/(8.9 \times 10^3\,\text{kg/m}^3)}$$
$$= 3.5 \times 10^3\,\text{m/s}$$

(b) For water, $v = \sqrt{B/\rho}$, and
$$v_{\text{H}_2\text{O}} = \sqrt{B/\rho} = \sqrt{(2.2 \times 10^9\,\text{N/m}^2)/(1.0 \times 10^3\,\text{kg/m}^3)}$$
$$= 1.5 \times 10^3\,\text{m/s}$$

(c) For air at 20°C, by Eq. 14.1, we have
$$v_{\text{air}} = (331 + 0.6\,T_{\text{C}})\,\text{m/s} = 331 + 0.6(20) = 343\,\text{m/s}$$

Follow-up Exercise. In this Example, how many times faster is the speed of sound (a) in copper than in water and (b) in copper than in air (at room temperature)?

A generally useful figure for the speed of sound in air is $\frac{1}{3}$ km/s (or $\frac{1}{5}$ mi/s). Using this figure, you can, for example, estimate how far away lightning has struck by counting the number of seconds between the time the flash is observed and the time the associated thunder is heard. Because of the very fast speed of light, the lightning flash is seen almost instantaneously. The sound waves of the thunder travel relatively slowly, at about $\frac{1}{3}$ km/s. For example, if the interval between

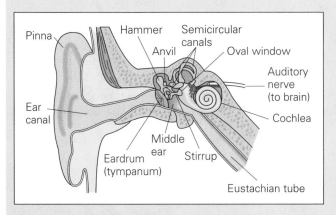

FIGURE 1 Anatomy of the human ear
The ear converts pressure variations in the air into electrical nerve impulses that are interpreted by the brain as sound. (See text for details.)

surface of the basilar membrane rest thousands of special receptor cells called *hair cells*.

The response of the basilar membrane to sounds of different frequency depends largely on resonance. Different sound frequencies transmitted by vibrations of the oval window to the fluid in the cochlea cause different sections of the basilar membrane to vibrate in resonance. As in the resonance of a violin string (Section 13.5), the resonant frequencies depend on mass and tension. The relevant mass in this case is partly that of the fluid in the ducts, which vibrates back and forth, and partly that of the membrane itself, which varies along its length (it is narrower near the window and broader near the tip). The tension is developed mainly by the bending

of the basilar fibers, which vary in length and rigidity in such a way that the membrane is most taut near the window and loosest near the tip. As we would expect, the broad, loose tip of the membrane vibrates most strongly in response to sounds of low frequency, while higher-frequency tones produce the most vibration in the narrower, tauter portion of the membrane nearest the window. Thus frequency information is translated into positional information (different regions of the basilar membrane).

The hair cells that rest on the basilar membrane are specialized nerve receptors. When a region of the basilar membrane is set into vibration by sound of a particular frequency, hair cells in that region are stimulated and nerve impulses are sent to the brain, where they are interpreted as sound. The brain translates this positional information (impulses from particular fibers originating in specific regions of the basilar membrane) back into frequency information (the subjective physiological sensation of pitch).

The sensation of loudness is determined by the amplitude of vibration of the basilar membrane. The greater the amplitude, the more strongly the hair cells are stimulated and the greater the number of impulses they send to the brain in a given period of time. The brain translates this information into degrees of loudness.

Incidentally, the middle ear is connected to the throat by the Eustachian tube, the end of which is normally closed. It opens during swallowing and yawning to permit air to enter and leave, so that internal and external pressures are equalized. You have probably experienced a "stopping up" of the ears with a sudden change in atmospheric pressure (for example, during rapid ascents or descents in elevators or airplanes). Swallowing opens the Eustachian tubes and relieves the excess pressure difference on the middle ear.

the two is measured to be 6 s (often by counting "one thousand one, one thousand two, . . ."), the lightning stroke was about 2 km away ($\frac{1}{3}$ km/s $\times$ 6 s).

You may also have noticed the delay of sound relative to light at a baseball game. If you're sitting in the outfield stands, you see the batter hit the ball before you hear the crack of the bat.

EXAMPLE 14.2 ■ SAFE OR OUT? YOU MAKE THE CALL

On a cool October afternoon (air temperature = 15°C), from your seat in the center-field stands 113 m from first base, you witness the play that will decide the 2001 World Series. You see the runner's foot touch the bag; half a second later, straining your ears, you hear the faint thud of the ball in the first baseman's glove. The umpire signals safe; half the fans boo loudly. As a student of physics, you make the call—did the ump blow it?

Solution. Listing the data, we have

Given: $d = 113$ m   *Find:* t (sound travel time)
$T = 15°C$

The visual observation takes place almost instantaneously, but the sound generated by the ball striking the glove travels more slowly than light and takes more time to reach you. With a temperature of 15°C, the speed of sound is (Eq. 14.1):

$$v = (331 + 0.6\,T_C)\text{ m/s} = 331 + 0.6(15°C) = 340\text{ m/s}$$

For a constant speed, the general distance–time relation is $d = vt$, so the time for the sound to travel the distance to your seat is

$$t = \frac{d}{v} = \frac{113 \text{ m}}{340 \text{ m/s}} = 0.332 \text{ s}$$

or just under $\frac{1}{3}$ of a second. Thus the runner actually reached the base about $\frac{1}{2}$ s $- \frac{1}{3}$ s $= \frac{1}{6}$ s before the ball arrived. The ump was right!

To show why it is justified to say that the visual observation takes place almost instantaneously, let's see how long it takes for light to travel from first base to your seat. For practical purposes, the speed of light in air is the same as that in a vacuum, $c = 3.00 \times 10^8$ m/s, which is on the order of 10^6 (a million) times greater than the speed of sound. Then, the time for light to travel 113 m is

$$t = \frac{d}{c} = \frac{113 \text{ m}}{3.00 \times 10^8 \text{ m/s}} = 3.77 \times 10^{-7} \text{ s}$$

or 0.377 μs—not a very long time.

Follow-up Exercise. On a day when the air temperature is 20°C, a hiker shouts and hears the echo from the vertical stone cliff face 5.00 s later. (a) How far away is the cliff? (b) If the air temperature were 15°C, what would be the difference in the times of hearing the echoes?

14.3 Sound Intensity

Objectives: **To be able to (a) define sound intensity and explain how it varies with distance from a point source, and (b) calculate sound intensities on the decibel scale.**

Wave motion involves the propagation of energy. The rate of the energy transfer is expressed in terms of **intensity**, which is the energy transported per unit time across a unit area. Since energy/time is power, intensity is power/area:

$$\text{intensity} = \frac{\text{energy/time}}{\text{area}} = \frac{\text{power}}{\text{area}}$$

The standard units for intensity (power/area) are watts per meter squared (W/m²).

In the definition of sound intensity we use the component of force perpendicular to the surface area in the direction of wave propagation (just as we used the component of force perpendicular to an area in the definition of pressure in Chapter 9). For example, consider a point source that sends out spherical sound waves, as shown in •Fig. 14.4. If there are no losses, the sound intensity at a distance R from the source is

$$I = \frac{P}{A} = \frac{P}{4\pi R^2} \tag{14.2}$$

where P is the power of the source and $4\pi R^2$ is the area of a sphere with a radius R, through which the sound energy passes perpendicularly.

The intensity for a point source is therefore *inversely proportional to the square of the distance from the source* (an inverse square relationship). Two intensities at different distances from a source of constant power may be compared using a ratio:

$$\frac{I_2}{I_1} = \frac{P/4\pi R_2^2}{P/4\pi R_1^2} = \frac{R_1^2}{R_2^2}$$

or

$$\frac{I_2}{I_1} = \left(\frac{R_1}{R_2}\right)^2 \tag{14.3}$$

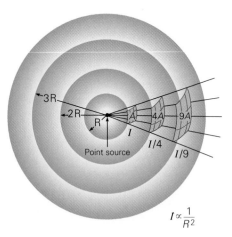

$$I \propto \frac{1}{R^2}$$

•**FIGURE 14.4 Intensity of a point source**
The energy emitted from a point source spreads out equally in all directions. Since intensity is power/area, $I = P/A = P/4\pi R^2$, where the area is that of a spherical surface. The intensity then decreases with the distance from the source as $1/R^2$.

Suppose that the distance from a point source is doubled; that is, $R_2 = 2R_1$, or $R_1/R_2 = \frac{1}{2}$. Then

$$\frac{I_2}{I_1} = \left(\frac{R_1}{R_2}\right)^2 = \left(\frac{1}{2}\right)^2 = \frac{1}{4}$$

and

$$I_2 = \frac{I_1}{4}$$

Since the intensity decreases by a factor of $1/R^2$, doubling the distance decreases the intensity to a quarter of its original value.

A good way to understand this inverse-square relationship intuitively is to look at the geometry of the situation. As you can see in Fig. 14.4, the greater the distance from the source, the larger the area over which a given amount of sound energy is spread, and thus the lower its intensity. (As an analogy, imagine having to paint two walls of different areas. If you had the same amount of paint to use on each, you'd obviously have to spread it more thinly over the larger wall.) Since this area increases as the square of the radius R, the intensity decreases accordingly—that is, as $1/R^2$.

Sound intensity is perceived by the ear as loudness. On the average, the human ear can detect sound waves (at 1 kHz) with an intensity as low as 10^{-12} W/m^2. This intensity (I_o) is referred to as the *threshold of hearing*. Thus, for us to hear a sound, it must not only have a frequency in the audible range, but also be of sufficient intensity. As the intensity is increased, the perceived sound becomes louder. At an intensity of 1.0 W/m^2, the sound is uncomfortably loud and may be painful to the ear. Thus, this intensity (I_p) is called the *threshold of pain*.

Note that the thresholds of pain and hearing differ by a factor of 10^{12}:

$$\frac{I_p}{I_o} = \frac{1.0 \text{ W/m}^2}{10^{-12} \text{ W/m}^2} = 10^{12}$$

Threshold of hearing:
$I_o = 10^{-12}$ W/m^2

Threshold of pain:
$I = 1.0$ W/m^2

That is, the intensity at the threshold of pain is a *trillion* times greater than that at the threshold of hearing. Within this enormous range, the perceived loudness is not directly proportional to the intensity. That is, if the intensity is doubled, the perceived loudness does not double. A doubling of perceived loudness corresponds approximately to an increase in intensity by a factor of 10. For example, a sound with an intensity of 10^{-5} W/m^2 would be perceived to be twice as loud as one with an intensity of 10^{-6} W/m^2 (the smaller the negative exponent, the larger the number).

It is convenient to compress the large range of sound intensities by using a logarithmic scale (base-10) to express intensity levels. The intensity level of a sound must be referenced to a standard intensity, which is taken to be that of the threshold of hearing, $I_o = 10^{-12}$ W/m^2. Then, for any intensity I, the intensity level is the log of the ratio of I to I_o, that is, log I/I_o. For example, if a sound has an intensity of $I = 10^{-6}$ W/m^2,

$$\log \frac{I}{I_o} = \log \frac{10^{-6} \text{ W/m}^2}{10^{-12} \text{ W/m}^2} = \log 10^6 = 6$$

(Recall that $\log_{10} 10^x = x$.) Thus, a sound with an intensity of 10^{-6} W/m^2 has an intensity level of 6 bel (B) on this scale. In this way, the intensity range from 10^{-12} W/m^2 to 1.0 W/m^2 is compressed into a scale of intensity levels running from 0 B to 12 B.

Note: The bel was named in honor of Alexander Graham Bell, the inventor of the telephone.

Sound Level Intensity—The Decibel

A finer intensity scale is obtained by using a smaller unit, the **decibel** (dB), which is a tenth of a bel. The 0–12 B range corresponds to 0–120 dB. In this case, the equation for the relative **sound intensity level**, or **decibel level** (β), is

$$\beta = 10 \log \frac{I}{I_o} \quad \text{where } I_o = 10^{-12} \text{ W/m}^2 \quad (14.4)$$

Sound level intensity in decibels

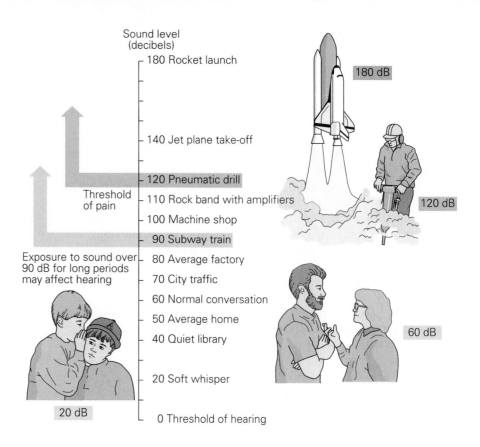

Sound level
(decibels)

- 180 Rocket launch

180 dB

- 140 Jet plane take-off

- 120 Pneumatic drill

120 dB

Threshold
of pain
- 110 Rock band with amplifiers
- 100 Machine shop
- 90 Subway train

Exposure to sound over
90 dB for long periods
may affect hearing
- 80 Average factory
- 70 City traffic
- 60 Normal conversation

60 dB

- 50 Average home
- 40 Quiet library

- 20 Soft whisper

20 dB

- 0 Threshold of hearing

•FIGURE 14.5 **Sound intensity
levels and the decibel scale**
The intensity levels of some
common sounds on the decibel
(dB) scale.

The decibel intensity scale and familiar sounds at some intensity levels are shown
in •Fig. 14.5. Sound intensities can have detrimental effects on hearing, and because of this the government has set occupational noise-exposure limits.

EXAMPLE 14.3 ■ SOUND INTENSITY LEVELS: USING LOGARITHMS

What are the intensity levels for sounds with intensities of (a) $10^{-12}\,\text{W/m}^2$ and
(b) $5.0 \times 10^{-6}\,\text{W/m}^2$?

Solution.

Given: (a) $I = 10^{-12}\,\text{W/m}^2$ *Find:* (a) β (sound level intensity)
         (b) $I = 5.0 \times 10^{-6}\,\text{W/m}^2$ (b) β

Using Eq. 14.4 we have,

(a)
$$\beta = 10 \log \frac{I}{I_o} = 10 \log \frac{10^{-12}\,\text{W/m}^2}{10^{-12}\,\text{W/m}^2}$$
$$= 10 \log 1 = 0\,\text{dB}$$

The intensity is the same as that at the threshold of hearing. (Recall that $\log 1 = 0$, since
$1 = 10^0$ and $\log 10^0 = 0$.)

(b)
$$\beta = 10 \log \frac{I}{I_o} = 10 \log \left(\frac{5.0 \times 10^{-6}\,\text{W/m}^2}{10^{-12}\,\text{W/m}^2} \right)$$
$$= 10 \log (5.0 \times 10^6) = 10(\log 5.0 + \log 10^6)$$
$$= 10(0.70 + 6.0) = 67\,\text{dB}$$

Follow-up Exercise. Note in this example that the intensity of $5.0 \times 10^{-6}\,\text{W/m}^2$ is
halfway between 10^{-6} and 10^{-5} (or 60 and 70 dB), yet this intensity does not correspond
to a midway 65 dB. (a) Why is this? (b) What intensity does correspond to 65 dB? (Compute to 3 significant figures.)

EXAMPLE 14.4 ■ INTENSITY LEVEL FACTORS: USING RATIOS

(a) What is the difference in the intensity level if the intensity of a sound is doubled? (b) By what factors does the intensity increase for intensity level *differences* of 10 dB and 20 dB?

Solution. Note in (b) that these values are differences, $\Delta\beta = \beta_2 - \beta_1$, not intensity levels. Then,

Given: (a) $I_2 = 2I_1$ **Find:** (a) $\Delta\beta$ (intensity level difference)
   (b) $\Delta\beta = 10$ dB   (b) I_2/I_1 (factors of increase)
      $\Delta\beta = 20$ dB

(a) The intensity is doubled, so $I_2/I_1 = 2$. This ratio can be used directly to find the intensity level *difference* because of the properties of logarithms. Recall that $\log a - \log b = \log a/b$. Using Eq. 14.4 and this relationship, we have for the difference $\Delta\beta = \beta_2 - \beta_1 = 10[\log(I_2/I_o) - \log(I_1/I_o)] = 10 \log(I_2/I_o)/(I_1/I_o) = 10 \log I_2/I_1$. Then,

$$\Delta\beta = 10 \log \frac{I_2}{I_1} = 10 \log 2 = 3 \text{ dB}$$

Thus, doubling the intensity increases the intensity level by 3 dB (for example, an increase from 55 dB to 58 dB).

(b) For a 10-dB difference,

$$\Delta\beta = 10 \text{ dB} = 10 \log \frac{I_2}{I_1}, \quad \text{and} \quad \log \frac{I_2}{I_1} = 1.0$$

Since $\log 10^1 = 1$,

$$\frac{I_2}{I_1} = 10^1 \quad \text{and} \quad I_2 = 10I_1$$

Similarly, for a 20-dB difference,

$$\Delta\beta = 20 \text{ dB} = 10 \log \frac{I_2}{I_1} \quad \text{and} \quad \log \frac{I_2}{I_1} = 2.0$$

Since $\log 10^2 = 2$,

$$\frac{I_2}{I_1} = 10^2 \quad \text{and} \quad I_2 = 100I_1$$

Thus, an intensity level difference of 10 dB corresponds to increasing (or decreasing) the intensity by a factor of 10. An intensity level difference of 20 dB corresponds to increasing (or decreasing) the intensity by a factor of 100. You should be able to guess the factor that corresponds to an intensity level difference of 30 dB. In general, the factor of the intensity change is $10^{\Delta B}$, where ΔB is the level difference in bels. Since 30 dB = 3 B and $10^3 = 1000$, the intensity changes by a factor of 1000 for an intensity level difference of 30 dB.

Follow-up Exercise. A $\Delta\beta$ of 20 dB and a $\Delta\beta$ of 30 dB correspond to intensity change factors of 100 and 1000, respectively. Does a $\Delta\beta$ of 25 dB correspond to an intensity change factor of 500? Explain.

EXAMPLE 14.5 ■ COMBINED SOUND LEVELS: ADDING INTENSITIES

Sitting at a sidewalk restaurant table, a friend talks to you in normal conversation (60 dB). At the same time, the intensity level of the street traffic reaching you is also 60 dB. What is the total intensity level of the combined sounds?

Solution. We have

Given: $\beta_1 = 60$ dB **Find:** Total β
   $\beta_2 = 60$ dB

It is tempting to simply add the two sound intensity levels together and say the total is 120 dB. However, this is not correct. Keep in mind that intensity levels are logarithmic functions, so they do not add like regular numbers.

Intensities (W/m^2) do add regularly, so let's find the intensities associated with the intensity levels:

$$\beta_1 = 60 \text{ dB} = 10 \log \frac{I_1}{I_o} = 10 \log \frac{I_1}{10^{-12} \text{ W/m}^2}$$

by inspection, we have

$$I_1 = 10^{-6} \text{ W/m}^2$$

Similarly, $I_2 = 10^{-6} \text{ W/m}^2$, since both intensity levels are 60 dB. So, the total intensity is

$$I_t = I_1 + I_2 = 1.0 \times 10^{-6} \text{ W/m}^2 + 1.0 \times 10^{-6} \text{ W/m}^2 = 2.0 \times 10^{-6} \text{ W/m}^2$$

Then, converting back to intensity level,

$$\beta = 10 \log \frac{I_t}{I_o} = 10 \left(\frac{2.0 \times 10^{-6} \text{ W/m}^2}{10^{-12} \text{ W/m}^2} \right) = 10 \log (2.0 \times 10^6) = 10(\log 2.0 + \log 10^6)$$

$$= 10(0.30 + 6.0) = 63 \text{ dB}$$

—a long way from 120 dB! Notice that the combined intensities doubled the value and the intensity level increased by 3 dB, in agreement with our finding in part (a) of the previous example.

Follow-up Exercise. In this Example, suppose the added noise gave a total that *tripled* the value of the conversation intensity. What would be the total combined intensity level in this case?

14.4 Sound Phenomena

Objectives: To be able to (a) explain sound reflection, refraction, and diffraction, and (b) distinguish between constructive and destructive interference.

Reflection, Refraction, and Diffraction

A sound wave can be reflected. An echo is a familiar example of sound *reflection*.

Sound *refraction* is less common, but you may have experienced this effect on a calm summer evening. Then it is sometimes possible to hear distant voices or other sounds that ordinarily would not be audible. This effect is due to the refraction, or bending, of the sound waves as they pass into a region where the air density is different. The effect is similar to what would happen if the sound passed into another medium.

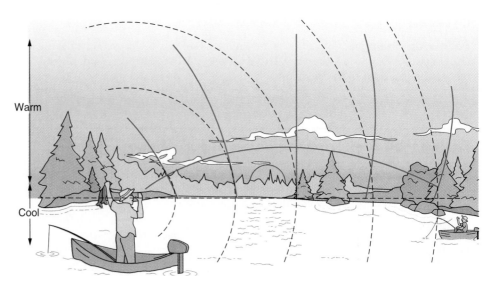

•**FIGURE 14.6 Sound refraction** Sound is refracted (bent) more in the overlying warmer air. This increases the intensity of the sound at a distance where it otherwise might not be heard.

The required conditions are a layer of cooler air near the ground with a layer of warmer air above it. These conditions occur frequently over bodies of water after sunset because of cooling (•Fig. 14.6). The sound waves spread out and upward from their source and would ordinarily be too faint to be heard by someone at a distance. But, as the waves pass into the overlying layer of warm air, they travel faster and are bent, or refracted, toward the distant person. This increases the intensity of the sound received by that person. If the intensity is above the threshold of hearing, the distant sound can be heard.

Sound is also *diffracted* or bent around corners, as you know from experience. Reflection, refraction, and diffraction of light are important phenomena that will be considered in more detail in Chapters 22 and 24.

Interference

Like waves of any kind, sound waves *interfere* when they meet. Suppose that two loudspeakers separated by some distance emit sound waves in phase at the same frequency (see •Fig. 14.7a). Consider the speakers to be point sources—the waves spread out spherically and interfere. The lines from a particular speaker represent wave crests (or condensations), and the troughs (or rarefactions) lie in the intervening white areas.

In particular regions of space, there will be constructive or destructive interferences. For example, if the waves meet in a region where they are exactly in phase (i.e., where two crests or two troughs coincide), there will be total **constructive interference**. This situation is illustrated in •Fig. 14.7b. Notice that the waves have the same motion at point C. Conversely, if the waves meet so that the crest of one coincides with the trough of the other (at point D), the two waves will cancel each other out (•Fig. 14.7c). The result will be total **destructive interference**.

It is convenient to describe the path lengths traveled by the waves in terms of wavelength (λ) to determine whether they arrive in phase. In such analysis we work with points of zero displacement rather than crests or troughs for convenience. Consider the waves arriving at point C in Fig. 14.7b. The path lengths in this case are AC = 4λ and BC = 3λ. The **phase difference** ($\Delta\theta$) is related to the **path difference** (ΔL) by the simple relationship:

$$\Delta\theta = \frac{2\pi}{\lambda}(\Delta L) \quad \textit{phase difference and path difference} \quad (14.5)$$

Since 2π rad is equivalent, in angular terms, to a full wave cycle or wavelength, multiplying the path difference by $2\pi/\lambda$ gives the phase difference in rad. For the example illustrated in Fig. 14.7b, we have

$$\Delta\theta = \frac{2\pi}{\lambda}(AC - BC) = \frac{2\pi}{\lambda}(4\lambda - 3\lambda) = 2\pi \text{ rad}$$

When $\Delta\theta = 2\pi$ rad, the waves are shifted by one wavelength. This is the same as $\Delta\theta = 0°$, so the waves are in phase. Thus, the waves interfere constructively in the point C region, increasing the intensity, or loudness, of the sound detected there.

From Eq. 14.5, it should be clear that the sound waves are in phase at any point where the path difference is zero or an integral multiple of the wavelength. That is,

$$\Delta L = n\lambda \quad (n = 0, 1, 2, 3, \ldots) \quad \begin{array}{l}\textit{condition for}\\\textit{constructive interference}\end{array} \quad (14.6)$$

A similar analysis for the situation in •Fig. 14.7c, where AD = $2\frac{3}{4}\lambda$ and BD = $2\frac{1}{4}\lambda$, gives

$$\Delta\theta = \frac{2\pi}{\lambda}[2(\tfrac{3}{4}\lambda) - 2(\tfrac{1}{4}\lambda)] = \pi \text{ rad}$$

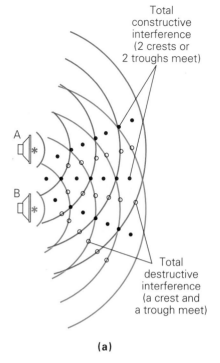

Total constructive interference (2 crests or 2 troughs meet)

Total destructive interference (a crest and a trough meet)

(a)

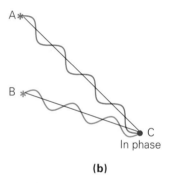

In phase

(b)

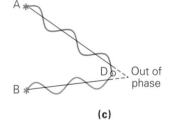

Out of phase

(c)

•**FIGURE 14.7 Interference** **(a)** Sound waves from two point sources spread out and interfere. At points where the waves arrive in phase (zero phase difference), such as point C in **(b)**, constructive interference occurs. At points where the waves arrive completely out of phase (phase difference of 180°), such as point D in **(c)**, destructive interference occurs. The phase difference at a particular point depends on the path lengths the waves travel to reach that point.

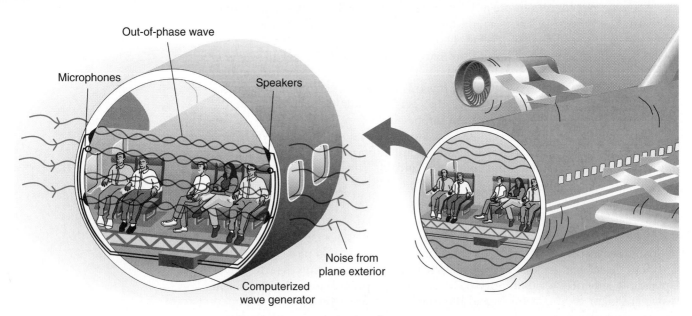

•FIGURE 14.8 **Destructive interference**
Destructive interference is used in a technique to reduce the cabin noise in commercial jet aircraft. Microphones route noise to a computer, which then drives speakers to produce sound waves 180° out of phase with the noise. The resulting destructive interference makes the cabin quieter.

or $\Delta\theta = 180°$. At point D, the waves are completely out of phase, and destructive interference occurs in this region.

Sound waves will be out of phase at any point where the path difference is an odd number of half-wavelengths ($\lambda/2$), or

$$\Delta L = m\left(\frac{\lambda}{2}\right) \quad (m = 1, 3, 5, \ldots) \qquad \text{condition for destructive interference} \qquad (14.7)$$

At these points, a softer or less intense sound will be heard or detected. If the amplitudes of the waves are equal, the destructive interference is total, and no sound is heard!

An application of destructive interference for sound waves is illustrated in •Fig. 14.8. Engine noise in commercial jet aircraft cabins can be distracting. In a new technique, microphones throughout the cabin pick up the various sounds; a computer analyzes them and generates sound waves that are 180° out of phase with the undesirable noise. Destructive interference then reduces the cabin noise.

EXAMPLE 14.6 ■ PUMP UP THE VOLUME: SOUND INTERFERENCE

At an open-air concert on a hot day (air temperature of 25°C), a person sits at a location that is 7.00 m and 9.10 m, respectively, from speakers at each side of the stage. A musician, warming up, plays a single 494-Hz tone. What does the listener hear? (Consider the speakers to be point sources.)

Solution. Since the sound waves interfere, what the person hears depends on the phase difference at that location. Hence, we have

Given: $d_1 = 7.00$ m and $d_2 = 9.10$ m   *Find:* $\Delta\theta$ (phase difference)
  $f = 494$ Hz
  $T = 25°C$

The phase difference between the waves arriving at the listener's location is found by expressing the path lengths in wavelengths. To do this, we first need to know the wavelength. Given the frequency, this can be found from the relationship $\lambda = v/f$ provided we know the value of the speed of sound v at the given temperature. We calculate v by means of Eq. 14.1:

$$v = 331 + 0.6T_C = 331 + 0.6(25) = 346 \text{ m/s}$$

The wavelength of the sound waves is then

$$\lambda = \frac{v}{f} = \frac{346 \text{ m/s}}{494 \text{ Hz}} = 0.700 \text{ m}$$

Thus, the distances in terms of wavelength are

$$d_1 = (7.00 \text{ m})\left(\frac{\lambda}{0.700 \text{ m}}\right) = 10.0\lambda$$

and

$$d_2 = (9.10 \text{ m})\left(\frac{\lambda}{0.700 \text{ m}}\right) = 13.0\lambda$$

The path difference is

$$\Delta L = d_2 - d_1 = 13.0\lambda - 10.0\lambda = 3.0\lambda$$

This is an integral number of wavelengths ($n = 3$), so constructive interference occurs; the sounds of the two speakers reinforce each other, and the listener hears an intense tone.

Follow-up Exercise. Suppose that in this Example the tone traveled to a person sitting 7.00 m and 8.75 m, respectively, from the two speakers. What would be the situation in this case?

Another interesting interference effect occurs when two tones of nearly the same frequency ($f_1 \approx f_2$) are sounded simultaneously. The ear senses pulsations in loudness known as **beats**. (The human ear can detect as many as 7 beats per second before they sound "smooth" or unpulsating.)

For example, suppose that two sinusoidal waves with the same amplitude have different frequencies. As these waves interfere, the total displacement at some location is (by the principle of superposition)

$$y = y_1 + y_2 = A \sin 2\pi f_1 t + A \sin 2\pi f_2 t$$

The trigonometric identity

$$\sin a + \sin b = 2 \cos\left(\frac{a - b}{2}\right) \sin\left(\frac{a + b}{2}\right)$$

allows the equation for y to be rewritten as

$$y = \left[2A \cos 2\pi\left(\frac{f_1 - f_2}{2}\right)t\right]\left[\sin 2\pi\left(\frac{f_1 + f_2}{2}\right)t\right] \tag{14.8}$$

•Figure 14.9 represents the resulting sound wave. The purple curve is a sinusoidal wave whose frequency is the average of the two frequencies producing it: $(f_1 + f_2)/2$. This wave is described by the sine term in Eq. 14.8. The amplitude of the wave varies sinusoidally, as shown by the black curve (known as an *envelope*). This variation in amplitude, expressed by the bracketed cosine term in Eq. 14.8, has a frequency of $(f_1 - f_2)/2$. The maximum amplitude is $2A$ (at those points where maxima of the two original tones interfere constructively).

What does this mean in terms of perception? A listener will hear a sound of frequency $(f_1 + f_2)/2$ that pulsates with a frequency of $(f_1 - f_2)/2$. You can see from Fig. 14.9 that a full cycle of the envelope representing this pulsation comprises two maximum amplitudes of the heard tone (red curve). This means that two maximum amplitudes of $2A$ are heard each cycle. Thus, the **beat frequency** (f_b) that is perceived is twice the envelope (cosine) frequency, or

$$f_b = |f_1 - f_2| \tag{14.9}$$

(The absolute value is taken because there cannot be a negative frequency.)

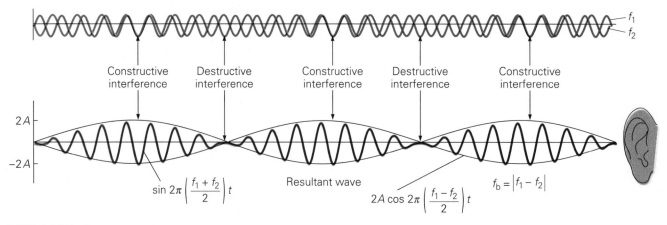

Constructive interference　Destructive interference　Constructive interference　Destructive interference　Constructive interference

$2A$

$-2A$

$\sin 2\pi \left(\dfrac{f_1 + f_2}{2} \right) t$

Resultant wave

$2A \cos 2\pi \left(\dfrac{f_1 - f_2}{2} \right) t$

$f_b = |f_1 - f_2|$

●FIGURE 14.9 Beats

Two traveling waves of equal amplitude and slightly different frequencies interfere and give rise to pulsating tones called beats. The beat frequency is given by $f_b = |f_1 - f_2|$.

Beats may be produced when tuning forks of nearly the same frequency are vibrating at the same time. For example, using forks with frequencies of 516 Hz and 513 Hz, the beat frequency is $f_b = 516$ Hz $-$ 513 Hz $= 3$ Hz, and 3 pulsations, or beats, are heard each second. Musicians tune two stringed instruments to the same note by adjusting the strings until the beats disappear ($f_1 = f_2$).

14.5 The Doppler Effect

Objectives: **To be able to (a) describe and explain the Doppler effect, and (b) give some examples of its occurrences and applications.**

If you stand along a highway and a car or truck approaches you with the horn blowing, the pitch (the perceived frequency) of the sound is higher as the vehicle approaches and lower as it recedes. You can also hear variations in the frequency of the motor noise when watching a racing car going around a track. A variation in the perceived sound frequency due to the motion of the sound source is an example of the **Doppler effect**.

As ●Fig. 14.10 shows, the sound waves emitted by a moving source tend to bunch up in front of the source and spread out in back. The Doppler shift in frequency can be found by assuming that the air is at rest in a reference frame such as that depicted in ●Fig. 14.11. The speed of sound in air is v and the speed of the moving source is v_s. The frequency of the sound produced by the source is f_s. In one period, $T = 1/f_s$, a wave crest moves a distance $d = vT = \lambda$. (The sound wave would travel this distance in still air in any case, whether or not the source were moving.) But, in one period, the source travels a distance $d_s = v_sT$ before emitting another wave crest. The distance between the successive wave crests is thus shortened to a wavelength λ':

$$\lambda' = d - d_s = vT - v_sT$$

$$= (v - v_s)T = \frac{v - v_s}{f_s}$$

The frequency heard by the observer (f_o) is related to the shortened wavelength by $f_o = v/\lambda'$, and substituting v/λ' gives

$$f_o = \frac{v}{\lambda'} = \left(\frac{v}{v - v_s} \right) f_s$$

The Austrian physicist Christian Doppler (1803–1853) first described what we now call the Doppler effect.

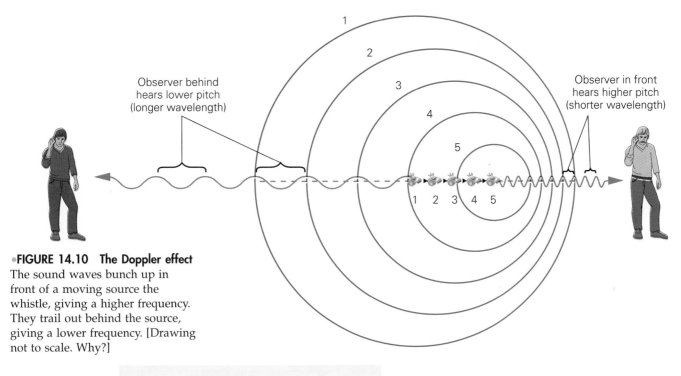

Observer behind
hears lower pitch
(longer wavelength)

Observer in front
hears higher pitch
(shorter wavelength)

•FIGURE 14.10 **The Doppler effect**
The sound waves bunch up in
front of a moving source the
whistle, giving a higher frequency.
They trail out behind the source,
giving a lower frequency. [Drawing
not to scale. Why?]

or

$$f_o = \left(\frac{1}{1 - \dfrac{v_s}{v}} \right) f_s$$

(14.10)

source moving toward a stationary observer
where v_s = speed of source
and v = speed of sound

Since $(1 - v_s/v)$ is less than 1, f_o is greater than f_s in this situation. For example,
suppose that the speed of the source is a tenth of the speed of sound: $v_s = v/10$,
or $v_s/v = \frac{1}{10}$. Then, by Eq. 14.10, $f_o = \frac{10}{9}f_s$.

Similarly, when the source is moving away from the observer ($\lambda' = d + d_s$),
the observed frequency is given by

$$f_o = \left(\frac{v}{v + v_s} \right) f_s = \left(\frac{1}{1 + \dfrac{v_s}{v}} \right) f_s$$

(14.11)

source moving away from a stationary observer

Here f_o is less than f_s. (Why?)

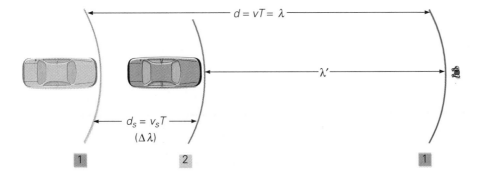

$d = vT = \lambda$

λ'

$d_s = v_s T$
$(\Delta \lambda)$

1   2   1

•FIGURE 14.11 **The Doppler effect
and wavelength**
Sound from a moving car's horn
travels a distance d in a time T.
During this time, the car (the
source) travels a distance d_s before
putting out a second pulse, thereby
shortening the observed
wavelength of the sound in the
approaching direction.

Combining Eqs. 14.10 and 14.11 yields a general equation for the observed frequency with a moving source and a stationary observer.

$$f_o = \left(\frac{v}{v \pm v_s} \right) f_s = \left(\frac{1}{1 \pm \dfrac{v_s}{v}} \right) f_s \tag{14.12}$$

$\begin{cases} - \text{ for source moving toward stationary observer} \\ + \text{ for source moving away from stationary observer} \end{cases}$

As you might expect, the Doppler effect also occurs with a moving observer and a stationary source. This situation is a bit different. As the observer moves toward the source, the distance between successive wave crests is the normal wavelength (or $\lambda = v/f_s$), but the measured wave speed is different. Relative to the approaching observer, the sound from the stationary source has a wave speed of $v' = v + v_o$, where v_o is the speed of the observer and v is the speed of sound in still air. (The observer moving toward the source is moving in a direction opposite that of the propagating waves, and thus meets more wave crests in a given time.) The observed frequency (f_o) is then (with $\lambda = v/f_s$):

$$f_o = \frac{v'}{\lambda} = \left(\frac{v + v_o}{v} \right) f_s$$

or

$$f_o = \left(1 + \frac{v_o}{v} \right) f_s$$

observer moving toward a stationary source   (14.13)
where v_o = speed of observer
and v = speed of sound

Similarly, for an observer moving away from a stationary source, the perceived wave speed is $v' = v - v_o$, and

$$f_o = \frac{v'}{\lambda} = \left(\frac{v - v_o}{v} \right) f_s$$

or

$$f_o = \left(1 - \frac{v_o}{v} \right) f_s$$

observer moving away from a stationary source   (14.14)

Equations 14.13 and 14.14 can be combined into another general equation for a moving observer and a stationary source.

$$f_o = \left(\frac{v \pm v_o}{v} \right) f_s = \left(1 \pm \frac{v_o}{v} \right) f_s \tag{14.15}$$

$\begin{cases} + \text{for observer moving toward stationary source} \\ - \text{for observer moving away from stationary source} \end{cases}$

There are also the cases when both the source and the observer are moving, either toward one another or away from one another. However, these will not be considered mathematically.

EXAMPLE 14.7 ■ ON THE ROAD AGAIN: THE DOPPLER EFFECT

As a truck traveling at 96 km/h approaches and passes a person standing along the highway, the driver sounds the horn. If the horn has a frequency of 400 Hz, what are the frequencies of the sound waves heard by the person (a) as the truck approaches and (b) after it has passed? (Assume that the speed of sound is 346 m/s.)

Solution.

Given: $v_s = 96$ km/h $= 27$ m/s *Find:* (a) f_o (observed frequency approaching)
  $f_s = 400$ Hz (b) f_o (moving away)
  $v = 346$ m/s

(a) Using Eq. 14.12 with a minus sign (source approaching stationary observer),

$$f_o = \left(\frac{v}{v - v_s}\right)f_s = \left(\frac{346 \text{ m/s}}{346 \text{ m/s} - 27 \text{ m/s}}\right)(400 \text{ Hz}) = 434 \text{ Hz}$$

(b) A plus sign is used when the source is moving away:

$$f_o = \left(\frac{v}{v + v_s}\right)f_s = \left(\frac{346 \text{ m/s}}{346 \text{ m/s} + 27 \text{ m/s}}\right)(400 \text{ Hz}) = 371 \text{ Hz}$$

Follow-up Exercise. Suppose that the observer in this Example was initially moving toward a stationary 400-Hz source with a speed of 96 km/h. What would be the observed frequencies in this case?

CONCEPTUAL EXAMPLE 14.8 ■ IT'S ALL RELATIVE: MOVING SOURCE AND MOVING OBSERVER

Suppose a sound source and an observer were moving away from one another in opposite directions, each at one half the speed of sound in air. Then, the observer would (a) receive sound with a frequency greater than the source frequency, (b) receive sound with a frequency less than the source frequency, (c) receive sound with the same frequency as the source frequency, (d) receive no sound from the source. *Clearly establish the reasoning and physical principle(s) used in determining your answer before checking it below. That is, **why** did you select your answer?*

Reasoning and Answer. As we know, when a source moves away from a stationary observer, the observed frequency is lower (Eq. 14.11). Similarly, when an observer moves away from a stationary source, the observed frequency is also lower (Eq. 14.14). With both source and observer moving away from each other in opposite directions, the combined effect would make the observed frequency even less, so the answer is not (a) or (c).

Looking at (b) and (d), it would appear that (b) is the correct answer, but we must logically eliminate (d) for completeness. It should be remembered that the speed of sound relative to the air is constant. Therefore, (d) would be correct *only if the observer is moving faster than the speed of sound* relative to the air. Since the observer is moving at only one half the speed of sound, (b) is the correct answer. Think about it this way. The sound from the source is moving at the speed of sound through the air toward the observer regardless of how fast the source is moving. The observer is moving at only one half the speed of sound through the air, so the sound from the source can easily reach the observer.

Follow-up Exercise. What would be the situation if the source and the observer were both traveling in the same direction with the same subsonic speed? (Subsonic, as opposed to supersonic, refers to a speed that is less than the speed of sound in air.)

PROBLEM-SOLVING HINT

You may find it difficult to remember whether a plus or minus sign is used in the general equations for the Doppler effect. Let your experience help you. For the commonly experienced case of being a stationary observer, the sound frequency increases when the source approaches, so the denominator in Eq. 14.12 must be smaller than the numerator. For this case, you use the minus sign. When the source is receding, the frequency is lower. The denominator in Eq. 14.12 must then be larger than the numerator, and you use the plus sign for this case. Similar thinking will help you choose a plus or minus sign for the numerator in Eq. 14.15.

The Doppler effect also occurs for light waves, although the formulas describing it are different from those given above. When a distant light source such as a star moves away from us, the frequency of the light we receive from it is lowered. That is, the light is shifted toward the red (longer wavelength) end of the spectrum, an effect known as a Doppler *red shift*. Similarly, the frequency of light from an object approaching us is increased—the light is shifted toward the blue (shorter wavelength) end of the spectrum, producing a Doppler *blue shift*. The magnitude of the frequency shift is related to the speed of the source.

The Doppler shift of light from astronomical objects is very useful to astronomers. The rotation of a planet, star, or other body can be established by looking at the Doppler shifts of light from opposite sides of the object—because of the rotation, one side is receding (and hence red-shifted) and the other approaching (blue-shifted). Similarly, the Doppler shifts of light from stars in different regions of our galaxy, the Milky Way, indicates that the galaxy is rotating.

Light from distant galaxies is also found to be red-shifted, which indicates that they are moving away from us. This has been taken as evidence of an expanding universe. However, by modern interpretations, this is not a true Doppler shift, but a shift in the wavelength of light produced by the expansion of the universe as a whole. The wavelength of light expands along with the universe, giving a *cosmological red shift* (see Chapter 26).

You have been subjected to a practical application of the Doppler effect if you have ever been caught speeding in your car by police radar, which uses reflected radio waves. (Radar stands for *r*adio *d*etecting *a*nd *r*anging and is similar to underwater sonar, which uses ultrasound.) If the radio waves are reflected from a parked car, the reflected waves return to the source with the same frequency. But for a car that is moving toward a patrol car, the reflected waves have a higher frequency, or are Doppler-shifted. Actually, there is a double Doppler shift. The moving car acts like a moving observer in receiving the wave (first Doppler shift), and in reflecting it the car acts like a moving source emitting a wave (second Doppler shift). The magnitudes of the shifts depend on the speed of the car. A computer quickly calculates this speed and displays it for the police officer.

Sonic Booms

Consider a jet plane that can travel at supersonic speeds. As the speed of a moving source of sound approaches the speed of sound, the waves ahead of the source come close together (•Fig. 14.12). When a plane is traveling at the speed of sound, the waves can't outrun it, and they pile up in front. At supersonic speeds, the waves overlap. This overlapping of a large number of waves produces many points of constructive interference, forming a large pressure ridge, or shock wave. This is sometimes called a bow wave because it is analogous to the wave produced by the bow of a boat moving through water at a speed greater than the speed of the water waves. (Fig. 14.12c).

From aircraft traveling at supersonic speed, the shock wave trails out to the sides and downward. When this pressure ridge passes over an observer on the ground, the large concentration of energy produces what is known as a **sonic boom.** There is really a double boom because shock waves are formed at both ends of the aircraft. Under certain conditions, the shock waves can break windows and cause other damage. (Sonic booms are no longer heard as frequently as in the past. Pilots are now instructed to fly supersonically only at high altitudes and away from populated areas.)

On a smaller scale, you have probably heard a "mini" sonic boom. The "crack" of a whip is actually a sonic boom created by the transonic speed of the whip's tip.

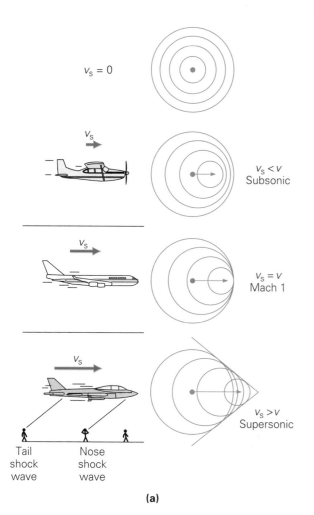

(a)

(b)

(c)

•**FIGURE 14.12 Bow waves and sonic booms**
(a) When an aircraft exceeds the speed of sound in air, the sound waves form a pressure ridge, or shock wave. As the trailing shock wave passes over the ground, observers hear a sonic boom—actually two booms, because shock waves are formed at the front and tail of the plane. **(b)** A bullet traveling at a speed of 500 m/s. Note the shock waves produced (and the turbulence behind the bullet). The image was made using interferometry with polarized light and a pulsed laser, with an exposure time of 20 ns. **(c)** A shock wave in water. Bow waves like the ones shown here are produced whenever a boat travels faster than the speed of the water waves that it creates.

A common misconception is that a sonic boom occurs only when the plane breaks the sound barrier. As an aircraft approaches the speed of sound, the pressure ridge in front of it is essentially a barrier that must be overcome with extra power. However, once supersonic speed is reached, this barrier is no longer there, and the shock wave, continuously created, trails behind the plane, producing booms along its ground path.

Ideally, the sound waves produced by a supersonic aircraft form a cone-shaped shock wave (•Fig. 14.13). The waves travel outward with a speed v, and the speed of the plane (the source) is v_s. Note from the figure that the angle between a line tangent to the spherical waves and the line along which the plane is moving is given by

$$\sin \theta = \frac{vt}{v_s t} = \frac{v}{v_s} \tag{14.16}$$

The inverse ratio of the speeds is called the **Mach number** (M):

$$M = \frac{v_s}{v} \tag{14.17}$$

If v equals v_s, the plane is flying at the speed of sound, and the Mach number is 1 (that is, $v_s/v = 1$). Therefore, a Mach number less than 1 indicates a subsonic speed, and a Mach number greater than 1 indicates a supersonic speed. In the latter case, the Mach number tells the speed of the aircraft in terms of a multiple of the speed of sound.

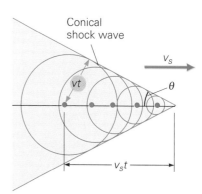

•**FIGURE 14.13 Shock wave cone and Mach number**
When the speed of the source (v_s) is greater than the speed of sound in air (v), the interfering spherical sound waves form a conical shock wave that appears as a V-shaped pressure ridge when viewed in 2 dimensions. The angle θ is given by $\sin \theta = v/v_s$, and the inverse ratio v_s/v is called the Mach number.

14.6 Musical Instruments and Sound Characteristics

Objective: To be able to explain some of the sound characteristics of musical instruments in physical terms.

Musical instruments provide good examples of standing waves and boundary conditions. On some stringed instruments, different notes are produced by varying the lengths of the strings using finger pressure (•Fig. 14.14). As you learned in Chapter 13, the natural frequencies of a stretched string (fixed at each end, as is the case for the strings on an instrument), are $f_n = nv/2L$, where $v = \sqrt{F/\mu}$. Initially adjusting the tension in a string tunes it to a particular (fundamental) frequency. Then the effective length of the string is varied by finger pressure.

Standing waves are also set up in wind instruments. For example, consider a pipe organ with fixed pipe lengths, which may be open or closed (•Fig. 14.15). An open pipe is open at both ends, and a closed pipe is closed at one end and open at the other (the antinode end). Analysis similar to that done in Chapter 13 for a stretched string with the proper boundary conditions shows that the natural frequencies for the pipes (where v is the speed of sound in air) are

•**FIGURE 14.14 A shorter vibrating string, a higher frequency** Different notes are produced on stringed instruments such as guitars, violins, or cellos by placing a finger on a string to change its effective, or vibrating, length.

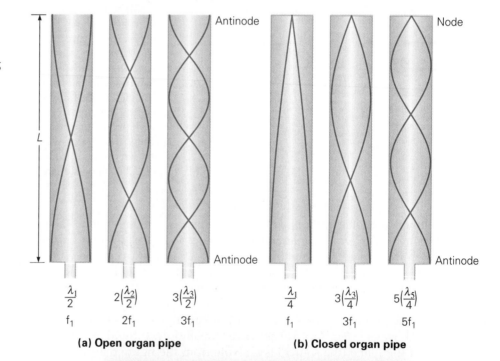

$\frac{\lambda_1}{2}$	$2\left(\frac{\lambda_2}{2}\right)$	$3\left(\frac{\lambda_3}{2}\right)$
f_1	$2f_1$	$3f_1$

(a) Open organ pipe

$\frac{\lambda_1}{4}$	$3\left(\frac{\lambda_3}{4}\right)$	$5\left(\frac{\lambda_5}{4}\right)$
f_1	$3f_1$	$5f_1$

(b) Closed organ pipe

•**FIGURE 14.15 Organ pipes** Longitudinal standing waves are formed in vibrating air columns in pipes (illustrated here with sinusoidal curves). **(a)** An open pipe has antinodes at both ends. **(b)** A closed pipe has a closed (node) end and an open (antinode) end. **(c)** A modern pipe organ. The pipes can be open or closed.

(c)

$$f_n = \frac{v}{\lambda_n} = n\left(\frac{v}{2L}\right) = nf_1 \quad n = 1, 2, 3, \ldots \tag{14.18}$$

natural frequencies for an open pipe

and

$$f_m = \frac{v}{\lambda_m} = m\left(\frac{v}{4L}\right) = mf_1 \quad m = 1, 3, 5, \ldots \tag{14.19}$$

natural frequencies for a closed pipe

Note that the natural frequencies depend on the length of a pipe. This is an important consideration in a pipe organ (Fig. 14.15c), particularly in selecting the dominant or fundamental frequency. (Pipe diameter is also a factor, but is not considered in this simple analysis.)

The same physical principles apply to wind and brass instruments. In all of these, the human breath is used to create standing waves in an open tube. Most such instruments allow the player to vary the effective length of the tube, and thus the pitch produced—either by opening and closing holes in the tube, as in woodwinds, or with the help of slides or valves that vary the actual length of tubing in which the air can resonate, as in most brasses ($\bullet$Fig. 14.16).

Recall that a musical note or tone is referenced to the fundamental vibrational frequency. In musical terms, the first overtone is the second harmonic, the second overtone is the third harmonic, and so on. Note that for a closed organ pipe (Eq. 14.19) the even harmonics are missing.

FIGURE 14.16 Wind instruments
The saxophone, like all wind instruments, is essentially an open tube. The effective length of the air column, and hence the pitch of the sound, is varied by opening and closing holes along the tube.

EXAMPLE 14.9 ■ PIPE DREAMS: FUNDAMENTAL FREQUENCY

A particular organ pipe has an open length of 0.653 m. Taking the speed of sound in air to be 345 m/s, what is the fundamental frequency of this pipe?

Solution.

Given: $L = 0.653$ m   **Find:** f_1 (fundamental frequency)
            $v = 345$ m/s (speed of sound)

We can use Eq. 14.18 directly, and with $n = 1$,

$$f_1 = \frac{v}{2L} = \frac{345 \text{ m/s}}{2(0.653 \text{ m})} = 264 \text{ Hz}$$

This is middle C (C_4).

Follow-up Exercise.   A closed organ pipe has a fundamental frequency of 256 Hz. What would be the frequency of its first audible overtone?

Perceived sounds are described by terms whose meanings are similar to those used to describe the physical properties of sound waves. Physically, a wave is generally characterized by intensity, frequency, and waveform (harmonics). The corresponding terms used to describe the sensations of the ear are loudness, pitch, and quality (or timbre). These general correlations are shown in Table 14.2. However, the correspondence is not perfect. The physical properties are objective and can be measured directly. The sensory effects are subjective and vary from person to person. (As an analogy, think of temperature as measured by a thermometer and by the sense of touch.)

Sound intensity and its measurement on the decibel scale were covered in Section 14.3. **Loudness** is related to intensity, but the human ear responds differently to sounds of different frequencies. For example, two tones of the same intensity (in W/m^2) but different frequencies may be judged by the ear to have different loudnesses.

TABLE 14.2   General Correlation between Perception and Physical Characteristics of Sound

Sensory Effect	Physical Wave Property
Loudness	Intensity
Pitch	Frequency
Quality (timbre)	Waveform (harmonics)

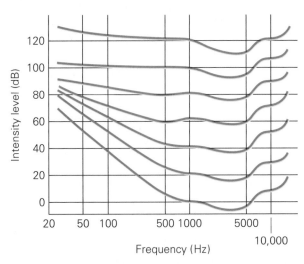

•FIGURE 14.17 Equal loudness contours
The curves indicate tones that are judged to be equally loud, though they have different frequencies and intensity levels. For example, on the next-to-lowest contour, a 1000-Hz tone at 20 dB sounds as loud as a 100-Hz tone at 50 dB. Note that the frequency scale is logarithmic to compress the large frequency range.

Frequency and **pitch** are often used synonymously, but again there is an objective–subjective difference. If the same low-frequency tone is sounded at two intensity levels, most people will say that the more intense sound has a lower pitch, or perceived frequency.

The curves in the graph of intensity level versus frequency shown in •Fig. 14.17 are called equal loudness contours. They join points representing intensity–frequency combinations that the average person judges to be equally loud. The top curve shows that the decibel level of the threshold of pain does not vary much from 120 dB, regardless of the frequency of the sound. However, the threshold of hearing, represented by the lowest contour, varies widely with frequency. For a tone with a frequency of 2000 Hz, the threshold of hearing is 0 dB. But a 20-Hz tone would have to have an intensity level of over 70 dB just to be heard (the extrapolated y-intercept of the lowest curve).

It is interesting to note the dips (or minima) in the curves. These indicate that the ear is most sensitive to sounds with frequencies around 4000 Hz and 12,000 Hz. Note that a tone with a frequency of 4000 Hz can be heard at intensity levels *below* 0 dB. The reason these minima occur is due to a resonance in a closed cavity in the auditory canal (similar to a closed pipe). The length of the cavity is such that it has a fundamental resonance frequency of about 4000 Hz, resulting in extra sensitivity. As a closed cavity, the next natural frequency is the third harmonic (see Eq. 14.19), and this is three times the fundamental frequency, or about 12,000 Hz.

The **quality** of a tone is the characteristic that enables it to be distinguished from another of basically the same intensity and frequency. Tone quality depends on the waveform, specifically, on the number of harmonics (overtones) present and their relative intensities (•Fig. 14.18). The tone of a voice depends in large part on the vocal resonance cavities. One person can sing a tone with the same basic frequency and intensity as another, but different combinations of overtones give the two voices different qualities.

The notes of a musical scale correspond to certain frequencies; for example, middle C (C_4) has a frequency of 262 Hz. When a note is played on an instrument, its assigned frequency is that of the first harmonic, the fundamental frequency. This frequency is dominant over the accompanying overtones that determine the sound quality of the instrument. Recall from Chapter 13 that the overtones produced depend on how an instrument is played. Whether a violin string is plucked or bowed, for example, can be discerned from the quality of identical notes.

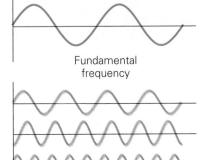

Fundamental
frequency

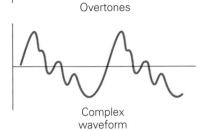

Overtones

Complex
waveform

(a)

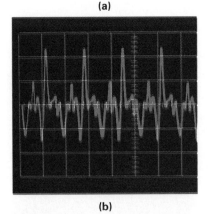

(b)

•FIGURE 14.18 Waveform and quality
(a) The superposition of sounds of different frequencies and amplitudes gives a complex waveform. The overtones determine the quality of the sound. (b) The waveform of a violin tone is electronically displayed on an oscilloscope.

Chapter Review

Important Terms

sound waves   447
audible region   447
sound frequency spectrum   447
infrasonic region   447
ultrasonic region   447
intensity   452
decibel (dB)   453

sound intensity level   453
constructive interference   457
destructive interference   457
phase difference   457
path difference   457
beats   459
beat frequency   459

Doppler effect   460
sonic boom   464
Mach number   465
loudness   467
pitch   468
quality   468

Important Concepts

- The sound frequency spectrum can be divided into three frequency regions: infrasonic ($f < 20$ Hz), audible (20 Hz $< f <$ 20 kHz), and ultrasonic ($f > 20$ kHz).
- The speed of sound in a medium depends on the elasticity of the medium and its density. In general, $v_{solids} > v_{liquids} > v_{gases}$.
- The intensity of a point source is inversely proportional to the square of the distance from the source.
- The sound intensity level is a logarithmic function and is expressed in decibels (dB).

- Sound wave interference of two point sources depends on phase difference as related to path difference. Sound waves that arrive at a point in phase reinforce each other (constructive interference); sound waves that arrive at a point out of phase tend to cancel each other (destructive interference).
- The Doppler effect depends on the velocities of the sound source and the observer relative to still air. When the relative motion of the source and observer is toward each other, the observed pitch increases; when the relative motion of source and observer is away from each other, the observed pitch decreases.

Important Equations

Speed of Sound (in m/s):

$$v = (331 + 0.6T_C) \text{ m/s} \qquad (14.1)$$

Intensity of a Point Source:

$$I = \frac{P}{4\pi R^2} \quad \text{and} \quad \frac{I_2}{I_1} = \left(\frac{R_1}{R_2}\right)^2 \qquad (14.2\text{–}3)$$

Intensity Level (in dB):

$$\beta = 10 \log \frac{I}{I_o} \quad \text{where } I_o = 10^{-12} \text{ W/m}^2 \qquad (14.4)$$

Phase Difference (where ΔL is path difference):

$$\Delta\theta = \frac{2\pi}{\lambda}(\Delta L) \qquad (14.5)$$

Condition for Constructive Interference:

$$\Delta L = n\lambda \quad (n = 0, 1, 2, 3, \ldots) \qquad (14.6)$$

Condition for Destructive Interference:

$$\Delta L = m\left(\frac{\lambda}{2}\right) \quad (m = 1, 3, 5, \ldots) \qquad (14.7)$$

Beat Frequency:

$$f_b = |f_1 - f_2| \qquad (14.9)$$

Doppler Effect:

$$f_o = \left(\frac{v}{v \pm v_s}\right)f_s = \left(\frac{1}{1 \pm \dfrac{v_s}{v}}\right)f_s \qquad (14.12)$$

where v_s = speed of source
and v = speed of sound

$\begin{cases} -\text{for source moving toward stationary observer} \\ +\text{for source moving away from stationary observer} \end{cases}$

$$f_o = \left(\frac{v \pm v_o}{v}\right)f_s = \left(1 \pm \frac{v_o}{v}\right)f_s \qquad (14.15)$$

where v_o = speed of observer
and v = speed of sound

$\begin{cases} +\text{for observer moving toward stationary source} \\ -\text{for observer moving away from stationary source} \end{cases}$

Mach Number:

$$M = \frac{v_s}{v} \qquad (14.17)$$

Natural Frequencies of Open Organ Pipe:

$$f_n = n\left(\frac{v}{2L}\right) = nf_1 \quad (n = 1, 2, 3, \ldots) \qquad (14.18)$$

Natural Frequencies of Closed Organ Pipe:

$$f_m = m\left(\frac{v}{4L}\right) = mf_1 \quad (m = 1, 3, 5, \ldots) \qquad (14.19)$$

Exercises

14.1 Sound Waves

14.2 The Speed of Sound

1 A sound wave with a frequency of 15 Hz is in what region of the sound spectrum: (a) audible, (b) infrasonic, (c) ultrasonic, (d) supersonic?

2 A sound wave in a solid (a) is longitudinal, (b) is transverse, (c) has longitudinal and transverse components, (d) travels slower than when the material is in the liquid phase.

3 The speed of sound is generally greatest in (a) solids, (b) liquids, (c) gases, (d) vacuum.

4 The speed of sound in air (a) is about 1/3 km/s, (b) is about 1/5 mi/s, (c) depends on temperature, (d) all of these.

5 Suggest a possible explanation of why some flying insects produce buzzing sounds and some do not.

6 At a concert, notes are played on a bass violin and a flute, which are equidistant from you. Which note do you hear first? Explain.

7 When water is poured over ice cubes in a glass, cracking sounds may be heard. What causes these?

8 The speed of sound in air is temperature dependent. What effect does humidity have, if any?

9 The wave speed in a liquid is given by $v = \sqrt{B/\rho}$, where B is the bulk modulus and ρ is the density. Show that this equation is dimensionally correct. What about $v = \sqrt{Y/\rho}$ for a solid?

10 ■ What is the speed of sound in air at (a) 15°C and (b) 20°C?

11 ■ The thunder from a lightning flash is heard by an observer 6.0 s after she sees the flash. What is the approximate distance to the lightning strike in (a) kilometers and (b) miles?

12 ■ What is the air temperature if the speed of sound is (a) 346 m/s and (b) 340 m/s?

13 ■■ Particles that are about 3.0×10^{-2} cm in diameter are to be scrubbed loose in an aqueous ultrasonic cleaning bath. Above what frequency should the bath be operated to produce wavelengths of this size and smaller?

14 ■■ Brass is an alloy of copper and zinc. Does the addition of zinc to copper cause an increase or decrease in the speed of sound in brass rods as compared to copper rods? If so, by what factor?

15 ■■ A tuning fork vibrates at a frequency of 512 Hz. What is the wavelength of the sound coming from the fork when the air temperature is (a) 0°C and (b) 20°C?

16 ■■ The speed of sound in steel is about 4.5 km/s. A steel rail is struck with a hammer, and there is an observer 0.30 km away with one ear to the rail. (a) How much time will elapse from the time the sound is heard through the rail to the time it is heard through the air? Assume that the air temperature is 20°C and there is no wind blowing. (b) How much time would elapse if the wind were blowing toward the observer at 36 km/h from where the rail was struck?

17 ■■ At a baseball game on a cool day (air temperature of 16°C), a fan hears the crack of the bat 0.40 s after observing a batter hit a ball. How far is the fan from home plate?

18 ■■ A 2000-Hz tone is sounded at times when the air temperature is 20°C and 10°C. What is the percent change in the wavelength of the tone in going from the higher to the lower temperature?

19 ■■ A person holds a rifle horizontally and fires at a target. The bullet has a muzzle velocity of 460 m/s, and the person hears the bullet strike the target 2.00 s after firing it. The air temperature is 72°F. What is the distance to the target?

20 ■■ A hiker gives a shout and hears the echo reflected from a rock wall 3.40 s later. If the air temperature is 10°C, how far is the hiker from the wall? (Assume no wind.)

21 ■■ The driver of a truck parked on a mountain road gives the air horn a quick blast toward a rock cliff 0.760 km away. If he hears the echo 4.50 s later, what is the approximate air temperature at this mountain elevation?

22 ■■■ An engineer who uses units of feet and degrees Fahrenheit wants an equation for the speed of sound in air. Rewrite Eq. 14.1 to meet this need.

23 ■■■ Sound propagating through air at 20°C passes through a vertical cold front into air that is 4.0°C. If the sound has a frequency of 2400 Hz, by what percentage does its wavelength change in crossing the boundary?

14.3 Sound Intensity

24 If the air temperature increases, would the sound intensity from a constant output point source (a) increase, (b) decrease, or (c) remain unchanged?

25 The decibel scale is referenced to a standard intensity of (a) 1.0 W/m², (b) 10^{-12} W/m², (c) normal conversation, (d) the threshold of pain.

26 The Richter scale used to measure the intensity levels of earthquakes is a logarithmic scale, as is the decibel scale. Why are such scales used?

27 Can there be negative decibel levels, such as -10 dB? If so, what would these mean?

28 ■ By how many times must the distance from a point source be increased to *reduce* the sound intensity by ⅓?

29 ■ Find the intensity levels in decibels for sounds with intensities of (a) 10^{-3} W/m², (b) 10^{-5} W/m², and (c) 10^{-14} W/m².

30 ■ What is the intensity of a sound that has an intensity level of (a) 60 dB and (b) 110 dB?

31 ■ If the intensity of one sound is 10^{-4} W/m² and the intensity of another is 10^{-2} W/m², what is the difference in their intensity levels?

32 ■■ The takeoff and landing noise levels for some common commercial jet aircraft are given in •Table 14.3. What are the lowest and highest intensities for (a) take-off and (b) landing?

TABLE 14.3 Takeoff and Landing Noise Levels for Some Common Commercial Jet Aircraft*

Aircraft	Takeoff noise (dB)	Landing noise (dB)
737	85.7–97.7	99.8–105.3
747	89.5–110.0	103.8–107.8
DC-10	98.4–103.0	103.8–106.6
L-1011	95.9–99.3	101.4–102.8

*Noise level readings taken from 650 ft. Range depends on the aircraft model and the type of engine used.

33 ■■ What would be the intensity level of a 23-dB sound after being amplified (a) 10 thousand times, and (b) a million times? (c) How about a billion times?

34 ■■ A tape player has a signal-to-noise background ratio of 53 dB. How many times larger is the intensity of the signal than that of the background noise?

35 ■■ Sound from a loudspeaker, which approximates a point source, is measured to have a sound level intensity of 70 dB at a distance of 3.0 m from the speaker. What is the approximate sound power emitted by the speaker?

36 ■■ The sound intensity levels for a machine shop and a quiet library are 90 dB and 40 dB, respectively

(a) How many times greater is the intensity of the sound in the machine shop than that in the office? (b) What is each intensity?

37 ■■ Two identical balloons burst simultaneously, producing a sound with an intensity level of 100 dB. What would the intensity level have been if only one balloon had burst?

38 ■■ At a rock concert, the average sound intensity level for a person in a front-row seat is 110 dB for a single band. If all the bands scheduled to play produce sound of that same intensity, how many of them would have to play simultaneously for the sound level to be at or above the threshold of pain?

39 ■■ At a distance of 10.0 m from a point source, the intensity level is measured to be 70 dB. At what distance from the source will the intensity be 40 dB?

40 ■■ At a 4th of July celebration, a firecracker explodes as illustrated in •Fig. 14.19. Considering the firecracker to be a point source, what are the intensities heard by observers at B, C, and D, relative to that for the observer at A?

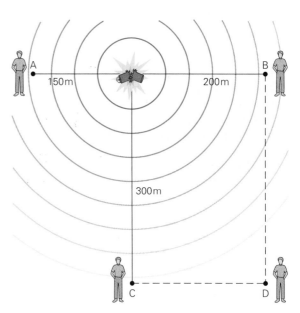

•**FIGURE 14.19 A big bang**
See Exercise 40.

41 ■■ An orchestra plays a movement pianissimo (very softly) at an average intensity of 7.5×10^{-6} W/m² and another movement is played fortissimo (very loudly) at 2.5×10^{-4} W/m². What is the difference in the average sound levels for the movements?

42 ■■ A person has lost 20 dB of the normal range of hearing response. How many times must the intensity of a sound with an intensity level of 70 dB be amplified by a hearing aid to allow the person to hear it at a normal intensity?

43 ■■ A person standing 4.0 m from a wall shouts so that the sound strikes the wall with an intensity of 2.5×10^{-4} W/m^2. Assuming that the wall absorbs 20% of the incident energy and reflects the rest, what is the sound intensity level just before and after the sound is reflected?

44 ■■ An engine-powered lawnmower is rated at 95 dB. (a) What is the sound intensity for this particular mower? (b) How many times more intense is the sound of this mower than that of an electric-powered mower rated at 83 dB?

45 ■■ Show that $10 \log I/I_0$ is $\frac{1}{10}$ of a bel.

46 ■■■ A 1000-Hz tone issuing from a loudspeaker has an intensity level of 100 dB at a distance of 2.5 m. If the speaker is assumed to be a point source, how far from the speaker will the sound have intensity levels of (a) 60 dB and (b) just barely enough to be heard?

47 ■■■ A flying bee produces a buzzing sound that is just barely audible to a person 3.0 m away. How many bees would have to be buzzing at that distance to produce a sound with an intensity level of 40 dB?

14.4 Sound Phenomena
14.5 The Doppler Effect

48 Interference plays a major role in (a) refraction, (b) diffraction, (c) beats, (d) the Doppler effect.

49 The Doppler effect is used in (a) radar, (b) sonar, (c) determining galactic red shifts, (d) all of these.

50 Do interference beats have anything to do with the beat of music? Explain.

51 Is there a Doppler effect if a sound source and an observer are moving (a) with the same velocity or (b) at right angles? (c) What would be the effect if a moving source accelerated toward a stationary observer?

52 How fast would a "jet fish" have to swim to create an aquatic sonic boom?

53 Two sound waves with the same wavelength, 0.50 m, arrive at a point after having traveled (a) 2.50 m and 3.75 m and (b) 3.25 m and 8.25 m. What type of interference occurs in each case?

54 ■ Two adjacent point sources, source A and source B, are in front of an observer and emit identical 600-Hz tones. To what closest distance behind source B would source A have to be moved for the observer to hear no sound? (Assume that the air temperature is 20°C.)

55 ■ A violinist and a pianist simultaneously sound notes with frequencies of 436 Hz and 440 Hz, respectively. What beat frequency will be heard by the musicians?

56 ■ A violinist tuning an instrument to a piano note of 264 Hz detects three beats per second. What are the possible frequencies of the violin tone?

57 ■ What is the beat frequency of two tones that have equal amplitudes and frequencies of 256 Hz and 260 Hz? How could the beat frequency be made zero?

58 ■■ Two identical strings on different cellos are tuned to the 440-Hz A note. The peg holding one of the strings slips, so its tension is decreased by 1.5%. What is the beat frequency heard when the strings are then played together?

59 ■■ While standing near a railroad crossing, a person hears a train horn. The frequency emitted by the horn is 400 Hz. If the train is traveling at 90.0 km/h and the air temperature is 25°C, what is the frequency heard by the bystander (a) when the train is approaching and (b) when it has passed by?

60 ■■ What is the frequency heard by a person driving 50 km/h toward a blowing factory whistle ($f = 800$ Hz) if the air temperature is 0°C?

61 ■■ A bystander hears a siren vary in frequency from 476 Hz to 404 Hz as a fire truck approaches, passes by, and moves away on a straight street. What is the speed of the truck? (Take the speed of sound in air to be 343 m/s.)

62 ■■ How fast, in km/h, must a sound source be moving toward you to make the observed frequency 3.0% greater than the true frequency? (Assume that the speed of sound is 340 m/s.)

63 ■■ What is the half-angle of the shock wave of a jet aircraft just as it breaks the sound barrier?

64 ■■ The supersonic transport (SST) *Concorde* on transatlantic flights flies at a speed of Mach 1.5. What is the half-angle of the conical shock wave formed by the *Concorde* at this speed?

65 ■■ The half-angle of the conical shock wave formed by a supersonic jet is 30°. What are (a) the Mach number of the aircraft and (b) the actual speed of the aircraft if the air temperature is 0°C?

66 ■■■ Two point-source loudspeakers are a certain distance apart and a person stands 12.0 m in front of one of them on a line perpendicular to the base line of the speakers. If the speakers emit identical 1000-Hz tones, what is their minimum nonzero separation so the observer hears no sound? (Let the speed of sound be exactly 340 m/s.)

67 ■■■ A stationary source directs an 800-Hz sound wave toward an approaching object moving with a speed of 25.0 m/s. What is the frequency shift of the reflected wave if the air temperature is 20°C? (*Hint*: There are two Doppler shifts here. Why?)

68 ■■■ A stationary submerged submarine tracks an approaching submarine with sonar. A short pulse of ultrasound with a frequency of 400 kHz is sent toward the sub and returns 10 s later with an observed frequency of 412 kHz. (a) What is the speed of the approaching sub? Take the speed of sound in seawater to be 1200 m/s. [*Hint:* The ultrasound received by the moving sub is Doppler-shifted, and the moving sub acts as a moving source of sound with this shifted frequency. That is, the stationary sub receives a reflected pulse that is doubly Doppler-shifted.] (b) What is the approximate range of the approaching sub at the time of reflection?

14.6 Musical Instruments and Sound Characteristics

69 The human ear can best hear tones at (a) 1000 Hz, (b) 4000 Hz, (c) 6000 Hz, (d) any frequency

70 The quality of sound depends on its (a) waveform, (b) frequency, (c) speed, (d) intensity.

71 (a) After a snowfall, why does it seem particularly quiet? (b) Why do sounds in empty rooms sound hollow? (c) Why do people's voices sound fuller or richer when they sing in the shower?

72 The first three natural frequencies of an organ pipe are 136 Hz, 408 Hz, and 680 Hz. (a) Is the pipe an open or closed one? (b) Taking the speed of sound in air to be 340 m/s, find the length of the pipe.

73 Physically, why aren't the frets on a guitar evenly spaced?

74 It is possible for an open and a closed organ pipe of the same length to produce notes of the same frequency? Justify your answer.

75 ■ A closed organ pipe has a fundamental frequency of 528 Hz (a C note) at room temperature. What is the fundamental frequency of the pipe when the temperature is 0°C?

76 ■■ An open organ pipe has a length of 0.75 m. What would be the length of a closed organ pipe whose third harmonic ($m = 3$) is the same as the fundamental frequency of the open pipe?

77 ■■ A closed organ pipe has a length of 0.800 m. At room temperature, what are the frequencies of (a) the second harmonic and (b) the third harmonic?

78 ■■ An open organ pipe and a closed organ pipe both have lengths of 0.52 m at 20°C. What is the fundamental frequency of each pipe?

79 ■■ Suppose the pipes in Exercise 77 were at 10°C. Would the change in the fundamental frequency be the same for both pipes? Justify your answer. (Is there

another effect that might play a role here? Explain.)

80 ■■ An organ pipe has a fundamental frequency of 340 Hz in air at 0°C. When filled with another gas at the same temperature, the fundamental frequency is 992 Hz. What is the gas?

81 ■■ Show that the general equation for the Doppler effect for a moving source and a moving observer is given by

$$f_o = f_s \left(\frac{v \pm v_o}{v \pm v_s} \right)$$

using the sign convention of Eqs. 14.12 and 14.15.

82 ■■ A fire engine travels at a speed of 90 km/h with its siren emitting sound at a frequency of 500 Hz. What is the frequency heard by a passenger in a car traveling at 65 km/h in the opposite direction to the fire engine, (a) approaching it and (b) moving away from it? [*Hint:* See Exercise 67 and take the speed of sound to be 354 m/s (a hot day).]

83 ■■ A tuning fork with a frequency of 440 Hz is held above a resonance tube partially filled with water. Assuming that the speed of sound in air is 342 m/s, for what heights of the air column will resonances occur?

84 ■■■ A closed organ pipe is filled with helium. The pipe has a fundamental frequency of 660 Hz in air at 0°C. What is the fundamental frequency with the helium?

Additional Exercises

85 A jet flies at a speed of Mach 2.0. What is the half-angle of the conical shock wave formed by the aircraft?

86 Two sources of 440-Hz tones are located 6.97 m and 8.90 m from an observation point. If the air temperature that day is 15°C, how do the waves interfere at the point?

87 Derive Eqs. 14.18 and 14.19.

88 How long does it take sound to travel 3.5 km in air if the temperature is 30°C?

89 If a person standing 30.0 m from a 550-Hz point source walks 5.0 m toward the source, by what factor does the sound intensity change?

90 The value of the speed of sound in air at 100°C is listed in Table 14.1 as 387 m/s. Is this the value given by Eq. 14.1? If not, explain.

91 The speed of sound in air for normal *environmental* temperatures to a good approximation is $v = 331 + 0.6T_C$ m/s (Eq 14.1). Yet, a better approximation is given by $v = (331)\sqrt{1 + (T_C/273)}$ m/s. Show that this approximation is justified. [*Hint:* Consider a binomial expansion, $(1 \pm x)^n = 1 \pm nx + n(n-1)x^2/2! \pm \dots$

92 What is the decrease in intensity if the distance from a point source is tripled?

93 Two people speak at the same time with intensity levels of 60.0 dB and 65.0 dB. What is the intensity level of the combined sounds heard by another person?

94 The intensity level of a sound is 90 dB at a certain distance from its source. How much energy falls on a 1.5-m^2 area in 5.0 s?

95 Show that the difference in the intensity levels for intensities I_2 and I_1 is given by 10 log I_2/I_1.

96 The note A (440 Hz) is sounded on a stringed musical instrument. The air temperature is 20°C. Approximately how many vibrations does the string make before the sound reaches a person 30 m away?

97 (Here's an old one.) A person drops a stone into a deep well and hears the splash from its hitting the water 3.16 s later. How deep is the well? (Assume the air temperature in the well to be 10°C.)

98 A stereo speaker is rated at 60 W of output at 1000 Hz. At the lower limit of the audible sound range, the output decreases by 1.5 dB. What is the power output of the speaker at the lower frequency?

99 One hunter sees another who is 1.5 km away fire his rifle (sees smoke come from the barrel). If the air temperature is 5°C, how long will it be until the first hunter hears the report of the shot? (Assume no wind.)

100 A factory whistle is heard by one worker at an intensity level of 60 dB and by another at 80 dB. How many times farther away from the whistle is the first worker?

101 If the room temperature (20 °C) of the air increases by 10 C°, what will be the percentage change in the speed of sound?

15 Electric Charge, Forces, and Fields

Few sights are more awe-inspiring than a bolt of lightning from a dark storm cloud. And very few processes in nature deliver such an enormous amount of energy to a small area in such a tiny fraction of a second. Yet, while everyone has seen lightning, not many people have actually experienced its tremendous power at close range. Though it usually makes the papers and TV news broadcasts when it happens, only a few hundred people are actually struck by lightning each year in the United States.

It might surprise you, therefore, to realize that you have almost certainly had an experience that is practically identical, from a physicist's point of view. Have you ever walked across a carpeted room and then gotten a tiny shock when you reached for a brass doorknob? Though the scale is dramatically different, the physical processes involved are essentially the same.

Ever since Ben Franklin flew his famous kite in a thunderstorm, if not before, it has been known that lightning and electricity are closely connected. In this chapter you will begin the study of a large and important area of physics: that dealing with electrical phenomena. Among other things, you'll learn what the lightning bolt and the spark from the doorknob have in common.

Electricity seems to have a flair for the dramatic. To most of us, the word conjures up striking images: a blaze of floodlights, crackling live wires, the skyline of a great city at night, or a flash of lightning. There is often a hint of anxiety in our associations, for we all know that electricity can be dangerous. We also know, however, that electricity can be tamed, even domesticated. In the home or the office, we have come to take it almost for granted. Indeed, the extent to which we depend on it becomes evident only when the power goes off unexpectedly, giving us a dramatic reminder of the role that it plays in our daily lives. Yet less than a century ago there were no power lines crossing the land, no electric lights or appliances—none of the seemingly endless electrical applications surrounding us today.

Less than 200 years ago, the discovery was made that electricity and magnetism are in fact related phenomena. Today, the electromagnetic force is recognized as being one of the four fundamental forces (along with gravity and the strong and weak nuclear forces discussed in Chapters 29 and 30). It is customary in introductory physics courses to consider the electrical part of the electromagnetic force before the magnetic part and only then to combine electricity and magnetism into electromagnetism. This will be the approach followed in this book.

15.1 Electric Charge

Objectives: To be able to **(a)** distinguish between the two types of electric charge, **(b)** state the force law that operates between charged objects, and **(c)** understand and use the law of charge conservation.

What is electricity? Perhaps the broadest answer is that electricity is a collective term describing phenomena associated with the interaction between *electrically charged* objects. We start our study of electricity by investigating electric forces between charged objects *at rest*. This area of physics is called electro*statics*, because all the charges remain at rest.

Like mass, **electric charge** is a fundamental property of matter. Electric charge is fundamentally associated with atomic particles, the electron and the proton. The simplistic solar system model of the atom, shown in •Fig. 15.1, likens its structure to planets orbiting the Sun. The electrons are viewed as orbiting a nucleus, a core containing protons and electrically neutral particles called neutrons. As we saw in Section 7.5, the centripetal force that keeps the planets in their orbits about the Sun is supplied by gravity. Similarly, the force that keeps the electrons orbiting the nucleus is supplied by electrical attraction. However, there is an important distinction between the gravitational and electric forces.

There is apparently only one type of mass and this gives rise to only attractive gravitational forces. Electric charge, on the other hand, comes in two types, distinguished by the labels positive (+) and negative (−). Protons carry a positive charge and electrons carry a negative charge. Different combinations of the two kinds of charges can produce *either* attractive *or* repulsive forces.

The directions of the electric forces when charges interact with one another are given by the following simple principle, sometimes known as the **law of charges** or the **charge–force law:**

Like charges repel, and unlike charges attract.

That is, two negatively charged particles or two positively charged particles experience mutually repulsive forces, whereas particles with opposite charges are mutually attracted (•Fig. 15.2). These forces are equal and opposite, and act on different objects, in keeping with Newton's action–reaction law for forces.

The charge of an electron and the charge of a proton are equal in magnitude, though opposite in sign. The charge on the electron (*e*) is taken as the fundamental

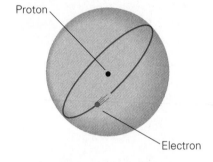

Proton

Electron

(a) Hydrogen atom

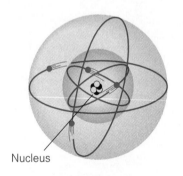

Nucleus

(b) Beryllium atom

•**FIGURE 15.1 Simplistic model for atoms**
The so-called solar system model of **(a)** a hydrogen atom and **(b)** a beryllium atom views the electrons as orbiting the positively charged nucleus, analogous to the planets orbiting the Sun. (The electronic structure of atoms is actually much more complicated than this.)

Note: Recall the discussion of Newton's third law in Section 4.5.

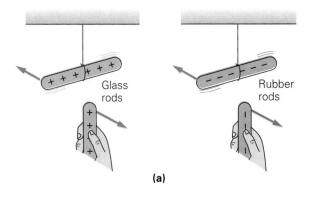
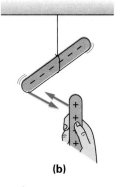

Glass rods

Rubber rods

(a)

(b)

•FIGURE 15.2 **The charge-force law, law of charges**
(a) Like charges repel. **(b)** Unlike charges attract.

unit of charge, since it is the smallest charge that has been observed in nature.* Thus the electric charge on an object is an integral multiple of fundamental charges (either positive or negative). Our general symbol for charge will be the letter q. So for the charge of an object we can write,

$$q = ne \tag{15.1}$$

where n is an integer. We sometimes say that charge is "quantized," which means that it occurs only in integral multiples of the fundamental electronic charge. Mass, on the other hand, is not quantized.

The unit of charge in the SI is the **coulomb** (**C**), named for the French physicist Charles A. de Coulomb (1736–1806), who discovered a relationship between electric force and charge (to be considered later in this chapter). The charges of the electron and the proton in terms of the coulomb are given in Table 15.1.

There are several terms frequently used when discussing electrical properties. Saying that an object has a **net charge** means that it has an excess of either positive or negative charges. As you will learn in the next section, excess charges are commonly produced by a transfer of electrons. For example, if an object has a net charge of $+1.60 \times 10^{-19}$ C, it could be that one of its atoms has lost an electron. This means that the atom has one proton whose charge is no longer canceled by an electron, so that the atom is no longer electrically neutral. (Such charged atoms are called *ions*.) The charge on the electron is a tiny fraction of a coulomb, and a coulomb's worth of net charge on an object is rarely seen in everyday situations. More typically we deal with net charges in the microcoulomb (μC, or 10^{-6} C) to picocoulomb (pC, or 10^{-12} C) range.

In dealing with any electrical phenomena, another important principle is **conservation of charge**:

> The net charge of an isolated system remains constant.

The net charge may be other than zero, but it remains constant. Suppose, for example, that a system consists initially of two electrically neutral objects, and electrons are transferred from one to the other. The object with added electrons will then have a net negative charge and the object with fewer electrons will have a net positive charge of equal magnitude. However, the net charge of the *system* is still zero. If we think about the universe as a whole, conservation of charge means that the net charge *of the universe* is constant, though no one knows what the numerical value of that net charge might actually be.

This principle does *not* mean that charged particles cannot be created or destroyed. Physicists have, in fact, been doing this since the early part of this cen-

---

*According to recent experiments, protons are made up of particles called quarks that carry charges of $\pm\frac{1}{3}$ and $\pm\frac{2}{3}$ of the electronic charge. There is experimental evidence for the existence of quarks within the nucleus, but not outside the nucleus in common electrical interactions. (See Chapter 30.)

TABLE 15.1	Particles and Electric Charge	
Particle	*Electric Charge*	*Mass*
Electron	-1.60×10^{-19} C	$m_e = 9.11 \times 10^{-31}$ kg
Proton	$+1.60 \times 10^{-19}$ C	$m_p = 1.672 \times 10^{-27}$ kg
Neutron*	0	$m_n = 1.674 \times 10^{-27}$ kg

*The neutron is an electrically neutral particle found in the nucleus of an atom.

tury, as you will learn in the chapters on modern physics. However, by conservation of charge, charged particles are created or destroyed only in pairs with equal and opposite charges. Like the other conservation laws of physics, the concept of charge conservation is called a law because no violation of it has ever been observed experimentally.

EXAMPLE 15.1 ■ ON THE CARPET: QUANTIZED CHARGE

(a) If you shuffle across a carpeted floor on a dry day and acquire a net charge of -2.0 μC (1 microcoulomb = 1 μC = 10^{-6} C), will you have a deficiency or excess of electrons? (b) How many missing or extra electrons will you have?

Solution. First, we express the charge in coulombs and list what we are to find.

Given: $q = -2.0 \ \mu C \left(\dfrac{10^{-6} \, C}{1 \, \mu C} \right) = -2.0 \times 10^{-6}$ C *Find:* (a) whether you have lost or gained electrons
(b) number of missing or excess electrons

(a) Since the sign of your net charge is negative and electrons carry a negative charge, you have acquired an excess of electrons.

(b) The net charge is made up of an integral number of electronic charges. Using Eq. 15.1,

$$n = \frac{q}{e} = \frac{-2.0 \times 10^{-6} \, C}{-1.60 \times 10^{-19} C/\text{electron}} = 1.3 \times 10^{13} \, \text{electrons}$$

As you can see, net charges usually involve huge numbers of electrons. This fact makes the addition of one electron, or even a million electrons, virtually impossible to detect. Since charge is quantized in extremely small packages it is possible to think of the electrons as flowing like a charged liquid and, in most cases, to ignore that the transferred charge is composed of individual particles. This is similar to our picture of a river as a moving continuous fluid, even though we know it is actually composed of many individual water molecules.

Follow-up Exercise. In this Example; (a) what type of charge does the carpet acquire? (b) How much charge does the carpet acquire? (c) Does it have an excess or deficiency of electrons? (d) How many?

15.2 Electrostatic Charging

Objectives: To be able to (a) distinguish between conductors and insulators, (b) explain the operation of the electroscope, and (c) distinguish among charging by conduction, induction, and polarization.

That there are two types of electric charges and attractive and repulsive forces can be demonstrated easily. Before learning how this is done, you need to be able to distinguish between electrical conductors and insulators. What distinguishes these broad groups of substances is their ability to conduct, or transmit, electric charge.

Some materials, particularly metals, are good **conductors** of electric charge. Others, such as glass, rubber, and most plastics, are **insulators**, or poor electrical conductors. A comparison of the relative magnitudes of the conductivities of some materials is given in •Fig. 15.3.

A general picture is that in conductors, valence electrons of atoms—the ones in the outermost orbits—are loosely bound. As a result, they can be easily removed from the atom and moved about in the conductor. That is, they are not permanently bound to a particular atom. (This electron mobility also plays a major role in thermal conduction, Section 11.4.) In insulators, on the other hand, the valence electrons are more tightly bound. Continuous conduction of electric charge is called an *electric current*, and will be considered in more detail in Chapter 17.

As Fig. 15.3 shows, there is also an intermediate class of materials called **semiconductors**. Their ability to conduct charge is much less than that of metals, though much greater than that of insulators. The conductivity of semiconductors can be adjusted by adding certain types of atomic impurities in varying concentrations. Semiconductors form the basis of the transistors, solid-state circuits, and chips that have become the backbone of the modern computer industry.

The electroscope is a device that can be used to demonstrate the characteristics of electric charge (•Fig. 15.4). In its simplest form, it consists of a metal rod with a metallic bulb at one end and a pair of hanging foil "leaves," usually made of gold, at the other end. This arrangement is insulated from its glass container by a rubber cork. When charged objects are brought close to the bulb, electrons in the bulb are either attracted or repelled, according to the charge–force law. For example, if a negatively charged rod is brought near the bulb, electrons are re-

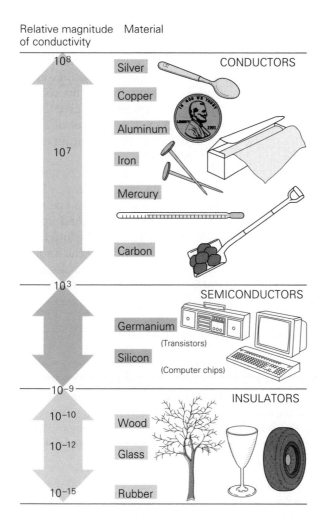

•FIGURE 15.3 Conductors, semiconductors, and insulators A comparison of the relative magnitudes of the electrical conductivities of various materials. (Not to scale.)

Bulb ⊕

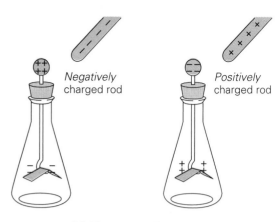

Negatively charged rod

Positively charged rod

(a) Neutral electroscope has charges evenly distributed; leaves are close together.

(b) Electrostatic forces cause leaves to diverge.

•FIGURE 15.4 The electroscope
An electroscope can be used to determine whether an object is electrically charged. When a charged object is brought near the bulb, the leaves diverge.

Note: From an external viewpoint you cannot tell whether the rubber rod gained negative charges or the fur gained positive charges. In other words, moving electrons to the rubber rod results in the same physical situation as moving positive charges to the fur. However, because the rubber is an insulator, and its electrons are therefore tightly bound, we might suspect that the fur lost electrons and the rubber gained them. In solids, the protons, being in the nuclei of the atoms, do not move. It is just a question of which material most easily loses electrons.

Note: Electrical ground refers to earth (hence "ground") or some other object that can receive or supply electrons without significantly changing its own electrical condition.

pelled, and the bulb is left with a positive charge. The electrons are conducted to the metal foil leaves, which diverge because of a mutually repulsive electric force between them (Fig. 15.4b). Similarly, if a positively charged rod is brought near the bulb, the leaves also diverge. (Can you explain why?)

Notice that the net charge of the electroscope remains zero in these instances because it is isolated (insulated)—only the *distribution* of charge is altered. However, it is possible to give an electroscope (and other objects) a nonzero net charge by electrostatic charging.

Charging by Friction

In general, **electrostatic charging** is a process by which an insulator or an insulated conductor receives a net charge. In one such charging process, when certain insulator materials are rubbed with cloth or fur, they become electrically charged. For example, if a hard rubber rod is rubbed with fur, the rod will acquire a net negative charge; and rubbing a glass rod with silk will give the rod a net positive charge. This is called **charging by friction**. The transfer of charge is due to the contact between the materials, and depends on the nature of the materials; the charges are not merely rubbed off by friction.

You have almost certainly experienced a result of frictional charging when, after walking across a carpet on a dry day, you get "zapped" by a spark when you reach for a metal object, such as a doorknob. This happens because you have become electrostatically charged, that is, you have picked up a net charge from the carpet. The charge produces an electric force great enough to *ionize* (free electrons from) the air molecules between your hand and the knob when your hand comes close to the metal knob. The resulting flow of charges gives rise to a spark discharge between hand and metal. This doesn't occur on humid days. With adequate humidity, a thin film of moisture on objects prevents the build-up of charge by conducting it away.

As with heat, when a charge moves in a conductor, it is common to talk about a flow. Like heat flow, the flow of electricity was once thought to be due to some type of fluid transfer. Ben Franklin proposed a single-fluid theory of electricity. He assumed that all bodies had some normal amount of "electrical fluid." When some of this was transferred, for example, by rubbing two bodies together, one body would then have an excess of it and the other a deficiency. Franklin indicated these conditions by plus and minus signs, respectively, which is the origin of the sign convention for charge.

Charging by Contact

Bringing a charged rod close to an electroscope will reveal that the rod is charged, but won't tell you *how* the rod is charged (positively or negatively). This distinction can be made, however, if the electroscope is first given a known type of charge.

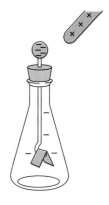

(a) Neutral electroscope is touched with negatively charged rod; charges are transferred to bulb.

(b) Electroscope has net negative charge.

(c) Negatively charged rod repels electrons; leaves diverge farther.

(d) Positively charged rod attracts electrons; leaves collapse.

•**FIGURE 15.5 Charging by contact or conduction**
Charge is transferred to the electroscope when the charged rod touches the bulb. Then, when an oppositely charged rod is brought near the bulb, the leaves collapse or come closer together.

For example, electrons can be transferred from one object to another if the two objects touch, as illustrated in •Fig. 15.5a for an electroscope bulb and a negatively charged rod. The electrons in the rod are mutually repelled by one another. Some will be able to "escape" and move onto the electroscope. In this case we say the electroscope has been **charged by contact** (•Fig. 15.5b). If a negatively charged rod is brought close to the now negatively charged electroscope, the leaves will diverge even further as more electrons are pushed onto them from the bulb (•Fig. 15.5c). An oppositely (positively) charged rod will cause the leaves to collapse by attracting some electrons up to the bulb and away from the leaves (•Fig. 15.5d).

Charging by Induction

Since it is electrons that are transferred in conductors, you might wonder how an electroscope can ever become positively charged. This can be done by **charging by induction** (•Fig. 15.6). Touching the bulb with a finger "grounds" the electroscope, that is, provides a path by which electrons can escape from the bulb. Then when a negatively charged rod is brought close to the bulb (but carefully, so as not to touch), it repels electrons from the bulb into the finger and down into the Earth. Removing the finger while the charged rod is kept nearby leaves the electroscope with a net positive charge, because the removed electrons have no way of flowing back once the finger is removed.

Polarization

Charging by contact and induction both involve a removal of charge from an object. However, an object can have some charge moved *within* it to give different regions of charge, yet keep a net charge of zero. In this case, induction brings about **polarization**, or separation of charge (•Fig. 15.7). Now you can understand why a balloon will stick to the wall or ceiling after being rubbed on someone's hair or sweater. The balloon is charged by friction, and the charged balloon induces an opposite charge on the surface, creating an attractive electric force.

Electrostatic charging can be annoying (even dangerous), but it can also be beneficial in a variety of practical applications. Clothes and papers often stick together because of static cling, and an electrostatic spark discharge can start a fire or cause an explosion in the presence of a flammable gas. On the other hand, the air we breathe is cleaner because of electrostatic precipitators used in smokestacks.

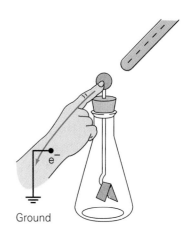

Ground

(a) Electrons are transferred to ground through hand.

(b) Electroscope is left positively charged.

•**FIGURE 15.6 Charging by induction**
(a) Touching the bulb provides a path to ground for charge transfer.
(b) When the finger is removed, the electroscope has a net charge.

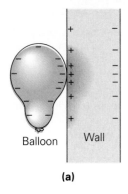

Balloon   Wall

(a)

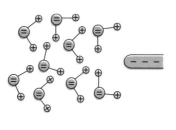

Polar molecules

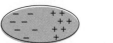

Nonpolar molecule

Induced molecular dipole

(b)

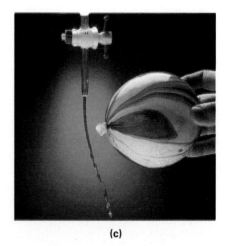

(c)

•**FIGURE 15.7   Polarization**
(a) When the balloons are charged by friction and placed in contact with the wall, an opposite charge is induced on the wall's surface, to which the balloons then stick by the force of electrostatic attraction. **(b)** Some molecules, such as those of water, are polar in nature—that is, they have separated regions of positive and negative charge. But even some molecules that are not normally polar can be polarized by the presence of a nearby charged object. The electric force induces a separation of charge, making the molecules into induced molecular dipoles. **(c)** A stream of water bends toward a charged balloon. No charge needs to be induced in the water molecules, because they are already natural dipoles. The charged balloon simply attracts the ends of the molecules that carry the opposite charge.

In these devices, electrical discharges cause the particles that are a byproduct of fuel combustion to acquire a charge. This particulate matter is then removed from the flue gases by means of electric force. On a smaller scale, electrostatic air cleaners are available for the home. Another, now almost indispensable, application is the electrostatic copier (see the Insight on p. 483).

15.3   Electric Force

Objectives: **To be able to (a) understand Coulomb's law, and (b) use it to calculate the electric force between two charged particles, and on one charge due to several other charges.**

The relative directions of the electric forces on mutually interacting charges are given by the charge–force law. However, what about the *magnitude* or strength of the electric force? This was investigated by Charles de Coulomb (after whom the unit of electric charge is named), using a delicate balance to measure the force. He found that the magnitude of the electric force between two point charges (q_1 and q_2) depended on the product of the charges and varied inversely as the square of the distance between them; that is $F \propto q_1 q_2 / r^2$. (Note that this is an inverse-square relationship, mathematically similar to the one for the force of gravity between two point masses, $F \propto m_1 m_2 / r^2$.)

Like Cavendish's measurements for the determination of the universal gravitational constant G, where $F = G m_1 m_2 / r^2$ (Section 7.4), Coulomb's measurements provided a constant of proportionality so that the electric force may be written in

Note: Compare Eq. 7.15.

Xerography, coined from the Greek words *xeros* (meaning "dry") and *graphein* (meaning "to write"), refers to a dry process by which almost any printed material can be copied. This process makes use of a photoconductor, which is a light-sensitive semiconductor, such as selenium. When kept in darkness, a photoconductor is a good insulator that can be electrostatically charged. However, when light strikes the material, it becomes conductive and the electric charge can be removed.

In transfer xerography, a photoconductor-coated plate, drum, or belt is electrostatically charged, and then receives a projected image of the page to be copied. The illuminated portions of the photoconductor coating become conducting and are discharged, but the areas corresponding to the dark print remain charged. Essentially, this creates an electric charge copy of the original sheet. Then the photoconductor copy comes into contact with a negatively charged powder called toner, or dry ink. The toner is attracted to and adheres on the charged regions. Paper is placed over the inked photoconductor and given a positive charge. The toner is then attracted to the paper and heating causes it to be permanently fused to the paper. All of this takes place very quickly—out comes your copy!

The laser printer used with computers is basically a xerographic machine. In this case, there is no "original copy" per se; the information to be printed is stored in the computer. In the printer a laser (Chapter 27) scans back and forth across a rotating, charged drum (Fig. 1). The laser beam passes through a device called a modulator, which allows the beam to reach

the drum or blocks it, according to the signals received from the computer. Wherever the fine light beam strikes the drum, that point is discharged, while unilluminated areas, corresponding to the letters, remain charged. In this way a charged image of what is to be printed is produced. The rest of the printing process then takes place as described previously.

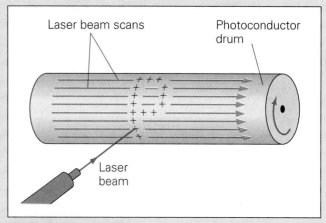

FIGURE 1   Laser printer
A computer-controlled laser scans the charged photoconductor drum, causing the charge to bleed off where the beam strikes. When the laser beam is turned off, charged regions remain, which can be reproduced as in the xerography process.

equation form. Thus, the magnitude of the electric force between two electric point charges is described by an equation called **Coulomb's law:**

> **Note:** Coulomb's law gives the electric force, but only between *point charges*.

$$F = \frac{kq_1q_2}{r^2} \qquad \text{(point charges only)} \qquad (15.2)$$

where r is the distance between the charges (Fig. 15.8a), and k is a constant:

$$k = 8.988 \times 10^9 \ \text{N·m}^2/\text{C}^2 \approx 9.00 \times 10^9 \ \text{N·m}^2/\text{C}^2$$

> **Note:** In calculations, we will take k to be *exact* at $9.00 \times 10^9 \ \text{N·m}^2/\text{C}^2$ for significant figure purposes.

There are equal and opposite forces on the charges (Fig. 15.8a). These forces are sometimes referred to as electro*static* forces, emphasizing the fact that Coulomb's law applies to fixed or *static* charges.

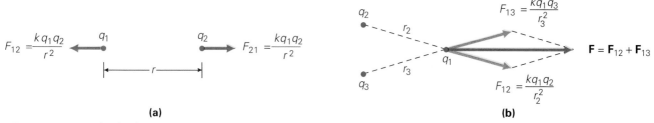

FIGURE 15.8   Coulomb's law
(a) The mutual electrostatic forces on two point charges are equal and opposite. **(b)** For a configuration of more than two charges, the force on a particular charge is the vector sum of the forces on it due to all the other charges.

In some instances, we are concerned with the force on a particular charge in a configuration of two or more charges. The net electric force on a particular charge is simply the vector sum of the forces on it due to all the other charges (•Fig. 15.8b).

CONCEPTUAL EXAMPLE 15.2 ■ FREE OF CHARGE: ELECTRIC FORCES

A rubber comb pulled through dry hair can acquire a net negative charge. It can then be used to pick up small pieces of uncharged paper. This seems to violate Coulomb's force law: since the paper has zero net charge, you might expect there to be no electric force. Explain how the attraction comes about. *Clearly establish the reasoning and physical principle(s) used in determining your answer before checking it below. That is, **how** did you arrive at your answer?*

Reasoning and Answer. When the charged comb is brought near the paper, the paper becomes polarized. The positive end of the paper is closer to the comb than the negative end. Since the electric force varies *inversely* with the square of the distance, the attraction between the comb and the positive end is greater than the repulsion between the comb and the negative end, and the vector sum of the forces on the paper is toward the comb. Instead of contradicting Coulomb's force law, this phenomenon depends crucially on it and illustrates clearly that the electrostatic forces gets weaker with greater distance!

Follow-up Exercise. Does this phenomenon tell you the *type* of charge on the comb? Why or why not?

EXAMPLE 15.3 ■ COULOMB'S LAW: REVIEWING VECTOR ADDITION

(a) Two point charges of $-1.0\ \mu C$ and $2.0\ \mu C$ are separated by a distance of 0.30 m, as illustrated in •Fig. 15.9a. What is the electrostatic force on each particle? (b) A configuration of three charges is shown in •Fig. 15.9b. What is the electrostatic force on q_3?

Solution. Listing the data and converting microcoulombs to coulombs, we have

Given: (a) $q_1 = -1.0\ \mu C \left(\dfrac{10^{-6}\ C}{\mu C}\right) = -1.0 \times 10^{-6}\ C$ *Find:* (a) F_{12} and F_{21}
(b) F_3

$q_2 = +2.0\ \mu C \left(\dfrac{10^{-6}\ C}{\mu C}\right) = +2.0 \times 10^{-6}\ C$

$r = 0.30\ m$

(b) Data given in Figure 15.9b. Convert charges to coulombs as in (a).

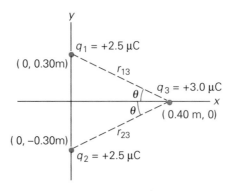

$q_1 = -1.0\ \mu C$ $q_2 = +2.0\ \mu C$

F_{12} F_{21}

0.30 m

(a)

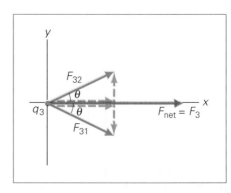

•**FIGURE 15.9 Coulomb's law and electrostatic forces**
See Example 15.3.

(b)

(a) Eq. 15.2 gives the magnitude of the force acting on each point charge.

$$F_{12} = F_{21} = F = \frac{kq_1q_2}{r^2} = \frac{(9.00 \times 10^9 \,\text{N·m}^2/\text{C}^2)(1.0 \times 10^{-6}\,\text{C})(2.0 \times 10^{-6}\,\text{C})}{(0.30\,\text{m})^2}$$

$$= 0.20\,\text{N}$$

Only the magnitude of the charges is used since Coulomb's law gives only the force magnitude. However, because the charges are unlike, we know that the force is attractive. By Newton's third law, F_{21} is equal to F_{12} but in the opposite direction (see the Problem-Solving Hint below).

(b) We must vectorially add the forces F_{31} and F_{32}. Since all the charges are positive, the forces are repulsive as shown in the vector diagram in Fig. 15.9b. Since $q_1 = q_2$ and the charges are equidistant from q_3, then F_{31} and F_{32} are of equal magnitude.

Note from the figure that $r_{31} = r_{32} = 0.50$ m. (Why?) With data from the figure we again use Eq. 15.2.

$$F_{31} = F_{32} = \frac{kq_2q_3}{r_{32}^2}$$

$$= \frac{(9.00 \times 10^9 \,\text{N·m}^2/\text{C}^2)(2.5 \times 10^{-6}\,\text{C})(3.0 \times 10^{-6}\,\text{C})}{(0.50\,\text{m})^2}$$

$$= 0.27\,\text{N}$$

Taking into account the directions of F_{31} and F_{32}, we see by symmetry that the y components of the vectors cancel. Thus F_3 is along the x axis:

$$F_3 = F_{31}\cos\theta + F_{32}\cos\theta = 2F_{32}\cos\theta$$

since $F_{31} = F_{32}$. The angle θ may be determined from the distance triangles; that is,

$$\tan\theta = \frac{y}{x} = \frac{0.30\,\text{m}}{0.40\,\text{m}} = 0.75, \quad \text{and} \quad \theta = \tan^{-1}(0.75) = 37°$$

So

$$F_3 = 2F_{32}\cos\theta = 2(0.27\,\text{N})\cos 37° = 0.43\,\text{N}$$

in the $+x$ direction.

Follow-up Exercise. (a) In part (b) of this Example, what is the force on q_3 if it is instead placed at the origin? (b) What is the force on q_3 if it is a very large distance out on the x axis? (*Hint:* Make a vector sketch first and the answers will follow without any trigonometry.)

PROBLEM-SOLVING HINT

The signs of the charges may be used explicitly in Eq. 15.2. If they are, a positive value for F indicates a repulsive force and a negative value an attractive force. However, it is usually less confusing to calculate the magnitude of the force using only the magnitude of the charges (as was done in Example 15.3) and apply the charge–force law to determine whether the force is attractive or repulsive. This latter approach will be the one we use.

The forces in Example 15.3 are not large by everyday standards. But on the atomic scale, even these tiny forces can produce huge accelerations on very light particles like electrons. Also, the inverse-square relationship can make a huge difference, since atomic and nuclear distances are so small. Consider the following nuclear example involving protons.

EXAMPLE 15.4 ■ FEAR AND LOATHING IN THE NUCLEUS: REPULSIVE ELECTROSTATIC FORCES

(a) What is the magnitude of the repulsive electrostatic force between two protons in a nucleus? Take the distance between the centers of nuclear protons to be 3.0×10^{-13} cm. (b) If these protons were released from rest, how would the magnitude of their initial acceleration compare to that of the acceleration due to gravity on the Earth's surface, g?

Solution. We know the charge on the proton (Table 15.1). Listing the known quantities, we have

Given: $r = 3.0 \times 10^{-13}$ cm $= 3.0 \times 10^{-15}$ m
$q_1 = q_2 = +1.60 \times 10^{-19}$ C

Find: (a) F (magnitude of force)
(b) a/g (magnitude of acceleration compared to g)

(a) Using Coulomb's law (Eq. 15.2), we have

$$F = \frac{kq_1q_2}{r^2} = \frac{(9.00 \times 10^9 \, \text{N·m}^2/\text{C}^2)(1.60 \times 10^{-19} \, \text{C})(1.60 \times 10^{-19} \, \text{C})}{(3.0 \times 10^{-15} \, \text{m})^2}$$

$$= 26 \, \text{N}$$

(This is about 6 lb of force, so we might expect a huge acceleration in part [b].)

(b) The acceleration this force acting alone on one proton would produce is given by Newton's second law

$$a = \frac{F_{\text{net}}}{m} = \frac{26 \, \text{N}}{1.67 \times 10^{-27} \, \text{kg}} = 1.6 \times 10^{28} \, \text{m/s}^2$$

Then

$$\frac{a}{g} = \frac{1.6 \times 10^{28} \, \text{m/s}^2}{9.8 \, \text{m/s}^2} = 1.6 \times 10^{27}$$

That is, $a = (1.6 \times 10^{27})g \approx 10^{27} \, g$! (A whole *lot* bigger.)

Large atoms contain many protons in their nuclei, so the repulsive force on any of these protons would be even larger, which should tend to cause the nucleus to fly apart. Since this doesn't generally happen, there must be a stronger attractive force holding the nucleus together. This is the nuclear (or strong) force, which will be discussed in Chapters 29 and 30.

Follow-up Exercise. Suppose you could place a free proton on the ground, and wished to place a second one directly above the first so that its weight would be exactly balanced by the electric repulsion between them. How far apart would the protons have to be?

Although there is a striking similarity between the mathematical form of the expressions for the electric and gravitational forces, there is a huge difference in their relative strengths, as is shown in the following example.

EXAMPLE 15.5 ■ INSIDE THE ATOM: ELECTRIC VERSUS GRAVITATIONAL FORCE

How do the magnitudes of the electric and gravitational forces between a proton and an electron compare? Express your answer as a ratio of electric force to gravitational force. In other words, how many times larger is the electric force than the gravitational force?

Solution. The charges and masses of the particles are known (Table 15.1), and we also know the values of the constants in the equations for the electric and gravitational forces. Thus, we have

15.4 Electric Field

Objectives: **To be able to (a) understand how the electric field vector is defined, (b) plot electric field lines for simple charge distributions, and (c) calculate electric fields due to several point charges.**

The electric force, like the gravitational force, is an action-at-a-distance force. In fact, we can see that the range of the electric force between two point charges is infinite, since $F_e \propto 1/r^2$, and so approaches zero only if r approaches infinity. Thus, a particular arrangement, or configuration, of charges can have an effect on an additional charge placed anywhere nearby (or possibly anywhere in space).

The idea of a force acting across space was a difficult one for early investigators, and the concept of a *force field*, or simply a *field*, was introduced. Conceptually, an *electric field* surrounds every arrangement of charges. Thus, the field represents the *physical effect* of the originating configuration of charges on the nearby space. We think of the field as representing what is different about the nearby space because the charges are there. This conception allows us to see other charges as interacting with the electric field rather than with the originating charges themselves.

The electric field is a *vector field* and can tell you what force a charge experiences at a particular position in space. The strength or magnitude of the electric field is expressed as *force per unit charge*. We think of mapping an electric field by placing a *test charge* at various locations and measuring the force per unit charge at points throughout space. With these values, the electric field can be plotted. Once you know the electric field pattern due to a charge or charge configuration throughout space, you can then ignore the charge(s) and talk solely in terms of the field. This often greatly facilitates calculations.

In investigating the electric field around a charge or configuration of charges, it is important for the test charge to be small enough that the force it exerts on the field-producing charge or charges is negligible. This will ensure that the in-

Note: We can think of the magnitude of the electric field as being useful in the same way that the price *per pound* is for food items. Knowing how much you want of an item, you can compute how much it will cost if you know the price *per pound*. Similarly, given the magnitude of a charge placed in an electric field, you can compute the force on it if you know the field strength in newtons *per coulomb*.

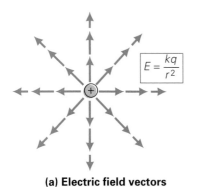

(a) Electric field vectors

$$E = \frac{kq}{r^2}$$

The closer together the lines of force, the stronger the field

(b) Electric field lines (lines of force)

• **FIGURE 15.10   Electric field**
(a) Electric field direction is determined using a positive test charge. Note that the magnitude of the field (lengths of vectors) becomes smaller as the distance from the source charge increases, reflecting the inverse square relationship characteristic of the field from a point charge. **(b)** The vectors are connected to give electric field lines, or lines of force.

troduction of the test charge does not change the field-producing charge arrangement. Since the direction of the force on the charge depends on whether it is positive or negative, this too must be specified in a field representation. By convention, a *positive* test charge (q_o) is used for measuring the field. The **electric field** is then defined as

$$\mathbf{E} = \frac{\mathbf{F}_{\text{on } q_o}}{q_o} \qquad \textit{electric field definition} \qquad (15.3)$$

SI unit of electric field: newton/coulomb (N/C)

E is a vector field and at any point has the direction of the force experienced by a *positive* (test) charge placed there. The SI unit for the electric field is the newton per coulomb (N/C).

The magnitude of the electric field (E), or the force per unit charge, at a distance of r meters from a point charge of q coulombs is then easily computed using Coulomb's force law:

$$E = \frac{F}{q_o} = \frac{1}{q_o}\left(\frac{kq_o q}{r^2}\right) = \frac{kq}{r^2}$$

That is,

$$E = \frac{kq}{r^2} \qquad \begin{array}{l}\textit{electric field} \\ \textit{due to point charge q}\end{array} \qquad (15.4)$$

The direction of the field is given by the charge–force law, with q_o taken as positive.

It is important to note that in the derivation of Eq. 15.4 q_o canceled out. This must always happen since the field is produced by the nearby arrangement of charges, *not* by the test charge. In the case of a point charge, the electric field is produced by q, and the expression should not have the test charge in it.

Some electric field vectors in the vicinity of a positive charge are illustrated in • Fig. 15.10. Note that the vectors point away from the positive charge, which is the direction of the force a positive test charge experiences, and that the magnitude of the vectors decreases with distance from the charge (an inverse-square relationship). For a *configuration* of charges, the total electric field at any point is the vector sum of the electric fields due to the individual charges. The vector character of the electric field is shown in the next two examples.

EXAMPLE 15.6 ■ ELECTRIC FIELDS IN ONE DIMENSION: ZERO NET FIELD

Two point charges are placed on the x axis as shown (• Fig. 15.11). Find the locations on the axis where the electric field is zero.

Solution. The electric field is the vector sum of the field from each of the point charges. Since both charges are positive, their fields point to the right at all points right of q_2 (the test charge gives the field direction as positive). Therefore they cannot cancel in that region. Similarly, to the left of q_1 both fields point to the left and cannot cancel. The only possibility is *between* the charges. Let us specify this location by x relative to q_1.

Given: $d = 0.60$ m (distance between charges)   **Find:** x [the location(s) of zero E]
   $q_1 = +1.5\ \mu C = +1.5 \times 10^{-6}$ C
   $q_2 = +6.0\ \mu C = +6.0 \times 10^{-6}$ C

where the charges were converted from microcoulombs to coulombs.

We can see that if a small positive test charge is placed on the axis between the charges, the two forces will oppose one another. Thus we will have cancellation of the two fields

---

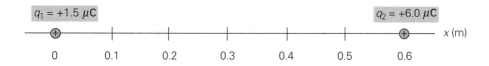

•FIGURE 15.11 **Electric field in one dimension**
See Example 15.6.

if their two magnitudes are equal. We use Eq. 15.4, the expression for the electric field from a point charge

$$E_1 = E_2 \quad \text{or} \quad \frac{kq_1}{x^2} = \frac{kq_2}{(d-x)^2}$$

Rearranging this expression so that the constant k cancels, we can write

$$\frac{1}{x^2} = \frac{(q_2/q_1)}{(d-x)^2}$$

and since $q_2/q_1 = 4$, we can take the square root of both sides:

$$\sqrt{\frac{1}{x^2}} = \sqrt{\frac{q_2/q_1}{(d-x)^2}} = \sqrt{\frac{4}{(d-x)^2}} \quad \text{or} \quad \frac{1}{x} = \frac{2}{d-x}$$

This is easily solved: $x = d/3 = 0.60 \text{ m}/3 = 0.20 \text{ m}$. The answer makes sense physically: because q_2 is the larger of the two positive charges, the only way to make the two fields equal in magnitude is to be closer to q_1.

Follow-up Example. Repeat this Example, changing the sign of the right-hand charge. How is this situation physically different?

EXAMPLE 15.7 ■ ELECTRIC FIELDS IN TWO DIMENSIONS: USING COMPONENTS

What is the electric field at the origin for the three-charge configuration shown in •Fig. 15.12?

Solution. From the figure, we have

Given: $q_1 = -1.00 \ \mu\text{C} = -1.00 \times 10^{-6} \text{ C}$ **Find:** **E** (total electric field at origin)
$q_2 = +2.00 \ \mu\text{C} = +2.00 \times 10^{-6} \text{ C}$
$q_3 = -1.50 \ \mu\text{C} = -1.50 \times 10^{-6} \text{ C}$
$r_1 = 3.50 \text{ m}$
$r_2 = 5.00 \text{ m}$
$r_3 = 4.00 \text{ m}$

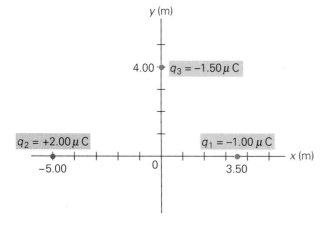

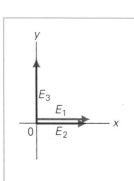

•FIGURE 15.12 **Finding the electric field**
See Example 15.7.

By the law of charges, a *positive* test charge q_o at the origin would tell us the direction of the electric field due to each charge. The directions are indicated in the figure inset. Then, using Eq. 15.4 to find the magnitude of each of the electric field vectors,

$$E_1 = \frac{kq_1}{r_1^2} = \frac{(9.00 \times 10^9 \, \text{N} \cdot \text{m}^2/\text{C}^2)(1.00 \times 1.0^{-6} \, \text{C})}{(3.50 \, \text{m})^2}$$
$$= 7.34 \times 10^2 \, \text{N/C}$$

$$E_2 = \frac{kq_2}{r_2^2} = \frac{(9.00 \times 10^9 \, \text{N} \cdot \text{m}^2/\text{C}^2)(2.00 \times 1.0^{-6} \, \text{C})}{(5.00 \, \text{m})^2}$$
$$= 7.20 \times 10^2 \, \text{N/C}$$

$$E_3 = \frac{kq_3}{r_3^2} = \frac{(9.00 \times 10^9 \, \text{N} \cdot \text{m}^2/\text{C}^2)(1.50 \times 1.0^{-6} \, \text{C})}{(4.00 \, \text{m})^2}$$
$$= 8.44 \times 10^2 \, \text{N/C}$$

The magnitudes of the x and y components of the total field are then

$$E_x = E_1 + E_2 = 7.34 \times 10^2 \, \text{N/C} + 7.20 \times 10^2 \, \text{N/C}$$
$$= 1.45 \times 10^3 \, \text{N/C}$$

$$E_y = E_3 = 8.44 \times 10^2 \, \text{N/C}$$

So, in component form,

$$\mathbf{E} = E_x \mathbf{x} + E_y \mathbf{y} = (1.45 \times 10^3 \, \text{N/C})\mathbf{x} + (8.44 \times 10^2 \, \text{N/C})\mathbf{y}$$

You should be able to show that in magnitude-angle form this is

$$E = 1.68 \times 10^3 \, \text{N/C at } \theta = 30.2° \quad \text{relative to the } +x \text{ axis}$$
$$\text{(in the first quadrant)}$$

Follow-up Exercise. In this Example, remove q_3 and find the electric field at its location due to the other two charges.

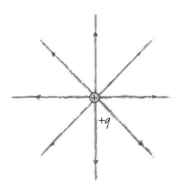

$+q$

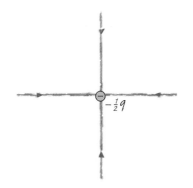

$-\frac{1}{2}q$

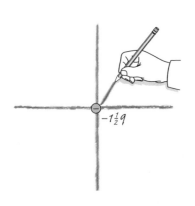

$-1\frac{1}{2}q$

How many lines should be drawn, and what should their direction be ?

Electric Lines of Force

There is a convenient way of graphically representing the electric field pattern in space through the use of **lines of force.** Consider the electric field vectors near a positive point charge, as in Fig. 15.10a. If we connect the vectors as in Fig. 15.10b, we are constructing the lines of force. Notice that the electric field is stronger nearer a charge, where the lines of force are closer together. Note, too, that at any point on a line of force, the electric field direction is along the tangent to the line. The lines of force also have arrows attached to them indicating the field direction. These are the general "rules" for sketching and interpreting electric field lines:

1. The closer together the lines of force, the stronger the electric field;
2. The direction of the electric field is tangent to the lines of force;
3. The electric field lines start at positive charges and end on negative charges;
4. The number of lines leaving or entering a charge is proportional to the magnitude of that charge.

These rules enables us to "map" the pattern of electric lines of force pattern from various charge configurations.

The construction of electric field vectors and then electric lines of force is an exercise in artistic vector addition. In •Fig. 15.13a, we show how the electric field pattern for the equal but opposite charges is mapped out. This arrangement of charges is called an **electric dipole** because it consists of two separate electric charges. One reason the study of electric dipoles is important is that they give us an approximate model for molecules that are permanently polarized, like the water molecule. Even though the net charge on the dipole is zero, it creates an electric field, because the charges are separated. If they were not, the charges would

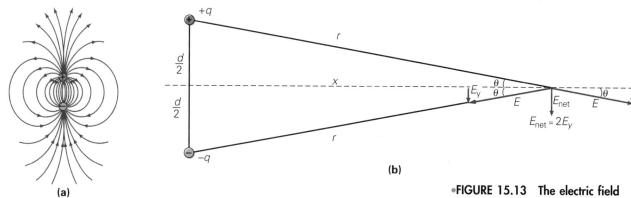

(a)

(b)

•FIGURE 15.13 **The electric field from a dipole**
(a) The electric field line pattern produced by a dipole. (b) The electric field on a dipole bisector (See Example 15.8).

cancel, as would the net field. In the following example, you can see that the field from a dipole, although not zero, certainly decreases much faster than that due to a point charge.

EXAMPLE 15.8 ■ FAST FADE: ELECTRIC FIELD DUE TO A DIPOLE

Using •Fig. 15.13b, show that the electric field far from a dipole on its perpendicular bisector axis is given by kqd/x^3.

Solution. We are given the two charges and the requirement of being "very distant":

Given: $q_1 = q$ *Find:* The electric field on the x axis
$\qquad\ q_2 = -q$
$\qquad\ x \gg d$

Since we are far from each charge, we know that $r = \sqrt{x^2 + (d/2)^2} \approx x$ since $d \ll x$. The net field is the vector sum of the two fields. The two fields have the same magnitude (Why?), each given by $E \approx kq/x^2$. By symmetry, the x components of these two fields cancel and the y components add to give a field pointing down. Since $\sin\theta = (d/2)/r$ from the geometry, the magnitude of the field is

$$E_{net} = 2E\sin\theta \approx 2(kq/x^2)\left(\frac{d/2}{r}\right) = \frac{kqd}{x^3} \propto \frac{1}{x^3}$$

At large distances the electric field decreases *more rapidly* than that of a point charge. If you looked back at the charges from the very distant field point, you would see the charges almost on top of one another, thereby almost canceling out. No wonder the net field decreases more rapidly!

Follow-up Exercise. Repeat the calculation in this Example after changing the negative charge to a positive one. Why is d *not* in your final answer, and why is it an inverse *square* decrease, unlike the dipole result?

Fig. •15.14 shows the construction of the electric field between two oppositely charged plates. Once we understand what the field looks like near one large plate, putting two together is easy.

The complete electric field pattern is shown for some charge configurations and charged conductors in •Fig. 15.15. Note that the electric field lines begin on positive charges and end on negative charges (or at infinity when there is no nearby negative charge). Also note that the lines do not cross, and that we choose the number of lines emanating from, or ending at, a charge, in proportion to the magnitude of that charge. Thus if we draw six lines *leaving* a point charge of $+2.0$ μC, we must show three lines *ending on* a nearby point charge of -1.0 μC in order to be consistent.

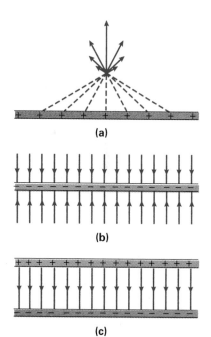

(a)

(b)

(c)

•FIGURE 15.14 **Electric field due to very large parallel plates**
(a) By symmetry, the electric field points up above the positive plate and down below it. (b) The field is oppositely oriented for the negative plate. (c) Superimposing the fields from both plates results in cancellation outside the plates and an approximately uniform field between them.

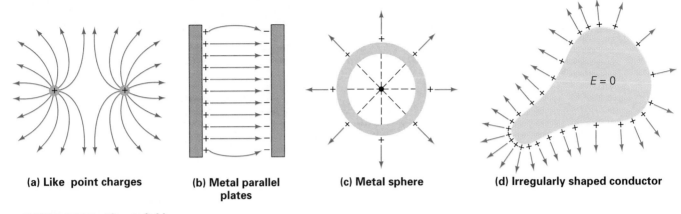

(a) Like point charges   **(b) Metal parallel plates**   **(c) Metal sphere**   **(d) Irregularly shaped conductor**

•**FIGURE 15.15   Electric fields**
Electric fields for various charge configurations: **(a)** Like point charges. **(b)** Oppositely-charged parallel plates. The field is relatively uniform between the plates. **(c)** A charged metal sphere. The electric field outside the sphere is as though all the charge on the sphere were concentrated at its center. The electric field inside the sphere is zero. **(d)** An irregularly shaped charged conductor. Charge tends to accumulate at sharp points, or locations of greatest curvature, producing large electric

15.5   Conductors and Electric Fields

Objectives: **To be able to (a) describe the electric field near the surface and in the interior of a conductor, (b) determine where the highest concentration of excess charge accumulates on a charged conductor, and (c) sketch the electric field line pattern outside a charged conductor.**

The electric fields associated with charged conductors (that are isolated or insulated) have several interesting properties. By definition, in a static situation, charges that remain at rest must experience no electric force, so it follows that

> The electric field is zero everywhere inside a charged conductor.

Also, as you might expect, excess charges on a conductor tend to get as far away from each other as possible since they are very mobile. Thus,

> Any *excess* charge on an isolated conductor resides entirely on the surface of the conductor.

Another property of static electric fields and conductors is that there cannot be any tangential component of the electric field at the conductor surface. If this were not true, charges could move along the surface, contrary to our assumption of a static situation. Thus it must be true that

> The electric field at the outer surface of a charged conductor is perpendicular to the surface.

Lastly, the excess charge on a conductor of irregular shape is most closely packed where the surface is highly curved, that is,

> Excess charge tends to accumulate at sharp points, or locations of greatest curvature, on charged conductors, and the places of highest charge accumulation are where the electric field from the conductor is the largest.

These results are illustrated in •Fig. 15.16. Note that these results are true only for good conductors, and even then only under static conditions. Electric fields can exist inside a nonconductor, for example, and even inside conductors when conditions vary over time.

To help understand *why* most of the charge accumulates in the highly curved surface regions, consider the forces acting *between* charges on the surface of the

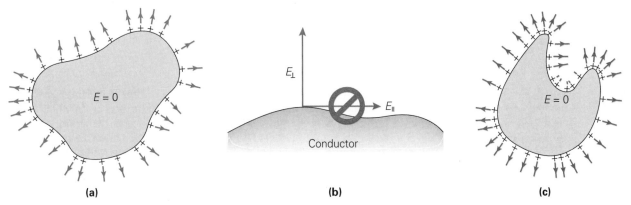

(a) **(b)** **(c)**

•FIGURE 15.16 Electric fields and conductors
(a) The electric field is zero inside a conductor. Any excess charge resides entirely on the conductor's surface. **(b)** Under static conditions, the electric field must not have a component tangential to the conductor's surface. **(c)** For an irregularly shaped conductor, the excess charge accumulates in the regions of highest curvature. The electric field near the surface is perpendicular to that surface and strongest where the charge is densest.

conductor. Where the surface is slightly curved, these forces will be directed nearly parallel to the surface (they would be parallel for a completely flat surface). The charges will spread out until the parallel forces in opposite directions cancel out. At a sharp end, the intercharge forces will be directed more nearly perpendicular to the surface. Here there will be less tendency for the charges to move parallel to the surface, resulting in a concentration of charge. (Draw some surfaces and force vectors to prove this to yourself.) This is shown in •Fig. 15.17.

Let's look more closely at some of these results for a spherical conductor. Suppose that a metal sphere is given a charge $+Q$, as shown in Fig. 15.15b. Because of the mutually repelling forces between them, the like charges will be distributed evenly over the sphere and will eventually come to rest on its outside surface. At this point, the conductor is in electrostatic equilibrium. Because of symmetry, the electric field outside a charged sphere is as though all the excess charge on the sphere were concentrated at its center; that is, $E = kQ/r^2$, where r is greater than the radius of the sphere. Inside the actual charged sphere, the electric field is zero. (Notice the analogy to the gravitational case, in which all the mass of a uniform sphere can be thought of as concentrated at its center.)

Furthermore, we can represent the electric field just outside the sphere of radius R ($r \approx R$) by rewriting Eq. 15.4, multiplying both the numerator and denominator by 4π to give

$$E_{surface} = 4\pi k(Q/4\pi R^2)$$

But the expression $Q/4\pi R^2$ is simply the *surface charge density* (total surface charge divided by surface area of a sphere in C/m²) on the surface of the sphere, while $4\pi k$ is a constant. Thus, we see that if you place the same excess charge on spheres of different size, the larger spheres will have smaller surface electric field strengths. This is because the surface charge density is less on the larger spheres, since they have more surface area. This special case illustrates the general result of conductor surface electric fields being highest where the radius of curvature (R) is smallest (that is, the areas with the greatest curvature).

An interesting situation occurs if there is a large concentration of charge on a conductor with a sharp point. The electric field strength in that region may be large enough to ionize molecules of air (that is, pull electrons off the molecules) and a spark discharge may occur. More charge can be placed on a gently curved conductor, such as a sphere, before a spark discharge will occur. The concentration of charge at the sharp point of a conductor is one reason for the effectiveness of lightning rods. (See the Insight on p. 494.) As an example of some of these charged conductor properties, consider the following classic electrostatic experiment.

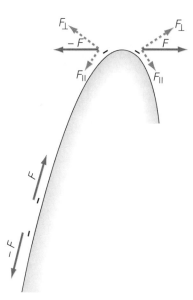

•FIGURE 15.17 Concentration of charge on a curved surface
On a flat surface, the repulsive forces between excess charges will be parallel to the surface and will tend to push the charges apart. On a sharply curved surface, by contrast, these forces will be directed at an angle to the surface. Their components parallel to the surface will be smaller, allowing more charge to concentrate in such areas.

Insight | Lightning and Lightning Rods

We are all familiar with the violent release of electrical energy in the form of lightning. Although it is a common occurrence, we still have a lot to learn about the formation of lightning. We do know that during the development of a cumulonimbus or storm cloud, a separation of charge occurs. The cloud acquires regions of different charge, with the bottom of the cloud generally negatively charged. As a result, an opposite charge is induced on the surface of the Earth (Fig. 1a). Eventually, lightning may reduce this charge difference by ionizing the air, allowing a flow of charge between cloud and ground. However, air is a good insulator and the electric field must be quite strong for this ionization to occur.

How the separation of charge takes place in a cloud is not fully understood, but it must be associated somehow with the rapid vertical movement of air and moisture within storm clouds. Water is a polar molecule—it has regions of charge, and under certain circumstances water molecules can break apart to produce positively charged and negatively charged ions. It is thought that some ionization may occur as a result of frictional forces between water droplets. However, a more plausible theory describes the separation as taking place during ice pellet formation. It has been shown experimentally that as water droplets freeze, positively charged ions are concentrated in the colder, outer regions of the droplets, whereas negatively charged ions are concentrated in the warmer, interior regions. Thus, a freezing droplet has a positively charged outer ice shell and a negatively charged liquid interior.

As the interior of a droplet begins to freeze, it expands and shatters the outer shell. This gives rise to positively charged ice fragments, which are carried upward by the internal cloud turbulence. This occurs on a large scale, and the remaining, relatively heavy droplets with their negative charge eventually settle to the base of the cloud.

Most lightning occurs entirely within a cloud (intracloud discharges) where it cannot be seen directly. However, familiar visible discharges do take place between clouds (cloud-to-cloud discharges) and between a cloud and the Earth (cloud-to-ground discharges). Pictures of cloud-to-ground discharges taken with special high-speed cameras reveal a nearly invisible downward ionization path. This occurs in a series of jumps or steps and so is called a *stepped leader*. As the leader nears the ground, positively charged ions in the form of a *streamer* rise from trees, tall buildings, or the ground to meet it.

When a streamer and leader make contact, the electrons along the leader channel begin to flow downward. The initial flow is near the ground, and as it continues, electrons positioned successively higher begin to migrate downward. Hence, the path of electron flow is continuously extended upward in what is called a *return stroke*. The surge of charge flow in the return stroke causes the conductive path to be illuminated, producing the bright flash seen by the eye and recorded in time-exposure photographs of lightning (Fig. 1b). Most lightning flashes have a duration of less than 0.50 s. Usually after the initial discharge, ionization again takes place along the original channel and another return stroke occurs. Typical lightning events have 3 or 4 return strokes.

Ben Franklin is often said to have been the first to demonstrate the electrical nature of lightning. In 1750 he suggested an experiment using a metal rod on a tall building. However, a Frenchman named d'Alibard set up the experiment and drew

Note: Compare Section 7.5.

CONCEPTUAL EXAMPLE 15.9 ■ AN ICE PAIL EXPERIMENT

A positively charged rod is held inside an isolated metal container which has uncharged electroscopes conductively attached to its inside and outside surfaces (•Fig. 15.18a). With the charged rod held inside the container, (a) neither electroscope would show a deflection, (b) only the outside-connected electroscope would be deflected, (c) only the inside-connected electroscope would be deflected, (d) both electroscopes would be deflected. *Clearly establish the reasoning and physical principle(s) used in determining your answer before checking it below. That is, **why** did you select your answer?*

Reasoning and Answer. The positively charged rod would attract negative charges, causing the inside of the metal container to become negatively charged. The outside would thus acquire a positive charge. Hence, both electroscopes would be charged (though with opposite signs) and would show deflections, so the answer is (d). A similar experiment was performed by the nineteenth-century English scientist Michael Faraday using ice pails, so this setup is often called Faraday's ice pail experiment.

Follow-up Exercise. Suppose that the positively charged rod *touched* the metal container as illustrated in •Fig. 15.18b. What would be the effect on the electroscopes?

sparks from a rod during a thunderstorm. Franklin later performed a similar experiment with a kite he flew during a thunderstorm. He drew sparks to his knuckles from a key hanging at the end of a conductive kite cord. Ben was extremely lucky that he wasn't electrocuted. *Under no circumstances should you try to duplicate this experiment.* On the average, lightning kills 200 people a year in the United States and injures another 550.

A practical outcome of Franklin's work with lightning was the lightning rod, which was described in *Poor Richard's Almanac* in 1753. It consists simply of a pointed metal rod connected by a wire to a metal rod that has been driven into the Earth, or grounded (Fig. 1c). Franklin wrote that the rod "ei-

ther prevents the stroke from the cloud or, if the stroke is made, conducts the stroke to Earth with safety of the building."

The latter is the general principle of operation of a lightning rod. The elevated rod intercepts the downward ionized stepped leader from a cloud, harmlessly discharging it to the ground before it reaches the structure or makes contact with an upward streamer. This prevents the formation of the damaging electrical surge associated with the return stroke.

(b)   (c)

FIGURE 1   Lightning and lightning rods
(a) Cloud polarization induces a charge on the Earth's surface.
(b) When the field becomes large enough, an electrical discharge results, which we call lightning. See the text for details. **(c)** A lightning rod or arrestor provides a path to ground so as to prevent damage.

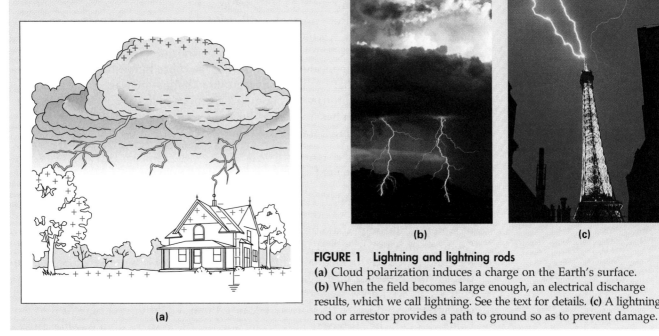

(a)   (b)

•**FIGURE 15.18   An ice pail experiment**
(a) See Example 15.9. **(b)** See Follow-up Exercise.

*15.6   Gauss's Law for Electric Fields: A Qualitative Approach

**Objectives: To be able to (a) state the physical basis of Gauss's Law, and
(b) use the law to make qualitative predictions.**

A fundamental law for electric fields was discovered by Karl Friedrich Gauss (1777–1855), a German mathematician. In its mathematical form, this relationship can be used to calculate electric field strengths in cases of high symmetry. However, even a purely qualitative account of Gauss's law can teach us much interesting physics.

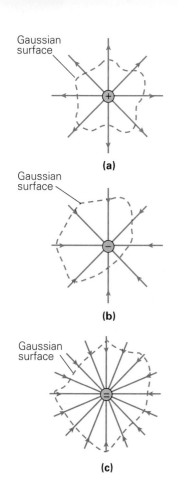

(a)

(b)

(c)

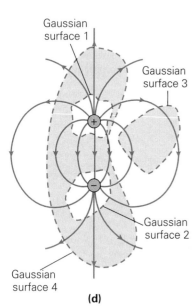

(d)

•**FIGURE 15.19 Various Gaussian surfaces and lines of force**
(a) Surrounding a single positive point charge. (b) Surrounding a single negative point charge. (c) Surrounding a larger negative point charge. (d) Four different surfaces surrounding various parts of an electric dipole.

Consider the single positive electric charge in •Fig. 15.19a. Let us picture an imaginary closed surface surrounding it. (Such surfaces are called **Gaussian surfaces.**) Now, let us designate outward-pointing electric field lines as positive and inward-pointing field lines as negative. If we count the lines of both types and total them up (that is, subtract the number of negative lines from the number of positive ones),we would naturally find that the total is positive (since in this simple case there are *only* positive lines). That is, there is a certain net number of outward-pointing electric field lines passing through the surface. Similarly, for a negative charge, as shown in •Fig. 15.19b, the count would yield a negative total—a certain net number of inward-pointing lines passing through the surface. You can easily see that this would be true for *any* surface surrounding the charge. If we double the magnitude of the negative charge (•Fig. 15.19c), we see that our negative field line count would also double.

•Figure 15.19d shows a dipole with four different imaginary surfaces. It is easy to see that Surface 1, which encloses a net positive charge, has a positive field line count, and Surface 2, which encloses a net negative charge, has a negative field line count, just as in Figs. 15.7a and b. The more interesting cases are Surfaces 3 and 4. Note that both include zero net charge—Surface 3 because it includes no charges and Surface 4 because it includes equal and opposite charges. And we find that each has a net field line count of zero.

These situations illustrate the fundamental physical idea of **Gauss's law*:**

> The net number of electric field lines passing through an imaginary surface is proportional to the amount of net charge enclosed within that surface.

An everyday analogy may help you to understand this principle. If you surround a lawn sprinkler with an imaginary surface, you will find that there is a net flow of water out through that surface—clear evidence that what you have inside is a "source" of water (disregarding the hidden pipe bringing water into the sprinkler underground). In an analogous way, a net outward-pointing electric field indicates the presence of a net positive charge inside the surface, since positive charges, as we know, are "sources" of the electric field. Similarly, a drain inside our surface would give itself away because there would be a net inward flow of water through the surface (again neglecting the drain carrying the water away). This analogy is illustrated in •Fig. 15.20. The following example illustrates the power of Gauss's law, even qualitatively.

CONCEPTUAL EXAMPLE 15.10 ■ CHARGED CONDUCTORS REVISITED: GAUSS'S LAW

A net charge Q is placed on an arbitrarily shaped conductor. Use the qualitative version of Gauss's law to prove that all the charge must lie on the conductor's surface under electrostatic conditions. (This situation is illustrated in •Fig. 15.21.) *Clearly establish the reasoning and physical principle(s) used in determining your answer before checking it below. That is,* **how did you arrive at your answer?**

Reasoning and Answer. Since the situation is static equilibrium, there can be no electric field inside the volume of the conductor; otherwise, the almost-free electrons would be moving around. Now for our Gaussian surface, let us take one that follows the shape of the actual conductor, but is just barely inside the actual surface. Since there are no electric field lines inside the conductor, there are no electric field lines passing through our imaginary surface in either direction. So the net result is zero electric field lines penetrating the Gaussian surface.

But, by Gauss's law, the net number of field lines is proportional to the amount of charge enclosed inside the surface. Thus, there must be no net charge within the sur-

---

*Strictly speaking, this is Gauss's law for electric fields. There is also a version of Gauss's law for magnetic fields which we will not discuss in this text.

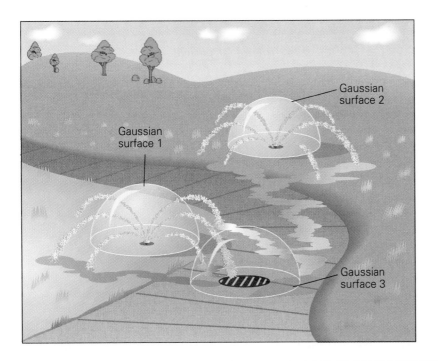

•FIGURE 15.20 Water analogy to Gauss's law
A net outward flow of water indicates a source of water inside the surface. A net inward flow of water indicates a water drain inside the surface.

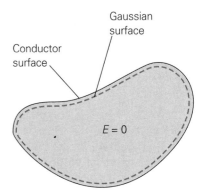

FIGURE 15.21 Gauss's law: Excess charge on a conductor
See Conceptual Example 15.10.

face! Since our imaginary surface can be made as close as we wish to the true surface, it follows that the excess charge, if it cannot be inside the volume of the conductor, must be on the outside. This is a very elegant proof of what we determined before by considering intercharge forces and repulsion.

Follow-up Exercise. In this Example, if the net charge on the conductor is negative what would be the sign of the net number of lines through a Gaussian surface that completely encloses the conductor, excess charge and all? Explain your reasoning.

Chapter Review

Important Terms

electric charge   476
law of charges or charge-force
  law   476
coulomb (C)   477
net charge   477
conservation of charge   477
conductors   479

insulators   479
semiconductors   479
electrostatic charging   480
charging by friction   480
charging by contact   481
charging by induction   481
polarization   481

Coulomb's law   483
electric field   488
electric lines of force   490
electric dipole   490
Gaussian surface   496
Gauss's law   496

Important Concepts

• Electric charge is the property of an object that determines its electrical behavior: the electric force it can exert and the electric force it can experience.

• Opposite charges attract and like charges repel one another.

• The law of conservation of charge states that the net charge on an isolated system remains constant.

• Coulomb's law gives the force between two point charges.

• The electric field is a vector field that describes how nearby charges modify the space around them. It is defined as the electric force per unit positive charge.

• Electric lines of force are the imaginary lines formed by connecting electric field vectors. Their closeness and direction indicate the magnitude and direction of the electric field at any point.

*• A Gaussian surface is an imaginary closed surface. The net number of lines of force passing through it indicates the amount and sign of net charge enclosed by it.

Important Equations

Quantization of Electric Charge
($|e| = 1.60 \times 10^{-19}$ C):

$$q = ne \qquad (15.1)$$

Coulomb's Law ($k \approx 9.0 \times 10^9$ N·m²/C²):

$$F = \frac{kq_1q_2}{r^2}$$

(two point charges, magnitude only) $\qquad (15.2)$

Electric Field (definition):

$$\mathbf{E} = \frac{\mathbf{F}_{\text{on } q_\text{o}}}{q_\text{o}} \qquad (15.3)$$

Electric Field Due to a Point Charge, q:

$$E = \frac{kq}{r^2} \quad \text{(magnitude only)} \qquad (15.4)$$

Exercises*

15.1 Electric Charge

1 A combination of three electrons and two protons would have a net charge of (a) +1, (b) −1, (c) +1.6 × 10^{-19} C, (d) −1.6 × 10^{-19} C.

2 The directions of the interacting electric forces on two charges is given by (a) the conservation of charge, (b) the charge–force law, (c) the magnitude of the charges, (d) none of these.

3 (a) How do we know that there are two types of electric charge? (b) What would be the effect of designating the charge on the electron as positive and the charge on the proton as negative?

4 An electrically neutral object can be given a net charge by several means. Does this violate the conservation of charge? Explain.

5 ■ What net electric charge would one million protons have?

6 ■ A glass rod rubbed with silk acquires a charge of +8.0 × 10^{-10} C. (a) What is the charge on the silk? (b) How many electrons have been transferred to the silk?

7 ■ A rubber rod rubbed with fur acquires a charge of −2.4 × 10^{-9} C. (a) What is the charge on the fur? (b) How much mass is transferred to the rod?

8 ■ In walking across a carpet, a person acquires a net negative charge of −5.0 μC. How many excess electrons does the person have?

9 ■■ An alpha particle is the nucleus of a helium atom with no electrons. What is the charge on an alpha particle?

*Take k to be *exact* at 9.00×10^9 N·m²/C² and e to be *exact* at 1.60×10^{-19} C for significant figure purposes.

10 ■■ What is the total charge on 1.0 mol of Ca^{2+} ions (calcium atoms that have lost two electrons apiece)?

15.2 Electrostatic Charging

11 A rubber rod is rubbed with fur. The fur is quickly brought near the bulb of an uncharged electroscope. What is the sign of the charge on the leaves of the electroscope?

12 When a positively charged rod is brought near the bulb of a negatively charged electroscope, the leaves (a) diverge farther, (b) collapse, (c) remain unchanged.

13 Electrostatic charging may be done by (a) friction, (b) contact, (c) induction, (d) all of these.

14 ■ How could an electroscope be negatively charged by induction? How could you prove it was negatively charged?

15 ■■ Two metal spheres mounted on insulated supports are in contact. How could both spheres be given a net electric charge without touching them? Would the spheres be charged positively or negatively?

15.3 Electric Force

16 The magnitude of the electric force between two point charges is given by (a) the charge–force law, (b) conservation of charge, (c) Coulomb's law, (d) both (a) and (b).

17 Compared to that of the electric force, the strength of the gravitational force between two protons is (a) about the same, (b) somewhat larger, (c) very much larger, (d) very much smaller.

18 ■ An electron is a certain distance from a proton. How would the electric force be affected if the electron were moved (a) half that distance toward the proton, and (b) three times that distance away from the proton?

19 ■ Two identical point charges are at a fixed distance from one another. How would the electric force be affected if (a) one of their charges was doubled and the other was halved, (b) both their charges were doubled, and (c) one of the charges was doubled, and the other was unchanged?

20 ■ We often experience the effect of the gravitational force, for example, in picking up a heavy object, but we don't normally experience the electric force. Explain why.

21 ■ Why can astronomers safely ignore the electric force when calculating planetary orbits around the Sun?

22 ■ In a certain organic molecule the nuclei of two carbon atoms are separated by a distance of 0.25 nm. What is the magnitude of the electric repulsion between them?

23 ■ An electron and a proton are separated by 10 cm. (a) What is the magnitude of the force on the electron? (b) What is the net force on the system?

24 ■ Two charges originally separated by 30 cm are moved farther apart, until the force between them has decreased by a factor of 10. How far apart are the charges then?

25 ■ Two charges are separated to a distance of 100 cm causing the electric force between them to decrease by exactly a factor of 5. What was their initial separation distance?

26 ■■ Two charges are attracted by a force of 25 N when separated by 20 cm. What is the force between the charges when the distance between them is 90 cm?

27 ■■ Two charges of $-1.0 \ \mu C$ are placed at opposite ends of a meterstick. Where on the meter stick could (a) a free electron and (b) a free proton be placed on the meterstick and be in electrostatic equilibrium?

28 ■■ Charges of $-1.0 \ \mu C$ and $+1.0 \ \mu C$ are placed at opposite ends of a meterstick. Where could (a) a free electron and (b) a free proton be placed and be in electrostatic equilibrium?

29 ■■ Two charges, q_1, and q_2, are located at the origin and at (0.50 m, 0), respectively. Where on the x axis must a third charge, q_3, of arbitrary sign be placed to be in electrostatic equilibrium if (a) q_1 and q_2 are like charges of equal magnitude, (b) q_1 and q_2 are unlike charges of equal magnitude, and (c) $q_1 = +3.0 \ \mu C$ and $q_2 = -7.0 \ \mu C$?

30 ■■■ The electron and proton in a hydrogen atom are separated on the average by a distance of 5.3×10^{-11} m (•Fig. 15.22). Assuming the orbit of the electron to be circular, (a) what is the electric force on the electron? (b) What is the electron's orbital speed? (c) What is the magnitude of the electron's centripetal acceleration in units of g?

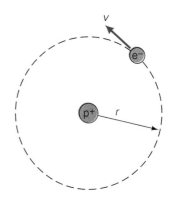

•**FIGURE 15.22 Hydrogen atom**
See Exercises 30 and 31.

31 ■■■ Compute the magnitude of the (a) gravitational force and (b) the electrical force between the electron and proton in the hydrogen atom (Fig. 15.22).

32 ■■■ Three charges are located at the corners of an equilateral triangle, as depicted in •Fig. 15.23. What are the magnitude and the direction of the force on q_1?

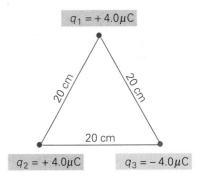

•**FIGURE 15.23 A charge triangle**
See Exercises 32, 51, and 52.

33 ■■■ Four charges are located at the corners of a square as illustrated in •Fig. 15.24. What are the magnitude and the direction of the force on (a) q_2 and (b) q_4?

34 ■■■ Two 0.10-g pith balls are suspended from the same point by threads 30 cm long. (Pith is a light insulating material once used to make helmets worn in tropical climates.) When the balls are given equal charges, they come to rest 18 cm apart as shown in •Fig. 15.25. What is the magnitude of the charge on each ball? (Neglect the mass of the thread.)

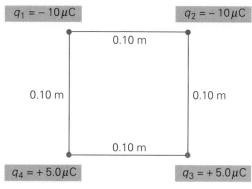

•**FIGURE 15.24 A charge rectangle**
See Exercises 33, 53, 56, 81, and 83.

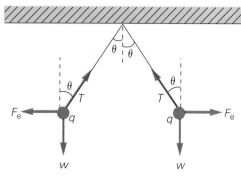

•**FIGURE 15.25 Repelling pith balls**
See Exercise 34.

15.4 Electric Field

35 The electric field due to a negative charge (a) varies as $1/r$, (b) points toward the charge, (c) has a finite range, (d) is the same as that of a positive charge.

36 The units of electric field are (a) C, (b) N/C, (c) N, (d) J.

37 At a point in space an electron feels an electric force vertically upward. The direction of the electric field at that point is (a) down, (b) up, (c) zero, (d) undetermined by the data.

38 How is the relative magnitude of the electric field in different regions determined from a field vector diagram?

39 How can the relative magnitudes of the field in different regions be determined from an electric field line diagram?

40 ■ Why can electric field lines never cross?

41 ■ A positive charge is inside an isolated metal sphere as shown in •Fig. 15.26. Describe the situation in terms of the electric field and the charge on the sphere. How would the situation change if the charge were negative?

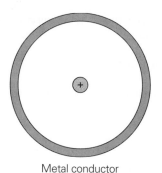

Metal conductor

•**FIGURE 15.26 A point charge inside a thick metal spherical shell**
See Exercise 41.

42 ■ Explain how a region may be shielded from electrical influences (electric fields). (Do you think gravitational shielding is possible? Explain.)

43 ■ Sketch the electric field between a positively charged flat plate and an equal but oppositely charged pointed object near the plate and pointing perpendicular to its plane.

44 ■ An isolated electron experiences an electric force of 6.4×10^{-14} N. What is the magnitude of the electric field at the electron's location?

45 ■ What are the magnitude and the direction of a vertically oriented electric field that would just support the weight of an electron?

46 ■ What is the magnitude of the electric field at a point 0.25 cm away from a point charge of $+4.0 \ \mu C$?

47 ■■ Two charges, $-4.0 \ \mu C$ and $-5.0 \ \mu C$, are separated by a distance of 20 cm. What is the electric field halfway between the charges?

48 ■■ Two charges, $-3.0 \ \mu C$ and $-4.0 \ \mu C$, are located at $(-0.50 \text{ m}, 0)$ and $(0.75 \text{ m}, 0)$, respectively. (a) Where on the x axis is the electric field zero? (b) Is there a position (or positions) where the electric field has only a y component? Explain.

49 ■■ Three charges, $+2.5 \ \mu C$, $-4.8 \ \mu C$, and $-6.3 \ \mu C$, are located at $(-0.20 \text{ m}, 0.15 \text{ m})$, $(0.50 \text{ m}, -0.35 \text{ m})$, and $(-0.42 \text{ m}, -0.32 \text{ m})$ respectively. What is the electric field at the origin?

50 ■■ Two charges of $+4.0 \ \mu C$ and $+9.0 \ \mu C$ are 30 cm apart. Where on the line joining the charges is the electric field zero?

51 ■■ What is the electric field at the center of the triangle in Fig. 15.23?

52 ■■ Compute the electric field at a point midway between charges q_1 and q_2 in Fig. 15.23.

53 ■■ What is the electric field at the center of the square in Fig. 15.24?

54 ■■ A particle with a mass of 1.0×10^{-5} kg and a charge of $+2.0$ μC is released in a uniform electric field of 12 N/C. (a) How far does it travel in 0.50 s? (b) What is its speed at that point?

55 ■■ A thin, hollow spherical conductor with a radius of 0.20 m has a uniform charge density of -1.0×10^{-8} C/m^2 on its surface. What is the electric field (a) inside the sphere, (b) at the surface of the sphere, and (c) 0.80 m from the center of the sphere?

56 ■■■ Compute the electric field at a point 4.0 cm from q_2 along a line running toward q_3 in Fig. 15.24.

57 ■■■ Two equal but opposite charges form a dipole as shown in •Fig. 15.27. (a) What is the electric field direction and magnitude at point P? (b) Compute your answer (in terms of k, q, d, and x) using exact vector addition and (c) show that it gives the distant field approximation of kqd/x^3 (see Example 15.8)

•**FIGURE 15.27   Electric dipole field**
See Exercise 57.

15.5   Conductors and Electric Fields

58 In electrostatic equilibrium, the electric field just below the surface of a charged conductor is (a) the same value as the field just above the surface (b) zero (c) kq/R^2

59 An uncharged thin metal slab is placed in an external electric field which points horizontally to the right. If the slab is oriented so that its large area side is perpendicular to the field, what is the sign of the charge on its left side?

60 In Exercise 59, what is the electric field inside the slab? (a) zero (b) same value as the original external field (c) somewhere between zero and the original external field value.

61 ■ A solid conducting sphere is surrounded by a thick spherical conducting shell. A point charge of Q is initially placed at the center of the inner sphere. After equilibrium how much charge is on (a) the interior of the solid sphere, (b) the surface of the solid sphere, (c) the inner surface of the shell, and (d) the outer surface of the shell?

62 ■ In Exercise 61, what is the electric field direction (a) in the interior of the solid sphere, (b) between the sphere and the shell, (c) inside the shell, and (d) outside the shell?

63 ■■ In Exercise 61, write expressions for the electric field magnitude (a) in the interior of the solid sphere, (b) between the sphere and the shell, (c) inside the shell, and (d) outside the shell. Your answer should be in terms of Q, r (the distance from the center of the sphere), and k.

64 ■■ A flat triangular piece of metal with rounded corners has a net positive charge on it. Sketch how the charge distributes itself on the surface.

65 ■■ In Exercise 64, add the electric field lines near the surface of the metal (including direction) to your sketch.

66 ■■ How much work does it take to move a small test charge around inside the volume of a charged conductor?

67 ■■■ Approximate a metal needle as a long cylinder with a very pointed but slightly rounded end, and sketch the charge distribution and outside electric field lines if the needle has an excess of electrons on it.

*15.6   Gauss's Law for Electric Fields: A Qualitative Approach

68 What would be the net number of electric field lines passing through a Gaussian surface that was located totally within a set of oppositely charged parallel plates.

69 A Gaussian surface surrounds an object with a net charge of -5.0 μC. Which of the following is true? (a) There will be more electric field lines pointing outward than inward. (b) There will be more electric field lines point inward than outward. (c) The net number of field lines through the surface is zero.

70 ■ The same Gaussian surface is used to separately surround two charged objects. The net number of field lines penetrating the surface is the same in both cases, but oppositely directed. What can you say about the net charges on the two objects?

71 ■■ If a Gaussian surface has 16 field lines leaving it when it surrounds a point charge of $+10.0$ μC and 75 fields lines leaving it when it surrounds an unknown point charge, what is the amount of charge on the unknown?

72 ■■ If 8 field lines are counted leaving a Gaussian surface when it completely surrounds the positive end of an electric dipole, what is the count if it surrounds just the other end?

73 ■■ Suppose a Gaussian surface encloses a positive point charge which has 6 field lines leaving it, and also a negative point charge with twice the magnitude of charge as the positive one. What is the net number of field lines passing through the Gaussian surface?

74 ■■ If there is a net number of electric field lines pointing out on a Gaussian surface, does that necessarily mean there are no negative charges on the interior?

Additional Exercises

75 An electron is placed in a constant electric field of 2.5×10^3 N/C directed along the $+y$ axis. (a) What is the force on the electron? (b) If the electron is released, how much kinetic energy does it have after moving 10 cm?

76 How many electrons are required to make up a net charge of -0.50 μC?

77 Two charges, -3.0 μC and -5.0 μC, are 0.40 m apart. (a) Where should a charge of -1.0 μC be placed to put it in electrostatic equilibrium? (b) Where should a charge of $+1.0$ μC be placed to be in electrostatic equilibrium?

78 An electron is released from rest at the origin in a uniform electric field that has a magnitude of 450 N/C in the $+x$ direction. (a) How long will it take the electron to travel 1.0 m? (b) What will its position be (in xy coordinates) at half that time?

79 A proton is traveling horizontally to the right with a speed of 2.0×10^5 m/s. (a) What electric field (magnitude and direction) does it take to stop it in 4.5 cm? (b) How long will it take to stop?

80 A charge of $+6.0$ μC at the origin of a set of coordinate axes experiences a force of 1.8 N in the $+y$ direction due to another charge located 0.30 m away. What are the magnitude and the position of the other charge?

81 Compute the electric field at a point midway between charges q_1 and q_4 in Fig. 15.24.

82 A solid spherical conductor (uncharged) is placed in a uniform electric field pointing to the right. Sketch the charges on the conductor and the electric field after electrostatic conditions are attained.

83 If a Gaussian surface completely surrounds the charge arrangement in Fig. 15.24, what is the sign of the net number of lines passing through the surface?

84 In the dipole electric field calculation (Example 15.8) the product of the magnitude of the charge on one end of the dipole, times the distance of separation, appears as qd in the numerator. Explain why the electric field on the perpendicular axis to the dipole goes to zero if d is zero.

85 For an electric dipole, the product qd is called the *dipole moment* and is made into a vector, called **p**, by assuming it points from the negative end of the dipole toward the positive end (see •Fig. 15.28a). With a sketch (•Fig. 15.28b), show that a dipole placed in a uniform field will tend to rotate until its dipole moment "lines up with the field."

86 What happens to the electric dipole in Exercise 85 if the field is not uniform (see •Fig. 15.28c), assuming the dipole is free to move?

87 A pair of oppositely charged parallel plates has a uniform metal slab inserted parallel to the plates. Sketch what the resulting electric field looks like.

88 Using symmetry arguments in a sketch, show that the electric lines of force outside (but close to) a long straight uniformly charged wire "stick out" perpendicularly to the wire in a pattern reminiscent of a test tube scrubbing brush. Sketch how the field looks from the side view of the wire, and also head on.

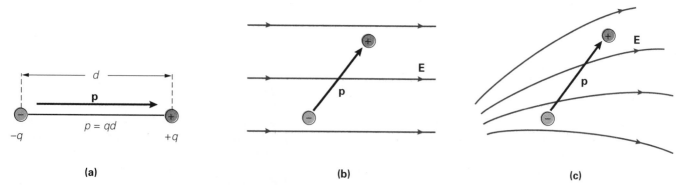

(a)   **(b)**   **(c)**

•**FIGURE 15.28 Dipole moment, dipoles in electric fields**
(a) The dipole moment vector has a magnitude of qd and points from the negative to the positive end. **(b)** An electric dipole placed in a uniform electric field. **(c)** An electric dipole placed in a nonuniform electric field. See Exercises 85 and 86.

16 Electric Potential, Energy, and Capacitance

Perhaps one of the reasons that the study of electricity sometimes seems harder than, say, mechanics is that electricity is not very apparent to our senses. You have seen objects sliding or falling all your life, so you have some sense of what to expect of them. Even liquids and gases behave in ways that are basically familiar to us from everyday experience. But electricity is invisible, inaudible, odorless. Indeed, when we do become aware of electricity, it usually means that something is wrong—often dangerously so. The crackling of shorted wires, the smell of burning insulation on an overloaded coil, the blue flash of an arcing circuit, the warning tingle indicating you are touching something that is "live" and shouldn't be—all of these signal trouble.

Yet this isn't always true. The girl in the photo above is clearly experiencing the effects of electricity. In fact, she is charged to a potential of several thousand volts! You probably know that household circuits, which carry only 120 volts, can give you a nasty and potentially dangerous shock. Yet the girl doesn't seem to be in any trouble or danger. What's going on? You'll find the explanation of this and many other electrical phenomena in this chapter.

503

16.1 Electric Potential Energy and Potential Difference

Objectives: To be able to **(a) understand the concept of electric potential difference ("voltage") and its relationship to electric potential energy,** and **(b) calculate electric potential differences and electric potential energy.**

In Chapter 15 we analyzed electrical effects using vectors, that is, electric field vectors and electric lines of force. Recall that in mechanics, we first used Newton's laws with free-body diagrams and vectors. Then, in search of a simpler approach to many situations, we introduced quantities such as work, kinetic energy, and potential energy. With these concepts, we could solve many problems using the work–energy theorem and the principle of energy conservation. We will find it very useful, both conceptually and from a problem-solving standpoint, to extend these energy methods to the study of electric fields.

Electric Potential Energy Difference

Let's start with one of the simplest electric field patterns, the field between two large, oppositely charged parallel plates. Near the center of the plates the field is uniform in magnitude and direction (•Fig. 16.1a). Suppose we take a small positive charge, q_o, and move it at constant velocity against the electric field, from the negative plate (A) to the positive plate (B). An external force with the same magnitude as the electric force, $F = q_o E$, is required. The work done by the external force is positive (force and displacement in the same direction) and equal to $F_e d \cos 0° = +F_e d$ or $q_o E d$. At the same time negative work is done *by the field* (force and displacement in opposite directions), equal to $F_e d \cos 180° = -F_e d$ or $-q_o E d$. The net work on the charge is zero, so it experiences no kinetic energy change, in keeping with the work–energy theorem.

However, if the test charge is released, it will accelerate back toward the negative plate, gaining kinetic energy. Therefore, we can see that by moving the charge from A to B, the external force increases the system's **electric potential energy** by an amount equal to the work done:

$$\Delta U_e = U_B - U_A = q_o E d$$

It is important to realize that, although at first glance we may associate this increase with the charge alone, this is not the case. It is really a potential energy gain *of the whole system*—the charge *and* the plates.

The gravitational analogy to the parallel plate electric field is the Earth's gravitational field near the surface. When an object is raised a vertical distance h, the change in its potential energy is positive and equal to the work done by the external force (•Fig. 16.1b):

$$\Delta U_g = U_B - U_A = mgh$$

The gravitational potential energy change can be qualitatively envisioned as the stretching of an invisible "gravitational spring" connecting the mass with the Earth. The value of the potential energy at A is arbitrary and we can assign it any value we desire. Similarly, in the parallel plate case, the value of the potential energy at A or B (or any point between) is also arbitrary. It is common (though

•FIGURE 16.1 Potential energy changes in uniform electric and gravitational fields
(a) Moving a positive charge against the electric field takes positive work and increases the electric potential energy. **(b)** Moving a mass against the gravitational field takes positive work and increases the gravitational potential energy.

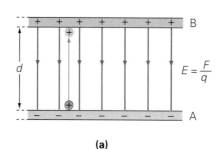

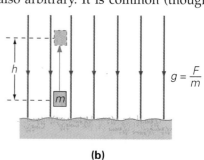

(a)

(b)

not required) to designate the zero point for electrical potential energy as the negative plate. However, in both cases the only measurable physical quantity is the potential energy *difference* relative to some (arbitrary) reference point.

Electric Potential Difference

In dealing with electric forces we eliminated the dependence on the test charge by defining the electric field as the *electric force per unit charge*. Then, knowing the electric field, we could determine the force on *any* point charge placed at that location. Similarly, we define the **electric potential difference** as the change in potential energy *per unit positive test charge*. This definition can also be written in terms of work:

$$\Delta V = \frac{\Delta U}{q_0} = \frac{W}{q_0} \qquad \text{\textit{potential difference}} \atop \text{\textit{definition}} \qquad (16.1)$$

SI unit of electric potential difference: joule/coulomb (J/C) or volt (V)

The SI unit for potential difference is the joule per coulomb. This has been renamed the **volt** (V), in honor of Alessandro Volta, an 18th century Italian scientist. Thus, 1 V = 1 J/C. Potential difference is sometimes called **voltage**, and the symbol for potential difference is commonly changed from ΔV to just V, as we will do later in this chapter.

Potential difference, although based on the potential energy difference, is *not* the same thing. Since potential difference is defined per unit charge, it does not depend on the amount of charge moved, a very useful property. To illustrate this point, let us calculate the potential difference for the case of a uniform field between parallel plates:

$$\Delta V = \frac{\Delta U}{q_0} = \frac{q_0 Ed}{q_0} = Ed \qquad \text{\textit{potential difference}} \atop \text{\textit{(parallel plates)}} \qquad (16.2)$$

Note: Eq. 16.2 is for parallel plates only.

You can see that the amount of charge moved cancels out and the result depends only on the charged plate characteristics, that is, the field they produce (E) and their separation (d). We say that the positively charged plate is at a *higher* **electric potential** than the negatively charged one by an amount ΔV.

Note: We commonly assign a value of zero for the electric potential of the negative plate, but this is arbitrary. Only potential *differences* are meaningful.

It may strike you as odd that we define electric potential *difference* without first defining electric potential. Although this may seem backwards, there is a good reason for it. Electric potential difference is a physically meaningful quantity that we can actually measure and use in our calculations. The value of the electric potential, by contrast, can't be measured in any absolute way—it depends entirely on our choice of a reference point.

To illustrate this idea, recall that when we studied mechanical forms of potential energy associated with springs and gravitation in Chapter 5, we found ourselves dealing with *changes* in potential energy. *Values* of potential energy could be specified, but only with respect to an arbitrary reference point. Thus, in the case of gravity, we sometimes defined the zero point as the Earth's surface for convenience. However, we could just as easily define the zero point as an infinite distance from Earth, as was done in Figure 7.19.

Note: Review Section 5.4

The same is true of electric potential energy—and therefore of electric potential, since this is just the potential energy per unit charge. We may want to define the potential as zero at the negative plate of a pair of charged parallel plates. But it is sometimes more convenient to locate the zero point at infinity, as we shall see shortly for the case of a point charge. Either way, the *difference* between two given points will be unaffected. Suppose, for example, that with one choice of the zero value, point A is at a potential of 100 V and point B at 200 V. With a different zero point, the potential at A might be 1100 V, but the potential at B would then be 1200 V. Or we might designate a zero point that would give A a potential of −500 V; the potential at B would then be −400 V. The *difference* would always be +100 V, however regardless of where we chose to assign the zero point.

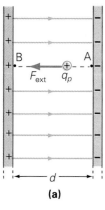

(a)

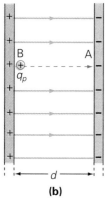

(b)

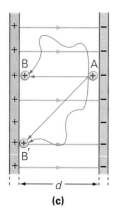

(c)

• **FIGURE 16.2 Accelerating a charge**
(a) Moving a proton from the negative to the positive plate increases the potential energy of the system. **(b)** When the proton is released from the positive plate, it accelerates back toward the negative plate, gaining kinetic energy at the expense of the electric potential energy. **(c)** The work done to move a proton between any two points in an electrostatic field, such as points A and B is independent of the path. No work is done to move the proton between B and B'. See Example 16.1

Thus, *V* cannot be considered a unique physical quantity; only *changes* in *V*, which we will call *voltage*, are meaningful. Notice in the following Example that we implicitly assign zero potential to the negative plate, thereby assuring that the positive plate has a positive value of *V*. After studying the Example, ask yourself what the potential of the positive plate would be if the negative plate were assigned a potential of +5.0 V, and how this would affect the results of the Example.

EXAMPLE 16.1 ■ MOVING A PROTON: PARALLEL PLATES AND POTENTIAL DIFFERENCE

A proton is moved from the negative to the positive plate of a parallel plate arrangement in • Fig. 16.2a. The plates are 1.50 cm apart and the field is uniform with a magnitude of 1500 N/C. (a) What is the potential energy gain? (b) What is the potential difference between the plates? (c) What is the potential difference between the negative plate and a point midway between the plates? (d) If the proton is released from rest at the positive plate (• Fig. 16.2b), what speed will it have as it hits the negative plate?

Solution. We are given *E*, and since we are moving a proton, we know its mass and charge.

Given: $E = 1500 \text{ N/C}$ *Find:* (a) ΔU (potential energy gain)
$q_p = +1.60 \times 10^{-19} \text{ C}$ (b) ΔV (plate potential difference)
$m_p = 1.67 \times 10^{-27} \text{ kg}$ (c) ΔV_{mp} (potential difference:
$d = 1.50 \times 10^{-2} \text{ m}$ negative plate to midpoint)
                                             (d) v (speed of proton at
                                                     negative plate)

(a) The potential energy gain is the work done by the external force acting on the proton:

$$\Delta U = (qE)d = (1.60 \times 10^{-19}\text{C})(1500 \text{ N/C})(1.50 \times 10^{-2}\text{ m})$$
$$= 3.60 \times 10^{-18}\text{ J}$$

(b) The potential difference, or plate voltage, is the potential energy change per unit charge, as expressed in Eq. 16.1:

$$\Delta V = \frac{\Delta U}{q} = \frac{3.60 \times 10^{-18}\text{ J}}{1.60 \times 10^{-19}\text{ C}}$$
$$= +22.5 \text{ V}$$

We conclude that the positive plate is 22.5 volts higher in potential than the negative plate.

(c) Since the potential energy gain at the midpoint, ΔU_{mp}, is proportional to the distance moved, the potential difference, ΔV_{mp}, between the negative plate and the midpoint is half of the full difference, or about 11.3 V.

(d) Conservation of energy requires that the proton must gain in kinetic energy what it loses in potential energy. Since $K_o = 0$, $\Delta K = K - K_o = K$, we can write

$$\Delta K = |\Delta U|$$
$$K = \tfrac{1}{2}m_p v^2 = |\Delta U|$$
$$v = \sqrt{\frac{2|\Delta U|}{m_p}} = \sqrt{\frac{2(3.60 \times 10^{-18}\text{ J})}{1.67 \times 10^{-27}\text{ kg}}} = 6.57 \times 10^4 \text{ m/s}$$

Even though the amount of kinetic energy gained is very small on an everyday basis, the proton acquires considerable speed (about 0.02% the speed of light) because of its extremely small mass.

Follow-up Exercise. How would your answers change in this Example if an alpha particle were moved instead of a proton? (An alpha particle, the nucleus of a helium atom, has a charge of +2*e* and a mass approximately four times that of a single proton.)

Notice in the preceding Example that the potential energy gain and the potential difference are independent of the path (●Fig. 16.2c). This is true of all conservative forces, of which the electrostatic force is one. The work done in moving the proton from A to B is always the same regardless of the route taken. It is also the same as the work done in moving it from A to B', since movement perpendicular to the electric field requires no work. (Why?)

Consider how our energy discussion would differ if we had used an electron instead of a proton. The electron, being negatively charged ($q_e = -1.60 \times 10^{-19}$ C), would be attracted to B, so the external force would have to act in a direction *opposite* the displacement to keep the electron moving at a constant velocity (that is, to prevent the electron from accelerating). It would thus do *negative* work on the electron, so the potential energy of the electron would be *decreased*:

$$\Delta U_{electron} = q_e \Delta V = U_B - U_A < 0.$$

In this case, the electron would be attracted to the positive plate, which is at a higher potential than the negative plate. If allowed to move freely, it would "fall" toward the region of higher potential, just as the proton "fell" toward the region of lower potential, losing potential energy and (in the absence of any external force) gaining kinetic energy. This is a general result for charges in electric fields:

> Positive charges, when released, tend to move to regions of low potential, and negative charges tend to move to regions of high potential.

Note: Potential differences are decided with positive charges. Electrons experience the same potential difference but the opposite potential *energy* change. To determine whether potential increases or decreases, decide whether an external force does positive or negative work on a positive test charge.

EXAMPLE 16.2 ■ ZERO TO 1860 MILES PER SECOND IN $2\frac{1}{2}$ mm: ACCELERATING AN ELECTRON

It is desired to accelerate an electron to 1.00% the speed of light in a space of 2.50 mm, between a pair of large horizontal parallel plates. If the top plate is positively charged, (a) what plate voltage is required? (b) What is the magnitude and direction of the electric field? (Disregard the effects of gravity—why?)

Solution. One percent the speed of light is $(0.0100)(3.00 \times 10^8 \text{ m/s}) = 3.00 \times 10^6$ m/s, and since it is an electron, we know its mass and charge.

Given: $q_e = -1.60 \times 10^{-19}$ C   *Find:*   (a) ΔV (plate voltage)
$m_e = 9.11 \times 10^{-31}$ kg   (b) E (magnitude and direction)
$d = 2.50 \times 10^{-3}$ m

(a) The electron must be released from the bottom plate (why?). The gain in kinetic energy is

$$\tfrac{1}{2}m_e v_e^2 = \tfrac{1}{2}(9.11 \times 10^{-31} \text{ kg})(3.00 \times 10^6 \text{ m/s})^2 = 4.10 \times 10^{-18} \text{ J}$$

This gain in kinetic energy comes at the expense of electric potential energy, thus $\Delta U = -\Delta K = -4.10 \times 10^{-18}$ J. Thus the voltage (potential difference) across the plates is

$$\Delta V = \frac{\Delta U}{q_e} = \frac{-4.10 \times 10^{-18} \text{ J}}{-1.60 \times 10^{-19} \text{ C}} = 25.6 \text{ V}$$

(Can you explain why this is less than the potential difference required to accelerate a proton to the same speed?)

(b) The electric field direction will be vertically downward, starting on the positively charged top plate and ending on the negatively charged lower plate. Its magnitude is given by the parallel plate expression, Eq. 16.2:

$$E = \frac{\Delta V}{d} = \frac{25.6 \text{ V}}{2.50 \times 10^{-3} \text{ m}} = 1.03 \times 10^4 \text{ V/m}$$

Follow-up Exercise. Use energy methods to determine the direction and magnitude of the uniform electric field required to stop a proton initially traveling in the $+x$ direction at 1.00% the speed of light in the space of 50 cm.

Insight   The Van de Graaff Generator

A method of generating large electrostatic potentials was developed by the American physicist Robert Van de Graaff around 1930. A diagram illustrating the operation of a simple Van de Graaff generator is given in Fig. 1a. Such generators are commonly used for classroom demonstrations and can develop potential differences of more than 50,000 V.

A continuous, motor-driven, insulating (rubber) belt runs vertically around two pulleys. When the motor is turned on, the rubber belt begins to move and frictional contact with the lower pulley transfers electrons from the pulley to the belt. The negatively charged belt moves upward to the upper pulley, where electrons flow from the belt to the upper pulley. With continuous running, charges build up on both pulleys.

After a short time, the built-up charge is sufficient to ionize the air in the vicinity of the metal electrodes, or combs (C_1 and C_2 in the figure). When this occurs at the upper electrode (C_2), electrons are transferred to the metal sphere. When the positive charge on the lower pulley is sufficient to cause an ionizing discharge, electrons are transferred from the grounded lower electrode (C_1) to the lower pulley. The net effect is that electrons are transferred from the ground to the metal sphere, leaving the sphere positively charged. Since the electric field is virtually zero inside the charged sphere (cf. Fig. 15.15b), electrons are continuously added to it and a build-up of negative charge results, giving a high electric potential. (Another method of charging the belt is through electric arcing, or corona discharge, from a high-voltage source.)

An electric potential between 50,000 V and 100,000 V is great enough to cause a so-called corona discharge through the ionization of air molecules around the sphere, and small electric arcs can be seen (and heard) between the sphere and a nearby object (Fig. 1b). In a demonstration of this, an instructor may arc the discharge to his or her knuckles. Another common demonstration is to have an insulated person touch the generator. The person becomes negatively charged, and repulsive forces cause his or her hair to stand on end (see the chapter-opening photo). In effect, the person becomes an electroscope, with hair serving as the leaves.

A voltage of 50,000–100,000 V may sound impressive, but there is no shock hazard for a healthy person because of the small amount of charge involved. The corona discharge can be suppressed and higher voltages achieved by surrounding the sphere with a gas that has a greater ionization potential than air. Because of this possibility of producing large voltages, Van de Graaff generators are used to accelerate charged particles in particle accelerators (Chapter 30).

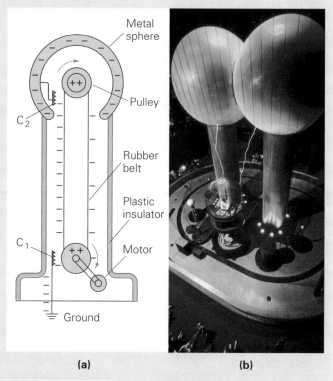

(a)   **(b)**

FIGURE 1   The Van de Graaff generator
(a) A charged, continuously moving belt transports electrons to the metal sphere, giving rise to large potential differences between the sphere and ground. (See text for description.) **(b)** The discharges from these large demonstration Van de Graaff generators in the Boston Science Museum are a good deal more conspicuous. They reach the ground through the wire cage in which the operator sits.

Potential Difference Due to Point Charges

For nonuniform electric fields, the potential difference between two points is determined by applying the definition given in Eq. 16.1. That is, we calculate the work per unit charge in moving a positive charge between the two points at constant speed. However, when the field strength varies with location, the mathematics is no longer simple because the work is done by a varying force. The only nonuniform field we will consider in any detail is that due to a fixed point charge (•Fig. 16.3). We give (without deriving) the result for the potential difference between two points at distances r_A and r_B from a point charge (q):

$$\Delta V = \frac{kq}{r_B} - \frac{kq}{r_A} \qquad \begin{array}{l}\textit{potential difference} \\ \textit{(point charge)}\end{array} \qquad (16.3)$$

In Fig. 16.3, the point charge is positive. We see that if B is closer to the point charge than A, the potential difference is positive: B is at a higher potential than A. This is because it takes *positive* external work to move a positive test charge from A to B. Thus the potential increases as we consider locations nearer to a positive charge. Again, the potential difference is the same regardless of the path. The work done along Path II from A to B is the same as along the direct route, Path I.

If the point charge were negative, B would be at a lower potential than A. Thus, in going from A to B, a positive charge would lose potential energy. In summary, the potential changes according to the following general rules:

> **Electric potential increases as we consider locations closer to positive charges or farther from negative charges.**

> **Electric potential decreases as we consider locations farther from positive charges or nearer to negative charges.**

Sometimes, the potential at very large distance from a point charge is defined to be zero (which is convenient, although not required). In this case, the *electric potential* at a distance r from the charge is

$$V = \frac{kq}{r} \qquad \begin{array}{l}\textit{potential due to} \\ \textit{point charge}\end{array} \qquad (16.4)$$

You should keep in mind, however, that only electric potential *differences* are important, as the next Example illustrates.

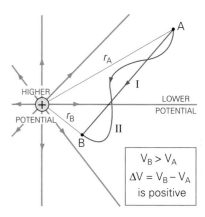

$$V_B > V_A$$
$$\Delta V = V_B - V_A$$
is positive

•FIGURE 16.3 Electric field and potential due to a point charge Electric potential increases as you move closer to a positive charge. Thus, B is at a higher potential than A, and $\Delta V = V_B - V_A$ is positive.

Note: Eq. 16.4 is for point charges only.

EXAMPLE 16.3 ■ THE HYDROGEN ATOM REVISITED: POTENTIAL DIFFERENCES NEAR A PROTON

In the Bohr model of the hydrogen atom (see Chapter 27), the electron in orbit around the proton can exist only in certain circular orbits. The smallest has a radius of 0.0529 nm and the next largest has a radius of 0.212 nm. (a) What is the potential difference between the two orbits? (b) Which orbit is at the higher potential?

Solution. The electron orbits in the field of a proton, and we know its charge. First converting the radii into meters, we have:

Given: $q_p = +1.60 \times 10^{-19}$ C $\qquad\qquad$ *Find:* (a) ΔV (potential difference)
$\qquad\quad r_1 = 0.0529$ nm $= 5.29 \times 10^{-11}$ m $\qquad\qquad$ (b) Orbit at higher potential
$\qquad\quad r_2 = 0.212$ nm $= 2.12 \times 10^{-10}$ m

(a) The difference in potential is given by Eq. 16.3:

$$\Delta V = \frac{kq_p}{r_1} - \frac{kq_p}{r_2} = kq_p\left(\frac{1}{r_1} - \frac{1}{r_2}\right)$$

$$= \left(9.00 \times 10^9 \, \frac{\text{N} \cdot \text{m}^2}{\text{C}^2}\right)(+1.60 \times 10^{-19} \, \text{C})\left(\frac{1}{5.29 \times 10^{-11} \, \text{m}} - \frac{1}{2.12 \times 10^{-10} \, \text{m}}\right)$$

$$= +20.4 \, \text{V}$$

(b) The smaller orbit is at the higher potential, because it is closer to the positively charged proton.

Follow-up Exercise. In this Example, (a) what is the change in the potential energy if the electron is moved from the smallest orbit very far away from the proton (to "ionize" the atom). (b) Does the electron move to a region of higher or lower electrical potential? (c) Does the potential energy of the system increase or decrease, and if so, by how much?

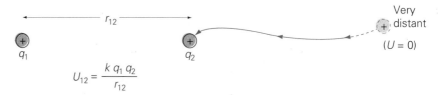

$$U_{12} = \frac{k\,q_1\,q_2}{r_{12}}$$

•**FIGURE 16.4 Mutual electric potential energy of two point charges**
If we move a positive charge from a very great distance to a distance r_{12} from another charge, there is an increase in potential energy because positive work must be done to bring the mutually repelling charges closer together.

Electric Potential Energy of Various Charge Configurations

In our discussion of gravitational energy, we considered the gravitational potential energy of systems of masses. The expressions for electric force and gravitational force are mathematically similar, and so are those for the corresponding potential energies, except for the use of charge in place of mass and the two signs associated with charge. In the gravitational case of two masses, the mutual gravitational potential energy was negative because the force is always attractive. For electric potential energy, the result can be either positive or negative because the electric force can be either attractive or repulsive.

If we start with a single positive point charge fixed in space, and move a positive charge toward it from a very large distance to a distance r_{12} (•Fig. 16.4), the work done is positive and the system of two charges gains electric potential energy:

$$\Delta U = q_2 \Delta V = q_2\left(\frac{kq_1}{r_{12}} - \frac{kq_1}{\infty}\right) = \frac{kq_1q_2}{r_{12}}$$

Although we can choose our zero electric potential energy at any arbitrary point, it is natural and convenient to think of the electric potential energy as being zero at very large distances. We can then write

$$U_{12} = \frac{kq_1q_2}{r_{12}} \qquad \begin{array}{l}\textit{potential energy} \\ \textit{(two charges)}\end{array} \qquad (16.5)$$

For *unlike* charges, the electrical force is attractive. In that case, by analogy with an attractive gravitational force, the electrostatic potential energy is said to be negative. For *like* charges, the electrical force is repulsive, and the potential energy is positive. Thus if the two charges are of the same sign, it takes positive work to move them together, and upon release, they will move apart, gaining kinetic energy as they lose potential energy. Conversely, it takes positive work to increase the separation of two opposite charges, such as the proton and the electron in a hydrogen atom.

The signs of the charges can be used to keep things straight mathematically, as shown in the following example. Remember that energy is a scalar quantity, and in the case of a static configuration of charges, the *total* potential energy is the algebraic sum of the mutual potential energies of all pairs of charges. For a configuration of many point charges, the total potential energy (U_t) is

$$U_t = U_{12} + U_{23} + U_{13} + \cdots \qquad (16.6)$$

EXAMPLE 16.4 ■ A CHARGE TRIANGLE: CALCULATING ELECTROSTATIC POTENTIAL ENERGY

What is the total electrostatic potential energy of the charge configuration shown in •Fig. 16.5?

Solution. From the figure, we have

Given: $q_1 = +2.0\ \mu\text{C} = +2.0 \times 10^{-6}\ \text{C}$ **Find:** U_t (total potential energy)
$q_2 = -2.0\ \mu\text{C} = -2.0 \times 10^{-6}\ \text{C}$
$q_3 = +1.0\ \mu\text{C} = +1.0 \times 10^{-6}\ \text{C}$
$r_{12} = r_{13} = r_{23} = 0.30\ \text{m}$

The total potential energy is the algebraic sum of the mutual potential energies of all pairs of charges. Here, with three charges, Eq. 16.6 gives

$$U_t = U_{12} + U_{23} + U_{13} = k\left(\frac{q_1 q_2}{r_{12}} + \frac{q_2 q_3}{r_{23}} + \frac{q_1 q_3}{r_{13}}\right)$$

$$= (9.00 \times 10^9\ \text{N·m}^2/\text{C}^2)\left[\frac{(+2.0 \times 10^{-6}\ \text{C})(-2.0 \times 10^{-6}\ \text{C})}{0.30\ \text{m}}\right.$$

$$+ \frac{(-2.0 \times 10^{-6}\ \text{C})(+1.0 \times 10^{-6}\ \text{C})}{0.30\ \text{m}}$$

$$\left. + \frac{(+2.0 \times 10^{-6}\ \text{C})(+1.0 \times 10^{-6}\ \text{C})}{0.30\ \text{m}}\right]$$

$$= -0.12\ \text{J}$$

The minus sign in the answer means that the charges in this configuration have a lower potential energy than they would if they were at a great distance from one another (where the potential energy is defined as zero). Thus, if they were released from rest, they would fall together, because this would lower their potential energy still further (a larger negative value).

Follow-up Exercise. (a) In this Example, if q_2 were $+2.0\ \mu\text{C}$, how would U_t change? (b) Explain the significance of the sign in your answer to (a). [Hint: If these charges were released from their fixed locations, would they all fly apart or fall together?]

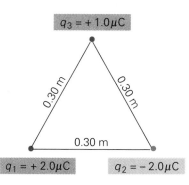

•**FIGURE 16.5 Electrostatic potential energy**
A charge configuration has electric potential energy due to the work done in bringing the charges together from very great distances. This energy can be easily computed. See Example 16.4.

Electric potential energy is a particularly useful quantity when no nonconservative forces are involved. Suppose that two positively charged particles are initially at rest near one another. If we release them, they will fly apart, gaining kinetic energy at the expense of the system's electric potential energy. The situation is analogous to that of two masses connected by a compressed spring. Consider the following Example.

EXAMPLE 16.5 ■ RELEASING THE ELECTRICAL "SPRING": CONSERVATION OF ENERGY INCLUDING ELECTRIC POTENTIAL ENERGY

Two positively charged particles of equal mass are initially at rest near one another (see •Fig. 16.6). If they are released, find their speeds after they have separated by twice their initial separation distance, using the values given in the figure.

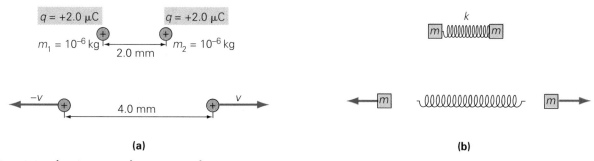

(a) **(b)**

•**FIGURE 16.6 Electric potential energy into kinetic energy**
(a) Two charges released from rest will gain kinetic energy. See Example 16.5. (b) A spring analogy for this situation.

Solution. From the figure, we have

Given: $q_1 = q_2 = q = +2.0\ \mu C = +2.0 \times 10^{-6}\ C$ *Find:* v (final speed of
$r_1 = 2.0\ mm = 2.0 \times 10^{-3}\ m$ each charge)
$r_2 = 2r_1 = 4.0 \times 10^{-3}\ m$
$m_1 = m_2 = 1.0\ mg = 10^{-6}\ kg$

Because no external forces act on the system, conservation of momentum requires that the two particles must have the same speed after their release, since they have the same mass. (Can you draw the momentum diagram, showing the initial momentum and the final momentum?) Let's designate their common speed as v and their common mass as m. Since there are no nonconservative forces acting, mechanical energy is also conserved:

$$K_i + U_i = K_f + U_f$$

$$0 + \frac{kq^2}{r_1} = \tfrac{1}{2}mv^2 + \tfrac{1}{2}mv^2 + \frac{kq^2}{2r_1}$$

Rearranging this equation and putting in the numbers,

$$\frac{kq^2}{2r_1} = mv^2$$

$$v = \sqrt{\frac{kq^2}{2mr_1}} = \sqrt{\frac{(9.00 \times 10^9\ N\cdot m/C^2)(2.0 \times 10^{-6}\ C)^2}{2(10^{-6}\ kg)(2.0 \times 10^{-3}\ m)}}$$

$$= 3.0 \times 10^3\ m/s$$

Follow-up Exercise. In this Example, what is the speed of each object after they are a very large distance from one another?

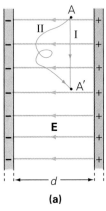

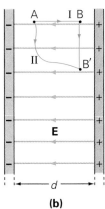

●FIGURE 16.7 Equipotential surfaces between parallel plates (a) The equipotential surfaces are planes parallel to the plates. The work done in moving a charge is zero as long as you start and stop on the same equipotential (compare path I to II). **(b)** Once you move to a higher potential (A to B), you can stay on that new equipotential by moving perpendicularly to the electric field (B to B'). The change in potential is independent of path, since the same change occurs via path I as via path II.

16.2 Equipotential Surfaces and the Electric Field

Objectives: To be able to **(a)** explain what is meant by an equipotential surface, **(b)** sketch equipotential surfaces for simple charge configurations, and **(c)** explain the relationship between equipotential surfaces and electric fields.

Suppose that a positive charge is moving at constant speed perpendicularly to an electric field, as shown in ●Fig. 16.7. As the charge moves from A to A' along path I between parallel charged plates, no work is done on it by the electric field. The angle between the force due to the field and the displacement is 90°, so the work done is $W = Fd\cos\theta = Fd\cos 90° = 0$. If no work is done, then the potential energy of the charge does not change between A and A', so $\Delta U_{AA'} = 0$. Thus, we can conclude that these two points—and *all* other points on path I—are at the same potential:

$$\Delta V_{AA'} = \frac{\Delta U_{AA'}}{q} = 0 \quad \text{or} \quad V_A = V_{A'}$$

Clearly this is also true for all points on the *plane* containing path I. This plane, on which the value of the electric potential is constant, is an example of an **equipotential surface** (sometimes referred to simply as an *equipotential*). The word equipotential means "same potential." An equipotential shape may not be a plane, nor is it necessarily parallel to the plates or charges that created the field, as is true in the special parallel plate case.

Since it takes no work to move a charge along an equipotential surface, then it must be generally true that

equipotential surfaces are always at right angles to the electric field.

Moreover, since the electric field is a conservative field, the work done on the charge will be the same whether we take path I, path II, or any other path from A to A' (Fig. 16.7a). The exact route is irrelevant. If the charge returns to the same equipotential surface, the net work done to move it is zero.

If we move the charge at right angles (against **E**) to the equipotential we just found, the electric potential energy, and hence the electric potential, will increase. When we reach point B, if we then travel perpendicularly to the electric field lines, we will find that we are on another equipotential surface, but one of a higher potential. If we moved the charge from A to B' (Fig. 16.7b) the work required would be the same as in moving from A to B (why?).

To help understand the concept of an equipotential surface, consider a gravitational analogy. If we call the gravitational potential energy zero at ground level, and raise an object a height h (from A to B in •Fig. 16.8), the work required by an external agent is mgh and is positive (why?). If we then move horizontally, the potential energy value does not change. Thus, surfaces of constant gravitational potential energy are planes parallel to the Earth's surface. Topographic maps, which map the land contours by plotting lines of constant elevation (usually relative to sea level), are actually maps of constant gravitational potential (•Fig. 16.9). Thus, topographic maps could be called maps of gravitational equipotential surfaces.

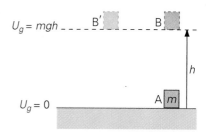

•**FIGURE 16.8   Gravitational potential energy**
Raising an object in a uniform gravitational field results in an increase in potential energy.

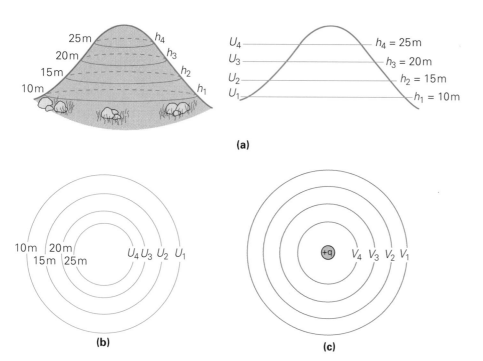

(a)

(b)

(c)

•**FIGURE 16.9   Topographic maps —a gravitational analogy for equipotential surfaces**
(a) A symmetrical hill with slices at different elevations. Each slice is a plane of constant gravitational potential. **(b)** The topographic map based on the slices in (a). The contours, where the planes intersect the surface, represent gravitational equipotentials. **(c)** Electrical equipotentials around a point charge.

To reinforce the idea of equipotential surfaces, consider the following Example.

EXAMPLE 16.6 ■ MAPPING ELECTRICAL EQUIPOTENTIAL SURFACES: PARALLEL PLATES

A pair of parallel plates (•Fig. 16.10), separated by 20.0 mm, is charged fully by a 12.0-V battery. (a) How far apart are two equipotential planes that differ in potential by 0.200 V? Does this information tell you *where* between the plates these planes are located? (b) Assigning zero potential to the negative plate, where is the equipotential surface with a voltage of +4.50 V located?

Solution.

Given: $d = 20.0 \text{ mm} = 2.00 \times 10^{-2} \text{ m}$ *Find:*   (a) Δx (distance between
$\quad\quad\quad V = 12.0 \text{ V}$ $\quad\quad\quad\quad\quad\quad$ potential planes)
$\quad\quad\quad \Delta V = 0.200 \text{ V}$ $\quad\quad\quad\quad\quad\quad$ (b) x (location of equipotential
$\quad\quad\quad\quad\quad\quad\quad\quad\quad\quad\quad\quad\quad\quad\quad$ surface)

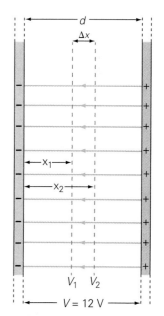

•**FIGURE 16.10   Equipotentials in a uniform electric field**
The equipotentials are planes parallel to the charged plates. The change in potential is related to the electric field strength and the distance moved. See Example 16.6.

(a) We know that the difference in the potentials of the two surfaces is 0.200 V. Think of moving a positive charge away from the negative surface (designated zero potential). The potential at two different distances is the external work done divided by the charge moved, or

$$V_1 = Ex_1 \quad \text{and} \quad V_2 = Ex_2$$

We know that for this case, $E = V/d$ and we are interested only in differences:

$$V_2 - V_1 = Ex_2 - Ex_1 = E\Delta x = \left(\frac{V}{d}\right)\Delta x$$

or

$$\Delta x = \left(\frac{V_2 - V_1}{V}\right)d = \left(\frac{0.200 \text{ V}}{12.0 \text{ V}}\right) \times 2.00 \times 10^{-2}\text{ m}$$

$$= 3.33 \times 10^{-4}\text{ m } (= 0.333 \text{ mm})$$

This answer does *not* provide us with location. *Any* equipotentials separated by this distance will have a difference of 0.200 V in potential. Relative to surface 1, there is a surface located 0.333 mm farther from the negative plate that is higher in potential by 0.200 V, and one located 0.333 mm closer to the negative plate that is lower in potential by 0.200 V.

(b) Since the gain in potential for the parallel plate case varies linearly with the distance from the negative plate (assigned a potential value of zero volts), we have for the voltage at any distance, x,

$$V_x = Ex = \left(\frac{V}{d}\right)x$$

or

$$x = \left(\frac{V_x}{V}\right)d = \left(\frac{4.5 \text{ V}}{12.0 \text{ V}}\right) \times 2.00 \times 10^{-2}\text{ m}$$

$$= 7.5 \times 10^{-3}\text{ m } (=0.75 \text{ cm}) \text{ from the negative plate}$$

Follow-up Exercise. (a) In this Example, what would be the potential difference, or voltage, between equipotential surfaces located 1.0 cm and 1.4 cm from the positive plate, respectively? (b) Which one is at a higher potential? (c) What is the magnitude and direction of the electric field between the plates?

It is extremely useful to know how to sketch equipotential surfaces, because they are intimately related to the electric field. In the Learn by Drawing box on p. 516 we summarize the method used for sketching equipotentials, given an electric field pattern, and conversely for sketching the electric field when equipotential sufaces are given.

For the special case of the uniform electric field between parallel plates (•Fig. 16.11), as was shown in Example 16.6, the difference in potential (ΔV) between any two equipotential planes separated by a distance Δx is

$$\Delta V = V_2 - V_1 = E\Delta x$$

Thus if you started on surface 1 and moved *perpendicularly* away to surface 2 you would measure a potential increase that depended solely on the electric field strength and the distance traveled. Notice that for a given travel distance Δx, this is the *maximum* gain in potential you could achieve. To help you understand this point, think of taking one step in any direction starting from surface 1. The way to maximize the increase in potential would be to step onto surface 2. A step in any other direction, not perpendicular to surface 1, would yield a smaller change in potential because you would end up on an equipotential surface closer to the one you started on (see, for example, the three steps, each Δx long, in Fig. 16.11).

By finding the *direction of the maximum potential change*, we are finding the direction *opposite* that of **E**. In one dimension, using plus and minus to indicate directions (here + means away from the negative plate), the expression for the electric field (including direction) in terms of the potential is

$$E = -\left(\frac{\Delta V}{\Delta x}\right)_{max} \qquad \textit{electric field from potential} \qquad (16.7)$$

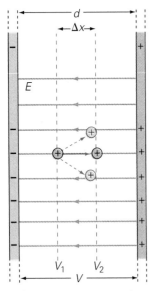

•FIGURE 16.11 Relationship between the potential change and E The electric field direction is that of maximum decrease in potential, or opposite the direction of maximum increase in potential. Its magnitude is given by the rate at which the potential changes per meter.

Here the minus sign means that **E** is in the direction opposite that in which V increases most rapidly, or in the direction in which V decreases the most rapidly.

As you can see, the units for electric field are volts per meter (V/m). In Chapter 15 we saw that E can be expressed in units of newtons per coulomb (N/C). You should be able to show, through unit analysis, that these two units are equivalent.

Note: See Section 15.4.

As mentioned above, the potential difference, rather than the electric field, is usually specified for practical applications. For example, a D-cell flashlight battery has a terminal voltage of 1.5 V, that is, a potential difference of 1.5 V between its terminals.

The concept of electric potential provides a convenient unit of energy that is used in calculating the kinetic energy of charged particles that have been accelerated through potential differences. An **electron volt** (eV) is defined as the kinetic energy acquired by an electron accelerated through a potential difference of exactly 1 V. Because the electron's gain in kinetic energy is equal to its loss in electric potential energy, we can write

$$\Delta K = -\Delta U = -(e\Delta V) = -(-1.60 \times 10^{-19} \text{ C})(1.00 \text{ V})$$
$$= 1.60 \times 10^{-19} \text{ J}$$

This provides us with an equivalence statement between the electron volt and the joule to three significant figures:

$$1 \text{ eV} = 1.60 \times 10^{-19} \text{ J}$$

The eV turns out to be a typical energy on the atomic scale. Thus it is much more convenient to express atomic energies in terms of electron volts than joules. The energy of *any* charged particle accelerated through *any* potential difference can be expressed in electron volts. The loss of electric potential energy (a negative change) equals the gain in kinetic energy (a positive change), thus their magnitudes (absolute values) are the same. For example, if an electron is accelerated from rest through a potential difference of 1000 V, its gain in kinetic energy (K) is

Note: Electron volt is a unit of energy. Volt is not. Do not confuse the two!

$$K = |e\Delta V| = (1 \text{ e})(1000 \text{ V}) = 1000 \text{ eV} = 1 \text{ keV}$$

The "keV" is the abbreviaton for the *kilo*electron volt.

Although an electron volt is conceptually defined using an electron, a particle with any integral number of units of electronic charge, positive or negative, can have its energy expressed in terms of it. Thus, if a particle with a charge of $+2e$, such as an alpha particle (a helium nucleus, consisting of two protons and two neutrons), were accelerated through this voltage, it would gain a kinetic energy of $K = (2e)(1000 \text{ V}) = 2000 \text{ eV} = 2 \text{ keV}$. Note how easy it is to compute the kinetic energy of an accelerated charged particle if you work in eV.

Larger units for energy are sometimes needed. For example, in nuclear and elementary particle physics it is not uncommon to find particles with energies of megaelectron volts (MeV), and gigaelectron volts (GeV). (1 MeV $= 10^6$ eV, and 1 GeV $= 10^9$ eV).*

You should be aware that the electron volt is *not* an SI unit. Thus in using energies to calculate speeds of a particle for example, you should first convert from electron volts back to joules. To calculate the speed of an electron after it has been accelerated from rest through 10.0 V, you would first convert its kinetic energy (10.0 eV) to joules: $K = (10.0 \text{ eV})(1.60 \times 10^{-19} \text{ J/eV}) = 1.60 \times 10^{-18}$ J. Then, using its mass in kg, its speed is in m/s:

$$v = \sqrt{2K/m} = \sqrt{2(1.60 \times 10^{-18} \text{ J})/(9.11 \times 10^{-31} \text{ kg})} = 1.87 \times 10^6 \text{ m/s}.$$

---

*At one time, a billion electron volts was referred to as BeV, but this was abandoned because confusion arose. In some countries, such as Great Britain and Germany, a billion means 10^{12} (which Americans call a trillion).

Graphical Relationship between Electric Field Lines and Equipotentials

Since it takes no work to move a charge along an equipotential, we know that equipotential surfaces and electric field lines must be perpendicular to one another. We also know that the electric field at any location points in the direction in which the potential decreases most rapidly with distance and has a magnitude equal to the change in potential per meter. We can take advantage of these facts to construct equipotentials for a given pattern of electric field lines. The reverse is also true: given the equipotentials, we can sketch the electric field lines, and if we also know the numerical values of the equipotentials, we can estimate the strength and direction of the field.

A couple of examples should provide some graphical insight into the close connection between equipotentials and fields. First, consider Fig. 1, in which you are given the electric field lines and want to determine the general shape of the equipotentials. Pick any point, such as A, and begin moving at right angles to the electric field at that point, as indicated by the electric field lines. (Remember that the field lines are essentially a map of the electric field vector **E** from point to point.) Keep moving, always maintaining this same perpendicular orientation to the electric field. Between lines you may have to approximate, but always look ahead to the next field line so that you can approach and cross it at a right angle. To find another equipotential, start at another point, such as B, and proceed the same way. Sketch as many equipotentials as you need to map out the area of interest. Fig. 1 shows the result of sketching four such equipotentials, from A (at the highest potential value—can you tell why?) to D (at the lowest potential).

Suppose that you are given the equipotentials instead of the field lines, as in Fig. 2. The electric field lines point in the direction of decreasing V and are perpendicular to the equipotential surfaces. Thus, to map the field, simply start at any point and move in such a way that your path intersects each equipotential surface that it encounters at a right angle. If you start at point A in Fig. 2, the field line shown results. Similarly, starting at points B, C, and D gives rise to additional field lines that suggest the complete electric field pattern; you need only add the arrows in the direction of decreasing potential.

Suppose that you wanted to estimate the magnitude of **E** at some point P, knowing the values of the equipotentials 1.0 cm on either side of it as shown in Fig. 3. From this information you can easily tell that the field points in the direction from A to B (why?). Its approximate magnitude is.

$$|E| = \frac{\Delta V}{\Delta x} = \frac{(1000 \text{ V} - 950 \text{ V})}{2.0 \times 10^{-2} \text{ m}}$$
$$= 2.5 \times 10^3 \text{ V/m}$$

Can you tell the direction of E at P?

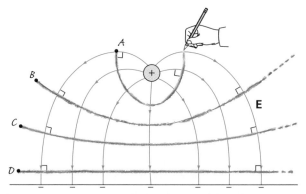

FIGURE 1   Sketching equipotentials from electric field lines
Given the electric field pattern, pick a point in the region of interest and move so that your path is always perpendicular to the next field line. Keep your path as smooth as possible, since kinks and loops in the equipotentials are not allowed. (Why?) To map a higher potential value, move opposite the direction of the electric field (why?) and repeat the process.

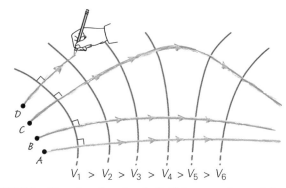

FIGURE 2   Mapping the electric field from equipotentials
Start at a convenient point and trace a line that crosses each equipotential at a right angle. Repeat the process as often as needed to reveal the field pattern, adding arrows to indicate the direction of the field lines from high to low potential.

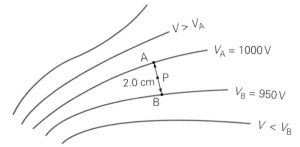

FIGURE 3   Estimating the magnitude of the electric field
The magnitude of the potential change per meter at any point gives the strength of the electric field at that point.

16.3 Capacitance

Objectives: To be able to (a) define capacitance and explain what it means physically, and (b) calculate the charge, voltage, electric field, and energy storage for parallel plate capacitors.

The electric field that exists between charged conductors stores electrical energy. Consider the parallel metal plates shown in •Fig. 16.12. Such an arrangement of conductors is a typical example of a **capacitor**. (In theory, any two nearby conductors can be thought of as constituting a capacitor.) Work must be done to transfer charge from one plate to the other. Suppose that one electron is moved from one plate to the other. This requires some external work, since the net positive charge left behind would attract this electron back. Once this is done, transferring the second electron would be more difficult because it would be repelled by the original electron *and* attracted by a double positive charge left behind. Thus, to separate the charges requires more and more work as more and more charge accumulates on the plates. (This is analogous to pumping up a tire with a hand pump: the more air there is in the tire, the harder the pumping gets, and the more work you have to do per stroke.)

The work needed to charge parallel plates may be very quickly done by a battery. The battery actually removes electrons from the positive plate and transfers or "pumps" electrons through the wire over to the negative plate, doing work on them in the process. The result is a separation of charge and the creation of an electric field in the capacitor. The battery charges the capacitor until the potential difference between the plates is equal to the terminal voltage of the battery. This process typically takes only microseconds of time. The result is a relatively uniform, static electric field. When the capacitor is disconnected from the battery, it remains a kind of reservoir of electrical energy. In accordance with the conservation of energy, we can consider the work done by the battery to be stored as potential energy in the electric field. This stored energy can then be used to do work.

In most everyday capacitors, the potential difference across the plates is proportional to the charge Q. (Here Q means the magnitude of the charge on *either* of the plates, since energy is stored even though the net charge on the capacitor is zero.)

$$Q \propto V$$

We can make this into an equation by introducing a constant of proportionality, C:

$$Q = CV \quad \text{or} \quad C = \frac{Q}{V} \qquad \textit{definition of capacitance} \qquad (16.8)$$

SI unit of capacitance: coulomb/volt, or farad (F)

Note: Capacitors store energy by means of an electric field.

Note: The net charges on the plates are $+Q$ and $-Q$, but it is customary to refer in general to the charge Q on a capacitor.

Note: From here on, we use V instead of ΔV as our notation for potential difference, or *voltage*. It is still, however, a *difference*.

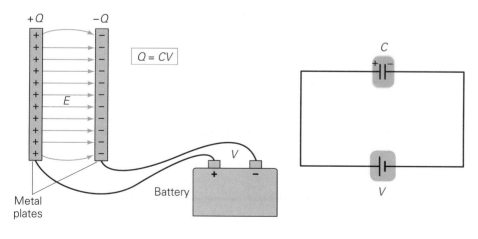

(a) Parallel-plate capacitor   **(b) Schematic circuit diagram**

•**FIGURE 16.12 Capacitor and circuit diagram**
(a) Two parallel metal plates are charged by a battery to a charge $Q = CV$, where C is the capacitance. Work is done in charging the capacitor, and energy is stored in the electric field. **(b)** This circuit diagram shows the symbols used for a battery and a capacitor. The longer line of the battery symbol is taken as the positive terminal and the short line as the negative terminal. The symbol for a capacitor is similar, but with lines of equal length.

This constant of proportionality C is called the **capacitance** and expresses the charge stored per volt or $C = Q/V$. Capacitance is a measure of the *per* volt charge capacity of any arrangement. When we say a capacitor has a "large capacitance" we mean that it has been constructed so as to hold more charge on its plates *per volt* than one of "smaller capacitance." In other words, if you connected the same battery to two different capacitors, thus providing identical voltages across their plates, the one with the larger capacitance would store more charge, and thus more energy. The units of capacitance are coulombs per volt (C/V), and this combined unit is called a **farad** (F). The farad is a rather large unit (see the following example), so the microfarad (1 μF = 10^{-6} F) or picofarad (1 pF = 10^{-12} F) is frequently used.

Capacitance depends *only* on the geometry (size, shape, spacing) of the plate arrangement (and the material between the plates, as we shall soon see). For parallel plates of equal surface area (A, the area of *one* of the plates), which are large compared to the distance between them (d), the capacitance is given by

$$C = \frac{\epsilon_{\mathrm{o}} A}{d} \qquad \textit{(parallel plates only)} \qquad (16.9)$$

Here ϵ_{o} is a constant called the *permittivity of free space* (vacuum), which has a value of approximately 8.85×10^{-12} C^2/N·m^2. ϵ_{o} is a fundamental constant and is related to the constant in Coulomb's law by $k = 1/4\pi\epsilon_{\mathrm{o}}$.

EXAMPLE 16.7 ■ WILL A 1.0-F PARALLEL PLATE CAPACITOR FIT INTO YOUR GARAGE? HOW LARGE IS A FARAD?

What would be the plate area of a 1.0-F parallel-plate capacitor if the plate separation were 1.0 mm?

Solution.

Given: $C = 1.0$ F   *Find:* A (area of one of the plates)
$d = 1.0$ mm $= 1.0 \times 10^{-3}$ m

Solving Eq. 16.9 for the area gives

$$A = \frac{Cd}{\epsilon_{\mathrm{o}}} = \frac{(1.0\,\mathrm{F})(1.0 \times 10^{-3}\,\mathrm{m})}{8.85 \times 10^{-12}\,\mathrm{C}^2/\mathrm{N{\cdot}m}^2}$$
$$= 1.1 \times 10^{8}\,\mathrm{m}^2$$

This is about 100 km^2 (about 40 mi^2), or a square 10 km (over 6 mi) on a side. This will definitely not fit in most garages! 1.0 F is therefore a very large value of capacitance!

Follow-up Exercise. In this Example, what would the spacing between the plates have to be if you wanted the capacitor to have a plate area roughly the area of the roof of a car? (Assume a rectangular area 3.0 m × 4.0 m.) Compare your answer to a typical atomic diameter of 10^{-9} to 10^{-10} m. Do you think it is feasible to build this?

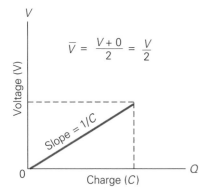

•FIGURE 16.13 Voltage versus charge
A plot of voltage versus charge for a charging capacitor is a straight line with a slope of $1/C$ (since $Q = CV$). The average voltage is $\overline{V} = V/2$, and the total work done is equivalent to transferring the charge through $\overline{V}$. Thus $U = W = \frac{1}{2}QV$, which is the area under the curve.

The expression for the energy stored in a charged capacitor may be obtained through graphical analysis, since both Q and V vary during the charging process. A plot of voltage versus charge for charging a capacitor gives a straight line with a slope of $1/C$ since $V = (1/C)Q$ (•Fig. 16.13). The graph represents the charging of an initially uncharged capacitor ($V_{\mathrm{o}} = 0$) to some final voltage (V). The total work done is equivalent to transferring the charge through the average voltage $\overline{V}$, and, since the voltage varies linearly with the charge, we can compute the average simply as

$$\overline{V} = \frac{V_{\mathrm{final}} + V_{\mathrm{initial}}}{2} = \frac{V + 0}{2} = \frac{V}{2}$$

Thus, the energy stored in the capacitor (or the work done by the battery to store that energy) is expressed as

$$U = W = Q\overline{V} = \tfrac{1}{2}QV$$

Since $Q = CV$, this equation may be written in several equivalent forms:

$$U = \tfrac{1}{2}QV = \frac{Q^2}{2C} = \tfrac{1}{2}CV^2 \qquad \text{\textit{energy storage in a capacitor}} \qquad (16.10)$$

The form $U = \tfrac{1}{2}CV^2$ is usually the most practical, since the capacitance and the applied voltage are often known. Rarely in everyday applications is the charge on a capacitor measured directly.

Note: Compare this to the analysis of the work done in compressing a spring, Section 5.2.

Note: Practice using the various forms of capacitor energy. In a pinch, you need recall only one and the definition of capacitance, $C = Q/V$.

EXAMPLE 16.8 ■ HOW MUCH WORK? ENERGY STORAGE IN A CAPACITOR

A small capacitor (6.0 μF) is charged by a 12-V car battery. (a) How much energy is stored in the capacitor? (b) How much stored energy did the battery lose? (c) If all of the capacitor's energy could be used to lift an 80-kg man, how far vertically would he be lifted?

Solution.

Given: $C = 6.0\ \mu\text{F} = 6.0 \times 10^{-6}\ \text{F}$ **Find:**   (a) U (energy stored in capacitor)
$V = 12\ \text{V}$ (b) ΔU_b (battery potential energy loss)
$m = 80\ \text{kg}$ (c) h (height man would be lifted)

(a)   The most convenient form of Eq. 16.10 to use to compute the capacitor's energy is

$$U = \tfrac{1}{2}CV^2 = \tfrac{1}{2}(6.0 \times 10^{-6}\,\text{F})(12\ \text{V})^2$$
$$= 4.3 \times 10^{-4}\,\text{J}$$

(b)   By energy conservation, if there are no losses in the wires, the battery loses whatever the capacitor gains. (We will learn later that charges flowing through wires can generate heat and therefore lose energy.) Hence

$$\Delta U_b = -4.3 \times 10^{-4}\ \text{J}$$

(c)   Lifting the man would increase his *gravitational* potential energy at the expense of the capacitor's *electric* potential energy. In principle, the capacitor could be used to run a very efficient motor and lift the man. Recall that the increase in gravitational potential will be mgh, so we can find h:

$$mgh = 4.3 \times 10^{-4}\ \text{J}$$
$$h = \frac{4.3 \times 10^{-4}\,\text{J}}{(80\ \text{kg})(9.8\ \text{m/s}^2)} = 5.5 \times 10^{-7}\ \text{m}$$

This small answer tells us just how little energy a typical capacitor stores under typical conditions. This height is only about one thousand times the diameter of an atom and would not be measurable with ordinary instruments!

Follow-up Exercise.   (a) How much charge is on the plates of the capacitor in this Example? (b) What voltage would be required to change the capacitor so that its stored energy could be used to raise the man 1.0 mm?

16.4 Dielectrics

Objectives: To understand (a) what a dielectric is, and (b) how it affects the physical properties of a capacitor.

In most capacitors, a sheet of insulating material, such as paper or plastic, is placed between the plates. This insulating material is called a **dielectric** and serves several purposes. It keeps the plates from coming into contact, which would allow the electrons to flow back onto the positive plate, thereby neutralizing the charge

TABLE 16.1 Dielectric Constants for Some Materials	
Material	Dielectric Constant (K)
Vacuum	1.00000
Air	1.00054
Paper	3.5
Polyethylene	2.3
Polystyrene	2.6
Teflon	2.1
Glass	4.6
Silicone oil	2.6

on the capacitor and energy stored. It also allows flexible plates of aluminum foil to be rolled into a cylinder, giving the capacitor a practical size. Finally, it increases the energy storage *capacity* of the capacitor, and therefore under the right conditions, the energy stored in the capacitor. This capability varies with materials and is characterized by the **dielectric constant** (K). Values of the dielectric constant for some materials are given in Table 16.1.

How a dielectric affects the electrical properties of a capacitor is illustrated in •Fig. 16.14. The capacitor is fully charged, then carefully disconnected from the battery, and a dielectric is inserted. In the dielectric material, work is done on molecular dipoles by the existing electric field as they experience electrical forces that tend to align them with that field. (The molecular polarization may be permanent or temporarily induced by the electric field. In either case the effect is the same.) Work is also done on the dielectric slab as a whole, since the charged plates attract it into the space between them. Both of these effects decrease the potential energy stored in the capacitor *as long as the charge on the plates stays the same*. The dielectric creates its own "reverse" electric field lines (Fig. 16.14c) and effectively partially cancels the field between the plates.

This means that the *net* field between the plates, and the voltage across the plates, are both reduced from their vacuum values. If we designate the vacuum values by E_o and V_o, then the dielectric constant of the material is defined as the ratio of the voltage with the material in place divided by the vacuum voltage, or, since for parallel plates V is proportional to E (Eq. 16.2, $V = Ed$), we have

$$K = \frac{V_o}{V} = \frac{E_o}{E} \qquad (16.11)$$

This definition is designed so that K is always greater than 1. (Can you show this?) The dielectric constant can most easily be determined by measuring the two voltages. (We will see how a voltmeter works in Chapter 18.) Since the battery was disconnected and the capacitor isolated, the charge on the plates, Q, is unaffected, and the new value of the capacitance is

$$C = \frac{Q}{V} = \frac{Q_o}{V_o/K} = K\frac{Q_o}{V_o} \quad \text{or} \quad C = KC_o \qquad (16.12)$$

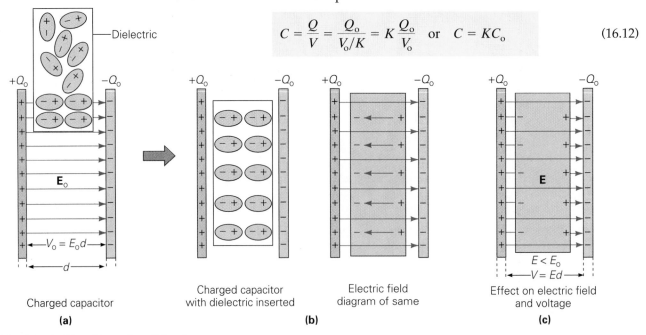

•FIGURE 16.14 The effects of a dielectric
(a) A dielectric with randomly oriented permanent molecular dipoles is inserted between the plates of an isolated charged capacitor. The capacitor does work on the dielectric slab as it is inserted due to the attractive forces between its charges and those induced on the dielectric surfaces. We thus expect, if the battery has been disconnected, the energy storage to decrease. **(b)** If the dielectric is put into the capacitor, the dipoles are oriented (or partially oriented) with the field, giving rise to an opposing electric field. The dipole field effectively cancels some of the field due to the plate charges. **(c)** The net effects are decreases in the electric field and voltage by a factor of $1/K$, where K is the dielectric constant of the material.

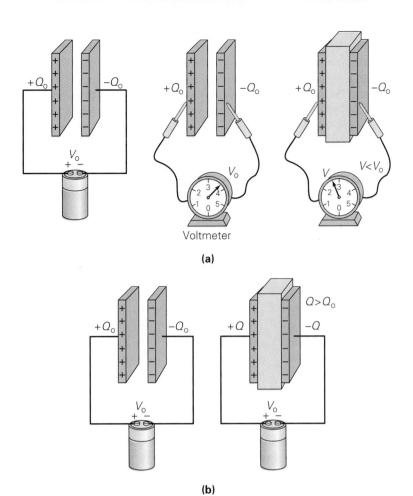

FIGURE 16.15 Dielectric and capacitance
(a) A parallel-plate capacitor in air is charged by a battery to a charge Q_o and a voltage V_o (left). If the battery is disconnected and the potential across the capacitor measured by a voltmeter, a reading of V_o is obtained (center). But if a dielectric is inserted between the capacitor plates, the voltage drops to $V = V_o/K$ (right), so the amount of energy stored is less. **(b)** A capacitor is charged as in part (a), but the battery is left connected. When a dielectric is inserted into the capacitor, the voltage is maintained at V_o by the battery, but the amount of charge that the capacitor accepts increases to: $Q = KQ_o$, so the amount of energy stored is greater. Both of these observations reflect the increase in capacitance produced by the dielectric.

We have seen that inserting a dielectric into an isolated capacitor results in a larger capacitance (the same charge stored at a lower voltage). But what about energy storage? Since there is no energy input (the battery is disconnected) and since the capacitor does work on the dielectric, the stored energy drops by a factor of K:

$$U = \frac{Q^2}{2C} = \frac{Q_o^2}{2KC_o} = \frac{Q_o^2/2C_o}{K} = \frac{U_o}{K} < U_o \qquad \text{(battery disconnected)}$$

If, on the other hand, the dielectric is inserted *and the battery remains connected*, the original voltage is maintained and the battery must supply more charge, and therefore do work (•Fig. 16.15). We expect the energy storage to go up in this situation. Here it is the charge on the plates that increases by the factor of K, $Q = KQ_o$. Once again the capacitance increases, since more charge is now stored at the same voltage $C = Q/V = K(Q_o/V_o) = KC_o$. Thus, *a dielectric increases the capacitance by a factor of K regardless of the conditions under which the dielectric is inserted.* However, in this case (constant V), as we have seen, the energy storage of the capacitor goes up at the expense of the battery

$$U = \tfrac{1}{2}CV^2 = \tfrac{1}{2}KC_oV_o^2 = K(\tfrac{1}{2}C_oV_o^2) = KU_o \approx U_o \qquad \text{(battery connected)}$$

For a parallel-plate capacitor with a dielectric, the capacitance is given by

$$C = KC_o = \frac{K\epsilon_o A}{d} \qquad (16.13)$$

This is sometimes written as $C = \epsilon A/d$, where $\epsilon = K\epsilon_o$ is the **dielectric permittivity** and is always greater than ϵ_o, the permittivity of the vacuum (free space).

An example of the use of a capacitor for electrical energy storage is shown in •Fig. 16.16.

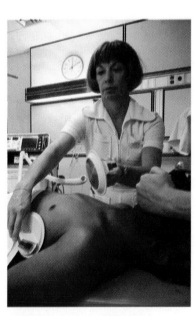

FIGURE 16.16 Defibrillator
A burst of electric current from a defibrillator can often restore normal heartbeat in persons who have suffered cardiac arrest. Capacitors are used to store the electrical energy that the device depends on.

EXAMPLE 16.9 ■ AIR VS PLASTIC: DIELECTRIC EFFECT ON CHARGE AND ENERGY

A parallel-plate capacitor having a plate area of 0.70 m² and a plate separation of 1.0 mm is connected to a source with a voltage of 50 V. Find the capacitance, the charge on the plates, and the energy of the capacitor (a) when there is air between the plates and (b) when the capacitor contains polystyrene (see Table 16.1).

Solution.

Given: $A = 0.70 \text{ m}^2$
$d = 1.0 \text{ mm} = 1.0 \times 10^{-3} \text{ m}$
$V = 50 \text{ V}$
$K = 2.6$

Find: for (a) and (b)
C (capacitance)
Q (charge)
U (energy)

(a) For air, K is essentially 1.0, and

$$C_\text{o} = \frac{\epsilon_\text{o} A}{d} = \frac{(8.85 \times 10^{-12} \,\text{C}^2/\text{N·m}^2)(0.70 \text{ m}^2)}{1.0 \times 10^{-3} \text{ m}}$$

$$= 6.2 \times 10^{-9} \,\text{F} = 0.0062 \,\mu\text{F} \ (=6.2 \text{ nF})$$

Then

$$Q_\text{o} = C_\text{o} V = (6.2 \times 10^{-9} \text{ F})(50 \text{ V}) = 3.1 \times 10^{-7} \text{ C} = 0.31 \,\mu\text{C}$$

and

$$U_\text{o} = \tfrac{1}{2} C_\text{o} V^2 = \tfrac{1}{2}(6.2 \times 10^{-9} \text{ F})(50 \text{ V})^2 = 7.8 \times 10^{-6} \text{ J}$$

(b) With a dielectric introduced, the capacitance increases. Since the voltage across the plates remains constant (why?), the charge on the capacitor must increase, and thus the energy stored increases.

$$C = K C_\text{o} = (2.6)(0.0062 \,\mu\text{F}) = 0.016 \,\mu\text{F}$$

$$Q = K Q_\text{o} = (2.6)(0.31 \,\mu\text{C}) = 0.81 \,\mu\text{C}$$

and

$$U = \tfrac{1}{2} Q V = \tfrac{1}{2} Q V_\text{o}$$
$$= K (\tfrac{1}{2} Q_\text{o} V_\text{o}) = K U_\text{o}$$
$$= (2.6)(7.8 \times 10^{-6} \text{ J}) = 2.0 \times 10^{-5} \text{ J}$$

Note that the energy is increased by a factor of 2.6, the dielectric constant, as expected. The battery had to deliver energy—it had to "pump" more charge to the capacitor which took work on its part.

Follow-up Exercise. Answer the questions in part (b) of this Example assuming that the battery is disconnected before the dielectric is inserted.

16.5 Capacitors in Series and Parallel

Objectives: **To be able to (a) find the equivalent capacitance of capacitors connected in series and in parallel, (b) calculate the charges, voltages, and energy storage of individual capacitors in series and parallel configurations, and (c) analyze capacitor networks that include both series and parallel arrangements.**

Capacitors can be connected in two basic ways: in series or in parallel. In series, the capacitors are connected head to tail, so to speak (•Fig. 16.17a). When they are connected in parallel, all the leads on one side of the capacitors have a common connection (all the tails hooked together, and all the heads hooked together, as shown in •Fig. 16.17b).

Capacitors in Series

When capacitors are wired in series, the charge (magnitude Q) is the same on all the plates

$$Q = Q_1 = Q_2 = Q_3 = \ldots$$

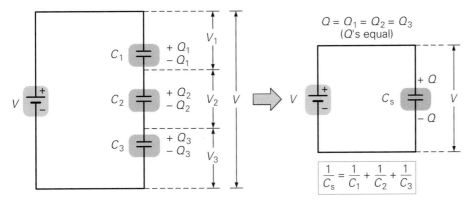

(a) Capacitors in series

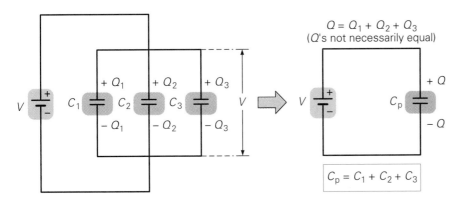

(b) Capacitors in parallel

•**FIGURE 16.17 Capacitors in series and parallel**
(a) Capacitors connected in series all have the same charge, and the sum of the voltage drops is equal to the voltage of the battery. The total capacitance is equivalent to C_s. **(b)** When capacitors are connected in parallel, the voltage drops across the capacitors are the same, and the total charge is equal to the sum of the charges on the individual capacitors. The total capacitance is equivalent to C_p.

To see this, refer to •Fig. 16.18. Note that only plates A and F are actually connected to the battery. Since the plates labeled B and C are isolated, the total charge on them must always be zero. Thus, if the battery puts a charge of $+Q$ on plate A, then $-Q$ is induced on B at the expense of plate C, which acquires a charge of $+Q$. This charge in turn induces $-Q$ on D, and so on down the line.

As we have seen, "voltage drop" is just another name for "change in electrical potential energy per unit charge." Thus, when we add up all the capacitor voltage drops (see Fig 16.17a), going from high to low potential, we get the same value as the drop across the terminals of the battery. (Can you explain why this must be true?) Thus, the sum of the individual voltage drops across all the capacitors is equal to the voltage of the source:

$$V = V_1 + V_2 + V_3 + \cdots$$

The individual voltages are related to the individual charges by

$$V_1 = \frac{Q}{C_1}, \quad V_2 = \frac{Q}{C_2}, \quad V_3 = \frac{Q}{C_3}, \cdots$$

We define the **equivalent series capacitance**, C_s, as the value of a single capacitor that could replace the series combination. Since the combination of capacitors stores a charge of Q at a voltage of V, we have, $C_s = Q/V$ or $V = Q/C_s$. Substituting these expressions into the previous equation and canceling the common Q's, we have

$$\frac{Q}{C_s} = \frac{Q}{C_1} + \frac{Q}{C_2} + \frac{Q}{C_3} + \cdots$$

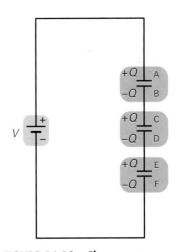

•**FIGURE 16.18 Charges on capacitors in series**
Plates B and C together had zero net charge to start. When the battery placed $+Q$ on plate A, this induced $-Q$ on B; thus C must have acquired $+Q$ for the BC combination to remain neutral. Continuing the argument through the string, we see that all the charges must be the same.

and

$$\frac{1}{C_s} = \frac{1}{C_1} + \frac{1}{C_2} + \frac{1}{C_3} + \cdots \qquad \textit{equivalent series} \atop \textit{capacitance} \qquad (16.14)$$

It is interesting to note that the value of C_s is always smaller than the smallest capacitance in the combination. The charge stored by the series is $Q = C_i V_i$ (where the subscript i refers to any one capacitor in the string). Since $V_i < V$, this arrangement stores *less* charge than any individual capacitor would if it were connected by itself to the same battery, $Q = C_i V$ (Fig. 16.17a).

It makes *physical* sense that in series the smallest capacitance will get the largest voltage, since $V = Q/C$, and all the Q's are the same. A small value of C means less charge stored per volt. Since the charges are the same in series, the smaller capacitors require a larger fraction of the total voltage.

Capacitors in Parallel

With a parallel hook-up (Fig. 16.17b), the voltages across the capacitors are the same (can you explain why in terms of energy?) with each individual voltage equal to that of the battery:

$$V = V_1 = V_2 = V_3 = \cdots$$

The total charge stored is equal to the sum of the charges on the individual capacitors (Fig 16.17b). We expect the equivalent capacitance in parallel to be larger than the largest one, because more charge per volt can be stored in this way than if any one of the capacitors were connected to the battery by itself.

$$Q_t = Q_1 + Q_2 + Q_3 + \cdots$$

The individual charges are given by $Q_1 = C_1 V$, $Q_2 = C_2 V$, . . . A capacitor with the **equivalent parallel capacitance**, C_p, would have to hold this total charge when connected to the battery, so $C_p = Q_t/V$ or $Q_t = C_p V$. Substituting these expressions into the previous equation and canceling the common V, we have

$$C_p V = C_1 V + C_2 V + C_3 V + \cdots$$

and

$$C_p = C_1 + C_2 + C_3 + \cdots \qquad \textit{equivalent parallel} \atop \textit{capacitance} \qquad (16.15)$$

In this case, the equivalent capacitance C_p is simply the sum of the individual capacitances. The equivalent capacitance in parallel is larger than the largest individual capacitance, as expected. Since the capacitors in parallel have the same voltage, the largest capacitance will store the most charge.

Example 16.10 ■ Which Stores More Charge? Capacitors in Series and Parallel

You have two capacitors, one 2.50 μF and the other 5.00 μF. What is the charge on each and the total charge stored if they are connected across a 12.0-V battery (a) in series, and (b) in parallel.

Solution. We make use of the equivalent capacitance relationships just derived. Listing the data, we have

Given: $C_1 = 2.50 \ \mu\text{F} = 2.50 \times 10^{-6}$ F
$C_2 = 5.00 \ \mu\text{F} = 5.00 \times 10^{-6}$ F
$V = 12.0$ V

Find: (a) Q on each capacitor in series and Q_t (total charge)
(b) Q on each capacitor in parallel

(a) In series the total (equivalent) capacitance is given by $1/C_s = 1/C_1 + 1/C_2$. Thus,

$$\frac{1}{C_s} = \frac{1}{2.50 \times 10^{-6}\,\text{F}} + \frac{1}{5.00 \times 10^{-6}\,\text{F}} = \frac{3}{5.00 \times 10^{-6}\,\text{F}}$$

and thus

$$C_s = 1.67 \times 10^{-6} \text{ F} \; (= 1.67 \; \mu\text{F})$$

Since the charge on each capacitor is the same in series and the same as the total, we have

$$Q_t = Q_1 = Q_2 = C_s V = (1.67 \times 10^{-6} \text{ F})(12.0 \text{ V}) = 2.00 \times 10^{-5} \text{ C}$$

(b) Here we use the equivalent capacitance in parallel relationship, $C_p = C_1 + C_2 = 2.50 \times 10^{-6} \text{ F} + 5.00 \times 10^{-6} \text{ F} = 7.50 \times 10^{-6} \text{ F}$. Therefore,

$$Q_t = C_p V = (7.50 \times 10^{-6} \text{ F})(12.0 \text{ V}) = 9.00 \times 10^{-5} \text{ C}$$

In parallel, each capacitor has the full 12.0 V across it, hence

$$Q_1 = C_1 V = (2.50 \times 10^{-6} \text{ F})(12.0 \text{ V}) = 3.00 \times 10^{-5} \text{ C}$$
$$Q_2 = C_2 V = (5.00 \times 10^{-6} \text{ F})(12.0 \text{ V}) = 6.00 \times 10^{-5} \text{ C}$$

Notice that the total stored charge is the sum of the two charges, as we would expect in parallel.

Follow-up Exercise. In part (a) of this Example, what is the voltage across each capacitor and how much energy is stored in each capacitor?

EXAMPLE 16.11 ■ ONE STEP AT A TIME: CAPACITORS IN SERIES–PARALLEL COMBINATION

Three capacitors are connected in a circuit as shown in ●Fig. 16.19a. What is the voltage across each capacitor?

Solution.

Given: Values of capacitance and   *Find:* V_1, V_2, and V_3
voltage from figure   (voltages across capacitors)

The voltage across each capacitor could be found from $V = Q/C$ if the charge on each capacitor were known. The total charge transferred from the battery in charging the capacitors is found by reducing the series–parallel combination to a single equivalent capacitance. Starting with the parallel combination,

$$C_p = C_1 + C_2 = 0.10 \; \mu\text{F} + 0.20 \; \mu\text{F} = 0.30 \; \mu\text{F}$$

and the grouping is reduced as shown in ●Fig. 16.19b. Note that the equivalent parallel capacitance is larger than that of the largest capacitor. For the series combination of C_p and C_3,

$$\frac{1}{C_s} = \frac{1}{C_3} + \frac{1}{C_p} = \frac{1}{0.60 \; \mu\text{F}} + \frac{1}{0.30 \; \mu\text{F}} = \frac{1}{0.60 \; \mu\text{F}} + \frac{2}{0.60 \; \mu\text{F}} = \frac{1}{0.20 \; \mu\text{F}}$$

and

$$C_s = 0.20 \; \mu\text{F}$$

(Don't forget to take the inverse of $1/C_s$.) Note that C_s is less than the least of the capacitances. Using the total equivalent capacitance, we can find the charge on the equivalent capacitance

$$Q = C_s V = (0.20 \times 10^{-6} \text{ F})(12 \text{ V}) = 2.4 \times 10^{-6} \text{ C} = 2.4 \; \mu\text{C}$$

This is the charge on C_3, since it is in series with the battery, and the voltage across C_3 is

$$V_3 = \frac{Q}{C_3} = \frac{2.4 \; \mu\text{C}}{0.60 \; \mu\text{F}} = 4.0 \text{ V}$$

The sum of the voltage drops across the capacitors equals the voltage gain across the terminals of the battery. Notice that it does not matter if you add the voltage drops by going across C_1 and then across C_3, or going across C_2 and then across C_3, since the drops across C_1 and C_2 are the same. Thus we have $V = V_{12} + V_3 = 12$ V (see Fig. 16.19).

$$V_{12} = V - V_3 = 12 \text{ V} - 4.0 \text{ V} = 8.0 \text{ V}$$

There are 8.0 V across both C_1 and C_2 since they are connected in parallel.

Follow-up Exercise. In this Example, find (a) the charge stored on each capacitor, and (b) the energy stored in each.

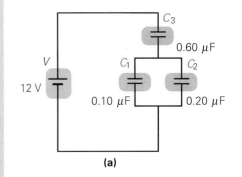

(a)

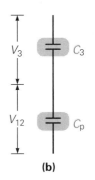

(b)

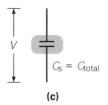

(c)

●**FIGURE 16.19 Circuit reduction** By combining capacitances, the capacitor combination may be reduced to a single equivalent capacitance. See Example 16.11.

Chapter Review

Important Terms

electric potential energy   504
electric potential difference   505
volt (V)   505
voltage   505
electric potential   505
equipotential surfaces   512

electron volt (eV)   515
capacitor   517
capacitance   518
farad (F)   518
dielectric   519
dielectric constant (*K*)   520

dielectric permittivity   521
equivalent series capacitance   523
equivalent parallel capacitance   524

Important Concepts

- The electric potential difference between two points is the work per unit positive charge done by an external force in moving charge between those two points, or the change in electric potential energy per unit positive charge.

- Voltage is synonymous with electric potential *difference*.

- Equipotential surfaces (equipotentials) are surfaces on which a charge has a constant electric potential energy. They are perpendicular to the electric field at all points. It takes no work to move a charge along an equipotential.

- An electron volt (eV) is the kinetic energy gained by an electron accelerated from rest through a potential difference of 1 V.

- A capacitor stores charge, and therefore electric energy, in the form of an electric field. Capacitance is a quantitative measure of how effective a capacitor is in storing charge.

- A dielectric is any nonconducting material capable of being partially polarized when placed in an electric field. When inserted between the plates of a capacitor, a dielectric raises the capacitance by a factor *K*, the dielectric constant.

- The equivalent series capacitance is always less than that of the smallest capacitor of the series combination.

- The equivalent parallel capacitance is always larger than that of the largest capacitor in the parallel combination.

Important Equations

Electric Potential Difference (voltage) (definition):

$$\Delta V = V_B - V_A = \frac{\Delta U}{q_o} \qquad (16.1)$$

Electric Potential Difference due to a Point Charge:

$$\Delta V = \frac{kq}{r_B} - \frac{kq}{r_A} \qquad (16.3)$$

Electric Potential Energy of a Configuration of Point Charges:

$$U_t = U_{12} + U_{23} + U_{13} + \cdots \qquad (16.6)$$

Relationship between Potential and Electric Field:

$$E = -\left(\frac{\Delta V}{\Delta x}\right)_{max} \qquad (16.7)$$

Capacitance (definition):

$$Q = CV \text{ or } C = \frac{Q}{V} \qquad (16.8)$$

Capacitance of a Parallel-Plate Capacitor (in air):

$$C = \frac{\epsilon_o A}{d} \qquad (16.9)$$

(where $\epsilon_o = 8.85 \times 10^{-12} \text{ C}^2/\text{N·m}^2$)

Energy in a Charged Capacitor:

$$U = \tfrac{1}{2}QV = \frac{Q^2}{2C} = \tfrac{1}{2}CV^2 \qquad (16.10)$$

Dielectric Effect on Capacitance:

$$C = KC_o \qquad (16.12)$$

Equivalent Capacitance for Capacitors in Series:

$$\frac{1}{C_s} = \frac{1}{C_1} + \frac{1}{C_2} + \frac{1}{C_3} + \cdots \qquad (16.14)$$

Equivalent Capacitance for Capacitors in Parallel:

$$C_p = C_1 + C_2 + C_3 + \cdots \qquad (16.15)$$

Exercises

16.1 Electric Potential Energy and Potential Difference

1   The SI unit of electric potential difference is (a) joule, (b) newton, (c) newton·meter, (d) joule per coulomb.

2   What is the difference (a) between electrostatic potential energy and electric potential, and (b) between electric potential difference and voltage?

3   The electrostatic potential energy of two point charges (a) is inversely proportional to their separation dis-

tance, (b) is a vector quantity, (c) is always positive, (d) has units of N/C.

4 When an electron approaches a proton, its electric potential energy is (a) increasing (b) decreasing (c) remaining constant.

5 In terms of the way positive charges react to differences in electrical potential, describe why they slow down as they approach other positive charges.

6 In an electric dipole field, where is the electric potential the greatest? (a) near the positive charge, (b) near the negative charge, (c) midway between the two charges.

7 In a charged parallel plate situation, where does a proton have the least electric potential energy? (a) near the positive plate, (b) near the negative plate, (c) midway between the plates.

8 For charged parallel plates, where does an electron have the least electric potential energy? (a) near the positive plate, (b) near the negative plate, (c) midway between the plates.

9 ■ A pair of parallel plates is charged by a 24-V battery. How much work does it take to move a charge of +4.0 μC from the positive to the negative plate?

10 ■ If it takes $+1.6 \times 10^{-5}$ J to move a positive charge between two charged parallel plates (a) what is the magnitude of the charge if the plates were connected to a 6.0-V battery? (b) Taking the charge to be negative, how was it moved, from negative to positive or positive to negative plate?

11 ■ What is the magnitude and direction of the electric field between two charged parallel plates in Exercise 10 if the plate separation is 2.0 mm?

12 ■ If an electron is accelerated by a uniform electric field (1000 V/m) pointing vertically up, use Newton's laws to determine its velocity after it moves 0.50 cm from rest.

13 ■ (a) Repeat Exercise 12, but find the speed using energy methods. Get the direction to your answer by considering electric potential. (b) Does it gain or lose potential energy?

14 ■ A deuteron which consists of a proton and neutron (and thus has the charge of a proton but approximately twice its mass) is accelerated by a potential difference of 10 kV. How fast is it moving if it started from rest?

15 ■ An alpha particle consists of two protons and two neutrons (and thus has a charge of $+2e$ and a mass of approximately four protons) is moving at 0.50% the speed of light to the right. It encounters a uniform electric field head-on and comes to rest in 2.0 m. (a) How

much electric potential energy will it gain? (b) Determine the magnitude and direction of the required electric field.

16 ■ (a) What is the difference in potential between two points, one at 10 cm, and the other at 20 cm from a charge of -5.5 μC? (b) Which one is at a higher potential?

17 ■■ (a) How far from a $+0.60$ μC charge would a point be if it had a value of electric potential of 10 kV? (b) If the point were moved to four times that distance, how much of a potential change would occur? Is it an increase or a decrease?

18 ■■ In the Bohr model of the hydrogen atom, the electron can exist only in circular orbits of a certain radius. One of them is 0.21 nm and another is 0.48 nm. (a) Determine the potential difference between these two orbits. (b) Which one is at a higher potential?

19 ■■ In Exercise 18, by how much does the potential energy of the atom change if the electron goes (a) from the lower to the higher orbit, (b) from the higher orbit to the lower orbit, and (c) from the higher orbit to a very large distance.

20 ■■ How much work does it take to completely separate two charges (each -1.4 μC) if they were initially 4.0 mm apart?

21 ■■ In Exercise 20, if the two charges are released at their initial separation distance, how much kinetic energy would each have when they were very distant from one another?

22 ■■ It takes $+5.5$ J of work to move two charges from a large distance to 1.0 cm from one another. If the charges have the same magnitude, (a) how large is each charge and (b) what can you tell about the signs of the charge?

23 ■■ A charge of -4.0 μC is initially 40 cm from a fixed charge of -6.0 μC and is then moved to a position 90 cm from the fixed charge. (a) What is the change in the mutual potential energy of the charges? (b) Does this change depend on the path through which the charge is moved?

24 ■■ How much work (magnitude) is done in moving an electron along two legs of an equilateral triangle with sides 0.25 m long in an electric field of 15 V/m parallel to the other side of the triangle? (Take the electron as being initially at one end of the side of the triangle parallel to the electric field.)

25 ■■ Compute the energy necessary to bring together the charges in the configuration shown in ●Fig. 16.20.

26 ■■ Compute the energy necessary to bring together the charges in the configuration shown in ●Fig. 16.21.

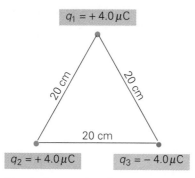

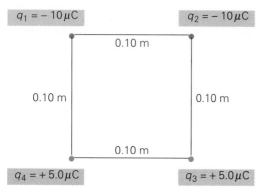

FIGURE 16.20   A charge triangle
See Exercises 25, 27, and 28.

FIGURE 16.21   A charge rectangle
See Exercises 26, 31, 32, and 34.

27 ■■ What is the electric potential at the center of the triangle in Fig. 16.20?

28 ■■ Compute the electric potential at a point midway between q_2 and q_3 in Fig. 16.20.

29 ■■■ Two electrons that are initially at rest 10 cm apart are released. What will be their speed when they are 20 cm apart?

30 ■■■ A proton moving directly toward another fixed proton has a speed of 0.50 m/s when the two are 1.0 m apart. How close to the stationary proton will the moving proton be when it stops and reverses direction?

31 ■■■ What is the electric potential at the center of the square in Fig. 16.21?

32 ■■■ Compute the electric potential at a point midway between q_1 and q_4 in Fig. 16.21.

33 ■■■ In a television picture tube, electrons are accelerated from rest through a potential difference of 10 kV in an electron gun. (a) What is the "muzzle velocity" of the electrons emerging from the gun? (b) If the gun is directed at a screen 35 cm away, how long does it take the electrons to reach the screen?

34 ■■■ Compute the electric potential at a point 4.0 cm from q_4 along the line joining it to q_1 in Fig. 16.21.

16.2   Equipotential Surfaces and the Electric Field

35 Equipotential surfaces are those surfaces on which (a) there is no change in potential, (b) the electric field is zero, (c) the potential is zero.

36 Equipotential surfaces are (a) parallel to the electric field, (b) perpendicular to the electric field, (c) at any angle with respect to the electric field.

37 Sketch the topographic map you would expect near a flat mesa that suddenly becomes a very sharp cliff. Label the gravitational equipotentials as to relative height and potential value. Show how to predict, from the map, which way a ball would roll if released from the top of the cliff.

38 Sketch the topographic map you would expect as you walk away from the ocean up a gently sloping uniform beach area. Label the gravitational equipotentials as to relative height and potential value. Show how to predict, from the map, which way a ball would accelerate if initially rolled up the beach away from the water.

39 Explain why two equipotential surfaces cannot intersect.

40 Suppose that you start with a charge at rest on an equipotential surface, move it *off* that surface, then return it to the same surface and bring it to rest. Is it still true that the net work done on that charge is zero? Explain.

41 The equipotential surfaces inside two charged parallel plates are approximately what geometrical shape?

42 In between charged parallel plates, which equipotentials have a higher potential: (a) the ones near the positive plate, (b) the ones near the negative plate, (c) the ones near the middle?

43 What geometrical shape are the equipotential surfaces outside a point charge?

44 For a positive point charge, if you go from one equipotential surface to a smaller one, what happens to the potential?

45 What geometrical shape are the equipotential surfaces near a long uniformly charged wire?

46 Show that the N/C is the same as the V/m.

47 Explain why an electron volt is a unit of energy. Which is larger, a GeV or an MeV?

48 At a given point on an equipotential surface, the electric field points directly to the (a) next highest equipotential, (b) the next lowest equipotential, (c) parallel to the equipotential surface?

49 Can electric field lines ever cross? Explain using the result of Exercise 39.

50 Can the potential at a point be zero, yet there exist a non-zero electric field at that point? Explain.

51 ■ A uniform electric field (10 kV/m) exists pointing vertically upward. How far apart are the equipotential planes that differ by 10 V?

52 ■ In Exercise 51, if the ground is called zero potential, how far above the ground is the equipotential surface corresponding to 7.0 kV?

53 ■ Determine the value of the potential 2.5 mm from the negative plate of a pair of parallel plates separated by 8.5 mm and connected to a 24-V battery.

54 ■ In Exercise 53, where, relative to the positive plate, is the point with a potential value of 20.0 V?

55 ■ For a point charge of +3.50 μC what is the radius of the equipotential surface that is at a potential of 2.50 kV?

56 ■ For a point charge of +3.50 μC what is the radius of the equipotential surface that is at a potential of 2.40 kV?

57 ■ Using your results from Exercises 55 and 56, calculate the amount of work (in eV) it would take to move an electron from the closer to the more distant equipotential.

58 ■ The potential difference involved in a typical lightning discharge may be up to 100 million volts. What would be the gain in kinetic energy of an electron after moving through this potential difference in (a) eV and (b) joules? (Assume no collisions.)

59 ■ In a typical Van de Graaff linear accelerator, protons can be accelerated through a potential difference of 20 MV. What is their kinetic energy if they started from rest in (a) eV, (b) keV, (c) MeV, (d) GeV, (e) joules?

60 ■■ In Exercise 59, how do your answers change if a doubly charged (+2e) alpha particle is accelerated instead? (Recall that an alpha particle consists of two neutrons and two protons.)

61 ■■ In Exercises 59 and 60, compute the speed of the proton and alpha particle on being accelerated through the potential.

62 ■■ What Van de Graaff voltage is required to produce a beam of protons (initially at rest) with a kinetic energy of (a) 3.5 keV, (b) 4.1 MeV, (c) 6.0 eV, and (d) 8.0 × 10^{-13} J?

63 ■■ Repeat Exercise 62 for electrons instead of protons.

64 ■■■ A proton is moved 15 cm on a path parallel to the field lines of a uniform electric field of 2.0 × 10^5

V/m. (a) What are the possible changes in potential? Consider both cases, of moving the proton with and against the field. (b) What is the change in potential *energy* in electron volts? (c) How much work would be done if the proton were moved perpendicularly to the electric field?

65 ■■■ Two large parallel plates are separated by 3.0 cm and connected to a 12-V battery. Starting at the negative plate and moving 1.0 cm toward the positive plate at a 45° angle, what value of potential would be reached, assuming the negative plate was defined as zero potential?

66 ■■■ In Exercise 65, what would be the value of the potential if you then moved 0.50 cm parallel to the plates?

16.3 Capacitance

67 Capacitance has units of (a) farads, (b) joules, (c) coulomb/volt, (d) both (a) and (c).

68 As can be seen from the formulas given in this chapter, the capacitance and the energy-storing capability of a parallel-plate capacitor can be increased by reducing the plate separation distance. This fact seems to imply that any amount of capacitance or energy storage could be had. Why is this not the case?

69 If the plates of an isolated parallel plate capacitor are moved farther apart, does the energy storage increase, decrease, or remain the same?

70 ■ Show that $\epsilon_o A/d$ has units of farads.

71 ■ How much charge flows through a 12-V battery when a 1.0-μF capacitor is connected across its terminals?

72 ■ What is the capacitance of a capacitor that acquires a charge of 0.40 μC when connected to a 250-V source?

73 ■■ A 12-V battery is connected to a parallel-plate air capacitor with plate areas of 0.20 m^2 and a plate separation of 5.0 mm. (a) What is the resulting charge on the capacitor? (b) How much energy is stored in the capacitor?

74 ■■ If the capacitor in Exercise 73 has its plate separation changed to 10.0 mm after being disconnected from the battery, how do your answers change?

75 ■■ If the capacitor in Exercise 73 has its plate separation changed to 10 mm with the battery remaining connected, how do your answers change?

76 ■■■ Assume that a high-tech fully charged 1.0-F capacitor is able to light a small 0.50-W bulb at full power for 5.0 s before it quits. What was the terminal voltage of the battery that charged the capacitor?

16.4 Dielectrics

77 Putting a dielectric in a charged, parallel-plate capacitor (a) decreases the capacitance, (b) decreases the voltage, (c) increases the charge, (d) causes a discharge because the dielectric is a conductor.

78 The dielectric constant of a conductor is said to be infinite. Explain why.

79 Give several reasons why a conductor would not be a good choice as a capacitor dielectric.

80 Explain why a stream of water is deflected, or bent, when an electrically charged object is brought close to it. (Try this yourself with a plastic pen, hard rubber comb, or balloon.)

81 ■ A capacitor has a capacitance of 50 pF, which increases to 175 pF with a dielectric material between its plates. What is the dielectric constant of the material?

82 ■ A 50-pF capacitor is immersed in silicone oil ($K = 2.6$) while connected to a 24-V battery. What will be the charge on it and the amount of stored energy?

83 ■ How much *more* energy is stored at a given voltage in a capacitor with a paper dielectric ($K = 3.5$) than in a capacitor with the same dimensions containing a polyethylene dielectric ($K = 2.3$)?

84 ■ A parallel-plate capacitor has a capacitance of 1.5 μF with air between the plates. The capacitor is connected to a 12-V battery and charged. The battery is then removed. When a dielectric is placed between the plates, a potential difference of 5.0 V is measured across the plates. What is the dielectric constant of the material?

85 ■■ A parallel-plate capacitor has rectangular plates with dimensions of 6.0 cm × 8.0 cm. If the plates are separated by a sheet of Teflon ($K = 2.1$) 1.5 mm thick, how much energy is stored in the capacitor when it is connected to a 12-V battery?

16.5 Capacitors in Series and Parallel

86 Capacitors in series have the same (a) voltage, (b) charge, (c) energy storage.

87 Capacitors in parallel have the same (a) voltage, (b) charge, (c) energy storage.

88 Under what conditions would two capacitors in series have the same voltage?

89 Under what conditions would two capacitors in parallel have the same energy storage?

90 ■ What is the equivalent capacitance of two capacitors of 0.60 μF and 0.80 μF when they are connected (a) in series and (b) in parallel?

91 ■ When a series combination of two uncharged capacitors is connected to a 12-V battery, 173 μJ of energy is drawn from the battery. If one of the capacitors has a capacitance of 4.0 μF, what is the capacitance of the other?

92 ■■ For the arrangement of three capacitors in •Fig. 16.22, with $C_2 = 0.20$ μF and $C_3 = 0.30$ μF, what value of C_1 will give a total equivalent capacitance of 1.70 μF?

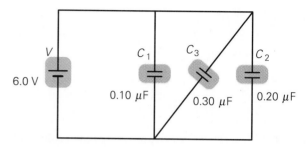

•**FIGURE 16.22 A capacitor triad**
See Exercises 92 and 95.

93 ■■ Three capacitors of 0.25 μF each are connected in parallel to a 12-V battery. (a) What is the charge on each capacitor? (b) How much charge is drawn from the battery?

94 ■■ Two capacitors, C_1 and C_2, are connected in parallel, and this combination is connected in series with another capacitor, C_3. If $C_1 = C_3 = 1.0$ μF and the total equivalent capacitance is 0.75 μF, what is the capacitance of C_2?

95 ■■■ What is the charge on each of the capacitors in the circuit in •Fig. 16.22?

96 ■■■ What are the maximum and minimum equivalent capacitances that can be obtained by combinations of three capacitors of 1.5 μF, 2.0 μF, and 3.0 μF?

97 ■■■ Four capacitors are connected in a circuit as illustrated in •Fig. 16.23. Find (a) the charge on and (b) the voltage difference across each of the capacitors.

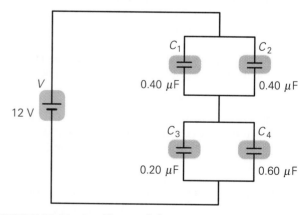

•**FIGURE 16.23 Double parallel in series**
See Exercise 97.

Additional Exercises

98 It is determined that a proton, when moved 10 mm straight up, moves in the direction in which the electric potential is decreasing the most rapidly. (a) If the change in potential is 200 V, estimate the direction and magnitude of the electric field. (b) If the ground is 10 cm away, what is the value of the potential there?

99 How much work is required to move an electron in from a large distance to the midpoint between the two poles of an electric dipole? [Hint: The work is independent of the path; choose your path "conveniently."]

100 An electric dipole consists of two point charges, $\pm q$, separated by a distance d. How much work (in terms of k, q and d) did it take to bring these charges together from a large distance?

101 Sketch the equipotential surfaces and the electric field line pattern outside a uniformly (negatively) charged long wire. Be sure to label the surfaces as to relative potential value and indicate the electric field direction.

102 Is there such a thing as an equipotential *volume*? If so, give a concrete example. [Hint: This means there is no work required to move a charge around a volume.]

103 Use energy methods to determine the minimum electric field strength needed to stop a proton moving at 3.7×10^5 m/s in the space of 5.0 cm.

104 A helium atom with one electron already removed (a positive helium ion) consists of a single orbiting electron and a nucleus of two protons. If the electron is in its minimum orbital radius of 0.027 nm, what is the potential energy of the system? Give your answer in eV.

105 Suppose that in Figure 16.19 the three capacitors have the following values: $C_1 = 0.15$ μF, $C_2 = 0.25$ μF, and $C_3 = 0.30$ μF. (a) What is the equivalent capacitance of this arrangement? (b) How much charge will be drawn from the battery? (c) What is the voltage across each capacitor?

106 How much work is required to bring two charges of $+2.5$ μC and $+4.5$ μC that are separated by an infinite distance to a distance of 0.50 m?

107 Two horizontal, conductive parallel plates are separated by a distance of 1.5 cm. What voltage across the plates would be required to suspend (a) an electron or (b) a proton in midair between them? (c) What would be the signs of the charges on the plates in each case?

108 A 1.0-F parallel-plate capacitor with a distance of 0.50 mm between the plates is to be built. (a) What is the required area of the plates without a dielectric between them? (b) What is the required area of the plates if a polystyrene film ($K = 2.6$) 0.50 mm thick is used between them?

109 The electric field between two parallel plates separated by 3.0 cm has a magnitude of 5.5×10^4 V/m. If the electron is released from the negative plate, how much kinetic energy has it gained when it reaches the positive plate?

17 Electric Current and Resistance

When you think of electricity, probably the first thing that comes to mind is electric lights, and with good reason. Electric light bulbs are among the earliest devices designed to use electricity. They are almost certainly the commonest, used in everything from your pocket flashlight to the landing lights on a 747, from Christmas tree lights to advertising signs 10 stories high. And electric bulbs would seem to be among the easiest electrical devices to understand. People who would never dream of trying to explain the operation of their stereo amplifier probably wouldn't hesitate to tell you how a 60-watt bulb works. Most people would say that something known as electricity flows through the wires and causes the filament of the bulb to glow. Perhaps they also realize that the electric current doesn't itself cause the filament to emit light—it merely heats the filament, which becomes luminous because it is hot.

All this is true as far as it goes—but if you stop to think about it, this explanation, like so many others that we lean on in everyday life, raises more questions than it answers. What is this "electricity" that actually passes through the wires? Where is it and what is it doing when the light is off? What causes it to "flow" when we flick the switch? Why does its passage heat the filament to such a high temperature, but not the wires connected to the bulb?

Such questions are basic ones for understanding, not just light bulbs, but electric current in general. In this chapter you'll find the answers. In the process, you'll learn the basic principles that explain the operation of dozens of common appliances, all of which "run on electricity."

Objectives: To be able to (a) summarize the basic features of a battery, and
(b) explain how a battery produces a direct current in a circuit.

After studying electric force and energy in Chapters 15 and 16, you can probably guess what is required to produce an *electric current*, or a flow of charge. A flow of mass may occur when there is a gravitational potential energy *difference*. For example, the water in a stream flows downhill, from an area of high to an area of lower gravitational potential. So, too, spontaneous heat flow occurs only when there is a temperature *difference*, as when a cold object comes in contact with a warm one. Similarly, a flow of electric charge is dependent on a *difference* in electric potential, or what we call, in everyday language, "voltage."

In solid conductors, particularly metals, the outer electrons of atoms are relatively free to move, while the positively charged nuclei are fixed in the lattice structure of the material. (In liquid conductors and charged gases called plasmas, both positive and negative ions can move, as well as electrons.) Energy is required to move electric charge. This electric energy is produced, or generated, through the conversion of other forms of energy, giving rise to a potential difference or voltage.

A **battery** is a device that converts chemical energy into electrical energy. The Italian scientist Allesandro Volta (1745–1827) is credited with constructing one of the first practical batteries. Basically, a battery consists of two *electrodes* in an *electrolyte*, a solution that conducts electricity (•Fig. 17.1). With the appropriate electrodes and electrolyte (Volta used zinc and copper electrodes in dilute sulfuric acid), a potential difference develops across the electrodes as a result of chemical action. As you may recall from chemistry, different metals dissolve in acids at different rates. When they dissolve, they move into solution as positively charged atoms (ions), leaving behind electrons. The result is a solution of acid and positive ions of both metal types. On average, both electrodes will have an excess of electrons, but because of differences in the chemical properties of the metals, one will have a larger excess.

When a complete circuit is formed, for example by connecting a light bulb and wires as shown, electrons from the more negative electrode (B in the figure) move through the wire and bulb to the less negative electrode (A). The result is a flow of electrons, or an electric current. As the current moves through the filament of the bulb, it can heat it to a sufficient temperature for it to give off visible light (glow).

At this point, you might think things would stop, but that is not the case. As electrode A receives more electrons than normal, it attracts the A ions from solution. (A chemical barrier can be placed in the solution to prevent the A and B ions from mixing.) The A ions reattach to electrons and "plate out" onto the A electrode as neutral atoms once again. Meanwhile, at the B electrode, the electrons are leaving. To generate more electrons, more B ions are released into solution. Thus, negative electrons flow from B to A through the wire and bulb while positive ions flow from B to A in the solution. Eventually this chemical action will cease when the B electrode is entirely dissolved into solution. This process results in an overall loss of chemical potential energy from the battery.

The electrode that has the larger number of excess electrons is named the **cathode**, and designated the negative (−) terminal of the battery. The other electrode, although in reality still having an excess negative charge, has a smaller excess than the cathode, and thus is designated the positive (+) terminal. It is called the **anode**. A color code that assigns red to the anode and black to the cathode is sometimes used—for example, on battery jumper cables for automobiles.

Because of the difference in charge on the battery terminals, a potential difference exists between them. The anode, having less negative charge, is at a higher potential than the cathode. The potential difference across the terminals of a bat-

Note: The term voltage in this text is used to mean "difference in electric potential."

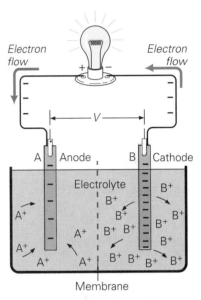

•**FIGURE 17.1 The chemical battery or cell**
Chemical processes involving the electrolyte and two unlike metal electrodes cause ions of both metals to dissolve into the solution. Since they dissolve at different rates, and thus leave different quantities of excess electrons behind, one electrode (the cathode) becomes more negatively charged than the other (the anode). Thus the anode is at a higher potential than the cathode. The electric potential, or voltage, developed across the electrodes can cause a current, or a flow of charge (electrons) in the wire. There is also a current in the solution as positive ions migrate, completing the circuit. (A membrane is necessary to prevent mixing of ion types.) By convention, the anode is designated the positive terminal and the cathode the negative terminal.

tery *when it is not connected* to a circuit is called the **electromotive force (emf)**. This is a somewhat misleading name because the electromotive force is not a force but a voltage difference, and, as such, is measured in volts. The emf ($\mathcal{E}$) is the maximum possible potential difference between the battery terminals (•Fig. 17.2a). When a battery is connected to a circuit and charge flows, the voltage across the battery is slightly less than the emf because of internal resistance. The operating voltage (V) of a battery is called its **terminal voltage** (•Fig. 17.2b). In this chapter, we will be chiefly concerned with terminal voltage.

The internal resistance r of the battery in Fig. 17.2b is represented explicitly in the circuit diagram. This resistance or opposition to charge flow results from the resistance of the materials from which the battery is made. Internal resistances are typically small, and the terminal voltage of a battery is usually only slightly less than its emf. However, when a battery supplies a large current to a circuit, the terminal voltage V may drop appreciably below the maximum emf value. As you will learn in our discussion of electricity, the emf for the circuit in Fig. 17.2b is given by $\mathcal{E} = Ir + IR = Ir + V$, or $V = \mathcal{E} - Ir$ where I is the current. Note that as I goes to zero, V approaches $\mathcal{E}$. Thus you can see that the emf is not a true indication of the state of the battery. Even if $\mathcal{E}$ is the expected terminal voltage, when we ask the battery to deliver current (and thus energy) to a light bulb, for example, the terminal voltage can be considerably less than $\mathcal{E}$, especially if the internal resistance is high, a common fault of used batteries.

As long as the internal chemical action maintains a potential difference across its terminals, current is supplied to a circuit as in Fig. 17.2b. The battery is said to deliver current to a circuit. Alternatively, we say that the circuit (or its components) draws current from the battery. Because electrons can flow only in one direction in such a circuit, from the negative ($-$) terminal (cathode) to the positive ($+$) terminal (anode), this is called **direct current (dc)**. Notice the circuit diagram in the figure. The symbol for a battery is two parallel lines, the longer of which represents the positive ($+$) terminal and the shorter the negative ($-$) terminal. A circuit element, such as the lamp, may be generally represented by the symbol ($-\!\wedge\!\wedge\!\wedge\!-$), which stands for a resistance (R). (Electrical resistance—resistance to

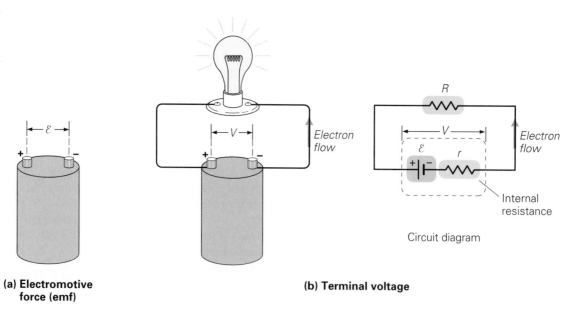

(a) **Electromotive force (emf)**

(b) **Terminal voltage**

•**FIGURE 17.2 Electromotive force (emf) and terminal voltage**
(a) The emf of a battery is the maximum potential difference across its terminals. This occurs when the battery is not connected to an external circuit. **(b)** Because of internal resistance (r), the terminal voltage when the battery is in operation is less than the emf. As will be seen, $V = \mathcal{E} - Ir$, where I is the current through the battery.

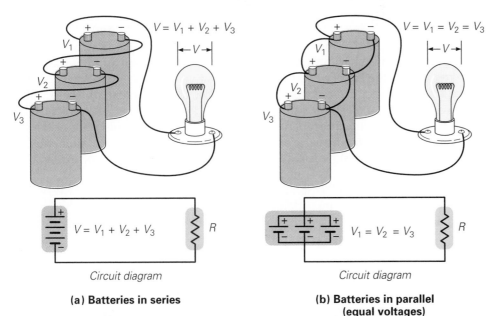

$$V = V_1 + V_2 + V_3$$

$$V = V_1 = V_2 = V_3$$

Circuit diagram

(a) Batteries in series

Circuit diagram

**(b) Batteries in parallel
(equal voltages)**

•**FIGURE 17.3** **Batteries in series
and in parallel**
(a) When batteries are connected in
series, their voltages add and the
voltage across the resistance R is
the sum of the voltages. **(b)** When
batteries of the same voltage are
connected in parallel, the voltage
across the resistance is the same, as
if only a single battery were
present, and each battery supplies
a fraction of the total current.

charge flow—will be considered in detail in later sections of this chapter. Here,
we merely introduce the circuit symbol.)

There are a wide variety of batteries now in use. One of the most common is
the 12-V automobile battery. It actually consists of six 2-V cells connected in series. That is, the positive terminal of each cell is connected to the negative terminal of the next cell (shown for three cells in •Fig. 17.3a). When batteries or cells
are connected in this fashion, their voltages add.*

If cells are connected in parallel, all their positive terminals have a common
connection, as do their negative terminals (•Fig. 17.3b). When identical batteries
are connected in this way, the potential difference is the same across all of them,
and each one supplies a fraction of the current to the circuit (for three batteries
with equal voltages, each one supplies 1/3 of the current). Parallel connections of
batteries are familiar to people who have had their car jump started. In that situation, the weak (high r) battery is connected in parallel to a healthy (low r) battery, which delivers most of the current to get the car started. If sources of different voltages are connected in parallel, the situation is more complex.

Other sources of voltage, such as generators and photocells, will be considered in later chapters. One of the more common is the so-called "power supply"
that plugs into the wall socket and converts the alternating voltage delivered by
the power company into direct (and usually adjustable) voltage. It can be thought
of as a variable battery.

17.2 Current and Drift Velocity

**Objectives: To be able to (a) define electric current, (b) distinguish between
electron flow and conventional current, and (c) explain the
concept of drift velocity.**

A battery or some other voltage source connected to a continuous conducting path
forms a **complete circuit**. In ordinary applications, *a sustained electric current requires a voltage source and a complete circuit*. As shown in •Fig. 17.4, a circuit may
have a switch (symbolized by S), that is used to open or close the circuit. When
the switch is open, the current or circuit component is turned off; when the switch
is closed, there is current in the circuit.

*Chemical energy is converted to electrical energy in a chemical *cell*. The term *battery* generally
refers to a "battery" of cells.

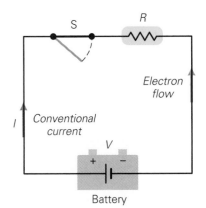

•**FIGURE 17.4** **Conventional
current**
For historical reasons, circuit
analysis is usually done using
conventional current. Conventional
current is in the direction in which
positive charges would flow, or
opposite to the actual electron flow.

Direct Current

Since it is electrons that move in the wires of a circuit, the net flow of charge is in the direction in which an electron experiences a force—away from the negative terminal of the battery and toward the positive terminal. (Recall the charge–force law, or the fact that electrons flow from low to high potential). Historically, however, circuit analysis has generally been done in terms of **conventional current**, which is in the direction in which positive charges would flow, or the direction opposite to the actual electron flow (see Fig. 17.4). (Some situations in which positive charge flow is really responsible for the current—for example in semiconductor mechanisms—are discussed later in the text.)

Conventional current: in direction opposite that of electron flow.

In general we say electric current is a flow of charge. Quantitatively, the **electric current** (I) in a wire is defined as the time rate of flow of net charge. If a net charge q passes a given point in a wire in time t at a constant rate of flow (•Fig. 17.5), the current is given by

$$I = \frac{q}{t} \qquad \textit{definition of electric current} \qquad (17.1)$$

SI unit of current: coulomb/s (C/s) or ampere (A)

As can be seen from the defining equation, the unit of current is coulombs per second (C/s). This combination of units is called an **ampere (A)** in honor of the French physicist Andre Ampère (1775–1836), an early investigator of electrical and magnetic phenomena. The ampere is commonly shortened to "amp," for example, a current of 10 A is read as "ten amps." Small currents are expressed in milliamperes (mA, 10^{-3} A) or microamperes (μA, or 10^{-6} A), which are shortened to milliamps and microamps. In a typical household circuit, it is not unusual for several amps of current to flow.

EXAMPLE 17.1 ■ COUNTING ELECTRONS: CURRENT AND CHARGE

There is a steady current of 0.50 A in a circuit for 2.0 min. How much charge passes by a given location in one of the connecting wires during this time? How many electrons does this represent?

Solution. We are told the current (rate) and time elapsed. We will use the definition of current.

Given: $I = 0.50$ A $\qquad\qquad$ *Find:* q (amount of charge)
$\qquad\quad t = 2.0$ min $= 1.2 \times 10^2$ s

By Eq. 17.1, $I = q/t$, so the charge is given by $q = It$.
$$q = It = (0.50 \text{ A})(1.2 \times 10^2 \text{ s}) = (0.50 \text{ C/s})(1.2 \times 10^2 \text{ s}) = 60 \text{ C}$$

The amount of charge q is made up of n electrons or electronic charges, that is, $q = ne$ (Eq. 15.1). Solving for n using the magnitude of the charge on the electron, we have

$$n = \frac{q}{e} = \frac{60 \text{ C}}{1.6 \times 10^{-19} \text{ C/electron}} = 3.8 \times 10^{20} \text{ electrons}$$

This may seem like a large number of electrons, but keep in mind that in an aluminum wire, for example, each atom has 13 electrons. From chemical principles, we know that there are about 6×10^{23} aluminum atoms in every 27 g of aluminum. Thus the percentage of electrons involved in the current flow is only a small fraction of the total number of electrons in the wire.

Follow-up Exercise. Many laboratory instruments can measure currents in the nanoampere range. How long would it take for one coulomb of charge to flow through a wire carrying a current of 1.0 nA?

Cross-sectional area A

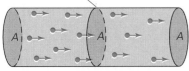

•**FIGURE 17.5 Electric current**
Electric current (I) in a wire is defined as the net amount of charge (q) that passes through a cross-sectional area of the wire per time, that is, $I = q/t$. I has the units of amperes (A), or "amps" for short.

Drift Velocity

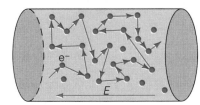

Although charge flow is frequently mentioned, you should be aware that electric charge does not flow in a conductor in the same way that water flows through a pipe. In the absence of a potential difference, the free electrons in a metal wire move about randomly at very high speeds, colliding many times per second with the atoms of the lattice. As a result, there is no net flow of charge one way or the other in the wire. (On the average, the same amounts of charge pass through a cross-sectional area in opposite directions in a given time.) When a potential difference is applied across the ends of the wire (for example, by a battery), this creates an electrical field in one direction and thus a net flow of charge in that direction. This does *not* mean electrons are moving directly from one end of the wire to the other. They still move about in all directions because they collide with the atoms of the conductor (•Fig. 17.6), but there is an added component in one direction to their velocities. The overall result is that their motions are less random, and more of them move toward the positive terminal of the battery than away from it.

The net electron flow is characterized by an average velocity called the **drift velocity**, which is much smaller than the random velocities of the electrons themselves. The magnitude of the drift velocity is on the order of 0.10 cm/s. The net movement of electrons at this speed is relatively slow. A quick calculation using this drift velocity would show that on the average it would take an electron about 17 minutes to travel 1 m along a wire. Yet a lamp comes on almost instantaneously when you turn the switch (close the circuit), and the electronic signals carrying telephone conversations travel almost instantaneously over miles of wire.

Evidently, *something* must be moving faster than the net electron motion. This is the electric field. When a potential difference is applied, the associated electric field that drives the electrons in their net motion travels along the conductor at a speed close to the speed of light (on the order of 10^8 m/s). The electric field therefore influences the motion of electrons throughout the conductor almost instantaneously. Thus there is current everywhere in the circuit almost simultaneously. You don't have to wait for electrons to "get there" from somewhere else. The electrons already in the light bulb filament begin to move almost immediately as they feel the electric force, thus delivering energy, and creating light.

•FIGURE 17.6 Drift velocity Because of collisions with the atoms of the conductor, electron motion is random. However, when the conductor is connected in a circuit, there is a relatively small net motion in the direction opposite the electric field in the wire (or toward the positive high potential terminal of the battery or other voltage source). This net motion is characterized by a drift velocity.

17.3 Resistance and Ohm's Law

Objectives: To be able to (a) define electrical resistance and explain what is meant by an ohmic resistor, (b) summarize the factors that determine resistance, and (c) calculate the effect of these factors in simple situations.

Given that an applied voltage (potential difference) across the two ends causes an electric current in a conductor, how much charge flows for a given voltage? As you might expect, the current is usually proportional to the voltage. That is, the greater the voltage, the greater the current, a relationship that can be written mathematically as

$$I \propto V$$

This is analogous to more water flowing through a spigot in a tank when the level of fluid in the tank is higher.

However, another factor besides voltage affects current flow. Just as internal friction (viscosity) affects fluid flow, the internal resistance of materials affects the flow of electric charge. Any object that offers resistance to the flow of charge is called a *resistor* and represented by the resistor symbol in circuit diagrams. In the pages that follow you will find this symbol used for a wide range of "resistors," from the cylindrical color-coded ones on printed circuit boards to electrical devices such as hairdryers and light bulbs.

Note: Remember that V really stands for ΔV.

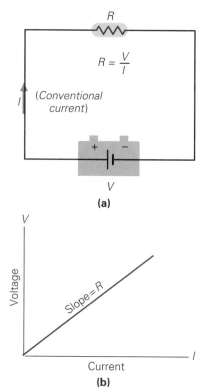

$$R = \frac{V}{I}$$

(Conventional current)

I

(a)

Voltage — *V*

Slope = *R*

Current — *I*

(b)

• FIGURE 17.7   Resistance definition and Ohm's law

(a) In principle, any circuit element's resistance is determined by dividing the voltage across it by the resulting current flow through it. **(b)** If the element obeys Ohm's law (or is "ohmic") then a plot of voltage versus current gives a straight line with a slope equal to *R*. This is not usually the case, as even ohmic resistors are only approximately so over a range of voltages.

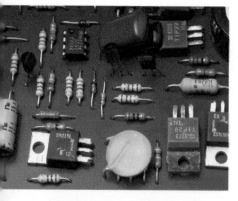

• FIGURE 17.8   Resistors

Various types of resistors (cylinders) of different values on a circuit board. The resistance (in ohms) of a particular resistor is indicated by a 4-band color code.

Resistance (R) is defined as the ratio of the voltage to the resulting current. This makes physical sense, because if we need to apply a huge voltage to a resistor and still get only a small current, we would say that it has a high resistance! Mathematically, its resistance is

$$R = \frac{V}{I} \quad \text{or} \quad V = IR \qquad \textit{definition of electrical resistance} \qquad (17.2)$$

SI unit of resistance: volt/ampere (V/A), or ohm (Ω)

For many materials, the resistance is constant, or at least approximately so over a range of voltages. A resistor that has constant resistance is said to obey **Ohm's law**, or to be ohmic. This law was named after Georg Ohm (1789–1851), a German physicist who investigated the relationship between current and voltage and found that there were cases when the resistance did not change with voltage. The units for resistance are volts per ampere (V/A). This combined unit is called an **ohm (Ω)**. A plot of voltage versus current for an ohmic resistance gives a straight line with a slope equal to its resistance, *R* (• Fig. 17.7). A common and practical form of Ohm's law is *V* = *IR* where *R* is constant.

Ohm's law is not a fundamental law in the sense that Newton's law of gravitation or the first and second laws of thermodynamics are. There is no law that states that materials *must* have constant resistance. Indeed, many of our technological advances in computers and communication are based on semiconductors, which can have highly *nonlinear* voltage–current behaviors. Rather, Ohm's law is an extremely useful generalization that applies to a wide range of materials, particularly metals. Unless we specify otherwise, we will assume the resistors are ohmic. But you should keep in mind that many everyday appliances, such as light bulbs, have variable resistances. The tungsten filaments of these bulbs have more resistance at their high operating temperatures than at room temperature.

EXAMPLE 17.2 ■ A CAR BATTERY: ELECTRICAL RESISTANCE

When a 12-V automobile battery is connected across an unknown resistor, there is a current of 2.5 mA in the circuit. What is the value of the resistor?

Solution.   Here we can simply apply the definition of resistance.

Given: $V = 12$ V   *Find:* R (resistance)
         $I = 2.5$ mA $= 2.5 \times 10^{-3}$ A

Resistors are used in circuits to limit current, and are manufactured in a wide variety of resistance values (see • Fig. 17.8). In this case, using Eq. 17.2,

$$R = \frac{V}{I} = \frac{12 \text{ V}}{2.5 \times 10^{-3} \text{ A}} = 4.8 \times 10^{3} \, \Omega$$

That is, the resistor is a 4.8-kΩ (kilohm) resistor. Resistors in the kΩ and MΩ (megohm) ranges are quite common.

Follow-up Exercise.   Estimate the resistance of the following common household appliances, assuming 120 V is applied across them: (a) a hairblower through which flows 10 A, (b) a 60-W bulb that carries a current of 0.50 A, and (c) a corded electric razor that carries a current of 0.10 A.

Factors That Influence Resistance

On the atomic level, resistance arises from collisions of electrons with lattice atoms or ions making up a material. Thus, resistance is a material property; that is, it depends on the type of material. The resistance of a particular conductor of uniform cross section, such as a length of wire, depends on several factors (•Fig. 17.9). The factors directly affecting resistance are

1. Type of material
2. Length
3. Cross-sectional area
4. Temperature

As you might expect, the resistance of an object (such as an everyday strand of wire) is directly proportional to its length (L) and inversely proportional to its cross-sectional area (A):

$$R \propto \frac{L}{A}$$

For example, a uniform metal wire 4 m long offers twice as much resistance as a similar wire 2 m long. But a wire with a cross-sectional area of 0.06 cm^2 has only half the resistance of one with an area of 0.03 cm^2. These geometrical conditions are analogous to those for liquid flow in a pipe. The longer the pipe, the more resistance (drag) there is; but the larger the diameter (or cross-sectional area) of the pipe, the greater the amount of liquid it can carry.

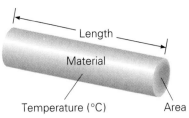

•**FIGURE 17.9 Resistance factors**
Factors directly affecting the electrical resistance of uniform conductor are the type of material (ρ), the length (L), the cross-sectional area (A), and the temperature.

Resistivity and Conductivity

The resistance of a material also depends on intrinsic properties (the kind of atoms that make it up), which are characterized by the **resistivity** (ρ). This acts as a constant of proportionality for the relationship of resistance to length and area:

$$R = \rho \frac{L}{A} \quad \text{or} \quad \rho = \frac{RA}{L} \tag{17.3}$$

SI unit of resistivity: $\Omega \cdot \text{m}$

As can be seen from Eq. 17.3, ρ has units of ohm·meters ($\Omega \cdot \text{m}$), that is, $\rho = RA/L$ and its units are $\Omega \cdot \text{m}^2/\text{m}$ or $\Omega \cdot \text{m}$. Resistivity is an atomic property of the material, not the geometry of the actual resistor. Thus it is of practical importance, since it allows calculation of the resistance of any cylindrical resistor or wire using Eq. 17.3.

The resistivities of some common conductors (and insulators) are given in Table 17.1. The values given are for a particular temperature (20°C, or room tem-

Note: Be careful not to confuse resistivity with mass density, which has the same symbol (ρ).

Resistivity—a material property.

TABLE 17.1	Resistivities (at 20°C) and Temperature Coefficients of Resistivity for Various Materials*				
	ρ ($\Omega \cdot$m)	α (C$^{\circ -1}$)		ρ ($\Omega \cdot$m)	α (C$^{\circ -1}$)
Conductors			*Semiconductors*		
Aluminum	2.82×10^{-8}	4.29×10^{-3}	Carbon	3.6×10^{-5}	-5.0×10^{-4}
Copper	1.70×10^{-8}	6.80×10^{-3}	Germanium	4.6×10^{-1}	-5.0×10^{-2}
Iron	10×10^{-8}	6.51×10^{-3}	Silicon	2.5×10^{2}	-7.0×10^{-2}
Mercury	98.4×10^{-8}	0.89×10^{-3}			
Nichrome alloy of nickel	100×10^{-8}	0.40×10^{-3}	*Insulators*		
and chromium			Glass	10^{12}	
Nickel	7.8×10^{-8}	6.0×10^{-3}	Rubber	10^{15}	
Platinum	10×10^{-8}	3.93×10^{-3}	Wood	10^{10}	
Silver	1.59×10^{-8}	6.1×10^{-3}			
Tungsten	5.6×10^{-8}	4.5×10^{-3}			

*Values for semiconductors are general values, and resistivities for insulators are typical orders of magnitude.

perature), because resistivity is somewhat temperature-dependent. For most metallic conductors, the resistivity increases as the temperature rises.

A quantity that is sometimes mentioned is the electrical conductivity of a material. As you might guess, the **conductivity** (σ) and the resistivity (ρ) are inversely proportional.

Conductivity—the reciprocal of resistivity.

$$\sigma = \frac{1}{\rho} \qquad (17.4)$$

The unit of conductivity is $(\Omega \cdot m)^{-1}$. You can choose to describe resistance by either resistivity or conductivity, since one is simply the inverse of the other.

EXAMPLE 17.3 ■ A LONGER CORD: RESISTANCE OF A WIRE

A 1.5-m insulated extension cord, which actually consists of two wires to make a complete circuit, is used to connect a lamp to an outlet. If the copper wire of the cord has a diameter of 2.3 mm (equivalent to American Wire Gauge, or AWG, No. 12), what is the resistance of *one* of the wires at room temperature?

Solution. After looking up the resistivity of room temperature in Table 17.1 we have enough information to do the calculation, assuming the wire to be cylindrical in shape.

Given: $L = 1.5\,m$ *Find:* R (resistance)
            $d = 2.3\,mm = 2.3 \times 10^{-3}\,m$
            $\rho_{Cu} = 1.70 \times 10^{-8}\,\Omega \cdot m$

The cross-sectional area of the wire is

$$A = \pi r^2 = \frac{\pi d^2}{4} = \frac{\pi(2.3 \times 10^{-3}\,m)^2}{4} = 4.2 \times 10^{-6}\,m^2$$

Then, using Eq. 17.4,

$$R = \frac{\rho L}{A} = \frac{(1.70 \times 10^{-8}\,\Omega \cdot m)(1.5\,m)}{4.2 \times 10^{-6}\,m^2}$$

$$= 6.1 \times 10^{-3}\,\Omega$$

This result demonstrates why the resistances of the connecting wires in an electrical circuit can usually be neglected. The resistances of the circuit *components* are generally much larger than this. We can thus ignore the resistances of the connecting wires unless the circuit components have comparable resistances.

Follow-up Exercise. Old-fashioned light bulbs used to be constructed from carbon filaments. Estimate the length of the cylindrical carbon filament with a diameter of 0.50 mm that would have a total resistance of 150 Ω at room temperature.

Temperature dependence of resistivity and resistance.

The temperature dependence of resistivity is nearly linear if the temperature change is not too great. An expression similar to the one for thermal linear expansion (Chapter 10) may be written for this relationship. That is, the resistivity (ρ) at a temperature T after a temperature change $\Delta T = T - T_o$ is given by

Note: Compare Eq. 10.9, Section 10.4.

$$\rho = \rho_o(1 + \alpha \Delta T) \qquad (17.5)$$

Here α is a constant (over a small temperature range) and is called the **temperature coefficient of resistivity**, and ρ_o is a reference resistivity at T_o (usually 20°C or 0°C). Eq. 17.5 may be rewritten as:

$$\Delta \rho = \rho_o \alpha \Delta T \qquad (17.6)$$

where $\Delta \rho = \rho - \rho_o$ is the change in resistivity which occurs over a given change in temperature (ΔT). Since the ratio $\Delta \rho / \rho_o$ is dimensionless, α must have the units of $C^{\circ-1}$ (or $1/C^\circ$). Physically, α is the fractional change in resistivity per degree. The temperature coefficients of resistivity for some materials are listed in Table 17.1. Our discussions will assume that these coefficients are constant over the temperature ranges used in the examples and problems.

Since resistance is directly proportional to the resistivity, we can use this fact together with Eqs. 17.3, 17.5, and 17.6 to derive an expression for the resistance of a conductor of uniform cross section as a function of temperature:

$$R = R_o(1 + \alpha \Delta T) \quad \text{or} \quad \Delta R = R_o \alpha \Delta T \qquad (17.7)$$

where $\Delta R = R - R_o$. The variation of resistance with temperature provides a means of measuring temperature in the form of an electrical resistance thermometer.

EXAMPLE 17.4 ■ AN ELECTRICAL THERMOMETER: VARIATION IN RESISTANCE WITH TEMPERATURE

A platinum wire has a resistance of 0.50 Ω at 0°C. It is placed into a water bath where its resistance jumps to 0.60 Ω. What is the temperature of the bath? (Assume α is constant over the temperature range.)

Solution. We need to look up the temperature coefficient of resistivity for platinum in Table 17.1. Once we have that it is a simple matter to solve for the bath temperature.

Given: $T_o = 0°C$ **Find:** T (temperature of the bath)
         $R_o = 0.50 \, \Omega$
         $R = 0.60 \, \Omega$
         $\alpha_{Pt} = 3.93 \times 10^{-3} \, C^{\circ-1}$ (Table 17.1)

The ratio $\Delta R/R_o$ is the fractional change in the initial resistance R_o (at 0°C). Using Eq. 17.7 with the values given for α, T, and T_o,

$$\frac{\Delta R}{R_o} = \alpha(T - T_o) = \alpha \Delta T$$

$$\Delta T = \frac{\Delta R}{\alpha R_o} = \frac{(R - R_o)}{\alpha R_o} = \frac{(0.60 - 0.50) \, \Omega}{(3.93 \times 10^{-3} \, C^{\circ-1})(0.50 \, \Omega)}$$

$$= 51 \, C^\circ$$

Thus the bath is at $T = T_o + \Delta T = 51°C$. Would the resistance change have been the same if the wire had been made of a different material and immersed in the same bath? (See the Follow-up Exercise.)

Follow-up Exercise. In this Example, if the material had been copper with $R_o = 0.50 \, \Omega$, what would its resistance be at the bath temperature? What conclusions can you draw about whether you would choose a material with a high or low temperature coefficient of resistivity to make a sensitive "resistance" thermometer?

Carbon and other semiconductors have negative temperature coefficients of resistivity (see Table 17.1). This means that the resistance of a semiconductor decreases with increasing temperature or increases with decreasing temperature. Many materials have positive temperature coefficients of resistivity, and their resistances decrease with decreasing temperature. You might be wondering how far electrical resistance can be reduced by lowering the temperature. In certain cases, the resistance can be reduced to zero—not just close to zero, but, as accurately as we can measure, to exactly zero! This condition of **superconductivity**, which may lead to technological applications of great importance, is discussed in the Insight on p. 542.

Insight   Superconductivity

The electrical resistance of metals and alloys generally decreases with decreasing temperature. At relatively low temperatures, some materials exhibit what is called **superconductivity**; that is, the electrical resistance vanishes, or goes to zero.

Superconductivity was discovered in 1911 by Heike Kamerlingh Onnes, a Dutch physicist. The phenomenon was first observed in solid mercury at a temperature of about 4 K (−269°C). The mercury was cooled to this temperature using liquid helium. The boiling point of helium (the temperature at which it condenses to a liquid) is about −267°C at 1 atm. Lead also exhibits superconductivity when cooled to such a temperature.

An electrical current established in a superconducting loop should persist indefinitely with no resistive losses. Currents introduced in superconducting loops have been observed to remain constant for several years. In 1957, a theory was presented by the American physicists John Bardeen, Leon Cooper, and Robert Schrieffer in an attempt to explain various aspects of metallic superconductors. Known as the BCS theory for obvious reasons, it presents a quantum-mechanical model in which electrons are viewed as waves traveling through a material. According to the theory, lattice vibrations and imperfections, which scatter electrons and give rise to resistance and energy loss in metals under normal conditions, have no effect on the electrons in superconductors. In the absence of this electron scattering, the resistance is zero, and a current can persist as long as the metal is in a superconducting state.

Keeping a superconductor at the low temperatures of liquid helium is somewhat difficult and costly. Thus, there has been an ongoing search for materials that become superconducting at higher temperatures. Other superconducting metals and alloys were found for which the critical temperature was about 18 K (−255°C). In 1973, a material with a critical temperature of 23 K (−250°C) was discovered.

In 1986, there was a major breakthrough—a new class of superconductors was discovered. These were the so-called ceramic alloys of rare earth elements, such as lanthanum and yttrium. These superconductors were prepared by grinding the mixture of metallic elements together and heating it to a high temperature to produce a ceramic material. For example, one such mixture consisted of barium, yttrium, and copper oxide. The critical temperature for these ceramic mixtures was about 57 K (−216°C).

In 1987, a ceramic material with a critical temperature of 98 K (−175°C) was discovered. This was a major advance because it meant that superconductivity could be obtained using liquid nitrogen, which has a boiling point of 77 K (−196°C). (See Fig. 1.) Liquid nitrogen is relatively plentiful (nitrogen is the chief constituent of air) and inexpensive—it costs cents per liter compared to dollars per liter for liquid helium.

There have been more recent reports of higher critical temperatures, and even suggestions that certain copper oxide

FIGURE 1   Magnetic levitation
A magnetic cube levitates above a piece of "high-temperature" superconducting material that is cooled with liquid nitrogen. When the material becomes cold enough to superconduct, it acquires certain magnetic properties that cause a magnet to be repelled and levitate.

compounds may lose their resistance to electrical current at room temperature and above. As scientists study the new superconductors, better understanding will be gained, and no doubt the critical temperature will rise. One problem is the difficulty of confirming a superconducting state in small samples or regions of samples. Another is that the BCS theory does not appear to apply to the new high-temperature superconductors. The mechanism for their superconductivity is not well understood.

There are many possible applications for superconductors. One is superconducting magnets. The strength of an electromagnet depends on the magnitude of the current in the windings (Chapter 19). If there were no resistance, there would be greater current and no losses. Used in motors or engines, superconducting electromagnets would provide more power. (Superconducting magnets cooled by liquid helium have been used in ships' engines and particle accelerators for some time now.) Such magnets could be used to levitate and propel trains and electric cars. Another application of superconductors might be underground transmission cables with no resistive losses. However, among the many technological problems still to be overcome is how to form wires out of ceramic materials (which are generally brittle). The application most likely to be realized soon will be in making faster computer circuits.

Imagine what it would mean if someone discovered a material that was superconducting at refrigerator or room temperature. The absence of electrical resistance opens many possibilities. You're likely to hear more about superconductor applications in the near future.

Objectives: To be able to (a) define electric power, (b) calculate the power delivery of simple electric circuits, and (c) explain joule heating and its significance.

When a sustained current exists in a circuit, the electrons are given energy by the voltage source, such as a battery. As these charge carriers pass through circuit components, they collide with the atoms of the material (experience resistance) and lose energy because of these collisions (Chapter 6). The energy transferred in the collisions can result in a temperature increase of the component. In this way electrical energy can be transformed, at least partially, into thermal energy.

As you know, electrical energy in general is commonly converted into various forms for everyday use: not only heat (as in electric stoves), but also light (as in light bulbs) and mechanical motion, or work (as in electric drills). Overall, however, all of the energy given to the charge carriers by the battery must be lost in the circuit. That is, a charge carrier traversing the circuit must lose all the electric potential energy it gained from the battery when it returns to the negative terminal of the battery.

The energy gained by a charge q from a voltage source having a terminal voltage difference V is qV. This is the work done by the source on the charge, $W = qV$. Over time, the average *rate* at which energy is delivered to the external circuit by the battery (called the **electric power**, $\overline{P}$) is given by

$$\overline{P} = \frac{W}{t} = \frac{qV}{t}$$

If the current and voltage are steady with time, then the average power is the same as the power at any instant. However, if the voltage varies with time, so will the current, and we can compute only the average power. If we assume steady currents for the rest of this chapter, then $I = q/t$ (Eq. 17.1), and we can rewrite the power equation in terms of the two commonly measured quantities, voltage and current:

$$P = IV$$

As you learned in Chapter 5, the SI unit of power is the watt (W).

Since at any voltage the resistance is $R = V/I$, power can be rewritten in two other convenient forms:

$$P = IV = \left(\frac{V}{R}\right)V = \frac{V^2}{R} \quad \text{and} \quad P = IV = I(IR) = I^2R$$

Thus there are three equivalent expressions for electric power calculations:

$$P = IV = \frac{V^2}{R} = I^2R \qquad electric\ power \qquad (17.8)$$

Usually you choose the one that is most convenient to the problem at hand.

Joule Heating

The thermal energy expended in a current-carrying resistor is sometimes referred to as **joule heat,** or I^2R **losses (I squared R).** In many instances, joule heating is an undesirable side effect: for example, in electrical transmission lines. However, in other uses, the conversion of electrical energy to thermal energy is the main object. Heating applications include the heating elements (burners) of electric stoves, hair dryers, immersion heaters, and toasters. A novel application of joule heat in battery testers is discussed in the Insight on page 544.

Note: Recall that temperature is a measure of the average translational kinetic energy per molecule of a sample.

Electric power: three equivalent expressions.

Insight   Joule Heat and Battery Testing

You may have noticed and used the individual battery testers on battery packages (Fig. 1). A battery is placed between the terminals of the tester and a rising yellow stripe indicates the condition of the battery. Joule heat is one of the ingredients of the operation.

The tester consists of a polyester strip with a printed circuit on one side and a liquid crystal on the other. (Many organic substances exhibit an "intermediate" phase of matter reminiscent of both a solid and a liquid, hence the name liquid crystal. These substances have unique optical properties that are affected by heat and voltage. See the Insight on LCDs—liquid crystal displays—in Chapter 24.)

The printed circuit is a combination of silver and graphite that appears as a dark film. It is designed so as to produce graduated resistances. For example, for the AA battery tester, the circuit film is tapered, being narrow at the bottom and wide at the top. The resistance per unit length ($R/L = \rho/A$) varies accordingly, with the greatest resistance at the narrow (small A) bottom.

You test a battery by inserting it into the back of the package so as to make contact with the ends of the tapered resistor. The printed circuit strip then carries a current I, the magnitude of which depends on the condition of the battery, and joule heat raises the temperature of the resistor. Since $P = I^2R$, the heating for a given current is greatest at the bottom, where the resistance is highest, and least at the top, where the resistance is lowest (Fig. 2).

The heating causes a reaction in the liquid crystal, which results in the thermometer effect on the display side of the tester. The black liquid crystal becomes transparent at a critical temperature. The "color change" occurs when the underlying yellow strip becomes visible through the transparent overcoat. The scale rates the battery as "good" if it supplies enough current to heat the complete liquid crystal strip above the critical temperature, and indicates "replace" if the critical heating occurs only at the bottom of the strip.

FIGURE 1   Battery tester
The test battery is inserted into the back of the package and contact is made by pressing with the fingers. The rising yellow stripe indicates the condition of the battery—the longer the stripe, the better the battery.

FIGURE 2   Resistance and joule heat
As seen in the back of the test package, a printed circuit provides a calibrated resistance. In this AA battery tester, the circuit film is tapered, being narrow at the bottom and wide at the top. It thus has the greatest resistance at the bottom.

TABLE 17.2	Typical Power and Current Requirements for Household Appliances (120 V)				
Appliance	*Power*	*Current*	*Appliance*	*Power*	*Current*
Air conditioner, room	1500 W	12.5 A	Microwave oven	625 W	5.2 A
Air conditioning, central	5000 W	41.7 A*	Radio-cassette player	14 W	0.12 A
Blender	800 W	6.7 A	Refrigerator, frost-free	500 W	4.2 A
Coffee maker	1625 W	13.5 A	Stove, top burners	6000 W	50.0 A*
Dishwasher	1200 W	10.0 A	Stove, oven	4500 W	37.5 A*
Electric blanket	180 W	1.5 A	Television, color	100 W	0.83 A
Hair dryer	1200 W	10.0 A	Toaster	950 W	7.9 A
Heater, portable	1500 W	12.5 A	Water heater	4500 W	37.5 A*

*High-power appliances such as these are typically wired to 240 V house supply to reduce the current to half these values (see Section 18.5).

Electric light bulbs are rated directly in watts (power), for example, 100 W or 60 W (•Fig. 17.10a). In such bulbs, electrical energy excites the atoms of the filament, and the excited atoms then radiate away the extra energy as a combination of light and heat. The greater the wattage of a bulb, the greater the energy consumption per unit time (joules per second). Incandescent lamps are relatively inefficient as light sources—only about 5% of the electrical energy is converted to visible light. Most of the energy produced is invisible infrared radiation and heat (Sections 11.4 and 20.4).

Electrical appliances are tagged or stamped with their power ratings. Either the voltage and power requirements or the voltage and current requirements are given (•Fig. 17.10b). In either case, the current, power, and effective resistance may be found using Eqs. 17.8 and 17.9. The power requirements of some household appliances are given in Table 17.2 on p. 544. You should keep in mind that lamps and appliances generally operate on alternating current, which will be discussed in Chapters 20 and 21. However, the current–voltage equations for alternating current have the same form as those for direct current as long as we are calculating average quantities.

EXAMPLE 17.5 ■ A LIGHT BULB: POWER AND CURRENT

A 60-W light bulb operates on 120-V household voltage. [Although in reality this voltage varies with time at 60 Hz, 120 V is the *average* voltage. We will assume it to be a constant value here.] (a) How much average current does the light bulb draw? (b) What is the resistance of the bulb at its operating temperature?

Solution. We can use the known power ratings and the voltage to determine the current flow, and then the voltage and the current to calculate the resistance as follows.

Given: $P = 60$ W   *Find:* (a) I (average current)
         $V = 120$ V   (b) R (resistance)

(a) The current may be conveniently found using $P = IV$:

$$I = \frac{P}{V} = \frac{60 \text{ W}}{120 \text{ V}} = \frac{60 \text{ J/s}}{120 \text{ J/C}} = 0.50 \text{ C/s} = 0.50 \text{ A}$$

(b) Using the definition of resistance, we have

$$R = \frac{V}{I} = \frac{120 \text{ V}}{0.50 \text{ A}} = 2.4 \times 10^2 \text{ } \Omega$$

Alternatively, if you had not been asked for the current, you could have jumped directly to the resistance as $R = V^2/P = (120 \text{ V})^2/60 \text{ W} = 2.4 \times 10^2 \text{ } \Omega$.

Follow-up Exercise. In this Example, suppose there was a brownout in which the household voltages in your neighborhood fell to 90 V. Assuming the light bulb to have a constant resistance independent of temperature, to what would its power output drop? In actuality, the drop would be less than your answer would indicate. What does this tell you about how the resistance of tungsten (the filament material) varies with temperature?

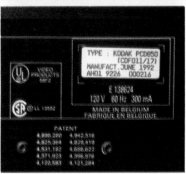

•FIGURE 17.10 Power ratings
(a) Light bulbs are rated directly in watts. This 60-W bulb uses 60 J of energy each second. (b) Appliance ratings list either voltage and power or voltage and current. In either case, the current, power, and effective resistance may be found. Here, one appliance is rated at 120 V and 18 W and the other at 120 V and 300 mA. Can you compute the current for the former and the power required for the latter?

CONCEPTUAL EXAMPLE 17.6 ■ AN ELECTRIC HEATER: RESISTANCE AND JOULE HEAT

An electric bathroom heater is constructed using a coil of wire. For efficient operation, the heater coil should have relatively (a) low resistance, (b) high resistance, or (c) the resistance doesn't matter. *Clearly establish the reasoning and physical principle(s) used in determining your answer before checking it below. That is, **why** did you select your answer?*

Reasoning and Answer. This problem is simplified by choosing the most convenient form of Eq. 17.9. If we use the form $P = I^2R$, we might be fooled into thinking that *increasing* the resistance would increase the power consumption, and thus the joule heating. However, note that power also increases with the square of the current—and current, being inversely proportional to resistance ($I = V/R$), increases with *lower* resistance.

We might conclude (correctly) that because of the I^2 dependence, this second effect will outweigh the first. However, a more direct approach is to use the form of Eq. 17.9 that eliminates the current (which is variable here) and gives the power in terms of the resistance and the voltage (which we can assume to be household voltage and therefore fixed): $P = V^2/R$. From this we can see clearly that, for a circuit with fixed voltage, power is inversely proportional to resistance. Thus, to promote joule heating, the heating element should have a relatively low resistance, and the answer is (a).

Follow-up Exercise. Two heater coils of equal length are made of the same material, but one uses narrower gauge wire than the other. If the wire in one coil has a diameter only 80% that of the other, how do their power outputs compare when they are connected to the same voltage source? Express the relationship in the form of a ratio (thin to thick wire).

EXAMPLE 17.7 ■ A POTENTIALLY DANGEROUS REPAIR: JOULE HEAT REVISITED

A hair dryer is rated at 1200 W for 115 V (average, assumed constant here). The uniform wire filament breaks near one end, and the owner repairs it by removing one section near the break and reconnecting it. The filament then has a length 10.0% shorter than its original length. What effect will this have on the heater's power output?

Solution. We expect the resistance to be less and therefore the current and power to increase. Listing the numerical values, we have

Given: $P_o = 1200$ W   *Find:* P (new power output)
$\quad\quad\quad V = 115$ V

Because 10.0% of the filament's length is gone, it will have 90.0% of its original resistance (R_o) after being reconnected, or $R = 0.900\,R_o$. This means that the current will increase, since the applied voltage is the same ($V = V_o$). We can write this as $V = IR = I_oR_o = V_o$ and therefore determine that the current increases by 11% since

$$I = \left(\frac{R_o}{R}\right)I_o = \left(\frac{1}{0.900}\right)I_o = (1.11)I_o$$

Under the initial conditions, the current was $I_o = P_o/V = 1200$ W/115 V = 10.4 A. Then the new current will be

$$I = (1.11)I_o = (1.11)(10.4\ \text{A}) = 11.6\ \text{A}$$

The power output could be computed directly from $P = IV$ or by computing the resistance and then using Eq. 17.9. By the latter method, the original resistance was

$$R_o = \frac{V}{I_o} = \frac{115\ \text{V}}{10.4\ \text{A}} = 11.1\ \Omega$$

The new resistance is therefore

$$R = (0.900)R_o = (0.900)(11.1\ \Omega) = 9.99\ \Omega$$

For the repaired dryer, then,

$$P = I^2R = (11.6\ \text{A})^2(9.99\ \Omega) = 1.34 \times 10^3\ \text{W}$$

The power output of the dryer has been increased by 150 W. *You should not attempt such a repair job.* With reduced resistance and increased current, the filament could overheat and possibly start a fire.

Follow-up Exercise. Suppose that because of a parts shortage, the owner of the hairdryer in this Example replaced the coil with one of the same length and material but made with wire that was 10.0% thicker. How would this affect the power of the dryer? Would such a repair create a potentially dangerous situation? Explain.

People often complain about their electric bills. But what do we actually buy from the electric or power company? As you may know, we pay for electricity measured in **kilowatt·hours (kWh).** Let's look at our defining equation of work and power to see what this is a unit of. Recall that power is the time rate of doing work, $P = W/t$ or $W = Pt$, so work has the units of watt·second (power × time). Converting this to a larger derived unit of kilowatt-hour, we see that the kWh is a unit of work (or energy), equivalent to 3.6 megajoules:

$$1 \text{ kWh} = (1000 \text{ W})(3600 \text{ s}) = (1000 \text{ J/s})(3600 \text{ s}) = 3.6 \times 10^6 \text{ J}$$

Thus, we pay the "power" company for electrical energy that we use to do work. We don't really pay for "power" because that is only the rate. (Similarly we do not pay the water company for the rate at which they deliver water, only for the total number of gallons delivered.) Let's now take a look at the cost of electricity.

EXAMPLE 17.8 ■ THE PRICE OF FRESHNESS: ELECTRIC POWER COST

If a frost-free refrigerator runs 15% of the time, how much does it cost to operate per month if the power company charges 11¢ per kilowatt-hour? (Assume 30 days in a month.)

Solution. We will work in kWh because the joule is a tiny unit when talking about practical electrical energy delivery.

Given: Frost-free refrigerator   *Find:* Operating cost per month
(see Table 17.2)

Since the refrigerator operates 15% of the time, in one day it runs

$$t = (0.15)(24 \text{ h}) = 3.6 \text{ h}$$

Taking the power requirement of the refrigerator to be 500 W (see Table 17.2), the electrical work done, or the average energy expended *per day* is, since $P = W/t$,

$$W = Pt = (500 \text{ W})(3.6 \text{ h}) = 1800 \text{ Wh} = 1.8 \text{ kWh}$$

Then

$$\left(\frac{1.8 \text{ kWh}}{\text{day}}\right)\left(\frac{\$0.11}{\text{kWh}}\right) = \frac{\$0.20}{\text{day}} \text{ or } 20¢ \text{ per day}$$

For a 30-day month, we have

$$\left(\frac{\$0.20}{\text{day}}\right)\left(\frac{30 \text{ day}}{\text{month}}\right) = \$6.00 \text{ per month}$$

Follow-up Exercise. Compare the cost (to the nearest penny) to run the refrigerator in this Example to the cost of leaving a 50-W porch light on every night of a 30-day month. (Assume the nights are 12 hours long and the rate charged by the power company is the same.)

The cost of electricity varies around the country. On the average, it ranges from about 8¢ to 18¢ per kilowatt·hour. Do you know the price of electricity in your locality? Check an electric bill to find out. The Insight on electrical costs (p. 548) gives a comparison you might find interesting.

Electrical Efficiency

In our electrical society, use of and demand for electricity are continually increasing. About 25% of the electricity generated in the United States goes into lighting (•Fig. 17.11). This is roughly equivalent to the output of 100 electrical generating (power) plants. Refrigerators consume about 7% of the electricity produced in the United States (the output of about 25 power plants).

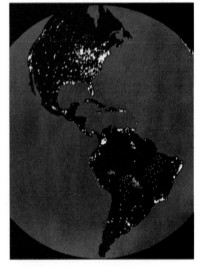

•**FIGURE 17.11 All lit up**
A satellite image of the Americas at night. Can you identify the major population centers in the United States and elsewhere? The red spots across part of South America indicate large-scale burning of vegetation. The small yellow spot in Central America shows burning gas flares at oil production sites. At the top right edge you can just glimpse a few of the white city lights of Europe. The image was recorded by a visible/infrared system.

The Cost of Portable Electricity

As mentioned following Example 17.8, the cost of household electricity ranges from about 8¢ to 18¢ per kilowatt-hour, depending on where one lives. The example also shows that electrical energy prices are still a bargain, considering the work that electricity does and the convenience it provides.

As noted earlier, batteries are widely and increasingly used to supply electrical energy. They power toys, radios, TV remote controls, and a variety of other portable devices. Have you ever thought about the cost of this "portable" electricity, as compared to that of household electricity? Let's take a look and make a general cost comparison. To do this, we need some data on batteries.

For one thing, we need to know the battery cost. This varies a great deal with location and quantity purchased. Focusing on just the AA-cell and the D-cell, a brief survey in the author's (JDW) area showed average costs to be about those listed in Table 1.

TABLE 1	Battery Data		
	Cost	Rating	Voltage
AA-cell	$0.95	2450 mA-h = 2.45 A·h	1.5 V
D-cell	$1.15	14,250 mA-h = 14.24 A·h	1.5 V

Then, we need to know how much "electricity" we get out of a battery. Batteries are rated in ampere (amp)·hours (A·h). This allows you to figure the lifetime of a battery for a given current (rate) output—for example, that needed to operate a flashlight: battery life (h) = amp·h rating/current output. Thus, the greater the current output, the shorter the battery life, as one would expect. The amp·hour ratings are not usually listed on batteries or their packages, but they are available from manufacturers, and some typical values are given in Table 1. (The ratings were furnished in units of mA·h.)

Now, notice that amp·hour is current × time (It), and recall that $q = It$. Hence, the battery rating is a measure of the charge supplied over the chemical lifetime of a battery. Converting the ratings to amp·s, which is equivalent to one coulomb (C), we have

AA–cell: $q = It = 2.45\,\text{A}\cdot\text{h}(3.60 \times 10^3\,\text{s/h}) = 8.82 \times 10^3\,\text{C}$
D–cell: $q = It = 14.25\,\text{A}\cdot\text{h}(3.60 \times 10^3\,\text{s/h}) = 5.13 \times 10^4\,\text{C}$

The energy supplied by a battery is given by $W = qV$, so the energy outputs over the lifetimes of the batteries are

AA–cell: $qV = (8.82 \times 10^3\,\text{C})(1.5\,\text{V})$

$$= 1.3 \times 10^4\,\text{J}\left(\frac{1\,\text{kWh}}{3.6 \times 10^6\,\text{J}}\right) = 0.0036\,\text{kWh}$$

D–cell: $qV = (5.13 \times 10^4\,\text{C})(1.5\,\text{V})$

$$= 7.7 \times 10^4\,\text{J}\left(\frac{1\,\text{kWh}}{3.6 \times 10^6\,\text{J}}\right) = 0.021\,\text{kWh}$$

where the conversions to kWh are made for comparison purposes. The cost of portable (battery) electricity is then about

AA–cell: price/kWh = $0.95/ 0.0036 kWh = $260 per kWh
D–cell: price kWh = $1.15/ 0.021 kWh = $55 per kWh

Compared to $0.08–$0.18/kWh for household electrical energy costs, we pay a great deal for battery convenience.

This huge consumption of electricity has prompted the federal and many state governments to set minimum efficiency limits for refrigerators, freezers, air conditioners, water heaters, dishwashers, heat pumps, and so forth (•Fig. 17.12). It is estimated that these new standards have saved over 3×10^{10} kilowatt·hours of electric energy.

Also, more efficient fluorescent lighting is being developed. The most efficient fluorescent lamp now in use consumes about 25–30% less energy than the average fluorescent lamp and roughly 75% less energy than incandescent lamps with an equivalent illumination output. Researchers are using new techniques in an effort to develop fluorescent lamps with even better efficiencies.

EXAMPLE 17.9 ■ WHAT WE CAN SAVE: INCREASING ELECTRICAL EFFICIENCY

Most modern power plants produce electricity at a rate of about 1.0 GW (electric power output). Estimate how many fewer power plants (running continuously) the state of California would need if all its households switched from the 500-W refrigerators of Example 17.8 to more efficient 400-W refrigerators. (Assume 6 million homes with an average of 1.2 refrigerators per home.)

Solution. We have seen in Example 17.8 that the 500-W refrigerators need 1.8 kWh per day to run. For the state this would be

$$\left(\frac{1.8\ \text{kWh/day}}{\text{refrig}}\right)(6 \times 10^6\ \text{homes})\left(\frac{1.2\ \text{refrig}}{\text{home}}\right) = 1.3 \times 10^7\ \frac{\text{kWh}}{\text{day}}$$

The more energy efficient refrigerators would use only 80% of this, or approximately 1.0×10^7 kWh/day. The difference is the rate at which electric energy is saved or 2.6×10^6 kWh/day. One 1.0 GW power plant produces

$$(1.0\ \text{GW/plant})\left(24\ \frac{\text{h}}{\text{day}}\right) = (10^6\ \text{kW/plant})\left(24\ \frac{\text{h}}{\text{day}}\right) = 2.4 \times 10^7\ \frac{(\text{kWh/plant})}{\text{day}}$$

So the replacement refrigerators would save about

$$\frac{2.6 \times 10^6\ \text{kWh/day}}{2.4 \times 10^7\ \text{kWh/plant-day}} = 0.11\ \text{plants}$$

or about 1/9 the output of a typical modern plant. Note that this saving would result from a change in just a single domestic appliance. Developing (and using!) more efficient electrical appliances is one way to avoid having to build new electric generating plants as population and electrical usage increase.

Follow-up Exercise. Electric and gas water heaters are often said to be equally efficient—typically, about 95%. In reality, while gas water heaters are capable of 95% efficiency, it might be more accurate to describe electric water heaters as only about 33% efficient, even though about 95% of the electrical energy they consume is transferred to the water in the form of heat. Explain. [*Hint: Where does the energy for an electric water heater come from? Recall the discussion of electrical generation in Section 12.4 and Carnot efficiency in Section 12.5*]

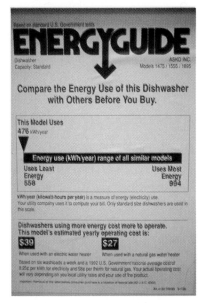

•**FIGURE 17.12 Energy guide**
Consumers are made aware of appliance efficiencies in terms of the average yearly cost of operation. Sometimes the yearly cost is given for different kilowatt-hour (kWh) rates, which vary around the country.

Chapter Review

Important Terms

battery   533
cathode   533
anode   533
electromotive force (emf)   534
terminal voltage   534
direct current (dc)   534
complete circuit   535
conventional current   536

electric current   536
ampere (A)   536
drift velocity   537
resistance   538
Ohm's law   538
ohm (Ω)   538
resistivity   539
conductivity   540

temperature coefficient
   of resistivity   540
superconductivity   541
electric power   543
joule heat (I^2R) loss   543
kilowatt·hour (kWh)   547

Important Concepts

• Electric current is the net rate at which charge flows past a given point.

• The electromotive force of a battery (or any power supply) is the voltage difference between the two terminals when there is no current.

• Terminal voltage is the actual voltage difference across the terminals of a battery or power supply when there is a current. It is usually less than the emf because of internal resistance.

• The direction of conventional current is that in which positive charge would move. In metals, the actual current is carried by electrons moving in the opposite direction.

• The electrical resistance of a circuit element is defined as the voltage difference across the element divided by the resulting current flow ($R = V/I$).

• Ohm's law is said to apply to a circuit element if that element exhibits a constant electrical resistance. Ohm's law is commonly written $V = IR$ where R is constant.

• The overall electrical resistance of a circuit element depends on its geometry (cross-sectional area and length) as well as the intrinsic electrical resistive properties of its material (resistivity). For most metallic conductors, resistivity increases with increasing temperature.

• Electric power is the rate at which work is done or energy is transferred. The power delivered to a circuit element depends on its resistance and the voltage difference across it.

• Joule heat is the heat energy expended in current-carrying circuit elements.

Important Equations

Electric Current (definition):

$$I = \frac{q}{t} \qquad (17.1)$$

Electrical Resistance (definition):

$$R = \frac{V}{I} \qquad (17.2)$$

Ohm's Law:

$$V = IR \ (R = \text{constant})$$

Resistivity:

$$\rho = \frac{RA}{L} \qquad (17.3)$$

Conductivity:

$$\sigma = \frac{1}{\rho} \qquad (17.4)$$

Temperature Coefficient of Resistivity (α constant):

$$\rho = \rho_o(1 + \alpha\Delta T) \ \text{ or } \ \Delta\rho = \rho_o\alpha\Delta T \qquad (17.6)$$
$$(\text{where } \Delta\rho = \rho - \rho_o)$$

Temperature-dependence of Resistance (α constant):

$$R = R_o(1 + \alpha\Delta T) \ \text{ or } \ \Delta R = R_o\alpha\Delta T \qquad (17.7)$$
$$(\text{where } \Delta R = R - R_o)$$

Electric Power:

$$P = IV = \frac{V^2}{R} = I^2R \qquad (17.8)$$

Exercises

Unless otherwise indicated, in this chapter assume all batteries have negligible internal resistance.

17.1 Batteries and Direct Current

1 When a battery is placed into a complete circuit, the voltage difference across its terminal is known as its (a) emf (b) terminal voltage (c) power (d) all of these.

2 Batteries A and B are both classified as 6-V batteries. When they are both connected (separately one at a time) to the same circuit element, forming a complete circuit, the terminal voltage of A is measured to be considerably less than B. What does this tell us about the batteries? (a) B is younger than A, (b) A has a lower internal resistance than B, (c) B has a lower internal resistance than A, (d) their emfs are really different.

3 When three 6-V batteries are connected in parallel, the output voltage of the combination is (a) 6 V, (b) 2 V, (c) 18 V, (d) none of these.

4 Explain clearly why a battery cannot be made using the same two metals for electrodes.

5 In the text we mentioned that our hypothetical battery required a chemical membrane to prevent the two ions from each electrode from mixing. Why is this necessary?

6 Is the electromotive force really a force? Explain.

7 A battery has a small internal resistance r (•Fig. 17.13). Explain why the emf of a battery is the terminal voltage for an open circuit condition.

8 Is it possible for a 12-V car battery to measure 12-V for its emf and yet not work when put into a circuit it was designed for? If so, explain.

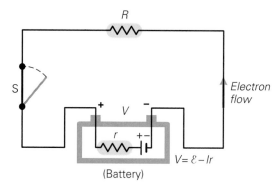

•**FIGURE 17.13 Emf and terminal voltage**
See Exercises 7 and 27.

9 ■ (a) Two 3.0-V dry cells are connected in series. What is the total voltage of the combination? (b) What would be the total voltage if the cells were connected in parallel?

10 ■ What is the voltage across six 1.5-V batteries when they are connected (a) in series and (b) in parallel?

11 ■■ Two 12-V batteries and one 24-V battery are connected in series. (a) What is the voltage across the whole arrangement? (b) What arrangement of these three batteries would give a total voltage of 24 V?

12 ■■ Given three batteries with voltages of 1.0 V, 3.0 V, and 12 V, how many different voltages could be obtained using one or more of the batteries connected in series, and what are these voltages?

17.2 Current and Drift Velocity

13 The unit of electrical current is (a) C, (b) C/s, (c) A, (d) both (b) and (c).

14 ■ A net charge of 30 C passes through a cross-sectional area of a wire in 2.0 min. What is the current carried by the wire?

15 ■ How long would it take for a net charge of 2.5 C to pass through a cross-sectional area of wire so as to produce a steady current of 5.0 mA?

16 ■ A small toy car draws 0.50 mA of current from a 3.0-V nicad battery. In 10 minutes of operation, (a) how much net charge flows in the circuit and (b) how much energy is lost by the battery?

17 ■ A car motor draws 50 A when starting up. If the start-up time is 1.5 s, what is the net charge passing through a spot in the circuit in this time?

18 ■■ A net charge of 20 C passes through a wire with a cross-sectional area of 0.30 cm² in 1.25 min; a net charge of 30 C passes through another wire with a cross-sectional area of 0.450 cm² in 1.52 min. Which wire carries more current and how much more?

19 ■■ A battery is rated at 60 A·h at 3.0 A. Thus the life of the battery is 20 h (60 A·h/3.0 A) when fully charged. How much charge will the battery deliver in this time?

20 ■■ What is the net number of electrons that move past a point in a wire carrying 500 A of current in 4.0 min?

21 ■■ In 0.7 min, if 6.7 C of electrons flow to the right past a point at the same time 8.3 C of protons flow to the left, what is the direction and magnitude of the current past that point?

22 ■■ In a linear accelerator, a 9.5 μA current of protons hits the target. (a) How many protons per second hit the target? (b) What is the energy delivered to the target each second if the protons have a kinetic energy of 20 MeV?

17.3 Resistance and Ohm's Law*

23 The unit of resistance is the (a) volt per ampere, (b) watt, (c) volt

24 For an ohmic resistor, current and resistance (a) vary with temperature, (b) are directly proportional, (c) are independent of voltage, (d) none of these.

25 Electrical conductivity (σ) is the reciprocal of the resistivity, (b) has units of $\Omega \cdot m$, (c) is equal to $1/R$, (d) does not vary with temperature.

26 If the voltage (V) were plotted versus current (I) for two ohmic conductors with different resistances on the same graph, how could you tell the less resistive one?

*Assume that the temperature coefficients of resistivity given in Table 17.1 apply over large temperature ranges.

27 ■ A battery labeled as 12.0-V is measured to supply 1.90 A to a 6.00-W resistor (Fig. 17.13). (a) What is the terminal voltage of the battery? (b) What is its internal resistance?

28 ■ What is the emf of a battery with an internal resistance of 0.15 Ω if it delivers 1.5 A to an externally connected 5.0 W resistor?

29 ■ (a) If a connecting wire in a circuit is replaced with one of the same material that is twice as long and has twice the cross-sectional area, how will the current in the wire be affected? Assume that the potential difference remains constant. (b) How will the current be affected if the new wire has the same length but half the diameter of the old one?

30 ■ In a semiconductor the number of conducting charges can vary drastically with temperature? What does a negative temperature coefficient of resistivity for a semiconductor imply about how the number of charge carriers varies with temperature in such a material?

31 ■ Some states allow the use of aluminum wire in houses in place of copper. If you wanted the resistance of your wires to be the same as copper, how would the thickness of your aluminum wires have to compare with those of the equivalent copper ones?

32 ■ How much current is drawn from a 12-V battery when a 150-Ω resistor is connected across its terminals?

33 ■ What voltage must a battery have to produce 0.50 A of current through a 2.0-Ω resistor?

34 ■ A circuit component with a resistance of 14 Ω draws 1.5 A of current when connected to a dc power supply. What is the voltage of the power supply?

35 ■■ A copper wire is 0.6 m long and has a diameter of 0.10 cm. What is the resistance of the wire?

36 ■■ A 100-Ω resistor is placed in a circuit with two identical batteries connected in series. If it draws 1.2 A, what is the terminal voltage of each battery?

37 ■■ An automobile starter is connected to a 12-V battery. If the starter has an effective resistance of 2.5 Ω, what is the rate at which electrons drift past a point in the circuit?

38 ■■ A certain material is formed into a long rod with a square cross section 0.50 cm on a side. When a voltage of 100 V is applied across a length of the rod 2.0 m long, 5.0 A of current flows. What are (a) the resistivity and (b) the conductivity of the material? Is it a conductor, insulator, or semiconductor?

39 ■■ Two copper wires have equal cross-sectional areas and lengths of 2.0 m and 0.50 m. What is the ratio of the current flow through the shorter to that through the longer if they are connected to the same power supply?

40 ■■ Two copper wires have equal lengths, but the diameter of one is triple that of the other. What is the ratio of the resistance of the thicker to that of the thinner wire?

41 ■■ The wire in a heating element of an electric stove burner has an effective length of 0.75 m and a cross-sectional area of 2.0×10^{-6} m^2. If the wire is made of iron, what is its resistance when operating at a temperature of 380°C?

42 ■■ What is the percentage variation of the resistivity of copper over the range from room temperature (20°C) to 100°C?

43 ■■ A copper wire has a resistance of 25 mΩ at 20°C. When the wire is carrying a current, joule heat causes its temperature to increase by 27 C°. What is the change in the wire's resistance?

44 ■■ What is the fractional change in the resistance of an aluminum wire 2.0 m long over a temperature range from the freezing point to the boiling point of water?

45 ■■ When a resistor is connected to a 12-V source, it draws 185 mA of current. When the same resistor is connected to a 90-V source, it draws 1.38 A of current. Is the resistor ohmic? Justify your answer mathematically.

46 ■■ What length of wire is required to make a 20.0 Ω resistor by winding AWG (American Wire Gauge) No. 10 nichrome wire in a coil? (AWG No. 10 wire has a diameter of 2.588 mm.)

47 ■■ A 1.0-m length of copper wire with a diameter of 2.0 mm is stretched out; its length increases by 25% while its cross-sectional area decreases but remains uniform. Is the resistance of the wire the same before and after? Justify your answer mathematically.

48 ■■ A particular application requires a 20-m length of aluminum wire to have a resistance of 1.0 mΩ. What diameter should the wire have?

49 ■■ A 20-m length of 1.0-mm-diameter nichrome wire is wound in a coil and connected to a 1.5-V flashlight battery. How much current will initially flow in the coil?

50 ■■■ A carbon resistor has a resistance of 150 Ω at room temperature. In a circuit with a voltage of 12 V across it, its operational temperature is 90°C. What is the difference in current through the resistor at 90°C and at room temperature when the circuit is initially closed? Which is larger?

51 ■■■ A rod of silicon is used in a circuit that carries a current of 0.50 A with a potential difference of 6.0 V at 20°C. If the temperature of the rod is increased to 25°C, how much current flows through it?

52 ■■■ Pieces of carbon and copper have uniform cross sections and the same resistance at room temperature.

(a) If the temperature of each piece is increased by 10.0 C°, which will have the greater resistance? (b) What is the ratio of the resistances?

53 ■■■ An electrical resistance thermometer made of platinum has a resistance of 5.0 Ω at 20°C and is in a circuit with a 1.5-V battery. What is the change in the current in the circuit when the thermometer is heated to 2020°C?

17.4 Electric Power

54 Electric power has units of (a) A$^2 \cdot \Omega$, (b) J/s, (c) V$^2/\Omega$, (d) all of these.

55 Show that the watt per amp is the same thing as the volt by using SI units for all the quantities.

56 If the voltage across an ohmic resistor is doubled, the power expended in the resistor (a) increases by a factor of 2, (b) increases by a factor of 4, (c) decreases by one-half, (d) none of these.

57 Assuming your hair blower/dryer obeys Ohm's law, what would happen if you plugged it directly in the 240-V outlets in Europe if it is made to be used in the 120-V outlets of the United States?

58 Most light bulb filaments are made of tungsten and are approximately the same length. What would be different about the filament in a 60-W bulb compared to a 40-W bulb?

59 ■ An appliance is rated for 2.5 A at 120 V. How much electric power does it require?

60 ■ How much power will be expended in a 10-kΩ resistor when it is connected across a 120-V voltage source?

61 ■ A hair dryer is rated for 1200 W at 120 V. (a) How much current does it use? (b) What is its resistance?

62 ■ Show that (volt)^2/ohm has SI units of power (watt).

63 ■ An electric heater is designed to produce 50 kW of heat when connected to a 240-V source. What must the resistance of the heater be?

64 ■■ How long would the heater in Exercise 63 take to heat 50 gallons of water from 20°C to 80°C assuming it is 90% efficient?

65 ■■ An electric toy rated at 2.5 W is operated by four 1.5-V batteries connected in series. (a) How much current does the toy draw? (b) What is its effective resistance?

66 ■■ A welding machine draws 18 A of current at 240 V. (a) How much energy does the machine use each second? (b) What is its effective resistance?

67 ■■ An electric water heater automatically operates for 2.0 h each day. (a) If the cost of electricity (to the nearest dollar) is 12¢/kWh, what is the cost of operating the heater during a 30-day month? (b) What is the effective resistance of a typical water heater? [*Hint:* See Table 17.2.]

68 ■■ What resistor should be used to generate 10 kJ of heat per minute when connected to a 120-V source?

69 ■■ A 1500-W hair dryer is used for 10 min every day. If the cost of electricity is 12¢/kWh, what is the cost (to the nearest dollar) of using the dryer for a year (365 days)?

70 ■■ A 240-V air conditioner unit draws 15 A of current. If it operates for a period of 20 min, (a) how much energy in kWh does it use? (b) If the cost of electricity is 12¢/kWh, what is the cost (to the nearest penny) of operating the unit for this time?

71 ■■ Two resistors of 100 Ω and 25 kΩ are rated for maximum wattages of 1.5 W and 0.25 W, respectively. What is the maximum voltage that can be safely applied to each resistor?

72 ■■ A wire 5.0 m long and 3.0 mm in diameter has a resistance of 100 Ω. A potential difference of 15 V is applied across the wire. Find (a) the current in the wire, (b) the resistivity of its material, and (c) the rate at which heat is being produced in it.

73 ■■ A coil of tungsten wire initially dissipates 500 W of power when connected to a voltage source. In a short time, the temperature of the coil increases by 150 C° because of joule heat. What is the corresponding change in the power?

74 ■■ A 0.50-kΩ resistor is in a circuit with three 12-V batteries. What is the joule heat loss of the resistor (a) if the batteries are connected in series and (b) if the batteries are connected in parallel?

75 ■■ A 10-W light bulb in a refrigerator operates on 120-V household voltage. What is the resistance of this bulb?

76 ■■ A 5.5-kW water heater operates on 240 V. (a) Should the heater circuit have a 20-A or a 30-A circuit breaker? (A circuit breaker is a safety device that opens the circuit at its rated current.) (b) Assuming 85% efficiency, how long will the heater take to heat the water in a 55-gal tank from room temperature to 80°C?

77 ■■ A student uses an immersion heater to heat a cup of water (300 mL) from 20°C to 80°C for tea. If it is 75% efficient and takes 2.5 min, what is the resistance of the heater? (Assume 120-V household voltage.)

78 ■■ By what factor is the power expended in a 150-Ω ohmic resistor increased when the voltage across it is increased from 15 V to 60 V?

79 ■■ In an electrical brownout, a light bulb's output drops from its usual 60 W at 120 V operating voltage. If the voltage was cut in half and the power dropped to 20 W during the brownout, what is the ratio of the bulb's resistance at full power to the resistance during the brownout?

80 ■■ A water pump is needed to deliver useful work (emptying a flooded basement) at a rate of 2.00 kW. If it is wired to a 240-V source and the manufacturer claims it is 84% efficient, (a) how much current does it draw, and (b) what is its resistance?

81 ■■ What is the efficiency of a 350-W motor that draws 4.0 A from a 120-V line?

82 ■■■ An immersion heater with a resistance of 9.0 Ω operates on 120 V. The heater is placed in a cup of water (0.25 L) at room temperature (20°C) to heat it up to make coffee. Assuming that no heat loss occurs, how long will the heater have to operate to bring the water to a boil?

83 ■■■ How long will the heater in Exercise 82 have to operate to boil all the water away after reaching the boiling point?

84 ■■■ A 500-W heating element made of nichrome wire whose diameter is 4.0 mm operates on 120 V. (a) What is the length of the wire? (b) What power would be dissipated if the voltage were reduced to 105 V?

85 ■■■ Find the total monthly (30-day) electric bill (to the nearest dollar) for the following household appliance usage if the utility rate is 9.5¢/kWh: central air conditioning runs 30% of the time; a blender is used 0.50 h/month; a dishwasher is used 8.0 h/month; a microwave oven is used 15 min/day; a frostfree refrigerator runs 15% of the time; a stove (range and oven) is used a total of 10 h/month; and a color television is operated 120 h/month. (Use the information given in Table 17.2.)

Additional Exercises

86 In electronics, the "characteristic curve" of a circuit element is a graph of the current, I, on the y axis, versus the voltage applied across it, V, on the x axis. (a) Show that for an ohmic resistor this is a straight line and the slope is the reciprocal of the resistance. (b) Sketch the curves (on the same set of axes) for a semiconductor material.

87 A resistance thermometer made of platinum shows a 25% change in resistance over a certain temperature range. What is the change in temperature for this range?

88 An iron conductor with a resistance of 1.75 Ω at 20°C is connected to a 12-V source. If the conductor is placed in an oven and heated to 300°C, what is the change in the current flowing through it?

89 An electric heater that operates on 120 V is rated at 1500 W. (a) How much current does the heater draw? (b) What is the resistance of its heating element?

90 If the price of electricity is 12¢/kWh, how much would it cost (to the nearest penny) to operate ten 75-W light bulbs for 6 hours?

91 A 20-Ω, 5.0-W resistor is used in a circuit. (a) What is the maximum voltage that should be applied across the resistor? (b) Could the resistor safely carry a current of 0.60 A?

92 An aluminum wire has a mass of 10 g and a length of 50 cm. Find the current in the wire when it is connected to a 10-V source.

93 A toaster oven is rated for 1600 W at 120 V. When the filament ribbon of the oven burns out, the owner makes a repair that reduces the length of the filament by 10%. How does this affect the power output of the oven, and what is the new value?

94 The tungsten filament of an incandescent lamp has a resistance of 200 Ω at room temperature. What will the filament's resistance be at an operating temperature of 1600°C?

95 Pieces of aluminum and copper wire are identical in length and diameter. At some temperature, one of the wires will have the same resistance as the other has at room temperature. What is the temperature? (Is there more than one temperature?)

96 Find the resistance of a piece of copper wire that is 75 cm long and has a diameter of 2.0 mm.

97 An electric iron with a 14-Ω heating element operates on 120 V. How many kWh does the iron convert to heat in one hour?

98 When a wire 1.0 mm in diameter and 2.0 m in length is connected to a 3.00-V source, a current of 11.8 A flows. (a) What is the resistivity of the wire? (b) What material is the wire made of?

99 A 100-Ω resistor is rated at 0.25 W. (a) What is the maximum current the resistor can carry? (b) What is the maximum operating voltage that should be applied across the resistor?

100 A wire carries a current of 75 mA. How many electrons have to pass through a cross-sectional area of the wire in 2.0 s to produce this current?

101 When a 1.0-m length of wire is connected to a 6.0-V source, it carries 0.48 A of current. Calculate the current in a 3.0-m length of the same type of wire connected to the same voltage source.

102 What is the resistance of a 50-km length of aluminum cable with an effective diameter of 1.0 cm?

103 A current of 0.10 A exists in a single resistor in a circuit with a 40-V voltage source. (a) What is the resistance of the resistor? (b) How much power is dissipated by the resistor? (c) How much energy is dissipated in 2.0 min?

104 Find the ratio of the resistance of a copper wire to the resistance of an aluminum wire (a) if the wires have the same length and diameter and (b) if the copper wire is twice as long as the aluminum wire and has a diameter that is half that of the aluminum wire.

105 A portable radio draws 150 mA of current from a battery. If the radio is played for a half-hour, what is the net number of electrons that move through it?

106 A copper bus bar used to conduct large amounts of current at a power station is 1.5 m long and 8.0 cm by 4.0 cm in cross section. What voltage difference across the ends of the bar will cause a current of 3000 A to flow through it?

107 Why is electric power delivered over long distances at high voltages when we know they can be very dangerous?

108 •Figure 17.14 shows charge carriers having a charge q and moving with a speed v_d (drift speed) in a conductor of cross-sectional area A. Let n be the number of free charge carriers per unit volume. (a) Show that the total charge (ΔQ) free to move in the volume element shown is given by $\Delta Q = (nAx)q$. (b) Show that the current in the conductor is given by $I = nqv_dA$.

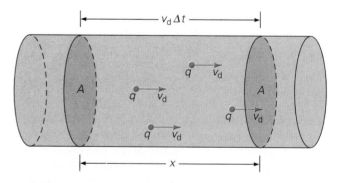

•**FIGURE 17.14 Total charge and current**
See Exercises 108 and 109.

109 A copper wire with a cross-sectional area of 13.3 mm^2 (AWG No. 6) carries a current of 1.2 A. If there are 8.5×10^{22} free electrons per cubic centimeter, what is the drift velocity of the electrons? [*Hint:* See Exercise 108.]

18 Basic Electric Circuits

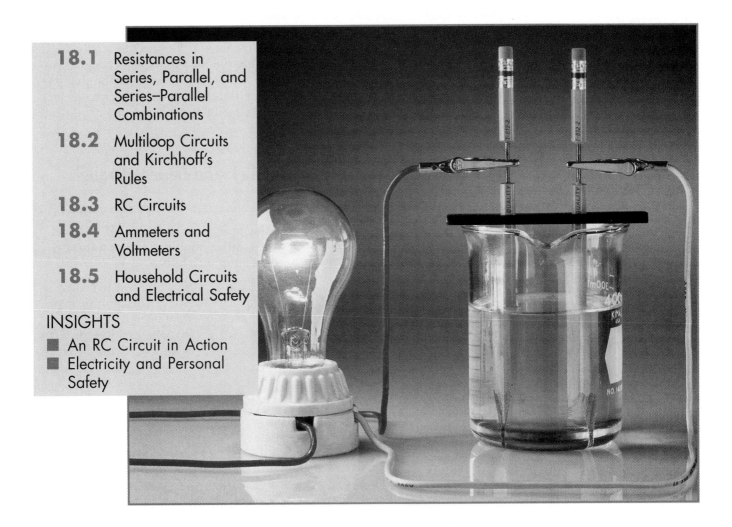

The term "electric circuit" probably brings to mind an image of wires. But metal wires are not the only thing that can conduct electricity, as the photo above shows. Since the bulb is lit, we know the circuit is complete. We can conclude, therefore, that the "lead" in a pencil (actually a form of carbon called graphite) is a good conductor of electricity. The same must be true for the liquid in the beaker—in this case, a solution of ordinary table salt. (The results would be very different if the electrodes consisted of the wooden part of the pencil or the liquid were salad oil.)

Electric circuits are of many kinds and can be designed for many purposes, from boiling water for coffee to lighting a Christmas tree. The one in the photo above is intended as a demonstration, but circuits of this type do have practical applications in the laboratory and in industry. They are used in synthesizing or purifying chemical substances, and in electroplating metals (for example, making silver plate). In this chapter you'll find out more about various kinds of electric circuits, including those that provide our household electric power, and learn how their behavior can be described in quantitative terms.

Armed with the principles learned in Chapters 15, 16, and 17, you are now ready to analyze some relatively simple circuits. This will give you an appreciation of how electricity actually works.

Circuit analysis most often deals with voltage, current, and power requirements. You will find that the way in which two or more components are connected in a circuit typically affects the voltage across those components, the current through them, and the power they consume. A circuit may be analyzed theoretically before actually being hooked up. The analysis might show that the circuit would not function properly as designed, or that there could be a safety problem (such as overheating due to joule heat).

Circuit diagrams are used to visualize circuits in order to understand their functioning. A few of these diagrams were included in figures in Chapter 17. In actuality, the wires of the circuit are not ordinarily placed in the rectangular pattern of a circuit diagram. The rectangular form is simply a convention that provides a neater presentation and easier visualization of the circuit arrangement.

We'll begin our analysis of circuits by looking at arrangements of resistive elements.

18.1 Resistances in Series, Parallel, and Series–Parallel Combinations

Objectives: To be able to (a) determine the equivalent resistance of resistors in series, parallel, and series–parallel combinations, and (b) use equivalent resistances to solve simple circuits.

The resistance symbols in a circuit diagram can represent any resistive element —a commercial resistor, a light bulb, an appliance, and so on. This discussion will consider all of these to be ohmic (that is, to have a constant operating resistance), unless otherwise stated. Also, the total resistance of all the connecting wires in a circuit will be considered to be negligible in comparison to those of the circuit elements.

Resistors in Series

In analyzing a circuit, you need to keep in mind that the sum of the voltages around a circuit loop (that is, the gains and losses with $+$ and $-$ signs, respectively) is zero. For example, for the circuit in •Fig. 18.1a the sum of the voltage drops (V_i) across the three circuit components, or resistances, equals the voltage

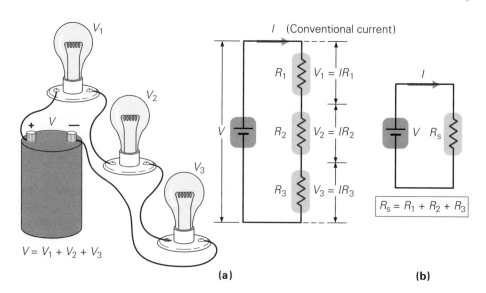

•**FIGURE 18.1 Resistors in series** (a) When resistors are connected in series, the current through each of them is the same. Also, the sum of the voltage drops across the resistors is equal to the voltage of the battery. (b) The equivalent resistance R_s of the resistors in series.

(a)

(b)

"rise" of the battery. In general, there will be a string of resistances, not necessarily just three, and we would write

$$V = \Sigma_i V_i = \Sigma_i I R_i \qquad (18.1)$$

or

$$V - \Sigma_i I R_i = 0$$

where the i summation is over the individual resistances. (The subscript is often omitted as being understood.)

The fact that the sum of the voltages around the circuit loop is zero is essentially a statement of the conservation of energy. We saw in Chapter 17 that for the charge carriers traversing the circuit, the energy given to them by a voltage source must be lost in the circuit. That is, they must have the same electric potential (conventionally designated zero) on returning to the negative terminal as they had when starting. Otherwise, the charge carriers would gain energy with each round trip of the circuit, which would violate the law of the conservation of energy.

The three resistances or resistors in Fig. 18.1 are in series, or connected end-to-end, so to speak. *When resistors are connected in series, the current is the same through all of them,* as required by the conservation of charge. Otherwise, there would be a build-up of charge somewhere along the way. The electrons entering one region of the wire repel the ones ahead of them, and the whole "stream" moves along without charge pile-up. •Fig. 18.2 shows the analogous flow of water along a smooth streambed punctuated by a series of rapids. The rapids can be thought of as gravitational "resistors," with the rocks and pebbles impeding the flow of water. The water current through each course of rapids is the same, and the sum of the potential energies lost in each is equal to the work that would have to be done to bring the water back to the top of the hill.

The total voltage drop around a circuit is equal to the sum of the individual voltage drops across the resistors. If we label their common current I, Eq. 18.1 can be written explicitly for three resistors (see Fig. 18.1):

$$\begin{aligned} V &= V_1 + V_2 + V_3 \\ &= IR_1 + IR_2 + IR_3 \\ &= I(R_1 + R_2 + R_3) \end{aligned}$$

If we wanted to replace the three resistors by a single resistor R_s (called the **equivalent series resistance**) and maintain the same current, we would need $V = IR_s$ or $R_s = V/I$. Hence the three resistors in series have an equivalent resistance R_s:

$$R_s = R_1 + R_2 + R_3$$

That is, *the equivalent resistance of resistors in series is simply the sum of the individual resistances.* The three resistors of Fig. 18.1 could be replaced with a single resistor with a resistance value of R_s without having any effect on the current or voltage source. For example, if each of the resistors in Fig. 18.1 had a value of

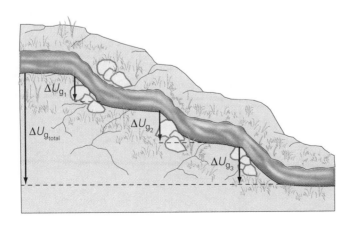

•**FIGURE 18.2 Water flow analogy to resistors in series**
Even though a different amount of gravitational potential energy is lost as the water flows through each rapids, the current (flow per unit time) is the same in each. The total loss of gravitational potential energy is the sum of the three losses. (To make this a complete circuit, some external agent, such as a pump, would have to do work to return the water to the top of the hill, restoring its original gravitational potential energy.)

10 Ω, then R_s would be 30 Ω. This makes physical sense—more resistors in a row, more overall resistance. *Note that the equivalent series resistance is larger than that of the largest resistor in the series.*

This result may be extended to any number of resistors in series:

$$R_s = R_1 + R_2 + R_3 + \cdots = \Sigma_i R_i \qquad \text{(equivalent series resistance)} \qquad (18.2)$$

QUESTION: Suppose that one of the light bulbs in the circuit in Fig. 18.1a blew out. What would happen?

ANSWER: All of the bulbs would go out because the circuit would no longer be complete. If the filament of one of the bulbs is broken or blows out, the circuit is open. An open circuit is like a drawbridge stuck in the up position: no traffic can flow across it. You can think of an open circuit as equivalent to an infinite resistance; thus the current through it is zero.

Resistors in Parallel

Resistors in parallel are wired so that voltage across them must be the same. That is, all of their "heads" (or high potential sides) are connected together and the same for their low potential sides.

Another basic way of connecting resistors in a circuit is in parallel (•Fig. 18.3a). In this case, all the resistors have common connections—that is, all the leads on one side of the resistors are attached together and the same for the leads on the other side. *When resistors are connected in parallel across a battery, the voltage drop across each resistor is the same* (and in this case equal to the voltage of the battery). However, the current from the battery divides among the different paths, as shown in Fig. 18.3a. Think of how the flow of charge must divide when it comes to a branch point in the circuit, like traffic on a highway that divides into two different forks (•Fig. 18.4a). Since charge does not pile up at the junctions, the total current flowing out of the battery breaks up into smaller currents whose sum equals the total current. Specifically, for three resistors in parallel, we have

$$I = I_1 + I_2 + I_3$$

If the individual resistances are equal, the current divides equally. But in general the resistances are not equal, so the current divides among the resistors proportionately—that is, the greatest current flows through the path of least resistance.

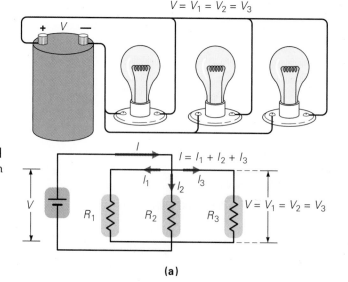

•**FIGURE 18.3 Resistors in parallel**
(a) When resistors are connected in parallel, the voltage drop across each of the resistors is the same. Note that the current from the battery divides proportionally among the resistors. **(b)** The equivalent resistance R_p of the resistors in parallel.

$$\frac{1}{R_p} = \frac{1}{R_1} + \frac{1}{R_2} + \frac{1}{R_3}$$

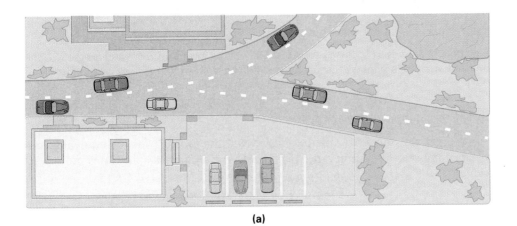

(a)

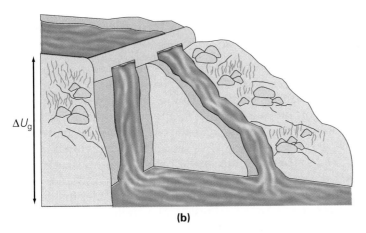

ΔU_g

(b)

•FIGURE 18.4 Parallel analogies
(a) When a road forks, the total number of cars entering the two branches each second is equal to the number arriving at the fork each second. (If this were not so, the traffic would slow and begin to back up.) **(b)** When water flows from behind a dam at a given height, the amount of gravitational potential energy that it loses in falling to the streambed below is the same regardless of the path by which it descends.

We define the **equivalent parallel resistance,** R_p, as the value of a single resistor that could replace these three and maintain the same current flow through the battery. Thus, $R_p = V/I$. The voltage drop (V) across each resistor is the same. As an analogy, imagine two separate paths for water, each leading from the top of a dam to the bottom. You can readily see that the water loses the same amount of potential energy regardless of the path it takes (•Fig. 18.4b). Thus for each individual resistor we can write $I_1 = V/R_1$. Then,

$$I = I_1 + I_2 + I_3$$
$$= \frac{V}{R_1} + \frac{V}{R_2} + \frac{V}{R_3}$$
$$= V\left(\frac{1}{R_1} + \frac{1}{R_2} + \frac{1}{R_3}\right) = \frac{V}{R_p}$$

Thus, the equivalent resistance R_p of three resistors in parallel is given (in reciprocal form) by

$$\frac{1}{R_p} = \frac{1}{R_1} + \frac{1}{R_2} + \frac{1}{R_3}$$

That is, the reciprocal of the equivalent resistance is equal to the sum of the reciprocals of individual resistors connected in parallel. The three resistors could be replaced with a single resistor with a resistance value of R_p (see Fig. 18.3b).

This result may also be generalized to include any number of resistors in parallel.

$$\frac{1}{R_p} = \frac{1}{R_1} + \frac{1}{R_2} + \frac{1}{R_3} + \cdots = \Sigma_i \frac{1}{R_i} \qquad \begin{array}{l} \textit{equivalent} \\ \textit{parallel resistance} \end{array} \qquad (18.3)$$

Equation 18.3 gives $1/R_p$. Don't forget to take the reciprocal of this value to get R_p. The units are not ohms until you invert for the final answer. (If you always carry units along in your calculations, you will be less likely to make this mistake.)

If there are only two resistors in parallel, the formula for the equivalent resistance may be written in a nonreciprocal form, which is sometimes more convenient.

$$\frac{1}{R_p} = \frac{1}{R_1} + \frac{1}{R_2} = \frac{R_1 + R_2}{R_1 R_2}$$

or

$$R_p = \frac{R_1 R_2}{R_1 + R_2} \qquad \text{(equivalent resistance, two resistors in parallel)} \qquad (18\text{-}3a)$$

An interesting fact to note is that *the equivalent resistance of resistors in parallel is always less than that of the smallest individual resistance.* You can easily demonstrate this for two resistors in parallel by putting some values in the equation. More generally, you can see that $\sum_i 1/R_i$, the sum of the reciprocals of the individual resistances, must be greater than the reciprocal of any individual resistance: $1/R_p = \sum_i 1/R_i > 1/R_{min}$, where R_{min} is the smallest individual resistance. Therefore R_p must be *less than* even the smallest individual resistance.

Note: R_p is smaller than the least resistor in parallel.

Physically, the reason that the overall or equivalent resistance becomes smaller when you wire resistors in parallel is as follows. Consider first just a battery and one resistor. Now connect a second resistor in parallel to the first. The original resistor keeps the same current because it has the same voltage across it as before. However, the new resistor now carries a current also. Therefore the *total* current into the junction (and thus out of the battery) has increased! The conclusion is that we now have more current for the same voltage, so the overall resistance of the external circuit must have dropped.

Notice that this argument does not depend on the *value* of the added resistor. If the original were $1\ \Omega$ and the new one $1\ M\Omega$, the equivalent resistance would be just slightly *less* than $1\ \Omega$! The extra resistor would allow only an added tiny trickle of current, but this would still be enough to produce a decrease in the overall circuit resistance. Think again of letting water out of a dam. If you punch one hole, you get a certain current, driven by the pressure of the water behind the dam. If you then punch even the tiniest second hole, the total water flow is increased. In effect, the overall resistance of the dam to water flow has been reduced. *Thus, series connections provide a way to increase resistance, and parallel connections a way to reduce resistance.*

EXAMPLE 18.1 ■ CONNECTIONS COUNT: RESISTORS IN SERIES AND IN PARALLEL

What is the equivalent resistance of three resistors ($1.0\ \Omega$, $2.0\ \Omega$, and $3.0\ \Omega$) when they are connected (a) in series (Fig. 18.1a), and (b) in parallel (Fig. 18.3a)? (c) What current will be delivered from a 12-V battery for each of these arrangements?

Solution. Listing the data, we have

Given: $R_1 = 1.0\ \Omega$
$R_2 = 2.0\ \Omega$
$R_3 = 3.0\ \Omega$
$V = 12\ V$

Find: (a) R_s (series resistance)
(b) R_p (parallel resistance)
(c) I (current for each case)

(a) The equivalent series resistance is simply the sum of the resistances of the three resistors (Eq. 18.2):

$$R_s = R_1 + R_2 + R_3 = 1.0 \ \Omega + 2.0 \ \Omega + 3.0 \ \Omega = 6.0 \ \Omega$$

(b) For resistors in parallel, Eq. 18.3 applies.

$$\frac{1}{R_p} = \frac{1}{R_1} + \frac{1}{R_2} + \frac{1}{R_3} = \frac{1}{1.0 \ \Omega} + \frac{1}{2.0 \ \Omega} + \frac{1}{3.0 \ \Omega}$$

With a common denominator,

$$\frac{1}{R_p} = \frac{6.0}{6.0 \ \Omega} + \frac{3.0}{6.0 \ \Omega} + \frac{2.0}{6.0 \ \Omega} = \frac{11}{6.0 \ \Omega}$$

or

$$R_p = \frac{6.0 \ \Omega}{11} = 0.55 \ \Omega$$

Note that R_p is found by inverting the value of $1/R_p$, and it is less than the smallest resistance. If you do not get such a result in parallel (for example, if the equivalent parallel resistance had come out to be 1.5 Ω, which is larger than one of the individual resistors), then you must have made a mistake!

(c) Knowing the equivalent resistance for the series arrangement gives the current directly:

$$I = \frac{V}{R_s} = \frac{12 \ V}{6.0 \ \Omega} = 2.0 \ A$$

It is interesting and instructive to consider also the individual voltages. In series, the fractional share of the total voltage is in proportion to the resistance of each resistor. That is, the largest resistors get the largest voltage drops. This makes sense—since the currents are all the same, *the larger resistors require the most voltage*. We can see that here by calculating the voltage drop across each resistor:

$$V_1 = IR_1 = (2.0 \ A)(1.0 \ \Omega) = 2.0 \ V$$
$$V_2 = IR_2 = (2.0 \ A)(2.0 \ \Omega) = 4.0 \ V$$
$$V_3 = IR_3 = (2.0 \ A)(3.0 \ \Omega) = 6.0 \ V$$

The sum of the voltage drops thus equals the battery voltage, as must be the case.

Similarly, for the parallel arrangement, the current is

$$I = \frac{V}{R_p} = \frac{12 \ V}{0.55 \ \Omega} = 22 \ A$$

Note that the current for the parallel combination is large relative to that for the series combination. (Why?) Also note that the smallest resistance gets most of the current, giving rise to the old adage "the current takes the path of least resistance." This makes sense because we know that in parallel, resistors all have the same voltage, so the smallest resistor will have the largest current. Here the currents through the three resistors are

$$I_1 = \frac{V}{R_1} = \frac{12 \ V}{1.0 \ \Omega} = 12 \ A$$

$$I_2 = \frac{V}{R_2} = \frac{12 \ V}{2.0 \ \Omega} = 6.0 \ A$$

$$I_3 = \frac{V}{R_3} = \frac{12 \ V}{3.0 \ \Omega} = 4.0 \ A$$

Follow-up Exercise. Calculate the power dissipated in each resistor and the total power dissipated in all the resistors for each arrangement in this Example. What generalizations can you come to? For instance, which resistor gets the most power when the resistors are wired in series? In parallel? Does the series arrangement require more or less power?

As a wiring application, consider strings of Christmas tree lights. In old strings of lights with large bulbs, when one bulb blew out, all the others on that string also went out, leaving you to hunt for the faulty bulb—and you had a real prob-

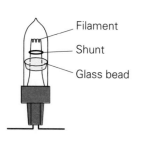

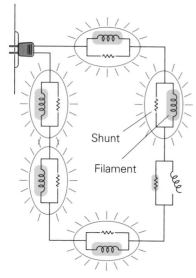

lem if more than one bulb blew out at the same time. With newer strings having smaller bulbs, one or more bulbs may burn out, but the others remain lit. Are the bulbs now wired in parallel? No, this would make the total equivalent resistance very small, giving a large current in the circuit, which is not desirable. Also, more costly wire would be required.

Instead, an insulated jumper or "shunt" is wired in parallel with each bulb filament (•Fig. 18.5). When a bulb is in operation, the shunt does not conduct because it is insulated from the filament wires. When the filament breaks and the bulb "burns out," there is momentarily no current in the string. The voltage difference across the open circuit at the broken filament will then be the full 120-V household voltage—the high side of the string at 120 V and the low side at 0 V. This causes a sparking that burns off the shunt insulation material. Making contact with the filament wires, the shunt completes the circuit and the rest of the lights in the string continue to glow. (The shunt, a wire with very little resistance compared to a bulb filament, is indicated by the small symbol in Fig. 18.5.)

CONCEPTUAL EXAMPLE 18.2 ■ OH, TANNENBAUM! CHRISTMAS TREE LIGHTS REVISITED

In a string of Christmas tree lights composed of bulbs with jumper shunts, if the filament of one bulb burns out, will the other bulbs (a) glow a little more brightly, (b) glow a little more dimly, or (c) be unaffected? *Clearly establish the reasoning and physical principle(s) used in determining your answer before checking it below. That is, **why** did you select your answer?*

Reasoning and Answer. If one bulb filament burns out and its shunt completes the circuit, there will be less total resistance in the circuit because the shunt resistance is much less than the filament resistance. (Note that the filaments of the good bulbs and the shunt of the burned-out bulb are in series, so the resistances add.)

With less total resistance, there will be more current in the circuit and the bulbs will all glow a little brighter, so the answer is (a). For example, suppose a string of lights had 18 identical bulbs. Then the voltage drop across each would be $120\ \text{V}/18 = 6.7\ \text{V}$. (The voltage drops across the bulbs must add up to the wall voltage, $V_t = 120\ \text{V}$.) If one bulb is burned out, the voltage across each lighted bulb would be $V = V_t/17 = 120\ \text{V}/17 = 7.1\ \text{V}$. Hence, the current through each bulb increases.

Follow-up Exercise. Suppose that the shunt resistors in Christmas tree bulbs were wired in *parallel* with the filaments so as always to be part of the circuit—that is, to conduct current at *all* times rather than only when the filament burned out. If one bulb in a string of such lights burned out, would the result be different from that described above? Explain.

Series–Parallel Resistor Combinations

Resistors may be connected in a circuit in a variety of series–parallel combinations. As shown in •Fig. 18.6, circuits with only one voltage source can usually be reduced or collapsed into a single equivalent loop, containing just the voltage source and one equivalent resistance, by applying the above results.

A general procedure for analyzing circuits with different series–parallel combinations of resistors is to find the voltage drops across and the currents through the various resistors as follows:

1. Starting with the resistor combination farthest from the voltage source, find the equivalent series and parallel resistances.

2. Reduce the circuit until there is a single loop with one total equivalent resistance.

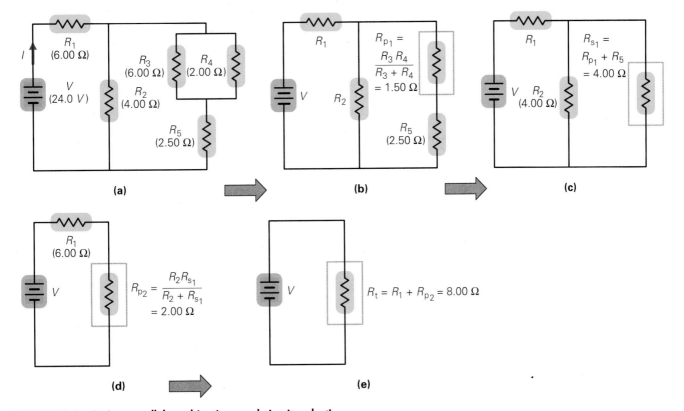

•**FIGURE 18.6 Series–parallel combinations and circuit reduction**
Reducing series combinations and parallel combinations to equivalent resistances reduces the circuit with one voltage source to a single loop with a single equivalent resistance. Example 18.3.

3. Find the current delivered to the reduced circuit using $I = V/R$.
4. Expand the circuit by reversing the reduction steps, and use the current for the reduced circuit to find the currents and voltage drops for the resistors in each step.

The following example illustrates this procedure. Analyze the various steps carefully in terms of the series and parallel relationships you have learned previously.

EXAMPLE 18.3 ■ SAME VOLTAGE OR SAME CURRENT? SERIES–PARALLEL
COMBINATION OF RESISTORS

(a) What is the voltage drop across and the current in each of the resistors R_1 through R_5 in Fig. 18.6a? (b) How much power is dissipated in R_4?

Solution. In order to verify voltage drops, we will calculate to three significant figures in the solution.

Given: Values in figure (three significant figures)   *Find:* (a) V and I for all the resistors
(b) P_4

(a) The parallel combination at the right side of the circuit diagram (farthest from the battery) is first reduced to the equivalent resistance R_{p_1} (see Fig. 18.6b):

$$R_{p_1} = \frac{R_3 R_4}{R_3 + R_4} = \frac{(6.00\ \Omega)(2.00\ \Omega)}{6.00\ \Omega + 2.00\ \Omega} = 1.50\ \Omega$$

This leaves a series combination along that side, which is reduced to R_{s_1} (Fig. 18.6c):

$$R_{s_1} = R_{p_1} + R_5 = 1.50\ \Omega + 2.50\ \Omega = 4.00\ \Omega$$

Then, R_2 and R_{s_1} are in parallel and can be reduced to R_{p_2} (Fig. 18.6d):

$$R_{p_2} = \frac{R_2 R_{s_1}}{R_2 + R_{s_1}} = \frac{(4.00\ \Omega)(4.00\ \Omega)}{4.00\ \Omega + 4.00\ \Omega} = 2.00\ \Omega$$

This leaves two resistances in series, which can be combined to give the *total* equivalent resistance (R_t) of the circuit (Fig. 18.6e):

$$R_t = R_1 + R_{p_2} = 6.00\ \Omega + 2.00\ \Omega = 8.00\ \Omega$$

Then, from the definition of resistance, the battery delivers a current of

$$I = \frac{V}{R_t} = \frac{24.0\ \text{V}}{8.00\ \Omega} = 3.00\ \text{A}$$

This *is* the current through R_1 and R_{p_2} since they are in series (Fig. 18.6d). Their voltage drops are

$$V_1 = IR_1 = (3.00\ \text{A})(6.00\ \Omega) = 18.0\ \text{V}$$

and

$$V_{p_2} = IR_{p_2} = (3.00\ \text{A})(2.00\ \Omega) = 6.00\ \text{V}$$

Since R_{p_2} is made up of R_2 and R_{s_1} (Fig. 18.6c), there is a 6.00 V drop across each of these resistors. With $R_2 = R_{s_1}$, the current divides equally:

$$I_2 = 1.50\ \text{A} \quad \text{and} \quad I_{s_1} = 1.50\ \text{A}$$

Then I_{s_1} is the current through R_{p_1} and R_5 in series (Fig. 18.6b). Their voltage drops are therefore

$$V_{p_1} = I_s R_{p_1} = (1.50\ \text{A})(1.50\ \Omega) = 2.25\ \text{V}$$
$$V_5 = I_s R_5 = (1.50\ \text{A})(2.50\ \Omega) = 3.75\ \text{V}$$

Note that these in fact do add up to 6.00 V exactly.

Finally, the voltage drop across R_3 and R_4 is the same as V_{p_1}:

$$V_3 = V_4 = 2.25\ \text{V}$$

The current of 1.50 A (I_{s_1}) divides at the R_3–R_4 junction.

$$I_3 = \frac{V_3}{R_3} = \frac{2.25\ \text{V}}{6.00\ \Omega} = 0.38\ \text{A}$$

$$I_4 = \frac{V_4}{R_4} = \frac{2.25\ \text{V}}{2.00\ \Omega} = 1.13\ \text{A}$$

Note that $I_3 + I_4 = I_{s_1}$ within rounding errors.

To see how the current divides proportionally between R_3 and R_4, consider these equations:

$$V_3 = V_4 \quad \text{and} \quad I_3 R_3 = I_4 R_4$$

or

$$I_3 = \left(\frac{R_4}{R_3}\right)I_4 = \left(\frac{2.00\ \Omega}{6.00\ \Omega}\right)I_4 = \frac{I_4}{3}$$

That is, I_3 is $\frac{1}{3}$ of I_4. That means that when the current I_{s_1} divides at the junction, $\frac{1}{4}$ of it goes through R_3 and $\frac{3}{4}$ goes through R_4, in inverse proportion to their relative resistances. For R_3, the relative resistance expressed as a fraction of the total resistance is $R_3/(R_3 + R_4) = \frac{6}{8} = \frac{3}{4}$, and for R_4 it is $R_4/(R_3 + R_4) = \frac{2}{8} = \frac{1}{4}$.

(b) The power expended in R_4 is

$$P_4 = I_4 V_4 = (1.13\ \text{A})(2.25\ \text{V}) = 2.54\ \text{W}$$

Circuit analysis like that done in Example 18.3 may seem involved, but it is simply repeated applications of equations for equivalents of series and parallel resistor combinations, along with the basic relationship $V = IR$.

Follow-up Exercise. In this Example, verify that the total power expended in all of the resistors is the same as the power output of the battery.

18.2 Multiloop Circuits and Kirchhoff's Rules

Objectives: To be able to (a) understand the physical principles that underlie Kirchhoff's circuit rules, and (b) apply these rules in the analysis of actual circuits.

Simple series–parallel circuits with a single voltage source that can be reduced to a single loop can be analyzed simply, as you learned in the preceding section. However, in general, circuits may contain several loops, with each one having voltage sources and/or resistances (collectively referred to as circuit elements or components). A simple multiloop circuit is shown in •Fig. 18.7a. Some combinations of resistors may be replaced by equivalent resistances (•Fig. 18.7b), but this circuit can be reduced only so far by using the procedures illustrated in Section 18.1.

A general method for analyzing multiloop circuits is by applying **Kirchhoff's rules**. These rules embody the conservation of charge and the conservation of energy. (Although they were not stated specifically, Kirchhoff's rules were applied to the simple circuits analyzed in Section 18.1.) Before presenting these rules, it is useful to introduce some terminology that will help us to describe complex circuits:

Kirchhoff's rules were developed by the German physicist Gustav Kirchhoff (1824–1887).

- A point in a circuit at which three or more connecting wires are joined together is called a **junction** or **node**—for example, point A in Fig. 18.7b. Current either divides or merges at a junction.
- A path connecting two junctions is called a **branch**. A branch may contain one or more circuit elements.

Kirchhoff's Junction Theorem

Kirchhoff's first rule, or **junction theorem**, is that the algebraic sum of the currents at any junction is zero.

$$\Sigma_i I_i = 0 \qquad \text{\textit{sum of currents at junction}} \qquad (18.4)$$

This simply means that the sum of the currents going into a junction (taken as positive) is equal to the sum of the currents leaving the junction (taken as negative), or that charge is conserved. For the junction at A in Fig. 18.7b, the algebraic sum of the currents is $I_1 - I_2 - I_3 = 0$, or

$$I_1 = I_2 + I_3$$
current in = current out

(This rule was applied in analyzing parallel resistances in Section 18.1.)

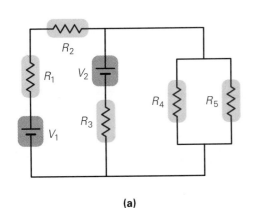

(a)

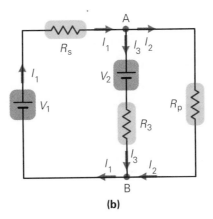

(b)

•**FIGURE 18.7 Multiloop circuit** In general, a circuit that contains voltage sources in more than one loop may not be able to be further reduced by series and parallel reductions. However, some reductions within each loop may be possible, such as from **(a)** to **(b)** here. At a circuit junction, where three or more wires come together, the current divides or currents come together, as at junctions A and B in part (b), respectively. The path between two junctions is called a branch. There are three branches in diagram (b), that is, three different paths between junctions A and B.

Of course, you cannot tell whether a particular current is directed into or out of a junction simply by looking at a multiloop circuit diagram. You have to *assume* current directions at a particular junction. If these assumptions are wrong, you will find out later from the opposite sign in your mathematical results. A negative current will indicate the wrong choice of direction. (You will see this in Example 18.5.)

Kirchhoff's Loop Theorem

Kirchhoff's second rule, or **loop theorem**, is that the algebraic sum of the potential differences (or voltages) across all of the elements of any *closed* loop is zero.

$$\Sigma_i V_i = 0 \qquad \text{sum of voltages} \atop \text{around loop} \qquad (18.5)$$

This means that the sum of the voltage rises equals the sum of the voltage drops around a closed loop, which must be true if energy is conserved. (This rule was used in analyzing series resistances in Section 18.1.)

Since traversing a circuit loop in either a clockwise or a counterclockwise sense will give rise to either a voltage rise or a voltage drop across each circuit element, respectively, it is important to establish a sign convention for voltage (electric potential *change*). This book uses the one illustrated in •Fig. 18.8. The voltage across a battery is taken to be positive (a voltage rise) if the loop is traversed toward the positive terminal (in the direction of conventional current from the battery), as shown in Fig. 18.8a. The voltage across a battery is taken to be negative if the loop is traversed in the opposite direction, that is, toward the negative terminal. (Note that the assigned branch currents have nothing to do with determining the sign of the voltage across a battery. The sign of the voltage depends only on the direction in which we choose to traverse the loop.)

The voltage across a resistor is taken to be negative (a drop) if the loop is traversed in the direction of the assigned current in that branch (Fig. 18.8b) and positive if the loop is traversed in the opposite direction. This sign convention allows you to go around a loop either clockwise or counterclockwise. Either way, Eq. 18.5 remains mathematically the same.

A graphical interpretation of Kirchhoff's loop theorem that you may find helpful is presented in the Learn by Drawing feature on pp. 568-569.

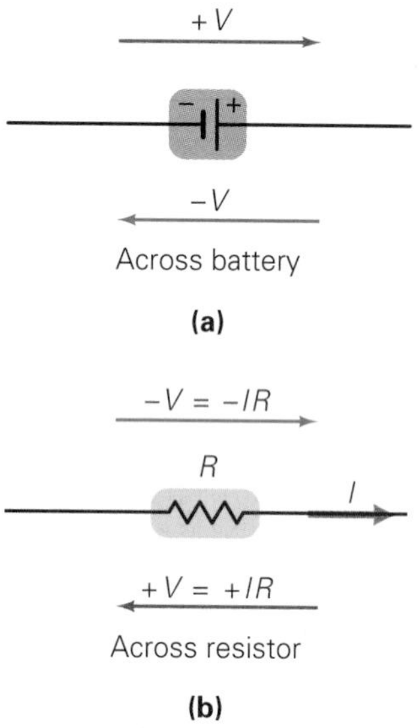

•**FIGURE 18.8   Sign convention for Kirchhoff's rules**
(a) When Kirchhoff's rules are applied in going around a circuit loop, the voltage across a battery is taken to be positive ($+$) if a battery is traversed from the negative to the positive terminal and negative ($-$) if the battery is traversed going from the positive to the negative terminal. **(b)** The voltage across a resistance is taken to be negative ($-$) if the resistance is traversed in the direction of the assigned branch current and positive ($+$) if the resistance is traversed in the direction opposite that of the assigned branch current.

EXAMPLE 18.4 ■ A SIMPLE CIRCUIT: USING KIRCHHOFF'S RULES

Two resistors ($R_1 = 6.00\ \Omega$ and $R_2 = 3.00\ \Omega$) are connected in parallel across a 12.0-V battery. Use Kirchhoff's rules in this simple case to calculate (a) the current through each resistor and the battery, and (b) the equivalent resistance of the pair.

Given: $R_1 = 6.00\ \Omega$ **Find:**   (a) current through R_1, R_2, and battery
           $R_2 = 3.00\ \Omega$ (b) equivalent resistance of R_1 and R_2
           $V = 12.0\ \text{V}$

Solution.

(a) The total current (I) through the battery divides between the two resistors, so we can use the junction theorem to write;

$$I = I_1 + I_2$$

Using the loop theorem to travel across the battery (from negative to positive terminals) and each resistor in turn (in the direction of the current), we have two equations

$$+V - I_1 R_1 = 0$$

and

$$+V - I_2 R_2 = 0$$

As we have already seen, the full battery voltage appears across each resistor, so the individual currents are

$$I_1 = \frac{V}{R_1} = \frac{12.0 \text{ V}}{6.00 \text{ }\Omega} = 2.00 \text{ A}$$

and

$$I_2 = \frac{V}{R_2} = \frac{12.0 \text{ V}}{3.00 \text{ }\Omega} = 4.00 \text{ A}$$

Thus the current through the battery will be $I = I_1 + I_2 = 2.00\text{A} + 4.00\text{A} = 6.00$ A.

(b) In effect, there is one equivalent resistance hooked across the battery and it draws 6.00 A when the 12.0 V is applied. Hence it has a resistance of

$$R_p = \frac{V}{I} = \frac{12.0 \text{ V}}{6.00 \text{ A}} = 2.00 \text{ }\Omega$$

Follow-up Exercise. Suppose that the resistors in this Example are connected in series across the battery rather than in parallel. Use Kirchhoff's rules to determine (a) the current through each, (b) the voltage across each, and (c) the power expended by each.

Application of Kirchhoff's Rules

Example 18.4 is a relatively simple application of Kirchhoff's rules. More complicated, multiloop circuits require a more structured approach. In this book we will use the following general steps in applying Kirchhoff's rules:

1. Assign a current and current direction for each branch in the circuit. This is done most conveniently at junctions.
2. Indicate the loops and the arbitrarily chosen directions in which they are to be traversed (see • Fig. 18.9). Every branch *must* be in at least one loop.
3. Apply Kirchhoff's first rule and write the equations for the currents, one for each junction that gives a different equation. (In general, this gives a set of equations that includes all branch currents.)
4. Traverse the number of loops necessary to include all branches. In traversing a loop, apply Kirchhoff's second rule and write the equations using the adopted sign convention.

Steps 3 and 4 give a set of N equations with N unknowns (the number of currents N), which may be solved for the currents. If more loops are traversed than necessary, you will have redundant equations. Only the number of loops that includes each branch once is needed.

This procedure may seem complicated, but it's really straightforward, as the following Example shows.

EXAMPLE 18.5 ■ BRANCH CURRENTS: USING KIRCHHOFF'S RULES

For the circuit diagrammed in Fig. 18.9, find the branch currents.

Given: See values in Fig. 18.9   *Find:* The three branch currents

Solution. The branch currents and their directions as well as the loops have been arbitrarily assigned in the diagram. Note that there is a current in every branch and that every branch is in at least one loop. (Some branches are in more than one loop, but that is acceptable.)

Applying Kirchhoff's first rule at the left hand node gives

$$I_1 = I_2 + I_3 \qquad (1)$$

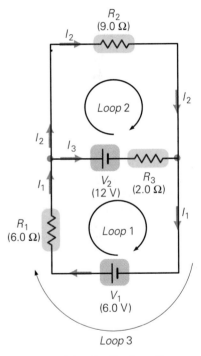

• **FIGURE 18.9   Application of Kirchhoff's rules**
To analyze a circuit like the one shown, assign a current and current direction for each branch in the circuit (this is most conveniently done at junctions). Then write equations for each independent junction (using Kirchhoff's first rule) and for as many loops as needed to include every branch (using Kirchhoff's second rule). Be careful to observe sign conventions. See Example 18.5.

LEARN BY DRAWING

Kirchhoff Plots: A Graphical Interpretation of Kirchhoff's Loop Theorem

Kirchhoff's loop theorem is usually stated in mathematical terms. However, there is a clever geometrical interpretation of it that may help you to develop a better insight into its meaning. This graphical approach allows you to visualize the potential changes in a circuit, either to anticipate the results of mathematical analysis or to confirm the order of magnitude of your results. In relatively simple circuits, it can give you much insight into the numerical answers.

The idea is to make a three-dimensional plot out of the circuit wiring diagram. The wires and elements of the circuit form the basis for the x–y plane, or "floor" of the diagram. Plotted perpendicularly to this plane, along the z axis, is the value of the electric potential, with an appropriate choice for zero voltage. Such a diagram is called a *Kirchhoff plot*.

The rules for constructing a Kirchhoff plot are simple: start at a known potential value and form a complete loop, finishing at the starting location. The fact that you must come back to the same starting location means that all the rises must be balanced out by the drops. This requirement is the geometrical expression of the conservation of energy, as embodied mathematically in Kirchhoff's loop theorem.

Thus, if the potential increases (say, in traversing a battery from cathode to anode), show a rise in the z direction. The rise, of course, is just the terminal voltage of the battery. Similarly, if the potential decreases (for example, traversing a resistor in the direction of the current), show a drop. If possible, try to make the sizes of the rises and drops (the voltages) to scale. That is, if there is a large rise in potential (such as you would have across a high-voltage battery), then make that rise large in proportion to the others on the diagram.

You may find it helpful if, before attempting a mathematical analysis of a circuit, you use this graphical method to preplan your attack and to qualitatively estimate the results (for example, which resistor gets most of the voltage, etc.). For elaborate circuits, this plotting may be too complicated for practical use. Nevertheless, it is always good to keep this concept in mind, as it illustrates the fundamental idea behind the loop theorem.

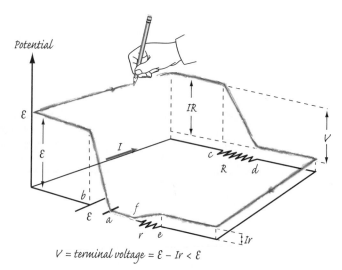

$V = \text{terminal voltage} = \mathcal{E} - Ir < \mathcal{E}$

FIGURE 1 Kirchhoff plots: A graphical problem solving strategy
The schematic of the circuit is laid out in the x–y plane and the potential plotted perpendicularly along the z axis. Usually, the zero of the potential is taken to be the negative terminal of the battery. A direction for current flow is assigned, and the value of the potential is plotted around the circuit, following the rules for potential gains and losses, Note that the wires are assumed resistanceless. (How can you tell?) The Kirchhoff plot shows a rise in potential when traversing the battery from cathode to anode, a drop in potential across the external resistor, and a final (smaller) drop across the internal resistance of the battery.

For the other node, you could write $I_2 + I_3 = I_1$ (currents in = currents out), but this is the same equation. (If there were another branch, it would have a different current equation for its junctions.)

Going around loop 1 as indicated in the figure and applying Kirchhoff's second rule with the sign convention gives

$$+V_1 - I_1R_1 - V_2 - I_3R_3 = 0$$

Then, putting in the numerical values from the figures gives

$$+6 - I_1(6) - 12 - I_3(2) = 0$$

and

$$6I_1 + 2I_3 = -6 \quad \text{or} \quad 3I_1 + I_3 = -3 \qquad (2)$$

(Units are omitted and resistances are written to one significant figure, both for convenience.)

Similarly, for loop 2,

$$+V_2 - I_2R_2 + I_3R_3 = 0$$

As an example of this method, consider the circuit diagrammed in Figure 1: a battery with internal resistance r wired to a single external resistor R. The current will flow from the anode to the cathode through the external resistor as shown. We start, as usual, by calling the potential of the battery cathode zero. (This is conventional, but not necessary; the assignment is yours, and in circuits with several batteries you will have to decide among several equally plausible choices.) Then, traversing the circuit in the direction of the arrows, we must show a rise in potential going from the cathode to the anode. Next, follow the current through the wires until it meets the external resistor. Notice that we show the wires as lines parallel to the x–y plane; that is, we indicate no voltage drop across them. (Why?)

When we reach the resistor, there must be a significant drop in potential. Note, however, that we cannot make the drop across the external resistor equal in magnitude to the rise across the battery. The potential does not go all the way to zero because there must be some voltage drop left to produce a current through the internal resistance of the battery. Thus, the terminal voltage of the battery, V, is less than its emf (the rise between points a and b), and the current in the external resistor is less than it would be if the battery had no internal resistance.

Figure 2 shows a more complicated situation: two resistors in series, and that combination in parallel with a third resistor. All three resistors have the same resistance, R. For simplicity, we assume that the battery here has negligible internal resistance. Starting at point a, there is a rise in potential corresponding to the battery voltage. Then, following the loop through the single resistor, there is a potential drop of equal magnitude across the resistor. However, if we follow the loop with the two resistors, we see that each gets only half the total voltage drop; thus we would expect each to carry only half the current of the single resistor. We know from the formulas developed in Section 18.1 that in parallel, the largest resistance carries the least current, but this geometrical approach can

help you to develop your intuition and allow you to anticipate the numerical results. Here, since all the resistances are equal, we would expect one-third of the total current through the two resistors and two-thirds through the single resistor. (Can you show this mathematically?)

As an exercise, you might wish to try redrawing Figure 2 as it would appear if the two resistors in series had resistances of $0.5R$ and $1.5R$. Can you predict what changes would occur in the actual circuit? Which resistor now has the largest voltage across it, and how do the currents in the branches compare? Analyze the circuit mathematically and see if your expectations are confirmed.

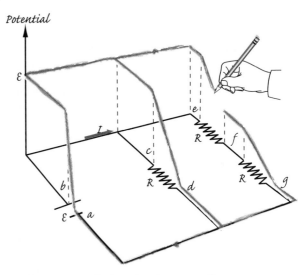

FIGURE 2 Kirchhoff plot of a more complex circuit
Try varying the values of the three resistors and imagine how the plot would change. Then analyze the circuit mathematically and see whether your plots allowed you to anticipate the voltages and currents in the various circuit elements.

and
$$12 - I_2(9) + I_3(2) = 0$$

Thus,
$$9I_2 - 2I_3 = 12 \tag{3}$$

Equations 1, 2, and 3 form a set of three equations with three unknowns. Physically, the problem is solved. The rest is math. You can solve for the I's in several ways. For example, first substitute from Eq. 1 into Eq. 2 to eliminate I_1:
$$3(I_2 + I_3) + I_3 = -3$$

This simplifies to
$$3I_2 + 4I_3 = -3 \tag{4}$$

Then, substituting from Eq. 4 into Eq. 3 eliminates I_2:
$$9(-1 - \tfrac{4}{3}I_3) - 2I_3 = 12$$

That is,
$$-14I_3 = 21 \quad \text{or} \quad I_3 = -1.5 \text{ A}$$

The minus sign on the results tells you that the wrong direction was assumed for I_3.
Putting the value of I_3 into Eq. 4 gives I_2:

$$3I_2 = 4(-1.5) = -3$$
$$I_2 = 1.0 \text{ A}$$

Then, by Eq. 1,

$$I_1 = I_2 + I_3 = 1.0 \text{ A} - 1.5 \text{ A} = -0.5 \text{ A}$$

The minus sign here indicates that I_1 was also assigned the wrong direction. So, at the upper junction in Fig. 18.9, I_3 really goes into the junction and I_1 and I_2 go out. (You might have suspected this to be the case because of the larger 12-V battery in the I_3 branch.)

Note that loop 3 was not used in this analysis. The equation for this loop would be redundant, giving four equations and three unknowns. However, loop 3 could have been used with either loop 1 or loop 2 in solving the problem. (Why?)

This application of Kirchhoff's rules is called the *branch current method*. A similar loop current method, which some consider to be simpler mathematically, is described in Exercise 51.

Follow-up Exercise. In this Example, show explicitly that the result for traversing loop 3 in the direction shown is just the sum of the equations obtained from traversing loop 1 and loop 2 separately. Thus, although it appears that there are four equations (three loop equations and one junction equation) and only three unknowns (the currents), there are, in reality three equations (two independent loop equations and the junction equation) and three unknowns.

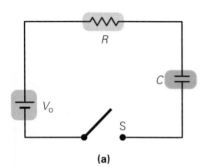

(a)

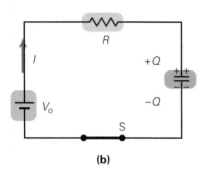

(b)

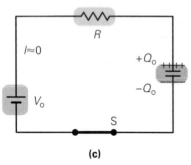

(c)

FIGURE 18.10 A series RC circuit
(a) Even though the circuit is open because of the space between the plates of the capacitor, **(b)** there is a current in the circuit when the switch is closed until the capacitor is charged to its maximum value. The rate of charging (and discharging) depends on the product of the values of the resistance and capacitance, which is called the time constant (τ) of the circuit: $\tau = RC$. **(c)** For times much larger than RC, the current is essentially zero and the capacitor is essentially fully charged.

18.3 RC Circuits

Objective: To be able to **(a) understand the charging and discharging of a capacitor through a resistor, and (b) calculate the current and voltage at specific times during these processes.**

The previous sections dealt with circuits having constant currents. In some dc circuits, the current may vary with time. This is the case with **RC circuits**, which have a resistor (R), or other resistive components, and a capacitor (C) in series.

Charging a Capacitor through a Resistor

The charging of an initially uncharged capacitor is depicted in • Fig. 18.10. You may be quick to notice that even after the switch is closed, the circuit is still open, or incomplete, because of the separation between the plates of the capacitor. This is true; nevertheless, when the switch is closed, charge does flow for a short time while the capacitor is charging.

The maximum amount of charge (Q_o) built up on the capacitor depends on the capacitance (C) and the voltage of the battery (V_o). Recall from Chapter 16 that this charge is $Q_o = CV_o$. Immediately after the switch is closed, at $t = 0$, there is an initial current in the circuit, $I_o = V_o/R$, through the resistor. Also recall that as charge accumulates on the capacitor's plates, it takes more work to add charge because of the repulsion between like charges. Eventually, the capacitor is charged to the maximum, and the current diminishes to zero.

The resistance helps determine how fast the capacitor is charged, since the larger its value, the greater the resistance to charge flow. The capacitance also influences the speed of charging—it takes longer to charge a larger capacitor. The voltage across the capacitor is found to vary exponentially with time according to the equation

$$V_C = V_o(1 - e^{-t/RC}) \qquad \text{(18.6)}$$

charging capacitor voltage in an RC circuit

where e has an approximate value of 2.718. (An irrational number, e is the base of the system of *natural logarithms*.) A graph of V_C versus t and a graph of I versus t are presented in •Fig. 18.11. The current varies with time according to the equation

$$I = I_o e^{-t/RC} \qquad (18.7)$$

The current decreases exponentially with time. Note that the current is greatest initially. (Why?)

According to Eq. 18.6, it would theoretically take an infinite amount of time for the capacitor in an RC circuit to become fully charged (to reach the battery voltage, V_o). However, in practice, such a capacitor charges and discharges in relatively short times. It is customary to use a special value to express the charging and discharging times. This **time constant** (τ) for an RC circuit is

$$\tau = RC \qquad (18.8)$$

At a time equal to the time constant, $t = \tau = RC$, the voltage across the charging capacitor is

$$V = V_o(1 - e^{-t/\tau}) = V_o(1 - e^{-\tau/\tau}) = V_o(1 - e^{-1})$$

That is,

$$V \approx 0.63 V_o$$

After one time constant, the voltage across the capacitor is at 63% of its maximum (or the capacitor is 63% charged, since $Q = CV$). Note that at that time the current has decayed to 37% of its initial maximum value (I_o).

At the end of two time constants, $t = 2\tau = 2RC$, the capacitor is charged to more than 86% of its maximum value, and so on. For practical purposes, the capacitor is considered to be fully charged after only a few time constants.

Discharging a Capacitor through a Resistor

When a fully charged capacitor is *discharged* through a resistance, the voltage across the capacitor decays exponentially with time (as does the current):

$$V_C = V_o e^{-t/RC} = V_o e^{-t/\tau} \qquad \begin{array}{l}\textit{discharging capacitor voltage}\\ \textit{in an RC circuit}\end{array} \qquad (18.9)$$

The voltage is said to "decay" exponentially. In one time constant, the voltage across the capacitor falls to 37% of its original value (see •Fig. 18.12). The current in the circuit decays exponentially, following Eq. 18.7.

EXAMPLE 18.6 ■ CHARGING A CAPACITOR: TIME CONSTANT

The capacitance and resistance in the RC circuit in Fig. 18.10 are 6.0 μF and 0.25 MΩ, respectively, and the battery has a 12 V terminal voltage. (a) What is the voltage across the capacitor at one time constant after the switch is closed if it was initially uncharged? (b) What is the voltage across the capacitor and its charge at $t = 5.0$ s?

Solution. Listing the data and applying the charging capacitor expressions:

Given: $C = 6.0 \ \mu\text{F} = 6.0 \times 10^{-6} \text{ F}$ *Find:*   (a) V (voltage at $t = \tau$)
$R = 0.25 \text{ M}\Omega = 2.5 \times 10^5 \ \Omega$ (b) V and Q at $t = 5.0$ s
$V = 12 \text{ V}$ (voltage and charge)

(a) In one time constant, the capacitor is charged to 63% of its maximum charge, and thus is at 63% of its maximum voltage.

$$V_C = 0.63 V_o = (0.63)(12 \text{ V}) = 7.6 \text{ V}$$

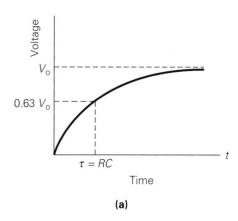

(a)

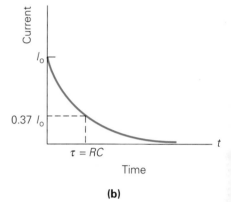

(b)

•**FIGURE 18.11 Capacitor charging**
(a) In a series RC circuit, the voltage across the capacitor increases nonlinearly with time, reaching 63% of its maximum voltage (V_o) in one time constant, $t = \tau = RC$. **(b)** The current in the circuit is initially a maximum ($I_o = V_o/R$) and decays exponentially with time, falling to 37% of its initial value in one time constant.

•**FIGURE 18.12 Capacitor discharging**
When a charged capacitor is discharged through a resistance, the voltage across the capacitor (and the current in the circuit) decays exponentially with time, falling to 37% of its initial value in one time constant, $t = \tau = RC$.

(b) When a specific time is given, you also need to know the time constant for the circuit, which in this case is

$$\tau = RC = (2.5 \times 10^5 \ \Omega)(6.0 \times 10^{-6} \ \text{F}) = 1.5 \ \text{s}$$

Then, for $t = 5.0$ s, using Eq. 18.6,

$$V_C = V_o(1 - e^{-t/\tau}) = (12 \ \text{V})(1 - e^{-5.0\text{s}/1.5\text{s}})$$
$$= (12 \ \text{V})(1 - 0.036) = (12 \ \text{V})(0.964)$$
$$= 11.6 \ \text{V}$$

(Significant figure rules would round this to 12 V, but we carry three figures here to make the point that the voltage is not exactly 12 V.)

Note that after $t = 5.0 \approx 3.3\tau$, the voltage across the capacitor is more than 96% of its maximum value.

Expressing τ as RC and multiplying both sides of Eq. 18.6 by C gives

$$CV_C = CV_o(1 - e^{-t/RC})$$

Since for a capacitor we know that the charge is proportional to the voltage, or $Q = CV$, we have

$$Q = Q_o(1 - e^{-t/RC})$$

and we see that the charge varies with time just as the voltage does.

The maximum charge (Q_o) on the capacitor is

$$Q_o = CV_o = (6.0 \times 10^{-6} \ \text{F})(12 \ \text{V}) = 7.2 \times 10^{-5} \ \text{C}$$

At $t = 5.0$ s $\approx 3.3\tau$, the charge on the capacitor is 96.4% of the maximum charge (the same percentage as for the voltage):

$$Q = (0.964)Q_o = (0.964)(7.2 \times 10^{-5} \ \text{C}) = 6.9 \times 10^{-5} \ \text{C}$$

Follow-up Exercise. In this Example, (a) what is the maximum energy storage in the capacitor? (b) After 5.0 s, compare the energy storage in the capacitor to the maximum possible, expressing your answer as a percentage. Is it 96.4%, like the voltage? Explain.

An application of an RC circuit is diagrammed in •Fig. 18.13a. This is called a blinker circuit (or, more impressively, a neon-tube relaxation oscillator). The resistor and capacitor are in series, and a small neon tube is connected across (in parallel with) the capacitor. (These neon tubes are about the size of miniature Christmas tree lights.)

When the circuit is closed, the voltage across the capacitor (and the neon tube) rises from 0 to V_b, which is the breakdown voltage of the neon gas in the tube (about 80 V). At that voltage, the gas is ionized and begins to conduct electricity, and the tube lights up. When the tube is in a conducting state, the capacitor discharges through it, and the voltage falls rapidly (Figure 18.13b). When the voltage drops below V_m, called the maintaining voltage, the tube discharge can no longer be sustained, and the tube stops conducting. The capacitor then begins charging again, the voltage rises from V_m to V_b, and the cycle repeats. The continual repetition of this cycle causes the tube to blink on and off. The period of the oscillator, or the time between flashes, depends on the RC time constant.

Other applications of RC circuits are discussed in the Insight feature on p. 573.

18.4 Ammeters and Voltmeters

Objectives: To understand **(a)** how galvanometers are used as ammeters and voltmeters, **(b)** how multirange versions of these devices are constructed, and **(c)** how they are connected to measure current and voltage in real circuits.

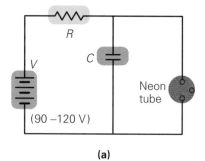

(a)

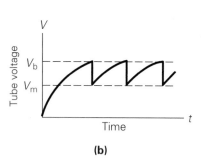

(b)

•**FIGURE 18.13 Neon-tube relaxation oscillator circuit**
(a) When a neon tube is connected across the capacitor in a series RC circuit having the proper voltage source, the voltage across the tube (and capacitor) will relax, or oscillate, with time. As result, the tube periodically flashes or blinks. **(b)** A graph of voltage versus time shows the oscillating effect.

Insight   An RC Circuit in Action

To install shelving or cabinets, or even to hang a picture securely, it is usually necessary to locate the studs in the walls of your house. Studs are the vertical wooden elements to which plasterboard or paneling is nailed, and they can be very elusive. The traditional way of finding them, by tapping on the wall, generally leads to more frustration than success. Commercially available stud finders employ pivoted magnets, which move when they pass over a nail head in the stud behind the wall, but these too are far from infallible.

An electronic device called Studsensor™, much easier to use, relies on capacitive sensing. The capacitor is in a printed circuit and forms part of a simple RC charging network. However, for this application the metallic plates of the capacitor are arranged end to end rather than facing one another as in a parallel-plate capacitor (Fig. 1). In this configuration the

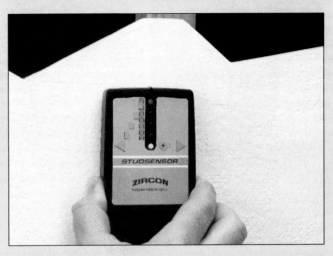

FIGURE 2   Studsensor™
In this cutaway view of the Studsensor in use, the lower green light indicates that the unit is calibrated. Vertical red lights signal the approach to a stud and the top red light indicates that the sensor is over a stud. Some models have both light and sound indicators.

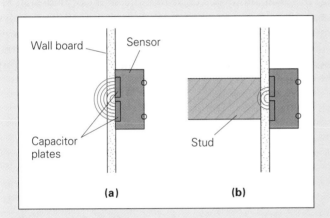

FIGURE 1   RC sensing
(a) The fringe electric field of the capacitor plates permeates the wall board, which acts as a dielectric. **(b)** When over a stud, the capacitance changes. The sensor detects this change as a change in an RC circuit.

fringe electric field extends a considerable distance beyond the plane of the plates, and any nearby material acts as a dielectric in this field.

The unit is initially calibrated by placing it against the wall and balancing the charging time of the RC circuit against a fixed reference. When the sensor containing the capacitor plates is moved over a stud, the character of the dielectric changes, and the capacitance of the plates increases. The greater the capacitance, the longer the charging time of the circuit relative to the calibration standard. The changes in charging time are sensed electronically and conveyed to the user by lighted indicators (Fig. 2).

As the names imply, an **ammeter** measures current flow through circuit elements (in amps) and a **voltmeter** measures voltage differences across circuit elements (in volts). A basic component of both of these meters is a **galvanometer** (•Fig. 18.14). The galvanometer operates on magnetic principles that will be covered in Chapter 19. In this chapter, it will simply be considered to be a circuit element having an internal resistance (r) and whose needle deflection is directly proportional to the current through it.

Ammeters

The resistance of a galvanometer coil (r) is relatively small, since the coil consists of metal wire. In effect, a galvanometer measures current, but because of the small coil resistance, only currents in the microamp range can be measured without burning out the coil. An ammeter that can be used to measure larger currents therefore must have a small *shunt resistor* in parallel with a galvanometer to take

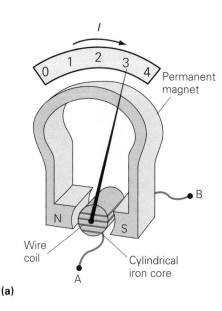

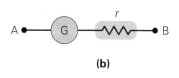

(b)

•**FIGURE 18.14 The galvanometer**
(a) A galvanometer is a current-sensitive device whose needle deflection is proportional to the current through its coil. **(b)** The circuit symbol is a circle containing a G. The internal resistance (r) of the meter is indicated explicitly.

(a)

most of the current (•Fig. 18.15). It should be clear that the shunt provides an alternate path by which part of a large current (I) can bypass the galvanometer ($I = I_g + I_s$, at the left junction in the figure). The shunt and the galvanometer resistance are in parallel; thus, to carry most of the current the shunt must be the smaller of the two. Because the voltages across the galvanometer and the shunt resistor are equal, we can write

$$V_g = V_s$$

or

$$I_g r = I_s R_s$$

Using the Kirchhoff's junction rule, we see that $I = I_g + I_s$, so the previous equation can be rewritten as

$$I_g r = (I - I_g)R_s$$

and

$$I_g = \frac{IR_s}{r + R_s} \qquad \begin{array}{l}\textit{galvanometer current} \\ \textit{for dc ammeter}\end{array} \qquad (18.10)$$

This equation allows you to select the proper shunt resistance for a given current range and galvanometer. The following Example illustrates how to do this.

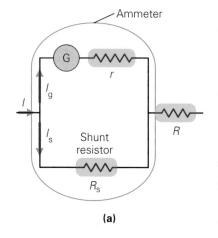

(a)

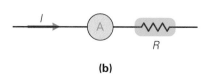

(b)

•**FIGURE 18.15 dc ammeter**
(a) A galvanometer in parallel with a shunt resistor (R_s) is an ammeter capable of measuring various current ranges, depending on the value of R_s. **(b)** The circuit symbol for an ammeter is a circle with an A inside it.

EXAMPLE 18.7 ■ GALVANOMETER AS AMMETER: CHOOSING A SHUNT RESISTOR

A galvanometer that can safely carry a maximum coil current of 200 μA (called the full-scale sensitivity) has a coil resistance of 50 Ω. It is to be used in an ammeter designed to read currents up to 3.0 A (at full scale). What is the required shunt resistance? (Disregard significant figures.)

Solution. We recognize that the galvanometer can only carry a small current. Most of the 3.0 A will have to be shunted (diverted) through the shunt resistance, so we expect its value to be much smaller than the coil resistance.

Given: $I_g = 200 \ \mu A = 2.00 \times 10^{-4}$ A   *Find:* R_s (shunt resistance)
           $r = 50 \ \Omega$
      $I_{max} = 3.0$ A

A full-scale reading of 3.0 A means that when a current of 3.0 A enters the ammeter, I_g should be 200 μA. Solving Eq. 18.10 for the shunt resistance and inserting the known values gives

$$R_s = \frac{I_g r}{I_{max} - I_g}$$

$$= \frac{(2.0 \times 10^{-4}\,\text{A})(50\,\Omega)}{3.0\,\text{A} - 2.00 \times 10^{-4}\,\text{A}}$$

$$= 3.3 \times 10^{-3}\Omega = 3.3\,\text{m}\Omega$$

Note the small size of the shunt resistance compared to the coil resistance r (50 Ω), as expected. This disparity allows most of the current (2.9998 A at full scale) to pass through the shunt resistor branch. The shunt resistor is made of a material that does not burn out as readily as the thin wire of the galvanometer coil. The ammeter will read currents linearly up to 3.0 A. For example, if a current of 1.5 A flowed into the ammeter, there would be a current of 100 μA in the coil of the galvanometer, which would give a half-scale reading.

Follow-up Exercise. When an ammeter is used to measure the current through a circuit element, it is connected in series so that the current through the element also must flow through the ammeter. Since the ammeter has resistance, wouldn't the insertion of the ammeter in series with the circuit reduce the current from its true value? Explain your reasoning

Voltmeters

A voltmeter that is capable of reading voltages higher than the microvolt range is constructed by connecting a large *multiplier resistor* in series with a galvanometer (•Fig. 18.16). Because the voltmeter has a large internal resistance due to the multiplier resistor, it draws little current from the main circuit. When connected *across* a circuit element, the voltmeter (or galvanometer branch) experiences a voltage drop of $V = V_g + V_m$, and most of this drop is across the multiplier resistor rather than the galvanometer coil. By Kirchhoff's loop theorem,

$$V = V_g + V_m$$

$$= I_g r + I_g R_m = I_g(r + R_m)$$

and

$$I_g = \frac{V}{r + R_m} \qquad \begin{array}{l}\textit{galvanometer current} \\ \textit{for dc voltmeter}\end{array} \qquad (18.11)$$

The voltage V is also the potential difference across the circuit element having a resistance R because of the parallel connection (see Fig. 18.16b). If the galvanometer scale is calibrated in volts (instead of amps), the voltage drop across that circuit element can be read from the meter.

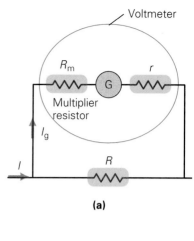

(a)

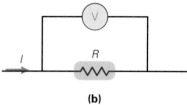

(b)

•**FIGURE 18.16 dc voltmeter**
(a) A galvanometer in series with a multiplier resistor (R_m) is a voltmeter capable of measuring various voltage ranges, depending on the value of R_m. **(b)** The circuit symbol for a voltmeter is a circle with a V inside it. Here R is the resistor whose voltage is being measured.

Note: Ammeters are galvanometers with *small resistors in parallel*. Voltmeters are galvanometers with *large resistors in series*.

EXAMPLE 18.8 ■ GALVANOMETER AS VOLTMETER: CHOOSING A MULTIPLIER RESISTOR

Suppose that the galvanometer described in Example 18.7 is to be used instead in a voltmeter with a full-scale reading of 3.0 V. What is the required value of the multiplier resistor?

Solution. To turn the galvanometer into a voltmeter, we must replace the shunt resistor by a multiplier resistor in series with the coil.

Given: $I_g = 200\,\mu\text{A} = 2.00 \times 10^{-4}\,\text{A}$ *Find:* R_m (multiplier resistance)
(from Example 18.6)
$r = 50\,\Omega$ (from Example 18.6)
$V = 3.0\,\text{V}$

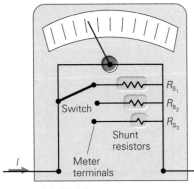

(a) Multirange ammeter

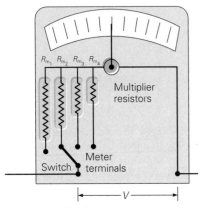

(b) Multirange voltmeter

(c)

• FIGURE 18.17 Multirange meters
(a) An ammeter or **(b)** a voltmeter can be used to measure several ranges of current and voltage by switching among different shunt and multiplier resistors, respectively. (Instead of a switch, there may be an exterior terminal for each range.) **(c)** Both functions may be combined into a single multimeter. Multimeters are available with electronic digital readouts.

Eq. 18.11 can be used to find R_m.

$$R_m = \frac{V - I_g r}{I_g}$$

$$= \frac{3.0\ \text{V} - (2.00 \times 10^{-4}\ \text{A})(50\ \Omega)}{2.00 \times 10^{-4}\ \text{A}}$$

$$= 1.5 \times 10^4\ \Omega = 15\ \text{k}\Omega$$

Follow-up Exercise. The voltmeter in this Example is used to measure the voltage of a resistor in a circuit. 3.0 A flows through the resistor (1.0 Ω) *before* the voltmeter is connected. Assuming the total *incoming* current (*I* in Fig. 18.16b) remains the same after the voltmeter is connected, calculate the current in the galvanometer.

For versatility, ammeters and voltmeters may be multiranged. This is accomplished by providing the user with a choice of several different shunt or multiplier resistors (•Fig. 18.17). Such meters may also be combined into *multimeters* that can measure voltage, current, and often resistance as well. Electronic digital multimeters are now common (Fig. 18.17c). In place of mechanical galvanometers, these use electronic circuits that analyze the voltage drops across internal resistors.

18.5 Household Circuits and Electrical Safety

Objectives: To understand (a) how household circuits are wired, and (b) the underlying principles that govern electrical safety devices.

Although household circuits generally use alternating current, which has not yet been discussed, they include practical applications of some of the principles already studied.

For example, would you expect the elements in a household circuit (lamps, appliances, and so on) to be connected in series or in parallel? From the discussion of Christmas tree lights in Section 18.1, it should be apparent that household elements must be connected in parallel. For example, when the bulb in a lamp blows out, other elements in the circuit continue to work. This would not be the case for a series circuit. Moreover, household appliances and lamps are generally rated for 120 V. If these elements were connected in series, the voltage drops across them would add up to a *total* of 120 V. None of the circuit elements would have adequate voltage, and the more that were connected in series, the less voltage each would have.

Power is supplied to a house by a three-wire system (•Fig. 18.18). There is a potential difference of 240 V between the two hot, or high-potential, wires, and each of these has a 120-V potential difference with respect to the ground. The third wire is grounded at the point where the wires enter the house, usually by a metal rod driven into the ground. This wire is defined as the zero potential and is called the ground, or neutral, wire.

The potential difference of 120 V needed for most household appliances is obtained by connecting them between the grounded wire and either of the high-potential wires: $\Delta V = 120\ \text{V} - 0\ \text{V} = 120\ \text{V}$ or $\Delta V = 0\ \text{V} - (-120\ \text{V}) = 120\ \text{V}$. (See Fig. 18.18.) Even though the grounded wire has a zero potential, it is a *current-carrying* wire because it is part of the complete circuit. Large appliances such as central air conditioners, ovens, and hot water heaters need 240 V for operation, and this is obtained by connecting them between the two hot wires: $\Delta V = 120\ \text{V} - (-120\ \text{V}) = 240\ \text{V}$.

As you learned in Chapter 17, the amount of current an appliance draws (uses) may be given on a rating tag but can also be determined from the power rating (using $P = IV$). For example, a stereo rated at 180 W at 120 V would draw

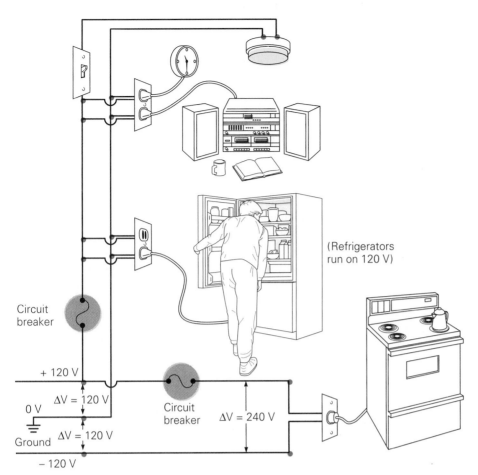

•FIGURE 18.18 Household wiring
A 120-V circuit is obtained by connecting between the +120-V (or −120-V) line and the ground line. A potential difference of 240 V for large appliances such as electric stoves, central air conditioners, and hot water heaters is obtained by connecting between the +120-V line and the −120-V line.

Circuit breaker

+ 120 V

0 V

$\Delta V = 120$ V

Ground

$\Delta V = 120$ V

− 120 V

Circuit breaker

$\Delta V = 240$ V

(Refrigerators run on 120 V)

an average current of 1.5 A ($I = P/V$). There are limitations to the number of elements that can be put in a circuit and the total amount of current drawn. In particular, the joule heat (or I^2R loss) must be considered. Generally, the more elements (resistances) connected in parallel, the smaller the equivalent resistance. In any case, the equivalent resistance is smaller than the resistance of the smallest element. As we saw in Example 17.7 in the previous chapter, for a constant voltage the joule heat produced is inversely proportional to the resistance of the circuit: $P = V^2/R$. Thus, by adding too many current-carrying elements it is possible to overload a household circuit so that it draws too much current and produces too much heat *in the wires*. This heat in the circuit wires could melt the insulation and start a fire.

Overloading is prevented by limiting the current in a circuit by means of two types of devices: fuses and circuit breakers. **Fuses** (•Fig. 18.19) are common in older homes. An Edison-base fuse has threads like those on the base of a light bulb (see Fig. 18.19b). Inside the fuse is a metal strip that melts because of joule heat when the current is larger than the rated value (which might be 15 A). The melting of the strip opens the circuit, and electrically everything comes to a halt.

There are several problems associated with fuses. One problem with Edison-base fuses in particular is that they are interchangeable. For example, a 30-A fuse can be put in a circuit that is rated for 15 A. This would allow a current flow twice as great as that for which the circuit was designed. Such a problem is avoided with another type of fuse, called a Type-S fuse (Fig. 18.19b). A nonremovable adapter is installed in the socket for a fuse; the adapter is specific for a particular rated value (the fuses and adapters for different ratings have different threads). That is, a 30-A Type-S fuse will not screw into a 15-A Type-S adapter, for example, so the wrong fuse can't be used.

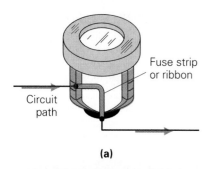

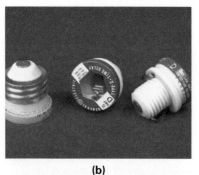

Fuse strip or ribbon

Circuit path

(a)

(b)

(c)

•**FIGURE 18.19 Fuses**
(a) A fuse contains a metallic strip, or ribbon, that melts when the current exceeds a rated value. This opens the circuit and prevents overheating. **(b)** Edison-base fuses (left) have threads similar to those on light bulbs; fuses with different ampere ratings can be interchanged. Type-S fuses (right) have different threads for different ratings that fit specific adapters in the fuse sockets and so cannot be interchanged. **(c)** A type of small fuse commonly found in electrical equipment. Note that the fuse on the right is "blown."

Circuit breakers are now used exclusively in wiring new homes. One type of circuit breaker, shown in •Fig. 18.20, uses a bimetallic strip (see Chapter 10). As the current through the strip increases, it becomes warmer and bends. At the rated current value, the strip will be bent sufficiently to cause the circuit to open mechanically. The strip quickly cools, so the breaker may be reset. However, when a fuse blows or a circuit breaker trips, this is an indication that the circuit is drawing or attempting to draw too much current. *You should always investigate to find and correct the problem before replacing the fuse or resetting the circuit breaker.*

Switches, fuses, and circuit breakers are placed in the hot (high-potential) side of the line. They would work in the grounded side as a heating overload sensor—but even when the switch was open, the fuse blown, or the breaker tripped, the circuit and any elements in it would still be connected to a high potential, which could be dangerous if a person made electrical contact. Even with fuses or circuit breakers in the hot side of the line, there is still a possibility of getting an electrical shock from a defective appliance that has a metal casing, such as a hand drill. A wire may come loose inside and make contact with the casing, which would then be "hot," or at a high potential. As illustrated in •Fig. 18.21, a person could provide a path to the ground and become part of the circuit, thus receiving a shock.

To prevent this from happening, a third, dedicated grounding wire is added to the circuit that grounds the metal casing (•Fig. 18.22). Note that this wire is not normally a current-carrying wire. If a hot wire comes into contact with the casing, the circuit is completed to this grounded wire. The fuse is blown or the circuit breaker tripped as most of the current follows the third wire to the ground (a low-resistance path) rather than flowing through you (a high-resistance path).

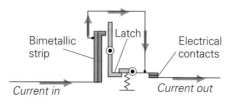

Bimetallic strip   Latch   Electrical contacts

Current in   *Current out*

Thermal trip   *Circuit broken*

After current overload

(a)

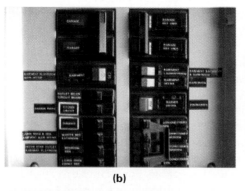

(b)   **(c)**

•**FIGURE 18.20 Circuit breaker**
(a) A diagram of a thermal trip element. With increased current and joule heating, the element bends until it opens the circuit at some preset current value. Magnetic trip elements are also used. **(b)** A bank of typical household circuit breakers. (Notice the numbers on the switches; what do you think they represent?) **(c)** These heavy-duty circuit breakers at an electrical substation must be capable of cutting off current almost instantaneously in power lines operating at very high voltages.

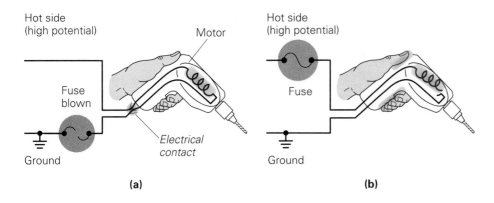

(a) **(b)**

•**FIGURE 18.21 Electrical safety**
(a) Switches and fuses or circuit breakers should always be wired in the hot side of the line, *not* in the grounded side as shown here. If these elements are wired in the grounded side, the line is at a high potential even when the fuse is blown or a switch is open.
(b) Even if the fuse or circuit breaker is in the hot side of the line, a potentially dangerous situation exists. If an internal wire comes into contact with the metal casing of an appliance or power tool, a person touching the casing, which is at high voltage, can get a shock.

On three-prong **grounded plugs**, the large round prong connects with the dedicated grounding wire. Adapters can be used between a three-prong plug and a two-prong socket. Such an adapter has a grounding lug or grounding wire (•Fig 18.23a). This should be fastened to the receptacle box by the plate-fastening screw or some other means. The receptacle box is grounded by means of the grounding wire. If the adapter lug or wire is not connected, the system is left unprotected, which defeats the purpose of the dedicated grounding safety feature.

You have probably noticed that there is another type of plug, a two-prong plug that only fits in the socket one way because one prong is larger than the other and one of the slits of the receptacle is also larger (•Fig. 18.23b). This is called a **polarized plug**. *Polarizing* in the electrical sense is a method of identifying the hot and grounded sides of the line so that particular connections may be made.

Such polarized plugs and sockets are an older safety feature. Wall receptacles are wired so that the small slit connects to the hot side and the large slit connects to the neutral, or ground, side. Having the hot side identified in this way makes two safeguards possible. First, the manufacturer of an electrical appliance can design it so that the switch is always in the hot side of the line. Thus, all of the wiring of the appliance beyond the switch is safely neutral when the switch is open and the appliance off. Moreover, the casing of an appliance is connected by the manufacturer to the ground side by means of a polarized plug. Should a hot wire inside the appliance come loose and contact the metal casing, the effect would be similar to that with a dedicated grounding system. The hot side of the line would be shorted to the ground, which would blow a fuse or trip a circuit breaker.

However, the three-wire dedicated grounding system is preferred over polarized plugs and sockets. There could be a wiring error in an appliance or receptacle, which might cause a problem even with a polarized plug. (What would the problem be?) Another type of electrical safety device, the ground fault interrupter, is discussed in Chapter 20.

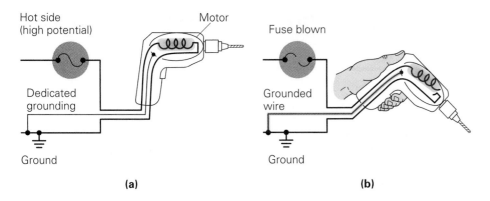

(a) **(b)**

•**FIGURE 18.22 Dedicated grounding**
(a) For electrical safety, a third wire is connected from an appliance or power tool to ground. This dedicated grounding wire normally carries no current (as opposed to the grounded wire of the circuit). **(b)** If a loose internal wire comes into contact with the grounded casing, the shorting to ground through the dedicated grounding wire blows the protective fuse or circuit breaker, and the appliance is not at a high potential as in Fig. 18.21b.

•**FIGURE 18.23** Plugging into ground

(a) A three-prong plug and adapter for a two-prong socket. Note the grounding lug on the adapter. This should be connected to the plate-fastening screw on the grounded receptacle box—otherwise, the safety feature is lost. **(b)** A polarized plug. The differently sized prongs permit prewired identification of the high and ground sides of the line.

(a)   (b)

Insight Electricity and Personal Safety

Safety precautions are necessary to prevent injuries when people work with and use electricity. Electricity conductors (such as wires) are coated with insulating materials so that they can be handled safely. However, when a person comes into contact with a charged conductor, a potential difference may exist across part of the body. A bird can sit on a high-voltage line without any problem because the bird and the line are at the same potential, and there is no potential difference or circuit path. But if a person carrying an aluminum (conducting) ladder touches it to an electrical line, a potential difference exists from the line to the ground, and the ladder and the person are part of the circuit.

The extent of personal injury in such a case depends on the amount of electric current that flows through the body and on the circuit path. The current is given by

$$I = \frac{V}{R_{body}}$$

where R_{body} is the resistance of the body. Thus, for a given voltage, the current depends on the body resistance.

Body resistance varies. If the skin is dry, the resistance may be 0.50 MΩ (0.50×10^6 Ω) or more. For a potential difference of 120 V, there would be a current of

$$I = \frac{V}{R_{body}} = \frac{120 \text{ V}}{0.50 \times 10^6 \, \Omega} = 0.24 \times 10^{-3} \text{A} = 0.24 \text{ mA}$$

This current is almost too weak to be felt.

Suppose, however, that the skin is wet with perspiration.

Then R_{body} is about 5.0 kΩ (5.0×10^3 Ω), and the current is

$$I = \frac{V}{R_{body}} = \frac{120 \text{ V}}{5.0 \times 10^3 \, \Omega} = 24 \times 10^{-3} \text{A} = 24 \text{ mA}$$

which could be very dangerous.

A basic precaution to take is to avoid coming into contact with an electrical conductor that might cause a potential difference to exist across the body or part of it. The effect of such contact depends on the path of the current. If that path is from the little finger to the thumb on one hand, probably only a burn would result from a large current. However, if the path is from hand to hand through the chest, the effect can be much worse. Some of the possible effects of this circuit path are given in Table 1.

Injury results because the current interferes with muscle functions and/or causes burns. Muscle functions are regulated by electrical impulses through the nerves, and these can be influenced by currents from external voltages. Muscle reaction and pain can occur from a current of a few milliamps. At about 10 mA, muscle paralysis can prevent a person from releasing the conductor. At about 20 mA, contraction of the chest muscles occurs, which may cause breathing to be impaired or to stop. Death can occur in a few minutes. At 100 mA, there are rapid uncoordinated movements of the heart muscles (called ventricular fibrillation), which prevent the proper pumping action.

Working safely with electricity requires a knowledge of fundamental electrical principles *and* common sense. Electricity must be treated with respect.

TABLE 1	Effects of Electric Current on the Human Body*
Current (approximate)	*Effect*
1.0 mA (0.001 A)	Mild shock or heating
10 mA (0.01 A)	Paralysis of motor muscles
20 mA (0.02 A)	Paralysis of chest muscles, causing respiratory arrest; fatal in a few minutes
100 mA (0.1 A)	Ventricular fibrillation, preventing coordination of the heart's beating; fatal in a few seconds
1000 mA (1 A)	Serious burns; fatal almost instantly

*The effect on the human body of a given amount of current depends on a variety of conditions. This table gives only general and relative descriptions.

Chapter Review

Important Terms

equivalent series resistances   557
equivalent parallel resistance   559
junction (node)   565
branch   565
Kirchhoff's first rule (junction theorem)   565

Kirchhoff's second rule (loop theorem)   566
RC circuit   570
time constant   571
ammeter (dc)   573
voltmeter (dc)   573
galvanometer   573

fuse   577
circuit breaker   578
grounded plug   579
polarized plug   579

Important Concepts

- When resistors are wired in series, the current through all of them must be the same. The equivalent resistance of resistors in series is always greater than that of the largest individual resistance.

- When resistors are wired in parallel, the voltage across all of them must be the same. The equivalent resistance of resistors in parallel is always less than that of the smallest individual resistance.

- Kirchhoff's junction theorem states that the total current into a given junction equals the total current out of that junction (conservation of electric charge).

- Kirchhoff's loop theorem says that in traversing a complete circuit loop, the algebraic sum of the voltage gains and losses is zero, or the sum of the volt-

age gains equals the sum of the voltage losses (conservation of energy in an electric circuit).

- The time constant for an RC circuit is a "characteristic time" by which we measure the capacitor's charge and discharge.

- An ammeter is a device for measuring current; it consists of a galvanometer and a shunt resistance in parallel. Ammeters are connected in series with the circuit element carrying the current to be measured.

- A voltmeter is a device for measuring voltage; it consists of a galvanometer and a multiplier resistor wired in series. Voltmeters are connected in parallel with the circuit element experiencing the potential difference to be measured.

Important Equations

Equivalent Series Resistance:
$$R_s = R_1 + R_2 + R_3 + \cdots = \Sigma_i R_i \qquad (18.2)$$

Equivalent Parallel Resistance:
$$\frac{1}{R_p} = \frac{1}{R_1} + \frac{1}{R_2} + \frac{1}{R_3} + \cdots = \Sigma_i \frac{1}{R_i} \qquad (18.3)$$

Equivalent Parallel Resistance (two resistors):
$$R_p = \frac{R_1 R_2}{R_1 + R_2} \qquad (18.3a)$$

Kirchhoff's Rules:

(1) $\Sigma_i I_i = 0$ *(junction theorem)*   (18.4)

(2) $\Sigma_i V_i = 0$ *(loop theorem)*   (18.5)

Charging Voltage across a Capacitor in an RC Circuit:
$$V_C = V_o(1 - e^{-t/RC}) \qquad (18.6)$$

Current (Charging and Discharging) in an RC Circuit:
$$I = I_o e^{-t/RC} \qquad (18.7)$$

Time Constant for an RC Circuit:
$$\tau = RC \qquad (18.8)$$

Discharging Voltage across a Capacitor in an RC Circuit:
$$V_C = V_o e^{-t/RC} = V_o e^{-t/\tau} \qquad (18.9)$$

Ammeter Current:
$$I_g = \frac{IR_s}{r + R_s} \qquad (18.10)$$

Voltmeter Current:
$$I_g = \frac{V}{r + R_m} \qquad (18.11)$$

Exercises

Assume all resistors are ohmic unless otherwise stated.

18.1 Resistances in Series, Parallel, and Series–Parallel Combinations

1 The same amount of current must flow through each resistor in a circuit when they are connected in (a) series, (b) parallel, (c) series–parallel, (d) parallel–series.

2 The potential difference across each resistor in a circuit must be the same when they are connected in (a) series, (b) parallel, (c) series–parallel, (d) parallel–series.

3 Are the voltage drops across resistors in series generally the same? If not, could they be so in certain cases?

4 Are the currents in resistors in parallel generally the same? If not, could they be so in certain cases?

5 ■ Three resistors that have values of 10 Ω, 5 Ω, and 2 Ω are connected in series to a battery. Which gets the most delivered power and why?

6 ■ Three resistors that have values of 10 Ω, 5 Ω, and 2 Ω are connected in parallel to a battery. Which gets the most delivered power and why?

7 ■ Three resistors that have values of 20 Ω, 40 Ω, and 60 Ω are to be connected. (a) What is the maximum equivalent resistance? (b) What is the minimum equivalent resistance?

8 ■ Two identical resistors (R) are connected in parallel and then wired in series to a 40-Ω resistor. If the total equivalent resistance is 55 Ω, what is the value of R?

9 ■ Three 4.0-Ω resistors can be connected in different ways to produce different equivalent resistances. Draw a circuit diagram showing each way, and find the total resistance in each case.

10 ■ Three resistors with values of 5.0 Ω, 10 Ω, and 15 Ω are connected in series in a circuit with a 9.0-V battery. (a) What is the total equivalent resistance? (b) What is the current through each resistor? (c) What is the rate at which energy is dissipated in the 15-Ω resistor?

11 ■ Find the equivalent resistances for all possible combinations of two or more of the three resistors in Exercise 10.

12 ■■ If you had only an infinite supply of 3 Ω resistors, what would be the simplest way to make an equivalent resistance of 4.5 Ω ?

13 ■■ A length of wire with a resistance of 27 μΩ is cut into three equal segments. The segments are then wrapped together to form a conductor a third as long as the original wire. What is the resistance of the shortened conductor?

14 ■■ You are given four resistors with values of 2.5 Ω, 3.5 Ω, 4.5 Ω, and 5.5 Ω. Is it possible to connect all of the resistors in a combination so as to produce an effective total resistance of 5.0 Ω? If so, describe the arrangement.

15 ■■ Three resistors with values of 2.0 Ω, 4.0 Ω, and 6.0 Ω are connected in series and put in a circuit with a 12-V battery. (a) How much current is delivered to the circuit by the battery? (b) What is the current through each resistor? (c) How much power is dissipated by each resistor? (d) How does this compare with the power dissipated by the total equivalent resistance?

16 ■■ Suppose that the resistors in Exercise 15 are connected in parallel. (a) How much current is drawn from the battery? (b) How much current flows through each resistor? (c) How much power is dissipated by each resistor? (d) How does this compare with the power dissipated by the total equivalent resistance?

17 ■■ Two 8.0-Ω resistors are connected in parallel, and two 4.0-Ω resistors are connected in parallel. These combinations are then connected in series and put in a circuit with a 12-V battery. What are the current through and the voltage across each resistor?

18 ■■ What is the equivalent resistance for the resistors in •Fig. 18.24?

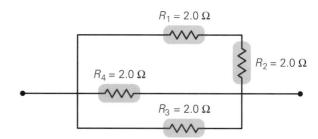

•**FIGURE 18.24 Series–parallel combination**
See Exercises 18 and 28.

19 ■■ What is the equivalent resistance between points A and B in •Fig. 18.25?

20 ■■ Find the equivalent resistance for the arrangement of resistors shown in •Fig. 18.26.

21 ■■ Two 60-W and one 100-W light bulbs are connected across (in parallel with) a 120-V source. (a) What is the current through each bulb? (b) How much power is dissipated by each bulb?

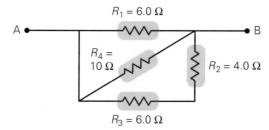

•FIGURE 18.25 Series–parallel combination
See Exercises 19 and 30.

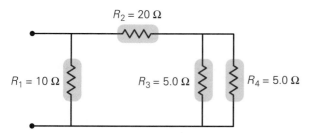

•FIGURE 18.26 Series–parallel combination
See Exercise 20.

22 ■■ Several 60-W light bulbs are connected in parallel with a 120-V source and the very last one blows a 15-A fuse in the circuit. (a) Sketch the schematic circuit being sure to show the fuse in relation to the bulbs. (b) How many light bulbs are in the circuit (including the very last one)?

23 ■■ Use three 50-Ω resistors and a 120-V source in a circuit in any arrangement. (a) What is the maximum power for the circuit? Sketch the arrangement. (b) What is the minimum power? Sketch the arrangement.

24 ■■ Find the current and the voltage drop for the 10-Ω resistor in •Fig. 18.27.

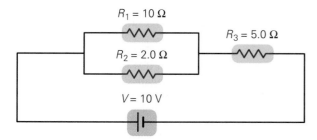

•FIGURE 18.27 Current and voltage drop for a resistor
See Exercise 24.

25 ■■ A three-way light bulb can produce 50 W, 100 W, or 150 W of power at 120 V. (a) Find the current for each power setting. (b) Find the resistance for each power setting.

26 ■■ For the circuit shown in •Fig. 18.28, find (a) the current through each resistor, (b) the voltage across each resistor, and (c) the total power dissipated.

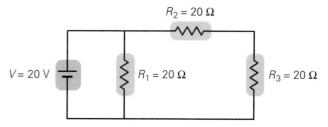

•FIGURE 18.28 Circuit reduction
See Exercises 26 and 46.

27 ■■ A 240-V circuit has a circuit breaker rated to trip at 20 A. How many 30-Ω resistors could be connected in parallel in the circuit and not trip the breaker?

28 ■■■ Suppose that the resistor arrangement in Fig. 18.24 is connected to a 12-V battery. (a) What will the current through each resistor be? (b) What will the voltage drop across each resistor be? (c) What will the total power dissipated be?

29 ■■■ Three 100-Ω immersible resistors and a 120-V source are used to heat 0.50 kg of water initially at 10°C as rapidly as possible. (Assume that there is no heat loss to the atmosphere, cup, etc.) (a) How much time is required to heat the water to 100°C? (b) How much additional time is required to boil away all of the water?

30 ■■■ The terminals of a 6.0-V battery are connected to points A and B in Fig. 18.25. (a) How much current flows through each resistor? (b) How much power is dissipated by each resistor? (c) Compare the sum of the individual power dissipations to the power dissipation of the equivalent resistance for the circuit.

31 ■■■ Light bulbs with the wattage ratings shown in •Fig. 18.29 are connected in a circuit. (a) What current does the voltage source deliver to the circuit? (b) Find the power dissipated by each bulb. (Assume normal operating resistances of the bulbs to be constant at any voltage.)

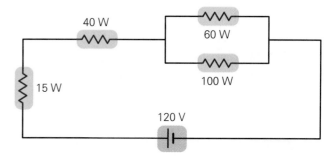

•FIGURE 18.29 Watt's up?
See Exercise 31.

32 ■■■ For the circuit in •Fig. 18.30, find (a) the current through each resistor and (b) the voltage drop across each resistor.

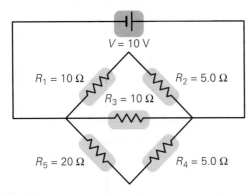

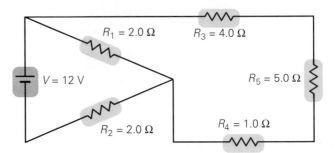

•**FIGURE 18.30 Resistors and currents**
See Exercise 32.

33 ■■■ What is the total power dissipated in the circuit shown in •Fig. 18.31?

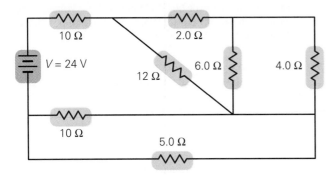

•**FIGURE 18.31 Power dissipation**
See Exercise 33.

34 ■■■ What is the equivalent resistance of the arrangement in •Fig. 18.32?

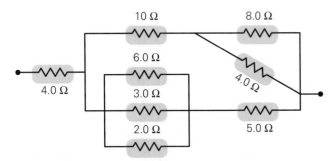

•**FIGURE 18.32 Equivalent resistance replacement**
See Exercise 34.

35 ■■■ Find the current through each of the resistors in the circuit in •Fig. 18.33.

36 ■■■ The circuit shown in •Fig. 18.34 is called a Wheatstone bridge after Sir Charles Wheatstone (1802–1875) and is used to make accurate measurements of resistance without the errors of ammeter–voltmeter measurements (see Exercises 77 and 78). The resistances R_1, R_2, and R_s are all known, and R_x is the unknown resistance to be measured; R_s is variable and is adjusted until the bridge circuit is balanced, when the galvanome-

•**FIGURE 18.33 How much current?**
See Exercise 35.

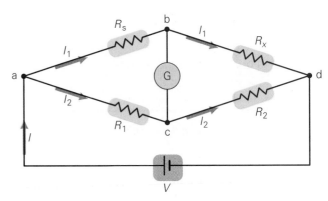

•**FIGURE 18.34 Wheatstone bridge**
See Exercise 36.

ter (G) shows a zero reading (no current through the galvanometer branch). Show that when the bridge is balanced R_x is given by

$$R_x = \left(\frac{R_2}{R_1}\right)R_s$$

18.2 Multiloop Circuits and Kirchhoff's Rules

37 A multiloop circuit has more than one (a) junction, (b) branch, (c) current, (d) all of these.

38 By our sign convention, if a resistance is traversed in the direction opposite to that of the conventional current, (a) the current is negative, (b) the current is positive, (c) the potential difference (voltage) is negative, (d) the potential difference (voltage) is positive.

39 If you traversed a circuit and came upon a capacitor and you went from negative to positive plate, the potential difference (voltage) would be classified as (a) positive, (b) negative, (c) either.

40 ■ Traverse loop 3 *opposite* the direction in Fig. 18.9, and show that the resulting equation is the same as if you followed the arrow direction.

41 ■ For the circuit in Fig. 18.9, reverse the directions of the interior loops and show that equations equivalent to those in Example 18.5 are obtained.

42 ■ An ideal battery (emf $\mathscr{E}$ and zero internal resistance) is connected to a single external resistor (R). Assume the wires have a small but not negligible resistance, R'. Use Kirchhoff's loop theorem to show that the voltage across the resistor is less than $\mathscr{E}$ and given by $V = \mathscr{E}R/(R + R')$.

43 ■ Two resistors, R_1 and R_2, are connected in series across a battery with terminal voltage V. Use Kirchhoff's rules to show that (a) the current each carries, (b) the voltage across each, and (c) the power expended in each (joule heat) is the same regardless of the order in which they are connected.

44 ■■ Two batteries (terminal voltages 10 V and 4 V) are connected with positive terminals together. A 12-Ω resistor is wired between their negative terminals. (a) Use Kirchhoff's loop theorem to find the current in the circuit and the power dissipated in the resistor. (b) Compare this result to the power output of each battery.

45 ■■ Find the current in each resistor in •Fig. 18.35 using Kirchhoff's rules.

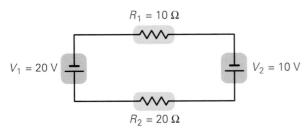

•**FIGURE 18.35 Single-loop circuit**
See Exercise 45.

46 ■■ Apply Kirchhoff's rules to the circuit in Fig. 18.28 to find the current through each resistor.

47 ■■ Apply Kirchhoff's rules to the circuit in •Fig. 18.36, and find (a) the current in each resistor and (b) the rate at which energy is being dissipated in the 8.0-Ω resistor.

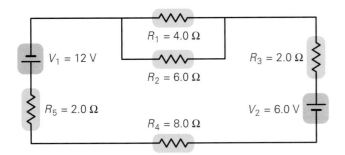

•**FIGURE 18.36 A loop in a loop**
See Exercise 47.

48 ■■ Find the current through each resistor in the circuit in •Fig. 18.37.

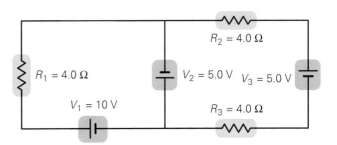

•**FIGURE 18.37 Double-loop circuit**
See Exercise 48.

49 ■■ Find the currents in the circuit branches in •Fig. 18.38.

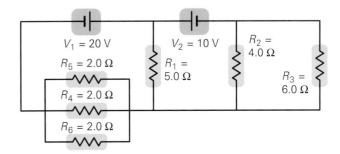

•**FIGURE 18.38 How many loops?**
See Exercise 49.

50 ■■■ For the multiloop circuit shown in •Fig. 18.39, what is the current through each branch?

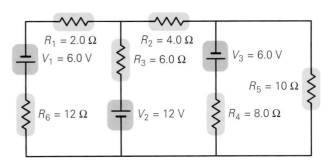

•**FIGURE 18.39 Triple-loop circuit**
See Exercise 50.

51 ■■■ A procedure called the loop current method is considered by some to be mathematically simpler than Kirchhoff's branch current method. In the loop current method, a particular current is assigned to each loop (in •Fig. 18.40, I_A and I_B). Kirchhoff's second rule is then used to write an equation for each loop, incorporating the sign convention. The equations are solved for the loop currents, and then Kirchhoff's first rule is applied to each junction in the circuit to find the branch currents by comparison with the loop currents. Show that this loop current method gives the same currents for the circuit in Fig. 18.40 as the branch current method. (Remember that all the loop currents in a branch must be taken into account when applying the loop and junction theorems.)

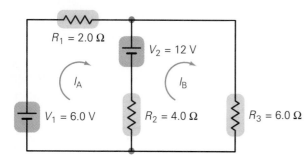

•**FIGURE 18.40** **The loop-current method**
See Exercise 51.

18.3 RC Circuits

52 In charging a capacitor in series with a resistance, in one time constant (a) the capacitor is 63% charged, (b) the voltage across the capacitor is 37% of its maximum value, (c) the current in the circuit decreases to 37% of its maximum value, (d) both (a) and (c).

53 When charged by a battery, an RC circuit has its maximum current (a) at the beginning of the charging process, (b) near the middle of the charging process, (c) at the end of the charging process, (d) after one time constant.

54 When discharging a capacitor through a resistor, an RC circuit has its maximum current (a) at the beginning of the discharging process, (b) near the middle of the discharging process, (c) at the end of the discharging process, (d) after one time constant.

55 ■ A single resistor (R) and an uncharged capacitor (C) are wired in series with a battery (terminal voltage V) and an open switch. At time $t = 0$, the switch is closed and the capacitor begins to charge. What is the voltage across the resistor and across the capacitor, expressed as fractions of V (to two significant figures), (a) just after the switch is closed, (b) after one time constant has elapsed, and (c) after many times constants?

56 ■ A single resistor (R) and a fully charged capacitor (C) are wired in series with a switch that has been open for a long time. The capacitor had been charged by a battery with terminal voltage V for a long time before being wired into this circuit. At time $t = 0$, the switch is closed and the capacitor begins to discharge through the resistor. What is the voltage across the resistor and across the capacitor, expressed as fractions of V (to two significant figures), (a) just after the switch is thrown, (b) after one time constant has elapsed, and (c) after many times constants?

57 ■ Why doesn't the voltage across the neon tube in an oscillator circuit fall below the maintaining voltage and go to zero?

58 ■ A capacitor in a single-loop RC circuit is charged to 63% of its final voltage in 1.5 s. Find (a) the time constant for the circuit and (b) the percentage of its final voltage after 15 s. (Use five significant figures to show how close to 100% this is.)

59 ■ For an RC circuit with a 2.5 MΩ resistance, what capacitance will give a time constant of 1.0 s?

60 ■■ An RC circuit has a time constant of 2.25 s. The capacitor is in the microfarad range and has the same numerical prefix as the resistor. What are the values of the capacitor and the resistor?

61 ■■ How many time constants will it take for an initially charged capacitor to be discharged to half of its initial voltage?

62 ■■ A 1.00 μF capacitor in series with a resistor is connected to a 12.0-V source and charged. (a) What resistance is necessary to cause the capacitor to have only 37% of its initial charge 1.50 s after starting to discharge? (b) What is the voltage across the capacitor at $t = 3\tau$ when it is charging?

63 ■■ An RC circuit with $C = 40$ μF and $R = 6.0$ Ω has a 24-V source. With the capacitor initially uncharged, an open switch in the circuit is closed. (a) What is the potential difference across the 6.0-Ω resistor immediately afterward? (b) What is the potential difference across the capacitor at that time? (c) What is the current in the resistor at that time?

64 ■■ (a) For the circuit in Exercise 63 after the switch has been closed for $t = 4\tau$, what is the charge on the capacitor? (b) After a long time, what are the voltages across the capacitor and the resistor?

65 ■■■ An RC circuit with a resistance of 5.0 MΩ and a capacitance of 0.40 μF is connected to a 12-V source. If the capacitor is initially uncharged, what is the change in voltage across it between $t = 2\tau$ and $t = 4\tau$?

66 ■■■ A 3.0-MΩ resistor is connected in series with a 0.28-μF capacitor. If this arrangement is hooked across four 1.5-V batteries connected in series, (a) what is the initial voltage across the capacitor and the initial current in the circuit? (b) How much charge is on the capacitor after 4.0 s?

18.4 Ammeters and Voltmeters

67 A voltmeter has a (a) large shunt resistance, (b) large multiplier resistance, (c) small shunt resistance, (d) small multiplier resistance.

68 To measure the current through a circuit element, an ammeter is connected (a) in series with the element, (b) in parallel with the element, (c) between the high potential side of the element and ground, (d) none of these.

69 What would happen if (a) an ammeter were connected in parallel in a circuit, and (b) a voltmeter were connected in series?

70 ■ A galvanometer with a full-scale sensitivity of 2000 μA has a coil resistance of 100 Ω. If it is to be used in an ammeter with a full-scale reading of 3.0 A, what is the necessary shunt resistance?

71 ■ If the galvanometer in Exercise 70 were to be used in a voltmeter with a full-scale reading of 1.5 V, what would the required value of the multiplier resistor be?

72 ■ A galvanometer with a full-scale sensitivity of 600 μA and a coil resistance of 50 Ω is to be used in an ammeter designed to read 5.0 A at full scale. What is the required value of the shunt resistor?

73 ■■ A galvanometer has a coil resistance of 20 Ω. A current of 200 μA deflects the needle through 10 divisions full-scale. What is needed to convert the galvanometer to a full-scale 10-V voltmeter?

74 ■■ An ammeter has a resistance of 1.0 mΩ. Find the current in the ammeter when it is properly connected to a 10-Ω resistor and a 6.0-V source. (Express your answer to five significant figures to show how different your answer is from 0.60 A.)

75 ■■ A voltmeter has a resistance of 20 kΩ. What is the current through the meter when it is connected across a 10-Ω resistor that is hooked to a 6.0-V source?

76 ■■■ An ammeter and a voltmeter can be used to measure the value of a resistor in a circuit. Suppose that the ammeter is connected in series with the resistor, and the voltmeter is placed across the resistor *only*. (a) Show that the value of the resistance when measured like this is given by

$$R = \frac{V}{I - (V/R_V)}$$

where I is the current measured by the ammeter, V is the voltage measured by the voltmeter, and R_v is the resistance of the voltmeter. [*Hint:* Draw a circuit diagram to help analyze the problem.] (b) Use this result to explain why the "perfect" voltmeter would have infinite resistance.

77 ■■■ An ammeter and a voltmeter are used to measure the value of a resistor in a circuit. The ammeter is connected in series with the resistor, and the voltmeter is placed across *both* the ammeter and the resistor. (a) Show that the value of the resistance when measured like this is given by

$$R = (V/I) - R_a$$

where V is the voltage measured by the voltmeter, I is the current measured by the ammeter, and R_a is the resistance of the ammeter. [*Hint:* Draw a circuit diagram to help analyze the problem.] (b) Use this result to explain why the "perfect" ammeter has zero resistance.

18.5 Household Circuits and Electrical Safety

78 The so-called neutral or ground wire in household wiring (a) is a current-carrying wire, (b) is at a voltage difference of 240 V from a "hot" wire, (c) carries no current, (d) is not part of house circuits.

79 A dedicated grounding wire (a) is the basis for the polarized plug, (b) is necessary for a circuit breaker, (c) normally carries no current, (d) none of these.

80 ■ What is wrong with the circuit in ●Fig. 18.41 in terms of electrical safety, and why?

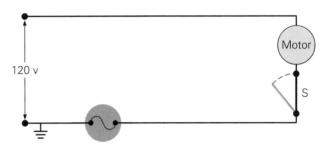

●**FIGURE 18.41 A safety problem?**
See Exercise 80.

81 ■ Bodily injury depends on the magnitude of the current and its path, yet you commonly see signs that warn "Danger: High Voltage" (●Fig. 18.42). Shouldn't this refer to high current? Explain.

●**FIGURE 18.42 Danger—High Voltage**
Shouldn't it be high current? See Exercise 81.

82 ■ Explain why it is perfectly safe for birds to perch with both feet on the same high voltage wire, even if the insulation is worn through.

83 ■ After a collision with a power pole, you are fine but trapped in a car with a high voltage line (frayed insulation) in contact with the hood of your car. Is it safer to step off the car one foot at a time, or to jump, both feet leaving the car at the same time? Explain your reasoning.

84 ■ Most electrical codes require the metal case of an electric clothes dryer to have a wire running from the case to a nearby faucet (or metal piece of plumbing). Explain why this is the case.

Additional Exercises

85 Five 30-Ω resistors are connected in parallel. What is the total equivalent resistance?

86 Two 1.0-kΩ resistors are connected in parallel. That arrangement is connected in series to another 1.0-kΩ resistor in a circuit containing a battery. If each resistor is rated at 0.25 W, find the maximum voltage of the battery that can be safely used in the circuit.

87 A 4.0-Ω resistor and a 6.0-Ω resistor are connected in series. A third resistor is connected in parallel with the 6.0-Ω resistor. This gives a total equivalent resistance of 7.0 Ω for the whole arrangement. What is the value of the third resistor?

88 Find the current through each resistor in the circuit in •Fig. 18.43.

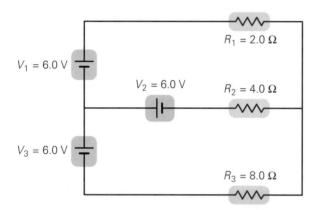

•**FIGURE 18.43** Kirchhoff's rules
See Exercise 88.

89 Resistors of 20 Ω and 15 Ω are connected in parallel in a circuit with a 9.0-V battery. How much energy is dissipated each second by the resistors?

90 Four resistors are connected to a 90-V source as shown in •Fig. 18.44. (a) Which resistor(s) dissipate the most power and how much? (b) What is the total power dissipated in the circuit?

91 Four resistors are connected in a circuit with a 110-V source, as shown in •Fig. 18.45. (a) What is the current through each resistor? (b) How much power is dissipated by each resistor?

92 Nine equal resistors, each of value R, are connected in a "ladder" fashion as shown in •Fig. 18.46. What is the effective resistance of this network between points A and B?

93 In Exercise 92, if a 12.0-V battery is connected from point A to point B, how much current flows through each resistor?

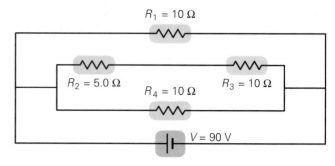

•**FIGURE 18.44** How much power is dissipated?
See Exercise 90.

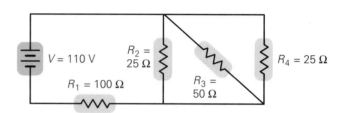

•**FIGURE 18.45** Joule heat losses
See Exercise 91.

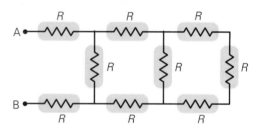

•**FIGURE 18.46** A resistance ladder
See Exercise 92.

94 Show that the unit of the time constant is the second.

95 An RC circuit has a time constant of 0.75 s. If the capacitor has a capacitance of 0.25 μF, what is the value of the resistance?

96 A resistor of 6.0 Ω and a capacitance of 0.20 μF is connected *in parallel* to a 12-V source. Initially the switch on the battery branch is open and the capacitor uncharged. The switch is then closed. After a long time, (a) what is the current through the resistor? (b) The current through the battery? (c) The voltage across the capacitor? (d) The charge on the capacitor?

97 The time between flashes of the neon tube, or the period of the oscillator circuit, in Fig. 18.13 is the time it takes for the voltage to rise from V_m to V_b. Show that the period of such an oscillator is given by

$$T = t_b - t_m = RC \ln\left(\frac{V_o - V_m}{V_o - V_b}\right)$$

where V_o is the maximum voltage, or the voltage of the battery in the circuit.

98 •Fig. 18.47 shows the workings of a potentiometer which provides a very accurate way of determining emf. It consists of three batteries, an ammeter, and several resistances including a uniform long wire which may be "tapped" for a specific fraction of its total resistance. $\mathcal{E}_o$ is called the working battery, #1 is a battery with a very well-known emf, and #2 is the unknown battery. The switch S is thrown toward battery #1 and the point T (for "tapped") is moved along the resistor until the ammeter reads zero. Let us call the resistance of this R_1. This procedure is repeated with the switch thrown toward battery #2 and the point T is moved to T′ until the ammeter again reads zero. Let us call the resistance of this R_2. Show that

$$\mathcal{E}_2 = \frac{R_2}{R_1}\mathcal{E}_1$$

99 Two 75-W light bulbs are connected to a 110-V power source. What is the current in each bulb and the total power dissipated if the bulbs are connected (a) in series and (b) in parallel?

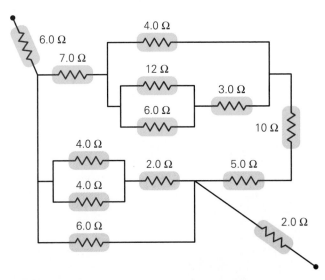

•**FIGURE 18.48 Connect it to a voltage source**
See Exercise 100.

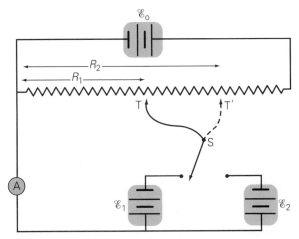

•**FIGURE 18.47 The potentiometer**
See Exercise 98.

100 If a 120-V source were connected across the leads (open ends) of the circuit shown in •Fig. 18.48, how much current would be drawn from it?

101 You have three light bulbs with power ratings of 40, 60, and 100 W. Assuming they are ohmic (which they are not) and that they are wired in *parallel* to a 120-V source, compute (a) the voltage across each, (b) the current flow through each, (c) the power delivered to each, (d) the total current delivered to the group, and (e) the total power delivered to the group. (f) Redo your answers assuming the 100-W bulb is turned off.

102 Assume you have the same bulbs as in Exercise 101, but they are wired in series. Again assuming they are ohmic, compute (a) the voltage across each, (b) the current flow through each, (c) the power delivered to each, (d) the total current delivered to the group, and (e) the total power delivered to the group. (f) Redo your answers assuming the 100-W bulb is turned off.

103 If a combination of three 30-Ω resistors dissipates 3.2 W when connected to a 12-V battery, how are the resistors connected in the circuit?

104 A battery has three cells, each with an internal resistance of 0.020 Ω and an emf of 1.50 V. The battery is connected in parallel with a 10.0-Ω resistor. (a) Determine the voltage across the resistor. (b) How much current passes through each cell? (The cells in a battery are connected in series.)

105 A galvanometer with an internal resistance of 50 Ω and a full-scale sensitivity of 200 μA is used to construct a multivoltmeter. What values of multiplier resistors allow for a three full-scale voltage readings of 10 V, 50 V, and 250 V? (See Fig. 18.17b.)

106 A galvanometer with an internal resistance of 100 Ω and a full-scale sensitivity of 100 μA is used to construct a multiammeter. What values of shunt resistors allow for three full-scale current readings of 0.30 A, 3.0 A and 30 A? (See Fig. 18.17a.)

19 Magnetism

When we think of magnetism, we tend to think of attraction. We know that you can pick things up with a magnet—at least, certain kinds of things. Most of us have probably encountered magnetic latches that hold cabinet doors shut, or used magnets to stick notes to the refrigerator. We're less likely to think of repulsion. Yet, wherever there are magnetic attractive forces, there must also be magnetic repulsive forces—and they can be just as useful.

The photo above shows an interesting example. At first glance, this looks like a perfectly ordinary locomotive. But—where are the wheels? In fact, this isn't a conventional train but a high-speed magnetically-levitated train. This means that the train doesn't touch the "rails" at all: it "floats" above them, sup-

ported by repulsive forces that are the product of powerful magnets. The advantages are obvious—with no wheels, there is no rolling friction, no bearings to lubricate—in fact, very few moving parts of any kind. Magnetic forces are also used to accelerate and decelerate the train.

But where do the magnetic forces come from? You probably realize that it wouldn't be feasible to rely on permanent magnets of the kind that you find on your refrigerator door. Maglev trains, like so many other important modern magnetic devices, use electromagnets, which create a magnetic field by means of an electric current. In this chapter you'll explore the close connection between electricity and magnetism, which underlies applications as diverse as high-speed trains and TV picture tubes.

Children, and many adults, are fascinated by magnetism. Whether we have used a compass to guide ourselves through the woods or merely attached a shopping list to the refrigerator with a magnet, few of us have not experienced this strange force at first hand. But magnets, produced commercially in great quantities today, were once quite scarce—they existed only as natural magnets, magnetized stones found in nature.

For centuries, the attractive properties of magnets were attributed to supernatural forces. We now associate magnetism with electricity (electromagnetism) as both turn out to be aspects of a single fundamental force or interaction (the electromagnetic force). We have put electromagnetism to use in motors, generators, radios, telephones, and so many other familiar applications that ours might be called an electromagnetic society. The development of high-temperature superconducting magnets (Chapter 17) may soon open the way for practical exploitation of many more devices that are now found only in the laboratory or as experimental prototypes—for example, high-speed trains levitated and propelled by magnetic fields.

Although electricity and magnetism are manifestations of the same fundamental force, it is instructive first to consider them individually and then put them together, so to speak. This and the next chapter will investigate magnetism and its intimate relationship to electricity.

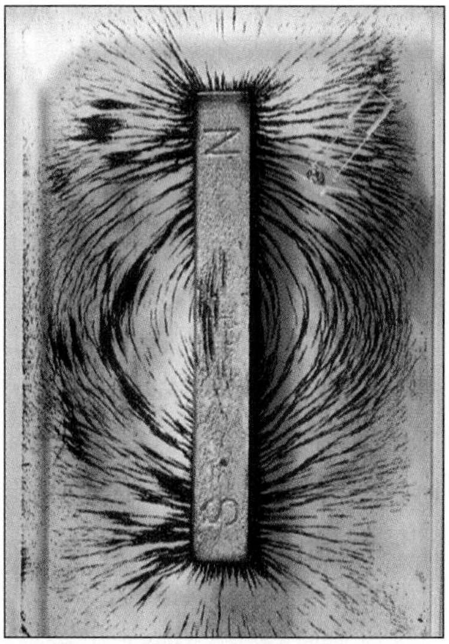

19.1 Magnets and Magnetic Poles

Objectives: To be able to (a) state the force rule between magnetic poles, and (b) explain how the magnetic field is defined.

One of the first things anyone notices in examining a common bar magnet is that it has two "centers" of force called poles, one at or near each end of the magnet (•Fig. 19.1). Rather than being distinguished as positive and negative, these poles are called north (N) and south (S). This terminology comes from the early use of the magnetic compass. The north pole of a compass magnet is the *north-seeking* end—that is, the end that tends to point *north* on Earth. With two bar magnets, we notice a pattern of attraction and a repulsion between their various ends: Each end is attracted to one end of a second magnet and repelled by the other.

The attraction and repulsion between poles of magnets is similar to the behavior of like and unlike electric charges. That is, analogous to the charge–force law, is the **pole–force law**, or **law of poles**:

> Like magnetic poles repel, and unlike magnetic poles attract.

That is, north and south poles (N–S) attract each other, and north and north poles (N–N) or south and south (S–S) poles repel each other (•Fig. 19.2).

Magnetic poles have always been observed to occur in pairs, never singly. Two opposite poles form what is called a magnetic *dipole*. You might think you could break a bar magnet in half and get two single isolated poles. However, you would find that the pieces are two shorter magnets, *each with a north and a south pole*. Although an isolated, single magnetic pole (a magnetic monopole) has been postulated to exist, it has yet to be found experimentally.

•**FIGURE 19.1 Bar magnet**
The iron filings indicate the poles or centers of force of a common bar magnet. The compass direction designates these poles as north (N) and south (S). (See Fig. 19.3.)

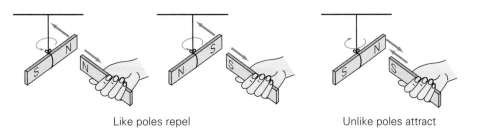

Like poles repel   Unlike poles attract

•**FIGURE 19.2 The law of poles**
Like poles (N–N or S–S) repel, and unlike poles (N–S) attract.

•FIGURE 19.3 Magnetic fields
Magnetic field lines may be outlined using iron filing or a compass. The filings behave like tiny compasses and line up with the field. The closer together the field lines, the stronger the magnetic field. **(a)** The magnetic field of a single bar magnet traced using a compass. **(b)** Iron filing pattern for the magnetic field between unlike poles. **(c)** Iron filing for the magnetic field between like poles. Notice how the field lines run between the poles in (b) and diverge in (c).

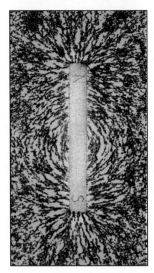

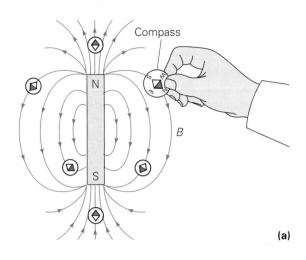

(a)

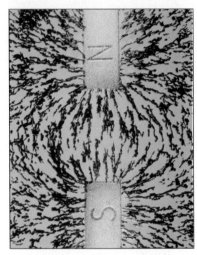

(b)

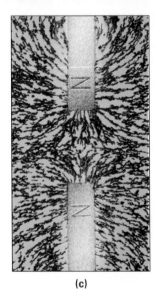

(c)

The fact that a magnet always has two poles may make you wonder about the nature of magnetism. Although it appears to have many similarities to electricity, it does not have the same fundamental nature. Even the source of magnetism, although ultimately derived from electric charge, is different. As will be discussed shortly, magnetism is produced by *electric charges in motion*, such as electric currents, orbiting atomic electrons, and so on. Stationary electric charges produce only electric fields, not magnetic ones. This understanding of the source of magnetism and some knowledge of magnetic materials (to be covered in a later section) will allow you to see why pieces of a magnet are themselves dipole magnets.

Defining the Magnetic Field (**B**)

From your knowledge of electrostatics, you may be tempted to think of a pole as a magnetic "charge" and to wonder if the magnetic force can be expressed in a form similar to Coulomb's law for electric charges, given in Chapter 15. In fact, such a law was developed by Coulomb using the magnetic pole strengths in place of the electric charges. However, this law is now rarely used because it doesn't match present understanding of magnetism and because it is more convenient to work with magnetic fields.

In Section 15.4, you learned how convenient it is to describe the interaction of electrically charged objects in terms of electric fields. There is an electric field surrounding any electric charge, and this is represented by electric field lines, or lines of force. Mathematically, the electric field is defined as the force per unit charge, $\mathbf{E} = \mathbf{F}/q$. Similarly, it is convenient to describe magnetic interactions in terms of magnetic fields. Hence, we consider the magnetic field that surrounds a magnet. Like an electric field, a **magnetic field** is a vector quantity and is represented by the symbol **B**. The pattern of the magnetic field lines surrounding a magnet can be visually demonstrated by sprinkling iron filings over a magnet covered with a piece of paper or glass (•Fig. 19.1). As you will learn, the iron filings become induced magnets and line up with the field much as compass needles line up with the Earth's magnetic field.

To describe any vector field, we must specify both the magnitude, or strength, and direction of the field at various points. The direction of a magnetic field (which we refer to simply as a "*B* field") is defined using another north magnetic pole:

> The direction of a magnetic field (**B**) at any location is in the direction that the north pole of a compass would point if a compass were at that location.

This definition provides another method of mapping a magnetic field by placing a small compass near a magnet. The torque on the compass needle (a small magnet) due to the magnetic forces on the two opposite poles of the needle causes the needle to rotate and line up in the direction of the existing field. If the compass is moved in the direction in which the needle points, the path of the needle traces out a field line, as illustrated in •Fig. 19.3a.

The field outside a bar magnet, by the pole-force law, points away from the north pole of a magnet and toward the south pole. As with an electric field, the closer together the field lines, the stronger the field. Note how this is indicated by the concentration of iron filings in the pole regions in •Fig. 19.3 b and c.

Since the magnetic north pole of a compass can be used to map out a magnetic field, you might think that **B** would be the magnetic force per unit pole (as **E** is the electric force per unit charge). The magnetic field was once thought of in this manner. However, the magnetic field is now defined in terms of the force on a moving electric charge, as will be discussed in the next section.

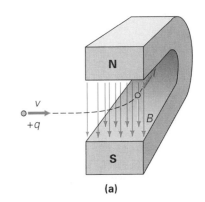

(a)

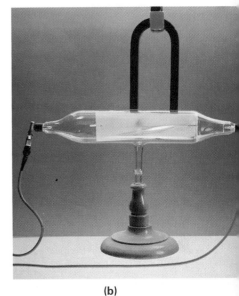

(b)

•**FIGURE 19.4 Force on a moving charged particle**
(a) When a charged particle enters a magnetic field, it experiences a force that is evident because the charge is deflected from its original path. (b) The electron beam in a cathode ray tube (made visible by fluorescent paper) is normally horizontal between the end electrodes, but is deflected here because of the magnet.

19.2 Electromagnetism, Magnetic Forces, and the Source of Magnetic Fields

Objectives: To be able to (a) determine the direction and magnitude of the magnetic force exerted by a magnetic field on a moving charged particle, and (b) calculate the direction and magnitude of the magnetic field produced by an electric current.

Experiments indicate that one of the important quantities in determining the *magnetic* force on a particle is the *electric* charge of that particle. That is, there seems to be some connection between *electrical* properties of objects and how they respond to *magnetic* fields. This connection is referred to as an *electromagnetic effect* and the study of interactions between electrically charged particles and magnetic fields is called **electromagnetism**.

A specific example of such an electromagnetic interaction occurs when a particle with a positive charge (q) moving with a constant velocity enters a region with a uniform magnetic field (for example, between two closely placed pole faces of a large permanent magnet) in a path such that the velocity and the magnetic field are at right angles (•Fig. 19.4a). When the charge enters the field, it is deflected into a curved path, which, if carefully measured, turns out to be a circular arc.

From the study of dynamics, you know that this circular motion must be caused by a force perpendicular to the particle's velocity. But what gives rise to this force? No electric field is present (other than that of the charge itself). The force of gravity is too weak to cause this deflection and, furthermore, as we learned in our study of two-dimensional motion, it would deflect the particle *downward* rather than sideways and into a parabolic arc, not a circular one. Evidently, the force is due to the interaction of the moving charge and the magnetic field. That is, *an electrically charged particle moving in a magnetic field may experience a force.*

Varying the charge, its speed, and the magnitude of the magnetic field experimentally shows that the magnitude of the deflecting force is directly proportional to each of these quantities. That is, $F \propto qvB$. In equation form, for the special case in which the velocity vector (**v**) of the charge is perpendicular to the magnetic field (**B**), we can write

$$F = qvB \qquad \text{(only for } \mathbf{v} \text{ perpendicular to } \mathbf{B}) \qquad (19.1)$$

This gives an expression for the strength (magnitude) of the magnetic field in terms of familiar quantities:

$$B = \frac{F}{qv} \qquad \text{(only for } \mathbf{v} \text{ perpendicular to } \mathbf{B}) \qquad (19.2)$$

SI unit of magnetic field: N/A·m, or tesla (T)

Note: Equations 19.1 and 19.2 are true only if the velocity is perpendicular to the magnetic field.

That is, B is the magnetic force per unit moving charge per unit velocity. The units are chosen so that there is no constant of proportionality needed in the preceding two equations. From Eq. 19.2, you can see that a magnetic field has SI units of N/(C·m/s) or N/A·m where the ampere $A = C/s$. This combination of units is given the name **tesla (T)**. Thus $1\text{ T} = 1\text{ N/A·m}$. The magnetic field is sometimes given in webers per square meter (Wb/m^2): $1\text{ Wb/m}^2 = 1\text{ T}$. (The weber, a unit of magnetic *flux,* is discussed in Chapter 20.)

If the direction of the charged particle's velocity is not perpendicular to the magnetic field, the magnitude of the magnetic force on the particle is not given by Eq. 19.1. It has been found that the magnitude of this force depends on the angle (θ) between the velocity and the field vectors—more specifically, on the sine of that angle ($\sin\theta$). That is, the force is zero when **v** and **B** are parallel and a maximum when those two vectors are perpendicular. In general, the magnitude of the force is

$$F = qvB\sin\theta \qquad \begin{array}{l}\textit{magnetic force on}\\ \textit{a charged particle}\end{array} \qquad (19.3)$$

When **v** and **B** are perpendicular ($\theta = 90°$), the equation reduces to Eq. 19.1, $F = qvB\sin 90° = qvB$ (maximum force). However, when the **v** and **B** vectors are parallel ($\theta = 0°$ or $180°$), the force on the moving charge is zero, $F = qvB\sin 0° = 0$. You should be able to prove for yourself that in Eq. 19.3, $B\sin\theta$ is the magnitude of the component of **B** perpendicular to the velocity.

The Right-Hand Force Rule

The *direction* of the magnetic force on a positively charged particle is determined by considering the velocity and field vectors. This method is stated in the form of a **right-hand force rule** (•Fig. 19.5a):

> When the fingers of the right hand are pointed in the direction of **v** and then turned or curled toward the vector **B**, the extended thumb points in the direction of **F** for a *positive* charge. (For a negative charge, the force is in the opposite direction.)

You might imagine the fingers of the right hand to be physically turning or pushing the **v** vector into **B** so that both are aligned in the same direction. The magnetic

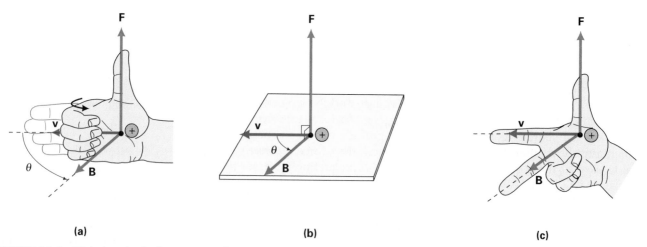

(a)　　　　　　　　　　(b)　　　　　　　　　　(c)

•**FIGURE 19.5 Right-hand rule for magnetic force**
(a) When the fingers of the right hand are pointed in the direction of **v** and then turned or curled toward the direction of **B**, the extended thumb points in the direction of the force **F** on a *positive* charge. **(b)** The force is always perpendicular to the plane of **B** and **v**, or always perpendicular to the direction of the particle's motion. **(c)** Another method. When the extended forefinger of the right hand points in the direction of **v** and the middle finger points in the direction of **B**, the extended right thumb points in the direction of **F**.

force is always *perpendicular to the imaginary plane* formed by the vectors **v** and **B** (•Fig. 19.5b). If the moving charge is negative, the right thumb points in the direction opposite that of the force. (You can simply assume that the negative charge is positive, apply the right-hand rule, and then reverse the direction given by the rule.)

Some people prefer a right-hand "three finger" rule as illustrated in •Fig. 19.5c: With the forefinger of the right hand pointing in the direction of **v** and the middle finger flexed in the direction of **B**, the extended right thumb points in the direction of **F**. Use whichever method you find easier.

CONCEPTUAL EXAMPLE 19.1 ■ EVEN FOR LEFTIES: USING THE RIGHT-HAND RULE

In a linear accelerator, a beam of protons travels horizontally northward. In order to have the beam deflected eastward by a uniform magnetic field which it enters, the field would have to be directed in which direction? (a) vertically downward, (b) west, (c) vertically upward, (d) south. *Clearly establish the reasoning and physical principle(s) used in determining your answer before checking it below. That is, **why** did you select your answer?*

Reasoning and Answer. Since the force on a moving charged particle is perpendicular to the plane of **v** and **B**, the magnetic field could not be in the horizontal plane, which eliminates (b) and (d). The field must be either vertically up or down. Using the right-hand rule for a positively charged particle to see if **B** is downward (answer a), you should find that the initial force would then be to the west. Hence, the answer is (c) for an easterly deflection. Prove this to yourself using the right-hand rule.

Follow-up Exercise. In what direction would a beam of *electrons* have to travel to experience a force in the same *direction* from the same magnetic field (vertically upward) as in this Example?

Note: B fields into the plane of the page are designated by ×. If the field points out of the plane, we use a dot •.

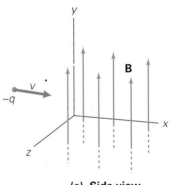

(a) Side view

EXAMPLE 19.2 ■ MOVING IN CIRCLES: FORCE ON A MOVING CHARGE

A negatively charged particle with a charge of -5.0×10^{-4} C moves at a speed of 1.0×10^2 m/s in the +x direction toward a uniform magnetic field of 2.0 T in the +y direction (•Fig. 19.6a). (a) What is the force on the particle just as it enters the magnetic field? (b) Describe the path of the particle while it is in the field.

Solution. We expect a circular trajectory, since the velocity and magnetic field are perpendicular.

Given: $q = -5.0 \times 10^{-4}$ C   *Find:*   (a) **F** (magnetic force on the particle)
        $v = 1.0 \times 10^2$ m/s (+x)   (b) Path of the particle in the field
        $B = 2.0$ T (+y)

(a) Equation 19.3 can be used to find the magnitude of the force:

$$F = qvB \sin \theta = (5.0 \times 10^{-4} \text{ C})(1.0 \times 10^2 \text{ m/s})(2.0 \text{ T})(\sin 90°) = 0.10 \text{ N}$$

The direction of the force just as the particle enters the magnetic field is in the −z direction in Fig. 19.6a. By the right-hand rule, the force on a positive charge would be in the +z direction. But since the charge is negative, the force is in the opposite direction.

(b) Since the magnetic force is always perpendicular to the velocity of the particle, it supplies the required centripetal force that causes the particle to move in a circular path (•Fig. 19.6b). The radius of the path (r) may be found using Newton's second law:

$$F = ma_c$$

That is, in this case, the centripetal force is supplied by the magnetic force, so we can write

$$qvB = \frac{mv^2}{r}$$

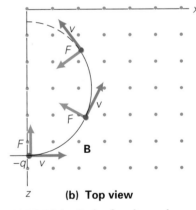

(b) Top view

•**FIGURE 19.6 Path of a charged particle in a magnetic field** (a) A charged particle entering a uniform magnetic field will be deflected—here toward the *xy* plane by the right-hand rule, because the charge is negative. **(b)** In the field, the force is always perpendicular to the particle's velocity. The particle moves in a circular path if the field is constant and the particle enters perpendicularly to the field direction. Example 19.2.

and

$$r = \frac{mv}{qB}$$

If the mass of the particle were known, the radius of its circular path could be computed easily.

Note that since there is no tangential acceleration, only the direction of the particle's motion changes—its speed, and thus its kinetic energy, remains constant. Looked at from another point of view, the fact that the magnetic force on the particle is always perpendicular to the direction of motion means that no work is done by the force. Thus the particle's kinetic energy cannot change.

Follow-up Exercise. In this Example, if the particle had been a proton instead, (a) in what direction would it be deflected, and (b) what would be the radius of its circular path if its speed were 1.0×10^6 m/s?

Insight MHD—Magnetohydrodynamics

Magnetohydrodynamics refers to the use of magnetic fields (*magneto*) in a liquid (*hydro*) propulsion system. Such systems are now being developed for use in sea-going ships and for other applications. For ships, the idea is to use MHD thrusters instead of the conventional propeller drive for propulsion. Arrays of thrusters would be used on each side of a ship like jet engines on an aircraft, or perhaps around the hull of a submarine.

The principle of an MHD thruster is illustrated in Fig. 1. The thruster uses the simple $F = qvB$ relationship in producing a jet of seawater. A powerful magnet surrounds a thruster tube through which seawater flows. The magnet is fashioned so as to produce a magnetic field perpendicular to the tube's length (see Fig. 1). Inside the tube, electrodes produce a current across the tube, composed of ions from the dissolved salts in seawater (Video Demonstration 11). Hence, we have charged particles moving in a magnetic field, which you know gives rise to a force on the particles. By the right-hand rule, the force is along the length of the tube, so seawater is ejected from the thruster. By Newton's third law, this jet action would cause a boat to move in the opposite direction (again analogous to a jet aircraft).

The intense magnetic fields needed for an MHD system are produced by superconducting magnets. Recall from Chapter 17 that certain materials become superconducting, or lose all electrical resistance, at some low critical temperature. With no resistance, large currents are possible without joule heat (I^2R losses), and large magnetic field can be produced ($B \sim I$).

MHD propulsion systems are currently in the development stage, and at present can only propel small boats very slowly. It is predicted, however, that some day MHD will be able to propel ships and submarines at speeds in excess of 160 km/h (100 mi/h). The speed of propeller-driven vessels is limited by what is known as cavitation. At higher propeller speeds, areas of low pressure form in front of the blades, causing the water to vaporize and form bubbles or cavities. This reduces the propeller's efficiency, and at very high speeds can cause damage. Because MHD systems have no propellers, they are not subject to this limitation. For the same reason they are relatively quiet —there is no propeller noise or noise resulting from cavitation. Another advantage of MHD is that, because the MHD thrusters have no mechanical moving parts, they should require less maintenance than conventional propulsion systems.

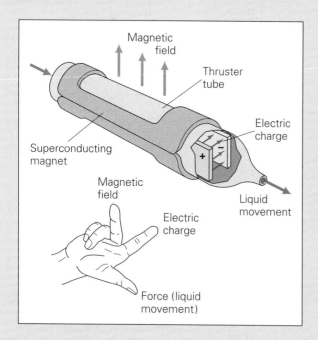

FIGURE 1 MHD thruster
The thruster uses an $F = qvB$ interaction to produce a jet of sea water. See text for description.

Electric Currents, Magnetic Fields, and the Right-Hand Source Rule

Electric and magnetic phenomena, although not exactly alike, are closely related. In fact, *magnetic* fields are produced by *electric currents* (moving charges) and we have seen they can exert forces on *moving electric charges*. This fact that magnetic fields were produced by electric currents was discovered by the Danish physicist Hans Christian Oersted in 1820. He noted that an electric current could produce a deflection of a compass needle. This can be demonstrated using an arrangement like that in •Fig. 19.7. When the circuit is open and there is no current in it, the compass needle points in the northerly direction due to the Earth's magnetic field (to be discussed in Section 19.6). However, when the switch is closed and there is current in the circuit, the compass needle is deflected, indicating that another magnetic field (other than that of the Earth) is affecting the needle. When the switch is opened, the compass needle goes back to pointing north.

Developing expressions for the magnitude of the magnetic field around a current-carrying wire requires mathematics beyond the scope of this book. This section will simply present the resulting formulas for the magnitude of the **B** field for several arrangements that are used in many practical applications.

Magnetic Field around a Long, Straight, Current-Carrying Wire. At a perpendicular distance d from a long, straight wire carrying a current I (•Fig. 19.8), the magnitude of **B** is

$$B = \frac{\mu_o I}{2\pi d} \qquad \text{(long straight wire)} \qquad (19.4)$$

where $\mu_o = 4\pi \times 10^{-7}$ T·m/A or Wb/A·m) is a constant called the *magnetic permeability of free space*. The field lines are closed circles around the wire.

Note in Fig. 19.8 that the direction of **B** is given by another right-hand rule which we will call the **right-hand source rule:**

> If a current-carrying wire is grasped with the right hand with the extended thumb pointing in the direction of the conventional current, the curled fingers indicate the circular sense of the magnetic field.

Magnetic Field at the Center of a Circular Current-Carrying Wire Loop. At the *center* of a circular loop of wire of radius r carrying a current I (•Fig. 19.9), the magnitude of **B** is

$$B = \frac{\mu_o I}{2r} \qquad \text{(center of circular loop)} \qquad (19.5)$$

Remember that this equation is valid *only* at the center of the loop. The direction of **B** is given by the right-hand source rule, and is perpendicular to the plane of the loop at its center (Fig. 19.9b). Notice that the overall field of the loop (Fig. 19.9c) is similar to that of a bar magnet (magnetic dipole).

Magnetic Field in a Current-Carrying Solenoid. A solenoid is constructed by winding a long wire in a tight coil, or helix (many circular loops, as shown in •Fig. 19.10). If the radius of the loops is small compared to the length (L) of the coil, **B** is parallel to the solenoid's longitudinal axis. If the solenoid has N turns (loops) and carries a current I, the magnitude of the magnetic field near the center is given by

$$B = \frac{\mu_o N I}{L} \qquad \text{(near center of a solenoid)} \qquad (19.6)$$

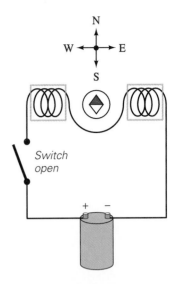

(a) No current

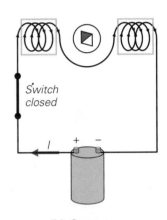

(b) Current

•**FIGURE 19.7 Electric current and magnetic field**
(a) With no current in the wire, the compass needle points north.
(b) With a current in the wire, the compass needle is deflected, indicating the presence of an additional magnetic field superimposed on that of the Earth. In this illustration, the strength of the additional field is roughly equal in magnitude to that of the Earth. How can you tell? (See Fig. 19.9.)

Note: The field vector is tangent to a field line at any point on the circle. The field lines, while typically curved, are true circles only near very long straight wires.

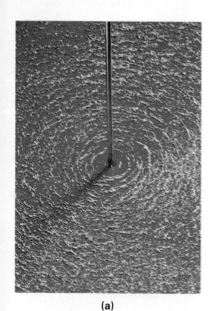

(a)

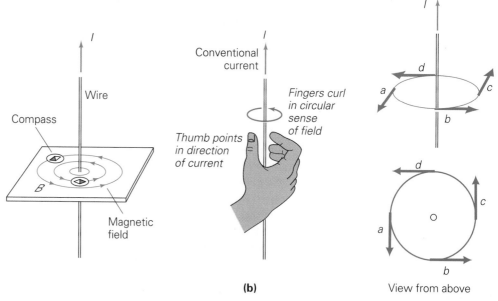

(b)

View from above

•**FIGURE 19.8   Magnetic field around a straight current-carrying wire**
(a) The field lines form concentric circles around the wire, as revealed by the iron filing pattern. **(b)** The circular sense of the field lines is given by a right-hand rule, and the magnetic field vector is tangent to the circular field line at any point.

•**FIGURE 19.9   Magnetic field at the center of a current-carrying loop**
(a) Iron filing pattern for a current-carrying loop. The magnetic field is perpendicular to the plane of the loop. **(b)** The direction of the field is given by the right-hand rule. With the thumb of the right hand in the direction of the conventional current, the curled fingers indicate the direction of **B**. **(c)** The overall magnetic field of a current-carrying circular loop is similar to that of a bar magnet.

The direction of **B** is given by the right-hand source rule as applied to one of the loops of the coil.

The expression $n = N/L$, or the number of turns *per meter*, is called the *linear turn density*. Using it allows Eq. 19.6 to be written in simpler form:

$$B = \mu_o n I \qquad \text{(center of a solenoid)} \qquad (19.7)$$

The longer the solenoid, the more uniform the magnetic field across the cross-sectional area within its coil. An ideal, or infinitely long, solenoid would have a uniform internal field. Of course, this ideal condition can only be approached in reality, but a long solenoid does produce a relatively uniform magnetic field. Once again, note how the pattern of the field lines for a loop (Fig. 19.9) and especially that for a solenoid (19.10) is similar to that of the magnetic field of a permanent bar magnet.

EXAMPLE 19.3 ■ NATIVE FIELDS: MAGNETIC FIELD FROM A CURRENT-CARRYING WIRE

The maximum household current in a wire is about 15 A. Assume that this current is in a west-to-east direction (•Fig. 19.11). What is the magnitude and direction of the magnetic field 1.0 cm directly below the wire?

Solution   We use the expression for the magnetic field due to a long straight wire. Listing the data, we have

(a)

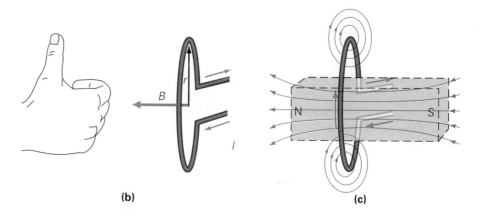

(b)   **(c)**

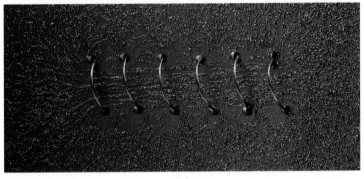

(a)

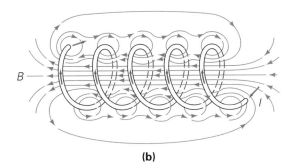

(b)

●FIGURE 19.10 Magnetic field of a solenoid
(a) The magnetic field of a current-carrying solenoid is fairly uniform near the central axis of the solenoid, as seen in this iron filing pattern. (b) The direction of the field may be determined by applying the right-hand rule to one of the loops.

Given: $I = 15$ A   *Find:* **B** (magnitude and direction)
           $d = 1.0$ cm $= 0.010$ m

Equation 19.4 can be used to find the magnitude of the field at the point 1.0 cm directly below the line:

$$B = \frac{\mu_o I}{2\pi d} = \frac{(4\pi \times 10^{-7}\,\text{T·m/A})(15\,\text{A})}{2\pi(0.010\,\text{m})} = 3.0 \times 10^{-4}\,\text{T}$$

By the right-hand source rule, you should be able to show that the magnetic field vector points northward. (Note that in Fig. 19.11 north is into the paper and south is outward.)

Follow-up Exercise. Estimate the electric current flowing in a long straight wire that would produce a magnetic field of the same strength as that in the Example at a distance of 1.0 m.

PROBLEM-SOLVING HINT

In Example 19.3, with $\mu_o = 4\pi \times 10^{-7}$ T·m/A, the 4π conveniently canceled. This is not the case in the following Example. Note that π may be left in symbol form unless a numerical answer is needed.

EXAMPLE 19.4 ■ WIRE VS SOLENOID: CONCENTRATING THE MAGNETIC FIELD

A solenoid is 0.30 m long with 10^3 turns per meter and carries a current of 5.0 A. What is the magnitude of the magnetic field through the center of this solenoid? Compare your result with the field near the single wire (notice that it carries *three times* the current in the solenoid in this Example!) in the previous Example, and comment.

Solution. The solenoid has an approximately uniform magnetic field if you stay away from the ends. In this case, we are near the center, and we have an expression for the field.

Given: $I = 5.0$ A   *Find:* B (magnitude of magnetic field)
           $n = 10^3$ turns/m

Here we are given the turn density, so using Eq. 19.7,

$$B = \mu_o n I = (4\pi \times 10^{-7}\,\text{T·m/A})(10^3\,\text{m}^{-1})(5.0\,\text{A}) = 2\pi \times 10^{-3}\,\text{T} \approx 6.3\,\text{mT}$$

Notice that this is about twenty times larger than the field near the typical household wire in Example 19.3. This is because the field in the solenoid is the vector sum of all the fields from each coil that makes up the solenoid.

Follow-up Exercise. In Example 19.4, if the current was reduced to 1.0 A, what turn density would be needed to create the same magnetic field near the center of the solenoid?

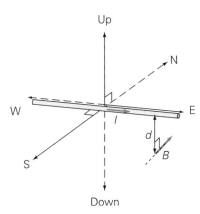

●FIGURE 19.11 Magnetic field
Finding the magnetic field produced by a straight current-carrying wire. Example 19.3.

19.3 Magnetic Materials

Objectives: **To be able to explain (a) how ferromagnetic materials enhance external magnetic fields, (b) how "permanent" magnets are produced, and (c) how "permanent" magnetism can be destroyed.**

Note: Review Fig. 15.1.

You may wonder why some materials are magnetic or easily magnetized, and others are not. You may also wonder how a bar magnet can create a magnetic field when there is no obvious current flowing. Knowing that a current produces a magnetic field and comparing the magnetic fields of a bar magnet and a current-carrying loop (see Figs. 19.1 and 19.9), you might begin to think that the magnetic field of the bar magnet is in some way due to *internal* currents or moving charges.

Recall that, according to the simplistic solar system model of the atom presented in Chapter 15, the electrons orbit the nucleus. They are thus charges in motion, and each orbiting electron is in effect a current loop that produces a magnetic field. This model pictures a material as having many internal atomic magnets. However, detailed analysis of atomic structure shows that the magnetic field produced by the orbiting atomic electrons is much smaller than what would be expected from this simplistic model. Also, the atoms are sometimes randomly arranged. As a result, the net magnetic effect due to orbiting electrons for most materials is either zero or very small.

What then is the source of the magnetism of magnetic materials? Modern theory says that magnetism is an atomic effect due to electron spin. The word "spin" is used because of an early idea that an additional contribution to the atomic magnetic field might come from an electron spinning on its axis. The classical picture likens a spinning electron to the Earth rotating on its axis. However, this is *not* the case. Electron spin is strictly a quantum-mechanical effect with no direct classical analog. Nonetheless, this picture of spinning electrons creating magnetic fields is one that is useful for qualitative thinking and reasoning. (We will revisit the topic of electron spin in relation to the magnetic field it creates later in this Chapter.)

In atoms with two or more electrons, the electrons usually pair up, in pairs whose spins are opposite. The magnetic fields then cancel each other, and the material is not magnetic. Aluminum is an example. However, in certain strongly magnetic materials, known as **ferromagnetic materials,** electron spins do not pair, or cancel, completely. As a result, there is a strong coupling or interaction between neighboring atoms, leading to the formation of large groups of atoms called **magnetic domains** in which many of the electron spins are aligned. There aren't very many ferromagnetic materials. The most common are iron, nickel, and cobalt (gadolinium and certain alloys are also ferromagnetic).

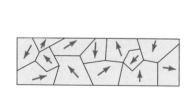

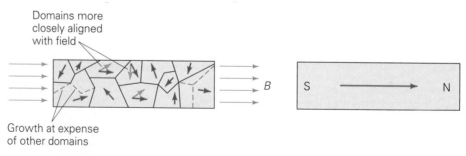

(a) No external magnetic field   **(b) With external magnetic field**   **(c) Resulting bar magnet**

•**FIGURE 19.12 Magnetic domains**
(a) With no external magnetic field, the magnetic domains of a ferromagnetic material are randomly oriented, and the material is unmagnetized. **(b)** In an external magnetic field, domains with orientations parallel to the field may grow at the expense of other domains, and the orientations of some domains may become more aligned with the field. **(c)** As a result, the material becomes magnetized, or exhibits magnetic properties.

In an unmagnetized ferromagnetic material, the domains are randomly oriented and there is no net magnetic effect (•Fig. 19.12a). But, when a ferromagnetic material is placed in an external magnetic field, two things may happen:

1. Domain boundaries change, and the domains with magnetic orientations parallel to the external field grow at the expense of the other domains.
2. The magnetic orientation of some domains may change slightly so as to be more aligned with the field (•Fig. 19.12b).

You now can understand why an unmagnetized piece of iron is attracted to a magnet and why iron filings line up with a magnetic field. Essentially, the pieces of iron become induced magnets.

Electromagnets and Magnetic Permeability

Ferromagnetic materials are used to make electromagnets, usually by wrapping a wire around an iron core (•Fig. 19.13). The current in the coils creates a magnetic field in the iron, which in turn creates its own field many times larger than the coil field. Turning the current on and off, the magnet (or magnetic field) may be turned on and off at will. The iron used in an electromagnet is called soft iron. It is treated so that when the external field is removed, the magnetic domains quickly become unaligned and the iron is unmagnetized. ("Soft" does not refer to the metal's mechanical hardness but to its magnetic properties.)

When an electromagnet is on (Fig. 19.13a), the iron core becomes magnetized and adds to the field of the solenoid. The total field is expressed as

$$B = \mu n I$$

Notice that this equation is similar to that for the magnetic field of a solenoid (Eq. 19.7), but contains μ instead of μ_o, the permeability of free space. Here, μ expresses the **magnetic permeability,** not of free space, but of the *core material*. The role permeability plays in magnetism is similar to that of the permittivity, ϵ, in electricity, discussed in Chapter 16. For magnetic materials,

$$\mu = K_m \mu_o \qquad (19.8)$$

Here K_m, called the *relative permeability*, is the magnetic analog of the dielectric constant K.

Both K and K_m are equal to one in a vacuum. Magnetic permeability is a useful material property. A core of a ferromagnetic material with a large permeability in an electromagnet can enhance its field thousands of times (compared to an air core).

Note that the strength of the field of an electromagnet depends on the current ($B = \mu n I$). Large currents produce large fields, but this field generation is accompanied by much greater joule heating (or $I^2 R$ losses). Large electromagnets need water-cooling coils to remove this heat. The problem may be alleviated by using a superconductor, which has zero resistance. However, superconducting magnets now in use also require tremendous cooling just to keep the conductors in the superconducting state, and this cooling also consumes energy and produces heat, as we saw in Section 12.4. Perhaps someday we will have electromagnets made with new high-temperature superconductors (Section 17.3), but currently these are ceramic materials which are not easily made into wires that can be wound in an electromagnet.

Iron that retains some magnetism after being in an external magnetic field is called "hard" iron and is used to make so-called permanent magnets. You may have noticed that a paper clip or a screwdriver blade becomes slightly magnetized after being near a magnet. Permanent magnets are produced by heating pieces of some ferromagnetic material in an oven and then cooling them in a strong magnetic field to get the maximum effect.

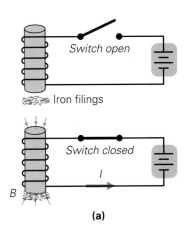

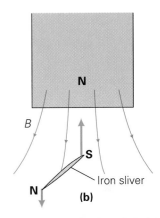

(b)

(c)

•**FIGURE 19.13 Electromagnet** (a) With no current in the circuit, there is no magnetic force. However, with a current in the coil, there is a magnetic field and the iron core becomes magnetized. (b) The lower end of the electromagnet in (a). Notice the sliver of iron is attracted to the end of the electromagnet. (c) An electromagnet in action picking up scrap metal.

A *permanent* magnet need not be really permanent, however—its magnetism can be lost. Hitting such a magnet with a hard object or dropping it on the floor may cause it to lose its domain alignment. Heating can also cause a loss of magnetism because the resulting increase in random motions of atoms tends to unalign the domains. Above a certain critical temperature, called the **Curie temperature** (or Curie point), domain coupling is destroyed by the increased thermal oscillations and a ferromagnetic material loses its ferromagnetism. The Curie temperature for iron is 770 °C. As you probably know, one of the worst things you can do to an audio or video magnetic tape cassette is to leave it on the dashboard of your car on a hot day. The thermal motion caused by the heat partially destroys the magnetic voice and/or video signal imprinted on the tape.

19.4 Magnetic Forces on Current-Carrying Wires

Objectives: **To be able to (a) calculate the strength and direction of the magnetic force on a current-carrying wire and the torque on a current-carrying loop, and (b) explain the concept of a magnetic moment for such a loop.**

An electric charge moving in a magnetic field experiences a force unless it is moving parallel to the field lines. Thus, a current-carrying wire in a magnetic field should also experience such a force because a current is made up of moving charges. That is, the magnetic forces on the moving charges should give a resultant force on the wire conducting them, and in fact this is found to be the case.

As you know, a conventional current can be thought of as positive charges moving in a straight wire, as depicted in •Fig. 19.14. In a time t, a charge (q) would move on the average through a length given by $L = vt$, where v is the drift velocity and is perpendicular to the B field. Since all the moving charges ($\Sigma_i q_i$) in this length of wire will experience a force due to the magnetic field, the magnitude of the total force on the length of wire, from Eq. 19.1, is

$$F = (\Sigma_i q_i)vB$$

Substituting L/t for v and rearranging gives

$$F = (\Sigma_i q_i)\left(\frac{L}{t}\right)B = \left(\frac{\Sigma_i q_i}{t}\right)LB$$

But $\Sigma_i q_i/t$ is the current (I), since it represents the charge passing through a cross-sectional area of the wire per unit time. Therefore,

$$F_{\max} = ILB \qquad (19.9)$$

Equation 19.9 gives the maximum force on the wire because the direction of the charges' motion (the current) is at an angle of 90° to the magnetic field. If the wire is not perpendicular to the field, but at some angle θ, then the force on the

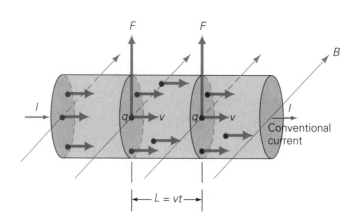

•**FIGURE 19.14 Force on a wire segment**
The diagram helps visualize and determine the force on a length of current-carrying wire in a magnetic field. Charge carriers are considered to be positive for convenience. See text for description.

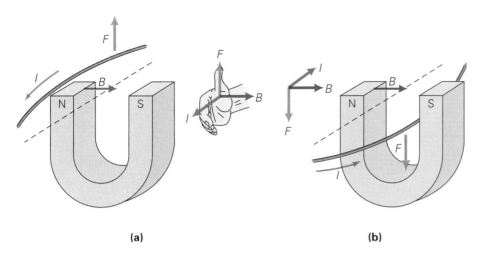

(a)   (b)

•FIGURE 19.15 **Force on a current-carrying wire**
The direction of the force is given by pointing the fingers of the right hand in the direction of the conventional current I and then turning or curling them toward **B**. The extended thumb points in the direction of **F**. The force is upward for the case shown in **(a)** and downward for the case shown in **(b)**.

wire will be less, as in the case of the force on a moving charge. In general, the force on a length of current-carrying wire is given by

$$F = ILB \sin \theta \qquad \begin{array}{l} \textit{magnetic force on} \\ \textit{a current-carrying wire} \end{array} \qquad (19.10)$$

Note that if the wire is parallel to the field, the force on it is zero (as is the force on a charge moving parallel to a magnetic field).

The direction of the magnetic force on a current-carrying wire is also given by a right-hand rule. The length of wire is treated as a vector **L** in the direction of the conventional current. If it is imagined to be turned by the fingers of the right hand toward **B**, the extended thumb points in the direction of **F**. This right-hand force rule (for a wire) is often stated as follows and is just an extension of the basic right-hand force rule on individual charges we learned previously.

> When the fingers of the right hand are pointed in the direction of the conventional current I and then turned or curled toward the vector **B**, the extended thumb points in the direction of the magnetic force on the wire.

Two examples of magnetic forces on current-carrying wires in a magnetic field are given in •Fig. 19.15. Apply the right-hand force rule in each case to see how the direction of the force is found.

EXAMPLE 19.5 ■ ATTRACTION OR REPULSION? MAGNETIC FORCE BETWEEN TWO PARALLEL WIRES

Two long parallel wires are separated by a distance d and carry conventional currents I_1 and I_2 in the same direction, as illustrated in •Fig. 19.16. (a) Derive an expression for the magnetic force per unit length on one of the wires due to the other and show that, in keeping with Newton's third law, the forces on the two wires are equal in magnitude. (b) Show that if the currents are parallel, these forces are mutually attractive, also in keeping with the third law that the forces of an action–reaction pair must be opposite in direction.

Solution. The symbols are given in Fig. 19.16.

(a) There is a force on each wire because of the magnetic field produced by the current in the other wire. By a right-hand rule, the magnetic field due to I_1 at the second wire is into the page, and its magnitude, from (Eq. 19.4), is

$$B_1 = \frac{\mu_o I_1}{2\pi d}$$

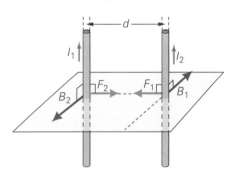

•FIGURE 19.16 **Mutual interaction**
For two parallel current-carrying wires, the magnetic field due to each current interacts with the other current, and there is a force on each wire. If the currents are in the same direction, the forces are attractive. See text for description and how this effect is used to define the ampere unit of current.

The direction of B_1 is perpendicular to the wire carrying I_2, so from Eq. 19.9 the magnitude of the *force per unit length* (F_2/L) on the wire is

$$\frac{F_2}{L} = I_2 B_1 = \frac{\mu_o I_1 I_2}{2\pi d}$$

The magnetic field due to I_2 also exerts a force per unit length on the wire carrying I_1, which by a similar analysis is given by

$$\frac{F_1}{L} = \frac{\mu_o I_2 I_1}{2\pi d}$$

This is clearly the same magnitude as the force on wire 2.

(b) By the right-hand force rule, each force is *toward* the other wire. Thus, the forces are equal and opposite, in keeping with Newton's third law. Therefore, two parallel wires carrying current in the same direction are attracted toward one another.

Follow-up Exercise. In a household wire cord, the centers of two wires are separated by 1.5 mm. Each carries a current of 5.0 A, but in opposite directions. (a) Calculate the magnetic force on 1.0 m of one of the wires. (b) Is the force between the wires attractive or repulsive?

The magnetic force between parallel wires in an arrangement like that analyzed in Example 19.5 is used to define the ampere. According to the National Institute of Standards and Technology (NIST, formerly the National Bureau of Standards),

> The ampere is defined as that current which, if maintained in each of two long parallel wires separated by a distance of 1 m in free space, would produce a magnetic force between the wires of exactly 2×10^{-7} N for each meter of wire.

Torque on a Current-Carrying Loop

Another important instance of magnetic force on a current is that on a current-carrying loop, for example, the rectangular loop in Fig. 19.17a. (The wires connecting the loop to a voltage source are not shown.) Suppose that the loop is free to rotate about an axis passing through opposite sides, as shown in the figure. There are no net forces or torques on the pivot sides of the loop (the sides through which the axis of rotation passes). When these sides are parallel to the field, the force on them is zero. (Why?) At any other angle to the field, the forces on them are equal and opposite in the plane of the loop (can you see why?), and so produce no torque. However, the equal and opposite forces on the other two sides of the loop (the sides parallel to the axis of rotation) *do* produce torques. These two torques produce a net torque that, in general, tends to rotate the loop. To see how this works, consider Fig. 19.17b, which shows a side view of Fig. 19.17a. The magnitude of the magnetic force on each of the nonpivoted sides, whose length is L, if given by

$$F = ILB$$

Note: Torque is discussed in Section 8.2.

since L and B are always at right angles to each other. Recall that the torque produced by a force is given by $\tau = r_\perp F$, where $r_\perp$ is the perpendicular lever arm from the axis of rotation to the line of the force. From Fig. 19.17b, we see that $r_\perp = \frac{1}{2}w \sin\theta$, where θ is the angle between the normal to the plane of the loop and the direction of the magnetic field. The total torque on the loop from both forces is

$$\begin{aligned}\tau &= r_\perp F + r_\perp F = (\tfrac{1}{2}w \sin\theta)F + (\tfrac{1}{2}w \sin\theta)F = wF\sin\theta \\ &= w(ILB)\sin\theta\end{aligned}$$

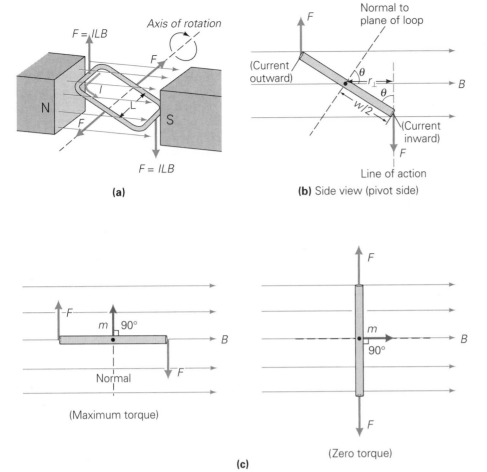

(a)

(b) Side view (pivot side)

(Maximum torque)

(Zero torque)

(c)

•FIGURE 19.17 Force and torque on a current-carrying pivoted loop (a) A current-carrying rectangular loop oriented in a magnetic field as shown here experiences a force on each of its sides. The forces on the sides parallel to the axis of rotation produce a torque that causes the loop to rotate. (b) A side view shows the geometry for determining the torque. See text for description. (c) Conditions of maximum and zero torque. The magnetic moment ($m=IA$) of the loop is determined by a right-hand rule, as discussed in the text. The loop will tend to align its magnetic moment with that of the external magnetic field.

Then, since wL is the area (A) of the loop, we may write for the magnitude of the torque on a pivoted, current-carrying loop,

$$\tau = IAB \sin \theta \qquad (19.11)$$

Torque on a current-carrying loop

Even though derived for a rectangular loop, Eq. 19.11 is valid for a flat loop of *any* shape.

The quantity IA is often referred to as the magnitude of the **magnetic moment** vector. The symbol for its magnitude is m. Thus we have $m = IA$. The magnetic moment vector *direction* is determined by circling the fingers of your right hand in the direction of the (conventional) current flow. Your thumb then gives the direction of the magnetic moment vector, **m**, as shown in Fig. 19.17c. Thus we can rewrite Eq. 19.11 in terms of the loop's magnetic moment

$$\tau = mB \sin \theta \qquad (19.12)$$

Note that if more than 1 loop makes up the coil—in general, N loops—the torque is N times that due to one loop (since they all have the same current), or $\tau = NIAB \sin \theta$.

Note: A *coil* consists of N loops of the same size, all carrying the same current, I, in series.

We see that the torque tends to line up the magnetic moment with the magnetic field direction. Thus, a loop in a magnetic field will rotate, or experience a torque, until $\sin \theta = 0$ (when the forces producing the torque are parallel to the plane of the loop, •Fig. 19.17c). This occurs when the plane of the loop is perpendicular to the field. If the loop is started at rest so that its magnetic moment makes some nonzero angle with the magnetic field direction, it will undergo an angular acceleration that will rotate it to the zero angle position. Rotational

m Electron "spin"
magnetic moment

(a)

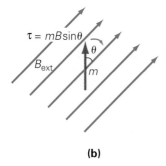

$\tau = mB\sin\theta$

B_{ext} θ m

(b)

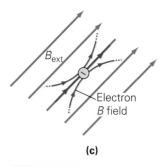

B_{ext}

Electron
B field

(c)

•**FIGURE 19.18 Magnetic moment of the electron**
(a) A spinning electron can be pictured as a spinning sphere of negative charge, having its own magnetic moment as shown.
(b) When the electron is placed in a magnetic field, its magnetic moment will tend to line up with that field. **(c)** When lined up with the external field, the electron's own field enhances the external field. In ferromagnets, many electrons doing this in unison produce a net field that is many times stronger than the external applied field.

inertia will cause it to coast through the equilibrium position (zero angle, zero torque) on to the other side. On that side, the torque will slow it down, stop it, then reaccelerate it back toward equilibrium. In the absence of frictional forces, this rotational oscillation will go on forever. In other words, the torque on the current loop is restoring and tends to cause the loop's magnetic moment to oscillate about the field direction.

There are many interesting applications of the concept of magnetic moment at the atomic level. One such arises when we consider a spinning electron as a series of concentric current loops (•Fig. 19.18a). Imagined in this way, the electron evidently must possess a magnetic moment. (Remember that this spinning picture is not to be taken literally, however; it is just a way of visualizing a quantum-mechanical concept.) When placed in an external magnetic field, this moment will tend to line up with the field (•Fig. 19.18b), thus enhancing the total field by producing a field of its own in the same direction (•Fig. 19.18c).

In addition to the magnetic moment (and field) produced by its spin, an atomic electron has a magnetic moment and associated field as a result of its orbital motion around the nucleus, as the following Example shows.

EXAMPLE 19.6 ■ THE HYDROGEN ATOM REVISITED: MAGNETIC MOMENT OF THE ELECTRON

An orbiting electron can create an *orbital* magnetic moment and a magnetic field. In the Bohr model of the hydrogen atom (discussed in Chapter 27), the single electron (in its smallest orbit) travels in a circle with a radius of 0.0529 nm and has a period of 1.5×10^{-16} s. (a) Derive a *general* expression for the orbital magnetic moment in terms of the orbit radius (r), the charge on the electron (e), and the orbital period (T). (b) What is the magnitude and direction (in comparison to its orbital rotation sense) of the orbital magnetic moment of the electron in the hydrogen atom?

Solution. We are given or know the following numerical data:

Given: $|e| = 1.60 \times 10^{-19}$ C   *Find:* (a) Expression for magnetic moment (m)
$T = 1.5 \times 10^{-16}$ s   (b) Magnetic moment of the electron
$r = 0.0529$ nm $= 5.29 \times 10^{-11}$ m   in orbit

(a) The time for one orbit is T. In the course of one orbit, a net charge of magnitude $q = |e|$ passes any given point. Here we let e represent the magnitude of the charge on the electron, so the equivalent current is $I = q/t = e/T$ and the orbital magnetic moment is

$$m = IA = \left(\frac{e}{T}\right)\pi r^2 = \frac{e\pi r^2}{T}$$

(b) Inserting the given data, we have

$$m = \frac{e\pi r^2}{T} = \frac{(1.60 \times 10^{-19} \text{ C})\pi(5.29 \times 10^{-11} \text{ m})^2}{1.5 \times 10^{-16} \text{ s}}$$
$$= 9.4 \times 10^{-24} \text{ A} \cdot \text{m}^2$$

If, for example, the electron is orbiting clockwise as we look down on its orbital plane, this is equivalent to a counterclockwise conventional current (since the electron is negatively charged). Curling the fingers of your right hand in the counterclockwise direction gives an orbital magnetic moment direction out of the plane of the orbit (thumb direction).

Follow-up Exercise. Compare the magnetic moment of the electron in this Example to that of an ordinary single wire loop having an area of 3.0 cm^2 and carrying a current of 0.50 A.

On an everyday scale, current-carrying loops, even ones with such limited motion, have important applications. Since the current is directly related to the torque, such a loop provides a means to measure current. Also, there are ways to keep a loop rotating, allowing continuous conversion of electrical energy to mechanical energy. This is the operating principle of motors. These and other applications of electromagnetism are considered in the following section.

19.5 Applications of Electromagnetism

Objectives: To be able to explain the operation of various instruments whose functions depend on electromagnetic interactions.

With the principles you have learned so far about electromagnetic interactions, you can understand the operation of several widely used scientific instruments.

The Galvanometer

In the discussion of ammeters and voltmeters in Chapter 18, the galvanometer was considered to be simply a circuit element that measured small currents. Now you are ready to see how it does this. As •Fig. 19.19a shows, a galvanometer consists of a coil of wire loops on an iron core that is pivoted between the pole faces of a permanent magnet. When a current exists in the coil, the coil experiences a torque. A small spring supplies a counter (restoring) torque, and in equilibrium a pointer indicates a deflection ϕ that is proportional to the current in the coil.

As you learned in Section 19.4, the torque on a single current-carrying loop in a magnetic field is given by $\tau = IAB \sin \theta$. The dependence of the torque on the angle θ between the field lines and a line normal to the plane of the loop would cause a problem in measuring current with a galvanometer coil. If the coil rotated from its position of maximum torque ($\theta = 90°$, or the B field perpendicular to the nonpivoted sides of the coil), the torque would become less, and the pointer deflection ϕ would not be proportional to the current alone. This problem is avoided by making the pole faces curved and wrapping the coil on a cylindrical iron core. The core tends to concentrate the field lines so that **B** is always perpendicular to the nonpivoted side of the coil (•Fig. 19.19b).

The magnetic force is then always perpendicular to the plane of the coil, and the torque does not vary with the angle θ. We assume that the galvanometer coil is made of N loops, *each* carrying a current I. Thus, the total torque on the coil is

$$\tau = NIAB \qquad (19.13)$$

The restoring torque of the spring (τ_s) in a galvanometer is given by a rotational form of Hooke's law ($F = kx$ for linear restoring force).

$$\tau_s = k\phi \qquad (19.14)$$

Here ϕ is the deflection angle as measured by the pointer. With a current in the coil, the coil and the pointer will rotate until the torque on the coil is balanced by that on the spring. In equilibrium, the magnitudes of the torques are equal, and by combining Eqs. 19.13 and 19.14 we obtain

$$\phi = \frac{NIAB}{k} \qquad (19.15)$$

Thus, the pointer deflection (ϕ) is directly proportional to the current (I).

A properly calibrated galvanometer can be used to measure small currents (in the microamp range). With an appropriate shunt resistor, a galvanometer forms the basis of an ammeter for measuring larger currents. Also, with an appropriate series multiplier resistor, the galvanometer forms a voltmeter.

The dc Motor

In general, *an electrical motor is a device that converts electrical energy into mechanical energy.* An example of such a conversion occurs with a galvanometer: A cur-

Note: See Section 18.4.

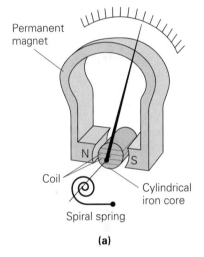

(a)

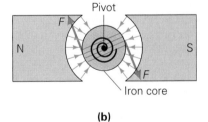

(b)

•**FIGURE 19.19 The galvanometer**
(a) The deflection of the needle is proportional to the current in the coil. A galvanometer can therefore be used to detect and measure currents. **(b)** A magnet with curved pole faces is used so that the field lines are always perpendicular to the core surface and the torque does not vary with angle.

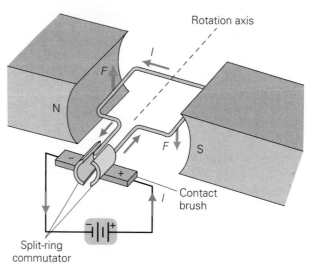

(a)

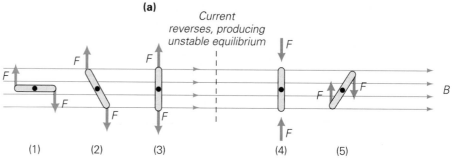

Current reverses, producing unstable equilibrium

(b) End view of loop showing loop rotation sequence

FIGURE 19.20 A dc motor
(a) A split-ring commutator reverses the polarity and current each half-cycle, so the coil rotates continuously. (b) An end view shows the forces on the coil and its orientation during a half-cycle. (For simplicity we show a single loop but it is actually a coil of *N* loops.)

rent causes the coil to rotate mechanically. However, a galvanometer is not considered to be a motor. A practical **dc motor** must have continuous rotation for continuous energy output.

A pivoted, current-carrying coil (*N* loops) in a magnetic field will rotate freely, but for only a half-cycle, or through a maximum angle of 180°. Recall that $\tau = NIAB \sin \theta$, and when the magnetic field is perpendicular to the plane of the coil ($\sin \theta = 0$), the torque is zero and the coil is in equilibrium.

To provide for continuous rotation, the current is reversed every half turn so that the torque forces are reversed. This is done by means of a split-ring commutator, an arrangement of two metal half-rings insulated from each other (•Fig. 19.20a). The ends of the wire that forms the coil are fixed to the half-rings, and the current is supplied to the coil through the commutator by means of contact brushes. Then, with one half-ring electrically positive (+) and the other negative (−), the coil and ring rotate. When they have gone through half a rotation, the half-rings come in contact with the opposite brushes. Their polarity is thus reversed, and the current in the coil is in the opposite direction. This changes the directions of the torque (•Fig. 19.20b). The torque is zero at the equilibrium position, but the coil is in unstable equilibrium. The coil has enough rotational momentum to continue through this point, and it rotates another half-cycle. This process repeats in continuous operation. For a real motor the rotating shaft is called the *armature*.

The Cathode Ray Tube (CRT)

The **cathode ray tube (CRT)** is a vacuum tube that is used in an oscilloscope (•Fig. 19.21a). Electrons are "boiled off" a hot filament in an electron gun and accelerated through a potential difference. Because they are emitted from the negative

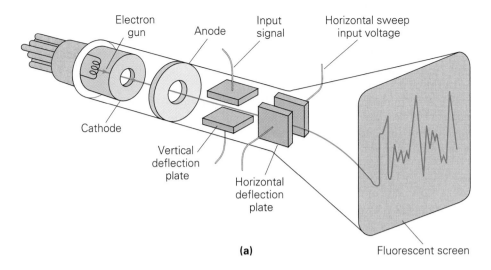

(a)

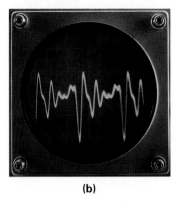

(b)

•**FIGURE 19.21 Cathode ray tube (CRT)**
(a) A beam for an electron gun is controlled by deflection plates. See text for description. **(b)** The motion of the deflected beam traces out a pattern on a fluorescent screen.

terminal, or cathode, the electrons are called cathode rays. (The positive terminal toward which the electrons are accelerated is the anode.) In 1897, the English physicist J. J. Thomson used a type of cathode ray tube to demonstrate the existence of the electron. Previously, it had not been known that the rays emitted from the hot filament of such a tube were electrons.

The electrons pass through a small hole in the anode, forming a beam. The beam passes through two sets of parallel plates, one for vertical deflections and the other for horizontal deflections. Input voltages (which usually vary with time) are placed across the plates, and the deflected beam makes patterns on a screen. (The fluorescent material coating the inside of the screen glows when struck by the electrons.) In normal operation, the horizontal input voltage from an internal voltage source sweeps the dot illuminated by the beam across the screen periodically at a constant rate. The vertical deflection is proportional to an external voltage input. Thus, the display on a CRT is a graph of the external voltage versus time (•Fig. 19.21b).

The picture tube of a television set is also a cathode ray tube (•Fig. 19.22). In this case, magnetic coils are usually used instead of plates to deflect the electron beam. The beam scans the fluorescent screen in a 525-line pattern in a fraction of a second. For a black-and-white TV, the signals transmitted from a video camera reproduce an image on the screen as a mosaic of light and dark dots, depending on whether or not the intermittent beam is on or off at a particular instant.

Producing the images of color TV is somewhat more involved. A common color picture tube has three beams, one for each of the primary colors (Chapter 25).

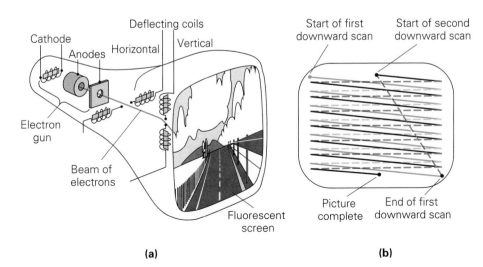

(a)   **(b)**

•**FIGURE 19.22 Television tube**
(a) A television picture tube is a CRT. **(b)** The beam scans every other line on the screen in one downward scan that takes $\frac{1}{60}$ s and then scans the lines in between in a second scan of $\frac{1}{60}$ s. This yields a complete picture of 525 lines in $\frac{1}{30}$ s.

Phosphor dots on the screen are arranged in groups of three (triads) with one dot for each of the primary colors. The excitation of the appropriate dots and the resulting combination of colors produce a color picture. There are also single-beam color picture tubes in which color dots are excited sequentially in rapid succession to produce the image on the screen.

The fact that a picture tube is a CRT can be demonstrated by bringing a magnet near a black-and-white TV screen. The magnetic field will deflect the electrons and distort the picture. *Do not try this on a color TV screen!* Internal metallic parts may become magnetized, causing the picture to be permanently distorted.

The Mass Spectrometer

Have you ever thought about how the mass of an atom or molecule is measured? Electric and magnetic fields provide a way in one form of a **mass spectrometer** (often called a mass spec for short). Actually, the masses of ions are measured since electric and magnetic fields have motional effects only on charged particles. (Recall that an ion is an atom or molecule with a net electric charge.)

Ions with a known charge ($+q$) are produced by heating a substance. A beam of ions introduced into the mass spec has a wide distribution of speeds. Ions with a particular velocity are selected by setting the values of the electric and magnetic fields between the plates of a *velocity selector* (•Fig. 19.23). For a positively charged ion, the electric field produces a downward force (magnitude $F = qE$), and the magnetic field produces an upward force (magnitude $F = qvB_1$). [Recall that a cross ($\times$) means into the page, and a dot ($\bullet$) means out of the page; you can visualize these symbols as though observing the feathered end and tip of an arrow, respectively. This is the general way of showing vectors that point into or out of the page.] If the beam is not deflected, the resultant force must be zero, so

$$qE = qvB_1$$

or

$$v = \frac{E}{B_1}$$

If E is written as V/d, since the voltage and the plate separation distance are the controllable or measured quantities,

$$v = \frac{V}{B_1 d} \tag{19.16}$$

Thus, a beam with a known speed can be selected by varying the values of V, B_1, and d.

The beam then passes through a slit into another magnetic field ($\mathbf{B}_2$), which is perpendicular to the direction of the beam. The force due to this magnetic field (magnitude $F = qvB_2$) is always perpendicular to the velocity of the ions, which

•**FIGURE 19.23 Principles of the mass spectrometer**
Ions pass through the velocity selector; those with a particular velocity enter a magnetic field (B_2). The particles are deflected, with the radius of the circular path depending on the mass of the particle. Paths of two different radii indicate that the beam contains particles of two different masses.

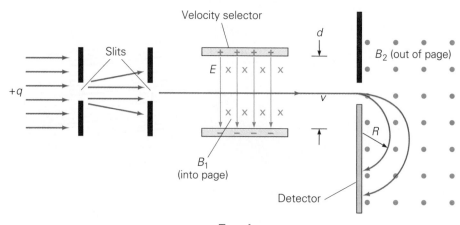

are deflected in a circular arc. The magnetic force supplies the centripetal force for this motion, and

$$\frac{mv^2}{R} = qvB_2$$

Using the expression for the selected ion velocity (Eq. 19.16),

$$m = \left(\frac{qdB_1B_2}{V}\right)R \qquad (19.17)$$

The quantity inside the parentheses is a constant, so the greater the mass of an ion, the greater the radius of the circular path. Two circular paths of different radii are shown in Fig. 19.23, which indicates that the beam contains ions of two different masses. If the radius of curvature R for an ion beam is measured (e.g., by recording the position of the beam with a detector), the mass of the ion can be calculated using the other known quantities in Eq. 19.17.

Mass spectrometers have many functions in modern laboratories (•Fig. 19.24). For example, they are used to help establish the age of ancient rocks and more recent human artifacts by measuring the relative abundance of different isotopes that they contain. (This method, called radioactive dating, is discussed in Section 29.3.) Among their other roles are tracking short-lived intermediates in studies of the biochemistry of living organisms; helping to determine the structure of large organic molecules; analyzing the composition of complex mixtures, such as a sample of smog-laden air or the output of a petroleum refinery; and detecting tiny amounts of impurities in metal alloys and semiconductors.

Another interesting application of the principles that underlie the magnetic mass spectrometer can be found in the momentum selector. This device is used extensively in accelerator laboratories throughout the world as a means of selecting particles having a desired momentum (and thus, in effect, of selecting the particle energy). For example, it is often essential to learn how a certain nuclear reaction depends on the energy of the bombarding particles. On a more practical note, in future cancer therapy using particle beams it may be important to control the kinetic energy of the particles very precisely in order to assure that they penetrate just to the tumor depth and not beyond. The following Example demonstrates how such control might be achieved.

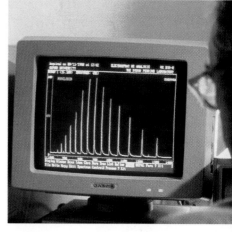

•FIGURE 19.24 **Mass spectrometer** Computer display of a mass spectrometer. The molecule being analyzed is myoglobin, a protein that stores oxygen in muscle tissue. Each of the peaks on the display represents the mass of a particular ionized fragment of the molecule. Such mass spectra can help to determine the composition and structure of large molecules. The mass spectrometer can also be used to identify tiny amounts of a particular molecule in a complex mixture.

EXAMPLE 19.7 ■ AROUND THE CORNER: MAGNETIC MOMENTUM SELECTOR

Consider positively charged particles (mass m, charge q) traveling down an evacuated beam tube and entering a uniform magnetic field, as shown in •Fig. 19.25. The particles initially move to the right with speed v, and B is directed into the page. Only those particles that reach point P (i.e., that travel in a quarter-circular arc of radius r) can proceed down the evacuated beam tube. (a) Show that the particles will be deflected as shown (upward). (b) Show that the emerging particles all have the same momentum (magnitude), given by qBr. (c) Calculate the strength of the field needed to deflect protons moving with a speed of 0.50% the speed of light if the radius of the quarter circle is to be 50 cm.

Solution. Remember that the magnetic force provides the required centripetal acceleration for the circular motion. Here the velocity and magnetic field are perpendicular, so the magnetic force is at its maximum. Listing the data and known quantities,

Given: $m = 1.67 \times 10^{-27}$ kg   *Find:* (a) direction of deflection
$q = 1.60 \times 10^{-19}$ C   (b) p (momentum of the selected
$v = 0.0050\,c = 1.5 \times 10^6$ m/s   particles)
$r = 50$ cm $= 0.50$ m   (c) B (magnetic field required for
                                                                    protons)

(a) Using the force right-hand rule, the velocity vector is to the right and the magnetic field into the paper. Thus, the force is up (toward the top of the page). The particle will be deflected upward, since it is positively charged.

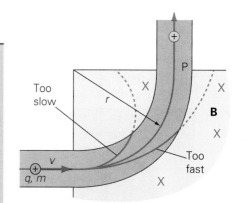

•FIGURE 19.25 **Momentum selector** By selecting only those particles traveling in a path of a particular radius, we are selecting only those particles having a specific linear momentum. Particles with a higher or a lower momentum follow paths that do not take them through the exit opening of the beam tube. See Example 19.7.

The Electronic Balance

Traditional laboratory balances measure mass by balancing the weight force of an unknown mass against that of a known mass. The newer, digital electronic balances (•Fig. 19.26a) work on a different principle. There is still a suspended beam with a pan on one end that holds the object to be weighed (or massed), but there is no known mass. The balancing downward force is supplied by a current-carrying coil of wire in the field of a permanent magnet (•Fig. 19.26b). The coil moves up and down in the cylindrical gap of the magnet, and the downward force is proportional to the current in the coil. The mass of the object in the pan is determined from the coil current that produces a force just sufficient to balance the beam.

The current required to produce balance is controlled automatically. This is done by means of light photosensing and a feedback loop. When the beam is horizontal or balanced, a knife-edge obstruction cuts off part of the light from a source that falls on a photosensitive "electric eye" (see Chapter 27). The resistance of the electric eye is a function of the amount of light falling on it. This resistance controls the current that an amplifier sends through the coil. For example, if the beam tilts so that the knife-edge raises and more light strikes the eye, the current in the coil is increased to counterbalance the tilting. In this manner, the beam is electronically maintained in nearly horizontal equilibrium. The current that keeps the beam in the horizontal position is read out on a digital ammeter that is calibrated in grams or milligrams instead of amperes.

•**FIGURE 19.26 Electronic balance**
(a) A digital electronic balance.
(b) Diagram of the principle of an electronic balance. The balance force is supplied by electromagnetism. See text for description.

(a)

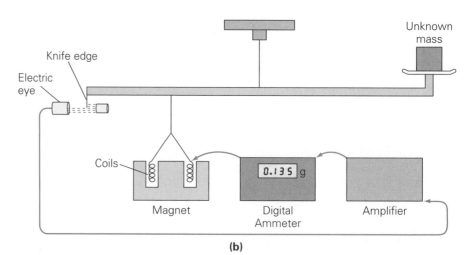

(b)

Objectives: To be able to (a) state some of the general characteristics of the Earth's magnetic field, (b) explain the theory about its possible source, and (c) discuss some of the ways in which it affects the Earth's local environment.

The magnetic field of the Earth has been used for centuries. In ancient times, navigators used lodestones or magnetized needles to show them where north was. An early study of magnetism was done by the English scientist Sir William Gilbert around 1600. In investigating the magnetic field of a specially cut spherical lodestone that simulated the Earth, he came to believe that the magnetic field was associated with the entire Earth, or that the Earth as a whole acted as a magnet. Gilbert thought that the field might be produced by a large body of permanently magnetized material within the Earth.

In fact, the Earth's magnetic field has approximately the configuration that would be produced by a large interior bar magnet, as •Fig. 19.27 shows. The magnitude of the horizontal component of the Earth's magnetic field at the magnetic equator is about 10^{-5} T, and the vertical component at the geomagnetic poles is about 10^{-4} T. It has been calculated that for a ferromagnetic material of maximum magnetization to produce this field, it would have to occupy about 0.01% of the Earth's volume.

The idea of a ferromagnet of this size within the Earth may not seem unreasonable at first, but this cannot be a correct model. The interior temperature of the Earth is well above the Curie temperatures of iron and nickel, the ferromagnetic materials believed to be the most abundant in the Earth's interior. For iron, this temperature is 770°C, which is attained at a depth of only 100 km. The temperature of the Earth is higher at greater depths. This means that such a permanent internal magnet is impossible.

The fact that an electric current produces a magnetic field has led scientists to speculate that the Earth's magnetic field is associated with motions in the liquid outer core. (Recall that the magnetic field of a loop of wire or a solenoid is similar to that of a bar magnet—compare Figs. 19.9 and 19.10 to Fig. 19.1.) These motions may be associated in some way with the Earth's rotation. Jupiter and Saturn have magnetic fields much larger than that of Earth. These planets are largely gaseous and rotate rapidly, with periods of 10–11 hours. Mercury and Venus, on the other hand, have very weak magnetic fields. These planets are more like Earth in density and rotate relatively slowly, with periods of 58 and 243 days, respectively.

Several theoretical models have been proposed to explain the Earth's magnetic field. For example, it has been suggested that it arises from currents associated with thermal convection cycles in the liquid outer core caused by heat from the inner core. But the details of the mechanism are still not clear.

The axis of the Earth's magnetic field does not lie along its rotational axis, which defines the geographic poles. That is, the north magnetic pole and the geographic North Pole do not coincide. The magnetic pole is about 1500 km (930 mi) south of the geographic North Pole (true north), displaced about 13° of latitude. (The south magnetic pole is displaced by even more from its geographic pole, so the magnetic axis is not a straight line through the center of the Earth.) A compass indicates the direction of magnetic north, not "true" or geographic north. The angular variation in these two directions is called the *magnetic declination* (•Fig. 19.28). As shown on the map in the figure, the magnetic declination varies for different locations. Knowing the variations in the magnetic declination is particularly important in navigation, as you can imagine.

The Earth's magnetic field shows a variety of fluctuations. For one thing, there is evidence that there have been magnetic pole reversals. That is, the Earth's magnetic poles have switched polarity at various times in the past, most recently about

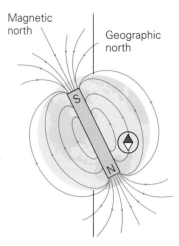

•**FIGURE 19.27 Geomagnetic field** The Earth's magnetic field is similar to that of a bar magnet. However, a permanent solid magnet could not exist within the Earth because of the high temperatures there. The field is believed to be associated with motions in the liquid outer core.

Magnetic north

Geographic north

•**FIGURE 19.28 Magnetic declination**

(a) The angular variation between magnetic north and "true" or geographic north is called the magnetic declination. **(b)** The magnetic declination varies with location. The map shows isogonic (same magnetic declination) lines for the conterminous United States. For locations on the 0° line, magnetic north is in the same direction as true (geographic) north. On either side of this line, a compass has an easterly or westerly variation. For example, on a 20°W line, a compass has a westerly declination of 20° (magnetic north is 20° west of true north).

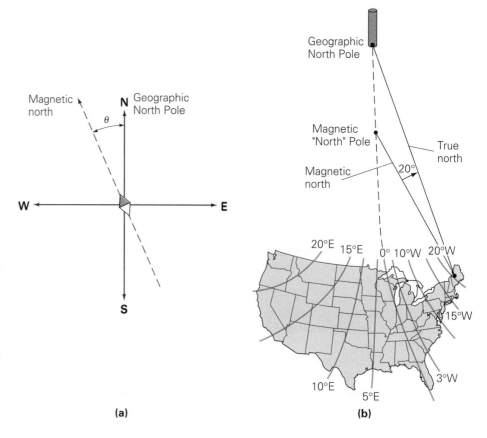

(a)

(b)

•**FIGURE 19.29 Magnetic confinement**

(a) A charged particle entering a uniform magnetic field at an angle moves in a spiraling path. **(b)** In a nonuniform, bulging magnetic field, particles spiral back and forth as though confined in a magnetic bottle. **(c)** Particles are trapped in the Earth's magnetic field, and the regions where they are concentrated are called Van Allen belts.

700,000 years ago. During a period of pole reversal, the south magnetic pole is near the south geographic pole. We are not certain why such pole reversals have occurred.

Even on a shorter time scale, the magnetic poles do not always remain in the same locations, but tend to "wander." The north magnetic pole, on which there are more data, moves about 1° of latitude (roughly 110 km or 70 mi) in a decade. For some unknown reason, it has moved consistently northward from its 1904 latitude position of 69°N, and westward, crossing the 100°W longitudinal meridian. In addition, there are sometimes dramatic daily shifts of as much as 80 km (50 mi), followed by a return to the starting positions. These are thought to be caused by charged particles from the Sun that reach the Earth's upper atmosphere and set up currents that disturb the magnetic field.

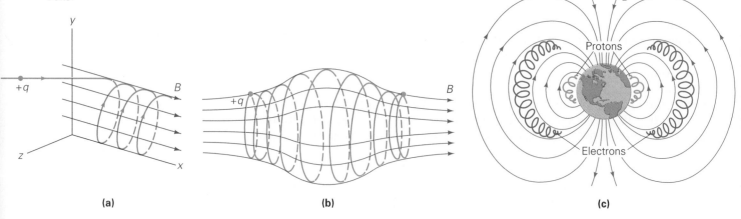

(a)

(b)

(c)

Charged particles from the Sun and cosmic rays entering the Earth's magnetic field give rise to other phenomena. A charged particle, entering a uniform magnetic field not perpendicular to the field, spirals in a helix (•Fig. 19.29a). This is because the component of the particle's velocity parallel to the field does not change. The motions of charged particles in a nonuniform field are quite complex. However, for a bulging field such as that depicted in •Fig. 19.29b, analysis shows that the particles spiral back and forth as though in a magnetic container, a so-called bottle. This is an important consideration in confining a plasma (a gas of charged particles) in fusion research (Chapter 30).

An analogous phenomenon occurs in the Earth's magnetic field, giving rise to regions where there are concentrations of charged particles. Two large donut-shaped regions at altitudes of several thousand kilometers are called the Van Allen belts (•Fig. 19.29c). It is in the lower Van Allen belt that light emissions called aurorae occur—the aurora borealis, or northern lights, in the Northern Hemisphere and the aurora australis, or southern lights, in the Southern Hemisphere. These eerie, flickering lights are most commonly observed in the Earth's polar regions, but have been seen at all latitudes (•Fig. 19.30).

It is believed that an aurora is created when charged solar particles are trapped in the Earth's magnetic field. Maximum aurora activity is noted to occur after a solar disturbance, such as a solar flare. Solar flares are violent magnetic storms on the Sun that spew out enormous quantities of charged particles and radiation. Trapped in the Earth's magnetic field, the charged particles are guided toward the polar regions. They excite or ionize oxygen and nitrogen atoms in the atmosphere. When the excited atoms return to their normal state and ions regain their normal number of electrons, light is emitted, producing the beautiful glow of the aurora.

•FIGURE 19.30 Aurora borealis—the northern lights
This spectacular aurora display is believed to be caused by energetic solar particles trapped in the Earth's magnetic field. The particles excite or ionize air atoms, and on de-excitation (or recombination), light is emitted. Aurorae are most commonly observed in the polar regions, where the solar particles are concentrated by the Earth's magnetic field.

Chapter Review

Important Terms

pole-force law or law of poles 591
magnetic field 592
electromagnetism 593
tesla (T) 594

right-hand force rule 596
right-hand source rule 597
ferromagnetic materials 600
magnetic domains 600
magnetic permeability 601

Curie temperature 602
magnetic moment 605
dc motor 608
cathode ray tube (CRT) 608
mass spectrometer 610

Important Concepts

- Opposite magnetic poles attract and like poles repel.
- The magnetic field is defined by the force it exerts on moving charges. The SI unit of the magnetic field is the tesla (T).
- The direction of the magnetic force exerted on charged particles moving in a magnetic field is given by the right-hand force rule.
- The direction of the magnetic field produced by a current or moving charge is given by the right-hand source rule.
- Ferromagnetic materials are those in which the electron spins lock together to create domains where the magnetic fields of individual electrons add constructively. The magnetic permeability of ferromagnetic materials is many times that of the vacuum, reflecting the fact that these materials enhance an external magnetic field.
- Above the Curie temperature the magnetic domains in a ferromagnet become thermally disordered and the material loses its "permanent" magnetic field.
- The direction of the magnetic moment of a current loop is specified by a right-hand rule. In an external magnetic field, the magnetic moment vector tries to align itself with the field direction.

Important Equations

Magnitude of the Magnetic Force on a Moving Charge (q):

$$F = qvB \sin \theta \qquad (19.3)$$

Directional right-hand force rule: When the fingers of the right hand are pointed in the direction of **v** and then turned or curled toward the vector **B**, the extended thumb points in the direction of the force **F** on a *positive* charge. (**F** is in the opposite direction for a negative charge.)

Magnitude of the Magnetic Field around a Long, Straight Current-Carrying Wire:

$$B = \frac{\mu_o I}{2\pi d} \qquad (19.4)$$

(where $\mu_o = 4\pi \times 10^{-7}$ T·m/A and is called the magnetic permeability of free space)

Magnitude of the Magnetic Field at the Center of a Circular Loop of Current-Carrying Wire:

$$B = \frac{\mu_o I}{2r} \qquad (19.5)$$

Magnitude of the Magnetic Field in a Solenoid (along the axis):

$$B = \frac{\mu_o NI}{L} \qquad (19.6)$$

or

$$B = \mu_o nI \quad \text{(where } n = N/L) \qquad (19.7)$$

Directional right-hand source rule: When a current-carrying wire is grasped with the right hand with the extended thumb pointing in the direction of the conventional current, the curled fingers indicate the circular sense of the magnetic field. (The field vector is tangent to the circular field line at any point on the circle.)

Magnetic Permeability:

$$\mu = K_m \mu_o \qquad (19.8)$$

Magnitude of Force on a Straight Current-Carrying Wire:

$$F = ILB \sin \theta \qquad (19.10)$$

Directional right-hand force rule for currents: When the fingers of the right hand are pointed in the direction of the conventional current I and then turned or curled toward the **B** vector, the extended thumb points in the direction of the force on the wire.

Magnitude of Torque on a Single Current-Carrying Loop:

$$\tau = IAB \sin \theta \qquad (19.11)$$

[The quantity IA is called the magnetic moment, m, of the loop: $m = IA$. For a coil of N loops, $\tau = NIAB \sin \theta$.]

Magnetic Velocity Selector:

$$v = \frac{E}{B_1} = \frac{V}{B_1 d} \qquad (19.16)$$

Particle Mass (mass spectrometer):

$$m = \left(\frac{qdB_1 B_2}{V}\right) R \qquad (19.17)$$

Exercises

19.1 Magnets and Magnetic Poles

1. When two bar magnets are near each other, how many possible repulsive magnetic force interactions are there between their poles? (a) 1, (b) 2, (c) 4, (d) the number depends on the orientation of the magnets.

2. If a compass is placed in a horizontal magnetic field, it will point (a) perpendicular to the field lines, (b) with the south pole in the opposite direction of the field, (c) at a nonzero angle $\theta < 90°$ to the field lines, (d) none of these.

3. If you look directly at the S pole of a bar magnet, the magnetic field points (a) to the right, (b) to the left, (c) away from you, (d) toward you.

4. Given two identical iron bars, one of which is a permanent magnet and the other unmagnetized, how could you tell the difference using only the two bars?

19.2 Electromagnetism, Magnetic Forces, and the Source of Magnetic Fields

5. A magnetic field (a) has units of N/A, (b) may affect stationary charges, (c) can produce a force on a moving charge.

6. A proton moves vertically upward in a uniform magnetic field and deflects to the right as you watch it. What is the magnetic field direction? (a) Directly away from you, (b) directly toward you, (c) to the right, (d) to the left.

7. If a negatively charged particle were moving downward along the right edge of this page, which way should a magnetic field be oriented so that the particle would initially be deflected to the left?

8. A proton moves towards you out of the plane of the page. What is the direction of the magnetic field it cre-

ates to the right of it as viewed by you. (a) toward the top of the page, (b) toward the bottom of the page, (c) right, (d) left, (e) zero.

9 ■ A magnetic field can be used to determine the sign of charge carriers. Consider a wide conducting strip in a magnetic field oriented as shown in •Fig. 19.31. The charge carriers are deflected by the magnetic force and accumulate on one side of the strip, giving rise to the measurable voltage across it. (This is known as the Hall effect.) If the sign of the charge carriers is unknown (they are either positive charges moving as indicated by the arrows in the figure, or negative charges moving in the opposite direction), how does the measured voltage polarity or sign allow this to be determined?

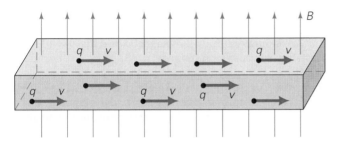

•FIGURE 19.31 The Hall effect
See Exercise 9.

10 ■ A positive charge of 0.25 C moves horizontally to the right at a speed of 2.0×10^2 m/s and enters a magnetic field of 0.40 T directed vertically downward. What is the magnitude of the force on the charge?

11 ■ A charge of 0.050 C moves vertically in a field of 0.080 T that is oriented horizontally. What speed must it have to experience a force of 10 N?

12 ■ A beam of protons is accelerated to a speed of 5.0×10^6 m/s in a particle accelerator and emerges horizontally from the accelerator into a uniform magnetic field. What **B** field will cancel the force of gravity and keep the beam moving exactly horizontally?

13 ■ An electron traveling at a uniform speed of 300 m/s in the $+x$ direction enters a uniform magnetic field and experiences a maximum force of 5.0×10^{-17} N in the $-y$ direction. What is the magnitude and direction of the magnetic field?

14 ■ An electron travels at a speed of 2.0×10^4 m/s through a uniform magnetic field whose magnitude is 1.2×10^{-3} T. What is the magnitude of the force on the electron if its velocity vector and the magnetic field vector (a) are perpendicular, (b) make an angle of 45° or (c) are parallel?

15 ■ What angle(s) does a particle's velocity vector have to make with the magnetic field direction for it to feel half the maximum possible magnetic force?

16 ■ A proton has a speed v near the equator. What will be the direction of the force experienced by a proton

due to the Earth's magnetic field if its velocity is directed (a) due south, (b) northwest, (c) upward?

17 ■ A horizontal proton beam in a particle accelerator is accelerated to a speed of 3.0×10^5 m/s. The beam enters a uniform magnetic field of 0.50 T oriented at an upward angle of 37° relative to the direction of the beam. (a) What is the initial acceleration of a proton in the beam? (b) If the beam were made up of electrons, what would be the difference in the force on the particles as the beam entered the magnetic field?

18 ■ A long, straight wire carries a current of 2.5 A. Find the magnitude of the magnetic field 25 cm from the wire.

19 ■ The magnetic field at the center of a 50-turn coil of radius 15 cm is 0.60 mT. Find the current in the coil.

20 ■ What current in a long, straight wire will produce a magnetic field with a magnitude of 10.0 μT at a perpendicular distance of 40.0 cm from the wire?

21 ■ Three particles enter a uniform magnetic field as shown in •Fig. 19.32a. Particles 1 and 3 have equal speeds and charges of the same magnitude. What can you say about (a) the charges of the particles and (b) their masses?

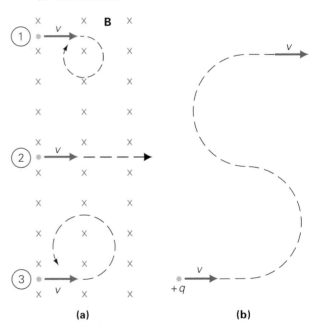

(a)   (b)

•FIGURE 19.32 Charges in motion
See Exercises 21 and 22.

22 ■ It is desired to deflect a positively charged particle in an "S" path as shown in •Fig. 19.32b using only magnetic fields. (a) Explain how this could be done. (b) How does the magnitude of the emerging velocity v compare to that of the initial velocity v_o?

23 ■ A solenoid 50.0 cm long with 1000 turns per meter has a central magnetic field of 1.5 mT. Find the current in the coil.

24 ■■ Two long, parallel wires separated by 50 cm carry currents of 4.0 A each in a horizontal direction. Find the magnetic field midway between the wires if the currents are (a) in the same direction and (b) in opposite directions.

25 ■■ A long, straight wire has a resistance of 2.0 Ω. What potential difference between its ends will produce a magnetic field of 1.0×10^{-4} T at a distance of 4.0 cm from the wire?

26 ■■ Two long, parallel wires separated by 0.20 m carry equal currents of 1.5 A in the same direction. Find the magnitude of the net magnetic field 0.15 m away from each wire on the side opposite the other wire.

27 ■■ Two long, parallel wires carry currents of 8.0 A and 2.0 A (•Fig. 19.33). (a) What is the magnitude of the magnetic field at the point midway between the wires? (b) Where on a line perpendicular to and joining the wires is the magnetic field zero?

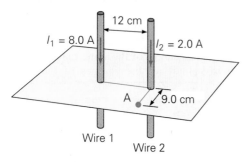

•**FIGURE 19.33 Parallel current-carrying wires**
See Exercises 27, 29, 30, 70, and 71.

28 ■■ A proton is accelerated to a speed of 4.0×10^6 m/s in a horizontal direction. It then enters a vertical magnetic field of 3.0 T. (a) Calculate the radius of the circular path of the particle. (b) Find the electric field that would cause the proton to move in a straight line.

29 ■■ Find the magnetic field (magnitude and direction) at point A in Fig. 19.33, which is located 9.0 cm away from wire 2 on a line perpendicular to the line joining the wires.

30 ■■ Suppose that the current in wire 1 in Fig. 19.33 were in the opposite direction. (a) What would be the magnetic field at a point midway between the wires? (b) Where on a line perpendicular to and joining the wires would the magnetic field be zero?

31 ■■ A circular loop of wire with a diameter of 12 cm is in the horizontal plane and carries a current of 3.6 A in a counterclockwise sense, as viewed from above. What is the magnetic field (magnitude and direction) at the center of the loop?

32 ■■ How much current must flow in a circular loop with a radius 10 cm to produce a magnetic field at its center that is approximately the same magnitude as the horizontal component of the Earth's magnetic field at the Equator?

33 ■■ A coil of four circular loops of radius 5.0 cm carries a current of 2.0 A in a clockwise sense, as viewed along a line perpendicular to the circles formed by the loops. What is the magnetic field at the center of the coil?

34 ■■ A circular loop of wire with a radius of 5.0 cm carries a current of 1.0 A. Another circular loop of wire is concentric with the first (has a common center with it) and carries a current of 2.0 A. The magnetic field at the center of the loops is measured to be zero. What is the radius of the second loop?

35 ■■ A solenoid 10 cm long is wound with 1000 turns of wire. How much current must flow through the windings to produce a magnetic field of 8.0×10^{-4} T at the solenoid's central axis?

36 ■■ An electron has a kinetic energy of 20 keV. It enters a uniform magnetic field of 3.5 T perpendicular to its line of motion. What is the radius of the resulting circular path?

37 ■■ A proton travels in a circular path with a radius of 60 mm in a uniform magnetic field of 12 μT. What is the kinetic energy of the proton?

38 ■■■ Two long, perpendicular wires carry currents of 15 A, as illustrated in •Fig. 19.34. What is the magnitude of the magnetic field at the midpoint of the line joining the wires?

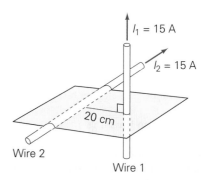

•**FIGURE 19.34 Perpendicular current-carrying wires**
See Exercises 38 and 57.

39 ■■■ Four wires running through the corners of a square with sides of length a as shown in •Fig. 19.35 carry equal currents I. Calculate the magnetic field at the center of the square in terms of these parameters.

40 ■■■ A solenoid is wound with 200 turns per cm. An outer layer of insulated wire with 180 turns per cm is wound over the first layer of wire. In operation, the inner coil carries a current of 5.0 A, and the outer coil carries a current of 7.5 A in the direction opposite to that of the current in the inner coil. What is the magnitude

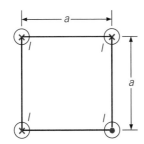

•**FIGURE 19.35 Current-carrying wires in a square array**
See Exercise 39.

of the magnetic field at the central axis of this doubly wound solenoid?

41 ■■■ A charged particle (charge q, mass m) moves in a horizontal plane at a right angle to a uniform vertical magnetic field. (a) What is the frequency (f) of the particle's circular motion in terms of q, B, and m? (This frequency is called the cyclotron frequency.) (b) Show that the time required for any charged particle to make one complete revolution is independent of its speed and radius. (c) Compute the path radius and frequency for an electron if $v = 10^5$ m/s and $B = 10^{-4}$ T.

42 ■■■ A proton enters a uniform magnetic field and completes a half circle, exiting in the exact opposite direction. (a) If the proton moves to the right and is deflected upward, what is the direction of the field? (b) How long does it take to turn around if it has an energy of 100 keV and the field is 0.10 T?

19.3 Magnetic Materials

43 The main source of magnetism in magnetic materials is from (a) electron orbits, (b) electron spin, (c) magnetic poles, (d) nuclear properties.

44 When a ferromagnetic material is placed in an external magnetic field, (a) the domain orientation may change slightly, (b) the domain boundaries may change, (c) new domains are created, (d) both (a) and (b).

45 Can a magnetic monopole be obtained by continually breaking a bar magnet in two? Explain.

46 ■ The magnitude of the magnetic moment (m) of a current loop is the product of the current (I) and the cross-sectional area of the loop (A): $m = IA$. Show that for an atomic electron in a circular orbit of radius r at an orbital speed v the orbital magnetic moment is $evr/2$.

47 ■ You are looking down on a circular current loop with the current circulating in the clockwise direction, what is the (a) direction of the magnetic field at the center of the loop, and (b) the direction of the loop's magnetic moment vector.

48 ■■ What is the magnetic moment of the electron in a hydrogen atom, which travels around the nuclear proton at a radius of 0.0529 nm? [Hint: Find the electron's speed by considering the centripetal force and see Exercise 46.]

49 ■■ What is the magnitude of the magnetic moment of a rectangular loop of wire that measures 10 cm by 15 cm and carries a current of 4.0 A?

50 ■■■ What is the magnetic field at the center of the circular orbit of the electron in a hydrogen atom (the orbital radius is 0.0529 nm)? [Hint: Find the electron's period by considering the centripetal force.]

51 ■■■ In Exercise 50, if you are looking down on the electron orbit and the electron orbits counterclockwise, (a) what is the direction of the magnetic field it produces at the site of the proton? (b) If the proton's magnetic moment (due to its spin) is oriented opposite to the field of the electron, what will it tend to do?

52 ■■■ A solenoid with 90 turns per cm has an iron core with a relative permeability of 2000. The solenoid carries a current of 1.6 A. (a) What is the magnetic field at the central axis of the solenoid? (b) How much greater is the magnetic field with the iron core than it would be without it?

19.4 Magnetic Forces on Current-Carrying Wires

53 A long straight, horizontal wire has a conventional current directed toward the north. In what direction would the needle point if a compass were placed above the wire: (a) east, (b) west, (c) south, (d) upward?

54 Hold up your left hand, palm to the right, the index and middle fingers forming a V. If the middle finger pointed in the direction of a magnetic field and the index finger in the direction of a conventional current in a straight wire, in what direction would the force on the wire be: (a) downward, (b) toward the back of the hand, (c) out of the palm, (d) none of these?

55 ■ Electric fields have equipotentials (Section 16.2), or paths, along which no work is done in moving an electric charge from one point to a different point. Are there analogous paths for magnetic fields? Explain.

56 ■ Compare the geometrical pattern of the electric field lines around an electric dipole and the magnetic field lines around a magnetic dipole (a bar magnet).

57 ■ Two straight wires are positioned at right angles to each other as in Fig. 19.34. If one wire carries a conventional current toward the north and the other a current toward the east, what is the direction of the force on each wire?

58 ■ A straight, horizontal segment of wire is 1.0 m long and carries a current of 5.0 A in the $+x$ direction in a magnetic field of 0.30 T that is directed vertically downward in the $-z$ direction. What is the force on the wire (magnitude and direction)?

59 ■ A 2.0-m length of straight wire carries a current of 20 A in a uniform magnetic field of 50 mT whose field lines make an angle of 37° with the direction of the conventional current. Find the force on the wire.

60 ■ Use a right-hand rule to find the direction of the conventional current in a wire in a magnetic field that results in the force on the wire shown for each case in •Fig. 19.36.

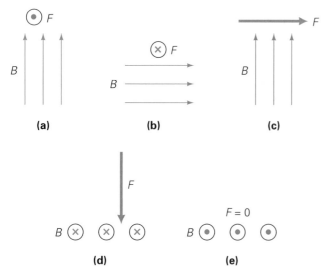

(a)   **(b)**   **(c)**

(d)   **(e)**

•**FIGURE 19.36   The right-hand rule**
See Exercise 60.

61 ■■ A straight wire 50 cm long conducts a current of 4.0 A directed vertically upward. If the wire experiences a force of 2.0×10^{-2} N in the eastward direction due to a magnetic field at right angles to its length, what is the magnitude and direction of the magnetic field?

62 ■■ A horizontal magnetic field of 1.0×10^{-2} T is at an angle of 30° to the direction of the current in a straight, horizontal wire 75 cm long. If the wire carries a current of 15 A, what is the magnitude of the force on the wire?

63 ■■ A wire carries a current of 10 A in the $+x$ direction in a uniform magnetic field of 0.40 T. Find the magnitude of the force per unit length and direction of the force on the wire if the magnetic field is directed in (a) the $+x$ direction, (b) the $+y$ direction, (c) the $+z$ direction, (d) the $-y$ direction, and (e) the $-z$ direction.

64 ■■ A straight wire 25 cm long is oriented vertically in a uniform horizontal magnetic field of 3.0 T in the $-x$ direction. What current (include direction) will cause the wire to experience a force of 5.0 N in the $+y$ direction?

65 ■■ A wire carries a current of 10 A in the $+x$ direction. Find the force per unit length on the wire if the magnetic field has components $B_x = 0.020$ T and $B_y = 0.040$ T.

66 ■■ A set of jumper cables used to start one car from another's battery is connected to the terminals of the cars' batteries. If 15 A of current flows in the cables when the one car is started and the cables are parallel and 15 cm apart, what is the force per unit length on the cables?

67 ■■ Find the force per unit length on each of two long, straight, parallel wires that are 24 cm apart when one carries a current of 2.0 A and the other a current of 4.0 A in the same direction.

68 ■■ Two long, straight, parallel wires 10 cm apart carry equal currents of 3.0 A in opposite directions. What is the force per unit length on the wires?

69 ■■ Two parallel wires are 4.0 m long and 18 cm apart and carry equal currents of 2.5 A in opposite directions. (a) Find the force on each wire. (b) Find the force on each wire if their currents are in the same direction.

70 ■■ What is the force per unit length on wire 1 in Fig. 19.33?

71 ■■ What is the force per unit length on wire 2 in Fig. 19.33?

72 ■■ Two long paralell wires are oriented so one is vertically above the other and free to move. The bottom one cannot move, and they have the same current. They each have a linear mass density of 1.5×10^{-3} kg/m. (a) How should the current directions in the two wires compare so that the top wire will be in equilibrium ("float")? (b) What is the current in each wire?

73 ■■ A wire is bent as shown in •Fig. 19.37 and placed in a magnetic field with a magnitude of 10 T in the indicated direction. Find the magnitude of the force on each segment of the wire if $x = 50$ cm and the wire carries a current of 5.0 A.

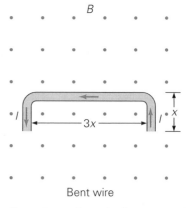

Bent wire

•**FIGURE 19.37   Current-carrying wire in a magnetic field**
See Exercise 73.

74 ■■ A nearly horizontal dc power line carries a current of 500 A directly eastward. If the only component of the Earth's magnetic field at the location is 5.0×10^{-5} T due north, what is the magnitude and direction of the magnetic force on a 75-m section of the line between two utility poles?

75 ■■ A rectangular loop of wire whose dimensions are 20 cm by 30 cm carries a current of 1.5 A. (a) What is the magnitude of the magnetic moment of the loop? (b) How should the loop be oriented in a uniform magnetic field to obtain the maximum torque on it?

76 ■■■ A 50.0-cm length of wire carries a current of 3.0 A in the $+x$ direction in a uniform magnetic field with components of $B_x = B_y = B_z = 100$ mT. Find the force on the wire.

77 ■■■ A rectangular wire loop of a cross-sectional area of 0.20 m² carries a current of 0.75 A. If the loop is free to rotate about an axis perpendicular to a uniform magnetic field of 3.0 T and the plane of the loop is initially at an angle of 30° to the direction of the magnetic field, what is the magnitude of the torque on the loop?

78 ■■■ A loop of wire with a magnetic moment of 1.6 A·m² is in a uniform magnetic field of 10.0 T. Find the magnitude of the torque on the loop if the magnetic field is (a) perpendicular to the plane of the loop and (b) at an angle of 30° to the plane of the loop.

19.5 Applications of Electromagnetism

79 For continuous operation, a dc motor must have (a) a split-ring commutator, (b) brushes, (c) current input, (d) all of these.

80 A mass spectrometer (a) can be used to determine the masses of atoms and molecules, (b) requires charged particles, (c) can be used to determine relative abundances of isotopes, (d) all of these.

81 ■ Explain the operation of the door bell and door chimes illustrated in •Fig. 19.38.

82 ■ An electron is accelerated horizontally from rest through a potential difference of 250 V in a CRT. The electron then passes through a perpendicular electric field of 0.10 V/m, while moving a horizontal distance of 12 cm. Find the electron's deflection from its original straight-line path.

83 ■ An ionized deuteron (a particle with a +1 charge) passes through a velocity selector whose magnetic and electric fields have magnitudes of 40 mT and 4.0 kV/m, respectively. Find the speed of the ion.

84 ■ A 3.5-keV proton passes in an undeflected straight line through a region of crossed electric and magnetic fields. If the magnetic field has a magnitude of 200 mT, what is the magnitude of the electric field?

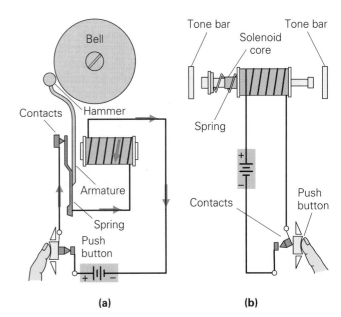

•**FIGURE 19.38 Electromagnetic applications**
(a) A doorbell and **(b)** door chimes both have electromagnets. See Exercise 81.

85 ■ In a velocity selector, a uniform magnetic field of 1.5 T from a large magnet and two parallel plates with a separation distance of 1.5 cm produce a perpendicular electric field. What voltage should be applied across the plates so that (a) a singly charged ion traveling at a speed of 8.0×10^4 m/s will pass through or (b) a doubly charged ion traveling at the same speed will pass through?

86 ■ A charged particle travels undeflected through crossed electric and magnetic fields whose magnitudes are 3000 N/C and 300 mT, respectively. Find the speed of the particle if it is (a) a proton or (b) an alpha particle. (An alpha particle is a helium nucleus, a positive ion with a double positive charge.)

87 ■■ An alpha particle (see Exercise 86) is accelerated in the $+x$ direction through a voltage of 1000 V. The particle then moves in an undeflected path between two oppositely charged parallel plates in a uniform magnetic field of 50 mT in the $+y$ direction. (a) If the plates are parallel to the xy plane, what is the magnitude of the electric field between them? (b) If the distance between the plates is 10 cm, what is the potential difference between them? (c) Which plate (top or bottom) is positively charged? Take the mass of the alpha particle to be 6.64×10^{-27} kg.

88 ■■ In a mass spectrometer, a singly charged ion having a particular velocity is selected using a magnetic field of 0.10 T perpendicular to an electric field of 1.0×10^3 V/m. The same magnetic field is used to deflect the ion, which moves in a circular path with a radius of 1.2 cm. What is the mass of the ion?

89 ■■■ In a mass spectrometer, a doubly charged ion having a particular velocity is selected using a magnetic

Exercises **621**

field of 100 mT perpendicular to an electric field of 1.0 KV/m. The same magnetic field is used to deflect the ion in a circlar path with a radius of 15 mm. Find (a) the mass of the ion and (b) the kinetic energy of the ion. (c) Does the kinetic energy of the ion increase in the circular path? Explain.

90 ■■■ In a "time-of-flight" mass spectrometer, a beam of protons and a beam of alpha particles (see Exercises 86 and 87) are accelerated through a potential difference of 500 kV. If the length of the flight tube is 1.25 m, what is the time difference for the different particles to arrive at the detector?

91 ■■■ In a momentum selector, a beam of protons enters a field and some of them make exactly a quarter circular arc with a radius of 0.50 m. If the magnetic field is oriented at right angles to their velocity, what is its magnitude if the exiting protons have a kinetic energy of 10 keV?

*19.6 The Earth's Magnetic Field

92 The Earth's magnetic field (a) has poles that coincide with the geographic poles, (b) is produced by internal ferromagnetic material, (c) reverses polarity every few hundred years, (d) none of these.

93 Aurorae occur (a) only in the Northern Hemisphere, (b) in the lower Van Allen belt, (c) because of pole reversals, (d) predominantly when there are no solar disturbances.

94 ■ What is the polarity of the magnetic pole near the Earth's geographic North Pole?

95 ■ Looking down on the Earth from above the North Pole, it would be spinning counterclockwise. What would the direction of the electric current inside the Earth have to be to create its magnetic field?

Additional Exercises

96 A proton is accelerated through a potential difference of 3.0 kV. It then enters a region between two parallel plates that are separated by 10 cm and have a potential difference of 250 V. Find the magnitude of the magnetic field (perpendicular to E) needed to allow the proton to pass between the plates undeflected.

97 How many turns should a solenoid 30 cm long have to produce a magnetic field of 5.0 mT at its axis when operating with a current of 5.0 A?

98 A solenoid 10 cm long has 3000 turns of wire and carries a current of 5.0 A. A 2000-turn coil of wire of the same length as the solenoid surrounds the solenoid and is concentric with it (shares a common center). If the outer coil carries a current of 10 A in the same direction as that of the current in the solenoid, find the magnetic field at the center of both coils.

99 A horizontal beam of electrons travels at a speed of 1.0×10^3 m/s along a north-to-south line in a discharge tube. What is the force on the electrons due to the downward vertical component of the Earth's magnetic field if it has a magnitude of 5.0×10^{-5} T at that location?

100 A proton enters a uniform magnetic field of 0.80 T in such a way that the proton follows a circular path with a radius of 4.6 cm. What is the kinetic energy of the proton in circular motion?

101 A current of 10 A is maintained through a square loop. The loop experiences a torque of 0.15 m·N about an axis through the loop's center and parallel to one of the sides. When it is placed in a magnetic field of 500 mT at an angle of 50° relative to its plane. Find the length of each side of the square.

102 A proton is accelerated from rest through a potential difference of 1.0 kV. It enters a uniform magnetic field of 4.5 mT that is perpendicular to the direction of its motion. (a) Find the radius of the circular path of the proton. (b) Calculate the period of revolution of the proton.

103 A beam of protons traveling north at a speed of 2.0×10^2 m/s passes through the space between two horizontal parallel plates, where a constant electric field and a constant magnetic field are at right angles to one another. If the electric field has a magnitude of 100 V/m and is directed upward from the bottom to the top plate, what must the magnitude and direction of the magnetic field be to allow the beam to pass undeflected? [*Hint:* Make a sketch of the situation.]

104 A horizontal power line carries a current of 2.5 A from west to east. What is the magnetic field 8.0 m below the line?

105 How fast should an electron travel at right angles to a magnetic field of 1.0×10^{-4} T so that the magnetic force just balances the gravitational force on it?

106 A particle with a charge of $+4.0 \times 10^{-8}$ C moves at a speed of 3.0×10^2 m/s through a magnetic field in the direction at which the magnetic force on the particle is maximum. If the force on the particle is 1.8×10^{-6} N, what is the magnitude of the magnetic field?

107 A current-carrying loop is in the form of a rectangle with dimensions of 20 cm by 30 cm. The loop carries a current of 10 A and is in a uniform magnetic field of 50 mT directed parallel to the plane of the loop. Find the torque on the loop.

108 How much current is in a straight wire 4.0 m long if the force on it is 0.040 N when it is placed in a 0.50-T magnetic field perpendicular to the wire?

20 Electromagnetic Induction

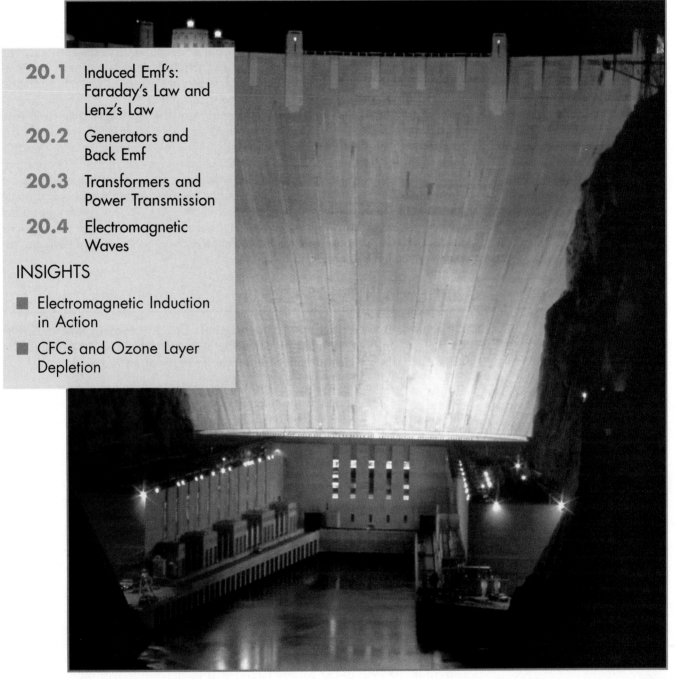

The great dams, with their hydroelectric power plants, are among the most impressive engineering feats of our civilization. At Hoover Dam (above) and others like it, one of the oldest and simplest power sources on Earth—falling water—is used to generate the electricity that lights our homes and drives our machines. In terms of basic physical principles, we can see that the gravitational potential energy of water trapped behind the dam is converted into kinetic energy (as the water is allowed to fall), and that some of this kinetic energy is then transformed into electric potential energy. But how? In this chapter you'll find the answer to this question. You'll also learn how electrical energy can be converted to yet another form of energy that we use in countless ways every day: electromagnetic radiation, including radio waves, X-rays, and light.

As we saw in Chapter 19, an electric current produces a magnetic field. But the mutual relationship of electricity and magnetism does not stop there. In this chapter we shall see that under certain conditions a magnetic field can be used to produce an electric current.

How is this done? Chapter 19 considered only constant magnetic fields. No current is induced in a loop of wire that is stationary in a constant magnetic field. However, if the magnetic field changes with time, or if the wire loop moves across or is rotated in the field, a current is induced in the wire.

The uses of this interrelationship of electricity and magnetism are legion. One example that will be familiar to most people is the playing of a cassette tape. The music you hear was encoded as tiny variations in a magnetic field. These variations produce electrical impulses, which are amplified and drive the speakers. The speakers, in turn, utilize electromagnetic interactions to translate the electrical impulses back into audible sound. Similar processes are involved when information is stored on or retrieved from a disk in your computer.

On a larger scale, consider the generation of the electricity that we use to power our cassette players, computers, and so many other devices that we use every day. If you stop for a moment and think how dependent we are on this form of energy, you will quickly realize that electric power is the basis of our modern civilization. A great deal of the energy that we extract from the environment is converted to electricity before being put to use. But regardless of the ultimate source of the energy—the burning of oil, coal, or gas, a nuclear reactor, or falling water—the actual conversion to electricity is accomplished by means of magnetic fields. This chapter examines the underlying electromagnetic principles that make such conversion possible.

20.1   Induced Emf's: Faraday's Law and Lenz's Law

Objectives: To be able to (a) define magnetic flux and explain how induced emfs are created by changing magnetic flux, and (b) calculate the magnitude and predict the polarity of an induced emf.

A magnet held stationary near a conducting wire loop does not induce a current in that loop (•Fig. 20.1a). If the magnet is moved toward the loop, however, as shown in •Fig. 20.1b, the deflection of the galvanometer needle indicates that there is a current in the loop during the motion. If the magnet is moved away from the loop, as shown in •Fig. 20.1c, the galvanometer needle is deflected in the opposite direction, which indicates a reversal of the current's direction.

Deflections of the galvanometer needle, indicating the presence of induced currents, also occur if the loop is moved toward or away from a stationary magnet. The effect, then, clearly depends on *relative* motion of the loop and magnet. The magnitude of the induced current is found to depend on the speed of the motion. However, there is a noteworthy exception to this general principle: If a loop is moved *parallel* to a uniform magnetic field, as shown in •Fig. 20.2, no current is induced in the loop.

Another way to induce a current in a stationary wire loop is to vary the *current* in another loop close to it. When the switch in the battery-powered circuit in •Fig. 20.3a is closed, the current in its loop goes from zero to a constant value in a short time. During this time, the magnetic field caused by this current and passing through both loops increases, and the galvanometer needle deflects. When the current in the loop in the battery circuit is at its maximum (constant) value, the resulting magnetic field is also constant, and the galvanometer reads zero. Similarly, when the switch is opened (•Fig. 20.3b), the current and the field decrease to zero, and the galvanometer deflects in the opposite direction. In both cases, the deflection and induced current occur only when the current and the magnetic field are *changing*.

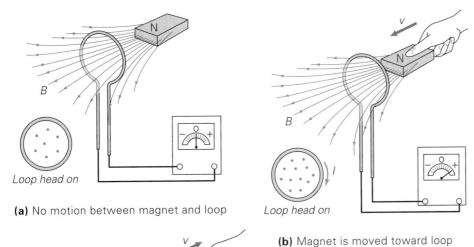

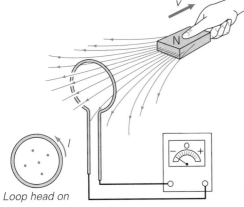

(a) No motion between magnet and loop

Loop head on

(b) Magnet is moved toward loop

(c) Magnet is moved away from loop

Loop head on

(d)

•FIGURE 20.1 Electromagnetic induction
(a) When there is no relative motion between the magnet and the wire loop, the number of field lines through the loop is constant, and the galvanometer needle shows no deflection. (b) Moving the magnet toward the loop increases the number of field lines passing through the loop, and an induced current is detected. (c) Moving the magnet away from the loop decreases the number of field lines passing through the loop. The induced current is in the opposite direction, as indicated by the needle deflection. (d) In an actual experiment of this type, a coil with a number of loops, or solenoid, is used to increase the effect.

The current induced in a loop is said to be set up by an induced electromotive force (emf) due to **electromagnetic induction**. Recall that an emf represents energy capable of moving charges around a circuit. For example, a battery is a chemical source of emf. In the case of a moving magnet and a stationary loop (Fig. 20.1), we say that an emf is *induced* in the loop, thereby causing the current. For the case of *two* stationary loops (Fig. 20.3), where a changing current in one circuit induces an emf in the other circuit, we speak of *mutual induction*.

Experiments on electromagnetic induction were done independently by Michael Faraday in England and Joseph Henry in the United States about 1830. Faraday

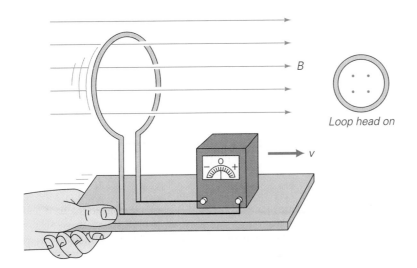

Loop head on

•FIGURE 20.2 Relative motion and no induction
When a loop is moved parallel to a uniform magnetic field, there is no change in the number of field lines passing through the loop and no induced current.

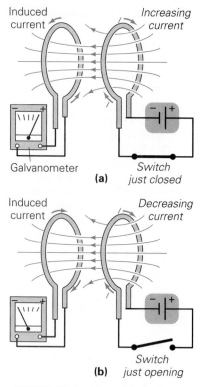

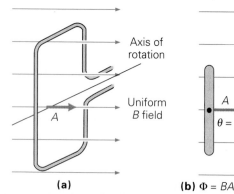

Galvanometer   **Switch**
(a)   **just closed**

Induced current   **Decreasing current**

Switch
(b)   **just opening**

•FIGURE 20.3 Mutual induction
(a) When the switch is closing in the right-loop circuit, the current buildup (typically over a few ms) produces a changing magnetic field that passes through the other loop, inducing a current in it. **(b)** Similarly, when the switch is opening, the magnetic field collapses and there is a decrease in the number of field lines through the second loop. The induced current in this loop is then in the opposite direction.

realized that the important factor in electromagnetic induction was the time rate of change of the number of magnetic field lines passing through the loop. That is,

> an induced emf is produced in a loop by changing the number of magnetic field lines passing through the plane of the loop.

(Note that this is the case for the situations in Figs. 20.1 and 20.3. Think about the situation in Fig. 20.2.)

Magnetic Flux

Since the induced emf or current in a loop depends on the change in the number of field lines through it, the ability to quantify the number of field lines through the loop at any time is useful. Consider a loop of wire in a uniform magnetic field **B**, as illustrated in •Fig. 20.4a. The number of field lines through the loop depends on its orientation relative to the **B** field. To describe this we use a vector **A** normal to the plane of the loop to represent its area with a directional sense. This is called the *area vector* and its magnitude is equal to the area of the loop. The orientation of the loop may then be described by the angle θ, which is the angle between **B** and **A**.

In general, a relative measure of the number of field lines passing through a particular area (the area within a loop, in our case) is given by the **magnetic flux** (Φ), which is defined as

$$\Phi = BA \cos \theta \qquad (20.1)$$

SI unit of magnetic flux: T·m², or weber (Wb)

A unit for magnetic flux is the weber (Wb). Recall from Section 19.2 that B has the units Wb/m² (or tesla, T). Thus the units of flux, $\Phi = BA$, are (Wb/m²)(m²) = Wb, which is T·m² in the SI.

The orientation of the loop to the magnetic field affects the number of field lines passing through it, and this is taken into account by including the cosine term in Eq. 20.1. Let us consider several possible orientations:

- If **B** and **A** are parallel ($\theta = 0°$), then the magnetic flux is a maximum: $\Phi_{max} = BA \cos 0° = +BA$, and the maximum number of field lines pass through the loop (•Fig. 20.4b).

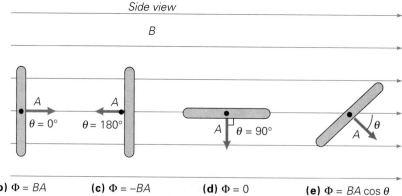

(a)

Axis of rotation

Uniform B field

Side view

B

A A
$\theta = 0°$ $\theta = 180°$

A $\theta = 90°$

θ
A

(b) $\Phi = BA$ **(c)** $\Phi = -BA$ **(d)** $\Phi = 0$ **(e)** $\Phi = BA \cos \theta$

•FIGURE 20.4 Magnetic flux
(a) Magnetic flux (Φ) is a relative measure of the number of field lines passing through an area (A). The area can be represented by a vector **A** perpendicular to the plane of the area. **(b)** When the plane of a rotating loop is perpendicular to the field and $\theta = 0°$, then $\Phi = \Phi_{max} = BA$. **(c)** When $\theta = 180°$, the magnetic flux has the same magnitude but is opposite in direction: $\Phi = -\Phi_{max} = -BA$. **(d)** When $\theta = 90°$, then $\Phi = 0$. **(e)** As the loop is rotated from an orientation perpendicular to the field to one more nearly parallel to the field, less area is open to the field lines and the flux decreases. In general, $\Phi = BA \cos \theta$.

- If **B** and **A** are opposite ($\theta = 180°$), then the magnitude of the magnetic flux is a maximum again, but of opposite sign: $\Phi_{180°} = BA \cos 180° = -BA = -\Phi_{max}$ (•Fig. 20.4c).
- If **B** and **A** are perpendicular, there are no field lines through the loop, and the flux is zero: $\Phi_{90°} = BA \cos 90° = 0$ (•Fig. 20.4d).
- For situations at intermediate angles (•Fig. 20.4e), $A \cos \theta$ is interpreted as the *effective* area of the loop open to field lines. Thus, in general, the flux is $\Phi = BA \cos \theta$. (Alternatively, we can think of $B \cos \theta$ as the perpendicular component of the *field* through the full area of the loop, A, as shown in •Fig. 20.5 Thus $\Phi = (B \cos \theta)A$, and the result is the same.)

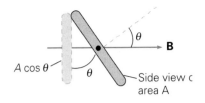

$\Phi = (A \cos \theta)B$

(a)

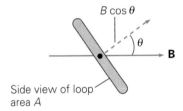

$\Phi = A (B \cos \theta)$

(b)

•**FIGURE 20.5 Magnetic flux through a loop: An alternative interpretation** Instead of defining the flux in terms of the magnetic field (B) passing through a reduced area ($A \cos \theta$), as in **(a)**, we can define it in terms of the perpendicular component of the magnetic field ($B \cos \theta$) passing through A, as in **(b)**. In either interpretation, $\Phi = BA \cos \theta$.

Faraday's Law of Induction and Lenz's Law

From his experiments, Faraday concluded that the emf induced in a coil of N loops depends on the time rate of change of the number of field lines through all the loops, or the *time rate of change of the magnetic flux* through all the loops. This is known as **Faraday's law of induction** and is mathematically expressed as

$$\mathcal{E} = -N \frac{\Delta \Phi}{\Delta t} \quad \text{induced emf} \tag{20.2}$$

where $\Delta \Phi$ is the change in flux through *one loop*. In a coil of N loops of wire, the total change of flux will be $N\Delta\Phi$. Note that $\mathcal{E}$ in Eq. 20.2 is an *average* value over the time interval Δt.

The minus sign is included in Eq. 20.2 to give an indication of the *polarity* or *direction* of the induced emf, which is found by considering the resulting induced current and its effect, according to **Lenz's law**:

> An induced emf gives rise to a current whose magnetic field opposes the change in magnetic flux that produced it.

This means that the magnetic field *due to the induced current* is in a direction that tends to keep the flux through the loop from changing. For example, if the flux increases in the $+x$ direction, the magnetic field due to the induced current will be in the $-x$ direction. This tends to cancel the increase in the flux, or *oppose the change*, since magnetic fields add vectorially. Essentially, the magnetic field due to the induced current tries to maintain the status quo value of the magnetic flux. This effect is sometimes described as "electromagnetic inertia," by analogy with the tendency of objects to resist changes in their velocity. In the long run, the induced current cannot maintain the magnetic flux unchanged. However, during the time the flux through the coil is changing, the induced magnetic field will tend to cancel the change.

The induced current direction is given by a right-hand rule: with the fingers of the right hand in the direction of the induced field, the extended thumb points in the direction of the conventional induced current. This is the right-hand rule used to find the direction of a magnetic field produced by a current (Chapter 19), applied in reverse, so to speak. Try applying Lenz's law to the loops in Fig. 20.3. (See also Demonstration 13.)

Lenz's law incorporates the conservation of energy. Suppose that the magnetic field due to an induced current *added* to the original field, that is, it increased the flux. This would give a greater induced emf and current, which would produce a greater magnetic field, which would give a greater induced current, and so on. This would be a something-for-nothing situation that would clearly be incompatible with the law of the conservation of energy—something that has never been observed.

DEMONSTRATION 12 ■ Magnetic Levitation and Lenz's Law

(a) A strong, lightweight magnet is taped to a strip of Manila folder and held above an aluminum disk on a turntable.

(b) When the disk rotates, the magnet levitates above it.

A demonstration of magnetic levitation—not with a superconductor, but with electromagnetic induction and induced currents.

When the disk rotates, it experiences a time-varying magnetic field because of its motion relative to the magnet, and induced currents (called eddy currents) are set up in the disk. By Lenz's law, the induced currents give rise to a magnetic field that opposes the change in flux experienced by the disk. These eddy currents experience a repulsive force due to the lightweight magnet's field, so the disk is pushed downward. By Newton's third law, the magnet experiences an upward reaction force. As a result, when the disk is rotating, the magnet "levitates" above it.

The levitation height depends on the tangential speed of the disk. This can be shown by moving the magnet along a radial axis. Levitation because of air currents is ruled out by using a similar strip of paper without a magnet. (The paper strip is necessary to prevent horizontal movement of the magnet.)

For the case of an emf induced in a loop by a moving magnet (Fig. 20.1), recall that a current-carrying loop has a magnetic field similar to that of a bar magnet. The induced current sets up a magnetic field whose effect is such that the loop acts like a bar magnet with a polarity that opposes the motion of the real bar magnet. Thus, there is opposition to the motion; work must be done to move the magnet.

Substituting the expression for the magnetic flux (Φ) given by Eq. 20.1 into Eq. 20.2 shows that a change in flux can result from changes other than just a change in magnetic field. In general,

$$\mathscr{E} = -\frac{N\Delta\Phi}{\Delta t} = -\frac{N\Delta(BA\cos\theta)}{\Delta t}$$

Hence we see that there are three quantities that can change with time to produce an induced emf: (1) the strength of the magnetic field B, (2) the area of the loop A, and (3) the angle θ. Expanding the equation so as to see (and label) these terms explicitly, we have

$$\mathscr{E} = -N\left[(A\cos\theta)\left(\frac{\Delta B}{\Delta t}\right) + (B\cos\theta)\left(\frac{\Delta A}{\Delta t}\right) + BA\left(\frac{\Delta(\cos\theta)}{\Delta t}\right)\right] \qquad (20.3)$$

$$(1) \qquad\qquad (2) \qquad\qquad (3)$$

1. The first term in the brackets represents the flux change due to a *time-varying magnetic field*, with the area (A) and orientation ($\cos\theta$) constant. A time-varying magnetic field is easily obtained using a time-varying current, or by moving a magnet in the space near a coil (or the coil in the space near the magnet).

2. The second term represents a flux change due to a *time-varying loop area*, with a constant magnetic field (**B**) and a constant orientation (cos θ). This could occur if a circular loop were being stretched or flattened in a plane, or if it had an adjustable circumference (imagine an adjustable loop around a balloon that was being blown up).

3. The third term represents a change in flux resulting from a *change in orientation of the loop with time*, with a constant magnetic field and a constant loop area. This occurs when a coil (made up of N loops) is rotated in a magnetic field. The change in the number of field lines through a single loop in this case is evident in the sequential views of a rotating loop in Fig. 20.4. Another way of looking at this is that there is a change in the *effective* area of the loop, or a change in the area that is open to the field lines.

The induced emf in a rotating loop will be considered in more detail in Section 20.2. The emfs resulting from the first two terms in Eq. 20.3 are analyzed in the following three examples. (Also, see the Insight on p. 630 for some applications of electromagnetic induction.)

CONCEPTUAL EXAMPLE 20.1 ■ FIELDS IN THE FIELDS: ELECTROMAGNETIC INDUCTION

In rural areas where electric power lines carry current to big cities, it is possible to generate small amounts of electricity from the lines by means of induction in a single coil. The overhead power lines carry relatively large currents that periodically reverse in direction 60 times per second. How would you orient the plane of the coil to produce maximum induced current if the power lines ran in a north-to-south direction? (a) Parallel to the ground, (b) perpendicular to the ground in the north–south direction, (c) perpendicular to the ground in the east–west direction, (d) the orientation of the coil would not affect the size of the induced current. *Clearly establish the reasoning and physical principle(s) used in determining your answer before checking it below. That is, **why** did you select your answer?*

Reasoning and Answer. Making a sketch will be helpful here. Magnetic field lines from long wires are circular, centered on the wire (see Fig. 19.8). The direction of the magnetic field at ground level will be parallel to the ground but will alternate in direction, from east to west and back to east again, every $\frac{1}{60}$ of a second. We can immediately eliminate answer (d), because we know that the flux *does* depend on the orientation of the loop relative to the field (see Eq. 20.1 and Fig. 20.4). Answer (a) cannot be correct, for in this orientation the coil would have *no* magnetic flux passing through it, and thus no emf. If the coil is oriented perpendicular to the ground but in the east–west direction, there would also be no magnetic flux through it, so answer (c) can also be ruled out. Hence the answer is (b). If the coil is oriented perpendicular to the ground with its plane in the north–south direction, the flux through it will vary from zero to its maximum value and back every 60 s, and the emf induced in the coil will be a maximum.

Follow-up Exercise. (a) Suggest some possible ways of increasing the amount of current produced by the arrangement discussed in this Example. (b) Explain clearly why this method would not work if the overhead wires carried a constant dc current.

EXAMPLE 20.2 ■ A LABORATORY HAZARD: INDUCED CURRENTS

If electrical instruments are left in the vicinity of an electromagnet when it is turned on or off, induced emfs (and the resulting currents) can cause damage to the wiring of the apparatus. Consider the following situation: The magnitude of a magnetic field passing perpendicularly through the 200-loop wire coil (radius 6.0 cm) of a galvanometer falls from its initial value of 2.0 T to zero in 0.010 s at a steady rate. (a) What emf is induced in the coil? (b) What is the direction and magnitude of the induced current if the resistance of the coil is 60 Ω?

Insight Electromagnetic Induction in Action

You have probably used applications of electromagnetic induction many times without knowing it. One of these is magnetic tape—either audio or video. In an audio tape recorder, a plastic tape coated with a film of iron oxide or chromium oxide runs past a recording head, which consists of a coil wound around an iron core with a gap (Fig. 1). Current in the coil produces a magnetic field in the gap which magnetizes the tape's film. The strength and direction of the magnetization in the film are determined by the gap field, which is determined by the current pulses that are generated by sound from a microphone.

To reproduce the sound, the tape is run by the same head or one similar to it. The changing flux due to the magnetization of the film on the tape induces an emf in the coil, matching the original voltage or current fluctuations. These are amplified and converted to sound in a speaker (which also commonly uses electromagnetic induction; see Exercise 4 and Fig. 20.24). A tape may be erased by passing it over the gap in a head that has a high-frequency alternating current (ac) input. The resulting magnetic field diminishes to zero as the tape moves away from the gap, leaving its film in a demagnetized condition.

Electromagnetic induction is also used in an electrical safety device called a ground fault circuit interrupter (GFCI). A GFCI may be plugged into a wall outlet or installed as part of a home circuit. It senses any problem in an electrical appliance plugged into the outlet or circuit and de-energizes the appliance, thus protecting individuals from electrical shock or injury.

The principle of the GFCI is illustrated in Fig. 2. It consists of a sensing coil and a circuit breaker. The coil is wrapped on an iron ring, through which pass the wires carrying current to and from the protected outlet or circuit. This arrangement is sometimes called a *differential transformer* because it can sense a difference in the currents carried by the two wires.

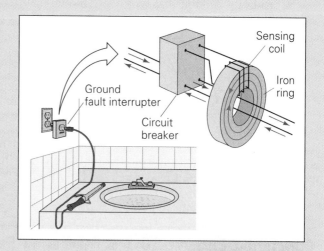

FIGURE 2 Ground fault circuit interrupter (GFCI) A safety device that quickly detects currents to ground and opens the circuit.

The opposite currents in the two wires produce opposite magnetic fields, which are concentrated in the iron ring. With alternating current, the directions of the fields are constantly changing. However, since the fields are always equal and opposite, the total field is normally zero. Hence, the net flux through the coil is zero, and no emf is induced in the sensing coil.

However, suppose that a wire breaks inside an appliance and touches its metal casing (Section 18.5), or a plugged-in curling iron falls into a sink full of water. If someone then touches the appliance or puts a hand in the water, some of the current will pass through the person's body to ground (a fault to ground). The currents in the wires passing through the ring are then not equal, giving rise to a nonzero magnetic flux. A changing flux causes an induced emf in the sensing coil and the resulting current trips the circuit breaker, opening the circuit.

This all happens in about 30 ms (30/1000 of a second) in response to an induced current of 4 to 6 mA. Notice that this is much less than the current that would normally trip a protective circuit breaker, which is usually set at 20 or 30 amps. The GFCI does not protect someone from receiving a shock, but it does limit the time the hazard exists and the potential for serious injury.

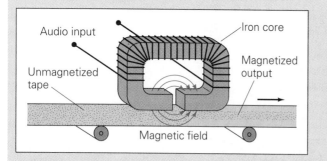

FIGURE 1 Tape recorder head
Current pulses in the coil on the iron core produce magnetic fields that magnetize the metallic coating on the tape as it passes by.

Solution. We note that the plane of the coil is perpendicular to the magnetic field, and that the angle of orientation stays constant, but the flux through the coil decreases due to the reduction in field strength. The data are

Given: $\theta = 0°$ **Find:** (a) $\mathscr{E}$ (induced emf)
$N = 200$ (b) I (induced current)
$r = 6.0 \text{ cm} = 0.060 \text{ m}$
$\Delta B = B_2 - B_1 = 0 \text{ T} - 2.0 \text{ T}$
$= -2.0 \text{ T (or Wb/m}^2)$
$\Delta t = 0.010 \text{ s}$
$R = 60 \ \Omega$

(a) Before the magnetic field starts changing and after it reaches zero (and stays constant), the induced emf and current will be zero. These quantities will be nonzero only for the 0.010 s during which the field is decreasing. Since the field decreases at a uniform rate, the induced emf and current will be constant during that period. Only the first term in Eq. 20.3 contributes, since A and θ are constant (for example, $\Delta A/\Delta t = 0$), and the induced emf is

$$\mathscr{E} = -N\frac{\Delta\Phi}{\Delta t} = -N(A\cos\theta)\left(\frac{\Delta B}{\Delta t}\right) = -N(\pi r^2)(\cos 0°)\left(\frac{\Delta B}{\Delta t}\right)$$

$$= -(200)\pi(0.060 \text{ m})^2\left(\frac{-2.0 \text{ T}}{0.010 \text{ s}}\right) = 4.5 \times 10^2 \text{ V}$$

If the field had decreased nonlinearly, then this answer would represent an average value.

(b) The magnetic field is decreasing. Thus, by Lenz's law, the induced current is in the direction that sets up a magnetic field in the *same* direction as the original field so as to add to it and thereby oppose the *decrease*. The magnitude of the induced current is given by the relationship between voltage, current, and resistance ($\mathscr{E} = IR$):

$$I = \frac{\mathscr{E}}{R} = \frac{4.5 \times 10^2 \text{ V}}{60 \ \Omega} = 7.5 \text{ A}$$

Follow-up Exercise. Suppose we have the same coil and orientation as in the Example. The field is held constant at 2.0 T, but the coil shrinks to half its radius in 0.50 s. Assume that the field direction is into the page ($\times$) and the plane of the coil is in the paper. Looking from above the plane of the paper, what is the direction and magnitude of the average induced current?

EXAMPLE 20.3 ■ MECHANICAL WORK INTO ELECTRICAL ENERGY: MOTIONAL EMF

A metal rod is pulled along a metal frame at a constant velocity, as illustrated in ●Fig. 20.6. A uniform magnetic field is normal to the plane of the frame. (a) What is the expression for the potential difference, or emf, developed in the rod–frame circuit in terms of B, L, and v? (b) What is the direction of the induced current?

Solution.

(a) In this situation, the magnetic field is constant and $\theta = 0°$, but the flux through the loop formed by the frame and the rod increases because its area increases. In a time Δt, the rod moves a distance $\Delta x = v\Delta t$. If L is the length of the rod, the change in the area of the loop is $\Delta A = L \Delta x = Lv\Delta t$. Therefore, the time rate of change (increase) of the area of the "moving" loop is given by

$$\frac{\Delta A}{\Delta t} = Lv$$

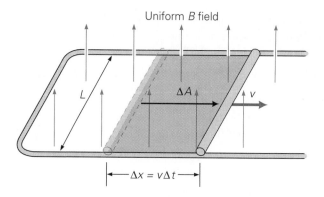

Uniform B field

• FIGURE 20.6 Motional emf
As the metal rod moves on the metal frame, the area of the rectangular loop varies with time. A current is induced in the loop as a result of the changing flux. See Example 20.3.

Thus, the emf that develops is

$$\mathscr{E} = -\frac{\Delta\Phi}{\Delta t} = -B\left(\frac{\Delta A}{\Delta t}\right) = -BLv$$

(b) The emf in the rod will set up an induced current in the loop. We know from Lenz's law that the magnetic field produced by this current will tend to oppose the change in flux. Since the moving rod causes an increase in the area of the loop, and thus an increase in the number of field lines through the loop in the upward direction, the induced field will be downward in the figure. Thus by the right-hand rule, the current *in the rod* will be directed toward the near side of the frame.

Follow-up Exercise. For the setup in this Example, estimate the maximum size of the emf that could be induced if the wire bar were 20 cm long and the magnetic field were taken to be 0.30 mT). Assume that the wire bar is pulled along at 10 cm/s.

The emf produced by the movement of the rod in Example 20.3 is known as *motional emf*. You might ask what would happen if the bar moved through the field on an *insulated* frame. With no conducting loop, and thus no complete circuit, there could be no sustained current. Yet the electrons in the bar would be moving through a magnetic field, and so would experience a force in a direction given by the right-hand force rule. So there would still be an induced emf, electrons would move in response to it, and the bar would quickly become polarized, with one end positive (from the deficiency of electrons) and the other negative (from the surplus of electrons). Once the bar was polarized, there would be no further movement of charge, but an induced emf would still exist. It is through this mechanism that a very small emf develops across the wings of an airplane in flight as they "cut through" the Earth's magnetic field lines.

20.2 Generators and Back Emf

Objectives: To be able to (a) understand the operation of electrical generators and calculate the emf produced by an ac generator at any time in its operating cycle, and (b) explain the origin of back emf and its effect on the behavior of motors.

As was pointed out in the preceding section, one method to induce an emf or current in a loop is through a change in the loop's orientation, or a change in its effective area. Note again in Fig. 20.4 that the number of field lines through the loop changes as it *rotates*; that is, the *effective* area of the loop (the area perpendicular to the field) is $A \cos\theta$. This way of producing a flux change is the principle of operation of a simple electrical generator.

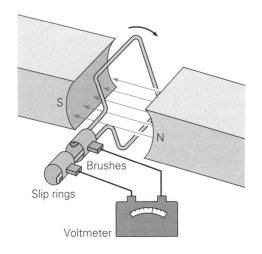

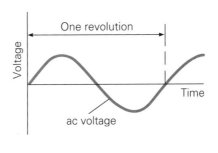

•FIGURE 20.7 A simple ac generator
The rotation of a loop of wire in a magnetic field produces a voltage output whose polarity reverses with each half-cycle. This alternating voltage gives rise to an alternating current.

Generators

A *generator* is a device that converts mechanical energy into electrical energy. Basically, the function of a generator is the reverse of that of a motor. In some instances, a generator may actually be run "backward" as a motor, and vice versa; however, this is not usually the case.

A battery supplies direct current (dc). That is, the polarity of the voltage does not change. Generators can produce either direct current or **alternating current (ac)**, for which the polarity of the voltage and the direction of the current periodically change. The electricity used in homes and industry is primarily ac.

An **ac generator** is sometimes called an *alternator*. The basic elements of a simple ac generator are shown in •Fig. 20.7. A wire loop is mechanically rotated in a magnetic field by some external means. The rotation of the loop (called an armature) causes the magnetic flux through it to change, and a current is induced in its wire. The ends of the wire loop are connected to an external circuit by means of slip rings and brushes. (In practice, generators have many loops, or windings, on their armatures.)

When the loop is rotated with a constant angular speed (ω), the angle (θ) between the magnetic field vector and the area vector of the loop (which is perpendicular to the plane of the loop) changes with time: $\theta = \omega t$ (assuming $\theta = 0°$ at $t = 0$). As a result, the cross-sectional area of the loop open to the magnetic field lines changes with time, and from Eq. 20.1 the flux at any time is

$$\Phi = BA \cos \theta = BA \cos \omega t$$

By Faraday's law, the induced emf for a rotating coil of N loops is then

$$\mathscr{E} = -N\frac{\Delta\Phi}{\Delta t} = -NBA\left(\frac{\Delta(\cos \omega t)}{\Delta t}\right)$$

It can be shown by mathematical methods beyond the scope of this book that as Δt approaches zero $[\Delta(\cos \omega t)]/\Delta t = -\omega \sin \omega t$. Thus, the instantaneous value of the emf is

$$\mathscr{E} = NBA\omega \sin \omega t$$

where $NBA\omega$ is the *maximum* value of the emf, which occurs when $\sin \omega t = \pm 1$. If we designate $NBA\omega$ as $\mathscr{E}_o$, we can write

$$\mathscr{E} = \mathscr{E}_o \sin \omega t \qquad (20.4) \qquad \textbf{Sinusoidal ac emf}$$

Since the value of the sine function varies between $+1$ and -1, the sign, or polarity, of the emf changes with time (•Fig. 20.8). Note from the figure that the emf has its maximum value when $\theta = 90°$ or $\theta = 270°$, that is, just when the area

Note: Motors are discussed in Section 19.5.

Note: Although redundant, the expression ac current is often used, as is ac voltage (usually abbreviated VAC, for volts ac).

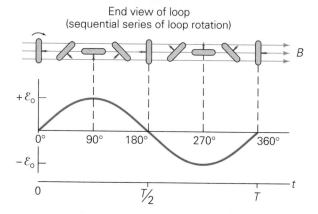

End view of loop
(sequential series of loop rotation)

B

$+\mathcal{E}_0$

0° 90° 180° 270° 360°

$-\mathcal{E}_0$

0 $T/2$ T t

• FIGURE 20.8 $\mathcal{E} = \mathcal{E}_0 \sin \omega t$
A graph of the sinusoidal generator output, together with an end view of the corresponding orientations of the loop during a cycle.

of the rotating loop becomes closed to field lines. This is because the *change* in flux is greatest at these points.

The direction of the current also periodically changes, which is why it is called *alternating* current. Since $\omega = 2\pi f$, Eq. 20.4 can be written as

$$\mathcal{E} = \mathcal{E}_0 \sin 2\pi ft \qquad (20.5)$$

where f is the rotational frequency of the generator's armature. The ac frequency in the United States is 60 Hz (or 60 cycles/s).

Keep in mind that Eqs. 20.4 and 20.5 give the instantaneous value of the emf and that $\mathcal{E}$ varies between $+\mathcal{E}_0$ and $-\mathcal{E}_0$ over one period. You will learn how to determine more practical time-averaged values for ac voltage and current in Chapter 21.

EXAMPLE 20.4 ■ AN AC GENERATOR: CALCULATING GENERATED EMF AT ANY TIME IN THE CYCLE

A 60-cycle ac generator (its coil rotates at 60 Hz) produces a maximum emf of 120 V. What is the value of the emf at $\frac{1}{120}$ s after it has a value of zero?

Solution. At 60 cycles per second, the armature takes $\frac{1}{60}$ s to make a full rotation. The time interval given is therefore equal to half a period of the armature.

Given: $\mathcal{E}_0 = 120$ V   *Find:* $\mathcal{E}$ (emf at $\frac{1}{120}$ s after it is zero)
$\quad\quad\quad f = 60$ Hz
$\quad\quad\quad t = \frac{1}{120}$ s

We can reason this without mathematics. A full cycle of the ac voltage takes $\frac{1}{60}$ s. In that time it would start at zero, reach a positive maximum, go through zero, reach a negative maximum and finally reach zero (where it started). Thus, since $\frac{1}{120}$ s is half a cycle, the voltage will be zero.

To see this mathematically, let's first check that Eq. 20.5 fits the conditions at $t = 0$. We can see that it does since

$$\mathcal{E} = \mathcal{E}_0 \sin 2\pi ft = (120 \text{ V})[\sin 2\pi(60 \text{ Hz})(0 \text{ s})] = 0$$

To find the ac voltage generated at any time after $t = 0$, we need only evaluate this equation at that time, specifically

$$\mathcal{E} = \mathcal{E}_0 \sin 2\pi ft$$
$$= (120 \text{ V})[\sin 2\pi(60 \text{ Hz})(\tfrac{1}{120} \text{ s})]$$
$$= (120 \text{ V})(\sin \pi) = 0$$

Follow-up Exercise. In this Example, what is the magnitude of the emf through the coils $\frac{1}{240}$ s after the emf has a value of zero?

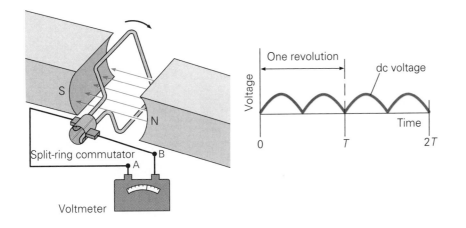

•**FIGURE 20.9 A simple dc generator**
The polarity of the induced emf reverses every half-cycle, but the split-ring commutator maintains the same polarity on terminals A and B. The output is a fluctuating dc (constant-polarity) voltage. (Compare this dc generator with the dc motor in Fig. 19.20.)

Electromagnetic induction can also be used to generate direct current. A simple **dc generator** is illustrated in •Fig. 20.9. Here a split-ring commutator is used to maintain the same polarity. One of the brushes is kept positively charged and the other negatively charged by an exchange of the ends of the armature just as the polarity changes. Note how this occurs from the figure. The output is a pulsating current, but because the polarity does not change, the current is in one direction, or direct. In practice, an armature has many windings, and when the pulses from a large number of loops are superimposed, the resultant direct current is steady and almost free of fluctuations. However, since alternating current is easily rectified, or converted to direct current, direct current is not usually generated in this manner.

A generator that uses a permanent magnet is referred to as a *magneto*. Magnetos are common in single-cylinder engines, such as those used to power small lawn mowers. (In the gasoline combustion engine of automobiles, the ignition voltage is usually supplied by an induction coil rather than a magneto.) The voltage output of a lawn mower's magneto is used to fire the spark plug that ignites the fuel mixture. In this application, permanent magnets mounted on a flywheel rotate about stationary armature coils. (The moving and stationary parts of a generator (or a motor) are called the rotor and stator, respectively.) The circuit is opened mechanically or by a solid-state device so that a built-up field collapses, producing a large induced voltage. About 15 kV is supplied to the spark plug at the proper time in the engine's cycle.

In most large-scale ac generators (power plants), the armature is stationary and magnets are revolved about it. The revolving magnetic field produces a time-varying flux through the coils of the armature and thus an ac output. The mechanical energy for the generator is supplied by a turbine (•Fig. 20.10). Turbines are powered by steam generated from the heat of combustion of fossil fuels, by nuclear reactions, or by water power (hydroelectricity). Thus the basic difference among the various types of power plants is the *source* of the energy that turns the turbines.

Back Emf

Motors also generate emfs. Like a generator, a motor has a rotating armature in a magnetic field. The induced emf in this case is called a **back emf,** $\mathcal{E}_b$ (or sometimes a "counter" emf), because its polarity is opposite to that of the line voltage and tends to reduce the current in the armature coils. (What would happen if this were not the case?)

For a motor with a coil of internal resistance R, the current it draws when in operation is given by

$$I = \frac{V - \mathcal{E}_b}{R}$$

(a)

(b)

•FIGURE 20.10 Electrical generation
The gravitational potential energy of water trapped behind a dam. **(a)** is used to turn the turbines that generate electric power **(b)**. (The dam shown in the first photo is the Glen Canyon dam on the Colorado River in Arizona; the turbines in the second photo are at Hoover Dam, also on the Colorado River.)

or

$$\mathscr{E}_b = V - IR \qquad\qquad back\ emf \qquad (20.6)$$

where V is the line voltage.

The back emf of a motor depends on the rotational speed of the armature and builds up from zero to some maximum value as the armature goes from rest to its normal operating speed. On startup, the back emf is zero, so the starting current is a maximum (Eq. 20.6). Ordinarily, a motor turns something; that is, it has a mechanical load. Without a load, the armature speed will increase until the back emf has built up to a point where it almost equals the line voltage, with just enough current in the coils to overcome friction and joule heat loss. Under normal load conditions, the back emf will be less than the line voltage. The larger the load, the slower the motor will rotate and the smaller the back emf will be. If a motor is overloaded and turns very slowly, the back emf may be reduced so much that the current becomes large enough to burn out the coils. Thus, the back emf is involved in the regulation of a motor's operation.

A back emf in a dc motor circuit can be represented in a circuit diagram as a battery with polarity opposite that of the driving voltage (•Fig. 20.11).

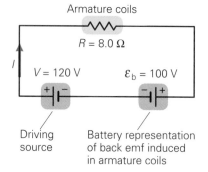

•FIGURE 20.11 Back emf
The back emf in the armature of a dc motor may be effectively represented as a battery with polarity opposite that of the driving source. See Example 20.5.

EXAMPLE 20.5 ■ GETTING UP TO SPEED: BACK EMF IN A DC MOTOR

A dc motor with a resistance in its windings of 8.0 Ω operates on 120 V. With a normal load, there is a back emf of 100 V when the motor reaches full speed (see Fig. 20.11). What are (a) the starting current drawn by the motor and (b) the armature current at operating speed under a normal load?

Solution. We assume the resistance of the winding is constant (ohmic) and list the given data

Given: $R = 8.0\ \Omega$ *Find:*   (a) I_s (starting current)
         $V = 120\ V$ (b) I_a (operational current)
         $\mathscr{E}_b = 100\ V$

(a) When the motor starts and its armature just begins to turn, the back emf is essentially zero. From Eq. 20.6, the current in the windings at that time is given by

$$I_s = \frac{V}{R} = \frac{120\ V}{8.0\ \Omega} = 15\ A$$

(b) At operating speed, the back emf opposes the driving voltage. The effective voltage is the difference in these two voltages (Eq. 20.6). Thus,

$$I_a = \frac{V - \mathscr{E}_b}{R} = \frac{120 \text{ V} - 100 \text{ V}}{8.0 \ \Omega} = 2.5 \text{ A}$$

Note that, with no back emf, the starting current of the motor is relatively large. (When a big motor, such as that of a central air conditioning unit for a whole building, starts up, you may notice the lights momentarily dim because of the large starting current that it draws.) Here again, there is danger of coil burnout at low operating speeds. In some instances, resistors are temporarily connected in series with a motor's coil to protect the windings from burning out because of large starting currents.

Follow-up Exercise. (a) In this Example, how much energy is required to start the motor, assuming it takes 10 s and the back emf is essentially zero in that time? (b) Compare this to the amount of energy required to keep it running for 10 s once it has reached its operating conditions.

Since motors and generators are opposites, so to speak, and a back emf develops in a motor, you may be wondering whether a back force develops in a generator. The answer is yes. When an operating generator is not connected to an external circuit, no current flows and there is no force on the coils of the armature due to the magnetic field. However, when the generator delivers power to a circuit and current flows in the coils (current-carrying wires in a magnetic field), there is a force that produces a *counter torque*, which opposes the rotation of the armature. As more current is drawn, the counter torque becomes greater, and a greater driving force is needed to turn the armature. Therefore, the higher the current output of a generator, the greater will be the energy expended (fuel consumed) in overcoming the counter torque.

20.3 Transformers and Power Transmission

Objectives: To be able to (a) explain transformer action in terms of Faraday's law, (b) calculate the output of step-up and step-down transformers, and (c) understand the importance of transformers in electric energy delivery systems.

Electricity is transmitted by power lines over long distances. Obviously, it is desirable to minimize the I^2R losses through these transmission lines. The resistance of a line is fixed, so reducing the I^2R losses means reducing the current. However, the power output of a generator is determined by its current and voltage outputs ($P = IV$), and for a fixed voltage (such as 120 V), a reduction in current would mean a reduced power output. It might appear that there is no way to reduce the current while maintaining the power supplied. Fortunately, however, electromagnetic induction can be applied to reduce power transmission losses by stepping up the voltage and reducing the current in just such a way that the delivered *power* is essentially unchanged. This is done with a device called a transformer.

As •Fig. 20.12 shows, a simple **transformer** consists of two coils of insulated wire wound on the same (closed) iron core. When ac voltage is applied to the input coil, or **primary coil**, the alternating current gives rise to an alternating magnetic flux that is concentrated in the iron core. The changing flux passes through the output coil, or **secondary coil**, inducing an alternating voltage and current in it.

The induced secondary voltage differs from the primary voltage depending on the ratio of the numbers of turns in the two coils. By Faraday's law, the secondary voltage is given by

$$V_s = -N_s \frac{\Delta \Phi}{\Delta t}$$

Note: For transformers, it is customary to use the term voltage rather than emf.

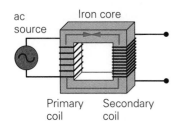

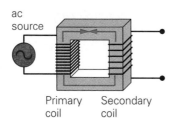

(a) Step-up transformer:
high-voltage
(low current) output

(b) Step-down transformer:
low-voltage
(high current) output

(c) Transformer
circuit symbol

•FIGURE 20.12 **Transformers**
(a) A step-up transformer has more turns in the secondary coil than in the primary. (b) A step-down transformer has more turns in the primary coil than in the secondary. (c) The circuit symbol for a transformer reflects its actual structure to some extent. (d) Workers inspect large transformers used in electrical power transmission systems (to be discussed shortly).

where N_s is the number of turns in the secondary coil. The changing flux in the primary coil produces a back emf equal to

$$V_p = -N_p \frac{\Delta \Phi}{\Delta t}$$

where N_p is the number of turns in the primary coil. If the resistance of the primary coil is neglected, this is equal to the external voltage applied to it. Then, forming a ratio gives

$$\frac{V_s}{V_p} = \frac{-N_s \Delta \Phi / \Delta t}{-N_p \Delta \Phi / \Delta t}$$

or

$$\frac{V_s}{V_p} = \frac{N_s}{N_p} \qquad (20.7)$$

Here it is assumed that the core concentrates the field so the flux through each coil is the same.

If the transformer is assumed to be 100% efficient (no energy losses), the power input is equal to the power output, and since $P = IV$,

$$I_p V_p = I_s V_s \qquad (20.8)$$

Although some energy is always lost, this is a good approximation, since a well-designed transformer may have an efficiency of more than 95%. (The sources of energy losses will be discussed shortly.) Then, using Eq. 20.7, the transformer currents and voltages are related to the turn ratio by the relationship

$$\frac{I_p}{I_s} = \frac{V_s}{V_p} = \frac{N_s}{N_p} \qquad (20.9)$$

With these equations, it is easy to see how a transformer affects the voltage and current. In terms of the output,

$$V_s = \left(\frac{N_s}{N_p}\right) V_p \quad \text{and} \quad I_s = \left(\frac{N_p}{N_s}\right) I_p \qquad (20.10)$$

That is, if the secondary coil has a greater number of windings than the primary does ($N_s > N_p$ or $N_s/N_p > 1$) as in Fig. 20.12a, the voltage is stepped up ($V_s > V_p$). This type of arrangement is called a **step-up transformer**. Notice, however, in this arrangement there is less current in the secondary than in the primary ($N_p/N_s < 1$ and $I_s < I_p$). For example, if the primary coil of a transformer has 50 turns and the secondary has 100 turns, $N_s/N_p = 2$ and $N_p/N_s = \frac{1}{2}$. Thus, a 220-V input at 10 A will be stepped up to a 440-V output at 5.0 A.

The opposite situation, in which the secondary coil has fewer turns than the primary, characterizes a **step-down transformer** (Fig. 20.12b). In this case, the voltage is

Step-up transformer: $N_s > N_p$ or $N_s/N_p > 1$ **(more turns on secondary than on primary)**

Step-down transformer: $N_p > N_s$ or $(N_s/N_p) < 1$ **(more turns on primary than on secondary)**

stepped down, or reduced, and the current is increased. A step-up transformer may be used as a step-down transformer by simply reversing the output and input connections. Also, it should be apparent that transformers operate only on ac (not on dc).

Note: The terms "step-up" and "step-down" refer to what happens to the primary or input *voltage*, not the current. The effect on the current is the reverse of the effect on the voltage.

EXAMPLE 20.6 ■ STEP-UP OR STEP-DOWN? TRANSFORMER ORIENTATION

A transformer has 50 turns on its primary and 100 turns on its secondary. (a) If the primary is connected to a 120-V source, what is the voltage output of the secondary? (b) If the transformer is operated in reverse, and the 120-V input is applied to the coil with 100 turns, what would be the voltage output?

Solution. Here the number of primary and secondary turns are given in part (a). In part (b), when operated in reverse, the secondary becomes the primary and vice versa, so we have

Given: (a) $N_p = 50$ **Find:** (a) V_s (secondary voltage outputs)
        $N_s = 100$ (b) V_s
        $V_p = 120$ V
    (b) $N_p = 100$
        $N_s = 50$
        $V_p = 120$ V

(a) The secondary voltage is given by Eq. 20.10. Note that this is a step-up transformer with $N_s > N_p$ and a turn ratio of $N_s/N_p = \frac{100}{50} = 2$. Then,

$$V_s = \left(\frac{N_s}{N_p}\right)V_p = (2)(120 \text{ V}) = 240 \text{ V}$$

Looking at the secondary current with $N_p/N_s = \frac{50}{100} = \frac{1}{2}$,

$$I_s = \left(\frac{N_p}{N_s}\right)I_p = \frac{1}{2}I_p$$

Thus we see that the voltage is stepped up or increased by a factor of 2, and the current is stepped down and is half as large.

(b) In this case, the turn ratio is $N_s/N_p = \frac{50}{150} = \frac{1}{2}$, and we have a step-down transformer with $N_p > N_s$. Then,

$$V_s = \left(\frac{N_s}{N_p}\right)V_p = (\tfrac{1}{2})(120 \text{ V}) = 60 \text{ V}$$

Again looking at the secondary current with $N_p/N_s = \frac{100}{50} = 2$,

$$I_s = \left(\frac{N_p}{N_s}\right)I_p = 2I_p$$

As you might expect, in this case, the voltage is stepped down and is half as large, while the current is stepped up by a factor of 2. (Since the currents were not given, the actual numerical values cannot be found.)

Follow-up Exercise. (a) In this Example, which connection would you utilize for European operation (at 240 V) of your American appliances? Explain your reasoning. (b) What would happen if you used the other connection by mistake? Show a numerical calculation for a 1200-W hair dryer designed for use in the United States and assumed ohmic.

The preceding equations apply to ideal transformers, but actual transformers do have some energy losses. First, there is always some flux leakage; that is, not all of the flux passes through the secondary coil. In some transformers, one of the insulated coils is wound over (on top of) the other rather than having the two on separate "legs" of a core. This helps avoid flux leakage and reduces the size.

Second, when ac current flows in the primary coil, the changing magnetic flux through the loops gives rise to an induced emf in that coil. This is called *self-induction*. By Lenz's law, the self-induced emf will oppose the change in cur-

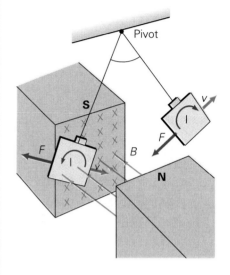

(a)

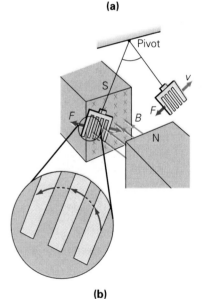

(b)

•FIGURE 20.13 Eddy currents
(a) Eddy currents are induced in a nonmagnetic conductive plate moving in a magnetic field. The induced currents oppose the change in flux, and there is a retarding force that opposes the motion. Note that the currents reverse as the plate swings through the field. **(b)** If the pendulum plate with the slits is used as shown here, the plate swings more freely.

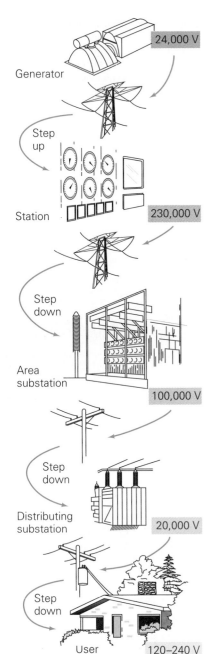

24,000 V

Generator

Step up

Station

230,000 V

Step down

Area substation

100,000 V

Step down

Distributing substation

20,000 V

Step down

User   120–240 V

• **FIGURE 20.14   Power transmission system**
A diagram of a typical electrical power distribution system.

rent and will thus limit the current (similar to a back emf in a motor). Remember that self-induction may be thought of as a kind of electromagnetic inertia—like the inertia of material bodies, it opposes change. Some energy is also lost due to the resistances of the coil wires (I^2R losses), but this is generally small.

A fourth cause of energy loss is **eddy currents** in the transformer core. A highly permeable material is used for the core to increase the density of the magnetic flux, but such materials are usually good conductors. The changing magnetic flux sets up swirling movements of charge, or eddy currents, in the core material, and these dissipate energy. (See Demonstration 13 earlier in the chapter for an eddy current effect.)

To reduce this effect, transformer cores are made of thin sheets of material (usually iron) laminated with an insulating glue between them. Because of the insulating layers between the sheets, the eddy currents are broken up, or confined to the sheets, which greatly reduces the loss due to them. Well-designed transformers generally have less than 5% internal energy loss.

An effect of eddy currents may be demonstrated by swinging a plate made of a nonmagnetic metal, such as aluminum, through a magnetic field, as illustrated in • Fig. 20.13a. Eddy currents are set up in the plate as a result of its motion in the field and the changing magnetic flux. By Lenz's law, the eddy currents set up opposing fluxes, or effectively opposite magnetic poles on the plate. This gives rise to a repulsive force that retards the swinging motion and brings the plate quickly to rest.

The breaking up of the eddy currents may be demonstrated using a plate with slits (• Fig. 20.13b). When this plate swings between the magnet's poles, it swings relatively freely. This is because the eddy currents are much reduced by the gaps (slits).

The damping effect of eddy currents is applied in the braking systems of rapid-transit cars that travel on rails. When an electromagnet on the car is turned on, it applies a magnetic field to a metal wheel or the rail. The repulsive force due to the induced eddy currents acts as a braking force. As the car slows, the eddy currents decrease, allowing a smooth braking action.

Power Transmission

For power transmission over long distances, transformers provide a means to increase (step up) the voltage and reduce the current of a generator's output in order to cut down the resistive I^2R losses. A schematic diagram of an ac power distribution system is shown in • Fig. 20.14. The voltage output of the generator is stepped up and transmitted over long distances to an area substation near the consumers. There the voltage is stepped down. There are further voltage step-downs at distributing substations and utility poles before 120–240 V are supplied to homes and businesses.

The following example illustrates the benefits of being able to step up the voltage (and step down the current) for electrical power transmission.

EXAMPLE 20.7 ■ CUTTING YOUR LOSSES: POWER TRANSMISSION AT HIGH VOLTAGE

A small hydroelectric power plant produces energy in the form of electric current at 10 A and a voltage of 440 V. The voltage is stepped up to 4400 V (by an ideal transformer) for transmission over 40 km of power line, which has a resistance of 0.50 Ω/km. [A power transmission line has two wires (for a complete circuit); however, for simplicity, you can take the length given to be the total wire length rather than doubling it.] (a) What percentage of the original energy would have been lost in transmission if the voltage had not been stepped up? (b) What percentage of the original energy is lost with the voltage stepped up?

Given: $V_p = 440$ V   *Find:* (a) Percentage energy loss
       $I_p = 10$ A   without voltage step-up
       $V_s = 4400$ V   (b) Percentage energy loss
       $R_o/L = 0.50 \ \Omega/\text{km}$ with voltage step-up
       $L = 40$ km

(a) The power output of the generator is

$$P = I_p V_p = (10 \text{ A})(440 \text{ V}) = 4400 \text{ W}$$

The resistance of 40 km of power line is

$$R = \left(\frac{R_o}{L}\right)L = \left(\frac{0.50 \ \Omega}{\text{km}}\right)(40 \text{ km}) = 20 \ \Omega$$

The energy loss rate (joules/s or watts) in transmitting a current of 10 A is

$$P_{\text{loss}} = I_p^2 R = (10 \text{ A})^2(20 \ \Omega) = 2000 \text{ W}$$

Thus, the percentage of the produced energy lost to joule heat in the wires is

$$\% \text{ loss} = \frac{P_{\text{loss}}}{P} \times 100\% = \frac{2000 \text{ W}}{4400 \text{ W}} \times 100\% = 45\%$$

(b) When the voltage is stepped up to 4400 V, the transmitted current is

$$I_s = \left(\frac{V_p}{V_s}\right)I_p = \left(\frac{440 \text{ V}}{4400 \text{ V}}\right)(10 \text{ A}) = 1.0 \text{ A}$$

(Note that the voltage was stepped up by a factor of 10 and the current is stepped down by the same factor.) The power loss in this case is

$$P_{\text{loss}} = I_s^2 R = (1.0 \text{ A})^2(20 \ \Omega) = 20 \text{ W}$$

and the percentage loss is reduced by a factor of 100 (since the current is reduced by a factor of 10 and the power depends on the *square* of the current)

$$\% \text{ loss} = \frac{P_{\text{loss}}}{P} \times 100\% = \frac{20 \text{ W}}{4400 \text{ W}} \times 100\% = 0.45\%$$

Follow-up Exercise.   Some heavy duty electrical appliances, such as water pumps, can be wired to 240 V or 120 V sockets. Their power rating is the same regardless of the voltage at which they run. (a) Explain the efficiency advantage of operating such appliances at the higher voltage. (b) Consider a 1 hp pump (774 W) and estimate the ratio of the power lost in the wires at 240 V to the loss at 120 V (assuming all resistances are ohmic).

Example 20.7 shows the advantage of using high-voltage, or high-tension, transmission lines (•Fig. 20.15a). However, there is a practical limit to the degree of voltage step-up. At very high voltages, the molecules in the air surrounding a power line may be ionized by the induced electric fields, forming a conducting path to nearby trees, buildings, or the ground. This is referred to as a leakage loss. Crews from electric companies are continually clearing foliage from near power lines, and long insulators are used to hold the high-voltage wires away from the metal towers (•Fig. 20.15b). Leakage losses are generally greater during wet weather because moist air is more easily ionized. Under certain conditions such as fog, you may see the arcing or corona discharge from a high-voltage line and/or hear the accompanying crackling noise.

20.4 Electromagnetic Waves

Objectives: To be able to **(a)** explain the physical nature, origin, and means of propagation of electromagnetic waves, and **(b)** describe the properties and uses of various types of electromagnetic waves.

Electromagnetic waves (or *radiation*) were considered as a means of heat transfer in Section 11.4. Now you are ready to understand more fully the production and

(a)

(b)

•**FIGURE 20.15   Long lines**
(a) Long-distance power lines carry electric power at high voltage across the plains of Wyoming. The use of high voltages in transmission reduces power losses. **(b)** Workers maintaining insulators at a power substation. Large insulators are needed to reduce leakage losses from high-voltage lines.

characteristics of electromagnetic radiation. As the name implies, these waves have both electric and magnetic properties, which may be described by quantities you have studied.

James Clerk Maxwell showed that four fundamental relationships could completely describe all observed electromagnetic phenomena. Maxwell also used this set of equations to predict the existence of waves of an electromagnetic nature. Because of his contributions, the set of equations is known as **Maxwell's equations,** although they were for the most part developed individually by other scientists (for example, Faraday discovered the law of induction).

Maxwell's equations

Essentially, Maxwell's equations combine the electric force and the magnetic force into a single electromagnetic force. The apparently separate fields are shown to be symmetrically related in the sense that either one can create the other. This symmetry is evident in the equations as presented in their advanced mathematical form. For our purposes, a qualitative description is sufficient:

Changing B produces E.
Changing E produces B.

A time-varying magnetic field produces an electric field.
A time-varying electric field produces a magnetic field.

The first statement is simply the field language version of the observation that, as we saw in Sec. 20.1, a changing magnetic flux gives rise to an induced emf and current in a wire. The second statement (the symmetrical twin of the first) is crucial to understanding the self-propagating characteristic of electromagnetic waves. This characteristic enables them to travel through a vacuum, whereas all other waves require a supporting medium.

According to Maxwell's theory, electromagnetic waves are produced by *accelerating* electric charges, such as an electron oscillating in simple harmonic motion. This could be one of the many electrons in the antenna of a radio transmitter, driven by an electrical oscillator with a frequency of about 10^6 Hz. As such an electron moves, it continually accelerates and decelerates, and radiates an electromagnetic wave (•Fig. 20.16). The continual oscillations of many such charges due to the alternating current in the transmitter produce time-varying electric and magnetic fields in the immediate vicinity of the antenna. The electric field (in the plane of the paper) continually changes direction, as does the magnetic field (into and out of the paper by a right-hand rule).

Both the electric and the magnetic fields carry energy and propagate outward with the speed of light (c in a vacuum, 3.00×10^8 m/s). Maxwell's equations show

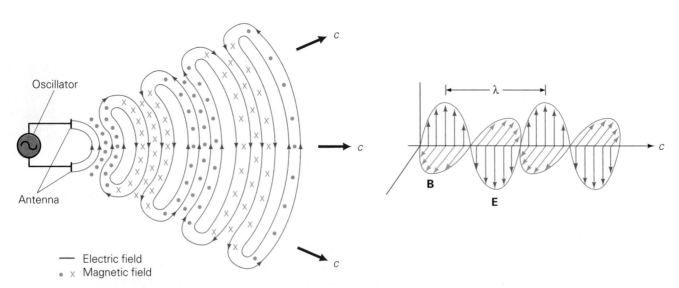

•**FIGURE 20.16 Source of electromagnetic waves**
Electromagnetic waves are produced by accelerating electric charges. The wave propagates outward with the electric field vectors (**E**) and magnetic field vectors (**B**) in phase, perpendicular to each other and to the direction of propagation.

that, except in the immediate vicinity of the source, the electromagnetic waves at a fixed instant of time are of the form shown in Fig. 20.16. The electric vector of the E field (**E**) is perpendicular to the magnetic vector of the B field (**B**), and each varies sinusoidally with time. Both **E** and **B** are in phase and perpendicular to the direction of propagation. Thus, electromagnetic waves are transverse waves but it is the *fields* that oscillate perpendicularly to the direction of propagation, not any material or medium.

EXAMPLE 20.8 ■ THE SPEED OF LIGHT (AND OTHER EM WAVES)

Two *Viking* space probes landed on the planet Mars in 1976 and sent radio and TV signals back to Earth. How much longer did it take for a signal to reach us when Mars was farthest from Earth than when it was closest? The average distances of Mars and Earth from the Sun are 229 million kilometers (142 million miles) and 150 million kilometers (93 million miles), respectively. Assume circular orbits with these average distances as radii.

Solution. Listing the data, we have

Given: $d_M = 229 \times 10^6$ km   *Find:* Δt (time difference)
          (average distance of Mars from Sun)
          $d_E = 150 \times 10^6$ km
          (average distance of Earth from Sun)

This is a simple time-distance calculation. To find the difference in the travel times of the radio signals, we first find the difference in the distances. The planets are farthest apart when they are on opposite sides of the Sun. At this time, they are separated by a distance $(d_M + d_E)$. The planets are closest when they are aligned on the same side of the Sun; their separation distance is then $(d_M - d_E)$. (Draw yourself a diagram to help visualize these relationships.)

The difference in the separation distances is then

$$\Delta d = (d_M + d_E) - (d_M - d_E) = 2d_E$$
$$= 2(150 \times 10^6 \text{ km}) = 3.0 \times 10^{11} \text{ m}$$

(Notice that the difference is twice the Earth's distance from the Sun, or the diameter of Earth's orbit, so the distance of Mars from the Sun was not actually needed. This illustrates the advantage of doing operations algebraically before putting in numerical values.)

Radio waves are electromagnetic waves and travel at "the speed of light," c. Since $\Delta d = c\Delta t$,

$$\Delta t = \frac{\Delta d}{c} = \frac{3.0 \times 10^{11} \text{ m}}{3.0 \times 10^8 \text{ m/s}} = 1.0 \times 10^3 \text{ s (or 16.7 min)}$$

Not too long to travel 186 million miles! (Note that this is twice the time it takes light to travel from the Sun to Earth, or twice the time it would take us to learn that the Sun had burned out or exploded. We say that the Earth is an average of 8.3 *light-minutes* from the Sun.)

Follow-up Exercise. A Russian–American joint venture to land a roving vehicle on the Martian surface is planned for 1999. Let's see why it might be important for it to carry its own decision-making computer on board. Suppose that a camera on the rover spots a huge cliff in the foreground, 50 m from the rover, and that the rover is moving at a steady 10 cm/s toward the cliff. (a) How long does it take for the video picture to get to the Earth and a directional change signal to be sent back to prevent it from going over the cliff? (Assume the decision is made and returned to Mars instantaneously from controllers on Earth and that the Earth is at its closest point to Mars.) (b) Can the rover be saved from certain disaster?

Radiation Pressure

An electromagnetic wave carries energy. Consequently, it can do work and can exert a force on a material it strikes. Consider light striking an electron at the surface of a material (•Fig. 20.17). The electric field of the electromagnetic wave does

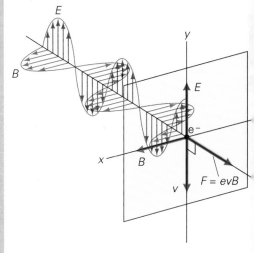

•FIGURE 20.17 Radiation pressure The electric field of an electromagnetic wave that strikes a surface acts on an electron, giving it a velocity (v). The magnetic field produces a force on the moving charge in the direction of propagation of the incident light. (Check this using the appropriate right-hand rule.)

work on the electron; assume that this gives the electron a velocity (v) as shown in the figure. Recall from Chapter 19 that a moving charged particle in a magnetic field experiences a force. As a result, there is a magnetic force on the electron due to the magnetic field component of the light wave. By the right-hand rule for the magnetic force on a moving charged particle, the force on the electron is in the direction shown in Fig. 20.17. That is, the electromagnetic wave produces a force on the electron in the direction in which the wave is propagating and therefore exerts a force on the material to which the electron is bound.

The radiation force per area is called the **radiation pressure**. Radiation pressure is negligible for most common situations, but can be of importance in some atmospheric and astronomical phenomena. For example, radiation pressure plays a key role in determining the direction in which the tail of a comet points. The sunlight delivers energy to the comet "head," which consists of frozen ices and dust. Some of this material is evaporated as the comet nears the Sun, and the evaporated particles are pushed away from the Sun by radiation pressure. Thus the tail points generally away from the Sun, whether the comet is approaching or leaving the Sun's vicinity.

Note: Even the light from your desk lamp exerts a very, very small force on your desk.

CONCEPTUAL EXAMPLE 20.9 ■ SAILING THE SEA OF SPACE: RADIATION PRESSURE IN ACTION

Some scientists have proposed that a relatively light spacecraft with a huge sail could be built and launched from a space station in Earth orbit. With little or no power of its own, it would use the pressure of sunlight to propel it to the outer planets. To get the maximum propulsive force, what kind of surface should the sails have? (a) Shiny and reflective. (b) Dark and absorptive. (c) The surface would make no difference.

Reasoning and Answer. At first glance, one might think that the answer is (c). However, as we have seen, radiation possesses energy and is capable of exerting force, and it clearly transfers momentum to whatever it strikes. Thus, we must think of the interaction between the radiation and the sail in terms of conservation of momentum, as discussed in Section 6.4. If the radiation is absorbed, the situation is analogous to a completely inelastic collision, and the sail will acquire all of the momentum (call it p) originally possessed by the radiation.

However, if the radiation is reflected, the situation is analogous to a completely elastic collision. Since the momentum of the radiation after the collision will be equal to its original momentum in magnitude but opposite in direction, its momentum will be reversed (from $+p$ to $-p$). To conserve momentum, the momentum transferred to the sail in this case must be twice as great ($2p$). Thus reflective sails would experience, on average, twice as much force as absorptive ones, and the answer is (a).

Follow-up Exercise. (a) Would the sail in this Example provide less or more acceleration as the interplanetary sailing ship moved farther from the Sun? (b) How could this change in acceleration be counteracted?

Types of Electromagnetic Waves

Electromagnetic waves are classified by ranges of frequencies or wavelengths in a spectrum. Frequency and wavelength are inversely related: $\lambda = c/f$. Thus, the greater the frequency, the shorter the wavelength, and vice versa. The electromagnetic spectrum is continuous, so the limits of the various ranges are approximate. Table 20.1 lists these frequency and wavelength ranges for the general types of electromagnetic waves (see also •Fig. 20.18).

Power Waves. Electromagnetic waves of 60-Hz frequency result from currents moving back and forth (alternating) in electrical circuits. As Table 20.1 indicates, these power waves have a wavelength of 5.0×10^6 m, or 5000 km (more than 3000 mi). Waves of this low frequency are of little practical use. They may occasionally

TABLE 20.1 Classification of Electromagnetic Waves

Type of Wave	Approximate Frequency Range (Hz)	Approximate Wavelength Range (m)	Source
Power waves	60	5×10^6	Electric currents
Radio waves			Electric circuits
AM	$0.53\text{--}1.7 \times 10^6$	570–186	
FM	$88\text{--}108 \times 10^6$	3.4–2.8	
TV	$54\text{--}890 \times 10^6$	5.6–0.34	
Microwaves	$10^9\text{--}10^{11}$	$10^{-1}\text{--}10^{-3}$	Special vacuum tubes
Infrared radiation	$10^{11}\text{--}10^{14}$	$10^{-3}\text{--}10^{-7}$	Warm and hot bodies
Visible light	$4.0\text{--}7.0 \times 10^{14}$	10^{-7}	Sun and lamps
Ultraviolet radiation	$10^{14}\text{--}10^{17}$	$10^{-7}\text{--}10^{-10}$	Very hot bodies and special lamps
X-rays	$10^{17}\text{--}10^{19}$	$10^{-10}\text{--}10^{-12}$	High-speed electron collisions
Gamma rays	above 10^{19}	below 10^{-12}	Nuclear reactions and processes in particle accelerators

produce a so-called 60-Hz hum on your stereo or introduce unwanted electrical noise in delicate instruments. More seriously, concerns have been expressed about possible health effects of these waves. Some early research tended to suggest that very low-frequency fields may have potentially harmful biological effects on cells and tissues. However recent surveys indicate that this may not be the case in everyday living. It is hoped that further investigation will tell.

Radio and TV Waves. Radio and TV waves are generally in the frequency range from 500 kHz to about 1000 MHz. The AM (amplitude modulated) band runs from 530 to 1710 kHz (1.71 MHz). Higher frequencies, up to 54 MHz, are used for "short wave" bands. TV bands range from 54 MHz to 890 MHz. The FM (frequency modulated) radio band runs from 88 to 108 MHz, which lies in a gap between channels 6 and 7 of the TV band. Cellular phones use radio waves to transmit voice communication in the ultrahigh frequency (UHF) band, with frequencies similar to those of radio waves used for television channels 13 and higher.

Early global communications used the "short wave" bands, as do amateur (ham) radio operators today. But how are the normally straight-line radio waves transmitted around the curvature of the Earth? This is accomplished by reflection from ionic layers in the upper atmosphere. Energetic particles from the Sun ionize gas molecules, giving rise to several ion layers. Certain of these layers reflect radio waves below a specific frequency. By "bouncing" radio waves off these layers, we can send radio transmissions beyond the horizon, to any region of the Earth.

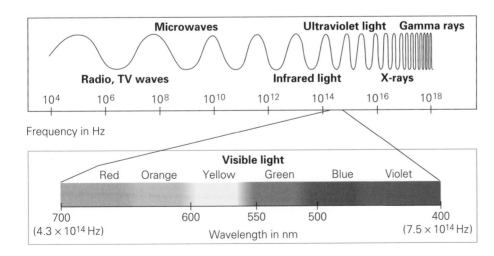

•FIGURE 20.18 The electromagnetic spectrum
The spectrum of frequencies or wavelengths is divided into regions, or ranges. Note that the visible region is a very small part of the total electromagnetic spectrum. The wavelengths are given in nanometers: 1 nm = 10^{-9} m. (Top illustrative wavelengths not to scale.)

•FIGURE 20.19　Global village
Radio and TV transmissions from around the world, relayed by orbiting satellites, can be picked up by backyard and rooftop antennas.

Note: Review the discussion of the Doppler shift in Section 14.4.

Note: See the Insight in Section 11.4.

•FIGURE 20.20　Slow down
Radar speed guns based on the Doppler effect employ radiation in the microwave region of the spectrum. Here the speed of cyclists is being monitored on a trail also used by hikers and joggers.

Such reflection of radio waves requires uniformity in the density of the ionic layers. When, from time to time, a solar disturbance produces a shower of energetic particles that upsets this uniformity, a communications "blackout" may occur as the radio waves are scattered in many different directions rather than reflected in straight lines. To avoid such disruptions, global communications have relied largely on transoceanic cables. Now we also have communications satellites, which can provide line-of-sight transmission to any point on the globe (•Fig. 20.19).

Microwaves. Microwaves, with frequencies in the gigahertz (GHz) range, are produced by special vacuum tubes (called klystrons and megatrons). Microwaves are used in communications and radar applications. In addition to its many roles in navigation and guidance, radar provides the basis for the speed guns used to time fast balls, tennis serves, motorists, and sometimes cyclists (•Fig. 20.20). When radar waves are reflected from a moving object, their wavelength is shifted by the Doppler effect (Chapter 14). The amount of the shift indicates the velocity of the object toward or away from the observer. Another very common use of microwaves today is in microwave ovens.

Infrared Radiation. The infrared region of the electromagnetic spectrum lies adjacent to the low-frequency or long-wavelength end of the visible spectrum. The frequency at which a warm body emits radiation depends on its temperature. Actually, such a body emits electromagnetic waves of many different frequencies, but the frequency of the maximum intensity characterizes the radiation. A body at about room temperature emits radiation in the far infrared region (meaning farthest from the visible region).

Recall from Chapter 11 that infrared radiation is sometimes referred to as heat rays. This is because water molecules, which are present in most materials, readily absorb infrared wavelengths. When they do, their random thermal motion is increased—that is, they heat up, and heat their surroundings. Infrared lamps are used in therapeutic applications and to keep food warm in cafeterias. Infrared radiation is also associated with maintaining the Earth's warmth or average temperature through the greenhouse effect. Incoming visible light (which passes relatively easily through the atmosphere) is absorbed by the Earth's surface and re-radiated as infrared (longer wavelength) radiation, which is trapped by greenhouse gases such as carbon dioxide and water vapor.

Visible Light. The visible region occupies a very small portion of the total electromagnetic spectrum. It runs from about 4×10^{14} Hz to about 7×10^{14} Hz, or a wavelength range of about 700–400 nm, respectively (Fig. 20.18). Only the radiation in this region can activate the receptors in our eyes. Visible light emitted or reflected from the objects around us provides us with much information about our world. Visible light will be considered in more detail in Chapters 22–25.

It is interesting to note that not all animals are sensitive to the same range of wavelengths. For example, snakes can detect infrared radiation, and the visible range of many insects extends well into the ultraviolet (see below). It is also interesting to note that the sensitivity range of our eyes conforms closely to the spectrum of wavelengths emitted by the Sun, both having maxima in the yellow–green region. If the Sun emitted mostly infrared radiation, for example, the surface of the Earth would appear very dark to us.

Ultraviolet Radiation. Beyond the violet end of the visible region lies the ultraviolet frequency range. Ultraviolet (or uv) radiation is produced by special lamps and very hot bodies. The Sun emits large amounts of ultraviolet radiation, but fortunately most of it received by the Earth is absorbed in the ozone (O_3) layer in the atmosphere at an altitude of about 40–50 km. Because the ozone layer plays a protective role, there is concern about its depletion by chlorofluorocarbon gases (such as Freon, once commonly used in refrigerators) that drift upward and react with the ozone. (See the Insight on p. 648.)

The small amount of ultraviolet radiation that reaches the Earth's surface from the Sun can cause human skin to burn or tan. Pigmentation in the skin acts as a protective mechanism against the penetration of ultraviolet radiation. The degree of penetration depends on the amount of a pigment called melanin in the skin and on the thickness of the skin's layers. All people (except albinos) have varying amounts of melanin in their skin. Exposure to sunlight induces the production of more melanin, causing the skin to tan. Overexposure, especially initially, may cause the skin to burn and turn red (sunburn). Healing thickens the outer skin layer, which then offers greater protection. Creams and lotions containing so-called sunscreens are also available to keep skin from sunburning. The molecules of chemicals in these preparations absorb some of the ultraviolet radiation before it reaches the skin. Physicians now urge people—especially those with fair skin, which contains little protective melanin —to use effective sunscreens or avoid bright sunlight altogether, because uv exposure greatly increases the risk of skin cancer, as mentioned in the Insight feature.

Exposure to sunlight is necessary for the natural production of vitamin D from compounds in the skin. This vitamin is essential for strong bones and teeth. In northern latitudes, where people's exposure to sunlight is relatively seasonal, diets need to be supplemented with synthetic vitamin D. You may have noticed that a vitamin D supplement is commonly provided in milk.

Most ultraviolet radiation is absorbed by certain molecules in ordinary glass. Therefore, you cannot get a tan or a sunburn through glass windows. Sunglasses are now labeled to indicate the uv protection standards they meet in shielding the eyes from this potentially harmful radiation. Welders wear special glass goggles or face masks to protect their eyes from the large amounts of ultraviolet radiation produced by the arcs of welding torches. Similarly, it is important to shield the eyes when using a sunlamp. The ultraviolet component of sunlight reflected from snow-covered surfaces can produce snowblindness in unprotected eyes.

"Black lights" used at discos emit radiation in the violet and near ultraviolet. The ultraviolet radiation causes fluorescent paints and dyes on signs, posters, and performers' clothes to glow with brilliant colors. *Fluorescence* is the process whereby a substance absorbs ultraviolet radiation and emits visible radiation. (The atomic mechanisms responsible are discussed in Chapter 27.) In fluorescent lamps, an electrically excited mercury vapor in the lamp emits ultraviolet radiation. The white material coating the inside of the tube absorbs the ultraviolet radiation and emits visible light. You may have noticed in the grocery store that many products, such as laundry detergents, are packaged in brightly colored boxes. In some instances, the ink is fluorescent, and appears brighter under the store's fluorescent lamps to get your attention. (A small amount of ultraviolet radiation is emitted from the lamps.)

X-rays. Beyond the ultraviolet region of the electromagnetic spectrum is the important X-ray region. We are all familiar with X-rays, primarily through medical applications. X-rays were discovered accidentally in 1895 by the German physicist Wilhelm Roentgen (1845–1923), when he noted the glow of a piece of fluorescent paper caused by some mysterious radiation coming from a cathode ray tube. Because of the apparent mystery involved, these were named x-radiation or X-rays.

The basic elements of an X-ray tube are shown in •Fig. 20.21. A potential difference of several thousand volts is applied across the electrodes in a sealed, evacuated tube. Electrons emitted from the negative electrode are accelerated toward the positive anode, which is called the target. When the electrons strike the target, they interact with the atomic electrons of its material, and the electrical repulsion abruptly decelerates them. This results in a loss of energy in the form of high-frequency X-rays. (Recall that accelerating charges radiate electromagnetic energy, and acceleration includes a *loss* of speed.) Similar processes take place in color television picture tubes, which use electron beams and high potential differences. Proper shielding is necessary to protect viewers from the resulting X-rays.

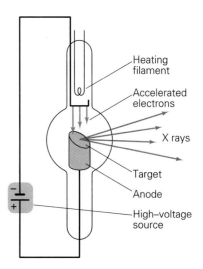

•**FIGURE 20.21 The X-ray tube** Electrons accelerated through a large potential difference strike a target electrode and interact with the atomic electrons of its material, causing the incoming electrons to be slowed down. Energy is emitted in the form of X-rays.

Insight   CFCs and Ozone Layer Depletion

We hear a lot about the ozone layer these days. What is it, and why should it concern us? Let's take a look.

At an altitude of about 30 km ($\approx$20 mi) in the atmosphere, there exists a concentration of ozone (O_3)—the "ozone layer." Here oxygen molecules (O_2) undergo dissociation by solar ultraviolet radiation ($O_2 + uv \rightarrow O + O$). The resulting atomic oxygen (O) combines with an oxygen molecule to form ozone ($O + O_2 \rightarrow O_3$). However, there is not a continual buildup of ozone. Ozone is very unstable in the presence of sunlight, and when it absorbs ultraviolet radiation, it is again dissociated into atomic and molecular oxygen. Moreover, an oxygen atom may react with an ozone molecule to form two ordinary oxygen molecules ($O + O_3 \rightarrow O_2 + O_2$), thus destroying the ozone. All these processes go on simultaneously in the ozone layer so that a natural balance between ozone production and ozone destruction is maintained.

The ozone layer is vitally important to the Earth's living creatures—ourselves included—because it absorbs most of the uv radiation in sunlight. It thus shields us from these energetic rays, which are potentially harmful to biological organisms. Not only can uv produce severe skin burns and eye damage, it is also a carcinogen, or cancer-causing agent. For example, uv exposure is thought to be responsible for most cases of melanoma, an extremely dangerous form of cancer that originates in the pigment-producing cells of the skin.

This protective blanket of ozone is now threatened by the products of human activity. The cause is the release into the atmosphere of a class of chemicals called chlorofluorocarbons (CFCs for short). CFCs are the most widely used refrigerants (a common commercial example is Freon). We depend on refrigeration to cool our homes, cars, and businesses, and use numerous refrigerators to keep food and beverages cool. But as gases, these CFCs often escape from refrigerator units. CFCs are also used as plastic foam blowing agents and in industrial solvents. They were once used as propellants in spray cans as well, but this use was discontinued in the 1970s after the effects of CFCs on the ozone layers were predicted.

The release of gaseous CFCs into the atmosphere is now a cause for worldwide concern. In the lower atmosphere, there is no known mechanism for their destruction, and the CFCs rise slowly through the atmosphere. In 20 to 30 years, they reach the altitude of the ozone layer. Here, the CFC molecules are broken apart by uv radiation and release reactive chlorine (Cl) atoms. These atoms in turn react with and destroy ozone molecules in a repeating cycle:

$$CFCs \overset{(uv)}{\rightarrow} Cl$$
$$Cl + O_3 \rightarrow ClO + O_2$$
$$ClO + O \rightarrow Cl + O_2$$

Notice that a chlorine atom is again available for reaction at the end of the sequence. It is estimated that these atoms may remain in the atmosphere for a year or more. During this time, a single Cl atom may destroy as many as 100,000 ozone molecules.

Attention was drawn to the fate of the ozone layer in the late 1980s when an ozone "hole" was discovered over Ant-arcica (Fig. 1). This prompted companies to seek environmentally safe substitutes for CFCs. Even so, there will be tradeoffs. Currently, the substitutes are much less efficient as refrigerants, and it is estimated that their universal adoption would cause a 3% increase in electrical use. This would both add to consumer cost and increase the amount of carbon dioxide (CO_2) released into the atmosphere, since most electricity is generated by burning fossil fuels. More CO_2 could in turn contribute to the greenhouse effect and global warming (Chapter 11).

It is estimated that for every 1% loss of the ozone layer, an additional 2% of the Sun's uv radiation will reach the Earth's surface. The direct effects of increased uv exposure on human communities are likely to include a higher incidence of melanoma. (In fact, a rise in the number of cases of this deadly cancer has already been observed in Australia and other southern regions where the erosion of the ozone layer has been most marked.) Eye damage, such as cataracts, is also likely to increase. On a global scale, stronger uv radiation could kill many of the oceanic microorganisms that play a key role in the food chain. And an increase in the amount of uv radiation reaching the Earth's surface could contribute to global surface warming, with its many poorly understood and potentially dangerous consequences (discussed in the Insight feature in Chapter 11).

The problem is real enough to have alarmed political leaders as well as scientists. In 1989, some 24 nations signed the *Montreal Protocol on Substances that Deplete the Ozone Layer*; and in 1990 more than 90 nations agreed to phase out CFCs by the year 2000.

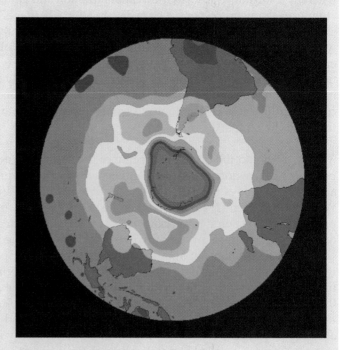

FIGURE 1   Ozone hole
The depletion of ozone over the South Pole. Purple shades indicate the regions of lowest ozone concentrations. The photo was taken by a TIROS weather satellite in October, 1995.

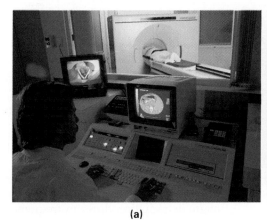

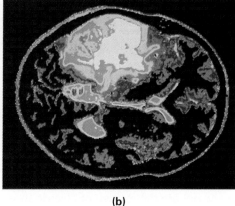

(a)   (b)

•**FIGURE 20.22 X-ray CT scan**
In an ordinary X-ray image, the entire thickness of the body is projected onto the film. Internal structures often overlap, making details hard to distinguish. In computerized tomography (from the Greek words *tomo*, "slice," and *graph*, "picture"), X-ray beams scan across a slice of the body. The transmitted radiation is recorded by a series of detectors and processed by a computer. Using information from multiple slices, the computer can construct a three-dimensional image. Any single slice can also be displayed for study, as on the monitor in the photo. CT scans typically provide physicians with much more information than can be obtained from a conventional X-ray. **(a)** A CAT scan in progress. **(b)** CAT image of a brain with a benign tumor.

As you will learn in Chapter 27, the energy of electromagnetic radiation depends on its frequency. High-frequency X-rays have very high energies and can cause cancer, skin burns, and other harmful effects. However, at low intensities, X-rays can be used with relative safety to view the internal structure of the human body and other opaque objects.* X-rays can pass through materials that are opaque to other types of radiation. The denser the material, the greater its absorption of X-rays and the less intense the transmitted radiation will be. For example, as X-rays pass through the human body, many more of them are absorbed by bone than by tissue. If the transmitted radiation is directed onto a photographic plate or film, the exposed areas show variations in intensity that form a picture of internal structures.

The combination of the computer with modern X-ray machines permits the formation of three-dimensional images by means of a technique called *computerized tomography*, or CT (•Fig. 20.22).

Gamma Rays. The electromagnetic waves of the upper frequency range of the known electromagnetic spectrum are called gamma rays (γ-rays). This high-frequency radiation is produced in nuclear reactions and in particle accelerators and can also be given off in certain types of nuclear radioactivity. Gamma rays will be discussed in Chapter 29.

Chapter Review

Important Terms

*Most health scientists believe that there is no safe "threshold" level for X-rays or other energetic radiation—that is, no level of exposure that is completely risk-free—and that some of the dangerous effects are cumulative over a lifetime. People should therefore avoid unnecessary medical X-rays or any other unwarranted exposure to "hard" radiation (Chapter 29).

Important Concepts

- Electromagnetic induction refers to the creation of induced emfs whenever the magnetic flux through a coil is changed. Faraday's law of induction states that the magnitude of the induced emf is equal to the time rate of change of the magnetic flux.
- Lenz's law states that when a change in magnetic flux induces an emf, the resulting current is in such a direction as to create a magnetic field that tends to oppose the change in flux.
- An ac current alternates in direction periodically.
- Generators (ac or dc) are the reverse of electric motors. When coils are spun in a magnetic field (or the field is spun while the coils are kept fixed), an emf is induced in the coils and used to create an electric current.
- Back emf is a reverse emf created by induction in motors when their armature is rotated in a mag-

netic field. The back emf partially cancels the voltage that drives the motor.
- A transformer is a device that changes the voltage supplied to it by means of induction. If the voltage is increased, the current is reduced, and vice versa.
- An electromagnetic wave consists of mutually perpendicular, time-varying electric and magnetic fields that propagate at a constant speed in vacuum ($c = 3.00 \times 10^8$ m/s). The different types of electromagnetic radiation differ in frequency, and thus in wavelength ($c = \lambda f$).
- Maxwell's equations are a set of four equations that describe all magnetic and electric field phenomena.
- Radiation carries energy and momentum and thus can exert force.

Important Equations

Magnetic Flux:

$$\Phi = BA \cos \theta \qquad (20.1)$$

Faraday's Law of Induction:

$$\mathscr{E} = -N\frac{\Delta\Phi}{\Delta t} \qquad (20.2)$$

$$= -N\left[\left(\frac{\Delta B}{\Delta t}\right)(A\cos\theta) + B\left(\frac{\Delta A}{\Delta t}\right)(\cos\theta) + BA\left(\frac{\Delta(\cos\theta)}{\Delta t}\right)\right]$$
$$(20.3)$$

Generator emf (where $\mathscr{E}_o = NBA\omega$):

$$\mathscr{E} = \mathscr{E}_o \sin \omega t = \mathscr{E}_o \sin 2\pi ft \qquad (20.4\text{–}5)$$

Back emf:

$$\mathscr{E}_b = V - IR \qquad (20.6)$$

Currents, voltages, and turn ratio for a transformer:

$$\frac{I_p}{I_s} = \frac{V_s}{V_p} = \frac{N_s}{N_p} \qquad (20.9)$$

Exercises

20.1 Induced Emf's: Faraday's Law and Lenz's Law

1   A unit of magnetic flux is (a) Wb, (b) T·m², (c) T·m/A, (d) both (a) and (b).

2   A magnetic flux change through a loop may be accomplished by the area of the coil and how many other parameters? (a) 2, (b) 3, (c) 4.

3   A bar magnet is dropped through a coil of wire as shown in ●Fig. 20.23. (a) Describe what is observed on the galvanometer by sketching a graph of $\mathscr{E}$ versus t. (b) Does the magnet fall freely? Explain.

4   ■ A telephone has both a speaker-transmitter and a receiver (●Fig. 20.24). The transmitter has a diaphragm coupled to a carbon chamber (called the button), which contains loosely packed granules of carbon. As the diaphragm vibrates because of incident sound waves, the pressure on the granules varies, causing them to be more or less closely packed. As a result, the resistance of the button changes. The receiver converts electrical impulses to sound. Applying principles of electricity and magnetism you have learned, explain the basic operation of the telephone.

5   ■ A circular loop of wire that encloses an area of 0.030 m² is in a uniform magnetic field of 0.25 T. What is the flux through the loop if its plane is (a) parallel to the field, (b) at an angle of 37° to the field, and (c) perpendicular to the field?

6   ■ A conductive loop enclosing an area of 2.0×10^{-3} m² is perpendicular to a uniform magnetic field of 6.0 T. If the field goes to zero in 0.0045 s, what is the magnitude of the average emf induced in the loop?

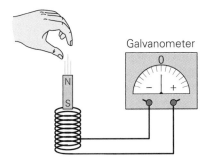

•FIGURE 20.23   A time-varying magnetic field
What will the galvanometer measure? See Exercise 3.

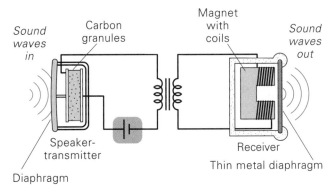

•FIGURE 20.24   Telephone operation
See Exercise 4.

7 ■ A circular loop with a radius of 20 cm is positioned in various orientations in a uniform 0.15-T magnetic field. Find the magnetic flux if the normal to the plane of the loop is (a) perpendicular to the magnetic field, (b) parallel to the magnetic field, and (c) at an angle of 40° to the magnetic field.

8 ■ A loop in the form of a right triangle with a side of 40.0 cm and a hypotenuse of 50.0 cm lies in a plane perpendicular to a uniform magnetic field of 550 mT. What is the flux through the loop?

9 ■ A square coil of wire with 10 turns is in a magnetic field of 0.25 T. The total flux through the coil is 0.50 T·m². Find the effective area of one turn if the field (a) is perpendicular to the plane of the coil and (b) makes an angle of 60° with the plane of the coil.

10 ■■ What is the magnetic flux (due to its own magnetic field) through the cross section of an ideal solenoid whose windings have a radius of 3.0 cm if the turn density is 250 turns/m and a current of 1.5 A flows through the wire?

11 ■■ A magnetic field perpendicular to the plane of a wire loop with an area of 0.40 m² decreases by 0.20 T in 10^{-3} s. What is the magnitude of the average value of the emf induced in the loop?

12 ■■ A square loop of wire with 50-cm sides experiences a uniform, perpendicular magnetic field of 100 mT. If the field goes to zero in 0.010 s, what is the magnitude of the average emf induced in the loop?

13 ■■ The magnetic flux through one turn of a 30-turn coil of wire is reduced from 35 Wb to 5.0 Wb in 0.10 s. The average induced current in the coil is 3.6×10^3 A. Find the total resistance of the wire.

14 ■■ If the magnetic flux through a single loop of wire increases by 30 T·m² and an average current of 40 A was induced in the wire (resistance 2.5 Ω), over what period of time was the flux increased?

15 ■■ During a time of 0.20 s, a coil of wire with 50 loops has an average induced emf of 9.0 V due to a changing magnetic field perpendicular to the plane of the coil. If the radius of the coil is 10 cm and the initial value of the magnetic field is 1.5 T, what is the final value of the field assuming the field decreased with time?

16 ■■ A single strand of wire of adjustable length is wound around the circumference of a round balloon that has a diameter of 20 cm. A uniform magnetic field with a magnitude of 0.15 T is perpendicular to the plane of the loop. If the balloon is blown up so its diameter and the diameter of the wire loop increase to 40 cm in 0.040 s, what is the magnitude of the average value of the emf induced in the loop?

17 ■■ The magnetic field perpendicular to the plane of a wire loop with an area of 0.10 m² changes with time as shown in •Fig. 20.25. What is the average emf induced in the loop for each time distinct interval on the graph (e.g., from 0 to 2.0 ms)?

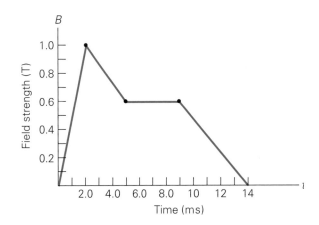

•FIGURE 20.25   Magnetic field versus time
See Exercise 17.

18 ■■ A metal rod, with its length oriented in the east–west direction, is moved at a uniform speed while held parallel to the ground. Starting at the Equator and heading due north, where would the most and least emf be generated in the rod?

19 ■■ A metal rod 20 cm long moves in a straight line at a speed of 4.0 m/s with its length parallel to a uniform magnetic field of 1.2 T. Find the resulting potential difference between the ends of the rod.

20 ■■ Suppose that the metal rod in Fig. 20.6 is 20 cm long and is moving at a speed of 10 m/s in a magnetic field of 3.0 T, but that the metal frame is covered with an insulating material. Find (a) the magnitude of the induced emf across the rod, and (b) the current in the rod.

21 ■■ The flux through a fixed loop of wire changes uniformly from +40 Wb to −20 Wb in 1.5 ms. (a) What is the significance of the negative flux? (b) What is the average induced emf in the loop?

22 ■■ A rectangular conductive loop measuring 20 cm by 30 cm has a resistance of 0.10 Ω. At what rate must a magnetic field perpendicular to the plane of the loop change with time to cause a current of 3.0 A to flow in the loop?

23 ■■■ A coil of wire with 10 turns (loops) and a cross-sectional area of 0.055 m² is placed in magnetic field of 1.8 T and oriented so the area is perpendicular to the field. The coil is then flipped by 90° in 0.25 s and ends up with the area parallel to the field. What is the magnitude of the average emf induced in the coil?

24 ■■■ A uniform magnetic field with a value of 1.5 T passes through a 0.40 m × 0.60 m rectangular loop at an angle of 60° relative to a normal to the plane of the loop. (a) If the field decreases to zero in 0.50 s, what is the average emf developed in the loop? (b) If the field remains constant at its initial value, at what rate would the loop have to shrink in size to generate the same average emf?

25 ■■■ A uniform magnetic field of 0.50 T permeates a double incline block as shown in •Fig. 20.26. Determine the magnetic flux through *each* surface of the block.

26 ■■■ A length of 20-gauge copper wire is formed into a circular loop with a radius of 20 cm. (20-gauge wire has a diameter of 0.8118 mm). A magnetic field perpendicular to the plane of the loop increases from zero to 7.0 mT in 0.25 s. Find the average electrical energy dissipated in the process.

27 ■■■ A metal airplane with a wing span of 30 m flies horizontally at a constant speed of 320 km/h in a region where the vertical component of the Earth's magnetic field is 5.0×10^{-5} T. What is the induced motional emf across its wing tips?

20.2 Generators and Back Emf

28 A split-ring commutator is used in (a) an ac generator, (b) a dc generator, (c) an alternator, (d) both (a) and (b).

29 The back emf of a motor depends on (a) the input voltage, (b) the input current, (c) the armature's rotational speed, (d) none of these.

30 What is the orientation of the armature loop in a simple ac generator when (a) the value of the emf is a maximum, (b) the value of the magnetic flux is a maximum? Explain why for each and also why they do not occur at the same orientation.

31 ■ A student has a bright idea for a generator: For the arrangement shown in •Fig. 20.27, the magnet is pulled down and released. With a highly elastic spring, the inventor thinks there should be a relatively continuous electrical output. What is wrong with this idea?

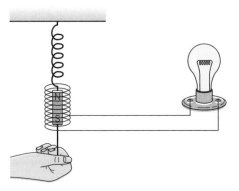

•**FIGURE 20.27 Inventive genius?**
See Exercise 31.

32 ■ An ac generator operates at a rotation frequency of 60 Hz. If the output voltage is a maximum (in magnitude) at $t = 0$ when is it next (a) a maximum (magnitude), (b) zero, (c) a minimum (maximum but negative, opposite direction), and (d) its initial value?

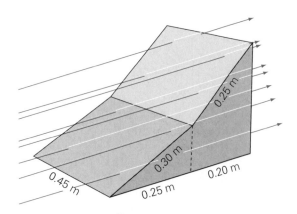

•**FIGURE 20.26 Magnetic flux**
See Exercise 25. (Drawing not to scale.)

33 ■ (a) What is the maximum value of the emf output from a simple ac generator having a single loop with an area of 90 cm² that rotates with a frequency of 60 Hz in a uniform magnetic field of 10^{-2} T? (b) What would the maximum value be if a coil of 15 such loops were used?

34 ■ A simple ac generator consists of a coil having 10 turns of wire, with each loop having an area of 50 cm². The coil rotates in a uniform magnetic field of 350 mT with a frequency of 60 Hz. (a) Write an equation showing how the emf of the generator varies as a function of time. (b) Compute the maximum emf.

35 ■ A 60-Hz sinusoidal ac voltage has a maximum value of 120 V. What is the value of the voltage at $\frac{1}{180}$ s after it has a value of zero? (Is there more than one such value? If so, list all possible answers. Explain why there might be more than one.)

36 ■■ A simple ac generator with a maximum emf value of 40 V is to be constructed, using loops of wire with a radius of 0.15 m, a rotational frequency of 60 Hz, and a magnetic field of 10^{-2} T. How many loops of wire will be needed?

37 ■■ A simple ac generator has a rotational frequency of 60 Hz. Assume zero emf at start-up. What percentage of the maximum emf is the instantaneous emf (a) at 0.50 s after start-up and (b) at $\frac{1}{360}$ s after the emf passes through zero in changing to a negative polarity?

38 ■■ The armature of a simple ac generator has 20 circular loops of wire with a radius of 10 cm. It is rotated with a frequency of 60 Hz in a uniform magnetic field of 800 mT. What is the maximum value of the emf induced in the loops, and how often is this value attained?

39 ■■ The armature of an ac generator has 100 turns, which are rectangular loops measuring 8.0 cm by 12 cm. The generator has a sinusoidal voltage output with an amplitude of 24 V. If the magnetic field of the generator is 250 mT, with what frequency does the armature turn?

40 ■■ Which has the greater voltage output—a generator with a loop area of 100 cm² rotating in a magnetic field of 20 mT at 60 Hz or a generator with a loop area of 75 cm² rotating in a magnetic field of 200 mT at 120 Hz? Justify your answer mathematically.

41 ■■ A motor has a resistance of 2.50 Ω and draws a current of 4.00 A when operating at its normal speed on a 115-V line. What is the back emf of the motor?

42 ■■ The starter motor in an automobile has a resistance of 0.40 Ω in its armature windings. The motor operates on 12 V and has a back emf of 10 V when running at normal operating speed. How much current does the motor draw (a) when running at its operating speed and (b) when initially starting up?

43 ■■ A 240-V dc motor with an armature whose resistance is 1.50 Ω draws a current of 16.0 A when running at its operating speed. (a) What is the back emf of the motor when it is operating normally? (b) What is the starting current? (Assume that there is no additional resistance.) (c) What series resistance would be required to limit the starting current to 25 A?

20.3 Transformers and Power Transmission

44 For a step-down transformer, which of the following is *not* correct? (a) $N_p > N_s$, (b) $N_s > N_p$, (c) energy is lost through eddy currents.

45 In order to reduce resistance losses, electric power is transmitted (a) at low voltage, (b) at high current, (c) using step-up transformers at the generating plant, (d) none of these.

46 ■ The voltage for ignition by a spark plug in an automobile is supplied by what is referred to as the coil, which is actually a pair of induction coils (•Fig. 20.28). Explain how the voltage of the car's battery (12 V) is raised to as high as 25 kV by this device. Explain the function of the distributor.

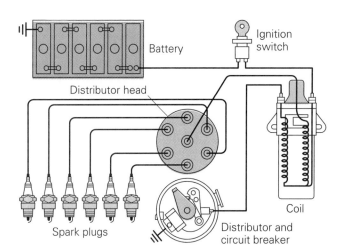

FIGURE 20.28 Auto ignition with coil
See Exercise 46.

47 ■ A transformer has 800 turns on its primary coil and 600 turns on its secondary. (a) What type of transformer is this? (b) If the input to the primary coil is 4.0 A at 120 V, what is the output of the secondary, assuming 100% efficiency?

48 ■ The number of turns on the secondary coil of an ideal transformer is 450 and on the primary it is 75. (a) What type of transformer is this? (b) What is the ratio of the current in the primary coil to the current in the secondary? (c) What is the ratio of the voltage in the primary coil to the voltage in the secondary?

49 ■ An ideal transformer steps up 8.0 V to 2000 V, and the 4000-turn secondary coil carries 2.0 A. (a) Find the number of turns on the primary coil. (b) Find the current in the primary.

50 ■ The primary coil of an ideal transformer has 720 turns, and the secondary coil has 180 turns. If 15 A flows in the primary with a voltage of 120 V, what are (a) the voltage and (b) the current output of the secondary?

51 ■ A transformer changes a 120-V input to a 3000-V output. Find the ratio of the number of turns on the primary coil to the number of turns on the secondary coil.

52 ■ The primary coil of an ideal transformer is connected to a 120-V source and draws 10 A. The secondary coil has 800 turns and a current of 4.0 A. (a) What is the voltage on the secondary? (b) How many turns are on the primary?

53 ■ An ideal transformer has 840 turns on its primary coil and 120 turns on its secondary. If the primary draws 1.50 A at 120 V, what are (a) the current and (b) the voltage output of the secondary?

54 ■■ An electric arc-welding machine requires 200 A of current. The transformer of the machine has 1200 turns on the primary coil, which draws 2.5 A at 240 V. (a) How many turns are on the secondary coil? (b) What is the voltage output of the secondary?

55 ■■ A 1.50-kV power line runs to a residential area, where a transformer with 1000 turns on its primary coil steps down the voltage to 240 V. How many turns are on the secondary coil of this transformer?

56 ■■ A circuit component operates on 20 V at 0.50 A. A transformer with 300 turns on its primary coil is used to convert 120-V household electricity to the proper voltage. (a) How many turns must the secondary coil have? (b) How much current flows in the primary?

57 ■■ A transformer in a door chime steps down the voltage from 120 V to 6.0 V and supplies a current of 0.50 A to the chime mechanism. (a) What is the turn ratio of the transformer? (b) What is the current input to the transformer?

58 ■■ An ac generator supplies 20 A at 440 V to a 10,000-V power line. If the step-up transformer has 132 turns on its primary coil, how many turns are on the secondary?

59 ■■ The electrical energy described in Exercise 58 is transmitted over an 80.0-km line ($R_o/L = 0.80$ Ω/km). (a) How many kilowatt-hours are saved in 5.00 h by stepping up the voltage? (b) At 10 cents per kWh, how much monetary savings (to the nearest dollar) is this to the consumer in a 30-day month assuming this happens continuously?

60 ■■ At an area substation, the power-line voltage is stepped down from 100,000 V to 20,000 V. If 1.0 MW of power is delivered to the 20,000-V circuit, what are the primary and secondary currents in the transformer?

61 ■■■ A voltage of 200,000 V in a transmission line is reduced to 100,000 V at an area substation, to 7200 V at a distributing substation, and finally to 240 V at a utility pole outside a house. (a) What turn ratio is required for each reduction step? (b) By what factor is the transmission line current stepped up in each voltage step-down and overall? (c) What is the overall factor by which the current is stepped up from transmission line to utility pole?

62 ■■■ An electric-generating plant produces electric energy at 50 A and 20 kV. The electricity is transmitted 25 km over transmission lines whose resistance is 1.2 Ω/km. (a) What would the power loss be if the energy is transmitted at 20 kV? (b) To what value should the output voltage of the generator be stepped up to decrease the energy loss by a factor of 15?

63 ■■■ Electrical power is transmitted through a 175-km power line with a resistance of 1.2 Ω/km. The generator output is 50 A at its operating voltage of 440 V. This voltage has a single step up for transmission to 44 kV. (a) How much power is lost as joule heat? (b) What is the necessary turn ratio of a transformer at the power delivery point that provides an output voltage of 220 V? (Neglect the voltage drop of the line.)

20.4 Electromagnetic Waves

64 Relative to the red end of the visible spectrum, the yellow and green regions have (a) lower frequencies, (b) longer wavelengths, (c) shorter wavelengths, (d) both (a) and (b).

65 Which of the following radiations has the highest frequency: (a) ultraviolet, (b) infrared, (c) X-rays, (d) microwaves?

66 On a cloudy summer day, you may work outside most of the day and feel relatively cool. Yet, that evening, you may discover that you have a bad sunburn. Why is this? Explain in terms of infrared and ultraviolet radiations.

67 ■ Find the frequencies of electromagnetic waves with wavelengths of (a) 2.0 m, (b) 25 m, (c) 75 m.

68 ■ The frequency ranges of the radio AM and FM bands and the TV band are given in the text. What are the corresponding wavelength ranges for these bands?

69 ■ A radar operator notes that a time of 0.24 ms elapses between the sending and return of a radar pulse. How far away was the object that reflected the pulse?

70 ■ How long does it take for a laser beam to travel from Earth to a reflector on the Moon and back? (Take the distance of the Moon to be 2.4×10^5 mi. This experiment was done when the Apollo flights of the early 1970s left laser reflectors on the lunar surface in order to accurately determine the Earth–Moon distance.)

71 ■ Two electromagnetic waves have the following wavelengths: (a) 800 nm and (b) 300 nm. To which spectral regions do these waves belong?

72 ■■ Orange light has a wavelength of 600 nm and green light a wavelength of 510 nm. What is the frequency difference of these colors?

73 ■■ A radio antenna made of wire is called a quarter wavelength antenna because of its length. If you were going to make such antennae for the AM and FM radio bands by using the mid frequency of each band, what lengths of wire would you use?

Additional Exercises

74 A 120-VAC motor has a resistance of 3.50 Ω. When operating at normal speed, the motor develops a back emf of 115 V. What are (a) the start-up current of the motor, and (b) the current when it is operating at normal speed?

75 An electric door bell operates on 4.5 V. If this voltage is obtained from standard 120-V household voltage, which of its windings has more turns and how many times more than the other?

76 A circular wire loop is in a uniform magnetic field of 1.5×10^{-2} T. The flux through the loop is 1.2×10^{-2} T·m². What is the radius of the loop if the normal to the plane of the loop makes an angle of 45° with the field?

77 The efficiency of a transformer is defined as the ratio of the power output to the power input: efficiency = $I_s V_s / I_p V_p$. Show that in terms of the ratios of currents and voltages given in Eq. 20.9 (for an ideal transformer), an efficiency of 100% is obtained. What does this imply?

78 A pivoted coil of wire is rotated 50 times per second in a uniform magnetic field. How often does the induced emf in the coil have a value of zero?

79 A 120-V dc motor draws a current of 6.0 A and has a back emf of 96 V at its operating speed. (a) What starting current is required by the motor? (Assume that there is no additional resistance.) (b) What series resistance would be required to limit the starting current to 15 A?

80 A circular wire loop with a diameter of 28 cm carries a current of 2.50 A. Assume that the magnitude of the magnetic field at the center of the loop is the same as

the magnetic field across the whole area. (a) What is the flux through the loop? (b) Suppose the current dropped to 0.50 A in 0.35 s, what average emf would be induced in the loop?

81 An ideal transformer has 120 turns on its primary coil and 840 turns on its secondary. If 14 A flows in the primary with a voltage of 120 V, what are (a) the voltage and (b) the current output of the secondary?

82 A loop of wire in the form of a square 1.0 m on a side is perpendicular to a uniform magnetic field of 0.75 T. What is the flux through the loop?

83 An ideal transformer has 80 turns on its primary coil and 360 turns on its secondary. If 3.0 A at 12 V is applied to the primary coil, what are (a) the current, (b) the voltage output, and (c) the power output of the secondary?

84 A square conductive loop 50 cm on a side has a resistance of 200 mΩ. Find the rate at which a magnetic field perpendicular to the plane of the loop must change with time to cause an average of 5.0 J/s to be expended in the loop.

85 A coil with 150 concentric loops of wire with a radius of 12 cm is used as the armature of a generator with a magnetic field of 0.80 T. With what frequency should the armature be rotated to make the polarity change every 0.010 s?

86 The transformer on a utility pole steps down the voltage from 20,000 V to 220 V for home usage. If a household circuit uses 6.6 kW of power, what are the primary and secondary currents in the transformer?

87 Gamma rays have typical wavelengths on the order of 3×10^{-15} m (nuclear diameter size). What is the frequency of typical gamma ray radiation?

88 In some countries, the voltage is delivered at 50 Hz instead of the 60 Hz as in the United States. (a) What is the rotation rate of the turbines that drive the electric generators at the plant? (b) Would different transformers have to be used if the step-up and step-down requirements of the delivery systems were to be the same? Explain carefully including ideal and nonideal transformers.

89 Radar operates at wavelengths of a few cm whereas FM radio operates at wavelengths on the order of meters. How do radar frequencies compare to those of the FM band on your radio?

90 Microwave ovens can have cold and hot spots due to standing electromagnetic waves analogous to standing wave nodes and antinodes in strings. If your microwave has cold spots (nodes) approximately every 5.0 cm, what frequency waves does it use to cook?

21 AC Circuits

D irect current circuits have many uses. Whenever you start your car in the morning or search for a lost object with a pocket flashlight, you are making use of direct current. It's even possible that this Montreal cafe has emergency lights on a dc backup circuit, designed to utilize batteries in the event of a power failure. But it's a safe bet that the lights we see in the photo operate on alternating current (ac). The electric power delivered to our homes and offices is also ac, and most of the devices we now use every day require this type of current.

There are several reasons for this. For one thing, almost all electric power is produced by generators using electromagnetic

induction, and such generators (as you saw in the last chapter) produce ac. Moreover, the voltage and current in ac circuits can be changed by transformers in a way that allows economical transmission of electric power over very long distances. (The electricity that lights these bulbs may have been produced by a hydroelectric station hundreds of miles away.) But perhaps the most important reason that ac is used so universally is that ac circuits are extremely versatile. The alternation of the current produces electromagnetic effects that can be exploited in a wide variety of devices. Every time you tune a radio to a favorite station, for example, you are taking advantage of a special property of ac circuits—one of many that you'll explore in this chapter.

Because alternating voltage and current are used in power transmission systems as well as industry and homes, ac circuits are very common. You might suppose that such circuits should be as simple and straightforward as dc circuits. However, the alternating current and the associated time-varying magnetic fields in such circuits add new dimensions to ac circuit analysis.

In common dc circuits, we are concerned only with ohmic resistances. There is ohmic resistance in ac circuits, too, but there are other factors that affect the flow of charge. Recall that in a dc circuit, once a capacitor has been fully charged it offers infinite resistance (since in effect it creates an open circuit). However, in an ac circuit, this is not the case. Alternating voltage continually charges and discharges a capacitor, so there is both current in the circuit and opposition to the current. Coils of wire also oppose an ac current through induction. An example, the back emf of a motor, was described in Chapter 20.

In this chapter, we will look at some of the basic principles of ac circuits. Forms of Ohm's law and expressions for power will be developed specifically for ac circuits. Finally, we will explore the phenomenon of resonance—the condition for maximum energy transfer in an ac circuit—and some of its important practical applications.

21.1 Resistance in an AC Circuit

Objectives: **To be able to (a) specify how voltage, current, and power vary with time in an ac circuit, and (b) explain how the rms and peak values of these quantities are related and how they are used in calculations.**

An ac circuit is one that contains an ac voltage source and one or more other circuit elements. The circuit diagram for an ac circuit with a single resistive element is shown in •Fig. 21.1. If the source output is assumed to be sinusoidal, as is the case for a simple generator (Section 20.2), the voltage across the resistor varies with time in accordance with the equation

$$V = V_o \sin \omega t = V_o \sin 2\pi f t \tag{21.1}$$

where ω is the angular frequency ($\omega = 2\pi f$). The voltage oscillates between values of $+V_o$ and $-V_o$, where V_o is the **peak voltage**.

AC Current and Power

Under ac conditions, the current through the resistor also oscillates—that is, its direction and magnitude changes. From the relationship between voltage, resistance, and current in a resistor, the current in the resistor can be expressed as a function of time:

$$I = \frac{V}{R} = \frac{V_o}{R} \sin 2\pi f t$$

The right hand side can be conveniently rewritten as

$$I = I_o \sin 2\pi f t \tag{21.2}$$

where the amplitude of the current is $I_o = V_o/R$ and is called the **peak current**.

Both current and voltage for an ac circuit are plotted versus time in the graph in •Fig. 21.2. Note that they are in step, or *in phase*, with each other, both reaching zero values and maxima at the same times. The current oscillates back and forth and has corresponding positive and negative values during each cycle. Thus, the summation of all the instantaneous current values (positive and negative) over one complete cycle is zero, and the *average current is zero*. Mathematically, this

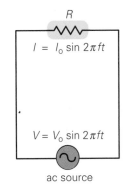

•FIGURE 21.1 **A purely resistive circuit**
The ac source delivers a sinusoidal voltage to the circuit, and the voltage across and current in the resistor are sinusoidal.

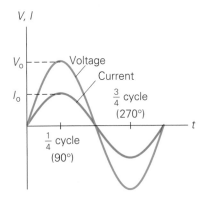

$$V = V_o \sin 2\pi f t$$
$$I = I_o \sin 2\pi f t$$

•FIGURE 21.2 **Voltage and current in phase**
In a purely resistive ac circuit, the voltage and current are in step, or in phase, with zero values and maxima occurring at the same respective times.

reflects the fact that the time average value of the sine function over one or more *complete* (360°) cycles is zero. Using angle brackets instead of overbars to denote a time average value, we can express this fact as $\langle \sin \theta \rangle = \langle \sin 2\pi ft \rangle = 0$. Similarly, $\langle \cos \theta \rangle = 0$.

The fact that the *average* current is zero, however, does not mean that there is no joule heating (I^2R losses). The dissipation of electrical energy depends on the collisions of electrons with the atoms of the material in which there is a current. These collisions occur when the electrons of an alternating current are going in either direction. Anytime there is current in a wire there will be joule heating, regardless of the direction of that current.

The instantaneous power is obtained using the instantaneous current (Eq. 21.2) and the power is a function of time because the current varies with time:

$$P = I^2R = I_o^2R \sin^2 2\pi ft \tag{21.3}$$

Even though the current changes sign each half-cycle, the square of the current, I^2, is always positive. Thus the average value of I^2R is *not* zero, even though the average value of I is zero. The average, or mean, value of I^2 is

$$\overline{I^2} = \langle I_o^2 \sin^2 2\pi ft \rangle = I_o^2 \langle \sin^2 2\pi ft \rangle$$

Using the trigonometric identity $\sin^2 \theta = \frac{1}{2}(1 - \cos 2\theta)$, we can write $\langle \sin^2 \theta \rangle = \frac{1}{2}(1 - \langle \cos 2\theta \rangle)$, and since $\langle \cos 2\theta \rangle = 0$ (just as $\langle \cos \theta \rangle = 0$), we have $\langle \sin^2 \theta \rangle = \frac{1}{2}$. Thus,

$$\overline{I^2} = I_o^2 \langle \sin^2 2\pi ft \rangle = \tfrac{1}{2}I_o^2 \tag{21.4}$$

The average power is therefore

$$\overline{P} = \overline{I^2}R = \tfrac{1}{2}I_o^2R \tag{21.5}$$

It is customary to write the ac power in the same form as the dc power ($P = I^2R$). To do this, a special value of current is used,

$$I_{rms} = \sqrt{\overline{I^2}} = \sqrt{\tfrac{1}{2}I_o^2} = \frac{I_o}{\sqrt{2}} = 0.707 I_o \tag{21.6}$$

where I_{rms} is called the **rms current**, or **effective current.** (Here rms stands for root-mean-square, in this case indicating the square root of the mean value of the square of the current.) Then with $I_{rms}^2 = (I_o/\sqrt{2})^2 = \tfrac{1}{2}I_o^2$, the effective or average power is

$$\overline{P} = \tfrac{1}{2}I_o^2R = I_{rms}^2 R \tag{21.7}$$

This power is the average power that is equivalent to the oscillating power averaged over time (see •Fig. 21.3).

AC Voltage

Since the peak values of voltage and current for a resistor are related by $V_o = I_o R$ a similar development can be used to show that the **rms voltage**, or **effective voltage**, is given by

$$V_{rms} = \frac{V_o}{\sqrt{2}} = 0.707 V_o \tag{21.8}$$

In previous chapters we have, in fact, calculated currents and voltages for ac cases. What we have *really* been finding are the rms values of these quantities. For resistors under ac conditions, then, we can use dc ideas as long as we realize we are solving for rms values. For alternating current, the relationship between current and voltage is

$$V_{rms} = I_{rms}R \tag{21.9}$$

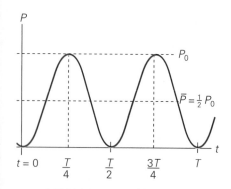

•**FIGURE 21.3 Power variation with time in a resistor**
Although both current and voltage oscillate in direction (sign), their product (power) is a positive oscillating quantity. Because of the way in which the rms values of current and voltage are defined, the average power is one-half the peak power, as you can see from the graph. Notice that the power reaches *two* peaks per cycle of current or voltage, thus the *power frequency* is double that of the current or voltage.

It is customary to measure and specify rms values for ac quantities. For example, the household line voltage of 120 V, as you may have guessed by now, is an rms value with a peak voltage of

$$V_o = \sqrt{2}\,V_{rms} = (1.414)(120 \text{ V}) = 170 \text{ V}$$

The various mean and peak values of ac current and voltage are shown in the graphs in •Fig. 21.4. Another voltage designation sometimes used is the peak-to-peak value which is the total voltage swing from the most negative voltage polarity in the cycle to the most positive. This is analogous to specifying twice the amplitude (the so-called double amplitude) for the total distance traveled by a harmonic oscillator as it travels from maximum compression to maximum extension:

$$V_{\text{p-p}} = 2V_o$$

For calculational significant figure purposes, we will assume the frequency of our voltages (for example, 60 Hz) to be exact.

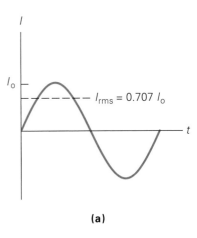

(a)

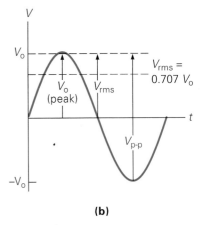

(b)

•FIGURE 21.4 **Root-mean-square (rms) current and voltage** The rms values of the current (a) and the voltage (b) are 0.707, or $1/\sqrt{2}$, times the peak or maximum values. The voltage is sometimes described by the peak-to-peak ($V_{\text{p-p}}$) value.

EXAMPLE 21.1 ■ A LIGHT BULB: RMS VALUES VS PEAK VALUES

A lamp with a 60-W bulb is plugged into a 120-V outlet. (a) What are the rms and peak currents through the lamp? (b) What is the resistance of the bulb? (Neglect the resistance of the lamp's wiring.)

Solution. We list the average power and the rms voltage of the source.

Given: $\bar{P} = 60$ W   **Find:** (a) I_{rms} and I_o (rms and peak currents)
$\quad\quad\quad V_{rms} = 120$ V   (b) R (resistance)

(a) The rms current is

$$I_{rms} = \frac{\bar{P}}{V_{rms}} = \frac{60 \text{ W}}{120 \text{ V}} = 0.50 \text{ A}$$

and the peak current is

$$I_o = \sqrt{2}\,I_{rm} = 1.41(0.50 \text{ A}) = 0.71 \text{ A}$$

(b) The resistance of the bulb is

$$R = \frac{V_{rms}}{I_{rms}} = \frac{120 \text{ V}}{0.50 \text{ A}} = 240 \ \Omega$$

Note that the resistance could also be obtained from the peak values (a V_o value of 170 V was computed above):

$$R = \frac{V_o}{I_o} = \frac{170 \text{ V}}{0.71 \text{ A}} = 240 \ \Omega$$

Follow-up Exercise. What would be the (a) rms current and (b) peak current in a 60-W light bulb in Great Britain, where the house rms voltage is 240 V at 50 Hz? (c) What would the resistance of a 60-W light bulb be in Great Britain compared to one designed for United States operation? Why are they so different?

CONCEPTUAL EXAMPLE 21.2 ■ ACROSS THE POND: BRITISH VS AMERICAN ELECTRICAL SYSTEMS

In European countries, the normal line voltage is 240 V. If a British tourist in the United States plugged in a hair dryer brought from home, you might expect the dryer to (a) not operate at all, (b) operate normally, (c) operate poorly, (d) burn out. *Clearly establish the reasoning and physical principle(s) used in determining your answer before checking it below. That is, **why** did you select your answer?*

•FIGURE 21.5 Converter and adapters
In foreign countries that have 240-V line voltages, a conversion to 120 V is needed to properly operate our normal appliances. Note the different types of plugs for different countries. The small plugs go into the foreign sockets and the converter prongs then fit into the back of a socket. Our standard two-prong plug fits into the converter which has a 120-V output.

Reasoning and Answer. British small appliances are designed to operate on 240 V. With half that voltage, an appliance would operate, but not as normally designed. With decreased voltage, there would be decreased current ($V = IR$) and greatly reduced joule heating ($P = V^2/R$). Assuming the same element resistance R, with half the voltage, there would be only one-fourth the power output. Thus, the heating element of the hair dryer might get warm, but would certainly not work to the user's expectations, so the answer is (c). Also, the decreased current might cause the blower motor to run slower than normal.

Fortunately, most people do not make this mistake, because the plugs and sockets are different (as are our 240-V plug and sockets different from the 120-V plug and sockets in the United States). There are several different types of plugs used in various countries. When on foreign travel with normal appliances, one should carry a converter/ adapter kit (•Fig. 21.5). It contains a selection of different plugs into which a voltage converter fits. The converter is a solid-state device that converts 240 V to 120 V for U.S. travelers and vice versa for tourists visiting the United States. (Note that some travel appliances are designed for dual voltage and will operate on either 120 V or 240 V.)

Follow-up Exercise. Foreign electrical systems generally operate at a frequency of 50 Hz—that is, 240 V, 50 Hz as compared to our 120 V, 60 Hz. Assuming the correct use of a voltage converter/adapter, what effect do you think this *frequency* difference might have on the hair dryer brought by the British tourist in the preceding Exercise? On his plug-in electric clock?

For ac circuits involving only resistors, the formulas derived for dc circuits can be used if the rms, or effective values, of the quantities are inserted. *For convenience and simplicity, the rms subscripts will be omitted from now on—but keep in mind that rms values are given for ac quantities unless otherwise specified.* Thus it is important to realize that we are working with only rms or peak quantities, not their values at specific times (instantaneous values).

Both alternating current and direct current are measured in amperes. But how is the ampere physically defined for alternating current? It cannot be derived from the mutual attraction of two parallel wires carrying ac current, as the dc ampere is defined (Section 19.4). An ac current changes direction with the source frequency (for example, 60 Hz), and the attractive force would average to zero. Thus, the ac ampere must be defined in terms of some property that is independent of the direction of the current. Joule heating is such a property, and *there is 1 ampere of alternating current in a circuit if the current produces the same average heating effect as 1 ampere of dc current would under the same conditions.*

21.2 Capacitive Reactance

Objectives: **To be able to (a) explain the behavior of capacitors in ac circuits, and (b) calculate their effect on ac current (capacitive reactance).**

Note: Review Section 16.3.

As we saw in Chapter 16, when a capacitor is connected to a voltage source in a dc circuit, current will flow for the short time required to charge the capacitor. As charge accumulates on the capacitor's plates, the voltage across them increases, opposing the current. That is, a capacitor in a dc circuit will limit or oppose the current as it charges. When the capacitor is fully charged, the current in the circuit falls to zero.

When a capacitor is in a circuit with an ac source, as shown in •Fig. 21.6a, it limits, or regulates, the current, but does not completely prevent the flow of charge. The capacitor is alternately charged and discharged as the current and voltage reverse each half-cycle.

Plots of ac current and voltage versus time for a circuit with a capacitor are shown in *Fig. 21.6b. Let's look at the changing conditions of the capacitor. We will choose as our starting point a time (t_o) when the voltage is at its maximum. This means that the capacitor is fully charged ($Q_o = CV_o$). Since the capacitor plates cannot accommodate any additional charge, there is no current in the circuit.

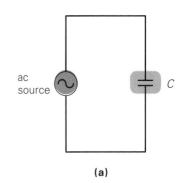

(a)

1. As the voltage starts to decrease, the capacitor begins to discharge, giving rise to a current in the circuit. The current reaches its maximum value as the voltage falls to zero and the capacitor plates are completely discharged.*

2. The voltage then reverses polarity and starts to increase. The capacitor becomes charged again, this time with the opposite polarity. With the plates initially uncharged, there is no opposition to the current (other than the resistance of the connecting wires), so the current is at a maximum. However, the charge accumulating on the capacitor plates inhibits the current, which therefore decreases, reaching zero when the voltage is again at a maximum and the capacitor fully charged.

    Note that at this point the charges will be the same as on the plates at the start, except reversed in sign, assuming no loss of energy in the wires (no resistance). So from an energy point of view, since no electrical energy is lost to heat, the charge on the capacitor must come back up to its initial value.

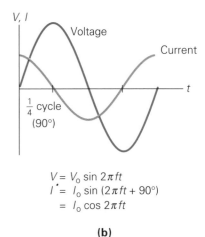

$$V = V_o \sin 2\pi ft$$
$$\dot{I} = I_o \sin (2\pi ft + 90°)$$
$$= I_o \cos 2\pi ft$$

(b)

During the next half cycle (steps 3 and 4) the same process is repeated, but with reversed polarities.

3. As the voltage again starts to decline, the capacitor loses its accumulated charge once more. The current is now in the opposite direction, attaining its maximum value as the voltage falls to zero.

4. The voltage rises, the capacitor becomes charged, and the current falls. At the end of this time the circuit has completed a full cycle—the voltage and current are exactly the same as they were at time t_o.

*FIGURE 21.6 A purely capacitive circuit

In a circuit with only capacitance, the current leads the voltage by 90° or $\frac{1}{4}$ cycle. (Notice that a current zero is 90° or a quarter cycle ahead of a voltage zero.)

Note that the current and voltage are *not* in step, or in phase, in this situation. The current reaches a maximum a quarter-cycle ahead of the voltage. The relationship between the current and the voltage is commonly stated in this way:

> In a purely capacitive ac circuit, the *current* leads the *voltage* by 90° or a quarter ($\frac{1}{4}$) cycle.

As in a dc circuit, there is opposition to the charging process in an ac circuit, but it is not totally limiting (as it is in a dc open-circuit condition). The quantitative measure of the impedance of a capacitor is referred to as its **capacitive reactance**. In an ac circuit, capacitive reactance, symbolized by X_C, is given by:

$$X_C = \frac{1}{2\pi fC} = \frac{1}{\omega C} \tag{21.10}$$

Capacitive reactance: opposition to current

where $\omega = 2\pi f$. Here C is the capacitance (in farads), and f is the frequency (in Hz). Like resistance, reactance is measured in ohms (Ω). Can you show that the ohm is equivalent to the second per farad?

*Note that although the current is depicted as negative in steps 1 and 2 of Fig. 21.6, this has no special significance. The positive and negative dimensions on the graph merely indicate the *direction* of the current and the *polarity* of the voltage. We could as easily have begun our analysis at step 3, with the voltage negative and the current positive.

As you can see, the reactance is inversely proportional to the capacitance (C) and the voltage frequency (f). Recall that capacitance is the charge per voltage ($C = Q/V$). For a particular voltage, therefore, the greater the capacitance, the more charge the capacitor can accommodate for a chosen frequency. Putting more charge on the plates requires a larger flow of charge, that is, a greater current, which means the opposition to the current, or the reactance, must be smaller.

The greater the frequency of the ac driving source, the shorter the charging time in each cycle. A shorter charging time means less charge accumulation on the plates, and therefore less opposition to the current. Thus, the capacitive reactance is inversely proportional to the frequency as well as to the capacitance. (Note that if $f = 0$, as is the case for dc, the capacitive reactance is infinite and there is no current flow, which corresponds to an open circuit.)

The capacitive reactance is related to the voltage across the capacitor and the current in the circuit by an equation that has the same form as $V = IR$ for pure resistances:

$$V = IX_C \qquad (21.11)$$

(Remember that V and I are rms values.)

EXAMPLE 21.3 ■ CURRENT FLOW UNDER AC CONDITIONS: CAPACITIVE REACTANCE

A 15.0-μF capacitor is connected to a 120-V, 60-Hz source. What are (a) the capacitive reactance and (b) the current (rms and peak) in the circuit?

Solution. As previously stated, we assume the 60 Hz frequency to be exact, thus our answers are to three significant figures.

Given: $C = 15.0 \ \mu\text{F} = 15.0 \times 10^{-6} \ \text{F}$ *Find:* (a) X_C (capacitive reactance)
        $V = 120 \ \text{V}$ (b) I (current—both rms
        $f = 60 \ \text{Hz (exactly)}$ and peak)

(a) The capacitive reactance is given by Eq. 21.10,

$$X_C = \frac{1}{2\pi f C} = \frac{1}{2\pi(60 \ \text{Hz})(15.0 \times 10^{-6} \ \text{F})} = 177 \ \Omega$$

(b) Then, using Eq. 21.11, the current is

$$I = \frac{V}{X_C} = \frac{120 \ \text{V}}{177 \ \Omega} = 0.678 \ \text{A}$$

Remember that this is the rms current. The instantaneous current will vary sinusoidally with time. Let us calculate the peak current

$$I_\text{o} = \sqrt{2}I = (1.41)(0.678 \ \text{A}) = 0.959 \ \text{A}$$

Thus the current oscillates between $+0.959$ A and -0.959 A, and is ahead of the voltage by $90°$.

Follow-up Exercise. In this Example, what voltage frequency would be needed to give the same current if the capacitance in the circuit were twice as large?

21.3 Inductive Reactance

Objectives: To be able to (a) explain what an inductor is, and (b) calculate the effect of inductors on ac current (inductive reactance).

Another important phenomenon in ac circuits is inductance. A coil of wire (an inductor) in a circuit with a time-varying current has a reverse voltage, or back emf, set up in it as a result of the changing magnetic flux produced by its own changing magnetic field. This is an example of self-induction. The induced emf opposes the change in flux (Lenz's law) and thus opposes the current flow in the circuit.

Note: Self-induction is introduced in Section 20.3; Lenz's law is given in Section 20.1

The self-induced emf is given by Faraday's law (Eq. 20.2): $\mathscr{E} = -N\Delta\Phi/\Delta t$. Here, since the geometry of the coil is fixed, N is constant. The time rate of change of the flux, $\Delta\Phi/\Delta t$, is proportional to the time rate of change of the current in the coil, $\Delta I/\Delta t$ (since it is the coil current that produces the magnetic field responsible for the changing flux in the coil). Thus, for the case of the fixed-geometry inductor we can write the back emf as

$$\mathscr{E} \propto -\frac{\Delta I}{\Delta t}$$

Writing this relationship in equation form with a constant of proportionality, the emf is

$$\mathscr{E} = -L\left(\frac{\Delta I}{\Delta t}\right) \tag{21.12}$$

Inductive reactance: opposition to current

where L is the inductance of the coil (more precisely, its self-inductance). As you can see from the equation, inductance has units of volt·second per ampere (V·s/A). This combination is called a henry (H). The smaller unit of millihenry (mH) is commonly used ($1 \text{ mH} = 10^{-3} \text{ H}$).

The impedance of an inductor to the current in an ac circuit depends on the inductance, and this is expressed quantitatively by its **inductive reactance** (X_L),

$$X_L = 2\pi f L = \omega L \tag{21.13}$$

where $\omega = 2\pi f$. Here f is the frequency of the driving source and L is the inductance. Like capacitive reactance, inductive reactance is measured in ohms (Ω), which can be shown to be equivalent to the units of henry per second.

Notice that the inductive reactance is directly proportional to both the inductance (L) of the coil and the frequency (f) of the voltage source. The inductance of a coil is a constant that depends on the number of turns, the coil's diameter, its length, and the material of the core (if any). The greater the inductance, the greater the opposition to current. The frequency of the ac driving source plays a role because the more rapidly the current in the coil changes, the greater will be the rate of change in its magnetic flux, and thus the greater the self-induced emf that opposes the current.

In terms of X_L, the equation for the voltage across an inductor has a mathematical form similar to that for resistors:

$$V = IX_L \tag{21.14}$$

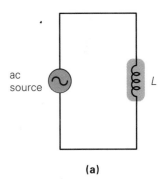

(a)

The circuit symbol for an inductor and the graphs of the voltage across the inductor and the current in the circuit are shown in •Fig. 21.7. For this kind of circuit, when the voltage is a maximum, the current is zero; and when the voltage goes to zero, the current is a maximum. The current lags a quarter-cycle behind the voltage, a relationship that is commonly expressed in the following way:

In a purely inductive ac circuit, the *voltage* leads the *current* by 90° or one quarter ($\frac{1}{4}$) cycle.

The phase relationships of current and voltage for purely inductive and purely capacitive circuits are opposite. A phase that may help you remember the relationship is

*EL*I the *IC*E man

With *E* representing voltage and *I* representing current, *EL*I indicates that with an inductance (*L*) the voltage leads the current (*I*). Similarly, *IC*E tells you that with a capacitance (*C*) the current leads the voltage.

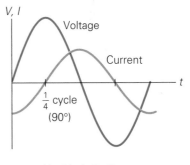

$V = V_0 \sin 2\pi ft$
$I = I_0 \sin(2\pi ft - 90°)$
$\quad = -I_0 \cos 2\pi ft$

(b)

•**FIGURE 21.7 A purely inductive circuit**
In a circuit with only inductance, the voltage leads the current by 90° or $\frac{1}{4}$ cycle. (Notice that a voltage zero is 90° or a quarter cycle ahead of a current zero.)

EXAMPLE 21.4 ■ IMPEDANCE WITHOUT RESISTANCE: INDUCTIVE REACTANCE

A 125-mH inductor is connected to a 120-V, 60-Hz source. What are (a) the inductive reactance and (b) the rms current in the circuit?

Solution. We list the given data and assume the frequency is exact as before.

Given: $L = 125 \text{ mH} = 125 \times 10^{-3} \text{ H}$ *Find:* (a) X_L (inductive reactance)
$V = 120 \text{ V}$ (b) I (current)
$f = 60 \text{ Hz (exact)}$

(a) The inductive reactance is given by Eq. 21.13
$$X_L = 2\pi f L = 2\pi (60 \text{ Hz})(125 \times 10^{-3} \text{ H}) = 47.1 \ \Omega$$

(b) Then, using Eq. 21.14 gives the rms current
$$I = \frac{V}{X_L} = \frac{120 \text{ V}}{47.1 \ \Omega} = 2.55 \text{ A}$$

Follow-up Exercise. (a) In this Example, what is the peak current? (b) What frequency of the voltage source would give the same current for an inductor with triple the inductance?

21.4 Impedance: RLC Circuits

Objectives: **To be able to (a) calculate currents and voltages when reactive circuit elements are present in ac circuits (series RC, RL, and RLC circuits), (b) use phase diagrams to calculate overall impedance and rms currents, and (c) understand and use the concept of the power factor in ac circuits.**

The previous sections considered purely capacitive and inductive circuits, in which only *nonresistive* opposition to current flow was found. However, it is impossible to have purely reactive circuits, since there is always some resistance, at minimum that from the connecting wires or the wire in an inductor. Therefore, real ac circuits usually have appreciable resistance, and the current is impeded by both resistances and reactances (capacitive and/or inductive). Analysis of some simple circuits will illustrate these effects.

Series RC Circuit

Suppose that an ac circuit consists of a voltage source and a resistance and a capacitive element connected in series, as illustrated in •Fig. 21.8a. The phase difference between the current and the voltage is different for each of these circuit elements. As a result, a special graphical method is used to find the effective opposition to the current flow in the circuit. This is conveniently found by means of a **phase diagram**, such as the one shown in Fig. 21.8b.

In a phase diagram, the resistance and reactance of the circuit are given vectorlike properties and their magnitudes are represented as arrows called **phasors**. On a set of *x–y* coordinate axes, the resistance is plotted on the positive *x* axis, since the voltage–current phase difference for a resistor is zero ($\phi = 0$). The capacitive reactance is plotted along the negative *y* axis, to reflect a phase difference of $-90°$. (A negative phase angle implies that the voltage lags behind the current, as is the case for a capacitor.)

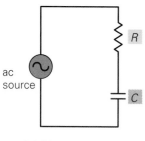

(a) Circuit diagram

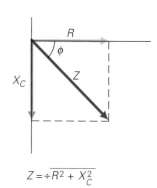

$$Z = \div\overline{R^2 + X_C^2}$$
$$\tan \phi = \frac{X_C}{R}$$

(b) Phase diagram

•**FIGURE 21.8 A series RC circuit**
In a series RC circuit **(a)**, the impedance Z is the phasor sum of the resistance R and the capacitive reactance X_C **(b)**.

The phasor sum is the effective or net opposition to the current, which we call the **impedance (Z)**. Phasors are added in the same way vectors are (as they must be because the effects are not in phase), so for the series RC circuit,

$$Z = \sqrt{R^2 + X_C^2} \tag{21.15}$$

The unit for the impedance is the ohm.

The Ohm's law relationship for the RC circuit in terms of impedance is

$$V = IZ \tag{21.16}$$

(Once again, remember current and voltage symbols here mean rms values.)

Impedance: effective opposition to current

EXAMPLE 21.5 ■ IS KIRCHHOFF'S LOOP THEOREM VALID FOR AC CIRCUITS? CAPACITIVE IMPEDANCE

A series RC circuit has a 100-Ω resistor and a 15.0-μF capacitor. (a) What is the current in the circuit when driven by a 120-V, 60-Hz source? (b) Compute the (rms) voltage across each circuit element and the two elements combined. Compare it to that of the voltage source. Is Kirchhoff's loop theorem satisfied?

Solution. We are working with three significant figures again because the frequency of 60 Hz is assumed to be exactly known.

Given: $R = 100\ \Omega$
$C = 15.0\ \mu\text{F} = 15.0 \times 10^{-6}\ \text{F}$
$V = 120\ \text{V}$
$f = 60\ \text{Hz}$

Find: (a) I (current)
(b) V_C (rms voltage across capacitor)
V_R (rms voltage across resistor)
$V_R + V_C$ (total rms voltage)

(a) First, the impedance is computed. From Example 21.3, we know the reactance for the capacitor is $X_C = 177\ \Omega$. Then using this directly in Eq. 21.15,

$$Z = \sqrt{R^2 + X_C^2} = \sqrt{(100\ \Omega)^2 + (177\ \Omega)^2} = 203\ \Omega$$

Since $V = IZ$, the current in the circuit is

$$I = \frac{V}{Z} = \frac{120\ \text{V}}{203\ \Omega} = 0.591\ \text{A}$$

(b) Since $V = IX$ in general ($V = IR$ for resistors), and the current is the same at all points at any given time in a series circuit, we have

$$V_R = IR = (0.591\ \text{A})(100\ \Omega) = 59.1\ \text{V}$$

and

$$V_C = IX_C = (0.591\ \text{A})(177\ \Omega) = 105\ \text{V}$$

The algebraic sum of these two rms voltages adds up to 164 V, which is *not* the same as the rms value of the voltage source. This does not, however, mean violation of Kirchhoff's loop theorem, which is based on conservation of energy per unit charge. *At any instant*, the source voltage does, in fact, equal the voltages across the capacitor and resistor *if you account for phase differences*. Thus we expect the rms values to agree if the combination of the voltages is calculated properly. The total rms voltage across both elements is found, not by simply adding them, but by recalling that the individual voltages are out of phase by ninety degrees. Thus the rms values must be added as in the Pythagorean theorem to obtain the total voltage:

$$V_{R+C} = \sqrt{V_R^2 + V_C^2} = \sqrt{(59.1\ \text{V})^2 + (105\ \text{V})^2} = 120\ \text{V}$$

When the individual voltage differences are combined properly, that is, taking into account the fact that they are out of phase and so never peak at the same time, we see that Kirchhoff's law *is* valid. The total voltage across both elements is equal (rms) to the voltage supplied by the source.

Follow-up Exercise. (a) How would the result in part (a) of this Example change if the circuit were driven by a voltage source with the same rms voltage but oscillating at 120 Hz? (b) Which circuit element, the resistor or the capacitor, is responsible for the change?

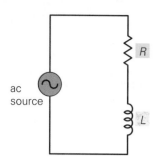

(a) Circuit diagram

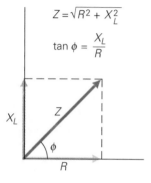

$$Z = \sqrt{R^2 + X_L^2}$$

$$\tan \phi = \frac{X_L}{R}$$

(b) Phase diagram

•**FIGURE 21.9   A series RL circuit**
In a series RL circuit **(a)**, the impedance Z is the phasor sum of the resistance R and the inductive reactance X_L **(b)**.

Series RL Circuit

The analysis of a series RL circuit (•Fig. 21.9) is similar to that of a series RC circuit. However, the inductive reactance is plotted along the positive y axis in the phase diagram to reflect a phase difference of $+90°$. (In this case, a positive phase angle implies that the voltage leads the current, as is the case for an inductor.)

The impedance can be seen to be

$$Z = \sqrt{R^2 + X_L^2} \tag{21.17}$$

as for an RC circuit, $V = IZ$.

Series RLC Circuit

An ac circuit may contain all three circuit elements—resistor, inductor, and capacitor—as in •Fig. 21.10. Again, phasor addition is used to determine the impedance by a component method. Summing the vertical components (the inductive and capacitive reactances) gives the total reactance ($X_L - X_C$, since X_C is defined as a negative phasor). The impedance is then the phasor sum of the resistance and the total reactance. From the phase diagram, you can see that the impedance is

$$Z = \sqrt{R^2 + (X_L - X_C)^2} = \sqrt{R^2 + \left(2\pi fL - \frac{1}{2\pi fC}\right)^2} \tag{21.18}$$

Again,

$$V = IZ \tag{21.19}$$

In this case, the **phase angle** (ϕ) between the voltage from the source and the current is given by

$$\tan \phi = \frac{X_L - X_C}{R} \qquad phase\ angle \tag{21.20}$$

Note that if X_L is greater than X_C (as is true in Fig. 21.9), the phase angle is positive ($+\phi$), and the circuit is said to be an **inductive circuit,** since the nonresistive part of the impedance is dominated by the inductor. If X_C is greater than X_L, the phase angle is negative ($-\phi$), and the circuit is said to be a **capacitive circuit** because the capacitor dominates the reactance.

A summary of impedances and phase angles for the three circuit elements and combinations considered is given in Table 21.1.

TABLE 21.1	Impedances and Phase Angles	
Circuit element(s)	*Impedance, Z (in Ω)*	*Phase angle, ϕ*
R	R	$0°$
C	X_C	$-90°$
L	X_L	$+90°$
RC	$\sqrt{R^2 + X_C^2}$	negative
RL	$\sqrt{R^2 + X_L^2}$	positive
RLC	$\sqrt{R^2 + (X_L - X_C)^2}$	positive if $X_L > X_C$ negative if $X_C > X_L$

EXAMPLE 21.6 ■ ALL TOGETHER NOW: IMPEDANCE IN AN RLC CIRCUIT

A series RLC circuit has a resistance of 25.0 Ω, a capacitance of 50.0 μF, and an inductance of 0.300 H. If the circuit is driven by a 120-V, 60-Hz source, what are (a) the impedance of the circuit, (b) the current in the circuit, and (c) the phase angle between the current and the voltage?

Solution.

Given: $R = 25.0 \ \Omega$
$C = 50.0 \ \mu\text{F} = 5.00 \times 10^{-5} \ \text{F}$
$L = 0.300 \ \text{H}$
$V = 120 \ \text{V}$
$f = 60 \ \text{Hz}$

Find: (a) Z (impedance)
(b) I (current)
(c) ϕ (phase angle)

(a) To find the impedance (Eq. 21.18), we first compute the reactances.

$$X_C = \frac{1}{2\pi f C} = \frac{1}{2\pi (60 \ \text{Hz})(5.00 \times 10^{-5} \text{F})} = 53.1 \ \Omega$$

and

$$X_L = 2\pi f L = 2\pi (60 \ \text{Hz})(0.300 \ \text{H}) = 113 \ \Omega$$

Then,

$$Z = \sqrt{R^2 + (X_L - X_C)^2}$$
$$= \sqrt{(25.0 \ \Omega)^2 + (113 \ \Omega - 53.1 \ \Omega)^2} = 64.9 \ \Omega$$

(b) Since $V = IZ$, we have

$$I = \frac{V}{Z} = \frac{120 \ \text{V}}{64.9 \ \Omega} = 1.85 \ \text{A}$$

(c) Solving $\tan \phi = (X_L - X_C)/R$ for the phase angle gives

$$\phi = \tan^{-1}\left(\frac{X_L - X_C}{R}\right) = \tan^{-1}\left(\frac{113 \ \Omega - 53.1 \ \Omega}{25.0 \ \Omega}\right) = +67.3°$$

Can you, once again, verify that the total rms voltage across all three circuit elements is equal to the rms voltage of the source?

Follow-up Exercise.   In an RLC circuit, how would the inductive and capacitive reactances compare if we wanted the current to be its maximum possible value?

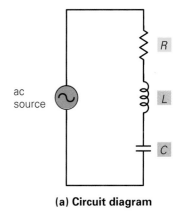

(a) Circuit diagram

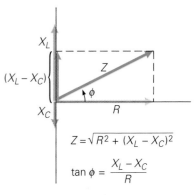

$$Z = \sqrt{R^2 + (X_L - X_C)^2}$$

$$\tan \phi = \frac{X_L - X_C}{R}$$

(b) Phase diagram

•**FIGURE 21.10   A series RLC circuit**
In a series RLC circuit **(a)**, the impedance Z is the phasor sum of the resistance R and the total reactance $(X_L - X_C)$, as shown in **(b)**.

You should now appreciate the importance of phasor diagrams in finding the impedance and the current in ac circuits. However, you may be wondering of what use the phase angle ϕ might be. This is answered by looking at the power loss for our RLC circuit, which illustrates another important benefit derived from using phasor diagrams.

Power Factor

The power, or energy dissipated per unit time, is an interesting feature of an RLC circuit. *There are no power losses associated with pure capacitances and inductances in an ac circuit.* Capacitors and inductors simply store energy and then give it back. Ideally, they have no resistance to produce joule heating. For example, during a half cycle, a capacitor in an ac circuit is charged, and energy is stored in the electric field between the plates. During the next half cycle, the capacitor discharges and returns the charge to the voltage source. Similarly, in an inductor (with negligible resistance in its coil wires), the energy is stored in the magnetic field as-

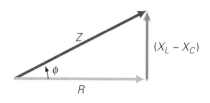

(a) Phasor triangle

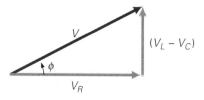

(b) Equivalent voltage triangle

•FIGURE 21.11 Phasor and voltage triangles
Since the voltages across the components of a series RLC circuit are given by $V = IZ$, $V_R = IR$, and $(V_L - V_C) = I(X_L - X_C)$, and the current is the same through all three, the phasor triangle **(a)** may be converted to a voltage triangle **(b)**. Note that $V_R = V \cos \phi$.

sociated with the current flowing through it which builds up, drops to zero, and then reverses during each cycle.*

Thus, *the only element that dissipates energy in an RLC series circuit is the resistive element.* As you learned earlier, the average power dissipated by a resistance R is $P = I^2R$ (where I is the rms value). The power can also be expressed in terms of the current and voltage, but the voltage in this case *must be that across the resistive element* (V_R), since this is the only dissipative element. That is, the power dissipated in a series RLC circuit is

$$P = P_R = IV_R$$

The voltage across the resistive element is conveniently found from a voltage triangle that corresponds to the phasor triangle (•Fig. 21.11). The voltages across the components of an RLC circuit are given by $V = IZ$, $V_R = IR$, $V_L = IX_L$, and $V_C = IX_C$. Combining the last two voltages, we may write $(V_L - V_C) = I(X_L - X_C)$. Hence, if we multiply each of the phasor triangle legs in Fig. 21.11a by the current I, we obtain an equivalent voltage triangle as in Fig. 21.11b. As you can see from this triangle, the voltage across the resistive element (V_R) depends on the phase angle.

$$V_R = V \cos \phi \qquad (21.21)$$

The term $\cos \phi$ is called the **power factor**, and from Fig. 21.11,

$$\cos \phi = \frac{R}{Z} \qquad power\ factor \qquad (21.22)$$

The power in terms of the power factor is

$$P = IV \cos \phi \qquad (21.23)$$

With power dissipated only in the resistance ($P = I^2R$) we can use Eq. 21.22 to express the average power in another form:

$$P = I^2Z \cos \phi \qquad (21.24)$$

Note that $\cos \phi$ varies from a maximum of $+1$ (when $\phi = 0$) to a minimum of 0 (when $\phi = \pi/2$). When $\phi = 0$, the circuit is said to be *completely resistive.* That is, there is maximum power dissipation (as though the circuit contained only a resistor). The power factor decreases as the phase angle increases in either direction [either $+\phi$ or $-\phi$, since $\cos(-\phi) = \cos \phi$], and the circuit becomes more inductive or capacitive. At $\phi = \pm 90°$, the circuit is *completely inductive or capacitive.* For these cases, it is as though the circuit contained only an inductor or a capacitor, respectively, and no power is dissipated ($P = IV \cos 90° = 0$). In actual practice, since there is always some resistance, it is impossible to be completely inductive or capacitive. It is possible, however, for an RLC circuit to be completely resistive even though it may contain both a capacitor and an inductor, as we shall see when we study resonance in Section 21.5.

EXAMPLE 21.7 ■ EXAMPLE 21.6 REVISITED: POWER FACTOR

How much power is dissipated in the circuit described in Example 21.6?

Solution. In Example 21.6, the circuit was found to have an impedance of $Z = 64.9\ \Omega$, and its resistance was $R = 25.0\ \Omega$. The power factor of the circuit is therefore

$$\cos \phi = \frac{R}{Z} = \frac{25.0\ \Omega}{64.9\ \Omega} = 0.385$$

---

*In Chapter 16, you saw that the energy stored in a capacitor was given by $U = \frac{1}{2}CV^2$. Similarly, it can be shown that the energy stored in an inductor is given by $U = \frac{1}{2}LI^2$, where L is the inductance and I is the current in the inductor.

$$P = IV \cos \phi = (1.85 \text{ A})(120 \text{ V})(0.385) = 85.5 \text{ W}$$

Note that this is quite different from the power that would be dissipated in a circuit without a capacitor and inductor—that is, with only a resistive element—or when the capacitive and inductive reactances cancel each other out and $\cos \phi = R/R = 1$.

Follow-up Exercise. In this Example, (a) what frequency *would* make the circuit completely resistive, and (b) what would be the power dissipated in the resistor then?

21.5 Circuit Resonance

Objectives: To be able to (a) understand the concept of resonance in ac circuits, and (b) calculate the resonance frequency of an RLC circuit.

When the power factor of an RLC circuit is equal to 1 ($\cos \phi = 1$), there is maximum power dissipation. For a particular voltage, the current in the circuit is then a maximum, since the impedance is a minimum. This occurs when the inductive and capacitive reactances effectively cancel each other. This can be made to happen in any RLC circuit by choosing the appropriate driving frequency.

The key to finding the correct frequency is to realize that the inductive and capacitive reactances are both frequency-dependent, so the impedance also depends on the frequency. From the expression for the impedance of an RLC circuit, $Z = \sqrt{R^2 + (X_L - X_C)^2}$, we can see that the impedance is a minimum when the inductive and capacitive reactances are equal in magnitude. This will occur at a particular value of the frequency, which we will designate f_o. We can find f_o by setting $X_L - X_C = 0$ or, using the expressions for the two reactances,

$$2\pi f_o L = \frac{1}{2\pi f_o C}$$

We can then solve this equation for the frequency:

$$f_o = \frac{1}{2\pi\sqrt{LC}} \qquad (21.25) \qquad \textbf{Resonance frequency}$$

Eq. 21.25 gives the frequency for the condition of minimum impedance, which is called the **resonance frequency**. A graph of reactance versus frequency is shown in •Fig. 21.12a. The X_C and X_L curves intersect at f_o where their values are equal ($X_L = X_C$).

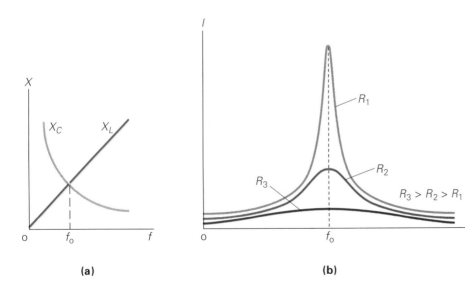

•FIGURE 21.12 Resonance frequency
(a) At the resonance frequency (f_o), the reactances are equal ($X_L = X_C$). On a graph of X versus f, this is the frequency at which the X_C and X_L curves intersect. (b) On a graph of I versus f, the current is a maximum at f_o. The curve becomes sharper and narrower as the resistance decreases.

Oscillator Circuits: Broadcasters of Electromagnetic Radiation

To generate the high-frequency electromagnetic waves used in radio communications and television (Fig. 1), electrons must be made to oscillate with high frequencies in electronic circuits. This may be accomplished through the use of resonant RLC circuits. Such circuits are called *oscillator circuits*, because they "oscillate" at resonant frequencies determined by their inductive and capacitive elements.

When the resistance R in an RLC circuit is very small, the circuit is essentially an LC circuit. Energy in such a circuit oscillates back and forth between the inductor and capacitor at a frequency f. This is the natural frequency of the circuit, and

is identical to its resonant frequency (Eq. 21.25). Some energy is dissipated by the small resistance. However, in an ideal LC circuit with no ohmic resistance, the oscillation would continue indefinitely.

To help understand this energy oscillation, consider an ideal LC parallel circuit as shown in Fig. 2a. (The resonance frequency for a parallel LC circuit is given by the same expression as for a series circuit.) Let's assume that the capacitor is initially charged and the circuit switch is then closed. The following sequence of events then occurs:

FIGURE 1 Broadcast antennas
These antennas are located atop the World Trade Center in New York City.

1. The capacitor would discharge instantaneously (since $RC = 0$) were it not for the necessity of the discharge current having to pass through the coil. When the switch is closed (at $t = 0$), the current through the coil is zero (Fig. 2a). As the current builds up, so does the magnetic field in the coil. By Lenz's law, the increasing magnetic field and change of flux through the coil induces an emf in such a direction as to oppose the current increase. Because of this opposition, the capacitor takes some time to discharge.

2. When the capacitor is (instantaneously) in a fully discharged state (Fig. 2b), the energy initially stored in it has been given to the inductor, where it was used to create a magnetic field. Recall that since $R = 0$, no energy has been lost because of joule heating. At this time (one-quarter of a cycle or period), the magnetic field in the coil has its maximum value and thus the current passing through the coil has its largest magnitude.

A Physical Explanation of Resonance

The physical explanation of resonance in an RLC circuit is worth exploring. First, recall that in a series circuit the current is the same at all times in all locations. Focusing on the resistor, we know that the resistor voltage is in phase with the current, so let us reference our phase relationships to the voltage across the resistor. For the capacitor, the voltage lags the current, and thus the voltage across the resistor, by ninety degrees. For the inductor, the voltage leads the current, and thus the voltage across the resistor, by ninety degrees. This means that *the capacitor and inductor voltages are always 180° out of phase, or of opposite polarity.* (This, of course, is embodied graphically in the phase diagram.)

Normally, the capacitive and inductive voltages are not of the same magnitude, and so do *not* cancel completely. Thus, we would generally find the voltage across the resistor to be less than the voltage of the source, and the power dissipated in the resistor (joule heating) to be less than the possible maximum. However, if the capacitive and inductive voltages do cancel exactly, then the full source voltage appears across the resistor, the power factor is 1, and we have maximum power output, or resonance conditions. Matching the two voltages means that $V_C + V_L = 0$, or $V_C = -V_L$, or $|V_C| = |V_L|$. But we know that $V_C = I_C X_C$ and $V_L = I_L X_L$, and in series circuits $I_C = I_L$. So resonance really is just the matching of the two impedances to make the voltages cancel: $X_L = X_C$. Since capacitive and in-

3. As the magnetic field collapses from its maximum value, an emf that tends to oppose the collapse is induced in the coil. This emf is in a direction that would tend to continue the current in the coil even as the current is decreasing (Lenz's law again). The polarity of the emf is now opposite what it was in step 1. Thus, current transports charge to the capacitor plates and charges the capacitor with a polarity the reverse of what it had initially.

4. When the capacitor is fully charged (with reverse polarity), it has the same energy that it had initially (Fig. 2c). This occurs halfway through the cycle, or half a period from the start. The magnetic field in the coil is now zero, as is the current in the circuit.

5. The capacitor then begins to discharge and the previous four steps are repeated, setting up an "oscillation" in the circuit.

For an ideal case, these oscillations would continue indefinitely. The oscillation of energy between a capacitor and inductor is analogous to the exchange of kinetic and potential energies for a mass oscillating on an ideal spring. (See the bar graphs next to the circuits in Fig. 1.)

In actual operation, a small ac voltage of the proper frequency applied to the circuit is sufficient to maintain strong oscillations between the electric field of the capacitor and the magnetic field of the inductor. A parallel LC circuit is sometimes called a tank circuit, which suggests its energy storage capability. Without voltage input, however, the oscillator would run down as a result of IR losses and radiation losses (as its mechanical counterpart would be slowed down by frictional losses).

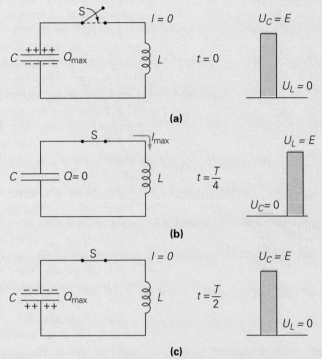

FIGURE 2 An oscillating LC circuit
If the resistance is negligible, this circuit will oscillate as described in the text. Half a cycle is shown. Essentially, energy is transferred back and forth between magnetic and electric potential energy. The oscillating electrons in the wire will give off electromagnetic radiation at the circuit's oscillating frequency, $f_o = 1/2\pi\sqrt{LC}$.

ductive reactances, unlike resistance, depend on the frequency, this requirement can be met by selecting the frequency at which this happens:

$$X_L = X_C$$

$$2\pi f_o L = \frac{1}{2\pi f_o C}$$

This is obviously the same result as Eq. 21.25.

CONCEPTUAL EXAMPLE 21.8 ■ IS RESONANCE POSSIBLE? RL AND RC CIRCUITS

Is it possible for resonance to occur in an RL or RC circuit? *Clearly establish the reasoning and physical principle(s) used in determining your answer before checking it below. That is, how did you arrive at your answer?*

Reasoning and Answer: With only one reactive element (capacitor *or* inductor) in each circuit, there is no possibility of two reactance voltages exactly cancelling, as we have seen can happen in an RLC circuit. Hence no resonance is possible.

Follow-up Exercise. What happens to the power dissipated in (a) an RL circuit, and (b) an RC circuit, as the frequency of the voltage source increases? Explain.

Resonance Applications

When a series RLC circuit is driven in resonance (at its resonance frequency), the impedance is a minimum, and the power transfer from the source to the circuit is a maximum. Since $X_L = X_C$, then $Z_{min} = R$, and the impedance is completely resistive. In terms of the power factor, $\cos \phi = R/Z = R/R = 1$; thus the power factor is a maximum and there is maximum power dissipation in the circuit.

A graph of current versus frequency is shown in Fig. 21.12b for several different resistances in an RLC circuit with fixed capacitance and inductance. Notice that the currents have maximum values at f_0. Also, notice how the curve becomes sharper and narrower as the resistance decreases.

Resonant circuits have a wide variety of applications, for example, in the tuning mechanism of a radio. Each radio station has an assigned broadcast frequency at which radio waves are transmitted. The oscillating electric and magnetic fields of the radio waves set electrons in the antenna into regular back-and-forth motion—that is, they produce an alternating current. (The process is the reverse of the production of electromagnetic waves by oscillating electrons in a transmitter.) Signals broadcast by many different stations generally reach a radio, but the receiver circuit selectively picks up only the one with a frequency at or near its resonant frequency. Most radios allow you to alter this resonant frequency to "tune in" different stations. Variable air capacitors in the tuning circuit have long been used for this purpose (Fig. 21.13). More compact variable capacitors in some smaller radios employ a polymer dielectric between thin plates. The polymer sheets help maintain the plate separation and increase the capacitance, allowing makers to use plates of smaller area. (Recall that $C = K\varepsilon_o A/d$.) In other radios, variable capacitors are replaced by solid-state devices.

Note: See Section 20.4, and the Insight on pp. 670–671.

Note: See Eq 16.9, Section 16.3.

CONCEPTUAL EXAMPLE 21.9 ■ AM vs FM: RESONANCE IN RADIO RECEPTION

When you switch from the AM band to the FM band in the design in some radios, you are effectively changing the capacitance of the receiving circuit. Is the capacitance (a) increased or (b) decreased when you make this change? *Clearly establish the reasoning and physical principle(s) used in determining your answer before checking it below. That is, why did you select your answer?*

Reasoning and Answer: Since FM stations broadcast at significantly higher frequencies than those of the AM band (see Table 20.1), we must increase the resonant frequency of the radio receiver to be able to receive a significant amount of power from them. To increase the resonant frequency of the receiver requires either reducing the inductance or the capacitance. Since we are told the capacitance is changed, we obviously must be switching to a lower capacitance than that used to tune in AM stations. Thus the answer is (b).

Follow-up Exercise. Based on the resonance curves shown in Fig. 21.12, can you explain how it might be possible to pick up two radio stations *simultaneously* on your radio? (You may have encountered this phenomenon, particularly between two large cities. Two stations are sometimes granted broadcasting licenses on closely spaced frequencies under the assumption that they are far enough apart so as never to be received by the same receiver. However, under certain atmospheric conditions, this may not be true.)

•FIGURE 21.13 Variable air capacitor
Rotating the movable plates between the fixed plates changes the overlap area and thus the capacitance. Such capacitors were common in tuning circuits in older radios.

EXAMPLE 21.10 ■ THAT CERTAIN FREQUENCY: RESONANCE IN AN RLC CIRCUIT

A series RLC circuit has a 50.0-Ω resistor, a 6.00 pF capacitor, and a 28.0-mH inductor. The circuit is connected to a wide-range, adjustable-frequency source with an output of 25.0 V. (a) What is the resonance frequency of the circuit? (b) How much current is in the circuit when it is in resonance? (c) What is the voltage across each of the circuit elements for this condition?

Solution. We list the values given:

Given: $R = 50.0 \ \Omega$
$C = 6.00 \ \text{pF} = 6.00 \times 10^{-9} \ \text{F}$
$L = 28.0 \ \text{mH} = 28.0 \times 10^{-3} \ \text{H}$
$V = 25.0 \ \text{V}$

Find: (a) f_o (resonance frequency)
(b) I (current)
(c) $V_R, \ V_C, \ V_L$ (voltages)

(a) By Eq. 21.25, we have for the resonance frequency

$$f_o = \frac{1}{2\pi\sqrt{LC}} = \frac{1}{2\pi\sqrt{(28.0 \times 10^{-3}\text{H})(6.00 \times 10^{-9}\text{F})}}$$

$$= 12.3 \times 10^3 \ \text{Hz} = 12.3 \ \text{kHz}$$

(b) At the resonance frequency, $X_L = X_C$, and $Z = R = 50.0 \ \Omega$. The current in the circuit is then

$$I = \frac{V}{Z} = \frac{V}{R} = \frac{25.0 \ \text{V}}{50.0 \ \Omega} = 0.500 \ \text{A}$$

(c) The reactances at f_o are

$$X_L = 2\pi f_o L = 2\pi(12.3 \times 10^3 \ \text{Hz})(28.0 \times 10^{-3} \ \text{H}) = 2.16 \times 10^3 \ \Omega = 2.16 \ \text{k}\Omega$$

and $X_C = X_L = 2.16 \ \text{k}\Omega$ at f_o. Then, using the relationship between voltage and current,

$$V_R = IR = (0.500 \ \text{A})(50.0 \ \Omega) = 25.0 \ \text{V}$$
$$V_L = IX_L = (0.500 \ \text{A})(2.16 \times 10^3 \ \Omega) = 1.08 \times 10^3 \ \text{V}$$
$$V_C = IX_C = (0.500 \ \text{A})(2.16 \times 10^3 \ \Omega) = 1.08 \times 10^3 \ \text{V}$$

Follow-up Exercise. In this Example, note that the voltages across the capacitor and inductor are much higher than the voltage source. (a) Can you explain where the additional voltage comes from? (b) Kirchhoff's loop theorem appears to be violated. Explain why this is not really the case.

Chapter Review

Important Terms

Important Concepts

- In an ac circuit, voltage and current vary sinusoidally with time. The rms (root mean square), or effective voltage and current are less than the peak, or maximum value that these quantities attain during a cycle of oscillation.

- Capacitors and inductors do not eliminate the current in an ac circuit but offer some opposition to it, termed capacitive and inductive reactance, respectively. Both of these quantities vary with the frequency of the circuit, but in opposite ways.

- In a purely capacitive ac circuit, the current leads the voltage by 90°. In a purely inductive circuit, the voltage leads the current by 90°.

- Impedance is the generalization of resistance to include not only resistance but also inductive and capacitive reactances.

- Phasors are vectorlike quantities that allow the voltages and impedances of capacitors, inductors, and resistors to be graphically represented under ac conditions.

- The power factor in an RLC circuit is a measure of how close the circuit is to expending the maximum power.
- The resonance frequency of an RLC circuit is the frequency at which the circuit dissipates maximum power. It is the frequency at which the inductive and capacitive reactances cancel one another completely.

Important Equations

Instantaneous Voltage in an ac Circuit:
$$V = V_o \sin 2\pi f t \qquad (21.1)$$

Instantaneous Current in an ac Circuit (resistivity):
(where $I_o = V_o / R$):
$$I = I_o \sin 2\pi f t \qquad (21.2)$$

Rms (or Effective) Current:
$$I_{rms} = \frac{I_o}{\sqrt{2}} = 0.707 \, I_o \qquad (21.6)$$

Rms (or Effective) Voltage:
$$V_{rms} = \frac{V_o}{\sqrt{2}} = 0.707 \, V_o \qquad (21.8)$$

Relationship between Voltage and Current in an ac Circuit (resistor only):
$$V_{rms} = I_{rms} R \qquad (21.9)$$

Effective Power in an ac Circuit:
$$\overline{P} = I_{rms}^2 R \qquad (21.7)$$

Capacitive Reactance and Relationship between Voltage and Current (capacitor only):
$$X_C = \frac{1}{2\pi f C} = \frac{1}{\omega C} \quad \text{and} \quad V = IX_C \quad (21.10\text{–}11)$$

Inductive Reactance and Relationship between Voltage and Current (inductor only):
$$X_L = 2\pi f L = \omega L \quad \text{and} \quad V = IX_L \quad (21.13\text{–}14)$$

Impedance for an RLC Circuit:
$$Z = \sqrt{R^2 + (X_L - X_C)^2} = \sqrt{R^2 + \left(2\pi f L - \frac{1}{2\pi f C}\right)^2} \quad (21.18)$$

Phase Angle Between Voltage and Current in an ac Circuit:
$$\tan \phi = \frac{X_L - X_C}{R} \qquad (21.20)$$

Relationship between rms Current and Voltage for RC, RL, and RLC Circuits:
$$V = IZ \qquad (21.19)$$

Power Factor for an RLC Circuit:
$$\cos \phi = \frac{R}{Z} \qquad (21.22)$$

Power in Terms of Power Factor:
$$P = IV \cos \phi \qquad (21.23)$$
or
$$P = I^2 Z \cos \phi \qquad (21.24)$$

Resonance Frequency for an RLC Circuit:
$$f_o = \frac{1}{2\pi \sqrt{LC}} \qquad (21.25)$$

Exercises

21.1 Resistance in an AC Circuit

1 Which of the following voltages has the greatest magnitude for a sinusoidal ac voltage: (a) V_o, (b) V_{rms}, (c) V_{p-p}?

2 The ac ampere is defined in terms of (a) Ohm's law, (b) the force between parallel, current-carrying wires, (c) joule heating, (d) the frequency of the ac current.

3 For the circuit in Fig. 21.1, why is the magnitude of V_o greater than that of I_o? Could it be the other way, with the magnitude of I_o greater than that of V_o? Explain.

4 If the average current in an ac circuit is zero, why isn't the average power zero?

5 ■ What are the peak voltages of a 120-VAC line and a 240-VAC line?

6 ■ An ac circuit has an rms current of 3.0 A. What is the peak current?

7 ■ The maximum (magnitude) potential difference across a resistor in an ac circuit is 156 V. Find the effective voltage.

8 ■ Based on the definition of the ac ampere, how much ac current must flow through a 6.0-Ω resistor to produce 15 J/s of joule heat?

9 ■ An ac circuit with a resistance of 5.0 Ω has an effective current of 0.75 A. (a) Find the rms voltage and peak voltage. (b) Find the average power dissipated by the resistance.

10 ■ A hair dryer rated at 1200 W is plugged into a 120-V outlet. (a) Find the effective current. (b) Find the peak current. (c) Find the resistance of the dryer.

11 ■■ The voltage across a resistor varies according to the following equation:

$$V = (120\ \text{V})(\sin 120\ t)$$

What are (a) the effective voltage across the resistor and (b) the period of the voltage cycle?

12 ■■ An ac voltage is applied to a 25-Ω resistor, and it dissipates 500 W of power. Find (a) the effective and peak currents and (b) the effective and peak voltages.

13 ■■ The current through a 60-Ω resistor is given by this equation:

$$I = (2.0\ \text{A}) \sin 380\ t$$

where t is in seconds as usual. (a) What is the frequency of the current? (b) What is the effective current? (c) How much power is dissipated by the resistor? (d) Write an equation expressing the voltage as a function of time. (e) Write an equation expressing the power as a function of time. (f) Show that the rms power obtained from (e) is the same as your answer to (c).

14 ■■ An ac voltage has an amplitude of 85 V and a frequency of 60 Hz. (a) If $V = 0$ at $t = 0$, what is the instantaneous voltage at $t = 2.0$ s? (b) What is the effective voltage?

15 ■■ An ac voltage has an rms value of 120 V. The voltage goes from zero to its maximum value in 1.5 ms. Write an expression for the voltage as a function of time.

16 ■■ A circuit has a current (in amps) given by $I = 8.0 \sin 40\pi t$ with an applied voltage (in volts) of $V = 60 \sin 40\pi t$. What is the average power delivered to the circuit?

17 ■■ The output of an ac generator has a negative peak voltage at $t = 0.075$ s and then the next positive peak voltage at $t = 0.125$ s. What is the operational frequency of the generator?

18 ■■ Find the effective and peak currents through a 40-W, 120-V light bulb.

19 ■■ A 50-kW heater is connected to a 240-V ac source. Find (a) the peak current and (b) the peak voltage.

20 ■■ The current and voltage outputs of an ac generator have peak values of 1.5 A and 12 V. What is the effective power output of the generator?

21 ■■■ A coil of 200 loops of wire with a radius of 10 cm is used as the armature of a simple generator with a magnetic field of 0.80 T. If the coil has a resistance of 2.0 Ω, how fast (angular speed in rad/s) must it be rotated to have an effective induced current of 2.7 A?

22 ■■■ Rotating coils are often used to measure magnetic fields. A coil of 40 loops with a radius of 7.5 cm rotates at 60 Hz about an axis along a diameter perpendicular to a uniform magnetic field. An *effective* voltage of 10 V,

as read from an ac voltmeter, is induced in the coil. What is the magnitude of the magnetic field?

21.2 Capacitive Reactance

21.3 Inductive Reactance

23 In a purely capacitive ac circuit, (a) the current and voltage are in phase, (b) the current leads the voltage, (c) the voltage lags the current.

24 The unit of capacitive reactance is (a) F, (b) (F·s)$^{-1}$, (c) Ω, (d) Ω·m.

25 In a purely inductive ac circuit, (a) the current leads the voltage, (b) the current and voltage are in phase, (c) the inductance is zero, (d) none of these.

26 The unit of inductance is (a) H/s, (b) H, (c) Ω, (d) Ω·s.

27 In an ac circuit, what can oppose current other than ohmic resistance and why?

28 Analyze the graph in Fig. 21.6b and explain why the current is zero when the voltage is a maximum, and vice versa. [*Hint:* $\Delta I/\Delta t$ is the slope of the curve for the current.]

29 ■ What capacitance is needed in a 60-Hz ac circuit to have a reactance of 45 Ω?

30 ■ Find the frequency at which a 25-μF capacitor will have a reactance of 25 Ω.

31 ■ A 2.0-μF capacitor is connected across a 60-Hz voltage source and a current of 2.0 mA is measured on an ammeter. What is the capacitive reactance of the circuit?

32 ■ A 50-mH inductor is connected in a circuit with a 120-V, 60-Hz source. (a) What is the inductive reactance of the circuit? (b) How much current is there in the circuit? (c) What is the phase angle between the current and the supplied voltage? (Assume negligible resistance.)

33 ■ How much current is delivered to a circuit with a 50-μF capacitor if it is connected to an ac generator with a 90-V, 60-Hz output?

34 ■ In Exercise 33, how much power is delivered to the circuit?

35 ■■ A variable capacitor in a circuit with a 120-V, 60-Hz source is initially at a value of 0.25 μF and is then increased to 0.40 μF. What is the percentage change in the current in the circuit?

36 ■■ An inductor has a reactance of 90 Ω in a 60-Hz ac circuit. What is its inductance?

37 ■■ Find the frequency at which a 450-mH inductor will have a reactance of 400 Ω.

38 ■■ With a 150-mH inductor in a circuit with a 60-Hz voltage source, a current of 1.6 A is measured on an ammeter. (a) What is the voltage of the source? (b) What is the phase angle between the current and that voltage?

39 ■■ What inductance will have the same reactance in a 120-V, 60-Hz circuit as a 10-μF capacitor does?

40 ■■ A circuit with a single inductor is connected to a 120-V, 60-Hz source. What is the value of the inductance if there is a current of 10 A in the circuit?

21.4 Impedance: RLC Circuits

21.5 Circuit Resonance

41 The impedance of an RLC circuit depends on (a) frequency, (b) inductance, (c) capacitance, (d) all of these.

42 If the capacitance is decreased in a series RLC circuit, (a) the capacitive reactance increases, (b) the inductive reactance increases, (c) the current remains constant, (d) the power factor stays constant.

43 Which of the following has a resonance frequency: (a) RC circuit, (b) RL circuit, (c) RLC circuit, (d) all of these?

44 When a series RLC circuit is driven at its resonance frequency, (a) energy is dissipated by only the resistive element, (b) the power factor is one, (c) there is maximum energy transfer to the circuit, (d) all of these.

45 If the resistance in an RLC circuit were zero and the circuit were driven with a variable frequency voltage source, what would a graph of current versus frequency look like? What would this mean physically?

46 ■■ A series RC circuit has a resistance of 200 Ω and a capacitance of 25 μF and is driven by a 120-V, 60-Hz source. (a) Find the capacitive reactance of the circuit. (b) How much current is drawn from the source?

47 ■■ To increase the impedance in the RC circuit in Exercise 46 by 20% while maintaining the 120-V source voltage, what would the required source frequency be?

48 ■■ In a series RC circuit with a 100-Ω resistor, what capacitance will give a current of 0.50 A if the circuit is driven by a 120-V, 60-Hz source?

49 ■■ What is the phase angle between the current and the source voltage in Exercise 48?

50 ■■ A coil in a 60-Hz circuit has a resistance of 100 Ω and an inductance of 0.45 H. Calculate (a) the coil's reactance and (b) the circuit's impedance.

51 ■■ A series RLC circuit has a resistance of 25 Ω, an inductance of 0.30 H, and a capacitance of 8.0 μF. At what frequency should the circuit be driven to have the maximum power transferred from the driving source?

52 ■■ In a series RLC circuit, for a particular driving frequency, $R = X_C = X_L = 40$ Ω. If the driving frequency is doubled, what will the impedance of the circuit be?

53 ■■ A resistor, an inductor, and a capacitor have values of 500 Ω, 500 mH, and 3.5 μF, respectively, and are connected in series to a 240-V, 60-Hz power supply. Is the circuit driven in resonance? If not, what values of resistance and inductance would be required for this to occur?

54 ■■ How much power is dissipated in the circuit described in Exercise 53?

55 ■■ Referring to •Fig. 21.14, find the currents supplied by the ac source for all possible connections.

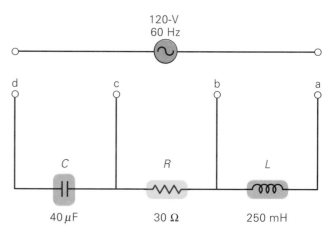

•**FIGURE 21.14 A series RLC circuit**
See Exercise 55.

56 ■■ A tuning circuit in a radio receiver has a fixed 0.50-mH inductance and a variable capacitor. If the circuit is tuned to a radio station broadcasting at 980 kHz on the AM dial, what is the capacitance of the capacitor?

57 ■■ What would be the range of the variable capacitor in Exercise 56 for tuning over the complete AM band? [*Hint:* See Table 20.1.]

58 ■■ A coil with a resistance of 30 Ω and an inductance of 0.15 H is connected to a 120-V, 60-Hz source. (a) How much current is in the circuit? (b) What is the phase angle between the current and the source voltage?

59 ■■ A series RLC circuit has a resistance of 25.0 Ω, an inductance of 0.450 H, and a capacitance of 5.00 μF. (a) What is the impedance with a driving frequency of 60 Hz? (b) Is the circuit driven in resonance? If not, find the resonance frequency of the circuit.

60 ■■ A circuit connected to a 220-V, 60-Hz power supply has the following components connected in series: a 10-Ω resistor, a coil with an inductive reactance of 120 Ω, and a capacitor with a reactance of 120 Ω. Compute the magnitude of the voltage across (a) the resistor, (b) the inductor, and (c) the capacitor.

61 ■■■ A series RLC circuit has a 25-Ω resistor, a 0.80-μF capacitor, and a 250-mH inductor. The circuit is connected to a variable-frequency source with a fixed rms output of 12 V. If the supplied frequency is set at the circuit's resonance frequency, what is the magnitude of the rms voltage across each of the circuit elements?

62 ■■■ An RL circuit operates on 120 V at 60 Hz and contains a 60.0-Ω resistor. If the wattage rating of the resistor is 10.0 W, what inductance should be used in the circuit so power is expended in the resistor at that rated limit?

63 ■■■ A series RLC circuit with a resistance of 400 Ω has capacitive and inductive reactances of 300 Ω and 500 Ω, respectively. (a) What is the power factor of the circuit? (b) If the circuit operates at 60 Hz, what additional capacitance should be connected with the original capacitance to give a power factor of one, and how should the capacitances be connected?

64 ■■■ A machine operates on 240 V at 60 Hz. In operation, it uses 1100 W of power, and the power factor is 0.75. Find the effective current in the circuit containing the machine.

65 ■■■ The current in a series RLC circuit is 4.0 A when it is connected to a 120-V, 60-Hz source. If the current leads the voltage by 37°, (a) what is the average power dissipated in the circuit? (b) What is the resistance of the circuit? (c) What is the value of $X_L - X_C$?

66 ■■■ A series RLC circuit has components with these values: $R = 50$ Ω, $L = 0.15$ H, and $C = 20$ μF. The circuit is driven by a 120-V, 60-Hz source. What is the power loss of the circuit as a percentage of its power loss when in resonance?

67 ■■■ If the values of all three circuit components in Exercise 66 were halved, what would be the change in the power loss?

Additional Exercises

68 A 60.0-W light bulb operates on 120-V household voltage. What are (a) the peak voltage across the bulb, and (b) the peak current in the circuit?

69 The rms current in an ac circuit is 0.25 A. What is the maximum instantaneous current?

70 A 120-V, 60-Hz source is connected across a 0.40-H inductor. (a) Find the current through the inductor. (b) Find the phase angle between the current and the supplied voltage.

71 A radio receiver circuit containing an inductor of 1.50 μH is tuned to an FM station at 98.9 MHz by adjusting a variable capacitor. What is the capacitance of the capacitor when the circuit is tuned to the station?

72 A series RL circuit has a resistance of 50 Ω and an inductance of 250 mH. The circuit is driven by a 120-V, 60-Hz source. (a) What is the inductive reactance of the circuit? (b) How much power is dissipated by the circuit?

73 Prove that $\langle \sin^2 \omega t \rangle = \frac{1}{2}$.

74 A circuit connected to a 110-V, 60-Hz source contains a 50-Ω resistor and a coil with an inductance of 100 mH. Find (a) the reactance of the coil, (b) the impedance of the circuit, (c) the current in the circuit, and (d) the power dissipated by the *coil*.

75 Calculate the phase angle between the current and the supplied voltage in Exercise 74.

76 A 1.0-μF capacitor is connected to a 120-V, 60-Hz source. (a) What is the capacitive reactance of the circuit? (b) How much current is there in the circuit? (c) What is the phase angle between the current and the supplied voltage?

77 The circuit in ●Fig. 21.15a is called a low-pass filter because a large current is delivered to the load (R_L) only by a low-frequency source. The circuit in ●Fig. 21.15b, on the other hand, is called a high-pass filter because a large current is delivered to the load only by a high-frequency source. Show mathematically why the circuits have these characteristics.

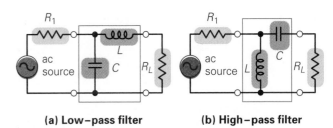

(a) Low–pass filter   (b) High–pass filter

●**FIGURE 21.15   Low-pass and high-pass filters**
See Exercise 77.

78 In a series RLC circuit with an inductance of 750 mH, what values of resistance and capacitance would give the circuit a resonance frequency of 60 Hz?

22 Geometrical Optics: Reflection and Refraction of Light

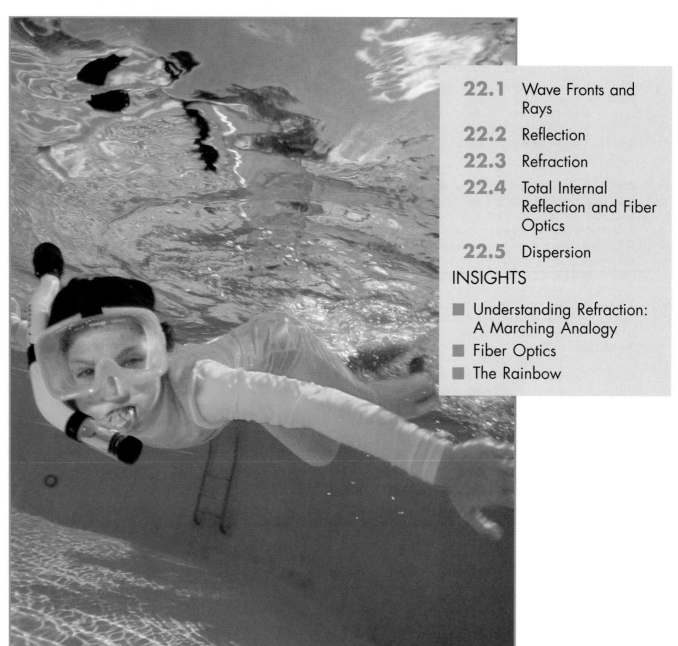

We live in a visual world, surrounded at every moment by images. How these images are created is something we take largely for granted—until we see something that we can't easily account for, as in the photo above. Nevertheless, although things are not always what they seem, simple physical principles can always explain why we see what we see. With their help, we can understand many intriguing everyday experiences: why a glass prism spreads out light into a spectrum of colors, the cause of the "wet spot" mirage seen on roads in the summer, how rainbows are created, why the legs of a person standing in a lake or swimming pool seem to lose a good deal of their length—and why the underwater swimmer in the photo is being shadowed by a phantom double.

Optics is the study of light and vision. Human vision, of course, requires light, specifically what is called visible light. The broad definition of light is any radiation in or near the visible region of the electromagnetic spectrum. This definition includes infrared and ultraviolet radiations. Similar optical properties are shared by all of these electromagnetic waves.

In this chapter we will investigate the basic optical phenomena of reflection and refraction. The principles that govern reflection explain the behavior of mirrors, while those that govern refraction explain the properties of lenses, such as those found in eyeglasses. With the aid of these and other optical principles, we can understand many optical phenomena that we experience every day. We will also explore some less familiar territory, including the fascinating field of fiber optics.

You may have noted that the title of the chapter is geometrical optics. What this means is that a simple geometrical approach, using straight lines and angles, can be used to investigate many aspects of reflection and refraction. For these purposes we do not need to be concerned with the physical nature of electromagnetic waves. The principles of geometrical optics will be introduced here and applied in greater detail in the study of mirrors and lenses in Chapter 23.

22.1 Wave Fronts and Rays

Objective: To be able to define and explain the concepts of wave fronts and rays.

Waves, electromagnetic or other, are conveniently described using wave fronts. A **wave front** is the line or surface defined by adjacent portions of a wave that are in phase. For example, if an arc is drawn along one of the crests of a circular water wave moving out from a point source, all the particles on the line will be in phase (•Fig. 22.1a). A line along a wave trough would work equally well (and would also form a circle). For a three-dimensional spherical wave, such as sound or light emitted from a point source, the wave front is a spherical surface rather than a circle.

At a distance far from the source, the curvature of a short segment of a circular or spherical wave front is small. Such a segment may be approximated as a linear wave front (in two dimensions) or a **plane wave front** (in three dimensions), just as we take the surface of the Earth to be locally flat (•Fig. 22.1b). A plane wave front may also be produced directly by a linear, elongated source such as a long bulb filament. In a uniform medium, wave fronts propagate outward from the source at a wave speed characteristic of the medium. For light in a vacuum, this speed is $c = 3.00 \times 10^8$ m/s.

The geometrical description of a wave using wave fronts tends to neglect the fact that the wave is actually sinusoidal, like those studied in Chapter 13. This simplification is carried a step further in the concept of a ray. As illustrated in Fig. 22.1, a line drawn perpendicular to a series of fronts and pointing in the direction of propagation is called a **ray**. Note that a ray is in the direction of the energy flow of a wave. A plane wave front is assumed to travel in a straight line in a medium in the direction of its rays. A beam of light may be represented by a group of parallel rays or simply as a single ray (•Fig. 22.2). The representation of light as rays is adequate and convenient for describing many optical phenomena.

The use of the geometrical representations of wave fronts and rays to explain phenomena such as the reflection and refraction of light is called **geometrical optics**. However, certain other phenomena, such as the interference of light, cannot be treated in this manner and must be explained in terms of actual wave characteristics. These phenomena will be considered in Chapter 24.

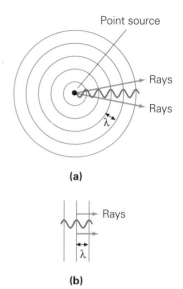

•**FIGURE 22.1 Wave fronts and rays**
A wave front is defined by adjacent points on a wave that are in phase, such as those along wave crests or troughs. **(a)** Near a point source, the wave fronts are circular (two-dimensional) or spherical (three-dimensional). **(b)** Far from a point source, the wave fronts are approximately linear or planar. A line perpendicular to a wave front in the direction of the wave's propagation is called a ray.

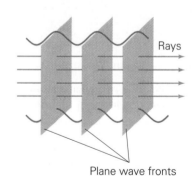

•**FIGURE 22.2 Light rays**
A plane wave front travels in a straight line in the direction of its rays. A beam of light may be represented by a group of parallel rays (or a single ray).

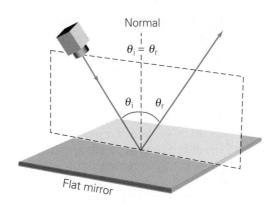

Normal

$\theta_i = \theta_r$

θ_i θ_r

Flat mirror

FIGURE 22.3 The law of reflection
According to the law of reflection, the angle of incidence (θ_i) is equal to the angle of reflection (θ_r). Note that the angles are measured relative to a normal (a line perpendicular to the reflecting surface). The normal and the incident and reflected rays lie all in the same plane.

22.2 Reflection

Objectives: To be able to (a) explain the law of reflection, and (b) distinguish between regular (specular) and irregular (diffuse) reflections.

The reflection of light is an optical phenomenon of enormous importance—if light were not reflected to our eyes by objects around us, we wouldn't see them at all. **Reflection** involves the absorption and re-emission of light by means of complex electromagnetic vibrations in the atoms of the reflecting medium. However, the phenomenon is easily described using rays.

A light ray incident on a surface is described by an **angle of incidence** (θ_i). This is measured relative to a line perpendicular, or normal, to the reflecting surface, a line commonly referred to as a *normal* (see •Fig. 22.3). Similarly, the reflected ray is described by an **angle of reflection** (θ_r). The relationship between these angles is given by the **law of reflection**:

The angle of incidence is equal to the angle of reflection:

$$\theta_i = \theta_r \qquad (22.1)$$

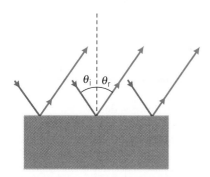

θ_i θ_r

(a) Regular, or specular reflection

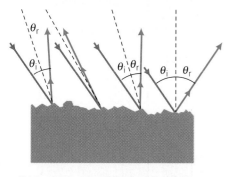

θ_r

θ_i θ_i θ_r θ_i θ_r

(b) Irregular, or diffuse reflection

FIGURE 22.4 Regular (specular) reflection and irregular (diffuse) reflection
(a) When a light beam is reflected from a smooth surface, and the reflected rays are parallel, the reflection is said to be regular or specular. **(b)** Reflected rays from a relatively rough surface, such as this page, are not parallel, and the reflection is said to be irregular or diffuse. (Note that the law of reflection applies to each individual ray.)

Another attribute of reflection is that the incident ray, the reflected ray, and the normal all lie in the same plane, which is sometimes called the plane of incidence.

When the reflecting surface is smooth, the reflected rays of a beam of light are parallel (•Fig. 22.4a). This is called **regular**, or **specular**, **reflection**. Reflection from a flat mirror is regular. If the reflecting surface is rough, on the other hand, the light rays are reflected in nonparallel directions because of the irregular nature of the surface (•Fig. 22.4b). This is termed **irregular**, or **diffuse**, **reflection**. The reflection of light from this page is an example of diffuse reflection.

Note in Fig. 22.4 that the law of reflection applies to both regular and diffuse reflection. However, the type of reflection involved determines whether we see images from a reflecting surface. In regular reflection, the reflected, parallel rays produce an image (•Fig. 22.5a). Diffuse reflection does not produce an image because the light is reflected in various directions (•Fig. 22.5b).

Experience with friction and direct investigations show that all surfaces are rough on a microscopic scale. What then determines whether reflection is specular or diffuse? In general, if the dimensions of the surface irregularities are greater than the wavelength of the light, the reflection is diffuse. Therefore, to make a good mirror, glass (with a metal coating) or metal must be polished until the surface irregularities are about the same size as the wavelength of light. Recall from Chapter 20 that the wavelength of visible light is on the order of 10^{-5} cm.

It is because of diffuse reflection that we are able to see illuminated objects, for example, the Moon. If the Moon's spherical surface were smooth, only the reflected sunlight from a small region would come to an observer on Earth, and

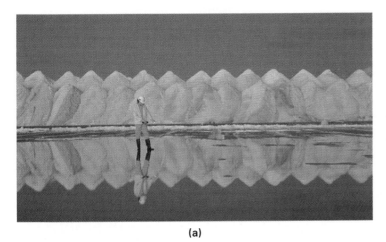

(a)

(b)

•**FIGURE 22.5 Reflections**
(a) Specular reflection from a smooth water surface produces an almost perfect mirror image of salt mounds at this Australian salt mine. (b) The bright white limestone in this Colorado mine obviously reflects much of the sunlight falling on it. Because its surface is not smooth on the microscopic level, however, it cannot produce a mirror image, but exhibits only diffuse reflection.

only a small illuminated area would be seen. Also, you can see the beam of light from a flashlight or spotlight because of diffuse reflection from dust and particles in the air.

CONCEPTUAL EXAMPLE 22.1 ■ ALL THAT GLITTERS. . . : REFLECTING ON REFLECTION

When you see the Sun or Moon over a lake or the ocean, you often observe a long swath of light, as shown in •Fig. 22.6, that is sometimes called a "glitter path." This reflection pattern is seen when (a) the surface of the water is disturbed by waves or ripples, (b) the reflection from the water surface is diffuse reflection, (c) the water surface is particularly smooth, as in Fig. 22.5a. *Clearly establish the reasoning and physical principle(s) used in determining your answer before checking it below. That is, **why** did you select your answer?*

Reasoning and Answer.   The fact that you see patches of light in the glitter path indicates that there is regular or specular reflection in these areas, so (b) is not correct. However, if the water had a smooth mirror surface as in Fig. 22.5a, we would simply see the Sun's or Moon's reflection as we observe it directly in the sky, so (c) is also ruled out. So by the process of elimination, the answer must be (a).

However, we should be able to justify this answer by explaining why it is correct. We can do this if we realize that because the water surface has waves, only certain segments of the surface are oriented so as to reflect the Sun's or Moon's image toward us at any instant. In effect, we have specular reflection, not from one big mirror surface, but from a series of little mirrors, like the facets of a glitter ball, with the reflecting facets changing almost randomly from moment to moment as the water surface undulates.

Follow-up Exercise.   When following an 18-wheeler, you may see a sign on the back of the truck stating: "If you can't see my mirror, I can't see you." What does this mean?

22.3   Refraction

Objectives: To be able to (a) explain refraction in terms of Snell's law and the index of refraction, and (b) give examples of refractive phenomena.

Refraction refers to the change in direction of a wave at a boundary where it passes from one medium into another. In general, when a wave is incident on a boundary between media, some of its energy is reflected and some is transmit-

•**FIGURE 22.6   A glitter path**
See Conceptual Example 22.1.

•FIGURE 22.7 Reflection and refraction

A beam of light is incident on a trapezoidal prism from the left. Part of the beam is reflected and part is refracted. The refracted beam is partially reflected and partially refracted at the bottom glass–air surface. (What happens to the reflected portion?)

ted. For example, when light traveling in air is incident on a transparent material such as glass (•Fig. 22.7), it is partially reflected and partially transmitted. But the direction in which the transmitted light is propagated is different from the direction of the incident light, so the light is said to have been refracted, or bent.

This phenomenon can be analyzed conveniently using a geometrical method developed by the Dutch physicist Christian Huygens (1629–1696). **Huygens' principle** for wave propagation is

> Every point on an advancing wave front can be considered to be a source of secondary waves, or wavelets, and the line or surface tangent to all these wavelets defines a new position of the wave front.

Huygens' principle is applied to incident and transmitted wave fronts at a media boundary as shown in •Fig. 22.8a. The wave speeds are different in the two media. In this case, $v_1 > v_2$. (The speed of light varies in different media, and in general, is less in denser media. Intuitively, you might expect the passage of light to take longer through a medium with more atoms per volume. And, in fact, the speed of light in water is about 75% of that in air or a vacuum.)

The distances the wave fronts travel in a time t are $v_1 t$ in medium 1 and $v_2 t$ in medium 2 (•Fig. 22.8b). As a result of the smaller wave speed in the second medium, the direction of the transmitted wave front is different from that of the incident wave front. The particles of the second medium are driven or set in motion by the incident wave disturbance, and thus the waves' frequency is the same in both media. However, the wavelengths are different because of the different wave speeds ($v = \lambda f$).

The change in the direction of wave propagation is described using the **angle of refraction**. In Fig. 22.8b, the angle of incidence is θ_1, and the angle of refraction is θ_2. From the geometry of two parallel rays, where d is the distance between the normals to the boundary at the points where the rays are incident,

$$\sin \theta_1 = \frac{v_1 t}{d} \quad \text{and} \quad \sin \theta_2 = \frac{v_2 t}{d}$$

Combining these two equations in ratio form gives

This law is named for the Dutch physicist Willebord Snell (1591–1626), who discovered it.

$$\frac{\sin \theta_1}{\sin \theta_2} = \frac{v_1}{v_2} \qquad \text{Snell's law} \qquad (22.2)$$

This expression is known as **Snell's law**.

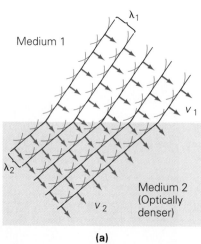
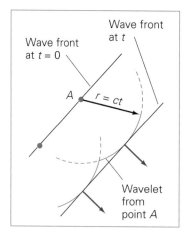
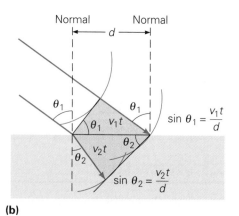

(a) **(b)**

•FIGURE 22.8 Huygens' principle
(a) Each point on a wave front is considered to be the source of a wavelet. The line or surface tangent to all these wavelets defines a new position of the wave front. When a wave enters an optically denser medium, its speed is less, and the wave fronts (or rays) are bent, or refracted. **(b)** This diagram shows the geometry for the derivation of Snell's law. See the text for a description.

Thus, light is refracted when passing from one medium into another because the speed of light is different in the two media (see the Insight on p. 687). The speed of light is greatest in a vacuum, and it is convenient to compare the speed of light in other media to this constant value (c). This is done by defining a ratio called the **index of refraction** (n).

$$n = \frac{c}{v} \frac{\text{(speed of light in a vacuum)}}{\text{(speed of light in a medium)}} \qquad (22.3)$$

Index of refraction: a ratio of speeds

As a ratio of speeds, the index of refraction is a unitless quantity. The indices of refraction of several substances are given in Table 22.1. Note that these values are for a specific wavelength of light. This is specified because v, and consequently n, are slightly different for different wavelengths ($n = c/v = c/\lambda_m f$, where λ_m is the wavelength of light in a particular medium). The values of n given in Table 22.1 will be used in examples and problems in this chapter for all wavelengths of light in the visible region, unless noted otherwise.

Remember that the frequency of light does not change when it enters another medium, but the wavelength of light in a material differs from the wavelength of that light in a vacuum, as can be easily shown:

$$n = \frac{c}{v} = \frac{\lambda f}{\lambda_m f}$$

or

$$n = \frac{\lambda}{\lambda_m} \qquad (22.4)$$

The wavelength of light in the medium is then $\lambda_m = \lambda/n$. Note that n is always greater than 1 because the speed of light in a vacuum is greater than the speed of light in any material ($c > v$). Therefore, $\lambda > \lambda_m$.

Note: When light is refracted
• its speed and wavelength are changed
• its frequency remains unchanged.

EXAMPLE 22.2 ■ THE SPEED OF LIGHT IN WATER: INDEX OF REFRACTION

What is the speed of light in water?

Solution. Obviously, there are some known quantities.

Given: $n = 1.33$ (from Table 22.1) $c = 3.00 \times 10^8$ m/s (known) *Find:* v (speed of light in H_2O)

Since $n = c/v$,

$$v = \frac{c}{n} = \frac{3.00 \times 10^8 \text{ m/s}}{1.33} = 2.26 \times 10^8 \text{ m/s}$$

Note that $1/n = v/c = 1/1.33 = 0.75$, so v is 75% of the speed of light in a vacuum.

Follow-up Exercise. The speed of light in a particular liquid is 2.40×10^8 m/s. What is the index of refraction of the liquid?

TABLE 22.1 Indices of Refraction (at $\lambda = 590$ nm)*	
Substance	*n*
Air	1.00029
Water	1.33
Ethyl alcohol	1.36
Fused quartz	1.46
Glycerine	1.47
Polystyrene	1.49
Oil (typical value)	1.50
Glass (depending on type)†	1.45–1.70
crown	1.52
flint	1.66
Zircon	1.92
Diamond	2.42

*A nanometer (nm) is 10^{-9} m.
†Crown glass is a soda–lime silicate glass and flint glass is a lead–alkali silicate glass. Flint glass is more dispersive than crown glass (see Section 22.5).

Hence we see that the index of refraction n is a measure of the speed of light in a transparent material, or technically, a measure of the *optical density* of a material. For example, the speed of light in water is less than that in air, so water is said to be optically denser than air. (Optical density does not correlate directly with mass density. In some instances, a material with a greater optical density than another will have a lower mass density.) So, the greater the index of refraction of a material, the greater its optical density and the smaller the speed of light in the material.

For practical purposes, the index of refraction is measured in air rather than in a vacuum, since the speed of light in air is very close to c, and

$$n_{air} \approx \frac{c}{c} = 1$$

(From Table 22.1, $n_{air} = 1.00029$.)

The index of refraction of a material may be determined experimentally using Snell's law.

$$\frac{\sin \theta_1}{\sin \theta_2} = \frac{v_1}{v_2} = \frac{c/n_1}{c/n_2} = \frac{n_2}{n_1}$$

or

$$n_1 \sin \theta_1 = n_2 \sin \theta_2 \qquad \begin{array}{l} \textit{Snells's law} \\ \textit{(another form)} \end{array} \qquad (22.5)$$

where n_1 and n_2 are the indices of refraction for the first and second media, respectively.

EXAMPLE 22.3 ■ ANGLE OF REFRACTION: SNELL'S LAW

Light in air is incident on a piece of crown glass at an angle of 37.0° (relative to the normal). What is the angle of refraction in the glass?

Solution. Listing the given quantities, we have

Given: $\theta_1 = 37.0°$ *Find:* θ_2 (angle of refraction)
$n_1 = 1.00$ (air)
$n_2 = 1.52$ (crown glass, Table 22.1)

Finding the angle of refraction is an easy matter using Snell's law in the form of Eq. 22.5:

$$\sin \theta_2 = \frac{n_1 \sin \theta_1}{n_2} = \frac{(1.00)(\sin 37.0°)}{1.52} = 0.396$$

and

$$\theta_2 = \sin^{-1}(0.396) = 23.3°$$

Follow-up Exercise. In the Example 22.2 Follow-up Exercise, it was stated that the speed of light in a liquid was 2.40×10^8 m/s. This would be difficult to measure directly. Instead, it was found experimentally that a beam of light entering the liquid from air at an angle of incidence of 37.0° had an angle of refraction of 28.8° in the liquid. What is the speed of light in the liquid determined from these data?

Note that Eq. 22.5 is a very practical form of Snell's law. If the first medium is air, $n_1 \approx 1$, and $n_2 \approx \sin \theta_1/\sin \theta_2$. Thus, only the angles of incidence and refraction need to be measured to determine the index of refraction of a material experimentally. If the index of refraction of a material is known, that value can be used in the practical form of Snell's law to find the angle of refraction for a given angle of incidence. Also note that the angle of refraction is inversely proportional to the index of refraction, $\sin \theta_2 \approx \sin \theta_1/n_2$. Hence, for a given angle of incidence, the greater the index of refraction, the smaller $\sin \theta_2$, and the smaller the angle of refraction θ_2.

More generally, the following relationships can be easily deduced from Eq. 22.5:

- If the second medium is more optically dense than the first medium ($n_2 > n_1$), the refracted ray is bent *toward* the normal ($\theta_2 < \theta_1$), as illustrated in ●Fig. 22.9a.
- If the second medium is less optically dense than the first medium ($n_2 < n_1$), the refracted ray is bent *away from* the normal ($\theta_2 > \theta_2$), as illustrated in ●Fig. 22.9b.

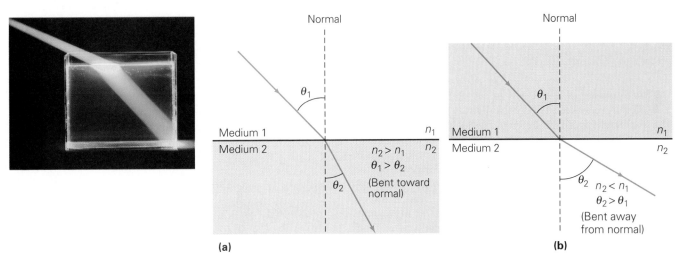

Normal

θ_1

Medium 1 n_1
Medium 2 n_2
$n_2 > n_1$
$\theta_1 > \theta_2$
(Bent toward
normal)
θ_2

(a)

Normal

θ_1

Medium 1 n_1
Medium 2 n_2
θ_2 $n_2 < n_1$
$\theta_2 > \theta_1$
(Bent away
from normal)

(b)

•**FIGURE 22.9   Index of refraction and ray deviation**
(a) When the second medium is more optically dense than the first ($n_2 > n_1$), the refracted ray is bent toward the normal, as in the case of light entering water from air. **(b)** When the second medium is less optically dense than the first ($n_2 < n_1$), the refracted ray is bent away from the normal. [Note that this is the case if the ray in part **(a)** is traced in reverse, going from medium 2 to medium 1.]

EXAMPLE 22.4 ■ A GLASS TABLE-TOP: MORE ABOUT REFRACTION

A beam of light traveling in air strikes the plate-glass top of a coffee table at an angle of incidence of 45° (•Fig. 22.10). The glass has an index of refraction of 1.5. (a) What is the angle of refraction for the light transmitted into the glass? (b) If the glass plate is 2.0 cm thick, what is the lateral (sideways) distance between the point where the ray enters the glass and the point where it emerges (x in the figure)? (c) Prove that the emergent beam is parallel to the incident beam, that is, that $\theta_4 = \theta_1$.

Solution.   Listing the data, we have

Given: $\theta_1 = 45°$ **Find:**   (a) θ_2 (angle of refraction)
$n_1 = 1.0$ (air)   (b) x (lateral distance)
$n_2 = 1.5$ (c) Show that $\theta_4 = \theta_1$
$y = 2.0$ cm

(a)   Using the practical form of Snell's law, $n_1 \sin \theta_1 = n_2 \sin \theta_2$, with $n_1 = 1.0$ gives

$$\sin \theta_2 = \frac{\sin \theta_1}{n_2} = \frac{\sin 45°}{1.5} = \frac{0.707}{1.5} = 0.47$$

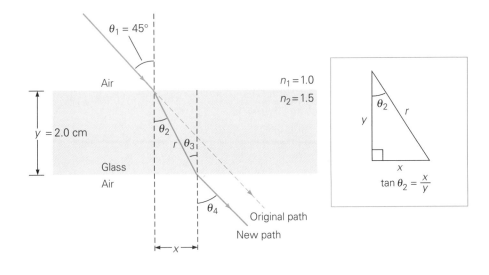

$\theta_1 = 45°$

Air

$n_1 = 1.0$
$n_2 = 1.5$

$y = 2.0$ cm

θ_2
r θ_3

Glass
Air

θ_4 Original path

New path

θ_2
y r

x

$\tan \theta_2 = \dfrac{x}{y}$

•**FIGURE 22.10   Double refraction**
The refracted ray travels laterally (sideways) a distance x in the glass, and the emergent ray is parallel to the original ray. See Example 22.3.

Insight

Understanding Refraction: A Marching Analogy

The refraction of light is not as easy to understand or visualize as reflection. We say that light is bent because waves have different speeds in different media. Intuitively, we might expect light to slow down when it enters a denser medium. The transmission of light through a transparent medium involves complex atomic absorption and emission processes, but it makes sense to suppose that these processes would take longer in a denser medium. This is indeed the case. The speed of light in water is 75% of that in air or in a vacuum, and in glass it is about 67% or less (depending on the type of glass). What is difficult to visualize is why light is bent or changes direction because of a change in speed.

To give some insight into this phenomenon, let's consider an analogy of a band marching across a field (Fig. 1). Part of the field is wet and muddy, and the marching column enters this region obliquely (at an angle of incidence). As the marchers in a row enter the wet, slippery region, they keep marching with the same cadence (frequency). However, slipping in the mud, the stride (wavelength) of the marchers is shorter, so they are slowed down.

The band members in the other part of the same row are still on dry ground and continue on with their original stride. The effect of the change in speed is a change in direction when the band enters the second medium. (A similar change in direction produced by changes in marching speeds is seen when a marching column turns a corner. The marchers nearest the corner deliberately shorten their stride and slow down, allowing those farther from the corner to swing around and complete their wider turn.) We might think of the marching

rows as wavefronts. As in refraction, the frequency (cadence) remains the same, but the wavelength, speed, and direction change (to keep a row aligned) on entering another medium.

FIGURE 1 Marching analogy for refraction
On obliquely entering a muddy field, the direction of a marching row is changed, analogous to the refraction of a wave front.

Thus,

$$\theta_2 = \sin^{-1}(0.47) = 28°$$

Note that the beam is bent toward the normal.

(b) You can see from the inset in Fig. 22.10 that the lateral distance (x) is related to θ_2 by $\tan \theta_2 = x/y$, and

$$x = y \tan \theta_2 = (2.0 \text{ cm})(\tan 28°) = (2.0 \text{ cm})(0.53) = 1.1 \text{ cm}$$

(c) If $\theta_1 = \theta_4$, then the emergent ray is parallel to the incident ray. Applying Snell's law to the beam at both surfaces gives

$$n_1 \sin \theta_1 = n_2 \sin \theta_2$$

and

$$n_2 \sin \theta_3 = n_1 \sin \theta_4$$

From the figure, we see that $\theta_2 = \theta_3$. Therefore,

$$n_1 \sin \theta_1 = n_1 \sin \theta_4$$

or

$$\theta_1 = \theta_4$$

Thus, the emergent beam is parallel to the incident beam but displaced laterally or sideways a distance x from the normal at the point of entry.

Follow-up Exercise. Show that the perpendicular distance between the original and emergent parallel ray paths is equal to $y \sin(\theta_1 - \theta_2)/\cos \theta_2$.

A beam of monochromatic (single frequency) light is directed at the side of a fish tank, so that the light passes from air to glass and from glass to water. The wavelength of the light in the water (a) is the same as that in air, (b) is independent of the wavelength in the glass, (c) does not change at the glass–water interface, (d) is shorter than that in the glass. *Clearly establish the reasoning and physical principle(s) used in determining your answer before checking it below. That is, **why** did you select your answer?*

Reasoning and Answer.   First, one needs to have an idea of the relative magnitudes of the optical densities of the materials or their indices of refraction. As can be seen from Table 22.1, $n_{glass} > n_{water} > n_{air}$.

Then, looking at the possible answers, we can eliminate (a) since the wavelength of light in a solid or liquid is less than that in vacuum or air, as shown previously (from Eq. 22.4, $\lambda_m = \lambda/n$). Similarly, since the indices of refraction of glass and water are different, the wavelength of the light changes at this interface, so (c) is not the answer. Noting that $n_{glass} > n_{water}$, we know that the wavelength is shorter in the more optically dense medium (glass), so (d) is eliminated.

This leaves only (b), which is the correct answer. Multiple choice questions may sometimes be answered correctly by such a process of elimination, but it is important to understand why a particular answer is correct—in this case, why the wavelength of light in water is independent of that in glass. Let's look at the interface of two general material media with indices n_2 and n_1. Forming a ratio,

$$\frac{n_2}{n_1} = \frac{c/v_2}{c/v_1} = \frac{v_1}{v_2} = \frac{\lambda_1 f}{\lambda_2 f} = \frac{\lambda_1}{\lambda_2}$$

or

$$\lambda_2 = \left(\frac{n_1}{n_2}\right)\lambda_1$$

Then, applying the wavelength-index conditions at the interfaces,

$$(\text{air} - \text{glass}) \quad \lambda_{glass} = \frac{\lambda_{air}}{n_{glass}}$$

$$(\text{glass} - \text{water}) \quad \lambda_{water} = \left(\frac{n_{glass}}{n_{water}}\right)\lambda_{glass}$$

and combining

$$\lambda_{water} = \left(\frac{n_{glass}}{n_{water}}\right)\frac{\lambda_{air}}{n_{glass}} = \frac{\lambda_{air}}{n_{water}}$$

We can see that the wavelength of the light in water is independent of that in glass, since the latter does not appear in the final equation. This is to be expected since the speed of light in a transparent medium, and thus the wavelength for a constant frequency, is characteristic of the material itself.

Follow-up Exercise.   A light source with a particular frequency in air is submersed in water in a fish tank. The beam travels in the water, through double plate glass at the side of the tank (each glass plate having a different n), and into air. In general, what are (a) the frequency and (b) the wavelength of the light in air?

A couple of other common examples of refraction are shown in •Fig. 22.11. The refraction of light by warm air near the surface of a road produces the "wet spot" mirage commonly seen on a road on a hot day (Fig. 22.11a). The illusion of water is actually produced by light from the sky diverted toward the viewer (Fig. 22.11b). We also "see" hot air rising from a hot road surface. Of course, you can-

(a)

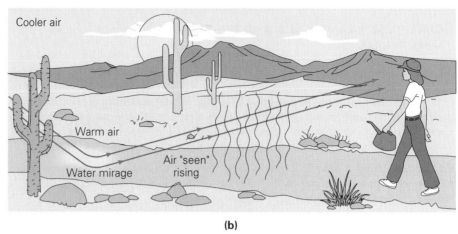

Cooler air

Warm air

Water mirage

Air "seen" rising

(b)

(c)

•FIGURE 22.11   Refraction in action
(a) A "wet spot" mirage. The road is actually dry; what looks like water is a refracted image of part of the sky. **(b)** The mirage is produced when light from the sky is refracted by warm air near the road surface. **(c)** Such refraction also enables us to perceive heated air as it rises. Although we do not see the air itself, objects viewed through the turbulent updrafts are distorted by refraction and seem to shimmer.

not really see the air, but gases have a property that makes it possible for you to sense them visually when heated. The index of refraction of a gas is proportional to its density, which is in turn inversely proportional to its temperature. Thus, you can perceive regions of air having different densities (temperatures) because the refraction of light as it passes through the rising air produces distortion of the image, which seems to shimmer as the air density fluctuates from moment to moment (Fig. 22.11c). (Ordinarily, the air near a road's surface has a uniform density, and you do not see these effects.)

Refraction of light in air gives rise to other effects. At night, the air through which starlight travels to your eyes is continually varying in density because of temperature variations and turbulence. The resulting refraction causes the stars' images to dance around and fluctuate in brightness—the familiar "twinkling." Atmospheric refraction also accounts for the fact that the Sun can be seen for a short time before it rises above or after it sinks below the horizon. (For a similar phenomenon see Exercise 23 at the end of the chapter.)

You may have experienced an effect of the refractive bending of light while trying to reach for something that is under water, for example, a fish or a coin (•Fig. 22.12a). You are used to light traveling in straight lines from objects to your eyes, but the light reaching your eyes from the submerged object is bent at the water surface. (Note in the figure that the ray is bent away from the normal.) The object appears to be closer to the surface than it actually is, and therefore

•FIGURE 22.12   Refraction and depth perception
(a) An object immersed in water appears closer to the surface than it actually is because of refraction. See Example 22.6. **(b)** Similarly, the legs of a person standing in water seem shortened.

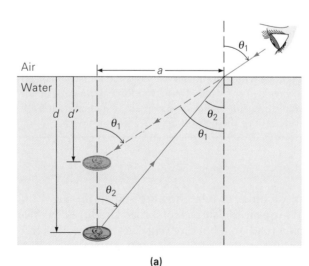

(a)

(a)

you tend to miss the object when reaching for it. For the same reason, the legs of a person standing in water often seem much shorter than their actual length (•Fig. 22.12b).

EXAMPLE 22.6 ■ A SUBMERGED COIN: REFRACTION AND DEPTH PERCEPTION

How is the apparent depth (d') of the coin in Fig. 22.12 related to the actual depth d and the angles of incidence and refraction?

Solution. From the figure, we see that the distance a is common to both d and d', and using trigonometry we may write

$$\tan \theta_1 = \frac{a}{d'} \quad \text{and} \quad \tan \theta_2 = \frac{a}{d}$$

Combining these equations to form a ratio,

$$\frac{d'}{d} = \frac{\tan \theta_2}{\tan \theta_1} \quad \text{or} \quad d' = \left(\frac{\tan \theta_2}{\tan \theta_1}\right) d$$

This expression can be simplified if we consider refraction for only small angles. Recall that for small angles ($\theta < 15°$), $\cos \theta \cong 1$ (actually $\cos 15° = 0.966$). Thus

$$\tan \theta = \frac{\sin \theta}{\cos \theta} \approx \sin \theta$$

The previous expression can therefore be approximated by

$$\frac{d'}{d} = \frac{\tan \theta_2}{\tan \theta_1} \approx \frac{\sin \theta_2}{\sin \theta_1}$$

But, by Snell's law, $\sin \theta_2 / \sin \theta_1 = 1/n$, and

$$d' \approx \frac{d}{n}$$

With $n = 1.33$ for water (Table 21.1),

$$d' \approx \frac{d}{1.33} = (0.752)d$$

and the apparent depth of the coin would be about $\frac{3}{4}$ of its actual depth.

Follow-up Exercise. Suppose that the angle of incidence in Fig. 22.12 were 30°, so that the small-angle approximation could not be used as in the Example. How do the apparent and actual depths compare in this case?

22.4 Total Internal Reflection and Fiber Optics

Objectives: **To be able to (a) describe internal reflection, and (b) give examples of fiber optic applications.**

An interesting phenomenon occurs when light traveling in one medium is incident on the boundary with another medium that is less optically dense, for example, when light goes *from* water *into* air. As you know, in such a case a transmitted beam will be bent away from the normal, and Snell's law tells you that the greater the angle of incidence, the greater the angle of refraction will be. That is, the more the angle of incidence increases, the farther the refracted ray diverges from the normal.

However, there is a limit. For a certain angle of incidence called the **critical angle** (θ_c), the angle of refraction is 90°, and the refracted ray is directed along the boundary. But what happens if the angle of incidence is even larger? If the angle of incidence is greater than the critical angle ($\theta_1 > \theta_c$), the light isn't trans-

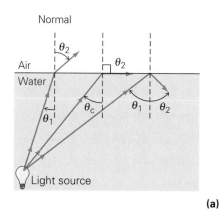

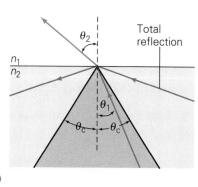

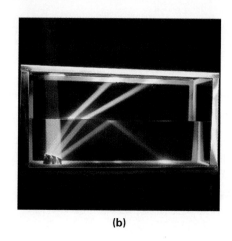

(b)

•FIGURE 22.13 Internal reflection
(a) When light enters a less optically dense medium, it is refracted away from the normal. At a critical angle (θ_c), the light is refracted along the interface (common boundary) of the media. At an angle greater than the critical angle ($\theta_1 > \theta_c$), there is total internal reflection. **(b)** Thus, total internal reflection occurs for incident angles outside a cone with an apex angle of $2\theta_c$. (Can you estimate the critical angle in the photograph?)

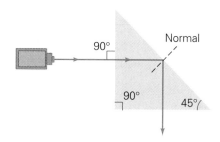

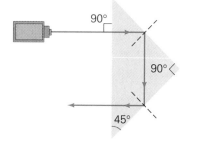

(a)

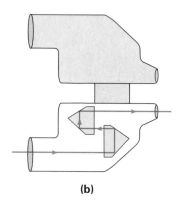

(b)

mitted but is internally reflected (•Fig. 22.13). This condition is called **total internal reflection** and the reflection process can be 100% efficient. Because of internal reflection, prisms can essentially be used as mirrors (•Fig. 22.14).

The critical angle for total internal reflection may be obtained using Snell's law. If $\theta_1 = \theta_c$ in the optically denser medium, $\theta_2 = 90°$, and

$$\frac{\sin \theta_1}{\sin \theta_2} = \frac{\sin \theta_c}{\sin 90°} = \frac{n_2}{n_1}$$

Since $\sin 90° = 1$,

$$\sin \theta_c = \frac{n_2}{n_1} \quad where\ n_1 > n_2 \tag{22.6}$$

If the second medium is air, $n_2 \approx 1$, and the critical angle for total internal reflection at the boundary from a medium into air is given by

$$\sin \theta_c = \frac{1}{n} \qquad \begin{array}{l} \textit{where n is the index of} \\ \textit{refraction of the medium} \\ \textit{(medium–air only)} \end{array} \tag{22.7}$$

EXAMPLE 22.7 ■ THE VIEW FROM THE POOL: CRITICAL ANGLE

(a) What is the critical angle for light traveling in water and incident on a water–air boundary? **(b)** If a diver submerged in a pool looked up at the water surface at an angle of $\theta < \theta_c$, what would she see? (Neglect any thermal or motional effects.)

Solution.

Given: $n = 1.33$ (from Table 22.1)   **Find:**   (a) θ_c (critical angle)
                                                                (b) View for $\theta < \theta_c$

•FIGURE 22.14 Internal reflection in a prism
(a) Because the critical angle of glass is less than 45°, prisms with 45° and 90° angles can be used to reflect light through 90° and 180°. **(b)** Internal reflection of light by prisms in binoculars make this instrument much shorter than a telescope. (See Chapter 25.)

(a) The critical angle can be found directly from Eq. 22.7.

$$\theta_c = \sin^{-1}\left(\frac{1}{n}\right) = \sin^{-1}\left(\frac{1}{1.33}\right) = 48.8°$$

(b) For viewing angles of $\theta < \theta_c$, the diver would have a conical view of things above the surface (see Fig. 22.13 and mentally trace the rays in reverse for light coming from all angles in three dimensions). That is, light coming from the above-water 180° panorama could be viewed in a cone with a half-angle of 48.8°. As a result, objects would appear distorted. Such panoramic views are seen in photographs made with "fish-eye" lenses (Fig. 22.15).

Follow-up Exercise. What would the diver see when looking up at the water surface at an angle of $\theta > \theta_c$?

Internal reflections enhance the brilliance of cut diamonds. The critical angle for a diamond–air surface is

$$\theta_c = \sin^{-1}\left(\frac{1}{n}\right) = \sin^{-1}\left(\frac{1}{2.42}\right) = 24.4°$$

A so-called brilliant-cut diamond has many facets, or faces (58 in all—33 on the upper face and 25 on the lower), and light entering the main and upper facets (or crown) above the critical angle is internally reflected in the diamond. It then emerges from the upper facets, giving rise to the diamond's brilliance (Fig. 22.16).

Fiber Optics

When a fountain is illuminated from below, the light is transmitted along the curved streams of water. This phenomenon was first demonstrated in 1870 by the British scientist John Tyndall (1820–1893), who showed how light was "conducted" along the curved path of a stream of water flowing from a hole in the side of a container. As you may have guessed, this phenomenon is observed because the light is internally reflected along the stream of water.

Internal reflection forms the basis of **fiber optics**, a relatively new and fascinating field centered on the use of transparent fibers to transmit light. Multiple internal reflections make it possible to "pipe" light along a transparent rod, even if the rod is curved (Fig. 22.17). Note from the figure that the smaller the diameter of the light pipe, the greater the number of internal reflections. In a small fiber, there can be as many as several hundred internal reflections per centimeter.

Internal reflection is an exceptionally efficient process. Optical fibers can be used to transmit light over very long distances with losses of only about 25% per kilometer. These losses are primarily due to fiber impurities, which scatter the light. Transparent materials have different degrees of transmission. Fibers are made out of certain plastics and special glass for maximum transmission, and the greatest efficiency is achieved with infrared radiation, for which there is less scattering. (See the Insight on p. 694.)

The comparatively greater efficiency of multiple internal reflections over multiple external reflections can be illustrated by considering a good reflecting surface, such as a plane mirror, which at best has a reflectivity of about 95%. Suppose two plane mirrors (common flat mirrors as in bathrooms) are placed facing and nearly parallel to each other. In this manner a beam from an object near one end and between the mirrors will be reflected back and forth, in a manner similar to the reflections in the light pipe in Fig. 22.17b.

After each reflection, the beam intensity would be 95% of the incident beam from the preceding reflection. The intensity I of the reflected beam after n reflections is given by

$$I = (0.95)^n I_o$$

FIGURE 22.15 Panoramic view A "fish-eye" lens provides an extremely wide-angle (but distorted) view. Shown here are replicas of Columbus's Nina, Pinta, and Santa Maria sailing into Miami.

(a)

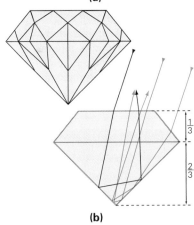

(b)

FIGURE 22.16 Diamond brilliance **(a)** Internal reflection gives rise to a diamond's brilliance. **(b)** The brilliant-cut diamond has 58 facets or faces. Light entering the upper facets is internally reflected and re-emerges through these facets, giving the diamond a sparkling brilliance. The depth proportions are critical. If a stone is too shallow or too deep, light will be lost through the lower facets.

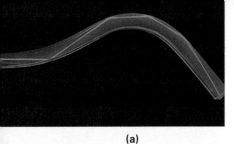

(a)

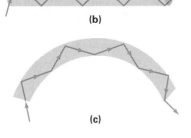

(b)

(c)

(d)

•**FIGURE 22.17   Light pipes**
(a) Total internal reflection in an optical fiber. (b) When light is incident on the end of a cylindrical form of transparent material so that the internal angle of incidence is greater than the critical angle of the material, the light is reflected down the length of the light pipe. (c) Light is also transmitted along curved light pipes by internal reflection. (d) As the diameter of the rod or fiber becomes smaller, the number of reflections per unit length becomes greater.

where I_o is the initial beam intensity before the first reflection. Then, after 14 reflections,

$$I = (0.95)^{14} I_o = 0.49 \, I_o$$

Thus, after only 14 reflections, the intensity is reduced to less than one-half (49%). For 100 reflections, $I = 0.006 \, I_o$, and there is only 0.6% of the initial intensity. Compare this to about 75% in optical fibers over a kilometer in length with *thousands and thousands* of reflections. An illustration of the reduction in intensity for multiple reflections in nearly parallel mirrors is shown in •Fig. 22.18.

Fibers whose diameters are about 10 μm (10^{-5} m) are grouped together in flexible bundles that are 4–10 mm in diameter and up to several meters in length, depending on the application (•Fig. 22.19). A fiber bundle with a cross-sectional area of 1 cm^2 can contain as many as 50,000 individual fibers. To prevent light from being transmitted between fibers in contact with each other, they are coated with a film.

22.5   Dispersion

Objective: To be able to explain dispersion and some of its effects.

Light of a single frequency is called monochromatic light (from the Greek, *mono* meaning "one" and *chroma* meaning "color"). Visible light that contains all the component frequencies, or colors, is termed white light. Sunlight is white light. When a beam of white light passes through a glass prism as shown in •Fig. 22.20a, it is spread out, or dispersed, into a spectrum of colors. This phenomenon led Newton to believe that sunlight is a mixture of component colors. When the beam enters the prism, the component colors, corresponding to different wavelengths, must be refracted at slightly different angles so that they spread out into a spectrum (•Fig. 22.20b).

The formation of a spectrum indicates that the index of refraction of glass must be slightly different for different wavelengths, and this is true of many transparent media (•Fig. 22.20c). The explanation has to do with the speed of light. In a vacuum, the speed of light c is the same for all wavelengths, but in a dispersive medium the speed of light is slightly different for different wavelengths. Since the

•**FIGURE 22.18   Multiple reflections**
Multiple reflections from nearly parallel mirrors.

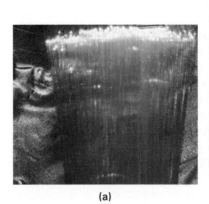

(a)

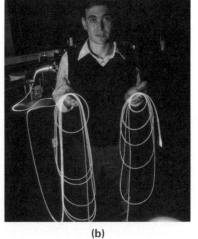

(b)

•**FIGURE 22.19   Fiber optic bundle**
Hundreds or even thousands of extremely thin fibers (a) are grouped together to make a bundle (b).

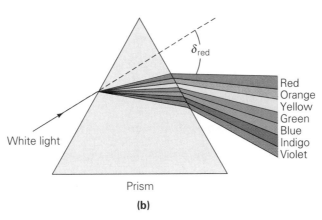

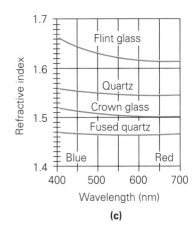

(a) (b) (c)

FIGURE 22.20 Dispersion
(a) White light is dispersed into a spectrum of colors by a glass prism. **(b)** In a dispersive medium, the index of refraction differs slightly for different wavelengths. Red light, with the longest wavelength, has the smallest index of refraction and is refracted least. The angle between the incident beam and a refracted ray is called the angle of deviation (δ) for that ray. (The angles are exaggerated for clarity. **(c)** Variation of the index of refraction with wavelength for some common transparent media.

index of refraction n of a medium is a function of the velocity of light in that medium ($n = c/v = c/\lambda f$), the index of refraction will then be different for different wavelengths. It follows from Snell's law that light of different wavelengths will be refracted at different angles.

We can summarize by saying that in a transparent material with different indices of refraction for different wavelengths of light, there is a separation of the wavelengths, and the material is said to exhibit **dispersion**. Dispersion varies with different media and may be neglected in many instances. Also, because the differences in the indices of refraction for different wavelengths are small, a representative value at some specified wavelength can be used for general purposes (see Table 22.1).

Note: You can remember the sequence of the colors of the visible spectrum (from the long-wavelength end to the short-wavelength end) using the name ROY G. BIV, which is an acronym for Red, Orange, Yellow, Green, Blue, Indigo, and Violet.

EXAMPLE 22.8 ■ FORMING A SPECTRUM: DISPERSION

A particular transparent material has an index of refraction of 1.4503 for the red end ($\lambda = 700$ nm) of the visible spectrum and an index of refraction of 1.4698 for the blue end ($\lambda = 400$ nm) of the spectrum. If white light is incident on a prism of this material as in Fig. 22.20b at an angle of 45°, what is the angular dispersion of the visible spectrum in the prism?

Solution.

Given: (red) $n_r = 1.4503$ for $\lambda_r = 700$ nm   *Find:* $\Delta\theta$ (angular dispersion)
(blue) $n_b = 1.4698$ for $\lambda_b = 400$ nm

To find the angular dispersion or spread of the visible spectrum in the prism, we compute the angles of refraction for each end component. Using Eq. 22.1 with $n_1 = 1.00$ (air),

$$\sin \theta_{2_r} = \frac{\sin \theta_1}{n_{2_r}} = \frac{\sin 45°}{1.4503} = 0.48756$$

and

$$\theta_{2_r} = 29.180°$$

Similarly,

$$\sin \theta_{2_b} = \frac{\sin \theta_1}{n_{2_b}} = \frac{\sin 45°}{1.4698} = 0.48109$$

Fiber Optics

Fiber optics can be used for purely decorative purposes, such as lamps, but a much more important application involves the piping of light to and images from hard-to-reach places. To do this, the ends of a fiber bundle are cut and polished to form a flexible *fiberscope*. A beam of light can be transmitted along the bundle to illuminate an area, even if the bundle is bent and twisted (Fig. 1). Equally important, an image or picture may be transmitted back by a fiberscope. Light travels through one set of fibers to illuminate an object and, after reflection, travels back in another set. The image has light and dark regions since each fiber has a circular cross section and transmits a dot. The overall image thus has a mosaic pattern, like a newspaper picture. The smaller the elements in the mosaic, the finer the detail. Thus, a fiberscope has a very large number of extremely fine fibers. A transmitted image can be magnified by a lens for viewing.

There are a number of interesting applications of fiber optics. For example, you're probably aware that many telephone transmissions are accomplished by means of fiber optics. Light signals, rather than electrical signals, are transmitted through optical telephone lines. Optical fibers have lower energy losses than current-carrying wires, particularly at higher frequencies, and can carry much more data. Also, these fibers are lighter than metal wires, have greater flexibility, and are not affected by electromagnetic disturbances (electric and magnetic fields) since they are made of materials that are electrical insulators.

Fiber optics has been widely applied in medicine. For example, endoscopes, or instruments used to view internal portions of the human body, previously consisted of lens systems in long narrow tubes. Some contained a dozen or more lenses and gave relatively poor images. Also, because the lenses had to be aligned in certain ways, the tubes had to have rigid sections, which limited the endoscope's maneuverability. Such an endoscope could be inserted down the throat into the stomach to observe the stomach lining. However, there would be blind spots due to the curvature of the stomach and the inflexibility of the instrument.

The flexibility of fiber bundles has eliminated this problem. Lenses are used at the end of the fiber bundles to focus the light, and a prism is used to change the direction for its return. The incident light is usually transmitted by an outer layer of fiber bundles, and the image is returned through a central core of fibers. Mechanical linkages allow maneuverability. The end of a fiber endoscope may be equipped with devices to obtain specimens of the viewed tissues for biopsy (diagnostic examination), or even to perform surgical procedures (Fig. 2). For example, you may have heard of arthroscopic surgery being performed on the knees of injured athletes. The arthroscope that is now routinely used for inspecting and repairing damaged joints is simply a fiber endoscope fitted with appropriate surgical implements.

A fiber-optic cardioscope used for direct observation of heart valves typically has a fiber bundle about 4 mm in diameter and 30 cm long. Such a cardioscope passes easily to the heart through the jugular vein, which is about 15 mm in diameter. To displace the blood and provide a clear field of view for observations and photographing, a transparent balloon at the tip of the cardioscope is inflated with saline (salt water) solution.

Another application of fiber optics is the coding and decoding of information. To make a "coded" image of a classified picture, for example, the component fibers of a bundle are deliberately misaligned and randomly interwoven. As a result, a transmitted image is jumbled and unrecognizable unless it is viewed through an identically interwoven bundle that "decodes" it.

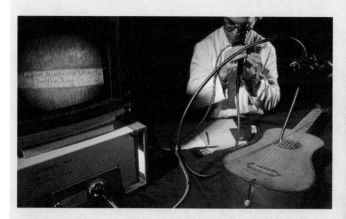

FIGURE 1 Fibroscopy
A technician uses a fiberscope to inspect the interior of a guitar. The label on the inside (image displayed on screen) reveals that it was made by Stradivarius in 1711.

FIGURE 2 Endoscopy
Surgeons use a fiber-optic endoscope to perform internal surgery.

and

$$\theta_{2_b} = 28.757°$$

So

$$\Delta\theta = \theta_{2_r} - \theta_{2_b} = 29.180° - 28.757° = 0.423°$$

Not much deviation, but as the light travels it spreads out and emerges from the prism, and the dispersion becomes evident (Fig. 22.20a).

Follow-up Exercise.   If green light exhibits an angular dispersion of 0.156° from the red light, what is the index of refraction for green light in the material?

A good example of a dispersive material is diamond, which is about five times more dispersive than glass. In addition to the brilliance resulting from internal reflections off many facets (Fig. 22.16), a cut diamond shows a play of colors, or "fire," resulting from the dispersion of the internally reflected light.

Another dramatic example of dispersion is the production of a rainbow, as discussed in the Insight on p. 696.

Chapter Review

Important Terms

Important Concepts

- Law of reflection: the angle of incidence equals the angle of reflection (as measured from the normal to the reflecting surface).

- The index of refraction of any medium is the ratio of the speed of light in a vacuum to its speed in that medium.

- The angle of refraction is given by Snell's law. If the second medium is more optically dense, the refracted ray is bent toward the normal; if less dense, bending is away from the normal.

- Light may be totally internally reflected if the second medium is less dense that the first, and if the angle of incidence exceeds the critical angle.

- Light is dispersed in some media because different wavelengths have slightly different indices of refraction.

Important Equations

Law of Reflection:

$$\theta_i = \theta_r \qquad (22.1)$$

Snell's Law:

$$\frac{\sin\theta_1}{\sin\theta_2} = \frac{v_1}{v_2} \quad \text{or} \quad n_1\sin\theta_1 = n_2\sin\theta_2 \qquad \begin{matrix}(22.2)\\(22.5)\end{matrix}$$

Index of Refraction:

$$n = \frac{c}{v} = \frac{\lambda}{\lambda_m} \qquad (22.3\text{–}4)$$

Critical Angle for Boundary between Two Materials (where $n_1 > n_2$):

$$\sin\theta_c = \frac{n_2}{n_1} \qquad (22.6)$$

Critical Angle for Material–Air Boundary: (where n is the index of refraction of the material)

$$\sin\theta_c = \frac{1}{n} \qquad (22.7)$$

The Rainbow

FIGURE 1   Double rainbow
Notice that the colors of the rainbows are reversed. In the lower, primary rainbow the colors run vertically from blue to red, whereas in the upper, secondary rainbow they run from red to blue.

We have all been fascinated by the beautiful array of colors of a rainbow (Fig. 1). With the optical principles learned in this chapter, we are now in a position to understand the formation of this spectacular display.

A rainbow is produced by refraction, dispersion, and internal reflection of light within water droplets. When millions of water droplets remain suspended in the air after a rainstorm, a multicolored arc is seen, whose colors run from violet along the lower part up the spectrum (in order of wavelength) to red along the upper. Below the arc, the light from the droplets combines to form a region of brightness. Occa-

sionally, more than one rainbow is seen; the main, or primary, rainbow is sometimes accompanied by a fainter and higher secondary rainbow.

The light that forms the primary rainbow is reflected once inside each water droplet (Fig. 2a). Being also refracted and dispersed, the light is spread out into a spectrum of colors. However, because of the conditions for refraction and internal reflection in water, the angles of deviation (between incoming and outgoing rays) for violet to red light lie within a narrow range of 40°–42° (Fig. 2b). This means that you can see a rainbow only when the Sun is behind you, so the dispersed light is reflected to you through these angles.

The less frequently seen secondary rainbow is caused by a double internal reflection (Fig. 2c). This results in an arc whose vertical color sequence is the inverse of the primary rainbow's. The angles of deviation in this case lie between 50.5° for red light and 54° for violet light.

We generally see only rainbow arcs, because the formation by water droplets is cut off at the ground. If you were on a cliff, you might possibly see a complete circular rainbow (Fig. 2c). Also, the higher the Sun is in the sky, the less of a rainbow you will be able to see from the ground. In fact, you won't see a primary rainbow if the Sun's altitude is greater than 42°. (Altitude in this case is the angle above the horizon.) The primary rainbow can still be seen from a height, however. As an observer's elevation increases, more of the arc becomes visible. It is common to see a completely circular rainbow from an airplane. You may also have seen a circular rainbow in the spray from a garden hose.

QUESTION: Can there be a third-order, or tertiary, rainbow?

ANSWER: Yes, a third-order rainbow resulting from three internal reflections in a water droplet is possible.

In a book entitled *Opticks*, Isaac Newton wrote, "The Light which passes through a drop of Rain after two Refractions,

Exercises*

22.1 Wave Fronts and Rays

1   A wave front is (a) always circular, (b) parallel to a ray, (c) described by a surface of equal phase, (d) none of these.

2   A ray (a) is perpendicular to the direction of energy flow, (b) is always parallel to other rays, (c) is perpendicular to a series of wave fronts, (d) illustrates the wave nature of light.

22.2 Reflection

3   For regular reflection, (a) the angle of incidence equals the angle of reflection, (b) the rays of a reflected beam are not parallel, (c) the incident ray, the reflected ray, and the normal all lie in the same plane, (d) both (a) and (c).

4   For diffuse reflection, (a) the angle of incidence equals the angle of reflection, (b) the rays of a reflected beam are not parallel, (c) the incident ray, the reflected ray, and the local normal all line in the same plane, (d) all of these.

*Assume angles to be exact.

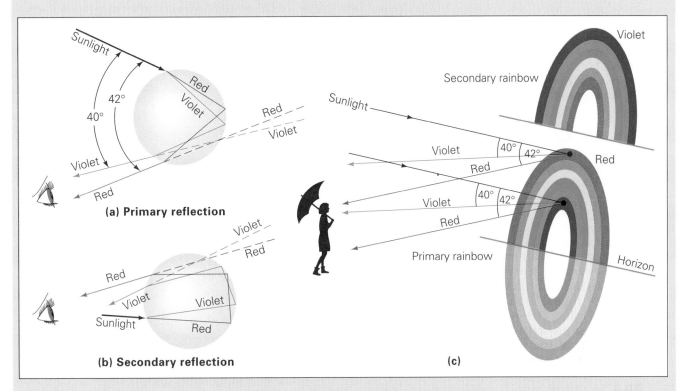

FIGURE 2 The rainbow
Rainbows are due to the refraction, dispersion, and internal reflection of sunlight. **(a)** A single internal reflection gives rise to the primary rainbow. **(b)** A double internal reflection produces a secondary rainbow. **(c)** In the primary rainbow, an observer sees red light at the top of the bow—that is, from the higher droplets—because red light is deviated most. (The other color components from these droplets pass above the observer's eyes.) Similarly, violet or blue is seen from the lower droplets.

and three or more Reflexions, is scarcely strong enough to cause a sensible rainbow." (By "sensible," he meant intense enough to be seen.) However, his friend Edmund Halley (for whom the famous comet was named) showed that the reason the tertiary rainbow is not seen is not because the reflected light lacks intensity, but rather because the tertiary arc is formed in the general direction of the Sun and cannot be seen against the brightness of the sky.

5 Explain why the sunbeams are visible in •Fig. 22.21.

6 ■ The angle of incidence of a light ray on a mirrored surface is 38°. What is the angle between the incident and reflected rays?

7 ■ A beam of light is incident on a plane mirror at an angle of 32° relative to the normal. What is the angle between the reflected rays and the surface of the mirror?

8 ■ A light ray incident on a plane mirror is at an angle of 65° relative to the mirror surface. At what angle is the reflected ray relative to the surface?

9 ■ Two people stand in a dark room, 3.0 m from a large plane mirror and 5.0 m apart. At what angle of incidence should one of them shine a flashlight on the mirror so the reflected beam directly strikes the other person?

•**FIGURE 22.21 Sunbeams**
See Exercise 5.

10 ■■ Standing 1.5 m in front of a plane mirror with your camera, you decide to take a picture of yourself. To what distance should the camera be manually focused?

11 ■■ A woman fixing the hair on the back of her head holds a hand (plane) mirror 30 cm in front of her face so as to look into a plane mirror on the bathroom wall behind her. If she is 90 cm from the wall mirror, approximately how far does the image of the back of her head appear behind her?

12 ■■ Two upright plane mirrors touch along one edge, where their planes make an angle of 70°. If a beam of light is directed onto one of the mirrors at an angle of incidence of 40° and is reflected onto the other mirror, what will be the angle of reflection of the beam from the second mirror?

13 ■■ Two plane mirrors are placed 50 cm apart with their mirrored surfaces parallel and facing each other. If the mirrors are 25 cm wide, at what angle should a beam of light be incident at one end of one mirror so that it just strikes the far end of the other?

14 ■■ If you hold a 30-cm square plane mirror 45 cm from your eyes and you can just see the full length of a 8.50-m flag pole behind you, how far are you from the pole? (*Hint:* A diagram is helpful here)

15 ■■■ Two plane mirrors M_1 and M_2 are placed together as illustrated in •Fig. 22.22. (a) If the angle α between the mirrors is 70° and the angle of incidence θ_{i_1} of a light ray incident on M_1 is 35°, what is the angle of reflection θ_{r_2} from M_2? (b) if $\alpha = 115°$ and $\theta_{i_1} = 60°$, what is θ_{r_2}?

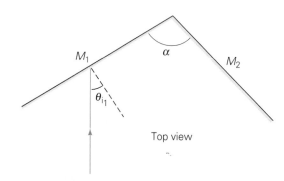

•**FIGURE 22.22 Plane mirrors together**
See Exercises 15, 16, and 17.

16 ■■■ For the plane mirrors in Fig. 22.22, what angles α and θ_{i_1} would allow a ray to be reflected back in the same direction from which it came (parallel to the incident ray)?

17 ■■■ Show that for two mirrors placed together at an angle α as in Fig. 22.22, a light ray successively reflected from both mirrors is deflected through an angle of 2α irrespective of the incident angle, where 2α is the obtuse angle between the incident and reflected rays.

22.3 Refraction

22.4 Total Internal Reflection and Fiber Optics

18 Light refracted at the boundary of two media (a) is bent toward the normal when entering the less optically dense medium, (b) is bent away from the normal when entering the more optically dense medium, (c) has the same frequency in both media, (d) always experiences a decrease in speed.

19 The index of refraction (a) is always greater than or equal to one, (b) is inversely proportional to the speed of light in a medium, (c) is inversely proportional to the wavelength of light in the medium, (d) all of these.

20 The critical angle for total internal reflection at a material–air boundary (a) is independent of the wavelength of the light in the medium, (b) is greater for a material with a smaller index of refraction, (c) may be greater than 90°, (d) none of these.

21 Total internal reflection will occur if (a) $\theta_1 < \theta_c$, (b) $n_1 = n_2$, (c) $n_2 > n_1$, (d) none of these.

22 Two refraction phenomena are shown in •Fig. 22.23. (a) Explain why the pencil appears almost severed. Illustrate with a ray diagram. (b) Explain why the setting Sun appears flattened.

•**FIGURE 22.23 Refraction effects**
See Exercise 22.

23 Explain the phenomenon illustrated in •Fig. 22.24. The pictures are taken with a camera on a tripod at a fixed angle. There is a penny in the container, but only its tip is seen initially. However, when water is added, more of the coin is seen. Why?

24 ■ The speed of light in a material is determined to be 2.22×10^8 m/s. What is the index of refraction of the material?

•FIGURE 22.24   You don't see it, then you do
See Exercise 23.

25 ■ Find the speed of light in diamond.

26 ■■ A beam of light enters water at an angle of 60° relative to the normal of the water surface. Find the angle of refraction.

27 ■■ A beam of light in air is incident on the surface of a slab of fused quartz. Part of the beam is transmitted into the quartz at an angle of 30° with a normal to the surface, and part is reflected. What is the angle of reflection?

28 ■■ Light passes from air into water. If the angle of refraction is 20°, what is the angle of incidence?

29 ■■ A beam of light is incident on a flat piece of polystyrene at an angle of 55° to a surface normal. What angle does the refracted ray make with the plane of the surface?

30 ■■ A beam of light traveling in air is incident on a transparent plastic material at an angle of incidence of 45°. The angle of refraction is measured to be 30°. What is the index of refraction of the material?

31 ■■ Is the speed of light greater in diamond or zircon? Express the difference as a percentage.

32 ■■ Monochromatic blue light having a frequency of 6.5×10^{14} Hz enters a piece of crown glass. What are the frequency and wavelength of the light in the glass?

33 ■■ Light passes through a piece of flint glass in 20 ps (picoseconds). Find the thickness of the glass.

34 ■■ Light passes from material A, which has an index of refraction of $\frac{4}{3}$, into material B, which has an index of refraction of $\frac{5}{4}$. Find the ratio of the speed of light in material B to the speed of light in material A.

35 ■■ If the angle of incidence in Exercise 34 is 30°, what is the angle of refraction?

36 ■■ A He–Ne (helium-neon) laser beam ($\lambda = 632.8$ nm in air) is directed through ethyl alcohol. What are the wavelength and frequency of the light in the alcohol?

37 ■■ A layer of water 30.0 mm thick floats on a layer of another liquid, which is 24.0 mm thick. The other liquid has an index of refraction of 1.50. How far below the water surface will the bottom of the container appear to be for a small angle of incidence?

38 ■■ A fish tank is made of glass with an index of refraction of 1.50. A person shines a light beam on the glass at an incident angle of 40° so as to see a fish inside. Is the fish illuminated? Justify your answer.

39 ■■ Prisms are sometimes used as optical components in place of mirrors. Find the minimum index of refraction for glass that is to be used in a 45°–90°–45° prism that is meant to change the direction of light by 100° on reflection.

40 ■■ A beam of light passes through a 45°–45°–90° prism as shown in •Fig. 22.25. (a) What can be said about the index of refraction of the prism for this to occur? (b) What if the prism were under water?

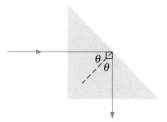

•FIGURE 22.25   Internal reflection in a prism
See Exercise 40.

41 ■■ A light ray in air is incident on a glass plate 10.0 cm thick at an angle of 40°. The glass has an index of refraction of 1.65. The emerging ray on the other side of the plate is parallel to the incident ray but laterally displaced. (a) How far is the emerging ray displaced relative to the normal? (b) What is the perpendicular distance between the original direction of the ray and the direction of the emerging ray?

42 ■■ A 45°–90°–45° prism is made of a material with an index of refraction of 1.85. Can the prism be used to deflect a beam of light by 90° (a) in air or (b) in water?

43 ■■ A person lying at poolside looks over the edge of the pool and sees a bottle cap on the bottom directly below, where the pool depth is 3.5 m. How far below the water surface does the bottle cap appear to be to the person?

44 ■■ What percentage of the actual depth is the apparent depth of an object submerged in water? (Assume that the observer is looking almost straight downward.)

45 ■■ How far will a beam of light travel in flint glass in the time it takes light to travel 25 cm in air?

46 ■■ The critical angle for a certain type of glass is determined to be 39° (at which no light is seen to emerge from a sample). What is the index of refraction of the glass?

47 ■■ A submerged diver shines a light toward the surface of the water at angles of incidence of 40° and 50°. Will a person on the shore see a beam of light emerging from the surface in either case? Justify your answer mathematically.

48 ■■ To a submerged diver looking upward through the water, the altitude of the Sun (the angle between Sun and horizon) appears to be 50°. What is the Sun's actual altitude?

49 ■■ At what angle must a diver submerged in a lake look toward the surface to see the setting Sun?

50 ■■ A coin lies on the bottom of a pool under 1.5 m of water and 0.90 m from the side wall (•Fig. 22.26). If a light beam is incident on the water surface at the wall, at what angle (θ) relative to the wall must the beam be directed so it will illuminate the coin? (See Fig. 21.24.)

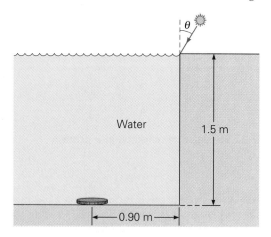

•**FIGURE 22.26 Find the coin**
See Exercise 50. (Drawing not to scale.)

51 ■■■ A crown-glass plate 2.5 cm thick is placed over a newspaper. How far beneath the top surface of the plate would the print appear to be to a person looking almost vertically downward through the plate?

52 ■■■ A circular jacuzzi has a diameter of 1.00 m and a depth of 1.00 m. When empty of water, a person standing a distance of 1.75 m from the near edge can just see the bottom of the opposite wall when looking at an angle of 45°. The jacuzzi is then filled completely with water. How far back from the near side must the person stand so as just to be able to see the bottom of the opposite wall over the near edge?

53 ■■■ An outdoor circular fish pond has a diameter of 4.0 m and a uniform full depth of 1.50 m. A fish halfway down in the pond and 0.50 m from the near side can just see the full height of a 1.8-m tall person. How far away from the edge of the pond is the person?

54 ■■■ A light beam traveling upward in a plastic material with an index of refraction of 1.60 is incident on an upper horizontal air interface at an angle of 45°. (a) Is the beam transmitted? (b) Suppose the upper surface of the plastic material is covered with a layer of liquid with an index of refraction of 1.20. What happens in this case?

55 ■■■ A cube of flint glass sits on a newspaper on a table. Is it possible to see the whole newpaper by looking into one of the vertical sides of the cube?

56 ■■■ Two glass prisms are placed together as shown in •Fig. 22.27. (a) If a beam of light strikes the face of one of the prisms at normal incidence as shown in the figure, at what angle θ does the beam emerge from the other prism? (b) At what angle of incidence would the beam be refracted along the prism interface?

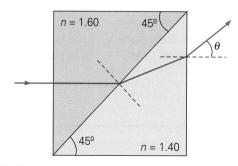

•**FIGURE 22.27 Joined prisms**
See Exercise 56.

22.5 Dispersion

57 Dispersion occurs for (a) monochromatic light, (b) polychromatic light, (c) both of these.

58 A transparent material (a) shows dispersion for $\theta_1 = 90°$, (b) has different n's for different λ's, (c) changes the frequency of a particular light wave, (d) none of these.

59 A prism disperses white light into a spectrum. Can a second prism be used to recombine the spectral components? Explain.

60 ■■ The index of refraction of crown glass is 1.515 for red light and 1.523 for blue light. Find the angle separating rays of the two colors in a piece of crown glass if their angle of incidence is 37°.

61 ■■ White light passes through a prism made of crown glass and strikes an air interface at an angle of 41.15°. Using the indices of refraction given in Exercise 60, describe what happens.

62 ■■ (a) If glass is dispersive, why don't we see a spectrum of colors when sunlight passes through a window pane? (b) Does dispersion occur for polychromatic light incident on a dispersive medium at an angle of 0°? Explain. (Are the speeds of the different colors of light the same in the medium?)

63 ■■ A beam of light with red and blue components having wavelengths of 670 nm and 425 nm, respectively, strikes a slab of fused quartz at an incident angle of 30°. It is observed that the different components are separated by an angle of 0.00131 rad on refraction. If the index of refraction for the red light is 1.4925, what is the index of refraction for the blue light?

64 ■■ Fused quartz is a dispersive medium with an index of refraction of 1.470 for blue light (400 nm in air) and 1.445 for red light (680 nm in air). A beam of light composed of these two colors is incident on a plate of fused quartz at an angle of 45°. (a) What is the angle of separation between the two components of the beam in the quartz? (b) What is the ratio of the wavelengths of the two components in the quartz?

65 ■■■ A beam of red light is incident on an equilateral prism as shown in ●Fig. 22.28. (a) If the index of refraction for red light is 1.400, at what angle θ does the beam emerge from the other face of the prism? (b) Suppose the incident beam were white light. What would be the angular separation of the red and blue components in the emergent beam if the index of refraction for blue light is 1.403? (c) What would be the angular separation if the index of refraction for blue light were 1.405?

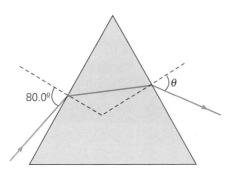

●**FIGURE 22.28 Prism revisited**
See Exercise 65

Additional Exercises

66 When looking into an opaque, empty container that is 15 cm deep at a viewing angle of 50° relative to the vertical side of the container, an observer sees nothing on the bottom. When the container is filled with water, the observer fully sees the coin on the bottom of and just beyond the side of the container (from the same viewing angle). See Fig. 22.24(b). How far is the coin from the side of the container?

67 Light travels from a material whose index of refraction is $n_1 = 2.0$ into another material whose index of refraction is $n_2 = 5/3$. The opposite parallel surface of the second material is exposed to the air. (a) Find the critical angle at which light will be reflected at the interface of the materials. (b) Find the angle at which the light will pass

through the interface and then be reflected at the opposite surface of the second material.

68 A light beam is incident on a mirror at an angle θ_i. The mirror is rotated through an angle ϕ about an axis in the plane of the mirror and perpendicular to the plane of incidence. The incident angle is then $\theta_i' = \theta_i - \phi$ (see ●Fig. 22.29). Show that the reflected ray rotates through an angle $\alpha = 2\phi$. (This is the principle of an optical lever, which is used to measure small angles of rotation.)

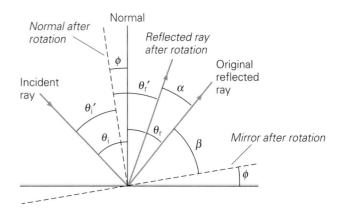

●**FIGURE 22.29 Rotation of reflected ray**
See Exercise 68.

69 Light strikes a surface at an angle of 40° relative to the surface. What is the angle of reflection?

70 A beam of light traveling in water strikes a glass surface at an angle of incidence of 43°. If the angle of refraction in the glass is 35°, what is the index of refraction of the glass?

71 Yellow-green light with a wavelength of 550 nm is incident on the surface of a flat piece of crown glass at an angle of 40°. (a) What is the angle of refraction of the light? (b) What is the speed of the light in the glass? (c) What is the wavelength of the light in the glass?

72 Light strikes water perpendicular to the surface. What is the angle of refraction?

73 Light passes from medium A into medium B at an angle of incidence of 30°. The index of refraction of A is 1.5 times that of B. (a) What is the angle of refraction? (b) What is the ratio of the speed of light in B to the speed of light in A? (c) What is the ratio of the frequency of the light in B to the frequency of light in A? (d) At what angle of incidence would the light be internally reflected?

74 A person in a boat shines a flashlight into the water of a clear lake at an angle of 12° and notes a large fish at an apparent depth of 4.5 m. (a) What are the x and y coordinates of the fish with the origin at the point where the light enters the water? (b) What is the angle of refraction?

23 Mirrors and Lenses

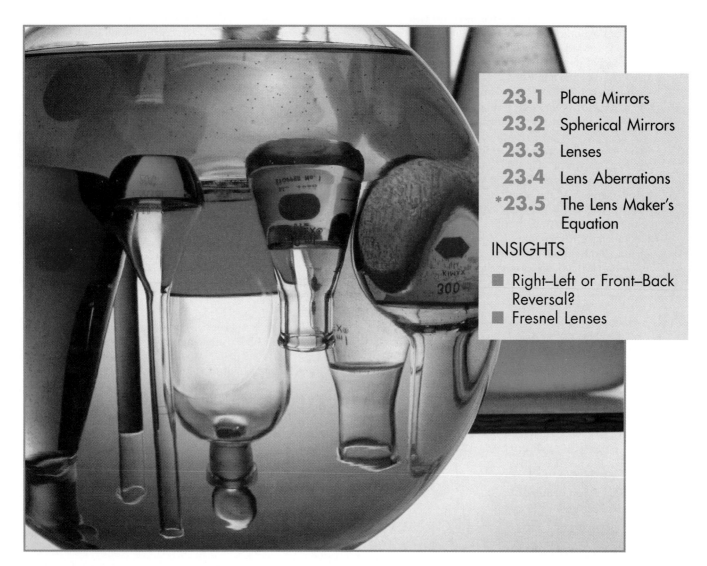

Think of what life would be like if there were no mirrors in bathrooms or on cars, and no one could get glasses. Imagine living in a world without optical images of any kind—no photographs, no movies, no TV. Think about how little we'd know about the universe if there were no telescopes with which to observe distant planets and stars; how little we'd know about biology and medicine if there were no microscopes with which to see cells and bacteria. We sometimes forget how dependent we are on mirrors and lenses.

The first mirror was probably the reflecting surface of a pool of water. Later, people discovered that polished metals and glass have reflective properties. They must also have noticed that when they looked at things through pieces of glass, the objects looked different, depending on the shape of the glass. In some cases, the objects appeared to be enlarged (magnified) or inverted, as in the photo above. In time, people learned to shape glass purposefully into lenses, opening the way for eventual development of the many optical devices we now take so much for granted.

In this chapter you'll learn how mirrors and lenses work. Among other things, you'll discover why the image in the photo above is upside down, whereas your image in an ordinary flat mirror is right-side up—but doesn't seem to comb its hair with the same hand you use!

The optical properties of mirrors and lenses are based on the principles of the reflection and refraction of light, introduced in Chapter 22. In this chapter, you will learn about these general properties by applying geometrical optics. Our emphasis will be on the formation of images by mirrors and lenses, and on the properties of those images. The practical uses of mirrors and lenses in optical instruments such as microscopes and telescopes will be discussed in Chapter 25.

23.1 Plane Mirrors

Objectives: To be able to (a) describe the characteristics of plane mirrors, and (b) explain apparent right–left reversals.

Mirrors are smooth reflecting surfaces, usually made of polished metal or glass that has been coated with some metallic substance. As you know, even an uncoated piece of glass such as a window pane can act as a mirror (see Demonstration 13 on page 706). However, when one side of a piece of glass is coated with a compound of tin, mercury, or silver, its reflectivity is increased, and light is not transmitted through the coating. A mirror may be front-coated or back-coated, depending on the application.

When you look directly into a mirror, you see the reflected images of yourself and objects around you. Usually, these images appear to be behind the mirror (on the other side of the surface). The geometry of a mirror's surface affects the size, orientation, and type of image.

A mirror with a flat surface is called a **plane mirror**. How images are formed by a plane mirror is illustrated by the ray diagram in •Fig. 23.1. As we know, an image appears to be behind or "inside" the mirror. This is because when the mirror reflects a ray from the object to the eye (Fig. 23.1a), it appears to us to originate from behind the mirror. Reflected rays from the top and bottom of an object are shown in Fig. 23.1b. In actuality, light rays coming from all points on the side of the object toward the mirror are reflected, and a complete image is observed.

The image formed in this way *appears* to be behind the mirror. Such an image is called a **virtual image**. Light rays appear to emanate from virtual images, but do not actually do so. However, spherical mirrors (discussed in the next section) can produce images from which light actually emanates. This type of image is called a **real image**. An example of a real image is one that you see on a movie screen.

Notice in Fig. 23.1b the positions or distances of the object and image from the mirror. Quite logically, the distance of an object from a mirror is called the *object distance* (d_o), and the distance its image appears to be behind the mirror is called the *image distance* (d_i). By geometry of similar triangles, it can be shown that $d_o = d_i$, which means that *the image formed by a plane mirror appears to be at a distance behind the mirror that is equal to the distance of the object in front of the mirror.*

We are interested in various characteristics of images. One of these is the size of an image compared to that of its object. This is expressed in terms of the **lateral magnification** (M)*, which is defined as a ratio:

$$M = \frac{\text{image height}}{\text{object height}} = \frac{h_i}{h_o} \qquad (23.1)$$

*Lateral, or height distance, magnification (M) is distinguished from angular, or angular distance, magnification (m) in Chapter 25. In this chapter, we will refer to M simply as the magnification, or magnification factor, for convenience.

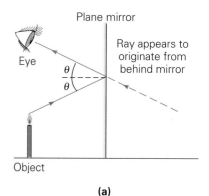

Plane mirror
Ray appears to originate from behind mirror
Eye
θ
θ
Object

(a)

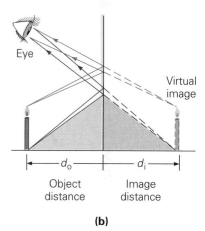

Eye
Virtual image
d_o
d_i
Object distance
Image distance

(b)

•**FIGURE 23.1 Image formed by a plane mirror**
(a) A ray from a point on the object is reflected in the mirror according to the law of reflection. **(b)** Rays from various points on the object produce an extended image. Because the two shaded triangles are similar, the image distance d_i (the distance of the image from the mirror) is equal to the object distance d_o. That is, the image is the same distance behind the mirror as the object is in front of the mirror. The rays appear to converge at the image position. In this case, the image is said to be virtual.

Right–Left or Front–Back Reversal?

A characteristic of a plane mirror that you have probably noticed is the *apparent* right–left reversal of images. If you stand in front of a mirror and raise your right hand, the mirror image will raise its left hand (Fig. 1a). Similarly, if you part your hair on one side, the part will appear on the other side of your image's head.

As shown in the ray diagram in Fig. 1b, this apparent reversal is consistent with the law of reflection. However, note the z axis. Here there is a front–back reversal. That is, the z′ axis (the reflected image of the z axis) has been reversed, so that it points

in the direction opposite that of the z axis, back toward the plane of the mirror. (Note that this is not the case for the x and y axes.) So, we see that a plane mirror image reproduces exactly all object points in the dimensions *parallel* to the mirror surface, but reverses the sequential order of things in the direction *perpendicular* to the mirror surface, giving rise to a front–back reversal.

It is this front–back reversal that produces the apparent right–left reversal. Note in Fig. 1a that with the right-hand palm facing the mirror, the image appears to be that of a left-hand palm. This is because the sequential order of the parts of the right hand

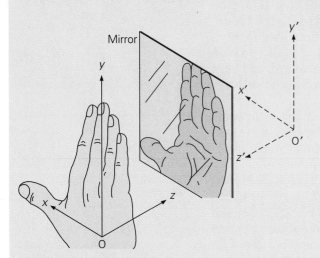
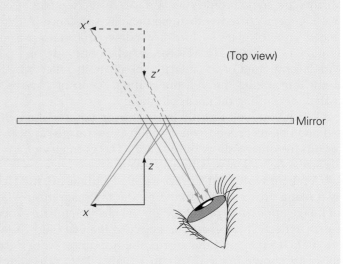

FIGURE 1 Symmetry reversal in a plane mirror
(a) An image in a plane mirror is said to show right–left reversal relative to the object. However, the object and image thumbs are both at the bottom of the mirror and point in the +x direction. There is a front-to-back reversal of the z–z′ axes inasmuch as they point toward each other. This doesn't occur for the x–x′ and y–y′ axes. **(b)** The front–back reversal of the z axis is consistent with the law of reflection. It is this reversal that gives rise to the apparent right–left reversal.

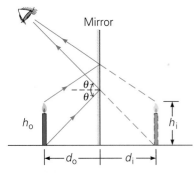

•FIGURE 23.2 Magnification
The lateral or height magnification is given by $M = h_i/h_o$. Writing this relationship as $h_i = Mh_o$, the M is then a magnification factor. For a plane mirror $M = 1$, which means that $h_i = h_o$ and the image is the same height as the object.

Referring to •Fig. 23.2, you should be able to show by similar triangles that $h_i = h_o$, so $M = 1$ for a plane mirror and there is no magnification. That is, you and your image in a plane mirror are the same size. Note that M is a magnification *factor*, since $h_i = Mh_o$.

We use a lighted candle as an object so as to allow us to address another image characteristic, its orientation—that is, whether it is upright or inverted with respect to the orientation of the object. (In sketching ray diagrams, an arrow is a convenient object for this purpose.) For a plane mirror, the image is always upright (or erect). This means that the image is oriented in the same vertical direction as the object. Note in Fig. 23.2, both the object and the image are oriented upward. (If the object and image were both oriented downward, the image would still be described as being upright to indicate that it has the same orientation as the object.)

With other types of mirrors, such as spherical mirrors (which we will consider shortly), it is possible to have inverted images. In this case, the vertical orientation of the image is opposite that of the object.

The main characteristics of an image formed by a plane mirror are summarized in Table 23.1. For another interesting plane mirror characteristic, see the Insight above.

have been reversed in the perpendicular, hand-thickness direction (that is, along the z axis). But it is also because we are symmetrical creatures, with left hands that are "mirror images" of our right hands! It's easier to understand this point if you imagine what you would see in the mirror if your left and right sides were as different as your head and your feet—for example, if you had a right hand but a left wing. Then it would be clear that even if you could step into the mirror and turn around, you would not match your mirror image. In fact, no movements or rotations can ever get you and your mirror image to coincide. (Even our highly symmetrical body plans can't entirely conceal this fact; for example, your image's heart isn't in the right place!)

When working with mirrors, it is good to keep in mind that right and left are directional *senses* (like clockwise and counterclockwise), rather than fixed directions referenced to a coordinate system. For example, consider the images shown in Fig. 2. Notice that the mirror does not reverse the sequence of the objects (letters) in the direction parallel to its surface, but does give a front–back reversal in the perpendicular direction, which is only noticeable for the nonsymmetric Ls. Also note that, with the letters positioned in this way, the front–back reversal of the L's appears not as a "right–left" reversal, but as a "top-to-bottom" reversal.

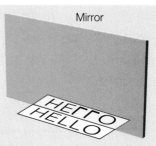

FIGURE 2 Hello!
The letters of HELLO on a piece of paper placed on a surface next to a vertical mirror form an image that shows no right–left reversal in the sequence of letters. However, the nonsymmetric Ls show a front-to-back reversal, which in this instance appears as a top-to-bottom reversal.

Finally, Fig. 3 shows a practical application of right–left reversal. But wait—this is left–right reversal! The lettering sequence has been reversed. What's going on? (See Exercise 6.)

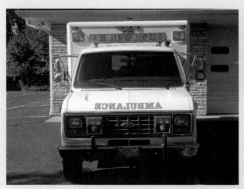

FIGURE 3 Practical right–left reversal
The letters of the word AMBULANCE are printed backwards and are reversed in sequence so that they appear in the proper orientation and order when seen in a rearview mirror.

TABLE 23.1   Characteristics of Images Formed by Plane Mirrors

The image distance is equal to the object distance ($d_i = d_o$). That is, the image appears to be as far behind the mirror as the object is in front.

The image is virtual, upright, and unmagnified ($M = 1$).

EXAMPLE 23.1 ■ ALL OF ME: MINIMUM MIRROR LENGTH

What is the minimum vertical length of a plane mirror needed for a person to be able to see a complete (head-to-toe) image?

Solution.   To determine this length, consider the situation shown in ●Fig. 23.3. With a mirror of minimum length, a ray from the top of the head would be reflected at the top of the mirror, and the ray from the feet would be reflected at the bottom of the mirror. The length L of the mirror is then the distance between the dashed lines perpendicular to the mirror at its top and bottom.

However, these lines are also the normals for the ray reflections. By the law of reflection, they bisect the angles between incident and reflected rays, that is, $\theta_i = \theta_r$. Then, because the large shaded triangles are similar, the length of the mirror from its bottom to a point even with the person's eyes is $h_1/2$, where h_1 is the person's height from the feet to the eyes. Similarly, the small upper length of the mirror is $h_2/2$ (the vertical distance between the eyes and the top of mirror). Thus,

$$L = \frac{h_1}{2} + \frac{h_2}{2} = \frac{h_1 + h_2}{2} = \frac{h}{2}$$

where h is the person's total height.

Hence, for a person to see his or her complete image in a plane mirror, the minimum height or vertical length of the mirror must be one-half the height of the person.

Follow-up Exercise. What effect does a person's distance from the mirror have on the minimum mirror length required to produce his or her complete image?

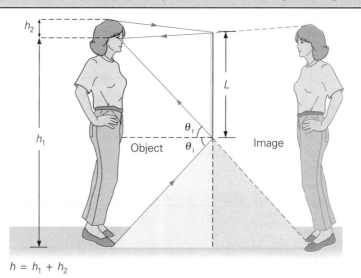

•**FIGURE 23.3 Seeing it all**
The minimum height or vertical length of a plane mirror needed for a person to see his or her complete (head-to-toe) image turns out to be one-half the person's height.

DEMONSTRATION 13 ■ A Candle Burning Underwater?

It would appear so, but you know this is not possible. It's really a matter of reflection and an image.

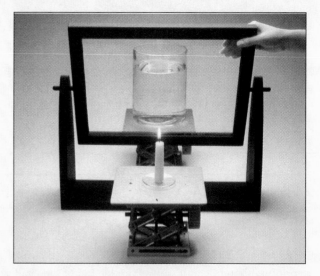

(a) The black frame holds a pane of glass, which acts as a plane mirror. The burning candle seen in the water is the image of the candle on the front stand. There is a container of water on a similar stand behind the glass, but no burning candle.

(b) The effect can be removed by tilting the glass—the image can no longer be seen from this viewing point. (Something to do with the law of reflection. What?)

23.2 Spherical Mirrors

Objectives: To be able to (a) distinguish between converging and diverging spherical mirrors, (b) describe images and their characteristics, and (c) determine these image characteristics using ray diagrams and the spherical mirror equation.

As the name implies, a **spherical mirror** is a reflecting surface with spherical geometry. As •Fig. 23.4 shows, if a portion of a sphere of radius R is sliced off along a plane, the severed section has the shape of a spherical mirror. Either the inside or outside of such a section can be reflective. For inside reflections, the mirror is a **concave mirror** (think of looking into a cave to remember the indented surface). For outside reflections, the mirror is a **convex mirror**.

The radial line through the center of the spherical mirror is called the *optic axis*, and it intersects the mirror surface at the *vertex* of the spherical section (Fig. 23.4). The point on the optic axis that corresponds to the center of the sphere of which the mirror forms a section is called the **center of curvature** (C). The distance between the vertex and the center of curvature is equal to the radius of the sphere and is called the **radius of curvature** (R).

When rays parallel to the optic axis are incident on a concave mirror, the reflected rays intersect at a common point called the **focal point** (F)*. As a result, a concave mirror is called a **converging mirror** (•Fig. 23.5a). Similarly, a beam parallel to the optic axis of a convex mirror diverges on reflection, as though the reflected rays came from a focal point behind the mirror's surface (Fig. 23.5b). Thus, a convex mirror is called a **diverging mirror**. An example of a diverging mirror is shown in •Fig. 23.6. In seeing diverging rays, the brain interprets or assumes there is an object from which the rays *appear* to diverge, even though there is none there. The true object is somewhere else.

<u>**Note:**</u>
Concave mirror = converging mirror
Convex mirror = diverging mirror

The distance from the vertex to the focal point of a spherical mirror is called the **focal length** (f). The focal length is related to the radius of curvature by this simple equation:

$$f = \frac{R}{2} \qquad \begin{array}{l} \textit{focal length,} \\ \textit{spherical mirror} \end{array} \qquad (23.2)$$

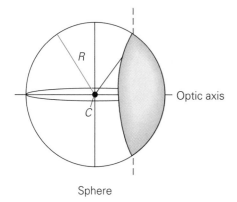

Sphere

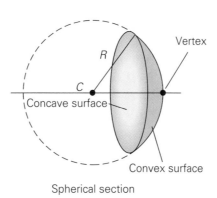

Spherical section

•**FIGURE 23.4 Spherical mirrors** A spherical mirror is a section of a sphere. Either the outside (convex) surface or the inside (concave) surface of the spherical section may be the reflecting surface.

*This is the case for the approximation where the width of the mirror or the illuminated area is small compared to the radius of curvature, that is, for small angles of reflection.

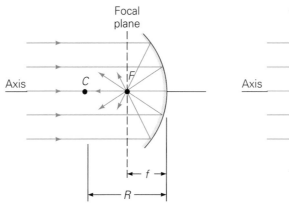

(a) Concave or converging mirror   **(b) Convex or diverging mirror**

•**FIGURE 23.5 Focal point**
(a) Rays parallel and close to the optic axis of a concave spherical converge at the focal point *F*. Off-axis parallel rays converge in the focal plane. **(b)** Rays parallel to the optic axis of a convex spherical mirror are reflected along paths as though they came from a focal point behind the mirror.

In some instances, spherical mirrors form virtual images that appear to be behind the surface just as images with plane mirrors do. However, in other instances, spherical mirrors form images that can be seen on a screen placed at certain positions. These are real images, formed by converging rays. The distinction between these two kinds of images is sometimes said to be that a real image can be formed on a screen, and a virtual image cannot. Stated another way, light actually emanates from a real image, but only appears to do so from a virtual image.

Ray Diagrams

The characteristics of images formed by spherical mirrors can be determined using geometrical optics. The method involves drawing rays emanating from one or more points on an object. The law of reflection applies ($\theta_i = \theta_r$), and certain key rays are defined with respect to the mirror's geometry as follows:

- A **parallel ray** is a ray that is incident along a path parallel to the optic axis and is reflected through the focal point (as are all rays near and parallel to the axis).

- A **chief ray**, or **radial ray**, is a ray that is incident through the center of curvature (*C*). Since it is incident normal to the mirror's surface, this ray is reflected back along its incident path, through *C*.

- A **focal ray** is a ray that passes through (or appears to go through) the focal point and is reflected parallel to the optic axis. (It is a mirror image, so to speak, of a parallel ray.)

These rays are illustrated in the ray diagrams in •Fig. 23.7 for concave and convex mirrors. It is customary to use the tip of the object (for example, the head of an arrow or the flame of a candle) as the origin of the rays. This makes it easy to see whether the image is upright or inverted. The corresponding point of the image is at the point of intersection of the rays. Also, the candle is arbitrarily taken to be upright with its base on the optic axis. Keep in mind, however, that properly traced rays from *any* point on the object can be used to find the image. Every point on a visible object acts as an emitter of light. For example, for a candle, the flame emits its own light, but every point on the wax stem reflects light.

•**FIGURE 23.6 Diverging mirror**
Note by reverse-ray tracing in Fig. 23.5b that a diverging (convex) spherical mirror gives an expanded field of view, as can be seen in this store-monitoring mirror.

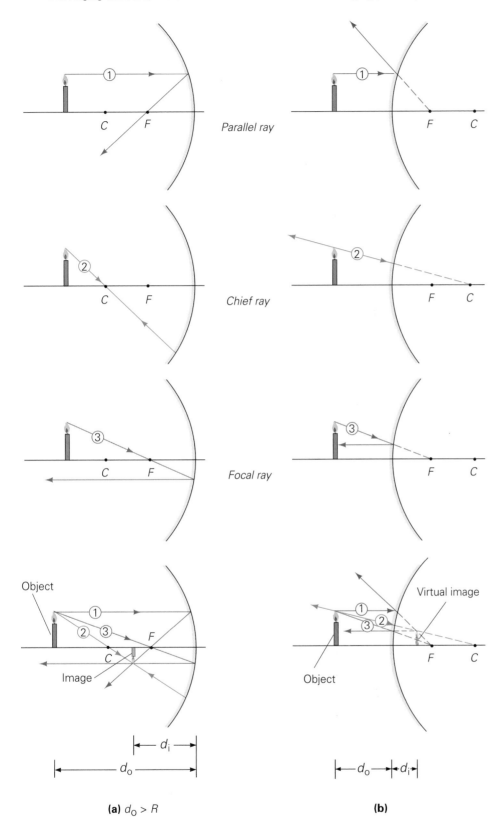

Converging (concave) mirror

Diverging (convex) mirror

Parallel ray

Chief ray

Focal ray

Object

Image

(a) $d_o > R$

Virtual image

Object

(b)

•FIGURE 23.7   Ray diagrams
Ray diagrams to find the images
for spherical mirrors can be drawn
using three rays: (1) A parallel ray
is reflected through the focal point
F for a converging mirror, and
appears to come from the internal
focal point for a diverging mirror.
(2) A chief ray is incident through
the center of curvature C and is
reflected back along its path of
incidence for a converging mirror,
and appears to be reflected from
the internal center of curvature for
a diverging mirror. (3) A focal ray
passing through the focal point F is
reflected parallel to the optic axis
for a converging mirror, and the
incident ray appears to pass
through the internal focal point for
a diverging mirror. **(a)** Ray
diagram for a converging mirror
for $d_o > R$. With the rays coming
from the tip of the object arrow,
their intersection defines the
location of the tip of the image
arrow relative to the optic axis.
(b) Ray diagram for a diverging
mirror. Here the image is virtual
and behind or inside the mirror.

Note in Fig. 23.7a that with a concave mirror and an object at a distance greater than the radius of curvature ($d_o > R$), a real image is formed. That is, the image may be seen on a screen (for example, a piece of white paper) positioned at a distance d_i from the mirror. The image has the characteristics of being real, inverted, and smaller than the object.

The rays reflected from a convex mirror diverge (Fig. 23.7b). Projecting the rays behind the mirror to find where they intersect indicates that the image is virtual. (The candle is drawn with its flame at the point of intersection.) This image is analogous to the virtual image formed by a plane mirror, and it cannot be projected on a screen. Since the reflected rays for an object at any distance from a convex mirror diverge, *a diverging mirror always forms a virtual image.*

For a converging spherical mirror, the characteristics of the image change with the distance of the object from the mirror. There are two points at which dramatic changes take place: C and F (the center of curvature and the focal point). These points divide the optic axis into three regions, as shown in •Fig. 23.8*: $d_o > R$, $R > d_o > f$, and $d_o < f$. Let's start with an object in the region farthest from the mirror ($d_o > R$) and move progressively toward the mirror.

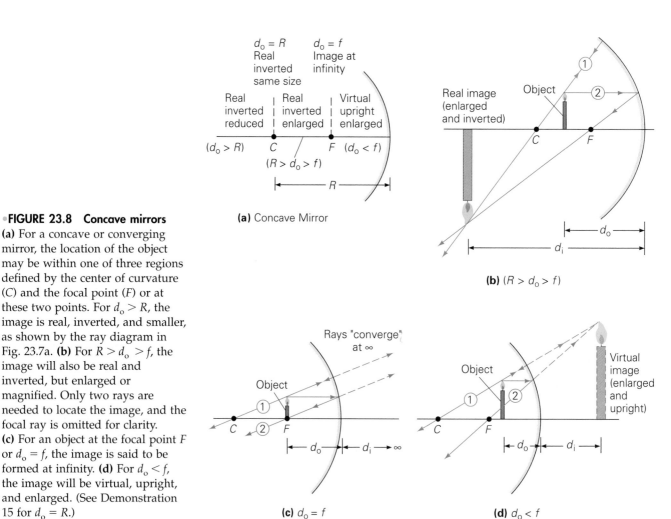

•**FIGURE 23.8 Concave mirrors**
(a) For a concave or converging mirror, the location of the object may be within one of three regions defined by the center of curvature (C) and the focal point (F) or at these two points. For $d_o > R$, the image is real, inverted, and smaller, as shown by the ray diagram in Fig. 23.7a. **(b)** For $R > d_o > f$, the image will also be real and inverted, but enlarged or magnified. Only two rays are needed to locate the image, and the focal ray is omitted for clarity. **(c)** For an object at the focal point F or $d_o = f$, the image is said to be formed at infinity. **(d)** For $d_o < f$, the image will be virtual, upright, and enlarged. (See Demonstration 15 for $d_o = R$.)

*Only two rays are needed to determine the image, and we will use only two, the parallel and chief rays, for illustration clarity. However, you are encouraged (or may be required) to draw the third ray in your diagrams as a check.

- The case of $d_o > R$ has already been dealt with in Fig. 23.7a.
- The case of $d_o = R$ is shown in Demonstration 14. The image is real, inverted, and the same size as the object.
- When $R > d_o > f$, we see from the ray diagram in Fig. 23.8b that an enlarged, inverted, real image is formed. As can be seen from the sequence, the image is magnified when the object is inside the center of curvature C.
- When $d_o = f$, so that the object is at the focal point (Fig. 23.8c), the reflected rays are parallel and the image is said to be formed at infinity. This expresses the idea that parallel lines converge at infinity (like parallel railroad tracks that appear to converge at a great distance). The focal point F is a special "crossover" point. When $d_o > f$, the image is real, and (as we shall see below) when $d_o < f$, the image is virtual. We can't actually see an image formed at infinity, but we say the image is formed there because of symmetry with the case represented in Fig. 23.8c—when an object is at "infinity," meaning that it is so far away that the rays emanating from it are essentially parallel, its image is formed at F. By reverse ray tracing, rays from an object at a great (taken as "infinite") distance from the mirror can be shown to be essentially parallel when they are near the mirror and to form an image on a screen aligned in the focal plane. This fact provides a method for determining the focal length of a concave mirror.
- When $d_o < f$, so that the object is inside the focal point (between the focal point and the mirror's surface), a virtual image is formed (Fig. 23.8d).

DEMONSTRATION 14 ■ A Candle Burning at Both Ends

Or is it? Notice that one flame is burning downward, which is rather strange. It's an illusion done with a spherical concave mirror.

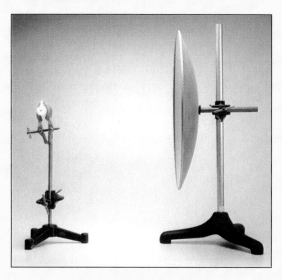

(a) When an object is at the center of curvature of a spherical concave mirror, a real image is formed that is inverted and the same size as the object, and the image distance is the same as the object distance. What is seen here is a horizontal candle burning at one end (flame up) and its overlapping image (flame down). Viewed at the same level, the inverted flame image appears to be at the opposite end of the candle.

(b) A side view showing the burning end of the horizontal candle in front of the spherical mirror.

The position and size of the image may be determined graphically by ray diagrams drawn to scale. However, these can be determined more quickly by analytical methods. The distances and focal length can be shown to be related by what is known as the **spherical mirror equation.**

$$\frac{1}{d_o} + \frac{1}{d_i} = \frac{1}{f} = \frac{2}{R} \qquad \textit{spherical mirror equation} \qquad (23.3)$$

Note that this equation can be written in terms of either the radius of curvature R or the focal distance f, since by Eq. 23.2, $f = R/2$.

Since the image distance d_i is often the quantity to be found, a convenient alternative form of this equation is

$$d_i = \frac{d_o f}{d_o - f} \qquad \begin{array}{l}\textit{spherical mirror equation} \\ \textit{in different form}\end{array} \qquad (23.4)$$

The **magnification factor** M can also be found analytically. This is expressed in terms of the image and object distances.

$$M = -\frac{d_i}{d_o} \qquad \begin{array}{l}\textit{magnification equation} \\ \textit{for a spherical mirror}\end{array} \qquad (23.5)$$

The minus sign is added as a sign convention to indicate the orientation of the image as given below. Hence, if $|M| > 1$, the image is magnified, or larger than the object. If $|M| < 1$, the image is reduced, or smaller than the object. (The origin of this equation is shown in the accompanying Box.)

*DERIVATION OF THE SPHERICAL MIRROR EQUATION

Students often wonder where various equations come from. Often, they appear in a textbook as if by magic. In many instances, however, their derivation is not excessively complicated. This is the case for the spherical mirror equation, which can be arrived at with the aid of a little trigonometry. Consider the ray diagram in •Fig 23.9. The object and image distances (d_o and d_i) and the heights of the object and image (h_o and h_i) are shown. Note that these lengths make up the bases and heights of triangles formed by the ray reflected at the vertex (V). These triangles ($O'VO$ and $I'VI$) are similar, since by the law of reflection their angles at V are equal. Hence, we can write

$$\frac{h_i}{h_o} = \frac{d_i}{d_o} \qquad (1)$$

The other (focal) ray through F also forms similar triangles, $O'FO$ and VFA. (Why are they similar?) Note that the bases of these triangles are $VF = f$ and $OF = d_o - f$. Then, if VA is taken to be h_i (in the approximation that the mirror is small compared to its radius),

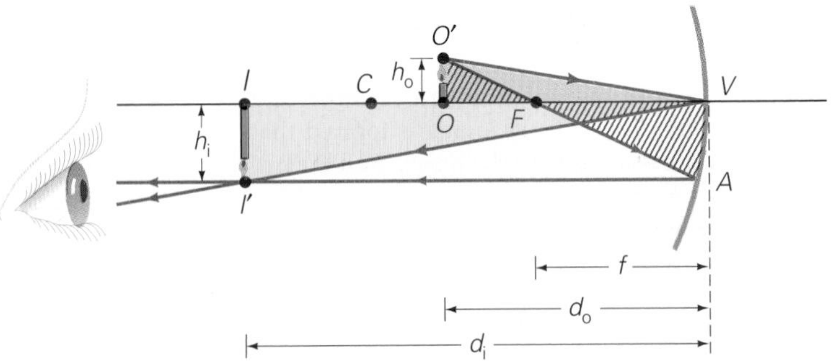

•**FIGURE 23.9 Spherical mirror equation**
The rays provide the geometry, through similar triangles, for the derivation of the spherical mirror equation. See text for description.

$$\frac{h_i}{h_o} = \frac{VF}{OF} = \frac{f}{d_o - f} \qquad (2)$$

Equating Eqs. 1 and 2, we have

$$\frac{d_i}{d_o} = \frac{f}{d_o - f} \qquad (3)$$

Multiplying both sides of this equation by d_o gives

$$d_i = \frac{d_o f}{d_o - f}$$

which is the spherical mirror equation in the form of Eq. 23.4.

We can go one step further and get the magnification factor M. Recall that the lateral magnification is the ratio of the height of the image (h_i) to the height of the object (h_o). That is, $M = h_i/h_o$ (Eq. 23.1). And, by Eq. 1 above,

$$M = \frac{h_i}{h_o} = -\frac{d_i}{d_o}$$

where the minus sign is inserted as a convention, as described in the text.

The signs on the various quantities are very important in the application of the preceding equations. This book uses the following sign conventions for single spherical mirrors:

Sign conventions for mirrors

- The focal length f (or R) is positive for a concave mirror and negative for a convex mirror.
- The object distance d_o is always taken to be positive.
- The image distance d_i is positive for a real image (formed on the same side of the mirror as the object) and negative for a virtual image (formed behind the mirror).
- The magnification M is positive for an upright image and negative for an inverted image.

This convention is summarized in Table 23.2.

The following examples show how the equations for a spherical mirror are used.

TABLE 23.2 Sign Convention for Spherical Mirrors

Concave mirror: f positive
Convex mirror: f negative
d_o always positive

d_i	Image	M	Image
+	Real	+	Upright
−	Virtual	−	Inverted

EXAMPLE 23.2 ■ WHAT KIND OF IMAGE? BEHAVIOR OF A CONCAVE MIRROR

A concave mirror has a radius of curvature of 30 cm. If an object is placed (a) 45 cm, (b) 20 cm, and (c) 10 cm from the mirror, where is the image formed and what are its characteristics? (Specify real or virtual, upright or inverted, and larger or smaller for each image.)

Solution.

Given: $R = 30$ cm   *Find:* d_i and image characteristics
         so $f = R/2 = 15$ cm   for given object distances
         (a) $d_o = 45$ cm
         (b) $d_o = 20$ cm
         (c) $d_o = 10$ cm

Note that these object distances correspond to the regions shown in Fig. 23.8a.

(a) In this case, the object distance is greater than the radius of curvature ($d_o > R$), and

$$\frac{1}{d_o} + \frac{1}{d_i} = \frac{1}{f}$$

or

$$\frac{1}{45} + \frac{1}{d_i} = \frac{1}{15}$$

where the units (cm) have been omitted for simplicity. Then

$$\frac{1}{d_i} = \frac{2}{45}$$

or

$$d_i = \frac{45}{2} = +22.5 \text{ cm}$$

and

$$M = -\frac{d_i}{d_o} = -\frac{22.5 \text{ cm}}{45 \text{ cm}} = -\frac{1}{2}$$

Thus, the image is real (positive d_i), inverted (negative M), and $\frac{1}{2}$ as large as the object ($|M| = \frac{1}{2}$).

(b) Here $R > d_o > f$, and the object is between the focal point and center of curvature. Using Eq. 23.4,

$$d_i = \frac{d_o f}{d_o - f} = \frac{(20 \text{ cm})(15 \text{ cm})}{20 \text{ cm} - 15 \text{ cm}} = 60 \text{ cm}$$

and

$$M = -\frac{d_i}{d_o} = -\frac{60 \text{ cm}}{20 \text{ cm}} = -3.0$$

In this case, the image is real, inverted, and three times the size of the object.

(c) For this case, $d_o < f$, and the object is inside the focal point. Then,

$$d_i = \frac{d_o f}{d_o - f} = \frac{(10 \text{ cm})(15 \text{ cm})}{10 \text{ cm} - 15 \text{ cm}} = -30 \text{ cm}$$

Then

$$M = -\frac{d_i}{d_o} = -\frac{(-30 \text{ cm})}{10 \text{ cm}} = +3.0$$

In this case, the image is virtual (negative d_i), upright (positive M), and three times the size of the object.

From the denominator of the right-hand side of the equation for d_i (Eq. 23.4), you can see that d_i will always be negative when d_o is less than f. Therefore, a virtual image is always formed for an object inside the focal point of a converging mirror.

(Draw representative ray diagrams for each of these cases to see that the image characteristics are correct.)

Follow-up Exercise. For the converging mirror in this Example, where is the image formed and what are its characteristics if the object is at 30 cm, or $d_o = R$?

PROBLEM-SOLVING HINT

When using the spherical mirror equations to find image characteristics, it is helpful to first make a quick sketch of the ray diagram for the situation. Doing this shows you the image characteristics and helps you avoid making mistakes when applying the sign convention. *The ray diagram and the mathematical solution must agree.*

EXAMPLE 23.3 ■ SIMILARITIES AND DIFFERENCES: BEHAVIOR OF A CONVEX MIRROR

An object is 30 cm in front of a diverging mirror that has a focal length of 10 cm. Where is the image and what are its characteristics?

Solution.

Given: $d_o = 30$ cm   *Find:* d_i and image characteristics
          $f = -10$ cm

Note that the focal length is negative for a convex mirror (see Table 23.2). Using Eq. 23.4,

$$d_i = \frac{d_o f}{d_o - f} = \frac{(30\text{ cm})(-10\text{ cm})}{30\text{ cm} - (-10\text{ cm})} = -7.5\text{ cm}$$

Then

$$M = -\frac{d_i}{d_o} = -\frac{(-7.5\text{ cm})}{30\text{ cm}} = +0.25$$

Thus, the image is virtual (negative d_o) and upright (positive M) and is 0.25 times $(\frac{1}{4})$ the size (height) of the object.

Follow-up Exercise.   As pointed out previously, a diverging mirror always forms a virtual image. What about the other characteristics of the image: its orientation and magnification? Can any general statements be made about them?

EXAMPLE 23.4 ■ IN THE DISTANCE: FINDING THE FOCAL LENGTH OF A CONCAVE MIRROR

Where is the image formed by a concave mirror if the object is at infinity? (An object at a great distance from a mirror, relative to the mirror's dimensions, may be considered to be at infinity.)

Solution.   With $d_o = \infty$, we have

$$\frac{1}{d_o} + \frac{1}{d_i} = \frac{1}{\infty} + \frac{1}{d_i} = \frac{1}{f} \quad \text{or} \quad d_i = f$$

Thus, the image is real ($+d_i$) and is formed at the focal point (or in the focal plane for an extended object).

This result provides an experimental means of determining the focal length of such a mirror. A screen (a piece of white paper) is adjusted until an image is formed on it. The screen is then in the focal plane of the mirror. (Does the magnification equation hold for this special case?)

Follow-up Exercise.   In the limit, a plane mirror may be thought of as a section of a sphere having an infinite radius of curvature ($R = \infty$). What do the spherical mirror equation and magnification factor tell you about the image characteristics of a plane mirror?

The descriptions of image characteristics for spherical mirrors are true only for objects near the optic axis, that is, only for small angles of reflection. If these conditions do not hold, the images will be blurred (out of focus) or distorted, because not all of the rays will converge in the same plane. As illustrated in •Fig. 23.10, incident parallel rays far from the optic axis do not converge at the focal point. The farther the incident ray is from the axis, the more distant is its reflected ray from the focal point. This effect is called **spherical aberration**.

Spherical aberration does not occur with a parabolic mirror. (As the name implies, a section through the center of a parabolic mirror has the form of a parabola.) All of the incident rays parallel to the optic axis of such a mirror have a common focal point. For this reason, parabolic mirrors are used in most astronomical telescopes, as we will see in the next chapter. However, these mirrors are more difficult to make than spherical mirrors (and therefore more expensive).

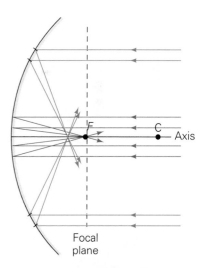

•**FIGURE 23.10 Spherical aberration for a mirror**
According to the small-angle approximation, rays parallel to and near the mirror's axis converge at the focal point. However, when parallel rays not near the axis are reflected, they converge in front of the focal point. This effect is called spherical aberration and gives rise to blurred images.

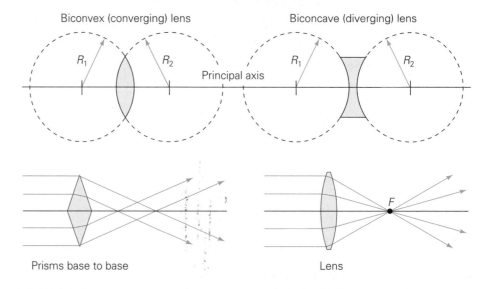

•FIGURE 23.11 Spherical lenses
Spherical lenses have surfaces defined by two spheres, and the surfaces may be either convex or concave. Biconvex and biconcave lenses are shown here. If $R_1 = R_2$, a lens is spherically symmetric.

Biconvex (converging) lens   Biconcave (diverging) lens

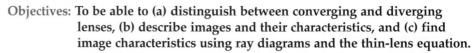

(a) Biconvex (converging) lens

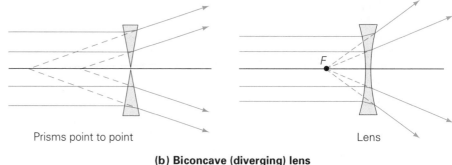

(b) Biconcave (diverging) lens

•FIGURE 23.12 Converging and diverging lenses
(a) A biconvex, or converging, lens may be approximated by two prisms placed base to base. For a thin biconvex lens, rays parallel to the axis converge at the focal point F. **(b)** A biconcave, or diverging, lens may be approximated by two prisms placed tip to tip. Rays parallel to the axis of a biconcave lens appear to diverge from a focal point on the incident side of the lens.

•FIGURE 23.13 Burning glass
A magnifying glass can be used to focus the Sun's rays to a point—with incendiary results.

23.3 Lenses

Objectives: To be able to (a) distinguish between converging and diverging lenses, (b) describe images and their characteristics, and (c) find image characteristics using ray diagrams and the thin-lens equation.

The word "lens" comes from the Latin word for lentil, a seed whose shape is similar to that of a common lens. An optical **lens** is made from some transparent material (most commonly glass but sometimes plastics or crystals). One or both surfaces usually have a spherical contour. Biconvex spherical lenses (both surfaces convex) and biconcave spherical lenses (both surfaces concave) are illustrated in •Fig. 23.11.

The properties of lenses are due to the refraction of light passing through them. When light rays pass through a lens, they are bent, or deviated from their original paths, according to the law of refraction. You studied refraction by prisms in Chapter 22. To analyze lens refraction, a biconvex lens may be approximated by two prisms placed base to base, as shown in •Fig. 23.12a. A biconvex lens is a **converging lens**: incident light rays parallel to the lens axis converge at a focal point on the opposite side of the lens. You may have focused the Sun's rays with a magnifying glass (a biconvex, or converging, lens) and have witnessed the concentration of radiant energy that results (•Fig. 23.13). The parallel rays coming from the Sun or some other distant object (at infinity) converge at the focal point. This fact provides a way for experimentally determining the focal length of a converging lens.

Conversely, a biconcave lens can be approximated by two prisms placed point to point (•Fig. 23.12b). A biconcave lens is a **diverging lens**: incident parallel rays emerge from the lens as though they emanated from a focal point on the incident side of the lens.

There are several types of converging and diverging lenses (•Fig. 23.14). Meniscus lenses are the type most commonly used for corrective eyeglasses. In general, a converging lens is thicker at its center than at its periphery, and a diverging lens is thinner at its center than at its periphery. This discussion will be limited to spherically symmetric biconvex and biconcave lenses, that is, ones for which both surfaces have the same radius of curvature.

When light passes through a lens, it is refracted and displaced laterally, as shown in Example 22.4 (Fig. 22.10). If a lens is thick, this displacement may be fairly large and can complicate the analysis of the lens's characteristics. This problem does not arise with thin lenses, for which the refractive displacement of transmitted light is negligible. This discussion will be limited to thin lenses.

Like a spherical mirror, a lens with spherical geometry has *for each lens surface* a center of curvature, a radius of curvature, a focal point, and a focal length. If each surface has the same radius of curvature, the focal points are at equal distances on either side of the lens. However, for a spherical lens, $f \neq R/2$, as it is for a spherical mirror (see Section 23.5). Usually, only the focal length of a lens is specified rather than the radius of curvature.

The general rules for drawing ray diagrams for lenses are similar to those for spherical mirrors, but obviously some modifications are necessary since light passes through a lens in either direction. Opposite sides of a lens are generally distinguished as the object and image sides. The object side is, of course, the side on which an object is positioned, and the image side is the opposite side of the lens (where a real image would be formed). The rays from a point on an object are drawn as follows:

- A **parallel ray** is a ray that is parallel to the lens axis on incidence and that after refraction either passes through the focal point on the image side of a converging lens, *or* appears to diverge from the focal point on the object side of a diverging lens.
- A **chief ray** is a ray that passes through the center of the lens and is undeviated.*
- A **focal ray** is a ray that passes through the focal point of the object side of a converging lens, *or* appears to pass through the focal point of the image side of a diverging lens, and after refraction is parallel to the lens axis.

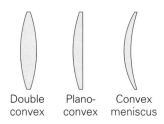

Double convex   Plano-convex   Convex meniscus

Converging lenses

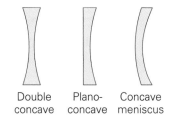

Double concave   Plano-concave   Concave meniscus

Diverging lenses

•**FIGURE 23.14 Lens shapes** The wide variety of lens shapes are generally categorized as converging or diverging. In general, a converging lens is thicker at its center than at the periphery, and a diverging lens is thinner at its center than at the periphery. This may be determined by feeling the lens with the finger and thumb.

DEMONSTRATION 15 ■ Inverted Image

A demonstration of how a cylindrical convex lens produces an inverted image.

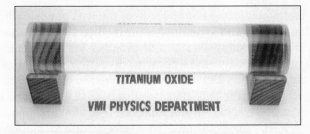

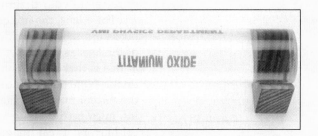

(a) Observe the words TITANIUM OXIDE as placed before a cylindrical plastic rod.
Note: A similar inversion occurs for spherical convex lenses.

(b) Viewed through the rod, the letters in TITANIUM are inverted, but the letters in OXIDE appear not to be. What is wrong, or is there something wrong?

*This incorporates the thin lens approximation (negligible lateral deviation).

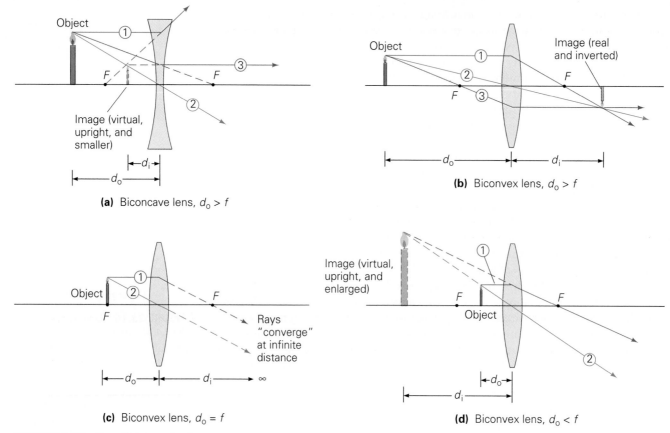

(a) Biconcave lens, $d_o > f$

(b) Biconvex lens, $d_o > f$

(c) Biconvex lens, $d_o = f$

(d) Biconvex lens, $d_o < f$

•**FIGURE 23.15 Ray diagrams for lenses**
As in ray diagrams for mirrors, parallel rays (1), chief rays (2), and focal rays (3) may be drawn to analyze lenses. These rays are shown for a diverging biconcave lens **(a)** and for a converging biconvex lens **(b)** when $d_o > f$. Diagrams **(c)** and **(d)** for a converging lens show the ray diagrams when $d_o = f$ and $d_o < f$, respectively. The focal rays have been omitted in these diagrams for clarity. For $d_o = f$, the image is formed at infinity, and for $d_o < f$, the image is virtual. (Note that for simplicity, the refraction of the rays is depicted as if it occurred at the center of each lens. In reality, it would occur at the air/glass and glass/air surfaces of each lens.)

These rays are shown in •Fig. 23.15 for converging and diverging lenses. Focal rays are drawn only in the initial diagrams. As with spherical mirrors, only two rays are needed to determine the image, and we will use the parallel and chief rays. (As in the case of mirrors, however, it is generally a good idea to include a third ray in your diagrams as a check.)

For a lens, the image is real when formed on the side of the lens opposite the object (on the image side, •Fig. 23.16a) and virtual when formed on the same side of the lens as the object (on the object side, •Fig. 23.16b). Another way of looking at this is that rays converge at a real image and appear to diverge from a virtual image. Note that for a diverging lens (Fig. 23.15a), the rays appear to diverge from the virtual image on the object side of the lens after being refracted. Like diverging mirrors, diverging lenses can form only virtual images. Also, as with spherical mirrors, a real image from a lens can be formed on a screen but a virtual image cannot.

Regions for the object distance with a converging lens could be similarly defined, as was done for a converging mirror in Fig. 23.11a. Here, an object distance of $d_o = 2f$ for a converging lens has a significance similar to that of $d_o = R$ for a converging mirror. However, the center of curvature for a converging lens does not have the same distinction as it does for a mirror, since $R \neq 2f$ for a lens. (See Demonstration 15 on the previous page.)

The image distances and characteristics for a lens can also be found analytically. The equations for thin symmetrical biconvex and biconcave lenses are identical to those for spherical mirrors. The **thin lens equation** is

$$\frac{1}{d_o} + \frac{1}{d_i} = \frac{1}{f} \quad \text{or} \quad d_i = \frac{d_o f}{d_o - f} \qquad \text{thin lens equation} \qquad (23.6)$$

The **magnification factor** is given by

$$M = -\frac{d_i}{d_o} \qquad \begin{array}{l} \textit{magnification equation} \\ \textit{for a thin lens} \end{array} \qquad (23.7)$$

(A thin lens is one for which the thickness of the lens is assumed negligible compared to its focal length.) The sign conventions for these thin lens equations are similar to those for spherical mirrors.

- The focal length (f) is positive for a converging lens (sometimes called a positive lens) and negative for a diverging lens (sometimes called a negative lens).
- The object distance (d_o) is always taken as positive for a single lens.
- The image distance (d_i) is positive for a real image (which is located on the image side of the lens) and negative for a virtual image (which is located on the object side of the lens).
- The magnification (M) is positive for an upright image and negative for an inverted image.

Just as when working with mirrors, it is helpful to sketch a ray diagram before working a lens problem analytically.

(a)

(b)

•**FIGURE 23.16 Images formed by a converging lens**
(a) When the distance of the object from the lens is greater than the focal length ($d_o > f$), an inverted, reduced, real image is formed on the opposite (image) side of the lens. **(b)** When the distance of the object from the lens is less than the focal length ($d_o < f$), an upright, virtual image is formed on the object side of the lens. It appears to be larger and more distant than the actual object, but cannot be projected on a screen.

EXAMPLE 23.5 ■ TWO IMAGES: BEHAVIOR OF A CONVERGING LENS

A biconvex lens has a focal length of 12 cm. Where is the image formed and what are its characteristics for an object (a) 18 cm from the lens and (b) 4 cm from the lens?

Solution.

Given: $f = 12$ cm
  (a) $d_o = 18$ cm
  (b) $d_o = 4$ cm (taken to be exact for simplicity)

Find: d_i and image characteristics for both parts

(a) Using the fractional form of Eq. 23.6,

$$\frac{1}{d_o} + \frac{1}{d_i} = \frac{1}{f}$$

or

$$\frac{1}{18} + \frac{1}{d_i} = \frac{1}{12}$$

With a common denominator,

$$\frac{2}{36} + \frac{1}{d_i} = \frac{3}{36}$$

Thus,

$$\frac{1}{d_i} = \frac{1}{36}$$

or

$$d_i = 36 \text{ cm}$$

Then

$$M = -\frac{d_i}{d_o} = -\frac{36}{18} = -2$$

The image is real (positive d_i), inverted ($-M$), and twice as tall as the object ($|M| = 2$).

(b) Using the alternative form of Eq. 23.6 for d_i gives

$$d_i = \frac{d_o f}{d_o - f} = \frac{(4 \text{ cm})(12 \text{ cm})}{4 \text{ cm} - 12 \text{ cm}} = -6 \text{ cm}$$

Then

$$M = -\frac{d_i}{d_o} = -\frac{(-6 \text{ cm})}{4 \text{ cm}} = +1.5$$

In this case, the image is virtual (on the object side of the lens), upright, and magnified by a factor of 1.5.

You should draw ray diagrams for both of these cases.

Follow-up Exercise. Suppose the lens in this Example were biconcave or diverging, with a focal length of 12 cm. What is the magnification of an object at $d_o = 4$ cm? (As you might expect, biconcave or diverging lenses, like diverging spherical mirrors, form only *upright, reduced, virtual* images. See Exercise 54.)

EXAMPLE 23.6 ■ TWO IMAGES REVISITED: FINDING THE TRANSITION POINT

As the object distance of a biconvex lens is varied, at what point does the real image go from being reduced to being magnified?

Solution. Recall that for a converging mirror, the real image of an object at $d_o = R$ is the same size as the object ($|M| = 1$). Also, for $d_o > R$, the image is smaller, and for $R > d_o > f$, the image is larger. For a biconvex lens, R does not have the same significance, since $R \neq 2f$. However, an object distance $d_o = 2f$, or a distance $2f$ from the lens, does have this significance for the size of the real image. Substituting $2f$ for d_o in Eq. 23.6 gives

$$\frac{1}{d_o} + \frac{1}{d_i} = \frac{1}{2f} + \frac{1}{d_i} = \frac{1}{f}$$

which can be solved for d_i, or by inspection $1/d_i = 1/2f$, and

$$d_i = 2f$$

Thus,

$$M = \frac{d_i}{d_o} = -\frac{2f}{2f} = -1$$

That is, when the object is at a distance of twice the focal length from the lens, a real, inverted image the same size as the object is formed at an equal distance on the opposite side of the lens.

It can be easily shown that for $d_o > 2f$, the image is smaller than the object, and for $2f > d_o > f$, the image is larger. (You can do this analytically or by quickly sketching two ray diagrams.)

Follow-up Exercise. What is the image magnification for an object placed at $d_o = 2f$ for a biconcave (diverging) lens?

CONCEPTUAL EXAMPLE 23.7 ■ HALF AN IMAGE?

A converging lens forms an image on a screen, as shown in •Fig. 23.17a. Then the lower half of the lens is blocked off, as shown in Fig. 23.17b. As a result, (a) only the top half

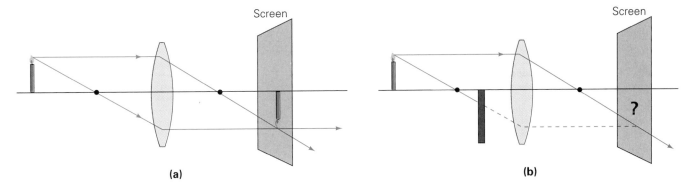

(a)

(b)

●FIGURE 23.17 Half a lens, half an image?
(a) A converging lens forms an image on a screen. (b) The lower half of the lens is blocked off. What happens to the image? See Example 23.7.

of the original image will be visible on the screen; (b) only the bottom half of the original image will be visible on the screen; (c) the entire image will be visible. *Clearly establish the reasoning and physical principle(s) used in determining your answer before checking it below. That is, **why** did you select your answer?*

Reasoning and Answer. At first thought, one might imagine that blocking off half of the lens would eliminate half of the image. However, rays from *every* point on the object pass through *all parts* of the lens. Thus, the upper half of the lens can form a total image (as could the lower half), so the answer is (c).

You might confirm this conclusion by drawing a chief ray in Fig. 23.17b. Or, you might use the scientific method and experiment—particularly if you wear eyeglasses. Block off the bottom part of your glasses, and you will find that you can still read through the top part (unless, of course, you wear bifocals).

Follow-up Exercise. Can you think of any property of the image that *would* be affected by blocking off half of the lens? Explain.

Combinations of Lenses

Many optical instruments such as microscopes and telescopes (Chapter 25) use a combination of lenses, or a compound lens system. When two or more lenses are used in combination, the overall image produced may be determined by considering the lenses individually in sequence. That is, the image formed by the first lens is the object for the second lens, and so on.

If the first lens produces an image in front of the second lens, that image is treated as a real object for the second lens (●Fig. 23.18a). If, however, the lenses are close enough together that the image from the first lens is not formed before the rays pass through the second lens (Fig. 23.18b), then a modification must be made in the sign convention. In this case, the image from the first lens is treated as a *virtual* object for the second lens, and the object distance for it is taken to be *negative* in the lens equation.

The total magnification (M_t) of a compound lens system is the product of the magnification factors (absolute values) of all the component lenses. For example, for a two-lens combination, as in Fig. 23.18,

$$M_t = |M_1| \times |M_2| \qquad (23.8)$$

You should be aware that not all lenses are spherical. A different type of lens is discussed in the Insight on p. 725.

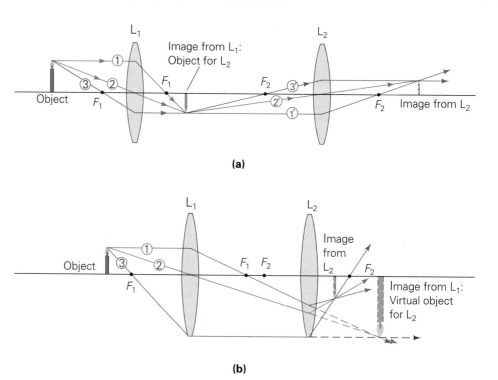

(a)

(b)

•**FIGURE 23.18 Lens combinations**
The final image produced by a
compound lens system may be
found by treating the image of one
lens as the object for the adjacent
lens. **(a)** If the image of the first
lens (L_1) is formed in front of the
second lens (L_2), the object for the
second lens is said to be real. (Note
that rays 1', 2', and 3' are the
parallel, chief, and focal rays for
L_2. They are *not* continuations of
rays 1, 2, and 3, the parallel, chief,
and focal ray for L_1.) **(b)** If the rays
pass through the second lens
before the image is formed, the
object for the second lens is said to
be virtual, and the object distance
for the second lens is taken to be
negative.

23.4 Lens Aberrations

**Objectives: To be able to (a) describe some common lens aberrations, and (b)
explain how they be can reduced or corrected.**

Lenses, like mirrors, also can have aberrations. Here are some common ones.

Spherical Aberration

The discussion of lenses thus far has focused on thin lenses, for which the lateral
deviation of light due to refraction is negligible. In general, with spherical mirrors and lenses, light rays parallel to and near the axis will be reflected or refracted to converge at the focal point. Converging lenses may, however, like spherical mirrors, show **spherical aberration**, the effect that occurs when parallel rays
passing through different regions of a lens do not come together on a common
focal plane. In general, rays close to the axis of a converging lens are refracted
less and come together at a point farther away from the lens than do rays passing through the periphery (•Fig. 23.19a). The place where the transmitted light
beam has the smallest cross section is called the *circle of least confusion*, and the
best (least distorted) image is formed at this location.

Spherical aberration can be minimized by using an aperture to reduce the effective area of the lens, so that only light rays near the axis are transmitted. Also,
combinations of converging and diverging lenses may be used. The aberration of
one lens can be compensated for (nullified) by the optical properties of another lens.

Chromatic Aberration

Chromatic aberration is an effect that occurs because the index of refraction of
the material making up a lens is not the same for all wavelengths (that is, the
material is dispersive). When white light is incident on a lens, the transmitted
rays of different wavelengths (colors) do not have a common focal point, and images of different colors are produced at different locations (•Fig. 23.19b). This dispersive aberration can be eliminated by using a compound lens system consist-

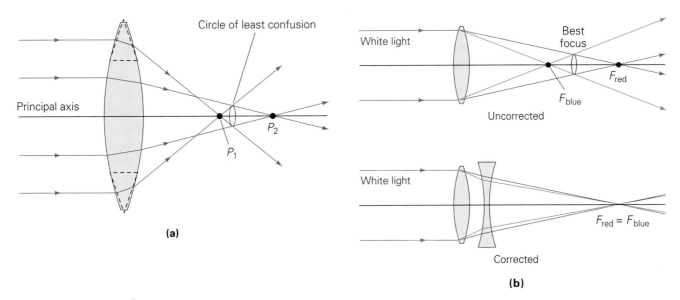

(a)

(b)

•**FIGURE 23.19 Lens aberrations**
(a) Spherical aberration. In general, rays closer to the axis of a lens are refracted less and come together at a point farther from the lens than do rays passing through the periphery of the lens. The smallest cross section of the transmitted beam is called the circle of least confusion, and the best (least distorted) image will be formed in its plane. **(b)** Chromatic aberration. Because of dispersion, different wavelengths (colors) of light are focused in different planes, which results in distortion of the overall image (top). This aberration may be corrected by using a converging lens and a diverging lens of different materials that compensate for each other's dispersion. The lens pair is called an achromatic doublet (bottom).

ing of lenses of different materials, such as crown glass and flint glass. The lenses are chosen so that the dispersion produced by one is compensated for by opposite dispersion produced by the other. With a properly constructed two-component lens system, called an *achromatic doublet* (achromatic means without color), the images of selected colors will coincide.

Astigmatism

A circular beam of light along the lens axis forms a circular illuminated area on the lens; and when incident on a converging lens, the parallel beam converges at the focal point. However, when a circular cone of light from an off-axis source falls on the convex spherical surface of a lens some distance away, the light forms an elliptical illuminated area on the lens. The rays entering along the major and minor axes of the ellipse then focus at different points after passing through the lens. This condition is called **astigmatism**.

With different focal points in different planes, the images in both planes are blurred. For example, the image of a point is no longer a point, but two separated short line images (blurred points). As with spherical aberration, the best image is formed somewhere between the images at the location of the circle of least confusion. Astigmatism can be reduced by reducing the effective area of the lens with an aperature, or by adding a cylindrical lens to compensate (similar to Fig. 23.19b).

*23.5 The Lens Maker's Equation

Objectives: To be able to **(a)** describe and apply the lens maker's equation, and **(b)** explain how its application differs from that of the thin-lens equation.

The biconvex and biconcave thin lenses considered so far in this chapter have been symmetric, with equal radii of curvature. However, there are converging and

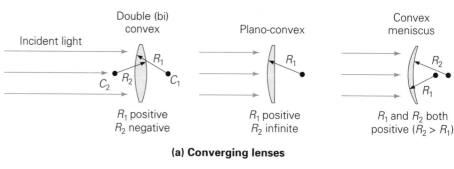

Double (bi)
convex

Plano-convex

Convex
meniscus

Incident light

R_1 positive
R_2 negative

R_1 positive
R_2 infinite

R_1 and R_2 both
positive $(R_2 > R_1)$

(a) Converging lenses

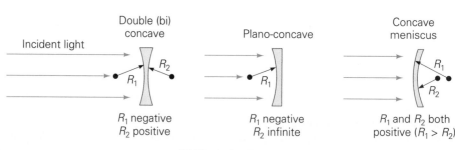

Double (bi)
concave

Plano-concave

Concave
meniscus

Incident light

R_1 negative
R_2 positive

R_1 negative
R_2 infinite

R_1 and R_2 both
positive $(R_1 > R_2)$

(b) Diverging lenses

•**FIGURE 23.20 Sign convention for the lens maker's equation** The radius of curvature is taken to be positive if it is on the emergent side of the lens and negative if it is on the incident side of the lens. A plane surface has an infinite radius of curvature (immaterial sign).

diverging lenses that have surfaces with different radii of curvature (see Fig. 23.14 and •Fig. 23.20). The focal lengths of such lenses are also important in optical analyses.

In general, the focal length of a thin lens in air ($n_{air} = 1$) is given by the so-called **lens maker's equation:**

$$\frac{1}{f} = (n - 1)\left(\frac{1}{R_1} - \frac{1}{R_2}\right) \qquad \textit{(for thin lens in air)} \qquad (23.9)$$

where n is the index of refraction of the lens material. If the lens is in a fluid other than air, then the term in parentheses becomes $[(n/n_m) - 1]$, where n_m is the index of refraction of the fluid medium. (This explains why some converging lenses become diverging when submerged in water. When $n_m > n$, then f is negative.)

For this equation, n is the index of refraction of the material, and R_1 and R_2 are the radii of curvature of the first (front side) and second (back side) lens surfaces, respectively (that is, the first surface is the one on which light from the object is first incident). The equation locates the focal point at which light passing through a lens will converge for a converging lens or the focal point from which transmitted light appears to diverge for a diverging lens.

The sign conventions for the radii of curvature in the lens maker's equation are included in Fig. 23.20 and summarized in Table 23.3. Note that the signs depend on identifying the location of the center of curvature (C) of a surface relative to either the side of the lens on which light is incident or the side from which it emerges.

The lens maker's equation can be used to show why the focal length of a biconvex glass lens is not equal to half the radius of curvature ($f \neq R/2$), as it is for a spherical mirror. For a biconvex lens, $|R_1| = |R_2| = R$, to have a focal length that was half of its radius of curvature would require an index of refraction of 2, which is greater than the values for known glasses (see Table 22.1).

TABLE 23.3 Sign Convention for Lens Maker's Equation

$+R$	when C is on the side of the lens from which light emerges or on the back side of lens.
$-R$	when C is on the side of the lens on which light is incident or on the front side of the lens.
$R = \infty$	for a plane (flat) surface
$+f$	converging (positive) lens
$-f$	diverging (negative) lens

Insight | Fresnel Lenses

To focus or to produce a large beam of parallel light rays, a sizable converging lens is necessary. The large mass of glass necessary to form such a lens is bulky and heavy; moreover, the thick lens absorbs some of the light, and is likely to show aberrations. A French physicist named Augustin Fresnel (Fre-nel'; 1788–1827) developed a solution to this problem for lenses used in lighthouses. Fresnel recognized that the refraction of light takes place at the surfaces of a lens. Hence, a lens could be made thinner—even flat—by removing glass from the interior, as long as this was done without changing the refracting properties of the surfaces.

This can be accomplished by cutting a series of concentric grooves in the surface of the lens (Fig. 1a). Note that the surface of each remaining curved segment is nearly parallel to the corresponding surface of the original lens. Together, the concentric segments refract light like the original biconvex lens (Fig. 1b). In effect, the lens has simply been slimmed down by the removal of unnecessary glass between the refracting surfaces.

A lens with such a series of concentric curved surfaces is called a Fresnel lens. Such lenses are widely used in overhead projectors and in beacons (Fig. 1c). A Fresnel lens is very thin and therefore much lighter than a conventional biconvex lens with the same optical properties. Also, Fresnel lenses are easily molded from plastic—often with one flat side (plano-convex) so that the lens can be attached to a glass surface.

One disadvantage of Fresnel lenses is that concentric circles are visible when an observer is looking through such a lens or when an image produced by one is projected on a screen as when using an overhead projector.

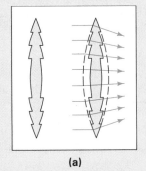

(a)

FIGURE 1 Fresnel lens
(a) The focusing action of a lens comes from refraction at its surfaces. It is therefore possible to reduce the thickness of a lens by cutting away glass in concentric grooves, leaving a set of curved surfaces with the same refractive properties as the lens from which they were derived. **(b)** A flat Fresnel lens with concentric curved surfaces magnifies like a biconvex converging lens. **(c)** An array of Fresnel lenses is used to produce focused beams in this Boston harbor light. (Fresnel lenses were in fact originally developed for use in lighthouses.)

(b)

(c)

EXAMPLE 23.8 ■ FINDING THE FOCAL LENGTH: THE LENS MAKER'S EQUATION

A thin plano-convex lens made of crown glass has an index of refraction of 1.50. If the radius of curvature for the convex surface is 20 cm, what is the focal length of the lens when light is incident on (a) the convex surface and (b) the plane surface?

Solution.

Given: $n = 1.50$ *Find:*   (a) f (focal lengths)
       (a) $R_1 = +20$ cm and $R_2 = \infty$ (b) f
       (b) $R_1 = \infty$ and $R_2 = -20$ cm

(a) The first surface in this case is the convex surface, so its radius of curvature is R_1. The center of curvature of R_1 is on the emergent side of the lens, so R_1 is positive by the sign convention (see Fig. 23.20). The plane surface is the second surface, and $R_2 = \infty$ (with no positive or negative designation). Then

$$\frac{1}{f} = (n - 1)\left(\frac{1}{R_1} - \frac{1}{R_2}\right) = (1.50 - 1)\left(\frac{1}{20\text{ cm}} - \frac{1}{\infty}\right) = \frac{0.5}{20\text{ cm}}$$

and

$$f = \frac{20\text{ cm}}{0.50} = 40\text{ cm}$$

Thus, the lens is converging (f is positive), and a beam parallel to the axis and incident on the convex surface will converge at a point 40 cm beyond the plane surface of the lens.

(b) In this case, the first surface is the plane surface, and the radius of curvature is negative for the convex surface since the center of curvature is on the incident side of the lens. Using the lens maker's equation with these conditions gives

$$\frac{1}{f} = (n - 1)\left(\frac{1}{R_1} - \frac{1}{R_2}\right) = (1.50 - 1)\left[\frac{1}{\infty} - \frac{1}{(-20\text{ cm})}\right] = \frac{0.50}{20\text{ cm}}$$

and

$$f = \frac{20\text{ cm}}{0.50} = 40\text{ cm}$$

The lens is converging, which you may find unexpected. Draw an off-axis parallel ray and show that convergence is the case when the ray is refracted away from the normal in emerging from the glass lens.

Follow-up Exercise. What would be the focal length of a crown glass biconvex lens with a radius of curvature of 20 cm for each surface?

Optometrists express the power (P) of a lens in **diopters** (D), which is simply the reciprocal of the focal length of the lens expressed in meters.

$$P\text{ (diopters)} = \frac{1}{f}\text{ (in meters)} \tag{23.10}$$

Note that the lens maker's equation (Eq. 23.9) gives a value in diopters ($1/f$) if the radii of curvature are expressed in meters.

Converging and diverging lenses are referred to as positive (+) and negative (−) lenses, respectively. Thus, if an optometrist prescribes a corrective lens with a power of +2 diopters, this means a converging lens with a focal length of

$$f = 1/P = 1/2 = 0.50\text{ m} = 50\text{ cm}$$

The greater the power of a lens in diopters, the shorter its focal length and the more converging or diverging it is.

EXAMPLE 23.9 ■ GRINDING A LENS: POWER IN DIOPTERS

A lens maker must fill a prescription for a corrective convexo-concave lens with a power of −2.1 diopters. The lens is to be ground from a glass blank ($n = 1.70$) having a convex front surface whose radius of curvature is 50 cm. To what radius of curvature should the second (back) surface of the lens be ground?

Solution.

Given: $\quad P = 1/f = -2.1\text{ D (m}^{-1})$ $\qquad\qquad$ *Find:* $\quad R_2$ (radius of curvature)
$\qquad\qquad n = 1.70$
$\qquad\qquad R_1 = 50\text{ cm} = 0.50\text{ m}$

Since the center of curvature of the convex surface is on the emergent side of the lens, R_1 is positive (see Fig. 23.20).

$$P = \frac{1}{f} = (n-1)\left(\frac{1}{R_1} - \frac{1}{R_2}\right)$$

or

$$-2.1 = (1.70 - 1)\left(\frac{1}{0.50 \text{ m}} - \frac{1}{R_2}\right)$$

Solving for R_2 gives

$$\frac{1}{R_2} = \frac{1}{0.50 \text{ m}} + \frac{2.1}{0.70 \text{ m}} = 5.0 \text{ m}^{-1}$$

and

$$R_2 = \frac{1}{5.0 \text{ m}^{-1}} = 0.20 \text{ m} = 20 \text{ cm}$$

A positive value for R_2 indicates that the center of curvature for the second surface is on the emergent (or back) side of the lens, as is that for R_1. With $R_1 > R_2$, the lens is diverging.

Follow-up Exercise. A biconcave lens is ground from a glass blank ($n = 1.5$) with a radius of curvature of 50 cm on each side. What is the power of the lens in diopters?

Chapter Review

Important Terms

plane mirror   703
virtual image   703
real image   703
lateral magnification   703
spherical mirror   707
concave (converging) mirror   707
convex (diverging) mirror   707
center of curvature   707
radius of curvature   707
focal point   707
focal length   707

parallel ray (mirror)   708
chief (radial) ray (mirror)   708
focal ray (mirror)   708
spherical mirror equation   712
magnification factor (spherical mirror)   712
spherical aberration (mirror)   715
lens   716
converging (biconvex) lens   716
diverging (biconcave) lens   716

parallel ray (lens)   717
chief ray (lens)   717
focal ray (lens)   717
thin lens equation   719
magnification factor (lens)   719
spherical aberration (lens)   722
chromatic aberration   722
astigmatism   723
*lens maker's equation   724
*diopters   726

Important Concepts

- Plane mirrors form virtual, upright, and unmagnified images.
- Spherical mirrors are either concave (converging) or convex (diverging). Diverging spherical mirrors always form upright, reduced, virtual images.
- Bispherical lenses are either convex (converging) or concave (diverging). Diverging spherical lenses always form upright, reduced, virtual images.

- In finding the location and determining the characteristics of the image formed by a mirror or lens analytically, an initial quick ray diagram sketch is helpful and recommended.
- The thin-lens equation relates focal length, object distance, and image distance.
- *The lens maker's equation is used to compute the grinding radii for the desired focal length of a lens.

Important Equations

Focal Length for Spherical Mirror:

$$f = \frac{R}{2} \qquad (23.2)$$

Spherical Mirror Equation:

$$\frac{1}{d_o} + \frac{1}{d_i} = \frac{1}{f} \quad \text{or} \quad d_i = \frac{d_o f}{d_o - f} \qquad (23.3\text{–}4)$$

Thin Lens Equation (where $f \neq R/2$):

$$\frac{1}{d_o} + \frac{1}{d_i} = \frac{1}{f} \quad \text{or} \quad d_i = \frac{d_o f}{d_o - f} \qquad (23.6)$$

Magnification Factor (spherical mirror and lens):

$$M = -\frac{d_i}{d_o} \qquad (23.5 \text{ and } 7)$$

Total Magnification with a Two-Lens System:
$$M_t = |M_1| \times |M_2| \qquad (23.8)$$

Sign Conventions for Spherical Mirrors and Lenses:

Concave mirror ⎤ f positive   Convex mirror ⎤ f negative
Biconvex lens ⎦ (converging)   Biconcave lens ⎦ (diverging)

d_o is always positive.

When d_i is positive, image is real; when d_i is negative, image is virtual.

When M is positive, image is upright; when M is negative, image is inverted.

Lens Maker's Equation:
$$\frac{1}{f} = (n - 1)\left(\frac{1}{R_1} - \frac{1}{R_2}\right) \qquad (23.9)$$

Lens Power in Diopters (where f is in meters):
$$P = \frac{1}{f} \qquad (23.10)$$

Sign Convention for Lens Maker's Equation:

$+R$ when C is on side of lens from which light emerges

$-R$ when C is on side of lens on which light is incident

$R = \infty$ for a plane (flat) surface

$+f$ converging (positive) lens

$-f$ diverging (negative) lens

Exercises

23.1 Plane Mirrors

1   A plane mirror (a) has a greater image distance than object distance, (b) produces a virtual, upright, unmagnified image, (c) changes the vertical orientation of an object, (d) reverses top and bottom.

2   A plane mirror (a) can be used to magnify, (b) produces both real and virtual images, (c) always produces a virtual image, (d) forms images by diffuse reflection.

3   (a) A transparent window pane can serve as a mirror, but this is usually observed only when it is dusk or dark outside. Why? (b) When looking at a reflecting window pane at night, you may see two similar images. Why? (c) One-way, half-silvered mirrors reflect on one side and can be seen through on the other. (This effect is sometimes used on sunglasses.) What is the principle used in making a one-way mirror? [*Hint*: At night a windowpane may be a one-way mirror.]

4   Day/night rearview mirrors are common in most cars. You flip a switch, tilting the mirror backwards, and the intensity and glare from headlights behind you at night are reduced (•Fig. 23.21). The mirror is actually wedge-shaped and is silvered on the back side. The effect has to do with front-surface and back-surface reflections. The unsilvered front surface reflects about 5% of the incident light, whereas the back-silvered surface reflects about 90% of the incident light. With this information, explain how the day/night mirror works.

5   When standing in front of a plane mirror, there is a "right–left" reversal. (a) Why is there not a top–bottom reversal of your body? (b) Could you effect an apparent top–bottom reversal by positioning your body differently?

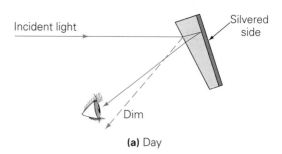

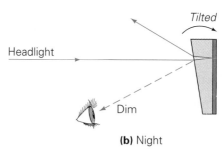

•FIGURE 23.21   Automobile day/night mirror
See Exercise 4.

6   Explain the sequence reversal in Fig. 3 of the Insight in this section. Was there any obvious front–back reversal?

7   ■ A person stands 3.0 m away from the reflecting surface of a plane mirror. (a) What is the apparent distance between the person and his or her image? (b) What are the image characteristics?

8   ■ An object 7.5 cm tall is placed 50 cm from a plane mirror. Find (a) the distance from the object to the image, (b) the height of the image, and (c) its magnification.

9 ■■ (a) If you move toward a plane mirror with a speed v, with what speed does your image move toward you relative to the ground? (b) If you stand stationary and the mirror is moved toward you with a speed v, with what speed does your image appear to move toward you?

10 ■■ A small dog sits in front of a plane mirror at a distance of 1.5 m. (a) Where is the dog's image? (b) If the dog jumps at the mirror with a speed of 0.50 m/s, how fast does it approach its image?

11 ■■ A woman 1.6 m tall stands 2.0 m in front of a plane mirror. (a) What is the minimum height the mirror must be to allow the woman to view her complete image from head to foot? Assume that her eyes are 10 cm below the top of her head. (b) What would be the minimum height of the mirror if she stood 4.0 m away?

12 ■■ Show that the spherical mirror equation and magnification equation give the correct image characteristics for a plane mirror.

13 ■■ A dance studio has plane mirrors on opposite walls, and multiple images are observed when a person stands between them. If a dancer stands 3.0 m from the mirror on the north wall and 5.0 m from the mirror on the south wall, what are the image distances for the first two images in both mirrors?

14 ■■ A student who is 1.4 m tall stands in front of a plane mirror that is tilted at an angle of 15° away from her. If the student's eyes are 12 cm below the top of her head, find the minimum height the mirror must be to allow her to see her complete image (head to foot) in it.

15 ■■ Find the minimum height of the mirror in Exercise 14 if it is tilted at an angle of 15° toward the student.

16 ■■■ Draw ray diagrams showing how three images of an object are formed in two plane mirrors at right angles as shown in •Fig. 23.22a. [*Hint:* Consider rays from each end of the object arrow in the drawing for each image.] Fig. 23.22b shows a similar situation from a different point of view that gives four images. Explain the extra image in this case?

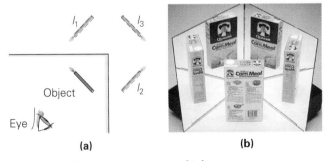

•**FIGURE 23.22   Two mirrors—multiple images**
See Exercise 16.

23.2 Spherical Mirrors

17 A spherical concave mirror with an object at its center of curvature will produce (a) a virtual image behind the mirror at $d_i = 2f$, (b) a virtual image in front of the mirror at $d_i = 2R$, (c) a real, upright, magnified image at $d_i = f$, (d) a real, inverted, unmagnified image at $d_i = R$.

18 Which one of the following statements concerning spherical mirrors is correct? (a) A converging mirror alone can produce an inverted virtual image. (b) A diverging mirror alone can produce an inverted virtual image. (c) A diverging mirror can produce an inverted real image. (d) A converging mirror can produce an upright real image.

19 (a) What is the purpose of using a dual mirror on a car or truck, such as the one shown in •Fig. 23.23? (b) Some rearview mirrors on the passenger side of automobiles have the written warning "OBJECTS IN MIRROR ARE CLOSER THAN THEY APPEAR." Explain why this is so. (c) Could a TV satellite dish as shown in Fig. 20.19 be considered a converging mirror? Explain.

•**FIGURE 23.23   Mirror applications**
See Exercise 19.

20 (a) If you look into a shiny spoon, you see an inverted image on one side and an upright image on the other. (Try it.) Why is this? (b) Could you see upright images on both sides? Explain.

21 (a) A 10-cm tall mirror for ladies' purses is advertised as a "Full-view mini mirror. See your full body in 10 cm." How can this be? (b) A popular novelty item consists of a concave mirror with a ball suspended at or slightly inside the center of curvature (see •Fig. 23.24). When the ball swings toward the mirror, its image grows larger and suddenly fills the whole mirror. The effect is that the image appears to be jumping out of the mirror. Explain what is happening.

22 ■■ A candle with a flame 1.5 cm tall is placed 5.0 cm from the front of a concave mirror. A virtual image is produced that is 10 cm from the vertex of the mirror. (a) Find the focal length and radius of curvature of the mirror. (b) How tall is the image of the flame?

•FIGURE 23.24 **Spherical mirror toy**
See Exercise 21.

23 ■■ An object 3.0 cm tall is placed 20 cm from the front of a concave mirror with a radius of curvature of 30 cm. Where is the image formed and how tall is it?

24 ■■ If the object in Exercise 23 is moved to a position 10 cm from the front of the mirror, what will the characteristics of the image be?

25 ■■ A concave mirror with a radius of curvature of 40 cm reflects sunlight. Find (a) the location and (b) the diameter of the Sun's image.

26 ■■ The image of an object 18 cm from a convex mirror is half the size of the object. What is the focal length of the mirror?

27 ■■ Using the spherical mirror equation and the magnification factor, show that for a concave mirror with $d_o < f$, the image of an object is always virtual, upright, and magnified.

28 ■■ An object 3.0 cm tall is placed at different locations in front of a concave mirror whose radius of curvature is 30 cm. Determine the location of the image and its characteristics when the object distance is (a) 40 cm, (b) 30 cm, (c) 20 cm, (d) 15 cm, and (e) 5.0 cm.

29 ■■ Draw a ray diagram for each of the cases in Exercise 28.

30 ■■ A concave shaving mirror is constructed so that a man at a distance of 20 cm from the mirror sees his image magnified 1.5 times. What is the radius of curvature of the mirror?

31 ■■ If an object is 15 cm in front of a convex mirror that has a focal length of 30 cm, how far behind the mirror will the image appear to an observer, and how tall will it be?

32 ■■ A concave mirror has a magnification of 3.0 for an object placed 50 cm in front of it. (a) What type of image is produced? (b) Find the radius of curvature of the mirror.

33 ■■ A child looks at a reflecting Christmas tree ball that has a diameter of 9.0 cm and sees an image of her face that is half the real size. How far is the child's face from the ornament?

34 ■■ A dentist holds a concave mirror whose radius of curvature is 40 mm at a distance of 15 mm from a tooth to view a small circular cavity 2.0 mm in diameter. How large is the image of the cavity?

35 ■■ A dentist uses a spherical mirror that produces an upright image that is magnified four times. What is the focal length of the mirror in terms of the object distance?

36 ■■ A 15-cm pencil is placed with its eraser on the optic axis of a concave mirror and its point directed upward at a distance of 20 cm in front of the mirror. The radius of curvature of the mirror is 30 cm. (a) Where is the image of the pencil formed, and what are its characteristics? (b) Draw a ray diagram for the situation where the pencil point is directed downward from the optic axis.

37 ■■ A pill bottle 4.5 cm tall is placed 12 cm from the front of a mirror. An upright image 9.0 cm tall is formed. What kind of mirror is it, and what is its radius of curvature?

38 ■■ A mirror at an amusement park shows anyone who stands 2.5 m from it an upright image three times the person's height. What is the mirror's radius of curvature?

39 ■■ (a) Sketch graphs of (1) d_i versus d_o and (2) M versus d_o for a converging mirror, for values of d_o from 0 to ∞. (b) Sketch similar graphs for a diverging mirror.

40 ■■■ The front surface of a wooden cube 5.0 cm on a side is placed a distance of 30 cm in front of a converging mirror having focal length of 20 cm. Where is the image of the front surface of the block located and what are its characteristics?

41 ■■■ A spherical section is mirrored on both sides. If the magnification of an object is +1.8 when the device is used as a concave mirror, what is the magnification of an object at the same distance in front of the convex side?

42 ■■■ A concave mirror has a radius of curvature of 20 cm. For what *two* object distances will the image have twice the height of the object?

43 ■■■ Show that $f = R/2$ for a converging spherical mirror. [*Hint:* Draw a ray diagram for the situation $R > d_o > f$ using all three types of rays, as in Fig. 23.7a. Look for similar triangles.]

44 ■■■ A convex mirror is used on the exterior of the passenger side of many cars. If the focal length of such a mirror is −40.0 cm, what will the location and height of

the image of a car that is 2.0 m tall be if the car is (a) 100 m behind and (b) 10.0 m behind? (See Exercise 19b.)

45 ■■■ Two students in a physics laboratory each have a concave mirror with the same radius of curvature, 40.0 cm. Each student places an object in front of a mirror. The image in both mirrors is three times the size of the object. However, when the students compare notes, they find that the object distances are not the same. Is this possible? If so, what are the object distances?

23.3 Lenses

23.4 Lens Aberrations

46 A diverging lens (a) must have at least one concave surface, (b) always produces a virtual image, (c) is thinner at the center than at the periphery, (d) all of these.

47 A virtual image will be formed by (a) a biconvex spherical lens with an object inside the focal point, (b) a biconvex spherical lens with the object at $d_o = 2f$, (c) a biconcave spherical lens with an object at $d_o = 2f$, (d) both (a) and (c).

48 A lens aberration that is caused by dispersion is called (a) spherical aberration, (b) chromatic aberration, (c) refractive aberration, (d) none of these.

49 How can the focal length be determined experimentally for (a) a concave mirror and (b) a biconvex lens?

50 ■■ Describe the image characteristics for the comparable regions shown in Fig. 23.8a for a mirror, along with $d_o = 2f$, for (a) a convex lens, and (b) a concave lens.

51 ■■ Find the location and magnification of the Sun's image with a convex lens whose focal length is 15 cm.

52 ■■ An object 4.0 cm tall is placed in front of a converging lens whose focal length is 22 cm. Where is the image formed and what are its characteristics if the object distance is (a) 15 cm and (b) 36 cm? Sketch ray diagrams for each.

53 ■■ An object is placed in front of a biconcave lens whose focal length is 18 cm. Where is the image located and what are its characteristics if the object distance is (a) 10 cm and (b) 25 cm? Sketch ray diagrams for each.

54 ■■ Using the thin lens equation and the magnification factor, show that for a spherical diverging lens, the image of a real object is always virutal, upright, and reduced.

55 ■■ A light source and a screen are separated by a distance d, and a converging lens with a focal length f can be placed anywhere between them. Determine whether sharp images will be formed on the screen (a) when $d < 4f$ and (b) when $d > 4f$.

56 ■■ A biconvex lens has a focal length of 0.12 m. Where on the lens axis should an object be placed in order to get (a) a real, enlarged image with a magnification of 2.5 and (b) a virtual, enlarged image with a magnification of 2.5?

57 ■■ An object 5.0 cm tall is 10 cm from a concave lens. The resulting image is 1/5 as large as the object. What is the focal length of the lens?

58 ■■ (a) Design a single-lens projector that will form a sharp image on a screen 4.0 m away with the transparent slides 6.0 cm from the lens. (b) If the object on a slide is 1.0 cm tall, how tall will the image on the screen be, and how should the slide be placed in the projector?

59 ■■ A single-lens camera (biconvex lens) is used to photograph a man 1.7 m tall who is standing 4.0 m from the camera. If the man's image fills the length of a frame of film (35 mm), what is the focal length of the lens?

60 ■■ An object is placed 40 cm from a screen. At what point between the object and the screen should a converging lens with a focal length of 10 cm be placed so it will produce a sharp image on the screen and what is its magnification?

61 ■■ A converging lens with a focal length of 20 cm is used to produce an image on a screen that is 2.0 m from the lens. How many times is the object magnified?

62 ■■ Using ●Fig. 23.25, derive (a) the thin lens equation and (b) the lens magnification equation. [Hint: Use similar triangles.]

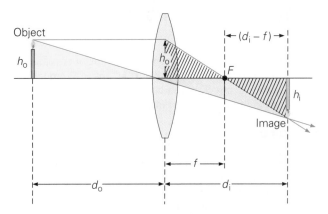

●FIGURE 23.25 The thin lens equation
The diagram shows the geometry for deriving the thin-lens equation (and magnification factor). Note that there are two sets of similar triangles. See Exercise 62.

63 ■■ (a) For a biconvex lens, what is the minimum distance between an object and its image if the image is

real? (b) What is the distance if the image is virtual? [*Hint:* Use ray diagrams and the thin lens equation.]

64   ■■ A stamp collector views a stamp with a magnifying glass (a convex lens). If she sees the stamp magnified by a factor of 2.5 when the glass is held 3.0 cm from it, what is the focal length of the lens?

65   ■■ (a) If a book is held 30 cm from an eyeglass lens with a focal length of −57 cm, where is the image of the print formed? (b) If an eyeglass lens with a focal length of +57 cm is used, where is the image formed?

66   ■■ For the arrangement shown in •Fig. 23.26, an object is placed 0.40 m in front of the converging lens, which has a focal length of 0.15 m. If the concave mirror has a focal length of 0.13 m, where is the final image formed and what are its characteristics?

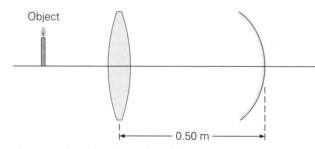

•**FIGURE 23.26   Lens-mirror combination**
See Exercise 66.

67   ■■ A simple camera uses a single converging lens with a focal length of 0.045 m. A picture is taken of a 26-cm tall physics book standing on a table at a distance of 1.5 m from the camera. If a sharp image is formed on the film, how far is the film from the lens? (b) What is the height of the image of the book on the film?

68   ■■ A converging mirror with a radius of curvature of 90 cm produces an upright, virtual image 45 cm tall and located 30 cm from the vertex of the mirror. What are the height and location of the object?

69   ■■■ The geometry of a compound microscope, which consists of two converging lens, is shown in •Fig. 23.27. (More detail on microscopes is given in Chapter 25.) The objective lens and the eyepiece lens have focal lengths of 2.8 mm and 3.3 cm, respectively. If an object is located 3.0 mm from the objective lens, where is the final image located?

70   ■■■ Two converging lenses, L_1 and L_2, have focal lengths of 30 cm and 20 cm, respectively. The lenses are placed 60 cm apart along the same axis, and an object is placed 50 cm from L_1 on the side opposite to L_2. Where is the image formed relative to L_2 and what are its characteristics?

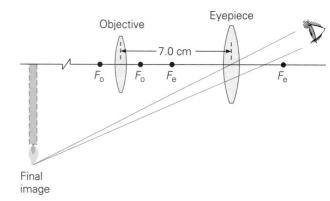

•**FIGURE 23.27   Compound microscope**
See Exercise 69.

71   ■■■ Show that for thin lenses with focal lengths f_1 and f_2 and in contact the effective focal length (f) is given by

$$\frac{1}{f} = \frac{1}{f_1} + \frac{1}{f_2}$$

*23.5 The Lens Maker's Equation

72   The lens maker's equation (a) expresses spherical aberration, (b) applies only to lenses with equal radii of curvature, (c) shows that $f \neq R/2$ for glass lenses, (d) is the same as the thin lens equation.

73   The power of a lens is expressed in units of (a) watts, (b) diopters, (c) m, (d) both (b) and (c).

74   According to the lens maker's equation, when a lens with $n = 1.60$ is immersed in water, (a) is there a change in the focal length of the lens? (b) What would be the case if the submerged lens had an index of refraction of 1.30?

75   ■ A symmetric biconvex lens has a focal length of 50 cm and is made of glass whose index of refraction is 1.5. What is the radius of curvature of both lens surfaces?

76   ■■ An optometrist prescribes a corrective lens with a power of +1.5 diopters. The lens maker will start with glass blank with an index of refraction of 1.6 and a convex front surface whose radius of curvature is 20 cm. To what radius of curvature should the other surface be ground?

77   ■■ A plastic plano-concave lens has a radius of curvature of 50 cm for its concave surface. If the index of refraction of the plastic is 1.35, what is the power of the lens?

78   ■■ A biconvex lens made of glass whose index of refraction is 1.5 has radii of curvature of 30 cm for one surface and 40 cm for the other. What is the focal length of the lens for light incident on each side?

Additional Exercises

79 A photographer uses a single-lens camera having a focal length of 60 mm to photograph a full moon. What will be the diameter of the Moon's image on the film? [*Note:* Data on the Moon may be found inside the back cover.]

80 A virtual image at a magnification of 0.50 is produced when an object is placed in front of a spherical mirror. (a) What type of mirror is it? (b) Find the radius of curvature of the mirror if the object is 7.0 cm from it.

81 A bottle 6.0 cm tall is located 75 cm from the concave surface of a mirror with a radius of curvature of 50 cm. Where is the image located and what are its characteristics?

82 A boy runs toward a plane mirror with a speed of 3.5 km/h. What is the speed of the boy's image in m/s relative to him?

83 A method of determining the focal length of a diverging lens is called autocollimation. As •Fig. 23.28 shows, first, a sharp image of a light source is projected on a screen by a converging lens. Second, the screen is replaced with a plane mirror. Third, a diverging lens is placed between the converging lens and the mirror. Light will then be reflected by the mirror back through the compound lens system, and an image will be formed on a screen near the light source. This image is made sharp by adjusting the distance between the diverging lens and the mirror. The distance at which the image is clearest is equal to the focal length of the lens. Explain why this is true.

84 Show that the magnification for objects near the optic axis of a convex mirror is given by $M = d_i/d_o$. [*Hint:* Use a ray diagram with rays reflected at the mirror's vertex.]

85 A biconvex lens produces a real, inverted image of an object that is magnified 2.5 times when the object is 20 cm from the lens. What is the focal length of the lens?

86 Two positive lenses, each having a power of 10 diopters, are placed 20 cm apart along the same axis. If an object is 60 cm from the first lens on the side opposite the second lens, where is the final image relative to the first lens, and what are its characteristics?

87 (a) Sketch graphs of (1) d_i versus d_o and (2) M versus d_o for a concave mirror, for values of d_o from 0 to ∞. (b) Sketch similar graphs for a convex mirror.

88 An object is 15 cm from a converging lens whose focal

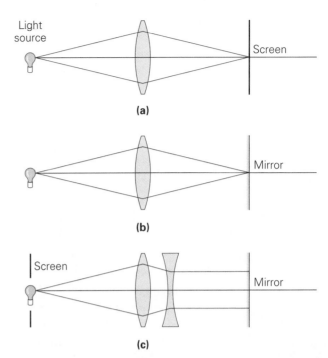

•FIGURE 23.28 Autocollimation
See Exercise 83.

length is 10 cm. On the opposite side of that lens at a distance of 60 cm is another converging lens with a focal length of 20 cm. Where is the final image formed, and what are its characteristics?

89 (a) Show that a ray parallel to the optic axis of a biconvex lens is refracted toward the axis at the incident surface and again at the exit surface. (b) Show that the refraction effect also holds for a biconcave lens, but with deflections away from the axis.

90 An object 3.0 cm tall is placed 15 cm from the front of a convex mirror whose focal length is 25 cm. Find the location of the image and its height.

91 A converging glass lens with an index of refraction of 1.62 has a focal length of 30 cm in air. What is the focal length when it is submerged in water?

92 The image of an object located 30 cm from a concave mirror is formed on a screen located 20 cm from the mirror. What is the mirror's radius of curvature?

93 A concave cosmetic mirror produces a virtual image 1.5 times the size of a person when his face is 20 cm from the mirror. (a) Draw a ray diagram of this situation. (b) What is the focal length of the mirror?

24 Physical Optics: The Wave Nature of Light

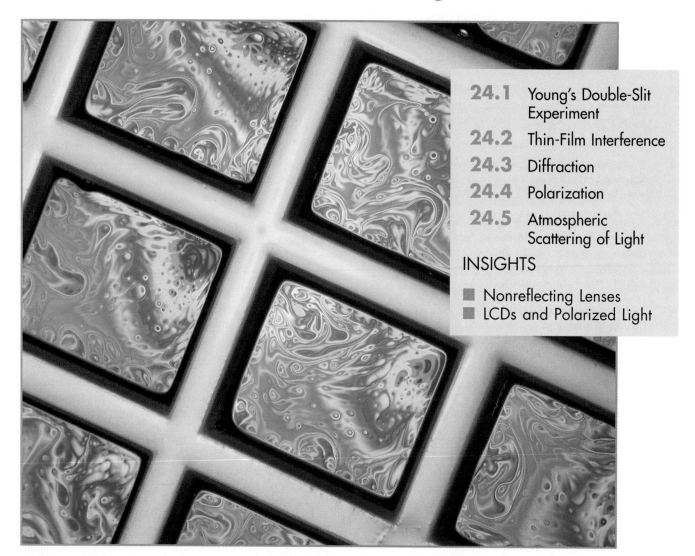

It's always intriguing to see brilliant colors produced by objects that we know don't ordinarily have any colors of their own. The glass of a prism, for example, clear and transparent by itself, nevertheless gives rise to a whole array of colors when white light passes, through it. "All the colors of the rainbow," we often say when describing the spectrum produced by a prism. Scientifically, it might be more correct to say that a rainbow contains "all the colors of the spectrum"—for a rainbow, as we saw in Chapter 22, is in fact a spectrum.

Of course, we know that prisms, like the water droplets that produce the rainbow, don't really create anything that isn't already there. They merely separate the different wavelengths that make up white light and spread them out for our pleasure and instruction. And if you recall some other common experiences, you'll quickly realize that neither a prism nor a sun shower is needed to produce a spectrum. You've probably noticed the rainbow sheen of colors reflected from the surface of a compact disc. Then, think of the colors that appear when a little motor oil is spilled on a puddle, or on wet pavement. And what about soap bubbles . . . ? In this chapter you'll learn how the colors in the photo above are created, and explore many other phenomena associated with the wave properties of light. Along the way, you'll find out why, although air is quite transparent and sunlight is yellow, our sky is blue and our sunsets are red.

The phenomena of reflection and refraction are conveniently analyzed using geometrical optics. Ray diagrams show what happens when light is reflected from a mirror or passed through a lens. However, some other phenomena involving light, such as the interference patterns in the chapter opening photo, cannot be adequately explained or described using rays, since this technique ignores the wave nature of light. Other such phenomena include diffraction and polarization.

Physical optics, or **wave optics**, takes into account those wave properties that geometrical optics ignores. The wave theory of light leads to satisfactory explanations of those phenomena that cannot be analyzed with rays. Thus, this chapter again considers waveforms.

24.1 Young's Double-Slit Experiment

Objectives: **To be able to (a) explain how Young's experiment demonstrated the wave nature of light, and (b) compute the wavelength of light from experimental results.**

It has been stated that light behaves like a wave, but no proof of this has been given. How would you go about demonstrating the wave nature of light? One method that involves the use of interference was first carried out in 1801 by the English scientist Thomas Young (1773–1829). **Young's double-slit experiment** not only demonstrated the wave nature of light but also allowed him to measure its wavelengths.

Recall from the discussion of wave interference in Sections 13.4 and 14.4 that superimposed waves may interfere constructively or destructively. Total constructive interference occurs when two crests are superimposed, and total destructive interference occurs when a crest and a trough of two identical waves are superimposed. Interference can be observed with water waves (•Fig. 24.1), for which constructive and destructive interference produce obvious interference patterns.

Note: Compare this figure with Figure 14.7a.

The interference of light waves is not as easily observed because of their relatively short wavelengths ($\approx 10^{-7}$ m). Also, stationary interference patterns are produced only with *coherent sources,* that is, sources that produce light waves having a constant phase relationship to one another. For example, for constructive interference to occur at some point, the waves meeting at that point must be in phase. As the waves meet, a crest must always overlap a crest and a trough must always overlap a trough. If a phase difference develops between the waves with time, the interference pattern changes, and a stable or stationary pattern will not be set up.

In an ordinary light source, the atoms are excited randomly, and the emitted light waves fluctuate in amplitude and frequency. Thus, light from two such sources is incoherent and does not produce a stationary interference pattern. Interference does occur, but the phase difference between the interfering waves changes so fast that the interference effects are not discernible.

To solve this problem, Young used light from a single source (sunlight) to illuminate two narrow, closely spaced slits (•Fig. 24.2a). The slits act as two sources, but light emerging from the slits is coherent because the slits merely separate the original beam into two parts. Any random changes in the light from the source will thus occur for the light passing through both slits, and the phase difference will be constant.

A popular theory, dating back to the ancient Greeks, considered light to be composed of "corpuscles" or particles. This theory survived relatively unquestioned for many centuries. Over the years, however, evidence to support a wave nature of light began to accumulate. Leonardo da Vinci (circa 1500), noting the similarity between sound echoes and the reflection of light, speculated that light might have a wave nature. By the 17th century a definite controversy existed over

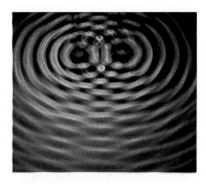

•FIGURE 24.1 Water wave interference The constructive and destructive interference of water waves from two coherent sources in a ripple tank produce interference patterns.

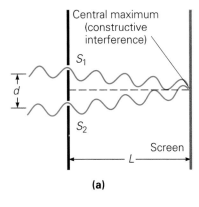

Central maximum
(constructive
interference)

S_1

d

S_2

Screen

L

(a)

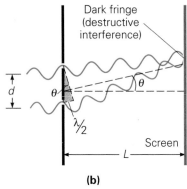

Dark fringe
(destructive
interference)

d

θ

θ

$\lambda/2$

Screen

L

(b)

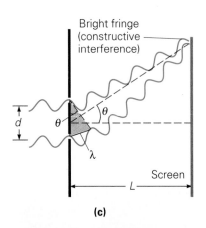

Bright fringe
(constructive
interference)

d

θ

θ

λ

Screen

L

(c)

•**FIGURE 24.3   Interference**
The interference that produces
bright or dark fringes depends on
the difference in the path lengths
of the light from the two slits. **(a)**
The path difference at the position
of the central maximum is zero, so
the waves arrive in phase and
interfere constructively. **(b)** At the
position of the first dark fringe, the
path difference is $\lambda/2$, and the
waves interfere destructively. **(c)** At
the position of the first bright
fringe, the path difference is λ, and
the interference is constructive.

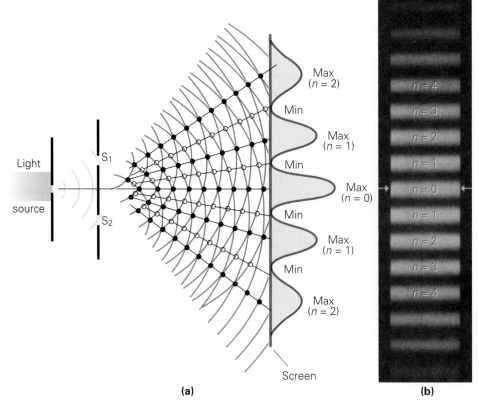

(a)

(b)

•**FIGURE 24.2   Double-slit interference**
(a) The coherent waves from two slits shown in blue (top slit) and red (bottom slit),
spread out and interfere, producing alternating maxima and minima, or bright and dark
fringes, on the screen. **(b)** An actual interference pattern. Note the symmetry of the
pattern about the central maximum ($n = 0$).

the nature of light. Isaac Newton favored a theory that treated light as a stream
of particles.

If light consisted of a stream of tiny particles, then in Young's experiment one
would expect to see two bright lines on a screen placed behind the slits. Instead,
Young observed a *series* of bright lines (•Fig. 24.2b), and he was able to explain
this pattern in terms of *wave interference*.

To help analyze what Young observed, let's imagine that light with a single
wavelength (monochromatic light) is used. Because of diffraction, or the bending
of the light around the corners of the slits, the waves spread out and interfere as
illustrated in the figure. Coming from two coherent "sources," the interfering
waves produce a stable interference pattern on the screen. The pattern consists of
a bright central maximum (•Fig. 24.3a) and a series of symmetrical dark (•Fig.
24.3b) and bright (•Fig. 24.3c) side fringes, which mark the positions at which destructive and constructive interference occurs. The bright fringes decrease in intensity on either side of this central maximum. Hence, the interference pattern
demonstrates the wave nature of light.

Measuring the wavelength of light requires looking at the geometry of Young's
experiment, as shown in •Fig. 24.4. Let the screen be a distance L from the slits,
and P be a point at the center of an arbitrary maximum, or bright side fringe. P
is located a distance y from the center of the central maximum and at an angle θ
relative to a normal line between the slits. The slits S_1 and S_2 are separated by a
distance d. Note that the light path from one slit to P is longer than the path from
the other slit to P. As the figure shows, the path difference (Δ) is

$$\Delta = d \sin \theta$$

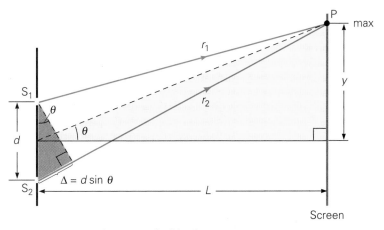

•FIGURE 24.4 Geometry of Young's double-slit experiment
The difference in the path lengths for light from the two slits traveling to a position P of
an arbitrary maximum, or bright fringe, is $r_2 - r_1 = \Delta$, which forms a side of the small
shaded triangle. Because the barrier with the slits is parallel to the screen, the angle
between r_2 and the barrier (at S_2, in the small shaded triangle) is equal to the angle
between r_2 and the screen. When L is very much greater than y, that angle becomes
almost identical to the angle between the screen and the dashed line, which is an angle
in the large shaded triangle. The two shaded triangles are then almost exactly similar,
and the angle at S_1 in the small triangle is almost exactly equal to θ. Thus, $\Delta = d \sin \theta$.
By the condition for constructive interference, $d \sin \theta = n\lambda$, where $n = 0, 1, 2, 3, \ldots$ is
the order of interference. (Drawing not to scale. Assume $y \ll L$.)

The fact that the angle in the small shaded triangle is equal to θ can be shown by
a simple geometrical argument involving similar triangles, as described in the
caption of Figure 24.4.

The relationship of the phase difference of two waves to their path difference
was discussed in Chapter 14 for the interference of sound waves. (There the path
difference was abbreviated ΔL.) The conditions for interference given for sound
waves hold for any sinusoidal waves, including light waves. Recall that constructive interference occurs at any point where the path difference between two
coherent, in-phase waves is an integral number of wavelengths. That is,

$$\Delta = n\lambda \quad \text{for } n = 0, 1, 2, 3, \ldots \qquad \begin{array}{l} \textit{condition for} \\ \textit{constructive interference} \end{array} \qquad (24.1)$$

Similarly, for destructive interference, the path difference is an odd number of
half-wavelengths.

$$\Delta = \frac{m\lambda}{2} \quad \text{for } m = 1, 3, 5, \ldots \qquad \begin{array}{l} \textit{condition for} \\ \textit{destructive interference} \end{array} \qquad (24.2)$$

Thus, in Fig. 24.4, point P will be the location of a bright fringe (constructive
interference) if

$$d \sin \theta = n\lambda \quad \text{for } n = 0, 1, 2, 3, \ldots \qquad (24.3)$$

where n is called the order number. The zeroth-order fringe ($n = 0$) corresponds
to the central maximum, the first-order fringe ($n = 1$) is the first bright fringe on
either side of the central maximum, and so on. As the path difference varies from
point to point, so does the phase difference and the resulting interference.

The wavelength can therefore be determined by measuring d and θ for a particular order bright fringe (other than the central maximum). The angle θ is related to the displacement of a fringe (y), which is conveniently measured from a
photograph of the interference pattern. If P is a point at the center of a bright

fringe, the distance from the center of the central maximum to that nth fringe is (see Fig. 24.4)

$$y = L \tan \theta = L\left(\frac{\sin \theta}{\cos \theta}\right) \tag{24.4}$$

If θ is small ($y \ll L$), $\cos \theta \approx 1$ and $\tan \theta$ may be approximated by $\sin \theta$. Therefore,

$$y \approx L \sin \theta \quad \text{or} \quad \sin \theta \approx \frac{y}{L} \tag{24.5}$$

Substituting this expression for $\sin \theta$ in Eq. 24.3 and then solving for y gives a good approximation of the distance of the nth bright fringe (y_n) from the central maximum on either side.

$$y_n \approx \frac{nL\lambda}{d} \quad \begin{array}{l} \textit{lateral distance to bright fringe} \\ \textit{(for small θ only)} \end{array} \tag{24.6}$$

Thus, the wavelength of the light is

Wavelength of light in Young's experiment from geometrical analyses

$$\lambda \approx \frac{y_n d}{nL} \quad \begin{array}{l} \text{for } n = 1, 2, 3, \ldots \\ \textit{(for small θ only)} \end{array} \tag{24.7}$$

A similar analysis gives the distances to the dark fringes (see Exercise 13).

From Eq. 24.3, we see that, except for the zeroth order fringe, $n = 0$ (the central maximum), the positions of the fringes depend on wavelength—different values of λ give different values of $\sin \theta$ and therefore of θ. Hence, when the experimenter uses white light, as Young did, the central fringe is white, but the other orders contain a spectrum of colors. By measuring the positions of the color fringes within a particular order, Young was able to determine the wavelengths of the colors of visible light and thus show that physiological perception of color is related to a physical quantity: the wavelength of light.

EXAMPLE 24.1 ■ MEASURING THE WAVELENGTH OF LIGHT: YOUNG'S DOUBLE SLIT

In a lab experiment, monochromatic light passes through two narrow slits that are 0.050 mm apart. The interference pattern is observed on a white wall 1.0 m from the slits, and the second-order bright fringe is measured as being 2.4 cm from the center of the central maximum. (a) What is the wavelength of the light? (b) What is the distance between the second-order and third-order bright fringes?

Solution. From the problem we have

Given: $d = 0.050$ mm   *Find:* (a) λ (wavelength)
 $L = 1.0$ m $= 10^3$ mm   (b) $y_3 - y_2$ (distance between
 $y_2 = 2.4$ cm $= 24$ mm $n = 2$ and $n = 3$)
 $n = 2$

(a) Using Eq. 24.7 with distances in millimeters for convenience gives

$$\lambda = \frac{y_n d}{nL} = \frac{(24 \text{ mm})(0.050 \text{ mm})}{2(10^3 \text{ mm})} = 6.0 \times 10^{-4} \text{ mm} = 6.0 \times 10^{-7} \text{ m}$$

This is 600 nm, which is the wavelength of yellow-orange light (see Fig. 20.18).

(b) With the wavelength, y_3 could be computed, and then the distance between the second-order and third-order fringes ($y_3 - y_2$) could be found. However, the bright fringes for a given wavelength of light are evenly spaced; in general, the distance between adjacent bright fringes is constant:

$$y_{n+1} - y_n = \frac{(n+1)L\lambda}{d} - \frac{nL\lambda}{d} = \frac{L\lambda}{d}$$

In this case,

$$\frac{L\lambda}{d} = \frac{(1.0 \text{ m})(6.0 \times 10^{-7}\text{ m})}{0.050 \times 10^{-3}\text{ m}} = 1.2 \times 10^{-2}\text{ m} = 1.2 \text{ cm}$$

Follow-up Exercise. Suppose white light were used instead of monochromatic light in this Example. What would be the separation distance of the red ($\lambda = 700$ nm) and blue ($\lambda = 400$ nm) components in the second-order fringe?

24.2 Thin-Film Interference

Objectives: **To be able to (a) describe how thin films produce colorful displays, and (b) give some examples of practical applications of thin-film interference.**

Have you ever wondered what causes the rainbowlike colors when light is reflected from a thin film of oil or a soap bubble? This effect is due to interference of light reflected from opposite surfaces of the film and may be readily understood in terms of wave interference.

First, however, you need to see how the phase of a light wave is affected by reflection. Recall from Chapter 13 that a wave pulse undergoes a 180° phase shift (change) when reflected from a rigid support and no phase shift when reflected from a free support (•Fig. 24.5). Similarly, as the figure shows, the phase change for the reflection of light waves at a boundary depends on the optical densities, or the indices of refraction, of the two materials.

- A light wave traveling in one medium and reflected from the boundary of a second medium whose index of refraction is greater than that of the first medium ($n_2 > n_1$) undergoes a 180° phase change.
- If the reflecting medium has the smaller index of refraction ($n_2 < n_1$), there is no phase change.

To understand why you see colors in an oil film (on water or on a wet road), consider the reflection of monochromatic light from a thin film, as illustrated in •Fig. 24.6. The path length of the wave in the film depends on the angle of incidence (why?), but for simplicity we will assume normal (perpendicular) incidence for the light, even though the rays are drawn at an angle in the figure for greater clarity of illustration.

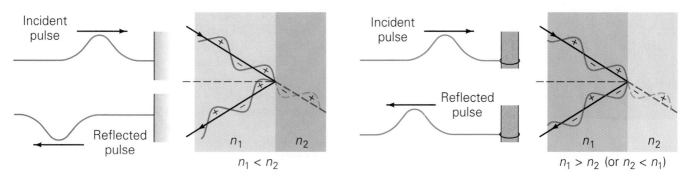

(a) Fixed end: 180° phase shift   **(b)** Free end: zero phase shift

•FIGURE 24.5 Reflection and phase shifts
The phase changes that light waves undergo on reflection are analogous to those for pulses in strings. **(a)** The phase of a pulse in a string is shifted by 180° on reflection from a fixed end, and so is the phase of a light wave when it is reflected from a more optically dense medium. **(b)** A pulse in a string has a phase shift of zero (is not shifted) when reflected from a free end. Analogously, a light wave is not phase-shifted when reflected from a less optically dense medium.

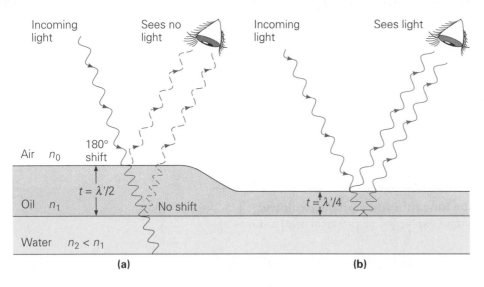

(a)   (b)   (c)

•**FIGURE 24.6 Thin-film interference**
For an oil film on water, there is a 180° phase shift for light reflected from the air–oil interface and a zero phase shift at the oil–water interface. **(a)** Destructive interference occurs if the oil film has a thickness of $\lambda'/2$ for normal incidence. (Waves displaced in drawing for clarity.) **(b)** Constructive interference occurs with a minimum film thickness of $\lambda'/4$. **(c)** Thin-film interference in an oil slick. Different film thicknesses give rise to the reflections of different colors.

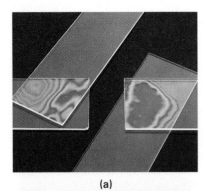

(a)

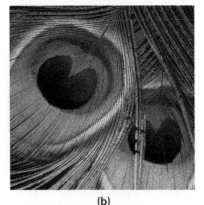

(b)

•**FIGURE 24.7 Thin-film interference**
(a) A thin air film between microscope slides gives rise to colorful patterns. **(b)** Multilayer interference in a peacock's feathers gives rise to bright colors. The brilliant throat colors of hummingbirds are produced in the same way.

The oil film has a greater index of refraction than air, and the light reflected from the air–oil interface undergoes a 180° phase shift. The transmitted waves pass through the oil film and are reflected at the oil–water interface. In general, the index of refraction of oil is greater than that of water (see Table 22.1); that is, $n_2 < n_1$, so a reflected wave in this instance is *not* phase shifted.

You might think that if the path length of the wave in the oil film ($2t$, twice the thickness, down and back) were an integral number of wavelengths [$2t = 2(\lambda'/2) = \lambda'$ in figure], then the waves reflected from the two surfaces would interfere constructively. But keep in mind that the wave reflected from the top surface undergoes a 180° phase shift. The reflected waves from the two surfaces are therefore *out of phase* for this film thickness, and they interfere destructively. This means that no light of this wavelength is reflected, only transmitted.

Similarly, if the path length in the film were an odd number of half-wavelengths [$2t = 2(\lambda'/4) = \lambda'/2$ in figure], the reflected waves would be in phase (as a result of a 180° phase shift of the incident wave), and they would interfere constructively. Light of this wavelength would be reflected from the oil film.

Keep in mind that the path length is expressed in terms of the wavelength of light *in the film*, λ'. Recall from Section 22.3 that the wavelength is different in different media. Here, $\lambda' = \lambda/n$, where λ' is the wavelength in the oil, λ the wavelength in air, and n the index of refraction of the oil. This may be seen from $n = c/v = f\lambda/f\lambda' = \lambda/\lambda'$.

Because oil and soap films generally have different thicknesses in different regions, particular wavelengths (colors) of white light interfere constructively in different regions and are reflected. As a result, a vivid display of various colors appears, which may change if the film thickness changes with time (see Demonstration 16). Thin-film interference may be seen if two glass slides are stuck together with an air film between them (•Fig. 24.7a). Also shown in the figure (•Fig. 24.7b) is an example of colorful interference we see in nature. The bright colors of a peacock are a result of layers of fibers in its feathers. Light reflected from successive layers interferes constructively, giving bright colors, even though the feather has no color pigment. Since the condition for constructive interference depends on the angle of incidence, the color pattern changes somewhat with the viewing angle and motion of the bird.

DEMONSTRATION 16 ■ Thin-Film Interference

A vivid display of colors from soap bubbles illuminated with white light from below.

(a) Interference of light reflected from the inner and outer soap-bubble surfaces produces the colors. As the film thickness increases from top to bottom, the spectrum of colors is repeated several times.

(b) Swirling colors may be produced by blowing gently tangent to the film.

From this analysis of thin-film interference, you can see that the word "destructive" in this context does not imply that energy is destroyed. Destructive interference is simply a description of a physical fact—that a light wave is not present at a particular location, but is somewhere else. When destructive interference occurs for light incident on the surfaces of a thin film, this tells you that the light is not reflected, but that the incident beam (energy) is transmitted. Similarly, the mathematical description of Young's double-slit experiment in the preceding section tells you that you *should not expect* to find light at the locations of the dark fringes. The light is at the bright fringes, and the interference analysis shows that the light is merely redistributed.

A practical application of thin-film interference is described in the Insight feature on p. 743 dealing with nonreflective coatings for lenses. The situation discussed there differs from that of the previous oil–water example, because glass has a greater index of refraction than the nonreflecting film. Consequently, phase shifts of incident light take place at both the film and glass surfaces. In such a case, the condition for destructive interference is that the path difference for normally incident light is $\lambda'/2$, where λ' is the wavelength of the light in the film. Thus, $2t = \lambda'/2$, and the film thickness that will produce the least reflection is

$$t = \frac{\lambda'}{4}$$

The wavelength of the light in the film (λ') is related to that in air (λ) by

$$\lambda' = \frac{\lambda}{n}$$

Substituting for λ' in the equation for t gives

$$t = \frac{\lambda}{4n} \quad \textit{minimum film thickness} \atop \textit{(for } n_2 > n_1\textit{)} \tag{24.8}$$

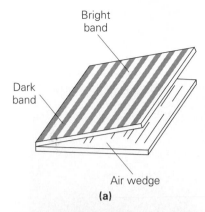

Bright band

Dark band

Air wedge

(a)

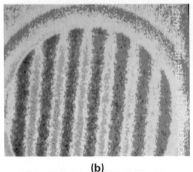

(b)

•**FIGURE 24.8 Optical flatness**
(a) An optical flat is used to check the smoothness of a reflecting surface. The flat is placed so that there is an air wedge between it and the surface. **(b)** If the surface is smooth, a regular or symmetric interference pattern is seen.

EXAMPLE 24.2 ■ NONREFLECTING COATINGS: THIN-FILM INTERFERENCE

A glass lens ($n = 1.60$) is coated with a thin transparent film of magnesium fluoride (MgF_2, $n = 1.38$) to make it nonreflecting. What is the minimum thickness of the film for the lens to be nonreflecting for incident light whose wavelength is 550 nm?

Solution.

Given: $n = 1.38$ (for film)   *Find:* t (film thickness)
            $\lambda = 550$ nm

Using Eq. 23.8, since n_2 (glass) $> n_1$ (film),

$$t = \frac{\lambda}{4n} = \frac{(550 \text{ nm})}{4(1.38)} = 99.6 \text{ nm}$$

which is quite thin ($\approx 10^{-5}$ cm).

Follow-up Exercise.   Suppose you wished to have the glass lens in this Example reflect rather than transmit the incident light through the lens. What would be the minimum film thickness for this case?

Optical Flats and Newton's Rings

The phenomenon of thin-film interference is used to check the smoothness and uniformity of optical components such as mirrors and lenses. **Optical flats** are made by grinding and polishing glass plates until they are as flat and smooth as possible. The degree of flatness may be checked by putting two such plates together at a slight angle, so that there is a thin air wedge between them (•Fig. 24.8). If the plates are smooth and flat, a regular interference pattern of bright and dark fringes, or bands, appears. This pattern is a result of the uniformly varying differences in path lengths between the plates. Any irregularity in the pattern indicates an irregularity in at least one of the plates. Once a good optical flat is verified, it can be used to check the flatness of a reflecting surface, such as that of a precision mirror.

A similar technique is used to check the smoothness and symmetry of lenses. When a curved lens is placed on an optical flat, the air wedge is a ring below the periphery of the circular lens (•Fig. 24.9a). The regular interference pattern in this case is a set of concentric bright and dark circular fringes (•Fig. 24.9b). They are called **Newton's rings**, after Isaac Newton, who first described this interference effect. Lens irregularities give rise to a distorted fringe pattern.

•**FIGURE 24.9 Newton's rings**
(a) A lens placed on an optical flat forms a ring-shaped air wedge, which gives rise to interference. The interference pattern is a set of concentric rings called Newton's rings. Lens irregularities produce a distorted pattern. **(b)** A Newton's ring pattern.

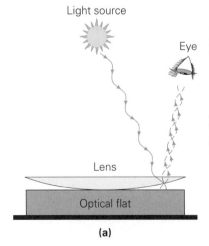

Light source

Eye

Lens

Optical flat

(a)

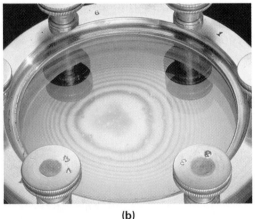

(b)

You have probably noticed the blue–purple tint of the coated optical lenses used in cameras and binoculars. The coating makes the lenses "nonreflecting." If a lens is nonreflecting, the incident light is almost totally transmitted. Complete transmission of light is desirable for the exposing of photographic film and for viewing objects with binoculars.

A lens is made nonreflecting by coating it with a thin film of a material that has an index of refraction between those of air and glass (Fig. 1). If the coating is a quarter-wavelength thick, the difference in path length between the reflected rays is $\lambda'/2$, where λ' is the wavelength of light in the coating. In this case, both reflected waves undergo a 180° phase shift, and they are out of phase for a path difference of $\lambda'/2$ and interfere destructively. That is, the incident light is transmitted, and the coated lens is nonreflecting.

FIGURE 2 Coated lenses
The nonreflective coating on binocular and camera lenses generally produces a characteristic bluish-purple hue.

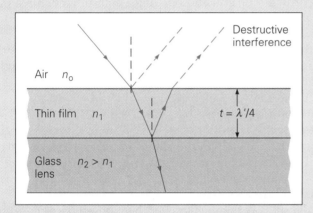

FIGURE 1 Nonreflective coating
For a thin film on a glass lens, there is a 180° phase shift at each interface when the index of refraction of the film is less than that of the glass. As a result, destructive interference occurs for a minimum film thickness of $\lambda'/4$, and the waves are transmitted rather than reflected, making the lens surface nonreflecting.

A quarter-wavelength thickness of film is, of course, specific for a particular wavelength of light. The thickness is usually chosen to be a quarter-wavelength of yellow–green light ($\lambda \approx 550$ nm), to which the human eye is most sensitive. The wavelengths at the red and blue ends of the visible region are reflected, giving the coated lens its bluish-purple tint (Fig. 2). Sometimes other quarter-wavelength thicknesses are chosen, giving rise to other hues, such as amber or reddish purple.

The coating on a nonreflecting lens actually serves a double purpose. It promotes nonreflection from the front of the lens and reduces back reflection. Some of the light transmitted through a lens is reflected from the back surface. This could be reflected again from the front surface of an uncoated lens, giving rise to poor images. However, the proper film thickness allows back reflections to be transmitted through the coating.

Nonreflective coatings are also applied to the surfaces of solar cells, which convert light into electrical energy (Chapter 27). Since the thickness of such a coating is wavelength-dependent, like that on a nonreflecting lens, not all of the light is transmitted. However, the reflective losses may be decreased from around 30% to 10%, making the cell more efficient.

24.3 Diffraction

Objectives: To be able to (a) define diffraction, and (b) give examples of diffractive effects.

In geometrical optics, light is represented by rays and pictured as traveling in straight lines. If this model represented the real nature of light, however, there would be no interference effects in Young's double-slit experiment. Instead, there would be only two bright slit images on the screen with a well-defined shadow area where no light enters. In fact, there are interference patterns, which means that the light must deviate from a straight-line path and enter the regions that would otherwise be in shadow. Huygens' principle requires that the waves spread out from the slits, and this deviation, or bending, of light is called **diffraction**. Diffraction generally occurs when waves pass through small openings or around

•FIGURE 24.10 Water wave diffraction
Diffraction of water waves can be observed at the seashore when conditions are suitable. See Fig. 13.17 for a laboratory demonstration of water wave diffraction.

sharp edges or corners. The diffraction of water waves is shown in •Fig. 24.10. (See also Fig. 13.17.)

As Fig. 13.17 showed, there are different degrees of bending or diffraction. The amount of diffraction depends on the wavelength of the wave and the size of the opening or object. In general, *the larger the wavelength compared to the size of the opening or object, the greater the diffraction*. This is illustrated in •Fig. 24.11. In Fig. 24.11a, the wavelength is smaller than the object ($\lambda < d$), and there is little diffraction—a large shadow zone behind the object is free of waves. (This corresponds to Fig. 13.17a for an opening. The waves pass through with little diffraction and the shadow zones are around the corners of the opening.) In Fig. 24.11b, with the wavelength and size of the object about equal ($\lambda \approx d$), there is some diffraction, but still a noticeable shadow zone. Finally, in Fig. 24.11c, with the wavelength much greater than the object ($\lambda > d$), there is much diffraction and little or no shadow zone. The wave bends around and passes on almost as if the object were not there.

The diffraction of sound (Chapter 14) is quite evident. Someone can talk to you from another room or around the corner of a building, and even in the absence of reflections, you can easily hear them. Audible sound waves have wavelengths of centimeters to meters. Thus, the wavelengths of sound are larger than or about the same size of ordinary objects and openings, and diffraction readily occurs for sound. Visible light, on the other hand, has wavelengths on the order of 10^{-7} m (Chapter 20), and diffraction phenomena for light waves often go unnoticed. However, careful observation will reveal that a shadow boundary is blurred or fuzzy. On close inspection, you can see that there is a pattern of bright and dark fringes (•Fig. 24.12). These interference patterns are evidence of the diffraction of the light around the edge of the object.

An illustrative example of diffraction is that with a single slit (•Fig. 24.13). Suppose that a single slit of width w is illuminated using monochromatic light whose wavelength (λ) is much smaller than the slit width. An interference pattern consisting of a bright central maximum and a symmetrical array of bright fringes (regions of constructive interference) on both sides is observed on a screen at a distance L from the slit (where $L \gg w$).

Because the width of the slit is very much greater than the wavelength of the light, the slit with light passing through cannot be treated as a point source of Huygens' wavelets. However, various points on the wave front passing through the slit may be considered to be such point sources. Then, the interference of those wavelets can be analyzed much as double-slit interference was analyzed earlier. The analysis will not be done here, but you will notice that the stated result is very similar in form to that for Young's double slit. However, for the single slit, dark fringes (regions of destructive interference) are analyzed, rather than bright

•FIGURE 24.11 Wavelength and object dimensions
In general, the larger the wavelength compared to the size of an object or opening, the greater the diffraction, as the illustrations show. Without diffraction, there is a shadow zone without waves behind the object.

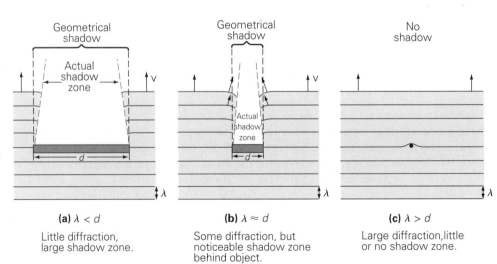

(a) $\lambda < d$
Little diffraction, large shadow zone.

(b) $\lambda \approx d$
Some diffraction, but noticeable shadow zone behind object.

(c) $\lambda > d$
Large diffraction, little or no shadow zone.

fringes. From the geometry, the condition for *destructive interference,* or interference minima (dark fringes) can be shown to be

$$w \sin \theta = m\lambda \quad \text{for } m = 1, 2, 3, \ldots \quad (24.9)$$

where θ is the angle of a particular minimum designated by $m = 1, 2, 3, \ldots$ on either side of the central bright fringe. (There is no $m = 0$. Why?)

Also, a small-angle approximation, $\sin \theta \approx y/L$, can be made in some instances in obtaining a formula for the lateral displacement of a particular interference minimum on the screen. This gives a good approximation of the distances of the dark fringes on either side of the center of the central maximum.

$$y_m = \frac{mL\lambda}{w} \quad \text{for } m = 1, 2, 3, \ldots \quad (24.10)$$

Note that Eq. 24.10 has the same form as Eq. 24.6, the equation for the displacement of bright fringes for double-slit interference. Here w is the width of the slit, and there d is the distance between the slits.

The qualitative predictions from Eq. 24.10 are quite interesting and instructive.

- For a given slit width (w), the greater the wavelength (λ), the wider the diffraction pattern.
- For a given wavelength (λ), the narrower the slit width (w), the wider the diffraction pattern.
- The width of the central maximum is twice the width of the side maxima.

Let's consider each of these.

As the slit is made narrower, the central maximum and the side fringes spread out and become larger (Fig. 24.13b). Equation 24.10 is not applicable to very small slit widths (because of the small-angle approximation). If the slit is decreased until it is the same order of magnitude as the wavelength of the light, the central maximum spreads out over the whole screen. That is, diffraction becomes dramatically evident when the slit width (or object) is about the same size as the wavelength. Diffraction effects are most easily observed when $\lambda/w \geq 1$, or $\lambda \geq w$.

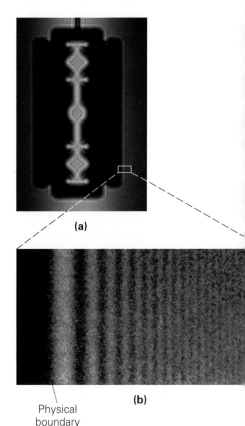

(a)

Physical
boundary

•**FIGURE 24.12 Diffraction in action**
(a) Diffraction patterns produced by a razor blade. **(b)** A close-up view of the fringes formed at the edge of the blade.

•**FIGURE 24.13 Single-slit diffraction**
(a) The diffraction of light by a single slit gives rise to an interference pattern consisting of a large bright central maximum and a symmetric array of side fringes. **(b)** The widths of the fringes depend on the width of the slit— the larger the slit width, the smaller the fringe widths.

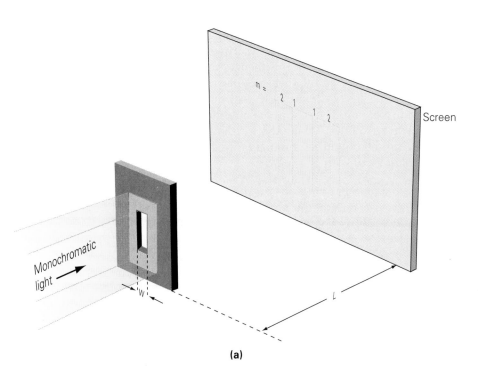

(a)

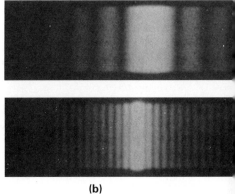

(b)

CONCEPTUAL EXAMPLE 24.3 ■ DIFFRACTION AND RADIO RECEPTION

Driving in the city, or in mountainous areas of the country, you have probably noticed that on certain broadcast bands the quality of your radio reception can vary sharply from place to place, with stations seeming to fade out and reappear. What do you think might cause this variation, and which band would you expect to be least affected by it? (a) Weather (162 MHz); (b) FM (88–108 MHz); (c) AM (525–1610 kHz). *Clearly establish the reasoning and physical principle(s) used in determining your answer before checking it below. That is, **why** did you select your answer?*

Reasoning and Answer.

Radio waves, like light, are electromagnetic waves, and so tend to travel in straight lines long distances from their sources. Although they are generally better than visible light at penetrating solid materials, they can still be blocked by objects in their path—especially if the objects are massive (like hills) or are made largely of metal (like buildings). However, because of diffraction, they can also "wrap around" obstacles or "fan out" as they pass through openings, just as sound waves do, provided their wavelength is at least roughly the size of the obstacle or opening. The longer the wavelength, the greater will be the diffraction, and so the less likely the waves are to be obstructed.

To determine which band benefits most by such diffraction, we need to find the wavelengths corresponding to the given frequencies, using the relationship $\lambda f = c$. When we do this, we find that AM radio waves, with $\lambda = 186$–571 m, are the longest of the three bands specified by a factor of about 100. We might conclude that weather and FM broadcasts would tend to be blocked by large objects such as hills or buildings, whereas AM broadcasts are likely to be diffracted around such objects or through the openings between them. Thus, the answer is (c).

Follow-up Exercise: During half-time at a football game, when a marching band faces you, you can easily hear the woodwind instruments (clarinet, saxophone, etc.) along with the brass instruments (trumpet, trombone, etc.). Yet, when the players turn and march away from you, the brass instruments sound muted but the woodwinds can still be heard quite well. Why is this?

Conversely, if the slit is made wider when using a particular wavelength of light, the diffraction pattern becomes narrower. The fringes move closer together and eventually become difficult to distinguish when λ is much smaller than w ($\lambda \ll w$). The pattern then appears as a fuzzy shadow around the central maximum, which is the illuminated image of the slit. This type of pattern is observed for the image produced by sunlight entering a dark room through a hole in a curtain. Such an observation led early experimenters to investigate the wave nature of light, and the acceptance of this theory was, in large part, due to the explanation of diffraction offered by physical optics.

Notice from Figs. 24.2 and 24.13 that the central maximum differs in width from the side maxima. The central maximum turns out to be twice as wide, which can be shown as follows. Taking the width of the central maximum to be the distance between the bounding minima on each side ($m = 1$), or a width of $2y_1$, we have from Eq. 24.10 with $y_1 = L\lambda/w$:

$$2y_1 = \frac{2L\lambda}{w} \qquad \textit{width of central maximum} \qquad (24.11)$$

Similarly, the width of the bright side fringes is given by

$$y_2 - y_1 = y_3 - y_2 = \frac{L\lambda}{w}$$

Or, in general,

$$y_{m+1} - y_m = \frac{L\lambda}{w} \qquad \textit{width of side fringes} \qquad (24.12)$$

Thus, the width of the central maximum is twice that of the side fringes.

EXAMPLE 24.4 ■ WIDTH OF THE CENTRAL MAXIMUM: SINGLE-SLIT DIFFRACTION

Monochromatic blue light ($\lambda = 425$ nm) passes through a slit whose width is 0.50 mm. What is the width of the central maximum on a screen located 1.0 m from the slit?

Solution.

Given: $\lambda = 425$ nm $= 4.25 \times 10^{-5}$ cm *Find:* $2y_1$ (width of
$w = 0.50$ mm $= 0.050$ cm central maximum)
$L = 1.0$ m $= 10^2$ cm

Equation 24.11 can be used directly.

$$2y_1 = \frac{2L\lambda}{w} = \frac{2(4.25 \times 10^{-5}\,\text{cm})(10^2\,\text{cm})}{0.050\,\text{cm}} = 0.17\,\text{cm}$$

Note that diffraction effects would not be readily observed in this case since $\lambda \ll w$.

Follow-up Exercise. By what factor would the width of the central maximum change if red light ($\lambda = 700$ nm) were used in this Example?

Diffraction Gratings

Bright and dark fringes result from diffraction and interference when light passes through a single slit or double slits. Fringe patterns increase as the number of slits is increased, and it is observed that the bright lines become sharper (narrower) and the dark lines broader. The intensity of the lines is less when the light has to pass through many narrow slits, but even so, the sharp lines are useful in optical analysis of light sources and other applications. As a result, arrangements of large numbers of parallel, closely spaced slits are fashioned in the form of **diffraction gratings** (•Fig. 24.14).

Diffraction gratings were first made of fine strands of wire. Their effects were similar to what may be seen by viewing a candle flame through a feather held

•FIGURE 24.14 **Diffraction grating** (a) A diffraction grating produces a sharply defined interference pattern. In each of the side fringes, components of different wavelengths are separated, since the deviation depends on the wavelength: $\theta = \sin^{-1}(n\lambda/d)$. (b) As a result, gratings are used in spectrometers to determine the wavelengths present in a beam of light by measuring their angles of diffraction and to separate the various wavelengths for further analysis.

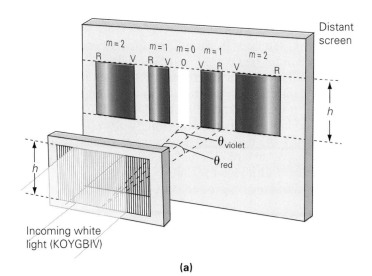

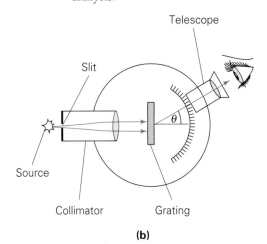

(a)

(b)

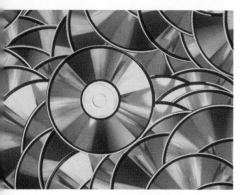

close to the eye. The closely spaced grooves of a CD can also act as a diffraction grating, giving rise to their familiar iridescent sheen of color (•Fig. 24.15). Better gratings have a large number of fine lines or grooves on glass or metal surfaces.

If light is transmitted through a grating, it is called a *transmission grating*. However, *reflection gratings* are more common. The colorful effect of a "reflection grating" that you might own is shown in Fig. 24.15. Commercial gratings are made by depositing a thin film of aluminum on an optically flat surface and then removing some of the reflecting metal by cutting regularly spaced, parallel lines. Precision diffraction gratings are made using two coherent laser beams, intersecting at an angle. The beams expose a layer of photosensitive material, which is then etched. The spacing of the grating lines is determined by the intersection angle of the beams. Precision gratings may have 30,000 lines per centimeter or more and are therefore expensive and difficult to fabricate. Most gratings used in laboratory instruments are replica gratings, which are plastic castings of high-precision master gratings.

Diffraction gratings are widely used in spectroscopy (the study of spectra), where they have almost replaced prisms. The forming of a spectrum and the measurement of wavelengths by means of a grating depend only on geometrical measurements such as lengths and/or angles. Wavelength determination using a prism, on the other hand, also depends on the dispersive characteristics of the glass or other material of which the prism is made.

It can be shown that the condition for interference maxima for a grating illuminated with monochromatic light is identical to that for a double slit. The expression is

Note: d is the distance between adjacent slits

$$d \sin \theta = n\lambda \quad \text{for } n = 0, 1, 2, 3, \ldots \qquad \begin{array}{c} \textit{interference maxima} \\ \textit{for a grating} \end{array} \qquad (24.13)$$

Here n is the order of interference and θ is the angle of deviation of a particular wavelength. The zeroth-order maximum corresponds to the central maximum of the diffraction pattern. The spacing between adjacent slits (d) is obtained from the number of lines per unit length of the grating, $d = 1/N$. For example, if $N = 5000$ lines/cm,

$$d = \frac{1}{N} = \frac{1}{5000} = 2.0 \times 10^{-4} \, \text{cm}$$

In contrast to a prism, which deviates red light least and violet light most, a diffraction grating produces the least angle of deviation for violet light (short λ), and the greatest for red light (long λ). Also, a prism disperses white light into a single spectrum. A diffraction grating, however, produces a number of spectra, one for each order other than $n = 0$. There is no deviation of the components of the light for the zeroth order (sin $\theta = 0$ for all wavelengths), so the central maximum is a white image. However, a spectrum is produced for each higher order.

The sharp spectra produced by diffraction gratings are used in instruments called spectrometers. In a spectrometer, materials are illuminated with light of various wavelengths to find which ones are strongly absorbed. The pattern of absorption helps to identify the material. The grating is rotated so that the sample is illuminated with a succession of different wavelengths. The wavelengths may also be in the infrared and ultraviolet regions.

The number of spectral orders produced by a grating depends on the wavelength of the light (which may be infrared, visible, or ultraviolet) and on the grating's spacing (d). From Eq. 24.13, since sin θ cannot exceed 1 (sin $\theta \leq 1$),

$$\sin \theta = \frac{n\lambda}{d} \leq 1$$

The order number is therefore limited as follows:

$$n \leq \frac{d}{\lambda}$$

(24.14) **Limit of spectral order**

(Why?)

EXAMPLE 24.5 ■ A DIFFRACTION GRATING: LINE SPACING AND SPECTRAL ORDERS

A particular diffraction grating produces an $n = 2$ spectral order at a deviation angle of 30° for light with a wavelength of 500 nm. (a) How many lines per centimeter does the grating have? (b) If the grating were illuminated with white light, how many orders of the *complete* visible spectrum would be produced?

Solution.

Given: $n = 2$
$\lambda = 500 \text{ nm} = 5.00 \times 10^{-7} \text{ m}$
$\theta = 30°$

Find: (a) N (lines/cm)
(b) Number of orders

(a) The grating spacing may be found using Eq. 24.13:

$$d = \frac{n\lambda}{\sin \theta} = \frac{2(5.00 \times 10^{-7} \text{ m})}{\sin 30°}$$

$$= 20.0 \times 10^{-7} \text{ m} = 2.00 \times 10^{-4} \text{ cm}$$

Then

$$N = \frac{1}{d} = \frac{1}{2.00 \times 10^{-4} \text{ cm}}$$

$$= 5000 \text{ lines/cm}$$

(b) The greatest angle of deviation for any order is for the longest wavelength, which is that of red light in the visible spectrum ($\lambda = 700 \text{ nm} = 7.00 \times 10^{-7}$ m). With a spacing of $d = 2.00 \times 10^{-4}$ cm $= 2.00 \times 10^{-6}$ m, by Eq. 24.14, the order numbers for the visible spectrum for this grating are limited to

$$n \leq \frac{d}{\lambda} = \frac{2.00 \times 10^{-6} \text{ m}}{7.00 \times 10^{-7} \text{ m}} = 2.86$$

Thus, for a grating with this spacing, only two complete orders are observed, or four complete visible spectra—two on each side of the central maximum. A large portion of the spectrum is seen in the third order, but this is limited to wavelengths of

$$\lambda \leq \frac{d}{n} = \frac{2.00 \times 10^{-6} \text{ m}}{3} = 6.70 \times 10^{-7} \text{ m} = 670 \text{ nm}$$

Thus, the red end of the visible spectrum is not seen completely in the third spectral order.

Follow-up Exercise. Suppose the grating in this Example had 10,000 lines/cm. How many orders of the complete visible spectrum would be observed for white light using this grating? (Before computing the answer, would you expect more or fewer orders to be seen?)

Note that it is possible for spectra produced by diffraction gratings to overlap at higher orders. That is, the angles of deviation for different orders may be the same for two different wavelengths. The spacing must be greater than or equal to the wavelength ($d \geq \lambda$) to have interference for the first spectral order. However, if the spacing is much greater than the wavelength ($d \gg \lambda$), the difference in the deviation angles for nearby wavelengths is quite small. This difference increases for higher orders, but then the overlapping of spectra becomes a problem. Thus, there is an optimal spacing of the lines on a diffraction grating for each spectral region of interest—infrared, visible, or ultraviolet. The spacing is typically chosen to be between 3λ and 6λ, where λ is the median (middle) wavelength of the spectral region.

X-ray Diffraction

The wavelength of any electromagnetic wave may be determined if a diffraction grating with the appropriate spacing is available. Diffraction was used to determine the wavelengths of X-rays early in this century. Experimental evidence indicated that the wavelengths of X-rays were probably around 10^{-8} cm, but it is impossible to construct a diffraction grating with this spacing. Around 1913, Max von Laue, a German physicist, suggested that the regular spacing of the atoms in a crystalline solid might make it act as a diffraction grating for X-rays, since the atomic spacing in crystals is on the order of 10^{-8} cm. X-rays were directed at crystals, and diffraction patterns were indeed observed.

•Figure 24.16a illustrates diffraction by the planes of atoms in a crystal such as sodium chloride. You can see that the path difference is $2d \sin \theta$, where d is the distance between the crystal's internal planes. Thus, the condition for constructive interference is

Bragg's Law

$$2d \sin \theta = n\lambda \quad \text{for } n = 1, 2, 3, \dots \tag{24.15}$$

This relationship is known as **Bragg's law**, after W. L. Bragg, the British physicist who first derived it.

The wavelengths of X-rays were experimentally determined by this means, and X-ray diffraction is now used to investigate the internal structure, not only of simple crystals, but of large, complex biological molecules such as proteins and DNA. Because of their short wavelengths, X-rays provide a diffraction "probe" for investigating the interatomic spacings of molecules.

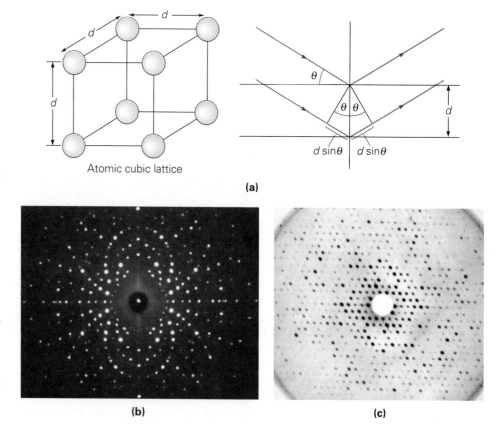

•**FIGURE 24.16 Crystal diffraction**
(a) The array of atoms in a crystal lattice structure acts like a diffraction grating, and X-rays are diffracted from the planes of atoms. With a lattice spacing of d, the path difference for the X-rays diffracted from adjacent planes is $2d \sin \theta$. **(b)** X-ray diffraction pattern of a crystal of potassium sulfate. By analyzing the geometry of such patterns, investigators can deduce the structure of the crystal and the position of its various atoms. **(c)** X-ray diffraction pattern of the protein hemoglobin that carries oxygen in the blood.

750   CHAPTER 24   Physical Optics: The Wave Nature of Light

24.4 Polarization

Objectives: To be able to (a) explain light polarization, and (b) give examples of polarization, both in the environment and in commercial applications.

When you think of polarized light, you may visualize polarizing (or Polaroid) sunglasses, since this is one of the more common applications of polarization. When something is polarized, this means that it has a preferential direction, or orientation (think of a polarized electrical plug, as described in Chapter 18). In terms of light waves, **polarization** refers to the orientation of the transverse oscillations.

Note: See Fig. 18.23

Recall that light is an electromagnetic wave with oscillating electric and magnetic field vectors (**E** and **B**, respectively) perpendicular to the direction of propagation (Chapter 20). Light from most sources consists of a large number of electromagnetic waves emitted by the atoms of the source. Each atom produces a wave with a particular orientation, corresponding to the direction of the atomic vibration. However, since electromagnetic waves are produced by numerous atoms, all orientations of the **E** and **B** fields are possible in the composite light emitted. When the field vectors are randomly oriented, the light is said to be *unpolarized*. This is commonly represented schematically in terms of the electric field vector as shown in •Fig. 24.17a.

Note: Refer to Figure 20.16.

If there is some preferential orientation of the field vectors, the light is said to be *partially polarized* (•Fig. 24.17b). If the field vectors oscillate in only *one* plane, the light is *plane polarized*, or *linearly polarized* (•Fig. 24.17c). Note that polarization is evidence that light is a transverse wave. Longitudinal waves, such as sound waves, cannot be polarized because there are no two-dimensional vibrations.

Light can be polarized in several ways. Polarization by reflection and double refraction will be discussed here. Polarization by scattering will be considered in Section 24.5.

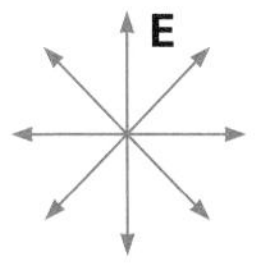

Polarization by Reflection

When a beam of unpolarized light strikes a smooth transparent medium such as glass, it is partially reflected and partially transmitted. The reflected light may be completely polarized, partially polarized, or unpolarized, depending on the angle of incidence. The unpolarized case occurs for 0° or normal incidence. As the angle of incidence is varied, it is found that the reflected light is partially polarized. The electric field components parallel to the surface are reflected more strongly, producing this partial polarization. However, at one particular angle of incidence, the reflected beam is completely polarized. At this angle, moreover, the refracted beam is partially polarized (•Fig. 24.18).

David Brewster (1781–1868), a Scottish physicist, found that the complete polarization of the reflected beam occurs when the reflected and refracted beams are 90° apart. The incident angle for this to occur is called the **polarizing angle** (θ_p) or the **Brewster angle**, and it is specific for a given material. As shown in Fig. 24.18, the reflected and refracted beams are 90° apart, and because $\theta_1 = \theta_p$ it is clear that

$$\theta_1 + 90° + \theta_2 = 180°$$

Then

$$\theta_1 + \theta_2 = 90°$$

or

$$\theta_2 = 90° - \theta_1$$

By Snell's law (Chapter 22), for incidence in air,

$$\frac{\sin \theta_1}{\sin \theta_2} = n$$

E

(a) Unpolarized

E

(b) Partially polarized

E

(c) Plane (linearly) polarized

•**FIGURE 24.17 Polarization** Polarization is represented by the orientation of the plane of vibration of the electric field vectors. **(a)** When the vectors are randomly oriented (as viewed along the direction of propagation), the light is unpolarized. **(b)** With preferential orientation of the vectors, the light is partially polarized. **(c)** When the vectors are in a plane, the light is plane polarized, or linearly polarized.

•FIGURE 24.18 Polarization by reflection

When the reflected and refracted components of a beam of light are 90º apart, the reflected component is linearly polarized, and the refracted component is partially polarized. This occurs when $\theta_1 = \theta_p = \tan^{-1} n$. With a stack of glass plates (or layers of thin film), there are multiple reflections which increase the intensity of the reflected polarized beam, and the refracted beam becomes more polarized.

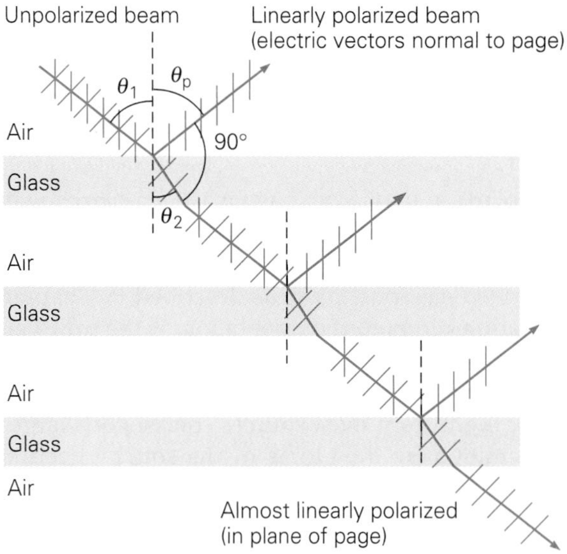

Unpolarized beam   Linearly polarized beam
                          (electric vectors normal to page)

Almost linearly polarized
(in plane of page)

In this case, $\sin \theta_2 = \sin (90° - \theta_1) = \cos \theta_1$. Thus,

$$\frac{\sin \theta_1}{\sin \theta_2} = \frac{\sin \theta_1}{\cos \theta_1} = \tan \theta_1 = n$$

With $\theta_1 = \theta_p$,

$$\tan \theta_p = n \quad \text{and} \quad \theta_p = \tan^{-1} n \qquad (24.16)$$

θ_p, polarizing or Brewster angle: incident angle for complete polarization of reflected light

(Since glass is dispersive, the Brewster angle also depends to some degree on the wavelength of the incident light.)

As shown in Fig. 24.18, in a stack of glass plates, the reflections from successive surfaces increase the intensity of the reflected polarized beam. Note that the refracted beam becomes more linearly polarized with successive refractions. This effect is sometimes referred to as *polarization by refraction*. In practical applications, several thin films of a transparent material are used instead of glass plates.

EXAMPLE 24.6 ■ SUNLIGHT ON A POND: POLARIZATION BY REFLECTION

Sunlight is reflected from the smooth surface of a pond of water. What is the Sun's altitude when the polarization of the reflected light is greatest?

Solution.   Here, we obtain the index of refraction of water from Table 22.1, and so we can write:

Given: $n = 1.33$ (Table 22.1)   *Find:* θ (altitude angle)
                                              (for greatest polarization)

The altitude is the angle between the Sun and the horizon, so the angle of incidence is the complementary angle (draw yourself a sketch), and

$$\theta = 90° - \theta_p$$

where θ_p is the polarizing angle for maximum or complete polarization.
    Then using Eq. 24.16,

$$\theta_p = \tan^{-1} n = \tan^{-1}(1.33) = 53°$$

So

$$\theta = 90° - \theta_p = 90° - 53° = 37°$$

Follow-up Exercise.   Light is incident on a flat, transparent material with an index of refraction of 1.38. At what angle of refraction would the transmitted light have the greatest polarization?

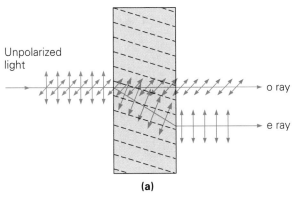

•FIGURE 24.19 Double refraction, or birefringence
(a) Unpolarized light incident normal to the surface of birefringent crystal and at an angle to a particular crystal direction is separated into two components. The ordinary (o) ray and the extraordinary (e) ray are plane polarized in mutually perpendicular directions. (b) Birefringence seen through a calcite crystal.

Unpolarized light

o ray

e ray

(a)

(b)

It is interesting to note that a rainbow, or light from a rainbow, is partially polarized. This occurs because the reflection angle inside a water droplet is somewhat close to the Brewster angle. The plane of polarization is tangential to the bow at each point.

Polarization by Double Refraction (Birefringence and Dichroism)

When monochromatic light travels in glass, its speed is the same in all directions and is characterized by a single index of refraction. Any material like this is said to be *isotropic,* meaning that it has the same characteristics in all directions. Some crystalline materials, such as quartz, calcite, and ice, are *anisotropic* with respect to the speed of light; that is, the speed of light is different in different directions within the material. Anisotropy gives rise to some unique optical properties, one of which is that the index of refraction may vary with the direction of propagation. Such materials are said to be doubly refracting, or to exhibit **birefringence**, and polarization is involved.

For example, a beam of unpolarized light incident on a birefringent crystal of calcite ($CaCO_3$, calcium carbonate) is illustrated in •Fig. 24.19. When the beam propagates at an angle to a particular crystal axis, it is doubly refracted and separated into two components, or rays. These two rays are linearly polarized in mutually perpendicular directions. One ray, called the *ordinary (o) ray,* passes through the crystal in an undeflected path and is characterized by an index of refraction (n_o). The second ray, called the *extraordinary (e) ray,* is refracted and is characterized by an index of refraction (n_e). The particular axis direction indicated by dashed lines in Fig. 24.19 is called the optic axis. Along this direction, $n_o = n_e$, and nothing extraordinary is noted about the transmitted light.

Some birefringent crystals, such as tourmaline, exhibit the interesting property of absorbing one of the polarized components more than the other. This property is called **dichroism.** If a dichroic crystal is sufficiently thick, the more strongly absorbed component may be completely absorbed. In that case, the emerging beam is plane polarized (•Fig. 24.20).

Another dichroic crystal is quinine sulfide periodide (commonly called herapathite, after W. Herapath, an English physician who discovered its polarizing properties in 1852). This crystal was of great practical importance in the development of modern polarizers. Around 1930, Edwin H. Land, an American scientist, found a way to align tiny, needle-shaped dichroic crystals in sheets of transparent celluloid. The result was a thin sheet of polarizing material that was given the commercial name Polaroid.

Better polarizing films have been developed using synthetic polymer materials instead of celluloid. During the manufacturing process, this kind of film is stretched in order to align the long molecular chains of the polymer. With proper treatment, the outer (valence) electrons of the molecules can move along the ori-

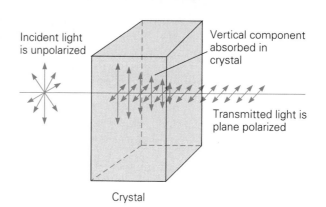

•**FIGURE 24.20 Selective absorption**
Dichroic birefringent crystals selectively absorb one polarized component more than the other. If the crystal is thick enough, the emerging beam is linearly polarized.

ented chains. As a result, light with **E** vectors parallel to the oriented chains is readily absorbed, but light with **E** vectors perpendicular to the chains is transmitted. The direction perpendicular to the orientation of the molecular chains is called the **transmission axis**, or the **polarization direction.** Thus, when unpolarized light falls on a polarizing sheet, the sheet acts as a polarizer and polarized light is transmitted (•Fig. 24.21).

The human eye cannot distinguish between polarized and unpolarized light. To tell whether or not light is polarized, we must use an analyzer, which may be a sheet of polarizing film. As shown in Fig. 24.21, if the transmission axis of an analyzer is parallel to the plane of polarization of polarized light, then there is maximum transmission. If the transmission axis of the analyzer is perpendicular to the plane of polarization, little light (ideally, none) will be transmitted. In general, the intensity of the transmitted light is given by $I = I_o \cos^2 \theta$, where θ is the angle between the transmission axes of the polarizer and analyzer. This expression is known as *Malus's law*, after its discoverer, French physicist E.L. Malus (1775–1812).

Now you can understand the principle of polarizing sunglasses. Light reflected from a smooth surface is partially polarized, as you learned earlier in this section. The direction of polarization is chiefly in the plane of the surface (see Fig. 24.18). Light reflected from water or snow may be so intense that it gives rise to visual glare (•Fig. 24.22). To reduce this, the polarizing lenses of glasses are oriented with their transmission axes vertical so that some of the partially polarized light from reflective surfaces is blocked or absorbed.

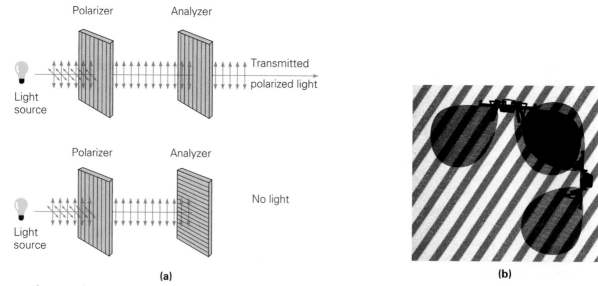

•**FIGURE 24.21 Polarizing sheets**
(a) When polarizing sheets are oriented so that their polarization directions are the same, the emerging light is polarized. The first sheet acts as a polarizer and the second as an analyzer. **(b)** When one of the sheets is rotated 90° and the polarization directions are perpendicular (crossed Polaroids), little light (ideally, none) is transmitted.

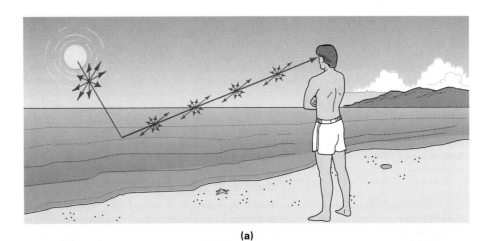

(a)

(b)

•**FIGURE 24.22 Glare reduction**
(a) Light reflected from a horizontal surface is partially polarized in the horizontal plane. When sunglasses are oriented so that their polarizing direction is vertical, the horizontally polarized component of such light is not transmitted. The intensity of the reflected light reaching the eye is diminished, so glare is reduced. **(b)** Polarizing filters for cameras exploit the same principle. The photo at right was taken with such a filter. Note the reduction in reflections from the store window.

Polarizing glasses whose lenses show different polarization directions are used to view some 3-D movies. The pictures are projected on the screen by two projectors that transmit slightly different images, photographed by cameras a short distance apart. The projected light is linearly polarized, but in directions that are mutually perpendicular. The lenses of the 3-D glasses also have polarization directions that are perpendicular. Thus, one eye sees the image from one projector and the other eye sees the image from the other projector. The brain interprets the slight difference in perspective of the two images as depth, or a third dimension, just as it does in normal vision.

Some transparent materials have the ability to rotate the direction of polarization of linearly polarized light. This property is called **optical activity** and is due to the molecular structure of the material (•Fig. 24.23). The rotation may be clockwise or counterclockwise, depending on the molecular orientation. Optically active molecules include those of certain proteins, amino acids, and sugars.

Glasses and plastics become optically active when under stress. The greatest rotation of the direction of polarization of the transmitted light occurs in the regions where the stress is the greatest. Viewing the stressed piece of material through crossed Polaroids allows the points of greatest stress to be identified. This determination is called optical stress analysis.

Another use of polarizing films is described in the Insight on p. 756.

LCDs and Polarized Light

LCDs (Liquid Crystal Displays) are commonplace on watches, calculators, gas pumps, and even some television screens. The name "liquid crystal" may seem self-contradictory. In general, when a crystalline solid melts, the resulting liquid no longer has an orderly atomic or molecular arrangement. Some organic compounds, however, pass through an intermediate state in which the molecules may rearrange somewhat but still maintain the overall order that is characteristic of a crystal.

A liquid crystal flows like a liquid, but its optical properties may depend on the order of its molecules. For example, certain liquid crystals with orderly arrangements are transparent. But the crystalline order can easily be disturbed by applied electrical forces. The disordered liquid then scatters light and is opaque.

A common type of LCD, called a twisted-nematic display, makes use of the effect of a liquid crystal on polarized light (Fig. 1). A display like those commonly used on wristwatches and calculators is made by sandwiching a layer of liquid crystal material between two glass plates that have fine parallel grooves, or channels, in their surfaces. One of the plates is then rotated or twisted 90°. The molecules in contact with the plates remain parallel to the grooves, and the result is a molecular orientation that rotates through 90° between the plates. In this configuration, the liquid crystal is optically active and will rotate the direction of polarization of linearly polarized light. The plates are then placed between crossed polarizing sheets and backed with a mirrored surface. Light entering and passing through the LCD is polarized, rotated 90°, reflected, and again rotated 90° by its components. After the return trip

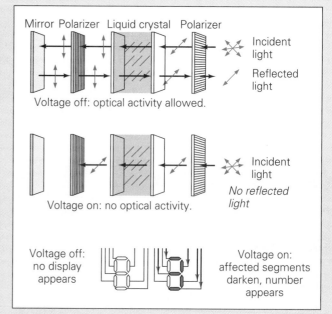

FIGURE 1   Liquid crystal display (LCD)
A twisted-nematic display is an application involving the optical activity of a liquid crystal and crossed Polaroids. When the crystalline order is disoriented by an electric field from an applied voltage, the liquid crystal loses its optical activity in that region, and light is not transmitted. Numerals and letters are formed by applying voltages to segments of a block display.

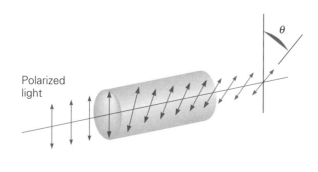

(a)

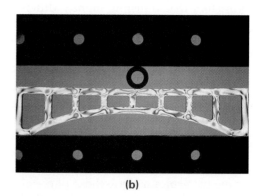

(b)

•**FIGURE 24.23   Optical activity**
(a) Some substances have the property of rotating the polarization direction of linearly polarized light. This ability depends on the molecular structure and is called optical activity. **(b)** Glasses and plastics become optically active under stress, and the points of greatest stress are apparent when the material is viewed through crossed Polaroids. Engineers can thus test plastic models of structural elements to see where the greatest stresses will occur when they are "loaded" (subjected to the forces that they will experience in actual use). Here a model of a suspension bridge strut is being analyzed.

FIGURE 2 Polarized light
The light from an LCD is polarized, as can be shown by using polarizing sunglasses as an analyzer.

molecules of the liquid crystal are polar, so an electric field can disorient them. Transparent, electrically conductive film coatings arranged in a seven-block pattern are put onto the glass plates. Each block, or display segment, has a separate electrical connection. When a voltage is applied across one or more of the segments, the electric field disorients the molecules of the liquid crystal in that area, and their optical activity is lost. (Note that the numerals 0 through 9 can all be formed by the segmented display.) The incident polarized light passing through the disoriented regions of the liquid crystal is absorbed by the second polarizer. Thus, these regions are opaque and appear dark. To produce images on small TV screens, the display segments of the LCD are coupled so that many of them may be energized with a single lead.

One of the major advantages of LCDs is their low power consumption. Other similar displays, such as those using red light-emitting diodes (LEDs), produce light themselves, using relatively large amounts of power. LCDs, on the other hand, produce no light, but instead use reflected light.

Some liquid crystals respond to temperature changes, and the orientation of the molecules affects the light-scattering properties of the crystal. Different wavelengths, or colors, of light are selectively scattered. The color of a liquid crystal can thus be an indication of the surrounding temperature, and liquid crystals are used to make thermometers.

Finally, there is a new and emerging display technology called the polymer dispersed liquid crystal (PDLC) display. This type of display consists of a liquid crystal and polymer mixture, which has the property of electrically controlled light scattering. That is, the transparency of the display can be varied from opaque to clear by changing the applied voltage. A major advantage of PDLC displays is that they do not require polarizers, making them more cost effective.

through the liquid crystal, the polarization direction of the light is the same as that of the initial polarizer. Thus, the light is transmitted and leaves the display unit. Because of the reflection and transmission the display appears to be a light color (usually light gray) when illuminated with white, unpolarized light.

The light from the bright regions of an LCD can readily be shown to be polarized using an analyzer. The whole display appears dark when the analyzer is properly oriented. You may have noticed this effect if you have ever tried to see the time on the LCD of a wristwatch while wearing polarizing sunglasses (Fig. 2).

The dark numbers or letters on an LCD are formed by applying an electric field to parts of the liquid crystal layer. The

24.5 Atmospheric Scattering of Light

Objectives: To be able to (a) define scattering, and (b) explain why the sky is blue and sunsets are red.

When light is incident on a suspension of particles, such as the molecules of air, some of the light may be absorbed and reradiated. This process is called **scattering**. The scattering of sunlight in the atmosphere produces some interesting effects. Among these are the polarization of sky light (that is, sunlight that has been scattered by the atmosphere), the blueness of the sky, and the redness of sunsets and sunrises.

Atmospheric scattering causes the sky light to be polarized. When unpolarized sunlight is incident on air molecules, the electric field of the light wave sets electrons of the molecules into vibration. The vibrations are complex, but the accelerating charges emit radiation, like the vibrating charges in the antenna of a radio broadcast station (see Section 20.4). The intensity of this emitted radiation is strongest along a line perpendicular to the oscillation. And, as illustrated in •Fig. 24.24, an observer viewing from an angle of 90° with respect to the direction of the sunlight will receive linearly polarized light because of the horizontal charge oscillations. The light also has a vertically polarized component. At other viewing angles, both components are present, and sky light seen through a polarizing filter appears partially polarized because of the stronger component.

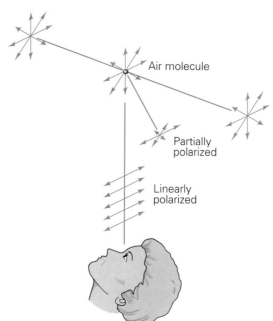

Unpolarized light

Air molecule

Partially polarized

Linearly polarized

•FIGURE 24.24 Polarization by scattering
When incident unpolarized sunlight is scattered by a molecule of gas of the air, the light in the plane perpendicular to the direction of the incident ray is linearly polarized. Light scattered at some arbitrary angle is partially polarized. An observer at a right angle to the direction of the incident sunlight receives plane-polarized light.

Since the scattering of light with the greatest degree of polarization is at a right angle to the direction of the Sun, at sunrise and sunset the scattered light from directly overhead has the greatest degree of polarization. The polarization of sky light may be observed by viewing the sky through a polarizing filter (or a polarizing sunglass lens) and rotating the filter. Light from different regions of the sky will be transmitted in different degrees, depending on its degree of polarization. It is believed that some insects, such as bees, use polarized sky light to determine navigational directions relative to the Sun.

Why the sky is blue

The scattering of sunlight by air molecules causes the sky to look blue. This is not a polarization effect but is caused by the selective absorption of light. As oscillators, air molecules have resonant frequencies in the ultraviolet region. Consequently, the sunlight is preferentially scattered, with light in the nearby blue end of the visible region scattered more than that in the red end.

For particles such as air molecules, which are much smaller than the wavelength of light, the scattering is found to be inversely proportional to the wavelength to the fourth power (that is, $1/\lambda^4$). This is called **Rayleigh scattering** after Lord Rayleigh (1842–1919), a British physicist who derived the wavelength relationship. This inverse relationship predicts that light of the shorter-wavelength, or blue, end of the spectrum will be scattered more than light of the longer-wavelength, or red, end. The scattered blue light is rescattered in the atmosphere and eventually is directed toward the ground. This is the sky light we see, so the sky appears blue.

EXAMPLE 24.7 ■ THE RED AND THE BLUE: RAYLEIGH SCATTERING

How much more is light at the blue end of the visible spectrum scattered by air molecules than light at the red end?

Solution. The Rayleigh scattering relationship is $S \propto 1/\lambda^4$, where S is the amount of scattering for a particular wavelength. Thus, you can form a ratio

$$\frac{S_b}{S_r} = \left(\frac{\lambda_r}{\lambda_b}\right)^4$$

where the subscripts stand for blue and red light. The blue end of the spectrum (violet light) has a wavelength of about $\lambda_b = 400$ nm and red light has a wavelength of about

$\lambda_r = 700$ nm. Putting these values into the ratio gives

$$\frac{S_b}{S_r} = \left(\frac{\lambda_r}{\lambda_b}\right)^4 = \left(\frac{700 \text{ nm}}{400 \text{ nm}}\right)^4 = 9.4 \quad \text{or} \quad S_b = 9.4S_r$$

Thus, blue light is scattered almost ten times as much as red light is.

Follow-up Exercise. What wavelength of light is scattered twice as much as red light? What color light is this?

You should keep in mind that all colors are present in sky light, but the dominant color is blue. (The sky doesn't appear violet because our eyes are more sensitive to blue and there is more blue than violet light in sunlight.) You may have noticed that the sky is more blue directly overhead than toward the horizon, and appears white just above the horizon (look for this effect the next time you are outside on a clear day). This is because there are relatively fewer air molecules or scatterers directly overhead than toward the horizon. Multiple scatterings occur in the denser air near the horizon, and the recombination of scattered light gives rise to the white appearance. (Analogously, if a drop or two of milk is added to a glass of water and the suspension is illuminated with intense white light, the scattered light has a bluish hue. But a glass of undiluted milk is white because of multiple scatterings.) In the same way, atmospheric pollution may impart a milky white appearance to most or all of the sky.

When the Sun is near the horizon, the sunlight travels a greater distance through the denser air near the Earth's surface. Since the light therefore undergoes a great deal of scattering, you might think that only the least scattered light, the red light, would reach observers on the Earth's surface. This would explain red sunsets. However, it has been shown that the dominant color of white light after only molecular scattering is orange. Thus, there must be other scattering that shifts the light from the setting (or rising) Sun toward the red (•Fig. 24.25).

Red sunsets have been found to result from the scattering of sunlight by atmospheric gases *and* small foreign particles. These particles are not necessary for the blueness of the sky, but are necessary for deep red sunsets and sunrises. (This is why spectacular red sunsets are observed in the months after a large volcanic eruption that puts tons of particulate matter into the atmosphere.) Red sunsets occur most often when there is a high-pressure air mass to the west, since the concentration of particles is generally greater in high-pressure air masses than in low-pressure air masses. Similarly, red sunrises occur most often when there is a high-pressure air mass to the east.

Why sunsets and sunrises are red

•FIGURE 24.25 Red sky at night A spectacular red sunset over a mountaintop observatory in Chile. The red sky results from the scattering of sunlight by atmospheric gases and small solid particles. A directly-observed reddening Sun is due to the scattering of the wavelengths toward the blue end of the spectrum in the direct line of sight.

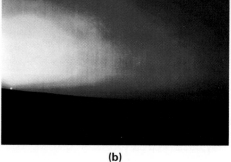

(a)   (b)

•**FIGURE 24.26 And now for something completely different**
(a) During the day, the sky of Mars is red, because carbon dioxide preferentially scatters red wavelengths. **(b)** Can you explain why Martian sunrises and sunsets have a bluish cast?

Now you can understand the meaning behind the old weather saying, "Red sky at night, sailors' delight. Red sky in the morning, sailors take warning." (See also the Biblical quote Matthew 16:1–4.) Fair weather generally accompanies high-pressure air masses because they are associated with reduced cloud formation. Most of the United States lies in the Westerlies wind zone, in which air masses generally move from west to east. A red sky at night is thus likely to indicate a fair-weather, high-pressure air mass to the west that will be coming your way. A red sky in the morning means the high-pressure air mass has passed, and poor weather may set in.

Red sunrises and sunsets are often made more spectacular by clouds. The clouds are colored pink by reflected red light. In general, clouds are white (when not shadowed) because the water droplets of which they are composed are not preferential scatterers and scatter visible light of all wavelengths (white light). Clouds do not affect the color of the incident light, but simply diffusely reflect the light from the Sun.

As a final note, how would you like a *red* sky? Then try Mars, the "red planet." The thin Martian atmosphere is about 95% carbon dioxide (CO_2). The CO_2 molecule is more massive than an oxygen (O_2) or nitrogen (N_2) molecule. As a result, CO_2 molecules have a lower resonant frequency (longer wavelength), and preferentially scatter the red end of the visible spectrum. Hence, the Martian sky is red during the day, adding to the true red of the iron oxide-laden ground (•Fig. 24.26).

And what of the color of sunrises and sunsets on Mars? Think about it. . . .

Chapter Review

Important Terms

Important Concepts

- Young's double-slit experiment provides evidence of the wave nature of light and a way to measure the wavelength of light ($\approx 10^{-7}$ m).

- Light reflected at a media boundary with $n_2 > n_1$ undergoes a 180° phase change. If $n_2 < n_1$, there is no phase change on reflection.

- Reflection phase changes at boundaries contribute to thin-film interference, which depends on film thickness.

- In general, the larger the wavelength compared to the size of an opening or object, the greater the diffraction.

- Polarization is the preferential orientation of the electromagnetic field vectors of light, and is evidence that light is a transverse wave. Light can be polarized by reflection, double refraction (birefringence), and scattering.

- The extent of Rayleigh scattering is inversely proportional to the fourth power of the wavelength. The blueness of Earth's sky results from the preferential scattering of sunlight by air molecules.

Important Equations for Review

Path Difference for Constructive Interference:

$$\Delta = n\lambda \quad \text{for } n = 0, 1, 2, 3, \dots \quad (24.1)$$

Path Difference for Destructive Interference:

$$\Delta = \frac{m\lambda}{2} \quad \text{for } m = 1, 3, 5, \dots \quad (24.2)$$

Bright Fringe Condition (double-slit interference):

$$d \sin\theta = n\lambda \quad \text{for } n = 0, 1, 2, 3, \dots \quad (24.3)$$

Wavelength Measurement (double-slit interference):

$$\lambda = \frac{y_n d}{nL} \quad \text{for } n = 1, 2, 3, \dots \quad (24.7)$$

Width of Bright Fringes (double-slit interference):

$$y_{n+1} - y_n = \frac{L\lambda}{d} \quad \text{(Example 24.1)}$$

Nonreflecting Film Thickness (minimum):

$$t = \frac{\lambda}{4n} \quad (24.8)$$

Dark Fringe Condition (single-slit diffraction):

$$w \sin\theta = m\lambda \quad \text{for } m = 1, 2, 3, \dots \quad (24.9)$$

Lateral Displacement of Dark Fringes (single-slit diffraction):

$$y_m = \frac{mL\lambda}{w} \quad \text{for } m = 1, 2, 3, \dots \quad (24.10)$$

Width of Bright Fringes (single-slit diffraction):

$$\Delta = y_{m+1} - y_m = \frac{L\lambda}{w} \quad (24.12)$$

Interference Maxima for a Diffraction Grating:

$$d \sin\theta = n\lambda \quad \text{for } n = 0, 1, 2, 3, \dots \quad (24.13)$$

where $d = 1/N$ and N is the number of lines per unit length.

Limit of the Order Number:

$$n \leq \frac{d}{\lambda} \quad (24.14)$$

Bragg's Law:

$$2d \sin\theta = n\lambda \quad \text{for } n = 1, 2, 3, \dots \quad (24.15)$$

Brewster (polarizing) Angle:

$$\tan\theta_p = n \quad \text{and} \quad \theta_p = \tan^{-1} n \quad (24.16)$$

Exercises

24.1 Young's Double-Slit Experiment

1   If in a Young's experiment using monochromatic light the screen distance is doubled, then which of the following would change: (a) d, (b) λ, (c) y_n, (d) none of these?

2   The formation of a stationary fringe pattern in Young's experiment depends on (a) diffraction, (b) interference, (c) coherent light, (d) all of these.

3   When white light is used in Young's experiment, bright fringes with spectrums of colors are seen. In a fringe spectrum, is the red end or the blue end closer to the central maximum?

4   What would be observed with the set-up for Young's experiment if the light passing through the slit were not coherent?

5   Television pictures are often seen to flutter when an airplane passes by (•Fig. 24.27). Explain a possible cause of this fluttering based on interference effects.

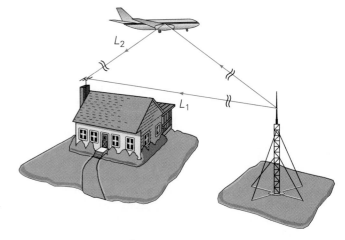

•**FIGURE 24.27 Interference**
See Exercise 5.

6   ■ What is the shortest path difference for two waves of equal frequency that arrive at a point with a phase difference of (a) $\pi/3$, (b) $\pi/2$, and (c) π? [*Hint:* See Eq. 14.5.]

7 ■ Two point sources of coherent sound are located at (2.0 m, 5.0 m) and (4.0 m, 3.0 m) in the xy plane. What is the wavelength of the sound waves at the point (15 m, 20 m) if (a) they interfere destructively to the first order and (b) they have a phase difference of 45°?

8 ■■ In the developments of Young's experiment and single-slit diffraction, a small-angle approximation was used to find the lateral displacements of the bright and dark fringes, respectively. How good is this approximation? For example, what is the percent error for $\theta = 10°$? (Percent error is the difference in values divided by the true value and expressed as a percent.)

9 ■■ An interference pattern is formed when light whose wavelength is 550 nm is incident on two parallel slits 50 μm apart. The second-order bright fringe is 3.0 cm from the center of the central maximum. How far from the slits is the screen on which the pattern is formed?

10 ■■ Monochromatic light illuminates two parallel slits that are 0.20 mm apart. The adjacent bright lines of the interference pattern on a screen 1.5 m away from the slits are 0.45 cm apart. What is the color of the light?

11 ■■ Blue light with a wavelength of 440 nm is used in a double-slit experiment where the slits are 0.50 mm apart. If the screen is 1.5 m from the slits, what is the angular separation of the first-order and third-order bright fringes?

12 ■■ Two parallel slits 1.0 mm apart are illuminated with monochromatic light whose wavelength is 640 nm, and an interference pattern is observed on a screen 1.50 m away. (a) What is the separation between adjacent interference maxima? (b) If the distance between the slits were increased to 1.5 mm, how would this affect the separation of the maxima?

13 ■■ Derive a relationship giving the locations of the dark fringes in Young's double-slit experiment. What is the distance between the dark fringes?

14 ■■ Monochromatic light passes through two narrow slits 0.25 mm apart and forms an interference pattern on a screen 1.5 m away. If light with a wavelength of 680 nm is used, what is the distance between the center of the central maximum and the center of the third-order bright fringe?

15 ■■ In a double-slit experiment using monochromatic light, a screen is placed 1.25 m from the slits, which have a separation distance of 0.0250 mm. The third-order bright fringe ($n = 3$) is measured to be 6.60 cm from the center of the central maximum. Find (a) the wavelength of the light, and (b) the lateral displacement of the second-order bright fringe ($n = 2$).

16 ■■ Yellow-orange light ($\lambda = 550$ nm) is used in a double-slit experiment with a slit separation distance of

1.75×10^{-4} m. With the screen located 2.00 m from the slits, determine (a) the angular separation in radians between the central maximum and the second-order bright fringe ($n = 2$) and (b) the lateral displacement of the fringe.

17 ■■ In a double-slit experiment with a screen at a distance of 1.50 m and using monochromatic light, the angle between the second-order bright fringe and the central maximum is measured to be 0.0230 rad. If the separation distance of the slits is 0.0350 mm, (a) what is the color of the light? (b) What is the lateral displacement of the fringe?

18 ■■■ What would be the effect on the interference fringes if the apparatus for Young's double-slit experiment were completely immersed in still water? (b) What would the displacement in Exercise 14 be if the entire system were immersed in still water?

19 ■■■ Light of two different wavelengths is used in a double-slit experiment. The location of the third-order bright fringe for yellow light ($\lambda = 600$ nm in air) coincides with the location of the fourth-order bright fringe for the other light. What is the wavelength of the other light?

24.2 Thin-Film Interference

20 For a thin film with $n_1 > n_o$ and $n_1 > n_2$, where n_1 is the index of refraction of the film, a film thickness for constructive interference of the reflected light is (a) $\lambda'/4$, (b) $\lambda'/2$, (c) λ', (d) both (a) and (b).

21 For a thin film with $n_0 < n_1 < n_2$, where n_1 is the index of refraction of the film, the minimum film thickness for destructive interference of the reflected light is (a) $\lambda'/4$, (b) $\lambda'/2$, (c) λ', (d) both (a) and (b).

22 What would be the effect of nondirect incidence on thin-film interference?

23 Suppose you wanted to make a reflecting surface using thin-film interference. How would you go about doing this?

24 ■ Light with a wavelength of 550 nm in air is normally incident on a glass plate ($n = 1.5$) whose thickness is 1.1×10^{-5} cm. (a) What is the thickness of the glass in terms of the wavelength in glass? (b) What is the phase difference in the beams reflected from each surface (that is, will they interfere constructively or destructively)?

25 ■ A lens with an index of refraction of 1.6 is to be coated with a material ($n = 1.4$) that will make it nonreflecting for red light ($\lambda = 700$ nm) normally incident on the lens. What is the minimum required thickness of the coating?

26 ■■ Magnesium fluoride ($n = 1.38$) is frequently used as a lens coating to make nonreflecting lenses. What is the difference in the minimum film thicknesses

required for maximum transmission of blue light ($\lambda = 400$ nm) and red light ($\lambda = 700$ nm)?

27   ■■ A solar cell is to have a nonreflecting coating of a transparent material whose index of refraction is 1.30. (a) What is the minimum thickness of the film for light with a wavelength of 550 nm? (b) Does the index of refraction of the adjacent material in the solar cell make a difference?

28   ■■ A thin layer of oil ($n = 1.5$) floats on water. Destructive interference is observed for light at two different locations with wavelengths of 480 nm and 600 nm. Find the two thicknesses of the oil film for normal incidence.

29   ■■ It is desired to make a nonreflecting lens for an infrared radiation detector. If the coating material has an index of refraction of 1.20, what should the film thickness be for radiation with a frequency of 3.75×10^{14} Hz?

30   ■■ Two parallel plates are separated by a small distance as illustrated in ●Fig. 24.28. If the top plate is illuminated with red light ($\lambda = 680$ nm), for what minimum separation distances will the light be (a) reflected back to the observer from air layer surfaces, and (b) transmitted through the plates? [$t = 0$ is *not* an answer for (b).]

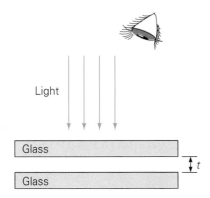

●FIGURE 24.28   Reflection or transmission?
See Exercise 30.

31   ■■ Show that the film thicknesses for a "nontransmissive" lens where n of the film is less than that of the lens material are given by

$$t = \frac{m\lambda}{2n}$$

where $m = 1, 2, 3, \ldots$, and n is the index of refraction of the film.

32   ■■■ An air wedge is shown in ●Fig. 24.29. If the top glass plate is illuminated with monochromatic light, describe the observed interference pattern. (a) Express the locations of the bright interference fringes in terms of wedge thicknesses measured from the apex of the wedge. (b) Show that the number of dark fringes is given by $m = 2t/\lambda_{air}$, where t is the maximum thickness of the wedge.

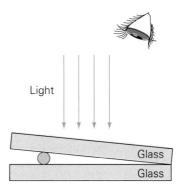

●FIGURE 24.29   Air wedge
See Exercises 32 and 33.

33   ■■■ The glass plates in Fig. 24.29 are separated by a thin, round filament. When the top plate is illuminated with light having a wavelength of 550 nm, it is observed that the filament lies directly below the sixth bright fringe. What is the diameter of the filament?

24.3 Diffraction

34   For single-slit diffraction, (a) all fringes are equal in width, (b) dark fringes occur for only odd integers, $m = 1, 3, 5, \ldots$, (c) for a given wavelength, the narrower the slit, the wider the fringes, (d) none of these.

35   As the number of lines per unit length of a diffraction grating increases, the number of spectral orders for a given wavelength (a) increases, (b) decreases, (c) remains unchanged.

36   (a) What does the equation for the width of the central maximum for single-slit diffraction predict when $w = \lambda$? Is the prediction correct? Explain. (b) Why is there no zeroth order ($n = 0$) in Bragg's law for crystal diffraction?

37   ■ A slit of width 0.20 mm is illuminated with monochromatic light having a wavelength of 480 nm, and a diffraction pattern is formed on a screen 1.0 m away from the slit. (a) What is the width of the central maximum? (b) What is the distance between the second-order and third-order bright fringes?

38   ■ A slit 0.025 mm wide is illuminated with red light ($\lambda = 680$ nm). How wide are (a) the central maximum and (b) the side maxima of the diffraction pattern formed on a screen 1.0 m from the slit?

39   ■■ Red light ($\lambda = 700$ nm) passes through a slit with a width of 1.0 mm. (a) Find the angular width and the linear width of the central maximum of the diffraction pattern on a screen 2.0 m from the slit. (b) Find the angular and linear widths if the slit is illuminated with blue light ($\lambda = 440$ nm).

40 ■■ How many fringes of interference maxima occur when monochromatic light with a wavelength of 560 nm illuminates a diffraction grating with 10,000 lines/cm?

41 ■■ In a particular diffraction pattern, the red component ($\lambda = 700$ nm) in the second-order spectrum is deviated at an angle of 20°. (a) How many lines per centimeter does the grating have? (b) If the grating is illuminated with white light, how many bright fringes of the complete visible spectrum are produced?

42 ■■ Find the locations of the blue ($\lambda = 420$ nm) and red ($\lambda = 680$ nm) components of the first-order and second-order spectra in a pattern formed on a screen 1.5 m from a diffraction grating with 7500 lines/cm.

43 ■■ White light whose components have wavelengths from 400 nm to 700 nm illuminates a diffraction grating with 4000 lines/cm. Do the first and second orders overlap? Justify your answer.

44 ■■ Visible white light traveling in air contains components with wavelengths from about 400 nm to 700 nm. If the white light illuminates a diffraction grating with 8000 lines/cm, what is the angular width of the first-order spectrum produced?

45 ■■ What is the angular width of the first-order spectrum in Exercise 44 if the whole system is immersed in water?

46 ■■ A Venetian blind is essentially a diffraction grating —not for visible light, but for waves with longer wavelengths. If the spacing between the slats of a blind is 2.5 cm, for what frequency and type of radiation is there a first-order maximum at an angle of 10°?

47 ■■ A certain crystal gives a deflection angle of 25° for the first-order diffraction of monochromatic X-rays with a frequency of 5.0×10^{17} Hz. What is the lattice spacing of the crystal?

48 ■■ A diffraction grating with 8000 lines per centimeter is illuminated with a beam of monochromatic red light from a He–Ne laser ($\lambda = 632.8$ nm). How many side maxima would be formed in the diffraction pattern, and at what angles would they be observed?

49 ■■ A single slit with a width of 0.25 mm is illuminated with monochromatic light ($\lambda = 680$ nm). A screen is placed 1.80 m from the slit to observe the fringe pattern. (a) What is the angle between the second dark fringe ($m = 2$) and the central maximum? (b) What is the lateral displacement of this dark fringe?

50 ■■ What is the width of the central maximum for the single slit arrangement in Exercise 49?

51 ■■■ Show that for a diffraction grating the violet ($\lambda = 400$ nm) portion of the third-order spectrum overlaps the yellow-orange ($\lambda = 600$ nm) portion of the second-order spectrum regardless of the grating's spacing.

52 ■■■ A teacher standing inside a doorway 1.0 m wide blows a whistle with a frequency of 1000 Hz to summon children from the playground. All the children come, except two boys playing near the swings, about 100 m away from the school building. Later the teacher begins to reprimand them, but the boys maintain rather convincingly that they did not hear the whistle. The swings are at an angle of 19.6° from a line normal to the doorway, and the speed of sound is 335 m/s. Could they be telling the truth? Justify your answer mathematically.

24.4 Polarization

53 Light may be polarized by (a) reflection, (b) refraction, (c) scattering, (d) all of these.

54 Brewster's angle depends on (a) the index of refraction of a material, (b) Bragg's law, (c) internal reflection, (d) interference.

55 Given two pairs of sunglasses, could you tell if one or both were polarizing?

56 How would you hold an analyzer to detect the polarization of a rainbow? Could you block out the polarized light completely? Explain.

57 Suppose that you held two polarizing sheets in front of you and looked through both of them. How many times would you see the sheets lighten and darken (a) if one of them were rotated through one complete rotation, (b) if both of them were rotated through one complete rotation at the same rate in opposite directions, (c) if both of them were rotated through one complete rotation at the same rate in the same direction, and (d) if both were rotated through one complete rotation in opposite directions, but one twice as fast as the other?

58 Sketch a graph of the Brewster angle versus the index of refraction.

59 ■ Some types of glass have a range of indices of refraction of about 1.4 to 1.7. What is the range of the polarizing angle for these glasses?

60 ■■ A light beam is incident on a glass plate ($n = 1.58$), and the reflected ray is completely polarized. What is the angle of refraction for the beam?

61 ■■ The critical angle for internal reflection in a certain medium is 55°. What is the Brewster angle for light externally incident on the medium?

62 ■■ The angle of incidence is adjusted so there is maximum linear polarization for the reflection of light from a transparent piece of plastic with $n = 1.22$. But some of the light is transmitted. What is the angle of refraction?

63 ■■ Sunlight is reflected off a glass plate window ($n = 1.55$). What would be the Sun's altitude (angle above the horizon) for the *reflected* light to be completely polarized?

64 ■■ The Brewster angle in Eq. 24.16 was derived for the incident light in air. Show that in general, $\theta_p = \tan^{-1} n$, where n is the relative index of refraction ($n = n_2/n_1$).

65 ■■ What angle of incidence is necessary for reflected light to have the maximum linear polarization at a water–crown glass interface?

66 ■■■ Find the Brewster angle for a piece of glass ($n = 1.60$) that is submerged in water.

67 ■■■ A plate of crown glass is covered with a layer of water. A beam of light traveling in air is incident on the water and partially transmitted. Is there any angle of incidence for which the light reflected from the water–glass interface will have maximum linear polarization? Justify your answer mathematically.

24.5 Atmospheric Scattering of Light

68 Which of the following colors is scattered the most in the atmosphere: (a) green, (b) yellow, (c) orange, (d) color makes no difference?

69 Red sunsets result from (a) denser air, (b) molecular scattering, (c) scattering by foreign particles, (d) all of these.

70 (a) Why does the sky not have a uniform blueness? (b) What color would an astronaut on the Moon see when looking at the sky or into space?

Additional Exercises

71 If the Brewster angle for a glass plate is 1.0 rad, what is the index of refraction of the glass?

72 In a double-slit experiment with slits that are 0.25 mm apart, the interference pattern is formed on a screen 1.0 m away from the slits. The third-order bright fringe is 0.60 cm from the center of the central maximum. (a) What is the frequency of the monochromatic light? (b) What is the distance between the second-order and third-order bright fringes?

73 If the slit width in a single-slit experiment were doubled, the distance to the screen reduced by a third, and the wavelength of the light changed from 600 nm to 450 nm, how would the width of the bright fringes be affected?

74 A glass plate has an index of refraction of 1.5. At what angle of incidence will light be reflected from the plate with the maximum linear polarization?

75 When illuminated with monochromatic light, a diffraction grating with 1000 lines/cm produces a first-order maximum 4.50 cm from the central maximum on a screen 1.00 m away. What is the wavelength of the light?

76 A thin air wedge between two flat glass plates forms bright and dark interference bands when illuminated with normally incident monochromatic light, as illustrated in Fig. 24.8. (a) Show that the thickness of the air wedge changes by $\lambda/2$ from one bright fringe to the next, where λ is the wavelength of the light. (b) What would the change in the wedge thickness between bright fringes be if the space were filled with a liquid with an index of refraction n?

77 A diffraction grating is designed to have the red end of the visible spectrum an angular distance of 7.5° from the center of the central maximum for the second order. How many lines per centimeter does the grating have?

78 Two parallel slits 0.75 mm apart are illuminated with monochromatic light whose wavelength is 480 nm. Find the angle between the center of the central maximum and the center of an adjacent bright fringe formed on a screen 1.2 m away from the slits.

79 Monochromatic light whose wavelength is 500 nm falls on two slits separated by a distance of 40 μm. What is the distance between the first-order and third-order bright fringes formed on a screen 1.0 m away from the slits?

80 A film on a lens is 1.0×10^{-7} m thick and is illuminated with white light. The index of refraction of the film is 1.4. For what wavelength of light will the lens be nonreflecting?

81 What is the spectral order limit for a diffraction grating with 9000 lines/cm when illuminated by white light?

82 In a double-slit experiment using monochromatic light, the angular separation between the central maximum and the second-order bright fringe is measured to be 0.16°. What is the wavelength of the light if the slit separation distance is 0.35 mm?

83 A camera lens is coated with a thin layer of a material that has an index of refraction of 1.35. This makes the lens nonreflecting for light with a wavelength of 450 nm (in air) that is normally incident on the lens. What is the thickness of the thinnest film in nm that will make the lens nonreflecting?

25 Optical Instruments

round 1609, Galileo constructed the first astronomical telescope and turned it on the heavens. He saw spots on the Sun, craters on the Moon, and four satellites circling Jupiter. About half a century later, a Dutch amateur lens grinder made a simple microscope and viewed microbes for the first time. It's interesting to think what our knowledge of the world around us would be like if these events had never happened. Perhaps some of our modern sciences would survive in recognizable form; there would still be chemistry, for example, though greatly impoverished. But astronomy would certainly not be one of them. Without telescopes, we would know little more of the astronomical universe than did ancient shepherds thousands of years ago. The stars and planets would remain nothing but mysterious points of light. We would have no idea of the size of our universe, or that it contains countless billions of stars and galaxies invisible to the naked eye.

The "eye" of the telescope in the photo above, situated on a mountaintop in Chile, is not a lens (as in Galileo's original instrument) but a parabolic mirror 3.5 meters in diameter, perfectly polished to within thousandths of a centimeter. Isaac Newton was one of the developers of this design, today used in all modern astronomical telescopes. But not all astronomical telescopes are equipped, as this one is, with computer controls that continually direct minute adjustments in the orientation of the mirror so as to achieve the clearest possible view of the heavens.

In this chapter you'll learn more about telescopes of this and other types, about microscopes, and about the factors that limit what we can see with these devices. You'll also find out more about the essential optical instrument without which all the others would be of little use: the human eye.

Optical instruments are used for a variety of purposes, but their basic function is to improve and extend our powers of visual observation. Indeed, vision is our chief means of acquiring information about the world. Without microscopes and telescopes to view objects that are not easily visible to the unaided eye, our knowledge of the world would be enormously impoverished. For example, biologists could not observe the cells of the human body or disease-causing bacteria. The science of astronomy, too, would be little advanced beyond the level attained by the ancient Greeks.

Although some optical instruments are extremely complex, they can generally be understood in terms of their basic components, usually mirrors and lenses. Without these reflective and refractive elements, our visual investigations would be severely limited. Nearly all large modern telescopes, for example, depend on mirrors like the one shown in the chapter-opening photo.

Mirrors and lenses were discussed in Chapter 23 and other optical phenomena in Chapter 24. You will find that the principles developed in those chapters can be easily applied to the study of optical instruments. This chapter first considers the optical properties of the human eye itself and then describes how optical instruments improve our view of the world.

25.1 The Human Eye

Objectives: **To be able to (a) describe the optical workings of the eye, and (b) explain some common vision defects and how they are corrected.**

The human eye is the most fundamental optical instrument, since without it the field of optics would not exist. The eye is analogous to a simple camera in several respects (•Fig. 25.1). A simple camera consists of a converging lens, which is used to focus images on light-sensitive film at the back of the camera's interior chamber. (Recall from Chapter 23 that for relatively distant objects, a converging lens produces a smaller, inverted, real image.) An adjustable diaphragm, or opening, and a shutter control the amount of light entering the camera.

Note: Image formation by a converging lens is discussed in Section 23.3; see Figure 23.15b.

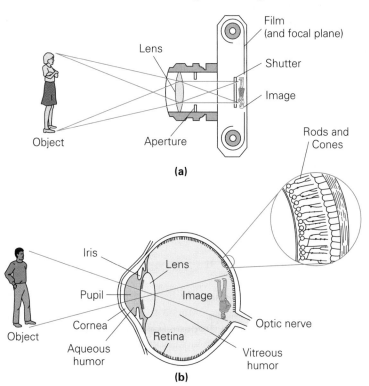

(a)

(b)

•**FIGURE 25.1 Camera and eye analogy**
In some aspects, a simple camera is similar to the human eye. An image is formed on the film in a camera and on the retina in the eye. (The complex refractive properties of the eye are not shown because multiple refractive media are involved. See the text for a description.)

The eye too has a converging lens that focuses images on the light-sensitive lining on the rear surface inside the eyeball. (Other transparent media in the eye also help to create the image by refracting incoming light.) The iris is a circular diaphragm that opens and closes to adjust the amount of light entering the eye. The eyelid might be thought of as a shutter; however, the shutter of the camera, which controls the exposure time, is generally opened for only a fraction of a second, and the eyelid is normally open for continuous exposure. The human nervous system performs a function analogous to that of a shutter in analyzing image signals from the eye at a rate of about 30 times per second. The eye might therefore be likened to a movie or videocamera, which exposes a similar number of frames (images) per second.

Although the optical functions of the eye are relatively simple, its physiological functions are quite complex. As Fig. 25.1b shows, the eyeball is a nearly spherical chamber (with an internal diameter of about 1.5 cm) filled with a jellylike substance called the *vitreous humor*. The eyeball has a white outer covering called the sclera, part of which is visible as the white of the eye. Light enters the eye through a curved, transparent tissue called the cornea and passes into a clear fluid known as the *aqueous humor*. Behind the cornea is a circular diaphragm, the iris, whose central hole is called the pupil. The iris contains the pigment that determines eye color. Through muscle action, the iris can change the area of the pupil (2 to 8 mm in diameter), thereby controlling the amount of light entering the eye.

Behind the iris is a **crystalline lens**, a converging lens composed of microscopic glassy fibers. When tension is exerted on the lens by attached muscles, the glassy fibers slide over each other, causing the shape of the lens to change and thus focusing the image properly, as described in more detail below.

On the back interior wall of the eyeball is a light-sensitive surface called the retina. From the **retina**, the optic nerve relays signals to the brain. The retina is composed of nerves and two types of light receptors, or photosensitive cells, called **rods** and **cones** because of their shapes. The rods are more sensitive to light and distinguish light from dark in low light intensities (twilight vision). The cone cells, on the other hand, can distinguish frequency ranges of sufficiently intense light, which the brain interprets as colors. The cells of the retina are present in large numbers (estimated at $125,000/mm^2$). Most of the cones are clustered together around a center (called the fovea centralis). The more numerous rods are outside this region and distributed nonuniformly over the retina.

Rods—dim light vision
Cones—color vision

The focusing adjustment of the eye is unlike that of a simple camera. A camera lens has a constant focal length, and the image distance is varied by moving the lens to produce sharp images on the film for different object distances. In the eye, the image distance is constant, and the focal length of the lens is varied to produce sharp images on the retina for different object distances.

Light passing through the eyeball travels through five different media with different indices of refraction: air [$n = 1.00$], the cornea [$n = 1.38$], the aqueous humor [$n = 1.33$], the crystalline lens [average $n = 1.40$], and the vitreous humor [$n = 1.34$]. The various refractions at the boundaries between these media all participate in the formation of an image. Most of the refraction of light occurs at the front surface of the cornea, while the crystalline lens makes the fine adjustments needed to focus the image. Nerve signals cause the muscles attached to the lens to change its radius of curvature, and thus the focal length of the lens. Thus, images of objects at different distances are focused sharply on the retina.

When the eye is focused on distant objects, the muscles are relaxed, and the crystalline lens has its thinnest shape and a power of about 20 D (diopters). Recall that the power (P) of a lens in diopters (D) is the reciprocal of its focal length *in meters*. When the eye is focused on closer objects, the lens becomes thicker, and the radius of curvature and focal length are decreased. The lens power may increase to 30 D, or even more in young children. The adjustment of the focal length

Note: This relationship is presented in Eq. 23.10, Section 23.5.

of the crystalline lens is called **accommodation**. (Look at a nearby object and then one in the distance and notice how fast accommodation takes place. It's practically instantaneous.)

The extremes of the range over which distinct vision (sharp focus) is possible are known as the far point and the near point. The **far point** is the greatest distance at which the normal eye can see objects clearly, and is taken to be infinity. The **near point** is the position closest to the eye at which objects can be seen clearly, and depends on the extent the lens can be deformed (thickened) by accommodation. The range of accommodation gradually diminishes with age as the crystalline lens loses its elasticity. That is, the near point gradually recedes with age. The approximate positions of the near point at various ages are listed in Table 25.1.

Children can see sharp images of objects that are within 10 cm (4 in.) of their eyes, and the crystalline lens of a normal young adult eye can be deformed to produce sharp images of objects as close as 12–15 cm (5–6 in.). However, at about the age of 40, the near point normally moves beyond 25 cm (10 in.). You may have noticed people over the age of 40 holding reading material at some distance from their eyes to bring it within the range of accommodation. When the print gets too small or the arm too short, corrective reading glasses are the solution. The recession of the near point with age is not considered an abnormal defect of vision, since it proceeds at about the same rate in all normal eyes.

Vision Defects

Speaking of the "normal" eye implies the existence of eyes with defective vision. That this is the case is apparent from the number of people who wear glasses or contact lenses. The eyes of many people cannot accommodate within the normal range of 25 cm to infinity. These people have one of the two most common visual defects: nearsightedness (myopia) and farsightedness (hyperopia). Both of these conditions can usually be corrected (•Fig. 25.2).

Nearsightedness (or *myopia*) is the condition of being able to see nearby objects clearly but not distant objects. That is, the far point is not infinity but some nearer point. When an object beyond the far point is viewed, the rays come to focus in front of the retina (Fig. 25.2b). As a result, the image on the retina is blurred, or out of focus. As the object is moved closer to the eye, its image moves back toward the retina. If the object is moved within the far point, a sharp image is seen.

Nearsightedness arises because the eyeball is too long or perhaps because the curvature of the cornea is too great. Whatever the reason, the images of distant objects are focused in front of the retina. Appropriate diverging lenses will correct this. Such a lens causes the rays to diverge, and the eye focuses the image farther back so that it falls on the retina.

Note: The eye sees clearly between its far point and near point.

TABLE 25.1
Approximate Near Points of the Normal Eye at Different Ages

Age (years)	Near Point (centimeters)
10	10
20	12
30	15
40	25
50	40
60	100

Nearsightedness: difficulty seeing distant objects.

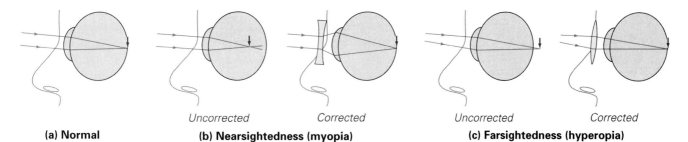

(a) Normal   (b) Nearsightedness (myopia)   (c) Farsightedness (hyperopia)

•**FIGURE 25.2 Nearsightedness and farsightedness**
(a) The normal eye produces sharp images on the retina for objects located between its near point and its far point. The image is real, inverted, and always smaller than the object (why?). Here, the object is a distant, upward-pointing arrow (not shown), and the light rays come from its tip. **(b)** In a nearsighted eye, the image of a *distant* object is focused *in front of* the retina. This defect is corrected by using a diverging lens. **(c)** In a farsighted eye, the image of a *nearby* object is focused *behind* the retina. This defect is corrected by using a converging lens. (The drawings are exaggerated.)

Farsightedness: difficulty seeing nearby objects.

Farsightedness (or *hyperopia*) is the condition of being able to see distant objects clearly but not nearby objects. That is, the near point is not at the normal position but at some point farther from the eye. The image of an object that is closer to the eye than the near point is formed behind the retina (Fig. 25.2c). Farsightedness arises because the eyeball is too short or perhaps because of insufficient curvature of the cornea. A similar farsighted condition (*presbyopia*) occurs with the normal receding of the near point with age, as discussed previously.

Farsightedness is usually corrected by appropriate converging lenses. Such a lens causes the rays to converge, and the eye is then able to focus the image on the retina. Converging lenses are also used to correct the farsightedness associated with the natural recession of the near point with age. Older people often must wear reading glasses, which have converging lenses.

EXAMPLE 25.1 ■ CORRECTING NEARSIGHTEDNESS: USE OF A DIVERGING LENS

A certain nearsighted person cannot see objects clearly when they are more than 78 cm from either eye. (a) What power must corrective lenses have if this person is to see distant objects clearly? (b) The person's near point is 23 cm. How is this affected when viewing through the corrective lenses? Assume that the lenses are in eyeglasses and are 3.0 cm in front of the eye. (See •Figure 25.3.)

Solution. Listing the given distances, we have

Given: $d_f = 78$ cm (far point)   *Find:* (a) P (in diopters)
$d_n = 23$ cm (near point)   (b) near point (with glasses)
$d = 3.0$ cm

Note: Review Examples 23.5 and 23.6.

(a) Here we have a lens problem similar to those in Chapter 23. Recall that the power of a lens is $P = 1/f$ (Eq. 23.10). We can use the thin-lens equation (Eq. 23.6) to find f or $1/f$ if we can determine the object and image distances, d_o and d_i.

The lens must effectively put the image of a distant object ($d_o = \infty$) at the far point d_f, which is 78 cm from the eye. The image, which acts as an object for the eye, is then within the range of accommodation. The image distance is *measured from the lens*, which is 3.0 cm from the eye, so $d_i = -75$ cm. ($d_i = d_f - d = 78$ cm $- 3.0$ cm $= 75$ cm; see the figure, which is not to scale.) A minus sign is used because the image is virtual, being on the object side of the lens. (You may recall from Chapter 23 that diverging lenses can form only virtual images.)

Note: Image formation by a diverging lens is discussed in Section 23.3; see Figure 23.15a.

Then, using the thin lens equation,

$$\frac{1}{f} = \frac{1}{d_o} + \frac{1}{d_i} = \frac{1}{\infty} - \frac{1}{75 \text{ cm}} = -\frac{1}{75 \text{ cm}} \quad \text{or} \quad f = -75 \text{ cm}$$

The minus sign indicates a diverging lens. From Eq. 23.10

$$P = \frac{1}{f} = \frac{1}{-0.75 \text{ m}} = -1.3 \text{ D}$$

A negative, or diverging, lens with a power of 1.3 D is needed.

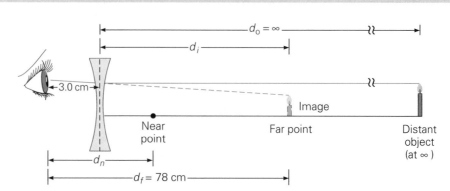

•**FIGURE 25.3 Correcting nearsightedness**
A diverging lens is used. See Example 25.1. (Drawing not to scale.)

(b) To determine the effect of wearing glasses on near point vision, we must find the shortest distance at which an object can still be seen clearly—i.e., the minimum d_o with glasses. Thus we need to calculate where the *object* must be so that its *image* formed by the corrective lens is at the near point (d_n), 23 cm from the eye. The image is then 23 cm − 3.0 cm = 20 cm *from the lens*, so $d_i = -20$ cm (a minus sign is used since the image is virtual). Then, with $f = -75$ cm,

$$d_o = \frac{d_i f}{d_i - f} = \frac{(-20 \text{ cm})(-75 \text{ cm})}{-20 \text{ cm} - (-75 \text{ cm})} = 27 \text{ cm}$$

This means that the near point for this person is 27 cm in front of the lenses when the glasses are being worn. If an object is closer than this, its image (which acts as an object for the eye) will be inside the near point and so will not be seen clearly.

Follow-up Exercise. Suppose a mistake was made in part (a) of this Example and a "corrective" lens of +1.3 D were used. What effect would this have?

If the near point is changed by corrective lenses as in Example 25.1 and this causes a problem, bifocal lenses can be used. Bifocals were invented by Ben Franklin, who glued two lenses together. They are now made by grinding lenses with different curvatures in two different regions. Both nearsightedness and farsightedness can be treated at the same time with bifocals. (There are also trifocals, with lenses having three different curvatures.)

EXAMPLE 25.2 ■ CORRECTING FARSIGHTEDNESS: USE OF A CONVERGING LENS

A farsighted person has a near point of 75 cm for one eye and a near point of 100 cm for the other. What powers should contact lenses have to allow the person to see an object clearly at a distance of 25 cm?

Solution.

Given: $d_{i_1} = -75 \text{ cm} = -0.75 \text{ m}$ *Find:* P_1 and P_2 (lens power
        $d_{i_2} = -100 \text{ cm} = -1.0 \text{ m}$ for each eye)
        $d_o = 25 \text{ cm} = 0.25 \text{ m}$

The eyes are distinguished as 1 and 2. Keep in mind that the optics of a person's eyes are usually different, and a different lens may be needed for each eye. In this case, each lens is to form an image at its eye's near point of an object that is at a distance (d_o) of 25 cm. This image will then act as an object within the eye's range of accommodation. This situation is illustrated in •Fig. 25.4 and corresponds to a person wearing reading glasses. (For the sake of clarity, the lens in the figure is not in contact with the eye.)

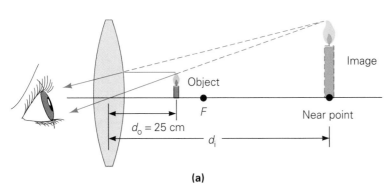

(a)

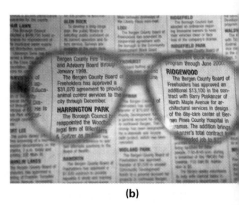

(b)

•**FIGURE 25.4 Reading glasses and correcting nearsightedness**
(a) When an object at the normal near point (25 cm) is viewed through reading glasses with converging lenses, the image is formed farther away but within the eye's range of accommodation (beyond the receded near point). **(b)** Small print as viewed through the lens of reading glasses. See Example 25.2.

The image distances are negative since the images are virtual. With contact lenses, the distance from the eye and the distance from the lens are taken to be the same. Then

$$P_1 = \frac{1}{f_1} = \frac{1}{d_{o_1}} + \frac{1}{d_{i_1}} = \frac{1}{0.25 \text{ m}} - \frac{1}{0.75 \text{ m}}$$

$$= \frac{2}{0.75 \text{ m}} = +2.7 \text{ D}$$

and

$$P_2 = \frac{1}{f_2} = \frac{1}{d_{o_2}} + \frac{1}{d_{i_2}} = \frac{1}{0.25 \text{ m}} - \frac{1}{1.0 \text{ m}}$$

$$= \frac{3}{1.0 \text{ m}} = +3.0 \text{ D}$$

Note that these are positive, or converging, lenses.

Follow-up Exercise. A mistake is made in grinding the corrective lenses in this Example—the left lens is ground to the prescription intended for the right eye, and vice versa. What effect would this have?

Another common defect of vision is **astigmatism**, which is usually due to a refractive surface, most usually the cornea or crystalline lens, being out of round (nonspherical). As a result, the eye has different focal lengths in different planes (•Fig. 25.5a). Points may appear as lines, and the image of a line may be distinct in one direction and blurred in another or blurred in both directions in the circle of least confusion (the location of the least distorted image). A test for astigmatism is given in •Fig. 25.5b.

Astigmatism may be corrected with lenses that have greater curvature in the plane in which the cornea or crystalline lens has deficient curvature (•Fig. 25.5c). Astigmatism is lessened in bright light because the pupil of the eye becomes smaller and the circle of least confusion is reduced.

QUESTION: What is meant by 20/20 vision?

ANSWER: Visual acuity is a measure of how vision is affected by object distance. This is commonly measured using a chart of letters placed at a given distance from the eyes. The result is usually expressed as a fraction: the *numerator* is the distance at which the test eye sees a standard symbol, such as the letter E, clearly, and the *denominator* is the distance at which the letter is seen clearly by a *normal* eye. A 20/20 (test/normal) rating, which is sometimes called "perfect" vision, means that at a distance of 20 ft, the eye being tested can see standard-sized letters as clearly as can a normal eye.

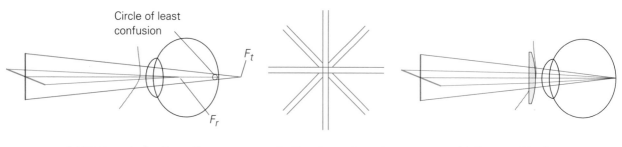

(a) Uncorrected astigmatism　　**(b) Test for astigmatism**　　**(c) Corrected by lens**

•**FIGURE 25.5 Astigmatism**
When one of the eye's refracting components is not spherical, the eye has different focal lengths in different planes. **(a)** The effect occurs because rays along the transverse and radial axes are focused at different points, F_t and F_r, respectively. **(b)** If the eye is astigmatic, some or all of the lines in this diagram will appear blurred. **(c)** Nonspherical lenses, such as plano-convex cylindrical lenses, are used to correct astigmatism.

A person's eyes may have differing visual acuity; for example, one eye may have 20/20 vision and the other 20/30. The latter means that the eye can read standard-sized letters from a chart at a distance of 20 ft and someone with normal vision could do so at 30 ft. That is, the eye in question has to be closer to the chart than a normal eye would to see the letters clearly. Similarly, 20/15 would indicate a person with a particularly sharp eye. The test eye could see an object clearly from a distance of 20 ft, whereas the normal eye would have to be a distance of 15 ft to do so. Thus we see that the greater the ratio or fraction, the better the visual acuity of the eye.

25.2 Microscopes

Objectives: To be able to (a) distinguish between lateral and angular magnification, and (b) describe simple and compound microscopes, and their magnifications.

Microscopes are used to magnify objects so that we can see more detail or see features that are normally indiscernible. Two basic types of microscope will be considered here.

The Magnifying Glass (Simple Microscope)

When we look at an object in the distance, it appears to be very small. As it is brought or comes closer to our eyes, it appears larger. How large an object appears depends on the size of the image on the retina. This may be related to the angle subtended by the object (●Fig. 25.6); the greater the angle, the larger the image.

When we want to examine detail, or look at something closely, we bring it close to our eyes so that it subtends a greater angle. For example, you may examine the detail of a figure in this book by bringing it closer to your eyes. The greatest amount of detail will be seen when the book is at your near point (assuming you are not wearing glasses). If your eyes were able to accommodate to shorter distances, an object brought very close to them would be further magnified. However, as you can easily prove by bringing this book very close to your eyes, images are blurred when objects are at distances inside the near point.

A **magnifying glass**, which is simply a single convex lens (sometimes called a simple microscope), allows a clear image to be formed of an object that is closer than the near point. In such a position, an object subtends a greater angle and therefore appears larger, or magnified. The lens produces a virtual image beyond the near point on which the eye focuses (●Fig. 25.7). If a hand-held magnifying glass is used, its position is usually adjusted until this image is seen clearly.

As illustrated in Fig. 25.7, the angle subtended by an object is much greater when a magnifying glass is used. The magnification of an object *viewed through a magnifying glass* is expressed in terms of this angle. This **angular magnification**, or magnifying power, is symbolized by m. The angular magnification is defined as the ratio of the angular size of the object viewed through the magnifying glass (θ) to the angular size of the object viewed without the magnifying glass (θ_o):

Note: Angular magnification is not the same as lateral magnification. Lateral magnification is discussed in Section 23.3 (see Eq. 23.7).

$$m = \frac{\theta}{\theta_o} \qquad \textit{angular magnification} \qquad (25.1)$$

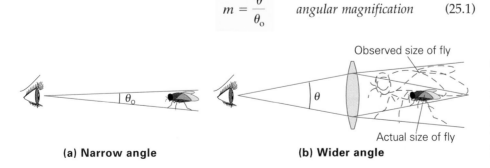

(a) Narrow angle   **(b) Wider angle**

Observed size of fly

θ

Actual size of fly

●**FIGURE 25.6 Magnification and angle**
(a) How large an object appears may be related to the angle subtended by the object. **(b)** The angle and the size of an object may be increased by using a converging lens.

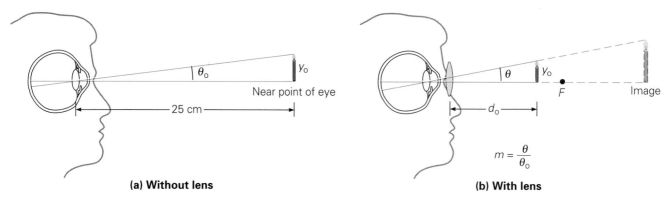

(a) Without lens   (b) With lens

FIGURE 25.7   Angular magnification
The angular magnification (m) of a lens is defined as the ratio of the angular size of an object viewed through the lens to the angular size of the object viewed without the lens, $m = \theta/\theta_o$.

(Note that this is not the same as M, the lateral magnification, which is a ratio of heights: $M = h_i/h_o$.)

The maximum angular magnification occurs when the image seen through the glass is at the eye's near point, that is, $d_i = -25$ cm, since this is as close as it can be clearly seen. (A value of 25 cm will be assumed to be typical for near point in this discussion. The minus sign is used because the image is virtual.) The corresponding object distance may be calculated from the thin lens equation, Eq. 23.6:

$$d_o = \frac{d_i f}{d_i - f} = \frac{(-25\ \text{cm})f}{-25\ \text{cm} - f}$$

or

$$d_o = \frac{25f}{25 + f} \qquad (25.2)$$

where f is in centimeters.

The angular sizes of the object are related to its height by (see Fig. 25.7)

$$\tan \theta_o = \frac{y_o}{25} \quad \text{and} \quad \tan \theta = \frac{y_o}{d_o}$$

Assuming that a small-angle approximation ($\tan \theta \approx \theta$) is valid gives

$$\theta_o \approx \frac{y_o}{25} \quad \text{and} \quad \theta \approx \frac{y_o}{d_o}$$

Then the maximum angular magnification may be expressed as

$$m = \frac{\theta}{\theta_o} = \frac{y_o/d_o}{y_o/25} = \frac{25}{d_o}$$

Substituting for d_o from Eq. 25.2 gives

$$m = \frac{25}{25f/(25 + f)}$$

which simplifies to

Magnifying glass magnification.

$$m = 1 + \frac{25\ \text{cm}}{f} \qquad \begin{array}{l} \textit{angular magnification} \\ \textit{for image at near point (25 cm)} \end{array} \qquad (25.3)$$

Thus, lenses with shorter focal lengths give greater angular magnifications.

In the derivation of Eq. 25.3, the object for the unaided eye was taken to be at the near point, as was the image viewed through the lens. Actually, the normal eye can focus on an image located anywhere between the near point and infinity. At the extreme where the image is at infinity, the eye is more relaxed (the muscles attached to the crystalline lens are relaxed, and the lens is thin). For the image to be at infinity, the object must be at the focal point of the lens. In this case,

$$\theta \approx \frac{y_o}{f}$$

and the angular magnification is simply

$$m = \frac{25 \text{ cm}}{f} \qquad \begin{array}{l} \textit{angular magnification} \\ \textit{for image at infinity} \end{array} \qquad (25.4)$$

Mathematically, it seems that the magnifying power may be increased to any desired value by using lenses with short enough focal lengths. Physically, however, lens aberrations limit the practical range of a single magnifying glass to about 3× or 4× (read as "three ex" and "four ex"), or a sharp image magnification of three or four times the size of the object when used normally.

EXAMPLE 25.3 ■ IT'S ELEMENTARY: ANGULAR MAGNIFICATION OF A MAGNIFYING GLASS

Sherlock Holmes uses a converging lens with a focal length of 12 cm to examine the fine detail of some cloth fibers found at the scene of a crime. (a) What is the maximum magnification given by the lens? (b) What is the magnification for relaxed eye viewing?

Solution.

Given: $f = 12$ cm   *Find:* (a) m (d_i = near point)
                                    (b) m ($d_i = \infty$)

(a) The maximum magnification occurs when the image formed by the lens is at the near point of the eye, which for Eq. 25.3 was taken to be 25 cm. Using that equation,

$$m = 1 + \frac{25 \text{ cm}}{f} = 1 + \frac{25 \text{ cm}}{12 \text{ cm}} = 3.1$$

(b) The eye is most relaxed when viewing distant objects. Eq. 25.4 gives the magnification for the image formed by the lens at infinity:

$$m = \frac{25 \text{ cm}}{f} = \frac{25 \text{ cm}}{12 \text{ cm}} = 2.1$$

Follow-up Exercise. Taking the maximum practical magnification of a magnifying glass to be 4× , which would have the longer focal length, a glass for near-point viewing or one for distant viewing, and how much longer?

The Compound Microscope

A compound microscope provides greater magnification than is attained with a single lens, or simple microscope. A basic **compound microscope** consists of a pair of converging lenses, each of which contributes to the magnification (•Fig. 25.8a). A converging lens having a relatively short focal length ($f_o < 1$ cm) is known as the **objective**. It produces a real, inverted, and enlarged image of an object positioned slightly beyond its focal point. The other lens, called the **eyepiece**, or **ocular**, has a longer focal length (f_e is a few centimeters) and is positioned so that the image formed by the objective falls just *inside* its focal point. This lens forms a magnified virtual image that is viewed by the observer. In essence, the objective acts as a projector, and the eyepiece is a simple magnifying glass used to view the projected image.

Note: It may be useful to review Section 23.3 and Fig. 23.16.

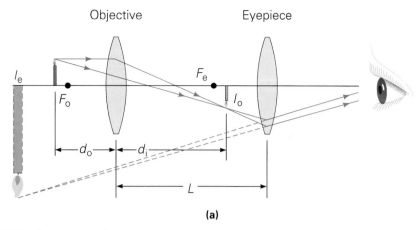

Objective   Eyepiece

(a)

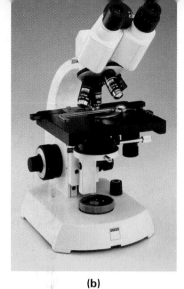

(b)

●**FIGURE 25.8   The compound microscope**
(a) In the optical system of a compound microscope, the real image formed by the objective falls just within the focal point of the eyepiece (F_e) and acts as an object for this lens. An observer looking through the eyepiece sees an enlarged image. **(b)** A compound microscope.

The total magnification (m_t) of a lens combination is the product of the magnifications produced by the two lenses. The image formed by the objective is greater in size than its object by a factor M_o equal to the lateral magnification ($M_o = d_i/d_o$, with the minus sign omitted). In Fig. 25.8a, note that the image distance for the objective lens is approximately equal to L, the distance between the lenses. That is, $d_i \approx L$. (The image I_o is formed by the objective just inside the focal point of the eyepiece, which has a very short focal length.) Also, since the object is very close to the focal point of the objective, $d_o \approx f_o$. With these approximations,

$$M_o \approx \frac{L}{f_o}$$

The angular magnification of the eyepiece for an object at its focal point is given by Eq. 25.4:

$$m_e = \frac{25 \text{ cm}}{f_e}$$

Since the object for the eyepiece (the image formed by the objective) is very near the focal point of the eyepiece, a good approximation is

$$m_t = M_o m_e = \left(\frac{L}{f_o}\right)\left(\frac{25 \text{ cm}}{f_e}\right)$$

Compound microscope magnification.

or

$$m_t = \frac{(25 \text{ cm})L}{f_o f_e} \quad \begin{array}{l} \textit{angular magnification of} \\ \textit{compound microscope} \end{array} \quad (25.5)$$

where f_o, f_e, and L are in centimeters.

EXAMPLE 25.4 ■ A COMPOUND MICROSCOPE: FINDING
MAGNIFICATION

A microscope has an objective with focal length of 10 mm and an eyepiece with a focal length of 4.0 cm. The lenses are positioned 20 cm apart in the barrel. Determine the approximate total magnification of the microscope.

Solution.

Given: $f_o = 10$ mm $= 1.0$ cm   ***Find:*** m_t (total magnification)
        $f_e = 4.0$ cm
        $L = 20$ cm

Using Eq. 25.5,

$$m_t = \frac{(25\text{ cm})L}{f_o f_e} = \frac{(25\text{ cm})(20\text{ cm})}{(1.0\text{ cm})(4.0\text{ cm})} = 125 \times$$

(Note the relatively short focal length of the objective.)

Follow-up Exercise. If the focal length of the eyepiece in this Example were doubled, how would this affect the length of the microscope for the same magnification? (Express the change as a percent.)

A modern compound microscope is shown in •Fig. 25.8b. Interchangeable eyepieces with magnifications from about 5× to over 100× are available. For standard microscopic work in biology or medical laboratories, 5× and 10× eyepieces are normally used. Microscopes are often equipped with rotating turrets, which usually contain three objectives for different magnifications, for example, 10×, 43×, and 97×. These objectives and the 5× and 10× eyepieces can be used in various combinations to provide magnifying powers from 50× to 970×. The maximum magnification with a compound microscope is about 2000×.

Opaque objects are usually illuminated with a light source placed above them. In many instances, the specimens to be viewed are transparent, such as glass slides containing cells or thin sections of tissues. These are illuminated with a light source beneath the microscope stage so that light passes through the specimen. A modern microscope is usually equipped with a light condenser (converging lens) and diaphragm below the stage, which are used to concentrate the light and control its intensity. A microscope may have an internal light source. The light is reflected into the condenser from a mirror. Older microscopes have two mirrors with reflecting surfaces: one a plane mirror for reflecting light from a high-intensity external source, and the other a concave mirror for converging low-intensity light such as sky light.

25.3 Telescopes

Objectives: **To be able to (a) distinguish between refractive and reflective telescopes, and (b) describe the advantages of each.**

Telescopes apply the optical principles of mirrors and lenses to improve our ability to see distant objects. Used for both terrestrial and astronomical observations, telescopes allow some objects to be viewed in greater detail and other fainter or more distant objects simply to be seen. Basically, there are two types of telescopes: refracting and reflecting. These depend on the gathering and converging of light by lenses and mirrors, respectively.

Refracting Telescope

The principle underlying one type of **refracting telescope** is similar to that of a compound microscope. The major components of this type of telescope are objective and eyepiece lenses, as illustrated in •Fig. 25.9. The objective is a large converging lens with a long focal length, and the movable eyepiece has a relatively short focal length. Rays from a distant object are essentially parallel and form an image (I_o) at the focal point (F_o) of the objective. This image acts as an object for the eyepiece, which is moved until the image lies just inside its focal point (F_e). A large, inverted, virtual image (I_e) is seen by an observer.

For relaxed viewing, the eyepiece is adjusted so that its image (I_e) is at infinity, which means that the objective image (I_o) is at the focal point of the eyepiece

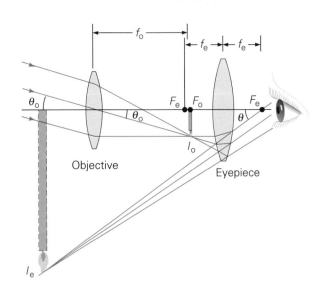

•**FIGURE 25.9 The refracting astronomical telescope**
In an astronomical telescope, rays from a distant object form an intermediate image at the focal point of the objective (F_o). The eyepiece is moved so that the image is at or slightly inside its focal point (F_e). An observer sees an enlarged image at infinity (shown at a finite distance here for illustration).

Refracting telescope magnification.

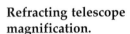

Note: Astronomical telescope—inverted image.

Galileo did not invent the telescope, but he built one after hearing about a Dutch instrument that could be used for distant viewing.

(f_e). As you can see in the figure, the distance between the lenses is then the sum of the focal lengths ($f_o + f_e$), which is the length of the telescope tube. The magnifying power of a telescope focused for the final image at infinity can be shown (see Exercise 52) to be

$$m = -\frac{f_o}{f_e} \quad \begin{array}{l} \textit{angular magnification} \\ \textit{of refracting telescope} \end{array} \qquad (25.6)$$

where the minus is inserted to indicate the image is inverted as in our lens sign convention in Section 23.3. Thus, to achieve the greatest magnification, the focal length of the objective should be made as great as possible and the focal length of the eyepiece as short as possible.

The telescope illustrated in Fig. 25.9 is called an **astronomical telescope**. Its final image is inverted, but this poses little problem to astronomers. (Why?) However, someone viewing an object on Earth through a telescope finds it more convenient to have an upright image. A telescope in which the final image is upright is called a **terrestrial telescope**. An upright final image can be obtained in several ways—two are illustrated in •Fig. 25.10.

In the telescope diagrammed in Fig. 25.10a, a diverging lens is used as an eyepiece. This type of terrestrial telescope is referred to as a Galilean telescope, because Galileo built one in 1609. The image of the objective is formed at the focal point of the eyepiece (F_e) on the side opposite the objective. Thus, the light rays pass through the eyepiece before the image is formed, and the image acts as a "virtual" object for the eyepiece (see Section 23.3). An observer sees a magnified, upright, virtual image. (Note that with a diverging lens and negative focal length ($-f_e$), Eq. 25.6 gives a $+m$, indicating an upright image.)

Galilean telescopes have several disadvantages, most notably very narrow fields of view and limited magnification. A better type of terrestrial telescope, illustrated in Fig. 25.10b, uses a third lens, called the erecting lens, between converging objective and eyepiece lenses. If the image is formed by the objective at a distance that is twice the focal length of the intermediate erecting lens ($2f_i$), then the lens merely inverts the image without magnification, and the telescope magnification is still given by Eq. 25.6.

However, to achieve the upright image in this way requires a greater telescope length. Using the intermediate erecting lens to invert the image increases the length of the telescope by four times the focal length of the erecting lens ($2f_i$ on each side). An erecting lens is used in a spyglass, which is telescoped to achieve the necessary length. The inconvenient length can be avoided by using internally

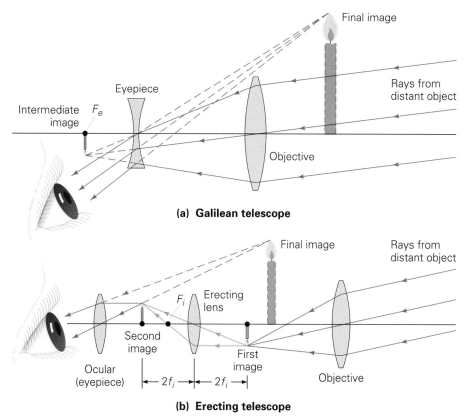

(a) **Galilean telescope**

(b) **Erecting telescope**

●**FIGURE 25.10   Terrestrial telescopes**
(a) A Galilean telescope uses a diverging lens as an eyepiece, producing upright images.
(b) Another way to produce upright images is to use a converging "erecting" lens between the objective and eyepiece in an astronomical-type telescope. This elongates the telescope, but the length may be shortened by using internally reflecting prisms.

reflecting prisms. This is the principle of prism binoculars, which are double telescopes—one for each eye (●Fig. 25.11).

EXAMPLE 25.5 ■ ASTRONOMICAL TELESCOPE—AND LONGER TERRESTRIAL TELESCOPE

An astronomical telescope has an objective lens with a focal length of 30 cm and an eyepiece with a focal length of 9.0 cm. (a) What is the magnification of the telescope? (b) If an erecting lens with a focal length of 7.5 cm is used to convert the telescope to a terrestrial type, what is the overall length of the telescope tube?

Solution.   Listing the data, we have

Given: $f_o = 30$ cm   *Find:*   (a) m (magnification)
            $f_e = 9.0$ cm   (b) L (length of
            $f_i = 7.5$ cm (intermediate   telescope tube)
                  erecting lens)

(a)   The magnification is given by Eq. 25.6:

$$m = -\frac{f_o}{f_e} = -\frac{30 \text{ cm}}{9.0 \text{ cm}} = -3.3\times$$

where the minus sign indicates that the final image is inverted.

(b)   Taking the length of the astronomical tube to be the distance between the lenses, this is the sum of their focal lengths.

$$L_1 = f_o + f_e = 30 \text{ cm} + 9.0 \text{ cm} = 39 \text{ cm}$$

Adding the intermediate erecting lens adds four times the focal length of this lens ($4f_i$) to the length of the scope (Fig. 25.10b). The overall length is then

$$L = L_1 + L_2 = 39 \text{ cm} + 4f_i = 39 \text{ cm} + 4(7.5 \text{ cm}) = 69 \text{ cm}$$

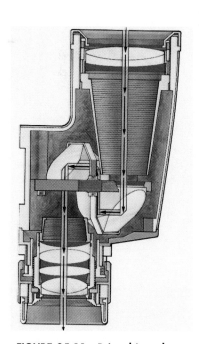

●**FIGURE 25.11   Prism binoculars**
A schematic cutaway view of one ocular showing the internal reflections in the prisms, which reduce the overall length.

Hence the telescope length is over two-thirds of a meter, with an upright image, but the same magnification. (Why?)

Follow-up Exercise. A terrestrial telescope 66 cm in length has an intermediate erecting lens with a focal length of 12 cm. It is desired to reduce the length to a more managable 50 cm. What is the focal length of an erecting lens that would do this?

CONCEPTUAL EXAMPLE 25.6 ■ CONSTRUCTING A TELESCOPE

A student is given two converging spherical lenses, one with a focal length of 5.0 cm and the other with a focal length of 20 cm. To construct a telescope to best view distant objects with these lenses, the student should hold the lenses (a) more than 25 cm apart, (b) less than 25 cm but more than 20 cm apart, (c) less than 20 cm but more than 5.0 cm apart, or (d) less than 5.0 cm apart. Also specify which lens should be used as the eyepiece. *Clearly establish the reasoning and physical principle(s) used in determining your answer before checking it below. That is, **why** did you select your answer?*

Reasoning and Answer. The only type of telescope that can be constructed with two converging lenses is an astronomical telescope. In this type of telescope, the lens with the longer focal length is used as an objective lens to produce a real image of a distant object. That image is then viewed with the lens with the shorter focal length (the eyepiece), used as a simple magnifier. If the object is at a great distance, a real image is formed by the objective lens in the focal plane of the lens. This image acts as the object for the eyepiece, which is positioned so that the image/object lies just inside its focal point so as to produce a large, inverted second image (Fig. 25.9).

This means that the two lenses must be *slightly* less than 25 cm apart. Thus, answer (a) is not correct. Answers (c) and (d) are also not correct because the eyepiece would be too close to the objective to get the large secondary image needed for optimal viewing of a distant object. In these cases, the rays would pass through the second lens before the image was formed, and a *reduced* image might be produced (see Section 23.3 and Fig. 23.17b). Thus, (b), with the objective image just inside the eyepiece focal point, is the correct answer.

Follow-up Exercise. A third converging lens with a focal length of 4.0 cm is used with the above two lenses to produce a terrestrial telescope in which the third lens does nothing more than invert the image. How should the lenses be positioned, and what distances should there be between them, for the final image to be of maximum size and upright?

Reflecting Telescope

For viewing the Sun, Moon, and nearby planets, magnification is necessary to see details. However, even with the highest feasible magnification, stars appear only as faint points of light. For distant stars and galaxies, it is more important to gather enough light so that the object can be seen at all. The intensity of light from a distant source is very weak. In many instances, such a source may be detected only when the light is gathered and focused on a photographic plate over a long period of time.

Recall from Section 14.3 that intensity is energy per *area* per unit time. Thus, more light can be gathered if the size of the objective is increased. This increases the distance at which the telescope can detect faint objects such as distant galaxies. (Recall that light intensity of a point source is inversely proportional to the *square* of the distance between the source and the observer.) However, producing a large lens involves difficulties associated with glass quality, grinding, and polishing. Compound lens systems are required to reduce aberrations, and a very large lens may sag under its own weight, producing further aberrations. The largest objective lens in use has a diameter of 40 in. (102 cm, or about 1 m) and is part of the refracting telescope of the Yerkes Observatory at Williams Bay, Wisconsin (Fig. 25.12).

These problems are reduced with a **reflecting telescope** utilizing a large, concave, front-surface parabolic mirror (Fig. 25.13). A parabolic mirror does not exhibit spherical aberration, and a mirror has no inherent chromatic aberration. High-quality glass is not needed, since the light is reflected by a mirrored front surface. And only one surface has to be ground, polished, and silvered.

As shown in Fig. 25.13a, light from a distant source entering the telescope tube is focused by the concave mirror. A small plane mirror is sometimes used to direct the converging rays to a more convenient viewing location (Fig. 25.13b). Such a mirror blocks only a small percentage of the incoming light. A third type of focus is shown in Fig. 25.13c.

The largest reflecting telescope in the United States, located at Hale Observatories on Palomar Mountain in California, has a mirror with a 200-in. (5.1-m) diameter (Fig. 25.14a). In the Hale design, the observer can actually sit in a cage at the prime focus of the telescope (Fig. 25.14b). The world's largest single mirror reflecting telescope has a reflector with a 6-m diameter and is located in the former Soviet Union (Fig. 25.14c).

Even though reflecting telescopes have advantages over refracting telescopes, they also have their own problems. Like a large lens, a large mirror may sag under its own weight, and the weight necessarily increases with the size of the mirror. The weight factor is also a cost factor for construction of a telescope, since the supporting elements for a heavier mirror must be more massive.

These problems are being addressed by new technologies. One approach is to use an array of small mirrors, coordinated so as to function as a single large mirror. Examples include the twin Keck telescopes at Mauna Kea in Hawaii

 FIGURE 25.12 Yerkes refracting telescope
The largest refracting telescope, at the Yerkes Observatory at Williams Bay, Wisconsin, has an objective lens with a diameter of 40 in. or 102 cm.

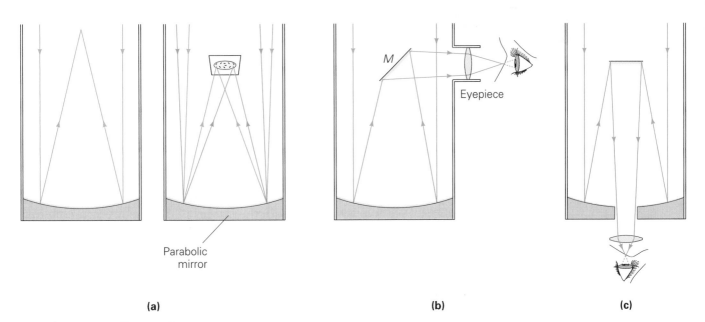

(a)   (b)   (c)

 FIGURE 25.13 Reflecting telescopes
A concave mirror may be used in a telescope to converge light to form an image of a distant object. The image may be at the prime focus as shown in **(a)**; or a small mirror and lens may be used to focus the image outside the telescope, as in **(b)**, called a Newtonian focus. **(c)** Another arrangement, called a Cassegrain focus, uses a mirror to form the image below the main mirror.

(a)

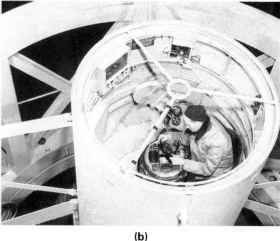

(b)

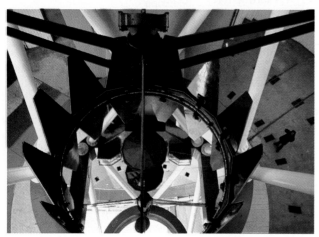

(c)

●**FIGURE 25.14 Reflecting telescopes**
(a) The Hale telescope on Palomar Mountain in California, with a mirror 200 in. (5.1 m) in diameter, is the largest single-mirror reflecting telescope in the United States. The mirror is at the bottom center, just above the heads of the people. **(b)** An astronomer in the observing capsule near the top of the telescope tube, at the prime focus. (Compare Figure 25.13a.) **(c)** The world's largest single-mirror reflecting telescope, located in the Caucasus Mountains of the former Soviet Union, has a mirror with a diameter of 6 m. (Note the size of the person standing on the observatory floor for comparison.) The dark flaps that surround the mirror fold down to protect it when not in use.

(●Fig. 25.15a). Each has a mirror consisting of 36 hexagonal segments that are computer-positioned to give the effect of a 10-m mirror. Similar methods are being used in the construction of many other new instruments.

On Earth's surface, we observe the heavens from the bottom of an ocean of air many kilometers deep. As we saw in Section 22.3, light is refracted as it passes through layers of air that differ in density. Temperature variations in the atmosphere produce convection currents and turbulence, which cause the air density to fluctuate continuously. As a result, the images of distant objects are blurred, distorted, or seem to shimmer as we observe them. To overcome this effect, many modern telescopes are being fitted with *active optics* or *adaptive optics* systems. Such instruments use tiny, computer-controlled adjustments in the precise shape and orientation of the telescope mirror during observations to compensate for atmospheric turbulence, as well as imperfections in the optical system caused by sagging, temperature changes, and the like. One such system is shown in the chapter opening photo. Another uses a laser beam (Chapter 27) to probe the atmosphere and obtain information about its motions from instant to instant (●Fig. 25.15b). A computer then continuously adjusts the configuration of the mirror to maintain optimum resolution.

(a)

(b)

•FIGURE 25.15 New types of telescopes
(a) The recently-completed Keck telescope on Mauna Kea, Hawaii, has a mirror consisting of 36 separate segments, positioned by computer to function like a single 10-m mirror. It is much cheaper and easier to construct a number of distortion-free small mirrors than a single equivalent large mirror. (b) In this experimental system developed by the U.S. Air Force, a laser beam is used to monitor atmospheric turbulence on a moment-by-moment basis. A computer adjusts the shape of the telescope mirror thousands of times per second to compensate for atmospheric distortions.

Another way of extending our view into space is to put telescopes into orbit about the Earth. Above the atmosphere, the view is unaffected by the twinkling effect of atmospheric turbulence and refraction, and there is no background problem from city lights. In 1990, the optical Hubble Space Telescope (HST) was launched into orbit. Even with a mirror of only 2.4 m, its privileged position was expected to allow the HST to produce images seven times clearer than those formed by Earthbound telescopes.

The HST's mirror was constructed in 1980, and apparently the optical system used to test it was faulty. The unfortunate result was spherical aberration (Section 23.4), which caused the images to be somewhat blurred. Even so, the HST was able to make many observations that are superior to those made from Earth's surface. In late 1993, astronauts performed repairs on the optical and other systems of the HST that improved its observation powers (•Fig. 25.16). Such missions to make repairs and to install updated equipment should make the HST a useful observatory well into the 21st century.

However, not all telescopes are optical in nature, as the Insight on p. 786 points out.

•FIGURE 25.16 Hubble Space Telescope (HST)
Late in 1993, astronauts from the Space Shuttle Endeavor visited the HST in orbit. They installed corrective equipment that compensated for many of the telescope's optical flaws and repaired or replaced other malfunctioning systems. Since that time the HST has produced a wealth of spectacular images and helped astronomers to make many important discoveries.

25.4 Diffraction and Resolution

Objectives: **To be able to (a) describe the relationship of diffraction and resolution, and (b) state and explain Rayleigh's criterion.**

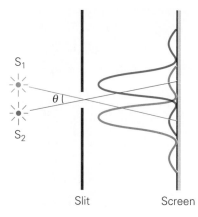

(a) Resolved

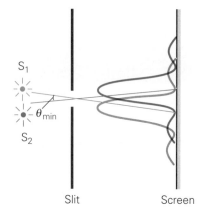

(b) Just resolved

•FIGURE 25.17   Resolution
Two light sources in front of a slit produce diffraction patterns. **(a)** When the angle subtended by the sources at the slit is large enough for the diffraction patterns to be distinguishable, the images are said to be resolved. **(b)** At smaller angles, the central maxima are closer together. At θ_{min} the central maximum of one pattern falls on the first dark fringe of the other, and the images are said to be just resolved. For smaller angles, the patterns are unresolved.

The diffraction of light places a limitation on our ability to distinguish objects that are close together when using microscopes or telescopes. This effect can be understood by considering two point sources located far from a narrow slit of width d (•Fig. 25.17). The sources could be distant stars, for example. In the absence of diffraction, two bright spots, or images, would be observed on a screen. As you know from Section 23.3, however, the slit diffracts the light, and each image consists of a central maximum with a pattern of weaker bright and dark fringes on either side. If the sources are close together, the two central maxima may overlap. Then the images cannot be distinguished, or are unresolved. For the images to be resolved, the central maxima must not overlap appreciably.

In general, images of two sources can be resolved if the central maximum of one falls at or beyond the first minimum (dark fringes) of the other. This generally accepted limiting condition for the **resolution** of two diffracted images was first proposed by Lord Rayleigh (1842–1919), a British physicist, and is known as the **Rayleigh criterion:**

> Two images are said to be just resolved when the central maximum of one image falls on the first minimum of the diffraction pattern of the other image.

The Rayleigh criterion may be expressed in terms of the angular separation (θ) of the sources (see Fig. 25.17). The first minimum ($m = 1$) for a single-slit diffraction pattern satisfies this relationship:

$$d \sin \theta = \lambda$$

or
$$\sin \theta = \frac{\lambda}{d}$$

This gives the minimum angular separation for two images to be just resolved according to the Rayleigh criterion. In general, for visible light, the wavelength is much smaller than the slit width ($\lambda \ll d$), so a good approximation is $\sin \theta \approx \theta$. The limiting, or minimum, angle of resolution (θ_{min}) for a slit of width d is then

$$\theta_{min} = \frac{\lambda}{d} \qquad \begin{array}{l} \textit{minimum angle of resolution} \\ \textit{(for a slit)} \end{array} \qquad (25.7)$$

(Note that θ_{min} is a pure number and therefore must be expressed in radians.) Thus, the images of two sources will be *distinctly* resolved if the angular separation of the sources is greater than λ/d.

The apertures (openings) of microscopes and telescopes are generally circular. Thus, there is a circular diffraction pattern around the central maximum, which is a bright circular disk (•Fig. 25.18). Detailed analysis for a circular aperture shows that the minimum angular separation for two objects for their images to be just resolved is

$$\theta_{min} = \frac{1.22\lambda}{D} \qquad \begin{array}{l} \textit{minimum angle of resolution} \\ \textit{(for a circular aperture)} \end{array} \qquad (25.8)$$

where D is the diameter of the aperture, and θ_{min} is in radians.

Equation 25.8 applies to the objective lens of a microscope or telescope, which may be considered to be a circular aperture for light. According to the equation, to make θ_{min} small so objects close together can be resolved, it is advantageous to make the aperture as large as possible. This is another reason for using large lenses (and mirrors) in telescopes, in addition to their greater light-gathering power.

EXAMPLE 25.7 ■ EYE AND TELESCOPE: EVALUATING RESOLUTION WITH THE RAYLEIGH CRITERION

Determine the minimum angle of resolution by the Rayleigh criterion for (a) the pupil of the eye (daytime diameter of about 4.0 mm), (b) the Yerkes Observatory 40-in. (102-cm) refracting telescope for visible light with a wavelength of 660 nm, and (c) a radio telescope of 25-m diameter for radiation with a wavelength of 21 cm.

Solution.

Given: (a) $D = 4.0 \text{ mm} = 4.0 \times 10^{-3} \text{ m}$
 (b) $D = 102 \text{ cm} = 1.02 \text{ m}$
  $\lambda = 660 \text{ nm} = 6.60 \times 10^{-7} \text{ m}$
 (c) $D = 25 \text{ m}$
  $\lambda = 21 \text{ cm} = 0.21 \text{ m}$

Find: (a) θ_{min} (minimum angles
 (b) θ_{min} of resolution)
 (c) θ_{min}

Eq. 25.8 applies to all three cases.

(a) For the eye,

$$\theta_{min} = \frac{1.22 \lambda}{D} = \frac{(1.22)(6.60 \times 10^{-7} \text{ m})}{4.0 \times 10^{-3} \text{ m}} = 2.0 \times 10^{-4} \text{ rad}$$

(b) For the light telescope,

$$\theta_{min} = \frac{(1.22)(6.60 \times 10^{-7} \text{ m})}{1.02 \text{ m}} = 7.9 \times 10^{-7} \text{ rad}$$

(c) For the radio telescope,

$$\theta_{min} = \frac{(1.22)(0.21 \text{ m})}{25 \text{ m}} = 0.010 \text{ rad}$$

The smaller the angular separation, the better the resolution. What do the results tell you?

Follow-up Exercise. Two of Jupiter's largest moons (it has 16 or more) are separated by a distance of 3.0×10^6 km at a time when the planet is 3.3×10^8 km from Earth. Would a person be able to resolve both moons with the unaided eye, assuming a night-time pupil diameter of 7.5 mm? (Consider light with a wavelength of 550 nm, to which the human eye is most sensitive and neglect atmospheric effects.)

For a microscope, it is convenient to specify the actual separation (s) between two point sources. Since the objects are usually near the focal point of the objective, to a good approximation

$$\theta_{min} = \frac{s}{f} \quad \text{or} \quad s = f\theta_{min}$$

where f is the focal length of the lens. (Here s is taken as the arc length subtended by θ_{min}, and $s = r\theta_{min} = f\theta_{min}$.) Then, using Eq. 25.8,

$$s = f\theta_{min} = \frac{1.22 \lambda f}{D} \quad \begin{array}{l} \textit{resolving power} \\ \textit{of a microscope} \end{array} \quad (25.9)$$

This minimum distance between two points whose images can be just resolved is called the **resolving power** of the lens. Practically, the resolving power of a microscope indicates the ability of the objective to distinguish fine detail in specimens' structures.

 Useful magnification refers to the magnification of resolvable details. Magnification beyond the limit of usefulness reveals no additional details of the object and so is called empty magnification. (For example, enlarging a newspaper picture is empty magnification—the image lacks fine detail, so nothing more is re-

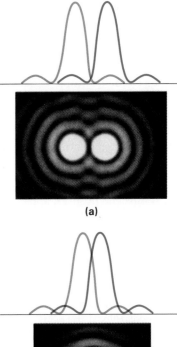

(a)

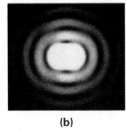

(b)

•FIGURE 25.18 Circular aperture resolution
(a) When the angular separation of two objects is large enough, the images are well resolved. (Compare Fig. 25.17a) **(b)** The minimum angular separation for two images to be just resolved is given by the Rayleigh criterion: the central maximum of the diffraction pattern of one image falls on the first minimum of the diffraction pattern of the other image. (Compare Fig. 25.17b) Objects with smaller angular separations produce images that overlap so much that they cannot be clearly distinguished.

Resolving power: minimum resolved distance between two points.

Insight Telescopes Using Nonvisible Radiation

The word "telescope" usually brings to mind visual observations. However, the visible region is a very small part of the electromagnetic spectrum, and celestial objects emit radiation of many other types, including radio waves. This fact was discovered accidentally in 1931 by an electrical engineer named Carl Jansky while he was working on the problem of static interference with intercontinental radio communications. Jansky found an annoying static hiss that came from a fixed direction in space, apparently from a celestial source. It was soon apparent that radio waves are another source of astronomical information, and radio telescopes were built.

A radio telescope operates similarly to a reflecting light telescope. A reflector with a large area collects and focuses the radio waves at a point where a detector picks up the signal (Fig. 1). The parabolic collector, called a dish, is covered with metal wire mesh or metal plates. Since the wavelengths of radio waves range from a few millimeters to several meters, wire mesh is a good reflecting surface for such waves.

Radio telescopes supplement optical telescopes and provide some definite advantages. Radio waves pass freely through the huge clouds of dust in our galaxy that hide a large part of it from visual observation. Also, radio waves easily penetrate the Earth's atmosphere, which reflects and scatters a large percentage of the incoming visible light (Fig. 2).

The Earth's atmosphere absorbs many wavelengths of electromagnetic radiation and completely blocks some from reaching the ground. Water vapor, a strong absorber of infrared radiation, is concentrated in the lower part of the at-

FIGURE 2 Ultraviolet astronomy from orbit
The atmosphere is largely opaque to infrared, ultraviolet, X-rays, and gamma rays. However, telescopes utilizing all of these wavelengths have made observations from above the atmosphere. This photo shows the Astro package, containing three ultraviolet telescopes, that was carried into orbit by the space shuttle in 1990 and 1995.

mosphere near the surface of the Earth. Thus, observations with infrared telescopes are made from high-flying aircraft or from orbiting spacecraft.

The first orbiting infrared observatory was launched in 1983. Not only are atmospheric interferences eliminated in space, but a telescope may be cooled to a very low temperature without becoming coated with condensed water vapor. Cooling the telescope helps eliminate interfering infrared radiation generated by the telescope itself. The orbiting telescope launched in 1983 was cooled with liquid helium to about 10 K; it carried out an infrared survey of the entire sky.

The atmosphere is virtually opaque to ultraviolet radiation, X-rays, and gamma rays from distant sources. Orbiting satellites with telescopes sensitive to these types of radiation have mapped out portions of the sky, and other surveys are planned.

At the altitudes reached by orbiting satellites, even observations in the visible region are not affected by air motion and temperature refraction. Perhaps in the not too distant future, a permanently staffed orbiting observatory carrying a variety of telescopes will help expand our knowledge of the universe.

FIGURE 1 Radio telescopes
Several of the dish antennae that make up the Very Large Array (VLA) radio telescope near Socorro, New Mexico. There are 27 movable dishes, each 25 m in diameter, forming the array along a Y-shaped railway network. The data from all the antennae are combined to produce a single radio image. In this way it is possible to attain a resolution equivalent to that of one giant radio dish.

vealed by enlargement.) As with a telescope, the useful magnification of a microscope is not limited by the power (focal lengths) of its lenses but by the aperture, that is, the size of the objective lens. However, making the objective lens larger leads to another limiting factor on resolution by a lens—aberrations. As you learned in Chapter 23, spherical and other aberrations give rise to blurred images. Compound lens systems can be used to reduce these effects significantly, so the limit of resolution is determined mainly by diffraction.

Note from Eq. 25.8 that higher resolution may be gained by using radiation of a shorter wavelength. Thus, a telescope with an objective of a given size will have greater resolution with violet light than with red light. For microscopes, it is possible to increase resolving power by shortening the wavelengths of the light used to create the image. This can be done with a specialized objective called an *oil immersion lens*. When such a lens is used, a drop of transparent oil fills the space between the objective and the specimen. Recall that the wavelength of light in the oil is $\lambda_m = \lambda/n$, where n is the index of refraction of the oil and λ the wavelength of light in air. With values of n of about 1.50 or higher, the wavelength is significantly reduced, and the resolving power increased proportionally.

Note: The relationship between wavelength and index of refraction is given in Section 22.3; see Eq. 22.4.

The wavelength can be further reduced by using ultraviolet light and photographic film. However, ordinary glass absorbs ultraviolet light, so the lenses of ultraviolet microscopes must be made of other materials, such as quartz.

The fact that the resolving power of a microscope can be increased by decreasing the wavelength of the incident light opened up new possibilities for seeing fine details after it was discovered that a beam of electrons has wave properties (Chapter 28). These beams have wavelengths that can be adjusted to be very much smaller than those of light in or near the visible region. Using a short-wavelength electron beam, an electron microscope can magnify up to 1000 times more than a visible-light microscope. X-rays with even shorter wavelengths than electron beams are used to attain even greater magnification and resolution.

25.5 Color and Optical Illusions

Objectives: To be able to (a) relate color vision and light, and (b) give examples of visual images and illusions.

In general, physical properties are fixed or absolute. For example, a particular radiation has a certain frequency or wavelength. However, visual perception of this radiation may vary from person to person. How we see radiation gives rise to the following interesting topics.

Color Vision

Color is perceived because of a physiological response to excitation by light of the cone receptors in the retina of the human eye. (Many animals have no cone cells and live in a black-and-white world.) The cones are sensitive to light with frequencies between 7.5×10^{14} Hz and 4.3×10^{14} Hz (wavelengths between 400 nm and 700 nm). Different frequencies of light are perceived by the brain as different colors. The association of a color with a particular frequency is subjective and may vary from person to person. As pitch is to sound and hearing, color is to light and vision.

Note: The relationship between pitch and frequency for sound is discussed in Sections 14.1 and 14.5.

Color vision is not well understood. One of the most widely accepted theories is that there are three different types of cones in the retina, responding to different parts of the visible spectrum, particularly in the red, green, and blue regions (•Fig. 25.19). Presumably, the types of cones absorb light of specific ranges of frequencies and functionally overlap one another to form combinations that are interpreted by the brain as various colors of the spectrum. For example, when red and green cones are stimulated equally by light of a particular frequency, the

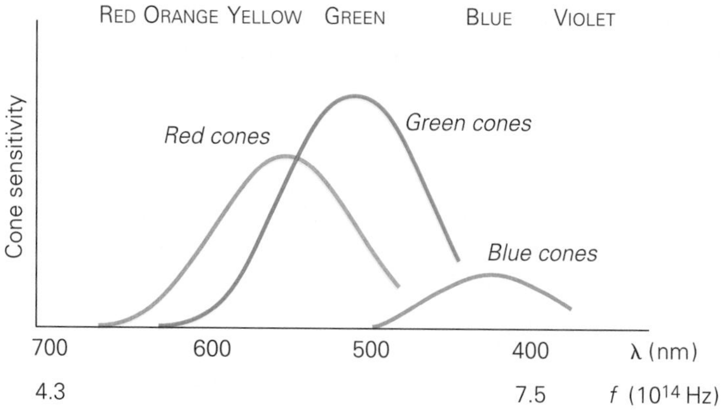

•FIGURE 25.19 **Sensitivity of cones**
Different types of cones in the
retina of the human eye respond to
different frequencies of light to
give three general color responses:
red, green, and blue.

brain interprets the color as yellow. But when the red cones are stimulated more
strongly than the green cones, the brain senses orange.

As Fig 25.19 shows, the human eye is not equally sensitive to all colors. Some
colors evoke a greater brightness response than others. The wavelength of maximum visual sensitivity is about 550 nm in the yellow-green region.

CONCEPTUAL EXAMPLE 25.8 ■ THE COLOR OF LIGHT: FREQUENCY OR WAVELENGTH?

What determines the color of light? (a) intensity, (b) frequency, (c) wavelength, (d)
both frequency and wavelength. *Clearly establish the reasoning and physical principle(s)
used in determining your answer before checking it below. That is, **why** did you select your
answer?*

Reasoning and Answer.   Clearly, (a) is not the answer; if it were, objects around
you would change color every time the light was dimmed or brightened in a room. The
other possible choices, however, may seem perplexing. You probably noticed in Figures
20.18 and 25.19 that both wavelengths and frequencies were specified for the various
colors. And in Chapter 24, we commonly talked about the color of light in terms of
wavelength—for example, blue light ($\lambda = 425$ nm). This practice is based on the relationship $\lambda f = c$, where f and λ have an exact relationship. Thus, you may think that you
don't have sufficient basis for deciding among the other possible answers. However,
you may possess more evidence than you realize, as the following thought experiment
shows.

Suppose you had an object that appeared red in color when observed in air, with
light coming from it having a wavelength of $\lambda = 700$ nm. Then you submerged the object in a swimming pool and jumped in to observe it from underwater. From Eq. 22.4,
with $n = 1.33$ for water, the wavelength of the light from the submerged object would be

$$\lambda' = \frac{\lambda}{n} = \frac{700 \text{ nm}}{1.33} = 526 \text{ nm}$$

If this were the case, underwater the red object would appear green! But this is not
what you observe in the real world. Obviously then, (c) is not correct, nor is (d). Therefore, the answer must be (b). But why frequency? Remember from Section 22.3 that frequency is the wave parameter that doesn't change when the same light travels in different media—wavelength and wave speed change, but not frequency (or color).

So, our color perception depends on frequency. When we refer to the wavelength
of a color, this is for the fixed speed of light in vacuum ($\lambda f = c$), which is commonly assumed in most situations involving the wavelength of light.

Follow-up Exercise.   Suppose a light source that appears blue in air ($\lambda = 450$ nm)
is viewed underwater. (a) What will be the observed speed, wavelength, and frequency
of the light from this object? (b) What color will it appear to the observer? (c) If color
perception depended on wavelength rather than frequency, what color would it appear?

Color blindness is believed to result when one type of cone is missing. This condition is genetically inheritable, and most color-blind people are males (about 4% of the male population). For example, if a man lacks red cones, he can see green through orange-red by means of his green cones. However, he is unable to distinguish among these colors satisfactorily because there are no red cones to provide contrast. A similar condition occurs if the green cones are missing. In either case, this is termed red–green color blindness because the person cannot effectively distinguish those colors.

The theory of color vision described above is based on the experimental fact that beams of varying intensities of red, green, and blue light will produce most colors, or hues. The red, blue, and green from which we interpret a full spectrum of colors are called the **additive primary colors** (or the additive primaries). When light beams of the additive primaries are projected on a white screen so they overlap, other colors will be produced, as illustrated in •Fig. 25.20. This is called the **additive method of color production** or mixing.

Triad dots consisting of three phosphors that emit the additive primary colors when excited are used in television picture tubes to produce colored images.

Note in Fig. 25.20 that the proper combination of the primary colors appears white to the eye. Also, many *pairs* of colors appear white to the eye when combined. The colors of such pairs are said to be **complementary colors**. For example, the complement of blue is yellow, that of red is cyan, and that of green is magenta. As you can see from the figure, the complementary color of a particular primary is the combination or sum of the other two primaries. Hence, the primary and its complement together appear white.

Edwin H. Land (the developer of Polaroid film) has shown that when the proper mixtures of only two wavelengths (colors) of light are passed through black and white transparencies (no color), they produce images of various colors. Land wrote, *"In this experiment we are forced to the astonishing conclusion that the rays are not in themselves color-making. Rather they are bearers of information that the eye uses to assign appropriate colors to various objects in an image."** From Land's experiments, it seems that information about colors, other than for the two wavelengths used, is developed in the brain via the cone cells. However, knowledge about color vision is far from complete.

Objects have color when they are illuminated with white light because they reflect (scatter) or transmit the light of the frequency of the color they appear to be. The other frequencies of the white light are absorbed. For example, when white light strikes a ripe red apple, only the waves in the red portion of the spectrum are reflected—all others (and thus all other colors) are absorbed. Similarly, when white light passes through a piece of transparent red glass, or a filter, only the red rays are transmitted. This occurs because the color pigments in the glass are selective absorbers.

Pigments are mixed to form various colors, such as in the production of paints and dyes. You are probably aware that mixing yellow and blue paints produces green. This is because the yellow pigment absorbs most of the wavelengths except those in the yellow and nearby regions of the visible spectrum, the blue pigment absorbs most of the wavelengths except those in the blue and nearby regions. The wavelengths in the intermediate green region, adjacent to both the yellow and blue wavelengths, are not strongly absorbed by either pigment, and therefore the mixture appears green. The same effect may be accomplished by passing white light through stacked yellow and blue filters. The light coming through both filters appears green.

Mixing pigments results in the subtraction of colors. The color that is perceived is created by whatever is *not* absorbed by the pigment—that is, not sub-

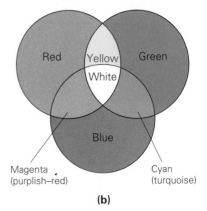

(a)

(b)

•**FIGURE 25.20 Additive method of color production**
When light beams of the primary colors (red, blue, and green) are projected onto a white screen, mixtures of them produce other colors. Varying the intensities of the beams allow most colors, or hues, to be produced.

Note: Recall from Chapter 22 that the colors of the visible spectrum may be remembered by ROY G. BIV.

*From "Experiments in Color Vision," by Edwin H. Land, in *Scientific American*, May 1959, pp. 84–99.

tracted from the original beam. This is the principle of what is called the **subtractive method of color production** or mixing. Three particular pigments—cyan, magenta, and yellow—are called the **subtractive primary pigments** (or subtractive primaries). Various combinations of two of the three subtractive primaries produce the three additive primary colors (red, blue, and green), as illustrated in ●Fig. 25.21. When the subtractive primaries are mixed in the proper proportions, the mixture appears black (all wavelengths are absorbed). Painters often refer to the subtractive primaries as red, yellow, and blue. They are loosely referring to magenta (purplish-red), yellow, and cyan ("true" blue). Mixing these paints in the proper proportions produces a broad spectrum of colors.

Note in Fig. 25.21 that the magenta pigment essentially subtracts the color green where it overlaps with cyan and yellow. As a result, magenta is sometimes referred to as "minus green." If a magenta filter were placed in front of a green light, no light would be transmitted. Similarly, cyan is called "minus red," and yellow is called "minus blue."

An example of subtractive color mixing is a photographer's use of a yellow filter to bring out white clouds on black and white film. This filter absorbs blue from the sky, darkening it relative to the clouds, which reflect white light. Hence, the contrast between the two is enhanced. What type of filter would you use to darken green vegetation on black and white film? To lighten it?

Visual Images and Illusions

Everything you see and therefore most of what you know depends on the information conveyed through the tiny pupils of your eyes. Thus, the better you understand the function of your eyes, the better you can interpret your visual observations. Even though you often hear people say "seeing is believing," you can't take this statement too literally. For example, from the discussion on converging lenses, you know that the images of objects formed on the retinas of your eyes are inverted. But you don't perceive the world as upside down—by

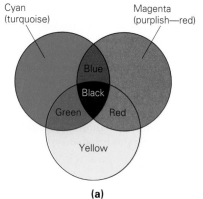

(a)

●**FIGURE 25.21 Subtractive method of color production**
(a) When the primary pigments (cyan, magenta, and yellow) are mixed, different colors are produced by subtractive absorption; for example, the mixing of yellow and magenta produces red. When all three pigments are mixed and all the wavelengths of visible light are absorbed, the mixture appears black. **(b)** Subtractive color mixing using filters. The principle is the same as in part (a). Each pigment selectively absorbs certain colors, removing them from the white light. The colors that remain are what we see.

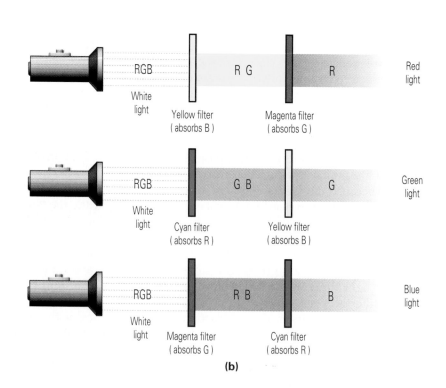

(b)

some means, the brain interprets the inverted image of the world as being right side up.

Knowledge of the eye will help you understand some of the things you see and do not see. For example, look intensely at the black cross in ●Fig. 25.22a with this book held at arm's length and your right eye closed. Then slowly bring the book toward you. At certain points, you will see the square and dot alternately disappear and reappear. Is there something wrong with your vision? No, the experiment demonstrates the eye's blind spot. This is the region where the optic nerve attaches to the retina and where there are no rods or cones. Consequently, there is no optical response. You don't ordinarily notice the blind spot because of movements of the eye and objects and because you have binocular vision.

The stimulation of one area of the retina may affect the sensations in an adjacent area, producing an illusion. For example, if you look at the pattern in ●Fig. 25.22b, you will probably see fleeting patches of gray where the white lines intersect between the black diamonds. Also, stimulation of a particular area of the retina may be "remembered" after your gaze has shifted to another object. Look intensely at the white dot in the black Y in ●Fig. 25.22c for about 30 seconds. Then transfer your gaze to the white dot between the E's. You will see a ghostly white Y between the E's. This is an example of an afterimage. Afterimages are commonly seen when you shift your eyes from a bright object to a dark background, for example, looking at a light in a darkened room and then looking away.

Keep in mind that there are many ways people can be visually fooled. For example, your depth perception is tricked in 3-D movies when objects give the illusion of jumping out of the screen. Other optical illusions are caused by the geometry of the situation in which they occur. Several common illusions are shown in the adjacent Learn by Drawing feature. How does your eye (or brain) interpret them?

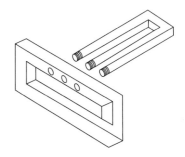

Is something dimensionally wrong here?

Parallel diagonal lines?

Big, bigger, biggest?

Which line at right connects with the one at left?

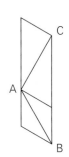

Is line AB shorter than line AC?

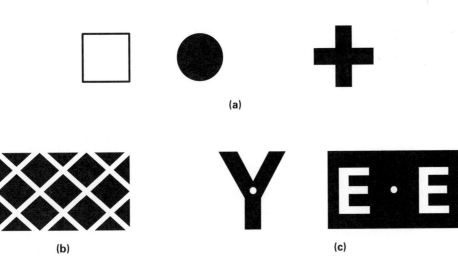

(a)

(b)   **(c)**

●**FIGURE 25.22 Visual images**
(a) To experience the blind spot, hold the book at arm's length and with the right eye closed, look intently at the black cross. Then bring the book slowly toward your face. The square and dot will alternately disappear and reappear. **(b)** Induced stimulation causes fleeting patches of gray to appear on the white areas between the points of the black diamonds. **(c)** Similarly, an image of a Y can be induced between the E's by looking intently at the white dot in the black Y, and then shifting your gaze to the white dot between the E's.

Chapter Review

Important Terms

crystalline lens   768
retina   768
rods   768
cones   768
accommodation   769
far point   769
near point   769
nearsightedness   769
farsightedness   770
astigmatism   772

magnifying glass (simple
  microscope)   773
angular magnification   773
compound microscope   775
objective   775
eyepiece (ocular)   775
refracting telescope   777
astronomical telescope   778
terrestrial telescope   778
reflecting telescope   781

resolution   784
Rayleigh criterion   784
resolving power   785
additive primary colors   789
additive method of color
  production   789
complementary colors   789
subtractive method of color
  production   790
subtractive primary pigments
  790

Important Concepts

- Nearsighted people cannot see distant objects clearly. Farsighted people cannot see nearby objects clearly. These conditions may be corrected by diverging and converging lenses, respectively.

- The magnification of a magnifying glass (or simple microscope) is expressed in terms of angular magnification (m), as distinguished from the lateral magnification (M, Chapter 23).

- The objective of a compound microscope has a relatively short focal length, and the eyepiece or ocular has a longer focal length. Both contribute to the magnification.

- A refracting telescope uses a converging lens to gather light, and a reflecting telescope uses a converging mirror. The image created by either one is magnified by the eyepiece.

- Diffraction places a limit on the ability to distinguish or resolve objects that are close together. Two images are said to be just resolved when the central maximum of one image falls on the first minimum of the diffraction pattern of the other image (Rayleigh criterion).

- Color is our brain's interpretation of the frequency of light.

Important Equations for Review

Angular Magnification:

$$m = \frac{\theta}{\theta_o} \qquad (25.1)$$

Magnification of a Magnifying Glass With Image at Near Point (25 cm):

$$m = 1 + \frac{25 \text{ cm}}{f} \qquad (25.3)$$

Magnification of a Magnifying Glass With Image at Infinity:

$$m = \frac{25 \text{ cm}}{f} \qquad (25.4)$$

Magnification of a Compound Microscope (with L and f's in centimeters):

$$M_t = M_o m_e = \frac{(25 \text{ cm})L}{f_o f_e} \qquad (25.5)$$

Magnification of a Refracting Telescope:

$$m = -\frac{f_o}{f_e} \qquad (25.6)$$

Resolution for a Slit:

$$\theta_{\min} = \frac{\lambda}{d} \qquad (25.7)$$

*Resolution for a Circular Aperture: (diameter **D**)*

$$\theta_{\min} = \frac{1.22 \, \lambda}{D} \qquad (25.8)$$

Resolving Power:

$$s = f\theta_{\min} = \frac{1.22 \, \lambda f}{D} \qquad (25.9)$$

Exercises

25.1 The Human Eye*

1 The rods of the retina (a) are responsible for 20/20 vision, (b) are responsible for twilight vision, (c) are responsible for color vision, (d) focus light.

2 A vision defect that commonly occurs due to normal aging is (a) astigmatism, (b) nearsightedness, (c) farsightedness, (d) accommodation.

3 The focal length of the crystalline lens of the human eye varies with muscle action. Discuss the shape (curvature) of the lens when looking at distant and close objects.

4 Which ratio indicates better vision, 20/15 or 15/20? Explain.

5 When using a flash camera, people in photographs often have "red eye." This is because of light reflected from the retina which is red because of blood vessels near the surface. Some cameras have an anti-red eye option, which when activated simply gives a quick flash before the longer picture-taking flash. Explain how this reduces red eye.

6 ■■ The far point of a certain nearsighted person is 300 cm. What type of lens with what focal length will allow this person to see distant objects clearly?

7 ■■ A certain farsighted person has a near point of 50 cm. What type and power of lens should an optometrist prescribe to enable the person to see objects clearly as close as 25 cm?

8 ■■ If the normal near point of the crystalline lens is taken to be 25 cm, what is the power of the lens? (Assume all refraction is done by the lens and that it is thin and spherical. Need more data? See text.)

9 ■■ A woman cannot see objects clearly when they are farther than 12.5 m away. (a) Is she nearsighted or farsighted? (b) What type of lens of what power (in diopters) will allow her to see distant objects clearly?

10 ■■ A farsighted professor can just see the print in a book clearly when holding the book at arm's length (0.80 m from the eyes). What type of lens with what focal length will allow the professor to read the text at the normal near point?

11 ■■ To correct a case of hyperopia, an optometrist prescribes positive contact lenses that effectively move the patient's near point from 100 cm to 25 cm. (a) What is the power of the lenses? (b) Will the person be able to

*Assume corrective lenses are in contact with the eye (contact lenses) unless otherwise stated.

see distant objects clearly when wearing the lenses or will she have to take them out?

12 ■■ A farsighted person with a near point of 1.15 m gets contact lenses and can then read a newspaper held at a distance of 25 cm. What is the power of the lenses? (Assume that the lenses are the same for both eyes.)

13 ■■ A nearsighted woman has an uncorrected far point of 200 cm. What type of lens of what power would correct this condition?

14 ■■ A farsighted man is unable to focus on objects nearer than 2.0 m. What type of lens of what power will allow him to focus on the print of a book held 25 cm from his eyes?

15 ■■ A certain myopic person has a far point of 150 cm. (a) What power must a lens have to allow him to see distant objects clearly? (b) If he is able to read print at 25 cm while wearing his glasses, is his near point less than 25 cm? (c) Give an indication of his age with normal recession of near point.

16 ■■ A middle-aged man starts to wear glasses with lenses of +2.0 D that allow him to read a book held as closely as 25 cm. Several years later, he finds that he must hold a book no closer than 33 cm to read it clearly using the same glasses, so he gets new glasses. What is the power of the new lenses? (Assume both lenses are the same.)

17 ■■ A nearsighted person has a far point located 7.5 m from one eye. (a) If a corrective lens is worn 2.0 cm from the eye, what would be the power of the lens necessary for the person to see distant objects? (b) What would be the power if a contact lens were used?

18 ■■■ A college professor can see objects clearly only if they are between 70 and 500 cm from her eyes. Her optometrist prescribes bifocals that enable her to see distant objects through the top half of the lenses and read students' papers at a distance of 25 cm through the lower half. What are the powers of the bifocal lenses? (Assume both lenses are the same.)

19 ■■■ On being examined by an optometrist, it is found that a person's near point has receded from 40 cm to 50 cm. What is the difference in the powers of the lenses prescribed so that the person would be able to read a newspaper at a distance of 25 cm? (Assume the same near point for both eyes.)

20 ■■■ Bifocal glasses are used to correct a person's nearsightedness and farsightedness (•Fig. 25.23). If the person's near points in the right and left eyes are 35.0 cm and 45.0 cm, respectively, and the far point is 220 cm for both eyes, what are the powers of the lenses pre-

Nearsightedness
correction

Farsightedness
correction

• **FIGURE 25.23 Bifocals**
See Exercise 20.

scribed for the glasses? (Assume the glasses are worn 3.00 cm from the eyes.)

21 ■■■ The eye of a nearsighted person cannot see objects farther away than 4.0 m. A contact lens is prescribed to correct the defect. (a) What type of lens of what power would allow the person to see distant objects in focus? (b) Show that treating the contact lens and the crystalline lens of the eye as a pair of lens in contact gives the same result.

25.2 Microscopes*

22 A magnifying glass (a) is a concave lens, (b) allows a clear image to be formed of an object that is closer than the near point, (c) magnifies by effectively increasing the angle the object subtends, (d) both (b) and (c).

23 A compound microscope has (a) unlimited magnification, (b) two lenses of the same focal length, (c) a diverging objective lens, (d) an eyepiece of relatively long focal length.

24 With an object at the focal point of a magnifying glass, the magnification is given by $m = 25$ cm$/f$ (Eq. 25.4). According to this, the magnification could be increased indefinitely by using lenses with shorter focal lengths. Why, then, have compound microscopes?

25 ■ A diamond cutter uses a jeweler's lens to examine a diamond. If the focal length of the lens is 16 cm, what is the maximum angular magnification of the diamond?

26 ■ Using the small-angle approximation, compare the angular sizes of a car 2.0 m in height when at distances of 500 m and 1025 m.

27 ■ An object is placed 10 cm in front of a converging lens with a focal length of 18 cm. What are (a) the lateral magnification and (b) the angular magnification?

28 ■ A physics student uses a converging lens with a focal length of 15 cm to read a small measurement scale. (a) What is the maximum magnification that can be obtained? (b) What is the magnification for viewing of the relaxed eye?

29 ■■ A student uses a magnifying glass to examine the details of a microcircuit in the lab. If the lens has a focal length of 8.0 cm and a virtual image is formed at the student's near point (25 cm), (a) how far from the circuit is the lens held, and (b) what is the magnification?

30 ■■ A detective looks at a fingerprint with a magnifying glass whose power is +2.5 D. What is the maximum magnification of the print?

31 ■■ A lens with a power of +10 D is used as a simple microscope. (a) How close can an object be brought to the eye to be examined through the lens? (b) What is the angular magnification at this point?

32 ■■ What is the maximum magnification of a magnifying glass with a power of +3.0 D for (a) a person with a near point of 25 cm and (b) a person with a near point of 10 cm?

33 ■■ The focal length of the objective lens of a compound microscope is 4.5 mm. The eyepiece has a focal length of 3.0 cm. If the distance between the lenses is 18 cm, what is the magnification of a viewed image?

34 ■■ A compound microscope has a 15-cm barrel and an ocular with a focal length of 8.0 mm. What power should the objective have to give a total magnification of 360×?

35 ■■ The tube of a compound microscope is 16 cm long. The focal length of the eyepiece is 3.0 cm, and the focal length of the objective is 0.45 cm. Find the total magnification of the microscope.

36 ■■ A compound microscope has an objective lens with a focal length of 0.50 cm and an ocular with a focal length of 3.25 cm. The separation distance between the lenses is 22 cm. For a person with a normal near point using the microscope, (a) what is the total magnification? (b) Compare this (as a percentage) with the magnification of the eyepiece alone as a simple magnifying glass.

37 ■■ A 1500× microscope has an objective whose focal length is 0.75 cm. If the tube of the microscope is 20 cm long, find the focal length of the eyepiece.

38 ■■ A specimen is 5.0 mm from the objective of a compound microscope, which has a power of +250 D. What must the magnifying power of the eyepiece be if the total magnification of the specimen is 100×?

*The normal near point should be taken as 25 cm unless otherwise specified.

39 ■■■ Referring to •Fig. 25.24, show that the magnifying power for a magnifying glass held at a distance d from the eye is given by

$$m = \left(\frac{25}{f}\right)\left(1 - \frac{d}{D}\right) + \frac{25}{D}$$

when the actual object is located at the near point (25 cm). [*Hint:* Use a small-angle approximation and note that $y_i/y_o = d_i/d_o$ by similar triangles.]

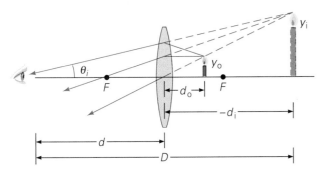

•FIGURE 25.24 Power of a magnifying glass
See Exercise 39.

40 ■■■ A magnifying glass with a focal length of 10 cm is held 4.0 cm from the eyes to view the small print of a book. What is the magnification if the magnifying glass is 5.0 cm from the book? [*Hint:* See Exercise 39.]

41 ■■■ The objective of a compound microscope has a focal length of 4.0 mm and is located 4.5 mm from a specimen. The specimen is viewed through an eyepiece ($f_e = 3.0$ cm) adjusted for minimum eye strain. (a) What is the magnification of the specimen? (b) How far apart are the lenses?

42 ■■■ A modern microscope is equipped with a turret having three objectives with focal lengths of 16 mm, 4.0 mm, and 1.6 mm and interchangeable eyepieces of 5.0× and 10×. A specimen is positioned so each objective produces an image 150 mm from the objective. What are the least and greatest magnifications possible?

25.3 Telescopes

43 An inverted image is produced by (a) a terrestrial telescope, (b) an astronomical telescope, (c) a Galilean telescope, (d) all of these.

44 Compared to large refracting telescopes, large reflecting telescopes have the advantage of (a) greater light-gathering capability, (b) freedom from chromatic aberration, (c) lower cost, (d) all of these.

45 ■ Find the magnification of a telescope whose objective has a focal length of 50 cm and whose eyepiece has a focal length of 2.7 cm.

46 ■ An astronomical telescope has an objective and an eyepiece whose focal lengths are 60 cm and 15 cm, respectively. What are (a) the magnifying power and (b) the length of the telescope?

47 ■■ A telescope has an ocular with a focal length of 10 mm. If the length of the tube is 1.5 m, what is the angular magnification of the telescope when it is focused for an object at infinity?

48 ■■ In a terrestrial telescope, an objective and eyepiece with focal lengths of 45 cm and 15 cm, respectively, are used. What should the focal length of the erecting lens be if the overall length of the telescope is to be 0.75 m?

49 ■■ An astronomical telescope has objective and eyepiece lenses with focal lengths of 87.5 cm and 7.50 mm, respectively. (a) What is the magnification of the telescope, and (b) what is its approximate length?

50 ■■ The three lenses of a terrestrial telescope have focal lengths of 40 cm, 20 cm, and 15 cm for the objective, erecting lens, and eyepiece, respectively. (a) What is the magnification of the telescope for an object at infinity? (b) What is the length of the telescope? (c) Does the erecting lens affect the magnification of the telescope? Explain.

51 ■■ Two astronomical telescopes have the following characteristics:

Telescope	Objective focal length (cm)	Eyepiece focal length (cm)	Objective diameter
1	90.0	0.84	75
2	85.0	0.77	60

Which telescope would you choose (a) for magnification? (b) For resolution?

52 ■■■ Referring to •Fig. 25.25, show that the angular magnification of a refracting telescope focused for the final image at infinity is $m = f_o/f_e$. (Since telescopes are designed for viewing distant objects, the angular size of an object viewed with the unaided eye is the angular size of the object at its actual location rather than at the near point, as is true for a microscope.)

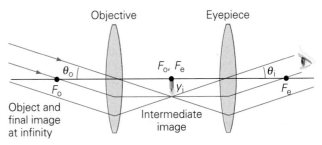

•FIGURE 25.25 Angular magnification of a refracting telescope
See Exercise 52.

25.4 Diffraction and Resolution

53 The images of two sources are said to be resolved when (a) the central maxima of the diffraction patterns fall on each other, (b) the first bright fringes of the diffraction patterns fall on each other, (c) the central maximum of one diffraction pattern falls on the first dark fringe of the other, (d) none of these.

54 For a telescope with a circular aperture, the minimum angle of resolution (a) is greater for red light than for blue light, (b) is independent of the frequency of the light, (c) is directly proportional to the radius of the aperture, (d) is independent of the area of the aperture.

55 The purpose of using oil-immersion lenses on microscopes is to reduce (a) the spherical aberration, (b) the useful magnification, (c) the chromatic aberration, (d) the wavelength of light so as to increase the resolving power.

56 ■ According to the Rayleigh criterion, what is the minimum angle of resolution for two point sources of red light ($\lambda = 680$ nm) in the diffraction pattern produced by a single slit with a width of 0.35 mm?

57 ■ The minimum angular separation of the images of two monochromatic point sources in a single-slit diffraction pattern is 0.0055 rad. If a slit width of 0.10 mm is used, what is the wavelength of the sources?

58 ■ What is the resolution limit due to diffraction (with no atmosphere) for the 40-in. (102-cm) Yerkes Observatory refracting telescope for light with a wavelength of 550 nm?

59 ■ What is the resolution due to diffraction (with no atmosphere) for the Hale telescope at Mount Palomar with its 200-in. diameter mirror for light with a wavelength of 550 nm? Compare this to the resolution limit for the Yerkes Observatory telescope.

60 ■■ The maximum diameter of the eye's pupil is about 8.0 mm. What is the minimum angle of separation for two, 550-nm sources from which light passes through the pupil of the eye?

61 ■■ Two moons of Jupiter are 4.0×10^8 km away from Earth and at a separation distance of 3.0×10^5 km. Based on the result of Exercise 60, would a person with normal vision be able to distinguish both moons with the unaided eye? (Assume the moons reflect sufficient light and their observation is not restricted by the planet and ignore Earth's atmosphere.)

62 ■■ The minimum angular separation of two stars emitting red light ($\lambda = 680$ nm) for a particular telescope is 5.6×10^{-6} rad. What is the diameter of the telescope's lens?

63 ■■ A refracting telescope with a lens whose diameter is 30.0 cm is used to view a binary star system that emits light in the visible region. (a) What is the minimum angular separation of the pair of stars for them to be barely resolved? (b) If the binary star is a distance of 6.00×10^{20} km from the Earth, what is the lateral separation between the stars of the pair? (Assume that a line joining the stars is perpendicular to our line of sight.)

64 ■■ The objective of a microscope is 2.50 cm in diameter and has a focal length of 30.0 mm. (a) If yellow light with a wavelength of 570 nm is used to illuminate a specimen, what is the minimum angular separation of two fine details of the specimen for them to be just resolved? (b) What is the resolving power of the lens?

65 ■■ For the microscope in Exercise 64, how great a difference is there in (a) the minimum angular separation and (b) the resolving power for blue light ($\lambda = 400$ nm) and red light ($\lambda = 700$ nm)?

66 ■■■ A microscope with an objective 1.20 cm in diameter is used to view a specimen using light from a mercury source with a wavelength of 546.1 nm. (a) What is the limiting angle of resolution? (b) What color of light in the visible spectrum would give the maximum limit of resolution? (c) If an oil-immersion lens were used ($n_{oil} = 1.40$), what would be the change (expressed as a percentage) in the resolving power?

25.5 Color and Optical Illusions

67 An additive primary color is (a) blue, (b) green, (c) red, (d) all of these.

68 A subtractive primary pigment is (a) cyan, (b) yellow, (c) magenta, (d) all of these.

69 Describe how the American flag would appear if it were illuminated with light of each of the primary colors.

70 Can white be obtained by the subtractive method of color production? Explain. It is sometimes said that black is the absence of all colors, or that a black object absorbs all incident light. If so, why do we see black objects?

71 Several beverages, such as root beer, when poured into a glass develops a "head" of foam. Why is the foam generally white, while the liquid is dark or colored?

72 White light is incident on two filters as shown in ●Fig. 25.26. Complete the color rays to show what color light emerges from the yellow filter.

Additional Exercises

73 A myopic person has glasses with a lens that corrects for a far point of 130 cm. What is the power of the lens?

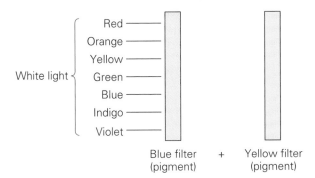

●**FIGURE 25.26 Color absorption**
See Exercise 72.

74 A man cannot see objects clearly when they are closer than 250 cm. (a) Is he nearsighted or farsighted? (b) What type of lens of what power (in diopters) will allow him to see objects clearly at 25 cm?

75 A farsighted person wears glasses whose lenses have a power of +1.25 D and is then able to see objects clearly as close as 25 cm. What is the person's near point?

76 A nearsighted person wears glasses whose lenses have a power of −0.15 D. What is the person's far point?

77 When viewing an object with a magnifying glass whose focal length is 20 cm, a person positions the lens so there is minimum eyestrain. What is the observed magnification?

78 Light from two monochromatic 550-nm point sources is incident on a slit that is 0.050 mm wide. What is the minimum angle of resolution in the diffraction pattern formed by the slit?

79 A compound microscope has an objective with a focal length of 4.00 mm and an eyepiece with a magnification of 10.0×. If the objective and eyepiece are 15.0 cm apart, what is the total magnification of the microscope?

80 A person using a magnifying glass with a focal length of 12 cm views an object at the focal point of the lens. What is the magnification?

81 A refracting telescope has an objective with a focal length of 50 cm and an eyepiece with a focal length of 15 mm. The telescope is used to view an object that is 10 cm high and located 50 m away. What is the apparent angular height of the object as viewed through the telescope?

82 An eyeglass lens with a power of +2.8 D allows a far-sighted person to read a book held at a distance of 25 cm from her eyes. At what distance must the person hold the book to read it without glasses?

83 A student views the details of a dollar bill with a magnifying glass, achieving its maximum magnification of 3×. What is the focal length of the magnifying glass?

84 A microscope whose tube is 15 cm long has objective and eyepiece lenses with focal lengths of 7.5 mm and 10 mm, respectively. What is the total magnification of the microscope?

85 The amount of light reaching the film in a camera depends on the lens aperture (the effective area) as controlled by the diaphragm. The f number is the ratio of the focal length of the lens to its effective diameter. For example, an f/8 setting means that the diameter of the aperture is 1/8 of the focal length of the lens. The lens setting is commonly referred to as the f-stop. (a) Determine how much light each of the following lens settings will admit to the camera compared to f/8: (1) f/3.2 and (2) f/16. (b) The exposure time of a camera is controlled by the shutter speed. If a photographer correctly uses a lens setting of f/8 with a film exposure time of 1/60 s, what *exposure* time should he use to get the same amount of light exposure if he sets the f-stop at f/5.6?

86 From a spacecraft in orbit 150 km above the Earth's surface, an astronaut wishes to observe her home town as she passes over it. What features will she be able to identify with the naked eye? (Hint: Estimate the diameter of the human iris.)

26 Relativity

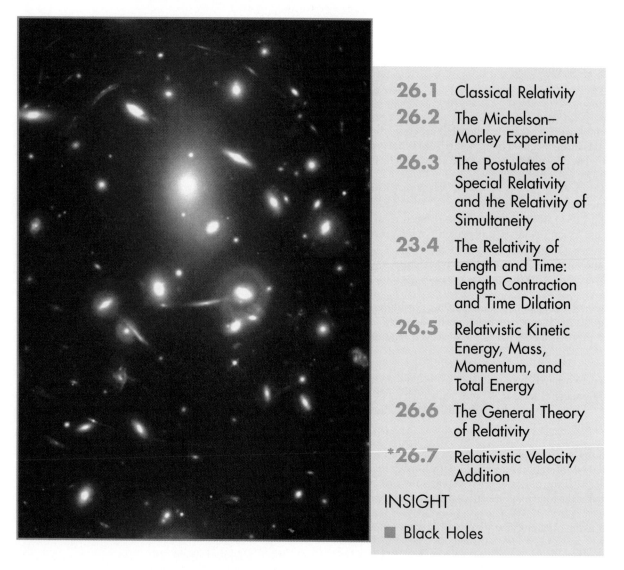

Y(ou might not think so at first glance, but this photograph tells us something very remarkable about the universe we inhabit. The bright shapes are galaxies, each consisting of many billions of stars—trillions, probably, for the larger ones. They are very far away from us—billions of light years (and recall that a light year is nearly 10 trillion kilometers). Mind-boggling numbers—yet these galaxies represent merely the *foreground* of the picture. The faint arcs, which make the photograph resemble a spider's web, represent light from galaxies far more distant still!

What is remarkable about these fine wisps of light, however, is not the billions of years that they have been traveling to reach us, but the path they have taken. We all know that light travels in straight lines—that is why the most accurate surveying instruments now use laser light. Yet the light from these galaxies has been warped by the gravitational field of the cluster of galaxies in the foreground, creating the filigree of arcs that we see in the photo.

The fact that light can be affected by gravity in this way was predicted by Albert Einstein as a consequence of his theory of relativity. In this chapter you'll find out what else this theory predicted—and how its predictions, even those that at first seemed most at odds with everyday experience, have been confirmed.

There was so much important scientific activity in so many areas at the beginning of the twentieth century that the study of modern physics could begin with any one of a variety of topics. We begin here with the fascinating and far-reaching subject of relativity, which was an outgrowth of nonclassical observations and thought.

Relativity originated from the analysis of physical phenomena when speeds approach that of light. This was not a subject of study in classical Newtonian kinematics and dynamics because objects with speeds of these magnitudes cannot ordinarily be observed. However, some particles in nature do travel at speeds comparable to the speed of light and others can now be accelerated to such speeds in particle accelerators.

When such speeds are obtained, measurements show changes in fundamental properties such as length, time, and mass. It is difficult to imagine observing metersticks becoming shorter, clocks running slower, or the masses of objects increasing—all because of traveling at very fast speeds—but this is what we observe. Indeed, relativity created a revolution in physics by causing us to rethink our basic understanding of space, time, and gravitation. It successfully challenged Newtonian concepts that had dominated scientific thinking for 250 years.

The impact of relativity has been especially great in those branches of science concerned with the extremes of physical reality—the subatomic realm (discussed in Chapters 29 and 30), where time intervals and distances are almost inconceivably small, and the cosmic realm, where time intervals and distances are almost unimaginably large. All modern theories about the birth, evolution, and ultimate fate of our universe are inextricably linked to our understanding of relativity.

In this chapter, you will learn how Einstein's theory of relativity explains the changes in length and time that we observe in rapidly moving objects, the equivalence of energy and mass, and the bending of light rays by gravitational fields—phenomena that appear very strange from the classical viewpoint. We will begin by exploring some of the deficiencies in classical concepts, and then describe how new ideas were developed.

26.1 Classical Relativity

Objectives: To be able to (a) summarize the concepts of classical relativity and relative velocities, (b) define inertial and noninertial reference frames, and (c) explain the reasoning behind the ether hypothesis.

Physics is concerned with the description of the world around us, and this endeavor depends on observations and measurements. We expect some aspects of nature to be consistent and unvarying. That is, the ground rules by which nature plays are consistent, and our descriptions of nature or physical principles do not change from observation to observation. We emphasize this fact by referring to such principles as "laws"—for example, the laws of motion. Not only have physical laws proved valid over time, but they are the same for all observers.

This means that a physical principle or law should be the same regardless of the observer's frame of reference. When we make a measurement or perform an experiment, we do so with reference to a particular frame or coordinate system, usually the laboratory, which is considered to be at rest. The experiment may also be observed by a person passing by (in motion). On comparing notes, the experimenter and the observer should find the results of the experiment or the physical principles involved to be the same in both frames of reference or coordinate systems. Basically, you can't change the laws of nature by the way you observe them. Measured quantities may vary, and descriptions may be different, but the laws that these quantities obey must be the same for all observers.

For example, suppose that you observe two cars traveling in the same direction on a straight road at speeds of 60 km/h and 90 km/h. Even though we rarely

Note: Review the discussion of relative velocities in Section 3.3.

Note: The relationship between Newton's first and second laws is discussed in Section 4.3.

Note: Newtonian relativity refers *only* to laws of mechanics.

Note: Maxwell's equations are discussed in Sec. 20.4

Note: More precisely, the speed of light has been measured as 2.99792458×10^8 m/s. We will assume c to be *exact* at 3.00×10^8 m/s for significant figure purposes.

say it, it is clear that these speeds are measured relative to your reference frame, that fixed to the ground. However, a person in the car traveling at 60 km/h observes the other car in front of it traveling at a speed of 30 km/h, *relative to her reference frame*—the car in which she is riding. (What does someone in the car traveling at 90 km/h observe?) In general, what one observes is *relative velocity*—that is, the velocity relative to the observer's frame of reference.

In measuring relative velocity, there seems to be no "true" rest frame. Any reference frame may be considered to be at rest *relative* to another. We can, however, make a distinction between what are called inertial and noninertial reference frames. In fact, an **inertial reference frame** is defined as one in which Newton's first law of motion holds. That is, in an inertial system, an *isolated* object (one on which there is no net force) is stationary or moves with a constant velocity. If Newton's first law holds, then his second law in its customary $\mathbf{F} = m\mathbf{a}$ form applies to the system.

A noninertial reference frame, on the other hand, is one in which an *isolated* object appears to be accelerating. Note, however, that it is actually the frame that is accelerating, not the object. If an isolated object is observed from an accelerated reference frame, then Newton's second law, $\mathbf{F} = m\mathbf{a}$, cannot be applied directly. One such everday example of a noninertial reference frame is an automobile. When you accelerate from a stop sign, a coffee cup perched on the dashboard may *appear*, when viewed from the car's noninertial reference frame, to accelerate backwards without being acted upon by any evident force. From the inertial frame of a sidewalk observer, however, the cup simply tends to stay put (in accordance with the first law) as the car accelerates out from under it. Here we will deal only with inertial reference frames, as we have done up to this point.

It follows that any reference frame moving with a constant velocity relative to an inertial reference frame is also an inertial frame. Given a constant relative velocity, no acceleration effects are introduced in comparing one frame to another. In such cases, $\mathbf{F} = m\mathbf{a}$ can be used by observers in *either* frame (with their own coordinate values) to analyze a dynamic situation, and both observers will come up with similar results. That is, the law will hold in both frames. Thus, there is no one inertial frame of reference that is preferred over another for the laws of mechanics. This is sometimes called the **principle of classical,** or **Newtonian, relativity:**

The laws of mechanics are the same in all inertial reference frames.

The Absolute Reference Frame: the Ether

In the late 1800s, with the development of the theories of electricity and magnetism, some serious questions arose. Maxwell's equations predicted light to be an electromagnetic wave that traveled with a speed of 3.00×10^8 m/s in vacuum. But *relative to what reference frame* does light have this speed? Classically, the speed of light measured from different frames of reference would be expected to differ—possibly to be even greater than 3.00×10^8 m/s if you were approaching the light wave—by simple vector addition of velocities.

Consider the situation illustrated in •Fig. 26.1. If a person in a reference frame moving in relation to another with a constant velocity $\mathbf{v}'$, threw a ball with a velocity $\mathbf{v}_b$, then the "stationary" observer would say that the ball had a velocity of $\mathbf{v} = \mathbf{v}' + \mathbf{v}_b$, relative to the stationary reference frame. For example, suppose a truck were moving with a speed of 20 m/s relative to the ground and a ball were thrown with a speed of 10 m/s relative to the truck in the direction of the truck's motion. Then the ball would have a speed of 20 m/s + 10 m/s = 30 m/s relative to the ground.

Now suppose that a person on the truck turned on a flashlight, projecting a beam of light. In this case, we have $\mathbf{v} = \mathbf{v}' + \mathbf{c}$, and the speed of light measured

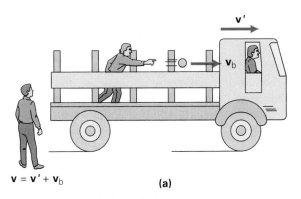
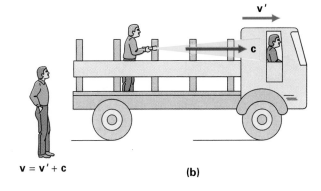

$\mathbf{v} = \mathbf{v}' + \mathbf{v_b}$ **(a)**

$\mathbf{v} = \mathbf{v}' + \mathbf{c}$ **(b)**

•**FIGURE 26.1 Relative velocity**
(a) According to a stationary observer on the ground, $\mathbf{v} = \mathbf{v}' + \mathbf{v_b}$. **(b)** Similarly, the velocity of light would classically be measured to be $\mathbf{v} = \mathbf{v}' + \mathbf{c}$, with a magnitude greater than c. (Velocity vectors not drawn to scale—why?)

by an observer on the ground would be expected to be greater than 3.00×10^8 m/s. Classically then, the measured speed of light is expected to have almost any value, depending on the observer's frame of reference.

It was reasoned, therefore, that the particular speed of $c = 3.00 \times 10^8$ m/s must be referenced to a particular frame, just as the speed of sound at room temperature, about 345 m/s, is referenced to the still air frame. Moreover, as a wave, light was thought to require some medium of transport, just as sound or water waves do. This was quite natural, since experience showed that *all* other physical disturbances (waves) must be transmitted through the stress or distortion of some medium. Since the Earth receives light from the Sun and from distant stars and galaxies, it was thought that this special light-transporting medium in which $c = 3.00 \times 10^8$ m/s must permeate all space. The medium was given the name *luminiferous ether* and referred to simply as **ether**.

The idea of an undetected ether became popular in the latter part of the nineteenth century. Maxwell, whose work laid the foundations for our understanding of electromagnetic waves, believed in the necessity of an etherlike substance:

> Whatever difficulties we may have in forming a consistent idea of the constitution of the ether, there can be no doubt that the interplanetary and interstellar spaces are not empty, but are occupied by a material substance or body which is certainly the largest, and probably the most uniform body of which we have any knowledge.

It would seem then that Maxwell's equations, which describe the propagation of light, did *not* satisfy the Newtonian relativity principle as did other physical laws. On the basis of the preceding discussion, a preferential or special inertial reference frame would seem to exist, one that could be considered absolutely at rest.

As another example of the apparent violation of the relativity principle by electromagnetic phenomena, consider two charges, stationary in the O reference frame and separated by a distance r, as illustrated in •Fig. 26.2. In this frame, an observer would measure a force of repulsion between the charges as given by Coulomb's law ($F = kq_1q_2/r^2$) and no magnetic force. (Why?)

Now suppose the charges were investigated by an observer in the O' frame moving with a velocity $\mathbf{v}$ relative to the O frame and perpendicularly to a line separating the charges. This observer would see the two charges approaching with a velocity $-\mathbf{v}$. The repulsive Coulomb force would be evident, as it was to the observer in the O frame. But, because a moving charge corresponds to a current, the observer in the O' system would also note attractive *magnetic* forces between the charges (similar to the forces for parallel current-carrying wires). In other words, observers in different inertial frames would measure *different* effects for the same situation, and the physical laws would not be the same in all iner-

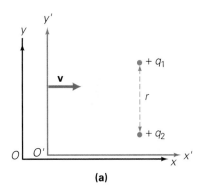

(a)

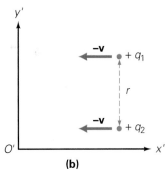

(b)

•**FIGURE 26.2 Nonrelativity**
(a) The repulsive forces between the stationary charges in the O reference frame are given by Coulomb's law: $F = kq_1q_2/r^2$.
(b) For an observer in the O' reference frame, moving toward the charges with a velocity $\mathbf{v}$, the charges would appear to be moving towards him. In addition to the repulsive Coulomb forces, there would thus be attractive forces due to the production of and interaction with magnetic fields (similar to the forces between parallel current-carrying wires).

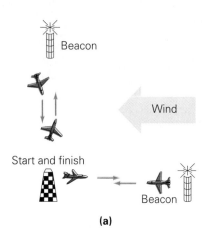

Beacon

Wind

Start and finish

Beacon

(a)

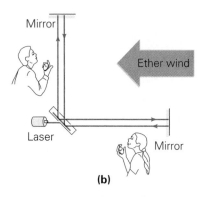

Mirror

Ether wind

Laser

Mirror

(b)

• **FIGURE 26.3 Ether wind detection**

(a) When two airplanes fly with equal air speeds over equal distances, the plane flying perpendicular to the wind direction takes less time to fly a round trip. **(b)** Classically, the ether wind should give rise to a similar difference in the times for light to be reflected back and forth.

Note: Interference is discussed in Sections 13.4, 14.4, and 24.1.

tial reference frames. Here again, the laws of electricity and magnetism (which govern the propagation of light) would seem *not* to satisfy the classical relativity principle.

This was the state of affairs toward the end of the nineteenth century when scientists set out to investigate whether they had come upon a new dimension of physics or perhaps a flaw in what were considered to be established principles. One of the first attempts was to test the ether theory. If it could be proved that the ether existed, then presumably a true absolute rest frame would finally be identified. This was the purpose of the famous Michelson–Morley experiment.

26.2 The Michelson–Morley Experiment

Objectives: **To be able to explain (a) the general concept and operation of the Michelson–Morley experiment, (b) its result, and (c) the effect on the ether concept.**

If the ether permeated all space, as was thought, the Earth itself would be surrounded by the medium. In its orbit around the Sun, the Earth presumably moved through the ether, like an airplane moving through still air. Relative to the airplane, there is air motion or wind, and similarly the orbiting Earth would experience an "ether wind."

If the ether wind existed, this would provide a means to detect the ether experimentally. Observers could carefully measure the speed of light on Earth in several directions, for example, parallel and perpendicular to the orbital velocity of the Earth, and compare the results. Because of the relative motion, the speeds should be different in different directions.

As an analogy, consider two airplanes flying at the same air speed and the same distances back and forth as illustrated in • Fig. 26.3a. The plane flying perpendicular to the wind must be directed slightly into the wind on both legs to stay on a straight-line course, an action that slows the plane somewhat. The other plane flies into the wind on the first leg and with the wind on the return leg. The velocities at which the planes fly relative to the ground are given by vector addition. It is a simple matter to show that the plane flying perpendicular to the wind takes less time to complete the two-leg trip than does the other plane.

A similar effect *should* be seen if the airplanes are replaced by light beams and the atmospheric wind by the ether wind (• Fig. 26.3b). However, this is more easily said than done. The average speed of the Earth in its orbit is about 30 km/s, compared to 300,000 km/s for the speed of light. In the late nineteenth century, the American physicist Albert A. Michelson devised a technique capable of detecting the required differences using the interference of light and designed an extremely sensitive **interferometer**.

The basic principle of Michelson's interferometer is shown in • Fig. 26.4. A monochromatic light beam is incident on a partially silvered glass plate (P), which acts as a beam splitter, dividing the incident beam into transmitted and reflected components. The beams travel to the plane mirrors (M_1 and M_2), and are reflected back toward P. There, part of each beam is reflected and transmitted to the detector screen (D). Another glass plate (P') is inserted in one beam, so both beams pass through an equal amount of glass.

Since the beams are from the same initial beam, they are coherent and, on combining, interfere according to their relative phase relationship. (Recall that the superposition principle can lead to constructive or destructive interference). This is determined by the difference in the path lengths of the beams.

Consider now the interferometer (and the Earth) traveling through the stationary ether with a velocity **v** and the interferometer arranged so that one of its perpendicular arms is parallel to **v** (• Fig. 26.5). The speed of the beam traveling

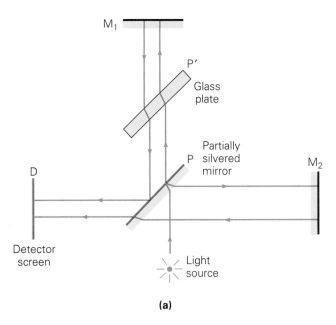

(a)

(b)

●**FIGURE 26.4 Michelson interferometer**
(a) If one of the arms of the interferometer were in the direction of the ether wind, the beams should arrive at the detector out of phase, and an interference pattern should be produced. (The light rays have been offset from one another for clarity.) See text for details. **(b)** A laboratory model of the Michelson interferometer.

to M_2 is $c - v$. On the return trip the relative velocity has a magnitude of $c + v$. The total time for the round trip is

$$t_2 = \frac{L_2}{c - v} + \frac{L_2}{c + v} = \frac{2L_2 c}{c^2 - v^2} \tag{26.1}$$

For the light beam traveling between P and M_1, the speed of light in the direction perpendicular to **v** is $v_\perp = \sqrt{c^2 - v^2}$ (as shown in the vector diagram in the figure). The speed is the same for the return path, so the time for the round trip on this arm is $t_1 = 2d/v_\perp$, or

$$t_1 = \frac{2L_1}{\sqrt{c^2 - v^2}} \tag{26.2}$$

Then, if $L_1 = L_2 = L$, we have with some rearranging of terms

$$\Delta t = t_2 - t_1 = \frac{2L}{c}\left[\frac{1}{1 - (v^2/c^2)} - \frac{1}{\sqrt{1 - (v^2/c^2)}}\right] \tag{26.3}$$

If there were a time difference, the beams would arrive at the detector out of phase, and an interference pattern would be observed. From the spacing of the interference fringes, one could determine the velocity of the light source relative to the ether. Notice, however, that the results depend on the *square* of the ratio of the speed of the Earth (relative to the ether) to the speed of light, a number almost assuredly very small, and especially small when squared. Thus this experimental setup needed to be, and was, very sensitive.

It is difficult to ensure experimentally that $L_1 = L_2$ to the required degree of accuracy. This problem may be resolved by rotating the apparatus by 90°, which

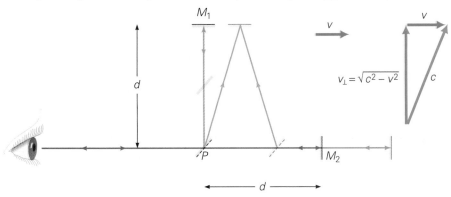

●**FIGURE 26.5 Interferometer light paths**
An interferometer moving through the stationary ether would be observed to have light paths as shown. See the text for description.

interchanges the arms. (With equal arms, the time difference, Δt, when the arms are interchanged becomes the negative of the value obtained using Eq. 26.3.) What would be observed then is a *shift* in the interference pattern or fringes.

Michelson performed this experiment with his colleague E. W. Morley in 1887. The interferometer was sensitive enough to detect the predicted fringe shift. But much to their surprise, Michelson and Morley found *no fringe shift at all*. Perhaps, they thought, the experiment had been done just at a time when the Earth was nearly at rest relative to the ether. To check this possibility, many measurements were made at different times (day and night, and at different seasons), always with the same null result. (Since then, essentially these same experiments have been repeated with better and better accuracy, and the result has never changed.)

Where then was the ether? Several hypotheses were suggested to explain the null result of the Michelson–Morley experiment. One suggested that the ether in the vicinity of the Earth was dragged along with the Earth, so the interferometer was at rest with respect to the ether (that is, $v = 0$). But because of the Earth's orbital motion, the viewing of a star perpendicular to the plane of the orbital plane throughout the year requires the telescope to be consistently tilted in the direction of the Earth's motion. This effect is called the *aberration of starlight* and was explained in 1725. The aberration requires that the Earth move with respect to the ether and *not* be stationary relative to it (as would be the case if the ether were dragged along).

An Irish physicist, George F. FitzGerald, proposed that as a result of the motion through the ether, perhaps the linear dimensions of objects in the direction of motion contracted or became smaller. If this occurred by a fractional amount of $\sqrt{1 - (v^2/c^2)}$ then the null result would be explained. Of course, this hypothesis cannot be experimentally investigated since the measurement instrument would also contract by the same fractional amount.

Another scientist, H. A. Lorentz, who also suggested this possibility, considered it in terms of possible changes in the electromagnetic forces between the atoms of a material due to the motion. This suggested length contraction, which is known as the Lorentz–FitzGerald contraction, might be termed the right idea, but for the wrong reason. In fact, such a length contraction was predicted by a theory proposed by Albert Einstein in 1905. Einstein's idea, however, had to do *not* with a mechanical shrinkage but rather with the fundamentals of the way we measure lengths and, indeed, the very nature of space and time. Einstein's special theory of relativity and some of its predictions, which forever changed the way physicists view space and time, are described in the following sections.

26.3 The Postulates of Special Relativity and the Relativity of Simultaneity

Objectives: **To be able to explain (a) how the two postulates of relativity imply the relativity of simultaneity, and (b) how the relativity of simultaneity leads to length contraction.**

The failure of the Michelson–Morley experiment to detect the ether left the scientific community in a quandary. The inconsistencies between Newtonian mechanics and electromagnetic theory remained unexplained. Many physicists were convinced that the experiments needed to be more accurate; they could not believe that light did not need a medium through which to propogate. These problems were resolved, however, by a theory introduced in 1905 by Albert Einstein (•Fig. 26.6). Interestingly, the Michelson–Morley experiment was apparently *not* a motivation for the development of his theory of relativity. Einstein later could not recall whether he had even known about the experiment at the time he formulated his ideas (while working as a clerk in the Swiss Patent Office!).

Einstein's key insight was based on his conception that the laws of mechanics should not be the only ones to obey the relativity principle. He reasoned that nature should be more symmetrical than that—*all* laws should be included under the relativity principle. In Einstein's view, the inconsistencies in electromagnetic theory were due to the assumption that an absolute rest frame (the ether) existed. His theory did away with the need for an ether by placing all laws of physics on an equal footing, thus eliminating any way of measuring the speed of an inertial reference frame.

The first of the two postulates on which relativity is based is thus an extension of the Newtonian principle of relativity, which applied only to the laws of mechanics. In Einstein's theory, the **principle of relativity** applies to *all* the laws of physics, including those of electricity and magnetism:

> Postulate I (principle of relativity): The laws of physics are the same in all inertial reference frames.

The first postulate implies that all inertial reference frames are equivalent, with physical laws being the same in all of them. There exists no type of experiment that can be performed within a reference frame that would enable an observer in that frame to detect its absolute motion. This means that *there is no such entity as an absolute reference frame* having some unique physical property that would distinguish it from other inertial reference frames. The first postulate seems quite reasonable, since one might expect the basic laws of nature to be the same for all inertial observers. We have no reason to think that nature would play favorites.

Einstein's second postulate involves the speed of light. As he saw it, the problem came from applying the vector addition of velocities to the velocity of light. With the appropriate relative velocity between inertial systems, the observed speed of light could have any value. With $(c + v)$, it could be greater than c; with $(c - v)$, it could be less than c; and with $(c - c)$, it could even reduce to zero! The last possibility was the source of Einstein's question, "What would I see if I rode a beam of light?" In the same frame as the electromagnetic wave, the electric and magnetic field vectors would be seen to vary in space but *not* with time. Wave motion is fundamental to waves, and such static fields were not consistent with electromagnetic theory.

To avoid such inconsistencies, Einstein formulated his second postulate, which can be simply stated as:

> Postulate II (constancy of the speed of light): The speed of light in vacuum has the same value in all inertial systems.

These two postulates form the basis of Einstein's **special theory of relativity**. The "special" designation is to indicate that it deals only with the particular case of inertial reference frames. The *general* theory of relativity, which will be discussed later, deals with noninertial or accelerating reference frames.

The second postulate is perhaps more difficult to accept than the first. It implies that two observers in different inertial reference frames would measure the speed of light to be c independent of the speed of the source and/or the observer. For example, if a person moving toward you with a constant velocity turned on a light, both you and the other person would measure the speed of light to be c, regardless of the magnitude of your relative velocity (•Fig. 26.7).

The second postulate is essential to the validity of the first, and consistent with the null result of the Michelson–Morley experiment. By doing away with the ideas of the ether and an absolute reference frame, Einstein was able to reconcile the apparent differences between mechanics and Maxwell's equations. Even so, the second postulate seemed to go against everyday common sense. But keep in mind that we have very little everyday experience in dealing with velocities at or near the speed of light! The ultimate test of any theory is provided by the scientific method. What does Einstein's theory predict, and has it been experimentally verified? As we shall see, the answer is a resounding "yes" in every experiment performed.

•**FIGURE 26.6 Einstein and Michelson**
A 1931 photo shows Einstein (center) with Michelson (left) during a meeting in Pasadena, California. At right is Robert A. Millikan, the recipient of the Nobel Prize in 1923 for his experimental determination of the value of the electronic charge.

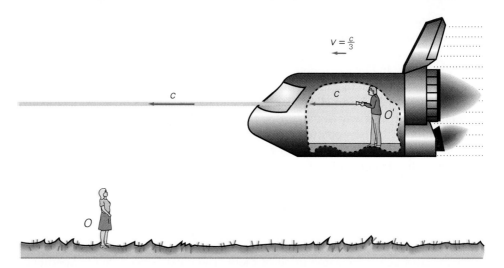

●**FIGURE 26.7 Constancy of the speed of light**
Two observers in different inertial frames measure the speed of the same beam of light. The observer in O' measures a speed of c inside the ship. Classically, we might expect O to measure a speed of $c + c/3 = 4c/3$ as the beam passes her, but she measures c relative to her reference frame, the ground.

The predictions of special relativity can be better understood by imagining simple situations to see what the theory predicts. Einstein did this himself in what he called *gedanken*, or thought, experiments. Let us begin with a series of famous Einstein gedanken experiments related to simultaneity and the fundamental idea of how we measure length and time. We will also look at some experimental evidence that supports the nonclassical predictions of the theory.

The Relativity of Simultaneity

In everyday life we think of two events that are simultaneous to one person as also being simultaneous to all other observers. What could be more obvious? Simultaneous events are those that occur "at the same time," and doesn't that mean the same thing to all observers? The answer is no—at least, not at high speeds!

To see this, let's think of an inertial frame (called O) in which two events are specifically designed to be simultaneous. For example, we could arrange for two small firecrackers (located at points A and B on the x axis) to explode when a switch controlling a voltage source, placed midway between them, is flipped to the "on" position (●Fig. 26.8a). Let us also equip the observer in this frame with a light receptor at point R, midway between the two firecrackers, capable of detecting whether two light flashes do, in fact, arrive at the same time. (We could, of course, place the light receptor at *any* location, but then the reception times would have to be corrected for time differences due to the different distances the light would have to travel. To avoid these unnecessary complications, we will place our simultaneity detector at the midpoint.)

Now let us actually perform the experiment by detonating the two firecrackers. Clearly, the light receptor would tell us that the two explosions went off simultaneously *in the O frame*. But consider the same two explosions as seen by another observer in a different intertial frame, O'. Viewed from O, the O' frame is moving to the right at a speed v relative to O. The observer in O' has equipped himself with a *series* of light receptors on his x' axis, because he is not sure which one will end up midway between the points where the firecrackers explode.

After the explosions, there are burn marks on both the x axis (at A and B) and x' axis (at A' and B'), as shown in Fig. 26.8b. He can use these marks to identify the particular O' light receptor (call it R') that was located midway between A' and B'. But when he reviews the data from this receptor, he finds that it did *not* register that the explosions occurred simultaneously! This fact does not cause the O observer to doubt his conclusion, however—he has a simple explanation of what happened. As he sees the situation, during the time it took for the light to get to the O' receptor, the receptor had moved some distance toward A' and away from B'. Consequently, the O' receptor recorded the flash from A' as arriving before that from B'.

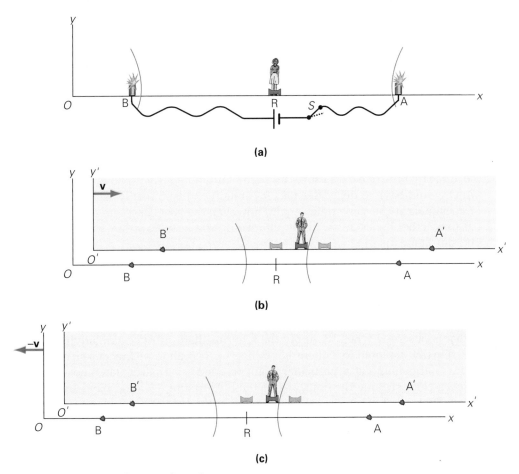

•FIGURE 26.8 The relativity of simultaneity
(a) An observer in reference frame O triggers two explosions (at points A and B) simultaneously. A light receptor R located midway between them records the two light signals as arriving at the same time. **(b)** An observer midway between the two marks, but in frame O', moving with respect to O, sees the burned marks made by the two explosions on the x' axis but sees A happen before B. **(c)** The situation as viewed from O'.

At this point you may be wondering which observer is correct. It's hard to find any objection to the conclusion reached by the observer in O—but isn't the observer in O' making a mistake? It should be obvious to him that he is moving with respect to the firecrackers. Why doesn't he realize this and take his motion into account? Since his light receptor was moving toward A and away from B, it should not surprise him that it recorded the flash from A before the flash from B. All he has to do is allow for this motion in his calculations, and he will conclude that the flashes "really" were simultaneous.

But if we reason in this way, we are ignoring the postulates of relativity! We are unconsciously assuming that when we look at this situation from O (as we are doing in •Fig. 26.8a and b), we are looking at what "really" happened from the vantage point of the frame that is "really" at rest. But, according to the first postulate, no reference frame is any more valid than any other, and none can be considered absolutely at rest. The observer in O' doesn't think of himself as moving—to him, the O frame is moving and his *own* frame is at rest. He would therefore see the firecrackers moving past him with a speed v (•Fig. 26.8c), but this motion would not affect his conclusions. If the explosions occurred at points A' and B', equally distant from R', but the flashes were recorded at different times, then the explosions could not have been simultaneous!

Notice that the second postulate also plays a crucial role here. If the speed of light were not constant for all observers, the interpretation of this experiment would be very different. If the speed of light were affected by the velocity of its source, the flash of light emitted by the explosion at A would travel faster with respect to the O' light receptor ($c + v$) than the flash from B ($c - v$), and would reach the receptor sooner. After adjusting for the different speeds of the light beams, the O' observer could conclude that the explosions were really simultaneous. But according to the second postulate, the velocity of light doesn't share the motion of its source. It is *always* measured as c, regardless of whether the source is moving toward the observer or away from him! So again, it is impossible to "explain away" the results from receptor R'.

You might wonder whether we could somehow arrange it for the O' observer to agree that the two explosions did occur simultaneously. To do this, the observer in O would have to delay firecracker A relative to B so that R' would receive the two signals at the same time. Again, however, the two observers would disagree as to whether the events were simultaneous—because now they would no longer be simultaneous in O!

What are we to make of this curious situation? In a nonrelativistic world, one of the observers would have to be wrong. But as we have seen, both observers performed the measurements correctly. Neither one did anything wrong procedurally, or used faulty instruments, or made any errors in logic. So we must conclude that both are correct! The results may seem paradoxical, but this is simply what is observed when we apply the two postulates of special relativity.

Finally, we note that there is nothing special about a firecracker explosion. Any "happening" at a particular point in space at a particular time—a karate kick, a clock striking the hour, a soap bubble bursting—would have done just as well. Such a happening is called an **event** in the language of relativity. Thus we can summarize our conclusions by saying that

> Events that are simultaneous in a particular inertial reference frame may *not* be simultaneous as measured in a different inertial frame.

So we must give up on the notion that simultaneity is an absolute concept. The nonintuitive results of length contraction and time dilation, discussed in the following section, are just two of the many relativistic results that follow directly from the two postulates and the relativity of simultaneity.

Notice that if the relative speed is slow, like those we are accustomed to in everyday life, then the lack of simultaneity is completely undetectable. This is why we come to the erroneous idea that simultaneity is absolute. This is not true, but at everyday speeds it is almost so. Most relativistic effects are like simultaneity—the departure from familiar experience is not apparent when the relative speeds between observers is much less than that of light. Since we rarely travel at speeds approaching that of light, it is hardly surprising that these generalizations seem strange to us. Keep in mind, however, that it is our everyday concepts such as the absoluteness of simultaneity that are *not* compatible with what is actually observed at high speeds!

CONCEPTUAL EXAMPLE 26.1 ■ AGREEING TO DISAGREE? THE RELATIVITY OF SIMULTANEITY

From Figure 26.8, estimate the relative speed of the two observers. If the relative speed were only 10 m/s, would there be better agreement on simultaneity? Why? *Clearly establish the reasoning and physical principle(s) used in determining your answer before checking it below. That is,* **how** *did you arrive at your answer?*

Reasoning and Answer: From the figure we can see that the O' frame has moved a third as far as the light wave has in the same time. Therefore, the relative speed is one third the speed of light. At 10 m/s, the positions of the two frames would not be noticeably different during the whole sequence; thus both observers would agree on simultaneity.

It should be clear that to measure the length of a moving object correctly, we must mark the positions of both ends *simultaneously*. However we have just seen that two inertial observers in relative motion will *disagree* on that concept. Thus, we expect that they will disagree on the length of the object. It also turns out to be true that they disagree on time intervals. These two effects, relativistic time dilation and length contraction, are the subjects of the next section.

26.4 The Relativity of Length and Time: Length Contraction and Time Dilation

Objectives: To be able to (a) understand the concepts of time dilation and length contraction, and (b) calculate the relationship between time intervals and lengths observed in different inertial frames.

Let us try yet another thought experiment, this time related to the measurement of time in different inertial frames. For the comparison of time intervals in moving inertial reference frames, let's use a special "relativistic" light clock as illustrated in •Fig. 26.9a. A tick or unit time interval on the clock corresponds to the time for a light pulse to make a round trip between the source and the mirror. It is assumed that two observers in the two frames, O and O', have identical light clocks and that the clocks run at the same rate when both are at rest relative to one another. *For an observer at rest with respect to one of these clocks,* the time interval for a light round trip is

$$\Delta t_o = \frac{2L}{c} \tag{26.4}$$

Now consider O' to be in motion with a constant velocity of magnitude v relative to O. For its clock (at rest in O') O' still measures the time interval as Δt_o. But, as seen by O, the clock in the O' system is moving, and the path of the light pulse forms the sides of the triangle, as shown in •Fig. 26.9b. Thus, O sees a longer light

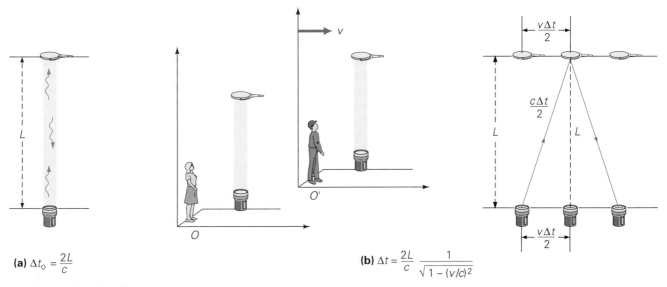

(a) $\Delta t_o = \dfrac{2L}{c}$

(b) $\Delta t = \dfrac{2L}{c} \dfrac{1}{\sqrt{1 - (v/c)^2}}$

•**FIGURE 26.9 Time dilation**
(a) A light clock that measures time in units of round-trip reflections of light. The time for light to travel up and back is $\Delta t_o = 2L/c$. **(b)** An observer in O measures a time interval of $\Delta t = (2L/c)\left[1/\sqrt{1 - (v/c)^2}\right]$ on the clock in the relatively moving O' frame, so the moving clock appears to be running more slowly. See the text for further description.

path for the clock in the moving system. From O's point of view, the geometry of the path of the light pulse is, by the triangle in the figure,

$$\frac{c^2 \Delta t^2}{2} = \frac{v^2 \Delta t^2}{2} + L^2$$

where Δt is the time interval measured by the observer in O.

It is now clear why O and O' will measure time intervals differently. If the light path in the clock moving relative to O's frame of reference, is longer, and the speed of light is the same for all observers, then the light in the moving clock must take a longer time to cover the path. Thus *from O's point of view*, the moving clock seems to be running slower—the ticks occur at a slower rate. We can calculate how much slower by solving the preceding equation for Δt:

$$\Delta t = \frac{2L}{c}\left[\frac{1}{\sqrt{1 - (v/c)^2}}\right] \tag{26.5}$$

But the same time interval (Δt_o) is measured by either observer on a clock at rest in his or her *own* frame and is given by Eq. 26.4. So the measured time intervals on the clock moving with respect to an inertial frame are different than the interval on a clock at rest. Combining the equations, we have

$$\Delta t = \frac{\Delta t_o}{\sqrt{1 - (v/c)^2}} \qquad \text{time dilation} \tag{26.6}$$

Since $\sqrt{1 - (v/c)^2}$ is less than 1 (why?), we have $\Delta t > \Delta t_o$, and O measures a longer time interval on the O' clock than does observer O' on the O' clock. This effect is called **time dilation**. With longer time between ticks, the O' clock appears to O to run more slowly than his or her own. Of course, the situation is relative, and O' would say that the clock in O's system ran slowly relative to the O' clock. Thus, because of the constancy of the speed of light, we have the qualitative statement of time dilation:

Time Dilation

> Moving clocks are measured to run more slowly than clocks at rest in the observer's own frame of reference.

This effect, like all relativistic effects, is only significant if the relative speeds are close to that of light.

Proper Time

To distinguish between the two time intervals, we refer to **proper time**. As with lengths, it is usually "proper" or normal to be at rest with respect to a clock when making a time interval measurement. In our development, the proper time can be seen to be Δt_o, or the time interval in the reference frame (O') in which the clock was at rest. Stated another way, the proper time interval between two events is that interval measured by an observer who is at rest relative to the events and sees those events occur *at the same location or point in space*. (What are the two events for the light pulse clock? Are they at the same location in O' for the O' clock?) In Fig. 26.9, observer in O sees the events by which the time interval of the O' clock is measured at different locations: because the clock is moving, the starting event (the light pulse leaving) occurs at a different location in O than the ending event (the light pulse returning). Thus Δt is *not* the proper time.

Many of the equations of special relativity can be written more simply if we represent the expression $1/\sqrt{1 - (v/c)^2}$ as a factor, gamma (γ):

$$\gamma = \frac{1}{\sqrt{1 - (v/c)^2}} \tag{26.7}$$

Gamma (γ) is always greater than or equal to 1. (When is it equal to 1?) Notice that as v approaches c, γ grows to infinity. Thus, if the mathematical results are in agreement with experiment, we expect that *relative speeds equal to or greater than that of light are not possible.*

The values of γ for several values of v expressed as fractions of c are given in Table 26.1. The values illustrate how the speed of an object must be an appreciable fraction of the speed of light for relativistic effects to be observed. The time dilation equation (Eq. 26.6) may then be written more compactly as

$$t = \frac{t_o}{\sqrt{1 - (v/c)^2}} = \gamma t_o \qquad (26.8)$$

Note: The proper time t_o is always less than the dilated time t.

where the deltas have been eliminated and it is understood that the t's represent time *intervals*. In words, the time interval measured on a moving clock is γ times the proper time. The proper time is thus always the shortest time interval between two events.

For example, suppose you were observing a clock that was in a system moving at a constant velocity (magnitude $0.600c$) relative to you. The gamma factor would be 1.25 (calculate or refer to Table 26.1). Because of the time dilation effect, when 20 minutes had elapsed on the moving clock, you would observe a time of $t = \gamma t_o = (1.25)(20 \text{ min}) = 25$ min to have elapsed on your clock. The 20 minutes is the proper time, since the events defining the 20-minute interval took place at the same location (that of the clock—for example, for readings at 8:00 and 8:20). As was pointed out earlier, the moving clock runs more slowly (20 minutes elapsed as opposed to 25 minutes on your clock) when viewed by an observer moving relative to it.

Finally, note that the time dilation effect cannot apply solely to our somewhat artificial light pulse clock. It must be true for *all* moving clocks (that is, anything that keeps to a rhythm or frequency, including the human heart). If this were not the case—if mechanical watches, for example, did not exhibit time dilation—then a watch and a light-pulse clock would run at different rates in a given inertial frame, and an observer in that frame would have a way of telling whether he or she was moving by making a comparison *solely within that frame*. Since that violates Postulate I, all moving clocks, regardless of their nature, must exhibit time dilation.

Let's take a look at an actual time dilation effect in nature.

EXAMPLE 26.2 ■ MUON DECAY VIEWED FROM THE GROUND: TIME DILATION VERIFIED BY EXPERIMENT

There are many subatomic elementary particles. One of these, called a muon, has the same charge as an electron but is about 200 times more massive. Muons are created in the Earth's atmosphere when cosmic rays (mostly protons) from outer space collide with the nuclei of the gas molecules of the air. The muons then approach the Earth with a speed near that of light (typically about $0.998c$).

However, muons are unstable and quickly decay into other particles. The average lifetime of a muon essentially at rest has been measured in the laboratory as 2.20×10^{-6} s. In this time, a muon would travel a distance of

$$d = v_o t = (0.998c)(2.20 \times 10^{-6} \text{ s})$$
$$= [(0.998)(3.00 \times 10^8 \text{ m/s})](2.20 \times 10^{-6} \text{ s}) = 659 \text{ m}$$

This is only 0.659 km, or less than half a mile. Since muons are created at altitudes of about 15–20 km, we would expect that none of them would reach the Earth's surface. However, an appreciable number of muons *do* reach the Earth. Explain this apparent paradox using time dilation ideas.

Solution. The paradox arises because the preceding calculation does *not* take time dilation into account. That is, a muon decays by its own clock, or in a time measured by a clock in its own reference frame (proper time, since in the rest frame of the muon, "birth" and "death" events take place at the same location).

To an observer on Earth, a clock in the moving muon reference frame would appear to run more slowly than a clock on Earth (•Fig. 26.10). So in a time $t_o = 2.20 \times 10^{-6}$ s on the muon "internal clock", the Earth clock reads a time interval that is greater by a factor of

TABLE 26.1 Some Values of

$$\gamma = \frac{1}{\sqrt{1 - (v/c)^2}}$$

v	γ
0	1.00
$0.100c$	1.01
$0.200c$	1.02
$0.300c$	1.05
$0.400c$	1.09
$0.500c$	1.15
$0.600c$	1.25
$0.700c$	1.40
$0.800c$	1.67
$0.900c$	2.29
$0.950c$	3.20
$0.990c$	7.09
$0.995c$	10.0
$0.999c$	22.4
c	∞

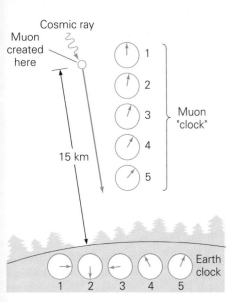

•FIGURE 26.10 Experimental evidence of time dilation
Muons are observed at the surface of the Earth as predicted by the theory of relativity. See Example 26.2.

$$\gamma = \frac{1}{\sqrt{1 - (v/c)^2}} = \frac{1}{\sqrt{1 - (0.998c/c)^2}} = 15.8$$

and with $t = \gamma t_o$, we have the muon traveling a distance of

$$d = vt = \gamma v t_o$$

$$= \gamma (0.660 \text{ km}) = (15.8)(0.659 \text{ km}) = 10.4 \text{ km}$$

Therefore, the theory of relativity predicts that the muons would travel a much greater distance in our frame than is predicted by classical physics. Since t_o is an *average* value, some muons do travel distances greater than 10.4 km and are observed at the Earth's surface.

Follow-up Exercise. In this Example, what speed would enable the muons to travel twice as far (20.8 km)? Would they have to travel twice as fast?

PROBLEM-SOLVING HINT

When working time dilation problems, it is important to identify the proper time interval t_o. To do this, first identify (1) the events that define a time interval and (2) a clock that measures it. Then, if the events occur in the same location in that reference frame, an observer at rest in the same frame (at rest relative to the clock) measures the proper time t_o. Since $t = \gamma t_o$ and $\gamma > 1$, the proper time interval t_o is always less than the dilated time interval t.

Length Contraction

As was mentioned earlier, the null result of the Michelson–Morley experiment could be explained if it were assumed that the length of the interferometer arm in the direction of motion contracted by a factor of $\sqrt{1 - (v/c)^2}$, the Lorentz–FitzGerald contraction. The same length contraction also turned out to be a consequence of the special theory of relativity. Its origin, however, is not some ether interaction with interatomic forces, but instead is explained by the two postulates of the relativity theory.

To measure the length of an object that is not at rest in our reference frame, we must mark both ends simultaneously. Consider again the two inertial reference frames in relative motion that we used in our discussion of simultaneity, and imagine a stick lying on the x axis at rest in O (•Fig. 26.11a). If the observer in O marks the ends simultaneously, the observer in O' will observe end A marked before B (why?). Thus for O' to make a correct length measurement, we must ask O to delay the marking of end A relative to B (•Fig. 26.11b). We can picture O setting off explosions that would make burn marks in both reference frames (on both x and x' axes), with the information about these events traveling out at the speed c in all directions as visible light. When the ends are marked correctly for O', all that needs to be done is subtract the two positions to get the length of the stick. Notice that it will be *less* than that measured by O.

Again we ask: "Which observer is correct?" The answer, of course, is that both are. *Both have measured the length correctly* in their own frames. Neither thinks that the other observer has done things correctly, but each is satisfied with his own measurement. The observer in O' would have measured the same length as O if he were *willing to overlook the fact that the burn marks were not made simultaneously*, as in Fig. 26.11a, but this would not have been the correct way to measure the length of the moving stick from the O' point of view. We see, then, that the lack of agreement on simultaneity leads to the following qualitative statement of length contraction:

The length of a moving object (in the direction of the relative motion of two inertial observers) is largest for the observer at rest with respect to the object and smaller for observers in motion with respect to it.

812 CHAPTER 26 Relativity

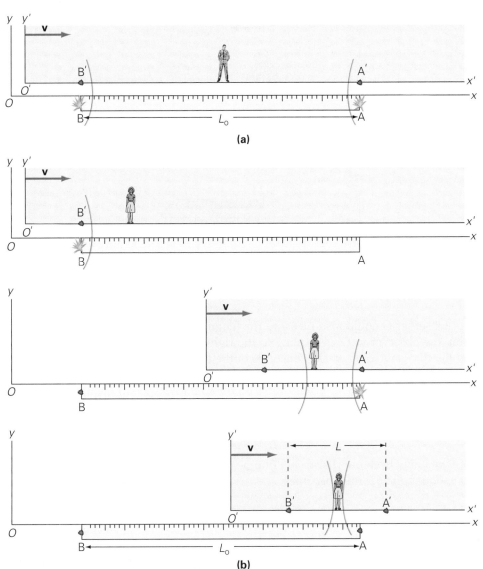

(a)

(b)

• FIGURE 26.11 **Measuring lengths correctly**
To correctly measure the length of a moving object, the ends must be marked simultaneously. **(a)** When *O* marks the ends simultaneously, *O'* does not agree, and observes A marked before B, resulting in too long a length from the *O'* view. **(b)** When *O* delays A relative to B by just the correct amount (how do we know from the sketch?), then *O'* measures the correct length from her point of view. The length measured by *O'* is less than the rest length measured by *O*.

Once again this effect is entirely negligible at speeds that are low compared to *c*.

The length of the stick or the distance between two points *as measured by the observer at rest with respect to the stick or the two points* is designated by L_0 and called the **proper length**. The proper length, sometimes known as the *rest length*, is the largest possible length. The term "proper" has nothing to do with the "correctness" of the measurement, because as we have seen, each observer is measuring correctly in his or her own frame.

Let us develop a mathematical expression, analogous to that for time dilation above, for length contraction. Consider a rod that is at rest in frame *O*. Consistent with our understanding of the term "proper," we see that the observer in frame *O* is going to be the proper length measurer for this rod. Thus we call the length of the rod that he measures L_0. Now this same rod is also viewed by an observer in *O'*, who is traveling with a constant speed *v* in the direction parallel to the stick (• Fig. 26.12). This observer also wishes to measure the length of the rod. She does this by using the clock she is holding to measure the interval between the times when the ends of the rod pass her. Since she is the one measuring the proper time interval in this case (why?), we call this time t_0. In her reference frame, the rod is moving to the left with speed *v*, so the length that she measures, *L*, is given by $L = vt_0$.

Proper Length

• FIGURE 26.12 Derivation of length contraction
The observer O measures the time it takes for the observer in O' to move past the ends of the rod. Similarly, the observer in O' measures the time it takes for the ends of the rod to pass her. The observer in O' is the *proper time measurer*, and thus measures the shortest possible time between these two events. The measured lengths of the rod are not the same. The observer in O is the *proper length measurer*, and so measures the longest possible length.

Observer O could also measure the length of the rod by the same means. To him, the observer in O' is moving to the right past the rod. If he notes the times on his clocks when O' passes the ends of the rod, he will measure a time interval t, and for him the length of the stick would be given by $L_o = vt$. (The v's are the same because the two observers agree on their relative speed.) Then, dividing one length by the other, we have

$$\frac{L}{L_o} = \frac{vt_o}{vt} = \frac{t_o}{t} \qquad (26.9)$$

But we know from our discussion of time dilation that $t = \gamma t_o$ (Eq. 26.8), or $t_o/t = 1/\gamma$. Equation 26.9 then becomes

Relativistic length contraction

$$L = \frac{L_o}{\gamma} = L_o\sqrt{1 - (v/c)^2} \qquad \begin{array}{l} length \\ contraction \end{array} \qquad (26.10)$$

Since γ is always greater than 1 with relative motion, we have $L < L_o$ or a **length contraction** by a factor of $1/\gamma$ or $\sqrt{1 - (v/c)^2}$, which is as expected from our previous qualititative discussion using simultanity.

Note: The proper length L_o is always greater than the contracted length L.

Thus the measurements of both time intervals and space intervals, that is, lengths and distances, are affected by relative motion. It should be remembered that the length contraction occurs only along the direction of motion. As a result, in a relativistic world, objects would appear to be contracted only in the dimensions that are in the direction of their motion.

EXAMPLE 26.3 ■ WARP SPEED? LENGTH CONTRACTION AND TIME DILATION

An observer sees a spaceship, measured as 100 m long when at rest, pass by in uniform motion with a speed of $0.500c$ (• Fig. 26.13). While the observer is watching the spaceship, a time of 2.00 s elapses on a clock on board the ship. (a) What would the observer measure the length of the moving spaceship to be? (b) What time interval elapses on the observer's clock for the 2.00 s interval measured on the ship's clock?

Solution. The 100 m is the proper length because it is measured when the spaceship is at rest with respect to the observer. The time interval of 2.00 s is the proper time interval because the beginning and end of the interval are measured by the same clock in the same location. Listing the given quantities, we have

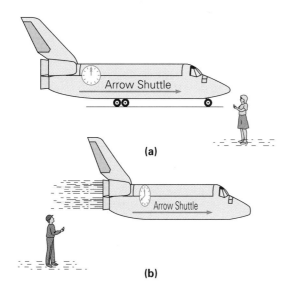

(a)

(b)

•FIGURE 26.13 **Length contraction and time dilation**
As a result of length contraction, moving objects are observed shorter, or contracted in the direction of motion, and moving clocks are observed to run more slowly because of time dilation. See Example 26.3.

Given: $L_o = 100$ m (proper length)   *Find:*   (a) L (contracted length)
$t_o = 2.00$ s (proper time)   (b) t (dilated time)
$v = 0.500c$

(a) By calculation or from Table 26.1, $\gamma = 1.15$ for $v = 0.500c$, and the length contraction is given by Eq. 26.10:

$$L = \frac{L_o}{\gamma} = \frac{100 \text{ m}}{1.15} = 87.0 \text{ m}$$

Thus the observer measures the length of the spaceship to be considerably shorter than its proper length.

(b) The time dilation is given by Eq. 26.8:

$$t = \gamma t_o = (1.15)(2.00 \text{ s}) = 2.30 \text{ s}$$

Follow-up Exercise. In this Example, calculate the time it takes the spaceship to pass a given point, at rest in the observer's reference frame, as seen by (a) a person in the spaceship, and (b) the observer watching the ship travel by. (c) Explain why these time intervals are not the same.

EXAMPLE 26.4 ■ WHAT THE MUON SAW: MUON DECAY REVISITED

We saw in Example 26.2 that more muons reach the Earth's surface than can be accounted for in nonrelativistic terms. Experimenters on Earth, who observe the muons moving at high speed, explain this phenomenon in terms of relativistic time dilation. How would a hypothetical observer on the muon explain the fact that the average muon makes it to the surface? Which explanation is correct?

Solution: In Example 26.2, we discussed a way of verifying time dilation using muon decay. A muon traveling at $v = 0.998c$ decays by its own clock (proper time) in $t_o = 2.20 \times 10^{-6}$ s. A muon would travel a distance of only 659 m in this time. Yet muons are observed on Earth, and so must travel more than 10 km, on the average, before decaying. This paradox was explained by a time dilation effect for an Earth observer, who sees the muon's clock running slow, thus enabling it to travel farther (before decay) than you would expect.

How does this situation look to a hypothetical observer riding with the muon? For such an observer, the muon clock will appear to be correct—and it registers a time interval that is *not* sufficient for the muon to reach the Earth's surface! How can this be reconciled with the observation of the Earth-bound observer? After all, we cannot have two observers disagreeing on the end result of the experiment. Either the muon made it to the surface or it did not. We have a signal from the Earth-based muon detector that a significant fraction of the muons do, indeed, make it to sea level.

The apparent paradox disappears when length contraction is taken into account. Then both observations can be seen to be consistent with the experimental results. The reasoning is as follows. For our imaginary observer riding on the muon, the muon clock reads correctly ($t_o = 2.20 \times 10^{-6}$ s), but the observed travel distance—a length—is shorter because of length contraction. With $\gamma = 15.8$ for $v = 0.998c$, a travel length of 10.0 km in the Earth frame (proper length L_o) is measured by the observer on the muon to be L,

$$L = \frac{L_o}{\gamma} = \frac{10.0 \text{ km}}{15.8} = 0.633 \text{ km} = 633 \text{ m}$$

To travel this distance would take a time of

$$t = \frac{L}{v} = \frac{L}{(0.998c)} = \frac{633 \text{ m}}{(0.998)(3.00 \times 10^8 \text{ m/s})} = 2.11 \times 10^{-6} \text{ s}$$

Note that this is roughly the average muon lifetime *in the muon's frame!* Thus, through relativistic considerations *both observers agree that the muons reach the Earth* (the experimental result). The Earth observer explains this result by saying that the muon clock is running slow (time dilation). The observer on the muon says, "No, my clock is fine—the distance we have to travel is considerably less than you claim" (length contraction). Who is correct? They both are. *The physical reasoning is different for different observers, but the end result is the same.*

Follow-up Exercise. Muons actually are created with a range of speeds in the upper atmosphere. What is the highest speed a muon could have and still *not* be detected 5.0 km from its creation point?

The Twin Paradox

Time dilation gave rise to another popular relativistic topic—the so-called **twin,** or **clock**, **paradox**. According to the special theory, a clock observed to be moving runs more slowly than, or not as fast as, one used by an inertial observer. For example, with $\gamma = 4$, for a proper time interval of 15 minutes, an hour would elapse on the observer's clock. Similarly, 1 year of proper time in the moving system takes 4 years in the observer's time frame. Since our heartbeat and age are measured by proper time, the question arises: Does an observer in one system age more quickly than a person in a system that is in motion relative to the first system?

One way to explore this question is to consider identical twins, one of whom goes on a high-speed space journey. The question then is: Will the space traveler come back younger than his Earth-bound twin? You might say that everything is relative and that the space twin sees the Earth clock run more slowly. He would claim that the Earth twin would age more slowly. They can't both be right.

The solution to the paradox lies in the fact that in leaving and returning to Earth, the space twin must experience *accelerations* and so is *not* always in an inertial reference frame. The stay-at-home twin does not feel the forces associated with speed and directional changes that the traveling twin experiences. Thus the two twins are individually "marked" and their experiences are *not* symmetrical. However, if the acceleration periods (at the start, at turnaround, and at return) occupy only a small part of the total trip time, the special theory can be applied to get an indication of what is likely to happen. The prediction is that the space-traveling twin does return younger than the Earth-bound twin. (Einstein's general theory, which considers accelerating systems, confirms this result.)

While traveling at a constant velocity on the outward and return trips, which occupy most of the total trip, both twins measure the same relative speed. However, the proper *length* of the trip is measured in the Earth twin's reference frame with fixed beginning and end points. The traveling twin thus measures a length contraction or shorter trip distance. Traveling at the same relative speed, the space twin travels a shorter time, and returns home younger than his Earth-bound twin *according to the space twin.* According to the Earth twin, the traveling twin's heart-beat (internal clock) runs slower so the space twin returns home younger than his Earth-bound twin also *according to the Earth twin!* Thus there is no disagreement, the traveler returns having aged less (fewer heartbeats) than the nontraveler. At extremely high speeds, it would even be possible for the traveler to come back and find many generations of Earthlings long since born and died.

The twin paradox has been experimentally verified using atomic clocks. In 1971, one of these clocks was flown around the world in jet planes. It was necessary to use these extremely accurate clocks to detect any effects because of the small γ factors, or low relative velocities. The time elapsed as recorded by the traveling clock was compared to that measured by a clock that stayed at home. These clocks were accurate enough to measure the predicted effect (with corrections made for accelerations, etc.), and time dilation was experimentally verified. As was stated in the reporting article, "These results provide an unambiguous empirical resolution of the famous clock 'paradox' with macroscopic clocks."*

EXAMPLE 26.5 ■ AD ASTRA: RELATIVITY AND SPACE TRAVEL

Assuming special relativity, ignoring the accelerations at the start, end, and turnaround points, and assuming that no time is spent at the destination, (a) find the speed at which a space explorer would need to travel in order to make the round trip to a nearby star 100 light-years away in only 20.0 years of traveler time. [The light-year (abbreviation ly) is defined as the distance light travels (in vacuum) in one year.] (b) How much time elapses for the people on Earth during this trip? Carry your answers to three significant figures to show the effect.

Solution. The 100 light-years is the proper length (one way) because it is with respect to the Earth. The ten years (one way) is measured by the traveling twin's clock (and heart) which is present for the start and end of the trip and therefore measures the proper time. Let us calculate a one-way trip. Listing the given quantities, we have

Given: $L_o = 100$ ly (proper length)
$t_o = 10.0$ y (one way proper time)

Find: (a) v (speed for the traveler)
(b) t (time elapsed on Earth)

(a) We can do the speed calculation using either time dilation (the Earth view) or length contraction (the traveler's view). Let's do the latter. The length to use then is the contracted version of the 100 light-years. According to the traveler, a one-way trip takes

$$t_o = \frac{\text{distance traveled}}{\text{speed}} = \frac{L}{v} = \frac{L_o\sqrt{1 - \left(\frac{v}{c}\right)^2}}{v}$$

This can be solved for v/c, giving

$$\frac{v}{c} = \frac{1}{\sqrt{1 + \left(\frac{ct_o}{L_o}\right)^2}} = \frac{1}{\sqrt{1 + \left(\frac{10 \text{ ly}}{100 \text{ ly}}\right)^2}} = 0.995$$

v is 99.5% the speed of light. Notice that we have used the fact that is the distance light travels in 10 years which is, by definition, 10 ly. We did not have to convert to meters because we found the solution in terms of symbols, without inserting numbers initially.

(b) The people on Earth would observe this traveler covering a total of 200 ly at $0.995c$, so the time taken would be 200 ly/$0.995c$ or about 201 y, compared to 20.0 years for the traveler. Clearly the traveler would come back to find that everyone on Earth when he left was deceased.

Follow-up Exercise. In this Example, show that the Earth observers calculate that 201 of their years will elapse during the trip, using time dilation considerations.

---

*Hafele and Keating, "Around the World Atomic Clocks: Relativistic Time Gains Observed," *Science*, vol. 117, no. 4044 (July 14, 1972), pp. 166–170.

26.5 Relativistic Kinetic Energy, Mass, Momentum, and Total Energy

Objectives: To be able to (a) understand the need for more general expressions for kinetic energy, mass, momentum, and total energy when objects move near the speed of light, and (b) use the relativistically correct expressions to calculate energy and momentum in particle interactions.

The ramifications of the theory of relativity are particularly important in particle physics, in which the speeds of the particles may be appreciable fractions of the speed of light. Such speeds are possible, for example, in modern particle accelerators. Many of our results from classical mechanics are incorrect when applied to high-speed particles. One of the most important quantities is kinetic energy. Does it have a form different from the expression that we used when we studied Newton's laws ($K = \frac{1}{2}mv^2$)? The answer is yes, as we will now see.

Relativistic Kinetic Energy and Mass

Einstein showed that kinetic energy at high speeds is indeed given by an expression different from the one we are used to. As in classical mechanics, it increases with speed, but in a different way. The **relativistic kinetic energy** expression is

$$K = \left[\frac{1}{\sqrt{1 - (v/c)^2}} - 1\right]m_{o}c^2 = (\gamma - 1)m_{o}c^2 \qquad \begin{matrix} \textit{relativistic} \\ \textit{kinetic energy} \end{matrix} \qquad (26.11)$$

Rest Mass

where m_{o} is called the **rest mass** or *proper mass* of the particle or object. The rest mass, like the proper length or time interval, is measured by an observer at rest with respect to the particle. Using mathematical techniques beyond the scope of this text, it can be shown that this expression becomes the more familiar $K = \frac{1}{2}mv^2$ when $v \ll c$.

Relativistic Mass

Yet another relativistic result is that the mass of the particle increases with speed according to the **relativistic mass** formula

$$m = \left[\frac{1}{\sqrt{1 - (v/c)^2}}\right]m_{o} = \gamma m_{o} \qquad \textit{relativistic mass} \qquad (26.12)$$

Here m is called the *relativistic mass* of the particle and is larger than the rest mass because the factor of γ is greater than 1.

According to Eqs. 26.11 and 26.12, as v approaches c, the kinetic energy and mass of an object approach infinity (•Fig. 26.14). A basic result of the special relativity is that no object can travel as fast as or faster than the speed of light. This may be seen from both equations. To accelerate an object to $v = c$ would require

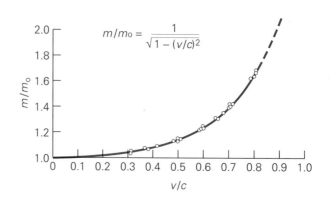

•FIGURE 26.14 Relativistic mass The data points on the curve are measured masses of accelerated particles. There is good agreement between the theoretical prediction for relativistic mass and the experimental data.

an infinite amount of energy, which makes c the maximum and unattainable speed limit. Time dilation and length contraction (Eqs. 26.5 and 26.10) also imply this. If a reference frame could travel with a speed c, then time intervals would become infinite (time would, in effect, stand still) and the lengths of objects would contract to nothing (in effect, they would vanish).

Suppose that we try to accelerate an object linearly by repeated brief applications of a force. If all objects are limited to speeds below c, the speed increase produced by each successive "push" must become less and less as the object's speed approaches c. Since an object's inertia, or resistance to change in velocity, is a measure of its mass, we can conclude that the particle's mass increases with speed.

There is complete agreement between experimental evidence and these expressions. Particle accelerators accelerate beams of charged particles to very high speeds using electric fields. Linear beams are deflected, or turned, by deflecting magnets. It is found that larger magnetic fields are needed to deflect high-speed particles (because of their relativistic mass increase) than would be needed if the nonrelativistic expressions for mass and kinetic energy were used.

Relativistic Momentum

In the theory of relativity, the definition of momentum must also be changed in order to maintain momentum as a conserved quantity. The basic conservation principle is valid if the **relativistic momentum** is defined as

$$\mathbf{p} = \frac{m_o \mathbf{v}}{\sqrt{1 - \left(\dfrac{v}{c}\right)^2}} = \gamma m_o \mathbf{v} = m\mathbf{v} \qquad (26.13) \qquad \textbf{Relativistic Momentum}$$

Notice momentum still retains its vector character in the direction of the particle's velocity.

Relativistic Total Energy and Rest Energy; the Equivalence of Mass and Energy

In classical mechanics, the total mechanical energy is $E = K + U$. When there is no potential energy, this reduces to $E = K$; that is, the total energy is simply the kinetic energy, *classically*. Equation 26.11, however, suggests a different way of looking at energy. The expression for kinetic energy given there can be expanded to take the form

$$K = (\gamma - 1)m_o c^2 = \gamma m_o c^2 - m_o c^2 = mc^2 - m_o c^2$$

or

$$mc^2 = K + m_o c^2 \qquad (26.14)$$

The right-hand side of this equation consists of a variable quantity (the kinetic energy K) plus a constant quantity ($m_o c^2$) that is a function of the body's rest mass. Mathematical techniques beyond the scope of this book can be used to show that mc^2 is the **relativistic total energy**, E. This means that even when a body is at rest, and thus has no kinetic energy ($K = 0$), it still has an energy of $m_o c^2$. This minimum energy that a body always possesses is called its **rest energy**, E_o:

$$E_o = m_o c^2 \qquad (26.15) \qquad \textbf{Rest Energy}$$

In the absence of potential energy, therefore, we can express the total energy as

$$E = K + E_o$$

or

$$E = K + m_o c^2 = mc^2 \qquad (26.16) \qquad \textbf{Relativistic Total Energy}$$

(When a particle has potential energy U, the total energy is $E = K + U + m_{\mathrm{o}}c^2$.)

A direct relationship between the total energy and the rest energy may be obtained by noting that the relativistic mass is given by $m = \gamma m_{\mathrm{o}}$ (Eq. 26.11). Multiplying both sides of the equation by c^2, we have

Relativistic Total Energy

$$E = \gamma E_{\mathrm{o}} \qquad (26.17)$$

and we have another gamma relationship. For a case with $v = 0$ (or $K = 0$) and $\gamma = 1$, we have $E = E_{\mathrm{o}}$, as noted above.

Mass-Energy Equivalence

Equation 26.16 shows that the relativistic mass is a direct measure of the total energy of the particle. This is Einstein's famous **mass-energy equivalence** formula, which points out that *mass is a form of energy.* Classically, when work is done on an object, its speed and energy increase. But, relativistically, the mass of an object also increases with increasing speed. Thus, the work goes not only into increasing the speed but also to increasing the mass. Moreover, now the particle has energy even at rest—its rest energy. Thus we can think of mass as another form of energy. Indeed, it is impossible in nuclear and particle physics to keep the law of conservation of energy if we do not adopt this idea. If the equivalence of mass and energy is adopted, however, and the expressions we have developed here are used, then total system energy is once again conserved.

The mass–energy equivalence does not mean that we can convert rest mass into useful energy at will. If we could, our energy problems would be solved, since we have a lot of mass on the Earth! Significant conversion of mass into other forms of energy, like heat, does take place on a limited scale in nuclear reactions (discussed in Chapter 30). For now, the important point is that any variation (increase or decrease) in the kinetic energy of a particle will give a variation in its relativistic mass.

The (rest) energy equivalent of a particle clearly depends on its rest mass, which varies from particle to particle ($E_{\mathrm{o}} = m_{\mathrm{o}}c^2$). For example, the rest mass of an electron in SI units is 9.109×10^{-31} kg, and the rest energy is (using four significant figures)

$$E_{\mathrm{o}} = m_{\mathrm{o}}c^2 = (9.109 \times 10^{-31} \text{ kg})(2.998 \times 10^8 \text{ m/s})^2 = 8.187 \times 10^{-14} \text{ J}$$

or

$$E_{\mathrm{o}} = (8.187 \times 10^{-14} \text{ J})\left(\frac{1 \text{ eV}}{1.602 \times 10^{-19} \text{ J}}\right) = 5.110 \times 10^5 \text{ eV} = 0.511 \text{ MeV}$$

In modern physics, particle energies are commonly expressed in units of electron volts because charged particles are often given energy by accelerating them through potential differences.

There is always the question of how you know whether you need to use the relativistically correct formulas or can "get away" with the classical ones. If the particle's speed is 10% the speed of light, then the error in using the nonrelativistic kinetic energy expression is less than 1% (the classical expression is low). At $v < 0.1c$, the kinetic energy is less than 0.5% of the rest (mass) energy. So a commonly accepted practice is to use this speed and kinetic energy region as a dividing line:

> For speeds below 10% the speed of light or kinetic energies less than 0.5% of the rest energy, the error in using the nonrelativistic formulas is less than 1% and it is usually acceptable to use the nonrelativistic expressions.

Thus, an electron with a kinetic energy of 50 eV would qualify as a nonrelativistic electron because 50 eV is 10,000 times smaller than its rest energy (only 0.01%). A proton with a kinetic energy of 0.511 MeV would also qualify as nonrelativistic because its rest energy is about 939 MeV. However, an electron with a kinetic energy of 0.511 MeV would be *highly* relativistic because its kinetic energy is the same as its rest energy. Thus, what counts is the kinetic energy relative to the rest energy of the particle, or the speed of the particle relative to that of light—even the dividing line is relative!

EXAMPLE 26.6 ■ A Speedy Electron: Accelerating Energy

(a) How much energy is required to give an electron initially at rest a speed of $0.900c$? (b) How much error would you have made if you had used the nonrelativistic expression?

Solution. We are told its speed and that it is an electron. Listing the data, we have:

Given: $v = 0.900c$ **Find:** (a) W (work required to get electron to $0.900c$)
$E_o = 0.511$ MeV (b) K (nonrelativistic kinetic energy)

(a) The energy needed is equal to the kinetic energy of the electron traveling at $0.900c$. However, we cannot correctly use the classical expression $K = \frac{1}{2}mv^2$ because the speed is close to that of light. Since the rest energy of an electron has been shown to be $E_o = m_oc^2 = 0.511$ MeV, a convenient form of the relativistic kinetic energy is given by Eq. 26.14:

$$K = m_oc^2\,(\gamma - 1) = E_o(\gamma - 1)$$

With $v = 0.900c$, by calculation or from Table 26.1, $\gamma = 2.29$, and

$$K = E_o(\gamma - 1) = (0.511 \text{ MeV})(2.29 - 1) = 0.659 \text{ MeV}$$

Thus, by the work-energy theorem, we know that 0.659 MeV of work is required to accelerate an electron to a speed of $0.900c$.

(b) From the discussion preceding this Example, we know it is foolish to use the nonrelativistic expression, but let's see what happens:

$$K_{\text{nonrel}} = \tfrac{1}{2}m_ov^2 = \tfrac{1}{2}m_oc^2\left(\frac{v}{c}\right)^2 = \tfrac{1}{2}(0.511 \text{ MeV})(0.900)^2 = 0.207 \text{ MeV}$$

As we suspected, the nonrelativistic kinetic energy expression produces results too low when the speed is near that of light.

Follow-up Exercise: In this Example, (a) what is the relativistically correct total energy of the electron? (b) What would we calculate the electron's total energy to be if we used the nonrelativistic formula?

EXAMPLE 26.7 ■ When One Plus One Doesn't Equal Two: Conservation of Relativistic Momentum and Energy

A particle of rest mass M_o, moving to the right with speed $v = 0.800c$, collides with and sticks to an identical particle, initially at rest. (a) What is the speed of the combined particles? (b) What is the rest mass of the combined particles? (c) What is the kinetic energy of the combined particles? [Parts (b) and (c) are to be answered in terms of M_o.]

Solution. In a collision we need to conserve both total energy and momentum.

Given: $v = 0.800c$ **Find:** (a) v' (speed after collision)
(b) M_o' (rest mass of combined particles)
(c) K' (kinetic energy of combined particles)

(a) The combined particle must be moving to the right after the collision to conserve the direction of the momentum, and the magnitude of the single particle's momentum beforehand must equal the magnitude of the combined particle's momentum afterward. We also know that the gamma factor for the incoming particle is

$$\gamma = 1/\sqrt{1 - (v/c)^2} = 1/\sqrt{1 - (0.800)^2} = 1.67.$$

Thus we can write momentum conservation as

$$p_i = p_f$$

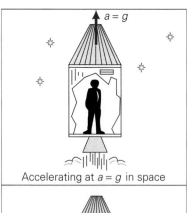

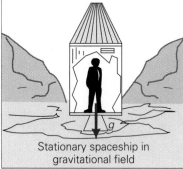

Accelerating at $a = g$ in space

Stationary spaceship in
gravitational field

(a)

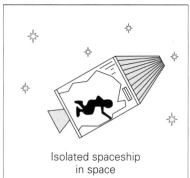

Isolated spaceship
in space

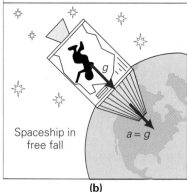

Spaceship in
free fall

**•FIGURE 26.15 The principle of
equivalence**
(a) The astronaut can perform no
experiment in a closed spaceship
that would determine whether he
was in a gravitational field or an
accelerating system. **(b)** Similarly,
an inertial frame without gravity
cannot be distinguished from free
fall in a gravitational field.

or

$$\gamma M_o v = \gamma' M'_o v'$$

or

$$1.67\, M_o(0.800c) = \gamma' M'_o v'$$

We use energy conservation and equate the total initial energy before the collision (which
includes the total energy of the moving particle and the rest energy of the "target") and
after the collision (which is the total energy of the combined particle).

$$E_i = E_f$$

or

$$1.67 M_o c^2 + M_o c^2 = \gamma' M'_o c^2$$

If we cancel the speed of light out of this last equation and divide it into the previous
result, we obtain

$$\frac{1.67\, M_o(0.800c)}{2.67 M_o} = \frac{\gamma' M'_o v'}{\gamma' M'_o} = v'$$

Thus $v' = 0.500c$.

(b)   From the speed afterward we have $\gamma' = 1/\sqrt{1 - (0.500)^2} = 1.16$. Then we can use
the energy equation above to solve for the rest mass of the combined particles:

$$2.67 M_o c^2 = 1.16 M'_o c^2$$

or

$$M'_o = 2.30 M_o.$$

(c)   For the kinetic energy afterward, we have

$$K' = (\gamma' - 1)M'_o c^2 = (0.16)(2.30 M_o c^2) = 0.368 M_o c^2$$

As a check on energy, the initial total energy is $2.67 M_o c^2$. The final total energy is $E' = \gamma' M'_o c^2 = (1.16)(2.30 M_o c^2) = 2.67 M_o c^2$ as expected.

Follow Up Exercise.   Repeat the calculations in this Example for two objects, each
traveling at a speed of $0.800c$, that collide head-on and stick together. Compare your
answers to those obtained from a nonrelativistic analysis.

26.6   The General Theory of Relativity

**Objectives: To be able to (a) explain the principle of equivalence, and (b)
specify some of the predictions of general relativity.**

Special relativity applies to inertial systems, but not to accelerating systems. Accelerating systems require an extremely complex theory, which was described by
Einstein in several papers published about 1915. Called the **general theory of relativity**, this surprisingly is essentially a gravitational theory.

The Equivalence Principle

An important aspect of the general theory is the **principle of equivalence:**

> An inertial reference frame in a uniform gravitational field is equivalent to a ref-
> erence frame in the absence of a gravitational field that has a constant accelera-
> tion with respect to that inertial frame.

What this basically means is that:

> No experiment performed in a closed system can distinguish between the effects
> of a gravitational field and the effects of an acceleration.

According to the principle of equivalence, an observer in an accelerating system
would find the effects of a gravitational field and those of an acceleration to be
equivalent or indistinguishable. Rotating frames are excluded, since rotational acceleration can be distinguished from gravitational acceleration by releasing an ob-

ject on a smooth, almost frictionless table. (Why?) Thus we stick with only linearly accelerating systems.

To better understand the equivalence principle, consider the situations illustrated in •Fig. 26.15. Imagine yourself as an astronaut in a closed spaceship. You drop a pencil and it accelerates to the floor. What does this mean? According to the equivalence principle, it could mean (a) that you are in a gravitational field, or (b) that you are in an accelerating system (Fig. 26.15a). You can't determine which. If the spaceship were in free space and accelerating with an acceleration a, or if it were stationary in a gravitational field with $g = -a$, the pencil would accelerate to the floor in either case. In your closed system (you are *not* allowed to look outside!), there is nothing you can do to determine whether the pencil's fall is a gravitational or an acceleration effect.

Suppose that the pencil does *not* begin to fall but instead remains suspended in midair next to you? This could mean that (a) you are in an inertial frame with no gravity, or (b) you are in free fall in a gravitational field. In the closed spaceship (Fig. 26.15b), you have no way of knowing whether you are in free space (negligible gravity) or accelerating toward the Earth in free fall. (Think of how a released pencil would appear to float to an observer in a closed, freely falling elevator.) Here, because the effects of gravitation and acceleration are equivalent, you have no way of telling whether both are present, or neither.

Light and Gravitation

The principle of equivalence of the general theory of relativity leads to an important prediction—that light is bent in a gravitational field. To see how this prediction arises, let's use our imaginations in the thought experiment illustrated in •Fig. 26.16. Suppose a beam of light traverses a spaceship that is accelerating rapidly. If the spaceship were stationary, light emitted at point A would arrive at point B. However, because the spaceship is accelerating during the finite time it takes for the light to traverse the ship, the light arrives at point C.

Now consider the situation as observed by a person on board the spaceship. From his point of view, he would observe the same effect of light emitted at A and arriving at C—*as though the light path were bent*. The acceleration of the rocket

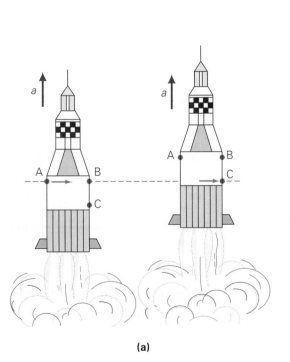

(a)

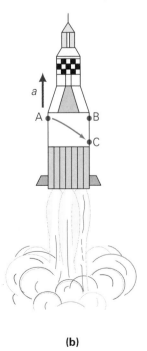

(b)

•FIGURE 26.16 Light bending (a) Light traversing an accelerating, closed rocket from point A arrives at point C. (b) In the accelerating system, the light would appear to be bent. Since an acceleration produces this effect, by the principle of equivalence, light should also be bent in a gravitational field.

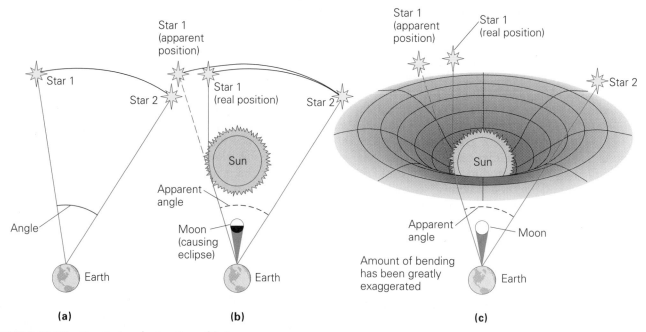

•FIGURE 26.17 Gravitational attraction of light
(a) Normally, two distant stars are observed to be a certain angular distance apart. **(b)** During a solar eclipse, the star behind the Sun can still be seen because of the effect of solar gravity on the starlight, which is evidenced by a larger measured angular separation. **(c)** General relativity views a gravitational field as a warping of space and time. A simplified analogy is the surface of a warped rubber diaphragm or sheet.

produces the effect, so by the equivalence principle we conclude that light path should also be bent in a gravitational field.

Of course, this effect must be very small, since we have observed no evidence of gravitational light bending in the Earth's gravitational field. However, this prediction of Einstein's theory was experimentally verified in 1919 during a solar eclipse (•Fig. 26.17). Distant stars appear to be motionless and are measured to have a constant angular distance separating them at night when they can be clearly seen. Light from one of the stars may pass near the Sun, which has a relatively strong gravitational field. However, any bending would not be observed on Earth because most of the starlight would be masked by the glare of the Sun itself and by the sunlight scattered in the atmosphere (which is why we can't see stars during the day).

But during a total solar eclipse, when the Moon comes between the Earth and the Sun, an observer in the Moon's shadow (the umbra) can see stars. If the light

•FIGURE 26.18 Gravitational lensing
(a) The bending of light by a massive object such as a galaxy or cluster of galaxies can give rise to multiple images of a more distant object. **(b)** The discovery of what appeared to be four images of the same quasar (the "Einstein cross") suggested the possibility of gravitational lensing. On investigation, a faint intervening galaxy was in fact found.

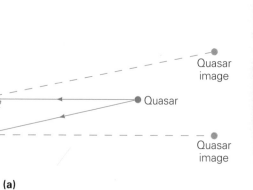

(a)

(b)

from a star passing near the Sun were bent, then the star would have an apparent location different from its actual position. As a result, the angular distance between stars would be measured to be slightly larger. Einstein's theory predicted the angular difference in the apparent and actual positions of the stars to be 1.75 seconds of arc, or an angle of about 0.00005°. The experimental angular distance was found to be 1.61 ± 0.30 seconds of arc.

General relativity views a gravitational field as a "warping" of space and time, as illustrated in Fig. 26.17c. A light beam follows the curvature of space–time like a ball rolling on a surface. The bending of light was verified during subsequent solar eclipses and also by signals from space probes. For example, signals from a Viking lander on the surface of Mars passed through the gravitational field of the Sun when Mars was on the far side of the Sun. The signals were observed to be delayed by about 100 μs. The delay was caused by the signals passing through the space–time warp, or "sink," of the Sun.

Gravitational Lensing.   Another effect of the gravitational bending of light is known as *gravitational lensing*. In the late 1970s, a double quasar was discovered. (A quasar is a powerful astronomical radio source.) The fact that it was a double quasar was not unusual, but everything about the two quasars seemed to be exactly the same, except that one was fainter than the other. It was suggested that perhaps there was only one quasar and that somewhere between it and the Earth, a massive object had deflected its electromagnetic radiation, producing multiple images. The subsequent detection of a faint galaxy between the two quasar images confirmed this hypothesis, and other examples have since been discovered (•Fig. 26.18).

Gravitational lenses can be considered to be a test of general relativity, but this concept has also given general relativity a new role in astronomy. By examining multiple images of a distant galaxy or quasar, their relative intensities and so on, astronomers can gain information about an intervening galaxy or cluster of galaxies whose gravitational field causes the bending of light.

Such an effect has been observed by the Hubble Space Telescope. On a photo of a remote galaxy, two mirror images of the structure were observed on opposite sides of the picture (•Fig. 26.19). These were believed to be caused by the gravitational lensing of an intervening cluster of galaxies containing much *dark matter:* matter that does not emit electromagnetic radiation, and so cannot be detected by regular observations. How much matter the universe contains is an important question for scientists. Is there enough matter for gravitational attraction to slow and

•FIGURE 26.19 **Gravitational lensing and dark matter** (a) Two similar images of a remote galaxy on either side of an intervening cluster of galaxies were observed by the Hubble Space Telescope. (b) The effect was thought to result from gravitational lensing by dark matter—matter that emits no electromagnetic radiation and so cannot be detected by ordinary observations. Gravitational lensing may offer a method to detect dark matter.

HUBBLE SPACE TELESCOPE VIEW

(a)

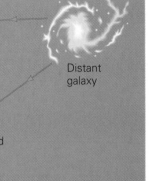

Photographic image

Intervening cluster of galaxies

Distant galaxy

Gravitational field bends light rays

Hubble Space Telescope

(b)

stop the Big Bang expansion of the universe? It is estimated that there might be 10 times more matter in the universe than is actually observed, but we do not know what this dark matter consists of, although several theories have been proposed. Gravitational lensing may provide a method to detect such dark matter.

Black Holes. The idea that gravity can affect light finds its most extreme application in the concept of a black hole. A **black hole** is generally considered to form from the gravitationally collapsed remnant of a star.* Such an object has a density so great and a gravitational field so intense that nothing can escape it. In terms of the space-time warp analogy presented above, a black hole would be a bottomless pit in the fabric of space–time. Even light can't escape the intense gravitational field of a black hole—hence the blackness.

An idea of the size of a black hole may be obtained by using the escape speed. Recall from Section 7.6 that the escape speed from a spherical body of mass M and radius R is given by

$$v_e = \sqrt{\frac{2GM}{R}} \qquad (26.18)$$

If light does not escape from a black hole, this means that the hole's escape speed must exceed the speed of light. The critical radius of a sphere around a black hole from which light will not be able to escape may be obtained by substituting $v_e = c$ in Eq. 26.18. Solving for R,

$$R = \frac{2GM}{c^2} \qquad (26.19)$$

Schwarzschild Radius

Note: A collapsing star becomes a black hole if it reaches the Schwarzschild radius, and collapses beyond this event horizon.

This is called the **Schwarzschild radius** after Karl Schwarzschild (1873–1916), a German astronomer who developed the concept. The boundary of a sphere of radius R defines what is called the **event horizon**. Any event occurring within this horizon is invisible to an observer outside, since light cannot escape.

It should be noted that the event horizon is not necessarily the radius of a black hole. It gives only the limiting distance within which light is unable to escape from a black hole. The mass inside would continue to collapse, making the black hole smaller than its event horizon.

EXAMPLE 26.8 ■ IF THE SUN WERE A BLACK HOLE: SCHWARZSCHILD RADIUS

Assuming our Sun collapsed to a black hole, what would be its Schwarzschild radius when it reached black hole status? ($M_S = 2.0 \times 10^{30}$ kg)

Solution.

Given: $M_S = 2.0 \times 10^{30}$ kg   *Find:* R (Schwarzschild
         $G = 6.67 \times 10^{-11}$ N·m²/kg² (known)   radius)
         $c = 3.00 \times 10^8$ m/s (known)

This is a straightforward calculation using Eq. 26.19. However, it should be mentioned that our Sun will not become a black hole. This is a fate that befalls only stars that are considerably more massive than the Sun.

$$R = \frac{2GM_S}{c^2} = \frac{2(6.67 \times 10^{-11}\,\text{N·m}^2/\text{kg}^2)(2.0 \times 10^{30}\,\text{kg})}{(3.00 \times 10^8\,\text{m/s})^2}$$

$$= 3.0 \times 10^3\,\text{m} = 3.0\,\text{km}$$

This is less than 2 miles! (Note the Sun currently has a radius of about 7.0×10^5 km)

---

*It is speculated that black holes may originate in other ways, such as the collapse of entire star clusters in the center of a galaxy or at the beginning of the Universe during the Big Bang. Some current theories (combining quantum mechanics and general relativity) hint that black holes actually might be able to emit material particles and radiation and thus "evaporate," but none have been discovered in this process. We will limit our discussion to stellar collapse.

Follow-up Exercise. Once black holes are formed, they continually draw in matter, thus expanding and increasing their Schwarzschild radius. How many times more massive would our Sun have to be in order for the Schwarzschild radius to extend out to the Earth?

If nothing, including radiation, escapes a black hole from inside its event horizon, how then might we observe or locate one? This is addressed in the Insight on p. 828.

Gravitational Red Shift. Another result of the general theory is that gravity affects time by causing it to slow down—the greater the gravitational field, the greater the slowing of time. There is evidence of this effect in what is called the *gravitational red shift* of light from the Sun. The gravitational field of the Sun is quite large, and if the slowing of time in a gravitational field is correct, the electronic vibrations in the Sun's atoms should be slower. As a result, light from these atoms would have a lower frequency and be shifted toward the red end of the visible spectrum.

The measurement of a solar gravitational red shift was unsuccessful for many years. This was because of a lack of understanding of the conditions at the solar surface. For example, violent motions with convection columns of rising hot gases and descending cooler gases give rise to Doppler shifts. It wasn't until 20 to 30 years ago, after a variety of effects were taken into account, that the gravitational red shifts of solar lines were measured. The gravitational red shift was also verified in the laboratory in the 1960s by using complicated nuclear techniques.

Another triumph for the general theory, also related to the slowing down of time in a gravitational field, involved the advance of the perihelion of the planet Mercury. For centuries it was known that Mercury's perihelion (the point in the planet's elliptical orbit closest to the Sun) does not remain constant each orbit, like those of the other planets. In the case of Mercury, the perihelion point moves "forward," in the direction of the planet's orbital motion. It was known that the small gravitational tugs of the outer eight planets would produce such an effect, but the detailed calculations predicted only half of the observed movement.

If Mercury's orbital period were increased (time slowed) because of the Sun's powerful gravitational field, this would allow the planet to orbit further each time to arrive at its perihelion. Calculations of this relativistic gravitational effect predicted the other half of the observed movement of the perihelion point, within the limits of experimental error. Mercury, of course, was a lucky choice because it is a planet with an obviously elliptical orbit (thus making its perihelion point easier to observe experimentally) as well as the closest of all the planets to the Sun, and thus the one most affected by the gravitational relativistic time effect.

Relativistic effects apply to all systems, no matter how complicated. Although these effects are most readily observed for atomic particles and light, they apply as well to real clocks, human beings, and so on. Although we don't commonly observe them, relativistic effects form the bases of some good science fiction and even some (not so good?) poetry, such as the following variation on a limerick that appeared in the British magazine *Punch* in 1923:

> A precocious student named Bright
> Could travel much faster than light;
> He departed one day
> In a relative way
> And arrived home the previous night.

Insight Black Holes

The idea of black holes has caught the public's fancy. Imagine—something so dense that neither light nor anything else can escape from it. So dense that a tablespoon full of its matter would probably weigh more than Mt. Everest. Black holes are becoming increasingly common in science fiction, but in actuality they are still to some extent theoretical. In the theory of stellar evolution, black holes result from the collapse of stars having much greater mass than the Sun. But experimental proof of the existence of a black hole is another matter.

The boundary of the "blackness" of the hole is the surface of the event horizon at its Schwarzschild radius (see the discussion in the text). The matter within can continue to contract, but the force of gravity at the event horizon would still be the same. (Why?) Thus, the boundary of a black hole is not a sphere of matter, but the radius at which the gravitational force is sufficiently strong to keep light from escaping. The actual radius of the matter of a black hole may be much less. What form matter takes inside a black hole is not known, and may never be known. If we ever determine the location of a black hole in space, any probe we might send to check this could never report back when it arrived. (Can you see why this is so?)

How then do you observe a black hole? Light passing nearby would be bent, but it is unlikely that we would ever observe this, considering the vastness of space. The most likely possibility comes from binary star systems, which are quite common. Cygnus X-1, the first X-ray source discovered in the constellation Cygnus, provides the best evidence so far for a black hole in our galaxy. (Many X-ray sources have been dis-

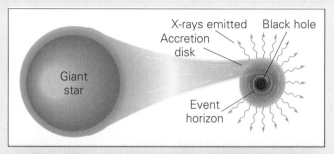

FIGURE 2 X-rays and black holes
Matter drawn from the other member of a binary star system forms a spiraling accretion disk around the hole. Matter falling into the disk is accelerated, and collisions give rise to the emission of X-rays.

covered through observations from rockets and satellites. Recall that X-rays do not penetrate the atmosphere.) If we look at the location of the Cygnus X-1 X-ray source in visible light (Fig. 1), we observe a giant star whose spectrum shows Doppler shifts, indicating an orbital motion. (Binary stars orbit about each other—more precisely, about their common center of mass.)

It is speculated that in binary star systems such as Cygnus X-1, one member has become a black hole. Matter drawn from the other member would create a so-called *accretion disk* of spiraling matter around the hole (Fig. 2). The matter falling into the disk would be accelerated and heated. Collisions and deceleration would produce X-rays, and the black hole would appear as an X-ray source.

A similar mechanism may explain the enormous energy output of *active galaxies* and *quasars*. It has been proposed that the centers of these brilliant objects, which produce vast amounts of radiation, might contain black holes with masses millions or even billions of times that of the Sun. Indeed, recent observations suggest that even many normal galaxies, including our own Milky Way galaxy, may harbor black holes in their cores.

Today, most physicists and astronomers probably believe that black holes do exist. But until definitive proof is obtained, there will continue to be a lot of speculation.

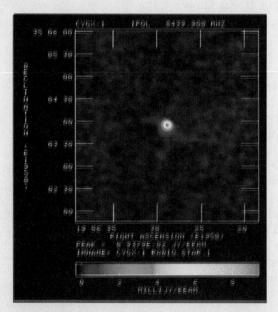

FIGURE 1 Cygnus X-1
The overexposed large dark spot at the center is a giant star believed to be a companion of the X-ray source and possibly a black hole, Cygnus X-1.

*26.7 Relativistic Velocity Addition

Objectives: To be able to (a) understand the necessity for a relativistic velocity addition formula, and (b) apply it to simple relative velocity calculations.

As we have seen, the postulates of special relativity affect our classical concepts of distance (and displacement) and time. Since velocity involves displacement and time, we might expect that measured relative velocities are also affected, and this is the case.

Recall from Section 26.1 that there was a problem with vector addition when light was involved. Ordinarily, if someone on a truck moving at 20 m/s threw a ball at 10 m/s in the direction of motion, an observer on the ground would say that the velocity of the ball was simply 20 m/s + 10 m/s = 30 m/s in the direction of the motion. However, if the ball is replaced by a beam of light (Fig. 26.1), such vector addition gives a speed greater than c, which violates the second postulate of relativity.

The same problem occurs with objects moving at appreciable fractions of the speed of light. Consider the rocket separation shown in ●Fig. 26.20. After separation, the jettisoned stage has a velocity **v** with respect to the Earth and the rocket has a velocity **u'** with respect to the jettisoned stage. Then, assuming straight line motion, by ordinary vector addition we would say that the velocity **u** of the rocket payload with respect to Earth was given by

Note: Review Section 6.6 and Fig. 6.22.

$$\mathbf{u} = \mathbf{v} + \mathbf{u'}$$

However, assuming relativistic speeds, suppose $v = 0.50c$ and $u' = 0.60c$. Then we would have $u = v + u' = 0.50c + 0.60c = 1.10c$. Thus, classical velocity addition says that an Earth observer would see the rocket traveling at a speed greater than the speed of light in vacuum. However, as we saw in Section 26.4, no object can travel faster than the speed of light.

Einstein recognized that since length and time are different with respect to different reference frames, classical velocity vector addition is not applicable in relativistic cases. Instead, he showed the correct formula for motion in a straight line to be

$$u = \frac{v + u'}{1 + \dfrac{vu'}{c^2}} \qquad \textit{(straight line only)} \qquad (26.20)$$

Relativistic velocity addition

where the velocities have the same meanings as above and sign notation is used to indicate velocity directions. Notice that the observed velocity u is *reduced* by a factor of $1/(1 + vu'/c^2)$ from that of the classical result.

It is important to clearly identify the velocities in working problems, so we list them again generally:

v = velocity of object 1 with respect to an inertial observer (e.g., on Earth)

u' = velocity of object 2 with respect to an object 1

u = velocity of object 2 with respect to an inertial observer (e.g., on Earth)

●**FIGURE 26.20 Relativistic velocity addition**
After jettisoning, the rocket has a velocity **u'** with respect to the jettisoned stage. The jettisoned stage has a velocity **v** with respect to the Earth. For relativistic velocities, the velocity **u** of the rocket with respect to the Earth is not given by ordinary vector velocity addition and must be obtained using a special formula. See text for description.

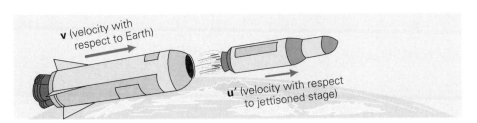

v (velocity with respect to Earth)

u' (velocity with respect to jettisoned stage)

Signs are used to indicate velocity directions. For example, if object 2 were approaching object 1, then we would have a negative value for u' in Eq. 26.20. Let's test Eq. 26.20 with some examples.

EXAMPLE 26.9 ■ FASTER THAN THE SPEED OF LIGHT? RELATIVISTIC VELOCITY ADDITION FOR OBJECTS

For the rocket separation in Fig. 26.20, let the velocities be $v = 0.50c$ and $u' = 0.60c$, as given in the text discussion. According to relativistic velocity addition, what is the velocity of the rocket nose cone as seen by an observer on Earth?

Solution. Listing the data,

Given: $v = 0.50c$ (velocity of jettisoned stage with respect to Earth)

$u' = 0.60c$ (velocity of rocket with respect to jettisoned stage)

Find: u (velocity of rocket with respect to Earth)

(The velocities are taken to be in the positive direction.) Using Eq. 26.20,

$$u = \frac{v + u'}{1 + \dfrac{vu'}{c^2}} = \frac{+0.50c + 0.60c}{1 + \dfrac{(+0.50c)(+0.60c)}{c^2}}$$

$$= \frac{+1.1c}{1.3} = +0.85c$$

Thus we have a speed less than c. (Note how it is convenient to carry c along in symbol form, which includes units.)

Follow-up Exercise. Repeat the calculations in this Example with both objects moving at 40% the speed of light. Classically, the answer is $0.80c$. Is the relativistically correct result the same? Explain mathematically.

EXAMPLE 26.10 ■ CAN LIGHT TRAVEL FASTER THAN LIGHT? RELATIVISTIC VELOCITY ADDITION FOR LIGHT

Suppose a spacecraft is moving in free space with a constant velocity of magnitude $0.25c$ relative to an observer. An astronaut on board the spacecraft sends out two light beams, one (a) in the direction of the craft's motion and the other (b) back toward the observer. What are the speeds of the light beams relative to the observer?

Solution. Here, if you apply the second postulate of special relativity, you already know the answers. But let's see if Eq. 26.20 gives us the correct answers.

Given: $v = 0.25c$
(a) $u' = c$ (in the direction of motion)
(b) $u' = -c$ (opposite the direction of motion)

Find: u (velocity of the light beam for both cases)

(a) Using Eq. 26.20, we have

$$u = \frac{v + u'}{1 + \dfrac{vu'}{c^2}} = \frac{+0.25c + c}{1 + \dfrac{(+0.25c)(c)}{c^2}}$$

$$= \frac{1.25c}{1.25} = +c$$

(b) Similarly, with $u' = -c$,

$$u = \frac{v + u'}{1 + \dfrac{vu'}{c^2}} = \frac{+0.25c - c}{1 + \dfrac{(+0.25c)(-c)}{c^2}}$$

$$= \frac{-0.75c}{0.75} = -c$$

Hence our relativistic velocity addition formula gives answers consistent with the special theory's second postulate—the constancy of the speed of light, as we expect, because the two postulates are what gave rise to all the relativistic formulas in this chapter!

Follow-up Exercise. Two rocket travelers approach one another, each traveling at $0.90c$ relative to a third inertial frame. Each shoots a light beam at the other. Calculate the speed that each measures for the other's light beam as it passes. [Hint: You know the answer from the postulates of special relativity.]

Chapter Review

Important Terms

relativity 799
inertial reference frame 800
principle of classical (Newtonian) relativity 800
ether 801
Michelson–Morley experiment 802
interferometer 802
principle of relativity 805
constancy of the speed of light 805

special theory of relativity 805
time dilation 809
proper time 810
proper length 813
length contraction 814
twin (clock) paradox 816
relativistic kinetic energy 818
rest mass 818
relativistic mass 818
relativistic momentum 819

relativistic total energy 819
rest energy 819
mass-energy equivalence 820
general theory of relativity 822
principle of equivalence 822
black hole 826
Schwarzschild radius 826
event horizon 826

Important Concepts

- An inertial (nonaccelerating) reference frame is one in which Newton's first law holds.

- The Principle of Relativity (Special Relativity Postulate I) states that all the laws of physics are the same in all inertial reference frames.

- The Constancy of the Speed of Light (Special Relativity Postulate II) states that the speed of light in a vacuum is the same for all inertial reference frames.

- Clocks moving relative to an observer in an inertial reference frame run more slowly than clocks at rest in the observer's frame (relativistic time dilation).

- Lengths of objects moving relative to an observer in an inertial reference frame are less (in the dimension of relative motion) than if the object were at rest in the observer's frame (relativistic length contraction).

- Relativistic mass, the mass of an object measured by an observer in motion with respect to it, is always greater than the object's proper mass, its mass measured in its rest frame.

- The relativistic mass–energy equivalence, $E = mc^2$, enables us to calculate an object's *total* energy based on its relativistic mass. Rest energy is the energy equivalent of an object's rest mass. An object has rest energy even if it has no kinetic or potential energy.

- The principle of equivalence, the guiding postulate of general relativity, states that no experiment performed in a closed system can distinguish between the effects of a gravitational field and the effects of an acceleration.

Important Equations

$$\gamma = \frac{1}{\sqrt{1 - (v/c)^2}} \qquad (26.7)$$

Time Dilation:

$$t = \gamma t_o = \frac{t_o}{\sqrt{1 - (v/c)^2}} \qquad (26.8)$$

Length Contraction:

$$L = \frac{L_o}{\gamma} = L_o\sqrt{1 - (v/c)^2} \qquad (26.10)$$

Relativistic Kinetic Energy:

$$K = \gamma m_o c^2 - m_o c^2 = E_o(\gamma - 1) \qquad (26.11)$$

Relativistic Mass:

$$m = \gamma m_o = \frac{m_o}{\sqrt{1 - (v/c)^2}} \qquad (26.12)$$

Relativistic Momentum:

$$\mathbf{p} = m\mathbf{v} = \gamma m_o \mathbf{v} \qquad (26.13)$$

Rest Energy:

$$E_o = m_o c^2 \qquad (26.15)$$

Relativistic Total Energy:

$$E = K + E_o \qquad (26.16)$$
$$E = K + m_o c^2 = mc^2$$

Total Energy and Rest Energy:

$$E = \gamma E_o \qquad (26.17)$$

Schwarzschild Radius:

$$R = \frac{2GM}{c^2} \qquad (26.19)$$

*Relativistic Velocity Addition:

$$u = \frac{v + u'}{1 + \dfrac{vu'}{c^2}} \qquad (26.20)$$

Exercises

Note: Assume c to be *exact* at 3.00×10^8 m/s. If a speed is given in terms of c it should be regarded as having the same number of significant figures as the numerical factor (e.g., $0.85c$ has two significant figures).

26.1 Classical Relativity

26.2 The Michelson–Morley Experiment

1   If an isolated object in a system exhibits a changing velocity, then (a) it is an inertial reference frame, (b) $\mathbf{F} = m\mathbf{a}$ applies to the system, (c) the laws of mechanics are the same in this reference frame as in all inertial frames, (d) none of these.

2   The principle of classical relativity (a) did not seem to apply to electricity and magnetism, (b) required a constant speed of light different from that predicted by Maxwell's equations, (c) implied that nothing could travel as fast as light.

3   The existence of the ether (a) would provide a special or absolute reference frame, (b) is necessary for light propagation, (c) made Maxwell's equations relativistically correct, (d) both (a) and (b).

4   The operation of the Michelson interferometer (a) was based on refraction, (b) depended on coherent light beams, (c) showed the expected fringe shift, (d) was based on the Lorentz–FitzGerald contraction.

5   The Michelson–Morley experiment (a) worked only at night, (b) showed the existence of the ether, (c) gave a null result, (d) none of these.

6   ■ We live on a rotating Earth and therefore are in an accelerating, noninertial system. How then can we apply Newton's laws of motion on Earth?

7   ■ How does the constancy of the speed of light explain the null result of the Michelson–Morley experiment?

8   ■ Car A is traveling eastward at 90 km/h. Car B is traveling with a speed of 65 km/h. Find the relative velocity of car B with respect to car A if car B is traveling (a) westward and (b) eastward.

9   ■ A person 1.50 km from your location fires a gun; you observe the muzzle flash. If a 10.0 m/s wind is blowing, how long will it take the sound to reach you if the wind is (a) toward you and (b) toward the other person? (The speed of sound is 345 m/s.)

10   ■ A small airplane has an air speed of 200 km/h (speed with respect to air). Find the airplane's ground speed if there is (a) a headwind of 35 km/h and (b) a tailwind of 25 km/h.

11   ■ A speedboat can travel 50 m/s in still water. If the boat is in a river that has a flow speed of 5.0 m/s, find the maximum and minimum values of the boat's speed relative to an observer on the riverbank.

12 ■■ A small boat can travel 0.45 m/s in still water. The boat heads directly across a 50-m-wide river, which has a flow speed of 0.15 m/s. (a) How long will it take the boat to reach the opposite shore? (b) How far downstream will the boat travel? (c) How would the boat have to be steered if it were to arrive at a point across the river directly opposite the departure point?

13 ■■■ The apparatus used for measuring the speed of light by the French scientist Fizeau in 1849 is illustrated in •Fig. 26.21. Teeth on a rotating disk periodically interrupt a beam of light. The light flashes travel to a plane mirror and are reflected back to an observer. Show that if the disk is rotated so that light passing through one tooth gap travels to the mirror and returns to be seen by the observer through the adjacent gap, the speed of light is given by $c = 2fNL$, where N is the number of gaps in the disk, f is the frequency of, or number of revolutions per second made by, the rotating disk, and L is the distance between the disk and the mirror.

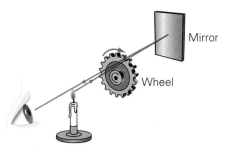

•**FIGURE 26.21   Fizeau's apparatus**
See Exercise 13.

14 ■■■ One proposed explanation of the null result of the Michelson–Morley experiment suggested that the interferometer arm in the direction of motion is shortened by a factor of $\sqrt{1 - (v/c)^2}$ (the Lorentz–FitzGerald contraction). Show how this contraction would give a null result. (This hypothesis was disproved by making the lengths of the arms of the interferometer unequal.)

15 ■■■ The swimmer analogy of the Michelson–Morley experiment is illustrated in •Fig. 26.22. Two swimmers start out from point A at the same time and swim with a constant speed c relative to the water. One swimmer swims directly across the river to point C and back to A, while the other swimmer swims to point B and back to A. Do the swimmers arrive back at A at the same time? Justify your answer mathematically.

26.3 The Postulates of Special Relativity and the Relativity of Simultaneity

16 The special theory of relativity (a) applies to inertial systems, (b) applies to noninertial systems, (c) applies to both inertial and noninertial systems, (d) disproves the constancy of the speed of light.

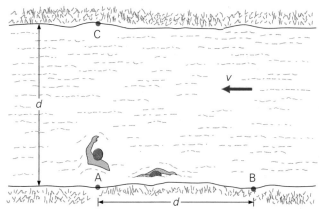

•**FIGURE 26.22   The swimmer analogy**
See Exercise 15.

17 The proper time interval between two events is (a) measured by a clock present at both events, (b) measured by any inertial reference frame, (c) dependent on the speed of the inertial frame measuring the interval.

18 ■ In a *gedanken* (thought) experiment, assume that you rode a beam of light coming from the face of a clock. How would time measured on the clock appear to you?

19 ■ In a *gedanken* experiment, you mark the ends of your spaceship simultaneously according to you, while also making marks on your friend's ship. He finds himself, coincidentally, in the exact middle of the marks in his reference frame. If he is traveling parallel to the length of your ship (from rear to front) at $0.5c$ he will observe (a) the marks made simultaneously, (b) the front mark before the rear mark, (c) the rear mark before the front.

20 ■ In the *gedanken* experiment in Exercise 19, in order for your friend to agree that the marks are made simultaneously, you must (a) make the marks simultaneously in your ship, (b) make the front mark before the rear mark, (c) make the rear mark before the front mark.

21 ■■ In a *gedanken* experiment (•Fig. 26.23a), two events in the same inertial reference frame (O) are related by cause and effect: (i) a gun at the origin fires a bullet along the x axis with a speed 300 m/s (assume that there are no gravitational or frictional forces). (ii) The bullet hits a target at $x = +300$ m. (a) What is the velocity of the inertial reference frame that would determine the proper time between these two events? (b) Use qualitative sketching arguments to show that these two events could not be viewed simultaneously by *any* inertial observer. [Thus we see that special relativity preserves the time ordering of two events if they could be cause and effect. Different observers may disagree on the time *interval* between the events, but not on their sequence—the firing of the gun will always precede the bullet hitting the target.]

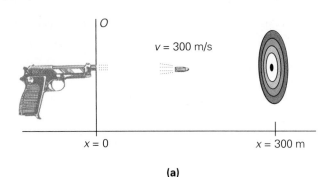

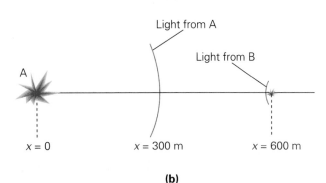

•FIGURE 26.23 Two thought experiments
See Exercises 21 and 22.

22 ■■ In a *gedanken* experiment (see •Fig. 26.23b), two events that cannot be related by cause and effect occur in the same inertial reference frame (O): (i) Strobe light A, at the origin of the x axis, flashes. (ii) 1.0 μs later, strobe light B, located at $x = +600$ m, flashes. B's flash clearly cannot be caused by A's flash (light travels only 300 m in the time between the two events). (a) Use a simple sketch to show that there exists another inertial reference frame, traveling at less than c, in which these two events would be observed to occur simultaneously. (b) What is the direction of the velocity of this reference frame? [Thus we see that special relativity does not preserve the time ordering of two events if they could not possibly be related by cause and effect. Different observers may disagree as to which event happened first, but this uncertainty has no significance for our understanding of physical laws because the two events are not causally related.]

26.4 The Relativity of Length and Time: Length Contraction and Time Dilation

23 An observer sees a friend passing by her in a rocket-ship with a uniform velocity that has a magnitude of an appreciable fraction of the speed of light. The observer knows her friend to be 5 ft 8 in. tall and he is standing so that his length is perpendicular to the relative velocity. To the observer, he will appear (a) taller, (b) shorter, (c) the same height (5 ft 8 in.)

24 ■ A person observes 5.0 min to elapse on her clock. How much time does she observe to elapse during this interval on a clock in another reference frame moving with a relative speed of 0.90c?

25 ■ A person has a pulse rate of 60 beats/min. What would the person's pulse rate be according to an observer moving with a speed of 0.85c relative to the person?

26 ■ A time of 3.0 min elapses on a clock, but an observer in another inertial reference frame measures 4 min to elapse. What is the relative speed of the reference frames of the clocks?

27 ■ A spaceship travels with a speed of $c/3$ relative to another inertial observer. What time change does a clock *on board* the spaceship show in 1.00 hour of elapsed time according to this other observer?

28 ■■ Alpha Centauri, the binary star closest to this solar system, is about 4.3 light years away. If a spaceship traveled this distance with a constant speed of 0.60c relative to Earth, (a) how much time would elapse on a clock on the spaceship? (b) How much time would elapse on an Earth-based clock?

29 ■■ One 25-year-old twin decides to take a round trip through space while the other remains on Earth. The traveling twin travels at a speed of 0.95c for 39 years according to Earth time. Assuming that special relativity applies, what are their ages when the traveling twin returns to Earth?

30 ■■ The proper lifetime of a muon is 2.20 μs. If the muon has a lifetime of 34.8 μs according to an observer on Earth, what is the speed of the muon, as a fraction of c?

31 ■■ An astronaut in a spaceship sets up a 25 g mass on a spring ($k = 0.28$ N/m) and starts it oscillating on a frictionless surface. If the spaceship travels with a constant velocity of 0.60c relative to Earth, what would be the period of the oscillator as measured by an observer on Earth?

32 ■■ A cylindrical-shaped spaceship has a length of 35.0 m and a diameter of 8.25 m. If it passes by the Earth with a relative speed of 2.44×10^8 m/s, traveling in the direction of its cylindrical axis, what are the dimensions of the ship measured by an Earth observer?

33 ■■ A "flying wedge" spaceship has a side view of a right triangle. At rest, the base and altitude (which form the 90° angle) measure 40.0 m and 15.0 m, respectively. As the ship moves by an observer on Earth with a speed of 0.900c, (a) what does she measure the area of the side of the ship to be? (b) What does she calculate the angle of the triangular "nose" to be?

34 ■■ How fast must a meterstick be moving relative to an observer so that he measures its length to be 50 cm?

35 ■■ How fast must a meterstick travel for its measured length to be (a) 1%, (b) 10%, and (c) 99% of its rest length?

36 ■■ The distance to a planet is 1.00 ly. How long does it take for a spaceship to reach the planet according to an inertial observer on the planet, ralative to which the speed of the spaceship is 0.800c?

37 ■■ A pole vaulter at the Relativistic Olympics sprints past you down the runway for a vault with a speed of 0.65c. When he is at rest, his pole is 7.0 m long. What does the length of the pole appear to be to you as he passes you? (What assumption do you need to make to arrive at your answer?)

38 ■■ What is the length contraction (ΔL) of an automobile 5.00 m long when it is traveling at 100 km/h? [Hint: $\sqrt{1 - x^2} \approx 1 - (x^2/2)$ for $x \ll 1$.]

39 ■■ An observer in a certain reference frame cannot understand why there is a problem in converting to the metric system because she notes the length of a meterstick in another moving reference frame to be the same length as her yardstick (parallel to the meter stick). Explain this situation quantitatively.

40 ■■■ Two reference frames move in relation to each other with a constant relative speed v in the x direction. Considering the O' frame to move in the $+ x$ direction relative to the O frame, (a) show that the coordinate systems are classically related by the transformation equations: $x = x' + vt$, $y = y'$, and $z = z'$. These are the so-called Galilean transformations, together with $t = t'$, since time classically is a universal quantity. (b) Taking into account the relativistic length contraction, show that the transformation equations are given by: $x' = \gamma(x - vt)$, $y' = y$, and $z' = z$. These relativistic transformations are called the Lorentz transformations. (Time is also affected, but the derivation of this transformation is beyond the scope of this book.)

26.5 Relativistic Kinetic Energy, Mass, Momentum, and Total Energy

41 The special theory of relativity shows relativistic effect(s) for the fundamental properties of (a) length, (b) mass, (c) time, (d) all of these.

42 The total energy for a relativistic moving particle is equal to (a) $E = \frac{1}{2}mv^2$, (b) $E = \gamma E_o$, (c) $E = K - m_o c^2$, (d) $E = m_o c^2$.

43 ■ If an electron has a kinetic energy of 2 keV, could the classical expression for kinetic energy be used to accurately compute its speed? What if the kinetic energy of the electron were 2 MeV?

44 ■ An electron travels at a speed of 0.500c. What is its total energy?

45 ■ An electron is accelerated from rest through a potential difference of 2.50 MV. Find the (a) speed, (b) kinetic energy, and (c) momentum of the electron.

46 ■ A home uses approximately 1.5×10^4 kWh of electricity per year. How much matter would have to be converted directly to energy to supply this amount of energy?

47 ■ How fast must an object travel for its mass to be (a) 1% more than its rest mass, and (b) 99% more than its rest mass?

48 ■ The United States uses over 3.0 trillion kWh of electricity annually. If 3.0 trillion kWh of electrical energy were supplied by nuclear generating plants, how much nuclear mass would have to be converted to energy, assuming a production efficiency of 35%?

49 ■ To travel to a nearby star, a spaceship is accelerated to 0.99c to take advantage of time dilation. If the ship has a rest mass of 3.0×10^6 kg, how much work must be done to get it up to speed from rest? Compare this to the annual electricity usage of the United States (see Exercise 48, 3.0 trillion kWh).

50 ■■ A person whose mass is 96 kg wishes to gain 24 kg relativistically with respect to another reference frame. (a) How fast would the person have to travel? (b) How would her height and width change if she stands so that her height is measured perpendicularly to the relative velocity direction? (c) How would her density change?

51 ■■ The kinetic energy of an electron is 60% of its total energy. Find the speed and momentum of the electron.

52 ■■ An electron has a total energy of 2.8 MeV. What is its momentum?

53 ■■ How much energy, in keV, is required to accelerate an electron from rest to 0.50c?

54 ■■ Sketch graphs of the (a) momentum, (b) energy, and (c) mass of an object as a function of speed.

55 ■■ A proton moves with a speed of 0.35c. What are its (a) total energy (b) kinetic energy, and (c) relativistic momentum?

56 ■■ A proton moving with a constant velocity has a total energy 2.5 times its rest energy. (a) What is the proton's speed? (b) What is its kinetic energy?

57 ■■ The Sun has a mass of 1.989×10^{30} kg and radiates at a rate of 3.827×10^{23} kW. (a) How much mass does the Sun lose in one hour? (b) How long will it take the Sun to lose one percent of its mass?

58 ■■ A beam of electrons is accelerated from rest to a speed of $0.950c$ in a particle accelerator. What are (a) the kinetic energy and (b) the total energy in MeV of an electron in the beam?

59 ■■ Phase changes require latent heat (Chapter 11). If one kilogram of ice at $0°C$ is converted to liquid at $0°C$, how much more mass would the liquid have as a result of absorbing energy?

60 ■■ A nickel has a mass of 5.00 g. If this mass could be completely converted to electric energy, how long would it keep a 60 W light bulb lit?

61 ■■■ Show that the *classical* kinetic energy and momentum of a particle can be expressed in terms of the rest energy of the particle as $K = \frac{1}{2}E_o \ (v^2/c^2)$ and $p = E_o v/c^2$.

62 ■■■ Using the relation $m^2c^4 = \gamma^2 m_o^2 c^4$, show that the total energy E and relativistic momentum p are related by $E^2 = p^2c^2 + (m_o c^2)^2$.

63 ■■■ The National Accelerator Laboratory at Batavia, Illinois, can accelerate particles to energies of giga-electron volts (1 GeV = 10^9 eV). (a) What is the speed of a proton with a kinetic energy of 1.00 GeV? ($m_o = 1.67 \times 10^{-27}$ kg.) (b) The planned Superconducting Super Collider (SSC) accelerator (which has been scrapped) would produce particles with kinetic energies of 40.0 GeV. What would be the speed of a proton in this case?

64 ■■■ At the Los Alamos Meson Physics Facility (LAMPF) protons are routinely accelerated to a kinetic energy of 600 MeV. (a) How do we know that 600 MeV must be kinetic energy and not total energy? (b) What is the speed of these protons? (c) What is their momentum?

26.6 The General Theory of Relativity

65 The general theory of relativity (a) is essentially a gravitational theory, (b) applies to rotating systems, (c) applies only to inertial systems, (d) refutes the principle of equivalence.

66 One of the predictions of the general theory is (a) mass-energy equivalence, (b) the constancy of the speed of light, (c) the twin paradox, (d) the bending of light in a gravitational field.

67 ■ Suppose a meterstick were dropped through the event horizon of a black hole. Describe what the effects might be on the meterstick.

68 ■■ If the Sun gravitationally collapsed to become a black hole, what would be its average density at the time it became a black hole? (See Example 26.8.)

69 ■■ Compare the radii of Earth's and Jupiter's event horizons if they contracted to black holes. ($M_E = 6.0 \times 10^{24}$ kg, $M_J = 318 \ M_E$)

70 ■■ A black hole has an event horizon of 5.00×10^3 m. (a) What is the mass of the "hole"? (b) Set a lower limit on the density of the black hole.

71 ■■ Assume that a collapsing star has a mass of 5.0 solar masses. (a) What would the radius of the collapsed star have to be for it to qualify as a black hole? (b) What would be the density of the black hole?

72 ■■ An apparatus like the one in •Fig. 26.24 was given to Albert Einstein on his 76th birthday by Eric M. Rogers, a physics professor at Princeton University. The goal is to get the ball into the cup without touching the ball. (Jiggling the pole up and down will not do it.) Einstein solved the puzzle immediately and then confirmed his answer with an experiment. How did Einstein get the ball into the cup? [*Hint:* He used a fundamental concept of general relativity.]*

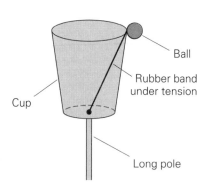

•**FIGURE 26.24 How to get the ball in the cup?** See Exercise 72.

*26.7 Relativistic Velocity Addition

73 ■ After jettisoning a stage, a rocket has a velocity of $+0.20c$ relative to the jettisoned stage. An observer on Earth sees the jettisoned stage moving with a velocity of $+7.5 \times 10^7$ m/s in the same direction as the rocket relative to her. What is the velocity of the rocket relative to the Earth observer?

74 ■ In moving away from a planet, a spacecraft fires a probe back toward the planet with a speed of $0.15c$ relative to the spacecraft. If the speed of the spacecraft is $0.40c$ relative to the planet, what is the velocity of the probe as seen by an observer on the planet?

*This problem is adapted from R. T. Weidner, *Physics.* Allyn and Bacon, 1985, p. 333.

75 ■■ A rocket launched outward from Earth has a speed of $0.100c$ relative to the Earth. It is directed toward an incoming meteor that may hit the planet. If the meteor moves directly toward the rocket with a speed of $0.250c$ relative to it, what is the velocity of the meteor as observed from Earth?

76 ■■■ In a linear particle accelerator, a proton with a speed of $0.800c$ moves in one direction and an electron moves with a speed of $0.900c$ in the opposite direction. Both speeds are relative to the laboratory frame of reference. (a) What is the velocity of the proton relative to the electron? (b) What is the velocity of the electron relative to the proton?

Additional Exercises

77 What is the energy equivalent of rest mass of a proton (in MeV)?

78 If the kinetic energy of a particle is increased by 10 keV, what is the corresponding increase in its relativistic mass?

79 A spaceship travels with a speed of 1.5×10^8 m/s relative to some observer in an inertial frame. How much time does a clock on board the spaceship appear to lose in one day according to that observer?

80 An electron travels at a speed of 9.5×10^7 m/s. What is its total energy?

81 A relativistic rocket is measured to be 50 m long, 2.5 m high, and 2.0 m wide. What are its dimensions to an inertial observer when the rocket travels at $0.75c$ in the direction of its length?

82 If the mass of 1.0 kg of coal (or any substance) could be completely converted into energy, how many kilowatt-hours of energy would be produced?

83 Imagine that you are moving with a speed of $0.80c$ past a person who is reading this chapter in the textbook. If he takes 30 min to read the chapter by his clock, how much time would be observed to elapse on your clock?

84 Each second the Sun gives off about 4.0×10^{26} J of radiant energy produced by converting mass to energy in nuclear reactions. (a) How many kilograms of mass are lost by the Sun each second? (b) If the Sun's mass is 2.000×10^{30} kg, at the rate calculated in part (a), how long in years will it be until the Sun's mass is reduced to 1.999×10^{30} kg?

85 A spaceship has a length of 150 m in its rest frame. An observer in another inertial frame measures the length of the spaceship as 110 m. What is the relative speed of the reference frames?

86 In the Michelson–Morley experiment, a difference in phase for the beams could be detected if the apparatus were rotated by 90°, since the interference pattern should change. Also, this would take care of the problem of assuming that the interferometer arm lengths are exactly equal, which should not be done arbitrarily. Show that the difference in the time differences for beam travel for the nonrotated and rotated conditions is

$$\Delta t - \Delta t' = \frac{2}{c}(L_1 + L_2)\left[\frac{1}{1 - (v^2/c^2)} - \frac{1}{\sqrt{1 - (v^2/c^2)}}\right]$$

where $\Delta t'$ is the time difference for the rotated condition and L_1 and L_2 are the lengths of the interferometer arms.

87 At typical nuclear power plants, refueling occurs every 18 months. Assuming that the plant has operated continuously since the last refueling, producing 1.2 GW of electrical power at an efficiency of 33%, how much less massive are the fuel rods at the end of the 18 months compared to the start. (Assume 30 day months.)

88 A proton, traveling at $0.90c$, has its path bent into a semicircular arc (radius 1.5 m) by a magnetic field at right angles to its velocity. What is the magnitude of the field (a) classically and (b) relativistically. (c) Explain why your answers are different.

27 Quantum Physics

For many years, the heroes and villains of science fiction routinely blasted away at each other with ray-guns, light sabers, energy beams, and the like. So when the laser was developed, it seemed at first as if real science was imitating fantasy. Lasers were associated with weaponry; indeed, the first thing many people heard about them was their possible use in the proposed "Star Wars" missile defense system.

Yet almost overnight, lasers started turning up in a variety of peaceful, everyday contexts—at light shows (like the one in the photo above), in bar-code scanners at store checkout counters, in computer printers. Today they hardly excite any special notice. No one is particularly surprised when a laser beam is bounced off a reflector on the Moon, or particularly apprehensive when told that they need to have a detached retina repaired by laser surgery. In fact, you probably own a laser yourself. (Think not? Do you have a CD player . . . ?)

As you'll see shortly, the laser is a practical application of principles that revolutionized physics when they were first developed in the early decades of this century. In this chapter, you'll learn how this "quantum revolution" took place, what was so radically new about it, and why some of its most basic concepts seemed so strange at the time—and still do today!

Einstein's theory of special relativity (Chapter 26) helped to resolve the problems that classical physics had in describing particles moving at speeds comparable to that of light. However, there were other troublesome areas in which classical theory did not agree with experimental results. As scientists explored the nature of the atom in the early years of the twentieth century, attempts to apply classical physics to explain phenomena on the atomic level often proved unsuccessful.

The period from about 1900 to 1930 was one of the most productive in the history of physics. Scientists devised new hypotheses based on nonclassical approaches, ushering in a profound revolution in our understanding of the physical world. Chief among these new theories was the idea that light is *quantized*, that it is made up of discrete amounts of energy. This hypothesis led to the formulation of a new set of principles and a new branch of physics, *quantum mechanics.*

Quantum mechanics introduced many seemingly strange, paradoxical concepts that nevertheless proved to be extremely powerful and far-reaching. Quantum theory forced physicists to reexamine the relationship between particles and waves. More specifically, it demonstrated that particles often exhibit wave properties and waves frequently behave like particles. It also showed that in the realm of the atom, explanations have to deal in *probabilities* rather than the precisely determined values of classical theory.

A detailed treatment of quantum mechanics is quite complex mathematically. However, a general overview of the important results is essential to an understanding of physics as we know it today. Thus, the important developments of "quantum" physics are presented in this chapter, and an introduction to quantum mechanics is provided in Chapter 28.

27.1 Quantization: Planck's Hypothesis

Objectives: To be able to (a) define blackbody radiation and use Wein's law
    to calculate the wavelength of maximum radiation, and
    (b) understand how Planck's hypothesis paved the way for
    quantum ideas.

One of the problems scientists were having at the end of the nineteenth century was how to explain the spectra of radiation emitted by hot objects. This is sometimes called **thermal radiation**. You learned in Chapter 11 that the total intensity of the emitted radiation is proportional to the fourth power of the absolute Kelvin temperature (T^4). Since the temperatures of all objects are above absolute zero, all objects emit thermal radiation.

Note: Review Eq. 11.5.

At normal temperatures, this radiation is in the infrared region and so is not detected visibly. At temperatures of about 1000 K, an object emits radiation in the long-wavelength end of the visible spectrum and has a reddish glow. A hot electric stove burner is a good example. Increased temperature causes the radiation and color to shift to a shorter wavelength, such as yellow-orange. Above a temperature of about 2000 K, an object glows with a yellowish-white color, like the glowing filament of a light bulb.

Analysis shows that, although we see only a particular color, there is a general continuous spectrum at all of these temperatures as illustrated in the graph in •Fig. 27.1a. The curves shown here do not correspond to a real radiating body; they are for an idealized blackbody. A **blackbody** is an ideal system that absorbs (and emits) all radiation that is incident on it.

Blackbody: an ideal absorber and emitter.

An absorbing blackbody may be approximated by a small hole leading to an internal cavity inside a block of material (see •Fig. 27.1b). Radiation falling on the hole enters the cavity and is reflected back and forth by the cavity walls. If the hole is very small in comparison to the surface area of the cavity, only a small amount of radiation will be reflected back out of the hole. Since nearly all radiation inci-

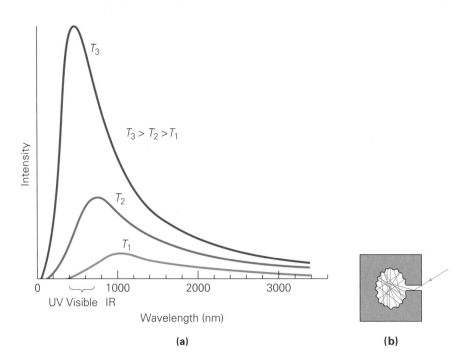

•**FIGURE 27.1 Thermal radiation**
(a) Intensity versus wavelength curves for the thermal radiation from an idealized blackbody at different temperatures. Notice how the location of maximum intensity shifts to a shorter wavelength with increasing temperature. **(b)** A blackbody may be approximated by a small hole leading to an interior cavity in a block of material.

Note: The concept of a blackbody was introduced in Section 11.4

Wien's displacement law— displacement of λ_{max} with temperature.

dent on the hole is absorbed and very little comes back out, the hole, viewed from the outside, is an excellent approximation to the surface of a blackbody.

Notice in the figure that as the temperature increases, more radiation is emitted at every wavelength and that the wavelength of the maximum-intensity component becomes shorter. This shift is found to obey a relationship, named **Wien's displacement law:**

$$\lambda_{max}\, T = \text{a constant} = 2.90 \times 10^{-3} \text{ m} \cdot \text{K} \qquad (27.1)$$

where λ_{max} is the wavelength of the radiation (in meters) at which maximum intensity occurs and T is the temperature of the body (in kelvins).

Wien's law can be used to determine the wavelength of the maximum spectral component if the temperature is known, or the temperature of the emitter if the wavelength of the strongest emission is known. Thus, it can be used to estimate the temperatures of stars from their radiations

---

EXAMPLE 27.1 ■ SOLAR COLORS: USING WIEN'S LAW

When we look at the Sun with the unaided eye, what we see is the gaseous photosphere from which radiation escapes. At the top of the photosphere, the temperature is 4500 K; at a depth of about 260 km, the temperature is 6800 K. What are the wavelengths of the radiation of maximum intensity for these temperatures?

Solution.

Given: $T_1 = 4500$ K   *Find:* λ_{max} (for different temperatures)
$T_2 = 6800$ K

The wavelengths are given by Wien's law (Eq. 27.1), so we have

$$\text{At top: } \lambda_{max} = \frac{2.90 \times 10^{-3}\,\text{m} \cdot \text{K}}{4500 \text{ K}} = 6.44 \times 10^{-7}\,\text{m} = 644 \text{ nm}$$

$$\text{At depth of 260 km: } \lambda_{max} = \frac{2.90 \times 10^{-3}\,\text{m} \cdot \text{K}}{6800 \text{ K}} = 4.26 \times 10^{-7}\,\text{m} = 426 \text{ nm}$$

Hence, at the Sun's surface, the emitted radiation of maximum intensity is in the orange region of the visible spectrum (Section 20.4). As the temperature increases with

Note: Review Fig. 20.18.

The Ultraviolet Catastrophe and the Planck Hypothesis

Classically, thermal radiation is pictured as resulting from the thermal agitation and acceleration of electric charges near the surface of an object. Electric charges in thermal agitation would undergo many different accelerations, so a continuous spectrum of emitted radiation would be expected. However, classical theoretical calculations based on the number of standing wave modes in a blackbody cavity predicted the intensity to be proportional to $1/\lambda^4$, that is, $I \propto 1/\lambda^4$.

At long wavelengths, the classical prediction agrees fairly well with experimental data. However, at shorter wavelengths, there is little correlation. Contrary to experimental observations (including Wien's law), classical theory predicts that the radiation intensity should increase as the wavelength goes to zero (•Fig. 27.2). This classical prediction is sometimes called the *ultraviolet catastrophe*—"ultraviolet" because the difficulty occurs for short wavelengths beyond the violet end of the visible spectrum and "catastrophe" because it predicts that the emitted intensity or energy will become infinitely large.

The failure of classical physics to explain the characteristics of thermal radiation led Max Planck, a German physicist, to reexamine the phenomenon. In 1900, Planck formulated a theory that correctly predicted the observed distribution of the blackbody radiation spectrum, but only by introducing a radical new idea. Planck showed that if it were assumed that the thermal oscillators could have only *discrete*, or particular, amounts of energy, rather than a continuous distribution of energies, or any arbitrary energy, then theory would agree with experiment.

These discrete amounts of energy of thermal oscillators were related to the frequency *f* of oscillation by

$$E_n = n(hf) \quad \text{for } n = 1, 2, 3, \ldots \tag{27.2}$$

that is, the energy is in integral multiples of hf. The symbol h represents a constant known (quite naturally) as **Planck's constant**, and has a value (to three significant figures) of

$$h = 6.63 \times 10^{-34} \text{ J·s}$$

This relationship is commonly called **Planck's hypothesis**. Rather than considering oscillator energy to be a continuous quantity as it is classically, by Planck's hypothesis the energy was *quantized*, or limited to discrete amounts. The smallest possible amount of oscillator energy, according to Eq. 27.2 with $n = 1$, is

$$E_1 = hf \tag{27.3}$$

All other permitted values of the energy are integral multiples of hf. The quantity hf is called a **quantum** of energy (from the Latin word *quantus*, meaning "how much"). Thus, Planck's hypothesis implied that each oscillator could absorb or emit energy *only in quantum amounts*.

Although the theoretical predictions agreed with experiment, Planck himself was not convinced of the validity of his quantum hypothesis. However, the concept of quantization was quickly used to explain other phenomena that could not be explained classically. His quantum hypothesis earned Planck the Nobel Prize in 1918.

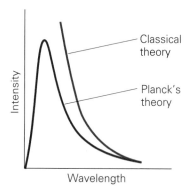

•**FIGURE 27.2   The ultraviolet catastrophe**
Classical theory predicts the intensity of thermal radiation to be proportional to $1/\lambda^4$, thus the intensity of the radiation should become infinitely large as the wavelength approaches zero. Planck's quantum theory, on the other hand, predicts the observed radiation distribution.

Note: As with other fundamental constants, we will take h to be *exact* at 6.63×10^{-34} J·s for significant figure purposes.

Planck's hypothesis—the birth of quantum physics.

Objectives: To be able to (a) describe the photoelectric effect, (b) explain how it can be understood by assuming that light energy is carried by particles, and (c) summarize the properties of photons.

The concept of the quantization of light was introduced in 1905 by Albert Einstein in a paper on light absorption and emission, interestingly enough at the same time he published his famous paper on the special theory of relativity! In contrast to Planck's hypothesis, which was made from a practical need to have a "theory" that fit the blackbody data, Einstein reasoned more fundamentally that energy quantization is a fundamental property of electromagnetic waves. He suggested that if the energy of the molecular oscillators in a hot substance is quantized, then it necessarily followed, to conserve energy, that *the emitted radiation should also be quantized*. He therefore boldly proposed that

> ...the radiant energy from a point source is not distributed continuously throughout an increasingly larger region, but, instead, this energy consists of a finite number of spatially localized energy quanta which, moving without subdividing, can only be absorbed and created in whole units.

A quantum or packet of light is referred to as a **photon,** and each photon has an energy

Energy of a photon of light with frequency f.

$$E = hf \qquad (27.4)$$

This idea suggests that light may sometimes behave as discrete quanta (plural of quantum), or "particles" of energy, rather than as waves. Equation 27.4 is thus the mathematical "connection" between the wave nature of light (a wave of frequency f) and the particle nature (photons each with an energy E). Given light of a certain frequency or wavelength, we can use this equation to calculate the amount of energy in each photon, or vice versa.

Einstein used this quantum concept to explain what is called the **photoelectric effect,** another area in which classical physical description was inadequate. Certain metallic materials are *photosensitive*. That is, when light strikes the surface of the material, electrons are emitted. The radiant energy supplies the work necessary to free the electrons from the material surface. Such materials are used to make photocells, as illustrated in •Fig. 27.3a. A voltage is maintained between the anode and the cathode. When light strikes the cathode which is the photosensitive material, electrons are emitted. These emitted *photoelectrons* complete the circuit, and a current is registered on the ammeter.

•FIGURE 27.3 The photoelectric effect and characteristic curves (a) Incident light on the photomaterial in a phototube (or cell) causes the emission of electrons, and a current flows in the circuit. The voltage applied to the tube is varied by means of the variable resistor. **(b)** As the plots of current versus voltage for two intensities of monochromatic light show, the current becomes saturated, or constant, as the voltage is increased. For negative voltages (by a reversal of the battery polarity), the current goes to zero at a particular stopping potential $-V_o$.

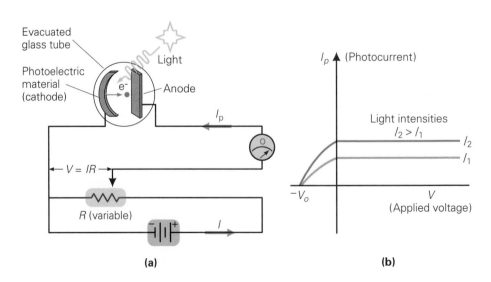

(a)

(b)

When a photocell is illuminated with monochromatic (single-frequency) light of different intensities, characteristic curves are obtained as shown in •Fig. 27.3b. The photocurrent increases as the voltage across the cell increases until a saturation current is reached. At this point, all of the emitted electrons are reaching the anode, and any further increase in voltage has no effect on the magnitude of the current.

As would be expected classically, the current is proportional to the intensity of the incident light—the greater the intensity ($I_2 > I_1$ in the figure), the more energy there is to free electrons.

A measure of the kinetic energy of the electrons being emitted from the photoelectric material can be gained by reversing the voltage across the electrodes. As the voltage is decreased and then made negative, the photocurrent decreases. Because of the retarding voltage, only the electrons with kinetic energy greater than $e|V|$ can make it to the negative plate and contribute to the current. (The kinetic energy is converted into electric potential energy as the electrons approach the negative plate.)

At some voltage V_o, called the **stopping potential**, no current will flow between the electrodes. The maximum kinetic energy (K_{max}) of the electrons emitted from the photomaterial is then related to the stopping potential by

$$K_{max} = eV_o \qquad (27.5)$$

since eV_o is the work needed to stop the most energetic electrons.

When the frequency of the incident light is varied, the maximum kinetic energy of the electrons is found to depend linearly on the frequency (•Fig. 27.4). No photoemission at all is observed to occur below a certain *cutoff frequency* f_o. Moreover, the electron emission begins instantaneously, with no observable time delay, when a photomaterial is illuminated by light with $f > f_o$, even if the light intensity is very low.

These four key characteristics of photoemission are summarized in Table 27.1 along with their correspondence with classical theoretical predictions. Note that only the first observation is predicted correctly by classical wave theory. Classically, the electric field of a light wave interacts with the electrons at the surface of the material and sets them into oscillation. The intensity (energy) of an oscillating electric field is proportional to the amplitude squared of the field vector, so the average kinetic energy of the electrons is proportional to the intensity of the light. Therefore, it is difficult to understand why the maximum kinetic energy of the photoelectrons is independent of the intensity, as is observed. Also, according to wave theory, photoemission should occur for any frequency of light, provided that the intensity is sufficient.

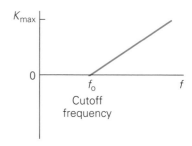

•**FIGURE 27.4 Maximum kinetic energy and frequency**
The maximum kinetic energy (K_{max}) of the photoelectrons is a linear function of the frequency of the incident light. Below a certain cutoff frequency f_o, no photoemission occurs.

TABLE 27.1	Photoelectric Effect Characteristics	
Characteristics		*Predicted by classical theory?*
1. The photocurrent is proportional to the intensity of the light.		Yes
2. The maximum kinetic energy of the emitted electrons is dependent on the frequency of the light but not on its intensity.		No
3. No photoemission occurs for light with a frequency below a certain cutoff frequency f_o regardless of its intensity.		No
4. A photocurrent is observed immediately when the light frequency is greater than f_o even if the light intensity is low.		No

The immediate emission of electrons raises an even more serious problem for classical theory. Classically, the time required for an electron to acquire enough energy to be freed and have a typically observed value of maximum kinetic energy is on the order of minutes for extremely low light intensities.

An explanation of the photoelectric effect was put forth by Einstein in 1905 using the quantum idea suggested by Planck in his theory of thermal oscillators. As was stated previously, he considered the energy of light to be in quantum bundles or photons, each with an energy $E = hf$. Thus, a light *wave* of frequency f could give energy to the electrons in a photomaterial only in discrete amounts equal to hf. In other words, we need to treat the electromagnetic energy flow (light) not as a wave, but as a particle in this particular experiment! An electron thus absorbs either the whole photon or nothing at all.

The electrons of a photomaterial are bound by attractive forces. Therefore, work must be done to free an electron, and some of the absorbed photon energy goes into doing this. Then, when a photon of energy hf is absorbed, by the conservation of energy we have

$$hf = K + \phi \tag{27.6}$$

where ϕ is the work or energy needed to free an electron from the material. Thus, *part of the energy of the absorbed photon goes into the work of freeing the electron, and the rest is carried off by the emitted electron as kinetic energy.*

The least tightly bound electron will have the maximum kinetic energy (K_{max}), and in this case the energy needed to free the electron is called the **work function** ϕ_o of the material. The energy conservation equation is then

Work function: the *minimum* energy needed to dislodge an electron.

$$\underset{\substack{\text{incident} \\ \text{photon} \\ \text{energy}}}{hf} \quad = \quad \underset{\substack{\text{maximum} \\ \text{kinetic} \\ \text{energy of} \\ \text{dislodged} \\ \text{electron}}}{K_{max}} \quad + \quad \underset{\substack{\text{minimum work} \\ \text{needed to} \\ \text{dislodge electron}}}{\phi_o} \tag{27.7}$$

Other electrons will require more energy than the minimum amount, and the kinetic energy of these electrons will be less than K_{max}.

EXAMPLE **27.2** ■ THE PHOTOELECTRIC EFFECT: ELECTRON SPEED AND STOPPING POTENTIAL

The work function of a particular metal is 2.00 eV. (a) If the metal is illuminated with monochromatic light having a wavelength of 550 nm, what will be the maximum speed of the emitted electrons? (b) What is the stopping potential?

Solution.

Given: $\phi_o = 2.00 \text{ eV} \times (1.6 \times 10^{-19} \text{ J/eV})$ *Find:*   (a) v_{max} (maximum speed)
            $= 3.20 \times 10^{-19} \text{ J}$ (b) V_o (stopping potential)
            $\lambda = 550 \text{ nm} = 5.50 \times 10^{-7} \text{ m}$

(a) The maximum speed is obtained from the maximum kinetic energy, which is given in Eq. 27.7. First computing the energy of a photon of light with the given wavelength, with $\lambda f = c$, we have

$$E = hf = \frac{hc}{\lambda} = \frac{(6.63 \times 10^{-34} \text{J·s})(3.00 \times 10^8 \text{m/s})}{5.50 \times 10^{-7} \text{m}}$$

$$= 3.62 \times 10^{-19} \text{ J}$$

Then

$$K_{max} = hf - \phi_o = 3.62 \times 10^{-19} \text{ J} - 3.20 \times 10^{-19} \text{ J}$$

$$= 4.2 \times 10^{-20} \text{ J} = \tfrac{1}{2} m v_{max}^2$$

where m is the mass of an electron and is a known quantity ($m = 9.11 \times 10^{-31}$ kg). Solving for v_{max}, we get

$$v_{max} = \sqrt{\frac{2K_{max}}{m}} = \sqrt{\frac{2(4.2 \times 10^{-20} \text{ J})}{9.11 \times 10^{-31} \text{ kg}}}$$

$$= 3.0 \times 10^6 \text{ m/s}$$

(b) The stopping potential is related to K_{max}, that is, $K_{max} = eV_o$, and

$$V_o = \frac{K_{max}}{e} = \frac{4.2 \times 10^{-20} \text{ J}}{1.60 \times 10^{-19} \text{ C}} = 0.26 \text{ V}$$

Follow-up Exercise. Suppose that, using light of a different wavelength, the stopping voltage is found to be 0.50 V. What is the wavelength of this light? Explain why this wavelength requires a different stopping voltage.

Einstein's quantum theory was consistent with experimental observation. Increasing the light intensity increases the number of photons and thus increases the number of electrons dislodged (the photocurrent). An increase in intensity does not change the energy of *individual* photons, however, which depends solely on the frequency of the light: $E = hf$. Therefore K_{max} is independent of intensity but linearly dependent on the frequency of the incident light as had been observed experimentally.

Einstein's theory also explained the existence of a limiting frequency. In the Einstein interpretation, the photon energy depends on frequency. Below a certain (cutoff) frequency, the photons do not have enough energy to dislodge electrons, so no current is observed. When $K_{max} = 0$, we can find the minimum cutoff frequency f_o from Eq. 27.7:

$$hf_o = K_{max} + \phi_o = 0 + \phi_o$$

or

$$f_o = \frac{\phi_o}{h} \qquad (27.8)$$

Cutoff frequency and work function.

In this case, the photon has just enough energy to free an electron from the material, but no extra energy to give it kinetic energy. (Notice that the graph in Fig. 27.4 provides a means to determine the value of h, which is the slope of the line.)

The frequency f_o is sometimes called the **threshold frequency**. Only light with a frequency above the threshold set by f_o will dislodge electrons. If the incident light has a frequency less than f_o, then no matter how many photons are available (that is, no matter how intense the light), the photons will not have enough energy to cause photoemission. No time delay in the emission of electrons would be expected in the quantum theory, since the energy is absorbed by an electron in a concentrated bundle instead of being continuously supplied and distributed over an area as in wave theory.

Note: A photon with energy hf_o will just dislodge an electron, but the electron will have no kinetic energy.

EXAMPLE 27.3 ■ THE PHOTOELECTRIC EFFECT: THRESHOLD FREQUENCY AND WAVELENGTH

What are the threshold frequency and corresponding wavelength of the metal described in Example 27.2?

Solution. We are given the work function, which is the minimum required photon energy to dislodge electrons. Listing the data, we have

Given: $\phi_o = 2.00$ eV
$\quad\quad\quad = 3.20 \times 10^{-19}$ J
(from Example 27.2)

Find: f_o (threshold or minimum frequency)
$\quad\quad\lambda_o$ (wavelength at threshold frequency)

The threshold (or cutoff) frequency is related directly to the work function, $\phi_o = hf_o$ (Eq. 27.8), so finding f_o is a straightforward calculation:

$$f_o = \frac{\phi_o}{h} = \frac{3.20 \times 10^{-19}\,\text{J}}{6.63 \times 10^{-34}\,\text{J·s}} = 4.83 \times 10^{14}\,\text{Hz}$$

as is finding the wavelength corresponding to the threshold frequency:

$$\lambda_o = \frac{c}{f_o} = \frac{3.00 \times 10^8\,\text{m/s}}{4.83 \times 10^{14}\,\text{Hz}} = 6.21 \times 10^{-7}\,\text{m} = 621\,\text{nm}$$

Any frequency lower than 4.83×10^{14} Hz or any wavelength longer than 621 nm will not yield photoelectrons. Notice that this wavelength lies in the red end of the visible region of the electromagnetic spectrum, so yellow light, for example, will dislodge electrons, but deeper red or infrared will not.

Follow-up Exercise. In this Example, what would the stopping voltage be if the frequency of the light were twice the cutoff frequency? Explain your reasoning.

PROBLEM-SOLVING HINT

In photon calculations, we are often given the wavelength of the light rather than the frequency, and need to find the corresponding photon energy. Instead of first calculating the frequency ($f = c/\lambda$), then the energy in joules ($E = hf$), and finally converting to electron-volts, we can do all this in one step. We simply use the relationship

$$E = hf = hc/\lambda$$

with hc expressed in eV·nm. The value of this useful constant is

$$hc = (6.63 \times 10^{-34}\,\text{J·s})(3.00 \times 10^8\,\text{m/s}) = 1.99 \times 10^{-25}\,\text{J·m}$$

$$= \frac{(1.99 \times 10^{-25}\,\text{J·m})(10^9\,\text{nm/m})}{1.60 \times 10^{-19}\,\text{J/eV}}$$

$$= 1.24 \times 10^3\,\text{eV·nm} \quad \text{(to 3 significant figures)}$$

This shortcut can save time and effort in working problems, and makes it very easy to estimate the photon energy associated with light of a given wavelength (or vice versa). Thus, if you are given orange light, for example, with a wavelength of 600 nm, you need only divide 1240 by 600 to find that each photon carries approximately 2 eV of energy:

$$E = \frac{hc}{\lambda} = \frac{1.24 \times 10^3\,\text{eV·nm}}{600\,\text{nm}} = 2.07\,\text{eV}$$

There are many applications of the photoelectric effect. The fact that the current produced by photocells is proportional to the light intensity makes them ideal for use in photographers' light meters. Photocells are also used in solar energy applications to convert sunlight to electricity. They are not yet efficient enough (about 20–25%) to produce electricity commercially, but this pollution-free method has great potential.

Another common application of the photocell is in the so-called "electric eye." Figure 27.5a illustrates the principle of the electric eye. As long as light strikes the photocell, current flows in the circuit. Interrupting or blocking the light opens the circuit in the relay (magnetic switch), which controls some device. A common application of the electric eye is to turn on street lights automatically at night. A safety application is shown in Fig. 27.5b. The threshold frequencies of some metals lie outside the visible region. Hence, nonvisible light (infrared or ultraviolet) may be used in such applications as burglar alarms or home protection systems.

Einstein's theory of the photoelectric effect represented a great success for the idea of quantization. It is interesting to note that Einstein won the Nobel Prize in physics in 1921 for his theory of the photoelectric effect rather than for his more famous theory of relativity (Chapter 26). Two years later, the quantum theories scored another triumph, which we will look at in the following section.

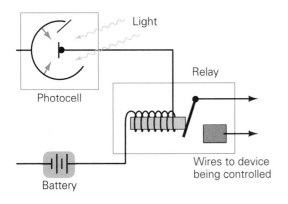

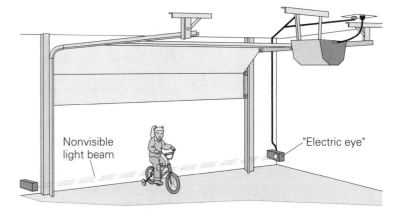

(a) Electric eye diagram

(b) Electric eye application

●**FIGURE 27.5 Photoelectric applications**
(a) A diagram of an electric eye circuit. When light strikes a phototube or cell, there is a current in the circuit. Interrupting the light beam opens the circuit in the relay or magnetic switch, which controls some device. **(b)** An electric eye may be used in an automatic garage door application. When the door starts downward, any interruption of the electric eye beam will cause the door to stop, protecting children or pets that may be passing under the descending door.

27.3 Quantum "Particles": The Compton Effect

Objectives: To be able to (a) understand how the photon model for light explains the scattering of light from various materials (the Compton effect), and (b) calculate scattered wavelengths in the Compton effect.

In 1923, the American physicist Arthur H. Compton (1892–1962) explained a phenomenon he observed in the scattering of X-rays from a graphite (carbon) block by considering the radiation to be composed of quanta. This explanation of the observed effect provided convincing evidence that light, and electromagnetic radiation in general, is, at least in certain types of experiments, composed of quanta, or "particles" of energy.

Compton had observed that when a beam of monochromatic X-rays was scattered by a material, the scattered radiation had a wavelength slightly longer than the wavelength of the incident beam. Also, the change in the wavelength was dependent on the angle through which the X-rays were scattered but not on the nature of the scattering material (see ●Fig. 27.6). This phenomenon came to be known as the **Compton effect.**

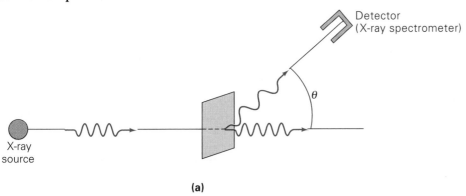

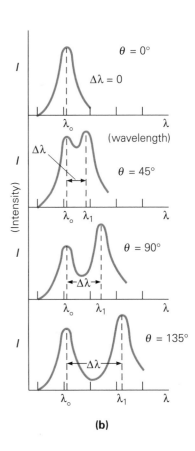

●**FIGURE 27.6 X-ray scattering**
(a) When X-rays of a single wavelength are scattered by a material, the wavelengths of the scattered X-rays are longer than the wavelength of the incident beam. **(b)** The wavelength shift depends on the scattering angle.

Classically, scattered radiation should have the same frequency or wavelength as the incident radiation. The electrons in the atoms of the scattering material absorb the radiation and oscillate at the same frequency as the incident wave. On emission, the scattered radiation should have the same frequency in all directions.

According to Einstein's photon theory, the energy of one quantum is directly proportional to the frequency of the light ($E = hf$), so a change in frequency or a shift in wavelength would indicate a change in photon energy. Here the scattered photons had *less* energy than the incident ones. The effect of the scattering angle on energy reminded Compton of scattering in the elastic collision of particles. Could the same principles apply in the scattering of quantum "particles"?

Pursuing this idea, Compton was able to explain the observed effect. He assumed that a quantum behaves like a particle or billiard ball in collision with other particles. He also considered only metals where he could assume that there were lots of essentially free electrons (the loosely bound ones responsible for the high electrical conductivity of metals). If an incident quantum collides with an electron initially at rest, the quantum loses some energy and momentum to the electron. Hence, he reasoned, the energy and frequency of the scattered quantum would be less ($E = hf$) and its wavelength is greater ($\lambda = c/f$). The collision is elastic, so both energy and momentum must be conserved. However, in this case it is the *relativistic* energy and momentum which are conserved. Applying these principles, Compton showed that the shift in the wavelength of the scattered quantum is given by

Note: Elastic collisions are discussed in Section 6.4.

Compton scattering

$$\Delta\lambda = \lambda_1 - \lambda_o = \lambda_C(1 - \cos\theta) \qquad (27.9)$$

where λ_o is the wavelength of the incident quantum and λ_1 that of the scattered quantum (see Exercise 77). The constant $\lambda_C = h/m_e c = 2.43 \times 10^{-12}$ m $= 2.43 \times 10^{-3}$ nm is called the *Compton wavelength* of the electron. The equation correctly predicted the observed wavelength shift, and Compton was awarded a Nobel Prize in 1927 for his work.

Notice that the maximum wavelength increase is $\Delta\lambda_{max} = 2\lambda_C = 0.00486$ nm which occurs when the photon is scattered directly backwards. (Can you use the elastic collision model to explain why?) Since this is the maximum possible wavelength *change*, clearly it will be negligible for incident wavelengths several thousand times this value or larger than several nm (strong UV). In other words, the Compton effect is experimentally negligible for ultraviolet light and longer wavelengths. It is only significant for X-ray and gamma ray scattering.

EXAMPLE 27.4 ■ X-RAY SCATTERING: THE COMPTON EFFECT

A monochromatic beam of X-rays with a wavelength of 1.35×10^{-10} m is scattered by a metal foil. By what percentage is the wavelength shifted for the scattered component observed at an angle of 90°?

Solution. We expect the change to be positive, indicating an increase in the scattered wavelength compared to the incident value. Listing the data, we have

Given: $\lambda_o = 1.35 \times 10^{-10}$ m 　　*Find:* 　% shift in wavelength
$\qquad\quad \theta = 90°$

The change or shift in the wavelength, $\Delta\lambda$, is given by Eq. 27.9, and the fractional change is $\Delta\lambda/\lambda_o$. Then

$$\frac{\Delta\lambda}{\lambda_o} = \frac{\lambda_C}{\lambda_o}(1 - \cos\theta)$$

$$= \frac{2.43 \times 10^{-12}\,\text{m}}{1.35 \times 10^{-10}\,\text{m}}(1 - \cos 90°) = 1.80 \times 10^{-2}$$

and

$$\frac{\Delta\lambda}{\lambda_{\mathrm{o}}} \times 100\% = 1.80\%$$

Follow-up Exercise. In this Example, (a) what would the fractional change be for backscattering (180°)? (b) Is it larger or smaller than the Example answer? Why is this so?

Einstein's and Compton's successes in explaining electromagnetic phenomena in terms of quanta left scientists with two apparently competing theories of electromagnetic radiation. Classically, the radiation is pictured as a continuous wave, and classical theory explains satisfactorily such wave-related phenomena as interference and diffraction. On the other hand, quantum theory was necessary to satisfactorily explain the photoelectric and Compton effects.

The two theories gave rise to a description that is called the **dual nature of light**. That is, light apparently behaves sometimes as a wave and at other times as photons or "particles." The relationship between the wave and particle nature of matter will be considered in the next chapter.

27.4 The Bohr Theory of the Hydrogen Atom

Objectives: **To be able to (a) understand how the Bohr model of the hydrogen atom explains atomic emission and absorption spectra, (b) calculate the energies and wavelengths of emitted photons for various transitions in hydrogen, and (c) understand how the generalized concept of atomic energy levels can be used to explain other atomic phenomena.**

In the 1800s, a great deal of experimental work was done with gas discharge tubes—for example, those containing hydrogen, neon, and mercury vapor (•Fig. 27.7a). (The common neon "lights," such as those in •Fig. 27.7b, are gas discharge tubes.) Normally, light from an incandescent source exhibits a *continuous spectrum*. However, when light emissions from these tubes were analyzed, discrete, or line, spectra were observed (•Fig. 27.8). Light coming directly from a tube gives a *bright-line* or **emission spectrum**, indicating that only certain specific wavelengths are present. In general, a discrete line spectrum is characteristic of the atoms or molecules of a particular material and is used to identify that material spectroscopically (Fig. 27.8a).

But atoms absorb light as well as emitting it. If white light is passed through a relatively cool gas, certain frequencies or wavelengths are found to be missing. The result is an **absorption spectrum**—a series of dark lines superimposed on a continuous spectrum (Fig. 27.8b). When the absorption and emission lines of a particular gas were compared, they were found to occur at the same frequencies. The reason for line spectra was not understood at the time, but they provided a clue to the electronic structure of the atoms.

Hydrogen, with a relatively simple visible spectrum, received much attention. Hydrogen is also the simplest atom, with only one electron and one proton. In the late nineteenth century, a Swiss physicist, J. J. Balmer, found an empirical formula that gives the wavelengths of the spectral lines of hydrogen in the visible region:

$$\frac{1}{\lambda} = R\left(\frac{1}{2^2} - \frac{1}{n^2}\right) \quad \text{for } n = 3, 4, 5, \ldots \tag{27.10}$$

where R is called the *Rydberg constant* and has a value of 1.097×10^{-2} nm^{-1}. The spectral lines of hydrogen in the visible region, called the **Balmer series**, were

(a)

(b)

•**FIGURE 27.7 Gas discharge tubes** The luminous glass tubes that we see everywhere today are actually gas discharge tubes, in which atoms of various gases emit light when electrically excited. Each gas radiates its own characteristic wavelengths. Not all "neon lights" actually contain neon, which glows with a red hue **(a)**; other gases are used to produce other colors **(b)**.

(a)

(b)

(c)

•**FIGURE 27.8 Spectra**
When a gas is excited by heat or electricity and the light that it emits is separated into
its component wavelengths with a prism or diffraction grating, the result is a bright-line,
or emission, spectrum. Each atom or molecule emits a characteristic pattern of discrete
wavelengths. The spectrum in **(a)** is that of barium; in **(b)**, that of calcium. When a
continuous spectrum consisting of all wavelengths is passed through a cool gas, a series
of dark lines is observed. Each line represents a "missing" wavelength—a particular
wavelength that has been absorbed by the gas. The wavelengths absorbed by any
substance are the same ones it emits when excited. The spectrum in **(c)** is that of the
Sun. Several prominent absorption lines can be seen, produced by the gases of the solar
atmosphere (Can you identify an element present in the Sun from these spectra?).

found to fit the formula, but it was not understood why. Similar formulas were
found to fit the spectral lines in the ultraviolet and infrared regions.

An explanation of the spectral lines was given in a theory of the hydrogen
atom put forth in 1913 by the Danish physicist Niels Bohr. Bohr assumed that the
hydrogen electron orbited the nuclear proton in a circular orbit (analogous to a
planet orbiting the Sun). The electrical Coulomb force (Chapter 15) supplies the
necessary centripetal force for the circular motion, and we may write

Note: Review Eq. 7.11 and Eq. 15.2.

$$F = \frac{mv^2}{r} = \frac{ke^2}{r^2} \tag{27.11}$$

where e is the magnitude of the charge of the proton and the electron, v the electron's orbital speed, and r the radius of the orbit (•Fig. 27.9).

The total energy of the electron is the sum of its kinetic and potential energies:

$$E = K + U = \tfrac{1}{2}mv^2 - \frac{ke^2}{r}$$

From Eq. 27.11, the kinetic energy may be written $\tfrac{1}{2}mv^2 = ke^2/2r$. Using this, the
total energy becomes

$$E = \frac{ke^2}{2r} - \frac{ke^2}{r} = -\frac{ke^2}{2r} \tag{27.12}$$

Notice that E is negative indicating that the system is bound, but that as the radius approaches infinity, E approaches zero. With $E = 0$, the electron would no
longer be bound to the proton, and the atom, having lost its electron, would be
ionized.

Thus far, only classical principles have been applied. But at this point in the
theory, Bohr made a radical assumption—radical in the sense that he introduced
a quantum concept. His assumption was that *the angular momentum of the electron
was quantized* and could have only discrete values that were integral multiples of
$h/2\pi$, where h is Planck's constant. In equation form,

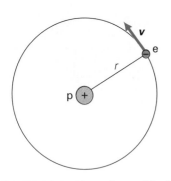

•**FIGURE 27.9 The Bohr model of
the hydrogen atom**
The electron of the hydrogen atom
is pictured as revolving around the
nuclear proton in a circular orbit,
similar to the way a planet orbits
the Sun.

$$mvr = n\left(\frac{h}{2\pi}\right) \quad \text{for } n = 1, 2, 3, 4, \ldots \tag{27.13}$$

The integer n is called a *quantum number*, specifically the **principal quantum number**. According to this assumption, the orbital speed of the electron (v) may be
found. It has only certain values given by

$$v = \frac{nh}{2\pi mr}$$

Putting this expression for v into Eq. 27.11 and solving for r, we find that

$$\frac{ke^2}{r^2} = \frac{mv^2}{r} = \frac{m(nh/2\pi mr)^2}{r}$$

and

$$r_n = \left(\frac{h^2}{4\pi^2 ke^2 m}\right) n^2 \quad \text{for } n = 1, 2, 3, 4, \ldots \qquad (27.14)$$

Here we use the subscript on r to indicate that for the electron orbits only certain radii are possible as determined by the principal quantum number n. The energy for a particular orbit can be found by substituting this expression for r into Eq. 27.12:

$$E_n = -\frac{ke^2}{2r_n} = -\frac{ke^2}{2}\left(\frac{4\pi^2 ke^2 m}{h^2 n^2}\right)$$

and

$$E_n = -\left(\frac{2\pi^2 k^2 e^4 m}{h^2}\right)\frac{1}{n^2} \quad \text{for } n = 1, 2, 3, 4, \ldots \qquad (27.15)$$

where the energy is also written as E_n to show its dependence on n. The quantities in the parentheses on the right-hand sides of Eqs. 27.14 and 27.15 are constants and are easily evaluated. The important results are

$$r_n = 0.0529 n^2 \text{ nm} \qquad (27.16)$$

$$E_n = \frac{-13.6}{n^2} \text{ eV} \quad \text{for } n = 1, 2, 3, 4, \ldots \qquad (27.17)$$

Principal quantum number, n: specifies orbit size and energy

where the radius is commonly expressed in nanometers and the energy in electron volts.

EXAMPLE 27.5 ■ THE FIRST EXCITED STATE: RADIUS AND ENERGY OF A BOHR ORBIT

Find the orbital radius and energy of an electron in a hydrogen atom characterized by the principal quantum number $n = 2$.

Solution. For $n = 2$, we make use of the numerical results given above

$$r_2 = 0.0529 n^2 \text{ nm} = 0.0529(2)^2 \text{ nm} = 0.212 \text{ nm}$$

and

$$E_2 = \frac{-13.6 \text{ eV}}{n^2} = \frac{-13.6 \text{ eV}}{2^2} = -3.40 \text{ eV}$$

Follow-up Exercise. In this Example, what is the speed of the orbiting electron?

However, there is still a classical problem with Bohr's theory. Classically, an accelerating electron radiates electromagnetic energy, and even in discrete circular orbits the electron would be centripetally accelerating. Thus, an orbiting electron would lose energy and spiral into the nucleus, as an Earth satellite might go into a decaying orbit because of frictional losses. This doesn't happen in the hydrogen atom, so Bohr had to make another nonclassical assumption. *He postulated that the hydrogen electron does not radiate energy when it is in a bound, discrete orbit but does so only when it makes a transition to another orbit.*

•FIGURE 27.10 Orbits and energy levels of the hydrogen electron
The Bohr theory predicts that the hydrogen electron can occupy only certain orbits having discrete radii. Each allowed orbit has a corresponding energy. These are conveniently displayed as energy levels. The lowest energy level ($n = 1$) is called the ground state, and those above ($n > 1$) are called excited states. The orbits are shown at left, with orbital radius plotted against the $1/r$ electrical potential of the proton. You can see that the electron in the ground state is at the bottom of a potential well analogous to the gravitational potential well of Figure 7.19. (Note that neither the radii nor the energy levels are drawn to scale—can you tell why?)

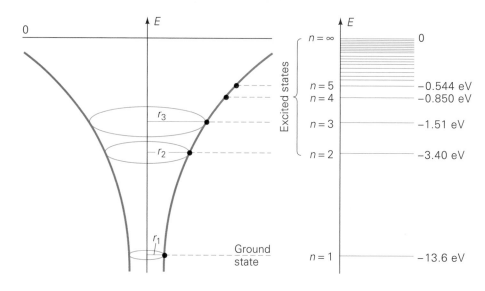

Note: Review the concept of potential energy wells presented in Sections 5.4 and 7.5 (see Figs. 5.14 and 7.18).

Ground state: $n = 1$
Excited states: $n \geq 2$

Energy Levels

The possible allowed orbits of the hydrogen electron are commonly expressed in terms of energy levels as illustrated in •Fig. 27.10. In this context, we simply refer to the electron as being in a particular energy level or state. The electron, being bound to the nuclear proton, is in a potential energy well, much like a gravitational potential well. The principal quantum number designates the energy level. The lowest energy level ($n = 1$) is called the **ground state**. The energy levels above the ground state are called **excited states**. For example, $n = 2$ is the first excited state and so on.

The electron is normally in the ground state and must be given energy to excite it, or to raise it up in the well to an excited state. The hydrogen electron can be excited only by discrete amounts. The energy levels are somewhat analogous to the rungs or steps of a ladder. Analogous to a person going up and down a real ladder, an electron goes up and down the energy ladder in discrete steps. Notice, however, that the energy rungs of the hydrogen atom are not evenly spaced.

If enough energy is absorbed to raise the electron to the top of the energy well so that the electron is no longer bound, the atom is ionized. For example, to raise a hydrogen electron from the ground state to $n = \infty$ requires 13.6 eV of energy. However, if the electron is already part way up the ladder (or higher up in the well)—that is, in an excited state—less energy is needed to get it to the top of the well. The energy of the electron in any state is E_n, and the energy needed to raise it to the top of the well is $-E_n$, which is called the **binding energy** of the electron. (Since E_n is negative—the electron is in a negative potential well—then $-E_n$ is positive, or the amount of energy needed to be *given* to the electron to raise it to the top of the well.)

An electron generally does not remain in an excited state for long; it decays, or makes a transition to a lower energy level, in a short time. The time an electron spends in an excited state is called the **lifetime** of the excited state. For many states, the lifetime is about 10^{-8} s. In making a transition to a lower state, energy is emitted as a photon of light (•Fig. 27.11). The energy of the photon is equal to the energy *difference* of the levels, that is,

$$\Delta E = E_{n_i} - E_{n_f} = \frac{-13.6}{n_i^2}\,\text{eV} - \left(\frac{-13.6}{n_f^2}\,\text{eV}\right)$$

or
$$\Delta E = 13.6\left(\frac{1}{n_f^2} - \frac{1}{n_i^2}\right) eV \qquad (27.18)$$

where the subscripts i and f refer to the initial and final states, respectively. Since $\Delta E = hf = hc/\lambda$, only photons of particular frequencies and wavelengths may be emitted. These correspond to the various discrete transitions.

The different values for the final principal quantum number refer to the energy level in which the electron starts or ends in the various emission or absorption series. The original Balmer series in the visible region corresponds to $n_f = 2$. There is only one series entirely in the ultraviolet range, the *Lyman* series, in which the transitions all end (in the case of emission) or start (in the case of absorption) with $n = 1$ (the ground state). There are many series entirely in the infrared region, most notably the *Paschen* series which ends or starts with the electron in the second excited state, $n = 3$.

Usually we are interested in the wavelength of the light (experimentally, that is what is commonly measured). Solving finally for the wavelength of the emitted light, λ, we get

$$\lambda = \frac{hc}{\Delta E} = \frac{1.24 \times 10^3}{\Delta E} nm \qquad (27.19)$$

where ΔE is expressed in electron volts and the constant hc has been replaced by its numerical value of 1.24×10^3 eV·nm (see the Problem-Solving Hint on p. 846).

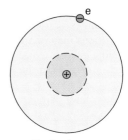

Excited atom

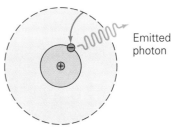

Emitted photon

De-excitation

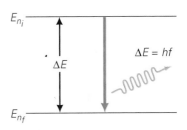

EXAMPLE 27.6 ■ INVESTIGATING THE BALMER SERIES: LINE WAVELENGTHS

What is the wavelength of the emitted light when excited electrons in hydrogen atoms make transitions from the $n = 3$ energy level to the $n = 2$ energy level?

Solution. We suspect that this will be the Balmer series since the final principal quantum number is two, in agreement with Balmer's original experimental findings (see Eq. 27.10)

Given: $n_i = 3$ *Find:* λ (wavelength of emitted light)
         $n_f = 2$

The energy of the emitted photon is

$$\Delta E = 13.6\left(\frac{1}{n_i^2} - \frac{1}{n_f^2}\right) eV = 13.6\left(\frac{1}{4} - \frac{1}{9}\right) eV = 1.89 \, eV$$

and, using Eq. 27.19 recalling that ΔE must be expressed in eV,

$$\lambda = \frac{1.24 \times 10^3}{\Delta E} nm = \frac{1.24 \times 10^3}{1.89} nm = 656 \, nm$$

which is light in the red portion of the visible spectrum.

Follow-up Exercise. Light of what wavelength would be just sufficient to ionize a hydrogen atom in its first excited state?

•FIGURE 27.11 Electron transitions and photon emission
When the electron in an excited atom decays or makes a transition to a lower orbit or energy level, a photon is emitted. The energy of the photon is equal to the energy difference in the energy levels.

We see that in the Bohr model, the hydrogen electron can make transitions only between discrete energy levels, thus emitting photons of discrete energies, or light of discrete wavelengths, producing spectral lines (•Fig. 27.12). The wavelengths of the various transitions were computed and found to correspond to the experimentally observed spectral lines.

•FIGURE 27.12 Hydrogen spectrum

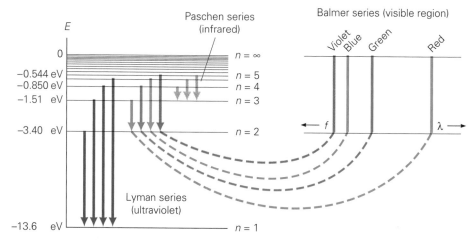

Paschen series (infrared)

Balmer series (visible region)

Lyman series (ultraviolet)

Transitions may occur between two or more energy levels as the excited electron returns to the ground state. Transitions to the *n* = 2 state give spectral lines that lie in the visible region of the spectrum and are called the Balmer series. Transitions to other levels give other series as shown.

(a) Visible illumination

(b) UV illumination

•FIGURE 27.13 Fluorescence
Many minerals will glow at visible wavelengths when illuminated by invisible ultraviolet light ("black light"). The visible light is produced when electrons excited by the UV fall back to lower energy levels in two or more smaller steps.

CONCEPTUAL EXAMPLE 27.7 ■ ENTIRELY VISIBLE? THE BALMER SERIES

We know that four wavelengths of the Balmer series are in the visible range. Are there more in the series, and if so, what type of light are they likely to be? (a) infrared, (b) visible, (c) ultraviolet. *Clearly establish the reasoning and physical principle(s) used in determining your answer before checking it below. That is,* **why** *did you select your answer?*

Reasoning and Answer.   The Balmer emission series ends with the electron in the first excited state (*n* = 2). We know that red light is emitted when the electron drops from the *n* = 3 level. The other three visible lines are emitted during the downward transitions from *n* = 4, 5, and 6. However, there is no reason why a transition cannot occur from *n* = 7 to *n* = 2. This is still in the Balmer series, but it involves a larger energy difference than the visible emissions. Hence the emitted photons will have larger energies, corresponding to smaller wavelengths. Wavelengths smaller than visible are likely to lie in the ultraviolet range, so the answer is (c). (This can be confirmed by direct calculation.)

Follow-up Exercise.   What type of light would be required to ionize a hydrogen atom in its second excited state? Explain.

Although Bohr's results gave agreement with the experiment for hydrogen as well as singly ionized helium and doubly ionized lithium (why?), it could not successfully handle multielectron atoms. Bohr's theory was incomplete in the sense that it patched new quantum ideas into a basically classical framework. The theory contains many correct ideas, but a complete description of the atom did not come until quantum mechanics was developed (Chapter 28).

Fluorescence.   Note in Fig. 27.12 that an electron excited to a higher energy level does not necessarily have to return to the ground state in a single jump. Often, several routes involving different sequences of transitions may be possible. For example, an electron in the *n* = 4 excited level could theoretically drop to either of the intermediate levels (*n* = 3 or *n* = 2)—or even to both in succession—before falling back to the ground state. This fact underlies the phenomenon of *fluorescence*, exhibited by a number of natural and synthetic substances.

In **fluorescence**, an electron that has been excited by absorbing a photon returns to the ground state in two or more steps, like a ball bouncing down a flight of stairs. At each step a photon is emitted. Since each drop represents a smaller energy transition than the original upward boost, the emitted photons must have a lower energy, and thus a longer wavelength, than the exciting photon. For example, many minerals are excited by ultraviolet (UV) light and fluoresce, or glow, in the visible region (•Fig. 27.13). A variety of living organisms, from corals to butterflies, manufacture fluorescent pigments that emit visible light.

Our familiar fluorescent lights make use of the same phenomenon. Fast-moving electrons, accelerated by the voltage across the tube, collide with atoms of gaseous mercury in the tube and excite them to higher energy levels. As the excited mercury atoms return to their normal energy levels they emit UV radiation, which in turn excites atoms in the fluorescent coating that lines the tube. It is this mineral that fluoresces, emitting visible light as described previously. A similar mechanism gives rise to the spectacular atmospheric light show known as the aurora (see Fig. 19.30).

27.5 A Quantum Success: The Laser

Objective: To understand some of the practical applications of the quantum hypothesis, in particular the laser.

The development of the laser was a major technological success derived from theoretical atomic physics. Although many scientific discoveries have been made accidentally, and many applications or inventions have come about by trial and error—Roentgen's discovery of X-rays and Edison's electric lamp are examples—the laser was developed on purpose. Using quantum physics, the principle of the **laser** was first predicted theoretically, then practically applied. (The word "laser" is an acronym and stands for *light amplification by stimulated emission of radiation*. Stimulated emission will be discussed shortly.)

This relatively new invention has found widespread usefulness. Laser light beams are used in medicine to control bleeding, to weld torn retinas, and to treat skin cancer. In industry, lasers are used to drill holes in materials, to weld, and to cut cloth. They are used in surveying, and laser light is used to carry telephone and television signals over fiber optic cables. There are laser printers, laser pickups in video and compact disc players, and lasers in supermarket checkouts.

The concept of atomic energy levels allows us to understand this important technological instrument. Bohr's model of the atom explains spectral absorption and emission lines in terms of quantum jumps between energy levels. However, other questions arise. Since there are usually several lower energy levels, what determines the energy level to which an excited electron will decay? Also, how long does an electron stay in an excited state before making a transition? These questions were answered by the use of quantum mechanics, a way of dealing with electrons as waves that is described in Chapter 28. The general results are that different transitions have different probabilities and, crucial to laser application, the time an electron stays in an excited state varies, depending on the atom and the state.

Usually, an excited electron makes a transition to a lower energy level almost immediately, remaining in an excited state for only about 10^{-8} s. However, the lifetimes of an excited electron in some energy levels are appreciable. An energy level in which an excited electron remains for some time is called a **metastable state**. *Phosphorescent* materials are examples of substances with metastable states. These materials are used on luminous watch dials, toys, and items that "glow in the dark." When a phosphorescent material is exposed to light, atomic electrons are excited to higher energy levels. Many of the excited electrons return to their normal state fairly soon, but there are metastable states in which electrons remain for seconds, minutes, even more than an hour. Consequently, the material, being made of billions of atoms, emits light or glows for some time (•Fig. 27.14).

A major consideration in laser operation is the emission process. As •Fig. 27.15 shows, spontaneous absorption and emission of radiation occur between two energy levels. That is, a photon is absorbed and a photon is emitted. However, there is another possible emission process called **stimulated emission**. Einstein proposed this process in 1919. If a photon with an energy equal to an allowed transition strikes an atom in an excited state, it may stimulate an electron to make a transition to a lower energy level, with the emission of another photon. Two photons with the same frequency and phase then go off in the same direction.

•FIGURE 27.14 Phosphorescence
When electrons in a phosphorescent material are excited, some of them do not immediately return to the ground state but remain in metastable states for periods of time. In this exhibit at the San Francisco Exploratorium, phosphorescent walls and floor continue to glow for about 30 seconds after being illuminated, retaining the shadows of children present when the phosphors were exposed to light.

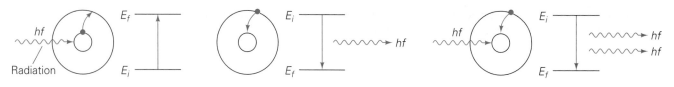

(a) Absorption **(b) Spontaneous emission** **(c) Stimulated emission**

•FIGURE 27.15 Photo absorption and emission
(a) In absorption, a photon is absorbed, and the electron is excited to a higher energy level. **(b)** After a short time, the electron spontaneously decays to a lower energy level with the emission of a photon. **(c)** If another photon with an energy equal to an allowed transition strikes an excited atom, stimulated emission occurs, and two photons with the same frequency and phase go off in the same direction as that of the incident photon.

Notice that stimulated emission is an amplification process—one photon in, two out. Of course, this is not a case of getting something for nothing, since the atom must be initially excited, and energy is needed to boost an electron to a higher state. This excitation process is somewhat analogous to pumping water to a roof-top reservoir for later use. However, stimulated emission does provide a way to amplify light.

Ordinarily, when light passes through a material, photons are more likely to be absorbed than to give rise to stimulated emission. This is because there are normally many more atoms in the ground state than in excited states.* However,

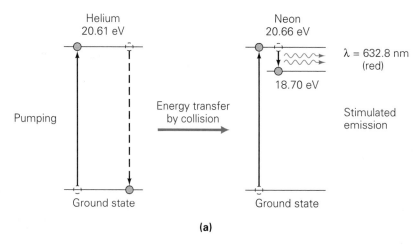

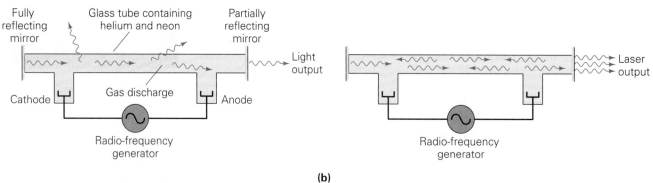

•FIGURE 27.16 The helium–neon laser
(a) Helium atoms are excited or pumped into an excited state. Energy is transferred to neon atoms by collisions and stimulated emission occurs (see text) with the emission of red light. **(b)** Reflections from the end mirrors of the laser tube set up an intense beam parallel to the axis of the tube.

*A multielectron atom is referred to as being in the ground state when its electrons occupy the lowest energy levels.

under specially arranged circumstances, it is possible to have more atoms in an excited state than in the ground state. This condition is known as a **population inversion**. In this case, there may be more stimulated emission than absorption, and we have amplification. *Population inversion and stimulated emission are two basic conditions necessary for laser action.*

There are several types of lasers, but the helium–neon (He–Ne) gas laser is most commonly used for classroom demonstrations and laboratory experiments. The characteristic reddish-pink light produced by the He–Ne laser ($\lambda = 632.8$ nm) can be observed in an optical scanning system at supermarket checkouts. The gas mixture is about 85% helium and 15% neon. Essentially, the helium is used for energizing and the neon for amplification. The gas mixture is subjected to a high dc voltage discharge rectified from a radio-frequency power supply, and helium atoms are excited (•Fig. 27.16a). This process is referred to as *pumping*. Energy is pumped into the system, and the helium atoms are pumped into an excited state (20.61 eV).

The excited state in the He atom has a relatively long lifetime (about 10^{-4} s) and has almost the same energy as an energy level in the Ne atom (20.66 eV). There is a good chance that before an excited He atom spontaneously emits a photon, it will collide with a Ne atom. When a collision occurs, energy is transferred to the Ne atom. Light is produced when the Ne electron drops down to a specific lower level. However, the lifetime of the 20.66-eV neon state is relatively long, and the delay causes a piling up of neon atoms in this state, leading to a population inversion.

The stimulated emission and amplification of the light emitted by the neon atoms are enhanced by reflections from mirrors placed at the end of the laser tube (•Fig. 27.16b). Some excited Ne atoms spontaneously emit photons in all directions, and these photons induce stimulated emissions. In stimulated emission, the two photons leave the atom in the same direction as that of the incident photon. Photons traveling in the direction of the tube axis are reflected back through the tube by the end mirrors. The photons, in reflecting back and forth, cause more stimulated emissions, and an intense, highly directional, coherent, monochromatic beam develops along the tube axis. Part of the beam emerges through one of the end mirrors, which is only partially silvered.

The monochromatic (single-frequency), coherent (same-phase), and directional properties of laser light give it unique properties (•Fig. 27.17). Light from sources like incandescent lamps is incoherent. The atoms emit randomly and at different frequencies (many different transitions). As a result, the light is out of phase or incoherent. Such beams spread out and become less intense. The properties of laser light allow the formation of a very tight beam, which with amplification can be very intense.

We hear about the possible development of "Star Wars" laser weapons that would shoot down missiles, and lasers powerful enough to induce nuclear fusion (Chapter 30). Such lasers must have large power outputs. Carbon dioxide gas lasers can have a continuous 25,000-W output, which is capable of cutting steel. Pulsed ruby lasers may deliver as much as 50 MW, but only for a short fraction of a second. The He–Ne gas lasers used in college classrooms and laboratories have typical outputs of a few milliwatts or less.

Some laser applications are shown in •Fig. 27.18. It is important to remember that, although laser beams are used to weld detached retinas in the eye, viewing a laser beam directly can be quite hazardous. The beam is focused on the retina in a very small spot. If the beam intensity and the viewing time are sufficient, the photosensitive cells of the retina may be burned and destroyed.

Another interesting application of laser light is the production of three-dimensional images in a process called **holography** (from the Greek word *holos*, meaning "whole"). The process does not use lenses as ordinary image-forming processes do, yet it recreates the original scene in three dimensions. The key to holography is the coherent property of laser light, which gives the light waves a definite spatial relationship to each other.

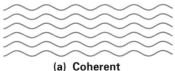

(a) Coherent

(b) Incoherent

•**FIGURE 27.17 Coherent light**
(a) Laser light is monochromatic (single frequency or color) and coherent, or in phase. **(b)** Light waves from sources in which atoms emit randomly are incoherent, or out of phase.

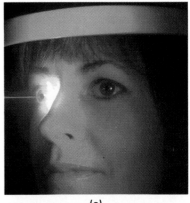

(a)

(b)

•**FIGURE 27.18 Laser applications**
(a) Laser eye surgery. Such procedures, which minimize bleeding and scarring, are used to repair detached retinas, treat glaucoma, correct retinal abnormalities produced by diabetes, and alleviate various other problems. **(b)** An infrared laser is used to cut through a steel plate.

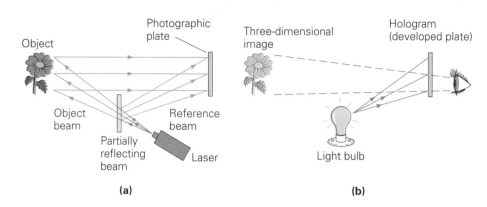

•FIGURE 27.19 Holography
(a) The coherent light from a laser is split into reference and object beams. The interference patterns of these beams are recorded on a photographic plate. **(b)** When the developed plate or hologram is illuminated, the viewer sees a three-dimensional image.

In the photographic process of making a *hologram*, an arrangement such as that illustrated in •Fig. 27.19 is used. Part of the light from the laser passes through a partial mirror to the object. The other part, or reference beam, is reflected to the film. The light incident on the object is reflected to the film, and it interferes with the reference beam. The film records the interference pattern of the two light beams, which essentially imprints on the film the information carried by the light wavefronts from the object.

When the film is developed, the interference pattern bears no resemblance to the object and appears as a meaningless pattern of light and dark areas. However, when the wavefront information is reconstructed by passing light through the film, a three-dimensional image is perceived (•Fig. 27.20). If part of the three-dimensional image is hidden from view, you can see it by moving your head to one side, just as you would to see a hidden part of a real object.

The technique of holography has many potential applications. Using coherent ultrasonic waves to form a hologram could provide a three-dimensional view of internal structures of the human body. Also, holograms made with X-ray lasers (yet to be developed) could provide better resolution for an in-depth view of microscopic structures. Holographic three-dimensional television may some day be commonplace.

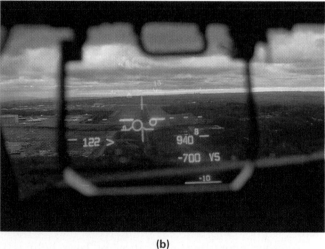

(a)   (b)

•FIGURE 27.20 Holographic images
(a) A three-dimensional hologram can be viewed from any angle, just as if it were a real object. **(b)** Holographic projection is used to display flight information at the pilot's eye level in an airplane cockpit.

Insight   The Compact Disc

The compact disc (CD) is a relatively new method of data storage, and most of us are familiar with it in terms of musical recordings. The CD system was introduced in 1980. There are now millions of CD players, and musical CD's outsell tapes and have nearly supplanted long-playing records. The disc itself is only 12 cm (5 in.) in diameter, but it can store more than 6 billion bits of information. This is the equivalent of more than 1000 floppy disks or over 275,000 pages of text that could be stored on a CD for display on a television monitor. For audio use, a CD can store 74 minutes of music, and the sound reproduction is virtually unaffected by dust, scratches, or fingerprints on the disc.

The information on the disc is in the form of raised areas called "pits," which are separated by flat areas called "land." The surface is coated with a thin layer of aluminum to reflect the laser beam that "reads" the information (Fig. 1). The pits are arranged in a spiral track like the grooves in a phonograph record, but the track is much narrower. The tracks on a CD are about 60 times closer together than the grooves on a long-playing phonograph record.

In the read-out system, a laser beam, which comes from a small semiconductor (solid state) laser, is applied from be-low the disc and focused on the aluminum coating of the track. The disc rotates at about 3.5 to 8 revolutions per second as the laser beam follows the spiral track. When the beam strikes a land area between two pits, it is reflected. When the beam spot overlaps a land area and a pit, the light reflected from the different areas interferes, causing fluctuations in the reflected beam. To make the fluctuations more distinct, the raised pit thickness is chosen to be one-quarter wavelength of the laser beam. Then, reflected light from land areas travels an additional path length of one-half wavelength, and destructive interference occurs when the two parts of the reflected beam combine (Section 24.1). As a result, there is less reflected intensity when the beam passes over the edge of a pit than when passing over a land area alone.

The reflected beam of varying intensity strikes a photodiode (solid state photocell), which reads the information. The fluctuations of the reflected light convey the coded information as a series of binary numbers (zeros and ones). A pit represents a binary 1 and a land area is read as a binary 0. The signals are electronically converted back into sound.

Research is underway to produce an erasable disc so that CD's can be used to record sound as well as play it back.

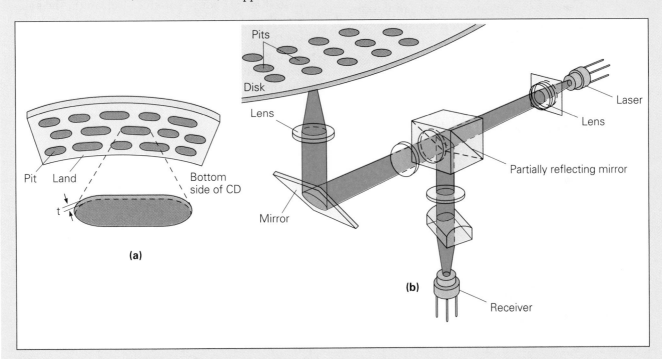

FIGURE 1   The compact disc
(a) The information on the disc is recorded in the form of raised areas called "pits," which are separated by flat areas called "lands." The pits or imprints are actually on the bottom of the disc. (b) The surface is coated with a thin layer of aluminum to reflect the laser beam that "reads" the information.

Chapter Review

Important Terms

thermal radiation   839
blackbody   839
Wien's displacement law   840
Planck's constant   841
Planck's hypothesis   841
quantum   841
photon   842
photoelectric effect   842
stopping potential   843
work function   844

threshold frequency   845
Compton effect   847
dual nature of light   849
Bohr theory of the hydrogen
   atom   849
emission spectrum   849
absorption spectrum   849
Balmer series   849
principal quantum number   850
ground state   852

excited states   852
binding energy   852
lifetime   852
fluorescence   854
laser   855
metastable state   855
stimulated emission   855
population inversion   857
holography   857

Important Concepts

- Thermal radiation, typically produced by hot objects, has a continuous spectrum.
- A blackbody is an ideal system that absorbs and emits all the radiation that falls on it.
- Wien's displacement law states that the wavelength of maximum intensity for radiation from a blackbody is inversely related to its temperature.
- Planck's constant (h) is the fundamental proportionality constant between energy and frequency for thermal oscillators and photons.
- The Compton effect is the increase in wavelength of light scattered by electrons or other charged particles.
- The dual nature of light means that light must be thought of as having both a particle and wave nature.

- The Bohr theory of the hydrogen atom treats the electron as a classical particle in circular orbit around the proton, held in orbit by the electric attraction force. The angular momentum of the electron is quantized in integral multiples of Planck's constant, h.
- The ground state of a hydrogen atom is the state of lowest energy for the atomic electron (in the smallest orbit, $n = 1$). Excited states ($n > 1$) have greater energies.
- Binding energy is the amount of energy required to completely free an electron from an atom, thus producing an ion.

Important Equations

Wien's Displacement Law:

$$\lambda_{max}T = 2.90 \times 10^{-3} \text{ m·K} \qquad (27.1)$$

Photon Energy:

$$E = hf \qquad (27.4)$$

(where Planck's constant is $h = 6.63 \times 10^{-34}$ J·s)

K_{max} and Stopping Potential in the Photoelectric Effect:

$$K_{max} = eV_o \qquad (27.5)$$

Energy Conservation in the Photoelectric Effect:

$$hf = K + \phi \qquad (27.6)$$

Energy Conservation for least bound electron in the Photoelectric Effect:

$$hf = K_{max} + \phi_o \qquad (27.7)$$

Work Function and Threshold Frequency in the Photoelectric Effect:

$$f_o = \frac{\phi_o}{h} \qquad (27.8)$$

Compton Equation (scattering from free electrons):

$$\Delta\lambda = \lambda_1 - \lambda_o = \lambda_C(1 - \cos\theta) \qquad (27.9)$$

(where $\lambda_C = h/m_e c = 2.43 \times 10^{-12}$ m = 0.00243 nm)

Bohr Theory Orbit Radius (for hydrogen atom):

$$r_n = 0.0529n^2 \text{ nm} \quad n = 1, 2, 3, \dots \qquad (27.16)$$

Bohr Theory Electron Energy (for hydrogen atom):

$$E_n = \frac{-13.6}{n^2} \text{ eV} \quad n = 1, 2, 3, \dots \qquad (27.17)$$

Bohr Theory Transition Energy (for hydrogen atom) (in eV):

$$\Delta E = 13.6\left(\frac{1}{n_f^2} - \frac{1}{n_i^2}\right) \text{eV} \qquad (27.18)$$

Bohr Theory Photon Wavelength (energies in eV):

$$\lambda = \frac{1.24 \times 10^3}{\Delta E} \text{ nm} \qquad (27.19)$$

Exercises

Note: Take *h* to be *exact* at 6.63×10^{-34} J·s for significant figure purposes and use $hc = 1.24 \times 10^3$ eV·nm (3 sf).

27.1 Quantization: Planck's Hypothesis

1 A blackbody (a) is an ideal system, (b) re-emits all radiation incident on it, (c) has a wavelength of maximum radiation intensity inversely proportional to its temperature, (d) all of these.

2 Planck's hypothesis (a) was the basis of Wien's law, (b) called for the intensity of blackbody radiation to be proportional to $1/\lambda^4$, (c) quantized the energy of thermal atomic oscillators, (d) predicted the ultraviolet catastrophe.

3 Does a blackbody at 200 K emit twice as much total radiation as when its temperature is 100 K? Explain.

4 ■ What is the wavelength and frequency of the most intense radiation component from a blackbody with a temperature of 0°C?

5 ■ Sketch a graph of temperature versus (a) frequency and (b) wavelength of the most intense radiation component of blackbody radiation.

6 ■ Find the approximate temperature of a star that emits light with a wavelength of maximum emission of 700 nm (deep red).

7 ■ Assuming the skin to be the same temperature as normal body temperature (37°C), what is the wavelength of the radiation component of maximum intensity emitted by our bodies? What region of the EM spectrum is this in?

8 ■ The walls of a blackbody cavity are at a temperature of 327°C. What is the frequency of the radiation of maximum intensity?

9 ■ What would be the approximate temperature of a blackbody if it appeared to be (a) yellow (as in a candle flame) and (b) blue (as in a bunsen burner flame)?

10 ■■ What would be the change in the frequency of the most intense spectral component of a blackbody that was initially at 200°C and was heated to 400°C?

11 ■■ What is the minimum quantized energy of a thermal oscillator producing radiation at λ_{max} in a blackbody with a temperature of 212°F?

12 ■■ The temperature of a blackbody is 1000 K. If the total intensity of the emitted radiation of 3.0 W were due *entirely* to the most intense frequency component,

how many quanta would be emitted per second per square meter*?

27.2 Quanta of Light: Photons and the Photoelectric Effect

13 In the photoelectric effect, classical theory predicted that (a) no photoemission can occur below a threshold frequency, (b) the photocurrent is proportional to the light intensity, (c) the kinetic energy of the emitted electron is dependent on frequency, but not on intensity, (d) a photocurrent is observed immediately.

14 A photocurrent is observed when (a) the light frequency is above the threshold frequency, (b) the energy of the photons is greater than the work function, (c) the light frequency is below the threshold frequency, (d) both (a) and (b).

15 If electromagnetic radiation is made up of quanta, why don't we detect the discrete packets of energy, for example, when listening to a radio?

16 From an ionizing (biological damage) viewpoint, is it more dangerous to stand in front of a very weak beam of X-ray radiation or a stronger beam of red light? How does the photon picture explain this apparent paradox?

17 ■ What is the energy of a quantum of light which has a frequency of 7.5×10^{14} Hz?

18 ■ A photon has an energy of 3.3×10^{-15} J. What type of radiation is this photon?

19 ■ Which has more energy and how many times more: a quantum of violet light or a quantum of red light from the ends of the visible spectrum?

20 ■ Ultraviolet light has a wavelength of 150 nm. How much energy does one of its photons have in (a) joules and (b) electron volts?

21 ■ Light with an intensity I produces a photocurrent I_p. If the intensity of the light is doubled, how is the current affected?

22 ■ Light with an intensity I produces a photoelectric emission, and the photoelectrons have a maximum kinetic energy of 5.0 eV. If the intensity of the light is tripled, what is the maximum kinetic energy of photoelectrons now?

23 ■■ Assume a 100-W light bulb gives off 2.50% of its energy as visible light. How many photons of visible light are given off in 1.00 minute? (Use an average visible wavelength of 550 nm.)

*(assuming each quantum has the minimum allowed energy)

24 ■■ A solar cell is 20% efficient and the intensity of sunlight received at the Earth's surface is 2.5×10^{12} J/km²·h. Assuming an average wavelength of 550 nm and a cell area of 1.0 cm², what is the maximum number of photons converted to electrical energy each second?

25 ■■ •Figure 27.21 shows a graph of stopping potential versus frequency for a photoelectric material. Determine (a) Planck's constant and (b) the work function of the material from the graph data.

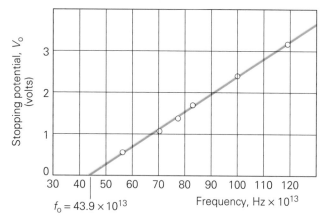

•**FIGURE 27.21** **Stopping potential versus frequency**
See Exercise 25.

26 ■■ A metal with a work function of 2.60 eV is illuminated with a beam of monochromatic light. It is found that the stopping potential for the emitted electrons is 2.80 V. What is the wavelength of the light in nanometers?

27 ■■ What is the longest wavelength of light that can cause the release of electrons from a metal that has a work function of 1.5 eV?

28 ■■ The threshold wavelength for emission from a metallic surface is 500 nm. Calculate the maximum speed of emitted photoelectrons for light having a wavelength of (a) 400 nm, (b) 500 nm, and (c) 600 nm.

29 ■■ In Exercise 28, what is the work function of the metal in eV?

30 ■■ The work function of a photoelectric material is 3.5 eV. If the material is illuminated with monochromatic light ($\lambda = 300$ nm), (a) find the stopping potential of emitted photoelectrons. (b) What is the cutoff frequency?

31 ■■ Blue light with a wavelength of 420 nm is observed to cause the emission of photoelectrons with a maximum kinetic energy of 1.00×10^{-19} J. Will red light ($\lambda = 700$ nm) cause the emission of photoelectrons from the material? Justify your answer mathematically.

32 ■■ Gold has a work function of 4.82 eV. If a gold surface is illuminated with ultraviolet light ($\lambda = 160$ nm), what is (a) the maximum kinetic energy (in eV) of the emitted photoelectrons and (b) the threshold frequency for gold?

33 ■■ Sodium and silver have work functions of 2.46 eV and 4.73 eV, respectively. (a) If the surfaces of the metals are both illuminated with the same monochromatic light, the photoelectrons from which metal will have the greater speed on emission? (b) What wavelengths (in nm) are required for photoelectrons to be emitted from each metal surface?

34 ■■■ When a certain photoelectric material is illuminated with red light ($\lambda = 680$ nm) and then blue light ($\lambda = 420$ nm), it is found that the photoelectrons resulting from the blue light have a maximum kinetic energy 1.8 times greater than those from red light. What is the work function of the material?

35 ■■■ When the surface of a particular material is illuminated with monochromatic light of various frequencies, the stopping potentials for the photoelectrons are determined to be:

Frequency (in Hz)

7.0×10^{15}	5.5×10^{15}	4.0×10^{15}	3.5×10^{15}

Stopping potential (in V)

12.6	10.3	4.10	1.97

Plot these data and determine the values of Planck's constant and the work function of the metal.

27.3 Quantum "Particles": The Compton Effect

36 The Compton effect was first observed using (a) visible light, (b) infrared radiation, (c) ultraviolet light, (d) X-rays.

37 The wavelength shift for Compton scattering is maximum when (a) the photon scattering angle is 90°, (b) the electron scattering angle is 90°, (c) the shift is equal to the Compton wavelength, (d) the incident photon is back scattered.

38 The maximum wavelength shift for Compton scattering from a free electron would compare how to the shift if the electron were bound to an atom? (a) it is larger if the electron is free, (b) it is larger if the electron is bound to an atom, (c) they are the same regardless.

39 In the center of the Sun, most of the electromagnetic energy is in the form of X-rays. By the time it reaches the surface, however, it is mostly in the visible range. Explain how many successive Compton scatterings can be a contributer to this wavelength change.

40 ■ Calculate the maximum wavelength shift for Compton scattering from a free electron.

41 ■ Show that the dimensions of Planck's constant are those of angular momentum.

42 ■ What is the change in wavelength when monochromatic X-rays are scattered by electrons through an angle of 20°?

43 ■■ A photon with an energy of 10 keV is scattered by a free electron. If the electron recoils with a kinetic energy of 5.0 keV, what is the wavelength of the scattered photon?

44 ■■ Show that the Compton wavelength has units of length and a value of 2.43×10^{-12} m = 2.43×10^{-3} nm.

45 ■■ X-rays scattered from a carbon target show a wavelength shift of 0.000711 nm. What is the scattering angle?

46 ■■ A monochromatic beam of X-rays with a wavelength of 2.80×10^{-10} m is scattered through a metal foil. What is the wavelength of the scattered X-rays observed at an angle of 45° from the direction of the incident beam?

47 ■■ X-rays with a wavelength of 0.0045 nm are used in a scattering experiment. If the X-rays are scattered through an angle of 53°, what is the wavelength of the scattered radiation?

48 ■■■ The Compton effect can occur for scattering from any charged particle, in particular it occurs for scattering from protons. (a) What is the value of the Compton wavelength for a proton? (b) Compare the maximum wavelength shifts for scattering by an electron and a proton by forming a ratio of electron value to that of the proton.

27.4 The Bohr Theory of the Hydrogen Atom

49 In his theory of the hydrogen atom, Bohr postulated the quantization of (a) energy, (b) centripetal acceleration, (c) light, (d) angular momentum.

50 An excited hydrogen atom emits light when its electron (a) makes a transition to a lower energy level, (b) is excited to a higher energy level, (c) is in the ground state.

51 Explain why the Bohr theory is applicable only to the hydrogen atom and hydrogen-like atoms, such as singly ionized helium, doubly ionized lithium, and other one-electron systems.

52 ■ Very accurate measurements of the wavelengths emitted by a hydrogen atom indicate that they are all slightly longer than expected from the Bohr theory. Explain how conservation of momentum ideas might help explain this. [Hint: Photons carry momentum and energy, both need to be conserved.]

53 ■ Find the energy required to excite a hydrogen electron from (a) the ground state to $n = 3$ and (b) $n = 2$ to $n = 3$.

54 ■ In Exercise 53, classify the type of light needed to create the transitions.

55 ■ Find the energy needed to ionize a hydrogen atom whose electron is in the (a) $n = 4$ state and (b) $n = 5$ state.

56 ■ What is the frequency of light that would excite the electron of a hydrogen atom from (a) $n = 2$ to $n = 5$ and (b) $n = 2$ to $n = \infty$?

57 ■ Use the Bohr theory to find the value of the Rydberg constant for the empirical formula (Eq. 27.10).

58 ■ Give the radius of the electron orbit of the hydrogen atom for each of the following states: (a) $n = 3$, (b) $n = 6$, (c) $n = 10$.

59 ■ Scientists are now beginning to study "large" atoms, that is, ones with orbits that are almost comprehensible to our everyday measurements. What excited state (give an approximate principal quantum number) are we talking about for hydrogen if we want the diameter of the orbit to be a micron (10^{-6} m)—that is, getting close to the size of a dust particle?

60 ■ Find the energy of the hydrogen electron for each of the following states: (a) $n = 3$, (b) $n = 6$, (c) $n = 10$.

61 ■■ Find the binding energy of the hydrogen electron for each of the following states: (a) $n = 3$, (b) $n = 5$, (c) $n = 10$.

62 ■■ The ionization energy of a hydrogen atom is 13.6 eV. Would the absorption of light having a frequency of 7.00×10^{15} Hz cause a hydrogen atom to be ionized? If so, what would be the kinetic energy of the emitted electron?

63 ■■ The hydrogen spectrum has a series of lines called the Brackett series, which results from transitions to the $n = 4$ energy level. What is the longest wavelength in this series and in what region of the EM spectrum does it lie?

64 ■■ A hydrogen electron in the ground state is excited to the $n = 5$ energy level. The electron makes a transition directly to the $n = 2$ level before returning to the ground state. (a) What are the wavelengths of the emitted photons? (b) Would any of the light emitted from a group of such hydrogen atoms be visible?

65 ■■ For which one of the following transitions of a hydrogen electron is the photon of greatest energy emitted: (a) $n = 5$ to $n = 3$, (b) $n = 6$ to $n = 2$, (c) $n = 2$ to $n = 1$? Justify your answer mathematically.

66 ■■ What is the binding energy for the electron in the ground state in the following hydrogen-like ions: (a) He^+; (b) Li^{2+}?

67 ■■ A hydrogen atom absorbs a photon of wavelength 434.1 nm. (a) How much energy did the atom absorb? (b) What were the initial and final states of the hydrogen electron?

68 ■■■ Show that the total energy of an electron in a hydrogen atom in an orbit with a radius of 0.0529 nm is -13.6 eV.

69 ■■■ In Excercise 68, (a) how much of the energy is potential and how much kinetic? (b) How do their magnitudes compare?

70 ■■■ Show that the speeds of an electron in the Bohr orbits are given (to two significant figures) by $v_n = (2.2 \times 10^6 \text{ m/s})/n$.

71 ■■■ A muon is a particle that has the same charge as an electron but a mass 207 times that of an electron. Using a muon to replace the electron in a hydrogen atom, a muonium atom is possible. For such an atom, ignoring any possible movement of the proton (in a muonium atom, the orbiting particle is 1/9 the mass of the proton whereas for an electron the ratio is a negligible 1/1840), (a) what would be the muon's energies and orbital radii for the $n = 1$ and $n = 2$ states? (b) What is the wavelength of the photon emitted for a muon transition from $n = 2$ to $n = 1$?

27.5 A Quantum Success: The Laser

72 Which of the following is *not* an essential for laser action? (a) population inversion, (b) phosphorescence, (c) pumping, (d) stimulated emission.

73 What do the letters *se* in the laser acronym stand for? Explain the process to which this term refers.

74 In what sense is a laser an "amplifier" of energy? Explain why this does not violate conservation of energy.

Additional Exercises

75 An FM radio station broadcasts at a frequency of 98.9 MHz and radiates 750 kW of power. How many quanta are radiated from the station's antenna each second?

76 A photon is a "massless" particle, i.e., it has no rest mass. Show that the momentum of a photon is given by $p = h/\lambda$. [*Hint:* Use the relativistic relationship $E^2 = p^2c^2 + (m_oc^2)^2$ which relates the total energy and momentum (Exercise 62, Chapter 26)].

77 Referring to ●Fig. 27.22, derive the equation for the wavelength shift in Compton scattering by applying the conservation of momentum and energy in an elastic collision. [*Hint:* Consider total relativistic energy. Also, analyze momentum classically with components as in Chapter 6, but remember that in the end $p = h/\lambda$ for a photon.]

78 Consider absorption of *visible light* by a hydrogen atom. Explain why it occurs only in a certain temperature (of the hydrogen gas) range. In other words, why does the gas not absorb visible light when it is either too cool or too hot? (By the way, the best temperature is about

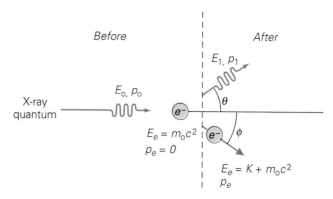

●**FIGURE 27.22 Compton scattering**
See Exercise 77.

10,000 K.) [Hint: In what state does the atom have to be to absorb *visible* light?]

79 What are the energies of the photons required to excite a hydrogen electron in the ground state to states (a) $n = 4$ and (b) $n = 8$?

80 How many quanta of red light ($\lambda = 700$ nm) would it take to have 1.0 J of energy?

81 What is the incident energy of each photon in a beam of monochromatic X-rays that are observed to have a wavelength of 4.5×10^{-10} m at a Compton scattering angle of 37°?

82 What is the frequency of the most intense spectral component from a blackbody at room temperature (20°C)?

83 The work function of a particular metal is 5.0 eV. (a) What is the frequency of light that causes electrons to be emitted with a maximum kinetic energy of 2.0×10^{-19} J? (b) What is the stopping potential of the electrons?

84 For what scattering angle would the wavelength shift for Compton scattering be 25 percent of the Compton wavelength?

85 How many transitions of the electron in a hydrogen atom result in the emission of light in the visible region of the spectrum (400–700 nm)?

86 Explain why, in the photoelectric effect, the photocurrent reduces *gradually* to zero as we turn up the retarding voltage. In other words, why does the current not stay at a large constant value until we reach the stopping potential, and then drop to zero? [Hint: See Fig. 27.3b, and also recall the energy structure of the electrons in the solid.]

87 Show that in any Bohr orbit in the hydrogen atom, the potential energy and kinetic energy of the electron are related by $U = -2K$.

28 Quantum Mechanics

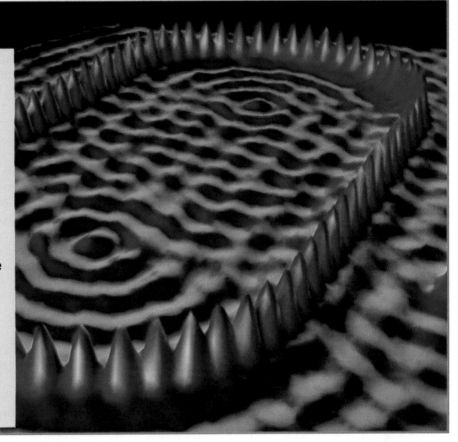

Just a few decades ago, if someone had claimed to have a photograph of an atom, they would have been laughed at, or perhaps sent for psychiatric care. Yet today, a device called the scanning tunneling microscope routinely produces images representing individual atoms. (In the picture above, the blue shapes represent iron atoms, neatly arranged on a copper surface.)

Of course, such "snapshots" have little in common with conventional photographs. For one thing, they are not made with light at all, but produced in a very different way. Their basis is a quantum-mechanical phenomenon called tunneling, which has no counterpart in our macroscopic world. Tunneling reflects two strange but very fundamental features of the subatomic realm: the probabilistic character of quantum processes and the wave nature of particles. In effect, tunneling explains how particles just may turn up in places where, according to classical notions, they have no right to be.

In this chapter you'll find out more about such intriguing aspects of the quantum world—including how a quantum-mechanical principle underlies the properties of all the chemical elements that make up the world around us.

> The opposite of a correct statement is a false statement. But the opposite of a profound truth may well be another profound truth.
>
> —Niels Bohr

The initial successes of the quantum theory were impressive, but they also created a perplexing situation. They seemed to show that light (that is, electromagnetic radiation) has a dual nature. On one hand, in many experiments, light exhibits classical wave behavior. On the other hand, explanations of certain phenomena require light to have a quantum or particle nature.

Around 1925, a new kind of physics based on the synthesis of wave and quantum ideas was introduced. This new theory, called **quantum mechanics**, attempts to combine the wave–particle duality into a single consistent description. It revolutionized scientific thought and provided the basis of our present understanding of microscopic (and submicroscopic) phenomena.

Quantum mechanics deals mainly with the minute aspects of the world of atoms. It has replaced the mechanistic view of the universe, in which all things moved according to exact natural laws, with a new concept of probability. All physical measurements are now accepted as being to some degree uncertain. Even so, when quantum mechanics is applied to macroscopic phenomena, it reproduces the results of classical physics. This means that there is no need to abandon the laws and principles of classical physics, since they provide a sufficiently accurate description of everyday phenomena. The situation is similar to that encountered in our discussion of relativity in Chapter 26. When objects are moving with speeds that are appreciable fractions of the speed of light, we must use the theory of relativity, but in less extreme situations, classical principles can still be applied.

In this chapter we will present some of the basic ideas of quantum mechanics to show how they describe waves and particles. We will also discuss some of the results made possible by the quantum-mechanical view of nature, such as the electron microscope.

28.1 Matter Waves: The de Broglie Hypothesis

Objectives: To be able to (a) explain de Broglie's hypothesis, (b) calculate the "wavelength" of a matter wave, and (c) specify under what circumstances the wave nature of matter will be observable.

Note: These relationships are discussed in Section 26.4.

Note: See Eq. 27.4 in Section 27.2.

Photon momentum

Since a photon travels at the speed of light, we treat it relativistically as a "massless" particle—that is, a particle having no rest mass m_o. If this were not the case, it would have infinite mass ($m = m_o/\sqrt{1 - (v/c)^2}$), and therefore infinite energy ($E = mc^2$). It can be shown that the total energy and relativistic momentum are related to each other by the equation $E^2 = p^2c^2 + (m_oc^2)^2$. (See Exercise 62, Chapter 26.) A photon without rest mass then has a total energy of $E = pc$, or a relativistic momentum of $p = E/c$. The energy may also be written $E = hf = hc/\lambda$, so the momentum of a photon is related to its wavelength by

$$p = \frac{E}{c} = \frac{hf}{c} = \frac{h}{\lambda} \qquad (photons\ only) \qquad (28.1)$$

Thus, the energy in electromagnetic waves of wavelength λ can be thought of as being carried by photon "particles" having momenta of h/λ.

Since nature exhibits a great deal of symmetry, the French physicist Louis de Broglie thought that there might be a wave–particle symmetry. That is, if light sometimes behaves like a particle, perhaps material particles, such as electrons, also had wave properties. In 1924, de Broglie put forth a hypothesis that a moving particle has a wave associated with it. He proposed that the wavelength of a material particle was related to the particle's momentum by an equation similar to that for a photon (Eq. 28.1). More specifically, the **de Broglie hypothesis** is

Whenever a particle has momentum of magnitude p, a wave is associated with it. This wave has a wavelength of

$$\lambda = \frac{h}{p} = \frac{h}{mv} \qquad (28.2)$$

de Broglie wavelength of a moving particle

These waves associated with moving particles were called **matter waves** or, more commonly, **de Broglie waves**, and were thought to somehow influence or "guide" particle motion. One might think of an electromagnetic wave as the de Broglie wave for a photon, but the de Broglie waves associated with particles such as electrons and protons are *not* electromagnetic waves. The de Broglie equation for the wavelength gives no clue as to what kind of wave is associated with a particle of matter (one that has nonzero rest mass).

Needless to say, de Broglie's hypothesis met with a great deal of skepticism. The idea that the motions of photons were somehow governed by the electromagnetic wave properties of light did not seem unreasonable. But to extend this idea and say that the motion of a particle *with mass* is somehow governed by the properties of an associated wave was difficult to accept. Moreover, there was no evidence at the time that particles exhibited any wave properties such as interference and diffraction.

In support of his hypothesis, de Broglie showed how it could give an interpretation of the quantization of the angular momentum in Bohr's theory of the hydrogen atom. Recall that in order to get agreement with the hydrogen atomic spectrum Bohr had to make the ad hoc hypothesis that the angular momentum of the electron was quantized in multiples of $h/2\pi$. De Broglie argued that for a *free* particle, the associated wave would have to be a *traveling* wave. However, the *bound* electron of a hydrogen atom travels repeatedly in discrete circular orbits according to the Bohr theory. The associated matter wave would then be expected to be a *standing* wave. That is, only an integral number of wavelengths could fit into a given orbit. The wave must meet itself, so to speak, with the same phase. That is, upon returning to the same spot, it must reinforce itself constructively, much as a standing wave in a circular string might do. This is a similar idea to the linear case for a standing wave in a stretched string fixed at both ends (see Fig. 28.1).

The circumference of a Bohr orbit of radius r_n is $2\pi r_n$, so this length would be equal to an integral number of wavelengths

$$2\pi r_n = n\lambda \quad (n = 1, 2, 3, \dots)$$

Note: The Bohr model is outlined in Section 27.4

Note: Compare the discussion of standing waves in Section 13.5.

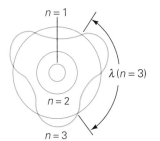

FIGURE 28.1   De Broglie waves and Bohr orbits
Similar to standing waves in a stretched string, de Broglie waves form circular standing waves on the circumferences of the Bohr orbits. Notice that the number of wavelengths in a particular orbit, shown here for $n = 3$, is equal to the principal quantum number of that orbit.

Note: See Eq. 8.14.

Using the de Broglie equation, $\lambda = h/mv$, for the wavelength, and recalling that the angular momentum for a circular orbit is $L = mvr$ we have

$$2\pi r_n = \frac{nh}{mv} \quad \text{or} \quad L_n = mvr_n = \frac{nh}{2\pi}$$

Thus, the angular momentum is quantized just as Bohr proposed. It was clear that de Broglie's hypothesis was consistent with Bohr's postulate for the hydrogen atom. For orbits other than those allowed by the Bohr theory, the de Broglie wave for the orbiting electron would *not* close on itself. That is, the circumference of a disallowed orbit would not be equal to an integral number of wavelengths. This would give rise to destructive interference, and the resulting wave would die out or have zero amplitude. This seems to imply that *the amplitude of the de Broglie wave must then be related in some way to the location of the electron*. This implication is, in fact, a fundamental cornerstone of quantum mechanics.

EXAMPLE 28.1 ■ SHOULD BALLPLAYERS WORRY ABOUT DIFFRACTION? DE BROGLIE WAVELENGTH

A pitcher throws a fast ball to his catcher at a speed of 40 m/s through a square strike-zone practice target—a sheet of canvas with an opening 50 cm on a side. If the mass of the ball is 0.15 kg, (a) what is the wavelength of the de Broglie wave associated with the moving ball? (b) Should the catcher be prepared for any diffraction effects as the ball passes through this opening?

Solution. We will use the de Broglie relationship to calculate the wavelength and then compare it to the size of the opening.

Given: $m = 0.15$ kg   *Find:*   (a) λ (de Broglie wavelength)
              $v = 40$ m/s   (b) Is significant diffraction likely?
              $d = 50$ cm $= 0.50$ m

(a) This is a straightforward calculation. Using the de Broglie equation (Eq. 28.2), we have

$$\lambda = \frac{h}{mv} = \frac{6.63 \times 10^{-34}\,\text{J·s}}{(0.15\,\text{kg})(40\,\text{m/s})} = 1.1 \times 10^{-34}\,\text{m}$$

(b) Recall from Section 24.3 that for a wave to significantly diffract through an opening, the wavelength must be roughly the same size as the opening. Since we have $\lambda = 1.1 \times 10^{-34}$ m $\ll 0.50$ m, this is very clearly *not* the case here. We should expect the ball to go straight through the opening into the catcher's mitt, as we observe.

Follow-up Exercise. In this Example, (a) how fast would the ball have to be thrown for diffractive effects to become important? (b) At that speed, how long would the ball take to travel the 20 m to the plate? (Compare your answer with the age of the universe, about 15 billion years. Do you think this movement would be noticeable?)

From the results of the previous Example, it is little wonder that we don't directly observe any effects of such matter waves in our everyday lives. Thus we cannot hope to detect the wave properties of ordinary objects. However, particles with very small masses traveling at relatively low speeds are another matter, as the following Example shows.

EXAMPLE 28.2 ■ A WHOLE DIFFERENT BALLGAME: DE BROGLIE WAVELENGTH OF AN ELECTRON

What is the de Broglie wavelength of the wave associated with an electron that has been accelerated from rest through a potential of 50.0 V?

Solution.

Given: $V = 50.0$ V   *Find:* λ (de Broglie wavelength)
$m = 9.11 \times 10^{-31}$ kg (known)

The work done in accelerating the electron is $W = eV$ and is equal to its kinetic energy gain ($\frac{1}{2}mv^2$). We can therefore calculate its speed as follows:

$$\tfrac{1}{2}\,mv^2 = eV \quad \text{or} \quad v = \sqrt{\frac{2eV}{m}}$$

so

$$v = \sqrt{\frac{2(1.60 \times 10^{-19}\,\text{C})(50.0\,\text{V})}{9.11 \times 10^{-31}\,\text{kg}}} = 4.19 \times 10^6\,\text{m/s}$$

Thus, the de Broglie wavelength is

$$\lambda = \frac{h}{p} = \frac{h}{mv} = \frac{6.63 \times 10^{-34}\,\text{J·s}}{(9.11 \times 10^{-31}\,\text{kg})(4.19 \times 10^6\,\text{m/s})}$$

$$= 1.74 \times 10^{-10}\,\text{m} = 0.174\,\text{nm}$$

Although wavelengths on the order of 10^{-10} m are extremely small, such waves can at least be detected. Slit widths on this order cannot be fabricated. However, recall from Chapter 24 that nature provides such slits in the form of crystal lattices.

Note: Diffraction from crystal lattices is discussed in Section 24.3.

Follow-up Exercise. (a) In this Example, how would the result change if the particle being accelerated were a proton? (Give a numerical answer.) (b) Would the de Broglie wave of a proton be more or less likely to diffract than that of an electron under the same conditions? Explain your reasoning.

PROBLEM-SOLVING HINT

In the previous Example, if we changed the accelerating voltage, we would have to go through the entire calculation of v and λ again, using the electron mass, electron charge, Planck's constant, and so on. Instead, it is very convenient to use the values of m, e, and h to derive a formula for an electron accelerated through a potential difference V. We begin by expressing the kinetic energy in terms of momentum. Since $p = mv$, the kinetic energy, $\frac{1}{2}mv^2$, can be written as $p^2/2m$. Then,

$$\frac{p^2}{2m} = eV \quad \text{or} \quad p = \sqrt{2meV}$$

Thus an alternate expression for the de Broglie wavelength is

$$\lambda = \frac{h}{p} = \frac{h}{\sqrt{2meV}} = \sqrt{\frac{h^2}{2meV}}$$

Inserting the most exact known values of h, e, and m and then rounding the result to 3 significant figures gives

$$\lambda = \sqrt{\frac{1.50}{V}} \times 10^{-9}\,\text{m} = \sqrt{\frac{1.50}{V}}\,\text{nm} \qquad \begin{array}{l}\textit{(nonrelativistic}\\ \textit{electrons only)}\end{array} \qquad (28.3)$$

where V is in volts. Thus, for $V = 50.0$ V,

$$\lambda = \sqrt{\frac{1.50}{50.0}} \times 10^{-9}\,\text{m} = 0.173\,\text{nm}$$

Note: An alternative de Broglie wavelength expression for electrons accelerated through a potential difference. V must be in volts, and this applies *only to nonrelativisitic electrons*.

(Note that this answer differs in the last digit from that found in Example 28.2, because there the values for h, e, and m were rounded to three significant figures *before* we performed the calculations.)

You may wish to derive a comparable formula for a proton, to use in solving similar problems. (What quantities would you have to change?)

In 1927, two physicists in the United States, C. J. Davisson and L. H. Germer, used a crystal to diffract a beam of electrons, thereby demonstrating a wavelike

Experimental proofs of the wave nature of particles.

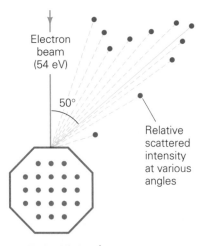

•FIGURE 28.2 The
Davisson–Germer experiment
When a beam of 54-eV electrons is
incident on the face of a nickel
crystal, a maximum in the
scattering intensity is observed at
an angle of 50º.

property of particles. A single crystal of nickel was used in the experiment. The
crystal was cut to expose a spacing of $d = 0.215$ nm between the lattice planes.
When a beam of electrons was directed normally on the crystal face, a maximum
in the intensity of the scattered electrons was observed at an angle of 50º relative
to the surface normal (•Fig. 28.2). The scattering was most intense for an accelerating potential of 54.0 V.

According to wave theory, the first-order maximum should be observed at an
angle given by

$$d \sin \theta = \lambda$$

This condition requires a wavelength of

$$\lambda = d \sin \theta = (0.215 \text{ nm}) \sin 50º = 0.165 \text{ nm}$$

Using Eq. 28.3 to determine the de Broglie wavelength of the electrons,

$$\lambda = \sqrt{\frac{1.50}{V}} \text{ nm} = \sqrt{\frac{1.50}{54.0}} \text{ nm} = 0.167 \text{ nm}$$

The agreement of the wavelengths was excellent within the limits of experimental accuracy, and the Davisson–Germer experiment gave convincing proof of the
validity of de Broglie's hypothesis.

Another experiment carried out by G. P. Thomson in Great Britain in the same
year added further proof. Thomson passed a beam of energetic electrons through
a thin metal foil. The diffraction pattern of the electrons was the same as that of
X-rays. A comparison of such patterns, as shown in •Fig. 28.3, leaves little doubt
that particles exhibit wavelike properties. An example of a practical application
of this fact is given in the Insight on p. 874.

28.2 The Schrödinger Wave Equation

**Objectives: To qualitatively understand (a) the reasoning that underlies the
Schrödinger wave equation, and (b) its use in finding particle
wave functions.**

De Broglie's hypothesis predicts that moving particles have associated waves that
somehow govern or describe their behavior. However, it does not tell us the *form*
of these waves, only their wavelengths. To have a useful theory, we need an equation that will give us the form of the matter waves, in particular the wave form
of a particle moving under the influence of a force, for example, an electron orbiting a proton under the influence of the electric force. Also, we must know how
these waves govern particle motion. In 1926, Erwin Schrödinger, an Austrian
physicist, presented his ideas on this subject. They included a general equation
that describes the de Broglie matter waves, how they travel, and the interpretation of the waves themselves.

As a wave, the de Broglie wave form is described by some mathematical function. (Recall how wave motion was described using equations in Section 13.2.) In
general, this function is commonly denoted by the symbol ψ (the Greek letter psi,
pronounced "sigh") and is called the **wave function**. It describes the wave as a
function of time and space. For example, the wave function for a (classical) traveling wave with wavelength λ and amplitude A is $\psi = A \sin [(2\pi x/\lambda) - \omega t]$.

Since the de Broglie wave governs a particle's motion, it would seem reasonable that the wave function was associated with the particle's energy, both kinetic and potential. Recall from the discussion of conservation of energy in Section 5.5 that for a conservative mechanical system the sum of the kinetic and
potential energies is a constant:

$$K + U = E = \text{a constant}$$

where the sum of the kinetic and potential energies equals the total mechanical

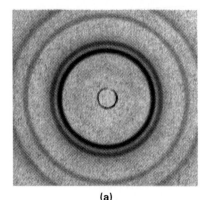

(a)

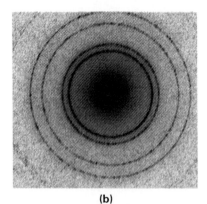

(b)

•FIGURE 28.3 Diffraction patterns
Comparison of an X-ray diffraction
pattern **(a)** and an electron
diffraction pattern **(b)** leaves little
doubt that electrons exhibit
wavelike properties.

energy, E. Schrödinger proposed a similar equation for the de Broglie matter waves involving the wave function ψ. This is known as **Schrödinger's wave equation**, and has the general form

$$(K + U)\,\psi = E\psi \qquad (28.4)$$

That is, the wave function is associated with the energy of a system. When applied to various situations with known values of K and U, complex mathematical equations result which are beyond the scope of this text. Using advanced mathematical procedures, the idea is to solve an equation for ψ and thus determine its form. But what is the physical significance of ψ?

In the early development of quantum mechanics (sometimes called wave mechanics), it was not clear how ψ should be interpreted. After much thought and investigation, it was concluded that ψ^2 (the wave function squared)* represents the probability of finding the particle at a certain position and time.

This interpretation involves the amplitude of ψ. Recall from Chapter 13 that the energy or intensity of a classical wave is proportional to the square of its amplitude. Similarly, the intensity of a light wave is proportional to E^2, where E is the electric field amplitude. Looking at this in terms of "particle" photons, the *intensity* of a light beam is proportional to the number (n) of photons in the beam, so $n \propto E^2$. That is, the number of photons is proportional to the square of the electric field amplitude of the wave.

Note: Review Eq. 13.5, Section 13.1

The wave function ψ is interpreted in an analogous manner. The wave function generally varies in magnitude in space and time. If ψ describes an electron beam, then ψ^2 will be proportional to the number of electrons that may be expected to be found in a small volume around a point at some time. However, when there is a single particle, ψ^2 represents the probability of finding the electron at the point. *Thus, in quantum mechanics the square of the wave function is proportional to the probability of finding a particle in space and time, or the* **probability density**.

Note: The *square* of the absolute value of the wave function solution to Schrödinger's equation, $|\psi|^2$, gives the probability of finding a particle at a location.

The interpretation of ψ^2 as the probability of finding a particle at a particular place altered the idea that the hydrogen electron could be found only in orbits at discrete distances from the nucleus, as described in the Bohr theory. When the Schrödinger equation was solved for the hydrogen atom, the radial wave functions for each energy level were found to have non-zero values at any radial distance from the nucleus. Thus, quantum mechanically, there is some finite probability (ψ^2) of finding the electron in a given energy level at almost *any* distance from the nucleus. For example, the relative probability that an electron in the ground state ($n = 1$) is at a given radius is shown in •Fig. 28.4a. The *maximum* probability coincides with the Bohr radius of 0.0529 nm, but it is possible (that is, there is a finite probability) that the electron could be at almost any distance from the nucleus, including inside the nucleus! The wave function exists for distances beyond 0.20 nm, but there is little chance (probability) of finding an electron in the ground state beyond this distance. The probability density distribution gives rise to the idea of an *electron cloud* around the nucleus (•Fig. 28.4b). The cloud density reflects the probability density that the electron is in a particular region.

Note: The Bohr model of the hydrogen atom is discussed in Section 27.4.

Thus, quantum mechanics uses the wave functions of particles to predict the probability of particle phenomena. This also applies to light and photon "particles." For example, the square of a classical electromagnetic wave function, which is the de Broglie wave for its photons, gives the probability of finding the photons at a location in space. In a single-slit diffraction experiment, we know that many photons will strike the region of the central maximum on a screen, but fewer photons will arrive in the regions of the first bright fringes, even fewer in the regions of the second bright fringes, none in the dark regions between them, and so on. The quantum mechanical probability distribution or density on the screen has the same relative distribution as the classical intensity or brightness.

Note: Single-slit diffraction is discussed in Section 24.3.

*More precisely, the absolute value of the wave function squared, $|\psi|^2$.

•FIGURE 28.4 Electron probability: the hydrogen atom
(a) The square of the wave function or the probability of finding the hydrogen electron in the ground state ($n = 1$) as a function of radial distance. The electron has the greatest probability of being found at distance of 0.0529 nm from the nuclear proton, which is the radius of the first Bohr orbit. **(b)** The probability distribution gives rise to the idea of an electron cloud around the nucleus. The cloud density reflects the probability

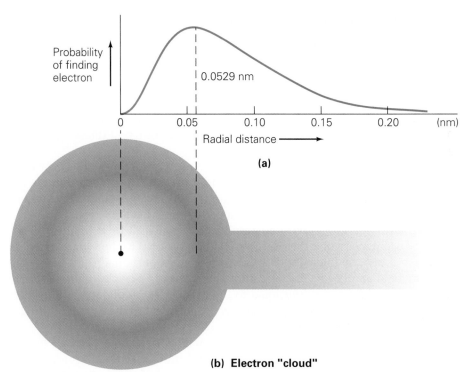

(a)

(b) Electron "cloud"

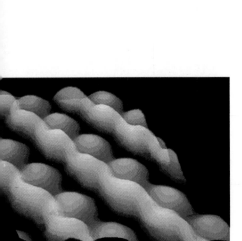

•FIGURE 28.5 Scanning tunneling microscopy
The scanning tunneling microscope, developed in the 1970s and 1980s, makes use of the quantum phenomenon called *vacuum tunneling*. The tip of a metal needle probe is moved across the contours of a metallized specimen surface. Applying a small voltage between the probe and the surface causes electrons to tunnel through the vacuum. The tunneling current is extremely sensitive to the separation of the needle tip and the surface. As a result, when the probe is scanned across the sample, surface features as small as atoms show up as variations in the tunneling current that can be processed to produce 3-dimensional images. This photograph shows atoms of the semiconductor gallium arsenide.

If this experiment could be done one photon at a time, *we could not predict exactly* to which region it would go. Its wave function would pass through both slits, and the two diffracted waves would interfere at the screen. We could, therefore, give only the *probability* that it would land in a particular spot on the screen. Thus, each photon lands at a spot, with a probability proportional to the square of its wave function at that point, which is the sum of the two diffracted waves from the two slits. Finally, after many such photons land, the classical two slit interference pattern (fringes) would build up.

Another quantum-mechanical result that runs counter to our everyday experiences with large objects is called *tunneling*. As we have seen, a basic theme of quantum mechanics is that a particle's location is governed by a probability wave, and the behavior of waves is different from that of particles. In classical physics, we know that there are regions where particles are forbidden because of energy considerations. These are regions where the particle's potential energy would be greater than its total energy. The particle is not allowed in such regions because it would have a negative kinetic energy ($E - U = K < 0$), which is impossible. In such situations we say that the particle's location is limited by a *potential energy barrier*.

In certain instances, however, quantum mechanics predicts a small, but finite, probability of the particle's wave function penetrating the barrier, and thus of the particle itself being found on the other side. We say that there is a certain probability of the particle "tunneling" through the barrier. Such tunneling is in fact observed to occur, and forms the basis of the scanning tunneling microscope (STM), which creates images with resolution on the order of the size of a single atom (•Fig. 28.5). (Such barrier penetration is also used to explain certain nuclear decay processes that classically should not occur, as we shall see in Chapter 29.)

28.3 Atomic Quantum Numbers and the Periodic Table

Objective: To understand the structure of the periodic table in terms of quantum mechanical electron orbits and the Pauli exclusion principle.

The Hydrogen Atom

When the Schrödinger equation was solved for the hydrogen atom, the results predicted the allowed energy levels to be the same as those in the Bohr theory (Section 27.4). Also, the energy values depended only on the principal quantum number n. However, in addition, the solution gave two other quantum numbers, which are designated as ℓ and m_ℓ. (See •Fig. 28.6 for a classical analog picture of the meaning of these quantum numbers.)

The quantum number ℓ is called the **orbital quantum number**. It is associated with the angular momentum of the electron due to its orbital motion. The ℓ quantum number has integer values for each n quantum number from zero up to $\ell = (n - 1)$. For example, if $n = 3$, the possible values of ℓ are 0, 1, and 2. The number of different ℓ values is then equal to n for a given principal quantum number. As just shown, for $n = 3$, there are three possible ℓ values. In the hydrogen atom, the energy of the orbit depends only on n, thus the different ℓ values all have the same energy, and are said to be *degenerate*.

The quantum number m_ℓ is called the **magnetic quantum number**. The term "magnetic" has an experimental origin. It was found that a strong magnetic field influences the energy of the hydrogen electron, as observed in its line spectrum. The quantum number m_l describes this "magnetic" effect, hence the name. In the absence of an external magnetic field, m_ℓ plays no role in determining the energy. Thus for a given ℓ value in the absence of an external magnetic field, there are even more different orbits with the same energy, thus we have additional degeneracy.

The magnetic quantum number m_ℓ is associated with the orientation of the angular momentum (a vector) in space. It is therefore associated with the ℓ quan-

ℓ = orbital quantum number

m_ℓ = magnetic quantum number

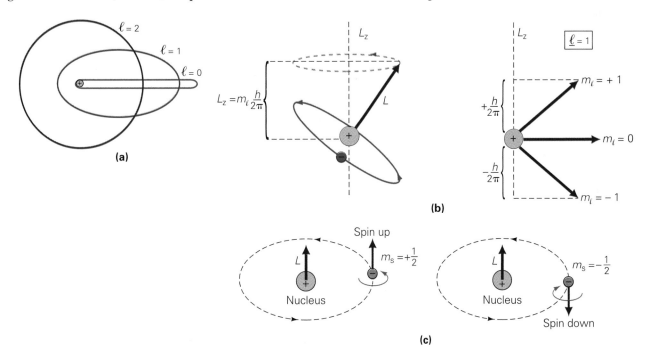

•FIGURE 28.6 Classical Interpretation of orbital quantum numbers.
(a) For planetary orbits, the energy depends only on the size (long axis) of the orbit. This provides a classical model for the subshell degeneracy. Here we plot the classical electron orbits around a proton for $n = 3$, showing the three subshells with different angular momentum but approximately the same energy. The circular orbit has the maximum angular momentum; the narrowest orbit would actually pass through the proton classically, and thus has zero angular momentum. **(b)** Here **L** represents the vector angular momentum about the nucleus associated with the orbit. Because the energy of an orbit is independent of its spatial orientation, all orbits with the same ℓ but different values m_ℓ have the same energy. There are $2\ell + 1$ of these possible orientations for a given ℓ. The value of m_ℓ tells the *component* of the angular momentum vector in a given direction, as shown. **(c)** The spin can be either up or down, depending on the rotational sense of the electron. However, electron spin is strictly a quantum mechanical property and cannot be identified with the physical spin of a macroscopic body.

Insight   The Electron Microscope

The de Broglie hypothesis helped to set the stage for the development of an important practical application of electron beams—the *electron microscope*. As we have seen from the discussion of the Davisson–Germer and Thomson experiments, the wave properties of the electron are not just conceptual or mathematical abstractions. Electrons can undergo diffraction, as do light waves. You might wonder whether this implies that electrons can also be focused like light waves. The answer is yes; and in fact, we can focus electron "waves" to form images, though the means employed are somewhat different from those used with light waves in a light microscope. Moreover, as you can see from Example 28.2, accelerated electrons have very short wavelengths. With such short wavelengths, it is possible to obtain greater magnification and finer resolution than any light microscope can provide. (Recall the discussion of the relationship of resolving power to wavelength in Section 25.4.) The resolving power of standard electron microscopes is on the order of a few nanometers.

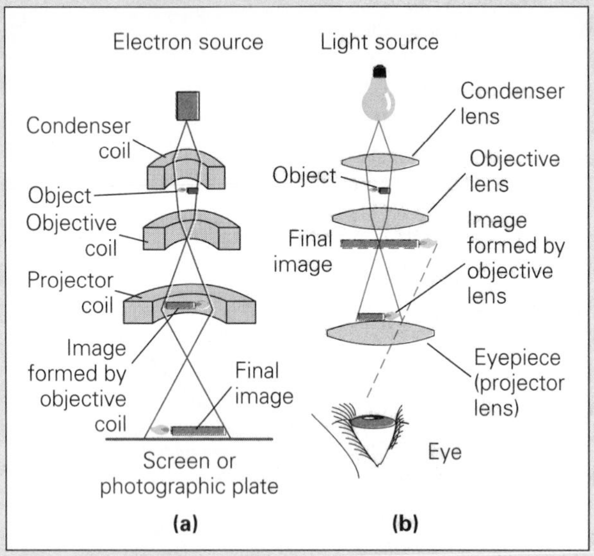

Technological developments made in the 1920s, in particular the focusing of electron beams by magnetic coils, permitted the construction of the first electron microscope in Germany in 1931. In a *transmission electron microscope*, an electron beam is directed onto a very thin specimen. Different numbers of electrons pass through different parts of the specimen depending on its structure. The transmitted beam is then brought to focus by a magnetic objective lens as illustrated in Fig. 1. The general components of electron and light microscopes are analogous, but an electron microscope is housed in a high-vacuum chamber so that the electrons are not deflected by air molecules. As a result, an electron microscope looks nothing like a light microscope (Fig. 2). Magnifications up to 100,000X can be achieved with an electron microscope (Fig. 3b), whereas a light microscope image (Fig. 3a) is limited to a magnification of about 2000X.

Another difference is that the final lens in an electron microscope, called the projector lens, has to project a real image onto a fluorescent screen or photographic film, since the eye cannot perceive an electron image directly. Specimens for transmission electron microscopy must be very thin. Special techniques allow the preparation of specimen sections as thin as 10 nm to 20 nm (only about 100 atoms thick).

The surfaces of thicker objects may be examined by the reflection of the electron beam from the surface. This is done with the more recently developed *scanning electron microscope*. A beam spot is scanned across the specimen by means of deflecting coils, much as is done in a television tube. Surface irregularities cause directional variations in the intensity of the reflected electrons, which gives contrast to the image. The specimens have to be coated with a thin layer of metal (such as gold or aluminum) to make them conducting. Otherwise, they would charge up nonuniformly from the electron beam and distort the image. Through such techniques, an electron microscope gives pictures with a remarkable three-dimensional quality such as those shown in Fig. 3c.

FIGURE 1 Electron and light microscopes
A comparison of the elements of **(a)** an electron microscope and **(b)** a compound light microscope. The light microscope is drawn upside down for a better comparison.

tum number. The m_ℓ quantum number has positive *and* negative integer values for each ℓ quantum number from zero up to $\pm \ell$. That is $m_\ell = 0, \pm 1, \pm 2, \pm 3, \ldots, \pm \ell$. For example, for $\ell = 3$, we have $m_\ell = -3, -2, -1, 0, +1, +2, +3$; or a total of $(2\ell + 1) = (2 \times 3 + 1) = 7$ values of m_ℓ.

But that's not all. There's one more quantum number for the hydrogen atom. Using high resolution spectrometers, it was found that each spectral line of hydrogen is in fact two very closely spaced lines. This is taken into account by a **spin quantum number**, m_s. It is associated with the intrinsic angular momentum of the electron. This property is called *electron spin*, and is sometimes described by analogy with the angular momentum of a spinning object in classical mechanics (Fig. 28.6). Once again, the energy of the orbit is, to a very high degree, independent of the spin quantum number, creating still further degeneracy in the hydrogen atom.

$m_s = $ **spin quantum number**

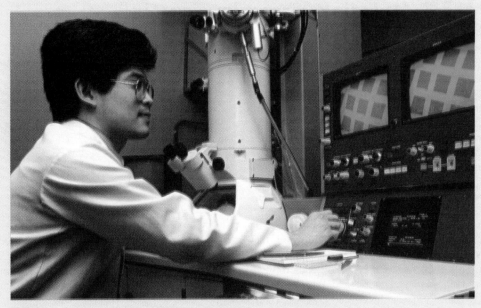

FIGURE 2 An electron microscope
The microscope is housed in the cylindrical vacuum chamber.

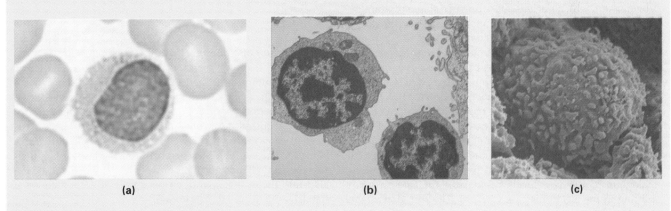

(a)   (b)   (c)

FIGURE 3 Lymphocytes (white blood cells)
Images produced by **(a)** a light microscope, **(b)** a transmission electron microscope (TEM), and **(c)** a scanning electron microscope (SEM).

If an electron were a spinning particle, it would have either angular momentum "spin up" or "spin down," or just two possible values. These are used to describe the so-called two-line spectral *fine structure*. However, electron spin is strictly a quantum mechanical property and cannot be accurately described by the classical analogy. Spin is an internal property of an electron that is characterized by the spin quantum number. The m_s quantum number for an electron has only two values, which are taken to be $m_s = \pm\frac{1}{2}$, for each value of m_ℓ. For example, for $\ell = 1$, there are three values of m_ℓ and each of these has two m_s values: for $m_\ell = -1$, $m_s = \pm\frac{1}{2}$; for $m_\ell = 0$, $m_s = \pm\frac{1}{2}$, and for $m_\ell = +1$, $m_s = \pm\frac{1}{2}$.

The four quantum numbers for the hydrogen atom are summarized in Table 28.1. It should be noted that, except in the case of the spin quantum number, the restrictions on their values came directly from mathematical theory. The value of the spin quantum number was dictated by experimental observations.

Magnetic Resonance Imaging (MRI)

You have probably heard of magnetic resonance imaging, or MRI. MRI, also known as NMR (for nuclear magnetic resonance), has become an increasingly common and important medical technique for visualizing parts of the human body (Fig. 1). It is based on a quantum mechanical spin phenomenon similar to one introduced in this chapter.

As we saw in Chapter 20, a current-carrying loop in a magnetic field experiences a torque, which tends to orient the loop in the field. Specifically, the torque tends to orient the magnetic moment of the loop parallel to the field (or the plane of the loop perpendicular to the field). Similarly, atomic electrons in orbit constitute tiny current loops and have magnetic moments. In this chapter we saw that there is a quantum mechanical property described as intrinsic electron angular momentum, or spin. This gives rise to a spin magnetic moment, and when atoms are placed in a magnetic field, the energy levels are split into two closely spaced levels that give rise to fine structure in spectra.

Nuclei also exhibit properties in magnetic fields that can be understood in terms of nuclear magnetic moments. Various nuclei have such magnetic moments, but to get a general understanding of MRI, we will consider the simplest, the hydrogen atom, with a nucleus consisting of a single proton. Hydrogen is also the most common nucleus used in medical applications, since it is the most abundant element in the human body. The spin angular momentum (or magnetic moment orientation) of the hydrogen nucleus can take on only two values, similar to that of electron spin. These values are commonly referred to as "spin up" and "spin down," in reference to the orientation of the magnetic moment in a magnetic field (Fig. 2a). With spin up, the magnetic moment is parallel to the field; with spin down, it is antiparallel to the field.

In terms of energy levels, in the presence of a magnetic field, there are two energy levels available for the nucleus, analogous to the two levels responsible for fine structure splitting in the electronic case (Fig. 2b). The spin down level has the higher energy, and the energy difference of the levels is proportional to the magnitude of the magnetic field: $\Delta E \propto B$. The transition to the higher level can be made through

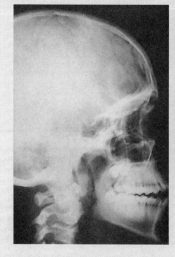
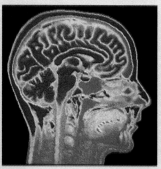

FIGURE 1 Diagnostic images **(a)** An X-ray of a human head. **(b)** A magnetic resonance image (MRI) of a human head. The amount of detail captured, especially in the soft tissues of the brain, makes such images very useful for medical diagnosis.

the absorption of a photon with this energy, $hf = \Delta E$.

In MRI imaging apparatus, the sample to be studied is placed in a uniform magnetic field **B**, the magnitude of which determines the energy (or frequency) of the photon needed to cause a transition: $hf = \Delta E \propto B$. The photons to do this are supplied by a pulsed beam of radio frequency (rf) radiation applied to the sample. If the frequency of the radiation corresponds to the energy difference between the levels (ΔE), then many nuclei will absorb photons from the rf beam and be excited into the upper, spin-down energy level. This frequency that drives such excitation is known as a "resonance" frequency, $f = \Delta E/h$; hence the name magnetic resonance imaging.

A diagram and photo of a typical MRI device are shown in Fig. 3. Large coils produce the magnetic field. (Notice the

TABLE 28.1	Hydrogen Atom Quantum Numbers		
Quantum Number		*Allowed Values*	*Number of Allowed Values*
Principal	n	1, 2, 3, . . . , ∞	No limit
Orbital angular momentum	ℓ	0, 1, 2, 3, . . . , $(n-1)$	n (for each n)
Orbital magnetic	m_ℓ	$0, \pm 1, \pm 2, \pm 3, \ldots, \pm \ell$	$2\ell + 1$ (for each ℓ)
Spin magnetic	m_s	$\pm\frac{1}{2}$	2 (for each m_ℓ)

Multielectron Atoms

The Schrödinger equation for atoms with more than one electron (multielectron atoms) cannot be solved exactly. However, a solution can be found, to a workable approximation, in which each electron occupies a state characterized by a set of

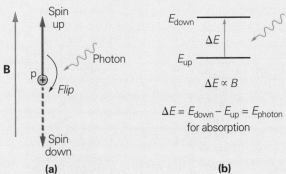

FIGURE 2 Nuclear spin
(a) In a uniform magnetic field, the spin angular momentum or magnetic moment of a hydrogen nucleus can have only two values called "spin up" and "spin down" in reference to the direction of the magnetic field. **(b)** This gives rise to two energy levels for the nucleus.

$$\Delta E \propto B$$

$$\Delta E = E_{down} - E_{up} = E_{photon}$$
for absorption

solenoid arrangement in Fig. 3a.) Other coils produce the rf signals that cause the nuclei to "flip their spin" or be excited from the lower to the upper energy level. The resulting absorption of energy is detected, either by this same rf coil or by another one. Emitted radiation coming from a return transition to the lower state may also be detected. The regions that produce the greatest absorption (or reemission) are those with the greatest concentration of the particular nucleus to which the apparatus is "tuned" by the choice of B and f. Images are produced by means of computerized tomography, similar to that used in X-ray CT scans (See Fig. 20.21). The result is a two- or three-dimensional image (Fig. 1b), which provides a great deal of diagnostic medical information.

Although a variety of atoms or nuclei exhibit nuclear magnetic resonance, most medical work is done with hydrogen atoms because of the varying water content of tissue. For example, muscle tissue has more water than fatty tissue, so there is a distinct contrast in radiation intensity. Fatty deposits in blood vessels are distinct from the tissue of the vessel walls. (The water-rich blood is not seen because it has moved to a new tomography plane by the time it reradiates.) A tumor with a water content different from that of the surrounding tissue would show up in an MRI image. MRI is also widely used in other fields, such as biology and biochemistry.

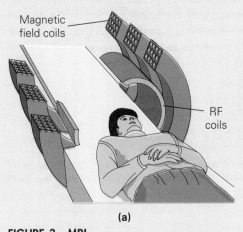
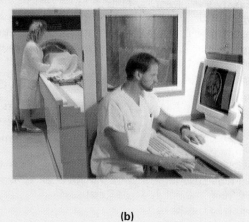

FIGURE 3 MRI
(a) A diagram and **(b)** a photograph of the apparatus used for magnetic resonance imaging.

hydrogen-type quantum numbers. As might be expected with repulsive forces between electrons, the description of electron energy in a multielectron atom is more complicated. For one thing, the energy depends not only on the principal quantum number n, but also on the orbital quantum number ℓ. This gives rise to a subdivision of energy levels. In a different language, in multielectron atoms, the degeneracies that we see in the hydrogen atom are removed, and the energy of an orbit depends on all four quantum numbers.

It is common to refer to the energy level of a given n as a **shell**, and to ℓ levels of that shell as **subshells**. That is, atomic electrons with the same n value are said to be in the same electron shell. Electrons with the same n and ℓ values are said to be in the same electron subshell.

The ℓ subshells can be designated by numbers, as discussed previously. However, it is common to use letters instead. The letters $s, p, d, f, g, \ldots$ correspond to

Note: n quantum numbers designate energy shells
  ℓ quantum numbers designate energy subshells.

TABLE 28.2 Subshell Designations

ℓ Value	Letter Designation
$\ell = 0$	s
$\ell = 1$	p
$\ell = 2$	d
$\ell = 3$	f
$\ell = 4$	g
$\ell = 5$	h
. . .	. . .

•FIGURE 28.7 Energy levels for a multielectron atom
(a) The shell–subshell (n–ℓ,) sequence shows that the energy levels are not evenly spaced and that the sequence of energy levels has numbers out of order. For example, note that the 4s level lies below the 3d level. The maximum number of electrons for a subshell, $2(2\ell + 1)$, is shown in parentheses on representative levels. **(b)** A convenient way to remember the energy level order for a multielectron atom is to list the n versus ℓ values as shown here. The diagonal lines then give the energy levels in ascending order.

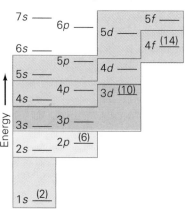

(a)

the values of $\ell = 0, 1, 2, 3, 4, \ldots$, respectively. (This designation derives from historical spectroscopic notation.) After f, the letters go alphabetically (Table 28.2). This letter designation helps avoid confusion with n numbers, as we shall see.

Since an electron's energy depends on both n and ℓ in multielectron atoms, both quantum numbers are used to label atomic energy levels (a shell and subshell). For distinction, we write n as a number, followed by the letter that stands for the value of ℓ. For example, 1s denotes an energy level with $n = 1$ and $\ell = 0$; 2p is for $n = 2$ and $\ell = 1$; 3d for $n = 3$ and $\ell = 2$, and so on.

Also, it is common to refer to the m_ℓ values as representing *orbitals*. For example, a 2p energy level has three orbitals corresponding to the m_ℓ values of -1, 0, and $+1$ (for the p subshell or $\ell = 1$). The spin quantum number also applies to electrons in multielectron atoms, but no special name is given to it.

As we saw in the Bohr theory, the energy levels for the hydrogen atoms are not evenly spaced, but run sequentially upward. However, in multielectron atoms, the numerical sequence of the energy levels has numbers out of order. The shell–subshell (n–ℓ notation) energy level sequence for a multielectron atom is shown in •Fig. 28.7a. Notice, for example, how the 4s level is below the 3p level. Such variations result in part from electrical forces between electrons in multielectron atoms. Also, the electrons in the outer orbits are shielded from the attractive force of the nucleus by the electrons in the inner orbits. A convenient way to remember the level order is given in •Fig. 28.7b.

The ground state for a multielectron atom is similar to that of the hydrogen atom, with the electron in the 1s or lowest energy level. For a multielectron atom, the ground state is the combination of energy levels with the lowest total energy. That is, the electrons are in the lowest possible energy levels. But to identify those levels, we must know how many electrons can occupy a particular energy level. For example, the lithium (Li) atom has three electrons. Can they all be in the 1s level? As we shall see, the answer is no.

The Exclusion Principle

Exactly how the electrons of a multielectron atom distribute themselves in the ground-state energy levels is governed by a principle set forth in 1928 by the Austrian physicist Wolfgang Pauli. The **Pauli exclusion principle** states

No two electrons in a multielectron atom can have the same set of quantum numbers (n, ℓ, m_ℓ, m_s). That is, no two electrons can be in the same quantum state.

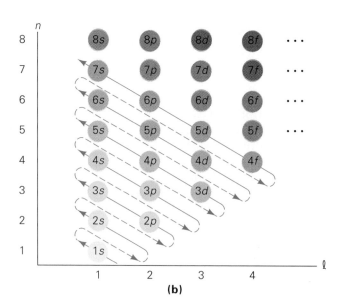

(b)

What this means is that each different set of quantum numbers (n, ℓ, m_ℓ, m_s) corresponds to a different quantum state that can be occupied by only one electron.

The limits on the quantum numbers set the limits on the number of states for a given energy level. For example, the $1s$ ($n = 1$, $\ell = 0$) can have only one m_ℓ value, $m_\ell = 0$, along with only two m_s values, $m_s = \pm\frac{1}{2}$. Thus, there are only two unique sets of quantum numbers (n, ℓ, m_ℓ, m_s) for the $1s$ level: $(1, 0, 0, +\frac{1}{2})$ and $(1, 0, 0, -\frac{1}{2})$, so only two electrons can be in the $1s$ level. When this is the case, we say *the shell is full*; all other electrons are excluded from it by Pauli's principle. Thus, for a Li atom, with three electrons, the third electron must occupy the next higher level ($2s$) when the atom is in the ground state. This is illustrated in •Fig. 28.8, along with the ground-state energy levels for some other atoms.

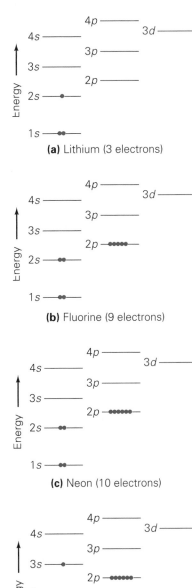

(a) Lithium (3 electrons)

(b) Fluorine (9 electrons)

(c) Neon (10 electrons)

(d) Sodium (11 electrons)

EXAMPLE 28.3 ■ HOW MANY STATES? THE QUANTUM SHELL GAME

How many possible sets of quantum numbers or electron states are there in (a) the $3p$ subshell and (b) the $4d$ subshell?

Solution. This is simply a job of following the quantum mechanical counting rules.

Given: (a) $3p$ level ($n = 3$, $\ell = 1$)   *Find:*   sets of quantum numbers
        (b) $4d$ level ($n = 4$, $\ell = 2$)   of electrons states

(a) For a particular subshell, it is the ℓ value that determines the number of states. Recall that there are $(2\ell + 1)$ possible m_ℓ values for a given ℓ. Thus, for $\ell = 1$, there are $[(2 \times 1) + 1] = 3$ values for m_ℓ. Each of these can have two m_s values ($\pm\frac{1}{2}$), making six different combinations of (n, ℓ, m_ℓ, m_s), or six states.

Notice that the number of possible states for a given value is then $2(2\ell + 1)$, since there are two possible "spin states" for each orbital state. Stated another way, if these all had the same energy we would have a degeneracy of $2(2\ell + 1)$ for each value of ℓ.

$$\text{Number of possible electron states for a given } \ell \text{ value} = 2(2\ell + 1)$$

(b) This part is now easy. The $4d$ level with $\ell = 2$ has

$$\text{Number of states} = 2(2\ell + 1) = 2[(2 \times 2) + 1] = 10$$

This development is summarized in Table 28.3. Notice that the number of states in all of the subshells in a given shell (given n) is $2n^2$. For example, for the $n = 2$ shell, the total number of states for its s and p subshells ($\ell = 0, 1$) is $2n^2 = 2(2)^2 = 8$. This means that up to eight electrons can be accommodated in the $n = 2$ shell.

Follow-up Exercise. In this Example, how many electrons could be accommodated in the $1s$ subshell if it were not for the spin quantum number?

Electron Configurations. We can build up the electron structure of atoms, so to speak, by putting an increasing number of electrons in the lower energy levels [hydrogen (H), 1 electron; helium (He), 2 electrons; lithium (Li), 3 electrons, etc.], as was done for four particular elements in Fig. 28.8. However, rather than draw diagrams each time we want to represent the electron arrangement in a ground-state atom, we can use a shorthand notation called the **electron configuration.**

In this notation, we write the energy levels in increasing order, and designate the number of electrons in each level with a superscript. For example, $3p^5$ means that a $3p$ subshell has five electrons:

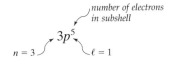

•**FIGURE 28.8 Filling subshells**
The electron subshell distributions for several unexcited atoms according to the Pauli exclusion principle. The s subshell can have a maximum of two electrons and the p subshell a maximum of six electrons.

Electron configuration: A shorthand notation for the electron quantum states occupied in a ground-state atom

TABLE 28.3 Possible Sets of Quantum Numbers and States

Electron Shell n	Subshell ℓ	Subshell Notation	Orbitals m_ℓ	Number of Orbitals (m_ℓ) in Subshell $(2\ell + 1)$	Number of States (m_s) in Subshell $2(2\ell + 1)$	Total Electron States for n Shell $2n^2$
1	0	1s	0	1	2	2
2	0	2s	0	1	2	8
	1	2p	1, 0, −1	3	6	
3	0	3s	0	1	2	
	1	3p	1, 0, −1	3	6	18
	2	3d	2, 1, 0, −1, −2	5	10	
4	0	4s	0	1	2	
	1	4p	1, 0, −1	3	6	32
	2	4d	2, 1, 0, −1, −2	5	10	
	3	4f	3, 2, 1, 0, −1, −2, −3	7	14	

The electron configurations for the atoms shown in Fig. 28.8 are

Li   (3 electrons) $1s^2 2s^1$

F   (9 electrons) $1s^2 2s^2 2p^5$

Ne   (10 electrons) $1s^2 2s^2 2p^6$

Na   (11 electrons) $1s^2 2s^2 2p^6 3s^1$

In writing an electron configuration, when one subshell is filled, you go on to the next higher one. Notice that the superscripts in a configuration add up to the total number of electrons in the atom.

The spacings between the energy levels are not equal, as can be seen from Figs. 28.7a or 28.8. In general, there are relatively large energy gaps between the s levels and the levels below them. The lower levels are p levels, with the exception of the lowest case—the $1s$ level below the $2s$ level (and there are no levels below the $1s$). The gaps between other levels—for example, between an s level and the p level above, or between the d and p levels—are smaller and do not differ greatly in energy.

The periodic large energy gaps in the levels may be represented by

$$1s^2 \,|\, 2s^2p^6 \,|\, 3s^2 3p^6 \,|\, 4s^2 3d^{10} 4p^6 \,|\, 5s^2 4d^{10} 5p^6 \,|\, 6s^2 4f^{14} 5d^{10} 6p^6 \,|\, \text{etc.}$$

(number of states)   (2)   (8)   (8)   (18)   (18)   (32)

where the vertical lines indicate the energy gaps. The states between the lines have similar energies. We refer to such sets of energy levels that have about the same energy as an **electron period.**

Electron periods: Sets of energy levels with about the same energy

These electron periods are the basis of the periodic table of elements. With your present knowledge of electron configurations, you are now in a position to understand the periodic table better than the person who originally developed it.

The Periodic Table of Elements

By 1860, over 60 chemical elements had been discovered. Several attempts had been made to classify the elements or put them into some orderly arrangement, but none proved to be very satisfactory. It had been noted in the early 1800s that the elements could be listed in such a way that similar chemical properties recurred periodically throughout the list. Following this idea, in 1869 a Russian chemist, Dmitri Mendeleev (pronounced men-duh-*lay*-eff), formulated an arrangement of the elements based on this periodic property. The modern version of his *periodic table of elements* is used today and can be seen on the walls of just about every science building, as well as in •Fig. 28.9.

Mendeleev arranged the known elements in rows, which are called **periods**, in order of increasing atomic masses. When he came to an element that had chemical properties similar to those of one of the previous elements, he went back and put this element below the previous similar one. In this manner, he formed both horizontal rows of elements, and vertical columns called **groups**, or families of elements with similar properties. The table was later rearranged in order of increasing atomic or proton number (the number of protons in an atom and the numbers at the top of the element boxes in Fig. 28.9) to resolve some inconsistencies. (Notice that if atomic masses were used, cobalt and nickel, atomic numbers 27 and 28, would fall in different groups.

With only 65 elements, there were vacant spaces in Mendeleev's table. The elements for these spaces were yet to be discovered. Because the missing elements were part of a sequence and had properties similar to those of other elements in a group, Mendeleev was able to predict their masses and chemical properties. Less than 20 years after Mendeleev devised his table, showing chemists what to look for to find the undiscovered elements, three of the missing elements were discovered.

Notice that the periodic table puts the elements into seven horizontal rows or periods, in order of increasing atomic or proton number. The first period has only two elements. Periods 2 and 3 have eight elements, and periods 4 and 5 have 18 elements. Recall that the *s*, *p*, *d*, and *f* subshells can contain a maximum of 2, 6, 10, and 14 electrons [$2(2\ell + 1)$], respectively. You should begin to see a correlation between these numbers and the arrangements of elements in the periodic table.

The periodicity of the periodic table can be understood in terms of the electron configurations of the atoms. For $n = 1$, the electrons are in one of two *s* states ($1s$); for $n = 2$, we have the last electrons going into the $2s$ and $2p$ states, which gives 10 electrons; and so on. Thus, for a given element, its period number is equal to the highest n shell containing electrons in the atom. Notice the electron configurations for the elements in Fig. 28.9. Also, compare the electron periods given earlier, as defined by energy gaps, and the periods in the periodic table (•Fig. 28.10). There is a one-to-one correlation, so the periodicity comes from energy level considerations in atoms.

Chemists refer to *representative elements*, which we see from Fig. 28.9 are those in which the last electron enters an *s* or *p* subshell. In *transition elements*, the last electron enters a *d* subshell; and in *inner transition elements*, the last electron enters an *f* subshell. So that the periodic table is not unmanageably wide, the *f* subshells are usually placed in two rows at the bottom of the table. Each row is given a name—the *lanthanide series* and the *actinide series*—based on where it is positioned within the period.

Finally, you can also understand why elements in vertical columns or groups have similar chemical properties. As you probably know from chemistry classes, the chemical properties of an atom, such as its ability to react and form compounds, depends on the outermost electrons in the atom—that is, the number of electrons in the outermost *unfilled* shell. It is these electrons, called *valence electrons*, that form chemical bonds. Because of the way the elements are arranged in the table, the outermost electron configurations of all the atoms in any one group are the same or very similar. The atoms would therefore be expected to have similar chemical properties, as indeed they do. For example, notice the first two groups at the left of the table. They have one and two outermost electrons in an *s* subshell, respectively. These elements are all highly reactive metals that form compounds having many similarities. The group at the far right, the noble gases, have completely filled subshells *and* come at the ends of electron periods, or just before a large energy gap. These gases are all chemically very nonreactive and form compounds only under very special conditions.

Periods: horizontal rows

Groups: vertical columns

PERIODS	GROUP I (s)	GROUP II (s)	Transition elements (d)										GROUP III (p)	GROUP IV (p)	GROUP V (p)	GROUP VI (p)	GROUP VII (p)	GROUP VIII (p)
1	1 H 1.01 $1s^1$																	2 He 4.00 $1s^2$
2	3 Li 6.94 $2s^1$	4 Be 9.01 $2s^2$											5 B 10.81 $2p^1$	6 C 12.01 $2p^2$	7 N 14.01 $2p^3$	8 O 16.00 $2p^4$	9 F 19.00 $2p^5$	10 Ne 20.18 $2p^6$
3	11 Na 22.99 $3s^1$	12 Mg 24.31 $3s^2$											13 Al 26.98 $3p^1$	14 Si 28.09 $3p^2$	15 P 30.97 $3p^3$	16 S 32.07 $3p^4$	17 Cl 35.45 $3p^5$	18 Ar 39.95 $3p^6$
4	19 K 39.10 $4s^1$	20 Ca 40.08 $4s^2$	21 Sc 44.96 $3d^14s^2$	22 Ti 47.88 $3d^24s^2$	23 V 50.94 $3d^34s^2$	24 Cr 52.00 $3d^54s^1$	25 Mn 54.94 $3d^54s^2$	26 Fe 55.85 $3d^64s^2$	27 Co 58.93 $3d^74s^2$	28 Ni 58.69 $3d^84s^2$	29 Cu 63.55 $3d^{10}4s^1$	30 Zn 65.39 $3d^{10}4s^2$	31 Ga 69.72 $4p^1$	32 Ge 72.61 $4p^2$	33 As 74.92 $4p^3$	34 Se 78.96 $4p^4$	35 Br 79.90 $4p^5$	36 Kr 83.80 $4p^6$
5	37 Rb 85.47 $5s^1$	38 Sr 87.62 $5s^2$	39 Y 88.96 $4d^15s^2$	40 Zr 91.22 $4d^25s^2$	41 Nb 92.91 $4d^45s^1$	42 Mo 95.94 $4d^55s^1$	43 Tc (98) $4d^55s^2$	44 Ru 101.07 $4d^75s^1$	45 Rh 102.91 $4d^85s^1$	46 Pd 106.42 $4d^{10}5s^6$	47 Ag 107.87 $4d^{10}5s^1$	48 Cd 112.41 $4d^{10}5s^2$	49 In 114.82 $5p^1$	50 Sn 118.71 $5p^2$	51 Sb 121.76 $5p^3$	52 Te 127.60 $5p^4$	53 I 126.90 $5p^5$	54 Xe 131.29 $5p^6$
6	55 Cs 132.91 $6s^1$	56 Ba 137.33 $6s^2$	57 La * 138.91 $5d^16s^2$	72 Hf 178.49 $5d^26s^2$	73 Ta 180.95 $5d^36s^2$	74 W 183.85 $5d^46s^2$	75 Re 186.21 $5d^56s^2$	76 Os 190.2 $5d^66s^2$	77 Ir 192.22 $5d^76s^2$	78 Pt 195.08 $5d^96s^1$	79 Au 196.97 $5d^{10}6s^1$	80 Hg 200.59 $5d^{10}6s^2$	81 Tl 204.36 $6p^1$	82 Pb 207.2 $6p^2$	83 Bi 208.98 $6p^3$	84 Po (209) $6p^4$	85 At (210) $6p^5$	86 Rn (222) $6p^6$
7	87 Fr (223) $7s^1$	88 Ra 226.03 $7s^2$	89 Ac † 227.03 $6d^17s^2$	104 (261) $6d^27s^2$	105 (262) $6d^37s^2$	106 (263) $6d^47s^2$	107 (262)	108 (265)	109 (266)	110 (269)	111 (272)							

Key: Atomic number — 26 Fe — Symbol; 58.85 — Atomic mass; $3d^64s^2$ — Outer electron configuration

f

*	58 Ce 140.12 $5d^14f^16s^2$	59 Pr 140.91 $4f^36s^2$	60 Nd 144.24 $4f^46s^2$	61 Pm (145) $4f^56s^2$	62 Sm 150.36 $4f^66s^2$	63 Eu 151.96 $4f^76s^2$	64 Gd 157.25 $5d^14f^76s^2$	65 Tb 158.93 $5d^14f^86s$	66 Dy 162.50 $4f^{10}6s^2$	67 Ho 164.93 $4f^{11}6s^2$	68 Er 167.26 $4f^{12}6s^2$	69 Tm 168.93 $4f^{13}6s^2$	70 Yb 173.04 $4f^{14}6s^2$	71 Lu 174.97 $5d^14f^{14}6s^2$	(Lanthanides)
†	90 Th 232.04 $6d^27s^2$	91 Pa 231.04 $5f^26d^17s^2$	92 U 238.03 $5f^36d^17s^2$	93 Np 237.05 $5f^46d^17s^2$	94 Pu (244) $5f^66d^07s^2$	95 Am (243) $5f^76d^07s^2$	96 Cm (247) $5f^76d^17s^2$	97 Bk (247) $5f^86d^17s^2$	98 Cf (251) $5f^{10}6d^07s^2$	99 Es (252) $5s^{11}6d^07s^2$	100 Fm (257) $5f^{12}6d^07s^2$	101 Md (258) $5f^{13}6d^07s^2$	102 No (259) $5f^{14}6d^07s^2$	103 Lr (260) $5f^{14}6d^17s^2$	(Actinides)

FIGURE 28.9 The periodic table

The elements are arranged in order of increasing atomic or proton number. Horizontal rows are called periods, and vertical columns are called groups. The elements in a group have similar chemical properties. Each atomic mass represents an average of that element's isotopes, weighted to reflect their relative abundance in our immediate environment. The masses have been rounded to two decimal places; more precise values can be found in Appendices IV and V. (A value in parentheses represents the mass number of the best-known or longest-lived isotope of an unstable element. Names for elements 104–109 are still being debated and elements 110 and 111 have not yet been named.)

Shell (last to be filled)	Subshells	Number of electrons in subshell, $2(2\ell + 1)$	Corresponding period in periodic table
n = 7	7p	6	
	6d	10	Period 7 (32 elements)
	5f	14	
	7s	2	
n = 6	6p	6	
	5d	10	Period 6 (32 elements)
	4f	14	
	6s	2	
n = 5	5p	6	
	4d	10	Period 5 (18 elements)
	5s	2	
n = 4	4p	6	
	3d	10	Period 4 (18 elements)
	4s	2	
n = 3	3p	6	Period 3 (8 elements)
	3s	2	
n = 2	2p	6	Period 2 (8 elements)
	2s	2	
n = 1	1s	2	Period 1 (2 elements)

•**FIGURE 28.10 Electron periods**
The periods of the periodic table are related to electron configurations. The last *n* shell to be filled is equal to the period number. The electron periods and the corresponding periods of the table are defined by energy gaps in the energy levels of the atoms.

28.4 The Heisenberg Uncertainty Principle

Objectives: To understand the inherent quantum mechanical limits on the accuracy of physical observations—in particular, those of location and speed, and energy and time.

Another important aspect of quantum mechanics has to do with measurement and accuracy. In classical mechanics, there is no limit to the accuracy of a measurement. Theoretically, by continual refinement of a measurement instrument and procedure, the accuracy could be improved to any degree so as to give *exact* values. This resulted in a *deterministic* view of nature. For example, if you know position and velocity of an object *exactly* at a particular time, you can determine where it will be in the future and where it was in the past (assuming you knew all the future and past forces).

However, quantum theory predicts otherwise and sets limits on the possible accuracy of measurements. This idea was introduced in 1927 by the German physicist Werner Heisenberg, who had developed another approach to quantum mechanics that complemented Schrödinger's wave theory. The **Heisenberg uncertainty principle** as applied to position and momentum (or velocity) may be stated as follows:

> It is impossible to know simultaneously an object's exact position and its momentum.

This concept is often illustrated with a simple thought experiment. Suppose that you wanted to measure the position and momentum (actually the velocity) of an electron. In order for you to "see," or locate, the electron, at least one photon must bounce off the electron and come to your eye, as illustrated in •Fig. 28.11. However, in the collision process, some of the photon's energy and momentum are transferred to the electron (as in the Compton effect, described in Section 27.3).

After the collision, the electron recoils. Thus, in the very process of trying to locate the position very accurately (trying to make the uncertainty of position Δx

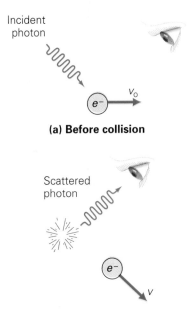

Incident photon

v_0

e^-

(a) Before collision

Scattered photon

e^-

v

(b) After collision

•**FIGURE 28.11 Measurement-induced uncertainty**
(a) To measure the position and momentum (or velocity) of an electron, at least one photon must collide with the electron and be scattered toward the eye. **(b)** In the collision process, energy and momentum are transferred to the electron, which induces uncertainty in the velocity.

very small), you induce more uncertainty into your knowledge of the electron's velocity or momentum ($\Delta p = m\,\Delta v$), since the process of determining its position sent the electron flying off. In the everyday macroscopic world, the uncertainty due to viewing an object would be negligible, because light does not appreciably alter the motion or position of an ordinary-sized object.

For our subatomic case, the position of an electron could be measured at best to an accuracy of about the wavelength λ of the incident light, that is, $\Delta x \approx \lambda$. The photon "particle" has a momentum of $p = h/\lambda$. Since we cannot tell how much of this momentum might be transferred during collision, the final momentum of the electron would have an uncertainty of $\Delta p \approx h/\lambda$.

The total uncertainty is given by the product of the individual uncertainties, and is at least,

$$(\Delta p)(\Delta x) \approx \left(\frac{h}{\lambda}\right)(\lambda) = h$$

This equation gives an estimate of the *minimum uncertainties* or maximum accuracies of *simultaneous* measurements of the momentum and position. In actuality, the uncertainties could be much worse depending on the amount of light (number of photons) used, the apparatus, and the technique. Through detailed theoretical calculations, Heisenberg found that, *at very best,*

Minimum uncertainties in momentum and position.

$$(\Delta p)(\Delta x) \geq \frac{h}{2\pi} \qquad (28.5)$$

That is, Heisenberg's uncertainties principle states that the product of the *minimum* uncertainties of position and momentum is on the order of Planck's constant divided by $2\pi\,(\approx 10^{-34}\,\text{J}\cdot\text{s})$. These are the minimum uncertainties, or the *best degree of accuracies* we can ever hope to achieve for *simultaneous* measurements. In the process of trying to locate the position of the particle accurately (that is, to make Δx small), the uncertainty in the momentum is made larger ($\Delta p \approx h/2\pi\,\Delta x$) and vice versa. To take the extreme case, if we could measure the exact location of a particle ($\Delta x \to 0$), we would have no idea at all about its momentum ($\Delta p \to \infty$). Thus, the measurement process itself limits the accuracy to which we can simultaneously measure position and momentum. In Heisenberg's words, "Since the measuring device has been constructed by the observer . . . we have to remember that what we observe is not nature in itself but nature exposed to our method of questioning."

EXAMPLE 28.4 ■ ELECTRON VERSUS BULLET: THE UNCERTAINTY PRINCIPLE

An electron and a 20-g bullet, both moving linearly, are measured to have equal speeds of 300 m/s to an accuracy of ± 0.010%. What is the minimum uncertainty in the position of each?

Solution. We expect the electron to be affected much more by the uncertainty principle because of its very small mass. Listing the quantities, we have

Given: $m_b = 20\,\text{g} = 0.020\,\text{kg}$
$m_e = 9.11 \times 10^{-31}\,\text{kg}$ (known)
$v = 300\,\text{m/s} \pm 0.010\%$

Find: Δx's (minimum uncertainties in position)

An uncertainty of 0.010% in velocity for both the electron and the bullet is

$$(300\,\text{m/s})(0.00010) = 0.030\,\text{m/s}\ (= 3.0\,\text{cm/s})$$

thus

$$v = 300\,\text{m/s} \pm 0.010\% = 300\,\text{m/s} \pm 0.030\,\text{m/s}$$

The *total* uncertainty in velocity is *twice* this amount because the measurements can be off by 0.010% above or below the actual values and

$$\Delta v = 0.060 \text{ m/s} \quad (= 6.0 \text{ cm/s})$$

Then, for the electron, the minimum uncertainty is

$$\Delta x = \frac{h}{2\pi\Delta p} = \frac{h}{2\pi m_e \Delta v}$$

$$= \frac{6.63 \times 10^{-34} \text{ J} \cdot \text{s}}{2\pi(9.11 \times 10^{-31} \text{ kg})(0.060 \text{ m/s})} = 0.0019 \text{ m} \quad (= 0.19 \text{ cm})$$

Similarly for the bullet, we can calculate its minimum position uncertainty

$$\Delta x = \frac{h}{2\pi m_b \Delta v} = \frac{6.63 \times 10^{-34} \text{ J} \cdot \text{s}}{2\pi(0.020 \text{ kg})(0.060 \text{ m/s})} = 8.8 \times 10^{-32} \text{ m}$$

Notice that the uncertainty in the position of the bullet (much smaller than nuclear diameters) is orders of magnitude less than that of the electron. The uncertainty for relatively massive objects traveling at ordinary speeds is practically negligible. However for electrons, the answer of 1.9 mm is very significant as this is many times larger than an atom.

Follow-up Exercise. In this Example, what would be the minimum uncertainty in the electron's speed in order for the minimum uncertainty in its positon to be 0.1 nm?

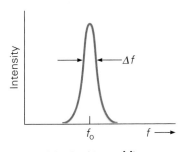

FIGURE 28.12 Natural line broadening
Because a measurement must be carried out in a time comparable to the lifetime (Δt) of an electron in an excited state, the energy is uncertain by an amount $\Delta E = h\Delta f$. The observed emission line has a width of Δf, rather than being a line of single frequency f_o.

Another form of the uncertainty principle relates energy and time. To understand this, consider the position of the electron in the previous thought experiment to be known with an uncertainty of $\Delta x \approx \lambda$. The photon used to detect the particle travels with a speed c, and it takes a time of $\Delta x/c \approx \lambda/c$ for this photon to traverse a distance equal to the uncertainty in the particle's position. Thus, the time when the particle is at the measured position is uncertain by about

$$\Delta t \approx \lambda/c$$

Since we can't tell whether the photon transfers some or all of its energy ($E = hf = hc/\lambda$) to the particle, the uncertainty in the energy is

$$\Delta E \approx \frac{hc}{\lambda}$$

Then the total uncertainty is at least

$$(\Delta E)(\Delta t) \approx \left(\frac{hc}{\lambda}\right)\left(\frac{\lambda}{c}\right) = h$$

As with the position–momentum relationship, at very best we have

$$(\Delta E)(\Delta t) \geq \frac{h}{2\pi} \qquad (28.6)$$

Minimum uncertainties in energy and time

This form of the uncertainty principle indicates that the energy of an object may be uncertain by an amount ΔE for a time $\Delta t \approx h/2\pi\Delta E$. During this time, the energy is uncertain and might not even be conserved. This is an important consideration in particle interactions, as you will learn in a later chapter.

Also, we cannot measure the energy of a particle exactly unless we take an infinite amount of time to do so. If a measurement of energy is carried out in a time Δt, then it must be uncertain by an amount ΔE. For example, the measurement of the frequency of a photon emitted by an atomic electron is a measurement of the energy associated with the transition from an excited state to the ground state. The measurement must be carried out in a time comparable to the time the electron is in the excited state—that is, the lifetime of the excited state. As a result, the observed emission line in the frequency spectrum has a finite energy width, since $\Delta E = h\Delta f$ (•Fig. 28.12).

Note: Spectral lines are discussed in Section 27.4.

This so-called *natural broadening* was ignored in Chapter 27, where spectral lines were considered to be exact frequencies (that is, to have no width at all). This is the same as assuming that the excited states of the Bohr atom have infinite lifetimes.

EXAMPLE 28.5 ■ NATURAL LINE BROADENING

An electron in an excited state has a lifetime of 1.00×10^{-8} s. (a) What is the minimum uncertainty in the energy of the photons emitted on de-excitation? (b) What is the magnitude of the natural broadening of the spectral line?

Solution.

Given: $\Delta t = 1.00 \times 10^{-8}$ s   *Find:* (a) ΔE (minimum energy uncertainty)
   (b) f (frequency broadening)

(a) The minimum uncertainty in the energy is given by Eq. 28.6:

$$\Delta E = \frac{h}{2\pi\Delta t} = \frac{6.63 \times 10^{-34}\,\text{J·s}}{2\pi(1.00 \times 10^{-8}\,\text{s})} = 1.06 \times 10^{-26}\,\text{J}$$

(b) The uncertainty, or broadening, of the frequency of the observed spectral line is then

$$\Delta f = \frac{\Delta E}{h} = \frac{1.06 \times 10^{-26}\,\text{J}}{6.63 \times 10^{-34}\,\text{J·s}} = 1.60 \times 10^{7}\,\text{Hz}$$

The uncertainty in the frequency of a spectral line is called the *natural line width*. The lifetimes of atomic electrons in their excited states are on the order of 10^{-8} s, and it might appear from the magnitude of the value of Δf that the spectral lines would be quite wide.

However, recall that the frequency of visible light is on the order of 10^{14} Hz. Expressed as a percentage, the uncertainty is about $10^7/10^{14}$ ($\times 100\%$) $= 10^{-5}\%$. Thus, the spectral line has a frequency width on the order of $10^{14} \pm 10^{-5}\%$, which is quite narrow.

Follow-up Exercise.   In this Example, what is the lifetime of the excited state if the atom emits photon with an energy of 2.5 eV $\pm$ $2.0 \times 10^{-6}\%$ when making the transition to the next lowest energy states?

●**FIGURE 28.13 Cloud-chamber photograph of pair production** In this false-color bubble-chamber photograph, a gamma-ray photon (not visible) interacts with an atomic nucleus to produce an electron and a positron (green and red spiral tracks at top). In the process, it also dislodges an orbital electron (the vertical green track). An external magnetic field causes the electron and positron to be deflected in paths of opposite curvature. A similar event is recorded in the bottom half of the photo. (Why do you suppose the paths of the particles created in this case show less deflection?)

28.5 Particles and Antiparticles

Objectives: To understand (a) the relationship between particles and antiparticles, and (b) the energy requirements for pair production.

The quantum mechanics of Schrödinger and Heisenberg were successful in explaining observations and in predicting new atomic phenomena. When the quantum theory was extended to include relativistic considerations by the British physicist Paul A. M. Dirac in 1928, something new and very different was predicted—a particle called the **positron**. The positron should have the same mass as the electron but should carry a *positive* charge. The oppositely charged positron is said to be the **antiparticle** of the electron.

The positron was first observed experimentally in 1932 by the American physicist C. D. Anderson in cloud chamber experiments with cosmic rays. The curvature of the particle tracks in a magnetic field showed two types of particles (●Fig. 28.13). The tracks indicated that both particles had the same mass, but their spiral curvatures were opposite. From the magnetic force relationship, $F = qvB$, this observation requires the particles to be oppositely charged. Thus, Anderson discovered the positron, a particle with a mass equal to that of the electron and with the same magnitude of electrical charge, but opposite in sign.

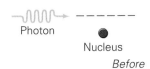
Photon

Nucleus

Before

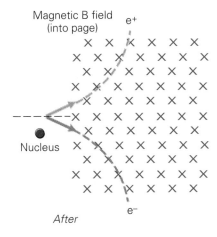

After

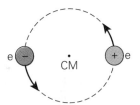

•**FIGURE 28.14 Pair production**
An electron and a positron are created when an energetic photon interacts with a nucleus.

By the conservation of charge, a positron can be created only with the simultaneous creation of an electron in a process called **pair production**. In Anderson's experiment, positrons were observed to be emitted from a thin lead plate exposed to cosmic rays from outer space, which contain highly energetic X-rays. Pair production occurs when an X-ray "photon" collides with a nucleus—the nuclei of the lead atoms of the plate in the Anderson experiment. In the collision process, the photon goes out of existence, and an "electron pair" (an electron and a positron) is created, as illustrated in •Fig. 28.14, in a conversion of energy into mass. By the conservation of energy,

$$hf = 2m_e c^2 + K_{e^-} + K_{e^+} + K_{nuc}$$

where hf is energy of the photon, $2m_e$ is the rest mass of the electron-positron pair, and the K's are the kinetic energies of the particles and the recoil nucleus. Because of its relatively large mass, the kinetic energy of the recoil nucleus can usually be considered negligible, and

$$hf \approx 2m_e c^2 + K_{e^-} + K_{e^+}$$

From the energy equation, we can see that a photon cannot create an electron pair unless

$$hf \geq 2m_e c^2 = 1.022 \text{ MeV} \qquad (28.7)$$

(Recall that the rest mass energy of an electron is $m_e c^2 = 0.511$ MeV.) This minimum energy is called the *threshold energy for pair production*.

But if positrons are created by cosmic rays, why are they not commonly found in nature? For example, why are they not evident in ordinary chemical processes? The answer is because positrons are taken out of existence by a process called **pair annihilation**. When an energetic positron appears in pair production, it loses kinetic energy in collisions as it passes through matter. Finally, moving at a low speed, it combines with an electron of the material and forms a hydrogenlike atom, called a *positronium atom*, in which a positron substitutes for a proton. The positronium atom is unstable and quickly decays ($\approx 10^{-10}$ s) into two 0.511-MeV photons (•Fig. 28.15). Pair annihilation is then a direct conversion of (rest) mass into electromagnetic energy, the inverse of pair production so to speak. These processes are striking examples of the mass–energy equivalence predicted by Einstein ($E = mc^2$), that is, mass is viewed as a form of energy.

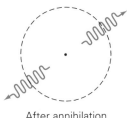

Before annihilation
(positronium atom)

After annihilation
(photons)

•**FIGURE 28.15 Pair annihilation**
The disappearance of a positronium atom is signaled by the appearance of two 0.511-MeV photons. Why do we not expect one photon with an energy of 1.022 MeV? (CM is center of mass.)

All subatomic particles have been found to have antiparticles, which are observed in cosmic rays from outer space and/or are produced in nuclear processes. For example, there is an antiproton with the same mass as a proton but with a negative charge. There are also antineutrons. In our environment, there is a preponderance of electrons, protons, and neutrons. When antiparticles are created, they quickly combine with their respective particles in annihilation processes.

It is conceivable that antiparticles predominate in some parts of the universe. If so, the atoms of the **antimatter** in this region would consist of negatively charged nuclei composed of antiprotons and antineutrons, surrounded by positively charged positrons (antielectrons). It would be difficult to distinguish a region of antimatter visibly, since it would appear the same as ordinary matter. The physical behavior of antimatter atoms would presumably be the same as those of ordinary matter. (Recall that the assignment of plus and minus signs to electric charges is an arbitrary convention.) However, if antimatter and ordinary matter came into contact, they would annihilate each other with an explosive release of energy.

Chapter Review

Important Terms

Important Concepts

- The de Broglie hypothesis assigns a wavelength to material particles, by analogy with the assignment of momentum to light.

- A quantum-mechanical wave function is the probability wave associated with a particle. The Schrödinger wave equation makes it possible to calculate the wave function in different situations. The probability density, the square of the wave function, gives the relative probability of finding a particle at a particular location.

- The orbital quantum number determines the allowed value of angular momentum for an electron orbit. The magnetic quantum number determines the orientation of the electron orbit with respect to a given axis. The spin quantum number is a purely quantum-mechanical concept that, in a classical analogy, indicates whether the spin angular momentum of an electron is up or down.

- The Pauli exclusion principle states that in a given atom, no two electrons may have exactly the same set of quantum numbers.

- The Heisenberg uncertainty principle states that you cannot simultaneously measure the position and momentum (or velocity) of a particle with zero uncertainty for each. The same holds true for the particle's energy and the period of time during which it is measured.

- Pair production refers to the creation of a particle and its antiparticle by high-energy photons. The reverse process is pair annihilation, in which a particle and antiparticle annihilate and their energy is converted into photon energy.

Important Equations

Momentum of a Photon:

$$p = \frac{E}{c} = \frac{hf}{c} = \frac{h}{\lambda} \qquad (28.1)$$

de Broglie Wavelength of a Moving Particle:

$$\lambda = \frac{h}{p} = \frac{h}{mv} \qquad (28.2)$$

Electron Wavelength When Accelerated Through Potential V (nonrelativistic):

$$\lambda = \sqrt{\frac{1.50}{V}} \times 10^{-9}\,\text{m} = \sqrt{\frac{1.50}{V}}\ \text{nm} \qquad (28.3)$$

Heisenberg Uncertainty Principle:

$$(\Delta p)(\Delta x) \geq \frac{h}{2\pi} \qquad (28.5)$$

$$(\Delta E)(\Delta t) \geq \frac{h}{2\pi} \qquad (28.6)$$

Condition for Electron Pair Production:

$$hf \geq 2m_e c^2 = 1.022\ \text{MeV} \qquad (28.7)$$

Exercises*

28.1 Matter Waves: The de Broglie Hypothesis

1 The momentum of a photon is (a) zero, (b) equal to c, (c) proportional to its frequency, (d) proportional to its wavelength, (e) given by the de Broglie hypothesis.

2 The Davisson–Germer experiment (a) was concerned with X-ray spectra, (b) verified the Heisenberg principle, (c) supported the Pauli exclusion principle, (d) demonstrated the wavelike properties of electrons.

3 ■ What is the de Broglie wavelength associated with photons of red light ($\lambda = 680$ nm)?

4 ■ What is the de Broglie wavelength of (a) an electron and (b) a proton, both moving with a speed of 100 m/s?

5 ■ Calculate the de Broglie wavelength of a 70-kg person running with a speed of 2.0 m/s.

6 ■■ A proton and an electron are accelerated from rest through a potential difference V. What is the ratio of the de Broglie wavelength of an electron to that of a proton (to two significant figures)?

7 ■■ An electron is accelerated from rest through a potential difference that gives it a matter wave with a wavelength of 0.100 nm. What is the potential difference?

8 ■■ Electrons are accelerated from rest through a potential of 250 kV. If the potential is increased to 600 kV, what is the ratio of the new deBroglie wavelength to the original?

9 ■■ Charged particles are accelerated through a potential difference V. By what factor would the de Broglie wavelength of the particles change if the voltage were doubled?

10 ■■ A proton traveling with a speed of 4.5×10^4 m/s is accelerated through a potential difference of 37 V. By what percentage does the de Broglie wavelength of the proton change?

*Note: Use Eq. 28.3 for λ where appropriate.

11 ■■ What is the energy of a beam of electrons that exhibits a first-order maximum at an angle of 30º when diffracted by a crystal grating with a spacing between the lattice planes of 0.15 nm?

12 ■■ What is the de Broglie wavelength of the Earth in its orbit about the Sun? (Assume a circular orbit.)

13 ■■ The resolution of a light microscope is directly proportional to the wavelength of the radiation it uses (Section 25.4). Similarly, for an electron microscope, the waves associated with the electrons set a limit on the resolution. (a) If the electrons in a microscope are accelerated through a potential difference of 1.0×10^5 V and the microscope has a circular aperture with a diameter of 0.10 mm, what is the minimum angular separation of two viewed objects? (b) Through what voltage difference would the electrons have to be accelerated to improve the angular resolution by a factor of two?

14 ■■ According to the Bohr theory of the hydrogen atom, the speed of the electron in the first Bohr orbit is 2.19×10^8 cm/s. (a) What is the wavelength of the matter wave associated with the electron? (b) How does this compare with the circumference of the first Bohr orbit?

15 ■■ It is desired to observe details whose size is of the order of 0.10 nm with an electron microscope. Through what potential must the electrons be accelerated so that they have a de Broglie wavelength of this order?

28.2 The Schrödinger Wave Equation

16 The wave function solution to the Schrödinger equation (a) can never be found, (b) is the probability of finding a particle, (c) functionally describes the de Broglie wave of a particle, (d) none of these.

17 The square of a particle's wave function is interpreted as being (a) the energy of the particle, (b) the probability of locating the particle, (c) the quantum number of a state, (d) the basis of the Pauli exclusion principle.

18 ■■■ A particle in a box is constrained to move in one dimension, like the bead on a wire illustrated in ●Fig. 28.16. Assuming no forces act on the particle in the interval $0 \leq x \leq L$ and that it hits a perfectly rigid wall, it may be thought of as a particle at the bottom of an infinite well with $U = 0$. (a) Show that the spatial wave function for the particle is $\psi_n = A \sin(n\pi x/L)$ for $n = 1, 2, 3, \ldots$, (b) Show that the kinetic energy of the particle is given by $K_n = n^2(h^2/8mL^2)$ where m is its mass. [Hint: (a) Consider boundary conditions like those of a standing wave. (b) Recall that $K = p^2/2m$ and that the wave function involves the de Broglie wavelength.]

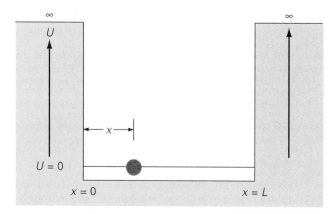

●**FIGURE 28.16 Particle in a box**
See Exercise 18.

19 ■■■ (a) Sketch the wave functions of the particle for the first three states in Exercise 18. (b) Where is the particle most likely to be found, and what is the probability of finding it there? Sketch the probability densities for these states.

20 ■■■ Sketch the form you would expect for the probability of the position of the hydrogen electron in the first three excited states as a function of the radial distance from the nucleus.

28.3 Atomic Quantum Numbers and the Periodic Table

21 The ℓ quantum number of the hydrogen atom (a) determines the total energy of a state, (b) is associated with the angular momentum of the electron, (c) is associated with the orientation of the angular momentum, (d) is associated with the electron spin.

22 The quantum number m_ℓ (a) determines the energy of the electron, (b) tells whether the electron is spinning up or down, (c) tells the orientation of the angular momentum of the electron, (d) all of these

23 The quantum number m_s (a) is purely a quantum mechanical concept, (b) arises from the orbital motion of an electron, (c) is due to actual electron spinning, (d) all of these.

24 Niels Bohr set forth a *correspondence principle*, which states that quantum mechanics and classical physics are in general agreement when the quantum numbers are very large. Discuss this principle in terms of the hydrogen atom.

25 What is the basis of the periodic table of elements in terms of quantum theory, and what do the elements in a particular group have in common?

26 ■ (a) How many possible sets of quantum numbers are there for $n = 2$ and $n = 3$ shells? (b) Explicitly write the (n, ℓ, m_ℓ, m_s) sets for these levels.

27 ■ How many possible sets of quantum numbers are there for the subshells (a) $\ell = 0$ and (b) $\ell = 3$?

28 ■ Which has more possible sets of quantum numbers associated with it, $n = 2$ or $\ell = 3$?

29 ■ An electron in a multielectron atom has a magnetic quantum number of $m_\ell = 3$. What are the minimum values that (a) ℓ, and (b) n could be for that electron?

30 ■■ Draw the ground-state energy level diagrams like those in Fig. 28.8 for (a) nitrogen, N, and (b) potassium, K.

31 ■■ Identify the atoms of each of the following ground-state electron configurations: (a) $1s^2 2s^2$, (b) $1s^2 2s^2 2p^3$, (c) $1s^2 2s^2 2p^6$, (d) $1s^2 2s^2 2p^6 3s^2 3p^4$.

32 ■■ Write the ground-state electron configurations for each of the following atoms: (a) boron, B, (b) calcium, Ca, (c) zinc, Zn, (d) tin, Sn.

33 ■■ Draw schematic diagrams for the electrons in the subshells of (a) sodium, Na, and (b) argon, Ar, atoms in the ground state.

34 ■■■ What would be the first two inert or noble gases if there was no electron spin?

35 ■■■ How would the electronic structure of lithium differ if electron spin has three possible orientations instead of just two?

28.4 The Heisenberg Uncertainty Principle

36 If the uncertainty in the position of a moving particle increases, (a) it may be located more exactly, (b) the uncertainty in its momentum decreases, (c) the uncertainty in its velocity increases, (d) none of these.

37 According to the uncertainty principle, to measure the exact energy of a particle requires (a) special equipment, (b) an infinite time, (c) uncertainty in the momentum, (d) none of these.

38 ■ What is the minimum uncertainty in the velocity of an electron that is known to be somewhere between 0.050 nm and 0.10 nm from a proton?

39 ■ What is the minimum uncertainty in the velocity of a 0.50-kg ball that is known to be at 1.0000 ± 0.0005 cm from the edge of a table?

40 ■ An electron travels with a speed of 3.60560 ± 0.00021 m/s. To what minimum uncertainty can its position be measured?

41 ■■ The energy of a 2.00-keV electron is known to $\pm 3.00\%$. How accurately can its position be measured?

42 ■■ Using a photoelectric gate, it is possible to determine the speed of a 0.15-kg ball rolling on a table within a minimum uncertainty of 5.0×10^{-4} m/s. What are the corresponding uncertainties in (a) the momentum and (b) the position of the ball?

43 ■■ If an excited state of an atom is known to have a lifetime of 1.0×10^{-7} s, what is the minimum error within which the energy of the state can be measured?

44 ■■ The energy of the first excited state of a hydrogen atom is measured to be -0.34 eV ± 0.0003 eV. What is the average lifetime for this state?

45 ■■ How much greater is the width of a spectral line due to natural broadening for a transition from an excited state with a lifetime of 10^{-12} s than for one from a state with a lifetime of 10^{-8} s?

28.5 Particles and Antiparticles

46 Pair production involves (a) the production of two electrons, (b) the production of two positrons, (c) a positronium atom, (d) a certain threshold energy.

47 Due to momentum considerations, pair annihilation *cannot* result in the emission of how many photons? (a) one, (b) two, (c) several.

48 It has been suggested in science fiction that matter and antimatter could be combined as a source of energy. (The starship Enterprise on *Star Trek* had antimatter engines with the antimatter being stored in antimatter "pods.") Speculate how or in what antimatter might be stored for such use. [*Hint:* See confinement methods for fusion in Chapter 30.]

49 ■■ A photon with a frequency of 1.25×10^{18} Hz collides with a nucleus. Will pair production occur? Justify your answer mathematically.

50 ■■ What is the frequency of the photons produced in electron pair annihilation?

51 ■■ What would be the required energy for a photon that could cause the production of a proton–antiproton pair?

52 ■■ A muon, or μ meson, has a negative charge like that of an electron, but its mass is 207 times greater. What would be the required energy for a photon that could cause the pair production of a meson and an antimeson?

Additional Exercises

53 If the spacing between the lattice planes of a particular crystal is 0.19 nm, what voltage is needed to accelerate electrons (from rest) into a beam that exhibits a first-order diffraction maximum at an angle of 50°?

54 With what accuracy would you have to measure the position of a moving electron so that its velocity is uncertain by 1.00×10^{-3} cm/s?

55 A gamma-ray photon has an energy of 7.5 MeV. What are (a) its momentum and (b) the wavelength of the associated de Broglie wave?

56 An electron is accelerated from rest through a potential of 5.00 kV $\pm 3.0\%$. What is the minimum uncertainty in the position of the electron after the acceleration?

57 Show that for a particle moving in one dimension between the fixed boundaries $-L \leq x \leq L$ the wave functions are given by $\psi_n = A \cos(n\pi x/2L)$ where $n = 1, 3, 5 \ldots$ and $\psi_n = A \sin(n\pi x/2L)$ where $n = 2, 4, 6 \ldots$

58 In Exercise 57, show that the allowed kinetic energies are four times smaller than those of Excercise 18. Why?

59 Where is the particle of Exercise 57 most likely to be found in its ground state?

60 What is the de Broglie wavelength for the matter wave associated with (a) a 250-g ball thrown at 30 m/s and (b) a 1800-kg automobile traveling at 80 km/h?

61 Is it possible for the number of quantum states for a given n quantum number to be equal to the number of states for a given ℓ quantum number? Justify your answer.

62 In proton–antiproton pair annihilation, assuming the protons initially at rest, what is (to two significant figures) the (a) energy and (b) frequency of each of the emitted photons?

63 Scientists can create "muonic" atoms by bombarding targets with beams of negative muons (see Exercise 52). These muons would be captured into similar orbitals (different energies and sizes, however) as electrons. Would they, in principle, be prohibited by the Pauli Exclusion Principle from sharing the lower orbitals that are filled with electrons? Explain.

64 Suppose a positronium atom were moving along the x axis when the annihilation happened. Sketch the emission paths of the two photons. Would they have to go off exactly back to back? Explain.

29 The Nucleus

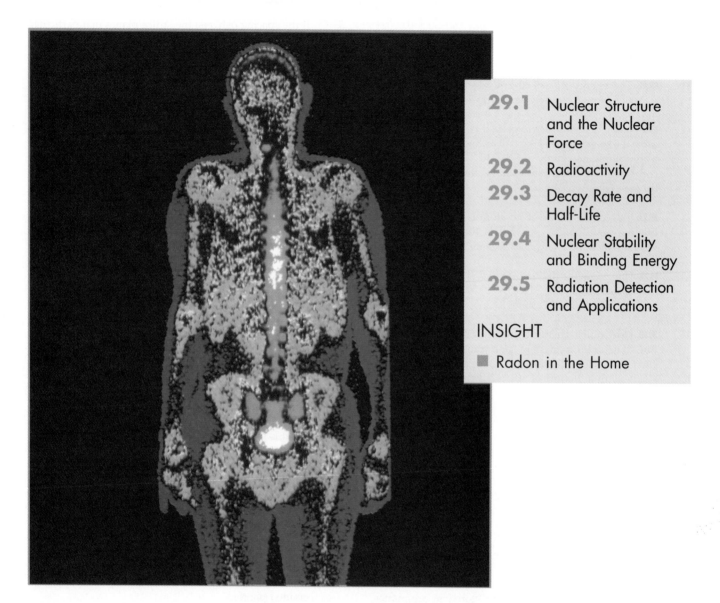

29.1 Nuclear Structure and the Nuclear Force

29.2 Radioactivity

29.3 Decay Rate and Half-Life

29.4 Nuclear Stability and Binding Energy

29.5 Radiation Detection and Applications

INSIGHT

■ Radon in the Home

Skeletons have always been used to evoke fear—for some reason, most of us don't like to see anything that reminds us of the inside of our bodies. But this high-tech skeletal image may be particularly frightening to many people, because it was created by radiation from a radioactive source. "Radiation," "radioactivity"—these are words that have their own special anxiety-producing associations. But while there may be good reason for this anxiety, we often overlook the other side of the picture—the many beneficial uses to which radiation can be put. Thus, everyone knows that exposure to high-energy radiation can cause cancer. Yet precisely the same sort of radia-

tion can be useful in the treatment of cancer, and even in the diagnosis of cancer.

The bone scan shown above is an example. The image was created by radiation released when unstable nuclei spontaneously break apart—a process we call radioactive decay. But why are some nuclei so much more unstable than others? What determines which nuclei are likely to be radioactive, the rate at which they break down, and what particles they emit when they do so? These are some of the questions you'll explore in this chapter. You'll also learn how radiation is detected and measured, as well as more about its dangers and uses.

Great advances have been made in nuclear physics since the introduction of the nuclear model of the atom in 1911. However, not everything is known about the nucleus. This central core of the atom presents a great challenge to the ingenuity of scientists, who would like to know its secrets and to apply its potential.

How does one study such a minute entity? Because of its unimaginably small size, all our knowledge of the nucleus hinges on indirect observations that are not fully explained by classical theory. Quantum mechanics, which was so helpful to our study of atomic physics, can be applied to the nucleus, but with some limitations, deriving mostly from the fact that we do not fully understand the force that holds the nucleus together.

Even so, a great deal can be learned about the nucleus by looking at the properties of various nuclei. One of the most revealing phenomena of the nucleus is *radioactivity*. Some nuclei are unstable and decay into nuclei of other elements with the telltale emission of detectable particles. These energetic particles, though they can be highly dangerous, also have many beneficial uses — for example, in treating cancer. In addition to such practical applications, the study of radioactivity and nuclear stability gives us some general ideas about the nature of the nucleus, the energy it possesses, and how this energy can be released. The release of nuclear energy has become one of our major energy sources. This will be considered in the next chapter. First, let's take a look at the nucleus itself.

29.1 Nuclear Structure and the Nuclear Force

Objectives: To be able to (a) distinguish between the Thompson and Rutherford–Bohr models of the atom, (b) specify some of the basic properties of the strong nuclear force, and (c) understand nuclear notation.

It is evident from the emission of electrons from heated filaments (thermionic emission) and the photoelectric effect that the atoms of certain materials, probably all materials, contain electrons. Since atoms are normally electrically neutral, they must also contain positive charge equal in magnitude to that of the total charge on all the electrons in an atom. Also, since the mass of an electron is small in comparison to the mass of even the lightest atoms, most of the mass appears to be associated with that positive charge.

Based on these observations, J. J. Thomson, a British physicist who had experimentally proven the existence of the electron in 1897, proposed a model of the atom. In the Thomson model, the negatively charged electrons were pictured as being uniformly distributed within a continuous sphere of positive charge. It was called the "plum pudding" model because the electrons in the positive charge were thought of as analogous to the raisins in a plum pudding. The region of positive charge was assumed to have a radius on the order of 10^{-8} cm, based on calculations from the bulk properties of matter.

Our modern model of atomic structure is quite different. This model pictures all of an atom's positive charges, and practically all of its mass, as concentrated in a central "nucleus," which is surrounded by the orbiting negatively charged electrons. The concept of an atomic nucleus was proposed by the British physicist Ernest Rutherford (1871–1937). Combined with Bohr's theory of orbiting electrons (Section 27.4), this idea led to the simplistic "solar system" or **Rutherford–Bohr model** of the atom.

Rutherford's insight came from the results of alpha particle scattering experiments performed in his laboratory around 1911. An alpha particle (α particle) is a doubly positively charged particle that is naturally emitted from some radioactive materials (to be discussed in more detail in Section 29.2). A beam of alpha

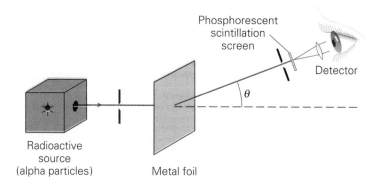

FIGURE 29.1 Rutherford's scattering experiment
A beam of alpha particles from a radioactive source is scattered by a thin foil, and the scattering is observed to be a function of the scattering angle θ. The detector was a scintillating phosphorescent screen.

Phosphorescent scintillation screen

Detector

Radioactive source (alpha particles)

Metal foil

particles from a radioactive source was directed at a thin gold foil "target" and the scattering angles of the particles were observed (•Fig. 29.1). Such experiments were designed to investigate the distribution of mass and electric charge in atoms.

An alpha particle is over 7000 times more massive than an electron. Thus, the Thomson model would predict only small deflections, the result of collisions with the light electrons as an alpha particle passed through a large gold atom (•Fig. 29.2). Surprisingly, however, Rutherford and his colleagues observed appreciable scattering angles. In some instances (about 1 in 8000), the alpha particles were actually found to be *back-scattered*, that is, scattered through angles greater than 90º (•Fig. 29.3).

Calculations showed that the probability of this taking place with a Thomson model of the atom and such a thin foil (10^{-4} cm) was minuscule—certainly much less than 1 in 8000. As Rutherford described the back-scattering, "It was almost as incredible as if you had fired a 15-inch shell at a piece of tissue paper and it came back and hit you."

The experimental results led Rutherford to the concept of an atomic nucleus:

> On consideration, I realized that this scattering backward must be the result of a single collision, and when I made calculations I saw that it was impossible to get anything of that order of magnitude unless you took a system in which the greater part of the mass of the atom was concentrated in a minute nucleus. It was then that I had the idea of an atom with a minute massive center carrying a charge.

If all of the positive charge of a target atom were concentrated in a very small region in the atom, then an alpha particle coming very close to this region would experience a large deflecting force. The mass of this nucleus of charge would be larger than that of the alpha particle, since most of the atomic mass is associated with the positive charge. Thus, back-scattering would be possible.

A simple calculation can give an idea of the size of the atomic nucleus. For a head-on collision, an alpha particle would attain its distance of closest approach

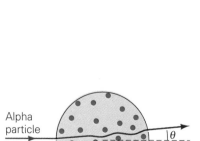

Alpha particle

FIGURE 29.2 The plum pudding model
With Thomson's plum pudding model of the atom, the alpha particles would be expected to be only slightly deflected by collisions with the electrons in the atom. The experimental results were quite different.

FIGURE 29.3 Rutherford scattering
A compact, dense atomic nucleus with a positive charge accounts for the observed scattering. An alpha particle in a head-on collision with the nucleus would be scattered directly backwards ($\theta = 180°$) after coming within a distance r_{min} of the nucleus. On this scale, the electron orbits (about the nucleus) are too large to show.

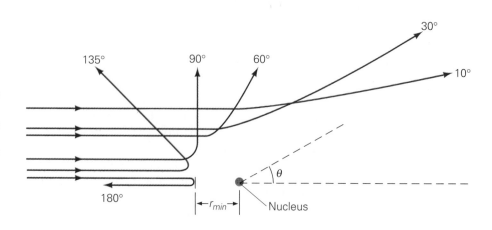

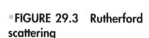

to the nucleus (r_{min} in Fig. 29.3). That is, the alpha particle approaching the nucleus would be stopped at a distance r_{min} by the repulsive Coulomb force and would be accelerated back along its original path. Assuming a spherical or point charge distribution, the electric potential is $V = kZe/r$, where Z is the *atomic number* or the number of protons in the nucleus and Ze is the total charge. The work done by the Coulomb force in stopping the alpha particle *is $W = qV = 2eV$*, since the alpha particle has a double positive electronic charge. This is equal to the initial kinetic energy of the alpha particle, that is,

$$\tfrac{1}{2} \, mv^2 = \frac{k(2e)Ze}{r_{min}}$$

and

$$r_{min} = \frac{4kZe^2}{mv^2} \tag{29.1}$$

By using the known value of the energy of the alpha particles from the particular source and other known values, r_{min} is found to be about 10^{-12} cm. This value is an *upper limit* for the nuclear radius, since the alpha particle does not reach the nucleus itself.

Although the nuclear model of the atom is useful, the nucleus is much more than simply a region of positive charge. For one thing, we now know that the atomic nucleus is composed of two types of particles—protons and neutrons—which are collectively referred to as **nucleons**. The nucleus of the common hydrogen atom is a single proton. Rutherford suggested that the hydrogen nucleus be given the name *proton* (from a Greek word meaning "first") after he became convinced that there was no positively charged particle in an atom lighter than the hydrogen nucleus. As you know from an earlier chapter, a neutron is an electrically neutral particle with a mass slightly greater than that of a proton. The existence of the neutron was not experimentally verified until 1932.

The Nuclear Force

Of the forces in the nucleus, we know there is an attractive gravitational force between nucleons (protons and neutrons). But in Chapter 15, we saw that the magnitude of the gravitational force is negligible in comparison with the mutually repulsive electrical force between positively charged protons. Taking only these repulsive forces into account, one might expect the nucleus to fly apart. Yet the nuclei of most atoms are stable, so there must be an additional force that holds the nucleus together. This strongly attractive force is called the **strong nuclear force**, usually referred to simply as the *nuclear force*.

Note: Review Example 15.4, Section 15.3.

The exact mathematical expression for the nuclear force is not known, and approximations indicate that it is extremely complicated. However, some general features of this force are as follows:

- The nuclear force is strongly attractive and much larger in relative magnitude than the electrostatic and gravitational forces.
- The nuclear force is very short-ranged; that is, a nucleon interacts only with its nearest neighbors, over distances on the order of 10^{-15} m.
- The nuclear force acts between any two nucleons within a short range, that is, between two protons, a proton and a neutron, or two neutrons.

Thus, nuclear protons in close proximity repel each other by the electric force but attract each other (and nearby neutrons) by the strong nuclear force. Having no electric charge, neutrons are attracted to nearby protons and neutrons only by the nuclear force (gravitational force being negligible).

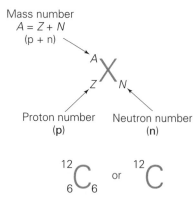

Mass number
$A = Z + N$
$(p + n)$

$$_{Z}^{A}X_{N}$$

Proton number   Neutron number
(p)   (n)

$$_{6}^{12}C_{6} \quad \text{or} \quad {}^{12}C$$

•FIGURE 29.4 Nuclear notation
A nucleus is represented by the chemical symbol of the particular element with the mass number A (number of protons and neutrons) as a left superscript and the proton (atomic) number Z as a left subscript. Sometimes the neutron number N is shown as a right subscript, but both Z and N are routinely omitted because the letter symbol tells you Z, and $N = A - Z$. This nuclear notation is shown for a nucleus of the most common isotope of carbon.

Nuclear Notation

To describe the nuclei of different atoms, it is convenient to use the notation illustrated in •Fig. 29.4. The chemical symbol of the element is used with subscripts and a superscript. The left subscript is the **atomic number** (Z), which indicates the number of protons in the nucleus. An alternative (and more descriptive) term, which we will generally use in this book, is **proton number**. For electrically neutral atoms, this is also the number of orbital electrons.

Recall that the number of protons in the nucleus of an atom defines the species of the atom—that is, the element to which the atom belongs. In the example in Fig. 29.4, the proton number $Z = 6$ indicates that this is a carbon nucleus. The proton number defines which chemical symbol is used. Electrons can be removed (or added) to an atom to form an ion, *but this does not change its species*. For example, a nitrogen atom with an electron removed, N^+, is still nitrogen—a nitrogen *ion*. The number of electrons can vary by ionization, but the proton number cannot vary without producing an atom of a different element. In that sense, adding the label Z is redundant because the chemical symbol tells us this; however we will usually include it so as not to have to refer to the periodic chart so often.

The left superscript on the chemical symbol is called the **mass number** (A)—the total number of protons and neutrons in the nucleus. Since protons and neutrons have approximately equal masses, the mass numbers of nuclei give a relative comparison of nuclear masses. For the example in Fig. 29.4, the carbon nucleus, the mass number is $A = 12$ because there are 6 protons and 6 neutrons. The number of neutrons, called the **neutron number** (N), is sometimes indicated by a subscript on the right side, but this is usually omitted because it can easily be determined from the other numbers: $N = A - Z$.

The atoms of an element, all of which have the same number of protons in their nuclei, may have different numbers of neutrons. For example, nuclei of different carbon atoms ($Z = 6$) may contain either 6, 7, or 8 neutrons and in nuclear notation would be written

$$_{6}^{12}C \qquad _{6}^{13}C \qquad _{6}^{14}C$$

Atoms whose nuclei have the same number of protons but different numbers of neutrons are called **isotopes**, in this case isotopes of carbon.

Isotopes are like members of a family; they all have the same Z number and the same surname, but the members of the family are distinct on the basis of the number of neutrons in their nuclei, and therefore their total mass. However, it should be noted that different isotopes of the same family have the same electronic structure and thus the same chemical properties. Isotopes are referred to by their mass numbers; for example, the previous isotopes of carbon are called carbon-12, carbon-13, and carbon-14. There are also other isotopes of carbon: ^{11}C, ^{15}C, and ^{16}C. A particular nuclear species or isotope of any element is called a **nuclide**. Thus, we have six nuclides, or isotopes, of carbon. Generally, only a few isotopes of a given species are stable. In this case, although it is relatively long-lived, carbon-14 is unstable. Only carbon-12 and carbon-13, in fact, are stable isotopes of carbon.

Another family of isotopes is that of hydrogen, which has three isotopes or nuclides: ^{1}H, ^{2}H, ^{3}H. These isotopes are not generally referred to as hydrogen-1, and so on, but are given special names. ^{1}H is called ordinary hydrogen or, simply, hydrogen. ^{2}H is called *deuterium*. Deuterium, sometimes known as heavy hydrogen, can combine with oxygen to form heavy water (commonly written D_2O). The third isotope of hydrogen, ^{3}H, is called *tritium*, which is unstable.

Isotopes of an element: same number of protons, different numbers of neutrons

Nuclide: a particular nuclear species or isotope

Objectives: To be able to (a) define the term "radioactivity," (b) distinguish among alpha, beta, and gamma decay, and (c) write nuclear decay equations.

Most elements have stable isotopes. It is the atoms of these stable nuclides with which we are most familiar in the environment. However, as we have mentioned, the nuclei of some isotopes are unstable and disintegrate spontaneously (decay) of their own accord, emitting energetic particles. Such isotopes are said to be *radioactive* or to exhibit **radioactivity**. For example, the tritium isotope of hydrogen just mentioned has a radioactive nucleus. The prefix "radio" refers to the emitted nuclear radiation, which in the modern context may be a particle or a wave. A radioactive isotope is "active" when it is emitting radiation. Of the nearly 1200 known unstable nuclides, only a small number occur naturally. The others are produced artificially (Chapter 30).

We know that radioactivity is completely unaffected by normal physical or chemical processes, such as heat and pressure or chemical reactions. Since chemical processes involve the outer electrons of atoms, the source of radioactivity must therefore lie deeper in the atom, that is, in the nucleus. This instability cannot be explained directly by a simple imbalance of attractive and repulsive forces in the nucleus because the nuclear disintegrations of a given isotope occur as a function of time at a fixed rate. Classically, one would expect identical nuclei to do the same thing at the same time. Therefore, radioactive decay processes suggest quantum mechanical probability effects.

The discovery of radioactivity is credited to the French scientist Henri Becquerel. In 1896, while studying the fluorescence of a uranium compound, Becquerel discovered that a photographic plate in the vicinity of a sample had been darkened when the compound had not been activated by exposure to light and was not fluorescing. Apparently the darkening was caused by some new type of radiation being emitted from the compound. In 1893, Pierre and Marie Curie announced the discovery of two radioactive elements, radium and polonium, which they had isolated from uranium pitchblende ore (•Fig. 29.5).

Experiment easily shows the radiation emitted by radioactive isotopes to be of three different kinds. When a radioisotope is placed in a chamber so that the emitted radiation passes through a magnetic field to a photographic plate (•Fig. 29.6), the radiations expose the plate, producing identifying spots. The positions of the spots show that some isotopes emit radiation that is deflected to the left; some, radiation that is deflected to the right; and some, radiation that is undeflected.

These spots are characteristic of what is known as alpha, beta, and gamma radiations. From the deflections of two of the types of radiation in the magnetic field, it is evident that positively charged particles are emitted from nuclei undergoing alpha decay and that negatively charged particles are emitted in beta decay. Also, the degree of deflection shows that alpha particles must be more massive than beta particles. The undeflected gamma radiation (gamma rays) is electrically neutral.

Investigations of the different radiations reveal that

- **Alpha particles** are doubly charged ($+2e$) particles containing two protons and two neutrons. They are identical to the nucleus of the helium atom (^4_2He).
- **Beta particles** are electrons.*
- **Gamma rays** are particles, or quanta, or photons, of electromagnetic energy.

•**FIGURE 29.5 The Curies**
Marie Sklodowska Curie (1867–1934) was born in Poland and studied in France. There she met and married Pierre Curie (1859–1906), who was a physicist well known for his work on crystals and magnetism. In 1903, Madame Curie (as she is commonly known) and Pierre shared the Nobel Prize in physics with Henri Becquerel for their work on radioactivity. Mme Curie was also awarded the Nobel prize in chemistry in 1911 for the discovery of radium and the study of its properties. The Curies are shown here on the cover of a 1904 magazine.

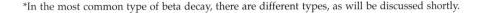

*In the most common type of beta decay, there are different types, as will be discussed shortly.

For a few radioactive elements, two spots are found on the film, indicating that some elements decay by two different modes.

Let's now look at what happens to a decaying nucleus in each of these three decay modes of radioactivity.

Alpha Decay

Alpha particle: a helium nucleus

When an alpha particle is ejected from a radioactive nucleus, the nucleus loses two protons and two neutrons, so the mass number (A) is decreased by four, that is, $\Delta A = -4$, and the proton number (Z) is decreased by two, $\Delta Z = -2$. Since the parent nucleus loses two protons, the resulting daughter nucleus must then be the nucleus of another element as defined by the proton number. (The original and resulting nuclei are commonly referred to as the parent and daughter nuclei, respectively.) Thus, the decay process is one of nuclear *transmutation* in which the nuclei of one element change into the nuclei of a lighter element.

An example of an isotope or nuclide that undergoes **alpha decay** is polonium-214. The decay process is represented in the form of a nuclear equation, similar to a chemical equation except that it refers to nuclei.

$$\underset{\text{polonium}}{{}^{214}_{84}\text{Po}} \rightarrow \underset{\text{lead}}{{}^{210}_{82}\text{Pb}} + \underset{\substack{\text{alpha particle} \\ \text{(helium nucleus)}}}{{}^{4}_{2}\text{He}}$$

Note that the totals of the mass numbers and the proton numbers are equal on each side of the equation: $(214 = 210 + 4)$ and $(84 = 82 + 2)$, respectively. This reflects the fact that *two conservation laws apply to all nuclear processes*. The first is the **conservation of nucleons**. That is, the total number of nucleons (A) remains constant in any process. The second is the familiar **conservation of charge**, when applied to nuclear reactions.

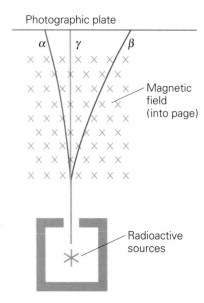

Photographic plate

α γ β

Magnetic field (into page)

Radioactive sources

• FIGURE 29.6 Nuclear radiation The radiations from radioactive sources can be distinguished by passing them through a magnetic field. Alpha and beta particles are deflected. Applying the right-hand force rule ($F = qvB$), we find that alpha particles are positively charged and beta particles are negatively charged. The radii of curvature (not to scale) allow the particles to be distinguished by mass. Gamma rays are not deflected, and so are uncharged, being in fact quanta of electromagnetic energy.

EXAMPLE 29.1 ■ URANIUM'S DAUGHTER: ALPHA DECAY

A ${}^{238}_{92}\text{U}$ nucleus undergoes alpha decay. What is the resulting daughter nucleus?

Solution. Since $\Delta Z = -2$ for alpha decay, the uranium-238 (${}^{238}\text{U}$) nucleus loses two protons, and the daughter nucleus has a proton number of $Z = 92 - 2 = 90$, which is the proton number of thorium (we look up the proton number in the periodic table, Fig. 28.9, if we don't know it from memory). The equation for the process is

$${}^{238}_{92}\text{U} \rightarrow {}^{234}_{90}\text{Th} + {}^{4}_{2}\text{He}$$

Sometimes the helium nucleus is written as just α or ${}^{4}_{2}\alpha$.

Follow-up Exercise. Using high-energy accelerators, it is possible to add an alpha particle to a nucleus, essentially the reverse of the reaction in this Example. Write this nuclear reaction and predict the resulting nucleus if an alpha particle is added to a carbon-12 nucleus in this fashion.

The energies of the alpha particles from radioactive sources are on the order of a few mega-electron volts (MeV). For example, the energy of the alpha particle emitted from the decay of ${}^{214}\text{Po}$ is about 7.7 MeV, and that from ${}^{238}\text{U}$ decay is about 4.2 MeV. Alpha particles from such radioactive sources were used in the scattering experiments that led to the Rutherford nuclear model (Section 29.1). Scattering experiments give some idea about the size of the nucleus, as you learned in the preceding section. They also shed light on its structure and the forces within it.

Outside the nucleus, the repulsive electric force increases as an alpha particle gets close to the nucleus. Inside the nucleus, however, the strongly attractive nuclear force dominates. These conditions are depicted in ●Fig. 29.7 in a graph of the electrical potential energy, U, as a function of r, the distance from the center of the nucleus. Outside the nucleus, U is proportional to, or "falls off," as $1/r$ since it represents the electrical potential. Within the nucleus, U is opposite in sign and has the form of a negative potential well because of the strongly attractive nuclear force. This could be the representation of a uranium-238 nucleus.

Consider alpha particles from a ^{214}Po source incident on a thin foil of ^{238}U. As illustrated in the figure, Rutherford scattering takes place. We say that this results from a potential barrier that the incident ^{214}Po alpha particles do not have enough energy to overcome. Instead, they are scattered away. If an incident particle did have enough energy to overcome, or cross, this Coulomb barrier, it would enter the nucleus. In this case, a nuclear reaction would result (to be discussed in Chapter 30). However, the ^{238}U nucleus itself undergoes alpha decay, emitting an alpha particle with an energy of 4.4 MeV. This fact appears to contradict the scattering experiment. How can these lower-energy alpha particles cross a potential barrier that higher-energy incident alpha particles cannot? Classically, this is impossible, since it violates the conservation of energy, and the inside of the potential barrier is referred to as a classically forbidden region. However, quantum mechanics offers an explanation.

Quantum mechanics predicts a finite probability of finding a particle in a classically forbidden region as a result of a wave function existing for this region. If the alpha particle exists as an entity in the nucleus, its wave function may tail off (in nonoscillatory behavior) while in the barrier region and make it through to the outside, where it then appears as a wave of much smaller amplitude (●Fig. 29.8). There would then be a finite probability (ψ^2, Section 28.2) of finding the alpha particle outside the nucleus. This is called **tunneling**, or **barrier penetration**, since the alpha particle has a small but nonzero chance or probability to tunnel through the barrier.

You might point out that there is still a violation of the conservation of energy. However, the uncertainty principle (Section 28.4) tells us that energy conservation can be violated by an amount ΔE for a time Δt. *Quantum mechanics thus allows the conservation of energy to be violated for brief periods, which may be long enough for the alpha particle to tunnel through the barrier.* The probability of finding an alpha particle outside the ^{238}U nucleus is extremely small, and this is reflected in the extremely slow decay rate of ^{238}U. However some other nuclei decay by alpha decay very rapidly. The height and width of the potential barrier determine the time or probability an alpha particle has to escape from the nucleus. Thus, the barrier would also control the decay rate (which will be considered in the next section).

Beta Decay

The emission of an electron (a beta particle) in a nuclear decay process might seem contradictory to the proton–neutron model of the nucleus. It should be noted that the electron emitted in beta decay is not an orbital electron, but is emitted from the nucleus, although it is *not* part of the original nucleus. In fact, the process of **beta decay** indicates that *the electron is created in the nucleus itself.* For example, the equation for carbon-14 beta decay is

$$\underset{\text{carbon}}{^{14}_{6}\text{C}} \rightarrow \underset{\text{nitrogen}}{^{14}_{7}\text{N}} + \underset{\substack{\text{beta particle} \\ \text{(electron)}}}{^{0}_{-1}\text{e}}$$

The parent carbon nucleus has six protons and eight neutrons, whereas the daughter nitrogen nucleus has seven protons and seven neutrons. Notice that the notation for the electron specifies a mass number of zero and a charge number of -1. This is in keeping with the conservation of nucleons and charge.

In this and similar beta decay processes, the neutron number of the parent nucleus decreases by one ($\Delta N = -1$), and the proton number of the daughter nu-

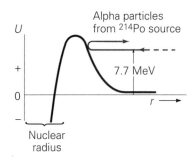

●**FIGURE 29.7 Nuclear potential barrier**
Alpha particles from radioactive polonium with energies of 7.7 MeV do not have enough energy to overcome the nuclear potential barrier of ^{238}U and are scattered.

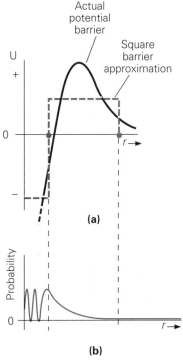

●**FIGURE 29.8 Tunneling effect, or barrier penetration**
(a) The potential barrier presented by a nucleus to an alpha particle inside can be approximated by a square barrier. **(b)** The square of the alpha particle wave function (the probability of finding it at a given location) is shown. Inside the nucleus it has a high likelihood of being found (large amplitude), but it can "tunnel" through the barrier and appear on the outside of the nucleus with a much smaller (small amplitude) but nonzero probability.

cleus increases by one ($\Delta Z = +1$). The mass number is unchanged. This indicates that, in essence, *a neutron within the nucleus decays into a proton* and an electron:

$$^{1}_{0}n \rightarrow ^{1}_{1}p + ^{\ 0}_{-1}e \quad (basic\ \beta^{-}\ decay)$$
$$\underset{neutron}{\qquad} \underset{proton}{\qquad} \underset{electron}{\qquad}$$

(This equation also turns out to describe the behavior of a *free* neutron, which is unstable when outside a nucleus.) In nuclear notation, the neutron has a mass number of 1 and a proton number of zero. (Why?)

Notice that this process enables an unstable nucleus to "kill two birds with one stone," so to speak. The carbon-14, for example, has one too many neutrons to be stable (the most massive stable isotope is carbon-13). The net result of beta decay is to decrease the neutron number and increase the proton number, as we shall see later both of which work in the direction of greater stability. In the carbon-14 example, nitrogen-14 is stable. We should also note that there is another elementary particle, called a neutrino, emitted in beta decay. For the sake of simplicity, it is not shown in the nuclear decay equations. Its place in beta decay will be discussed in Chapter 30.

There are actually two modes of beta decay, β^{-} and β^{+}, as well as a competing process called *electron capture*. The **β^{-} decay** involves the emission of an electron as above. Isotopes that decay by this means do so because they have too many neutrons compared to protons to be stable. However, unstable isotopes that undergo **β^{+} decay**, or *positron decay*, which involves the emission of a positron ($^{\ 0}_{+1}e$), have too many protons compared to neutrons. As a result, the net effect of β^{+} decay is to convert a proton into a neutron. An example of β^{+} decay is

$$^{15}_{8}O_{7} \rightarrow ^{15}_{7}N_{8} + ^{\ 0}_{+1}e$$
$$\underset{oxygen}{\qquad} \underset{nitrogen}{\qquad} \underset{positron}{\qquad}$$

Positron emission is also accompanied by a neutrino (of a different type than that associated with β^{-} decay), which we again omit until Chapter 30. As in β^{-} decay, the mass numbers of the parent and daughter nuclei do not change, but the proton number of the daughter nucleus in this case is one less than that of the parent nucleus. As we have guessed, this implies that fundamentally a proton disintegrates in the process into a neutron and a positron. (As with β^{-} decay, the positron is created during the decay and does not actually reside in the nucleus.)

$$^{1}_{1}p \rightarrow ^{1}_{0}n + ^{\ 0}_{+1}e \quad (basic\ \beta^{+}\ decay)$$
$$\underset{proton}{\qquad} \underset{neutron}{\qquad} \underset{positron}{\qquad}$$

The third related process, **electron capture (EC),** involves the absorption of one of the orbital electrons by a nucleus. The net result is a daughter nucleus the same as would have been produced by positron decay. Usually these two types of decay compete, since they produce the same, more stable daughter nucleus. This makes physical sense because the inward decay of an electron is equivalent electrostatically to the outward emission of a positron. In the language of quantum mechanics, there is a nonzero probability for both processes to happen to a given nucleus. What actually *does* happen is governed by the probabilistic rules of quantum mechanics. A specific example of electron capture is

$$^{\ 0}_{-1}e + ^{7}_{4}Be \rightarrow ^{7}_{3}Li$$
$$\underset{electron}{\qquad} \underset{beryllium}{\qquad} \underset{lithium}{\qquad}$$

As in β^{+} decay, a proton changes into a neutron, but no beta particle is emitted in the electron capture process. An electron in the innermost shell, called the K shell from spectroscopic notation, is usually captured, an event referred to as K-capture, although electrons in more distant orbits (for example L-capture) can occur with much less probability. Since there is no charged particle emission, the process is detected by observing the emission of characteristic X-rays produced when an electron from an outer shell moves into a K-shell vacancy. Because X-ray emission must take place *after* the K-capture, the X-rays are characteristic of the daughter nucleus, not the parent.

β^{-}-electron

β^{+}-positron

Note: Electron shells are discussed in Section 28.3.

Gamma Decay

In **gamma decay**, the nucleus emits a gamma (γ) ray, or a "particle" of electromagnetic energy. The emission of a gamma ray by a nucleus in an excited state is analogous to the emission of a photon by an excited atom. This may come about because of an energetic collision with another particle or, more commonly, because the nucleus is left in an excited state after a previous radioactive decay.

Gamma ray: a quantum or photon of electromagnetic energy

Nuclei are thought of as having energy levels like atomic electrons. However, the nuclear energy levels are much farther apart than those of an atom. The nuclear energy levels are on the order of kilo-electron volts (keV) and mega-electron volts (MeV) apart, compared to a few electron volts of difference between atomic energy levels. As a result, gamma rays are very energetic, having frequencies greater than those of X-rays, and thus much shorter wavelengths. As an example of gamma decay, consider the decay of nickel-61 from an excited nuclear state to one of lesser energy.

$$\underset{\substack{nickel \\ (excited)}}{^{61}_{28}\text{Ni}^*} \rightarrow \underset{nickel}{^{61}_{28}\text{Ni}} + \underset{gamma\ ray}{\gamma}$$

The asterisk indicates that the ^{61}Ni nucleus is in an excited state. The de-excitation process results in the emission of a gamma ray. Note that *in the gamma decay process, the mass and proton numbers do not change.* The daughter nucleus in this case is simply the parent nucleus with less energy.

As an example of gamma emission following beta decay, consider the following Example.

EXAMPLE 29.2 ■ TWO FOR ONE: BETA DECAY AND GAMMA EMISSION

Normal cesium found in nature consists entirely of the stable isotope $^{133}_{55}\text{Cs}$. An unstable isotope, $^{137}_{55}\text{Cs}$, is a common fragment of the fission process (for example, when the $^{235}_{92}\text{U}$ nucleus splits or "fissions" in a nuclear reactor fuel rod or during an atomic bomb explosion). When $^{137}_{55}\text{Cs}$ decays, its daughter nucleus is sometimes left in an excited state, giving off a gamma photon with an energy of 0.662 MeV to finally produce a stable nucleus. (a) Predict the beta decay mode for the $^{137}_{55}\text{Cs}$ nucleus and find the daughter product. (b) Write the gamma ray decay showing the daughter in its stable configuration.

Solution. We note that the $^{137}_{55}\text{Cs}$ fragment has too many neutrons to be stable. (How do we know? Compare to the stable ^{133}Cs.) The data are as follows

Given: Initial nucleus $^{137}_{55}\text{Cs}$ *Find:*   (a) Decay particle and daughter, decay reaction
                                                    (b) Reaction showing gamma decay to final stable state

(a) Since ^{137}Cs is overabundant in neutrons, we expect β^- decay to convert a neutron into a proton as we have seen previously. The proton number will increase by one; thus the daughter will be barium (see the periodic table, Figure 28.9). We write down the decay indicating the barium in an excited state ready to decay via gamma emission, and ignoring the emitted neutrino.

$$\underset{cesium}{^{137}_{55}\text{Cs}} \rightarrow \underset{\substack{barium \\ (excited)}}{^{137}_{56}\text{Ba}^*} + \underset{electron}{^{0}_{-1}\text{e}}$$

(b) The gamma emission of the barium is just

$$\underset{\substack{barium \\ (excited)}}{^{137}_{55}\text{Ba}^*} \rightarrow \underset{barium}{^{137}_{55}\text{Ba}} + \underset{\substack{0.662\ MeV \\ gamma\ ray}}{\gamma}$$

Radiation Penetration

The absorption or degree of penetration of nuclear radiations is an important consideration in applications such as radioisotope treatment of cancer and nuclear shielding (for example, around a nuclear reactor). Also, as we will see in a later section, the absorption of nuclear radiation is used to monitor and automatically control the thickness of metal and plastic sheets and films in fabrication processes.

The three types of radiation (alpha, beta, and gamma) are absorbed quite differently. The electrically charged alpha and beta particles interact with the atoms of a material and may produce ionizations along their paths. The greater the charge and the slower the particle, the greater is the energy transfer and ionization along the path, which determines the degree of penetration. Also, the penetration depends on the density of the material.

- Alpha particles are doubly charged, have a larger mass, and generally move more slowly. A few centimeters of air or a sheet of paper will usually completely stop or absorb them.
- Beta particles are singly charged and can travel a few meters in air or a few millimeters in aluminum before being stopped.
- Gamma rays are uncharged and are therefore more penetrating. A significant portion of a beam of high-energy gamma rays can penetrate a centimeter or more of a dense material such as lead. (Lead is commonly used as shielding for harmful X-rays, which are slightly less energetic than gamma rays.)

Radiation passing through matter can do considerable damage. In biological tissue, the radiation damage is chiefly due to ionizations in living cells (Section 29.5). We are exposed to normal background radiation from radioisotopes in the environment and cosmic radiation (radiation from outer space). The intensity of most such radiation is too low to be harmful. Another recent radiation concern involves airplanes. In the Chapter 25 Insight on telescopes using nonvisible radiation, it was mentioned that X-rays and gamma rays from distant sources do not penetrate the Earth's atmosphere. However, concern has been expressed that flight crews who spend many hours aboard high-flying jet aircraft may receive excessive exposure to such radiations.

Metals and other structural materials may become brittle and lose their strength when exposed to strong radiation, such as that in nuclear reactors used for electrical generation (Chapter 30) and the intense cosmic radiation that space vehicles are exposed to.

As pointed out earlier, of the nearly 1200 known unstable nuclides, only a small number occur naturally. Most of the radioactive nuclides found in nature occur as products of the decay series of heavy nuclei, that is, a series of continual radioactive decay into lighter elements. The ^{238}U decay series is given in •Fig. 29.9. It stops when the stable isotope ^{206}Pb is reached. Note that some nuclides in the series decay by two modes. As can be seen, radon (^{222}Rn) is part of this decay series. This radioactive gas has received a great deal of attention because it can accumulate in dangerous amounts in poorly ventilated buildings (see the Insight on p. 905).

Note: See also Figs. 29.23 (neptunium-237) and 29.24 (plutonium-239).

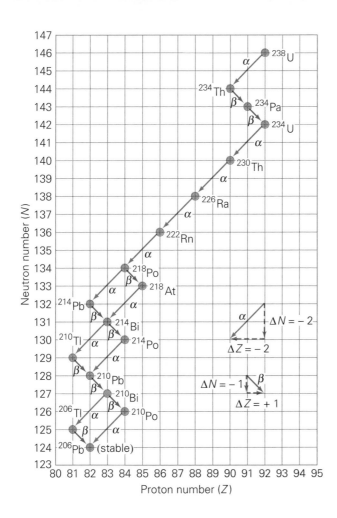

•FIGURE 29.9 Decay series for uranium-238
On this plot of N versus Z, a diagonal transition from right to left is an alpha decay process and a diagonal transition from left to right is a β^- decay process (how can you tell?). The decay series continues until a stable nucleus is reached.

29.3 Decay Rate and Half-Life

Objectives: To be able to (a) explain the concepts of activity, decay constant, and half life of a radioactive sample, and (b) use radioactive decay to find the age of objects.

The nuclei of a sample of a radioactive isotope do not decay all at once, but do so randomly at a characteristic rate that is unaffected by any external stimulus. No one can tell exactly when particular nuclei will decay. All that can be determined is how many nuclei in a sample will decay over a period of time.

The **activity** of a sample of radioactive isotope is defined as the number of nuclear disintegrations, or decays, per second. For a given amount of material, the activity decreases with time, as fewer and fewer radioactive nuclei remain. Each isotope has its own characteristic rate of decrease. Thus the rate at which the number of parent nuclei (N) decreases is proportional to the number present:

$$\frac{\Delta N}{\Delta t} \propto N$$

This can be written in equation form with a constant of proportionality.

$$\frac{\Delta N}{\Delta t} = -\lambda N \qquad (29.2)$$

The constant λ is called the **decay constant** and is different for different isotopes. The greater the decay constant, the greater the rate of decay and the more radioactive

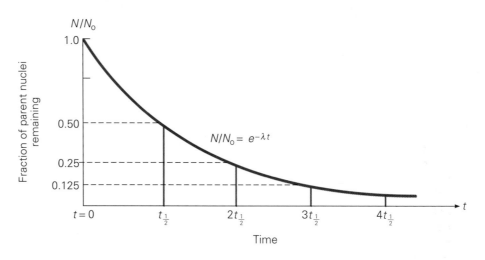

•FIGURE 29.10 Radioactive decay versus time
When the fraction of the remaining parent nuclei (N/N_o) in a radioactive sample is plotted as a function of time, an exponential decay curve is obtained, the shape or steepness of which depends on the decay constant λ.

Note: Activity = number of decays/second = $|\Delta N/\Delta t|$. We will omit the absolute value signs and write the activity as $\Delta N/\Delta t$, but keep in mind that we mean a *positive* number.

the isotope. The minus sign in the equation indicates that N is decreasing. The sample activity is defined as the *magnitude* of $\Delta N/\Delta t$, in decays per second without a minus sign. We will use $\Delta N/\Delta t$ for activity, but keep in mind that we mean a *positive* number. (See the usage in Example 29.3.)

The way in which the number of parent nuclei N decreases with time is illustrated in •Fig. 29.10. The curve and N are said to decay *exponentially* since the graph follows an exponential function $e^{-\lambda t}$ with time, where λ is the decay constant and $e \approx 2.718$ is the base of natural logarithms.

The number N of the undecayed parent nuclei at a particular time is given by

$$N = N_o e^{-\lambda t} \tag{29.3}$$

where N_o is the initial number of nuclei ($t = 0$).

The decay rate of an isotope is commonly expressed in terms of its half-life rather than the decay constant. The **half-life** ($t_{1/2}$) is defined as the time it takes for half of the radioactive nuclei in a sample to decay. This is the time corresponding to $N_o/2$ parent nuclei in Fig. 29.10, that is, $N/N_o = 1/2$. The activity also decreases to the fractional amount N/N_o in a time t, since it is proportional to the number of nuclei present. That is, the number of decays per second (or the activity or decay rate) itself decreases by half. The decay *rate* is what is usually measured to determine the half-life, rather than counting nuclei. In other words, one usually monitors the rate at which the decay particles are emitted until this rate decays by a certain fraction (not always the half-life, especially if the half-life is long).

For example, as may be seen from the graph in •Fig. 29.11, the half-life of strontium-90 (^{90}Sr) is 28 years. Another way of looking at what occurs in a half-

•FIGURE 29.11 Radioactive decay and amount of isotope
As illustrated here for strontium-90, after each half-life ($t_{1/2} = 28$ y), only half of the amount of ^{90}Sr present at the start of that time period remains, and the activity (decays/s) has decreased by one half. The other half of the sample has decayed into ^{90}Y via beta decay.

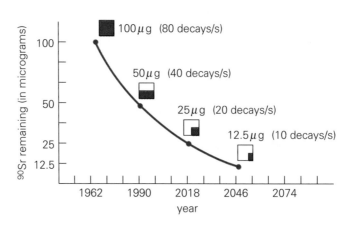

life is to consider the amount, or mass, of the parent material. As illustrated in Fig. 29.11, if one had 100 micrograms (μg) of ^{90}Sr initially, only 50 μg would remain at the end of 28 years. The other 50 μg would have beta decayed by the reaction

$$^{90}_{38}\text{Sr} \rightarrow {}^{90}_{39}\text{Y} + {}^{0}_{-1}\text{e}$$
$$\text{\textit{strontium} \quad \textit{yttrium} \quad \textit{electron}}$$

and the sample would contain both strontium and yttrium nuclei. After another 28 years, another half of the strontium nuclei would decay, leaving only 25 μg, and so on.

The half-lives of radioactive isotopes vary greatly, as may be seen from Table 29.1. Isotopes with very short half-lives are generally created in nuclear reactions, which will be considered in the next chapter. If these isotopes did exist when the Earth was formed, they would have long since decayed away. In fact, the elements technetium (Tc) and promethium (Pm) are not found on Earth (but can be produced in laboratories). On the other hand, the half-life of the naturally occurring ^{238}U isotope is 4.5 billion years, which is close to the estimated age of the Earth. Thus, we would expect about half of the original ^{238}U present when the Earth was formed to exist today.

The longer the half-life of an isotope, the more slowly it decays and the smaller the decay constant, λ. The half-life and the decay constant have an inverse relationship $t_{1/2} \propto 1/\lambda$. This can be seen from Eq. 29.3. The half-life is the time it takes for the number of nuclei present or the activity to decrease by half, that is, $N = N_o/2$. Thus,

$$\frac{N}{N_o} = e^{-\lambda t_{1/2}} = \frac{1}{2}$$

But

$$e^{-0.693} \approx \frac{1}{2}$$

TABLE 29.1	The Half-Lives of Some Radioactive Isotopes	
Nuclide	Primary Decay Mode	Half-Life
Beryllium-8 ($^{8}_{4}$Be)	α	1×10^{-16} s
Polonium-213 ($^{213}_{84}$Po)	α	4×10^{-16} s
Oxygen-19 ($^{19}_{8}$O)	β^-	27 s
Fluorine-17 ($^{17}_{9}$F)	β^+, EC	66 s
Polonium-218 ($^{218}_{84}$Po)	α, β^-	3.05 min
Technetium-104 ($^{104}_{43}$Te)	β^-	18 min
Krypton-76 ($^{76}_{36}$Kr)	EC	14.8 h
Magnesium-28 ($^{28}_{12}$Mg)	β^-	21 h
Iodine-123 ($^{123}_{53}$I)	EC	13.3 h
Radon-222 ($^{222}_{86}$Rn)	α	3.82 d
Cobalt-60 ($^{60}_{27}$Co)	β^-	5.3 y
Strontium-90 ($^{90}_{38}$Sr)	β^-	28 y
Radium-226 ($^{226}_{88}$Ra)	α	1600 y
Carbon-14 ($^{14}_{6}$C)	β^-	5730 y
Plutonium-239 ($^{239}_{94}$Pu)	α	2.4×10^4 y
Uranium-238 ($^{238}_{92}$U)	α	4.5×10^9 y
Rubidium-87 ($^{87}_{37}$Rb)	β^-	4.7×10^{10} y

so by comparison, to three significant figures,

Relationship between decay constant and half-life

$$t_{1/2} = \frac{0.693}{\lambda}$$

(29.4)

EXAMPLE 29.3 ■ BEFORE YOU SEND THE PATIENT HOME: HALF-LIFE AND ACTIVITY

The half-life of iodine-131 used in medical thyroid treatments is 8.0 days. At a certain time, an amount of ^{131}I containing 4.0×10^{14} nuclei is known to be in a patient's thyroid gland. (a) What will be the observed activity? (b) How many of the ^{131}I nuclei remain after 1.0 d?

Solution. We are given the elapsed time and the half life. Listing the data, we have

Given: $t_{1/2} = 8.0$ d
$= 6.9 \times 10^5$ s
$N_o = 4.0 \times 10^{22}$ nuclei
$t = 24$ h $= 8.6 \times 10^4$ s

Find: (a) $\Delta N/\Delta t$ (activity)
(b) N (number of undecayed nuclei)

(a) First, Eq. 29.4 is used to find the decay constant with $t_{1/2}$ in seconds:

$$\lambda = \frac{0.693}{t_{1/2}} = \frac{0.693}{6.9 \times 10^5 \text{s}} = 1.0 \times 10^{-6} \text{s}^{-1}$$

Then, using the expression for activity, we have the initial activity
$$|\Delta N/\Delta t| = \lambda N_o = (1.0 \times 10^{-6} \text{ s}^{-1})(4.0 \times 10^{14})$$
$$= 4.0 \times 10^8 \text{ decays/s}$$

(b) The actual number of parent nuclei present can be found from Eq. 29.3. With $t = 1$ day and $\lambda = 0.693/t_{1/2} = 0.693/8.0$ d $= 0.087$ d^{-1}, we have
$$N = N_o e^{-\lambda t} = (4.0 \times 10^{14} \text{ nuclei})e^{-(0.087 \text{ d}^{-1})(1.0 \text{ d})}$$
$$= (4.0 \times 10^{14} \text{ nuclei})e^{-0.087} = (4.0 \times 10^{14} \text{ nuclei})(0.917)$$
$$= 3.7 \times 10^{14} \text{ nuclei}$$

Note that for working with powers of e, most modern hand calculators have an e^x function button (sometimes the inverse of the ln x function). You should become familiar with it. Here we calculated $e^{-0.087} \approx 0.917$ to three significant figures.

Follow-up Exercise. Suppose that in this Example the attending physician would not allow the patient to be released from the hospital until the activity was one-eighth of its original level. (a) How long would the patient have to remain in observation? (b) In actual practice, the time is much shorter. Can you think of a possible biological reason?

Units of radioactivity

The strength of a radioactive sample or source may be specified at a given time by its activity. A common unit of radioactivity is named in honor of Pierre and Marie Curie. One **curie (Ci)** is defined as

$$1 \text{ Ci} \equiv 3.70 \times 10^{10} \text{ decays/s}$$

This is based on the activity of one gram of radium. The curie is the traditional unit for expressing radioactivity. However, the proper SI unit is the **becquerel (Bq),** which is simply defined as

$$1 \text{ Bq} \equiv 1 \text{ decay/s}$$

Therefore,

$$1 \text{ Ci} = 3.70 \times 10^{10} \text{ Bq}$$

Even with the emphasis on SI units, radioactive sources are commonly rated in curies. The curie is a relatively large unit, however, so the millicurie (mCi), the microcurie (μCi), and sometimes even smaller multiples are used. Teaching laboratories, for example, typically use samples with activities only on the order of microcuries. (Usually, if a sample has an activity of 1 Ci, it requires shielding or very short times of exposure.)

EXAMPLE 29.4 ■ GET A HALF-LIFE! DECLINING SOURCE STRENGTH

A ^{90}Sr beta source has an initial activity of 10.0 mCi. How many decays/s will be observed to take place at the end of 84 years, or three half-lives (approximately one human lifetime)?

Solution. We look up the half-life and use the fact that in each successive half-life, the activity decreases by a factor of one-half from what it was at the start of that interval.

Given: Initial activity = 10.0 mCi   *Find:* $\Delta N/\Delta t$ (activity,
                 t = 84.0 years   decays/s)
               $t_{1/2}$ = 28.0 years (from Table 29.1)

After three half-lives, the activity will be only $\frac{1}{8}$ as great ($\frac{1}{2} \times \frac{1}{2} \times \frac{1}{2} = \frac{1}{8}$), and the strength of the source will then be

$$\frac{\Delta N}{\Delta t} = 10.0 \text{ mCi} \times \frac{1}{8} = 1.25 \text{ mCi} = 1.25 \times 10^{-3} \text{ Ci}$$

and in terms of decays/s or Bq we have

$$\frac{\Delta N}{\Delta t} = (1.25 \times 10^{-3} \text{ Ci})\left(3.70 \times 10^{10} \frac{\text{decays/s}}{\text{Ci}}\right)$$

$$= 4.63 \times 10^7 \text{ decays/s}$$

$$= 4.63 \times 10^7 \text{ Bq}$$

Insight   Radon in the Home

You have probably heard about the widespread concern over radon gas accumulations in homes and other buildings. This is a relatively recent worry; a few years ago most people had never heard of radon.

Radon is a chemically inert gas belonging to the noble gas family, along with helium, neon, and others. However, it is radioactive and is part of the uranium-238 decay series (see Fig. 29.9). As such, it occurs naturally in the environment, the product of uranium deposits that exist in many areas. The amount of radon thus depends on the local geology.

Because the radon gas is chemically inert, it doesn't react with anything in the soil, and so is not immobilized in the ground. Also, the decay rate of radon is such that significant concentrations persist for days, and thus there is adequate time for it to seep up into homes. It enters through holes and cracks in the basement floor or walls or around pipes. Radon gas may also be dissolved in water, brought into the home with the water supply, and released on aeration.

What is the problem with this inert gas that is a natural part of our environment? When inhaled by humans, it can decay inside the lungs. Radon alpha decays into polonium-218. Polonium-218 is not a gas, nor is it chemically inert, but it is radioactive. It can lodge in the lungs where it too undergoes alpha decay, into lead-214. (Polonium-218 also beta decays—

see Fig. 29.9—but more than 99% of the nuclei decay by the alpha mode.)

As was discussed in the text, the penetration range of alpha particles is quite small. However, these particles can penetrate the cells of the lungs, causing damage and possibly inducing lung cancer (see Section 29.5).

Since radon occurs naturally in the environment, the next immediate questions are: How much radon in my home is too much, and how do I know how much radon I have? Let's answer the questions in reverse order. You can purchase sampling kits to test the level of radon in your home. As you might imagine, there is some controversy over the question of what level of radon is considered to be safe. The U.S. Environmental Protection Agency has set an activity standard of 4 picocuries (pCi)* per liter of air as an "action level." If your home shows a level of radon gas greater than this, you should consult the appropriate state agency for retesting and possible corrective actions, such as sealing cracks in the basement and around pipes coming through walls. A private water source may also need checking. Testing is particularly important in colder climates, where homes are insulated and sealed, and closed for long periods of time.

*A curie (Ci) is a unit of radioactivity discussed later in this chapter.

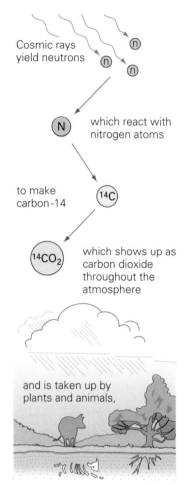

Cosmic rays yield neutrons

which react with nitrogen atoms

to make carbon-14

which shows up as carbon dioxide throughout the atmosphere

and is taken up by plants and animals,

but when an organism dies, no fresh carbon-14 replaces the carbon-14 decaying in its tissues, and the carbon-14 radioactivity decreases by half every 5730 years.

•FIGURE 29.12 Carbon-14 radioactive dating
The diagram illustrates the formation of carbon-14 in the atmosphere and its entry into the biosphere.

Note: The cosmic ray production of ^{14}C is an example of a nuclear reaction that induces a nuclear transmutation. Such reactions will be studied in more detail in Chapter 30.

Follow-up Exercise. In this Example, suppose that the campus radiation safety officer tells you that this sample will be able to go into a low-level waste disposal canister when its activity is approximately one-millionth of its current activity (in other words, when it drops from the present 10.0 mCi to 10.0 pCi, a factor of a million). Estimate, to one significant figure, how long you will have to keep it under its present cover? [*Hint:* Try your calculator. Two raised to approximately what power produces a million?]

Radioactive Dating

Because of their constant decay rates, radioactive isotopes can be used as nuclear clocks. As we have seen, the half-life of a radioactive isotope may be used to project how much of a given amount of material will exist in the future. Similarly, by using the half-life to project backward in time, scientists can determine the ages of objects containing radioactive isotopes. As you might surmise, some idea of the initial composition or initial amount of an isotope must be known.

To illustrate the principle of radioactive dating, let's take a look at how it is done with ^{14}C. **Carbon-14 dating** is used on materials that were once part of living things, or the remnants of objects made from or containing such materials (such as wood, bone, leather, or parchment). The process depends on the fact that living things—plants and animals (including yourself)—contain a known amount of radioactive ^{14}C. The concentration of carbon-14 is very small, about one ^{14}C atom for 7.2×10^{11} atoms of ordinary ^{12}C. Even so, this concentration cannot be due to an abundance of carbon-14 present when the Earth was formed, since a half-life of $t_{1/2} = 5730$ years for ^{14}C is brief in comparison to the estimated age of the Earth (over 4 billion years).

The observed concentration of ^{14}C is accounted for by its continuous production in the upper atmosphere. Cosmic rays from outer space cause reactions that produce neutrons (•Fig. 29.12). The neutrons are absorbed by the nuclei of the nitrogen atoms of the air which in turn decay by emitting a proton to produce ^{14}C by the reaction

$$^{14}_{7}N + {}^{1}_{0}n \rightarrow {}^{14}_{6}C + {}^{1}_{1}H$$

Recall that the carbon-14 then decays by beta decay ($^{14}_{6}C \rightarrow {}^{14}_{7}N + {}^{0}_{-1}e$). Although the intensity of incident cosmic rays may not be constant, the concentration of ^{14}C in the atmosphere is relatively constant because of atmospheric mixing and the fixed decay rate.

The ^{14}C is oxidized into carbon dioxide (CO_2), so a small fraction of the CO_2 molecules of the air are radioactive. Plants take in this radioactive CO_2 by photosynthesis, and animals ingest the material produced. As a result, the concentration of carbon-14 in living organic matter is the same as the concentration in the atmosphere, 1 part in 7.2×10^{11}. However, once an organism dies, the ^{14}C is *not* replenished, and the concentration decreases with radioactive decay (with $t_{1/2} = 5730$ years). *Measurement of the concentration of carbon-14 in dead matter relative to that in living things can then be used to establish when the organism died.*

Since radioactivity is generally measured in terms of activity, we must first know the ^{14}C activity in a living organism. This is derived in the following Example.

EXAMPLE 29.5 ■ WE THE LIVING: NATURAL CARBON-14 ACTIVITY

Determine the average ^{14}C activity, in decays per minute per gram of natural carbon, found in living organisms if the concentration of carbon-14 is the same as that in the atmosphere.

Solution. The relative concentration of carbon-14 in living organisms is one part in 7.2×10^{11} (given in text) or

$$\frac{^{14}C}{^{12}C} = \frac{1}{7.2 \times 10^{11}} = 1.4 \times 10^{-12}$$

Carbon has an atomic mass of 12.0, so the number of nuclei (atoms) in 1 g of carbon may be found by using Avogadro's number N_A (6.02×10^{23}) and the number of moles ($n = N/N_A$). With $n = 1.0$ g/(12 g/mole) $= \frac{1}{12}$ mole,

$$N = nN_A = (\tfrac{1}{12} \text{ mole})(6.02 \times 10^{23} \text{ nuclei/mole})$$
$$= 5.0 \times 10^{22} \text{ nuclei (per gram)}$$

Note: This relationship is introduced in Section 10.3.

The number of ^{14}C nuclei per gram is given by

$$N\left(\frac{^{14}C}{^{12}C}\right) = (5.0 \times 10^{22} \text{ nuclei/g})(1.4 \times 10^{-12})$$

$$= 7.0 \times 10^{10} \text{ } ^{14}C \text{ nuclei/g}$$

With a half-life of $t_{1/2} = (5730 \text{ y})(5.26 \times 10^5 \text{ min/y}) = 3.01 \times 10^9$ min, the decay constant is

$$\lambda = \frac{0.693}{t_{1/2}} = \frac{0.693}{3.01 \times 10^9 \text{ min}} = 2.30 \times 10^{-10} \text{ min}^{-1}$$

Then the activity, or the number of decays per minute per gram of carbon (to two significant figures), is given by

$$\Delta N/\Delta t = \lambda N = (2.30 \times 10^{-10} \text{ min}^{-1})(7.0 \times 10^{10})$$

$$= 16 \frac{\text{decays}}{\text{g}\cdot\text{min}}$$

Thus, if we determined that an artifact such as a bone or a piece of cloth had an activity of 8.0 counts/min *per gram* of carbon, the original living organism would have died about one half-life or about 5700 years ago. This would put the date of the artifact near 3700 B.C.

Follow-up Exercise. Suppose your instruments could measure carbon-14 beta emissions only down to 1.0 decays/g·min. How far back (to two significant figures) could you estimate the ages of dead organisms?

Now consider how the activity calculated in the preceding example can be used to date ancient organic finds.

Example 29.6 ■ Old Bones: Carbon-14 Dating

An old bone is unearthed in an archeological dig. Laboratory analysis determines that there are on the average 2.0 beta emissions per minute per gram of carbon in the bone. What is the approximate age of the bone?

Solution. Since we know the current activity, and the approximate initial activity (when the bone was part of a living animal constantly ingesting carbon-14) we can work backwards to determine the time elapsed.

Given: Activity = 2.0 decays/g·min   *Find:* Age of bone

Assuming that the organism had the normal concentration of carbon-14 when it died, at the time of death the carbon-14 activity would be 16 decays/g·min (Example 29.5). Afterward, the decay rate would decrease by one half for each half-life:

$$16 \xrightarrow{t_{1/2}} 8 \xrightarrow{t_{1/2}} 4 \xrightarrow{t_{1/2}} 2 \text{ decays}$$

So with the observed activity, the carbon-14 in the bone would have gone through three half-lives, or the bone would be three half-lives old. Thus with $t_{1/2} = 5730$ y (Table 29.1), (to two significant figures) we have

$$\text{Age} = 3.0 \text{ } t_{1/2} = (3.0)(5730 \text{ y}) = 1.7 \times 10^4 \text{ y} = 17 \text{ thousand years}$$

The limit of radioactive carbon dating depends on the ability to measure the very low activity after an appreciable number of half-lives. Current techniques give an age dating limit of about 40,000–50,000 years, depending on the size of the sample. After about nine half-lives, the radioactivity of a carbon sample has decreased so much that it is barely measurable (about 2 counts per gram per *hour*).

Another radioactive dating process uses lead-206 (^{206}Pb) and uranium-238 (^{238}U). This dating method is used extensively in geology because of the long half-life of ^{238}U. Lead-206 is the stable end isotope of the ^{238}U decay series (see Fig. 29.9). If a rock sample contains both of these isotopes, the lead is assumed to be a decay product of the uranium that was there when the rock was first formed. Thus, the ratio of ^{206}Pb$/^{238}$U is a measure of the geologic age of the rock.

29.4 Nuclear Stability and Binding Energy

Objectives: To be able to **(a)** state which proton and neutron number combinations result in stable nuclei, **(b)** explain the pairing effect and magic numbers in relation to nuclear stability, and **(c)** calculate nuclear binding energy.

Now that we have considered some of the properties of unstable isotopes, we turn our attention to the more common stable isotopes. Stable isotopes exist naturally for all elements having proton numbers from 1 to 83, except for those with $Z = 43$ (technetium) and $Z = 61$ (promethium), as was pointed out earlier. The nuclear interactions that give rise to nuclear stability are extremely complicated. However, by looking at some of the general properties of stable nuclei, it is possible to obtain some qualitative and quantitative criteria for nuclear stability. This will give us some general rules for determining whether or not a particular nuclide would be expected to be stable or unstable.

Nucleon Populations

One of the easiest things to consider is the relative number of protons and neutrons in stable nuclei. Nuclear stability must be related in some way to the dominance of either the repulsive Coulomb force between protons or the attractive nuclear force between nucleons, and this force dominance depends on the relative numbers, or ratio, of protons and neutrons.

For stable nuclei of low mass numbers for about $A < 40$, the nucleon ratio is approximately 1. That is, the number of protons and the number of neutrons are equal or nearly equal. For example, $^{4}_{2}$He, $^{12}_{6}$C, $^{23}_{11}$Na, and $^{27}_{13}$Al have the same or about the same number of protons as neutrons. For stable nuclei of higher mass numbers ($A > 40$), the number of neutrons exceeds the number of protons. The heavier the nuclei, the more the neutrons outnumber the protons.

This trend is illustrated in •Fig. 29.13, which shows a plot of the neutron number (N) versus proton number (Z) for stable nuclei. Notice how the heavier sta-

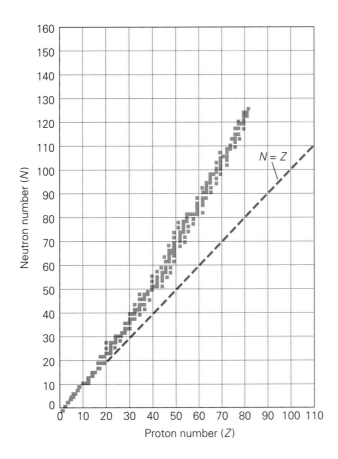

●FIGURE 29.13 A plot of N versus Z for stable nuclei
For nuclei with mass numbers $A < 40$ ($Z < 20$ and $N < 20$), the number of protons and the number of neutrons are equal or nearly equal. For nuclei with $A > 40$, the number of neutron exceeds the number of protons and the nuclei lie above the $N = Z$ line.

ble nuclei lie above the 45° or $N = Z$ line. Examples of heavy stable nuclei include $^{62}_{28}$Ni, $^{114}_{50}$Sn, $^{208}_{82}$Pb, and $^{209}_{83}$Bi. (Bismuth is the heaviest element that has a stable isotope.*) The extra neutrons of such heavy stable nuclei presumably allow the attractive forces among the nucleons to be greater than the repulsive forces between the numerous protons by acting as "spacers" between the protons, thus reducing the electrical repulsion.

Radioactive decay adjusts the proton and neutron numbers of an unstable isotope until a stable isotope on the neutron–proton stability curve in Fig. 29.13 is reached. Since alpha decay decreases the numbers of protons and neutrons by equal amounts, alpha decay alone would give nuclei with neutron populations that are larger than that of one of the stable isotopes on the curve. However, beta minus decay *following* alpha decay can lead to a stable combination, since the effect of beta minus decay is the loss of a neutron and the gain of a proton. Thus very heavy nuclei undergo a chain, or sequence, of alpha and beta decays until a stable nucleus is reached, as was illustrated in Fig. 29.9 for ^{238}U.

Pairing Effect

A fact that is not too evident from Fig. 29.13 is that many stable nuclei have even numbers of both protons and neutrons, and very few have odd numbers of both protons and neutrons. A survey of the stable isotopes (Table 29.2 shows that 168 stable nuclei have this even–even combination, 107 are even–odd or odd–even, and only four contain odd numbers of both protons and neutrons. These four are isotopes of the elements with the four lowest odd proton numbers: $^{2}_{1}$H, $^{6}_{3}$Li, $^{10}_{5}$B, and $^{14}_{7}$N.

TABLE 29.2 Pairing Effect of Stable Nuclei		
Proton Number	Neutron Number	Number of Stable Nuclei
Even	Even	168
Even	Odd	107
Odd	Even	
Odd	Odd	4

*Bismuth-209 does undergo alpha decay, but with a half-life of 2×10^{18} y; so for all practical purposes, it is stable.

The even and odd combinations seem to indicate that the protons and neutrons in stable nuclei above the very lightest elements tend to "pair up" (even–even). That is, two protons pair up and two neutrons pair up, but a proton and neutron are less likely to do so. Except for the very lightest elements, when there are odd numbers of nucleons (odd–odd), there is general instability. The instability is less if there is only one odd nucleon (odd–even or even–odd combination).

This so-called **pairing effect** gives a qualitative criterion for stability. For example, you would expect the aluminum isotope $^{27}_{13}$Al to probably be stable (even–odd), but not $^{26}_{13}$Al (odd–odd). This is actually the case.

The *general criteria for nuclear stability* can be summarized as follows:

Criteria for nuclear stability

1. All isotopes with proton number greater than 83 ($Z > 83$) are unstable.
2. (a) Most even–even nuclei are stable.
   (b) Many odd–even or even–odd nuclei are stable.
   (c) Only four odd–odd nuclei are stable ($^{2}_{1}$H, $^{6}_{3}$Li, $^{10}_{5}$B, and $^{14}_{7}$N).
3. (a) Stable nuclei with mass numbers less than 40 ($A < 40$) have approximately the same number of protons and neutrons.
   (b) Stable nuclei with mass numbers greater than 40 ($A > 40$) have more neutrons than protons.

EXAMPLE 29.7 ■ RUNNING DOWN THE CHECKLIST: NUCLEAR STABILITY

Is the sulfur isotope $^{38}_{16}$S likely to be stable?

Solution. Applying the general criteria:

1. *Satisfied.* Isotopes with $Z > 83$ are immediately known to be unstable. With $Z = 16$, this criterion is satisfied.
2. *Satisfied.* The isotope $^{38}_{16}$S$_{22}$ has an even–even nucleus and so has a good probability of being stable.
3. *Not satisfied.* $A < 40$, and Z ($= 16$) and N ($= 22$) are not approximately equal.

Therefore, the ^{38}S isotope is likely to be unstable. (The nucleus actually is unstable to beta minus decay since it is neutron-rich.)

Follow-up Exercise. (a) List some likely isotopes of copper ($Z = 29$). (b) Apply the criteria to see which ones should be stable and check your conclusions by consulting Appendix V.

Binding Energy

An important quantitative aspect of nuclear stability is the binding energy of the nucleons. This can be calculated by considering the sums of nuclear and electron masses and of their corresponding energies.

Since the masses of nuclei are so small in relation to the standard kilogram mass unit, another standard, the *unified* **atomic mass unit (u)** is used to measure nuclear masses. A *neutral atom* of carbon-12 is defined as having an exact value of 12.000000 u. Thus,

$$1 \text{ u} = 1.6606 \times 10^{-27} \text{ kg}$$

The various atomic particles then have masses in atomic mass units as shown in Table 29.3. The listed energy equivalents reflect Einstein's $E = m_0 c^2$ relationship.

TABLE 29.3   Particle Masses and Energy Equivalents

| Particle | Mass | | Energy (MeV) |
	u	kg	
	1	1.6606×10^{-27}	931.5
Electron	0.000548	9.1095×10^{-31}	0.511
Proton	1.00727	1.67265×10^{-27}	938.28
$^{1}_{1}$H atom	1.007825	1.67356×10^{-27}	938.79
Neutron	1.008665	1.67500×10^{-27}	939.57

Thus, a mass of 1 u has an energy equivalent of

$$m_o c^2 = (1.6606 \times 10^{-27} \text{ kg})(2.9977 \times 10^8 \text{ m/s})^2 = 1.4922 \times 10^{-5} \text{ J}$$

$$= \frac{1.4922 \times 10^{-5} \text{ J}}{1.602 \times 10^{-13} \text{ J/MeV}} = 931.5 \text{ MeV}$$

as given in the first entry of the table. We will use 931.5 MeV/u as a handy conversion factor to avoid having to multiply by c^2.

Note that the proton and the hydrogen atom ($^{1}_{1}$H) are listed separately, having different masses. The difference is the mass of the atomic electron. We deal almost exclusively with the masses of neutral atoms (nucleons plus Z electrons) rather than strictly with nuclei, since the atomic masses are what can be measured. Keep this in mind. We will soon be interested in very small mass differences, and the electronic mass is significant. We must balance out the masses of the electrons when we compare masses in nuclear processes.

Given this information, we can look at the effect of mass–energy relationships on nuclear stability. For example, if you compare the mass of a helium nucleus to the total mass of nucleons that make it up, a significant inequity emerges: a neutral helium atom has a mass of 4.002603 u. (Atomic masses of various isotopes are given in Appendix V.) The total mass of two protons (with two electrons, or actually two ^{1}H atoms) and two neutrons is, by addition,

$$2\, m(^{1}\text{H}) = 2.015650 \text{ u}$$
$$2\, m_n = \underline{2.017330 \text{ u}}$$
$$\text{total} \quad 4.032980 \text{ u}$$

Notice that this is greater than the mass of the helium atom (4.002603 u). The helium nucleus is then less massive than the sum of its parts by an amount

$$\Delta m = [2\, m(^{1}\text{H}) + 2\, m_n] - m(^{4}\text{He})$$
$$= 4.032980 \text{ u} - 4.002603 \text{ u} = 0.030377 \text{ u}$$

where the electronic mass cancels out, since the mass of a hydrogen atom was used for the proton. This mass difference, called the **mass defect**, has an energy equivalent of

$$(0.030377 \text{ u})(931.5 \text{ MeV/u}) = 28.30 \text{ MeV}$$

which is the **total binding energy** (E_b) of the nucleus—that is, the amount of energy released when the separate nucleons combined to form a nucleus. Then, in general, for any nucleus, the total binding energy is given by

$$E_b = (\Delta m)c^2 \qquad (29.5)$$

where Δm is the mass defect.

Another way of looking at the binding energy is that it is the energy that would be required to separate the constituent nucleons into free particles. This is

Binding energy: total energy needed to break the nucleus into its individual nucleons *or* the energy released when the nucleus is formed from its individual nucleons

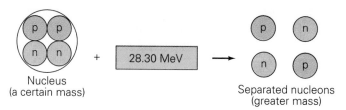

•FIGURE 29.14   Binding energy
As illustrated here, 28.30 MeV of work is required to separate a helium nucleus into free protons and neutrons. Conversely, if two protons and two neutrons could be combined to form a helium nucleus, 28.30 MeV of energy would be released. This is the binding energy of the nucleus.

illustrated in •Fig. 29.14 for the helium nucleus. Exactly 28.30 MeV of energy would be necessary to do the work of separating the nucleons. If the mass of a nucleus and the mass of its nucleons were the same, then the nucleons would not be bound together, and the nucleus would come apart without energy input.

We can gain some insight into the nature of the nuclear force by considering the *average binding energy per nucleon* for stable nuclei. This is simply the total binding energy of a particular nucleus divided by its number of nucleons, of E_b/A, where A is the mass number (number of protons plus number of neutrons). For example, for the helium nucleus (^{4}He), the average binding energy per nucleon is

$$\frac{E_b}{A} = \frac{28.30 \text{ MeV}}{4} = 7.075 \text{ MeV}$$

Compared to the binding energy of atomic electrons, for example, 13.6 eV for a hydrogen electron in the ground state, the nuclear binding energy is millions of times larger, which indicates a very strong binding force.

EXAMPLE 29.8 ■ THE STABLEST OF THE STABLE: BINDING ENERGY PER NUCLEON

Compute the average binding energy per nucleon for the iron-56 nucleus ($^{56}_{26}$Fe).

Solution.   We can find the atomic mass of the iron isotope in Appendix V, and we have the other needed masses from Table 29.3.

Given: $^{56}_{26}$F mass = 55.934939 u   *Find:* E_b/A (average binding energy
$^1_1$H mass = 1.007825 u   per nucleon)
$^1_0$n mass = 1.008665 u

(Notice that we use the mass of the hydrogen atom rather than the proton mass. Why?)
We then compute the mass defect or difference between the iron atom and its parts for $^{56}_{26}$Fe$_{30}$. The particles have a total mass of

$$26 \, m(^1\text{H}) = 26(1.007825 \text{ u}) = 26.203450 \text{ u}$$
$$30 \, m_n = 30(1.008665 \text{ u}) = \underline{30.259950 \text{ u}}$$
$$\text{total} \quad 56.463400 \text{ u}$$

and

$$\Delta m = [26 \, m(^1\text{H}) + 30 \, m_n] - m(^{56}\text{Fe})$$
$$= 56.463400 \text{ u} - 55.934939 \text{ u} = 0.528461 \text{ u}$$

The total binding energy is then

$$E_b = (\Delta m)c^2$$
$$= (0.528461 \text{ u})(931.5 \text{ MeV/u}) = 492.3 \text{ MeV}$$

where the conversion factor for MeV/u was used instead of making the c^2 calculation.

Our iron nuclide has 56 nucleons, so the average binding energy per nucleon is

$$\frac{E_b}{A} = \frac{492.3 \text{ MeV}}{56} = 8.791 \text{ MeV}$$

Follow-up Exercise. (a) To illustrate the pairing effect, compare the average binding energy per nucleon for ^{4}He (calculated previously to be 7.075 MeV) to that of ^{3}He. (Find atomic masses in Appendix V.) (b) How does your answer reflect the pairing idea?

If E_b/A is calculated for various nuclei and plotted versus mass number, the values generally lie along a curve as shown in •Fig. 29.15. The value of E_b/A rises rapidly with increasing A for light nuclei and starts to level off (around $A = 15$) at about 8.0 MeV, with a maximum value of about 8.8 MeV. The maximum of the curve occurs in the vicinity of iron, which has a very stable nucleus (as we have seen in Example 29.8). For $A > 60$, the E_b/A values decrease slowly for heavier nuclei, a finding indicating that the nucleons are less tightly bound.

The maximum in the curve gives an important indication that will be considered in the next chapter. If a large nucleus could be split or fissioned into two lighter nuclei, the nucleons would be more tightly bound, and energy would be released in the process—more than would be needed to induce the fission. (Recall how an *atom* gives up energy when one of its electrons becomes more tightly bound or makes a transition to a lower energy level.) Similarly, on the other side of the maximum, if we could fuse together two very light nuclei into a heavier nucleus, this would raise it up on the curve, and energy would also be released. Such fission and fusion processes are the sources of nuclear energy (Chapter 30).

In general, the E_b/A curve shows that, except for very light nuclei, the binding energy per nucleon does not change a great deal and is on the order of

$$E_b/A \approx 8 \text{ MeV}$$

Since E_b/A is relatively constant for most nuclei, an approximation is that

$$E_b \propto A \qquad \textit{(experimental approximation)}$$

In other words, the *total* binding energy is (approximately) proportional to the mass number.

This proportionality indicates a characteristic of the nuclear force that was mentioned earlier, one that makes it quite different from the electrical Coulomb force. Imagine that the attractive nuclear force acts among *all* nucleons (protons

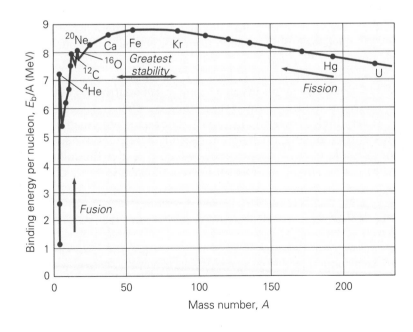

•FIGURE 29.15 A plot of binding energy per nucleon versus mass number
If the binding energy per nucleon (E_b/A) is plotted versus mass number (A), the curve has a maximum, which indicates that the nuclei in this region are on the average the most tightly bound together and have the greatest stability.

and neutrons) in a nucleus. Each pair of nucleons would then contribute to the total binding energy. In a nucleus containing A nucleons, there are $A(A - 1)/2$ pairs, so there would be $A(A - 1)/2$ contributions to the total binding energy. For nuclei for which $A \gg 1$ (heavier nuclei), then, $A(A - 1) \approx A^2$ and we would expect that

$$E_b \propto A^2 \qquad \text{(if nucleon–nucleon force is long range)}$$

But in fact, $E_b \propto A$ for most nuclei, which indicates that a given nucleon is not bound to all the other nucleons. This phenomenon, called *saturation*, implies that the nuclear forces act over a very short range.

Magic Numbers

In atoms we are familiar with the concept of a filled shell. Recall that at the end of every row on the periodic chart, there exists an inert gas (such as neon, Ne). An inert element consists of single atoms that tend not to form bonds with other atoms. We know that such atoms are relatively unreactive because their outer shells of electrons are full, up to the limit provided by the Pauli exclusion principle. Gaining or losing an electron would be very costly in terms of energy, and therefore unlikely to occur.

There is an analogous effect in the nucleus. Although the concepts of individual nucleon "orbits" and "filled shells" inside the nucleus are hard to visualize, experimental evidence (confirmed by intricate Schrödinger equation calculations) indicates the existence of "closed nuclear shells" when the number of protons *or* neutrons is 2, 8, 20, 28, 50, 82, or 126.

Solid evidence for the existence of such **magic numbers** is provided by the number of stable isotopes of various elements. If an element has a magic number of protons, it is also found to have an unusually high number of stable isotopes. Far away from a magic number, there may be only one or two (or even no) stable isotopes per element. Aluminum, for example, with 13 protons, has just one stable isotope, ^{27}Al. But tin, with $Z = 50$ (a magic number), has *ten* stable isotopes ranging from $N = 62$ to $N = 74$! (Not all N values are stable.) Nearby, indium, with $Z = 49$, has only two stable isotopes, and antimony, with $Z = 51$, also has only two.

The second piece of evidence is related to binding energies. Experiments can be done, for instance, with high-energy gamma ray photons, which knock out single nucleons from a nucleus (the so-called *photonuclear effect*) in a manner analogous to the photoelectric effect in metals. By carefully adjusting the photon energy, the binding energy of the last neutron or proton can be determined. It is found experimentally that a nuclide like tin requires about 2 MeV *more* photon energy to eject the least bound nucleon from its nucleus than a nuclide that is not at a magic number. *Thus, magic numbers are associated with extra large binding energies, another sure sign of higher stability.*

Other types of experiments involve trying to *add* neutrons to nuclei. It is found that with nuclei one neutron away from a magic number, the likelihood of adding a neutron is very high. However, the same experiments with magic neutron numbers show clearly that such nuclei do not readily accept more neutrons.

As an example of the magic number effect, consider the following calculations for removal of single neutrons from two different isotopes of calcium.

EXAMPLE 29.9 ■ EXPLORING MAGIC NUMBERS: THE PHOTONUCLEAR EFFECT

Compare the minimum photon energy needed to remove a neutron from $^{40}_{20}$Ca to that required for neutron removal from $^{40}_{18}$Ar.

[Here we chose two nuclei with even neutron numbers, so the pairing effect is about the same; only magic number effects will be significant. Thus, we are comparing neutron removal from a nucleus with a magic number of neutrons (20 for calcium) to neutron removal from a nucleus with a nonmagic number of neutrons (22 for argon). The necessary atomic masses (in u) not in Appendix V are supplied below, in addition to the ones you can look up in that appendix.]

Solution. It requires energy input to break loose a neutron in either case. Thus we expect the total mass (neutron plus product nucleus) after the removal to be more than the initial nucleus mass.

Given: ^{40}Ca mass = 39.962591 u   *Find:* Minimum energy needed to
^{39}Ca mass = 38.970719 u   remove a neutron from
^{40}Ar mass = 39.962383 u ^{40}Ca and ^{40}Ar
^{39}Ar mass = 38.964314 u
$^1_0$n mass = 1.008665 u

Essentially we are asked to compute the binding energy of the least bound neutron. For the ^{40}Ca case, the removal by gamma ray absorption can be written as

$$\gamma + ^{40}_{20}\text{Ca} \rightarrow ^{39}_{20}\text{Ca} + ^1_0\text{n}$$

The mass after breakup (at minimum energy, the neutron having almost no kinetic energy when it is released) can be found as follows:

$$^{39}_{20}\text{Ca} = 38.970719 \text{ u}$$
$$^1_0\text{n} = \underline{1.008665 \text{ u}}$$
$$\text{total} \quad 39.979384 \text{ u}$$

and

$$\Delta m = (^{39}\text{Ca} + ^1_0\text{n}) - ^{40}\text{Ca}$$
$$= 39.979384 \text{ u} - 39.962591 \text{ u} = 0.016793 \text{ u}$$

The binding energy for the neutron is then

$$E = (\Delta m)c^2$$
$$= (0.016793 \text{ u})(931.5 \text{ MeV/u}) = 15.64 \text{ MeV}$$

For the ^{40}Ar case, the removal by a gamma ray absorption can be written as

$$\gamma + ^{40}_{18}\text{Ar} \rightarrow ^{39}_{18}\text{Ar} + ^1_0\text{n}$$

The mass after breakup is

$$^{39}_{20}\text{Ar} = 38.964314 \text{ u}$$
$$^1_0\text{n} = \underline{1.008665 \text{ u}}$$
$$\text{total} \quad 39.972979 \text{ u}$$

and

$$\Delta m = (^{39}\text{Ar} + ^1_0\text{n}) - ^{40}\text{Ar}$$
$$= 39.972979 \text{ u} - 39.962383 \text{ u} = 0.010596 \text{ u}$$

The binding energy for the neutron is then

$$E = (\Delta m)c^2$$
$$= (0.010595 \text{ u})(931.5 \text{ MeV/u}) = 9.870 \text{ MeV}$$

Thus it takes about 6 MeV *more* energy to remove a paired neutron from a pair that fills a closed nuclear shell (a magic number) than if that pair does not complete a shell.

Follow-up Exercise. (a) Write down the reactions for gamma ray removal of a proton from ^{12}C and a neutron from ^{12}C and calculate the minimum energy required for each removal. (The necessary masses are given in Appendix V.) (b) In ^{12}C all the protons are paired, as are the neutrons. There is no magic number effect. Can you explain why removal of the proton takes less energy than removal of the neutron?

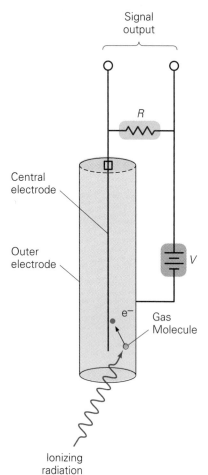

Signal output

R

Central electrode

Outer electrode

e⁻

Gas Molecule

Ionizing radiation

V

•**FIGURE 29.16   The Geiger counter**
A diagram showing the basic elements of a Geiger counter. See the text for a description.

Early detection was done with a scintillating phosphorescent screen that produced tiny flashes of light bright enough to be seen in the dark with the aid of a magnifying glass. Such a screen was used in the Rutherford scattering experiments with slow and tiring visual counting. It has been said that whenever Rutherford himself did the actual counting, he would "damn" vigorously for a few minutes and then have an assistant take over. More than a million scintillations or flashes were counted in the overall experiment.

29.5   Radiation Detection and Applications

Objectives: To (a) gain insight into the operating principles of various nuclear radiation detectors, (b) investigate the medical and biological effects of radiation exposure, and (c) study some of the practical uses and applications of radiation.

Detecting Radiation

Since the products of radioactive decay cannot be detected directly by our senses, detection must be done by some indirect means. We have already mentioned the detection of nuclear radiation by photographic film. People who work with radioactive materials and X-rays usually wear film badges, which indicate cumulative exposure to radiation by the degree of darkening of the film when it is developed. However, more immediate and quantitative methods of detection are desirable, and a variety of instruments have been developed for this purpose.

These **radiation detectors** are based on the ionization or excitation of atoms by the passage of energetic particles through matter. Alpha and beta particles are electrically charged particles, and they transfer energy to atoms along their paths by Coulomb electrical interactions, removing electrons and creating ions. Gamma rays although they have no electric charge, can also produce ionization. The gamma ray photons may remove electrons from atoms by the photoelectric effect or by Compton scattering (Section 27.3), which transfers energy from a photon to the electron it interacts with. Alternatively, they may *produce* electrons by pair-production (Section 28.5), depending on the energy of the photon. The energetic particles produced by these interactions are what is "detected" by a radiation detector.

One of the most common radiation detectors is the *Geiger counter*, which was developed chiefly by Hans Geiger, a student and then colleague of Lord Rutherford. The principle of the Geiger counter is illustrated in •Fig. 29.16. A voltage of 800–1000 V is applied across the wire electrode and metal tube of the Geiger tube. The tube contains a gas (such as argon) at low pressure. When an ionizing particle enters the tube through a thin window at one end, it ionizes a few atoms of the gas. The freed electrons are attracted and accelerated toward the positive wire anode. On their way, they strike and ionize other gas atoms. This process multiplies, and an "avalanche" discharge results that produces a current pulse. The pulse is amplified and sent to an electronic counter that counts the pulses or the number of particles detected. The pulses may also be used to drive a loudspeaker so that particle detection is heard as an audible click.

A major disadvantage of the Geiger counter is its relatively long "dead time." This is the recovery time required between successive electrical discharges in the tube (why?). The dead time of a Geiger tube is about 200 μs, which limits the counting rate to a few hundred counts per second. If particles are incident on the tube at a faster rate, not all of them will be counted.

Another method of detection, one of the oldest, has a much shorter dead time. This is used in the *scintillation counter* (•Fig. 29.17). In this counter, atoms of a phosphor material (such as sodium iodide, NaI) are excited by an incident particle, and visible light is emitted when the atoms return to their ground state. The light pulse is converted to an electrical pulse by a photoelectric material. This is amplified in a *photomultiplier tube*, which consists of a series of successively higher potential electrodes. The photoelectrons are accelerated toward the first electrode and acquire sufficient energy to cause several secondary electrons from ionization to be emitted when they strike the electrode. This process continues, and relatively weak scintillations are converted into sizable electrical pulses, which are counted electronically.

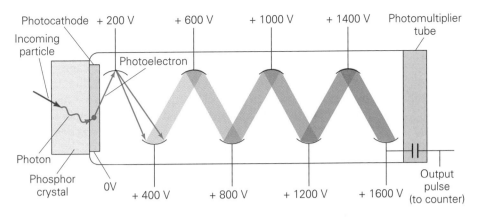

•FIGURE 29.17 The scintillation counter

A photon emitted by a phosphor atom excited by an incoming particle causes the emission of a photoelectron. Accelerated through a potential difference in a photo-multiplier tube, the photoelectrons free secondary electrons when they collide with successive electrodes at higher potentials. After several steps, a relatively weak scintillation is converted into a measurable electric pulse.

The dead time of a scintillation counter is a few microseconds. Also, the magnitude of the photomultiplier current pulse is proportional to the number of photons generated by the incident particle, which in turn is proportional to its energy. Thus, the magnitude of the counted pulses gives a measure of the energy of the incident particle. Counters with this feature are called *proportional counters*.

Semiconductor materials provide a relatively new method of radiation detection. In a *solid state* or *semiconductor detector*, charged particles passing through semiconductor material produce electron-hole pairs (solid state ionization, so to speak). With a voltage across the material, the electron-hole pairs give rise to electric signals, which may be amplified and counted. Solid state detectors are very fast—that is, they are capable of very high counting rates.

Counting methods are used to determine the number of particles produced by decaying nuclei of a radioactive isotope. Other methods of detection allow the tracks of charged particles to be seen or recorded visually. Those most commonly used are the cloud chamber, the bubble chamber, and the spark chamber. In the first two, vapors and liquids are supercooled and superheated, respectively, by suddenly varying the volume and pressure. The *cloud chamber* was developed early in this century by C.T.R. Wilson, a British atmospheric physicist. In the chamber, supercooled vapor condenses into droplets on ionized molecules along the path of an energetic particle. When the chamber is illuminated, the droplets scatter the light, making the path visible (•Fig. 29.18).

The *bubble chamber*, which was invented by the American physicist D.A. Glazer in 1952, uses a similar principle. A reduction in pressure causes a liquid to be superheated and able to boil. Ions produced along the path of an energetic particle become sites for bubble formation and a trail of bubbles is created. When a magnetic field is applied across the chamber, charged particles are deflected, and the energy of a particle can be calculated from the radius of curvature of the particle's path measured from a photograph. Since the bubble chamber uses a liquid, commonly liquid hydrogen, the density of atoms in it is much greater than in the vapor of a cloud chamber. Thus tracks are more readily observable, and bubble chambers have displaced cloud chambers.

Gamma rays do not leave visible tracks in a bubble chamber. However, their presence can be detected indirectly, for example, by electrons that have been Compton scattered and by the creation of electron–positron pairs.

The path of a charged particle is registered by a series of sparks in a *spark chamber*. Basically, a charged particle passes between a pair of electrodes with a high potential difference which are immersed in an inert (noble) gas. The charged particle causes the ionization of gas molecules, giving rise to a visible spark or flash between the electrodes as the released electrons travel to the positive electrode. A spark chamber is merely an array of such electrode pairs in the form of parallel plates or wires. A series of sparks, which can be photographed, marks the particle's path.

•FIGURE 29.18 Cloud chamber tracks

The circular track in this cloud chamber photograph was made by a positron in a strong magnetic field. (Can you explain the approximately circular path of the particle in terms of the orientation of the magnetic field relative to the positron velocity?)

Biological Effects and Medical Applications of Radiation

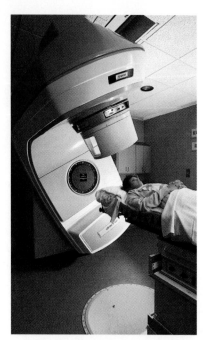

•FIGURE 29.19 Nuclear medicine A particle accelerator used in cancer treatment.

In medicine, nuclear radiation has both advantages and disadvantages. It can be used beneficially in the diagnosis and treatment of some diseases but can also be potentially harmful if it is not properly handled and administered. Nuclear radiation and X-rays can penetrate human tissue without pain or any other sensation. However, early investigators quickly learned that large doses or repeated small doses led to red skin, lesions, and other conditions.

The chief hazard of radiation is damage to living cells, primarily due to ionization. Ions, particularly complex ions or radicals produced by ionizing radiation, may be highly reactive (for example, a hydroxyl ion OH^- from water). These interfere with the normal chemical operations of the cell. If enough cells are damaged or killed by radiation, cell reproduction might not be fast enough, and the irradiated tissue could eventually die.

In other instances, there may be damage to a chromosome in the cell nucleus (genetic damage or mutation). If the affected cells are sperm or egg cells (or their precursors), any children that they produce may suffer from various birth defects. If the damaged cells are ordinary body cells, they may become cancerous, losing their normal form and reproducing in a rapid and uncontrolled manner. Such transformed cells grow at the expense of the surrounding tissues, producing a malignant tumor. Skin cancer and leukemia (cancer of the white blood cells, or leukocytes) often result from excessive radiation exposure. The human cells most susceptible to radiation damage are those of the reproductive organs, bone marrow, and lymph nodes.

Radiation can cause cancer, but it can also be used to treat cancer. Localized doses of radiation are used to destroy cancer cells. The radiation may be gamma rays from a radioactive ^{60}Co source or X-rays from a machine. Other particles produced in particle accelerators are also used to treat cancer (•Fig. 29.19). The electrical charge and energy of the particles determine the penetrating power of the radiation. X-rays and gamma rays are deeply penetrating, but, generally, beta particles penetrate only a few millimeters into biological tissue, and alpha particles penetrate only a fraction of a millimeter. Localization is achieved by using tightly focused beams. Also, radioactive compounds may be administered into the body that are preferentially absorbed by certain tissues or organs.

Radiation Dosage. An important consideration in radiation therapy and radiation safety is the amount, or *dose*, of radiation. Several quantities are used to describe this in terms of *exposure*, *absorbed dose*, and *equivalent dose*. The earliest unit of dosage, the **roentgen** (R), was based on exposure and was defined in terms of the ionization produced in air. (One roentgen is the quantity of X-rays or gamma rays required to produce an ionization charge of 2.58×10^{-4} C/kg of air.)

The roentgen has been largely replaced by the rad (*radiation absorbed dose*), which is an absorbed dose unit. One **rad** is an absorbed dose of radiation of 10^{-2} J of energy per kilogram in any absorbing material. The SI unit for absorbed dose is the **gray** (Gy):

$$1 \text{ Gy} \equiv 1 \text{ J/kg} \equiv 100 \text{ rad}$$

The gray unit was named in honor of Louis Harold Gray, a British radiobiologist whose studies laid the foundation for measuring absorbed dose.

These *physical* units give the energy absorbed per mass, but it is helpful to have some means for measuring the *biological damage* produced by radiation because equal doses of different types of radiation produce different effects. For example, a relatively massive alpha particle with a charge of $+2e$ moves through tissue rather slowly with a great deal of Coulomb interaction. The ionizing collisions thus occur close together along a short penetration path, and more localized damage is done than by a fast-moving electron or gamma ray.

This effect, or effective dose, is measured in terms of the **rem** unit (*r*oentgen or *r*ad *e*quivalent *man*). The different degrees of effectiveness of different particles are characterized by the **relative biological effectiveness (RBE)** or *quality factor* (QF), which has been tabulated for the various particles in Table 29.4. The RBE is defined in terms of the number of rads of X-rays or gamma rays that produces the same biological damage as one rad of a given radiation. (Note in Table 29.4 that gamma rays therefore have an RBE of 1.)

The effective dose is then given by the product of the dose in rads and the appropriate RBE.

$$\text{effective dose (in rem)} = \text{dose (in rad)} \times \text{RBE} \qquad (29.6)$$

Thus, one rem of any type of radiation does approximately the same amount of biological damage. For example, a 20-rem effective dose of alpha particles does the same damage as a 20-rem dose of X-rays, but 20 rad of X-rays are needed compared to 1 rad of alpha particles.

The SI unit of absorbed dose is the gray, and the effective dose with this unit is called the **sievert (Sv)**:

$$\text{effective dose (in Sv)} = \text{dose (in Gy)} \times \text{RBE} \qquad (29.7)$$

Since 1 Gy ≡ 100 rad, it follows that 1 Sv ≡ 100 rem. A summary of the radiation units is given in Table 29.5.

It is difficult to set a maximum permissible radiation dosage, but the general standard for humans is an average dose of 5 rem/year after the age of 18 with no more than 3 rem in any 3-month period. In the United States, the normal average annual dose per capita is about 200 mrem (millirem). About 125 mrem comes from the natural background of cosmic rays and naturally occurring radioactive isotopes in the soil, building materials, and so on. The remainder is chiefly from diagnostic medical applications, mostly X-rays (•Fig. 29.20), and from miscellaneous sources such as television tubes.

Diagnostic Applications. Radioactive isotopes offer an important technique for diagnostic procedures. For example, a radioactive isotope behaves chemically like a stable isotope of the element, and can participate in its normal chemical reac-

TABLE 29.4 Typical Relative Biological Effectiveness (RBE) of Radiations	
Type	*RBE (or QF)*
X-rays and gamma rays	1
Beta particles	1.2
Slow neutrons	4
Fast neutrons and protons	10
Alpha particles	20

TABLE 29.5 Radiation Units	
Unit	*Basis*
roentgen (R)	1 R–the quantity of X-rays or gamma rays that produces an ionization charge of 2.58×10^{-4} C/kg of air.
rad (*r*adiation *a*bsorbed *d*ose)	1 rad—an absorbed dose of radiation of 10^{-2} J/kg.
gray (Gy)	SI absorbed dose unit. 1 Gy = 1 J/kg = 100 rad.
rem (*r*ad *e*quivalent *man*) effective dose (in rem) = dose (in rad) × RBE	Effective dose. Relative effectiveness depends on type of radiation and is characterized by RBE (relative biological effectiveness). See Table 29.4.
sievert (Sv) effective dose (in Sv) = dose (in Gy) × RBE	SI unit of effective dose. 1 Sv = 100 rem.

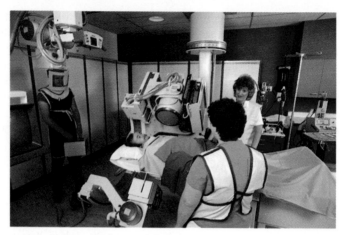

Though the radiation dosage represented by an individual diagnostic X-ray is generally relatively small, there is no threshold level for radiation exposure. That is, no exposure, no matter how low, can be guaranteed to be totally safe. Moreover, radiation effects are cumulative over time. For this reason, lead-lined aprons are used to protect parts of the patient's body that are not being photographed, as well as the technicians operating the X-ray equipment. It is especially important to shield the reproductive organs, since damage to reproductive cells could result in the production of children with birth defects, or even cause sterility.

(a)

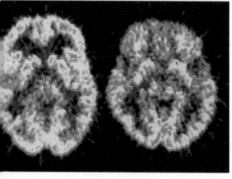

(b)

•FIGURE 29.21   PET
(a) A PET scanner monitors brain activity after administration of glucose containing radioactive isotopes. Note the array of detectors for the gamma radiation produced when an emitted positron is annihilated. (b) PET scans of a normal brain (left) and the brain of a schizophrenic patient (right).

tions. This makes it possible to label or tag molecules with radioisotopes, which can then be used as tracers.

By using tracers, many body functions can be studied simply by monitoring the location and activity of the labeled molecules as they are absorbed during body processes. For example, the activity of the thyroid gland in hormone production can be determined by monitoring its iodine uptake using radioactive iodine-123. This isotope emits gamma rays and has a half-life of 13.3 hours. Iodine is required by the body, and when ingested it collects in the thyroid gland near the throat. The uptake of radioactive iodine can be monitored by a detector to see if there is an abnormality.

Similarly, radioactive solutions of iodine and gold are quickly absorbed by the liver. This can be done safely because the isotopes have short half-lives. One of the most commonly used diagnostic tracers is technetium-99 (^{99}Tc). It has a convenient half-life of 6 hours and combines with a large variety of compounds. Detectors outside the body can scan over a region and record the activity so that a complete activity image may be reconstructed.

It is possible to image gamma ray activity in a single plane, or "slice," through the body. A gamma detector is moved around the patient to measure the emission intensity from many angles. A complete image can then be constructed by using computer-assisted tomography as in X-ray CT. This is referred to as single-photon emission tomography (SPET). Another technique, positron emission tomography (PET), uses tracers that are positron emitters such as ^{11}C and ^{15}O. When a positron is emitted, it is quickly annihilated, and two gamma rays are produced. Recall that the photons have equal energies and fly off in nearly opposite directions so as to conserve momentum (Section 28.5) The gamma rays are recorded in coincidence by a ring of detectors surrounding the patient (•Fig. 29.21)

Domestic and Industrial Applications of Radiation

The major application of radioactivity in the home is the smoke detector. In the smoke detector, a weak radioactive source ionizes the air molecules, setting up a small current in the detector circuit. If smoke enters the detector, the ions there become attached to the smoke particles. This causes a reduction in the current because the heavier, charged smoke particles move more slowly. The current difference is electronically sensed, and an alarm is triggered (•Fig. 29.22).

Industry also makes good use of radioactive isotopes. Radioactive tracers are used to determine flow rates in pipes, to detect leaks, and to study corrosion and wear. Also, it is possible to radioactivate certain compounds at a particular stage in a process by irradiating them with particles, generally neutrons. This technique is called **neutron activation analysis**, and is also an important method of identifying elements in a sample. Until its development, the chief methods of identification were chemical and spectral analyses. In both of these methods, a fairly large amount of a sample has to be destroyed in the analysis procedure. As a result, a sample may not be large enough for analysis, or small traces of elements in a sample may go undetected. Neutron activation analysis has the advantage over these methods on both scores: Only minute samples are needed, and the method can detect even tiny amounts of an element.

A neutron activation process is as follows:

$$\text{(source)} \quad {}^{252}_{98}\text{Cf} \rightarrow {}^{251}_{98}\text{Cf} + {}^{1}_{0}\text{n}$$

then

$$\underset{\substack{\text{(excited} \\ \text{nucleus)}}}{{}^{1}_{0}\text{n} + {}^{14}_{7}\text{N} \rightarrow {}^{15}_{7}\text{N}^*} \rightarrow \underset{\substack{\text{(gamma} \\ \text{ray)}}}{{}^{15}_{7}\text{N}} + \gamma$$

Here, neutrons from a californium-252 source bombard nitrogen-14, which absorbs a neutron and becomes an excited nitrogen-15 nucleus, which decays with the emission of a gamma ray with distinctive energy. Other neutron-activated nuclei may decay by other modes, such as beta decay.

The nitrogen activation process is used to screen for bombs in airport luggage. Virtually all explosives contain nitrogen. By using neutron activation and analyzing the energy and pattern of any gamma ray emission coming from a suitcase, the possibility of an explosive device being in the suitcase can be checked. Of course, other materials in the suitcase may contain nitrogen too, so manual checks are made to confirm any suspicious findings.

Finally, the federal government recently gave permission for the use of gamma radiation in the processing of poultry. The radiation kills bacteria and helps preserve the food and in no way makes the nuclei in the food radioactive. There are recent concerns with this process from food health professionals, because even though the gamma emission cannot make the poultry meat radioactive, it can change some of the *chemical* bonding through ionization effects. This has prompted concerns about whether these radiations can affect the chemical structure of the meat enough to warrant further study.

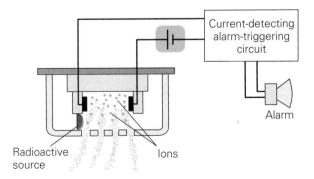

•FIGURE 29.22   Smoke detector In the common smoke detector, a weak radioactive source ionizes the air and sets up a small current. Smoke particles in the detector reduce the current, causing an alarm to sound.

Chapter Review

Important Terms

Rutherford–Bohr model   893
nucleon   895
strong nuclear force   895
atomic number (Z)   896
proton (atomic) number   896
mass number   896
neutron number   896
isotope   896
nuclide   896
radioactivity   897
alpha particle   897
beta particle   897
gamma ray   897
alpha decay   898

conservation of nucleons   898
conservation of charge   898
tunneling   899
barrier penetration   899
beta (β^-) decay   899
β^+ decay   900
electron capture   900
gamma decay   901
activity   903
decay constant   903
half-life   904
curie (Ci)   907
becquerel (Bq)   907
carbon-14 dating   908

pairing effect   912
atomic mass unit (u)   912
mass defect   913
total binding energy   913
magic number   916
radiation detectors   918
roentgen (R)   920
rad   920
gray (Gy)   920
rem   921
relative biological effectiveness
  (RBE)   921
sievert (Sv)   921
neutron activation analysis   923

Important Concepts

- The Rutherford–Bohr model of the atom is essentially the planetary model, with negative electrons orbiting the positively charged nucleus in which almost all the mass of the atom is located.

- The strong nuclear force is the short range attractive force between nucleons which is responsible for holding the nucleus together against the repulsive electric force of its protons.

- Isotopes of an element are nuclides that differ only in the number of neutrons they contain.

- Radioactivity refers to any type of nuclear instability that results in the emission of particles.

- Nuclei may undergo radioactive decay by the emission of an alpha particle, or helium nucleus (α de-

cay); a beta particle, or electron (β^- decay); a positron (β^+ decay); or a gamma ray, a high-energy photon of electromagnetic radiation (γ decay). Nuclei can also change by capturing one of the atom's orbital electrons (electron capture).

- The half-life of an isotope is the time required for the number of undecayed nuclei in a sample (and hence the radioactive decay rate or activity of the sample) to fall to one-half of its initial value.

- Total binding energy is the minimum amount of energy needed to pull a given nucleus apart into its constituent nucleons.

Important Equations

Decay Rate of a Radioisotope:

$$\frac{\Delta N}{\Delta t} = -\lambda N \qquad (29.2)$$

(1 Ci = 3.70×10^{10} decays/s = 3.70×10^{10} Bq)

Activity of a radioisotope:

$$|\Delta N/\Delta t| = \lambda N$$

Number of Undecayed Nuclei:

$$N = N_o e^{-\lambda t} \qquad (29.3)$$

Half-life and Decay Constant:

$$t_{1/2} = \frac{0.693}{\lambda} \qquad (29.4)$$

Total Binding Energy:

$$E_b = (\Delta m)c^2 \qquad (29.5)$$

Effective Dose:

$$\text{Dose (in rem)} = \text{dose (in rad)} \times \text{RBE} \qquad (29.6)$$
$$\text{Dose (in Sv)} = \text{dose (in Gy)} \times \text{RBE} \qquad (29.7)$$

Exercises

29.1 Nuclear Structure and the Nuclear Force

1 The concept of the atomic nucleus resulted from the work of (a) Bohr, (b) Thompson, (c) Rutherford, (d) Geiger.

2 Carbon-12 (a) has the same number of protons and neutrons, (b) is a nuclide, (c) is an isotope, (d) all of these.

3 The strong nuclear force (a) binds the orbital electrons to the atomic nucleus, (b) has a longer range than the electrostatic force, (c) acts only between identical particles, (d) overcomes the force of repulsion between protons in the nucleus.

4 ■ Calcium (atomic number 20) has 14 isotopes (some unstable but long-lived enough to investigate). Write the nuclear notations for the isotopes of calcium with mass numbers from 40 to 46. (The complete family of isotopes has mass numbers from 37 to 50.)

5 ■ Two isotopes of magnesium are ^{24}Mg and ^{25}Mg. What is the number of protons, neutrons, and electrons in each if (a) the atom is electrically neutral, (b) the atom has a −2 charge, and (c) the atom has a +1 charge?

6 ■ (a) Write the nuclear notations for the isotopes of hydrogen using the symbols H, D, and T. (b) Write the chemical symbols for six different forms of water using these symbols. (c) Identify the radioactive forms.

7 ■ For the isotopes of uranium (all unstable) that have mass numbers from 232 through 239, what are the numbers of protons, neutrons, and electrons in a neutral atom of each isotope?

8 ■ Oxygen has three stable isotopes with 8, 9, and 10 neutrons, respectively. Write the nuclear notations for these isotopes.

9 ■ An isotope of potassium has the same number of neutrons as the nuclide argon-40. Write the nuclear notation for the potassium isotope.

10 ■ Tin ($_{50}$Sn) has 23 isotopes that have mass numbers from 108 to 130 (ten of which are stable). Write the nuclear notations for the isotopes of tin with mass numbers from 115 to 120.

11 ■■ A theoretical result for the approximate nuclear radius (R) is $R = R_o A^{1/3}$, where $R_o = 1.2 \times 10^{-15}$ m and A is the mass number of the nucleus. (a) Find the nuclear radii of atoms of the noble gases: He, Ne, Ar, Kr, Xe, and Rn. (b) How does the order of magnitude of your answers compare to the approximate size of an atom, roughly 0.1 nm?

29.2 Radioactivity

12 The conservation of nucleons and the conservation of charge apply to (a) only alpha decay, (b) only beta decay, (c) only gamma decay, (d) all nuclear processes.

13 The neutron number is not conserved in (a) alpha decay, (b) beta decay, (c) gamma decay, (d) none of these.

14 ■ Write the nuclear equations expressing (a) the beta− decay of $^{60}_{27}$Co, and (b) the alpha decay of $^{226}_{88}$Ra.

15 ■ Write the nuclear equations for (a) the alpha decay of neptunium-237, (b) the β^- a decay of phosphorus-32, (c) the β^+ decay of cobalt-56, (d) electron capture in cobalt-56, and (e) the gamma decay of potassium-42.

16 ■ Tritium is radioactive. (a) Would you expect it to β^+ or β^- decay? (b) What is the daughter nucleus? Is it stable? [*Hint*: Write the nuclear equation for each.]

17 ■ Bismuth-210 can decay by alpha decay followed by beta decay, ending as lead-206. Write the nuclear equations for these sequential decay processes.

18 ■ Radon-222 ($^{222}_{86}$Rn), a radioactive gas, undergoes alpha decay. (a) What is the daughter nucleus? Write the nuclear equation for the decay process. (b) The daughter nucleus decays by either alpha decay or beta decay. What is the granddaughter nucleus for each of these processes?

19 ■ A lead-209 nucleus results from both alpha–beta sequential decays and beta–alpha sequential decays. What was the grandparent nucleus? (Show this for both decay routes by writing the nuclear equations for these decay processes.)

20 ■■ Complete the following nuclear decay equations:
(a) $^{8}_{4}$Be → $^{4}_{2}$He + _____
(b) $^{240}_{94}$Po → $^{97}_{38}$Sr + $^{139}_{56}$Ba + _____
(c) $^{47}_{21}$Sc* → $^{47}_{21}$Sc + _____
(d) $^{29}_{11}$Na → $^{0}_{-1}$e + _____

21 ■■ Complete the following nuclear equations:
(a) $^{238}_{92}$U → $^{234}_{90}$Th + _____
(b) $^{40}_{19}$K → $^{40}_{20}$Ca + _____
(c) $^{236}_{92}$U → $^{131}_{53}$I + 3^{1}_{0}n + _____
(d) $^{23}_{11}$Na + γ → _____
(e) $^{22}_{11}$Na + _____ → $^{22}_{10}$Ne

22 ■■■ Actinium-227 ($^{227}_{89}$Ac) decays by alpha decay or by beta decay and is part of a decay sequence like the one in Fig. 29.9. Each daughter nucleus then decays to radium-223 ($^{223}_{88}$Ra) Which subsequently decays to polonium-215 ($^{215}_{84}$Po). Write the nuclear equations for the decay process in the decay series from ^{227}Ac to ^{215}Po.

23 ■■■ The decay series for neptunium-237 is shown in •Fig. 29.23. (a) Identify the decay modes and each of the nuclei in the sequence. (b) Tell why each nucleus is likely to decay.

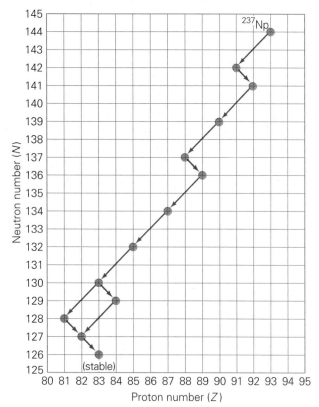

•FIGURE 29.23 Neptunium-237 decay series
See Exercise 23.

29.3 Decay Rate and Half-Life

24 After one half-life, a sample of a particular radioactive material (a) is half as massive, (b) has its half-life reduced by one half, (c) is no longer radioactive, (d) has its activity reduced by one half.

25 In three half-lives, the activity of a radioactive sample will have *decreased* by what percent? (a) 33.3%, (b) 50.0%, (c) 66.7%, (d) 87.5%.

26 What physical or chemical properties affect the decay rate or half-life of a radioactive isotope?

27 One radioactive nuclide has a decay constant that is half that of another. If they start with the same number of undecayed nuclei will twice as many of its nuclei decay in a given time? Explain.

28 What is the (a) half-life and (b) decay constant for a stable isotope?

29 ■ A particular radioactive sample is observed to undergo 1.25×10^6 decays/s. What is the activity of the sample in (a) curies and (b) becquerels?

30 ■ At present, a laboratory radioactive beta source has an activity of 40 mCi. (a) What is the present decay rate in decays per second? (b) Assuming that one beta particle is emitted per decay, how many are emitted per minute?

31 ■■ The half-life of a radioactive isotope is 1 h. What fraction of a sample of this isotope would be left after (a) 2 h and (b) 1 day?

32 ■■ A sample of technetium-104 has an activity of 10 mCi. Estimate the activity of the sample after one hour ($t_{1/2} = 18$ min).

33 ■■ What period of time is required for a sample of radioactive tritium ($^{3}_{1}$H) to lose 80.0% of its activity? Tritium has a half-life of 12.3 years.

34 ■■ Which has the greater decay constant, $^{28}_{12}$Mg or $^{104}_{32}$Tc? By what factor is it greater?

35 ■■ If an amount of ^{123}I is introduced into a patient in a medical diagnostic procedure, what percentage of the sample remains after 2.00 hours if all of the ^{123}I is retained in the patient's thyroid gland (answer to three significant figures)?

36 ■■ A sample of old bone is found to have 4.0 beta decays/min for each gram of carbon. Approximately how old is the bone?

37 ■■ Show that the number N of radioactive nuclei remaining in a sample after n half-lives is given by

$$N = \frac{N_{o}}{2^{n}} = \left(\frac{1}{2}\right)^{n} N_{o}$$

where N_{o} is the initial number of nuclei.

38 ■■ Some ancient writings on parchment are found sealed in a jar in a cave. If carbon-14 dating shows the parchment to be 28,650 years old, what percentage of the carbon-14 atoms still remain in the sample compared to the number when the parchment was made?

39 ■■ The activity of an unknown beta source is observed to decrease by 87.5% in 54 min. What is the radioactive isotope?

40 ■■ For a sample of cobalt-60, how long would it take (to the nearest whole year) for its activity to reduce to 10% of the original activity?

41 ■■ A soil sample contains 40 μg of ^{90}Sr. Approximately how much ^{90}Sr will be in the sample 100 years from now?

42 ■■ (a) What is the decay constant of fluorine-17? (b) How long will it take for the activity of a sample of ^{17}F to decrease to 12.5 percent of its initial value ($t_{1/2} = 66.0$ s)?

43 ■■ Francium-223 ($^{223}_{87}$Fr) has a half-life of 21.8 min. (a) If there is initially a 25.0-mg sample of this isotope, how many nuclei are present? (b) How many nuclei will be present 1 h and 49 min later?

44 ■■■ A basement room containing radon gas is sealed air-tight. If there are 7.50×10^{10} radon ($t_{1/2} = 3.82$ days) atoms trapped in the room at the time, (a) estimate how many radon atoms would remain in the room at the end of 30.0 days. (b) Radon undergoes alpha decay. Is there the same number of daughter nuclei as decayed parent nuclei at the end of 30 days? Explain.

45 ■■■ An old artifact is found to contain 250 g of carbon and has an activity of 475 decays per minute. What is the approximate age of the artifact to the nearest thousand years?

46 ■■■ The recoverable reserves of high-grade uranium-238 ore (high-grade ore contains about 10 kg of ^{238}U$_3$O$_8$ per ton) in the United States are estimated to be about 500,000 tons. Neglecting any geological changes, what mass of ^{238}U existed in this high-grade ore when the Earth was formed about 4.6 billion years ago? [*Hint:* See Appendix V.]

47 ■■■ In 1898, Pierre and Marie Curie isolated about 10 mg of radium-226 from 8 tons of uranium ore. If this sample had been placed in a museum, how much of the radium would remain in the year 2000?

48 ■■■ Nitrogen-13, with a half-life of 10 minutes, decays with the emission of positrons. (a) Write the nuclear equation for this decay process. (b) If a pure sample of ^{13}N has a mass of 0.0015 kg, what is the activity in 35 minutes? (c) What percentage of the sample is ^{13}N at this time?

29.4 Nuclear Stability and Binding Energy

49 For nuclei with mass number greater than 40, which of the following answers is correct? (a) The number of protons is approximately equal to the number of neutrons, (b) the number of protons exceeds the number of neutrons, (c) all are stable up to $z = 92$, (d) none of these.

50 The average binding energy per nucleon of the daughter nucleus in a decay process is (a) greater than, (b) less than, (c) equal to that of the parent nucleus.

51 ■ Determine whether each of the three isotopes of hydrogen is likely to be stable or unstable.

52 ■ Using Appendix V, list those stable nuclei that are "doubly-magic." That is, make a list of what are likely to have magic numbers for *both* neutrons and protons.

53 ■ From which one of the following pairs of nuclei would you expect it to be easier to remove a neutron? State your reasoning for your choice in each case. (a) $^{16}_{8}$O or $^{17}_{8}$O, (b) $^{40}_{20}$Ca or $^{42}_{20}$Ca, (c) $^{10}_{5}$B or $^{11}_{5}$B, (d) $^{208}_{82}$Pb or $^{209}_{83}$Bi.

54 ■ Write in nuclear notation the most abundant stable isotope of the nuclide with a proton magic number of 28. (See Appendix V.)

55 ■ Only two isotopes of Sb (antimony, $Z = 51$) are stable. Pick the two stable isotopes from the following: (a) ^{120}Sb, (b) ^{121}Sb, (c) ^{122}Sb, (d) ^{123}Sb, (e) ^{124}Sb.

56 ■ Determine whether the following isotopes are likely to be unstable: (a) ^{23}Na, (b) ^{50}V, (c) ^{209}Bi, (d) ^{209}Po, (e) ^{44}Ca.

57 ■ The total binding energy of $^{2}_{1}$H is 2.224 MeV. Use this to compute the mass of a ^{2}H nucleus from the known mass of a proton and a neutron.

58 ■■ Use Avogadro's number (see Section 10.3), $N_A = 6.02 \times 10^{23}$ atoms/mole, to show that 1 u = 1.66×10^{-27} kg. (Recall a ^{12}C atom has a mass of exactly 12 u.)

59 ■■ The mass of $^{3}_{2}$He is 3.016029 u. (a) What is the total binding energy of this nucleus? (b) What is the average binding energy per nucleon?

60 ■■ The mass of $^{9}_{4}$Be is 9.012183 u. What is E_b/A for this nucleus?

61 ■■ Which isotope of hydrogen has the greatest average binding energy per nucleon, deuterium or tritium? (Justify your answer mathematically.)

62 ■■ Near high neutron areas, say a nuclear reactor, neutrons will be absorbed by protons (hydrogen nuclei in water molecules) and give off a characteristic energy gamma ray in the process. What is the energy of the gamma ray (to three significant figures)?

63 ■■ How much energy (to four significant figures) would be required to completely separate all the nucleons of an oxygen-16 nucleus, the atom of which has a mass of 15.994915 u?

64 ■■ Determine the average binding energy per nucleon of sulfur-32 ($m = 31.972072$ u).

65 ■■ If an alpha particle could be removed intact from an aluminum-27 nucleus ($m = 26.981541$ u), a sodium-23 nucleus ($m = 22.989770$ u) would remain. How much energy would be required to do this?

66 ■■ On the average, are nucleons more tightly bound in an ^{27}Al nucleus or in a ^{23}Na nucleus?

67 ■■ The atomic mass of $^{238}_{92}$U is 238.050786 u. Find the average binding energy per nucleon for this isotope.

68 ■■■ The total binding energy of $^6_3$Li is 32.0 MeV. What is the mass (in u) of ^{6}Li nucleus?

69 ■■■ The mass of $^8_4$Be is 8.005305 u. (a) Which is less, the total mass of two alpha particles or the mass of the ^{8}Be nucleus? (b) Which is greater, the total binding energy of ^{8}Be nucleus or the total binding energy of two alpha particles? (c) On this basis alone do you expect the ^{8}Be nucleus to decay spontaneously into two alpha particles?

29.5 Radiation Detection and Applications

70 The detector with the greatest dead time is the (a) Geiger counter, (b) scintillation counter, (c) solid state detector, (d) all have the same dead time.

71 A unit of radiation dosage is the (a) rad, (b) gray, (c) sievert, (d) all of these.

72 A basic assumption of radiocarbon dating is that the cosmic ray intensity has been generally constant for the last 40,000 years or so. Suppose it were found that the intensity was much less 10,000 years ago than it is today. How would this affect carbon-14 dating?

73 A patient receives a 2.0-rad dose of radiation from X-rays and a 2.0 rad dose from an alpha source. (a) Which has the greater biological effectiveness? (b) What are the doses in rems?

74 ■ A technician working at a nuclear reactor facility is exposed to a slow neutron radiation and receives a dose of 1.25 rad. (a) How much energy is absorbed by 200 g of the worker's tissue? (b) Was the maximum permissible radiation dosage exceeded?

75 ■ A person working with nuclear isotopes for a 2-month period receives a 0.5-rad dose from a gamma source, a 0.3-rad dose from a slow-neutron source, and a 0.1-rad dose from an alpha source. Was the maximum permissible radiation dosage exceeded?

76 ■ In a diagnostic procedure, a patient in a hospital ingests 80 mCi of gold-198 ($t_{1/2}$ = 2.7 days). What is the activity at the end of two weeks if none is eliminated from the body by biological functions?

77 ■■ Neutron activation analysis was performed on small pieces of hair that had been taken from the exiled Napoleon after he died on the island of St. Helena in 1821. The samples were found to contain abnormally high levels of arsenic, which supported a theory that his death was not due to natural causes. If this evidence was derived through beta emissions coming from arsenic-76 nuclei, what was the arsenic isotope present *in the hair* and what was the final nucleus after the beta decay?

Additional Exercises

78 A fossil specimen is believed to be about 18,000 years old. Explain how this could be confirmed.

79 The binding energy per nucleon for $^4_2$H is 7.075 MeV. (a) What is the total binding energy? (b) What is the mass of a ^{4}He nucleus?

80 Starting with uranium-234 ($^{234}_{92}$U), there is a decay sequence of four alpha decays and two beta minus decays. What is the resulting nucleus of the sequence?

81 A sample of ^{215}Bi, which beta decays ($t_{1/2}$ = 2.4 min), contains Avogadro's number of nuclei. How many bismuth nuclei are present after (a) 10 min and (b) 1.0 h? (c) What are the activities in curies and becquerels at these times, and is this a realistic laboratory radioactive source?

82 An old bone yields, on the average, 1.00 beta emission per minute per gram of carbon. About how old is the bone?

83 Complete the following nuclear decay equations:

(a) $^{47}_{21}$Sc → $^{47}_{22}$Ti + _____

(b) $^{226}_{88}$Ra → $^4_2$He + _____

(c) $^{237}_{93}$Np → $^0_{-1}$e + _____

(d) $^{210}_{84}$Po* → $^{210}_{84}$Po + _____

(e) $^{11}_6$C → $^{11}_5$B + _____

84 •Figure 29.24 shows the decay series for plutonium-239. (a) Identify each of the nuclei in this decay process. (b) Compare the totals of the half-lives before and after the ^{227}Ac nuclei.

85 The value of E_b/A for $^{238}_{92}$U is 7.58 MeV. (a) What is the total binding energy of this nucleus? (b) What is its mass?

86 What are the equivalent rest energies of (a) a carbon-12 atom and (b) an alpha particle?

87 Determine which of the following isotopes are likely to be stable (a) ^{20}C, (b) ^{100}Sn, (c) ^{6}Li, (d) ^{9}Be, (e) ^{22}Na.

88 Thorium-229 ($^{229}_{90}$Th) undergoes alpha decay. (a) What is the daughter nucleus of this process? Write the decay equation. (b) This daughter nucleus undergoes beta decay. What is the resulting granddaughter nucleus? Write the decay equation.

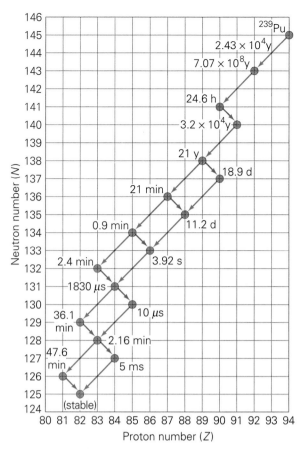

FIGURE 29.24 Plutonium-239 decay series
See Exercise 84.

89 Bismuth-214 ($^{214}_{83}$Bi) decays by alpha decay or beta decay. Identify the daughter nucleus in each case by writing the decay equations.

90 Gold-198 beta decays with a half-life of 2.7 days. (a) What is the activity of a sample of 1.0 g of pure $^{198}_{79}$Au? (b) What is the activity of the sample of the end of 1 week?

91 Using the nuclear radius in Exercise 11, [$R = R_o A^{1/3}$, where $R_o = 1.2 \times 10^{-15}$ m and A is the mass number of the nucleus], and assuming nuclei are spherical (they are approximately so in most cases), (a) show that the average nucleon density in a nucleus is 1.4×10^{44} nucleons/m^3 and (b) estimate the nuclear density in kilograms per cubic meter.

92 A handy unit for nuclear dimensions is the fermi (fm), named after the famous Italian–American physicist Enrico Fermi. The conversion factor between the fermi and the meter is 1 fm = 10^{-15} m. [Recently the trend is to call the fermi by its appropriate SI prefix, the femtometer, but in either case, the abbreviation is fm.] Referring to Exercise 91, what is (a) the average number of nucleons per cubic *fermi* and (b) the mass density in kilograms per cubic *fermi*?

93 (a) Compare the average *nuclear* density results of Exercise 91 to the average *atomic* density of, say, a hydrogen atom with a radius of 0.0529 nm assuming the atom to be spherical. (b) Why are they so different if the nuclear mass and atomic mass are essentially the same? (c) Does your number for hydrogen agree with the density of hydrogen gas at normal temperatures and pressures?

94 $^7_4$Be is unstable with a half-life of about 53.3 days, yet we find it on harvested vegetables and lying on the soil because it is produced when cosmic rays split nitrogen atoms in the upper atmosphere. The $^7_4$Be thus "floats" down to sea level depending on atmospheric conditions (rain, wind, etc.). (a) What are the decay possibilities for $^7_4$Be? (b) Look up the actual decay mechanism in Appendix V and write it out in equation form. (c) What is the daughter nucleus and is it stable?

30 Nuclear Reactions and Elementary Particles

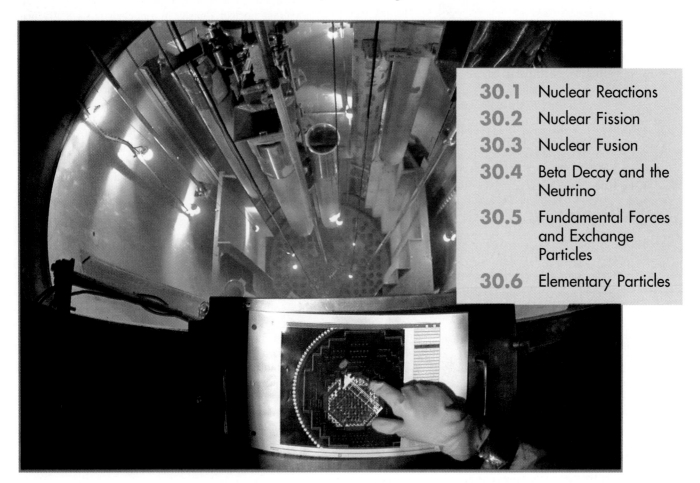

30.1 Nuclear Reactions

30.2 Nuclear Fission

30.3 Nuclear Fusion

30.4 Beta Decay and the Neutrino

30.5 Fundamental Forces and Exchange Particles

30.6 Elementary Particles

Today, more than half a century after the first nuclear reactor was built, these devices remain the center of intense controversy. Some people see them as playing an essential part in solving the world's energy problems. To others, they embody all that is wrong and dangerous about modern technology.

In fact, the reactor shown above is a perfect symbol of the ongoing debate about the safety of nuclear power generation. It is located at the Three Mile Island plant in Pennsylvania—the site of one of the largest and most widely publicized nuclear accidents to date. In this photo, you are looking down into the heart of the reactor itself. (The screen in the foreground is a touch-sensitive computer display.) The eerie glow illuminating

the shielding water in the containment structure results in part from Cerenkov radiation, produced by subatomic particles traveling faster than the speed of light in a transparent medium. You know from Chapter 26 that nothing travels faster than c, the speed of light *in vacuum;* but in a transparent medium the speed of light is less than c, and so can be exceeded by the highly energetic products of nuclear reactions.

In this chapter, you'll learn more about the nuclear fission reactions that occur in reactors here on Earth. You'll also learn how scientists are trying to harness even more powerful fusion reactions, which occur within the Sun and other stars, and could some day provide a safe and virtually inexhaustible supply of clean energy for our planet.

Thus far, we have considered radioactive processes in which the nucleus of one element decays into the nucleus of another with the emission of energetic particles. As might be expected, scientists wondered whether the reverse was possible. That is, could the nucleus of one element be bombarded with energetic particles and thus be converted into the nucleus of another element with the absorption of a particle—reverse radioactivity, so to speak?

Indeed, such processes are possible, and physicists have learned how to manipulate nuclear particles to create one isotope from another by nuclear reactions. Such reactions have made available the many artificial radioactive isotopes that do not occur naturally.

Knowledge of nuclear reactions has also made it possible to release energy from the nucleus. The many atomic and hydrogen bombs that have been produced since 1945 are stark reminders of the devastating power that can be released from the nucleus. On the more peaceful side, nuclear reactors are now an important source of energy for generating electricity.

In this chapter, we will look at these topics and at elementary particles. Investigations have shown that a variety of particles other than the proton and neutron are associated with the nucleus. The discovery of these particles has given some insight into the still puzzling world of the nucleus and what holds it together.

30.1 Nuclear Reactions

Objectives: To be able to (a) write nuclear reaction equations using charge and nucleon conservation, and (b) understand the concepts of Q value and threshold energy and utilize them in the analysis of nuclear reactions.

In chemical reactions, substances react with each other to form compounds. Since an ordinary chemical reaction involves only the atomic electrons, the atoms of the reactants do not lose their nuclear identities in the process. On the other hand, in **nuclear reactions**, nuclei are converted into the nuclei of other isotopes, which in general are nuclides of completely different elements. From the study of the nucleus, scientists have learned to initiate nuclear reactions by bombarding nuclei with energetic particles.

The first induced nuclear reaction was produced by Lord Rutherford in 1919. Nitrogen was bombarded with alpha particles from a natural source (bismuth-214). The occasional particles that came from the reactions were identified as protons. Rutherford reasoned that an alpha particle colliding with a nitrogen nucleus must sometimes induce a reaction that produces a proton. As a result, the nitrogen nucleus is *artificially* transmuted into an oxygen nucleus:

$$\underset{\substack{nitrogen \\ (14.003074\ u)}}{^{14}_{7}\text{N}} \quad + \quad \underset{\substack{alpha\ particle \\ (4.002603\ u)}}{^{4}_{2}\text{He}} \quad \rightarrow \quad \underset{\substack{oxygen \\ (16.999133\ u)}}{^{17}_{8}\text{O}} \quad + \quad \underset{\substack{proton \\ (1.007825\ u)}}{^{1}_{1}\text{H}}$$

where the atomic masses are given for a later purpose.

It turns out that this type of reaction and many others like it actually form a temporary intermediate "compound" nucleus in a highly excited state. For example, the preceding reaction can be written more precisely as

$$^{14}_{7}\text{N} + ^{4}_{2}\text{He} \rightarrow (^{18}_{9}\text{F}^*) \rightarrow ^{17}_{8}\text{O} + ^{1}_{1}\text{H}$$

The fluorine nucleus, $^{18}_{9}\text{F}$, formed in an excited state, is indicated by the usual asterisk. A compound nucleus typically rids itself of excess energy by ejecting a particle or particles. If a compound nucleus does occur, it only lasts a very short time. Hence the compound nucleus is commonly omitted from the equation for the nuclear reaction.

Think of the implication of Rutherford's discovery. One element had been converted into another! This was the age-old dream of the alchemists, although

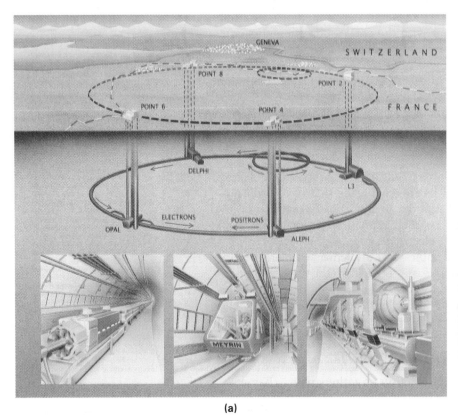

(a)

(b)

•**FIGURE 30.1   Particle accelerators**
(a) The Large Electron-Positron (LEP) Collider on the border between France and Switzerland, near Geneva, has a circumference of 26.7 km. The electrons and positrons that it accelerates in opposite directions to speeds approaching that of light collide with an energy of over 100 billion electron volts (100 GeV). **(b)** A 4500-ton particle detector at the Fermi National Accelerator Laboratory in Batavia, Illinois. Such devices are used to measure the trajectories and energies of subatomic particles, created in the nuclear reactions that occur when a beam of accelerated particles smashes into a target.

their main concern was changing common metals, such as mercury and lead, into gold. Even this transmutation can be done today through nuclear reactions. Large machines called **particle accelerators** use electric and magnetic fields to accelerate charged particles to very high energies (•Fig. 30.1). When the particles strike target nuclei, they initiate nuclear reactions. Different reactions require different particle energies. One nuclear reaction that is initiated when protons strike nuclei of mercury is

$$\underset{\substack{mercury \\ (199.968321\ u)}}{^{200}_{80}\text{Hg}} \quad + \quad \underset{\substack{proton \\ (1.007825\ u)}}{^{1}_{1}\text{H}} \quad \rightarrow \quad \underset{\substack{gold \\ (196.96656\ u)}}{^{197}_{79}\text{Au}} \quad + \quad \underset{\substack{alpha\ particle \\ (4.002603\ u)}}{^{4}_{2}\text{He}}$$

In this reaction, mercury is converted into gold, so it would seem that modern physics has fulfilled the alchemists' dream. However, making such small amounts of gold in an accelerator costs far more than the gold is worth.

Reactions like those written above have the general form

$$A + a \rightarrow B + b$$

where the uppercase letters represent the nuclei and the lowercase letters represent the particles. Such reactions are often written in a shorthand notation like this:

$$A(a, b)B$$

For example, in this form, the two previous reactions are written as

$$^{14}\text{N}(\alpha, p)\ ^{17}\text{O} \quad \text{and} \quad ^{200}\text{Hg}(p, \alpha)^{197}\text{Au}$$

In general, the periodic table lists over 100 elements, but only 90 elements occur naturally on Earth. Elements with proton numbers greater than uranium ($Z = 92$), as well as technetium ($Z = 43$) and promethium ($Z = 61$), are created artificially by nuclear reactions. The name technetium comes from the Greek word *technetos*, meaning "artificial," and technetium was the first unknown element to be created by artificial means. Elements up to $Z = 109$ have been confirmed.

There are some nuclear theorists who believe that there may be an "island of stability" at the next highest magic numbers (beyond 126) which might lead to a superheavy but stable element consisting of a nucleus with over 300 nucleons! To find such nuclei is the aim of massive "colliding beam" reaction experiments. Here high energy nuclei from the heavy end of the periodic chart are slammed into one another in hopes of leaping over the unstable nuclei we know about (just above $Z = 82$) and reaching the next magic number plateau.

Conservation of Mass–Energy and the Q Value

In every nuclear reaction, the total relativistic energy is conserved ($E = K + m_o c^2$, as shown in Chapter 26). Consider the reaction by which nitrogen is converted into oxygen, $^{14}N(\alpha, p)^{17}O$. By the conservation of total relativistic energy,

$$(K_N + m_N c^2) + (K_\alpha + m_\alpha c^2) = (K_O + m_O c^2) + (K_p + m_p c^2)$$

where the subscripts refer to the kinetic energy and mass of a particular particle and the respective masses are rest masses. Rearranging the equation, we have

$$K_O + K_p - K_N - K_\alpha = (m_N + m_\alpha - m_O - m_p)c^2 \tag{30.1}$$

The **Q value** of the reaction is defined as

$$Q = \Delta K = (K_O + K_p) - (K_N + K_\alpha) \tag{30.2}$$

Q value: energy released or absorbed in a reaction

The Q value is a measure of the total energy released or absorbed in a reaction, that is, the kinetic energy of the products of the reaction minus the kinetic energy of the reactants of the reaction. In terms of final and initial kinetic energies, we see that $Q = K_f - K_i$. The Q value is an important quantity in nuclear reactions. However, the kinetic energies of the reactants and products of the reaction do not have to be measured, since by Eqs. 30.1 and 30.2,

$$Q = (m_N + m_\alpha - m_O - m_p)c^2 \tag{30.3}$$

Similarly, in terms of a general reaction of the form A + a → B + b, or A(a, b) B,

$$Q = (m_A + m_a - m_B - m_b)c^2 = (\Delta m)c^2 \tag{30.4}$$

Thus, the Q value is given by the difference in the rest energies of the reactants and the products of a reaction. Note that this means that mass energy is generally converted into kinetic energy or vice versa. The amount of mass involved in these transformations, Δm, can be positive or negative. If mass is gained during the reaction, note the Q is negative; if mass is lost, Q is positive. In the former case, we would expect a net conversion of kinetic energy into mass, and in the latter, conversion of mass into extra kinetic energy. Clearly if Q is negative, the reaction *requires* a certain amount of kinetic energy before the mass of the products can be attained, that is, before there is enough kinetic energy to make the more massive product nuclei. To see these ideas more clearly, let us look at Lord Rutherford's original reaction in some detail.

Using the masses given in the parentheses under the reactants and products in the previous equation for $^{14}N(\alpha,p)^{17}O$ reaction, we obtain

$$Q = (m_N + m_\alpha - m_O - m_p)c^2$$

$$= [(14.003074 \text{ u} + 4.002603 \text{ u}) - (16.999133 \text{ u} + 1.007825 \text{ u})]c^2$$

$$= (-0.001281 \text{ u})c^2$$

or

$$Q = (-0.001281 \text{ u})(931.5 \text{ MeV/u}) = -1.193 \text{ MeV}$$

The negative Q value indicates that energy was absorbed in the reaction. When Q is less than zero (negative), the reaction is said to be **endoergic** (or endother-

Endoergic: energy in

mic). That is, energy (*ergic*) must be put into (*endo*) the reaction for it to proceed. In endoergic reactions, the kinetic energy of the reacting particles is at least partially converted into mass.

When the Q value of a reaction is positive ($Q > 0$), energy is released, and the reaction is said to be **exoergic** (or exothermic). That is, energy is produced by (*exo*) the reaction. In this case, mass is converted into energy and is carried away as kinetic energy of the products of the reaction.

Exoergic: energy out

EXAMPLE 30.1 ■ A POSSIBLE ENERGY SOURCE? Q VALUE OF A REACTION

Is the following reaction endoergic or exoergic?

$$\underset{\substack{\text{deuteron} \\ (2.014102\ u)}}{^{2}_{1}\text{H}} \quad + \quad \underset{\substack{\text{deuteron} \\ (2.014102\ u)}}{^{2}_{1}\text{H}} \quad \rightarrow \quad \underset{\substack{\text{helium} \\ (3.016029\ u)}}{^{3}_{2}\text{He}} \quad + \quad \underset{\substack{\text{neutron} \\ (1.008665\ u)}}{^{1}_{0}\text{n}}$$

Solution. The nature of the reaction is determined by the Q value: endoergic, $Q < 0$; exoergic, $Q > 0$. The Q value is given by Eq. 30.4, so we first calculate the mass difference (Δm) of the reaction using the given masses:

$$\Delta m = 2m_{\text{D}} - m_{\text{He}} - m_{\text{n}}$$

$$= 2(2.014102\ \text{u}) - 3.016029\ \text{u} - 1.008665\ \text{u} = 0.00351$$

Since the mass difference is positive, total mass has been lost and kinetic energy has been gained. Thus the reaction is exoergic. The value for Q can be obtained from the relationship $Q = (\Delta m)c^2$ or, more simply,

$$Q = (0.00351\ \text{u})(931.5\ \text{MeV/u}) = +3.27\ \text{MeV}$$

Follow-up Exercise. Show that the following reaction is endoergic and calculate its Q-value to four significant figures:

$$\underset{\substack{\text{carbon} \\ (12.000000\ u)}}{^{12}_{6}\text{C}} \quad + \quad \underset{\substack{\text{helium} \\ (4.002603\ u)}}{^{4}_{2}\text{He}} \quad \rightarrow \quad \underset{\substack{\text{helium} \\ (3.016029\ u)}}{^{3}_{2}\text{He}} \quad + \quad \underset{\substack{\text{carbon} \\ (13.003355\ u)}}{^{13}_{6}\text{C}}$$

PROBLEM-SOLVING HINT

Note that in the preceding Example the Q value was computed directly from the mass difference using the mass-energy conversion factor derived in Chapter 29 (see Table 29.3). This method eliminates using c^2 numerically and gives the energy in MeV.

The Q value of radioactive decay is always positive ($Q > 0$) because energetic particles are emitted spontaneously. This is sometimes called the *disintegration energy*. The meanings of the positive and negative Q values are summarized in Table 30.1.

When the Q value of a reaction is negative, or the mass of the products is greater than the mass of the reactants, then a certain amount of kinetic energy is converted into rest mass. Such reactions will not occur unless enough kinetic energy is initially available. One might think that if a particle incident on a station-

---

TABLE 30.1   Interpretation of Q Values

Q value	Effect
Positive ($Q > 0$)	Exoergic, mass converted into energy (mass of reactants greater than mass of products)
Negative ($Q < 0$)	Endoergic, energy converted into mass (mass of products greater than mass of reactants)

---

ary nucleus had kinetic energy equal in value to the Q of the reaction ($K = |Q|$),* then the reaction would occur, since this energy is equivalent to the mass gain. However, if all the kinetic energy were converted to mass, there would be none left over, so the particles would be at rest after the reaction (all $v = 0$). But this means that momentum could not be conserved, as it must be: the incoming particle had momentum, but the products wouldn't.

Hence, for the products of the reaction to have the same total momentum as the incident particle, the kinetic energy of the incident particle must be *greater* than $|Q|$. Just how much greater depends on the particular reaction, as we see next.

The minimum kinetic energy (K_{min}) that an incident particle needs in order to have to initiate an endoergic reaction is called the **threshold energy**. For classical (nonrelativistic) collisions, the threshold energy can be shown to be

$$K_{min} = \left(1 + \frac{m_a}{M_A}\right)|Q| \qquad \begin{array}{l}\text{(stationary}\\ \text{target only)}\end{array} \qquad (30.5)$$

Threshold energy for an endoergic reaction

where m_a and M_A are the masses of the incident particle and the *stationary* target nucleus, respectively. Since the expression in parentheses in this equation must be greater than 1, it follows that $K_{min} > |Q|$.

EXAMPLE 30.2 ■ NITROGEN INTO OXYGEN: THRESHOLD ENERGY

What is the threshold energy for the reaction $^{14}\text{N}(\alpha, p)^{17}\text{O}$?

Solution. The masses and the Q value for this reaction were given previously; thus we have

Given: $m_a = m_\alpha = 4.002603$ u   **Find:** K_{min} (threshold energy)
$M_A = m_N = 14.003074$ u
$Q = -1.193$ MeV
$|Q| = 1.193$ MeV

Using Eq. 30.5, we can write

$$K_{min} = \left(1 + \frac{m_\alpha}{M_N}\right)|-Q| = \left(1 + \frac{4.002603 \text{ u}}{14.003074 \text{ u}}\right)(1.193 \text{ MeV}) = 1.534 \text{ MeV}$$

Note that K_{min} is significantly larger than $|Q|$.

Follow-up Exercise. In this Example, how much of the incoming kinetic energy went into increasing the mass of the system and how much remained, in the final state, as kinetic energy? Explain your reasoning.

Reaction Cross Sections

Due to the probabilistic nature of quantum mechanics, more than one reaction is usually possible when a particle collides with a nucleus. As we have seen for an endoergic reaction, the incident particle must have a minimum kinetic energy to initiate a particular reaction. When a particle has more kinetic energy than the threshold energies of the *several* possible reactions, any of the reactions may occur. A measure of the probability that a particular reaction will occur is called the **cross section** for that reaction. If we knew the exact expression for the nuclear force and the form of the nuclear structure, we would, in principle, be able to calculate the cross section for each possible reaction. Even though this

---

*The kinetic energy is written as the absolute value of Q, that is, $|Q|$, because the kinetic energy cannot be negative. The sign of Q arises from the mass defect and merely indicates the gain or loss of energy to mass. Thus if Q is negative we want K to be positive, hence the use of absolute value.

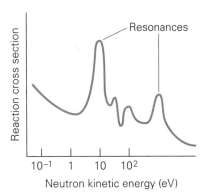

FIGURE 30.2 Reaction cross section
A typical graph of neutron reaction cross section versus energy. The peaks where the probabilities of reactions are greatest are called resonances.

Note: Review Figs. 29.7 and 29.8.

can be done (with fair success) today using high speed computers, it is clearly beyond the scope of this text. Thus, we restrict ourselves to using experimentally determined cross sections.

The cross section of a particular reaction depends on many factors. Usually an important factor is the kinetic energy of the incident particle. That is, the cross section usually depends on the kinetic energy of the initiating particle, sometimes very dramatically. For positively charged incident particles, such as protons and alpha particles, the Coulomb barrier of the nucleus influences the reaction cross section. Thus the probability of a given reaction initiated by charged particles generally increases as the kinetic energy of the incident particle increases. Since reactions do not occur for relatively low-energy charged particles, there must be barrier penetration (from the outside in). The increase in the cross section with particle energy may be thought of as being due to more energetic particles having a greater probability of tunneling through the Coulomb barrier, which is narrower at higher incoming energies.

Being electrically neutral, neutrons are unaffected by the Coulomb barrier. As a result, the cross section for a given reaction may be quite large, even for low-energy neutrons, in particular for reactions such as ^{27}Al(n, γ) ^{28}Al. Reactions such as these are called *neutron capture reactions*, and can be identified by the characteristic gamma ray energies emitted by the nucleus thus formed. For low energies, neutron reaction cross sections are often found to be proportional to $1/v$, where v is the speed of the neutron. That is, the probability of a reaction appears to be proportional to the time a neutron spends in the vicinity of a nucleus (classically, the time to transit the nucleus is $t \approx D/v$ where D is the nuclear diameter, or $t \propto 1/v$). As the neutron energy increases, the cross section varies a great deal, as ●Fig. 30.2 shows. The peaks in the curve are referred to as *resonances*. The resonances are associated with nuclear energy levels in the nucleus being formed. That is, if the energy of the neutron is "just right" to bring the final nucleus to one of its preferred energy levels, or states, there is a high probability for neutron absorption to occur.

30.2 Nuclear Fission

Objectives: To understand (a) what determines which nuclei are fissionable, (b) the nature and cause of a nuclear chain reaction, and (c) the basic principles involved in the operation of nuclear reactors.

In early attempts to make heavier elements artificially, uranium, the heaviest element known at the time, was bombarded with particles, mainly neutrons. An unexpected result was that the uranium nuclei sometimes broke apart or "split" into fragments. In the later 1930s, these fragments were identified as the nuclei of lighter elements. The process was dubbed "fission" after biological fission (the dividing of living cells).

Nuclear fission: "splitting" of the nucleus

Thus, in a **fission reaction**, a heavy nucleus divides into two lighter nuclei with the emission of two or more neutrons. Energy is emitted in the process, being carried off primarily by the neutrons and fission fragments. Some heavy nuclei undergo *spontaneous fission*, but at very slow rates. However, fission may be *induced*, and this is the important process in energy production. For example, when a ^{235}U nucleus absorbs an incident neutron, it may fission according to the reaction

$$^{235}_{92}\text{U} + {}^{1}_{0}\text{n} \rightarrow ({}^{236}_{92}\text{U}^*) \rightarrow {}^{140}_{54}\text{Xe} + {}^{94}_{38}\text{Sr} + 2({}^{1}_{0}\text{n})$$

The capture of a neutron results in the formation of an excited uranium-236 nucleus (●Fig. 30.3).

According to the so-called liquid drop model, the ^{236}U nucleus undergoes vi-

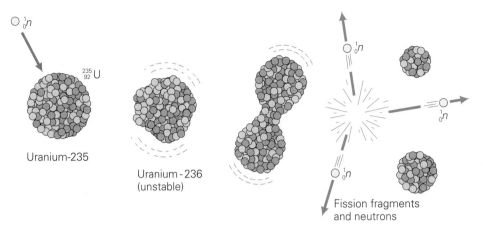

• **FIGURE 30.3 Fission**

When an incident neutron is absorbed by a fissionable nucleus, such as ^{235}U, the unstable nucleus (^{236}U) undergoes violent oscillations and breaks apart like a liquid drop, emitting two or more neutrons.

olent oscillations and becomes distorted like a liquid drop. The separation of the nucleons into different parts of the "drop" weakens the nuclear force, and the repulsive electrical force of the protons between the parts causes the nucleus to split or fission. Some isotopes, however, can capture neutrons *without* fissioning.

The preceding reaction is only one of a variety of ways in which uranium-235, the fissionable isotope of uranium, can fission. Others include (compound nuclei omitted)

$$_0^1n + {}_{92}^{235}U \rightarrow {}_{56}^{141}Ba + {}_{36}^{92}Kr + 3({}_0^1n)$$

$$_0^1n + {}_{92}^{235}U \rightarrow {}_{60}^{150}Nd + {}_{32}^{81}Ge + 5({}_0^1n)$$

Only certain nuclei undergo fission, and the probability of a fission reaction for a fissionable isotope depends on the energy of the incident neutrons. For example, the greatest cross sections for fission reactions for ^{235}U and ^{239}Pu are for "slow" neutrons with energies less than 1 eV. However, for ^{232}Th, the greatest cross section is for "fast" neutrons with energies of 1 MeV or more.

Rather than calculating the energy released in an exoergic fission reaction, we can obtain an estimate of its magnitude from the E_b/A curve for stable nuclei (Fig. 29.15). When a nucleus with a large mass number A, such as uranium, splits into two nuclei, it is in effect moving up along the downward-sloping tail of the E_b/A curve to a more stable state. As a result, the average binding energy per nucleon increases from a value of about 7.8 MeV to a value of approximately 8.8 MeV. Thus, the energy liberated is about 1 MeV per nucleon. In the first of the preceding fission reactions, there are $140 + 94 = 234$ nucleons in the fission products, so the accompanying energy release is approximately

$$\frac{1\,\text{MeV}}{\text{nucleon}} \times 234\,\text{nucleons} \approx 234\,\text{MeV}$$

At first glance, this might not seem like much energy, since 200 MeV is only about 3×10^{-11} J. By way of comparison, when you pick up your textbook from the desk, you do about 5 J of work or expend 5 J of energy. In fact, 200 MeV is less than 0.1 percent of the rest energy of the uranium-235 nucleus, which is approximately (235 nucleons)(939 Mev/nucleon) $= 2.2 \times 10^5$ MeV. Nevertheless, on a relative basis, it is many times larger than the energy release from chemical reactions, such as the burning of oil or coal. The energy released in an exothermic chemical process is only of a *few hundred electron volts per atom*. And you must keep in mind that there are billions and billions of nuclei in even a tiny sample of fissionable material. We simply need to have enough nuclei fissioning to produce practical amounts of energy.

This is accomplished by means of a **chain reaction**. For example, suppose a ^{235}U nucleus fissions with the release of two neutrons (• Fig. 30.4). Ideally, the neutrons may then initiate two more fission reactions, a process that results in the

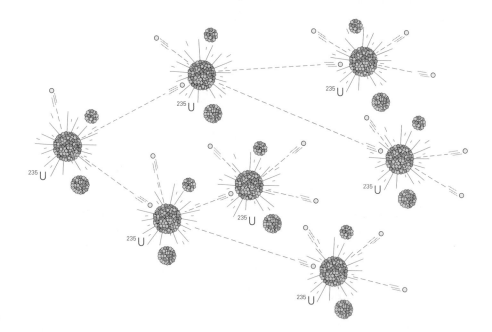

•FIGURE 30.4 Fission chain reaction
The neutrons that result from one fission event may initiate other fission reactions, and so on. When enough fissionable material is present, the sequence of reactions is self-sustaining (a chain reaction).

Note: A chain reaction requires a critical mass.

availability of four neutrons. These neutrons may initiate more reactions, and the process multiplies, the number of neutrons doubling with each generation. When this occurs uniformly with time, we have exponential growth.

To have a sustained chain reaction, there must be an adequate quantity of fissionable material. The minimum mass required to produce a chain reaction is called the **critical mass**. In effect, this means that there is enough fissionable material that one neutron from each fission event, on the average, goes on to fission another nucleus. With more neutrons, the chain and energy output would grow.

Several factors determine the critical mass. Most evident is the amount of fissionable material. If the quantity of material is small, many neutrons will escape from the sample before inducing a fission event, and the reaction will die out. Also, the nuclei of other isotopes in the sample may absorb neutrons, thereby limiting the chain reaction. As a result, the purity of the fissionable isotope affects the critical mass.

In general, natural uranium is made up of the isotopes ^{238}U and ^{235}U. The concentration of the fissionable ^{235}U isotope is only about 0.7%. The remaining 99.3% is ^{238}U, which may absorb neutrons for its own nuclear reaction other than fission, thereby preventing those neutrons from contributing to a chain reaction of ^{235}U fission. To have more fissionable ^{235}U nuclei in a sample and reduce the critical mass, the ^{235}U is concentrated. This enrichment varies from 3–5% ^{235}U for reactor-grade material used in electrical generation to over 99% for weapons-grade material. This is an important difference, since it is desirable that an energy-generating nuclear reactor *not* explode like an atomic bomb nor utilize fuel capable of being directly made into a bomb. For example, theft of reactor-grade fuel would still require significant refining to bring it up to weapons grade material.

Chain reactions take place almost instantaneously. If such a reaction proceeds uncontrolled in a fissionable sample of critical mass, the quick and enormous release of energy from the enormous number of fissioning nuclei causes an explosion. (See Demonstration 17.) This is the principle of the atomic bomb. In such a bomb, several subcritical pieces of fuel are suddenly imploded to form a critical mass. The chain reaction occurs in a very short time, thereby releasing an enormous amount of energy—a bomb. Clearly for the practical production of nuclear energy, the chain reaction process must somehow be controlled. This is done in nuclear reactors as we shall now see.

DEMONSTRATION 17 ■ A Simulated Chain Reaction

A simulation of how neutrons from fission reactions induce reactions in other nuclei so that the process grows in a chain reaction.

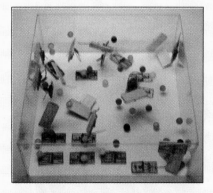

(a) Here, the "nuclei" are mouse-traps loaded with a single Super Ball (neutron) that will be emitted when a reaction occurs. A Plexiglas enclosure ensures a "critical mass," or that the balls (neutrons) are reflected instead of "escaping."

(b) The reaction begins.

(c) The reaction grows, with an increasing number of "fissions" and release of energy.

Nuclear Reactors

Currently, the only type of practical nuclear reactor is based on the fission chain reaction. Research reactors designed primarily to produce artifical isotopes exist, but we will focus on the type of reactor that is commonly used to generate electricity. The technology and initial design for this is over 40 years old, the first nuclear reactor for the purpose of electrical generation having gone into operation in 1957 in Shippingport, Pennsylvania. A typical design for the components of a nuclear reactor is shown in •Fig. 30.5. There are four key elements to a reactor: fuel rods, coolant, control rods, and moderator.

Tubes packed with pellets of uranium oxide form the **fuel rods**, which are located in the reactor core. A typical commercial reactor contains fuel rods bundled in fuel assemblies of approximately 200 rods each. A fuel rod assembly is constructed so that a coolant can flow around the rods to remove the energy emitted from the fission chain reaction. The reactors used in the United States are light water reactors, which means that ordinary water is used as a **coolant**. However, the hydrogen nuclei of ordinary water have a tendency to capture neutrons. This neutron absorption reaction can be written as

$$\,^1_0n + \,^1_1H \rightarrow \,^2_1H + \gamma$$

This removes neutrons from the chain reaction, so enriched uranium with 3–5% ^{235}U must used to achieve a critical mass using light water as a coolant. A nuclear reactor can run for 3 or 4 years before having to be refueled.

There are other types of reactors than light water ones. For example, many Canadian reactors traditionally have run using heavy water, that is, D_2O. The advantage here is that the deuterium does not absorb neutrons readily, and therefore we can use fuel of a considerably lower enrichment. (In the language of quantum mechanics, the reaction cross section is considerably lower.) Of course, the price to pay is the necessity of separating heavy water (recall that deuterium, although stable, constitutes less than 0.02% of normally occurring hydrogen).

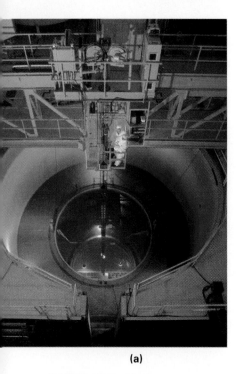

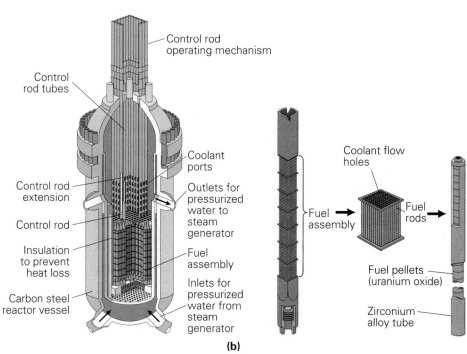

(a)

(b)

•FIGURE 30.5 Nuclear reactor
(a) The interior of a reactor during refueling. **(b)** A schematic diagram of a reactor showing a fuel rod assembly and a fuel rod.

Control rods: absorb neutrons and control chain reaction

The chain reaction and energy output of a reactor are controlled by means of boron or cadmium **control rods**, which can be inserted into or withdrawn from the reactor core. Cadmium and boron have a very high cross section for absorbing neutrons. By inserting the control rods between the fuel rod assemblies, neutrons are absorbed and removed from the chain reaction. The control rods can be adjusted so that energy is released at a relatively steady rate. The object is to create a sustainable fission chain reaction in which the average fission produces only one more fission. Anything even slightly greater than one, and the reaction can get out of control.

In practice, the energy production is monitored electronically and the control rods are operated by computers that push them in if the power level rises and pull them out if it begins to fall. Thus the operation of the reactor can be stablized using this *feedback* technique. When it comes time to refuel, or in an emergency, the control rods are fully inserted, and enough neutrons are removed to curtail the chain reaction and shut down the reactor. However, even with the chain reaction shut down, water must continue to circulate in order to prevent huge heat build-up due to the slow decay of the radioactive fission products in the fuel rods. If not, damage to the fuel rods can result, as discussed later in the chapter.

Moderator: slows neutrons to increase reaction cross section for uranium-235

The water flowing through the fuel rod assemblies acts not only as a coolant, but also as a **moderator**. The fission cross section of ^{235}U is largest for slow neutrons (kinetic energies less than 1 eV). The neutrons emitted from a fission reaction, however, are classified as fast neutrons with average energies of 2 MeV. These are slowed down, or their speed is moderated, by collisions. The hydrogen atoms in water are very effective in slowing down neutrons because their masses are nearly equal to the neutron mass. Recall from Chapter 6 that there is maximum energy transfer in a head-on collision of particles of equal mass. Not all collisions are head-on, but it takes only about 20 collisions to moderate fast neutrons to energies of less than 1 eV.

Note: This result was obtained in Section 6.4.

As we have seen before, an alternative coolant is heavy water (D$_2$O, where D stands for deuterium). It is not as efficient a moderator as light water (why?), however it does eliminate the need for enriched uranium because deuterium does not readily absorb neutrons.

Other materials, such as graphite (carbon), may be used as moderators. Because of their relatively heavy nuclei, not as much energy is transferred per collision. On the average, about 120 collisions with carbon atoms are needed to produce slow, or thermal, neutrons. (A thermal neutron is one that has acquired the average thermal speed of the atoms of the moderator.)

Although not the best moderator, carbon in the form of graphite also permits a chain reaction to occur for natural (unenriched) uranium. The first experimental proof that a chain reaction was feasible was accomplished in such a reactor in 1942 by a team of scientists working with Enrico Fermi on the World War II Manhattan Project. The reactor was called a "pile" because it essentially consisted of a pile of graphite blocks. (A carbon reactor was involved in the 1986 accident at Chernobyl in the former Soviet Union, discussed later.)

The Breeder Reactor. In general, the ^{238}U in a reactor goes along for the ride, so to speak. However, ^{238}U has an appreciable cross section for *fast* neutrons. Not all the neutrons are moderated, so there are some ^{238}U reactions, including its conversion into ^{239}Pu via successive beta decay (how would we predict this?):

$$\underset{(fast)}{^{1}_{0}n} + {}^{238}_{92}U \rightarrow {}^{239}_{92}U^* \rightarrow {}^{239}_{93}Np + {}^{0}_{-1}e$$
$$\hookrightarrow {}^{239}_{94}Pu + {}^{0}_{-1}e$$

Plutonium-239, with a half-life of 24,000 years, *is fissionable* and serves as additional fuel that prolongs the time to reactor refueling. It is possible to promote this conversion of ^{238}U in a reactor by reduced moderation. When, on the average, one or more neutrons from ^{235}U fission are absorbed by ^{238}U nuclei to produce ^{239}Pu, then the same amount or more of fissionable fuel is produced (^{239}Pu) as is consumed (^{235}U). This is the principle of the **breeder reactor**, which produces more fissionable fuel than it consumes.

Note: A breeder reactor doesn't produce something from nothing. It merely uses the kinetic energy of the neutrons, wasted in a non-breeder reactor, to create ^{239}Pu.

Of course, to utilize this fuel in other reactors or nuclear weapons, we must remove the fuel rods and chemically separate the plutonium from the uranium and fission fragments. This process is not trivial, involving the application of chemical methods to radioactive material. In addition, plutonium itself is chemically very reactive, so it is likely to contaminate any living tissue with which it comes in contact. It is a natural alpha emitter, and minute amounts of plutonium breathed into lung tissue have been known to cause lung cancer. In the United States, the spent fuel rods from typical power reactors contain significant quantities of fissionable plutonium. However, even though the government intends to provide a central warehouse for recycling these fuel rods, none has been built as of this writing, and spent fuel rods are stored on-site in borated water pools. (Why boron?—and why underwater?)

Developmental work on the breeder reactor in the United States was essentially stopped in the 1970s. France went on to develop an operational breeder reactor which provides nuclear fuel for its reactors. France and other nations are highly dependent on nuclear energy, as we will see later.

Electrical Generation. The components of a typical pressurized water reactor used in the United States are illustrated in •Fig. 30.6. The heat generated by the controlled chain reaction is carried away by the water moderator surrounding the rods in the fuel assembly. The water in the reactor is pressurized to between 100 and 200 atmospheres. This is done so that it can be heated to temperatures over 300°C for more efficient heat removal. (Recall that the boiling point of water increases with pressure.) The hot water is pumped to a heat exchanger, where the energy is transferred to the water of a steam generator. Notice that the reactor coolant and the exchanger water are in separate, closed systems. (Why do you think this is?)

Note: The relationship between pressure and boiling point is discussed in Section 11.53.

•FIGURE 30.6 Pressurized water reactor

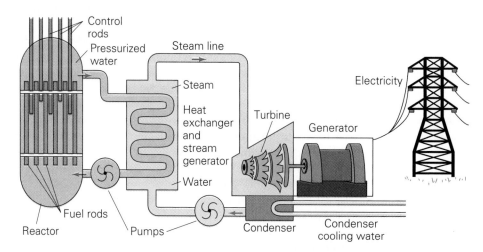

The diagram shows the components of a pressurized water reactor. The heat energy from the reactor core is carried away by the circulating water moderator. The water in the reactor is pressurized so that it can be heated to high temperatures for more efficient heat removal. The energy is used to generate steam, which drives the turbine that turns the generator.

High-pressure steam then turns a turbine that operates a generator, as would be the case for a coal- or oil-fired plant. The steam is cooled and condensed after turning the turbine.

Nuclear Reactor Safety Concerns

Nuclear energy is used to generate a substantial amount of the electricity in the world (•Fig. 30.7a). Some twenty-five countries now produce nuclear-generated electricity, and twelve more plan to do so in the 1990s. More than 430 nuclear reactor units are in operation throughout the world, with about 110 operating units in the United States (•Fig. 30.7b). With the increasing number of nuclear-generating facilities comes the fear of a possible nuclear accident and the re-

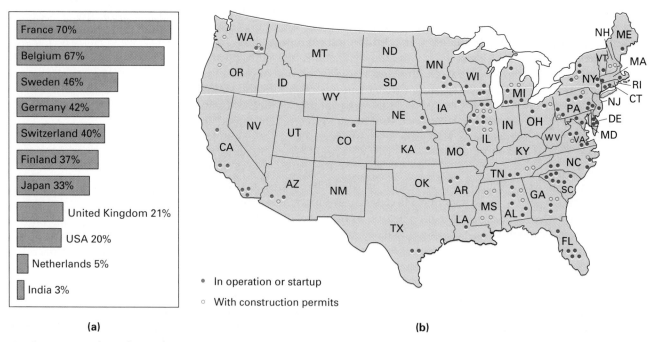

•FIGURE 30.7 Nuclear electrical generation

(a) A comparison of the percentages of nuclear electrical generation in various countries. Keep in mind that 20% of the electrical generation in a large country such as the United States represents more nuclear reactors than 67% of the electrical generation in a relatively small country such as Belgium. (b) Nuclear plant locations in the United States.

lease of radioactive materials into the environment. We now hear such terms as **LOCA** (loss-of-coolant accident) and **meltdown**.

Should the coolant of a light water reactor be lost or stop flowing through the core, the reactor would shut down. In such a reactor, the chain reaction will stop when the coolant is lost because the coolant is also the moderator—without it, there are no longer any thermal neutrons to cause successive fissions. However, decay reactions among the fission fragments, some of which have half-lives of hundreds of years, would continue. Loss of the coolant property of the light water might then allow the fuel rods to become hot enough (several thousand degrees Celsius) for the cladding (the outside covering) to melt and fracture. Once the fuel rods and cladding began to melt, the hot, fissioning mass could fall into the water on the floor of the containment vessel and cause a steam and hydrogen explosion that might rupture the walls of the vessel and allow radioactive fragments into the environment. Even if the walls were not breached, it is possible that the hot "melt" could burn through the floor of the containment building into the underground environment, eventually perhaps reaching ground level and the atmosphere (the so-called China syndrome).

A partial meltdown occurred at the Three Mile Island (TMI) generating plant near Middletown, Pennyslvania, in 1979. This was a LOCA in which fortunately only a relatively small amount of radioactive steam was vented into the atmosphere, although inside one reactor vessel, electronic robots have discovered heavy damage to the fuel rods. This vessel has had most of the radioactive liquid removed and is sealed for a long time.

The nuclear accident at Chernobyl in April 1986 was a meltdown, caused by human error but magnified by the inherent instability that results from the use of carbon as a moderator. The reactor involved was a graphite reactor, with some 1660 fuel assemblies encased in 1700 tons of graphite blocks (•Fig. 30.8a). When the cooling was inadvertently removed, the chain reaction went out of control—something that could not happen in a light water reactor—producing a tremendous rise in temperature. Chemical explosions blew the top off the building, and when the fuel rods melted down, the graphite blocks burned like a massive coal pile, spewing radioactive smoke into the air as well. The radioactive smoke was carried over much of Europe by the wind and also north over the pole into Canada and the United States, where significant amounts of core fission fragments (such as ^{131}I, ^{90}Sr, and ^{137}Cs) were measured on the ground and in the air.

In western Russian and eastern Europe, there was (and still is) concern over possible effects on the health of exposed human populations. For example, we have seen that iodine is taken up by the thyroid. Although small quantities of radioactive iodine are safely used for diagnostic purposes, large amounts can cause thyroid cancer. Only a few dozen people died as an immediate result of the accident in the region near the plant, but it is feared that thousands may eventually die of cancer related to the Chernobyl radiation. A decade after the accident, large areas of formerly productive land in Belorus and Ukraine remain indefinitely closed to agricultural use or even to human habitation. Meanwhile, medical and environmental studies continue, not only on the people who live in that region but also on the plants and livestock.

Finally, even if nuclear reactors operate safely (and the safety record, particularly in the United States, is a very good one), there is always the problem with their leftovers, or radioactive waste. The by-products of the fission reactions are radioactive and have long half-lives. As a result, nuclear waste will be a continuing problem for years. The United States, for example, is still deciding where and how to seal, bury, and guard the nuclear waste for many generations to come (•Fig. 30.8b). Even when the storage places are built, there will still be the highly dangerous problem of getting the radioactive fuel from the power plants to storage.

(a)

(b)

•FIGURE 30.8 Nuclear safety
(a) An aerial photo showing damage to the reactor at Chernobyl in the former USSR. This accident released large amounts of radioactive materials into the environment, with very dire consequences. (b) Experimental equipment being installed in an underground cavern to help test its suitability for the storage of radioactive nuclear wastes. The safety of such storage in the long term is still being studied and debated.

30.3 Nuclear Fusion

Objectives: To be able to (a) explain the fundamental difference between fusion and fission, (b) calculate energy releases in fusion reactions, and (c) understand how fusion might eventually produce useful electric energy for the world.

The other important type of reaction for the production of nuclear energy is fusion. In a **fusion reaction**, light nuclei fuse together to form a heavier nucleus, releasing energy in the process ($Q > 0$). A simple fusion reaction—the fusion of two deuterium nuclei (^2_1H), sometimes called a D–D reaction—was given in Example 30.1. This reaction, as we saw in the example, releases 3.27 MeV of energy per fusion.

Another example is the fusion of deuterium and tritium (a D–T reaction):

$$^2_1H \quad + \quad ^3_1H \quad \rightarrow \quad ^4_2He \quad + \quad ^1_0n$$

(2.014102 u)   *(3.016049 u)*   *(4.002603 u)*   *(1.008665 u)*

Computing the Q value for this reaction, it is found that there is a release of $Q = 17.6$ MeV per fusion.

Notice that the energy releases from these fusion reactions are very small in comparison to the more than 200 MeV energy release from a fission reaction. However, think about samples of hydrogen and uranium with equal masses. A given mass of hydrogen isotopes has many, many more nuclei than an equivalent mass of a heavy fissionable isotope. The fusion of hydrogen gives almost three times the energy released from the fissioning of uranium. (About 0.4% of the hydrogen mass is converted to energy; recall that for uranium the figure is only about 0.1%)

We might say that our lives are dependent on nuclear fusion. Fusion is the source of energy for stars, including the Sun. In the initial stages of a star's life, there is hydrogen burning, or the fusion of hydrogen into helium. One such sequence of fusion reactions that is believed to be possible for solar energy output is as follows:

$$^1_1H + ^1_1H \rightarrow ^2_1H + ^0_{+1}e + \nu$$

then protons and deuterons fuse

$$^1_1H + ^2_1H \rightarrow ^3_2He + \gamma$$

and finally the ^3He nuclei fuse

$$^3_2He + ^3_2He \rightarrow ^4_2He + ^1_1H + ^1_1H$$

The net effect of this sequence, called the *proton–proton cycle*, is the combining of four protons (^1_1H) to form one helium (^4_2He) nucleus plus two positrons ($^0_{+1}e$), two gamma rays (γ), and two neutrinos (ν) (a particle to be discussed in the next section) and the release of energy:

$$4(^1_1H) \rightarrow ^4_2He + 2(^0_{+1}e) + 2\gamma + 2\nu + \text{energy} \quad (Q = 24.7 \text{ MeV})$$

In a star, it is believed that as the gamma rays (Compton) scatter off many nuclei on their way to the surface, they gradually suffer a reduction in energy until, for example in our Sun, they each carry only a few electron-volts of energy. In other words, by the time they reach the surface, they have become visible light. In our Sun, fusion of hydrogen into helium occurs in the central 10% of the Sun's mass only. Even so, it has been going on for about five billion years and should continue, approximately as it is, for another five billion years. After that, it is believed in its next stage of evolution the central core will fuse helium into heavier elements like lithium and carbon. Consider the enormous number of fusion reactions generated by the Sun in the next Example!

EXAMPLE 30.3 ▪ STILL GOING? THE FUSION POWER OF THE SUN

Sunlight falls on the Earth's upper atmosphere at the rate of 1.4×10^3 W/m². This energy is first generated near the Sun's center by the fusion of hydrogen into helium (proton-proton cycle). Assume that the net result of the fusion process is conversion of four protons into a helium nucleus. Assuming we are 150 million km from the Sun and that only 10% of the hydrogen is available for fusion, calculate (a) the mass loss of the Sun per second, (b) the mass of hydrogen lost per second, and (c) the expected lifetime of the Sun. Assume the proton-proton cycle operates and look up the masses involved. Take the mass of the Sun to be 2.0×10^{30} kg and use atomic masses. Neglect electron imbalances. (Why? See solution to part b.)

Solution. We are given the data we need.

Given: $R = 1.50 \times 10^8$ km $= 1.5 \times 10^{11}$ m *Find:* (a) $\Delta m/\Delta t$ (overall
$\qquad\quad M_{Sun} = 2.0 \times 10^{30}$ kg $\qquad\qquad\qquad\qquad\qquad\qquad$ mass loss rate)
$\qquad\quad P_{Sun}/A = 1.4 \times 10^3$ W/m^2 $\qquad\qquad$ (b) $\Delta m_H/\Delta t$ (hydrogen
$\qquad\qquad\qquad\qquad\qquad\qquad\qquad\qquad\qquad\qquad\qquad$ mass loss rate)
$\qquad\qquad\qquad\qquad\qquad\qquad\qquad\qquad\qquad$ (c) Δt (Sun's lifetime)

(a) First we need to calculate the overall power output of the Sun. To do this, imagine a spherical surface surrounding the Sun with a radius equal to that of the Earth's orbit. The surface area of such a sphere (which will intercept all of the Sun's light energy) is

$$A = 4\pi R^2 = 4\pi (1.5 \times 10^{11} \text{ m})^2 = 2.8 \times 10^{23} \text{ m}^2$$

The amount of solar energy that falls on each square meter of this sphere is what we have been given. Thus the total power output of the Sun is

$$P_{Sun} = (1.4 \times 10^3 \text{ W/m}^2)(2.8 \times 10^{23} \text{ m}^2) = 4.0 \times 10^{26} \text{ W}$$

This means that the Sun is putting out 4.0×10^{26} J of light energy each second. To find the equivalent mass loss rate, we first convert the energy from joules to MeV (Sec. 16.2):

$$\frac{4.0 \times 10^{26} \text{ J/s}}{1.60 \times 10^{-13} \text{ J/MeV}} = 2.5 \times 10^{39} \text{ MeV/s}$$

Then, using the mass-energy relationships found in the previous chapter (see Table 29.3), we can find the mass loss rate (in kg/s):

$$\frac{(2.5 \times 10^{39} \text{ MeV/s})(1.66 \times 10^{-27} \text{ kg/u})}{931.5 \text{ MeV/u}} = 4.5 \times 10^9 \text{ kg/s}$$

Thus, the Sun's rate of mass loss, $\Delta m/\Delta t$, is 4.5×10^9 kg/s.

(b) This mass loss is attributable to many fusions per second, each one losing a miniscule amount of mass. Each fusion will lose

$$\Delta m \text{ per fusion} = 4m_H - m_{He}$$

$$= 4(1.007825 \text{ u}) - 4.002603 \text{ u}$$

$$= (0.028697 \text{ u})(1.66 \times 10^{-27} \text{ kg/u})$$

$$= 4.8 \times 10^{-29} \text{ kg/fusion}$$

We can use this to calculate the number of fusions occurring per second:

$$\text{Number of fusions per second} = \frac{4.5 \times 10^9 \text{ kg/s}}{4.8 \times 10^{-29} \text{ kg/fusion}} = 9.4 \times 10^{37} \text{ fusion/s}$$

Each fusion takes four protons, hence the amount of *hydrogen* being converted per second is

$$\Delta m_H/\Delta t = (9.4 \times 10^{37} \text{ fusions/s})(4)(1.67 \times 10^{-27} \text{ kg of H/fusion})$$

$$= 6.3 \times 10^{11} \text{ kg of H lost/s}$$

(c) At this rate, 10% of the Sun's hydrogen supply would last about

$$\Delta t = \frac{(0.1)(2.0 \times 10^{30} \text{ kg})}{6.3 \times 10^{11} \text{ kg/s}} = 3.2 \times 10^{17} \text{ s} \approx 10 \text{ billion years}$$

Since we believe the Sun to have formed into its current evolutionary stage about 5 billion years ago, it is approximately middle-aged.

Follow-up Exercise. In this Example, how many kilograms of helium are being produced in the Sun per second (to two significant figures)? Explain your reasoning.

Fusion as a Source of Energy Produced on the Earth

In several ways, fusion appears to be the ideal energy source of the future. Enough deuterium exists in the oceans, in the form of heavy water, to supply our needs for centuries. Controlled fusion does not depend on a chain reaction, as we shall see, so there is less danger of loss of control and the release of radioactive material through a meltdown. Also, the light nuclei of fusion products have relatively short half-lives. For example, tritium has a half-life of 12.3 years, as compared to hundreds or thousands of years for fission by-products. In addition, tritium is only a beta emitter and beta particles can be absorbed by a meter of air or a thin layer of aluminum foil.

However, there are many unresolved technical problems in the production of fusion energy for practical use. A primary problem is that very high temperatures are needed to *initiate* fusion reactions. This reflects the energy needed to overcome the Coulomb repulsion of the fusing nuclei. Temperatures on the order of *millions of degrees* are needed to initiate *thermonuclear reactions*, as they are called. Because of this, practical fusion reactors have not been achieved. The problem is in confining sufficient energy in a reaction region to maintain the necessary high temperatures. Uncontrolled fusion has been demonstrated in the form of the hydrogen (H) bomb. In this case, the fusion reaction was initiated by an initial implosion caused by a small atomic or fission bomb. This provided the necessary density and temperatures to begin the fusion process.

At such high temperatures, electrons are stripped from their nuclei, and a gas of positively charged ions and free, negatively charged electrons is obtained. Such a gas of charged particles is called a **plasma**. Plasmas have a number of special physical properties. As a result, they are sometimes referred to as the fourth phase of matter, a term used in 1879 by William Crookes, an English chemist, who generated plasmas in gas-discharge or Crookes tubes.

The problem of plasma confinement is being approached in two ways: magnetic confinement and inertial confinement. Since a plasma is a gas of charged particles, it can be controlled and manipulated by using electric and magnetic fields. In **magnetic confinement**, magnetic fields are used to hold the plasma in a confined space, a so-called magnetic bottle (see Fig. 19.29).

Electric fields are used to produce electric currents in the plasma which raise its temperature. Temperatures of 100 million kelvins have been achieved in a design that uses a donut-shape called a *tokamak* (•Fig. 30.9).

Plasma: a fourth phase of matter

•FIGURE 30.9 **Magnetic confinement**
Magnetic confinement is one method by which controlled nuclear fusion might be achieved. **(a)** Tokamak configuration showing the B field generated by external currents. The magnetic field confines the plasma in the ring. **(b)** Photo of the Princeton Tokamak Fusion Test Reactor (TFTR).

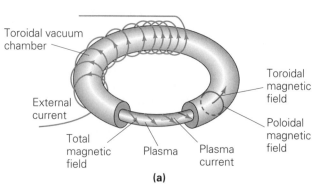

Toroidal vacuum chamber

External current

Total magnetic field

Plasma

Plasma current

Toroidal magnetic field

Poloidal magnetic field

(a)

(b)

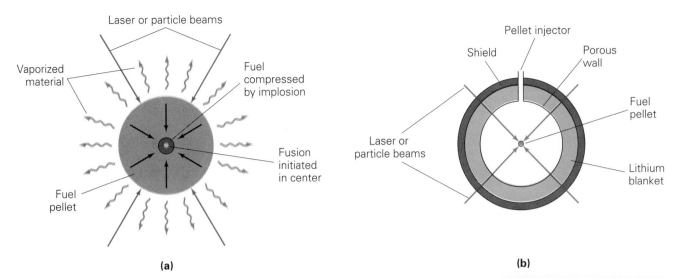

(a)

(b)

However, to initiate fusion, it is not enough to have a very high temperature; there are also requirements on the plasma density and confinement time. The trick is to put all these together in a fusion reactor. The generation of 1.7 MW of power for less than a second in a magnetically confined plasma, achieved in 1991, represents the best result obtained so far.

Inertial confinement depends on an implosion technique similar to that used in the ^{239}Pu bomb mentioned earlier. Hydrogen fuel pellets would be either dropped or positioned in a reactor chamber (•Fig. 30.10). Pulses of laser, electron, or ion beams would then be used to implode the pellet, producing compression and high temperatures. Fusion would occur if the pellet stayed together for a sufficient time, which depends on its inertia (hence the name, inertial confinement). At this time, lasers and particle beams are not powerful enough to sustain fusion by this means.

However, research continues. The technological problems of controlled thermonuclear fusion are enormous, but so are the benefits. Practical energy production from fusion is not expected until well into the next century, if at all.

30.4 Beta Decay and the Neutrino

Objectives: To be able to (a) explain why the neutrino is necessary to account for observed beta decay data, (b) specify some of the physical properties of neutrinos, and (c) write complete beta decay equations.

As described in Section 29.2, beta decay appears to be a straightforward process. That is, certain radioactive nuclei naturally decay and emit an electron or a positron in the process. Examples of β^- and β^+ reactions are

$$\underset{(14.003242\,u)}{^{14}_{6}\text{C}} \quad \rightarrow \quad \underset{(14.003074\,u)}{^{14}_{7}\text{N}} \quad + \quad ^{\ 0}_{-1}\text{e}$$

$$\underset{(13.005739\,u)}{^{13}_{7}\text{N}} \quad \rightarrow \quad \underset{(13.003355\,u)}{^{13}_{6}\text{C}} \quad + \quad ^{0}_{+1}\text{e}$$

However, more than this is going on in these beta decay processes. When they are analyzed in detail, there seem to be some violations of physical principles. One of the difficulties arises with energy. The energy released in the above β^- process may be calculated from the mass difference in the nuclei:*

$$\Delta m = 14.003242 \text{ u} - 14.003074 \text{ u} = 0.000168 \text{ u}$$

*Using the atomic mass of ^{14}N takes into account the emitted electron, since the daughter atom in beta decay would have only six electrons like the parent ^{14}C atom.

(c)

•**FIGURE 30.10 Inertial confinement**
Another method by which controlled nuclear fusion might be achieved is inertial confinement. **(a)** An illustration of fuel pellet implosion. The compression is enhanced by the vaporization of the outer shell material of the pellet. **(b)** A schematic diagram of a possible fusion reactor cavity. The lithium blanket captures neutrons from the fusion reactions, producing tritium, which may be used as a fuel. **(c)** The world's most powerful laser, Nova, created a brief burst of fusion in a tiny fuel capsule. In the 50 trillionths of a second it lasted, the reaction gave off ten trillion neutrons.

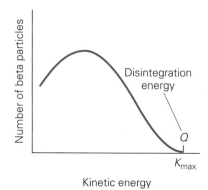

Number of beta particles

Disintegration
energy

Q

K_{max}

Kinetic energy

FIGURE 30.11   Beta ray spectrum
For a typical beta decay process,
most beta particles are emitted
with $K < Q$, leaving energy
unaccounted for.

and

$$E = (0.000168 \text{ u})(931.5 \text{ MeV/u}) = 0.156 \text{ MeV}$$

Thus, the decay reaction has a Q value or disintegration energy of 0.156 MeV. The electron, being a light particle, would be expected to carry away virtually all of this energy or to have a $K_{max} = Q$. However, this is not what happens.

When the kinetic energies of the electrons from a given beta decay process are determined, a *continuous spectrum* of energies is observed up to a cutoff point of $K_{max} \approx Q$ as shown in •Fig. 30.11. (See Conceptual Example 30.4, which shows why this is only approximate.) That is, electrons are emitted that have less kinetic energy than expected; in fact, most of the beta particles have $K < Q$. This would appear to be a violation of the conservation of energy, since not all the energy is accounted for.

Nor is this the only difficulty. Observations of individual beta decay processes show that the emitted electron and the daughter nucleus from a stationary parent nucleus do not always leave the disintegration site in opposite directions. Thus, we also have an apparent violation of the conservation of linear momentum.

To top it off, there is also an apparent violation of the conservation of angular momentum. Nucleons and electrons have intrinsic (spin) quantum numbers of $\frac{1}{2}$. For a nucleus with an even number of nucleons (such as ^{14}C), the total angular momentum quantum number will be an integer, since an even number of spin $\frac{1}{2}$ particles is present. When an electron is created in the decay process, an odd number of spin $\frac{1}{2}$ particles is present, and the total angular momentum quantum number after the decay will be a half-integer. Therefore, the sums of the total angular momentum quantum numbers would not be equal before and after spontaneous decay.

What, then, is the problem? We would hope that our conservation laws are not invalid. The other alternative is that these apparent violations are telling us something about nature that we do not know or do not yet recognize. The difficulties would be resolved if we assumed that, in addition to the electron, another unobserved particle of intrinsic spin $\frac{1}{2}$ were created and emitted in beta decay.

This explanation and the existence of such a particle were first proposed in 1930 by Wolfgang Pauli. The particle was christened the **neutrino** (meaning "little neutral one"). The name reflected the supposition that in order for charge to be conserved in beta decay, the new particle had to be electrically neutral. The fact that the neutrino had not been observed suggested that it must interact very weakly with matter. Later, scientists discovered that the neutrino interacts with matter by a second nuclear force, much weaker than the strong force we have studied. This interaction was called the *weak interaction*. Alternatively, we talk about the existence of a **weak nuclear force**. (This force is discussed in the next section.)

The observations of beta decay suggest that the neutrino has zero rest mass and therefore travels with the speed of light like a photon. It also has linear momentum p with a total relativistic energy $E = pc$ and has an intrinsic spin quantum number of $\frac{1}{2}$. In 1956, a particle with these properties was detected experimentally, and the existence of the neutrino was established. Thus, we add the neutrino to the list of subatomic particles. The nuclear equations for the previous beta decay reactions become

$$^{14}_{6}\text{C} \rightarrow ^{14}_{7}\text{N} + ^{0}_{-1}\text{e} + \bar{\nu}_e$$

and

$$^{13}_{7}\text{N} \rightarrow ^{13}_{6}\text{C} + ^{0}_{+1}\text{e} + \nu_e$$

where the Greek letter ν (nu) is the symbol used for the neutrino. The symbol with a bar over it represents an antineutrino. This bar method is a common way of indicating an antiparticle. There are two different neutrinos associated with common beta decay. In general, a neutrino is emitted in β^+ decay, and an antineutrino is emitted in β^- decay. The e subscript identifies the neutrinos as associated with beta decay. As you will learn in a later section, there are other types of neutrinos.

Note that the antineutrino is associated with the decay of a neutron:

$$n \rightarrow p + e^- + \bar{\nu}_e$$

The neutrino ν_e is associated with the decay of a proton into a neutron and a positron. A free proton cannot decay by positron emission because of the conservation of energy (rest masses). But because of binding energy effects, this does occur in the nucleus in β^+ decay with the emission of a neutrino. We will discuss neutrinos and their relationship to the weak nuclear force in a later section.

CONCEPTUAL EXAMPLE 30.4 ■ HAVING IT ALL? MAXIMUM KINETIC ENERGY IN BETA DECAY

Consider the beta decay of an unstable nucleus initially at rest. As a specific example, take the one we have previously mentioned in detail, the classic $^{14}_{6}C \rightarrow {}^{14}_{7}N + {}^{0}_{-1}e + \bar{\nu}_e$. We have said that Q represents the amount of energy released when the decay happens. Is it possible for the maximum kinetic energy of the emitted beta particle to be *exactly* equal to Q? *Clearly establish the reasoning and physical principle(s) used in determining your answer before checking it below. That is, **how** did you arrive at your answer?*

Reasoning and Answer. In beta decay, as in all interactions, we expect total energy and momentum to be conserved. Here energy is conserved, and since we consider mass just a form of energy, its loss shows up in the gain of kinetic energy for the product particles. Since there is no momentum to start with, we also expect there to be no overall momentum after the decay is over. Even in the simplest case, if the daughter nucleus does not recoil, this means that the neutrino must go off in a direction exactly opposite that of the beta particle in order to preserve zero momentum. Thus, even in the most favorable case, the beta particle cannot get all of Q as its kinetic energy. Some amount (however small) must go to either the neutrino or the daughter nucleus (or, in general, to both) in order to cancel the momentum of the beta particle. Thus we see that for the beta particle, $K_{max} < Q$.

Follow-up Exercise. A typical energy spectrum for emitted beta particles is shown in Figure 30.11. What this figure indicates is that the most likely energy for the beta particles is somewhere between the maximum allowable and zero. What would the decay product trajectories look like after the decay if the emitted beta particle had almost no kinetic energy? (There is a small probability that this will happen.)

30.5 Fundamental Forces and Exchange Particles

Objectives: To be able to (a) understand the quantum mechanical picture of forces, and (b) classify the various forces according to their strengths, properties, ranges, and virtual particles.

All interactions take place by means of forces. The forces involved in everyday activities are complicated because of the large numbers of atoms or molecules that make up ordinary objects. Frictional forces, the forces holding this book together, and so on, are actually due to the electromagnetic forces between atoms. Looking at the fundamental interactions between particles makes things simpler. On this level, there are only four known **fundamental forces**: the *gravitational force*, the *electromagnetic force*, the *strong nuclear force*, and the *weak nuclear force*.

The most familiar of these four are the gravitational and electromagnetic forces. Gravity acts between all particles, while the electromagnetic force is restricted to charged particles. Both forces have an inverse square relationship for the separation distance between interacting particles, and an infinite range.

Classically, the action-at-a-distance of a force is described by using the concept of a field. For example, a charged particle is considered to be surrounded by an electric field through which it *interacts* with other charged particles. Quantum mechanics, on the other hand, provides an alternative description of how forces

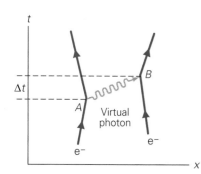

•FIGURE 30.12 Feynman diagram for an electron–electron interaction The interacting electrons undergo a change in energy and momentum due to the exchange of a virtual photon, which is created at A and absorbed at B in a time (Δt) that is consistent with the uncertainty principle.

are transmitted. This process is pictured as an exchange of particles, analogous to you and another person interacting by tossing a ball back and forth.

The creation of such particles would classically violate the conservation of energy. However, within the limitations of the uncertainty principle, a particle can be created for a *short time* with no extra energy. For ordinary time intervals, the conservation of energy is obeyed. But for *extremely* short time intervals, the uncertainty principle *permits* a large *uncertainty* in energy ($\Delta E \Delta t \geq h/2\pi$), so energy conservation may be briefly violated. A particle created in such a manner and time is called a **virtual particle.**

The fundamental forces are considered to be carried by virtual **exchange particles.** The exchange particle for each force is different in mass. The greater the mass of the particle, the greater the amount of energy required to create it and the shorter time it exists. Since a massive particle can exist for only a short time, the range of the interaction for the associated force would be small. That is, *the range of a force associated with an exchange particle is inversely proportional to the mass of that exchange particle.*

The exchange particle for the electromagnetic force is a (virtual) **photon.** As a "massless" particle, it has an infinite range, as does the electromagnetic force. The idea of a particle exchange is represented by the Feynman diagram of the collision of two electrons in •Fig. 30.12. Such space–time diagrams are named after American scientist and Nobel Prize winner Richard Feynman (1918–1988), who used them to analyze electrodynamic interactions. The important points are the vertices of the diagram. One electron may be considered to create a virtual photon at point A and the other electron to absorb it at point B. Each of the two interacting particles undergoes a change in energy and momentum by virtue of the exchange photon and force interaction.

The Strong Nuclear Force and Mesons

Using the idea of exchange particles, Japanese physicist H. Yukawa in 1935 proposed that the short-range strong nuclear force between two nucleons is associated with an exchange particle called the **meson.** An estimate of the mass of this particle may be gained from the uncertainty principle. If a nucleon were to create a meson, the conservation of energy would have to be violated by an amount of energy at least as great as the meson rest mass or

$$\Delta E = (\Delta m)c^2 = m_{\mathrm{m}}c^2$$

where m_{m} is the rest mass of the meson.

By the uncertainty principle, the meson would be absorbed in the exchange process in a time on the order of

$$\Delta t = \frac{h}{2\pi \Delta E} = \frac{h}{2\pi m_{\mathrm{m}}c^2}$$

In this time, the meson could travel a distance not greater than

$$R = c\Delta t = \frac{h}{2\pi m_{\mathrm{m}}c} \tag{30.6}$$

Taking this distance to be the range of the nuclear force ($R \approx 1.4 \times 10^{-15}$ m) and solving for m_{m} gives

$$m_{\mathrm{m}} \approx 274 m_{\mathrm{e}}$$

where m_{e} is the electron rest mass. Thus, if the meson existed, it would be expected to have a rest mass about 270 times that of an electron.

Of course, virtual mesons of an exchange process cannot be directly observed. But if sufficient energy were supplied to colliding nucleons, real mesons might be created through the energy made available in the collision process. The real mesons could then leave the nucleus and be detected. At the time of Yukawa's prediction, there were no known particles with masses between that of the electron (m_e) and that of the proton ($m_p = 1836m_e$).

In 1936, Yukawa's prediction seemed to come true when a new particle with a mass of about 200 m_e was discovered in cosmic rays. Originally called the μ meson, and now just **muon**, the particle was shown to have two charged varieties, positive and negative of electronic magnitude, with a rest mass of

$$m_{\mu^\pm} = 207m_e$$

However, further investigations showed that the muon did not behave like the particle of Yukawa's theory. In particular, the interaction of muons with matter (nuclei) was very weak. The muons from cosmic radiation could penetrate a large mass of material as evidenced by their detection in deep mines.

This situation was a source of controversy and confusion for several years. But in 1947, two more charged particles of the appropriate mass range were discovered in cosmic radiation. These particles were called π mesons (primary mesons), and now more commonly **pions**, together with a less massive electrically neutral π meson. Measurement showed the masses of the pions to be

$$m_{\pi^\pm} = 247m_e \quad \text{and} \quad m_{\pi^\circ} = 264m_e$$

Moreover, the pion interacted strongly with matter. The pions fulfilled the requirements of Yukawa's theory, and it has been generally accepted that this meson is the particle primarily responsible for the transmission (or mediation) of the *strong* nuclear force. The Feynman diagrams of nucleon–nucleon interactions are illustrated in •Fig. 30.13.

Free mesons are unstable. For example, the π^+ particle quickly decays (in about 10^{-8} s) into a muon:

$$\pi^+ \rightarrow \mu^+ + \nu_\mu$$

Here ν_μ is a μ neutrino which differs from the electron neutrino produced in beta decay. The muons can also decay into positrons and electrons with the emission of both types of neutrinos, for example, the positive muon decay would be written as

$$\mu^+ \rightarrow e^+ + \nu_e + \bar{\nu}_\mu$$

The Weak Nuclear Force and the W particle

The discrepancies in beta decay discussed in the preceding section led to another discovery. Electrons and neutrinos are emitted from unstable nuclei, but there was evidence that they did not exist *inside* the nucleus. Enrico Fermi proposed that these particles did, in fact, *not* exist before being emitted but were *instantaneously created* in the decaying radioactive nucleus. For β^- decay, this would mean that a single neutron was in some way *transmuted* into three particles:

$$n \rightarrow p^+ + e^- + \bar{\nu}_e$$

That a neutron could do this was confirmed by the observation of *free* neutrons. Free neutrons disintegrate after a few minutes into a proton, an electron, and a neutrino. But the question arose as to what force could cause a neutron to disintegrate in this manner. Since the neutron was being observed outside the nucleus (free), none of the known forces, including the strong nuclear force, was applicable. It was (correctly) reasoned that some other force must be acting in beta decay. Decay rate measurements indicated that the force was extraordinarily weak—weaker than the electromagnetic force, but still much stronger than the gravitational force. Thus, the **weak nuclear force** was discovered.

It was originally thought that the weak interaction was localized, without any range at all. However, we now know that the weak force has a range of about 10^{-17} m. In terms of an exchange particle, this means that the virtual carriers must be much more massive than the pions of the strong nuclear force. The virtual ex-

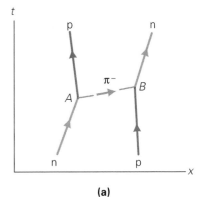

(a)

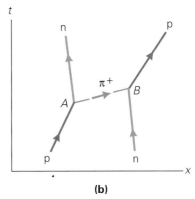

(b)

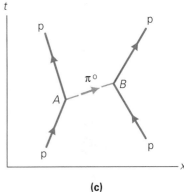

(c)

•FIGURE 30.13 **Feynman diagrams for nucleon–nucleon interactions**
Diagrams for nucleon–nucleon interactions through the exchange of virtual pions. **(a)** n–p reaction, **(b)** p–n reaction, **(c)** p–p reaction.

TABLE 30.2 Fundamental Forces

Force	Relative Strength	Action Distance	Exchange Particle	Particles with Interaction
Strong nuclear	1	Short range ($\approx 10^{-15}$ m)	Pion (π meson)	Hadrons*
Electromagnetic	10^{-3}	Inverse square (infinite)	Photon	Electrically charged
Weak nuclear	10^{-8}	Extremely short range ($\approx 10^{-17}$ m)	W particle†	All
Gravitational	10^{-45}	Inverse square (infinite)	Graviton	All

*Hadrons are elementary particles (see Section 30.6).

†Actually, three particles are involved: W^+, W^-, and Z. These are described in Sections 30.5 and 30.6.

W particle: the exchange particle for the weak nuclear force

change particles for the weak nuclear force were called **W particles**. W (weak) particles have masses about 100 times that of a proton, which explains the extremely short range and weakness of the force. The existence of the W particle was confirmed in the 1980s when accelerators were built with enough energy to create the first real W particles.

The weak force is the only force that acts on neutrinos, which explains why they are so difficult to detect. Research has shown that the weak force is involved in the transmutation of other subatomic particles. In general, the weak force is limited to transmuting the identities of particles within the nucleus. The only way it manifests its existence in the outside world is through the emitted neutrinos. One highly noticeable but infrequent announcement of the weak force at work is during the explosion of a supernova. The collapse of the core of an aging star gives rise to a huge release of neutrinos. The weak force of the neutrinos blasts the outer layers of the star into space in a cataclysmic explosion of a supernova. In a relatively "nearby" supernova (a mere 1.5×10^{18} km distant), observed in 1987, a burst of neutrinos was in fact detected at essentially the same time as the light flash. Since both signals arrived after traveling extremely long distances, this result can be used to set an upper limit on the rest mass of the neutrino. If it had rest mass, its speed would be less than that of light and it would arrive later.

Gravity and Gravitons

Graviton: the exchange particle for the gravitational force

Finally, let's not forget the gravitational force. The exchange particles of the gravitational force are called **gravitons**. There is still no firm evidence of the existence of this massless particle. Several searches have been made to detect the graviton or gravity waves, but the relative weakness of its interaction makes it a very elusive particle. A comparison of the relative strengths of the fundamental forces is given in Table 30.2.

30.6 Elementary Particles

Objectives: To be able to (a) classify the various elementary particles into families, and (b) understand the different properties of the various families of elementary particles.

The fundamental particles, the building blocks of atoms, are referred to as **elementary particles**. Simplicity reigned when it was thought that an atom was an indivisible particle and thus *the* elementary particle. However, scientists now know of a variety of subatomic particles, many of which have not been mentioned here. Indeed, scientists are working to simplify and reduce a long list of elementary particles, since some may be variations of one type of particle.

There are several systems for classifying elementary particles on the basis of their various properties. One classification uses the distinction of nuclear force interactions. Particles that experience or interact by the strong nuclear force are called **hadrons**. These include the proton, neutron, and pion. Other particles, which interact via the weak nuclear force but not the strong force, are called **leptons** ("light ones"). The lepton family includes the electron, muon, and their neutrinos.

Hadrons: strong nuclear force interactions
Leptons: weak nuclear force interactions

Note in Table 30.2 that gravity and the weak force are the only forces acting between *all* particles. But the force of gravity is so weak in elementary particle interactions that it can, in general, be neglected. This leaves the weak force as the only measurable force that affects all particles.

Let's take a brief look at the lepton and hadron families.

Leptons

The most familiar lepton is the electron. It is apparently the only member of the lepton family that exists naturally in atoms. There is no evidence of any internal structure, and, measured to be smaller than 10^{-17} m in size, the electron is presently considered to be a point particle.

Muons were first observed in cosmic rays. They are electrically charged and about 200 times heavier than an electron. Since they appear not to have any internal structure, they are sometimes called heavy electrons. Muons are unstable and decay in about 10^{-6} s. This is the decay reaction for the negative muon:

$$\mu^- \rightarrow e^- + \bar{\nu}_e + \nu_\mu$$

Recall that muon decay was used as evidence for relativistic time dilation in Chapter 26.

Note: Muon decay was treated in Examples 26.2 and 26.4.

A third charged lepton has been discovered. Known as a tau (τ^-) particle or **tauon**, it has a mass twice that of a proton. The electron, muon, and tauon are all negatively charged and appear to have no internal structure. Their antiparticles are positively charged.

The only other known leptons are neutrinos, which are present in cosmic rays, are emitted by the Sun, and also in some radioactive decays. Neutrinos appear to have no rest mass and to travel at the speed of light. They feel neither the electromagnetic force nor the strong force and so pass through matter as if it weren't there. Neutrinos are so weakly interacting that most neutrinos striking the Earth pass right through it.

Neutrinos come in several varieties. The electron neutrino (ν_e) and muon neutrino (ν_μ) are well documented, and it is believed that a tau neutrino (ν_τ) also exists. There is an antineutrino for each of these types.

This completes the list of leptons. With a total of six leptons *plus antiparticles*, there are twelve different leptons in all. Current theories predict that there should be no others.

Hadrons

Another category of elementary particles is called hadrons. All hadrons interact by the strong force, the weak force, and gravity. Also, some are electrically charged. The best known hadrons are nucleons: the proton and the neutron. All others are short-lived and decay via the weak force or more rapidly under the influence of the strong force.

The number of hadrons suggests that they are perhaps composites of other elementary particles. Some help came to sorting out the hadron "zoo" in 1963 when Murray Gell-Mann and George Zweig of Caltech put forth the quark theory. It suggested that **quarks** were elementary charged particles that made up hadrons. They could combine only in two possible ways, either in trios or in quark–antiquark pairs. Combinations of three quarks produce relatively heavy hadrons called **baryons** ("heavy ones"), which include the proton and neutron. Quark–antiquark pairs form the lighter particles mesons.

TABLE 30.3	Types of Quarks	
Name	*Symbol*	*Charge*
Up	u	$+\frac{2}{3}e$
Down	d	$-\frac{1}{3}e$
Strange	s	$-\frac{1}{3}e$
Charm	c	$+\frac{2}{3}e$
Top (Truth)	t	$+\frac{2}{3}e$
Bottom (Beauty)	b	$-\frac{1}{3}e$

To account for the hadrons that were known at that time, the theory proposed three types or "flavors" of quarks. These were given the names "up" (u), "down" (d), and "strange" (s). In addition, the quarks carried fractional electronic charges. The u, d, and s quarks had charges of $+\frac{2}{3}e$, $-\frac{1}{3}e$, and $-\frac{1}{3}e$, respectively, with antiquarks (designated as usual with overbars, e.g. $\bar{u}$) having charges of opposite signs (for example, a charge of $\bar{u} = -\frac{2}{3}e$). Thus, the quark combinations for the proton and neutron are uud and udd, respectively. A positive pion (π^+) is a ud combination (can you show the charge comes out $+e$?).

Despite much experimental effort, no isolated quark has been observed, and they are not believed to exist freely outside of the nucleus. This would explain why we do not observe fractional electronic charges in nature. Quarks can also exist in excited states, similar to the excited states of an atom. It is thought that many of the observed hadrons might be excited states of certain combinations of quarks.

Quarks interact by the strong force, but they are also subject to the weak force. A weak force acting on a quark changes its flavor and gives rise to the decay of hadrons.

The discovery of new elementary particles in the 1970s led to the addition of more quark flavors, which were called "charm" (c), "top" (or "truth," t), and "bottom" (or "beauty," b). A summary of the quark flavors is given in Table 30.3. There is an oppositely charged antiquark for each quark.

Since quarks interact, it is postulated that they too have interacting exchange particles. The exchange particle for quarks has been dubbed the **gluon**. Gluons bind or "glue" hadrons together, thus replacing the pion as the hadron exchange particle.

The quark theory was further extended in terms of a force field. To give the strong force a field representation, each quark is said to possess the analog of electric charge that is the source of the gluon field. Instead of calling this property "charge," it was called "color," with no relationship to ordinary color. Each quark can come in one of three colors: red, green, and blue. There are corresponding anticolors for antiquarks.

When a quark emits or absorbs a gluon, it changes color. The effect is to change the identity of a quark—for example, a blue quark to a red quark. In this respect, the strong force resembles the weak force (for which there is a change of one particle into another with the exchange of a W particle).

Actually, the quark color scheme was developed after a major discovery was made concerning the weak force. Like Maxwell's combining the electric force and the magnetic force into a single electromagnetic force, the weak force and the electromagnetic force were combined, or shown to be two parts of a single **electroweak force**. This came about as the result of a theory put forth by Sheldon Glashow, Abdus Salam, and Steven Weinberg, for which they received a Nobel Prize in 1979. In this theory, weakly interacting particles such as electrons and neutrinos carry a weak charge that gives rise to a weak force field. The exchange particles for the weak force interaction are the heavy, electrically charged W^+ and W^- particles, along with a neutral Z particle for reactions in which there is no change in or transfer of charge. Two weak interactions are illustrated in •Fig. 30.14.

In 1983, the existence of the Z was confirmed, and the four fundamental forces were reduced to three. Of course, scientists would like to reduce the list even further. A theory that would merge the strong nuclear force and the electroweak force into a single unified force is the so-called *grand unified theory* (GUT). Some two dozen exchange particles are required, including the 12 already known particles.

Should the three fundamental forces be reduced to two by the GUT or a similar theory, there is the hope of a further reduction with the idea that all forces are part of a single *superforce*. The combining of a grand unified force with gravity is a real challenge. Although the three components of a grand force may be represented as force fields in space and time, our current view is that gravity *is* space and time.

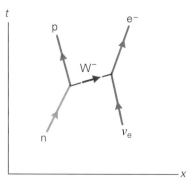

(a)

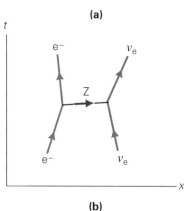

(b)

•**FIGURE 30.14 Weak force interactions**
The Feynman diagrams for **(a)** the decay of a neutron into a proton and an electron through the exchange of a W^- particle and **(b)** the scattering of a neutrino by an electron through the exchange of a neutral Z particle.

Chapter Review

Important Terms

nuclear reactions   931
particle accelerators   932
Q value   933
endoergic   933
exoergic   934
threshold energy   935
cross section   935
fission reaction   936
chain reaction   937
critical mass   938
nuclear reactor   939
fuel rods   939
coolant   939
control rods   940

moderator   940
breeder reactor   941
LOCA   943
meltdown   943
fusion reaction   944
plasma   946
magnetic confinement   946
inertial confinement   947
neutrino   948
fundamental forces   949
virtual particle   950
exchange particle   950
photon   950
meson   950

muon   951
pion   951
weak nuclear force   951
W particle   952
graviton   952
elementary particles   952
hadrons   953
leptons   953
tauon   953
quarks   953
baryons   953
gluon   954
electroweak force   954

Important Concepts

- The *Q*-value of a reaction or decay represents the energy released or absorbed in the process; it appears as a change in the total rest mass of the system.

- In fission, an unstable heavy nucleus decays by splitting into two fragments and several neutrons, which together have less total mass than the original nucleus; energy is thus released.

- A chain reaction occurs when the neutrons released from one fission trigger other fissions, which trigger further fissions, etc.

- In fusion, two light nuclei fuse together, producing a nucleus with less total mass than the original nuclei; energy is thus released.

- A plasma, produced when matter is completely ionized, is a mixture of positive ions and free electrons.

- Beta decay produces a daughter nucleus, an electron or positron, and an antineutrino or neutrino.

- Exchange particles are virtual particles believed to be associated with various forces. The pi meson or pion is the exchange particle primarily responsible for the strong nuclear force.

- The weak nuclear force, transmitted by the W particle, is primarily responsible for beta decay and the instability of the neutron.

- Hadrons are the family of elementary particles involved with the strong force. The nucleons are part of the hadron family.

- Leptons are the family of elementary particles involved with the weak force. Electrons, muons, and neutrinos are part of the lepton family.

- Quarks are elementary, fractionally charged particles that make up hadrons such as protons.

- The electroweak force is the name given to the unified electromagnetic and weak forces.

Important Equations

Q Value:

$$Q = (m_A + m_a - m_B - m_b)c^2 = (\Delta m)c^2 \qquad (30.4)$$

Threshold Energy:

$$K_{min} = \left(1 + \frac{m_a}{M_A}\right)|Q| \qquad (30.5)$$

Range of Exchange Particle:

$$R = c\Delta t = \frac{h}{2\pi m_m c} \qquad (30.6)$$

Exercises

30.1 Nuclear Reactions

1 The Q value of a reaction (a) is given by the difference in the kinetic energies of the products and reactants, (b) is equal to the mass–energy difference between the initial and final systems, (c) may be positive or negative, (d) all of these.

2 To initiate an endoergic reaction, a particle incident on a stationary nucleus must have (a) a minimum disintegration energy, (b) a kinetic energy equal to the Q value, (c) a kinetic energy less than a certain threshold energy, (d) a kinetic energy larger than the Q value which depends on m_a/M_A.

3 How does the threshold energy of an incident particle in a nuclear reaction vary with the mass of the target nuclei? (Sketch a graph if you can.) What is the interpretation of $Q = 0$?

4 ■ Complete the following nuclear reactions:
(a) $^{1}_{0}n + ^{40}_{18}Ar \rightarrow$ _____ $+ ^{0}_{-1}e$
(b) $^{1}_{0}n + ^{235}_{92}U \rightarrow ^{98}_{40}Zr +$ _____ $+ 3(^{1}_{0}n)$
(c) $^{1}_{0}n + ^{235}_{92}U \rightarrow ^{133}_{51}Sb + ^{99}_{41}Nb +$ _____
(d) _____ (α, p) $^{17}_{8}O$
(e) $^{10}_{5}B$ (_____, $\alpha)^{7}_{3}Li$

5 ■ Write the compound nuclei for the reactions in Exercise 4.

6 ■ Complete the following nuclear reactions:
(a) $^{13}_{6}C + ^{1}_{1}H \rightarrow \gamma +$ _____
(b) $^{10}_{5}B + ^{4}_{2}He \rightarrow ^{12}_{6}C +$ _____
(c) $^{27}_{13}Al$ (α, n) _____
(d) $^{14}_{7}N(\alpha, p)$ _____
(e) $^{13}_{6}C(p, \alpha)$ _____

7 ■ Give the compound nuclei for the reactions in Exercise 6.

8 ■ Lead-211 undergoes beta decay. Find the energy released in the decay reaction.

9 ■ Determine what the daughter nuclei in the following decay equations might be and whether the reactions occur spontaneously:
(a) $^{27}_{13}Al \rightarrow$ _____ $+ ^{0}_{-1}e$
(b) $^{226}_{88}Ra \rightarrow$ _____ $+ ^{4}_{2}He$
(c) $^{197}_{79}Au \rightarrow$ _____ $+ ^{4}_{2}He$

10 ■ Show that the Q value for the D–T reaction given in Section 30.3 is 17.6 MeV.

11 ■ Find the Q value for the alpha decay of uranium-238:

$$^{238}_{92}U \rightarrow ^{234}_{90}Th + ^{4}_{2}He$$
$$(238.050786\ u) \quad (234.043583\ u) \quad (4.002603\ u)$$

Would you expect the Q to be positive or negative?

12 ■■ Find the threshold energy for the following reaction:

$$^{16}_{8}O + ^{1}_{0}n \rightarrow ^{13}_{6}C + ^{4}_{2}He$$
$$(15.994915\ u) \quad (1.008665\ u) \quad (13.003355\ u) \quad (4.002603\ u)$$

13 ■■ Determine the minimum kinetic energy of an incident alpha particle that will initiate the following reaction:

$$^{14}_{7}N + ^{4}_{2}He \rightarrow ^{17}_{8}O + ^{1}_{1}H$$
$$(14.003074\ u) \quad (4.002603\ u) \quad (16.999131\ u) \quad (1.007825\ u)$$

14 ■■ Is the reaction ^{200}Hg (p, $\alpha)^{197}Au$ endoergic or exoergic? Prove your answer. (See the equation in the chapter for mass values.)

15 ■■ Is the following reaction endoergic or exoergic? Prove your answer.

$$^{7}_{3}Li + ^{1}_{1}H \rightarrow ^{4}_{2}He + ^{4}_{2}He$$
$$(7.016005\ u) \quad (1.007825\ u) \quad (4.002603\ u) \quad (4.002603\ u)$$

16 ■■ Determine the Q value of the following reaction:

$$^{9}_{4}Be + ^{4}_{2}He \rightarrow ^{12}_{6}C + ^{1}_{0}n$$
$$(9.012183\ u) \quad (4.002603\ u) \quad (12.000000\ u) \quad (1.008665\ u)$$

17 ■■ Find the threshold energy for the following reaction:

$$^{3}_{2}He + ^{1}_{0}n \rightarrow ^{2}_{1}H + ^{2}_{1}H$$
$$(3.016029\ u) \quad (1.008665\ u) \quad (2.014102\ u) \quad (2.014102\ u)$$

18 ■■ Find the threshold energy of the following reaction:

$$^{13}_{6}C + ^{1}_{1}H \rightarrow ^{1}_{0}n + ^{13}_{7}N$$
$$(13.003355\ u) \quad (1.007825\ u) \quad (1.008665\ u) \quad (13.005739\ u)$$

19 ■■ What is the minimum kinetic energy of a proton that will initiate the reaction $^{3}_{1}H$ (p,d) $^{2}_{1}H$? (d stands for a deuterium *nucleus* called the *deuteron*.)

20 ■■ ^{226}Ra decays and emits a 4.706-MeV alpha particle. Find the kinetic energy of the recoiling daughter nucleus from the decay of a stationary radium-226 nucleus.

21 ■■■ The same type of incident particle is used for two endoergic reactions. In one reaction, the mass of the target nucleus is 25 times greater, and in the other reaction 40 times greater. If the Q value of the first reaction is twice that of the second, which has the greater threshold energy and how many times greater?

22 ■■■ Consider n (where n is an integer ≥ 1) ceramic pie plates of radius R randomly fixed on a rectangular wall with dimensions L and W. If you threw a very small (point) object at the wall, what would be the percent probability of hitting a pie plate in terms of these parameters? [*Hint:* Think in terms of area and your "reaction" cross section.]

23 ■■■ As you might have guessed, the units of cross section in SI are m^2. When neutron resonances were first discovered, they were much, much larger than any cross section previously measured and someone said they were "as big as a barn." Hence a more common unit for cross section is the barn (b) defined as 1 b $\equiv$ 10^{-28} m^2. Using the empirical equation for the radius of a nucleus (Eq. 29.1), $R = R_o A^{1/3}$, where $R_o \approx 1.2 \times 10^{-15}$ m = 1.2 fm, estimate the *geometrical* cross section (in barns) for (a) carbon-12, (b) iron-56, (c) lead-208, and (d) uranium-238.

24 ■■■ Assume that the average kinetic energy of ions in a plasma is given by the equation for the kinetic energy of the atoms in an ideal gas ($\frac{1}{2}mv^2 = \frac{3}{2}k_B T$) and that fusion occurs when the ions approach each other within a distance of the upper limit for the nuclear diameter ($R = 10^{-12}$ cm). Calculate the temperature required for fusion of two deuterium ions. (Boltzmann's constant is $k_B = 1.38 \times 10^{-23}$ J/K.)

30.2 and 30.3 Nuclear Fission and Fusion

25 Nuclear fission (a) is endoergic, (b) occurs only for uranium-235, (c) releases about 500 MeV of energy per fission, (d) requires a critical mass for a sustained reaction.

26 A nuclear reactor (a) can operate on natural (unenriched) uranium, (b) has its chain reaction controlled by neutron-absorbing materials, (c) can be partially controlled by the amount of moderator, (d) all of these.

27 A nuclear fusion reaction (a) has a negative Q value, (b) may occur spontaneously, (c) is an example of "splitting" the atom, (d) releases less than 50 MeV of energy per fusion process.

28 Controlled fusion requires (a) no critical mass, (b) confinement, (c) formation of a plasma, (d) all of these.

29 The energy produced in fission reactions is carried off as kinetic energies of the products. How is this converted to heat in a nuclear reactor?

30 ■ Find the approximate energy released in the following fission reactions: (a) $^{235}_{92}$U + $^{1}_{0}$n → with the release of 5 neutrons and (b) $^{235}_{94}$Pu + $^{1}_{0}$n → with the release of 3 neutrons.

31 ■ Calculate the amounts of energy released in the following fusion reactions: (a) $^{2}_{1}$H + $^{2}_{1}$H → $^{3}_{2}$He + $^{1}_{0}$n and (b) $^{3}_{2}$He + $^{3}_{2}$He → $^{4}_{2}$He + 2^{1}_{1}H.

32 ■ Find the Q values for the (a) H–D reaction and (b) He–He reaction in the proton–proton cycle in Section 30.3.

33 ■■ Using water as a moderator works well because the proton and neutron have nearly the same mass. For a head-on elastic collision, we might expect a neutron to lose all of its kinetic energy in one collision, whereas for an "almost miss" we might expect it to lose essentially none. Assume that on average the neutron loses one-third of its kinetic energy each collision. Estimate how many collisions are needed to reduce a 20-MeV neutron to one with a kinetic energy of only 0.02 eV (approximately "thermal").

30.4 Beta Decay and the Neutrino

34 In the absence of a neutrino, what is not conserved in beta decay: (a) energy, (b) linear momentum, (c) angular momentum, (d) all of these?

35 A neutrino interacts with matter by (a) an electrical interaction, (b) a strong interaction, (c) a weak interaction, (d) both (b) and (c).

36 Why is it so difficult to detect neutrinos experimentally?

37 ■ A neutrino created in a beta decay process has an energy of 2.65 MeV. What is the de Broglie wavelength of the neutrino?

38 ■■ In Exercise 37, what is the momentum of the beta particle *plus* the daughter nucleus? Include the direction and your reasoning.

39 ■■ In Excercise 37, if the disintegration energy was 3.51 MeV, (a) what would be the maximum possible kinetic energy of the beta particle? (b) What would its momentum (include direction) be in this case? (c) What would the kinetic energy and momentum of the daughter nucleus be in this case?

40 ■■ Show that the disintegration energy for β^- decay is
$$Q = (m_P - m_D - m_e)c^2 = (M_P - M_D)c^2$$
where the m's represent the masses of the parent and daughter *nuclei* and the M's represent the masses of the neutral *atoms*.

41 ■■ What is the maximum kinetic energy of the electron emitted when a ^{12}B nucleus beta decays into a ^{12}C nucleus? (See Exercise 40.)

42 ■■ The kinetic energy of an electron emitted from a ^{32}P nucleus that beta decays into a ^{32}S nucleus is observed to be 1.00 MeV. What is the energy of the accompanying neutrino of the decay process? Neglect the recoil energy of the daughter. (See Exercise 40.)

43 ■■ Show that the disintegration energy for β^+ decay is

$$Q = (m_P - m_D - m_e)c^2 = (M_P - M_D - 2m_e)c^2$$

where the m's represent the masses of the parent and daughter *nuclei* and the M's represent the masses of the neutral *atoms*.

44 ■■■ The kinetic energy of a positron emitted from the β^+ decay of a ^{13}N nucleus into a ^{13}C nucleus is measured to be 1.190 MeV. What is the energy of the accompanying neutrino in the process? (Neglect the recoil energy of the daughter nucleus and see Exercise 43.)

45 ■■■ On the basis of the Q values given in Exercises 40 and 43, what are the mass requirements of the parent atoms if β^- and β^+ processes are going to be energetically possible?

30.5 Fundamental Forces and Exchange Particles

46 Virtual particles (a) form virtual images, (b) exist only in a time for the violation of the conservation of energy permitted by the uncertainty principle, (c) make up positrons, (d) can be observed in exchange processes.

47 The exchange particle for the strong nuclear force is the (a) pion, (b) W particle, (c) muon, (d) positron.

48 If virtual exchange particles are unobservable by themselves, how is their existence verified?

49 ■ Draw the Feynman diagrams for Compton scattering of a photon by an electron.

50 ■ Draw the Feynman diagram for electron–positron pair annihilation.

51 ■ Assuming the range of the nuclear force to be on the order of the Bohr radius (about 0.0529 nm), predict the mass of the exchange particle.

52 ■■ In a reaction process, a high-speed proton collides with a nucleus and travels a distance of 5.0×10^{-16} m on the average before the reaction takes place. What type of interaction is this and during what time period does the interaction take place?

53 ■■ By what minimum energy is the conservation of energy "violated" during a neutral pi meson exchange process?

54 ■■ How long is the conservation of energy "violated" in a neutral pi meson exchange process?

55 ■■ A W particle in a weak interaction is found to have an energy of 100 MeV. What is the approximate range for the weak interaction with this particle?

30.6 Elementary Particles

56 Particles that interact by the strong nuclear force are called (a) muons, (b) hadrons, (c) W particles, (d) leptons.

57 Quarks make up which of the following particles? (a) baryons, (b) muons, (c) Z particles, (d) all of these.

58 What is meant by quark flavor and color? Can these be changed? Explain.

59 With so many types of hadrons, why aren't fractional electronic charges observed?

Additional Exercises

60 Complete the following nuclear reactions:
(a) $^6_3$Li + $^1_1$H $\rightarrow$ $^3_2$He + _____
(b) $^{58}_{28}$Ni + $^2_1$H $\rightarrow$ $^{59}_{28}$Ni + _____
(c) $^{235}_{92}$U + $^1_0$n $\rightarrow$ $^{138}_{54}$Xe + 5^1_0n + _____
(d) $^9_4$Be(α, n) _____
(e) $^{16}_8$O(n, p) _____

61 Give the compound nuclei for the reactions in Exercise 60.

62 Compute the Q values (to three significant figures) of the following fusion reactions: (a) $^2_1$H + $^3_1$H $\rightarrow$ $^4_2$He + $^1_0$n and (b) $^2_1$H + $^2_1$H $\rightarrow$ $^3_1$H + $^1_1$H.

63 Determine the threshold energy of the following reaction:

$$\begin{array}{ccccccc} ^{16}_8\text{O} & + & ^1_0\text{n} & \rightarrow & ^{13}_6\text{C} & + & ^4_2\text{He} \\ (15.994915\ u) & & (1.008665\ u) & & (13.003355\ u) & & (4.002603\ u) \end{array}$$

64 Polonium-210 undergoes alpha decay. Find the energy released in the decay reaction.

65 Show that the Q value for electron capture is given by

$$Q = (m_P + m_e - m_D)c^2 = (M_P - M_D)c^2$$

where the m's represent the masses of the parent and daughter *nuclei* and the M's represent the masses of the neutral *atoms*.

66 (a) In the electron capture process of a ^{7}Be nucleus converting into a ^{7}Li nucleus, what is the energy of the emitted neutrino (neglect daughter recoil)? (See Exercise 65.) (b) Is it energetically possible for ^{7}Be to β^+ decay into ^{7}Li?

67 Complete the following nuclear reactions:
(a) $^{19}_9$F + $^1_0$n $\rightarrow$ $^1_1$H + _____
(b) $^{13}_7$N + $^1_0$n $\rightarrow$ $^4_2$He + _____
(c) $^{92}_{40}$Zr(p, α) _____
(d) $^3_2$He(γ, p) _____

Appendices

APPENDIX I Mathematical Relationships

ALGEBRAIC RELATIONSHIPS

$(a + b)^2 = a^2 + 2ab + b^2$
$(a - b)^2 = a^2 - 2ab + b^2$
$(a^2 - b^2) = (a + b)(a - b)$

Quadratic Formula

if $ax^2 + bx + c = 0$,

$$x = \frac{-b \pm \sqrt{b^2 - 4ac}}{2a}$$

Powers and Exponents

$x^0 = 1$

$x^1 = x$ $\qquad\qquad x^{-1} = \dfrac{1}{x}$

$x^2 = x \cdot x = x^2$ $\quad x^{-2} = \dfrac{1}{x^2}$ $\quad x^{\frac{1}{2}} = \sqrt{x}$

$x^3 = x \cdot x \cdot x = x^3$ $\quad x^{-3} = \dfrac{1}{x^3}$ $\quad x^{\frac{1}{3}} = \sqrt[3]{x}$

etc. $\qquad\qquad$ etc. $\qquad\qquad$ etc.

$x^a \cdot x^b = x^{(a+b)}$
$x^a / x^b = x^{(a-b)}$
$(x^a)^b = x^{ab}$

Logarithms

if $x = a^n$, then $n = \log_a x$

$\log xy = \log x + \log y$
$\log x/y = \log x - \log y$
$\log x^y = y \log x$

common logarithms: base 10
(assumed when abbreviation "log" is used unless another base is specified)

$\log 10^x = x$

natural logarithms: base $e = 2.71828 \ldots$ (abbreviated "ln")

$\ln e^x = x$
$\log x = 0.43429 \ln x$
$\ln x = 2.3026 \log x$

GEOMETRIC AND TRIGONOMETRIC RELATIONSHIPS

Areas and Volumes of Some Common Shapes

Circle: $\quad A = \pi r^2 = \dfrac{\pi d^2}{4}$ $\quad$ (area)

$\qquad\qquad c = 2\pi r = \pi d$ $\quad$ (circumference)

Triangle: $\quad A = \frac{1}{2}ab$

Sphere: $\quad A = 4\pi r^2$
$\qquad\qquad V = \frac{4}{3}\pi r^3$

Cylinder: $\quad A = \pi r^2$ $\quad$ (end)
$\qquad\qquad A = 2\pi rh$ $\quad$ (body)
$\qquad\qquad A = 2(\pi r^2) + 2\pi rh$ $\quad$ (total)
$\qquad\qquad V = \pi r^2 h$

Definitions of Trigonometric Functions

$\sin \theta = \dfrac{y}{r}$ $\qquad \cos \theta = \dfrac{x}{r}$ $\qquad \tan \theta = \dfrac{\sin \theta}{\cos \theta} = \dfrac{y}{x}$

$\theta°$ (rad)	$\sin \theta$	$\cos \theta$	$\tan \theta$
0° (0)	0	1	0
30° ($\pi/6$)	0.500	0.866	0.577
45° ($\pi/4$)	0.707	0.707	1.00
60° ($\pi/3$)	0.866	0.500	1.73
90° ($\pi/2$)	1	0	$\rightarrow \infty$

See trigonometric tables for other angles.

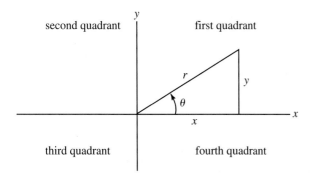

For very small angles:

$$\cos \theta \approx 1 \qquad \sin \theta \approx \theta \text{ (radians)}$$

$$\tan \theta = \frac{\sin\theta}{\cos\theta} \equiv \theta \text{ (radians)}$$

The sign of a trigonometric function depends on the quadrant, or the signs of x and y; for example, in the second quadrant $(-x, y)$, $-x/r = \cos \theta$ and $y/r = \sin \theta$. The sign can also be assigned using the reduction formulas.

Reduction Formulas

	(θ in second quadrant)	(θ in third quadrant)	(θ in fourth quadrant)
$\sin \theta =$	$\cos(\theta - 90°) =$	$-\sin(\theta - 180°) =$	$-\cos(\theta - 270°)$
$\cos \theta =$	$-\sin(\theta - 90°) =$	$-\cos(\theta - 180°) =$	$\sin(\theta - 270°)$

Fundamental Identities

$$\sin^2 \theta + \cos^2 \theta = 1$$
$$\sin 2\theta = 2 \sin \theta \cos \theta$$
$$\cos 2\theta = \cos^2 \theta - \sin^2 \theta = 2 \cos^2 \theta - 1 = 1 - 2 \sin^2 \theta$$
$$\sin^2 \theta = \tfrac{1}{2}(1 - \cos 2\theta)$$
$$\cos^2 \theta = \tfrac{1}{2}(1 + \cos 2\theta)$$

For half-angle ($\theta/2$) identities, replace θ with $\theta/2$, for example:

$$\sin^2 \theta/2 = \tfrac{1}{2}(1 - \cos \theta) \qquad \cos^2 \theta/2 = \tfrac{1}{2}(1 + \cos \theta)$$

$$\sin(\alpha \pm \beta) = \sin \alpha \cos \beta \pm \cos \alpha \sin \beta$$
$$\cos(\alpha \pm \beta) = \cos \alpha \cos \beta \mp \sin \alpha \sin \beta$$
$$\tan(\alpha \pm \beta) = \frac{\tan \alpha \pm \tan \beta}{1 \mp \tan \alpha \tan \beta}$$

Law of Cosines

For a triangle with angles X, Y, and Z and opposite sides x, y, and z, respectively:

$$x^2 = y^2 + z^2 - 2yz \cos X$$

For $X = 90°$, this reduces to the Pythagorean theorem:

$$x^2 = y^2 + z^2$$

Trigonometric Tables

| Angle | | | | | Angle | | | | |
Degrees	Radians	Sine	Cosine	Tangent	Degrees	Radians	Sine	Cosine	Tangent
0°	.000	0.000	1.000	0.000					
1°	.017	.018	1.000	.018	46°	0.803	0.719	0.695	1.036
2°	.035	.035	0.999	.035	47°	.820	.731	.682	1.072
3°	.052	.052	.999	.052	48°	.838	.743	.669	1.111
4°	.070	.070	.998	.070	49°	.855	.755	.656	1.150
5°	.087	.087	.996	.088	50°	.873	.766	.643	1.192
6°	.105	.105	.995	.105	51°	.890	.777	.629	1.235
7°	.122	.122	.993	.123	52°	.908	.788	.616	1.280
8°	.140	.139	.990	.141	53°	.925	.799	.602	1.327
9°	.157	.156	.988	.158	54°	.942	.809	.588	1.376
10°	.175	.174	.985	.176	55°	.960	.819	.574	1.428
11°	.192	.191	.982	.194	56°	.977	.829	.559	1.483
12°	.209	.208	.978	.213	57°	.995	.839	.545	1.540
13°	.227	.225	.974	.231	58°	1.012	.848	.530	1.600
14°	.244	.242	.970	.249	59°	1.030	.857	.515	1.664
15°	.262	.259	.966	.268	60°	1.047	.866	.500	1.732
16°	.279	.276	.961	.287	61°	1.065	.875	.485	1.804
17°	.297	.292	.956	.306	62°	1.082	.883	.470	1.881
18°	.314	.309	.951	.325	63°	1.100	.891	.454	1.963
19°	.332	.326	.946	.344	64°	1.117	.899	.438	2.050
20°	.349	.342	.940	.364	65°	1.134	.906	.423	2.145
21°	.367	.358	.934	.384	66°	1.152	.914	.407	2.246
22°	.384	.375	.927	.404	67°	1.169	.921	.391	2.356
23°	.401	.391	.921	.425	68°	1.187	.927	.375	2.475
24°	.419	.407	.914	.445	69°	1.204	.934	.358	2.605
25°	.436	.423	.906	.466	70°	1.222	.940	.342	2.747
26°	.454	.438	.899	.488	71°	1.239	.946	.326	2.904
27°	.471	.454	.891	.510	72°	1.257	.951	.309	3.078
28°	.489	.470	.883	.532	73°	1.274	.956	.292	3.271
29°	.506	.485	.875	.554	74°	1.292	.961	.276	3.487
30°	.524	.500	.866	.577	75°	1.309	.966	.259	3.732
31°	.541	.515	.857	.601	76°	1.326	.970	.242	4.011
32°	.559	.530	.848	.625	77°	1.344	.974	.225	4.331
33°	.576	.545	.839	.649	78°	1.361	.978	.208	4.705
34°	.593	.559	.829	.675	79°	1.379	.982	.191	5.145
35°	.611	.574	.819	.700	80°	1.396	.985	.174	5.671
36°	.628	.588	.809	.727	81°	1.414	.988	.156	6.314
37°	.646	.602	.799	.754	82°	1.431	.990	.139	7.115
38°	.663	.616	.788	.781	83°	1.449	.993	.122	8.144
39°	.681	.629	.777	.810	84°	1.466	.995	.105	9.514
40°	.698	.643	.766	.839	85°	1.484	.996	.087	11.43
41°	.716	.656	.755	.869	86°	1.501	.998	.070	14.30
42°	.733	.669	.743	.900	87°	1.518	.999	.052	19.08
43°	.751	.682	.731	.933	88°	1.536	.999	.035	28.64
44°	.768	.695	.719	.966	89°	1.553	1.000	.017	57.29
45°	.785	.707	.707	1.000	90°	1.571	1.000	.000	∞

The basic assumptions are as follows:

1. The molecules of a pure gas all have the same mass (m) and are in continuous and completely random motion. (The mass of each molecule is so small that the effect of gravity on it is negligible.)

2. The gas molecules are separated by large distances and occupy a volume that is negligible compared to these distances; that is, they are point particles.

3. The molecules exert no forces on each other except when they collide.

4. The collisions of the molecules with one another and the walls of the container are perfectly elastic.

The magnitude of the force exerted on the wall of the container by a gas molecule colliding with it is $F = \Delta p / \Delta t$. Assuming that the direction of the velocity (v_x) is normal to the wall, the magnitude of the average force is

$$F = \frac{\Delta(mv)}{\Delta t} = \frac{mv_x - (-mv_x)}{\Delta t} = \frac{2mv_x}{\Delta t} \qquad (1)$$

After striking one wall of the container, which for convenience is assumed to be cubical with sides of dimensions L, the molecule recoils in a straight line. Suppose that the molecule reaches the opposite wall without colliding with any other molecules along the way. The molecule then travels the distance L in a time equal to L/v_x. After the collision with that wall, again assuming no collisions on the return trip, the round trip will take $\Delta t = 2L/v_x$. Thus, the number of collisions per unit time a molecule makes with a particular wall is $v_x/2L$, and the average force of the wall from successive collisions is

$$F = \frac{2mv_x}{\Delta t} = \frac{2mv_x}{2L/v_x} = \frac{mv_x^2}{L} \qquad (2)$$

The random motions of the many molecules produce a relatively constant force on the walls, and the pressure (p) is the total force on a wall divided by its area:

$$p = \frac{\Sigma F_i}{L^2} = \frac{m(v_{x_1}^2 + v_{x_2}^2 + v_{x_3}^2 + \cdots)}{L^3} \qquad (3)$$

The subscripts refer to individual molecules.

The average of the squares of the speeds is given by

$$\overline{v_x^2} = \frac{v_{x_1}^2 + v_{x_2}^2 + v_{x_3}^2 + \cdots}{N}$$

where N is the number of molecules in the container. In terms of this average, Eq. 3 may be written

$$p = \frac{Nm\overline{v_x^2}}{L^3} \qquad (4)$$

However, the molecules' motions occur with equal frequency along any one of the three axes, so $\overline{v_x^2} = \overline{v_y^2} = \overline{v_z^2}$, and $\overline{v^2} = \overline{v_x^2} + \overline{v_y^2} + \overline{v_z^2} = 3\overline{v_x^2}$. Then,

$$\sqrt{\overline{v^2}} = \overline{v}$$

where $\overline{v}$ is called the root-mean-square (rms) speed, and we write $(\overline{v})^2 = \overline{v^2}$. Substituting this result into Eq. 4 along with V for L^3 (since L^3 is the volume of the cubical container) gives

$$p = \frac{Nm\overline{v}^2}{3V} \qquad \text{or} \quad pV = \tfrac{1}{3}Nm\overline{v}^2 \qquad (5)$$

This result is true even though collisions between molecules were ignored. Statistically, these collisions average out, so the number of collisions with each wall is as described. This result is also independent of the shape of the container. A cube merely simplifies the derivation.

Combining this result with the empirical perfect gas law gives

$$pV = NkT = \tfrac{1}{3}Nm\overline{v}^2$$

The average kinetic energy per gas molecule is thus proportional to the absolute temperature of the gas:

$$\tfrac{1}{2}m\overline{v}^2 = \tfrac{3}{2}kT \qquad (6)$$

The collision time is negligible compared with the time between collisions. Some kinetic energy will be momentarily converted to potential energy during a collision; however, this potential energy can be ignored because each molecule spends a negligible amount of time in collisions. Therefore, by this approximation, the total kinetic energy is the internal energy of the gas, and the internal energy of a perfect gas is directly proportional to its absolute temperature.

APPENDIX III Planetary Data

Name	Equatorial Radius (km)	Mass (Compared to Earth's)*	Mean Density (× 10³ kg/m³)	Surface Gravity (Compared to Earth's)	Semimajor Axis × 10⁶ km	Semimajor Axis AU	Period Years	Period Days	Eccentricity	Inclination to Ecliptic
Mercury	2,439	0.0553	5.43	0.378	57.9	0.3871	0.24084	87.96	0.2056	7°00′26″
Venus	6,052	0.8150	5.24	0.894	108.2	0.7233	0.61515	224.68	0.0068	3°23′40″
Earth	6,378.140	1	5.515	1	149.6	1	1.00004	365.25	0.0167	0°00′14″
Mars	3,397.2	0.1074	3.93	0.379	227.9	1.5237	1.8808	686.95	0.0934	1°51′09″
Jupiter	71,398	317.89	1.36	2.54	778.3	5.2028	11.862	4,337	0.0483	1°18′29″
Saturn	60,000	95.17	0.71	1.07	1427.0	9.5388	29.456	10,760	0.0560	2°29′17″
Uranus	26,145	14.56	1.30	0.8	2871.0	19.1914	84.07	30,700	0.0461	0°48′26″
Neptune	24,300	17.24	1.8	1.2	4497.1	30.0611	164.81	60,200	0.0100	1°46′27″
Pluto	1,500–1,800	0.02	0.5–0.8	~0.03	5913.5	39.5294	248.53	90,780	0.2484	17°09′03″

*Planet's mass/Earth's mass, where $M_E = 6.0 \times 10^{24}$ kg.

APPENDIX IV Alphabetical Listing of the Chemical Elements

Element	Symbol	Atomic Number (Proton Number)	Atomic Mass	Element	Symbol	Atomic Number (Proton Number)	Atomic Mass	Element	Symbol	Atomic Number (Proton Number)	Atomic Mass
Actinium	Ac	89	227.0278	Curium	Cm	96	(247)	Lawrencium	Lr	103	(260)
Aluminum	Al	13	26.98154	Dysprosium	Dy	66	162.50	Lead	Pb	82	207.2
Americium	Am	95	(243)[a]	Einsteinium	Es	99	(252)	Lithium	Li	3	6.941
Antimony	Sb	51	121.757	Erbium	Er	68	167.26	Lutetium	Lu	71	174.967
Argon	Ar	18	39.948	Europium	Eu	63	151.96	Magnesium	Mg	12	24.305
Arsenic	As	33	74.9216	Fernium	Fm	100	(257)	Manganese	Mn	25	54.9380
Astatine	At	85	(210)	Fluorine	F	9	18.998403	Mendelevium	Md	101	(258)
Barium	Ba	56	137.33	Francium	Fr	87	(223)	Mercury	Hg	80	200.59
Berkelium	Bk	97	(247)	Gadolinium	Gd	64	157.25	Molybdenum	Mo	42	95.94
Beryllium	Be	4	9.01218	Gallium	Ga	31	69.72	Neodymium	Nd	60	144.24
Bismuth	Bi	83	208.9804	Germanium	Ge	32	72.561	Neon	Ne	10	20.1797
Boron	B	5	10.81	Gold	Au	79	196.9665	Neptunium	Np	93	237.048
Bromine	Br	35	79.904	Hafnium	Hf	72	178.49	Nickel	Ni	28	58.69
Cadmium	Cd	48	112.41	Hahnium	Ha	105	(262)	Niobium	Nb	41	92.9064
Calcium	Ca	20	40.078	Helium	He	2	4.00260	Nitrogen	N	7	14.0067
Californium	Cf	98	(251)	Holmium	Ho	67	164.9304	Nobelium	No	102	(259)
Carbon	C	6	12.011	Hydrogen	H	1	1.00794	Osmium	Os	76	190.2
Cerium	Ce	58	140.12	Indium	In	49	114.82	Oxygen	O	8	15.9994
Cesium	Cs	55	132.9054	Iodine	I	53	126.9045	Palladium	Pd	46	106.42
Chlorine	Cl	17	35.453	Iridium	Ir	77	192.22	Phosphorus	P	15	30.97376
Chromium	Cr	24	51.996	Iron	Fe	26	55.847	Platinum	Pt	78	195.08
Cobalt	Co	27	58.9332	Krypton	Kr	36	83.80	Plutonium	Pu	94	(244)
Copper	Cu	29	63.546	Lanthanum	La	57	138.9055	Polonium	Po	84	(209)

Element	Symbol	Atomic Number (Proton Number)	Atomic Mass	Element	Symbol	Atomic Number (Proton Number)	Atomic Mass	Element	Symbol	Atomic Number (Proton Number)	Atomic Mass
Potassium	K	19	39.0983	Silicon	Si	14	28.0855	Tungsten	W	74	183.85
Praseodymium	Pr	59	140.9077	Silver	Ag	47	107.8682	Uranium	U	92	238.0289
Promethium	Pm	61	(145)	Sodium	Na	11	22.98977	Vanadium	V	23	50.9415
Protactinium	Pa	91	231.0359	Strontium	Sr	38	87.62	Xenon	Xe	54	131.29
Radium	Ra	88	226.0254	Sulfur	S	16	32.066	Ytterbium	Yb	70	173.04
Radon	Rn	86	(222)	Tantalum	Ta	73	180.9479	Yttrium	Y	39	88.9059
Rhenium	Re	75	186.207	Technetium	Tc	43	(98)	Zinc	Zn	30	65.39
Rhodium	Rh	45	102.9055	Tellurium	Te	52	127.60	Zirconium	Zr	40	91.22
Rubidium	Rb	37	85.4678	Terbium	Tb	65	158.9254			106	(263)
Ruthenium	Ru	44	101.07	Thallium	Tl	81	204.383			107	(262)
Rutherfordium	Rf	104	(261)	Thorium	Th	90	232.0381			108	(265)
Samarium	Sm	62	150.36	Thulium	Tm	69	168.9342			109	(266)
Scandium	Sc	21	44.9559	Tin	Sn	50	118.710			110	(269)
Selenium	Se	34	78.96	Titanium	Ti	22	47.88			111	(272)

APPENDIX V Properties of Selected Isotopes

Atomic Number (Z)	Element	Symbol	Mass Number (A)	Atomic Mass*	Abundance (%) or Decay Mode† (if radioactive)	Half-Life (if radioactive)
0	(Neutron)	n	1	1.008665	β^-	10.6 min
1	Hydrogen	H	1	1.007825	99.985	
	Deuterium	D	2	2.014102	0.015	
	Tritium	T	3	3.016049	β^-	12.33 years
2	Helium	He	3	3.016029	0.00014	
			4	4.002603	≈100	
3	Lithium	Li	6	6.015123	7.5	
			7	7.016005	92.5	
4	Beryllium	Be	7	7.016930	EC, γ	53.3 days
			8	8.005305	2α	6.7×10^{-17} s
			9	9.012183	100	
5	Boron	B	10	10.012938	19.8	
			11	11.009305	80.2	
			12	12.014353	β^-	20.4 ms
6	Carbon	C	11	11.011433	β^+, EC	20.4 ms
			12	12.000000	98.89	
			13	13.003355	1.11	
			14	14.003242	β^-	5730 years
7	Nitrogen	N	13	13.005739	β^-	9.96 min
			14	14.003074	99.63	
			15	15.000109	0.37	
8	Oxygen	O	15	15.003065	β^+, EC	122 s
			16	15.994915	99.76	
			18	17.999159	0.204	
9	Fluorine	F	19	18.998403	100	

Atomic Number (Z)	Element	Symbol	Mass Number (A)	Atomic Mass*	Abundance (%) or Decay Mode† (if radioactive)	Half-Life (if radioactive)
10	Neon	Ne	20	19.992439	90.51	
			22	21.991384	9.22	
11	Sodium	Na	22	21.994435	β^+, EC, γ	2.602 years
			23	22.989770	100	
			24	23.990964	β^-, γ	15.0 h
12	Magnesium	Mg	24	23.985045	78.99	
13	Aluminum	Al	27	26.981541	100	
14	Silicon	Si	28	27.976928	92.23	
			31	30.975364	β^-, γ	2.62 h
15	Phosphorus	P	31	30.973763	100	
			32	31.973908	β^-	14.28 days
16	Sulfur	S	32	31.972072	95.0	
			35	34.969033	β^-	87.4 days
17	Chlorine	Cl	35	34.968853	75.77	
			37	36.965903	24.23	
18	Argon	Ar	40	39.962383	99.60	
19	Potassium	K	39	38.963708	93.26	
			40	39.964000	β^-, EC, γ, β^+	1.28×10^9 years
20	Calcium	Ca	40	39.962591	96.94	
24	Chromium	Cr	52	51.940510	83.79	
25	Manganese	Mn	55	54.938046	100	
26	Iron	Fe	56	55.934939	91.8	
27	Cobalt	Co	59	58.933198	100	
			60	59.933820	β^-, γ	5.271 years
28	Nickel	Ni	58	57.935347	68.3	
			60	59.930789	26.1	
			64	63.927968	0.91	
29	Copper	Cu	63	62.929599	69.2	
			64	63.929766	β^-, β^+	12.7 h
			65	64.927792	30.8	
30	Zinc	Zn	64	63.929145	48.6	
			66	65.926035	27.9	
33	Arsenic	As	75	74.921596	100	
35	Bromine	Br	79	78.918336	50.69	
36	Krypton	Kr	84	83.911506	57.0	
			89	88.917563	β^-	3.2 min
38	Strontium	Sr	86	85.909273	9.8	
			88	87.905625	82.6	
			90	89.907746	β^-	28.8 years
39	Yttrium	Y	89	89.905856	100	
43	Technetium	Tc	98	97.907210	β^-, γ	4.2×10^6 years
47	Silver	Ag	107	106.905095	51.83	
			109	108.904754	48.17	
48	Cadmium	Cd	114	113.903361	28.7	
49	Indium	In	115	114.90388	95.7; β^-	5.1×10^{14} years
50	Tin	Sn	120	119.902199	32.4	
53	Iodine	I	127	126.904477	100	
			131	130.906118	β^-, γ	8.04 days
54	Xenon	Xe	132	131.90415	26.9	
			136	135.90722	8.9	
55	Cesium	Cs	133	132.90543	100	

Atomic Number (Z)	Element	Symbol	Mass Number (A)	Atomic Mass*	Abundance (%) or Decay Mode† (if radioactive)	Half-Life (if radioactive)
56	Barium	Ba	137	136.90582	11.2	
			138	137.90524	71.7	
			144	143.92273	β^-	11.9 s
61	Promethium	Pm	145	144.91275	EC, α, γ	17.7 years
74	Tungsten (wolfram)	W	184	183.95095	30.7	
76	Osmium	Os	191	190.96094	β^-, γ	15.4 days
			192	191.96149	41.0	
78	Platinum	Pt	195	194.96479	33.8	
79	Gold	Au	197	196.96656	100	
80	Mercury	Hg	202	201.97063	29.8	
81	Thallium	Tl	205	204.97441	70.5	
			210	209.990069	β^-	1.3 min
82	Lead	Pb	204	203.973044	β^-, 1.48	1.4×10^{17} years
			206	205.97446	24.1	
			207	206.97589	22.1	
			208	207.97664	52.3	
			210	209.98418	α, β^-, γ	22.3 years
			211	210.98874	β^-, γ	36.1 min
			212	211.99188	β^-, γ	10.64 h
			214	213.99980	β^-, γ	26.8 min
83	Bismuth	Bi	209	208.98039	100	
			211	210.98726	α, β^-, γ	2.15 min
84	Polonium	Po	210	209.98286	α, γ	138.38 days
			214	213.99519	α, γ	164 μs
86	Radon	Rn	222	222.017574	α, β	3.8235 days
87	Francium	Fr	223	223.019734	α, β^-, γ	21.8 min
88	Radium	Ra	226	226.025406	α, γ	1.60×10^3 years
			228	228.031069	β^-	5.76 years
89	Actinium	Ac	227	227.027751	α, β^-, γ	21.773 years
90	Thorium	Th	228	228.02873	α, γ	1.9131 years
			232	232.038054	100; α, γ	1.41×10^{10} years
92	Uranium	U	232	232.03714	α, γ	72 years
			233	233.039629	α, γ	1.592×10^5 years
			235	235.043925	0.72; α, γ	7.038×10^8 years
			236	236.045563	α, γ	2.342×10^7 years
			238	238.050786	99.275; α, γ	4.468×10^9 years
			239	239.054291	β^-, γ	23.5 min
93	Neptunium	Np	239	239.052932	β^-, γ	2.35 days
94	Plutonium	Pu	239	239.052158	α, γ	2.41×10^4 years
95	Americium	Am	243	243.061374	α, γ	7.37×10^3 years
96	Curium	Cm	245	245.065487	α, γ	8.5×10^3 years
97	Berkelium	Bk	247	247.07003	α, γ	1.4×10^3 years
98	Californium	Cf	249	249.074849	α, γ	351 years
99	Einsteinium	Es	254	254.08802	α, γ, β^-	276 days
100	Fernium	Fm	253	253.08518	EC, α, γ	3.0 days
101	Mendelevium	Md	255	255.0911	EC, α	27 min
102	Nobelium	No	255	255.0933	EC, α	3.1 min
103	Lawrencium	Lr	257	257.0998	α	$\approx$35 s
104	Rutherfordium	Rf	261	261.1087	α	1.1 min
105	Hahnium	Ha	262	262.1138	α	0.7 min

* The masses given throughout this table are those for the neutral atom, including the Z electrons.
† EC stands for electron capture.

Answers to Follow-up Exercises

Chapter 1

1.1 1000 kg has a weight of 2,200 lb. Weight of metric ton is equivalent to the British *long* ton (2,200 lb), or 200 lb greater than the British *short* ton (2000 lb).

1.2 Yes, m = m.

1.3 (a) 50 mi/h [(0.447 m/s)/mi/h] = 22 m/s. (b) Since (1609 m/mi) = 1 and (1 h/3600 s) = 1, then multiplying the two factors gives 1. Then, trading denominators, (1609 m/3600 s)(h/mi) = 1. Thus, 0.447 m/s = 1 mi/h.

1.4 $1 \text{ m}^3 = 10^6 \text{ cm}^3$.

1.5 (a) To get the smallest numerical value, we must make the number in the numerator as small as possible and the number in the denominator as large as possible. It is clear from Example 1.5 that, since 1 mi = 1.61 km, the distance in the numerator should be in miles. Similarly, since there are about 4 L in 1 gal, the volume in the denominator should be in liters. Expressing fuel economy in mi/L will give the smallest numerical value. (b) To demonstrate the comparative numerical values, we convert one set of units to the other three. Arbitrarily selecting 1 mi/gal, on converting, we obtain 1 mi/gal = 1.61 km/gal = 0.264 mi/L = 0.425 km/L.

1.6 (a) $7.0 \times 10^5 \text{ kg}^2$ (b) 3.02×10^2 (no units).

1.7 (a) 23.70 (b) 22.09.

1.8 $2.3 \times 10^{-3} \text{ m}^3$ or $2.3 \times 10^3 \text{ cm}^3$.

1.9 16.9 m.

1.10 $1.30 \times 10^3 \text{ cm}^2$.

Chapter 2

2.1 (a) Not if there is motion; $\Delta x \neq 0$, then $\bar{s} = \Delta x/\Delta t \neq 0$. A zero speed would simply imply that $\Delta x \neq 0$, or the object is at rest. (b) 4.4×10^4 mi/h (44,000 mi/h—pretty fast).

2.2 33.3 mi/h.

2.3 No. The + and − signs indicate the vector directions with respect to the reference axis. If the velocity and acceleration are in *opposite* directions, a moving object will slow down. For example, suppose that a car traveling in the negative x direction ($-v_o$) experiences an acceleration in the positive x direction ($+a$). The car would slow down, so a positive acceleration ($+a$) can produce a deceleration. Similarly, if the velocity and acceleration are both in the *same* direction, the car will speed up. For example, if the car is initially traveling in the negative x direction ($-v_o$), a negative acceleration ($-a$) will speed it up in that direction.

2.4 9.0 m/s in the direction of the original motion.

2.5 96 m. (A lot quicker isn't it?)

2.6 0.904 s. (Positive root taken. Why?)

2.7 The speeds of the cars as a function of time are given by Eq. 2.7, $v = v_o + at$. With $v_o = 0$, this becomes $v = at$. Hence, speed is proportional to t (i.e., $v \propto t$); so if car B accelerates twice as long, it will have twice the speed. (No t^2 dependence here.)

2.8 20 ft. (So speed in a school zone can make a difference.)

2.9 1.16 s longer.

2.10 Time for bill to fall its length = 0.179 s. This is less than the average reaction time (0.192 s) computed in the Example, so most people cannot catch bill.

2.11 $y_u = y_d = 5.12$ m, as measured from reference $y = 0$ at release point.

2.12 As long as the ball is in flight, its position is given by Eq.

2.9′. We know that at $t = 2.88$ s, the ball has returned to the starting point ($y = 0$), and the negative t^2 term in the equation is equal in magnitude to the t term. For $t > 2.88$ s, the negative t^2 term dominates and negative y values are obtained, which means that the ball's position is below the starting or zero reference point.

Chapter 3

3.1 $v_x = -0.40$ m/s (in $-x$ direction), $v_y = 0.30$ m/s.

3.2 $x = 9.00$ m, $y = 12.6$ m (same).

3.3 Relative to the ground, or "fixed" reference frame.

3.4 $\mathbf{v} = 0 \, \mathbf{x} + 3.7 \text{ m/s } \mathbf{y}$.

3.5 $d \, (= 524.976 \text{ m}) = 525$ m. (Rounding differences.)

3.6 14.5° W of N.

3.7 $\mathbf{v} = 8.25 \text{ m/s } \mathbf{x} - 22.1 \text{ m/s } \mathbf{y}$.

3.8 $\mathbf{v} = 30.0$ m/s, $\theta = -37°$ (relative to x-axis). (v_{x_o} unchanged, $v_y = -v_{y_o}$.)

3.9 From Example 3.8, $t_{max} = 1.85$ s, and Δt *(after)* = 3.00 s − 1.85 s = 1.15 s. By symmetry, then, height is the same at 1.15 s *before* t_{max} or $t = 1.85$ s − 1.15 s = 0.70 s.

3.10 $\theta = 13°$.

3.11 To understand this effect, consider the velocity components of the player's motion. Note that for the vertical motion near the maximum height (Fig. 3.16b), the upward speed (v_y) is quite small, going to zero at the maximum height and then increasing from zero in the descent. During this time, the combination of the slow vertical motions and the constant horizontal motion gives the illusion that the player hangs in the air.

3.12 27 m.

3.13 0.694 s and 3.00 s (Note the slight difference from the 3.9 answer.)

Chapter 4

4.1 Could compute $\Delta v = v_4 - v_3$, but by definition for a constant acceleration of 1.09 m/s², the velocity changes 1.09 m/s each second.

4.2 11 lb.

4.3 8.3 N.

4.4 Neglecting air resistance, $F = w = mg = (0.750 \text{ kg})(9.80 \text{ m/s}^2) = 7.35$ N, (downward).

4.5 2000 N.

4.6 $a = \dfrac{(m_2 - m_1)g - f}{m_1 + m_2}$.

4.7 (a) $m_2 > 1.7$ kg (b) $\theta < 17°$.

4.8 Yes, $T \propto 1/\sin \theta$. Move supports closer together so angle (and $\sin \theta$) are greater, and T is reduced.

4.9 (a) No, the juggler and the balls would be in near free fall with a relative velocity of zero. If a ball in hand were tossed upward, it would not return to the hand; a ball that had been falling toward the hand would not reach it. (b) If the bottom of the cage were solid (or had paper), then the force created by the downward strokes of the wings would be transmitted via the air to the cage floor, and the scale would read the same. However, if the cage bottom were open so that the air could pass through, the scale would read less.

4.10 $\mu_s = (1.40)\mu_k$, for three cases in Table 4.1.

4.11 F varies with angle, with the angle for minimum applied force being around 33° in this case (greater forces are required for 20° and 50°). In general, the optimum angle depends on the coefficient of friction.

4.12 Taking F_1 and F_2 as the pulling and pushing forces, respectively, we have $N = mg - F_1 \sin \theta_1 + F_2 \sin \theta_2$, where the θs are the direction angles of the respective forces. Thus, the magnitude of the normal force depends on both the magnitudes of the forces and the angles at which they act. If $F_1 = F_2$ and $\theta_1 = \theta_2$, the vertical components of the applied forces cancel each other. Otherwise, the normal force could be affected. (It is possible for both the forces and the angles to be different *without* affecting the normal force. How?)

4.13 $mg \sin \theta - f_k = ma$, block accelerates down the plane. No, would have to measure acceleration.

4.14 Air resistance depends not only on speed, but also on size and shape. If the heavier ball were larger, it would have more exposed area to collide with air molecules, and the retarding force would build up faster. Depending on the size difference, the heavier ball might reach terminal velocity first and the lighter ball strike the ground first. Also, the balls might reach terminal velocity together. How would they strike the ground in this case?

Chapter 5

5.1 15 J.

5.2 Work is done initially in lifting the handles and load, but when moving forward at a constant height no work is done by the vertical component because that component is perpendicular to the motion.

5.3 No, this would require the force of kinetic friction to be greater than the gravitational component down the plane, so that the net force is up the plane.

5.4 0.64 J.

5.5 No, $W_2/W_1 = 4$, or 4 times as much.

5.6 Here we have $m_s = m_g/2$ as before. However, $v_s/v_g = (6.0 \text{ m/s})/(4.0 \text{ m/s}) = \frac{3}{2}$. Using a ratio as in the Example, $K_s/K_g = \frac{9}{8}$, and the safety still has more kinetic energy than the guard. (Answer could also be obtained from direct calculations of kinetic energies, but for a relative comparison, a ratio is usually quicker.)

5.7 $W_3/W_2 = 1.4$ or 40% larger. More work, but smaller percentage increase.

5.8 $\Delta K_T = 0$, $\Delta U_T = 0$.

5.9 9.9 m/s.

5.10 To determine whether the speed depends on the mass of a ball *in principle*, one needs to look at the physical relationships that apply to the situation. Here, energy is the major consideration, and as always, one should keep in mind the versatility of the conservation of energy in analyzing phenomena. The mechanical energy is conserved while the balls are in flight, so let's consider the initial energy (E_o) of a ball and its energy (E) just before striking the ground. By the conservation of energy,

$$E_o = E$$
$$\tfrac{1}{2}mv_o^2 + mgh = \tfrac{1}{2}mv^2$$

and
$$v = \sqrt{v_o^2 - 2gh}$$

As can be seen, the mass cancels and doesn't appear in the equation. Thus, speed is independent of mass. (Recall that objects or projectiles in free fall all fall with the same vertical acceleration g—see Section 2.5.)

5.11 0.025 m.

5.12 59%.

5.13 52%.

5.14 (a) Does the same work in twice the time, or half the work in the same time. (b) Does the same work in half the time, or twice the work in the same time.

5.15 (a) No. This would mean that all the energy input could be converted into useful work with no losses. Practically, friction and losses are always present. (b) The work output would be greater than the energy input; or you'd get more out than you put in. That is, energy would be created. (That would be nice, but it would violate the conservation of energy, one of the cornerstones of physics.)

Chapter 6

6.1 5.0 m/s. Yes, this is 18 km/h or 11 mi/h, a speed at which humans can run.

6.3 5.0 kg·m/s, 53° (relative to $-x$ axis in second quadrant). The answer was in Fig. 6.2. Did you see it?

6.4 (a) Yes, for the $m_1 - m_2 -$ Earth system. But with m_2 attached to the Earth, the mass of this part of the system would be vastly greater than that of m_2, so its change in velocity would be negligible. (b) Assuming the ball is thrown in the + direction: for thrower, $v_t = -0.50$ m/s; for catcher, $v_c = 0.48$ m/s. For ball: $p = 0$, $+25$ kg·m/s, $+1.2$ kg·m/s.

6.5 $x_1 = \dfrac{mv_{1_o}^2}{2}$, so $\dfrac{x_1}{x_2} = \dfrac{1}{2}\left(\dfrac{m_1}{m_2}\right)\left(\dfrac{v_{1_o}}{v_{2_o}}\right)^2$

$= \dfrac{1}{2}\left(\dfrac{1}{2}\right)(2)^2 = 1$, and $x_1 = x_2$.

6.6 (a) Yes, $K_o = K_1 + K_2$. (b) No, the y component of momentum is zero before collision, so must be zero afterwards; thus one puck must be below the $+x$ axis and one above.

6.7 60 m/s. Greater speed, longer drive, ideally. (There is also a directional consideration.)

6.8 No. All of the kinetic energy cannot be lost to make the dent. The momentum (velocity) after the collision cannot be zero, since it was not zero initially. Thus, the balls must be moving and have kinetic energy. This may also be seen from Eq. 6.11: $K_f/K_i = m_1/(m_1 + m_2)$, and K_f cannot be zero (unless m_1 is zero, which is not possible).

6.9 5.0 m.

6.10 No. If $v_1 = v_2$, then $m_1 - m_2 = 2m_1$, which requires $m_2 = -m_1$. Negative mass is not possible, so $m_1 - m_2$ is always less than $2m_1$.

6.11 All of the balls swing out to different degrees. With $m_1 > m_2$, the stationary ball (m_2) moves off with a greater speed after collision than the incoming, heavier ball (m_1), and the heavier ball's speed is reduced after collision, in accordance with Eq. 6.16 (see Fig. 6.13b). Hence, a "shot" of momentum is passed along the row of balls with equal mass (see Fig. 6.13a), and the end ball swings out with the same speed as was imparted to m_2. Then, the process is repeated; m_1, *now moving more slowly*, collides again with the initial ball in the row (m_2), and another, but smaller, shot of momentum is passed down the row. The new end ball in the row receives less kinetic energy than the one that swung out just a moment previously, and so doesn't swing as high. This process repeats itself instantaneously for each ball, with the observed result that all of the balls swing out to different degrees.

6.12 $M = 11$ kg, initially at origin.

6.13 $(X_{CM}, Y_{CM}) = (0.47$ m, 0.10 m). Same location as in Example, $\frac{2}{3}$ of the length of the bar from m_1. Note: The location of the CM does not depend on the frame of reference.

Chapter 7

7.1 1.61×10^3 m $= 1.61$ km (about a mile).

7.2 1% (still a pretty good approximation).

7.3 (a) 0.022 min or 1.3 s.   (b) To equalize the running distances. This is necessary because the curved sections of the track have different radii, and thus different lengths.

7.4 120 rpm.

7.5 (a) 86%.   (b) No; the astronaut still has mass, and thus, with gravity acting, has weight by definition. (See end of Section 7.6.)

7.6 106 rpm.

7.7 The string cannot be exactly horizontal, but must make some small downward angle to the horizontal so that there will be an upward component of the tension force to balance the ball's weight.

7.8 No, depends on mass: $F_c = \mu_s mg$.

7.9 No. Both masses have the same angular frequency or speed ω, and $a_c = r\omega^2$, so actually $a_c \propto r$. Remember, $v = 2\pi r/T$, and note that $v_2 > v_1$, with $a_c = v^2/r$.

7.10 The direction of both $\boldsymbol{\omega}$ and $\boldsymbol{\alpha}$ for the CD would then be downward, perpendicular to the plane of the CD.

7.11 The angular speed (ω) is the same for all riders. The tangential speed (v) is not the same for all, but varies, depending on a rider's radial distance from the center of the merry-go-round ($v = r\omega$). However, all riders at the same radial distance (on a circle with the same radius) have the same tangential speed.

7.12 0.0028 m/s². Large force, but relatively small acceleration.

7.13 (a) No, does not vary linearly. $\Delta U = 2.4 \times 10^9$ J, only a 9.1% increase.   (b) A smaller negative value means a greater energy in the Earth's negative potential well (higher up in the well).

7.14 This is the amount of *negative* work done by an external force or agent when the masses are brought together. The masses attract each other gravitationally, so without an external force they would accelerate toward each other and collide. An external agent must apply forces in opposite directions to prevent this and get the masses in the configuration, hence the work is negative. To separate the masses by infinite distances, an equal amount of positive work (against gravity) would have to be done.

7.15 3.3.

7.16 Because of air resistance, the spacecraft would lose more energy and continue its downward spiral. With this frictional effect, the spacecraft heats up. (Recall that meteors burn up in the atmosphere.) In case you're wondering, to keep a spacecraft from overheating, a material is used that evaporates from the outer surface tiles. This requires work, which is supplied by heat energy. Enough heat is carried off for the spacecraft to remain relatively cool. (See latent heat, Chapter 11.)

7.17 You wouldn't want to apply a forward thrust as in this Example, since the radius of the orbit of your spacecraft would increase and it would travel more slowly, falling farther behind. By using a reverse thrust (Example 7.16), you would move to a lower orbit and speed up. Once you had overtaken the station, you could apply an appropriate forward thrust to put the spacecraft back into the proper orbit for docking.

Chapter 8

8.1 The weights of the balls and the forearm produce torques that tend to cause rotation in the direction opposite that of the applied torque.

8.2 $T \propto 1/\sin\theta$, and as θ gets smaller, so does $\sin\theta$, and T increases. In the limit, $\sin\theta \to 0$, and $T \to \infty$.

8.3 $\sum\tau$: $Rx - m_1gx_1 - m_2gx_2 - m_3gx_3 = (200\text{ g})g(50\text{ cm}) - (25\text{ g})g(0\text{ cm}) - (75\text{ g})g(20\text{ cm}) - (100\text{ g})g(85\text{ cm}) = 0$, where $R = Mg$.

8.4 No. With f_{s_2}, the reaction force R would not generally be the same (f_{s_2} and R are perpendicular components of the force exerted on the ladder by the wall). In this case, we still have $R = f_{s_1}$, but $Ry - (m_1g)x_1 - (m_mg)x_m - f_{s_2}x_3 = 0$.

8.5 5 bricks.

8.6 (d) No (equal masses).   (e) Yes, with larger mass farther from axis of rotation, $I = 360$ kg·m².

8.7 The long pole (or your extended arms) increases the moment of inertia by placing more mass farther from the axis of rotation (the tightrope or rail). When the walker leans to the side, a gravitational torque tends to produce a rotation about the axis of rotation so as to cause a fall. However, with a greater rotational inertia (greater I), the walker has time to shift his or her body so that the center of gravity is again over the rope or rail, and thus again in (unstable) equilibrium. In fact, with very flexible poles, the CG may be below the wire, thus ensuring stability.

8.8 $t = 0.63$ s.

8.9 $\alpha = \dfrac{2mg - (2\tau_f/R)}{(2m + M)R}$; $\dfrac{N-N}{\text{kg·m}}$ or $\dfrac{N}{\text{kg·m}} = \dfrac{\text{kg·m/s}^2}{\text{kg·m}} = \dfrac{1}{s^2}$

8.10 (a) At the location where the ball would land, $x = L\cos 35°$. For $L = 1.0$ m, then $x = (1.0\text{ m})\cos 35° = 0.82$ m from hinge, or 0.18 m from the free end of the board.   (b) No. $ay = a\cos\theta = ra\cos\theta = 3rg\cos^2\theta/2L = 8.1$ m/s². This is sufficient to bring the cup down before the ball, since the cup has a shorter vertical distance to travel.

8.11 No. In general, different moments of inertia.

8.12 (a) 0.25 m.   (b) f_s, the force of *static* friction, acts at the point of contact, which is always instantaneously at rest, and so does no work. Some frictional work may be done due to rolling friction, but this is considered negligible for hard objects and surfaces.

8.13 $v_{CM} = 2.2$ m/s; or using a ratio, 1.4 times greater.

8.14 You already know the answer: 5.6 m/s. (Doesn't depend on the mass of the ball.)

Chapter 9

9.1 39 kg.

9.2 2.3×10^{-4} m³.

9.3 3.03×10^4 N (or 6.82×10^3 lb—about 3.4 tons!) This is roughly the force on your back right now. Our bodies don't collapse under atmospheric pressure because cells are filled with incompressible fluids (mostly water!), bone, and muscle, which react with an equal outward pressure (equal and opposite forces). As with forces, it is a pressure *difference* that gives rise to dynamic effects.

9.4 $d_o = \sqrt{\dfrac{F_o}{F_i}}\, d_i = \sqrt{\dfrac{1}{10}}\,(8.0\text{ cm}) = 2.5\text{ cm}.$

9.5 This flask would have the same pressure on the base. Here the sloping walls exert a downward force on the liquid, which makes up for the weight of the liquid "missing" from the truncated vertical column above the base.

9.6 10.3 m (about 34 ft—you can see why we don't use water barometers).

9.7 Greater, 7.7×10^2.

9.8 The object would sink, so the buoyant force is less than the object's weight. Hence, the scale would have a reading greater than 40 N. Note that with a greater density, the object would not be as large, and less water would be displaced.

9.9 11%.

9.10 The water level falls. When the block is in the boat, it is supported by buoyant forces; that is, it is essentially floating, and so displaces its own weight in water. When the block is in the water, however, it displaces only its own *volume* of water. Since the block is denser than water, less water is displaced in the second case than in the first.

9.11 The height of the capillary column decreases with increasing temperature (and vice versa).

9.12 69%.

9.13 No, $r_{oil} = (5.8)r_{water}$.

Chapter 10

10.1 (a) $-40°C$, and you should immediately know the answer to (b). This is the temperature at which the Fahrenheit and Celsius temperatures are numerically equal.

10.2 (a) $T_R = T_F + 460$, (b) $T_R = \frac{9}{5}T_C + 492$, (c) $T_R = \frac{9}{5}T_R$.

10.3 96°C.

10.4 7.53×10^{23} molecules.

10.5 99.999% (same as in Example 10.5).

10.6 50 C°.

10.7 It depends on the metal of the bar. If the thermal expansion coefficient (α) of the bar is less than that of iron, it will not expand as much and not be as long as the diameter of the circular ring after heating. However, if the bar's α is greater than that of iron, the bar will expand more than the ring and the ring will be distorted.

10.8 Basically, the situations would be reversed. In Example 10.8, faster cooling would be achieved by submerging the ice —the cooler water would be less dense and would rise, promoting mixing. For a lake with cooling at the surface, cooler, less-dense water would remain at the surface until minimum density was achieved. With further cooling, the denser water would sink and freezing would occur from the bottom up.

10.9 3%.

10.10 Using Eq. 10.16, with $R_N = R_O$, we have $R_O/R_N = \sqrt{(T_N/T_O)(m_O/m_N)} = 1$. Rather than solving explicitly, note that $m_O/m_N = 1.14$, so for the relationship to hold, $T_N/T_O = 1/1.14$ or $T_O = (1.14)T_N = (1.14)(293\text{ K}) = 334\text{ K}$ (or 61°C).

Chapter 11

11.1 21 m (about the height of a 5-story building).

11.2 -2.5×10^4 J.

11.3 40°C.

11.4 Mercury, with a temperature change 30 times greater than that of water.

11.5 The final temperature (T_f) would be less, so the computed value of c would be smaller.

11.6 2.68×10^6 J.

11.7 24°C.

11.8 42 J/s.

11.9 No; the air spaces provide good insulation because air is a poor conductor. The many small pockets of air between a person's body and the outer garment form an insulating layer that minimizes conduction and so retards the loss of body heat. (There is little convection in the small spaces.) Similarly, for a diver in a wet suit, a thin film of water acts as an insulating layer.

11.10 2.1×10^4 s (350 min or 5.8 h).

11.11 Drapes reduce heat loss by radiation transmission through the window and by keeping convection currents away from the glass.

Chapter 12

12.1 For the same reason as in the Example: adiabatic expansion. The air rushing out of the nozzle of the pressurized tire does work against the atmosphere, cools, and cools the valve. (Try it yourself—but don't let too much air out without having a pump handy.)

12.2 122°C.

12.3 -1.2×10^3 J/K.

12.4 This would require the average temperatures to be equal, which gives $T_i + 18°C = 24°C + T_i$, where T_i is the intermediate equilibrium temperature. But this equation obviously cannot be solved for T_i, so $\Delta S = 0$ is not possible.

12.5 We know that the entropy of the universe increases in *every* natural process, including the death of your amoeba. We know, too, that *local* decreases in entropy are possible, provided that they are "paid for" with greater increases in entropy elsewhere. Living things "purchase" such decreases by capturing and using energy to grow and increase the complexity of their organism (see the Insight on Life, Order, and the Second Law). However, when an organism dies, it loses the ability to utilize energy and can no longer maintain its highly ordered structure. It decays, and eventually its elements return to the environment ("dust to dust"). Thus the death of an amoeba represents a local, as well as a global, increase in entropy.

12.6 Temperature is a factor in heat transfer to the body, and for evaporation from the head, which increases with temperature. Humidity also plays a role: With high humidity, there would be less evaporation from the head.

12.7 500 J/cycle.

12..8 An increase of 7.5%.

Chapter 13

13.1 No, 75%.

13.2 0.061 J.

13.3 Note that $T = 1/f = 1/0.20$ Hz $= 5.0$ s. At $t = 0$, $x = A$ and the mass is at its positive amplitude position. At $t = 3.1$ s, the mass has traveled to its negative amplitude position (in

0.50 cycle) and back to $x = -0.11$ cm, approaching the equilibrium position ($x = 0$) in 0.62 cycles.

13.4 Select a convenient length L to give an easily measured period T. Time several small-angle oscillations to find an average period (so as to minimize timing errors if done by hand) and use Eq. 13.14 to compute g.

13.5 When descending or accelerating downward, $w' = m(a - g)$, and with $a = g$, there is *apparent* weightlessness or $g' = g - g = 0$ (apparent zero g). From Eq. 13.14, the period would be infinite (or the pendulum does not swing).

13.6 (a) 0.50 m   (b) No.

13.7 440 Hz.

13.8 You could tune the string to 264 Hz by tightening it, increasing the tension. Assuming any change in the cross-sectional area of the string to be negligible, we have $f_2 = f_1\sqrt{F_2/F_1}$, and $F_2 = (f_2/f_1)^2 F_1 = (264\ \text{Hz}/220\ \text{Hz})^2 F_1 = (1.44)F_1$. Hence, the tension force would have to be increased by a factor of 1.44.

Chapter 14

14.1 (a) 2.3   (b) 10.2.

14.2 (a) 860 m   (b) $\Delta t = 0.06$ s.

14.3 (a) The dB scale is logarithmic, not linear.   (b) 3.16×10^{-6} W/m^2.

14.4 No, $I_2 = 316\ I_1$.

14.5 65 dB.

14.6 Destructive interference. $\Delta L = 2.5\ \lambda = 5(\lambda/2)$, and $m = 5$. No sound would be heard if the waves from the speakers had equal amplitudes. Of course, during a concert the sound would not be single-frequency tones, but would have a variety of frequencies and amplitudes. Spectators at certain locations might not hear certain parts of the audible spectrum, but this probably wouldn't be noticed.

14.7 Same as approaching and receding frequencies in the Example, because of relative motion. (There may be differences due to rounding errors.)

14.8 With the source and the observer traveling in the same direction at the same speed, their relative velocity would be zero. That is, the observer would consider the source to be stationary relative to him or her. Since their speed is subsonic, the sound from the source would simply overtake the observer without a shift in frequency. Keep in mind that generally for motions involved in a Doppler shift, the word *toward* is associated with an *increase* in frequency, and *away* with a *decrease* in frequency. Here, the source and observer remain a constant distance apart. (What would be the case if the speeds were supersonic?)

14.9 768 Hz.

Chapter 15

15.1 (a) The rug acquires a positive charge. (b) $+2.0\ \mu$C. (c) The rug has a deficiency of electrons. (d) 1.3×10^{13} electrons.

15.2 No. Regardless of the sign of the charge on the comb, it will polarize the paper so that the closest end is opposite in sign and therefore will attract it.

15.3 (a) The forces cancel to zero. (b) $2F_{31}$ (pointing away from the origin), which approaches zero at very large distances.

15.4 12 cm.

15.5 $F_e/F_g = 4.16 \times 10^{42}$. The magnitude of the electric force is the same as before, but the gravitational force is much less because the mass of the electron is much smaller than that of the proton.

15.6 The only possibility is 0.60 m to the left of q_1. The fields cannot cancel between the two charges because they point in the same direction. They cannot cancel to the right of q_2 because in that region the field from q_2 is always larger than that from q_1.

15.7 $\mathbf{E} = (5.53 \times 10^2\ \text{N/C})\mathbf{x} + (3.44 \times 10^1\ \text{N/C})\mathbf{y}$.

15.8 $E \approx k(2q)/x^2$; d is not in the answer because at large distances the two charges appear to be a single point charge of $+2q$. Thus we expect $1/x^2$ dependence.

15.9 In effect, positive charge would be transferred from the rod to the pail and would neutralize the induced negative charge on the inside surface. (In reality, it's the electrons that move, and they neutralize the positive charge on the rod.) This leaves a net positive charge on the outside of the pail. Thus, the deflection of the outside-connected electroscope's leaves is unchanged, while the inside-connected electroscope's leaves collapse to a neutral state. Basically, the *excess* charge acquired by the metal pail appears only on the outside of the pail.

15.10 The net enclosed charge is negative. This indicates a net number of inward pointing electric field lines. In this case, there are only inward field lines.

Chapter 16

16.1 (a) Doubles to 7.20×10^{-18} J.   (b) Unchanged.   (c) Unchanged.   (d) Decreases by a factor of $1/\sqrt{2}$ to 4.64×10^4 m/s.

16.2 $\mathbf{E} = (9.39 \times 10^4\ \text{V/m})(-\mathbf{x})$.

16.3 (a) $+4.36 \times 10^{-18}$ J. (b) Lower potential. (c) U increases because positive work is done to remove the electron.

16.4 $U = +0.24$ J. They would fly apart if released.

16.5 4.24×10^3 m/s.

16.6 (a) 2.4 V.   (b) The surface closer to the positive plate is at a higher potential.   (c) 6.00×10^2 V/m pointing from positive to negative plate.

16.7 1.1×10^{-10} m, which is *smaller* than an atomic diameter!

16.8 (a) 7.2×10^{-5} C.   (b) 5.1×10^2 V.

16.9 $C = 0.016\ \mu$F; $Q = 0.31\ \mu$C; $U = 3.0 \times 10^{-6}$ J.

16.10 $V_{2.5} = 8.0$ V; $V_{5.0} = 4.0$ V; $U_{2.5} = 8.0 \times 10^{-5}$ J; $U_{5.0} = 4.0 \times 10^{-5}$ J.

16.11 (a) $Q_1 = 8.0 \times 10^{-7}$ C; $Q_2 = 1.6 \times 10^{-6}$ C; $Q_3 = 2.4 \times 10^{-6}$ C   (b) $U_1 = 3.2 \times 10^{-6}$ J; $U_2 = 6.4 \times 10^{-6}$ J; $U_3 = 4.8 \times 10^{-6}$ J.

Chapter 17

17.1 1.0×10^9 s.

17.2 (a) 12 Ω (b) $2.4 \times 10^2\ \Omega$ (c) $1.2 \times 10^3\ \Omega$.

17.3 0.82 m.

17.4 0.67 Ω; the larger the temperature coefficient of resistivity, the more sensitive the thermometer.

17.5 34 W; the resistance of tungsten drops as it cools.

17.6 $P_{\text{thin}}/P_{\text{thick}} = 0.64$.

17.7 The power increases by 21%, which is potentially dangerous.

17.8 $1.98.

17.9 Electric and gas water heaters can be considered equally efficient only if you ignore the ultimate source of the energy they use. Gas heaters derive their energy *directly* from a fuel, whereas the energy for an electric water heater is produced by an electrical generating plant, which converts stored potential energy (nuclear, coal, gas, oil, etc.) into electricity. In Chapter 12, we saw that even if power plants operated at maximum (Carnot) efficiency, they would be only about 50% efficient. In reality, they are typically about 35% efficient. Thus an electric heater could be no more than 35% efficient in terms of its ultimate energy source even if it were able to transfer 100% of the energy it receives to the water. If it is rated at 95% efficiency, it would really be only about 95% × 35% = 33% efficient. A gas heater, by contrast, loses no energy in transport to the home, and thus can utilize about 95% of the primary fuel energy for heating water. Thus a gas heater is overall about three times more efficient in terms of primary fuel usage.

Chapter 18

18.1 (a) In series: $P_1 = 4.0$ W, $P_2 = 8.0$ W, $P_3 = 12.0$ W, and $P_t = 24.0$ W. They all have the same current; the largest resistor has the most voltage and therefore the most power. (b) In parallel: $P_1 = 144$ W, $P_2 = 72$ W, $P_3 = 48$ W, and $P_t = 264$ W. They all have the same voltage; the smallest resistor has the most current and therefore the most power. The series arrangement requires considerably less total power.

18.2 If the shunt were wired in parallel with the filament, current would normally be divided between the two elements. If a bulb blew out, all of the current would then flow through the shunt. We know that a combination of elements wired in parallel must always have *less* resistance than any one of its individual components. Therefore the resistance of the bulb would be greater with only the shunt in the circuit. (In fact, it would most likely be considerably greater, because the shunt would probably be designed to have a relatively high resistance. If it were low, most of the current would ordinarily flow through the shunt rather than through the bulb, wasting electricity and producing dangerous heat.) Since the bulb is in series with the rest of the circuit, the total resistance of the string of bulbs would be slightly increased when a bulb blew out. There would thus be slightly less current in the entire string, so all the bulbs would glow a little less brightly than normal.

18.3 $P_1 = 54.0$ W, $P_2 = 9.00$ W, $P_3 = 0.84$ W, $P_4 = 2.54$ W, and $P_5 = 5.63$ W. The total power (output) is the sum of these numbers or $P_t = 72.0$ W. The power output of the battery is $P_b = I_b V_b = (3.0$ A$)(24.0$ V$) = 72.0$ W. The two agree, as expected from energy conservation.

18.4 (a) $I_1 = I_2 = 1.33$ A. (b) $V_1 = 8.0$ V and $V_2 = 4.0$ V. (c) $P_1 = 10.6$ W and $P_2 = 5.3$ W.

18.5 The equation from Loop 1 (clockwise): $V_1 - I_1 R_1 - V_2 - I_3 R_3 = 0$. The equation from Loop 2 (clockwise): $V_2 - I_2 R_2 + I_3 R_3 = 0$. The equation for Loop 3 (around the outside, clockwise): $V_1 - I_1 R_1 - I_2 R_2 = 0$. This last equation is not a new equation, but just the sum of the Loop 1 and Loop 2 equations!

18.6 (a) $U_{max} = 4.32 \times 10^{-4}$ J. (b) $U_{5.0\,s}/U_{max} = (0.964)^2 = 0.929$. The stored energy after 5.0 s have elapsed is less than 96.4% because it varies as the *square* of the voltage.

18.7 In principle, insertion of the ammeter in series with the element of interest would reduce the current you are trying to measure. However, all ammeters are designed with very small resistance, so that this effect is negligible.

18.8 0.199 mA.

Chapter 19

19.1 South.

19.2 (a) Out of the page. (b) 5.2×10^{-3} m.

19.3 1500 A.

19.4 5.0×10^3 turns/m.

19.5 (a) 3.3×10^{-3} N (b) Repulsive.

19.6 1.5×10^{-4} A·m^2, which is many times greater than that of the hydrogen atom.

19.7 (a) 0.25 m (b) 0.75 m.

Chapter 20

20.1 (a) Increase the number of loops that make up the coil or the area of the coil. (b) There would be no *change* in magnetic flux through the plane of the coil and therefore no induced emf.

20.2 A clockwise current of 0.11 A (as viewed from above the plane of the paper) would be induced.

20.3 6.0×10^{-6} V.

20.4 120 V.

20.5 (a) 1.8×10^4 J. (b) 5.0×10^2 J, or about 36 times less energy than during starting.

20.6 (a) Step-down, because you want to lower the voltage from 240 V to 120 V. (b) The dryer would operate at 4800 W until it burned out, since $P_{240}/P_{120} = (V_{240}/V_{120})^2 = 2^2 = 4$ if R is assumed constant.

20.7 (a) Less energy would be lost in the wires to the pump in the form of joule heating. Operating at higher voltage enables the use of lower currents, thus reducing $I^2 R$ losses. (b) Assuming constant resistance, it will only be one-fourth of the loss, since $P_{240}/P_{120} = (I^2 R)_{240}/(I^2 R)_{120} = (I_{240}/I_{120})^2 = (\frac{1}{2})^2 = \frac{1}{4}$.

20.8 (a) 5.3×10^2 s. (b) In that time the rover would have traveled 53 m and be over the cliff, so it cannot be saved.

20.9 (a) Since the intensity of the sunlight (W/m^2) would diminish as the craft got further from the Sun, the force on the sail, and hence the acceleration of the craft, would decrease. (b) Increase the area of the sail.

Chapter 21

21.1 (a) 0.25 A (b) 0.35 A (c) 960 Ω in Great Britain and 240 Ω in the U.S. The one designed for operation in Great Britain needs *four* times resistance because the voltage is *twice* as large as in the U.S., and power depends on the *square* of the voltage.

21.2 If the motor of the dryer and the clock operate off the frequency of the voltage they would both run slower.

21.3 30 Hz.

21.4 (a) 3.61 A (b) 20 Hz.

21.5 (a) 0.90 A (b) The capacitor is responsible, since the resistance of a resistor does not depend on the frequency.

21.6 Maximum current means minimum Z, since $I = V/Z$. To do this we make the capacitive reactance equal to the inductive reactance so that $Z = R$, which is that minimum value.

21.7 (a) 41.1 Hz (b) 576 W.

21.8 (a) As frequency increases, inductive reactance increases, Z increases, I decreases, and therefore P decreases. (b) As frequency increases, capacitive reactance decreases, Z decreases, I increases, and therefore P increases.

21.9 If the two frequencies are close together, so that their resonance curves are very close, you may be right at resonance for one station, thereby picking it up very well, but also be picking up a fairly large signal (though not its maximum) from the nearby station.

21.10 (a) The capacitor and inductor have large amounts of energy stored in them, on average, and thus have large voltages across them.   (b) We have to remember that the voltages across the capacitor and the inductor are exactly out of phase and cancel; thus Kirchhoff's loop rule still works. That is, if you sum the voltages around the loop, taking into account phases, you have a gain of 25.0 V at the voltage source, a loss of 25.0 V at the resistor, and the capacitor and inductor voltages cancel out. Thus all voltage gains and drops cancel.

Chapter 22

22.1 Light travels in straight lines, and by the law of reflection, with $\theta_1 = \theta_r$, if you can see someone in a mirror, that person is able to see you. Conversely, if you can't see the trucker's mirror, then he or she won't be able to see your image in that mirror.

22.2 $n = 1.25$.

22.3 By Snell's law, $n_2 = 1.25$, and $v = c/n_2 = 2.40 \times 10^8$ m/s.

22.4 Draw a line that starts from the point where the ray emerges from the glass and strikes the original path (the dashed line) at a right angle. Call this segment d, and call the angle at the point of entry that subtends this segment ϕ. Then, $\phi = \theta_1 - \theta_2$, and $d = r \sin \phi = r \sin (\theta_1 - \theta_2)$. From the triangle in the Fig. 22.10 inset, $r = y/\cos \theta_2$; thus $d = y \sin (\theta_1 - \theta_2)/\cos \theta_2$.

22.5 (a) The frequency of the light is unchanged in the different media, and so has the same frequency as that of the source.   (b) The wavelength in air is independent of the water and glass media, as can be shown by adding another step (medium) to the Example solution. By reverse analysis, $\lambda_{air} = n_{water}\lambda_{water} = (c/v_{water})\lambda_{water} = c/f$. Thus, the wavelength in air is simply c/f.

22.6 $d' = (0.703)d$.

22.7 Because of total internal reflections, the diver sees the reflection of something on the sides and/or bottom of the pool. (Use reverse ray tracing.)

22.8 $n = 1.4574$.

Chapter 23

23.1 No effect. Note that the solution to the Example does not include the distance. The geometry of the situation is the same regardless of the distance from the mirror.

23.2 $d_i = 30$ cm, real, inverted, and same size. Note from the Example results (a) and (b) that when the object is closer to the mirror than is the center of curvature, the image becomes magnified.

23.3 Image always virtual, upright, and reduced.

23.4 $d_i = -d_o$ and $M = +1$. Virtual (behind mirror, same distance as object is in front), upright and same size.

23.5 $|M| = 0.75$.

23.6 $|M| = \frac{1}{3}$. (Recall that the image is always reduced with a diverging lens.)

23.7 Blocking off half of the lens would cut the *amount* of light focused at the image plane in half, so the resulting image would be less bright.

23.8 $f = 20$ cm.

23.9 $P = -2.0$ D.

Chapter 24

24.1 $\Delta y = y_r - y_b = 1.20 \times 10^{-2}$ m.

24.2 Twice as thick, $t = 199$ nm.

24.3 In brass instruments, the sound comes from a relatively large, flared opening. There is thus little diffraction, so most of the energy is radiated in the forward direction. In woodwind instruments, the sound comes from tone holes along the column of the instrument. These holes are small compared to the wavelength of the sound, so there is appreciable diffraction. As a result, the sound is radiated in nearly all directions, even backwards.

24.4 Would increase by a factor of 1.65.

24.5 Fewer; $n \leq 1.43$, so $n = 0, 1$.

24.6 $\theta_2 = 36°$.

24.7 $\lambda = 588$ nm (yellow).

Chapter 25

25.1 Wouldn't work. Real image formed on person's side of lens ($d_i = +75$ cm).

25.2 For an object at $d_o = 25$ cm, the image for eye 1 would be formed at 1.0 m; this is beyond the near point for that eye, so object could be seen clearly. The image for eye 2 would be formed at 0.77 m; this is inside the near point for that eye, so the object would not be seen clearly.

25.3 Glass for near-point viewing would have longer focal length, 20 cm longer.

25.4 Length doubles.

25.5 $f_i = 8.0$ cm.

25.6 Erecting lens (of focal length f_e) should go between objective and eyepiece, positioned a distance of $2f_e$ from the image formed by the objective, which acts as an object. Erecting lens then produces an inverted image of the same size at $2f_e$ on opposite side of the lens, which acts as an object for the eyepiece. Use of the erecting lens lengthens telescope by $4f_e$.

25.7 Yes, $\theta_{min} < \theta$ (angle subtending moon separation).

25.8 (a) $v = 2.26 \times 10^8$ m/s, $\lambda' = 338$ nm, $f' = 6.67 \times 10^{14}$ Hz. (b) blue (unchanged).   (c) Wavelength of 338 nm lies in the nonvisible ultraviolet region so object would not be seen, or would appear black.

Chapter 26

26.1 Since there is no relative motion *in the direction of the events*, observers in the two frames do *not* disagree on their simultaneity. If they are marked simultaneously in one frame, they are simultaneous in the other.

26.2 $0.9995c$. They can't travel twice as fast to travel twice as far because they are limited by the speed of light. However, by traveling faster they can increase the time dilation effect and travel farther.

26.3 (a) 6.67×10^{-7} s.   (b) 5.80×10^{-7} s.   (c) The observer uses the same clock (at the given point) to measure the time interval and thus measures the proper time, which is the shortest interval. The person in the spaceship will use two different clocks in the ship, located at the front and rear. Thus that time interval is not the proper interval.

26.4 $0.991\, c$ or slower.

26.5 From the Earth viewpoint, the proper time for the round trip is measured by the traveler as 20.0 y. (Why is that the proper time interval?) Applying the time dilation factor, γ, using the speed calculated in the Example results in 201 y as measured from the Earth, in agreement with the result in part (b) of the Example.

26.6 (a) 1.17 MeV. (b) 0.207 MeV.

26.7 $\frac{10}{3}M_o$. Classically we expect the answer to be $2M_o$. However, relativistically we know that some of the kinetic energy that was available before the collision has been converted into rest mass. We therefore expect an answer greater than $2M_o$.

26.8 Its mass would have to be larger than its present mass by a factor of 5.0×10^7.

26.9 The relativistic answer is $0.69c$ which is less than the classical answer of $0.8c$. In this situation the relativistic answer will always be less than the classical answer. The relativistic equation has the classical answer in the numerator ($v + u' = 0.80c$), but has a denominator [$1 + vu'/c^2 = 1 + (0.40c)(0.40c)/c^2 = 1.16$] that is always larger than 1.

26.10 Either the second postulate or putting the numbers into the relativistic velocity addition formula gives the speed of light for the light beam regardless of the which traveler is measuring.

Chapter 27

27.1 (a) 967 nm, which is in the IR, but the star should emit enough red light to seem reddish. (b) 297 nm, which is in the near UV, but the star should emit enough violet and blue light to seem bluish. (It actually emits enough light of all colors to appear bluish-white.)

27.2 497 nm. The wavelength is shorter, so each photon carries more energy. The most energetic electrons thus have more kinetic energy, so a higher stopping voltage is required.

27.3 2.0 V. The frequency in the Example is the minimum frequency, providing photons with just barely enough energy to eject the least bound electrons (2.0 eV). If you double the frequency you double the photon energy to 4.0 eV. Thus the most energetic electrons have 4.0 eV − 2.0 eV, or 2.0 eV of kinetic energy, and require 2.0 V to stop them.

27.4 (a) +0.0360 or +3.60%. (b) It is a larger change (increase) than in the Example because there is more momentum lost to the electron when the photon is back-scattered. Thus the photon loses more momentum and the scattered photon has a longer wavelength.

27.5 1.09×10^6 m/s.

27.6 365 nm, which is in the UV.

27.7 821 nm, which is in the IR.

Chapter 28

28.1 (a) 8.8×10^{-33} m/s. (b) 7.1×10^{16} billion years or 4.8×10^{15} times the age of the universe, so not noticeable.

28.2 (a) It would be less because the mass of the proton is significantly more than the electron:

$$\lambda_p = 4.06 \times 10^{-12} \text{ m} = 4.06 \times 10^{-3} \text{ nm.}$$

(b) It would be less likely to diffract because the wavelength is smaller than before.

28.3 One.

28.4 1.2×10^6 m/s.

28.5 6.6×10^{-11} s.

Chapter 29

29.1 $^{12}_{6}\text{C} + ^{4}_{2}\text{He} \rightarrow ^{16}_{8}\text{O}$.

29.2 $^{22}_{11}\text{Na} \rightarrow ^{22}_{10}\text{Ne} + ^{0}_{+1}\text{e}$; the daughter is $^{22}_{10}\text{Ne}$.

29.3 (a) 24 d (b) Bodily functions help to remove it before it decays.

29.4 About 20 half lives, since $2^{20} \approx 1.1 \times 10^6$, or about 600 years.

29.5 It takes four half lives to get to this activity level, so you could measure back to 2.3×10^4 years.

29.6 It takes four half lives to decay to one-sixteenth the original activity, so projecting backwards in time to when the two isotopes were present in equal amounts gives 5.1 billion years, which is close to the age of the Earth by other estimates. (Can you see that we expect our answer to be an *over*estimate anyway? Why?)

29.7 (a) Since copper has $A > 40$, we would expect the proton number to be slightly less than the neutron number. So a reasonable shopping list would be: $^{61}_{29}\text{Cu}$, $^{62}_{29}\text{Cu}$, $^{63}_{29}\text{Cu}$, $^{64}_{29}\text{Cu}$, $^{65}_{29}\text{Cu}$, and $^{66}_{29}\text{Cu}$. (b) The odd–odd criterion eliminates the second, and fourth, and sixth isotopes in the list. We might therefore expect that copper would have three stable isotopes, $^{61}_{29}\text{Cu}$, $^{63}_{29}\text{Cu}$, and $^{65}_{29}\text{Cu}$. (c) Appendix V confirms that the last two of these are in fact stable but that $^{61}_{29}\text{Cu}$ is not. Apparently $N = 32$ is not sufficiently larger than $Z = 29$ in this region of the periodic table to give a stable isotope with these numbers. Of course, you have no way of knowing this ahead of time; all three guesses are quite plausible.

29.8 (a) The average binding energy per nucleon for ^{4}He is 7.075 MeV/nucleon. For ^{3}He it is 2.573 MeV/nucleon. (b) We would expect ^{3}He to be less stable (less binding energy per nucleon) because it has one unpaired neutron, whereas both the neutrons and protons in ^{4}He are completely paired.

29.9 (a) Proton removal produces an isotope of boron: $^{12}_{6}\text{C} + \gamma \rightarrow ^{11}_{5}\text{B} + ^{1}_{1}\text{H}$; neutron removal produces an isotope of carbon: $^{12}_{6}\text{C} + \gamma \rightarrow ^{11}_{6}\text{C} + ^{1}_{0}\text{n}$. Proton removal takes a minimum energy of 15.96 MeV, whereas neutron removal takes at least 18.72 MeV. (b) The proton has a positive charge and is repelled by the other protons in the nucleus, making it easier to remove.

Chapter 30

30.1 Since the total mass afterwards is greater than that before, it is endoergic. $Q = -15.63$ MeV.

30.2 1.193 MeV went toward increasing the total system mass. This is the energy equivalent of the mass increase. However, the minimum kinetic energy to initiate this reaction is 1.534 MeV; hence the difference, 0.341 MeV, must show up as extra kinetic energy in the final state in order to conserve momentum. If we used 1.193 MeV for the incident kinetic energy, the product nuclei would be created at rest, thereby *not* conserving momentum. Hence the reaction would not occur.

30.3 6.3×10^{11} kg of helium is produced each second. To two significant figures, this is the same as the mass *loss* of hydrogen. However, we know that the He mass gain must be less than the hydrogen mass loss because it is the difference that is converted into light energy. The mass lost is too small to be seen if we calculate to only two significant figures.

30.4 Since the emitted beta particle has little kinetic energy, it also carries little momentum. Thus to conserve momentum, the neutrino and the daughter nucleus would have to carry equal momentum in opposite directions. Thus we would expect the neutrino and the daughter nucleus to go off in opposite directions.

Answers to Odd-Numbered Exercises

Note to the student: The answers given here were reached by solving the problems step-by-step and rounding the result at each step. If you work the problems fully and then round only your result, your correct answer may differ slightly from the answer you find here. Variations may be due either to rounding or to calculator differences.

Chapter 1

1. (b)
3. (a) yes; **(b)** no; **(c)** yes
5. a duodecimal system might have a 12¢ dime and a 12 "dime" dollar or 144¢ dollar
7. (d)
9. yes
13. yes; $m^3 = m^3$
17. yes; $m/s = m/s - m/s$
19. yes; $[L]^2 = [L]^2 + [L]^2$
21. (a) dimensionless; **(b)** 1/m; **(c)** m^2
23. no
25. (a) kg·m/s²; **(b)** yes
27. (c)
29. cm
31. (a) 3.28×10^3 ft; **(b)** 656 ft
33. no
35. 2.72 m; 200 kg; 19 in.
37. about 4.2 lb
39. (a) 2 L by 0.11 L; **(b)** 27 mL more in 500 mL
41. (a) 10.0 km/L; **(b)** $854.44
43. 16 oz = 454 g; 12 fl. oz = 355 mL
45. 2.06 m; 95.5 kg
47. (a) 91.5 m by 48.8 m; **(b)** 27.9 cm to 28.6 cm
49. (a) 26 m³; **(b)** 9.2×10^2 ft³
51. (a) 62.3 lb/ft³; **(b)** 8.33 lb
53. (a)
55. 5.06 cm; 5.06×10^{-1} dm; 5.06×10^{-2} m
57. no; only one doubtful digit; best 25.48 cm
59. (a) 1.0 m; **(b)** 8.0 cm; **(c)** 16 kg; **(d)** 1.5×10^{-2} μs
61. (a) 10.1 m; **(b)** 775 km; **(c)** 2.55×10^{-3} kg; **(d)** 9.30×10^7 mi
63. 20 cm²
65. 470 cm²
67. (a) 2.0 kg·m/s; **(b)** 2.1 kg·m/s, rounding difference
69. (c)
71. 5.5×10^3 kg/m³
73. (a) no; 4.5×10^3 lb; **(b)** yes; 6.00 ft **(c)** yes; 5.0 min; **(d)** yes; 90 km/h; **(e)** no; 11.0 in. and 20 in.; **(f)** no; 216 lb and 20 in.
75. 8.72×10^{-3} cm/page
77. (a) 57%; **(b)** 8.4%
79. same area for both; 1.3 cm²
81. 0.41 h or 25 min
83. 2.6×10^9 times
85. 46 cm²
87. 4.9×10^2 cm³

89. (a) 1.8×10^3 cm²; **(b)** 0.18 m²
(c) 1.2×10^3 kg/m³
91. 1 fathom = 0.183 dam; 1 furlong = 0.201 km
93. not reasonable (56 mi/h)
95. 4.48×10^3 cm³

Chapter 2

1. (c)
3. no; different coordinates for same location
5. no final position can be given; may be 0 to 500 m
7. (d)
9. no, vertical line has infinite slope or speed
11. 28 km; yes
13. (a) 0.50 m/s; **(b)** 8.3 min
15. (a) 7.5 m/s; **(b)** zero, displacement zero
17. 10.3 s
19. (a) 1.0 m/s; 0; 1.3 m/s; 2.8 m/s; 0, 1.0 m/s; **(b)** 1.0 m/s; 0; 1.3 m/s; −2.8 m/s; 0; 1.0 m/s; **(c)** 1.0 m/s; 0; 0; −2.8 m/s; **(d)** −0.89 m/s
21. (a) 74.6 mi/h; 160.3 mi/h; **(b)** 115%
23. (a) 185 km/h; **(b)** 115 mi/h
25. 12.5 s; 56.3 m (relative to first runner)
27. (c)
29. v_0
31. 1.74 m/s²
33. −2.2 m/s each s
35. 5.0 s
37. (a) $a_{0-1} = 0$; $a_{1-3} = 4.0$ m/s²; $a_{3-8} = -4.0$ m/s²; $a_{8-9} = 8.0$ m/s²; $a_{9-11} = 0$; **(b)** constant velocity of −4.0 m/s
39. (d)
41. no; would accelerate an object having $-v$ or $v_o = 0$
43. 7.7 s
45. (a) 81.4 km/h; **(b)** 0.794 s
47. 3.09 s and 13.7 s
49. -1.8×10^5 m/s²
51. (a) 4.0 s; **(b)** 12 m north
53. (a) $a = -8.0$ m/s²

Initial velocity	Stopping distance
36 km/h (22 mi/h)	15 m (49 ft)
72 km/h (45 mi/h)	50 m (160 ft)
90 km/h (56 mi/h)	78 m (260 ft)

(b) $a = -4.0$ m/s²

Initial velocity	Stopping distance
36 km/h (22 mi/h)	8.8 m (29 ft)
72 km/h (45 mi/h)	30 m (98 ft)
90 km/h (56 mi/h)	45 m (150 ft)

55. 75 m/s
57. 96 m
59. $x - x_o = \dfrac{v_o + v}{2}(t - t_o)$;
$v = v_0 + a(t - t_o)$;
$x - x_o = v_o(t - t_o) + \frac{1}{2}a(t - t_o)^2$;
$v^2 = v_o^2 + 2a(x - x_o)$.
x stands for displacement in Eq. 2.9 and position here
61. 457 m from starting position
63. (d)
65. (a) linear, slope $-g$.

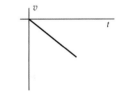

(b) parabola.

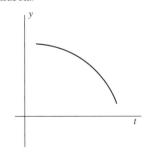

67. 78.4 m
69. 15.9 m
71. 15.3 m/s
73. (a) 5.6 m; **(b)** $t_{up} = 0.68$s; $t_{down} = 3.6$ s
75. (a) −44.4 m **(b)** 37.8 m/s downward
77. same; 9.80 m/s²
79. (a) $h = \frac{1}{2}gt_1 t_2$; **(b)** $v_o = \frac{1}{2}g(t_1 + t_2)$
81. 2.5 m/s²

83. (a)

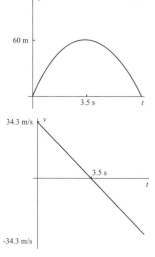

(b)

(c) For both motions the acceleration is a constant equal to $-g = -9.80\ \text{m/s}^2$

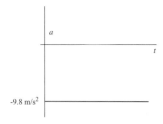

85. (a) 3.2 s;   **(b)** 0
87. (a) $t_M = 2.45\ t_E$;   **(b)** Moon: $y = 99.2$ m; $t = 22.0$ s; Earth: $y = 16.5$ m; $t = 3.67$ s
89. y (above top) = 1.49 m
91. -20.9 m/s;   **(b)** 2.87 s
93. (a) 18.5 m/s;   **(b)** 15.9 m
95. 45 mi/h
97. (a) 8.45 s;   **(b)** $x_m = 157$ m; $x_c = 132$ m;
(c) 13 m
99. 1.2×10^2 m
101. 7.1 m/s

103.

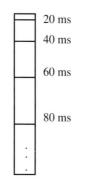

105. 5.00 s
107. (a) 60 mi/h;   **(b)** 27 m/s;   **(c)** same magnitude, 88 ft/s
109. (a) $\overline{v}_1 = 5.40$ km/h east; $\overline{v}_2 = 4.11$ km/h north;   **(b)** $\overline{s} = 4.58$ km/h;   **(c)** $\overline{s} = 5.40$ km/h; $\overline{v} = 0$

Chapter 3

1. (b)
3. (a) linear velocity increases or decreases;   **(b)** moves in a parabolic path;   **(c)** moves in a circle
5. $v_x = 7.0$ m/s; $v_y = 2.6$ m/s
7. (a) 70 m;   **(b)** 3.4 min, 2.6 min
9. (a) 8.4 m/s;   **(b)** 7.5 m/s
11. $x = 1.75$ m; $y = -1.75$ m
13. (a) 65.2 m;   **(b)** 86.5 m
15. $(x, y) = (10\ \text{m}, 6.0\ \text{m})$
17. (c)
19. $v_x = 5.00$ m/s; $v_y = 8.66$ m/s
21. yes

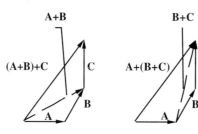

23.

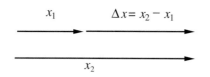

25. (a) -3.4 cm $\mathbf{x} - 2.9$ cm $\mathbf{y}$ **(b)** 4.5 cm, 63° (2nd quadrant);   **(c)** 4.0 cm $\mathbf{x} - 6.9$ cm $\mathbf{y}$
27. (a) 14.4 N $\mathbf{y}$;   **(b)** 1.1 N $\mathbf{x} + 12.7$ N $\mathbf{y}$
29. 4.1 km $\mathbf{x} + 102$ km $\mathbf{y}$
31. 5.0 m/s, 2.9° (3rd quadrant); $v = -5.0$ m/s $\mathbf{x} - 0.25$ m/s $\mathbf{y}$
33. (a) -1.9 m/s $\mathbf{x} + 4.9$ m/s $\mathbf{y}$;   **(b)** -5.2 m/s $\mathbf{y}$;   **(c)** -7.9 m/s $\mathbf{x} + 4.9$ m/s $\mathbf{y}$
35. -7.9 N $\mathbf{x} - 3.0$ N $\mathbf{y}$
37. (a) in the same direction;   **(b)** in opposite directions;   **(c)** at right angles;   **(d)** only if $\mathbf{F}_2 = 0$

39. $F_{\parallel} = (0.50)w$, $F_{\perp} = (0.87)w$
41. $\mathbf{d} = -4.3$ m $\mathbf{x} - 27$ m $\mathbf{y}$ or 27 m, 81° (3rd quadrant)
43. $\mathbf{d} = 3.5$ m $\mathbf{x} - 1.5$ m $\mathbf{y}$ or 3.8 m, 23° (4th quadrant)
45. $\mathbf{d} = 1.1$ m $\mathbf{x} + 1.2$ m $\mathbf{y}$
47. (a) 40 km/h;   **(b)** -50 km/h
49. (a) 18.6° upstream;
(b) 237 s = 3.96 min
53. (a) $y = 75.0$ m, $x = 6.25$ m;   **(b)** 16.7 m
55. 70 m/s or -20 m/s
57. 30 s
59. (b)
61. no; it is a parabolic path.
63. (a) 0.452 s;   **(b)** 0.565 m
65. 2.7×10^{-13} m
67. (a) 0.123 s;   **(b)** 1.10 m
69. 6.02 m/s
71. (a) 1.35 m;   **(b)** 20.4 m;   **(c)** kick the ball harder to increase v_o and/or increase the angle
73. $R_{35} = (1.1)\ R_{60}$
75. (a) the horizontal velocities of the cart and the ball are the same;   **(b)** 0.77 m;   **(c)** would not be caught
77. 40.9 /s, 11.9° above horizontal
79. 8.7 m/s
81. (a) 161 m;   **(b)** 160 m
83. (a) 4.8 m;   **(b)** 9.7 m
89. no; 16.7 s
91. (a) $v_x = 19.2$ m/s; $v_y = 16.1$ m/s
(b) $x = 48.0$ m; $y = 40.3$ m
93. (a) 32.6°;   **(b)** the target could not be hit
95. (a) 1.8 m/s $\mathbf{x} + 2.5$ m/s $\mathbf{y}$;
(b) $x = 3.6$ m; $y = 11$ m
97. (a) $y = v_{yo}t - \frac{1}{2}gt^2$

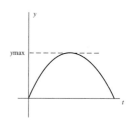

(b) $x = v_{xo}t$

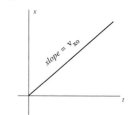

(c) $v_x = v_{xo}$

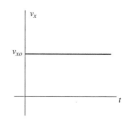

(d) $v_y = v_{yo} - gt$

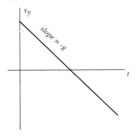

(e) $v = \sqrt{v_x^2 + v_y^2} = \sqrt{v_{xo}^2 + (v_{yo} - gt)^2}$

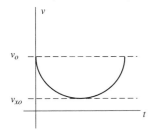

99. $(x, y, z) = (13.2\text{ m}, -2.20\text{ m}, 39.3\text{ m})$
101. (a) 9.14 m/s; **(b)** 12.3 m

Chapter 4

1. (d)
3. no
5. yes
7. (a) foward; **(b)** backwards
9. 1.5 N **y**, no
11. -7.6 N **x** $+ 0.64$ N **y**
13. (a)
15. extra acceleration
17. 6.75 N
19. 7.0×10^6 N
21. 20 m/s² in the direction of the force
23. 7.4 N
25. 2.73 m/s² in the direction of the force.
27. (a) 3.0 m/s²; **(b)** 2.7 m/s²
29. (a) 0.103 m/s²; **(b)** 400 N
31. 80 metric tons
33. yes; 1.8 m/s² downward
35. (a) 3.7×10^2 N; **(b)** 5.2×10^2 N;
(c) T increases as θ decreases
37. 10 m/s²
39. (a) 9.8 m/s²; 5.9º relative to boat's direction; **(b)** 6.3 m/s²; 9.1º relative to boat's direction
41. (a) 5.4 kg; **(b)** 5.2 kg
43. (a) 2.5 m/s² to right; **(b)** 2.0 m/s² to left
45. (a) 1.62 s; **(b)** 1.19 m
47. (a) 3.9 m/s²; **(b)** 4.9 m/s²
49. 1.8 m/s²
51. (d)
53. yes
55. the tension forces T's are action-reaction pair; the weights of the objects and the normal forces by the horizontal surface are not illustrated
57. three pairs of forces
59. 1.50×10^3 N
61. (d)
63. pull

65. $\tan \theta$
67. (a) no; **(b)** increasing or decreasing air resistance depends on the directions of the wind and motion
69. 2.7×10^2 N
71. 33° > 20º, not move
73. 30°; 0.58
75. (a) 11 kg; **(b)** 9.1 kg; **(c)** friction is too large for m_1 to move down
77. 53 m
79. 0.15
81. 0.064
83. 0.33
85. 0.25
87. 2.3 m/s² downward
89. 10 m/s²
91. (a) 12 m north; **(b)** 1.1 m south
93. 1.7°
95. 0.32 N
97. 5.2 m
99. 2.0 N **x** + 3.5 N **y**
101. 4.0 kg by 4.5 m ahead

Chapter 5

1. (d)
3. (a) while the weight is not moving, there is no displacement and therefore no work, regardles of the force exerted; **(b)** work is done by the force exerted by the weightlifter, acting in the direction of the displacement (upward); **(c)** as in (a), no work is done when there is no motion; **(d)** work is done on the falling barbell, but by the earth (gravity), not the weightlifter
5. positive on the way down and negative on the way up; no; maximum on the way up and down; minimum (zero) at top or bottom
7. 2.94×10^5 J
9. 6.65 m
11. 3.7 J
13. 1.47×10^5 J
15. 5.77 m
17. 2.7×10^3 J
19. (c)
21. 4.0×10^{-2} J
23. 0.21 J
25. (a) 15 kg; **(b)** 5.1 kg more
27. 0.50 J
29. 42.5 J
31. (d)
33. the same
35. (a) 3.8×10^5 J; **(b)** -3.8×10^5 J
37. -1.5×10^3 N
39. (a) 240 m; **(b)** 58 km/h
41. 2.7×10^{33} J
43. (d)
45. 0.21 m
47. 5.9 J
49. lowered 0.51 m
51. (a) -1.8×10^2 J; **(b)** 8.5×10^2 J; **(c)** 8.5×10^2 J
53. -1.0×10^2 J
55. (d)
57. chemical (food), potential (muscle), kinetic, and potential (gravity) energies
59. 1.7×10^9 J

61. (a) $K = 9.7$ J; $U = 4.7$ J; **(b)** $K = 10$ J
63. (a) 1.03 m; **(b)** 0.841 m; **(c)** 2.32 m/s
65. (a) 2.7 m/s; **(b)** 0.38 m; **(c)** 29º
67. (a) 0.22 m; **(b)** 0.19 m
69. (a) 2.6 m/s; **(b)** yes; start with $v_o > 2.6$ m/s
71. (b)
73. no; energy
75. yes; piece work based on work/time
77. 0.134 hp
79. (a) 9.9×10^5 J; **(b)** \$0.89
81. 5.7×10^{-5} W
83. (a) 1.6×10^{22} J; **(b)** absorbed by atmosphere and reflected back into space; **(c)** 3.2×10^5 J (50% received)
85. 6.7×10^2 J/s
87. 48.7%
89. 1.33×10^7 m/s
91. 7.4×10^2 J
93. (a) 9.0×10^4 J; **(b)** $W = 5.3 \times 10^5$ J
95. (a) 2.2×10^5 J; **(b)** -2.2×10^5 J
97. (a) 52 km/h = 14 m/s; **(b)** 37 km/h = 10 m/s
99. 17 m/s
101. no, because $E_{\text{bottom}} < E_{\text{top}}$
103. (a) 4.8×10^2 W; **(b)** 0.64 hp
105. 6.8×10^{-2} J

Chapter 6

1. (b)
3. air moves backwards and boat moves forward
5. (a) $(m$ kg)(25 m/s); **(b)** zero
7. 1.4×10^2 kg·m/s
9. 8.3×10^2 N
11. 5.88 kg·m/s in direction opposite v_0
13. (a) -1.8 kg·m/s; **(b)** -3.5 kg·m/s
15. 43 N
17. (a) 2.1 kg·m/s 45° below the $-x$ axis; **(b)** a collision does not have to occur; momentum would be the same
19. $\Delta p = -1.1$ kg·m/s **x** -2.8 kg·m/s **y**
21. (a) -491 kg·m/s; **(b)** yes; -519 kg·m/s
23. (a)
25. throw something or blow breath
27. 0.88 m/s westward
29. 3.5 m/s, 58° above $-x$ axis
31. 7.6 m/s, 12° above $+x$ axis
33. 0.78 m/s
35. 82.8 m
37. (c)
39. (a)
41. (a) drive: large impulse; chip shot: small impulse; **(b)** jab: small impulse; knock-out punch: large impulse; **(c)** bunting: small impulse; home-run swing: large impulse
43. momentum is a vector and kinetic energy is a scalar
45. uranium would have far less speed
47. 8.0×10^3 N
49. (a) -1.2 N·s; **(b)** -12 N
51. 14 N·s in direction opposite v_o

53. 1.4×10^3 lb

55. $\mathbf{v}_1 = -48$ cm/s; $\mathbf{v}_2 = 2.0$ cm/s

57. kinetic energy not conserved

59. 13 N upward

61. 7.6×10^5 N

63. $v_x' = mv/(m + M)$; $v_y' = M V/(m + M)$

65. $15.3°$; 0.57 m/s

67. (a) 28%; (b) 1.3×10^7 m/s

69. since $K_o \neq K$, collision is not elastic; primarily heat

73. (a) $v = v_o/90$; (b) 2.5×10^2 m/s (c) 99%

75. (b) $e_{elastic} = 1.0$; $e_{inelastic} = 0$

77. (d)

79. momentum zero, CM does not move

81. $x_{CM} = 0.98$ m from less massive sphere

83. (8.0 m, 0)

85. no change in (a), 0.01 m shift in (b)

87. $\mathbf{A} = 5.3$ m/s $\mathbf{x} + 2.7$ m/s $\mathbf{y}$

91. (a) 90 kg travels 3.2 m and 60 kg travels 4.8 m; (b) same distances

93. (a) 0.35 kg·m/s $\mathbf{x}$; (b) 9.2×10^{-2} kg·m/s $\mathbf{y}$

95. yes, $(4.3d, 3.8d)$

97. (a) 6.3×10^5 kg·m/s; (b) 7.8×10^4 N

99. (a) 36 km/h, $56°$ N of E; (b) 51% lost

101. 1.1×10^{12} m; it moves

103. 5.4×10^2 s

105. -8.2×10^3 kg·m/s

Chapter 7

1. (c)

3. (2 m, $45°$)

5. (a) 0.175 rad; (b) 4.71 rad; (c) 0.785 rad; (d) 7.85 rad

7. 4.7 cm

9. $115°$

13. 10.7 rad

15. 3.9×10^3 km

17. (a) 2.0×10^2 rad; (b) 60 m

19. (b)

21. perpendicular into the turntable

23. 0.070 rad/s

25. 2.4 rev/d

27. A is faster

29. (a) 2.1×10^{-2} rad/s; (b) 5.2 m/s

31. 0.120 s

33. (a) 11.1 rad/s; (b) 102 ft

35. (a) 1.16×10^{-5} s^{-1}; (b) 7.27×10^{-5} rad/s (c) $v_{45}/v_{eq} = 70.7\%$; $\omega_{45}/\omega_{eq} = 100\%$

37. (d)

39. not enough centripetal force on the water drops and so they fly out along a tangent

41. the inertia of our bodies have tendencies to keep moving forward along a straight line

43. 1.11 m/s^2

45. no; 1.33 m/s^2 > 1.25 m/s^2

47. (a) 3.0×10^{-26} N; (b) zero

49. $13.5°$

51. (a) 3.0×10^4 m/s = 6.7×10^5 mi/h (b) 6.0×10^{-3} m/s^2

53. (a) 2.9; (b) 0.93 mg

55. 1.01 m/s^2

57. (d)

59. 3.8 rad/s^2 into turntable

61. 1.4×10^{-3} rad/s^2

63. 7.5 rev

65. (a) 1.82 rad/s^2; (b) 28.7 rev

67. (a) 53 s; (b) 8.5 m/s^2 $\mathbf{r}$ + 1.4 m/s^2 $\mathbf{t}$

69. (b)

71. no

73. from the period and distance of the moon and gravitational acceleration on the surface of the Earth

75. 9.8 m/s^2

77. (a) 2.3×10^{20} N toward Sun; (b) 6.5×10^{20} N toward Sun

79. 3.4×10^5 m

81. 8.0×10^{-10} N, toward opposite corner

83. $F_{E-S} = (1.7 \times 10^2)F_{E-M}$

85. 3.7 m/s^2

87. 3.0 g's

89. -5.29×10^{-10} J

91. (a) 2.2×10^{-10} N $\mathbf{y}$; (b) 2.3×10^{-10} N

93. (d)

95. (a) to get more velocity relative to space (b) tangential speed of the Earth is higher in Florida

97. (a) 3.7×10^3 m/s; (b) 34%

99. (a) 3.0×10^{-19} s^2/m; (b) same

101. 1.53×10^9 m

103. (a) K_{500}; (b) U_{850}; (c) $K_{500} = (1.05)K_{850}$; $U_{850} = (1.05)U_{500}$

105. 1.89×10^{27} kg

107. Moon's gravitational attraction is greater on the water on the near side than on the Earth and produces one bulge for one tide; the attraction is greater on the Earth than on the water on the far side and so the Earth moves toward the moon and leaves the water behind for another bulge

109. (a) 4.9 m/s^2; (b) 8.7 m/s^2; (c) 7.0 m

111. 8.5 m

113. (a) 3.17×10^{-8} Hz; (b) 1.99×10^{-7} rad/s

115. 640 N

117. (a) 4.6×10^{-11} N towards 0.75 kg; (b) 1.8×10^{-10} N/kg; (c) 0.27 m; no

119. (a) $a_t = 2.5$ m/s^2; $a_c = 9.7$ m/s^2; (b) at the lowest point of the swing; $a_t = 0$

121. (a) none; (b) 1.8×10^{33} J

Chapter 8

1. (a)

3. yes

5. at the 9-o'clock position, the velocity is straight upward; so it is a "free-fall" with an initial upward velocity

7. (a) 0.30 m/s; (b) 0.60 m/s

9. yes

11. yes

13. 0.58 rotations

15. (a)

17. (a) directly over the base of support; (b) directly below and through the axis of rotation

19. cylinder or sphere on level surface

21. 2.3 m

23. (a) 83.3 cm; (b) 62.5 g

25. (a) 4.7 m·N; (b) 5.4 m·N; (c) 4.7 m·N

27. 1.8×10^3 N (> 400 lb)

29. (a) 88.2 N; (b) 10.5 kg

31. 5.08 J

33. yes

35. 1.2 m from left end

37. $m_2 = 0.20$ kg; $m_3 = 0.50$ kg; $m_4 = 0.40$ kg

39. (a)

41. depends on how mass is distributed about an axis

43. decrease moment of inertia

45. (a) in the direction of the force; (b) torque decreases

47. lengthen day

49. (a) 2.4 kg·m^2; (b) 0.27 kg·m^2; (c) 2.4 kg·m^2

51. (a) 5.0×10^{-3} kg·m^2; (b) 1.0×10^{-2} kg·m^2; (c) 5.0×10^{-3} kg·m^2; (d) 2.5×10^{-3} kg·m^2

53. 1.1×10^5 m·N

55. (a) 1.7×10^{-3} m·N; (b) 8.0×10^{-4} m·N

57. 5.4 rad/s

59. 3.6 N

61. cylinder goes higher by 7.1%

63. $\tan^{-1}(7\mu_s/2)$

65. (b)

67. 7.5×10^2 J

69. 5.8 m/s

71. 1.39×10^8 J; 1.54×10^6 W

73. (a) 29%; (b) 40%; (c) 50%

75. (a) 2.8×10^2 J; (b) 0.25 s

77. 3.4×10^{-3} J; the work is supplied by the motor

79. (a) $\sqrt{gR}$; (b) $h = (2.7)R$; (c) weightlessness

81. (c)

83. decrease the moment of inertia and therefore increase the angular velocity

85. the gravitational force is not through the axis of rotation; this generates a torque to change the angular momentum vector and to cause the top to precess; the clockwise or counterclockwise precession direction depends on the direction of the spinning of the top

87. 3.4 rad/s

89. (a) $v_p > v_a$; (b) 1.03

91. $L_{rev} = 2.6 \times 10^{34}$ m·N·s; $L_{rot} = 2.2 \times 10^{29}$ m·N·s

93. $I = (0.60)I_o$

95. (a) 4.3 rad/s; (b) $K = (1.1)K_o$

97. $b(v_o/v_{max})$

99. $\tan \theta = f_s/N = \mu_s$; (b) $30°$, yes

105. net torque is zero

107. (a) 2.5 cm; (b) 36 cm

109. (a) 1.3 m/s^2; (b) 3.4×10^{-2} m·N

111. $T_1 = 21$ N; $T_2 = 15$ N

113. smaller side; 1.3

Chapter 9

1. (c)

3. less strain for a given stress

5. N/m

7. (a) 1.9×10^5 N/m^2; (b) 2.5×10^5 N/m^2

9. 1.1×10^{11} N/m^2

11. 9.6×10^{10} N/m^2

13. bends toward brass

15. $\phi_{A1} = (1.5)\phi_{Cu}$

17. 1.3×10^4 N

19. (a) ethyl alcohol has the smallest B; **(b)** $\rho_w/\rho_e = 2.2$

21. 5.4×10^{-5}

23. none—same depth, same pressure

25. (a) when the liquid is poured from an unvented can, a partial vacuum develops inside and the pressure difference cause the pouring difficult; by opening the vent, you are allowing air to go into the can and the pressures are equalized so the liquid can be easily poured; **(b)** forcing the air out and reducing the pressure inside the dropper; with the dropper in a liquid, the liquid rises in the dropper due to atmospheric pressure; **(c)** when we inhale, the lungs expands so the pressure decreases; this rushes the air into the lungs; when we exhale, the lungs contract and the pressure increases; this pushes the air out

27. yes; the pressure taken from the calf is higher than that taken from the arm because of a height pressure difference

29. 33 cm^3

31. 74 g

33. 2.1×10^3 Pa

35. 6.39×10^{-4} m^2

37. air density decreases rapidly with altitude

39. 1.08×10^5 Pa

41. 34 cm

43. 0.50 N

45. 2.6 mm

47. (a)

49. no change; as the ice melts, the volume of the newly converted water decreases, however the ice which was initially above the water surface is now under the water; this compensated the decrease in volume; it does not matter whether the ice is hollow or not; both can be proved mathematically

51. no

53. 10,000 metric tons

55. no

57. 1.5×10^{-5} m^3; 6.0×10^3 kg/m^3

59. yes

61. 16.4 m

63. no

65. (d)

67. there are many more capillaries than arteries; the total area of the capillaries is greater than that of the arteries; so if A increases, v decreases

69. 7.4×10^{-2} N/m

71. 0.015 m

73. (a) water; **(b)** $h_w = (1.3)h_b$

75. h = -0.010 m

77. (a)

79. the speed of the air between the vehicles is greater and so the pressure is less; this causes a pressure difference which produces the force

81. the air speed is greater on top of the strip and so the pressure is less; this causes a pressure difference which lifts the strip.

83. the water flowing by decreases the pressure at the end of the tube away from the flask; this creates a pressure difference, and suction is created in the tube; as the eggs moves to one side, there is a change in the flow speed around the egg that creates an inward pressure that makes the egg back to midstream

85. 4.9×10^{-2} m^3/s

87. 91 s

89. 0.14 m^3/s

91. 5.7×10^3 Pa

95. (c)

99. 7.9×10^{-5} m^3/s

101. 0.22 m/s

103. $\phi = 90°$, h = 0

105. 1 part water to 0.62 parts alcohol, or 38% alcohol

107. 1.32×10^5 N or 2.96×10^4 lb

109. 3.0×10^3 N

111. 1.2×10^3 kg/m^3

113. 0.45 m

115. 2.2 m/s

117. (a) $v_s = 11$ cm/s; $v_n = 66$ cm/s **(b)** 1.2×10^3 Pa

119. (a) 3.3×10^4 N; **(b)** 3.8×10^4 N

Chapter 10

1. (a)

3. not necessarily; internal energy depends on mass in addition to temperature

5. let $T_F = T_C$; $(\frac{9}{5})T_C + 32 = T_C$; solve

7. (a) 302°F; **(b)** 90°F; **(c)** −13°C; **(d)** −460°F

9. (a) −45°C; **(b)** 375°F

11. 136°F; −128°F

13. (a) −101 F°; **(b)** 27 C°

15. (c)

17. (a) decreases; **(b)** constant

19. the pressure of the gas is held constant; so if the temperature increases, so does the volume and vice versa, therefore temperature is determined from volume

21. (a) 273 K; **(b)** 373 K; **(c)** 293 K; **(d)** 238 K

23. (a) 233 K; **(b)** 233 K

25. (a) 10,340°F; 5727°C; **(b)** 4.6%

27. (a) 2.2 moles; **(b)** 2.5 moles; **(c)** 3.0 moles; **(d)** 2.5 moles

29. 40°C

31. 33.4 lb/in.^{2}

33. 1.7 atm

35. 3.0%

37. 7.8×10^4 K, hotter than the Sun's surface; no

39. (d)

41. (a) ice moves upward; **(b)** ice moves downard; **(c)** Cu

43. decreases

45. 0.027 m

47. no

49. 4.8 C°

51. 8.4×10^{-3} m

53. 0.0027 cm

57. (a) larger; **(b)** 5.5×10^{-6} m^3

59. 0.24 gal

61. (b)

63. the gases diffuse through the porous membrane, but the helium gas diffuses faster because it has a smaller mass; eventually there will be equal concentrations of gases on both sides of the container

65. (a) 6.1×10^{-21} J; **(b)** 7.7×10^{-21} J

67. (a) 6.21×10^{-21} J; **(b)** 1.37×10^3 m/s

69. $v_{oxygen} = (1.22)v_{ozone}$

71. 1.25×10^3 J

73. 6.8% increase

75. 2.5:1

77. 0.99999 m

79. 6.8×10^{-4} m

81. 7.8 cm

83. 33 C°

85. (a) 3.3×10^{-2} moles; **(b)** 2.0×10^{22} molecules; **(c)** 0.94 g

87. (a) 18 F°; **(b)** 5.6 C°

89. $\alpha = 3.8 \times 10^{-6}$ C°$^{-1}$; yes, low coefficient of linear expression

91. 13.4 g/cm^3

95. 3.3 cm

97. 2.206×10^5 Pa

Chapter 11 ˙

1. (d)

3. 1.27×10^5 J

5. 5.86×10^3 W

7. 1 Btu = 1.06×10^3 J

9. (d)

11. water has higher specific heat and has more energy to lose

13. 700 J/kg·C°

15. 1.7×10^6 J

17. 21.5°C

19. 59.6 kcal

21. 77°C

23. 1.27 kg

25. 687 J/kg·C°

27. 3.72 min

29. (d)

31. because of different internal energy, different molecular structure and different intrinsic heat value; latent heats are also different for different substances because of different molecular structure, or bonds; the latent heat energy goes into breaking these bonds.

33. (a) under pressure, the boiling point of the coolant increases and the engine may be operated at a higher temperature for increased efficiency; if the cap is removed and the pressure is suddenly reduced, the coolant flash boils and can spurt out, causing burns; **(b)** at high altitudes, the pressure is lower, which decreases the boiling point of water; so it will take longer to cook the rice; this increases the amount of water evaporated; therefore more water is needed

35. 2.6×10^6 J more

37. 8.0×10^4 J

39. (a) 184 kcal; **(b)** −182 kcal

41. 0.45 L

43. 1.72×10^{-2} kg

45. 49 kcal

47. 9.7×10^{10} kcal; yes

49. 0.17 L

51. (d)

53. (a) this convects the heat from the hot soup to the cooler air; (b) no; the ice blocks air flow and cooling of the air (also, the air conditioner is less efficient and runs more, increasing electric cost.)

55. to increase the surface area for better conduction and radiation

57. the double-walled and partially evacuated container: conduction and convection; the mirrored interior: radiation

59. 63 W

61. 4.54×10^6 J

63. (a) 1.3×10^2 kcal/s; (b) 73 kg, no

65. (a) the greater the R-value, the greater the insulation value; (b) (1) 4.2 in.; (2) 51 in.

67. (a) 2.5×10^4 J/s; (b) 2.6×10^3 J/s; no

69. 2.8×10^6 J

71. (a) same; (b) $(1.3)P_o$; (c) $(16)P_o$

73. 9.3×10^{22} kcal/s

75. 3.9×10^5 J

77. 65°C

79. 2.6×10^2 kcal

81. 27°C

83. (a) 34 kcal; (b) 0.44 kg; (c) no, because ice is less dense than water

85. 4.0×10^2 m/s

87. Al, 4.4 kcal

89. $m_{Pb} = 6m_{Cu}$

Chapter 12

1. (c)

3. because $p = nRT/V$

5. (a) isothermal

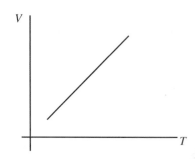

(b) isobaric

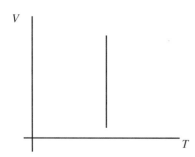

(c) isometric

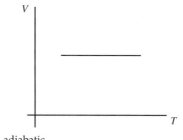

(d) adiabatic

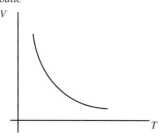

7. (a) on the liquid–vapor surface, water exists in two phases, liquid and vapor in equilibrium; on the solid–vapor surface, the solid phase and vapor phase coexist in equilibrium (b) at certain combinations of p, V, and T, all three phases of water coexist in equilibrium; the coordinates of these combinations are on the triple line

9. (c)

11. (a) a force is exerted to pump the air and does work on the system; after one cycle, the internal energy of the system returns to its original value; so heat must leave the system; this causes the body of the pump to heat up (b) the rapid "expansion" is almost adiabatic; work is done at the expense of the internal energy of the gas; this lowers the temperature and cools the valve

13. (a) added, 400 J; (b) $\Delta U = 0$ for a complete cycle, T is constant

15. 3.0×10^5 J

17. -5.7×10^4 J (out)

19. (a) zero; (b) $-p_1V_1$ (on the system) (c) $-p_1V_1$ (from the system)

21. 2.09×10^3 J

23. 3.6×10^4 J

25. (a) 1.0×10^5 J; 0 J; (b) -7.5×10^4 J (c) 7.5×10^4 J removed

27. (c)

29. energy created

31. 1.2×10^3 J/K

33. $\Delta S_i = -9.1 \times 10^2$ J/K; $\Delta S_s = 1.5 \times 10^3$ J/K; ice has more order

35. -332 J/K

37. 127°C

39. $+0.963$ J/K

41. -26 J/K

43. (a) 2.73×10^4 J/K; (b) isentropic

45. (a) 61.0 J/K; (b) -57.8 J/K; (c) 3.2 J/K

47. (b)

49. heat could be completely converted to work for single process such as isothermal process of an ideal gas

51. 1.20×10^3 J

53. 35%

55. (a) 200 J; (b) 75% or 600 J

57. (a) 67%; (b) not the same

59. (a) 6.5×10^8 J; (b) 28%

61. (a) 1.5×10^2 kcal; (b) 1.9×10^6 J

63. 500 kW

65. (a) 2.4×10^5 J; (b) -10 C°

69. (a)

71. (a) no; (b) $+Q_1$ and $+Q_4$; $-Q_2$ and $-Q_3$; T_1 and T_3 constants

73. 6.0×10^5 J

75. 9.2×10^3 J

77. 100 C° change in temperature

79. $\epsilon_{C_1} = 35\%$; $\epsilon_{C_2} = 67\%$

81. 20.1%

83. hi-temperature reservoir

85. 53%

87. (a) 40%; (b) cycle comes back to original entropy value; $\Delta S = 0$

89. (a) 64%; (b) a lot more energy is lost than ideally predicted

91. (a) 13; (b) no, $\epsilon_C = 11$

93. (a) 6.7%; (b) probably not; due to the poor return; fossil fuels are much better

95. 8.0 L

99. water: freezing point decreases with increasing pressure;

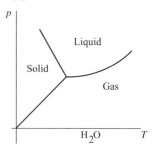

CO_2: freezing point increases with increasing pressure;

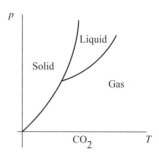

101. (a) 1.5×10^4 J; (b) 22°C

103. (b) Carnot efficiency

105. (a) 9.8×10^3 J; (b) 2.0×10^4 J

107. 3.10×10^3 J

Chapter 13

1. (b)

3. (a) four times as large (b) twice as large

5. $8A$

7. 0.025 s

9. left of $x = 0$ moving toward wall

13. (a) 10^{-12} s; (b) 63 m/s
15. 0.25 kg
17. 0.71 Hz
19. (b) 1.8×10^2 times more
21. (a) 2.5 m/s; (b) 2.5 m/s; (c) 2.7 m/s; equilibrium position
23. (a) 0.38 m; (b) 8.5×10^{-3} m
25. (d)
27. trace out the path on scolling paper
29. increase
31. (a) $x = A \sin \omega t$; (b) $x = A \cos \omega t$
33. 9.787 m/s^2
35. (a) 0.52%; (b) 2.0%; (c) 3.2%
37. (a) 4.8 cm; (b) 4.4 cm/s; (c) -1.2 cm/s^2
39. (a) 14 cm; (b) 89 cm/s; (c) 0.16 J
41. $y = (5.0$ cm$) \cos \pi t/4$; all the same
43. (a) $x = 0$; (b) $v = 94$ cm/s; (c) $a = 0$
45. $T_1 = \sqrt{2}\, T_2$
47. (a) 1.5×10^3 N/m; (b) 0.052π s or 0.16 s; (c) 1.5×10^2 N, to right (fully compressed)
51. (d)
53. (a) transverse and longitudinal; (b) longitudinal; (c) longitudinal
55. 20 m/s
57. 3×10^{-6} m
59. 1.7 cm to 17 m
61. 1.00×10^4 m
63. (a) 0.15 m; (b) 0.15 m/s
65. (a) 3.8×10^2 s; (b) yes; (c) 1.6×10^3 s; shear wave not transmitted through outer liquid core
67. (a) 0.20 s; (b) 2.0 s
69. (d)
71. only waveform, energy is not destroyed
73. (c)
75. clamp an object at one end (node) and make the other end an anti-node; the object could be an antenna, a metal rod, etc
77. (a) 6.0 m; (b) 2.0 m
79. (a) 8.49 m/s; (b) $f_n = 0.425\, n$ Hz; $n = 1, 2, 3, \ldots$
81. (a) yes; (b) no
83. (a) 3rd harmonics of 3.0 m = 1st harmonics of 1.0 m and 6th harmonic of 3.0 m = 2nd harmonic of 1.0 m; (b) 3rd harmonic of 2.0 m = 4th harmonic of 1.5 m
85. 12 m
87. 0.22 kg; 0.055 kg; 0.024 kg; 0.014 kg
89. $E_2/E_1 = \frac{1}{6}$
91. 1.7 s; 0.60 Hz
93. 2.8×10^2 Hz; 1.2 m
95. 210 Hz
97. (a) 8.3 s; (b) yes, never reach zero
99. 1.1 cm/s
101. 3.0 s

Chapter 14

1. (b)
3. (a)
5. some insects produce sound that is higher than the human audible range (>20 kHz)
7. the thermal stress caused by the temperature difference (gradient) results in

the crack of the ice; this is accompanied by the release of energy including sound
11. (a) 2.0 km; (b) 1.3 mi
13. 5.0×10^6 Hz
15. (a) 0.646 m; (b) 0.670 m
17. 1.4×10^2 m
19. 394 m
21. 12°C
23. 2.8%
25. (b)
27. yes, an intensity below the threshold intensity
29. (a) 90 dB; (b) 70 dB; (c) -20 dB
31. 20 dB
33. (a) 63 dB; (b) 83 dB; (c) 113 dB
35. 1.1×10^{-3} W
37. 97 dB
39. 316 m
41. 15 dB
43. 84 dB and 83 dB
47. 10^4 bees
49. (d)
51. (a) no; (b) no; (c) increasing frequency
53. (a) destructive; (b) constructive
55. 4 Hz
57. 4 Hz; make frequencies equal
59. (a) 431 Hz; (b) 373 Hz
61. 28 m/s
63. 90°
65. (a) 2; (b) 662 m/s
67. 126 Hz
69. (b)
71. (a) snow absorbs sound so there is little reflection (b) there is less absorption; so the reflection dies out more slowly and therefore sounds hollow and echoing (c) sound is reflected by the shower walls and standing waves are set up, giving rise to more harmonics and therefore richer sound quality
73. the spacing (change in length) depends not only on the frequency differences Δf, but also on the values of the frequencies themselves; for different frequencies, ΔL is therefore different
75. 510 Hz
77. (a) f_2 does not exist—only odd harmonics; (b) 322 Hz
79. no, greater for open pipe; linear expansion
83. 0.194 m; 0.583 m; 0.972 m
85. 30°
89. $I_2 = (1.44)I_1$
93. 66.2 dB
97. 45 m
99. 4.5 s
101. 1.7%

Chapter 15

1. (d)
3. (a) because attractive and repulsive forces can be produced by different combinations of just two types of charges; (b) no effect
5. $+1.60 \times 10^{-13}$ C

7. (a) $+2.4 \times 10^{-9}$ C; (b) 1.4×10^{-20} kg
9. $+3.20 \times 10^{-19}$ C
11. positive
13. (d)
15. charge by polarization, bring a charged object up close to one; sphere would have opposite charges
17. (d)
19. (a) same; (b) 4 times as large; (c) 2 times as large
21. both the Sun and the planets are electrically neutral
23. (a) 2.3×10^{-26} N; (b) zero
25. 44.7 cm
27. (a) 50 cm; (b) 50 cm
29. (a) $x = -0.25$ m; (b) no location; (c) $x = -0.94$ m for either $\pm q_3$
31. 3.6×10^{-47} N; 8.2×10^{-8} N
33. (a) 96 N, 39° below $+x$ (b) 61 N, 84° above $-x$
35. (b)
37. (a)
39. by the relative density of the field lines
41. negative charge on the inside surface of the sphere and equal amount of positive charge on the outside surface of the sphere; the excess like charges will be distributed evenly over the surface of the sphere; the electric field outside the sphere behaves as though all the excess charge on the sphere were concentrated at its center; the electric field is zero inside the sphere; if the charge were negative, the field lines would all point toward the charge; the field inside would still be zero, and the field outside would still behave as though all the excess charge were concentrated at the center, even though the excess charge is located on the surface
43.

45. 5.6×10^{-11} N/C downward
47. 9.0×10^5 N/C toward the -5.0 μC charge
49. $\mathbf{E} = 2.2 \times 10^5$ N/C $\mathbf{x} - 4.1 \times 10^5$ N/C $\mathbf{y}$
51. 5.4×10^6 N/C toward -4.0 μC charge
53. 3.8×10^7 N/C in the $+y$ direction
55. (a) zero; (b) 1.1×10^3 N/C toward the center; (c) 71 N/C toward the center
57. (a) $-y$; (b) $\dfrac{kqd}{[(d/2)^2 + x^2]^{3/2}}$
59. negative
61. (a) zero; (b) $-Q$; (c) $+Q$; (d) $-Q$
63. (a) zero; (b) $\dfrac{kQ}{r^2}$; (c) zero; (d) $\dfrac{kQ}{r^2}$
65. field lines are perpendicular to the surface

67.

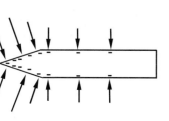

69. (b)
71. $+46.9\ \mu C$
73. net of 6 entering it or -6 lines
75. (a) 4.0×10^{-16} N in the $-y$ direction;
(b) 4.0×10^{-17} J
77. (a) 0.17 m from the $-3.0\ \mu C$ charge in between the charges; **(b)** 0.17 m from the $-3.0\ \mu C$ charge in between the charges
79. (a) 4.6×10^{3} N/C to the left;
(b) 4.5×10^{-7} s
81. (a) 5.9×10^{7} N/C at $87°$ above $+x$ axis
83. negative
85.

E
$+q$ F
F
$-q$

E
$-q$ $+q$
p

87. same field except zero inside slab, induced charges on slab create exactly opposite field to create zero field inside slab

Chapter 16
1. (d)
3. (a)
5. they slow down
7. (b)
9. -9.6×10^{-5} J
11. 3.0×10^{3} V/m from positive to negative
13. (a) 1.3×10^{6} m/s down; **(b)** it loses potential energy
15. (a) 7.5×10^{-15} J; **(b)** 1.2×10^{4} N/C to the left
17. (a) 0.54 m; **(b)** 7.5 kV, a decrease
19. (a) it gains 6.2×10^{-19} J; **(b)** it loses -6.2×10^{-19} J; **(c)** it gains 4.8×10^{-19} J
21. 2.2 J
23. (a) -0.30 J; **(b)** no
25. 0.72 J
27. 3.1×10^{5} V
29. 36 m/s
31. -1.3×10^{6} V
33. (a) 5.9×10^{7} m/s; **(b)** 5.9×10^{-9} s
35. (a)

37.

15.37
Mesa ——————

—————————
Ball rolling
—————————
—————————

Bottom of cliff ——————

39. a point in space can have only one potential value; if two equipotential surfaces intersect in space, the intersection will have two different potential values
41. planes, parallel to the plate surface
43. spheres centered on the charge
45. cylinders with their central axis coinciding with the wire
47. an electron volt is $e\ \Delta V = q\ \Delta V = \Delta U$; a GeV is larger than a MeV by 1000 times
49. no, equipotential surfaces cannot cross; so electric field lines cannot cross either
51. 1.0 mm
53. $+7.1$ V
55. 12.6 m
57. $+100$ eV
59. (a) 2.0×10^{7} eV; **(b)** 2.0×10^{4} keV; **(c)** 20 MeV; **(d)** 2.0×10^{-2} GeV; **(e)** 3.2×10^{-12} J
61. 6.2×10^{7} m/s (protons); 4.4×10^{7} m/s (alpha)
63. answers are the same and the direction is opposite
65. $+2.8$ V
67. (d)
69. it increases
71. 1.2×10^{-5} C
73. (a) 4.2×10^{-9} C; **(b)** 2.5×10^{-8} J
75. (a) halved to 2.1×10^{-9} C; **(b)** halved to 1.3×10^{-8} C
77. (b)
79. we cannot maintain a nonzero voltage on a conductor; charges will transport from the positive to the negative immediately
81. $K = 3.5$
83. 1.5 times as much
85. 4.3×10^{-9} J
87. (a)
89. equal in capacitance
91. 6.0 μF
93. (a) 3.0 μF; **(b)** 9.0 μF
95. 0.10 μF: 0.60 μC; 0.20 μF: 1.2 μC; 0.30 μF: 1.8 μC
97. (a) C_1: $Q = 2.4\ \mu C$; C_2: $Q = 2.4\ \mu C$; C_3: $Q = 1.2\ \mu C$; C_4: $Q = 3.6\ \mu C$; **(b)** 6.0 V for all capacitors
99. zero
101. electric field lines are pointing toward the wire.
103. 14 kV/m
105. (a) 0.17 μF; **(b)** 2.1 μC; **(c)** $V_1 = V_2 = 5.1$ V; $V_3 = 6.9$ V
107. (a) 8.4×10^{-13} V; **(b)** 1.5×10^{-9} V; **(c)** in (a), top plate is $+$; in (b) top plate is $-$
109. 2.6×10^{-16} J

Chapter 17
1. (b)
3. (a)
5. if they mix, the two electrodes are essentially the same and there will be no potential difference
7. there is no voltage drop on the internal resistor
9. (a) 6.0 V; **(b)** 3.0 V
11. (a) 48 V; **(b)** the 12-V batteries in series then combined in parallel with the 24-V battery
13. (d)
15. 8.3 min
17. 75 C
19. 2.2×10^{5} J
21. 0.36 A to left
23. (a)
25. (a)
27. (a) 11.4 V; **(b)** 0.32 Ω
29. (a) same; **(b)** one quarter the current
31. aluminum has to be 29% thicker than copper
33. 1.0 V
35. 1.3×10^{-2} Ω
37. 3.0×10^{19} electrons/s
39. the shorter carries 4 times the current
41. 0.13 Ω
43. 4.6 mΩ
45. $R = 65\ \Omega$ at both voltages; thus resistor is ohmic
47. it increases; $R_2/R_1 = 1.6$
49. 59 mA
51. 0.77 A
53. 0.27 A decrease
57. its power output would quadruple and it would overheat
59. 3.0×10^{2} W
61. (a) 10.0 A; **(b)** 12.0 Ω
63. 1.2 Ω
65. (a) 0.42 A; **(b)** 14 Ω
67. (a) \$32; **(b)** 3.2 Ω
69. \$11
71. 100 Ω: 12 V; 25 Ω: 79 V
73. decrease of 202 W
75. 1.4 kΩ
77. 21 Ω
79. $R_{120\ V}/R_{90\ V} = \frac{4}{3}$
81. 73%
83. 3.5×10^{2} s
85. \$120
87. 64 C°
89. (a) 12.5 A; **(b)** 9.60 Ω
91. (a) 10 V; **(b)** no
93. 1.8×10^{3} W
95. copper at 117°C or aluminum at -73°C
97. 1.0 kWh
99. (a) 5.0×10^{-2} A; **(b)** 5.0 V
101. 0.16 A
103. (a) 4.0×10^{2} Ω; **(b)** 4.0 W; **(c)** 4.8×10^{2} J
105. 1.7×10^{21} electrons
107. to reduce $I^2 R$ loses
109. 6.6×10^{-6} m/s

Chapter 18

1. (a)

3. no; only if all resistors are equal

5. 10 Ω

7. (a) 120 Ω; **(b)** 11 Ω

9. parallel: 1.3 Ω; series: 12 Ω; two in series-parallel: 2.7 Ω; two in parallel-series: 6.0 Ω

11. series: 15 Ω, 20 Ω, 25 Ω, 30 Ω; parallel: 3.3 Ω, 3.8 Ω, 6.0 Ω, 2.7 Ω; series-parallel: 11 Ω, 14 Ω, 18 Ω, 4.2 Ω, 6.7 Ω, 7.5 Ω

13. 3.0 $\mu\Omega$

15. (a) 1.0 A; **(b)** 1.0 A; **(c)** $P_{2\Omega} = 2.0$ W; $P_{4\Omega} = 4.0$ W; $P_{6\Omega} = 6.0$ W; **(d)** $P_{sum} = P_{total} = 12$ W

17. I (for all) = 1.0 A; $V_{8.0\Omega} = 8.0$ V; $V_{4.0\Omega} = 4.0$ V

19. 2.7 Ω

21. (a) 60 W: 0.50 A; 100 W: 0.83 A; **(b)** 60 W: 60 W; 100 W: 100 W

23. (a) 860 W; **(b)** 96 W

25. (a) 50 W: 0.42 A; 100 W: 0.83 A; 150 W: 1.25 A **(b)** 50 W: 290 Ω; 100 W: 144 Ω; 150 W: 96 Ω

27. two

29. (a) 4.3×10^2 s = 7.3 min; **(b)** 2.6×10^3 s = 44 min

31. (a) 0.085 A; **(b)** $P_{15} = 7.0$ W; $P_{40} = 2.6$ W; $P_{60} = 0.24$ W; $P_{100} = 0.41$ W

33. 35 W

35. $I_1 = 2.7$ A; $I_2 = I = 3.3$ A; $I_3 = I_4 = I_5 = 0.55$ A

37. (d)

39. (a)

45. $I_1 = 0.33$ A left; $I_2 = 0.33$ A right

47. (a) $I_1 = 0.75$ A left; $I_2 = 0.50$ A left; $I_3 = 1.25$ A up; $I_4 = 1.25$ A right; $I_5 = 1.25$ A down **(b)** 13 W

49. $I_1 = 3.23$ A down; $I_2 = 1.54$ A down; $I_3 = 1.02$ A down; $I_4 = I_5 = I_6 = 1.93$ A left

51. $I_1 = 0.273$ A left; $I_2 = 1.36$ A up; $I_3 = 1.09$ A down

53. (a)

55. (a) $V_R = V$, $V_C = 0$ **(b)** $V_R = 0.37 V$, $V_C = 0.63 V$ **(c)** $V_R = 0$, $V_C = V$

57. when tube ceases to conduct, voltage rises again

59. 0.40 μF

61. 0.693 τ

63. (a) 24 V; **(b)** 0; **(c)** 4.0 A

65. 1.4 V

67. (b)

69. (a) the galvanometer would burn out; **(b)** it would read the voltage of the source

71. 6.5×10^2 Ω

73. 50 kΩ

75. 0.30 mA

79. (c)

81. high voltage could produce harmful current

83. it is safer to jump

85. 6.0 Ω

87. 6.0 Ω

89. 9.5 J/s

91. (a) $I_1 = 1.0$ A; $I_2 = I_4 = 0.40$ A; $I_3 = 0.20$ A; **(b)** $P_1 = 100$ W; $P_2 = P_4 = 4.0$ W; $P_3 = 2.0$ W

93. 0.0293 A

95. 3.0 MΩ

99. (a) 0.34 A; 37 W; **(b)** 0.68 A; 150 W

101. (a) 120 V; **(b)** 0.33 A; 0.50 A; 0.83 A; **(c)** 40 W; 60 W; 100 W; **(d)** 1.67 A; **(e)** 200 W; **(f)** the current, voltage and power of the other two remain the same, the total current is reduced to 0.83 A and the total power is reduced to 100 W

103. two in parallel and in series with other resistor

105. $R_1 = 49.95$ kΩ; $R_2 = 200$ kΩ; $R_3 = 1.0$ MΩ

Chapter 19

1. (b)

3. (c)

5. (c)

7. into the page

9. the sign of the voltage drop across the strip would then indicate the type of charge

11. 2.5×10^3 m/s

13. 1.0 T in $-z$ direction

15. 30° or 150°

17. (a) 8.6×10^{12} m/s^2; **(b)** same magnitude, opposite direction

19. 2.9 A

21. (a) (1) negative charge; (2) no charge; (3) positive charge; **(b)** $m_3 > m_1$

23. 1.2 A

25. 40 V

27. (a) 2.0×10^{-5} T; **(b)** 9.6×10^{-2} m from wire 1

29. 1.4×10^{-5} T at 38° below a horizontal line to the left

31. 3.8×10^{-5} T, toward you (up)

33. 1.0×10^{-4} T, away from the observer

35. 6.4×10^{-2} A

37. 4.0×10^{-24} J

39. $B = \sqrt{2} \mu_o I / \pi a$, at 45° towards lower left wire

41. (a) $\dfrac{qB}{2\pi m}$; **(b)** $\dfrac{2\pi m}{qB}$; **(c)** 5.7×10^{-3} m; 2.8×10^6 Hz

43. (b)

45. no

47. (a) away from you; **(b)** away from you

49. 6.0×10^{-2} A·m^2

51. (a) away from you; **(b)** flip to line up with the field

53. (a)

55. yes

57. east for north wire and north for east wire

59. 1.2 N perpendicular to the plane of B and I

61. 1.0×10^{-2} T north to south

63. (a) 0; **(b)** 4.0 N/m in $+z$; **(c)** 4.0 N/m in $-y$; **(d)** 4.0 N/m in $-z$; **(e)** 4.0 N/m in $+y$

65. 0.40 N/m; $+z$

67. 6.7×10^{-6} N/m attractive

69. (a) 2.8×10^{-5} N repulsive; **(b)** 2.8×10^{-5} N attractive

71. 2.7×10^{-5} N toward the other wire

73. short: 25 N; long: 75 N

75. (a) 9.0×10^{-2} A·m^2; **(b)** with plane of coil parallel to **B**

77. 0.39 m·N

79. (d)

83. 1.0×10^5 m/s

85. (a) 1.8×10^3 V **(b)** same speed, independent of the charge

87. (a) 1.6×10^4 V/m; **(b)** 1.6×10^3 V **(c)** top is positive

89. (a) 4.8×10^{-26} kg; **(b)** 2.4×10^{-18} J; **(c)** no, work equals zero

91. 2.9×10^{-2} T

93. (b)

95. clockwise since the field points down at the north pole.

97. 2.4×10^2

99. 8.0×10^{-21} N west

101. 0.22 m

103. 5.0×10^{-1} T, east

105. 5.6×10^{-7} m/s

107. 3.0×10^{-2} m·N

Chapter 20

1. (d)

3. (a)

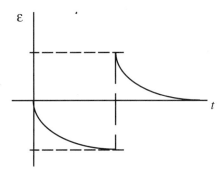

(b) no, by Lenz's law it is repelled moving toward the loop, and attracted as it leaves the loop

5. (a) 0; **(b)** 4.5×10^{-3} T·m^2; **(c)** 7.5×10^{-3} T·m^2

7. (a) 0; **(b)** 1.9×10^{-2} T·m^2; **(c)** 1.4×10^{-2} T·m^2

9. (a) 0.20 m^2; **(b)** 0.23 m^2

11. 80 V

13. 2.5 Ω

15. 0.35 T

17. (a) -50 V; **(b)** $+13$ V; **(c)** 0; **(d)** $+12$ V

19. zero

21. (a) final magnetic field is opposite the initial field; **(b)** 4.0×10^4 V

23. -4.0 V

25. lower: 0.037 T·m^2; upper: 0.034 T·m^2; back: 0.071 T·m^2; 0 for all other surfaces

27. 0.13 V

29. (c)

31. the magnet moving through the coils produces a current at the expense of the kinetic and potential energy

33. (a) 3.4×10^{-2} V; **(b)** 0.51 V

35. ±104 V, the initial voltage direction was not specified, so there are two possible directions to the answer

37. (a) 0; (b) 87%

39. 16 Hz

41. 105 V

43. (a) 216 V; (b) 160 A; (c) 8.1 Ω

45. (c)

47. (a) step-down; (b) 90 V; (c) 5.3 A

49. (a) 16 turns; (b) 5.0×10^2 A

51. $N_P/N_S = \frac{1}{25}$

53. (a) 10.5 A; (b) 17.1 V

55. 160 turns

57. (a) $N_S/N_P = \frac{1}{20}$; (b) 2.5×10^{-2} A

59. (a) 128 kWh; (b) $1840

61. (a) $N_S/N_P = \frac{1}{2}$, 1/13.9, $\frac{1}{30}$; (b) 2.0, 13.9, 30; (c) 833

63. (a) 53 W; (b) $N_P/N_S = 200$

65. (c)

67. (a) 1.5×10^8 Hz; (b) 1.2×10^7 Hz; (c) 4.0×10^6 Hz

69. 3.6×10^4 m

71. (a) 800 nm > 700 nm; infrared; (b) 300 nm < 400 nm; ultraviolet

73. AM: 67 m; FM: 0.77 m

75. primary; 27

77. 100%, ideal conditions

79. (a) 30 A; (b) 4.0 Ω

81. (a) 840 V; (b) 2.0 A

83. (a) 0.67 A; (b) 54 V; (c) 36 W

85. 50 Hz

87. 1×10^{23} Hz

89. radar frequencies are much higher because radar wavelengths are much shorter

Chapter 21

1. (c)

3. since $I = V_o/R$, V_o is greater than I_o if R is greater than 1 Ω; it could be the other way if R is less than 1 Ω

5. 170 V; 339 V

7. 110 V

9. (a) $V_{rms} = 3.8$ V; $V_o = 5.3$ V; (b) 2.8 W

11. (a) 85 V; (b) 5.2×10^{-2} s

13. (a) 60 Hz; (b) 1.4 A; (c) 1.2×10^2 W; (d) $V = (120$ V$)$ sin $380t$; (e) $P = (240$ W$)$ sin^2 $380t$; (f) $P = (240$ W$)$ <sin^2 $380t$> = $(240$ W$)/2 = 120$ W

15. $V = (170$ V$)$ sin $334\pi t$

17. 10 Hz

19. (a) 2.9×10^2 A; (b) 3.4×10^2 V

21. 1.5 rad/s

23. (b)

25. (d)

27. capacitor and inductor can oppose current flow

29. 5.9×10^{-5} F

31. 1.3×10^3 Ω

33. 1.7 A

35. an increase of 60%

37. 141 Hz

39. 0.70 H

41. (d)

43. (c)

45.

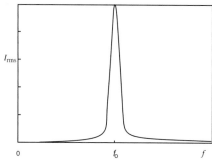

resonance at f_o

47. 35 Hz

49. −65°

51. 1.0×10^2 Hz

53. no; 2.0 H for any resistor

55. ab: 1.3 A; ac: 1.2 A; bc: 4.0 A; cd: 1.8 A; bd: 1.6 A; ad: 2.9 A

57. 1.7×10^{-11} F to 1.8×10^{-10} F

59. (a) 362 Ω; (b) no; 110 Hz

61. $V_R = 12$ V; $V_L = 2.7 \times 10^2$ V; $V_C = 2.7 \times 10^2$ V

63. (a) 0.894; (b) 13.3 μF

65. (a) 3.8×10^2 W; (b) 24 Ω; (c) −18 Ω

67. −80 W

69. 0.35 A

71. 1.7×10^{-12} F

75. 37°

77. in Fig. 21.14a, the inductor in series with R_L filters out the high frequency current and so only the low frequency current reaches R_L; in Fig. 21.14b, the capacitor in series with R_L filters out the low frequency current and so only the high frequency current reaches R_L.

Chapter 22

1. (c)

3. (d)

5. diffuse reflections by particulate matter

7. 58°

9. 40°

11. 1.50 m

13. 27°

15. (a) 35°; (b) 55°

19. (d)

21. (d)

23. light coming out of water is bent into the air with a larger angle of refraction and so it reaches our eyes

25. 1.24×10^8 m/s

27. 47°

29. 57°

31. 26% greater in zircon

33. 3.6×10^{-3} m

35. 32°

37. 38.6 mm

39. 1.56

41. (a) 4.2 cm; (b) 3.2 cm

43. 2.6 m

45. 15 cm

47. seen for 40°, not for 50°

49. 48.8°

51. 1.6 cm

53. 2.0 m

55. no

57. (b)

59. no; the light will be further dispersed by second prism

61. red is transmitted and blue is internally reflected

63. 1.4980

65. (a) 21.7°; (b) 0.22°; (c) 0.37°

67. (a) 56°; (b) 30°

69. 50°

71. (a) 25°; (b) 1.97×10^8 m/s; (c) 362 nm

73. (a) 49°; (b) 1.5; (c) 1; (d) 42°

Chapter 23

1. (b)

3. (a) during the day, light passes both ways through the pane, at night, there is little light coming through the pane, so the reflections are much more clear; (b) due to reflections on both sides of the pane glass, (c) this works on a combination of half-silvering and bright light on one side and dark on the other

5. (a) the left-right reversal is an apparent one; it is caused by the front-back reversal; (b) no

7. (a) 6.0 m; (b) upright, virtual, and same size

9. (a) v; (b) $2v$

11. (a) 0.80 m; (b) 0.80 m

13. 3.0 m and 13 m behind north mirror; 5 m and 11 m behind south mirror

15. 0.68 m

17. (d)

19. (a) plane mirror gives a large view of the area immediately around that side of the truck; small convex mirror gives a wide angle perspective of the road in back of both sides of the truck; (b) images formed by convex mirror are smaller than objects (so the image distances are smaller than object distances) and hence appear closer than they actually are; (c) yes, because it collects a large amount of radio waves and focuses them onto a small area

21. (a) the image is smaller than the object; (b) as the ball swings toward the mirror and approaches the focal point, the image enlarges; an enlarged image appears to be closer to our eyes and so it appears to move toward the observer and therefore produces the effect of appearing to jump out of the mirror as the ball swings through the focal point

23. $d_i = 60$ cm; $h_i = 9.0$ cm, image is real and inverted

25. (a) 20 cm; (b) 0.19 cm

29. (a)

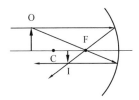

(b)

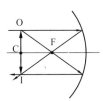

(c)

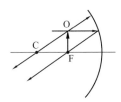

(d)

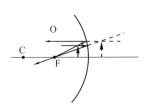

31. $d_i = -10$ cm; $h_i = (2/3)h_o$
33. 2.3 cm
35. $f = (4/3)d_o$
37. concave, 48 cm
39. (a)

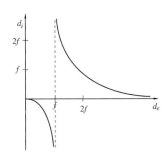

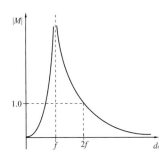

(b)

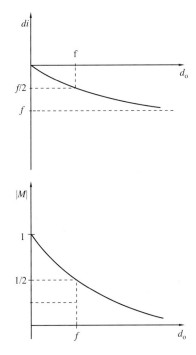

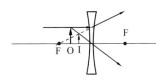

41. 0.69
45. yes; $d_o = 13.3$ cm; $d_o = 26.7$ cm
47. (d)
49. (a) distance from vortex to image of distant object; **(b)** distance from vortex to image of distant object
51. 15 cm; 1.0×10^{-12}
53. (a) $d_i = -6.4$ cm; $M = 0.64$;

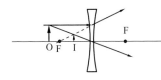

(b) $d_i = -10.5$ cm; $M = 0.42$

55. (a) no; **(b)** yes
57. −2.5 cm
59. 8.1×10^{-2} m
61. −9.0
63. (a) $d = 4f$; **(b)** approaches 0
65. (a) −20 cm; **(b)** −63 cm
67. (a) 4.6 cm; **(b)** 0.80 cm, inverted
69. virtual image 18 cm to left of eyepiece
73. (b)
75. 50 cm
77. −0.70 D
79. 0.55 mm
81. $d_i = 37.5$ cm, $h_i = 3.0$ cm, real and inverted
85. 14 cm
87. same as those in Exercise 23.39

89. (a)

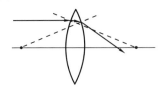

(b)

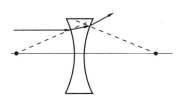

91. 85 cm
93. (a)

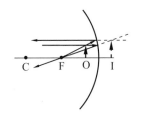

(b) 60 cm

Chapter 24

1. (c)
3. blue
5. the path difference will change; this change in path difference results in a change in the condition of interference
7. (a) 0.80 m; **(b)** 3.2 m
9. 1.4 m
11. 1.8×10^{-3} rad
13. $y_m = m\lambda L/(2d), m = 1, 3, 5, \ldots; \Delta y = L\lambda/d$
15. (a) 4.4×10^{-7} m; **(b)** 4.4×10^{-2} m
17. (a) 4.00×10^{-7} m, violet
(b) 3.45×10^{-2} m
19. 450 nm
21. (a) because both reflections have 180° phase shifts
23. coat a piece of glass with a material ($\lambda/2$ thick) whose index of refraction is between those of air and glass
25. 1.3×10^{-7} m
27. (a) 1.06×10^{-7} m; **(b)** yes, solar cell $n > $ film n
29. 1.67×10^{-7} m
33. 1.61×10^{-6} m
35. (b)
37. (a) 4.8×10^{-3} m; **(b)** 2.4×10^{-3} m
39. (a) 1.4×10^{-3} rad; 2.8×10^{-3} m;
(b) 8.8×10^{-4} rad; 1.8×10^{-4} m
41. (a) 2.44×10^3 lines/cm; **(b)** 11
43. do not overlap
45. 11.0°
47. 7.1×10^{-10} m
49. (a) $0.31° = 5.4 \times 10^{-3}$ rad;
(b) 9.8×10^{-3} m
53. (d)

55. looking through a lens of each pair, rotate one of the lens; if the intensity changes as the glasses are rotated, both pairs are polarized
57. (a) twice; (b) four times; (c) none; (d) six times
59. 54° to 60°
61. 51°
63. 57.2°
65. 48.8°
67. no
69. (d)
71. 1.6
73. $\Delta y = (0.25)\Delta y_o$
75. 450 nm
77. 9.3×10^2 lines/cm
79. 2.5×10^{-2} m
81. one red; two violet
83. 83.3 nm

Chapter 25

1. (b)
3. short focal length and small radius for close objects; long focal length and large radius for distance object
5. the pre-flash causes the iris to reduce down so that when the second flash comes momentarily, you don't have a wide opening through which you get the red-eye reflection from the retina
7. +2.0 D
9. (a) nearsighted; (b) −0.080 D
11. (a) +3.0 D; (b) take them out
13. −0.50 D
15. (a) −0.67 D; (b) yes; 21 cm; (c) 30–40 years old
17. (a) −0.13 D; (b) −0.13 D
19. $\Delta P = +0.50$ D
21. (a) −0.25 D; (b) −0.25 D
23. (d)
25. 2.6×
27. (a) 2.3×; (b) 2.5×
29. (a) 6.1 cm; (b) 4.1×
31. (a) 7.1 cm; (b) 3.5×
33. 330×
35. 300×
37. 0.44 cm
41. (a) 67×; (b) 6.6 cm
43. (b)
45. 19 ×
47. 150 ×
49. (a) 117 × ; (b) 88.3 cm
51. (a) second; (b) first
53. (c)
55. (d)
57. 5.5×10^{-7} m
59. 1.32×10^{-7} rad; Yerkes/Hale = 4.98
61. yes
63. (a) 1.63×10^{-6} rad; (b) 9.76×10^{17} m
65. (a) $\Delta\theta = 1.46 \times 10^{-5}$ rad; (b) $\Delta s = 4.39 \times 10^{-4}$ mm
67. (d)
69. with red light: red and white appear red; blue appears black; with green light: only white appears green; both red and blue appear black; with blue light: red appears black; white and blue appear blue

71. foam has very low material density and it can only absorb very little light or almost all light is reflected
73. −0.77 D
75. 0.36 m
77. 1.3×
79. 375×
81. 3.8°
83. 13 cm
85. (a) (1) 6.3; (2) 0.25; (b) 1/120 s

Chapter 26

1. (d)
3. (a)
5. (c)
7. since there was no fringe shift in the interference pattern, the speed of light is constant in all directions
9. (a) 4.23 s; (b) 4.48 s
11. 55 m/s and 45 m/s
15. no
17. (a)
19. (b) since your friend is traveling to the front and so he sees the front mark first
21. (a) 300 m/s
23. (c) her friend does appear the same height because the height is perpendicular to the velocity
25. 32 beats/min
27. 1.06 h
29. Earth twin is 64 y; traveling twin is 37 y
31. 2.4 s
33. (a) 131 m^2; (b) 40.7°
35. (a) 0.99c (0.99995c); (b) 0.44c; (c) 0.14c
37. 5.3 m
39. she is traveling at 0.40c in the direction of the lengths of the two sticks
41. (d)
43. yes; no
45. (a) 0.985c; (b) 2.50 MeV; (c) 1.56×10^{-21} kg·m/s
47. (a) 0.14c; (b) 0.86c
49. 1.6×10^{24} J or 150,000 times more!
51. (a) 0.92c; (b) 6.3×10^{-22} kg·m/s
53. 79 keV
55. (a) 1.0×10^3 MeV; (b) 63 MeV; (c) 1.9×10^{-19} kg·m/s
59. 3.7×10^{-12} kg
63. (a) 0.875c (b) 0.999c
65. (a)
67. the stick would be elongated by the gravity gradient (difference)
69. 8.9 mm and 2.8 m
71. (a) 1.5×10^4 m; (b) 7.2×10^{17} kg/m^3
73. 0.43c
75. 0.154c toward Earth
77. 939 MeV
79. 3.2 h
81. 33 m long, 2.5 m high, and 2.0 m wide
83. 50 min
85. 0.68c
87. 1.9 kg

Chapter 27

1. (d)
3. no

5. (a)

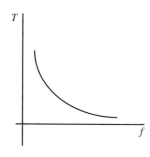

(b)

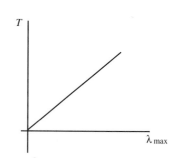

7. 9.4×10^{-6} m, infrared
9. 4.8×10^3 K; 7.3×10^3 K
11. 2.56×10^{-20} J
13. (b)
15. the packets of radio signal arrive in such quantity and so quickly that our ear cannot distinguish between discrete arrivals
17. 5.0×10^{-19} J
19. $E_v = (1.75)E_t$
21. doubled
23. 4.15×10^{20} quanta
25. (a) 6.7×10^{-34} J·s (b) 2.9×10^{-19} J
27. 8.3×10^{-7} m
29. 2.48 eV
31. no
33. (a) sodium; (b) $\lambda_{silver} < 262$ nm; $\lambda_{sodium} < 504$ nm
35. 6.7×10^{-34} J·s; 13 eV

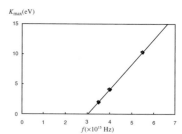

37. (d)
39. 200,000 scatterings
43. 0.25 nm
45. 45°
47. 0.0055 nm
49. (d)
51. the theory applies only on atoms with single electron
53. (a) 12.1 eV; (b) 1.89 eV
55. (a) 0.850 eV; (b) 0.544 eV
57. 1.095×10^{-2} nm^{-1}
59. $n \approx 100$

61. **(a)** 1.51 eV; **(b)** 0.544 eV; **(c)** 0.136 eV
63. 4.05×10^3 nm, infrared
65. **(c)**
67. **(a)** 2.86 eV; **(b)** $n_i = 2$ and $n_f = 5$
69. **(a)** potential is -27.2 eV and kinetic is $+13.6$ eV; **(b)** $|U| = 2K$
71. **(a)** $E = -2.82 \times 10^3$ eV; $E = -7.04 \times 10^2$ eV; $r = 2.56 \times 10^{-4}$ nm; $r = 1.02 \times 10^{-3}$ nm; **(b)** 0.586 nm
73. stimulated emission
75. 1.14×10^{31} quanta
79. **(a)** 12.8 eV; **(b)** 13.4 eV
81. 4.4×10^{-16} J
83. **(a)** 1.5×10^{15} Hz; **(b)** 1.3 V
85. four

Chapter 28

1. **(c)**
3. 680 nm
5. 4.7×10^{-36} m
7. 150 V
9. $\lambda_2/\lambda_1 = 1/\sqrt{2} \approx 0.71$
11. 2.7×10^2 eV
13. **(a)** 4.7×10^{-8} rad; **(b)** 4.0×10^5 V
15. 150 V
17. **(b)**
19. **(a)** wave functions

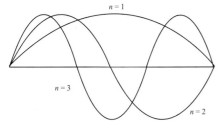

(b) probability

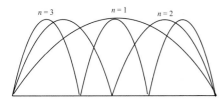

21. **(b)**
23. **(a)**
25. element groups according to the values of quantum numbers n and l; within a group, the elements have the same or very similar electronic configurations for the outmost electrons
27. **(a)** 2; **(b)** 14
29. **(a)** $l = 3$; **(b)** $n = 4$
31. **(a)** Be; **(b)** N; **(c)** Ne; **(d)** S
33. **(a)**

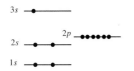

Na has 11 electrons

(b)

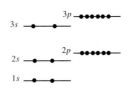

Ar has 18 electron

35. it would be $1s^3$
37. **(b)**
39. 2.1×10^{-29} m/s
41. 1.5×10^{-10} m
43. 1.1×10^{-27} J
45. 10^4
47. **(a)**
49. no
51. 1.9 GeV
53. 70 V
55. **(a)** 4.0×10^{-21} N·s **(b)** 1.7×10^{-13} m
59. $x = 0$
61. yes when $n = 1$ and $l = 0$
63. no, muons are not identical to electrons

Chapter 29

1. **(c)**
3. **(d)**
5. **(a)** ^{24}Mg: 12 p, 12 n, 12 e; ^{25}Mg: 12 p, 13 n, 12 e; **(b)** ^{24}Mg: 12 p, 12 n, 14 e; ^{25}Mg: 12 p, 13 n, 14 e; **(c)** ^{24}Mg: 12 p, 12 n, 11 e; ^{25}Mg: 12 p, 13 n, 11 e
7. 92 p and 92 e for each with n from 140 n to 147 n
9. $^{41}_{19}$K
11. **(a)** He: 1.9×10^{-15} m; Ne: 3.3×10^{-15} m; Ar: 4.1×10^{-15} m; Kr: 5.3×10^{-15} m; Xe: 6.1×10^{-15} m; Rn: 7.3×10^{-15} m; **(b)** they are roughly from ten thousand to fifty thousand times smaller
13. **(b)**
15. **(a)** $^{237}_{93}$Np $\rightarrow$ $^{233}_{91}$Pa $+ ^{4}_{2}$He; **(b)** $^{32}_{15}$P $\rightarrow$ $^{32}_{16}$S $+ ^{0}_{-1}$e; **(c)** $^{56}_{27}$Co $\rightarrow$ $^{56}_{26}$Fe $+ ^{0}_{+1}$e; **(d)** $^{56}_{27}$Co $+ ^{0}_{-1}$e $\rightarrow$ $^{56}_{26}$Fe; **(e)** $^{42}_{19}$K* $\rightarrow$ $^{42}_{19}$K $+ \gamma$
17. α: $^{210}_{83}$Bi $\rightarrow$ $^{206}_{81}$Tl $+ ^{4}_{2}$He; β: $^{206}_{81}$Tl $\rightarrow$ $^{206}_{82}$Pb $+ ^{0}_{-1}$e
19. $^{213}_{83}$Bi
21. **(a)** $^{4}_{2}$He; **(b)** $^{0}_{-1}$e; **(c)** $^{102}_{39}$Y; **(d)** $^{23}_{11}$Na*; **(e)** $^{0}_{-1}$e
23. **(a)** α to ^{233}Pa; β to ^{233}U; α to ^{229}Th; α to ^{225}Ra; β to ^{225}Ac; α to ^{221}Fr; α to ^{217}At; α to ^{213}Bi; α to ^{209}Tl or β to ^{213}Po; β to ^{209}Pb or α to ^{209}Pb; β to ^{209}Bi; **(b)** too many neutrons
25. **(d)**
27. no, decay is exponential
29. **(a)** 33.8 μCi **(b)** 1.25×10^6 Bq
31. **(a)** $\frac{1}{4}$; **(b)** $\frac{1}{2}^{24}$ or approximately 6×10^{-6} % of the original
33. 28.6 y
35. 90.1%
39. ^{104}Tc
41. 3.4 μg
43. **(a)** 6.75×10^{19} nuclei **(b)** 2.11×10^{18} nuclei
45. 17,000 y

47. 9.6 mg
49. **(d)**
51. ^{1}H and ^{2}H: stable; ^{3}H: unstable
53. **(a)** $^{17}_{8}$O because an unpaired neutron beyond a magic number; **(b)** $^{42}_{20}$Ca because of magic number difference making $^{42}_{20}$Ca more stable (both are paired); **(c)** $^{10}_{5}$B because of lack of pairing; **(d)** approximately the same as both have 126 neutrons (paired and magic) and different number of protons does not affect neutrons
55. **(b)** and **(d)**; others are odd–odd
57. 2.013553 u
59. **(a)** 7.72 MeV; **(b)** 2.57 MeV/nucleon
61. tritium
63. 127.6 MeV
65. 10.1 MeV
67. 7.57 MeV/nucleon
69. **(a)** two alphas; **(b)** two alphas; **(c)** yes
71. **(d)**
73. **(a)** alpha particles; **(b)** x-ray is 2.0 rem; alphas are 40 rem
75. yes
77. $^{76}_{33}$As; $^{76}_{32}$Ge
79. **(a)** 28.3 MeV; **(b)** 4.001501 u
81. **(a)** 3.4×10^{22} nuclei; **(b)** 1.8×10^{16} nuclei; **(c)** no, 9.7×10^{19} decays/min; 5.2×10^{15} decays/min
83. **(a)** $^{0}_{-1}$e; **(b)** $^{222}_{86}$Rn; **(c)** $^{237}_{94}$Pu; **(d)** γ; **(e)** $^{0}_{+1}$e
85. **(a)** 1.80×10^3 MeV **(b)** 237.997821 u
87. **(c)** and **(d)**
89. alpha: $^{210}_{81}$Tl; beta: $^{214}_{84}$Po
91. **(b)** 2.3×10^{17} kg/m^3
93. **(a)** the atomic density is 2.7×10^3 kg/m^3, which is about 10^{14} times less than the nuclear density; **(b)** this is because the mass is the same but averaged over a much larger volume; **(c)** the average atomic density is much higher than hydrogen gas (most gases are typically ≈ 2 kg/m^3) because the gas density includes mostly empty space between the gas molecules

Chapter 30

1. **(d)**
3.

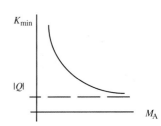

5. **(a)** $^{41}_{18}$Ar*; **(b)** $^{236}_{92}$U*; **(c)** $^{236}_{92}$U*; **(d)** $^{18}_{9}$F*; **(e)** $^{11}_{5}$B*
7. **(a)** $^{14}_{7}$N*; **(b)** $^{14}_{7}$N*; **(c)** $^{31}_{15}$P*; **(d)** $^{18}_{9}$F*; **(e)** $^{14}_{7}$N*
9. **(a)** $^{27}_{14}$Si; no; **(b)** $^{222}_{86}$Rn; yes; **(c)** $^{193}_{77}$Ir; yes
11. +4.28 MeV
13. 1.53 MeV

15. exoergic
17. 4.36 MeV
19. 5.38 MeV
21. $(K_1/K_2)_{min} = 2.03$
23. (a) 0.24 b;   (b) 0.66 b;   (c) 1.6 b;
(d) 1.7 b
25. (d)
27. (d)
29. the reaction products collide with the core materials and coolant and in the process produces heat which converts water to steam that powers a turbine
31. (a) 3.27 MeV;   (b) 12.9 MeV
33. 19
35. (c)
37. 4.69×10^{-13} m
39. (a) 0.86 MeV;   (b) 1.41×10^{-21} kg·m/s opposite neutrino direction   (c) both zero

41. 13.37 MeV
45. $\beta^-: M_P > M_d$; $\beta^+: M_P > M_d + 2m_e$
47. (a)
49.

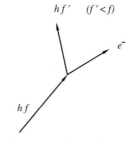

hf' $(f' < f)$

e^-

hf

51. 6.6×10^{-33} kg
53. 135 MeV
55. 1.98×10^{-15} m
57. (a)
59. all hadrons contain quarks and/or antiquarks; quars are believed to exist freely outside the nucleus
61. (a) $^7_4$Be*;   (b) $^{60}_{29}$Cu*;   (c) $^{236}_{92}$U*;
(d) $^{13}_6$C*;   (e) $^{17}_8$O*
63. 2.35 MeV
67. (a) $^{19}_8$O;   (b) $^{10}_5$B;   (c) $^{89}_{39}$Y;   (d) $^2_1$H

Index

Photo Credits

All Demonstration photos: Michael Freeman

Chapter 1   **CO.1** George Whiteley/Photo Researchers, Inc. **Fig. 1.2b** U.S. Department of Commerce   **Fig. 1.3c** U.S. Department of Commerce   **Fig. 1.5** Frank Labua/Simon & Schuster/PH College **Fig. 1.7a** Lander College   **Fig. 1.7b** Kip Peticolas/Fundamental Photographs   **Fig. 1.8** John Smith   **Fig. 1.15** Frank Labua/Simon & Schuster/PH College   **Fig. 1.16** Frank Labua/Simon & Schuster/PH College

Chapter 2   **CO.2** Carl Schneider/The Stock Connection   **Fig. 2.4** John Smith   **Fig. 2.6** Focus on Sports, Inc.   **Fig. 2.14b** James Sugar/Black Star   **IN, p. 49 Fig. 1** North Wind Picture Archives   **IN, p. 49 Fig. 2** J.M. Charles Rapho/Photo Researchers, Inc.   **Fig. 2.15** Frank Labua/Simon & Schuster/PH College

Chapter 3   **CO.3** C.E. Nagele/FPG International   **Fig. 3.12c** Photri, Inc.   **Fig. 3.15b** Richard Megna/Educational Development Center/Fundamental Photographs   **Fig. 3.17a** George Chan/Photo Researchers, Inc.   **Fig. 3.17b** David G. Curran/Rainbow   **Fig. 3.20** Focus on Sports, Inc.

Chapter 4   **CO.4** T.J. Florian/Rainbow   **Fig. 4.1** Bill Sanderson/Science Photo Laboratory/Photo Researchers, Inc.   **Fig. 4.4** John Smith   **Fig. 4.14** Tom Cogill   **Fig. 4.19a** Patrick Behare/Photo Researchers, Inc.   **Fig. 4.19b** Gabe Palmer/Mug Shots   **Fig. 4.25** Photo Researchers, Inc.   **Fig. 4.33** Focus on Sports, Inc.

Chapter 5   **CO.5** Asarco, Inc.   **Fig. 5.9a** Guntram Gerst/Peter Arnold, Inc.   **Fig. 5.9b** Michael Collier   **Fig. 5.11a** Ken Straiton/Photo Researchers, Inc.   **Fig. 5.11b** Joan Menschenfreund/The Stock Market   **Fig. 5.19** The Image Works   **Fig. 5.21** Robert Della-Piana/The Stock Market   **Fig. 5.22** Daemmrich/Stock Boston   **Fig. 5.26** David Lissy/Focus on Sports, Inc.

Chapter 6   **CO.6** John Cordes/Focus on Sports, Inc.   **Fig. 6.1a** Gary S. Settles/Photo Researchers, Inc.   **Fig. 6.1b** David G. Curran/Rainbow   **Fig. 6.1c** Stephen J. Krasemann/Photo Researchers, Inc.   **Fig. 6.7a** Omikron/Science Source/Photo Researchers, Inc.   **Fig. 6.9a** M. Hans/Vandystadt/Photo Researchers, Inc.   **Fig. 6.9b** Globus Brothers/The Stock Market   **IN, p. 180 Fig. 1** Ford Motor Company   **Fig. 6.10a** Ann Purcell/Photo Researchers, Inc.   **Fig. 6.10b** H.P. Martin/The Stock Market   **Fig. 6.14** Diane Schiumo/Fundamental Photographs   **Fig. 6.15a** Richard Megna/Fundamental Photographs   **Fig. 6.15b** Bruce Berman/The Stock Market   **Fig. 6.18** Paul Silverman/Fundamental Photographs   **Fig. 6.20** Focus on Sports, Inc.   **Fig. 6.22b** NASA Headquarters   **Fig. 6.24** Runk/Schoenberger/Grant Heilman Photography, Inc.

Chapter 7   **CO.7** Jeff Greenberg/Photo Reseachers, Inc.   **Fig. 7.6b** Tom Tracy/The Stock Market   **Fig. 7.10** Cris Priest/Science Photo Library/Photo Researchers, Inc.   **Fig. 7.12c** Lewis Portnoy/The Stock Market   **Fig. 7.13** Focus on Sports, Inc.   **IN, p. 223 Fig. 1** NASA Headquarters   **Fig. 7.23a** NASA/Science Source/Photo Researchers, Inc.   **Fig. 7.23c** Johan Elbers   **Fig. 7.23d** C. Seghers/Photo Researchers, Inc.   **IN, p. 234 Fig. 1** Photo Researchers, Inc.   **Fig. 7.29** NASA Headquarters   **Fig. 7.30** Frank Labua/Simon & Schuster/PH College   **Fig. 7.32** Alexander Lowry/Photo Reseachers, Inc.   **Fig. 7.36** Archive Photos   **Fig. 7.37** Rita Nannini/Photo Researchers, Inc.

Chapter 8   **CO.8** Jerry Wachter/Focus on Sports, Inc.   **Fig. 8.11a** Richard Hutchings/Photo Researchers, Inc.   **Fig. 8.11b** Dr. E.R. Degginger   **Fig. 8.11c** Jean-Marc Loubat/Agence Vandystadt/Photo Reseachers, Inc.   **Fig. 8.12b** Frank Labua/Simon & Schuster/PH College   **IN, p. 258 Fig. 1** California Institute of Technology   **Fig. 8.23a** Frank Labua/Simon & Schuster/PH College   **Fig. 8.23b** Frank Labua/Simon & Schuster/PH College   **Fig. 8.23c** NASA Headquarters   **Fig. 8.26** Jerry Wilson   **Fig. 8.36** Tony Salvino/The Image Works   **Fig. 8.38** Gerard Lacz/Natural History Photographic Agency   **Fig. 8.39** NASA Headquarters

Chapter 9   **CO.9** Focus on Sports, Inc.   **IN, p. 292 Fig. 1** Jonathan Watts/Science Photo Library/Photo Researchers, Inc.   **Fig. 9.12** NASA Headquarters   **Fig. 9.13** Balloon Excelsior, Inc.   **Fig. 9.15** Bill Curtsinger/Photo Researchers, Inc.   **Fig. 9.17a** Hermann Eisenbeiss/Photo Researchers, Inc.   **Fig. 9.18a** David Spears/Science Photo Library/Photo Researchers, Inc.   **Fig. 9.18b** Richard Steedman/The Stock Market   **Fig. 9.22b** Diane Schiumo/Fundamental Photographs   **IN, p. 317 Fig. 2** Blair Seitz/Photo Researchers, Inc.   **Fig. 9.28** Frank Labua/Simon and Schuster/PH College   **Fig. 9.31a** Tom Cogill   **Fig. 9.31b** John Smith   **Fig. 9.32** Tom Cogill   **Fig. 9.35** Tom Cogill

Chapter 10   **CO.10** Jim Corwin/Photo Researchers, Inc.   **Fig. 10.2c** Frank Labua/Simon & Schuster/PH College   **Fig. 10.3a** Leonard Lessin/Peter Arnold, Inc.   **Fig. 10.3b** Jay Boleman, *Physics: A Window on Our World, 3/e* 1995, Prentice Hall   **Fig. 10.5b** John Smith   **IN, p. 331 Fig. 2** Frank Labua/Simon & Schuster/PH College   **Fig. 10.6** Sinclair Stammers/Science Photo Library/Photo Researchers, Inc.   **Fig. 10.11a** Richard Choy/Peter Arnold, Inc.   **Fig. 10.11b** Jay Boleman, *Physics: A Window on Our World, 3/e* 1995, Prentice Hall   **Fig. 10.15** Paul Silverman/Fundamental Photographs   **Fig. 10.19** Tom Cogill

Chapter 11   **CO.11** Tom Cogill   **Fig. 11.2** Jerry Wilson   **Fig. 11.3** North Wind Picture Archives   **Fig. 11.5** John Smith   **Fig. 11.12** Frank Labau/Simon & Schuster/PH College   **Fig. 11.13c** Richard Lowenberg/Science Source/Photo Researchers, Inc.   **Fig. 11.17** Adam Hart-Davis/Science Photo Library/Photo Researchers, Inc.   **Fig. 11.18** John Smith   **Fig. 11.20** John Smith   **Fig. 11.21** Leonard Lessin/Peter Arnold, Inc.   **Fig. 11.24** Joe Sohm/Photo Researchers, Inc.

Chapter 12   **CO.12** Austrian National Tourist Office   **Fig. 12.8** Grant Heilman Photography, Inc.   **Fig. 12.10** Leonard Lessin/Peter Arnold, Inc.

Chapter 13   **CO.13** J.R. Berintenstein/Photo Researchers, Inc.   **Fig. 13.9** John Smith   **IN, p. 430 Fig. 1** Peter Menzel/Peter Arnold, Inc.   **IN, p. 430 Fig. 2** H. Yamaguchi/Sygma   **IN, p. 431 Fig. 4b** Russell D. Curtis/Photo Researchers, Inc.   **IN, p. 431 Fig. 4c** Russell D. Curtis/Photo Researchers, Inc.   **Fig. 13.14a** PSSC Physics, 2nd edition, 1965; Educational Development Center, Inc., Newton, MA.   **Fig. 13.17** Fundamental Photographs   **Fig. 13.18a** Richard Megna/Fundamental Photographs   **Fig. 13.19** Richard Megna/Fundamental Photographs   **Fig. 13.20** Torleif Svensson/The Stock Market   **Fig. 13.21** David Brownell/The Image Bank   **IN, p. 439 Fig. 1** Special Collection Division, University of Washington Libraries   **Fig. 13.26** CENCO

Chapter 14   **CO.14** Seitz/Photo Researchers, Inc.   **Fig. 14.1b** Leonard Lessin/Peter Arnold, Inc.   **Fig. 14.3** NASA Headquarters   **Fig. 14.12b** Philippe Plaily/Science Photo Library/Photo Researchers, Inc.   **Fig. 14.12c** Jim Kahnweiler/Positive Images   **Fig. 14.14** Joseph Nettis/Photo Researchers, Inc.   **Fig. 14.16** Michael Furman Photographer Ltd./The Stock Market   **Fig. 14.18b** Richard Megna/Fundamental Photographs

Chapter 15   **CO.15** Zefa/London/The Stock Market   **Fig. 15.7a** Jerry Wilson   **Fig. 15.7c** Charles D. Winters/Photo Researchers, Inc.   **IN, p. 495, Fig. 1b** Keith Kent/Peter Arnold, Inc.   **IN, p. 495, Fig. 1c** Philippe Wojazer/Reuters/Corbis-Bettmann

Chapter 16   **CO.16** Richard Megna/Fundamental Photographs   **IN, p. 508, Fig. 1b** Jerry Wilson   **Fig. 16.16** Spencer Grant/Photo Researchers, Inc.

Chapter 17 CO.17 Gord Handley **Fig. 17.8** T.J. Florian/Rainbow **IN, p. 542, Fig. 1** IBM Research **Fig. 17.10** Frank Labua/Simon & Schuster/PH College **IN, p. 545, Fig. 1 and Fig. 2** Frank Labua/Simon & Schuster/PH College **Fig. 17.11** NASA/Mark Marten/Photo Researchers, Inc. **Fig. 17.12** Frank Labua/Simon & Schuster/PH College

Chapter 18 CO.18 Richard Megna/Fundamental Photographs **IN, p. 573, Fig. 2** Zircon **Fig. 18.14a** CENCO **Fig. 18.17c** Richard Megna/Fundamental Photographs **Fig. 18.19b&c** Frank Labua/Simon & Schuster/PH College **Fig. 18.20b** Frank Labua/Simon & Schuster/PH College **Fig. 18.20c** U.S. Department of Energy/Science Photo Library/Photo Researchers **Fig. 18.23a&b** Frank Labua/Simon & Schuster/PH College **Fig. 18.42** M. Antman/The Image Works

Chapter 19 CO.19 Peter Menzel/Stock Boston **Fig. 19.1** CENCO **Fig. 19.3b&c** Richard Megna/Fundamental Photographs **Fig. 19.4** Richard Megna/Fundamental Photographs **Fig. 19.8a** Richard Megna/Fundamental Photographs **Fig. 19.9a** Richard Megna/Fundamental Photographs **Fig. 19.10a** Richard Megna/Fundamental Photographs **Fig. 19.13c** Grant Heilman Photography **Fig. 19.21b** Jerry Wilson **Fig. 19.24** James Holmes/Oxford Centre for Molecular Sciences/Science Photo/Photo Researchers, Inc. **Fig. 19.26a** CENCO **Fig. 19.30** Pekka Parviainen/Science Photo Library/Photo Researchers, Inc.

Chapter 20 CO.20 Charles O'Rear/Westlight **Fig. 20.1d** Richard Megna/Fundamental Photographs **Fig. 20.10a** U.S. Department of the Interior **Fig. 20.12d** Westinghouse Electric Corp. **Fig. 20.15a** Dan Guravich/Photo Researchers, Inc. **Fig. 20.15b** U.S. Department of Energy/Mark Marten/Photo Researchers, Inc. **Fig. 20.19** Peter Menzel/Stock Boston **Fig. 20.20** Gary Wagner/Stock Boston **IN, p. 648, Fig. 1** NOAA/Science Photo Library/Photo Researchers, Inc. **Fig. 20.22** Simon Fraser/Science Photo Library/Photo Researchers, Inc.

Chapter 21 CO.21 Thomas Kitchin/Tom Stack & Associates **Fig. 21.5** Frank Labua/Simon & Schuster/PH College **IN, p. 670, Fig. 1** Rip Griffith/Photo Researchers, Inc. **Fig. 21.13** Frank Labua/Simon & Schuster/PH College

Chapter 22 CO.22 Alise/Mort Pechter/The Stock Market **Fig. 22.5a** Peter M. Fisher/The Stock Market **Fig. 22.5b** U.S. Bureau of Mines **Fig. 22.6** Photri/The Stock Market **Fig. 22.7** Richard Megna/Fundamental Photographs **Fig. 22.9** Richard Megna/Fundamental Photographs **Fig. 22.11a** Runk/Schoenberger/Grant Heilman Photograph **Fig. 22.11b** Kent Wood/Photo Researchers, Inc. **Fig. 22.13c** DiMaggio/Kalish/Peter Arnold, Inc. **Fig. 22.15** W. Eastep/The Stock Market **Fig. 22.16a** Gemological Institute of America **Fig. 22.17a** CENCO **Fig. 22.18** Richard Megna/Fundamental Photographs **Fig. 22.19a** C. Falco/Photo Researchers, Inc. **Fig. 22.19b** Hank Morgan/Photo Researchers, Inc. **Fig. 22.20** David Parker/Science Photo Library/Science Source/Photo Researchers, Inc. **IN, p. 694 Fig. 1** P. Gontier/The Image Works **IN, p. 694 Fig. 2** SIU/Photo Reseachers, Inc. and Charles Lightdale/Photo Researchers, Inc. **IN, p. 696 Fig. 1** Doug Johnson/Science Photo Library/Photo Researchers, Inc. **Fig. 22.21** Martin Bond/Science Photo Library/Photo Researchers, Inc. **Fig. 22.23a** SIU/Photo Researchers, Inc. **Fig. 22.23b** J. Barry O'Rourke/The Stock Market **Fig. 22.24** Frank Labua/Simon & Schuster/PH College

Chapter 23 CO.23 Peticolas/Megna/Photo Researchers, Inc. **IN, p. 705 Fig. 3** Frank Labua/Simon & Schuster/PH College **Fig. 23.6** Paul Silverman/Fundamental Photographs **Fig. 23.13** Photo Researchers, Inc. **Fig. 23.16** John Smith **IN, p. 725 Fig. 1** Bohdan Hrynewych/Stock Boston; John Smith **Fig. 23.22** Michael Freeman **Fig. 23.23** Tom Tracy/The Stock Market **Fig. 23.24** John Smith

Chapter 24 CO.24 Adrienne Hart Davis/Science Photo Library/Photo Researchers, Inc. **Fig. 24.1** Richard Megna/Fundamental Photographs **Fig. 24.2b** From the "Atlas of Optical Phenomena," Michel Cagnet, Maurice Francon, Jean Claude Thrierr. © by Springer-Verlag OHG, Berlin, 1962. Published by Prentice Hall, Inc., Englewood Cliffs, NJ. **Fig. 24.7a** David Parker/Science Photo Library/Photo Researchers **Fig. 24.7b** Gregory G. Dimijian, M.D./Photo Researchers, Inc. **Fig. 24.8b** H.R. Bramaz/Peter Arold, Inc. **Fig. 24.9b** Ken Kay/Fundamental Photographs **IN, p. 743, Fig. 2** Kristen Brochmann/Fundamental Photographs **Fig. 24.10** Jay Boleman, *Physics: A Window on Our World, 3/e* 1995, Prentice Hall **Fig. 24.12** Ken Kay/Fundamental Photographs **Fig. 24.16** Prof. A.J. Stosick, University of Southern California **Fig. 24.19b** Dr. E.R. Degginger **Fig. 24.21b** Peter Aprahamian/Sharples Stress/Engineers Ltd./SPL/Photo Researchers, Inc. **Fig. 24.22** Nina Barnett **IN, p. 757, Fig. 2** Leonard Lessin/Peter Arnold, Inc. **Fig. 24.25** Roger Ressmeyer/Starlight **Fig. 24.26** Photri, Inc.

Chapter 25 CO.25 Roger Ressmeyer/Starlight Collection **Fig. 25.4b** Frank Labua/Simon & Schuster/PH College **Fig. 25.8b** Carl Zeiss, Inc., Thornwood, NY **Fig. 25.12** Yerkes Observatory Photographic Services **Fig. 25.14a&b** Hale Observatory **Fig. 25.14c** Science Photo Library/Photo Researchers, Inc. **Fig. 25.15** Roger Ressmeyer/Starlight Collection, A Division of Corbis **Fig. 25.16** NASA **IN, p. 786 Fig. 1** Tony Craddock/Science Photo Library/Photo Researchers, Inc.

Chapter 26 CO.26 David Malin/Royal Observatory, Edinburgh **Fig. 26.4b** PASCO Scientific **Fig. 26.6** Science Photo Library/Photo Researchers, Inc. **Fig. 26.18b** NASA **Fig. 26.19a** NASA **IN, p. 828, Figs. 1&2** David Hardy/Science Photo Library/Photo Researchers, Inc.

Chapter 27 CO.27 Sam C. Pierson, Jr./Photo Researchers, Inc. **Fig. 27.7** Larry Albright **Fig. 27.8** Wabash Instrument Corp./Fundamental Photographs **Fig. 27.13** Gary Retherford/Photo Reseachers, Inc. **Fig. 27.14** Dan McCoy/Rainbow **Fig. 27.18a** Will & Deni McIntyre/Photo Researchers, Inc. **Fig. 27.18b** Lawrence Livermore National Laboratory/Science Photo Library/Photo Researchers, Inc. **Fig. 27.20a** Hank Morgan/Rainbow **Fig. 27.20b** Chuch O'Rear/Westlight

Chapter 28 CO.28 IBM Corporation, Research Division, Almaden Research Center **Fig. 28.3a** Hank Morgan/Rainbow **Fig. 28.3b** Will & Deni McIntyre/Photo Researchers, Inc. **Fig. 28.5** IBM Research **IN, p. 875, Fig. 2** Sinclair Stammers/Science Photo Library/Photo Researchers, Inc. **IN, p. 875, Fig. 3a** Manfred Kage/Peter Arnold **Fig. 3b** Dr. R. Kessel/Peter Arnold; **Fig. 3c** David Scharf/Peter Arnold **IN, p. 876, Fig. 1a** Omikron/Science Source/Photo Researchers, Inc. **Fig. 1b** Mehau Kulyk/Science Photo Library/Photo Researchers, Inc. **IN, p. 877, Fig. 3b** Will & Deni McIntyre/Photo Researchers, Inc. **Fig. 28.13** Lawrence Berkeley/Science Photo Library/Photo Researchers, Inc.

Chapter 29 CO.29 Roger Tully/Tony Stone Images **Fig. 29.5** Mary Evans Picture Library/Photo Researchers, Inc. **Fig. 29.18** Lawrence Bewrkeley/Photo Researchers, Inc. **Fig. 29.19** Larry Mulvehill/Photo Researchers, Inc. **Fig. 29.20** Larry Mulvehill/Photo Researchers, Inc. **Fig. 29.21a** Dan McCoy/Rainbow **Fig. 29.21b** Monte S. Buchsbaum, M.D. Mount Sinai School of Medicine, New York, NY

Chapter 30 CO.30 Alexander Tsiaras/Stock Boston **Fig. 30.1a** George Retseck/George Retseck from Scientific American, July, 1990 **Fig. 30.1b** Fermilab Visual Media Services **Fig. 30.5a** Tom Tracy/The Stock Market **Fig. 30.8a** Igor Kostin/Imago/Sygma **Fig. 30.8b** U.S. Dept. of Energy/Science Photo Library/Photo Researchers, Inc. **Fig. 30.9b** Princeton University Plasma Physics Laboratory **Fig. 30.10c** Gary Stone/Lawrence Livermore National Laboratory

Jerry Wilson author photo, Jeremey Kwasney

Mathematical Symbols

$=$	is equal to
$\neq$	is not equal to
$\approx$	is approximately equal to
$\propto$	is proportional to
$>$	is greater than
$\geq$	is greater than or equal to
$\gg$	is much greater than
$<$	is less than
$\leq$	is less than or equal to
$\ll$	is much less than
$\bar{x}$	average value
Δx	change in x
$\lvert x \rvert$	absolute value of x
Σ	sum
∞	infinity

The Greek Alphabet

Alpha	A	α	Nu	N	ν
Beta	B	β	Xi	Ξ	ξ
Gamma	Γ	γ	Omnicron	O	o
Delta	Δ	δ	Pi	Π	π
Epsilon	E	ϵ	Rho	P	ρ
Zeta	Z	ζ	Sigma	Σ	σ
Eta	H	η	Tau	T	τ
Theta	Θ	θ	Upsilon	Y	υ
Iota	I	ι	Phi	Φ	ϕ
Kappa	K	κ	Chi	X	χ
Lambda	Λ	λ	Psi	Ψ	ψ
Mu	M	μ	Omega	Ω	ω

Quadratic Formula

If $ax^2 + bx + c = 0$, then

$$x = \frac{-b \pm \sqrt{b^2 - 4ac}}{2a}$$

Values of Some Useful Numbers

$\pi = 3.14159\ldots$ $\sqrt{2} = 1.41421$

$e = 2.71828\ldots$ $\sqrt{3} = 1.73205$

Trigonometric Relationships

Definitions of Trigonometric Functions

$$\sin\theta = \frac{y}{r} \qquad \cos\theta = \frac{x}{r} \qquad \tan\theta = \frac{\sin\theta}{\cos\theta} = \frac{y}{x}$$

θ° (rad)	$\sin\theta$	$\cos\theta$	$\tan\theta$
0° (0)	0	1	0
30° ($\pi/6$)	0.500	$\sqrt{3}/2 \approx 0.866$	$\sqrt{3}/3 \approx 0.577$
45° ($\pi/4$)	$\sqrt{2}/2 \approx 0.707$	$\sqrt{2}/2 \approx 0.707$	1.00
60° ($\pi/3$)	$\sqrt{3}/2 \approx 0.866$	0.500	$\sqrt{3} \approx 1.73$
90° ($\pi/2$)	1	0	∞

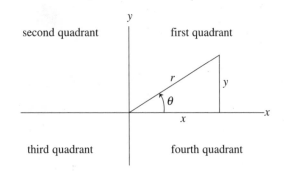